Profiles of

DRUG SUBSTANCES, EXCIPIENTS, AND RELATED METHODOLOGY

VOLUME 33

Profiles of DRUG SUBSTANCES, EXCIPIENTS, AND RELATED METHODOLOGY

VOLUME 33

Edited by

HARRY G. BRITTAIN
Center for Pharmaceutical Physics, Milford, New Jersey 08848

VOLUME 33 AUTHORED BY RICHARD J. PRANKERD

CRITICAL COMPILATION OF pK_A VALUES FOR PHARMACEUTICAL SUBSTANCES

RICHARD J. PRANKERD, PhD
Victorian College of Pharmacy, A Faculty of Monash University
Parkville VIC 3052, Australia

ELSEVIER

AMSTERDAM • BOSTON • HEIDELBERG • LONDON
NEW YORK • OXFORD • PARIS • SAN DIEGO
SAN FRANCISCO • SINGAPORE • SYDNEY • TOKYO
Academic Press is an imprint of Elsevier

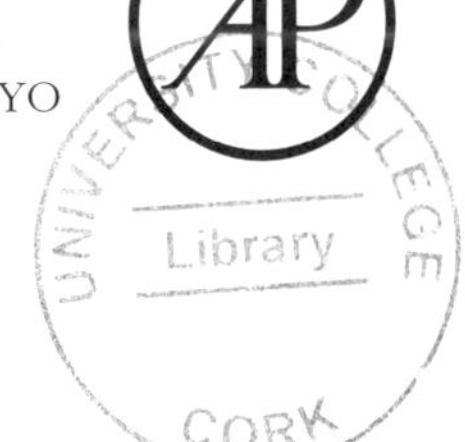

Academic Press is an imprint of Elsevier
84 Theobald's Road, London WC1X 8RR, UK
Radarweg 29, PO Box 211, 1000 AE Amsterdam, The Netherlands
Linacre House, Jordan Hill, Oxford OX2 8DP, UK
30 Corporate Drive, Suite 400, Burlington, MA 01803, USA
525 B Street, Suite 1900, San Diego, CA 92101-4495, USA

First edition 2007

ISBN: 978-0-12-260833-9
ISSN: 0099-5428 (Series)

Printed and bound in USA
07 08 09 10 11 10 9 8 7 6 5 4 3 2 1

CONTENTS

Foreword vii
Acknowledgments ix
Dedication xi
Preface xiii

Critical Compilation of pK_a Values for Pharmaceutical Substances

Richard J. Prankerd

1. Introduction 1
2. Evaluation of Methods for pK_a Determination 7
3. pK_a Values from Computer Programs 24
4. Summary of the Data Compilation 26

References 28

Data Compilations

Appendix A

Main List

pK_a Values Found with Significant Data Quality Information—Mainly Primary Literature 35

Appendix B

Supplementary List

pK_a Values Found with Little or No Data Quality Information—Mainly Secondary Literature 425

Index

Alphabetical Index of Drugs or Drug Related Compounds Found in pK_a Database Files 627

Subject Index 725

FOREWORD

It is against the backdrop of an ever increasing interest in the biological aspects of the pharmaceutical sciences that Richard Prankerd has undertaken the Herculean task of collating a database and reviewing the methodology and reliability of reported values for a fundamental physical chemical parameter that is often overlooked, or even taken for granted, in research studies and educational programs—this parameter is the almost ubiquitous pK_a. The theoretical and practical aspects of this important equilibrium constant have been the subject of significant reviews and treatises in the past. However, this is the first compilation in which the focus is a critical assessment of the reliability of reported pK_a values of compounds with particular relevance to the pharmaceutical and biomedical sciences.

Richard has systematically identified the relevant primary and secondary literature for nearly 3500 reported pK_a values for drugs and related or relevant compounds, and then assessed the reliability of these reported values using the IUPAC classification and guidelines.

This compendium provides an easy to read, excellent resource, and reference base for the pharmaceutical sciences and related disciplines that require a definitive source of information on drug-relevant pK_a values. In addition, a well-referenced introductory section presents an insightful summary of the varied practical and theoretical issues that research scientists should recognize in order to maximize the reliability of pK_a measurements undertaken in their laboratories.

William N. Charman

ACKNOWLEDGMENTS

The author acknowledges generous financial support from SmithKline Beecham (courtesy of Dr. Donald C. Monkhouse), during the first data collection phase of this project. In recent years, I have been much encouraged by my colleagues at the Victorian College of Pharmacy (Monash University), and by Dr. Harry Brittain, the Editor of *Analytical Profiles*, to see this project to completion. I am grateful to all for their support. In particular, I would like to express my appreciation to Professor Bill Charman for his valuable suggestions after reading the chapter, and for his Foreword. I extend thanks to Drs. K. McIntosh, M. Campbell, and D. McLennan of the Centre for Drug Candidate Optimization, Faculty of Pharmacy, Monash University, for their ready assistance with *in silico* pK_a predictions.

DEDICATION

To my father, Kenneth Prankerd (1921–2006), who gave me his curiosity gene.
To my mentor, Robert McKeown, who impressed upon me the need to be careful.

PREFACE

A compilation of this nature inevitably rests on the efforts of others. Earlier compilations of drug p*K*a values have been assembled by Ritschel [1], Smith and Rawlins [2], Speight [3], Hoover [4], Newton and Kluza [5], Craig, Hansch, and coworkers [6], Williams [7], and Delgado and Remers [8]. Inevitably, the more recent of these compilations are the larger, but most contain about 500–700 listings.

However, size is not the only criterion, and it is certainly not the most important. The quality of literature information is vital to those who use it. The reliability of literature data must be known, in order to be most useful. In addition to the above drug-specific p*K*a compilations, there are also the more general, but significantly critical IUPAC compilations of p*K*a values in water for weak organic acids [9], and for weak organic bases [10], the originals of which were later supplemented [11, 12]. The main purpose of the present compilation is to apply to the drug sciences p*K*a literature the same principles of critical assessment that marked the IUPAC-sponsored general compilations.

As suggested by the dedication, I was encouraged in my student years to do things carefully. Rob McKeown would say to me something like "Fifty years from now, there will always be better theories to interpret your data. But if you have measured the data to the best that the technique will allow, the only way anyone can get better understanding is to develop newer and more reliable techniques to collect the data." This principle has been a prime motivation for attempting this compilation and critical evaluation of the drug-relevant pK_a literature.

In the northern summer of 1988, discretionary funding from SmithKline Beecham allowed me to make a start on this project by hiring Jimy Gillette, a senior pharmacy student at the University of Florida, to begin data collection. Since then, I have checked through the hundreds of citations collected by Jimy, and the many more citations that I subsequently collected myself.

It is inevitable that a compilation of this size (which cannot claim to be exhaustive) will have errors of omission and commission. These are the fault of the author only. Readers are encouraged to draw these to the attention of either the author or the editor, so that information in future editions or supplements can be made more reliable and more complete.

Richard Prankerd
Victorian College of Pharmacy
381 Royal Parade
VIC 3052
Australia
richard.prankerd@vcp.monash.edu.au

REFERENCES

[1] W. Ritschel, pKa values and some clinical applications, in *Perspectives in Clinical Pharmacy* (eds. D. Francke and H. Whitney), 1st edn., Drug Intelligence Publications, Hamilton, IL, 1972, pp. 325–367.

[2] S. Smith and M. Rawlins, *Appendix C. Variability in human drug response*, 1st edn., Butterworths, London, 1973, pp. 154–165.

[3] T. Speight, *Avery's Drug Treatment*, 3rd edn., Publishing Sciences Group, Inc., Littleton, MA, 1987, pp. 1352–1380.

[4] J. Hoover, *Dispensing of Medication*, 8th edn., Mack Publishing, Easton, PA, 1976, p. 230, 247, 418–426, 468–634.

[5] D. Newton and R. Kluza, pKa values of medicinal compounds in pharmacy practice, *Drug Intell. Clin. Pharm*, 1978, **12**, 546–554.

[6] P. Craig, Compendium of Drugs, in *Comprehensive Medicinal Chemistry* (eds. C. Hansch, P. Sammes and J. Taylor), 1st edn., Pergamon Press, New York, 1990, pp. 237–965.

[7] D. Williams, Appendix A-1, in *Principles of Medicinal Chemistry* (eds. W. Foye, T. Lemke and D. Williams), 4th edn., Williams and Wilkins, Baltimore, MA, 1995, pp. 948–961.

[8] J. Delgado, and W. Remers, pKas of drugs and reference compounds, in *Wilson and Gisvold's Textbook of Organic Medicinal and Pharmaceutical Chemistry* (eds. J. Delgado and W. Remers), 8th edn., Lippincott-Raven, Philadelphia, PA, 2000.

[9] G. Kortum, W. Vogel, and K. Andrussow, *Dissociation Constants of Organic Acids in Aqueous Solution*, 1st edn., Butterworth, London, 1961.

[10] D. Perrin, *Dissociation Constants of Organic Bases in Aqueous Solution*, 1st edn., Butterworths, London, 1965.

[11] E. Serjeant and B. Dempsey, *Ionisation Constants of Organic Acids in Aqueous Solution*, 1st edn., Pergamon Press, Oxford, New York, 1979.

[12] D. Perrin, *Dissociation Constants of Weak Bases in Aqueous Solution*, 1st edn., Butterworths, London, 1972.

Critical Compilation of pK_a Values for Pharmaceutical Substances

Richard J. Prankerd*

Contents

1. Introduction 1
   1.1. The IUPAC pK_a compilations 2
   1.2. Limitations of current pharmaceuticals pK_a compilations 4
2. Evaluation of Methods for pK_a Determination 7
   2.1. Inherent precision 7
   2.2. Physicochemical factors controlling the accuracy and precision of pK_a values 8
   2.3. Validation of data quality 23
3. pK_a Values from Computer Programs 24
4. Summary of the Data Compilation 26
References 28

1. INTRODUCTION

There are numerous compilations of pK_a values'' in the physical chemistry literature, including several for pharmaceutically relevant organic weak acids and bases [1–8]. These are complemented by further compilations of pharmaceutically relevant physicochemical data such as partition coefficients, solubilities, and reaction rate constants [6]. At the same time, other pharmaceutically interesting phenomena have not yet received the attention they deserve, such as the detailed substrate specificity and kinetics of endogenous enzyme systems (e.g., esterases and phosphatases), which are relevant to the rational design of prodrugs and the predictability of their bioconversion to active drug [13].

Along with solubilities, partition coefficients, and reaction rates, pK_a values are the most important physicochemical properties of drugs and the excipients used

* Victorian College of Pharmacy, A Faculty of Monash University, Parkville VIC 3052, Australia

Profiles of Drug Substances, Excipients, and Related Methodology, Volume 33
ISSN 0099-5428, DOI: 10.1016/S0099-5428(07)33001-3

to formulate them into useful medicines. The determination of pK_a values is typically discussed either first or second (after solubility) in preformulation textbooks. The extent of ionization (overall state of charge) for a dissolved drug is a function of its intrinsic pK_a value(s) and the pH value of the solution [14]. The extent of ionization for a drug can control its solubility, dissolution rate, reaction kinetics, complexation with drug carriers (e.g., cyclodextrins), absorption across biological membranes, distribution to the site of action, renal elimination, metabolism, protein binding, or receptor interactions. Clearly, research in many aspects of the drug sciences requires knowledge and use of drug pK_a values. When an investigator chooses to make use of tabulated or compiled physicochemical constants, reliably assessed data is required to account best for physicochemical or biopharmaceutical results that are dependent on the relationship between pH and pK_a.

1.1. The IUPAC pK_a compilations

The critical assessment of data quality is one of the major features of the seminal group of pK_a compilations [9, 10] for weak organic acids and bases sponsored by the International Union of Pure and Applied Chemistry (IUPAC). These compilations feature assessment of data quality, based on aspects of pK_a measurement such as the mathematical definition chosen to calculate the value from the raw data, choice of the experimental method, and the degree to which technical refinements have been applied. The pK_a values are described in these compilations as "very reliable" (pK_a error $< \pm 0.005$), "reliable" (pK_a error ± 0.005 to ± 0.02), "approximate" (pK_a error ± 0.02 to ± 0.04), and "uncertain" (pK_a error $> \pm 0.04$). These error criteria may seem overly restrictive to some readers. However, it must always be remembered that pK_a values are logarithmic representations of the acid dissociation constant K_a. These error criteria, when applied to the dissociation constants that they represent, are substantially larger than first appears. For example, a decrease in pH of 0.3 unit (which many would regard as small) actually represents a twofold increase in hydronium ion activity. This increase in a_{H^+} may double a reaction rate, and thereby halve the corresponding shelf life. A variation in assigned pK_a of 0.05 unit (trivial to many) represents a change in the extent of ionization for a weak acid or base of 10.9% at a fixed pH. The resulting change in polarity may influence solubility or partitioning properties to a similar extent, and which approaches comparison with the experimental errors in careful pharmacokinetic or biopharmaceutical work. Attention was drawn to this issue with respect to partitioning behavior [15] many years ago. The variation in pK_a of $< \pm 0.005$ which is required to qualify a result for status of "very reliable" corresponds to an uncertainty in the dissociation constant, K_a, of $<1\%$, which seems reasonable for the most careful physicochemical work. It is most regrettable that very few authors in the pharmaceutical sciences [16] have ever seen fit to include comparisons of their data with the IUPAC reliability criteria.

Differences large enough to correspond to the "uncertain" classification may substantially alter aqueous solubilities or partition coefficients, and certainly can lead to gross errors in derived thermodynamic quantities where pK_a data are measured as a function of temperature. In one example, pK_a errors of 0.05 at most led to the derivation of ΔH^o and ΔS^o values for ionization of 5, 5-diethylbarbituric acid [17] that varied with temperature in the wrong direction, compared to the best data [18] in the literature. In other words, the temperature dependences of these quantities were shown [19] to have the wrong sign, because of apparently quite small errors in the pK_a values. In the present compilation, the criterion for "uncertain" has been widened to a pK_a error $> \pm 0.06$. This is equivalent to variability in the dissociation constant of about 15%.

The IUPAC dissociation constant compilations [9–12] are not focused on drug substances, although they do include some pharmaceuticals, such as morphine and other opiates, acetanilide, some barbiturates, vitamins, antibiotics, and alkaloids. In a sense, this is a limitation of these compilations. A set of cross-referencing indexes to these compilations would be a useful tool.

The IUPAC compilations are entirely confined to pK_a values of weak organic acids and bases measured in aqueous solutions only. As those compilations predated the Yasuda-Shedlovsky [20, 21] extrapolation procedure, any pK_a values derived from simple extrapolation of organic cosolvent composition to 100% water would have to be assigned "uncertain" reliability. The world of drug sciences has not been able to avoid the use of aqueous-organic cosolvent mixtures for pK_a determination, as well as numerous other purposes. This is due to the low aqueous solubilities often seen for drug-like molecules, which are often a function of the lipophilicity range needed to ensure adequate passive transfer across biological membranes. The apparent pK_a values that result from measurements in aqueous-organic solvent mixtures are generally not able to be precisely converted to the values that would result in water alone, due to: (a) the nonlinearity of direct relationships between apparent pK_a and solvent composition, especially for bases at low organic cosolvent content; and (b) the lack of values obtained in solvent mixtures with a sufficient number of different compositions (however, see Section 2.2.4). Wider application (and possible further refinement) of the Yasuda-Shedlovsky extrapolation for correlating apparent pK_a values with solvent composition [20, 21] may go some way toward solving this problem, although the method at present does not consistently lead to precisely the same results for compounds with sufficient water solubility to be used as validation controls. Hence, a compilation of drug pK_a values must, at the least, draw attention to the use of different cosolvent combinations, where this has occurred.

It has also been noted [22] that the use of pK_a values measured in aqueous media (or extrapolated to the same through the Yasuda-Shedlovsky procedure) are not ideal in accounting for enzyme-substrate interactions, where binding to an active site involves ionizable functional groups. The active sites of enzymes and receptors, along with biological membranes, tend to be less polar in nature than water, and thus the relevant pK_a value is more likely to be one that pertains to a partially aqueous solvent. Theoretical treatments for this situation were described [23] long ago by Bjerrum, Wynne-Jones, and Kirkwood and Westheimer, and further applied in the exceedingly careful physicochemical work of Ives and his school [24–26]. Hence, the use of partially aqueous cosolvent mixtures has a physicochemical–pharmacological legitimacy, although one that has so far received very little theoretical attention in the pharmaceutical literature compared to the mere convenience of keeping acidic or basic solutes in solution for the purpose of pK_a measurement. It must be noted, however, that pH meter calibration for use in partially aqueous solutions is more complex [20] than for purely aqueous solutions.

1.2. Limitations of current pharmaceuticals pK_a compilations

A major limitation of all current compilations of pharmaceutically relevant pK_a values is their incomplete coverage of the literature. First, there is a large number of new drugs (about 330) that has been commercialized since the early 1980s. Few of these have had pK_a values reported. Most compilations deal with older drugs, many of them no longer in therapeutic use. Rarely do any of the current compilations in the drug literature cite more than one value for each compound. For example, for glibenclamide, a secondary source cited only one of the three values reported in the original paper [27], without acknowledging the existence of the other two values. All three values were substantially different to each other, yet there was no apparent reason for citing one value over the others. Later work [28] suggested that one of the uncited values was the more accurate. Needless to say, it is recognized that a compilation goes out of date the moment that the first new piece of relevant data is entered into the literature.

A second limitation of existing compilations is overreliance on requoting the secondary literature, so that experimenters are unable to ascertain the quality of pK_a data that they wish to use. Also, there is the risk of errors creeping into data sets when they are copied and recopied. In the course of finding raw data for this compilation, it was often necessary to track sequentially through two or three secondary literature citations to find the original source, from which experimental validation information might be obtained. In one compilation, no literature references were given at all. Frequently, important details had been omitted in the transmission. For example, a pK_a value $= 5.93$ cited for atropine looked more like a pK_b value, until discovery of the primary reference [29] showed that the original data had been obtained from titrations in glacial acetic acid as solvent, rather than water. This vital fact was omitted in the secondary source. Occasionally, the pharmacology literature uses the symbol "pK_a" to mean the negative log of the affinity of a drug for a receptor, which is not at all the same as the acid–base behavior of the compound. The secondary pK_a compilations have sometimes quoted these affinity values as if they referred to acid–base equilibria. For procaine, three different pK_a values for the tertiary amine were found in seven secondary sources [5, 6, 8, 30–33], none of which had cited a primary reference. In a few cases, different values from the same primary source were cited by two separate secondary sources. One reason for the omission of key details such as temperature is that some data was originally published in hard-to-access foreign language journals and any information at all is only readily available through Chemical Abstracts.

It was disappointing to find a few extremely old and inaccurate data quoted in very recent secondary compilations, for example, one textbook cited a value for

5,5-dimethylbarbituric acid that had been first published [34] over 100 years ago, but which has also been requoted in other secondary literature [35]. The value given (pK_a = 7.1) is highly inaccurate and was superseded by a more reliable value (pK_a = 8.51) [36] over 25 years ago. Most compilations continue to list pK_a values for numerous other barbituric acid derivatives from a large study [37] published in 1940. That study had a flawed experimental design which resulted in low values for most compounds, when compared to validated, more recent studies [36, 38]. One published secondary compilation was acknowledged to be the result of a library data collection exercise given to undergraduate pharmacy students. The collected references were later checked randomly, but not exhaustively, by a graduate student. While that compilation is quite extensive (more than 700 total citations), ~68% were from the secondary literature. The compilation of the present database has as one of its objectives the citation of original sources wherever possible. Only one secondary source is regarded as sufficiently careful in its handling of the original data, and that is the series of compilations prepared under the auspices of IUPAC [9–12]. Even so, virtually all citations from the IUPAC compilations have been rechecked with the original literature, with one or two revisions necessary (e.g., see no. 1988 in the database).

A third limitation of all previous compilations is the failure to critically review data quality. It was regrettable to note a number of examples where the pK_b values for bases had been misquoted as if they were pK_a values due to failure to compare the numerical values with those expected for the functional groups to which they had been assigned. A further problem with reported pK_b values is that when secondary compilers did convert them to pK_a values (using pK_a + pK_b = pK_w) the value for pK_w at 25 °C (14.008) was always used, even when the temperature for the pK_b measurement ($\neq$ 25 °C) was available. This avoidable error led to discrepancies of several tenths of a pK_a unit over the commonly employed temperature range (15–37 °C) for single measurements.

There is also the important issue of data consistency. Many drugs and excipients have had their pK_a values measured by different methods with very good agreement, provided that all experimental conditions were maintained constant. Examples included atenolol [9.60 ± 0.04; potentiometric, partitioning ,and capillary electrophoresis (CE) studies in different laboratories], barbital (5,5-diethylbarbituric acid) (7.98 ± 0.01; electrometric, potentiometric, and spectrophotometric studies in different laboratories), benzoic acid (4.205 ± 0.015; electrometric, potentiometric, spectrophotometric, and conductance measurements in many separate studies), ephedrine (9.63 ± 0.05; five potentiometric or spectrophotometric studies in different laboratories), isonicotinic acid (pK_1 = 1.77 ± 0.07, pK_2 = 4.90 ± 0.06; five potentiometric or spectrophotometric studies in different laboratories), nicotinic acid (pK_1 = 2.07 ± 0.07, pK_2 = 4.79 ± 0.04; six potentiometric, spectrophotometric, or capillary zone electrophoresis studies in different laboratories), phenobarbital (5-ethyl-5-phenylbarbituric acid) (7.48 ± 0.02; potentiometric and spectrophotometric studies from several workers in multiple laboratories), nimesulide (6.51 ± 0.05; potentiometric and spectrophotometric studies in multiple laboratories), and chlorthalidone (9.35, potentiometric; 9.36, spectrophotometric/solubility-pH). The recent comparative studies of Takacs-Novak, Avdeef, and collaborators [20, 21, 39, 40] have gone some way toward increasing the stock of drug substances where agreement of replicated pK_a values to <0.05 has been found. Existing compilations of drug pK_a values almost invariably list only a single value for each compound. They do not comment on the conditions under which the value was obtained, or the quality of the data.

Conversely, there are many other drugs that have had their pK_a values reported more than once (in a few cases, eight or more times, e.g., ibuprofen, propranolol, and quinine), frequently by the same method, but often with discordant results. Examples include lidocaine (7.89 $\pm$ 0.07, from 6 independent potentiometric studies as long ago as 1948; but 7.18 from a recent conductance study), propranolol (11 studies giving values in the range 9.23–9.72), clofazimine (3 studies in the range 8.37–9.11), famotidine (6.76 by spectrophotometry, 6.89 by partitioning, and 6.98 by solubility–pH dependence), ibuprofen (8 studies giving values in the range 4.1–5.3), phenylbutazone (6 studies reporting values from 4.33 to 5.47), glibenclamide (5.3 by potentiometry, 6.3 by partitioning, and 6.8 by solubility–pH dependence), and also some of the fluoroquinolones. Some of these differences are due to differences in conditions, such as ionic strength or temperature, but others are experimental error. These variations are very rarely reported in previous secondary compilations.

Failure to take into account the effects of temperature, solvent composition, or ionic strength is usually responsible for differences between repeated measures of a drug pK_a value. This point was clearly made in a thorough report [41] on the pK_a values of numerous macrolide antibiotics. However, the large difference quoted above for lidocaine may be due to the failure of assumptions involved in an otherwise careful conductance study (pK_a = 7.18), the result of which is seriously at variance with six earlier potentiometric studies (pK_a = 7.89 $\pm$ 0.07, I = 0.00–0.15 M). It has been expressly stated [42] that the normal conductance method is unsuitable for acids with $K_a < 10^{-5}$ (i.e., pK_a > 5).

Many of the drug pK_a values recorded in earlier volumes of the monograph series entitled *Analytical Profiles of Drug Substances* (now titled *Analytical Profiles of Drugs and Excipients*) were reported with little or no experimental detail. As well, there may be no primary references, an unsupported reference to secondary literature, or only a personal communication to identify the source of the data. Experimental details such as temperature, ionic strength, or solvent composition are critical for the assessment of data quality. These omissions are a matter of regret, especially where the pK_a value has been reported only once. Fortunately, more recent volumes in this series have begun to address these deficiencies. The same reservations regarding lack of experimental information also apply to the values listed in other earlier compilations.

The overriding aim of the present database is an attempt to make available as many as possible of the original sources, conditions, and methods for pK_a values of pharmaceutical interest that are in the literature. Then the users of such data can judge for themselves, as much as is possible, the reliability of these numbers, without having to search out original publications, many of which are becoming increasingly difficult to access. It is surprising how many drug pK_a values rest on unconvincing experimental work. The present emphasis on high-throughput screening for potential therapeutic candidates means that there is an increasing demand for fundamental physicochemical quantities. These quantities are needed for the optimization of drug-like properties in lead compounds. This demand is presently being filled at least partly by computerized (*in silico*) predictions. As the quality of the outputs from such predictive programs depend on the quality of the input data used for algorithm development or neural network training purposes, it is best to use input data that is as reliable as possible.

2. EVALUATION OF METHODS FOR pK_a DETERMINATION

A number of methods have been used for the experimental measurement of pK_a values and closely associated quantities, such as pH. It is not the intent of this overview to describe in detail the theoretical and practical aspects of these methods, which have been satisfactorily described elsewhere [42–47]. Rather, the focus of the present work is to apply an understanding of these experimental methods to assessment of the quality of the resulting data. Research workers who make use of the data in this compilation can then have confidence that the values which they cite are as meaningful as possible.

2.1. Inherent precision

The pK_a compilations gathered under the auspices of the IUPAC sometimes reference experimental data of the highest quality. Measurement of such data employs sophisticated methods with the utmost refinements to both experimental techniques and calculational tools. Highly accurate methods are largely restricted to conductance methods [48, 49] (reliable to ± 0.0001 pK unit or better, for suitable compounds) and electrochemical cells (EMF method using hydrogen electrodes) without liquid junction potentials [50, 51] (reliable to ± 0.001 pK unit or better). These fundamental methods have been very rarely used to obtain ionization data for pharmaceutically relevant organic acids and bases. Most pK_a measurements on pharmaceutical substances are based on relationships between the measured solution pH and a measured physicochemical quantity such as added titrant concentration [47], solubility [47, 52, 53], spectrophotometric absorbance [54–57], partition coefficient, optical rotation, nuclear magnetic resonance data (chemical shifts, coupling constants), or fluorescence intensity. The use of pH values immediately limits the expected accuracy and precision of the measurements. As a reader of the published physical pharmacy literature and as a reviewer of submitted manuscripts, I am frequently amazed by the extent to which the limitations and pitfalls of pH measurements are ignored or forgotten. A cogent review [58] of the development of pH measurements up to the time of the work done at the former US National Bureau of Standards (NBS; now the National Institute of Science and Technology, NIST) has been given under the appropriate title of "Use and Abuse of pH Measurements."

The current definition of pH is clearly stated in a recent publication [59] from the IUPAC. This definition is largely based on work carried out in the 1950s and 1960s by Roger Bates and his collaborators at the NBS.

Although the definition of pH is well known [Eq. (1)], the limitations of this definition are not so widely recognized or applied in the daily practice of the pharmaceutical sciences:

$$\mathrm{pH} = -\log a_{\mathrm{H}^+} \tag{1}$$

where a_{H^+} is the hydronium ion (colloquially, hydrogen ion) activity. The hydronium ion activity is related to the hydronium ion concentration by a mean ionic activity coefficient, $f_\pm$, Eq. (2):

$$a_{\mathrm{H}^+} = [\mathrm{H}^+] f_\pm \tag{2}$$

The accuracy and usefulness of these relationships is dependent on the mathematical definition of $f_\pm$. This coefficient is estimated from various modifications of the Debye-Hückel equation [23, 44, 45] [Eqs. (13–17); Section 2.2.5]. The definition given by Eq. (1) is certified by British Standard no. 2586 (a revision of Br. Std. no. 1647) to be accurate to ±0.02 pH unit in the temperature region 0 to 95 °C. The British Standard [60] is stated [45] to be ''consistent in nearly all respects with the NBS conventional scale....''

The imprecision of ±0.02 in the definition of pH is made up of approximately equal contributions from two quite different sources. One source of uncertainty is the variability in the liquid junction potential [45, 59, 61] under different conditions of solution composition and dynamics. This is an intrinsic feature of the glass electrode–reference electrode combination used for pH measurements. This uncertainty does not arise in conductance measurements, and can be neglected in measurements of hydronium ion activity only by the use of reversible electrochemical (electrometric) cells that are constructed without liquid junction potentials. These approaches require considerable technical expertise and attention to detail in order to obtain results with maximal accuracy and precision.

The other source of uncertainty in the definition of pH arises from assumptions that must be made in the use of various modifications of the Debye-Hückel equation [44, 45, 59] which is used to estimate the mean ionic activity coefficient [Eqs. (13–17); Section 2.2.5]. The geometric mean ionic activity coefficient ($f_\pm = \sqrt{(f_+ f_-)}$) is used, as it is relatively easy to estimate with reasonable precision, whereas current methods for estimation of individual ionic activity coefficients still require significant computational effort, even for small inorganic ions [62]. The mean ionic activity coefficient [see Eqs. (13–17) in Section 2.2.5] is required to estimate hydronium ion concentrations from measured pH values. The hydronium ion concentrations are required, in turn, for calculations of the pK_a values.

2.2. Physicochemical factors controlling the accuracy and precision of pK_a values

Control of a number of physicochemical factors is critical to achieving maximal precision and accuracy in measured pK_a values. These include the choice of experimental method, pH meter calibration, temperature control, solvent composition, ionic strength, absence of atmospheric CO_2 contamination, estimation methods for activity coefficients, and chemical stability.

2.2.1. Choice of experimental method

Experimental methods differ in the quality of data which may reasonably be expected. The conductance method is potentially the most accurate and precise of all methods (±0.0001 pK_a unit or better), as demonstrated by the series of extremely careful pK_a-temperature dependence studies [24–26] reported by Ives and coworkers on several alkyl substituted cyanoacetic and malonic acids. These papers are models of care and attention to detail. However, the conductance method is only really suitable [42] for acids and conjugate acids with pK_a values of <5, or for very weak acids or bases as their conjugates. Very recent papers from Apelblat and his group [63–65] provide good examples of the use of the conductance method for pharmaceuticals. The electromotive force (EMF) method using reversible electrometric cells without liquid junction potentials is almost as good as the conductance method, with accuracies of ±0.001 pK_a unit or better. A number of important pK_a-temperature dependence studies [18, 66, 67] from the former US NBS used this method, and thereby laid the experimental basis for currently used pH calibration standards. Some NBS data, primarily buffers with pharmaceutical relevance, are cited in the tables forming Appendices A and B.

Methods based on pH measurements are very widely used, although they can only give results that are theoretically accurate to ±0.02 pK_a units at best. These are generally suitable for the needs of pharmaceutical sciences research.

The most common primary equations relating pK_a to pH and other solution properties are given in Eqs. (3–8):

Potentiometric titration (weak acids):

$$pK_a = pH + \log\left\{\frac{([Y] - [H^+] - [K^+] + [OH^-])}{([H^+] + [K^+] - [OH^-])}\right\} - \log y_{\pm(i)} \tag{3}$$

where [Y] is the total molar concentration of the protonated and deprotonated forms of the weak acid, [H^+] and [OH^-] are the molar concentrations of hydronium and hydroxyl ions [calculated from the measured pH value and the value for the water autoprotolysis constant (pK_w) at the temperature of measurement], [K^+] is the concentration of potassium ions added as standard potassium hydroxide during the potentiometric titration, and $y_{\pm(i)}$ is the mean ionic activity coefficient, as defined in Eqs. (13–17). Potassium hydroxide has generally been preferred over sodium hydroxide as the titrant for weak acids, as glass electrodes have always had negligible potassium ion error, although the effects of sodium ion concentrations on electrode response at high pH have been reduced through use of better materials. Inclusion of the corrections for [H^+], [OH^-], and $y_{\pm(i)}$ are essential for the best results by this approach.

Potentiometric titration (weak bases):

$$pK_a = pH + \log\left\{\frac{([OH^-] + [X^-] - [H^+])}{([Y] - [OH^-] - [X^-] + [H^+])}\right\} + \log y_{\pm(i)} \tag{4}$$

where [Y] is the total concentration of the protonated and deprotonated forms of the weak base, [X^-] is the concentration of counterions (usually chloride or perchlorate) added as standard strong acid titrant during the potentiometric titration and the other quantities are as described above for weak acids, Eq. (3). Inclusion of the corrections for $[H^+]$, $[OH^-]$, and $y_{\pm(i)}$ are essential for the best results.

Complete solution of Eqs. (3) and (4) requires an iterative approach, as the $[H^+]$ and $[OH^-]$ terms cannot be precisely calculated without an accurate estimate for $\log y_{\pm(i)}$, while assignment of the latter requires an accurate assessment of the total solution ionic strength, which includes contributions from the former terms. It is initially assumed that $\log y_{\pm(i)} = 0$, allowing initial estimates to be made of the hydronium and hydroxyl ion concentrations. These are then used to calculate an initial estimate for the ionic strength, which is substituted into the appropriate form of the Debye-Hückel equation [Eqs. (13–17)], giving a new estimate of the value for $\log y_{\pm(i)}$. The estimates are cycled iteratively until convergence takes place, typically requiring not more than five cycles for satisfactory results. It is important to evaluate all terms in order to obtain the most accurate final pK_a value.

Spectrophotometry:

$$\mathrm{p}K' = \mathrm{pH} + \log\left(\frac{[A_1 - A]}{[A - A_0]}\right) + \mathbf{A}|Z_+Z_-|\left(\frac{\sqrt{I}}{1+\sqrt{I}}\right) \tag{5}$$

where pK' is the apparent pK_a value at a fixed ionic strength; A_0, A_1, and A are the spectrophotometric absorbances for precisely equimolar concentrations of the solute when fully protonated, fully deprotonated, and partially protonated (at the corresponding pH value), respectively; Z_+ and Z_- are the overall charges on the cation (hydronium ion or protonated base) and anion (deprotonated acid or hydroxyl ion), respectively; $\mathbf{A}$ is a constant dependent only on the temperature and dielectric constant of the solvent and I is the solution ionic strength; and

$$\mathrm{p}K' = \mathrm{p}K_a + bI \tag{6}$$

where the thermodynamic pK_a value is found by plotting the apparent values (pK') as a function of I. In this approach, the Guggenheim modification [Eq. (17)] of the Debye-Hückel equation has been incorporated into Eq. (6), as this is the most precise for a wide range of I, and is relatively easy to set up experimentally.

Solubility–pH dependence:

$$\mathrm{p}K_a = \mathrm{pH} - \log\left(\frac{S - S_0}{S_0}\right)(\text{weak acids}) \tag{7}$$

where S is the solubility of the weak acid at a corresponding pH value, while S_0 is the intrinsic solubility of the protonated free acid:

$$\mathrm{p}K_a = \mathrm{pH} + \log\left(\frac{S - S_0}{S_0}\right)(\text{weak bases}) \tag{8}$$

where S is the solubility of the weak base at a corresponding pH value, while S_0 is the intrinsic solubility of the deprotonated free base. The equations for the solubility–pH dependence method, first proposed in 1945 by Krebs and Speakman [68] and later developed by Zimmermann [53], based on transformations amenable to linear least squares regression analysis, are very dependent on the accuracy of the experimental solubility-pH data, especially the S_0 value. It was later shown by Lewis [52] that the use of weighted linear least squares was successful in ameliorating the effects of a value for S_0 which lacks accuracy.

Equations (3–6) for the potentiometric and spectrophotometric methods will provide thermodynamic pK_a values. For the solubility–pH dependence method [Eqs. (7–8)], the values obtained are apparent values (pK_a'), which are relevant to the ionic strength (I) of the aqueous buffers used to fix the pH value for each solution. If the ionic strength of each buffer solution is controlled or assessed, then the apparent pK_a value can be corrected to a thermodynamic value, using an activity coefficient from one of the Debye-Hückel equations (Section 2.2.5). If the solubility–pH dependence is measured in several buffer systems, each with a different ionic strength, then the Guggenheim approach can be used to correct the result to zero ionic strength [Eq. (17)].

The most common methods used for pharmaceutical pK_a values are based on pH measurements, Eqs. (3–8). Thus, they cannot be interpreted with greater accuracy than ±0.02 pK_a unit [see the definition in Section 2.1, Eq. (1)]. This level of precision and accuracy should always be the aim in determining pK_a values for inclusion in the drug sciences literature. Potentiometric titrations [Eqs. (3–4)] are often performed with this level of accuracy, primarily for compounds with either a single ionization step or for multiple ionizations with >4 log units between the pK_a values. The careful use of precise pH meters (e.g., the series of Beckman Research models, or the corresponding Radiometer, Orion, or Metrohm instruments) for the determination of pH data means that reproducibility for replicate measurements may be rather better than ±0.02. In the author's experience, these instruments may be calibrated with a reproducibility of ±0.002 pH unit, which can be maintained (with proper temperature control and exclusion of CO_2) for at least 8 h. This does not imply accuracy of ±0.002 pH unit, which is not possible according to the current definition of pH. Spectrophotometric [Eq. (5)] and solubility–pH dependence [Eqs. (7–8)] methods are potentially capable of similar accuracy, but often do not give results better than ±0.05 pK_a unit, due to the inevitable inclusion of additional sources of error from the absorbance or concentration measurements.

A difficulty in the use of glass pH electrodes, even in aqueous solutions, but especially in partly aqueous solutions, is that they are relatively slow to respond to changes in hydronium ion activity. Equilibration times of at least 1 min are generally needed for each data point in a titration that might have a total of more than 20 points. This has two consequences: (a) potentiometric titrations are relatively time consuming; and thus (b) workers can become too impatient to wait for the electrode response to fully stabilize, leading to further measurement errors. Automated measurement, such as with Sirius® autotitration equipment, is able to measure about 25 pK_a values per day. Faster alternatives to the pH electrode are needed to rapidly measure pK_a values.

Besides pH issues, the only other sources of experimental error in potentiometric titrations are inaccuracies in the concentrations of the reagents, which can be minimized by careful weighing of properly stored materials with the highest purity, and volumetric errors, which can be compensated (with errors not exceeding $\pm 0.1\%$) by use of properly calibrated A Grade glassware, micrometer syringes (e.g., Agla or Hamilton brands), or autotitrators (e.g., Metrohm). Where multiple overlapping ionization steps occur, the titration end points cannot be determined analytically and must be estimated by some type of nonlinear regression analysis, for example, the four ionization constants [69] for desferrioxamine. Inevitably, some loss of precision will occur in such data fitting.

It has been stated [42] that the spectrophotometric method (or other spectroscopic methods) can be as precise as the EMF method based on electrochemical cells without liquid junction potentials. However, for this to be true, the pH term in Eq. (5) must not be measured with a pH meter, but replaced by a similar term based on the hydronium ion activities, $a_{H_3O^+}$), from cells without liquid junction potential. This approach was used in earlier years, for example, the work of Robinson [70] (see Vanillin, no. 1492 in the database) and of Bates and Schwarzenbach [71] on phenols. For the usual spectrophotometric method based on conventional pH measurements, additional factors which control overall accuracy include:

a. The differentiation between the spectra for the protonated and deprotonated forms. Where the protonated and deprotonated forms of the compound have very similar spectra, for example, aliphatic amines, the lack of dynamic range for *A* ensures there will be large errors in the measured absorbance ratios in Eq. (5)
b. Temperature or pH differences between the solutions in the spectrometer cuvette and in measured pH sample. The availability of miniaturized pH and temperature probes means that all absorbance and pH measurements should be performed in the spectrophotometer cuvette where possible
c. The potential for interference from a second ionization step (i.e., completeness of the isolation of spectra for the pure protonated and deprotonated forms from other ionization steps in the molecule). Orthogonal methods of data analysis have been applied in efforts to solve this problem. However, results suggest [72] that so far, the accuracy of the resulting pK_a values is not as good as desired. Another approach to this problem was employed in the spectrophotometric pK_a determinations for isonicotinic acid and 1,3-bis[(2-pyridyl)methyleneamino] guanidine [73] after a detailed analysis of the incurred errors.

It should be noted that the spectrophotometric method normally uses the changes in molecular absorbance with ionization (dA/dpH method) as the raw data for pK_a measurement. It is also possible to use changes in the maximum wavelength for absorption versus pH ($d\lambda/d$pH method), although this has been very rarely done, due to the requirement for measurement of a significant change in the dependent variable (see Ibuprofen, no. 662 in Appendix A)

For the solubility–pH dependence method [Eqs. (7–8)], other factors controlling accuracy (apart from the pH measurement) include: (a) the completeness of equilibration for the saturated solutions; (b) the care taken in separating the undissolved solid from the saturated solutions; and (c) the accuracy of the assay method used for the quantitative analysis of the saturated solutions. Completeness of equilibration is shown by constant total solubility as a function of time. The variation in "constant" solubility should be as small as possible, preferably not more than $\pm 1\%$, for good quality work. Separation of the excess solid from the saturated solution may be performed either by filtration or centrifugation. Both methods have their disadvantages. Filtration is more difficult to perform at temperatures that are increasingly remote from ambient. Filters should be presaturated with the analyte to counter the possibility of errors from adsorption, especially for very dilute solutions of poorly water-soluble compounds in their neutral (free acid or free base) forms. Conversely, access to a centrifuge with stable control over the desired temperature range is not always possible. The choice of separation method is a matter of experience and preference, modulated by the specific compound or series of compounds. Assay methods for saturated solutions must be properly validated and as accurate as possible, especially for the solution that defines the value for S_0, the solubility of the protonated free acid or deprotonated free base. This question has been addressed by Lewis [52].

Less commonly used measurement techniques include the pH dependence of partition coefficients [74], fluorescence spectra [75], nuclear magnetic resonance chemical shifts or coupling constants, HPLC or CE retention volumes [76, 77], and the dependence of reaction rates for ionizable substrates on pH (also called kinetic methods). Kinetic methods were amongst the earliest methods to be used for pK_a determination. In some cases, they may be the only feasible method, for example, extremely weak acids (p$K_a > 12$) without suitable absorption spectra. The difficulty with kinetic methods is that they may not actually measure the pK_a value for the substrate, but that of the reaction transition state. If the electronic configuration of the transition state is similar to that of the reactant (early transition state), then the kinetic pK_a may be quite close to the equilibrium value. However, if the transition state more nearly approximates the reaction products (late transition state), then the kinetic pK_a value may bear little resemblance to that for the reactant. This explanation might account for the lack of agreement between the first apparent kinetic (p$K' = 4.0$) and equilibrium (p$K' = 8.6$) pK_a values for hydrochlorothiazide at 60 °C [78]. Similar restrictions may be placed on the use of pK_a values from the pH dependence of fluorescence spectra, as these reflect the properties of the first excited state of the molecule rather than its ground state [75].

Any chemical property which has a sigmoidal pH-dependency could be used to measure apparent pK_a values, provided that the protonated and deprotonated forms have significantly different values for that property, Eq. (9):

$$\mathrm{p}K' = \mathrm{pH} + \log\left(\frac{[A_1 - A]}{[A - A_0]}\right) \tag{9}$$

where A_0 and A_1 represent the property for the protonated and deprotonated forms, respectively, while A represents the property for the partially protonated form at a specified pH value. Equation (9) is a simple restatement of the

Henderson-Hasselbalch equation. One such property is the retention time (or retention volume) of the compound on a reversed phase-high performance liquid chromatography (RP-HPLC) column, provided that a suitable means for detection of the compound exists. The retention time (volume) is a function of the pH of the buffered aqueous mobile phase and should be substantially smaller for the more polar ionized form, compared to the less polar unionized form. Judicious selection of the column packing type and column length needs to be exercised, as manipulation of retention times by the addition of organic mobile phase modifiers would make the resulting apparent pK_a values require further adjustment with a Yasuda-Shedlovsky correction (Section 2.2.4) to be applied for comparison with purely aqueous systems. Advantages of this approach to pK_a measurement, compared to more conventional methods (e.g., potentiometry, spectrophotometry, and solubility/pH techniques), are quite significant and include:

a. The need to use only very small amounts of the compound
b. The ability to use compounds that are not of high purity
c. The ability to use compounds that are not especially stable

In combination with modern micro-pH flow cell electrodes (volume as low as 50 μl of solution) placed close to the detector output of an HPLC system, pK_a values for multiple congeners can be assessed simultaneously, thus providing at least some way for experimentally coping with the vastly increased number of possible drug candidates arising from combinatorial chemistry libraries and high-throughput screening. This should allow diminished reliance on the admittedly less accurate linear free energy or artificial neural network (ANN) estimation methods for pK_a values of new compounds. As with all pK_a measurement techniques, the ability of the method to deal with compounds having multiple ionizing groups is a function of the separation in their pK_a values. It is unfortunate that one of the few reports of this method [79] gave pK_a values for standard substances (e.g., benzoic acid) that were significantly at variance with the best values in the literature, even after correcting for ionic strengths. Use could be made of the Yasuda-Shedlovsky procedure to account for the effects of partially aqueous mobile phases containing common organic modifiers such as methanol and acetonitrile. This requires careful calibration of the pH electrode [20].

A closely related approach uses CE coupled to mass spectrometry (MS) with volatile buffers as the carrier phase [76] or with high sensitivity amperometric detection [77]. CE methodology is favored for high-throughput pK_a screening, as it can elute a set number of compounds faster than a similar number of compounds on an HPLC column. By this means, the pK_a values for 50 compounds could be estimated in 150 min. The high sensitivity of MS detection is favored over UV (whether separation is by CE or HPLC) when the aqueous solubilities of the analytes are low. Furthermore, MS detection based on single ion currents can deal explicitly with problems such as rapid on-column degradation and incomplete peak resolution. A difficulty with mass spectral detection is that no explicit information is obtained regarding the site of ionization, for which UV detection has some advantage. Although mass spectra can be run in both positive and negative ion modes, signal response is not necessarily diagnostic of the site of protonation or deprotonation. Errors have occasionally been made in this way (e.g., see furosemide, no. 583; phenylbutazone, no. 1011; and phenytoin, no. 1103, in Appendix A of the compilation).

Careful comparisons with CE-derived pK_a data from other methods indicated good agreement only in some cases (e.g., see compilation data for atenolol, clioquinol, codeine, enalapril, ibuprofen, lidocaine, and phenylbutazone). Poorer agreement was seen in other cases (e.g., bumetanide, clofazimine, haloperidol, lisinopril, and oxytetracycline). Also, the reproducibility of the CE method [76, 77] was not as good as desired for a method based on pH measurements, with standard deviations for replicate pK_a values in the range 0.07–0.22. Samples for high-throughput screening are typically in the form of DMSO solutions, so it might be thought that the presence of this solvent contributes to reduced reproducibility. However, present literature [76] suggests that this is not so. Part of the reduced reproducibility may be due to acid–base solutions in the laminar flow regions near capillary walls behaving differently to bulk solutions. The response behavior of electrochemical HPLC detectors is known to change in the presence of laminar flow, so it seems reasonable that other electrolyte solution properties may also change under similar circumstances. Nevertheless, rapid analysis times combined with reasonably consistent results makes these approaches the most useful for present high-throughput screening of weakly acidic or basic compounds; the results are generally more accurate than those obtained from *in silico* predictions.

2.2.2. Calibration of pH meters with pH standards

Errors in calibration of pH meters used in pK_a determinations are directly reflected in the final results, whereas errors in other aspects of pK_a measurements are less intrusive. Suitable analytical grade standards for pH meter calibration are listed in Chapter 4 of Bates [45] and in Appendix 12.3 of Robinson and Stokes [44]. These include aqueous solutions of potassium hydrogen tartrate (saturated at 25 °C; pH = 3.557 at 25 °C), potassium hydrogen phthalate (0.05 M; pH = 4.008 at 25 °C), sodium tetraborate (0.01 M; pH = 9.180 at 25 °C), and calcium hydroxide (saturated at 25 °C; pH = 12.454 at 25 °C). These compounds have had their standard pH values [pH(S) values] assigned from EMF measurements in cells without liquid junction potentials (hydrogen electrodes). Standard prepackaged pH meter calibration solutions (e.g., pH 4, 7, and 9) are suitable for routine work, provided they are stored properly and that their temperature dependences are taken into account by the investigator, Table 1.1.

Rapid absorption of atmospheric CO_2 is a severe problem for solutions with pH > 9 and storage of these solutions in a CO_2-free environment is imperative for accuracy in calibration. For careful work, high pH standard solutions may be obtained as certified IUPAC standards, or prepared in small volumes from dry chemicals, using the highest grade material available. Solutions must be made with highly purified distilled water that has been boiled to remove dissolved CO_2, and then cooled under protection from the atmosphere. In the author's experience, the best quality water is prepared by refluxing Milli-Q or similar water with dilute alkaline permanganate for 30–40 min, followed by distillation in an all-glass still that is used only for water purification. Water prepared in this way has been

TABLE 1.1 Temperature dependence for the pH values of several standard solutions[a]

Temperature (°C)	Nominal pH 4 (red)	Nominal pH 7 (green)	Nominal pH 9 (yellow)
10	3.997	7.05	9.11
15	3.998	7.02	9.05
20	4.001	7.00	9.00
25	4.005	6.98	8.95
30	4.011	6.98	8.91
35	4.018	6.96	8.88
40	4.027	6.85	8.85

[a] IUPAC [59] has specified that primary pH standard solutions must have a temperature variance of <0.01 pH/°C.

found to give (compared to the source water): (a) lowered total background ion current when used as mobile phase for an HPLC-MS system; (b) lowered total background redox current when used as mobile phase for an HPLC system with electrochemical detection; and (c) lowered residual enthalpy of dilution when assessed by isothermal titration microcalorimetry. It can be stored under argon in borosilicate glass containers, but must be degassed by argon sparging if more than 24 h post-distillation.

These precautions are essential for the preparation and storage of standard solutions with pH values >7, as repeated exposure to atmospheric CO_2 will result in significant pH changes in a relatively short time. For example, 0.01M sodium tetraborate (borax) solutions (pH = 9.180 at 25 °C) will decrease by 0.02 pH unit after about two weeks if briefly exposed (4–5 min) to the atmosphere every 3–4 days, but otherwise stored under CO_2-free nitrogen (or argon). A pH decrease of this magnitude invalidates the use of this solution as a calibration standard. pH standard solutions have low buffer capacities due to their low concentrations (which are needed to minimize ionic strength effects on thermodynamic activity coefficients). They must be checked frequently against fresh solutions to guard against shifts in pH and replaced as soon as changes are noted, for example, a significant change in the Nernst slope. Storage of high pH standard solutions is enhanced if sealed containers are used, combined with positive nitrogen pressure displacement to dispense the solution immediately before use.

2.2.3. Temperature effects

Temperature dependence equations for pK_a and pK_w values show the importance of good temperature control during pH meter calibration as well as in measurements. To a greater or lesser extent, all acid–base reactions vary with temperature. The effect of temperature changes for equilibrium reactions are closely described by the Valentiner-Lannung equation [80–82]:

$$pK_a = B + \frac{A}{T} + C \log T \tag{10}$$

where A, B, and C can be represented by thermodynamic functions [80–82], for example, as in Eq. (11). The Valentiner-Lannung equation is an expansion of the van't Hoff isochore and was originally proposed to account for the solubility-temperature dependence and solution thermodynamics of inert gases in water. It has also been used to describe the temperature dependence of reaction rates [83–85] and aqueous solubilities of sparingly water-soluble drugs [86–88]. A van't Hoff plot of pK_a versus $1/T$ is curved, due to a finite value for the heat capacity change (ΔC_p) for ionization and thus, all coefficients in the Valentiner-Lannung equation are significantly different to zero:

$$pK_a = \left[\frac{\Delta C_p - \Delta S^o}{R \ln 10}\right] + \frac{\Delta H^o}{RT \ln 10} - \frac{\Delta C_p}{R} \log T \tag{11}$$

where ΔH° is the standard enthalpy change for ionization, ΔS° is the standard entropy change for ionization, and ΔC_p is the heat capacity change for ionization at constant pressure. The nonlinear van't Hoff nature of the pK_a-temperature relationship is not simply a consequence of the presence of ions in acid–base equilibria, for it is even shown by the solubilities of inert gases in water [80, 81]. Rather, it is a consequence of changes in hydrogen-bonded water structure, as well as changes in ion hydration that result from variations in the molecular charge densities that occur on proton transfer.

Equations (10) and (11) can also be expressed in exponential form, as well as in forms which use ΔG as the dependent variable rather than pK_a (from the van't Hoff isotherm, $\Delta G = -RT \ln K$). For equilibrium reactions in aqueous and other polar solutions, the ΔC_p value is expected to have a finite value, due to the significant changes in solvent structure which occur when ionization takes place. For some compounds, the ΔC_p value may have a large uncertainty and be not statistically different to zero, depending on the precision of the raw data (e.g., 5,5-di-isopropylbarbituric acid) [89]. In these cases, the pK_a-temperature dependence is satisfactorily described by the integrated van't Hoff equation [Eq. (10) without the $C \log T$ term]. This equation will give a linear van't Hoff plot of pK_a versus $1/T$.

As well as the inherent effects of temperature on the ionization of weakly acidic and basic functional groups, precise calculation of pK_a values from potentiometric titration requires the autoprotolysis constant for water, pK_w. This quantity has been shown by careful measurements [90, 91] to be very temperature dependent, with values ranging from 14.943_5 at 0 °C to 13.017_1 at 60 °C and 12.264 at 100 °C [92–94].

The temperature variance of pK_w means that the pH scale also changes as a function of temperature. A solution with neutral pH is one where $a_{H_3O^+} = a_{OH^-}$. So ingrained is the idea that pH = 7.0 is "the" neutral pH value, that many laboratory workers are unaware that this is only true at 24 °C, the temperature at which pK_w has a value of exactly 14.00, according to the data reported by Harned and Robinson.

TABLE 1.2 Values for pK_w as a function of temperature[a]

Temperature, °C (Kelvin)	pK_w	Temperature, °C (Kelvin)	pK_w
0.0 (273.15)	14.9435 (14.952)	35.0 (308.15)	13.6801 (13.690)
5.0 (278.15)	14.7338 (14.740)	40.0 (313.15)	13.5348 (13.545)
10.0 (283.15)	14.5346 (14.541)	45.0 (318.15)	13.3960 (13.407)
15.0 (288.15)	14.3463 (14.352)	50.0 (323.15)	13.2617 (13.276)
20.0 (293.15)	14.1669 (14.173)	55.0 (328.15)	13.1369 (13.150)
25.0 (298.15)	13.9965 (14.004)	60.0 (333.15)	13.0171 (13.031)
30.0 (303.15)	13.8330 (13.842)		

[a] The main values in Table 1.2 were taken from Harned and coworkers [90, 91] and Ramette [92]. The pK_w values in parentheses are values from the work of Covington [95].

At temperatures less than 24 °C, neutral pH is >7.0, while the converse is true at higher temperatures, for example, at 0 °C, neutral pH is 7.47, while at 60 °C, it is 6.51.

The effect of temperature on pK_a depends on the nature of the functional group. In particular, amine and imide pK_a values change significantly with temperature; typically $\Delta pK/\Delta T = 0.03$ deg^{-1} (amines) and 0.02 deg^{-1} (imides). Carboxylic acids display reduced dependence of pK_a on temperature ($\Delta pK/\Delta T =$ 0.003 deg^{-1} at worst) [96] and may pass through a minimum in the accessible temperature region (0–100 °C). The second pK_a of phosphoric acid passes through a minimum in water at 45 °C. To reduce unwanted thermal effects on the measured pK_a values to a negligible level, the experimental temperature should be controlled to ±0.01 °C, using a suitable thermostat. A recently calibrated thermometer (precision mercury-in-glass or platinum resistance) should be used for temperature measurement.

2.2.4. Solvent composition and polarity

Many drug substances are insufficiently water-soluble for their pK_a values to be determined with ease. This is especially so for potentiometric titration methods, where a drug concentration of 0.01 M is ideal (in terms of the magnitudes of resulting pH changes and the corrections needed for ionic strength effects, see Section 2.2.5), while 0.001 M represents an easily manageable lower limit. Below this limit, the pH changes that result from additions of titrant are so small as to be difficult to measure with precision, although automated instruments, such as the Sirius® autotitrators, have often been used at concentrations of <0.001 M. Although pK_a values for numerous poorly water-soluble drugs can be determined by either the spectrophotometric or solubility–pH dependence methods, some compounds do not have suitable chromophores, especially aliphatic amines. Investigators should make an effort to choose a method that will give the pK_a

value in water (or a closely allied relevant aqueous solvent, such as physiological saline), so that the measured value is as widely applicable as possible.

The solubility–pH dependence method should be used more often than it is, although it is not without its own difficulties. Its peculiar advantage, of course, is that the very low solubility which makes it difficult or impossible to assess a pK_a value by potentiometry or spectrophotometry is the very phenomenon that can be exploited in measuring the pK_a value, by studying the effects on solubility of pH changes. Provided that saturation of aqueous solutions over a wide range of pH can be achieved in reasonable time and a suitable method of quantitative analysis exists (with the required sensitivity and precision to measure the low solubility of the uncharged species, S_0), measurement of pK_a by this method is quite feasible. Workers should always check that saturation has in fact been achieved, by checking the measured concentration at 24 h intervals, until differences are purely statistical. The Noyes-Whitney equation for dissolution rates should remind workers that the poorer the aqueous solubility, the slower that saturation is attained. Other experimental issues include the propensity of many poorly water-soluble organic compounds to adsorb onto glass or plastic surfaces; and the difficulty of cleanly separating the saturated supernatant solution from the remaining undissolved solid. The experimenter must decide whether filtration or centrifugation is better, based on previous experience with similar compounds.

Where spectrophotometric or solubility methods are not feasible, organic cosolvents are often needed to obtain sufficiently high concentrations of the dissolved drug to get satisfactory titration curves. Some data in the older literature is of lesser reliability, due to precipitation of the less soluble neutral form of the analyte during titrations in aqueous media. Cosolvents that have been used in drug pK_a studies include methanol, ethanol, acetonitrile, *N,N*-dimethylformamide, *N,N*-dimethylacetamide, methylcellosolve, and 1,4-dioxane. Where this approach has been used in the past, it is imperative that investigators realize that the measured pH values and the reported pK_a values are "apparent" values, which may be very different to the thermodynamic values in water. The presence of organic cosolvents can have profound effects on pK_a values. For example, acetic acid (as a typical carboxylic acid) and aniline (as a typical aromatic amine) have the following pK_a' values at 25 °C, Table 1.3.

The sulfamic acid group of cyclamic acid behaves similarly to acetic acid, although the effect of ethanol on its apparent pK_a' values is not as extreme as is methanol on the carboxylic acid group. These values are only relevant to the conditions in which they were measured, unless valid interpolation or extrapolation methods can be used to estimate pK_a' values for different conditions. In particular, linear extrapolation of apparent pK_a' value to zero cosolvent content in older work is very error-prone. It is not generally appreciated that relationships between apparent pK_a' value and cosolvent content (or other functions of cosolvent content, such as dielectric constant), although linear over moderate to high concentrations of cosolvent, become progressively nonlinear in lower concentrations [45–47]. Worse, the solubility of the drug is at its lowest in solutions with low cosolvent content, and the level of error in apparent pK_a' measurements increases under the very conditions

TABLE 1.3 Effect of solvent composition on apparent pK_a' values

Compound	Methanol (vol%; pK_a') [97]					
Acetic acid	0; 4.76	20; 5.00	40; 5.31	60; 5.76	80; 6.43	–
Anilinium ion	0; 4.60	20; 4.45	40; 4.30	60; 4.12	80; 4.00	–
	Ethanol ((vol%); pK_a') [98]					
Cyclamic acid	0; 2.28	20; 2.48	40; 2.60	60; 2.90	80; 3.22	100; 3.45

where the error in extrapolations has the greatest influence on the estimated value for an aqueous solution.

In general, deprotonation of acids (e.g., carboxylates, phenols, and imides) is more susceptible to the effects of solvent polarity change than for bases (e.g., amines, pyridines) in their conjugate forms (Table 1.3). Deprotonation of a typical acid in water or a solvent mixture involves dissociation of a neutral species into an anion and a cation, thus leading to a large change in solvation of the two ionized species that result, whereas deprotonation of the conjugate acid form of a base simply involves transfer of a positively charged proton from one neutral base (i.e., the analyte) to another (e.g., water). The net charge does not change (although the charge density does) and the changes in solvation through charge–dipole interactions are less.

In recent years, the extrapolation procedure developed separately by Yasuda and Shedlovsky has been reported [20, 21, 99] to provide pK_a values from partially aqueous solutions that can sometimes closely approximate the values found in purely aqueous solutions. The Yasuda-Shedlovsky equation [Eq. (12)] can be used to correlate apparent pK_a' values in different solvent polarities (p_sK_a) to approximate the aqueous pK_a value by extrapolation as a reciprocal function of the dielectric constant (ε) of each of the cosolvent mixtures:

$$p_sK_a + \log\,[H_2O] = \frac{A}{\varepsilon} + B \tag{12}$$

A plot of the Yasuda-Shedlovsky equation generally is a straight line. The fitted coefficients A and B are then used to estimate the pK_a value in a 100% aqueous solution, for which $[H_2O] = 55.5$ molal and $\varepsilon = 78.3$. Successful use of this approach in cosolvent mixtures requires a complex pH electrode calibration procedure [20].

While the Yasuda-Shedlovsky procedure can give quite satisfactory results, it is not infallible, even in the hands of experts. Appendix A, quinine (no. 1223) gives a clear illustration of the difficulty in obtaining extrapolated results with good accuracy when using different organic solvent–water mixtures to provide adequate solubility for potentiometric measurements. The values for pK_{a1} and pK_{a2} for quinine (using data from the same source) ranged from 3.85 to 4.32 and 8.15 to 8.58 (respectively) when extrapolated by the Yasuda-Shedlovsky procedure from diverse solvents. If data from the worst performing solvent (dimethylacetamide; $pK_{a1} = 3.85$, $pK_{a2} = 8.15$) is excluded, then the range of values for pK_{a2} becomes an acceptable 0.07. However, the values for pK_{a1} still range from 4.07 to 4.32, and it is not immediately obvious why the extrapolated values always tend (at least in this case) to values lower than the best values in aqueous solution.

2.2.5. Ionic strength

The ionic strength of a solution generally alters the activity coefficients of all dissolved ionic species, thus influencing measured pK_a values. The Debye-Hückel equations [Eqs. (13–17)] are required to correct for solution ionic strength in order to obtain thermodynamic pK_a values:

Debye-Hückel equation:

$$-\log \gamma_{\pm} = A|z_+z_-|\frac{\sqrt{I}}{1 + Ba_0\sqrt{I}} \tag{13}$$

Debye-Hückel limiting law:

$$-\log \gamma_{\pm} = A|z_+z_-|\sqrt{I} \tag{14}$$

Güntelberg modification:

$$-\log \gamma_{\pm} = A|z_+z_-|\frac{\sqrt{I}}{1 + \sqrt{I}} \tag{15}$$

Davies modification:

$$-\log \gamma_{\pm} = A|z_+z_-|\frac{\sqrt{I}}{1 + \sqrt{I}} - 0.10I \tag{16}$$

Guggenheim modification:

$$-\log \gamma_{\pm} = A|z_+z_-|\frac{\sqrt{I}}{1 + \sqrt{I}} - \beta I \tag{17}$$

where A and B are constants that depend only on temperature and solvent dielectric constant, I is the ionic strength, a_0 is the so-called ion size parameter, β is an empirical constant, and z_+, z_- are the charges on the species in the ionization equilibrium reaction. Attempts have been made to estimate values for the ion size parameter, a_0, for several weak organic acids [100]. However, many studies make the assumption that the product Ba_0 in the Debye-Hückel equation [Eq. (13)] is equal to 1.00, and then apply either the Güntelberg or Guggenheim modifications. Elsewhere [47], it has been suggested that the product of Ba_0 should have a value of 1.6. The equations provide reasonable estimates for $\gamma_{\pm}$, provided that the ionic strength is not greater than a critical value. Extended forms of these equations, especially the Guggenheim modification, can be used for quite high ionic strengths (>1.0 M), especially where the β coefficient has been partitioned into separate ionic contributions.

The empirical constant β can be estimated from linear relationships between measured apparent pK_a' values and ionic strength. It is equated with 0.10 in the

Davies modification of the Debye-Hückel equation. However, this particular value was found to best account for the activity effects in solutions of multivalent metal cations such as calcium [44, 101]. Furthermore, it is only valid at 25 °C. It was subsequently increased to 0.15 for studies on the pK_a values of some phenols [70], where the larger value was found to give more consistent results [101]. It is debatable whether the value of 0.15 is applicable to solutions containing relatively large organic cations or anions such as ionized drug molecules. In some unpublished studies [102, 103] the value for β was found to be up to an order of magnitude greater than 0.15. Very recent studies on lactic acid ionization [104] demonstrated this point unequivocally.

2.2.6. Chemical stability

Stability of compounds during pK_a determinations can have deleterious effects on precision, through degradation reactions such as hydrolysis, oxidation, or photolysis. This is especially important during lengthy procedures, such as potentiometric titrations or equilibration for the solubility–pH dependence method. Oxidation and photolysis can often be excluded (or at least reduced to negligible proportions) by appropriate countermeasures, such as protection from light, degassing, use of inert gas sparging, inclusion of innocuous (inert) antioxidants, or chelating agents. However, hydrolysis of susceptible molecules is very difficult to stop in an aqueous environment. Although hydrolysis rates can be reduced by the use of organic cosolvents, the effect of such cosolvents on the pK_a values can be difficult to correct by extrapolation to zero cosolvent content. Conversely, spectrophotometric absorbance measurements can usually be completed fairly quickly after an aliquot of stock solution is diluted with the appropriate buffer. As the absorbance measurements can be followed for a period of time, and then extrapolated to zero time to correct for degradation during measurement, spectrophotometric methods may be preferred for moderately unstable compounds with pH-sensitive chromophores. This approach has been occasionally reported [105], with satisfactory results.

2.2.7. Exclusion of atmospheric CO_2

As carbonic acid is a weak acid with p$K_1 = 6.2$ [106], the presence of CO_2 in any pK_a measurement system may influence the results. If the pK_a value for the acid or conjugate acid is less than that of carbonic acid, little or no influence will occur. However, for ionizable systems where the pK_a value is greater than that of carbonic acid, the presence of dissolved CO_2 will have marked effects, giving apparent values that are lower than the true values. This is especially so where potentiometric titrations with sodium or potassium hydroxides are used. This is because the concentrations of protonated and deprotonated forms required for the Henderson-Hasselbalch equation [Eqs. (3), (4), and (9)] are calculated from the volumes of acid or base added to the titration system. The presence of dissolved CO_2 (in either the titrant or titrated solution) will alter the volume required. Spectrophotometric and other methods of pK_a measurement are free from the influences of dissolved CO_2, if the pH-dependent phenomenon is insensitive to its presence. Cases where this must be taken into account include drugs where the absorption spectra overlap with those of carbonates/bicarbonates (check wavelengths). However, dissolved CO_2 does alter ionic strengths, and hence, activity coefficients. Methods for removing and excluding dissolved CO_2 are described in detail by Albert and Serjeant [46, 47] and must be followed rigorously in order to be satisfactory. During titrations, CO_2 must also be rigorously excluded by use of a nitrogen or argon gas blanket (or hydrogen, for hydrogen electrodes). Covington *et al.* [107] have shown that gently passing the blanket gas through rather than over the titration solution has a significant advantage in that even trace amounts of CO_2 are removed by the former procedure, but not by the latter. This is also recommended by Albert and Serjeant. Only by carefully checking standard compounds with accurately known pK_a values (= 7 or greater) can a system for measuring new pK_a values be reliably tested for this source of interference.

2.3. Validation of data quality

The above factors must all be optimized in order to obtain reliable pK_a values, within the constraints of data based on pH measurements (accuracy of ± 0.02 pK_a unit at best). One of the key methods for assuring the reliability of pK_a measurements [46, 47] is to include concurrent measurements on compounds with accurately known pK_a values, as well as the unknown compounds. These reference compounds should be easily obtainable in a high state of purity, either commercially or through recrystallization from common solvents such as distilled or Milli-Q water, high purity alcohols or their mixtures with water. It is preferable for the reference compounds to have had their pK_a value(s) determined by more than one experimental method. Suitable compounds include benzoic acid (p$K_a = 4.205 \pm 0.015$ at 25 °C; available from British Drug Houses (United Kingdom) as compressed tablets of a high purity thermochemical standard) and 5,5-diethylbarbituric acid (p$K_{a1} = 7.980 \pm 0.002$ at 25 °C; this is easily recrystallized to high purity from ethanol-water, 40:60 v/v). Both of these compounds have reliably known pK_a values over a wide temperature range [9, 18]. In particular, the pK_a value for benzoic acid has been determined at 25 °C by conductance, electrometric, potentiometric, and spectrophotometric methods with agreement to ± 0.015 pK_a unit for all four methods. The pK_{a1} value for 5,5-diethylbarbituric acid has been determined by electrometric, potentiometric and spectrophotometric methods with even better agreement. This reference compound has the further advantage that errors due to atmospheric CO_2 absorption (into either the reaction solution or standardized KOH titrant) will influence the results. It is a severe test for sloppy technique. The pK_a value for benzoic acid, conversely, is insensitive to the presence of CO_2 and so can test for other instrumental, technical, or procedural faults. Whether reported new pK_a data measurements include comparisons with known data is a significant component in determining their reliability or validity. Very few reported pK_a values in the drug sciences literature meet this criterion, despite the advice [46, 47] of Albert and Serjeant. This point has also recently been expressed elsewhere [108], but little notice has been taken so far, either by experimentalists or the editors who publish their results.

3. pK_a VALUES FROM COMPUTER PROGRAMS

The estimation of pK_a values through predictive relationships has a long history, commencing with the Brønsted catalytic law, through the formalism of the Hammett and Taft relationships and their derivatives, then finally to predictions based on the application of ANNs. The Hammett and Taft relationships are described as structure-reactivity or linear-free energy relationships (LFERs), and are usually valid for a specific set of related compounds called a *reaction series*. At very best, these relationships give predicted values with a mean accuracy of about ± 0.1 log unit. There are now numerous commercial computer programs that can estimate the pK_a values for virtually any given structure, using a variety of predictive algorithms in combination with ANNs. One widely used package is from ACDLabs (Toronto, Ontario, Canada).

A reasonable question is to enquire about the reliability of the results of such estimations. Table 1.4 compares estimated values from ACD/p*K*a (ver. 7) with the most reliable values for a variety of randomly chosen compounds from the data compilation of the present work. It is clear from the data in Table 1.4 that sometimes the ACD/pK_a package predicts a value that is very close to the best literature values. This is especially true for some smaller molecules, for example, 4-aminobenzoic acid or 5,5-diethylbarbituric acid. However, other predictions may be quite unreliable. It is of concern that some of the deviations in Table 1.4 are larger than the estimated error allowed by the package in specific cases, for example, ebifuramin, camptothecin, or citric acid.

It is expected that predictions from such computer packages can be improved, the more that the package is "trained" by inclusion of additional measured values in its internal database for closely related compounds, although "overtraining" has also been known to occur. However, even training may not be sufficient. For example, Table 1.4 gives a close prediction for 5,5-diethylbarbituric acid (7.95), and it was found that the internal database contained the best measured value (7.980) from the literature. The package also gave an excellent prediction for the related 5,5-diphenylbarbituric acid (7.32). On the other hand, the database contained a value for lidocaine (7.90) that was within the range of reliable measured values, yet the package estimated a value (7.53) that was significantly at variance. Furthermore, although the package was able to provide an excellent estimated value for 5,5-diethylbarbituric acid, it could not do so for the very closely related cyclopentane-1′,5-spiro derivative (9.30, estimated; 8.83, measured). For this compound, the computed value was overestimated, when one would expect the opposite to be the case, given that changes in alkyl substitution generally do not alter pK_a values greatly. Similarly, for 5,5-dimethylbarbituric acid (7.95, estimated; 8.51, measured), the prediction was also inaccurate, but in the opposite direction.

It has been elsewhere suggested [109] that the predictive deficiencies of programs such as ACD/pK_a are due to their fragment-based approach. It was further suggested that an approach based on quantum mechanical methods would be more reliable. However, unpublished analysis (by the present author) of predicted values based on such quantum mechanical methods for a set of 40 carboxylic acids [109] indicated that many of them were still poorly estimated, with a mean deviation on the order of 0.3 log unit. The plot of predicted versus

TABLE 1.4 Estimated (ACD/pK_a) *versus* measured pK_a values

Compound	Estimated pK_a value(s)	Measured pK_a value(s)	Deviation(s) (estimated − measured)	Compound	Estimated pK_a value(s)	Measured pK_a value(s)	Deviation(s) (estimated − measured)
Acetaminophen	9.86 ± 0.13	9.63 ± 0.01	+0.23	Clioquinol	2.32 ± 0.30	2.96	−0.64
					7.23 ± 0.59	8.12	−0.89
Amantadine	10.76 ± 0.20	10.71 ± 0.01	+0.05	Clozapine	4.36 ± 0.30	3.58	+0.78
					6.19 ± 0.30	7.94	−1.75
4-Aminobenzoic acid	2.51 ± 0.10	2.501	+0.01	Cyclopentane-1′,	9.30 ± 0.20	8.83 ± 0.03	+0.47
	4.86 ± 0.10	4.874	−0.01	5-spirobarbituric acid			
4-Aminosalicylic acid	2.21 ± 0.10	1.78	+0.43	Diazepam	3.40 ± 0.10	3.42	−0.02
	3.58 ± 0.10	3.63	−0.05				
Ampicillin	2.44 ± 0.50	2.53 ± 0.004	−0.09	5,5-Diethylbarbituric acid	7.95 ± 0.20	7.98 ± 0.002	−0.03
	6.76 ± 0.29	7.24 ± 0.02	−0.48	(barbitone)			
Aspartame	3.71 ± 0.10	3.19 ± 0.01	+0.52	5,5-Dimethylbarbituric acid	7.95 ± 0.20	8.51 ± 0.02	−0.56
	7.70 ± 0.39	7.87 ± 0.02	−0.17				
Atenolol	9.17	9.60 ± 0.04	−0.43	5,5-Diphenylbarbituric acid	7.32 ± 0.20	7.30 ± 0.02	+0.02
Benzocaine	2.51	2.50 ± 0.04	−0.01	Diphenoxylate	7.63 ± 0.40	7.1	+0.53
Brucine	8.27 ± 0.20	8.28	−0.01	Ebifuramin	1.85 ± 0.50	5.24 ± 0.04	−3.39
Camptothecin	3.31 ± 0.40	1.18	+2.13	Ephedrine	9.38 ± 0.10	9.64 ± 0.03	−0.26
	11.02 ± 0.20	10.83	+0.19				
Cefroxadine	2.57 ± 0.50	3.30 ± 0.02	−0.73	Epinephrine (adrenaline)	9.16 ± 0.20	8.71 ± 0.04	+0.45
					9.60 ± 0.10	9.78 ± 0.12	−0.18
Chloral hydrate	10.54 ± 0.41	10.04	+0.50	Ethionamide	4.34	4.37	−0.03
Chlorcyclizine	2.24 ± 0.50	2.12 ± 0.04	+0.12	Lidocaine	8.53 ± 0.25	7.90 ± 0.05	+0.63
	7.89 ± 0.42	7.65 ± 0.04	+0.24				
8-Chlorotheophylline	4.57 ± 0.70	5.28	−0.71	Nitrazepam	3.19 ± 0.10	2.92 ± 0.07	+0.27
					11.4 ± 0.7	10.5 ± 0.1	+0.9
Chlorothiazide	5.95 ± 0.42	6.85	−0.90	Papaverine	6.32	6.38 ± 0.03	−0.06
	9.70 ± 0.20	9.45	+0.25				
Chlorzoxazone	9.44 ± 0.30	8.3	+1.14	Phenylbutazone	4.29 ± 0.60	4.53 ± 0.06	−0.24
Citric Acid	3.86 ± 0.23	3.128	+0.73	Propranolol	9.15	9.53 ± 0.04	−0.38
	4.63 ± 0.19	4.762	−0.13				
	5.48 ± 0.19	6.396	−0.916				
Clindamycin	8.74	7.77	0.97	Pyridoxine	8.37	8.95 ± 0.06	−0.58

observed pK_a values was linear with $R^2 = 0.9218$. Prediction of substituted aniline pK_a values was better with mean deviations on the order of 0.2 log unit. The linear predicted versus observed plot had $R^2 = 0.9739$. However, these predictions were for simple monofunctional acids or bases, not multifunctional drug substances.

The difficulty with all present computational models is that they attempt to predict pK_a values from the structural properties of a ***single thermodynamic state***, that is, the molecular form of the acid or base. However, pK_a values, as conventionally defined, are in direct proportion to ΔG^o, the free energy change for transition from the protonated state to the deprotonated state. For predictions of pK_a values for basic compounds such as amines, heteroaromatics, guanidines, amidines, and azomethines, the modeled structures are the electrically neutral deprotonated forms. For predictions of pK_a values for acidic compounds, such as carboxylic acids, phenols, imides and sulfonamides, the modeled structures are the neutral protonated forms. In both situations, the other state in the equilibrium reaction (the hydrated anion for acids; the hydrated cation for bases) is *a priori* ***assumed to be the same*** for all structures that have a specific functional group. Studies reported by McKeown *et al.* [89] have shown that this is not the case, at least for the 5,5-disubstituted barbituric acids. In particular, steric substituent effects ***cannot*** be assumed to be the same on both the protonated and deprotonated states, especially where charge separation occurs in the ionization of acids. Similarities in substitution effects on pK_a values suggest that the same conclusion applies [38] to several families of substituted mono- and dicarboxylic acids. Until predictive methods take into account the differential thermodynamic state origin of pK_a values, there will always be considerable uncertainty in the accuracies of the resulting computed values. This is irrespective of the methods and refinements of the computational procedure itself, whether by ANNs, molecular mechanics, semiempirical molecular orbital calculations or *ab initio* calculations.

Ultimately, the value of such *in silico* predicted pK_a data depends on the use(s) to be made of such numbers. They are generally satisfactory for use in a drug discovery program. If they are rough estimates of the extent of ionization needed to explain bioavailability or pharmacokinetic data, which also usually have experimental errors of more than a few percent, then these values will serve fairly well, although more accurate values would be an improvement. Conversely, if values are needed for use in conjunction with more precise physicochemical data, for example, solubilities or partition coefficients that can be measured with precision and accuracy in the 1–5% range, then deficiencies in the interpretation of such data are very likely to result from use of poorly defined pK_a values. And finally, values from such computational estimates are worthless when the objective is to further probe the relationships of molecular structure and physicochemical properties.

4. SUMMARY OF THE DATA COMPILATION

The database compilation following this introduction to some of the difficulties of pK_a measurement presents nearly 3500 pK_a values for drug and related substances that have been collected from the literature. A summary of the chief issues follows:

The reported pK_a measurements were assessed for the *quality of the data*, based on an examination of these factors:

- Experimental method
- Precision of temperature control
- Solvent composition
- pH meter calibration
- Exclusion of CO_2
- Use of thermodynamic activity corrections
- Validation of experimental results by comparison with gold standard compounds

For example, some experimental methods are more reliable than others. Methods based on conductance or electromotive cells without liquid junction potential are capable of very high accuracy and precision, although conductance methods are best applied to acids or conjugate acids with pK_a values in the range 2–5. Conversely, methods based on pH measurements are presently constrained to lower reliability (not better than ±0.02), due to the limits of the theoretical definition of pH. Titrimetric methods involving pH measurement are usually more accurate and precise than pH-based methods using spectrometric, partitioning, or solubility methods for discrimination between the protonated and deprotonated species because of the greater reliability of properly calibrated volumetric glassware or modern autotitrators. Where key experimental details, such as those listed above, were not reported in the original literature, then the reader cannot assume that these were in fact performed. The resulting data must then be held to have lower reliability. One of the most important of these is the last, the use of comparisons with measurements for compounds of known reliability. It is a matter of concern that while we inculcate our students with the importance of carefully validating HPLC assays of samples in biological matrices with the appropriate controls, we do not do the same for physicochemical measurements, for which significantly higher precision should be achieved, when appropriate care is taken.

The experimental data in the compilation were assessed according to the criteria established in the 1960s by the IUPAC for its compilations of dissociation constants for weak organic acids and bases. These criteria and the codes that denote them in the compilations are given in Table 1.5. In the present work, the cutoff for "uncertain" was increased to $> \pm 0.06$ pK unit, corresponding to an uncertainty in the dissociation constant, K_a, equivalent to $>15\%$.

TABLE 1.5 IUPAC reliability criteria for pK_a values of weak acids and bases

Classification	Code	Criterion as error in pK_a	Uncertainty in K_a
Very reliable	VR	$< \pm 0.005$	$\equiv <1\%$
Reliable	R	± 0.005 to ± 0.02	$\equiv \sim 1\%$
Approximate	A	± 0.02 to ± 0.04	$\equiv \sim 5\%$
Uncertain	U	$> \pm 0.04$	$\equiv >10\%$

Measured pK_a values for many drugs have been reported in the literature more than once. In some cases, there is very good agreement between repeated measures, while in others, there is very poor agreement (see below). The following were ranked as "R" = "Reliable" or "A" = "Approximate" under the criteria in Table 1.5, and were obtained from measurements using different methods and in different laboratories:

- Benzoic acid pK_a = 4.205 ± 0.015 ($n > 4$)
- 5,5-Diethylbarbituric acid pK_a = 7.980 ± 0.01 ($n = 3$)
- Nimesulide pK_a = 6.51 ± 0.05 ($n = 3$)
- Chlorthalidone pK_a = 9.35 ± 0.01 ($n = 2$)

Replicated pK_a data for numerous other drugs were in very poor agreement with each other, and were ranked as "U" = "Uncertain" or occasionally as "VU":

- Carbenoxolone pK_{a1} = 4.2–6.7 ($n = 3$)
- Glibenclamide pK_a = 5.3–6.8 ($n = 3$)
- Ibuprofen pK_a = 4.1–5.3 ($n = 8$)
- Lidocaine pK_a = 7.18–7.95 ($n = 7$)
- Phenylbutazone pK_a = 4.33–5.47 ($n = 8$)
- Propranolol pK_a = 9.23–9.72 ($n = 11$)

Altogether, ~74% of the pK_a values in the pharmaceutical sciences literature were found to be of "Uncertain" quality, based on the modified IUPAC criteria, whereas only 0.1% qualified as "Very Reliable"; these were almost all pharmaceutically relevant buffers that had been selected by the (former) US NBS (now the NIST) for evaluation as pH standards. "Reliable" values made up 0.33% of the total, while "Approximate" values comprised ~25%.

The compilation is divided into two sections, the larger section (Appendix A) comprising those drug or drug-related pK_a values for which the measurements were sufficiently well described for the data to be assessed for reliability. These were almost entirely taken from the primary literature. The smaller section (Appendix B) comprised those pK_a values for which little reliability data could be assessed, and were mostly from the secondary literature. The compounds are listed alphabetically, largely by common, rather than systematic name. Users unfamiliar with common drug names can use the molecular formula index that follows the database.

REFERENCES

[1] W. Ritschel, pKa values and some clinical applications, in *Perspectives in Clinical Pharmacy* (eds. D. Francke and H. Whitney), 1st edn., Drug Intelligence Publications, Hamilton, IL, 1972, pp. 325–367.

[2] S. Smith and M. Rawlins, *Appendix C. Variability in human drug response*, 1st edn., Butterworths, London, 1973, pp. 154–165.

[3] T. Speight, *Avery's Drug Treatment*, 3rd edn., Publishing Sciences Group, Inc., Littleton, MA, 1987, pp. 1352–1380.

[4] J. Hoover, *Dispensing of Medication*, 8th edn., Mack Publishing, Easton, PA, 1976, p. 230, 247, 418–426, 468–634.

[5] D. Newton and R. Kluza, pKa values of medicinal compounds in pharmacy practice, *Drug Intell. Clin. Pharm.*, 1978, **12**, 546–554.

[6] P. Craig, Compendium of Drugs, in *Comprehensive Medicinal Chemistry* (eds. C. Hansch, P. Sammes and J. Taylor), 1st edn., Pergamon Press, New York, 1990, pp. 237–965.

[7] D. Williams, Appendix A-1, in *Principles of Medicinal Chemistry* (eds. W. Foye, T. Lemke and D. Williams), 4th edn., Williams and Wilkins, Baltimore, MA, 1995, pp. 948–961.

[8] J. Delgado and W. Remers, pKas of Drugs and Reference Compounds, in *Wilson and Gisvold's Textbook of Organic Medicinal and Pharmaceutical Chemistry* (eds. J. Delgado and W. Remers), 8th edn., Lippincott-Raven, Philadelphia, PA, 2000.

[9] G. Kortum, W. Vogel and K. Andrussow, *Dissociation Constants of Organic Acids in Aqueous Solution*, 1st edn., Butterworth, London, 1961.

[10] D. Perrin, *Dissociation Constants of Organic Bases in Aqueous Solution*, 1st edn., Butterworths, London, 1965.

[11] E. Serjeant and B. Dempsey, *Ionisation Constants of Organic Acids in Aqueous Solution*, 1st edn., Pergamon Press, Oxford and New York, 1979.

[12] D. Perrin, *Dissociation Constants of Weak Bases in Aqueous Solution*, 1st edn., Butterworths, London, 1972.

[13] B. Liederer and R. Borchardt, Enzymes involved in the bioconversion of ester-based prodrugs, *J. Pharm. Sci.*, 2005, **95**(6), 1177–1195.

[14] A. Albert and E. Serjeant, *Appendix V. The Determination of Ionization Constants*, 2nd edn., Chapman and Hall, London, 1971.

[15] K. Murthy and G. Zografi, Oil-water partitioning of chlorpromazine and other phenothiazine derivatives using dodecane and n-octanol, *J. Pharm. Sci.*, 1970, **59**, 1281–1285.

[16] P. Seiler, The simultaneous determination of partition coefficient and acidity constant of a substance, *Eur. J. Med. Chem.*, 1974, **9**(6), 663–665.

[17] A. Briggs, J. Sawbridge, P. Tickle and J. Wilson, Thermodynamics of dissociation of some barbituric acids in aqueous solution, *J. Chem. Soc. (B)*, 1969, 802–805.

[18] G. Manov, K. Schuette and F. Kirk, Ionization constant of 5,5-diethylbarbituric acid from 0° to 60 °C, *J. Res. Natl. Bur. Stand.*, 1952, **48**(1), 84–91.

[19] R. Prankerd, Some physical factors and drug activity—Physical properties and biological activity in certain barbituric acid structures, University of Otago, Dunedin, NZ, Master of Pharmacy Thesis, 1977.

[20] A. Avdeef, K. Box, J. Comer, M. Gilges, M. Hadley, C. Hibbert, W. Patterson and K. Tam, PH-metric log P 11. pKa determination of water-insoluble drugs in organic solvent-water mixtures, *J. Pharm. Biomed. Anal.*, 1999, **20**, 631–641.

[21] K. Takacs-Novak, K. Box and A. Avdeef, Potentiometric pKa determination of water-insoluble compounds: Validation study in methanol/water mixtures, *Int. J. Pharm.*, 1997, **151**, 235–248.

[22] E. Canel, A. Gultepe, A. Dogan and E. Kihc, The determination of protonation constants of some amino acids and their esters by potentiometry in different media, *J. Solution Chem.*, 2006, **35**(1), 5–19.

[23] E. King, Medium effects, Ch. 10 in *Acid-Base Equilibria*, Vol 4:15 in *The International Encyclopaedia of Physical chemistry and Chemical Physics* (eds. E. Guggenheim, J. Mayer, F. Tompkins and R. Robinson), 1st edn., MacMillan, New York, 1965, pp. 269–279.

[24] F. Feates and D. Ives, The ionisation functions of cyanoacetic acid in relation to the structure of water and the hydration of ions and molecules, *J. Chem. Soc.*, 1956, 2798–2812.

[25] D. Ives and P. Marsden, The ionisation functions of diisopropylcyanoacetic acid in relation to hydration equilibria and the compensation law, *J. Chem. Soc.*, 1965, 649–676.

[26] D. Ives and P. Moseley, Derivation of thermodynamic functions of ionisation from acidic dissociation constants, *JCS Farad. Trans. I*, 1976, **72**, 1132–1143.

[27] P. Hadju, K. Kohler, F. Schmidt and H. Spingler, Physicalisch-chemische und analytische untersuchungen an HB 419, *Arzneim.-Forsch.*, 1969, **19**, 1381–1386.

[28] M. Crooks and K. Brown, The binding of sulphonylureas to serum albumin, *J. Pharm. Pharmacol.*, 1974, **26**, 305–311.

[29] T. Medwick, G. Kaplan and L. Weyer, Measurement of acidity and equilibria in glacial acetic acid with the glass calomel electrode system, *J. Pharm. Sci.*, 1969, **58**, 308–313.

[30] A. Martin, *Physical Pharmacy*, 3rd 1st edn., Lea and Febiger, Philadelphia, PA, 1983.
[31] L. Chatten, *Pharmaceutical Chemistry*, 1st edn., Marcel Dekker, New York, 1966.
[32] A. Al-Badr and M. Tayel, Procaine hydrochloride, in *Analytical Profiles of Drug Substances and Excipients* (ed. H. Brittain), 1st edn., Academic Press, New York, 1999, pp. 395–458.
[33] A. Moffat, *Clarke's Isolation and Identification of Drugs*, 2nd edn., The Pharmaceutical Press, London, 1986, p. 1223.
[34] J. Wood, The acidic constants of some ureides and uric acid derivatives, *J. Chem. Soc.*, 1906, **89**, 1831–1839.
[35] J. Kendall, Electrical conductivity and ionization constants of weak electrolytes in aqueous solution, in *International Critical Tables* (ed. E. Washburn), 1st edn., McGraw-Hill, New York, 1929, pp. 259–304.
[36] R. McKeown, First thermodynamic dissociation constants of 5,5-disubstituted barbituric acids in water at 25 °C. Part 1. 5,5-Dialkyl-, 5-alkenyl-5-alkyl-, 5-alkyl-5-aryl-, 5,5-dialkenyl-, 5,5-diaryl-, and 5,5-dihalogeno-barbituric acids, *J. Chem. Soc.* (Perkin II), 1980, 504–514.
[37] M. Krahl, The effect of variation in ionic strength and temperature on the apparent dissociation constants of thirty substituted barbituric acids, *J. Phys. Chem.*, 1940, **44**, 449–463.
[38] R. McKeown and R. Prankerd, First thermodynamic dissociation constants of 5,5-disubstituted barbituric acids in water at 25 C. Part 3. 5,5-Alkylenebarbituric Acids. A Comparison with 5,5-Dialkylbarbituric Acids, and with Mono- and Di-Carboxylic Acids, *J. Chem. Soc.* (Perkin II), 1981, 481–487.
[39] K. Takacs-Novak and A. Avdeef, Interlaboratory study of log P determination by shake-flask and potentiometric methods, *J. Pharm. Biomed. Anal.*, 1996, **14**, 1405–1413.
[40] K. Tam and K. Takacs-Novac, Multi-wavelength spectrophotometric determination of acid dissociation constants, *Anal. Chim. Acta*, 2001, **434**, 157–167.
[41] J. McFarland, C. Berger, S. Froshauer, S. Hayashi, S. Hecker, B. Jaynes, M. Jefson, B. Kamicker, C. Lipinski, K. Lundy, C. Reese and C. Vu, Quantitative structure-activity relationships among macrolide antibacterial agents: *In vitro* and *in vivo* potency against Pasteurella multocida, *J. Med. Chem.*, 1997, **40**, 1340–1346.
[42] E. King, Acidity constants from optical and magnetic measurements, Ch. 5 in *Acid-Base Equilibria*, Vol 4:15 in *The International Encyclopaedia of Physical Chemistry and Chemical Physics* (eds. E. Guggenheim, J. Mayer, F. Tompkins and R. Robinson), 1st edn., MacMillan, New York, 1965, pp. 90–115.
[43] J. Prue, *Ionic Equilibria*, 1st edn., MacMillan, New York, 1965.
[44] R. Robinson and R. Stokes, *Electrolyte Solutions*, 2nd Revised edn., Butterworths, London, 1965.
[45] R. Bates, *Determination of pH: Theory and Practice*, 2nd edn., Wiley, New York, 1973.
[46] A. Albert and E. Serjeant, *The Determination of Ionization Constants: A Laboratory Manual*, 2nd edn., Chapman and Hall, London, 1971.
[47] A. Albert and E. Serjeant, *The Determination of Ionization Constants: A Laboratory Manual*, 3rd edn., Chapman and Hall, London, 1984.
[48] D. MacInnes and T. Shedlovsky, The determination of the ionization constant of acetic acid, at 25°, from conductance measurements, *JACS*, 1932, **54**, 1429–1438.
[49] E. King, Acidity constants from conductance measurements, Ch. 2 in *Acid-Base Equilibria*, Vol 4:15 in *The International Encyclopaedia of Physical Chemistry and Chemical Physics* (eds. E. Guggenheim, J. Mayer, F. Tompkins and R. Robinson), 1st edn., MacMillan, New York, 1965, pp. 23–41.
[50] H. Harned and R. Ehlers, The dissociation constant of acetic acid from 0 to 35° centigrade, *JACS*, 1932, **54**, 1350–1357.
[51] E. King, Acidity constants from precise electromotive force measurements, Ch. 3 in *Acid-Base Equilibria*, Vol 4:15 in *The International Encyclopedia of Physical Chemistry and Chemical Physics* (eds. E. Guggenheim, J. Mayer, F. Tompkins and R. Robinson), 1st edn., MacMillan, New York, 1965, pp. 42–63.
[52] G. Lewis, Determination of dissociation constants of sparingly soluble compounds from solubility data, *Int. J. Pharm.*, 1984, **18**, 207–212.
[53] I. Zimmermann, Determination of pKa values from solubility data, *Int. J. Pharm.*, 1983, **13**(1), 57–65.

[54] L. Hammett, A. Dingwall and L. Flexser, The application of colorimetry in the ultraviolet to the determination of the strength of acids and bases, *JACS*, 1934, **56**, 2010.
[55] L. Flexser, L. Hammett and A. Dingwall, The determination of ionization by ultraviolet spectrophotometry: Its validity and its application to the measurement of strength of very weak bases, *JACS*, 1935, **57**, 2103–2115.
[56] L. Flexser and L. Hammett, The determination of ionization by ultraviolet spectrophotometry—Correction, *JACS*, 1938, **60**, 3097.
[57] H. von Halban and G. Kortum, The dissociation constants of weak and moderately strong electrolytes. I The dissociation constant of α-dinitrophenol and the range and validity of the limitation formula of Debye and Huckel, *Z. fur. Physik. Chem.*, 1934, **170**, 351–379.
[58] I. Feldman, Use and abuse of pH measurements, *Anal. Chem.*, 1956, **28**(12), 1859–1866.
[59] Anonymous, Measurement of pH. Definitions, Standards and Procedures (IUPAC pH Recommendations 2002), *Pure Appl. Chem.*, 2002, **74**(11), 2169–2200.
[60] Anonymous, *Specification for pH Scale*, British Standard 1647, British Standards Institution, London, 1961.
[61] D. Ives and G. Janz, *Reference Electrodes: Theory and Practice*, 1st edn., Academic Press, New York, 1961.
[62] C.-L. Lin and L.-S. Lee, A two-ionic-parameter approach for ion activity coefficients of aqueous electrolyte solutions, *Fluid Phase Equilib.*, 2003, **205**, 69–88.
[63] A. Apelblat, E. Manzurola and Z. Orekhova, Electrical conductance studies in aqueous solutions with ascorbate ions, *J. Solution Chem.*, 2006, **35**, 879–888.
[64] Z. Orekhova, M. Ben-Hamo, E. Manzurola and A. Apelblat, Electrical conductance and volumetric studies in aqueous solutions of nicotinic acid, *J. Solution Chem.*, 2005, **34**(6), 687–700.
[65] Z. Orekhova, Y. Sambira, E. Manzurola and A. Apelblat, Electrical conductance and volumetric studies in aqueous solutions of DL-pyroglutamic acid, *J. Solution Chem.*, 2005, **34**(7), 853–867.
[66] G. Manov, N. DeLollis and S. Acree, Ionization constant of boric acid and the pH of certain borax-chloride buffer solutions from 0° to 60° C, *J. Res. Natl. Bur. Stand.*, 1944, **33**, 287–306.
[67] R. Bates and S. Acree, pH values of certain phosphate-chloride mixtures, and the second dissociation constant of phosphoric acid from 0° to 60° C, *J. Res. Natl. Bur. Stand.*, 1943, **30**, 129–155.
[68] H. Krebs and J. Speakman, Dissolution constant, solubility and the pH value of the solvent, *J. Chem. Soc.*, 1945, 593–595.
[69] P. Ihnat and D. Robinson, Potentiometric determination of the thermodynamic ionization constants of deferoxamine, *J. Pharm. Sci.*, 1993, **82**, 110–112.
[70] R. Robinson and A. Kiang, The ionization constants of vanillin and two of its isomers, *Tr. Farad. Soc.*, 1955, **51**, 1398–1402.
[71] R. Bates and G. Schwarzenbach, Die Bestimmung thermodynamischer Aciditatskonstanten, *Helv. Chim. Acta*, 1954, **37**, 1069–1079.
[72] A. Wahbe, F. El-Yazbi, M. Barary and S. Sabri, Application of orthogonal functions to spectrophotometric analysis. Determination of dissociation constants, *Int. J. Pharm.*, 1993, **92**(1), 15–22.
[73] A. Asuero, M. Herrador and A. Camean, Spectrophotometric evaluation of acidity constants of diprotic acids: Errors involved as a consequence of an erroneous choice of the limit absorbances, *Anal. Lett.*, 1986, **19**, 1867–1880.
[74] N. Farraj, S. Davis, G. Parr and H. Stevens, Dissociation and partitioning of progabide and its degradation product, *Int. J. Pharm.*, 1988, **46**, 231–239.
[75] R. Kelly and S. Schulman, Proton transfer kinetics of electronically excited acids and bases, in *Molecular Luminescence Spectroscopy—Methods and Applications: Part 2* (ed. S. Schulman), 1st edn., Wiley-Interscience, New York, 1988, pp. 461–510.
[76] H. Wan, A. Holmen, Y. Wang, W. Lindberg, M. Englund, M. Nagard and R. Thompson, High-throughput screening of pKa values of pharmaceuticals by pressure-assisted capillary electrophoresis and mass spectrometry, *Rapid Commun. Mass Spectrom.*, 2003, **17**, 2639–2648.
[77] Q. Hu, G. Hu, T. Zhou and Y. Fang, Determination of dissociation constants of anthrocycline by capillary zone electrophoresis with amperometric detection, *J. Pharm. Biomed. Anal.*, 2003, **31**, 679–684.

[78] K. Connors, Hydrochlorothiazide, in *Chemical Stability of Pharmaceuticals—A Handbook for Pharmacists* (eds. K. Connors, G. Amidon and V. Stella), 2nd edn., Wiley-Interscience, New York, 1986, pp. 478–482.
[79] S. Unger, J. Cook and J. Hollenberg, Simple procedure for determining octanol-aqueous partition, distribution, and ionization coefficients by reversed phase high pressure liquid chromatography, *J. Pharm. Sci.*, 1978, **67**, 1364–1367.
[80] S. Valentiner, The solubility of the noble gases in water, *Z. fur. Physik.*, 1927, **42**, 253–264.
[81] A. Lannung, The solubilities of helium, neon and argon in water and some organic solvents, *JACS*, 1930, **52**, 67–80.
[82] D. Everett and W. Wynne-Jones, Thermodynamics of acid-base equilibria, *Trans. Farad. Soc.*, 1939, **35**, 1380–1401.
[83] M. Blandamer, R. Robertson and J. Scott, An examination of the parameters describing the dependence of rate constants on temperature for solvolysis of various organic esters in water and aqueous mixtures, *Can. J. Chem.*, 1980, **58**, 772–776.
[84] M. Blandamer, R. Robertson, J. Scott and A. Vrielink, Evidence for the incursion of intermediates in the hydrolysis of tertiary, secondary and primary substrates, *JACS*, 1980, **102**, 2585–2592.
[85] M. Blandamer, J. Burgess, P. Duce, R. Robertson and J. Scott, A re-examination of the effects of added solvent on the activation parameters for solvolysis of *t*-butyl chloride in water, *JCS Farad. Trans. I*, 1981, **77**, 1999–2008.
[86] D. Grant, M. Mehdizadeh, A.-L. Chow and J. Fairbrother, Non-linear van't Hoff solubility-temperature plots and their pharmaceutical interpretation, *Int. J. Pharm.*, 1984, **18**, 25–38.
[87] R. Prankerd and R. McKeown, Physico-chemical properties of barbituric acid derivatives. Part I. Solubility-temperature dependence for 5,5-disubstituted barbituric acids in aqueous solutions, *Int. J. Pharm.*, 1990, **62**(1), 37–52.
[88] R. Prankerd, Solid state properties of drugs. Part I. Estimation of heat capacities for fusion and thermodynamic functions for solution from aqueous solubility-temperature dependence measurements, *Int. J. Pharm.*, 1992, **84**(3), 233–244.
[89] R. McKeown, R. Prankerd and O. Wong, The Development of Drugs and Modern Medicines—The Beckett Symposium Proceedings, London, 1986, pp. 80–89.
[90] H. Harned and W. Hamer, The ionization constant of water and the dissociation of water in potassium chloride solutions from electromotive forces of cells without liquid junction, *JACS*, 1933, **55**, 2194–2205.
[91] H. Harned and R. Robinson, Temperature variation of the ionization constants of weak electrolytes, *Trans. Farad. Soc.*, 1940, **36**, 973–978.
[92] R. Ramette, On deducing the pK-temperature equation, *J. Chem. Educ.*, 1977, **54**(5), 280–283.
[93] R. Robinson and R. Stokes, *Appendix 12.2. Electrolyte Solutions*, 2nd Revised edn., Butterworths, London, 1971, p. 544.
[94] H. Harned and B. Owen, The Physical Chemistry of Electrolyte Solutions, in *The Physical Chemistry of Electrolyte Solutions* (eds. H. Harned and B. Owen), 3rd edn., Reinhold, New York, 1958, Ch. 15.
[95] A. Covington, M. Ferra and R. Robinson, Ionic product and enthalpy of ionization of water from electromotive force measurements, *J. Chem. Soc. Farad. Trans. I*, 1977, **73**, 1721–1730.
[96] L. Eberson, Acidity and hydrogen bonding of carboxyl groups, in *The Chemistry of Carboxylic Acids and Esters* (ed. S. Patai), 1st edn., Interscience Publishers, New York, 1969.
[97] R. Robinson and R. Stokes, *Appendix 12.1. Electrolyte Solutions*, 2nd Revised edn., Butterworths, London, 1971.
[98] J. Talmage, L. Chafetz and M. Elefant, Observation on the instability of cyclamate in hydro-alcoholic solution, *J. Pharm. Sci.*, 1968, **57**, 1073–1074.
[99] D. Newton, W. Murray and M. Lovell, pKa determination of benzhydrylpiperazine antihistamines in aqueous and aqueous methanol solutions, *J. Pharm. Sci.*, 1982, **71**(12), 1363–1366.
[100] J. Rubino, Electrostatic and non-electrostatic free energy contributions to acid dissociation constants in cosolvent-water mixtures, *Int. J. Pharm.*, 1988, **42**, 181–191.
[101] C. Davies, *Ion Association*, 1st edn., Butterworths, London, 1962.

[102] R. Prankerd, *Phenylbutazone. 4th Year Project Report, B. Pharm*, University of Otago, Dunedin, New Zealand, 1974.
[103] R. Prankerd, *A study of some physical properties and their relationships with the biological activities of barbituric acids*, Ph.D. Thesis, University of Otago, Dunedin, New Zealand, 1985.
[104] J. Partanen, P. Juusola and P. Minkkinen, Determination of stoichiometric dissociation constants of lactic acid in aqueous salt solutions at 291.15 and 298.15 K, *Fluid Phase Equilib.*, 2003, **204**, 245–266.
[105] L. Al-Razzak, A. Benedetti, W. Waugh and V. Stella, Chemical stability of pentostatin (NSC-218321), a cytotoxic and immunosuppressive agent, *Pharm. Res.*, 1990, **7**, 452–460.
[106] C. T. Flear, S. W. Roberts, S. Hayes, J. C. Stoddart and A. K. Covington, pK1′ and bicarbonate concentration in plasma, *Clin. Chem.*, 1987, **33**(1), 13–20.
[107] A. Covington, R. Robinson and M. Sarbar, Determination of carbonate in the presence of hydroxide. Part 2. Evidence for the existence of a novel species from first-derivative potentiometric titration curves, *Anal. Chim. Acta*, 1981, **130**, 93–102.
[108] S. Singh, N. Sharda and L. Mahajan, Spectrophotometric determination of pKa of nimesulide, *Int. J. Pharm.*, 1999, **176**, 261–264.
[109] U. Chaudry and P. Popelier, Estimation of pKa using quantum topological molecular similarity descriptors: Application to carboxylic acids, anilines and phenols, *J. Org. Chem.*, 2004, **69**(2), 233–241.

APPENDIX A

APPENDIX A. MAIN LIST

pK_a values found with significant data quality information—mainly primary literature

Reliability assessment (data quality) is based on information in the original source describing the method used (including evidence for calibration of pH meters; exclusion of CO_2 in determining pK_a values above 6.5), whether pK_a values for standard compounds were measured, presence of organic cosolvents, the presence or absence of corrections for $[H^+]$, $[OH^-]$ in potentiometric titrations, and use of mean ionic activity coefficients in the calculations. Considerable effort has been made to locate the original source for each measured pK_a value. Where only secondary sources have been located, data reliability cannot be assessed with confidence.

A small number of journal or other serial titles have been abbreviated in the tables. These include Analytical Profiles of Drug Substances (APDS) (this abbreviation is also used for the longer and more detailed recent titles of this series); *J. Am. Chem. Soc.* (JACS); *J. Chem. Soc.* (JCS); *J. Org. Chem.* (JOC); *J. Pharmacol. Expt. Ther.* (JPET); *J. Pharm. Pharmacol.* (JPP). Further abbreviations are found at the beginning of Appendix B.

No.	Name	pK_a value(s)	Data quality	Ionization type	Method	Conditions t°C; I or c M	Comments and Reference(s)
			See comments	($^+$H or $^-$H)		Solvent t (°C) Ionic strength (I) or analyte concentration (c) in molar (M) units	Data reliability cutoff points: R: Reliable = ± <0.02; A: Approx = ± 0.02 to ± 0.06; U: Uncertain = ± >0.06 These overall data reliability cutoffs apply when all other aspects of the pK_a values have been considered. Where key variables (temperature, ionic strength and solvent composition) have not been reported, or the value was obtained from a computer program, a value is automatically assessed as U: Uncertain. A few results have been classified as VR: Very Reliable, while a few others are VU: Very Uncertain.
1	Acetaminophen (paracetamol) ($C_8H_9NO_2$) HO—C6H4—NHAc	9.63 ± 0.01	A	−H	Potentiometric	H_2O t = 25.0 ± 0.1 I = 0.1 (NaCl) N_2 atmosphere	Takacs-Novak K and Avdeef A, Interlaboratory study of log P determination by shake-flask and potentiometric methods, *J. Pharm. Biomed. Anal.*, **14**, 1405–1413 (1996). "Titration in aqueous medium: Typically, 10 ml of 0.5 to 10 mM solutions of the samples were preacidified to pH 1.8–2.0 with 0.5 M HCl, and were then titrated alkalimetrically to some appropriate high pH (maximum 12.0). The titrations were carried out at 25.0 ± 0.1 °C, at constant ionic strength using NaCl, and under an inert gas atmosphere. The initial estimates of pK_a values were obtained from Bjerrum difference plots (n_H vs. pH) and then were refined by a weighted nonlinear least-squares procedure (Avdeef, 1992, 1993). For each molecule a minimum of three and occasionally five or more separate titrations were performed and the average pK_a values along with the standard deviations were calculated." "Titrations in solvent mixtures: A series of semi-aqueous solutions of the samples, containing 3–60% (w/w) methanol were titrated. From these titrations, the p_sK_a values (the apparent ionization constants in methanol-water solvent) were obtained, and the Yasuda-Shedlovsky procedure was applied to estimate the aqueous pK_a values (Avdeef, *et al.*, *Anal. Chem.*, **65**, 42–49 (1993)). The four-parameter procedure was used for electrode standardization in both aqueous and semiaqueous solutions."

2	Acetaminophen (paracetamol) ($C_8H_9NO_2$) HO–C_6H_4–NHAc	9.67 ± 0.08	U	–H	Potentiometric	H_2O $t = 25.0 \pm 0.1$ $I = 0.1$ (NaCl) N_2 atmosphere	Takacs-Novak K, Box KJ and Avdeef A, Potentiometric pK_a determination of water-insoluble compounds: Validation study in methanol/water mixtures, *Int. J. Pharm.*, 151, 235–248 (1997). "Titration in aqueous medium: Ten ml of 1 mM or 5 mM aqueous solutions of the samples were pre-acidified to pH 1.8–2.0 with 0.5 M HCl, and were then titrated alkalimetrically to some appropriate high pH (maximum 12.5). The titrations were carried out at 25.0 ± 0.1 °C, at I = 0.1 M ionic strength using NaCl, and under N_2 atmosphere. The initial estimates of pK_a values were obtained by difference plots (n_H vs. pH, where n_H is the average number of bound protons) and were then refined by a weighted non linear least-squares procedure (Avdeef, 1992, 1993). For each molecule a minimum of three and occasionally five or more separate titrations were performed and the average pK_a values along with the standard deviations were calculated." "Titrations in solvent mixtures: A series of 1 mM or 5 mM semiaqueous solutions of the samples, containing 3–70 wt% methanol were titrated under the same conditions as in aqueous titrations. For all the molecules of the validation set (group 1) measurements were carried out at six different R values ranging from 15 to 64 wt%. Titrations at each methanol/water mixture were repeated three times, and then the average of the p_sK_a values was calculated (Table 2). The Yasuda-Shedlovsky procedure was applied to estimate the aqueous pK_a values (Avdeef, *et al.*, *Anal. Chem.*, **65**, 42–49, (1993))." NB: pK_a = 9.67 ± 0.08 by extrapolation from 15.7–60.1%w/w aqueous MeOH. See other Avdeef papers in this reference for calibration procedure.
3	Acetaminophen (paracetamol)	9.55 ± 0.03	U	–H	Spectro	$t = 25$	Dobas I, Sterba V and Vecera M, Kinetics and mechanism of diazo coupling. X. The coupling kinetics of para-substituted phenol, *Collection of Czechoslovak Chem. Communications*, **34**(12), 3746–3754 (1969). Cited in Fairbrother JE, Acetaminophen, *APDS*, **3**, 1974, 27.
4	Acetaminophen (paracetamol)	9.35 ± 0.03	U	–H	Spectro	$t = 25$	Talukdar PB, Banerjee S and Sengupta SK, Intramolecular hydrogen bonding in 2-(substituted amino-) methyl-4-acetamidophenol, *J. Indian Chem. Soc.*, **47**(3) 267–272 (1970). Cited in Fairbrother JE, Acetaminophen, *APDS*, **3**, 1974, 27.
5	Acetaminophen (paracetamol)	9.75 (0.06)	U	–H	Spectro ($\lambda = 259$ nm)	H_2O $t = 20$	Wahbe AM, El-Yazbi FA, Barary MH and Sabri SM, Application of orthogonal functions to spectrophotometric analysis. Determination of dissociation constants, *Int. J. Pharm.*, **92**(1) 15–22 (1993); Wahbi AM, El-Yazbi FA, Hewala II and Awad AA, Use of

(continued)

No.	Name	pK_a value(s)	Data quality	Ionization type	Method	Conditions t°C; I or c M	Comments and Reference(s)
							ratios of orthogonal function coefficients for the determination of dissociation constants, *Pharmazie*, **48**, 422–425 (1993). NB: pH values were measured at 20 °C but it was not clear if the spectral data were obtained at this temperature. No details given for pH meter calibration or corrections for ionic strength. The orthogonal method is intended to correct for the effects of spectra which overlap for the protonated and deprotonated forms of the ionizing species. The error given for the pK_a value (in parentheses) is the "overall relative standard deviation", but this term is not defined. An alternative graphical method gave $pK_a = 9.65$.
6	Acetaminophen (paracetamol)	9.75	U	−H	CE/pH (+ve ion mode)	H_2O $t = 25$ $I = 0.025$	Wan H, Holmen AG, Wang Y, Lindberg W, Englund M, Nagard MB and Thompson RA, High-throughput screening of pK_a values of pharmaceuticals by pressure-assisted capillary electrophoresis and mass spectrometry, *Rapid Commun. Mass Spectrom.*, **17**, 2639–2648 (2003). NB: Reported a literature value of 9.56 (Box K, Bevan C, Comer J, Hill A, Allen R and Reynolds D, *Anal. Chem.*, **75**, 883–892 (2003)), and a predicted value (ACD Labs) of 9.86.
7	Acetaminophen (paracetamol)	GLpK_a: 9.45 ± 0.01	U	−H	Spectro	H_2O $t = 25$ $I = 0.15$ (KCl)	Tam KY and Takacs-Novac K, Multi-wavelength spectrophotometric determination of acid dissociation constants, *Anal. Chim. Acta*, **434**, 157–167 (2001).
		A&S: 9.58 ± 0.15	U	−H		Ar atmosphere	NB: See Clioquinol for details.
8	Acetanilide (C_8H_9NO) [structure: benzene ring–NHCOCH$_3$]	0.5	U	+H	Potentiometric	H_2O $t = 25 \pm 2$	Hall NF, The strength of organic bases in glacial acetic acid solution, *JACS*, **52**, 5115–5128 (1930). Cited in Perrin DD, Dissociation Constants of Organic Bases in Aqueous Solution, 1965, Butterworths, Lond (1965) No. 389. Ref. H11. Other values, also uncertain (U), are reported in Perrin.
9	Acetanilide (Antifebrin)	1.4	U	+H	Potentiometric	H_2O $t = 25$	Evstratova KI, Goncharova NA and Solomko VIa, Dissociation constants of weak organic bases in acetone, *Farmatsiya* (Moscow), **17**(4), 33–36 (1968). Abstract: The pK_a and pK_b values for some pharmaceutically important organic bases in water and 90% acetone solns. were calculated from emf. data measured in the Izmailov's arrangement (cf. I., CA 54: 114f). The results are (base, pK_a in H_2O, pK_a in aqueous acetone, pK_b in H_2O, and pK_b in aqueous acetone given): $MeNH_2$ 10.6, 10.4, 3.4, 9.9; Et_2NH 10.7, 9.4, 3.3, 10.9; Et_3N 10.7, 11.4, 3.3, 9.2; antifebrin 1.4, 4.4, 12.6, 15.9; phenacetin approx. 2.2, 3.5, ~11.8, 16.8; novocaine 8.85, 9.6, 5.15, 10.7; spasmolytine 7.7, 8.7, 6.3, 11.6; benzacine 7.8, 7.2, 6.2, 13.1; chloridine 5.6, 7.0, 8.4, 13.3; diethazine 7.0, 8.5, 7.0, 11.8; amizil 8.5, 8.2, 5.5, 12.2; sarcolysine II ~2.5, 5.9, ~11.5, 14.1; apressine 7.1, 4.7, 6.9, 15.6; promedol 8.4, 3.4, 5.6, 16.9;

No.	Compound	pKa			Method	Conditions	Comments and Reference
							dimedrol 8.2, 9.1, 5.8, 11.2; antipyrine ~2.2, 4.4, ~11.8, 15.9; bendazol 4.2, 4.4, 9.8, 15.9; Pyramidone 4.84, 4.9, 9.16, 15.4; acriquine 6.5, 8.9, 7.5, 11.4; Dionine 7.9, 9.0, 6.1, 11.3; caffeine 0.61, 4.4, 13.39, 15.9; methylcaffeine ~2.6, 4.0, ~11.4, 16.3; urotropine 4.9, 6.7, 9.1, 13.6; theophylline ~2.6, 4.4, ~11.5, 16.0; theobromine 0.11, 3.8, 13.89, 16.5; pyridine 5.31, 4.5. 8.69, 15.8; atropine 9.65, 9.9, 4.35, 10.4; pachycarpine 11.76, 5.5, 2.24, 14.8; salsoline 8.83, 10.1, 5.17, 10.2; salsolidine 9.11, 9.7, 4.89, 10.6; papaverine 5.9, 6.2, 8.1, 14.1; morphine 7.87, 8.7, 6.13, 11.6; codeine 7.95, 7.8, 6.05, 12.5; quinine 8.0, 9.3, 6.0, 11.0; platyphylline 8.1, 9.5, 5.95, 10.8; pilocarpine 6.85, 7.4, 7.15, 12.9; and ephedrine 9.66, 10.4, 4.34, 9.9.
10	Acetarsone ($C_8H_{10}AsNO_5$)	3.73 7.9 9.3	U U U	–H –H –H	Potentiometric	H_2O	Hiskey CF and Cantwell FF, Ultraviolet spectrum correlations with the conjugate acid-base species of acetarsone and arsthinol, *J. Pharm. Sci.*, **57**, 2105–2111 (1968). "...About 1.5 mmoles of the compound was suspended in a known volume of carbon dioxide-free water and titrated with ... 0.1 N sodium hydroxide. ... a vigorous stream of nitrogen was blown into the solution to prevent carbon dioxide uptake. ... Acetarsone has a very limited solubility in water .... Consequently, it was necessary in the initial stages of the titration to wait after each addition of alkali for the pH meter to drift to its final steady state values. During this period, more undissociated acetarsone went into solution until both the solubility and ionization equilibria were satisfied. This situation continued until about 80% of the first equivalent of alkali had been added, at which point all the dissolved acid was in solution and .... equilibrium was established quickly....
	[Structure: benzene ring with $AsO(OH)_2$, $NHCOCH_3$, OH]						
	Arsthinol	9.5 ± 0.1	U	–H			The solubility of undissociated acetarsone was determined ... to be 1.74×10^{-3} M in 0.1 N HCl. This value was used in estimating the pK_{a1} value, by ... fitting the potentiometric ... data to: $[H] - [OH] = (K_1S/H) - CB$ and $[H] - [OH] = \{mK_1/([H] + K_1)\} - CB$ where S is the solubility of undissociated acetarsone, CB is the added concentration of strong base and m is the formal concentration of acetarsone.... Separation of the pK_{a2} and pK_{a3} values was performed by a method developed by Linderstrom-Lang and modified by Bjerrum."
11	Acetazolamide ($C_4H_6N_4O_3S_2$) [Structure: 1,3,4-thiadiazole with CH_3CONH and SO_2NH_2]	7.2 8.8	U U	–H –H	Potentiometric	H_2O *t* unspecified *I* unspecified	Coleman JE, *Ann. Rev. Pharmacol.*, **15**, 238–240 (1975). NB: Reported the macroconstants and microconstants for acetazolamide by titration. $pK_1 = 7.20$; $pK_2 = 8.8$; $pK_{12} = 7.46$ (sulfa anion); $pK_{13} = 7.55$ (acetamide anion); $pK_{24} = 8.54$ (dianion); $pK_{34} = 8.45$ (dianion); $K_1 = K_{12} + K_{13}$; $1/K_2 = 1/K_{24} + 1/K_{34}$. Ref: Lindskog S. *In* CO_2: Chemical, Biochemical and Physiological Aspects. National Aeronautics and Space Administration, Special Pub. #SP-188, 157. Washington, DC, NASA. NB: More information is needed than a titration can provide to obtain the microconstants. Also gave chloroacetazolamide, $pK_1 = 5.7$ (chloroacetamide), $pK_2 = 8.4$ (sulfonamide).

(continued)

No.	Name	pK_a value(s)	Data quality	Ionization type	Method	Conditions t°C; I or c M	Comments and Reference(s)
12	Acetic acid ($C_2H_4O_2$) CH_3COOH	4.756	R	−H	Conductance Electrometric	H_2O $t = 25.0 \pm 0.1$ $I = 0.000$	MacInnes D and Shedlovsky T, The determination of the ionization constant of acetic acid, at 25 °C, from conductance measurements, *JACS*, **54**, 1429–1438 (1932); Harned H and Ehlers R, The dissociation constant of acetic acid from 0 to 35° centigrade, *JACS*, **54**, 1350–1357 (1932).
13	*N*-Acetylaminosalicylic acid ($C_9H_9NO_4$)	2.7	U	−H	^{13}C-NMR/pH	H_2O t unspecified I unspecified	Allgayer H, Sonnenbichler J, Kruis W, Paumgartner G, Determination of the pK_a values of 5-aminosalicylic acid and *N*-acetylaminosalicylic acid and comparison of the pH dependent lipid-water partition coefficients of sulphasalazine and its metabolites, *Arzneim.-Forsch.*, **35**(9), 1457–1459 (1985) "Extrapolated from d_4 methanol/H_2O mixtures or d_7 dimethylformamide/H_2O mixtures. . . . By a series of ^{13}C-NMR spectra at different pH (from pH 1.0 to 14 in one unit steps) the chemical shifts were recorded. The turning points gave the pK values of the different substituents of 5-ASA and AcASA, respectively. . . . the resonance of the substituted carbon atoms and the corresponding o- and p- positions were mainly shifted. . . . Thus, the following pK values were obtained: 5-ASA -COOH group: 3.0; -NH_3^+: 6.0; -OH: 13.9; AcASA-COOH: 2.7; -OH:12.9."
		12.9	U	−H			
14	α-Acetylmethadol (levomethadyl acetate) ($C_{23}H_{31}NO_2$)	8.3	U	+H	Potentiometric	H_2O $I = 0.1$ (NaCl)	Schanker LS, Shore PA, Brodie BB and Hagben CAM, Absorption of drugs from the stomach. I. The rat, *JPET*, **120**, 528–539 (1957).
15	6-Acetylmorphine ($C_{19}H_{21}NO_4$)	8.19	U	+H	Potentiometric	H_2O $t = 25.0 \pm 0.1$ $I = 0.15$ (KCl) under Ar	Avdeef A, Barrett DA, Shaw PN, Knaggs RD and Davis SS, Octanol-, chloroform-, and propylene glycol dipelargonat-water partitioning of morphine-6-glucuronide and other related opiates, *J. Med. Chem.*, **39**, 4377–4381 (1996). NB: See Morphine for details.
		9.55	U	−H			

16	Acetylpromazine (acepromazine) ($C_{19}H_{22}N_2OS$) 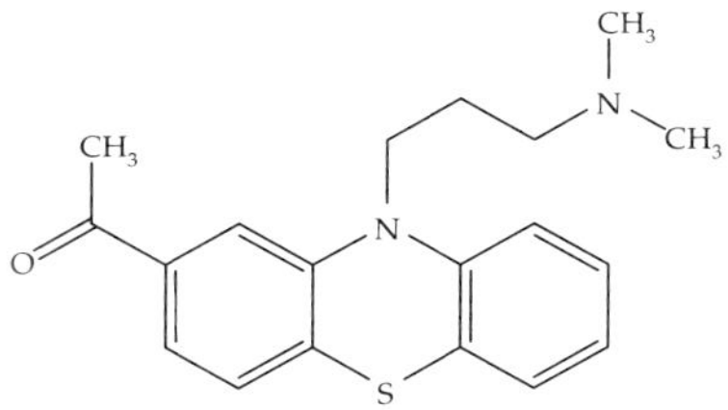	9.3	U	+H	soly-pH	H_2O $t = 25.0 \pm 0.01$ I undefined N_2 atmosphere	Liu S and Hurwitz A, The effect of micelle formation on solubility and pK_a determination of acetylpromazine maleate, *J. Colloid Interface Sci.*, **60**, 410–413 (1977). NB: I was undefined, but was kept low through adjustment of pH with additions of HCl or NaOH only.
17	Aconitine ($C_{34}H_{47}NO_{11}$) 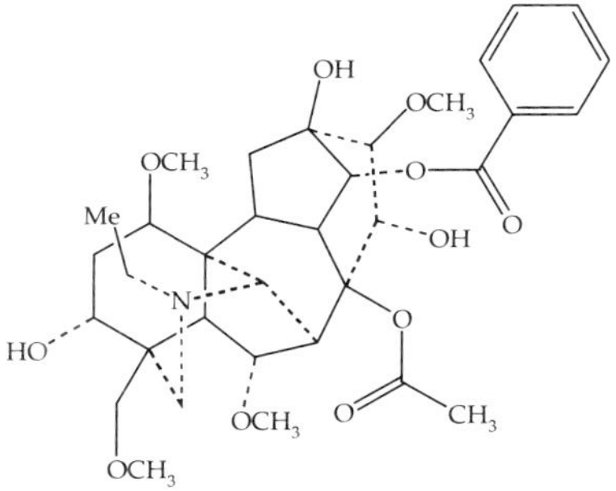	8.35	U	+H	Spectro	H_2O $t = 15$ $c = 0.004$ to 0.01	Kolthoff IM, The dissociation constants, solubility product and titration of alkaloids, *Biochem. Z.*, **162**, 289–353 (1925). Cited in Perrin Bases no. 2857 K47. Used the indicator method, where the pH of a solution of known molar ratio of protonated and deprotonated forms of the test substance is quantified by the color of a visual indicator of known pK_a value. Several different indicators are proposed, including methyl red, dimethyl yellow, methyl orange, or bromophenol blue. Applicable only where the compound is a univalent alkaloidal base, with $K_b > 5 \times 10^{-7}$. Data given for the following bases: pyridine, piperidine, pyrrole, quinoline, isoquinoline, piperazine, benzocaine, novocaine, coniine, piperine, arecoline, nicotine, atropine, tropacocaine, pseudotropine, cocaine, ecgonine, anhydroecgonine, benzoylecgonine, sparteine, pelletierine, quinine, quinidine, cinchonine, cinchonidine, cupreine, hydroquinine, strychnine, morphine, brucine, cytisine, papaverine, narcotine, narceine, thebaine, codeine, dionine, apomorphine, hydrastine, hydrastinine, berberine, pilocarpine, isopilocarpine, aconitine, colchicine, emetine, solanine, physostigmine, and cevadine.
18	Acridine derivatives				Spectro	H_2O T = room temperature (RT) I unspecified	Schulman SG, Naik DV, Capomacchia AC and Roy T, Electronic spectra and electronic structures of some antimicrobials derived from proflavine, *J. Pharm. Sci.*, **64**, 982–986 (1975).
	Acridine orange ($C_{17}H_{19}N_3$) 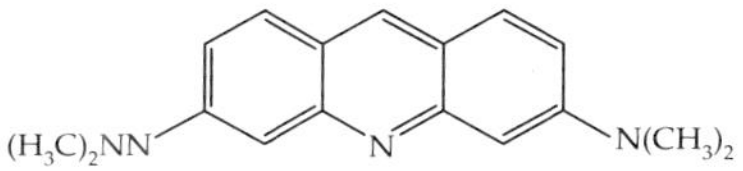	−3.3 0.2 10.1	U U U	+H +H +H			"The shifts in the absorption and fluorescence spectra of 3-aminoacridine, proflavine, acridine orange, and acridine yellow were employed to show that the singly charged cations, the predominant species at biological pH, exist in the ground state in the amino form. In the lowest excited singlet state, however, the monocations of the diaminoacridines have the imino structure, a conclusion supported by the relative ground and excited state pK_a values of the reactions of the monocation with H^+. The ground state amino structure has its positive charge concentrated at the heterocyclic nitrogen atom, a fact that is of primary importance in determining the geometry of binding to DNA."

(continued)

Appendix A *(continued)*

No.	Name	pK_a value(s)	Data quality	Ionization type	Method	Conditions t°C; I or c M	Comments and Reference(s)
							NB: No details were given of the pH meter calibration. Fluorometric measurements were used to assess the pK_a* values for the first excited state. Used previously described spectrophotometric titration procedures: Cappomacchia A, Casper J and Schulman SG, *J. Pharm. Sci.*, **63**, 1272–1276 (1974). Acridine yellow = 2,7-dimethyl-3,6-diaminoacridine; Proflavine = 3,6-diaminoacridine
	Acridine yellow ($C_{15}H_{15}N_3$)	−3.0 0.5 8.9	U U U	+H +H +H			
	Proflavine ($C_{13}H_{11}N_3$)	−2.7 0.3 9.5	U U U	+H +H +H			
	3-Aminoacridine ($C_{13}H_{10}N_2$)	−1.4 8.0	U U	+H +H			
19	Acridine derivatives				Spectro	H_2O t = RT I unspecified	Cappomacchia A, Casper J and Schulman SG, Valence tautomerism of singly protonated 9-aminoacridine and its implications for intercalative interactions with nucleic acids, *J. Pharm. Sci.*, **63**, 1272–1276 (1974).

No.	Compound	pK_a	Data quality	Ionization type	Method	Conditions	Comments and Reference(s)
	2-aminoacridine ($C_{13}H_{10}N_2$)	1.1 5.9 –	U U	+H +H −H			By measurements of fluorescence/pH, the excited state pK_a* values are: (see table below)
	9-aminoacridine ($C_{13}H_{10}N_2$)	−8.5 10.0 15.9	U U U	+H +H −H			
20	Acyclovir ($C_8H_{11}N_5O_3$)	2.41 ± 0.27 9.06 ± 0.88	U U	+H −H	partition	H_2O t = RT I = 0.35	Kristl A, Mrhar A and Kozjek F and Ionization properties of acyclovir and deoxyacyclovir, *Int. J. Pharm.*, **57**, 229–234 (1989); *ibid.* **99**, 79–82 (1993). "The ionization constant of N2-acetylacyclovir was determined and compared with those reported for acyclovir and desciclovir (deoxyacyclovir) with the aim of locating basic and acidic moieties in the molecules. For acyclovir, introducing a 2-hydroxyethoxymethyl group at position 9 of guanine lowered the basic strength of the molecule, whereas the acidic ionization constant remained unchanged. For desciclovir, basic strength was increased and acidic character was lost. For N2-acetylacyclovir, only acidic properties were present. Introduction of an acetyl group into the acyclovir molecule at the 2-NH_2 position hindered the basic character. It was concluded that the weak basic properties of this series are attributed to the imidazole moiety with the contribution of the 2-NH_2 group, while the acidic moiety is at the oxygen atom."
	4-Deoxyacyclovir ($C_8H_{13}N_5O_2$) N^2-Acetylacyclovir ($C_{10}H_{13}N_5O_4$)	3.63 ± 0.09 8.54 ± 0.03	U U	+H −H	partition Spectro	H_2O t = RT I = 0.25	

	+H	+H	−H
2-aminoacridine	−5.8	–	12.4
9-aminoacridine	−6.8	–	13.2

(continued)

No.	Name	pK_a value(s)	Data quality	Ionization type	Method	Conditions t°C; I or c M	Comments and Reference(s)
21	Acyclovir	2.34 9.23	U U	+H −H	Potentiometric	H_2O $t = 25$	Bergström CAS, Strafford M, Lazorova L, Avdeef A, Luthman K and Artursson P, Absorption classification of oral drugs based on molecular surface properties, *J. Med. Chem.*, **46**(4), 558–570 (2003). NB: From extrapolation of aqueous-methanol mixtures to 0% methanol.
		2.16 9.04	U U	+H −H	Potentiometric	H_2O $t = 37$ $I = 0.15$ (KCl)	Balon K, Riebesehl BU, Muller BW, Drug liposome partitioning as a tool for the prediction of human passive intestinal absorption, *Pharm. Res.*, **16**, 882–888 (1999).
22	Adenine ($C_5H_5N_5$) 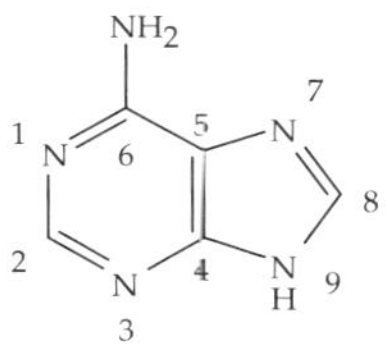	4.2 ± 0.1 10.1 ± 0.2	U U	+H (N1) −H (N9)	^{15}N-NMR	H_2O t undefined I undefined	Gonnella NC, Nakanishi H, Holtwick JB, Horowitz DS, Kanamori K, Leonard NJ and Roberts JD, Studies of tautomers and protonation of adenine and its derivatives by nitrogen-15 nuclear magnetic resonance spectroscopy, *JACS*, **105**, 2050–2055 (1983). NB: Apparent pK_a values were reported for two adenine derivatives in 66% aqueous DMF: 2-butyladenine, 4.3, 11.3; 8-butyladenine, 3.8, 11.7.
23	Adenosine ($C_{10}H_{13}N_5O_4$) 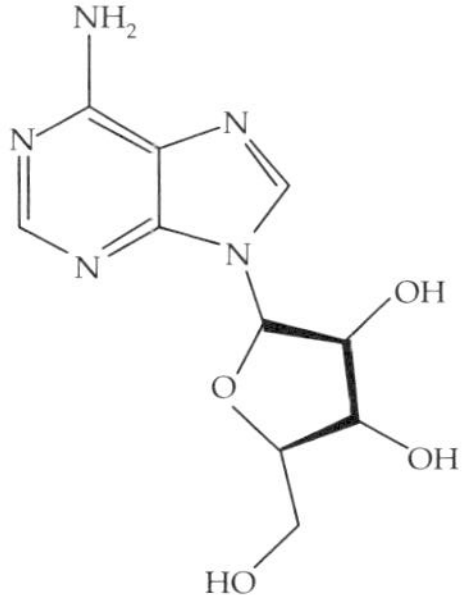	12.35 ± 0.03	A	−H	calorimetric titration	H_2O $t = 25$ $I = 0.00$	Izatt RM, Hansen LD, Rytting JH and Christensen JJ, Proton ionization from adenosine, *JACS*, **87**, 2760–2761 (1965). NB: Adenosine (0.01 M) (and several related compounds) were titrated with 0.6 M NaOH in a precision titration calorimeter. I was extrapolated to zero. The result was assigned to an average for ionization of the 2′ and 3′ hydroxyl groups of the ribose.

No.	Compound	pKa	Data quality	Ionization type	Method	Conditions	Comments and reference(s)
24	Adinazolam ($C_{19}H_{18}ClN_5$) 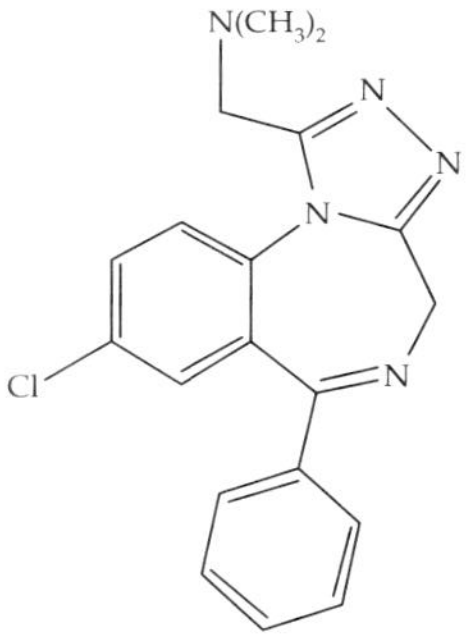	6.2	U	+H	NMR/pH	D_2O/DCl or D_2O/NaOD	Ogata M, Tahara T and Nishimura T, NMR spectroscopic characterization of adinazolam mesylate: pH-dependent structure change in aqueous solution and active methylene, *J. Pharm. Sci.*, **84**, 786–790 (1995).
		5.8 ± 0.1	U	+H	Potentiometric	H_2O	"The structural changes of adinazolam mesylate in aqueous solution under various pH conditions by NMR spectroscopy were investigated; by plotting of the signal integration and chemical shifts of the side chain, pK_a . . . for . . . the side chain amine were determined. Conformational analysis of the side chain was performed with deuterium-induced isotope effects on chemical shifts, nuclear Overhauser effects, relaxation times, and energy calculations." NB: The low value for the 3° amine pK_a was ascribed to interactions with the triazole ring.
25	Adrenaline						See Epinephrine
26	Albendazole sulphoxide ($C_{12}H_{15}N_3O_3S$) 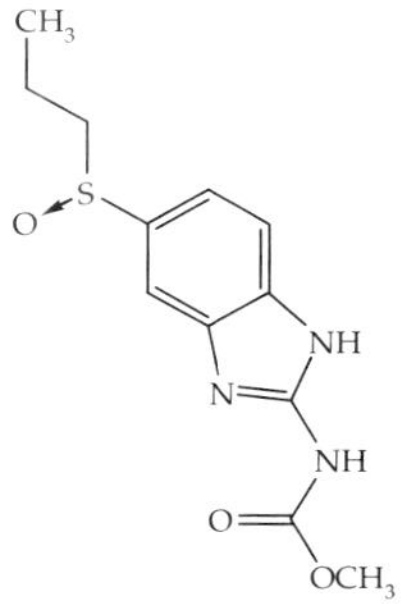	3.28 ± 0.01	A	+H	Potentiometric	H_2O $t = 25.0 \pm 0.1$ $I = 0.1$ (NaCl)	Takacs-Novak K, Box KJ and Avdeef A, Potentiometric pK_a determination of water-insoluble compounds: Validation study in methanol/water mixtures, *Int. J. Pharm.*, **151**, 235–248 (1997). NB: p$K_{a1} = 3.32 \pm 0.08$ and p$K_{a2} = 9.99 \pm 0.04$ by extrapolation from 15.2–52.8 %w/w aqueous MeOH. See Acetaminophen for details.
		9.93 ± 0.01	A	−H			
27	Albuterol (salbutamol) ($C_{13}H_{21}NO_3$) 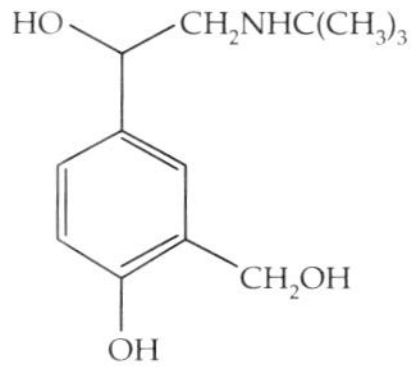	9.07	U	−H, +H	Spectro	H_2O $t = 25.0 \pm 0.05$ $I = 0.10$	Ijzerman AP, BultsmaT, Timmerman H and Zaagsma J, The ionization of β-adrenoceptor agonists: A method for unravelling ionization schemes, *J. Pharm. Pharmacol.*, **36**(1), 11–15 (1984). NB: Microscopic: 9.22 and 10.22. See Isoprenaline.
		10.37	U	+H, −H			

(continued)

No.	Name	pK_a value(s)	Data quality	Ionization type	Method	Conditions t°C; I or c M	Comments and Reference(s)
28	Alprenolol ($C_{15}H_{23}NO_2$) HO, $CH_2NHCH(CH_3)_2$, O, $CH_2CH{=}CH_2$	9.6	U	+H	Potentiometric		Barbato F, Caliendo G, LaRotonda MI, Morrica P, Silipo C and Vittoria A, Relationships between octanol-water partition data, chromatographic indices and their dependence on pH in a set of beta-adrenoceptor blocking agents, *Farmaco*, **45**, 647–663 (1990); Mannhold R, Dross KP and Reffer RF, Drug lipophilicity in QSAR practice: I. A comparison of experimental with calculative approaches, *Quant-Struct.-Act. Relat.*, **9**, 21–28 (1990). Cited in: Lombardo F, Obach RS, Shalaeva MY and Gao F, Prediction of human volume of distribution values for neutral and basic drugs. 2. Extended data set and leave-class-out statistics, *J. Med. Chem.*, **47**, 1242–1250 (2004) (refs 275, 277).
29	Amantadine ($C_{10}H_{17}N$) NH_2	10.71 ± 0.01	A	+H	Potentiometric	H_2O $t = 20.0$ $I = 0.000$	Perrin DD, Hawkins I, The dissociation constant of the 1-aminoadamantane cation, *Experientia*, **28**, 880 (1972). "Solutions 0.005 M in 1-aminoadamantane hydrochloride . . . were titrated potentiometrically with 1M carbonate-free potassium hydroxide under nitrogen, using the procedure of Albert and Serjeant[6]. The ionic strengths of the solutions were varied from 0.005 to 0.105 by adding potassium nitrate. Free hydroxyl ion concentrations were calculated from the measured pH values, using the ionic product of water and Davies equation[7] to approximate the required activity coefficients at the specified ionic strengths. At each temperature and ionic strength, 9 points were taken covering the range from 1/10 to 9/10 neutralization, and the pK_a' values calculated for these points were averaged. The maximum range in any set was within ±0.05 pH unit. The resulting 'practical' pK_a values for 20° and 37° are given in the Table.
		10.14 ± 0.02	A	+H		$t = 37.0$	

I (M)	0.005	0.010	0.015	0.030	0.055	0.105
$t = 20°$	10.74	10.77	10.79	10.80	10.84	10.92
$t = 37°$	10.12	10.21	10.23	10.26	10.30	10.34

Extrapolation to zero ionic strength afforded thermodynamic pK_a values of 10.71 ± 0.01 at 20° and 10.14 ± 0.02 at 37°."

NB: Differs in some details from the account presented in APDS 12 (below), but is clearly the original work. A good model to follow.

30	Amantadine	10.71 ± 0.01	A	+H	Potentiometric	H_2O $t = 20.0$ $I = 0.0$	Kirschbaum J and Amantadine, *APDS*, **12**, 1983, 1–31. "Solutions of recrystallized amantadine were titrated potentiometrically with 0.1M carbonate-free potassium hydroxide at various temperatures and concentrations (95). Extrapolation to zero ionic strength gave pK_a values of 10.71 ± 0.01 at 20°, 10.58 ± 0.02 at 25° and 10.14 ± 0.02 at 37°. A linear plot gave -d(pK_a)/dT = 0.034. These pK_a results are similar to such alkyl analogues as 2-amino-2-methylpropane (pK_a = 10.68 at 25°) and 3-amino-3-ethylpentane (pK_a = 10.59), and support the conclusion (96) that there is little strain in this alicyclic molecule.
		10.58 ± 0.02	A			$t = 25$	95. Perrin DD and Hawkins I, The dissociation constant of the 1-aminoadamantane cation, *Experientia*, **28**, 880 (1972).
		10.14 ± 0.02	A			$t = 37$	96. Korolev BA, Khardin AP, Radchenko SS, Novakov IA and Orlinson BS, Basicity and structure of mono- and diamino derivatives of adamantine, *Zhurnal Organicheskoi Khimii*, **14**(8), 1632–1634 (1978). CA 89:196776n." NB: The value at 25 °C was obtained by interpolation
31	Amdinocillin (mecillinam) ($C_{15}H_{23}N_3O_3S$) 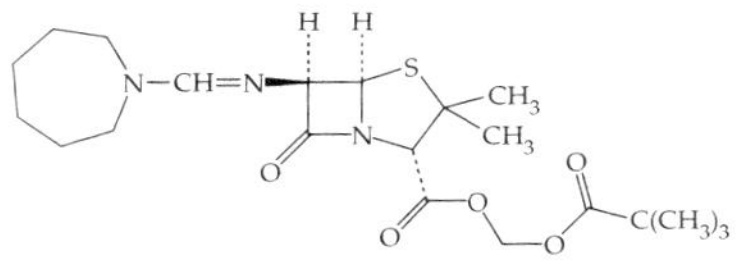	2.65	U	−H	Potentiometric	H_2O $t = 35$ $I = 1.0$	Bundgaard H, Aminolysis of the 6-β-amidinopenicillanic acid mecillinam. Imidazolone formation and intramolecular participation of the amidino side chain, *Acta Pharm. Suec.*, **14**(3), 267–278 (1977).
		8.79	U	+H			"The kinetics of reaction of mecillinam (I), with various primary amines in aqueous solution at 35 °C, were investigated. . . . The reactions with the amines are shown to be strongly facilitated by participation of the amidino side chain function (pK_a 8.8) in both its protonated and free base form."
32	Amiloride ($C_6H_8ClN_7O$) 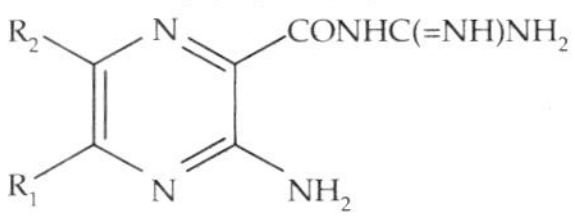				Potentiometric	H_2O $t = 24$ I unspecified	Bock MG, Schlegel HB and Smith GM, Theoretical estimation of pK_a values of pyrazinylguanidine derivatives, *JOC*, **46**(9), 1925–1927 (1981). Cited in Mazzo DJ, Amiloride hydrochloride, *APDS*, **15**, 1986, 1–34. "The dissociation constant of amiloride . . . from aqueous titration (18) indicates that amiloride is a moderately strong organic base with a pK_a of approximately 8.7 at 25 °C (amidino nitrogen). The pK_a of amiloride . . . has also been determined using gas-phase proton affinities, enthalpies of solution and semi-empirical calculations (25, 26). These theoretically derived pK_a values agree well with the experimentally determined value of pK_a = 8.7."
	R_1 R_2					[a] solvent was 30% aqueous EtOH	18. Rogers DH, Dept. Pharmaceutical Research and Development, Merck Sharp and Dohme Research Labs., West Point PA, internal communication
	NH_2 Cl	8.70	U	+H			25. Aue DH, Webb HM, Bowers MT, Liotta CL and Alexander CJ and Hopkins HP, Jr., A quantitative comparison of gas- and solution-phase basicities of substituted pyridines, *JACS*, **98**(3), 854–856 (1976)
	NH_2 H	9.30	U	+H			

(continued)

No.	Name	pK_a value(s)	Data quality	Ionization type	Method	Conditions t°C; I or c M	Comments and Reference(s)
	NH_2 S-Ph	9.00	U	+H			26. Bock MG, *et al.*, *JOC*, (1981) (above) NB: No further experimental details given.
	$N(CH_3)_2$ Cl	8.76	U	+H			
	NH_2 F	9.00	U	+H			
	NH_2 SCF_3	8.22	U	+H			
	CH_3O Cl	8.25	U	+H			
	SCH_3 Cl	8.05	U	+H			
	H Cl	7.10	U	+H			
	OH Cl	5.45	U	+H			
	SH Cl	4.00	U	+H			
	Cl Cl	6.60[a]	U	+H			
33	Amiloride	8.67	U	+H	Potentiometric	H_2O $t = 25$	Bergström CAS, Strafford M, Lazorova L, Avdeef A, Luthman K and Artursson P, Absorption classification of oral drugs based on molecular surface properties, *J. Med. Chem.*, **46**(4), 558–570 (2003). NB: From extrapolation of aqueous-methanol mixtures to 0% methanol.
		8.35	U	+H	Potentiometric	H_2O $t = 25$ $I = 0.15$ (KCl)	Balon K, Riebesehl BU and Muller BW, Drug liposome partitioning as a tool for the prediction of human passive intestinal absorption, *Pharm. Res.*, **16**, 882–888 (1999).
34	Amino acid esters				Potentiometric	H_2O $t = 25.0 \pm 0.1$ $I = 0.1$ (NaCl) N_2 atmosphere	Canel E, Gultepe A, Dogan A, Kihc E, The determination of protonation constants of some amino acids and their esters by potentiometry in different media, *J. Solution Chem.*, **35**(1), 5–19 (2006). "... the purity of substances used was measured by potentiometric titrations. ... A 0.10 mol.L^{-1} hydrochloric acid solution was prepared in water and standardized against sodium carbonate. Several 0.10 mol.L^{-1} sodium hydroxide solutions were prepared in 30, 50 and 70% (v/v) aqueous ethanol solutions and stored in glass bottles protected against the atmosphere. The base solutions were standardized ... by titration with hydrochloric acid. All potentiometric measurements were performed ... at (25.0 ± 0.1) °C under a nitrogen atmosphere. An Orion 940A Model pH-ionmeter fitted with a combination pH electrode containing a filling solution of 0.10 mol.L^{-1} NaCl was used for measuring the cell emf values ... The potentiometric cell was calibrated before each experiment so that the hydrogen ion concentration rather than the activity was measured. For all the solvent mixtures examined, reproducible values of autoprotolysis constants (Kapp) were calculated from several series of [H+] and [OH−] measurements at 0.10 mol.L^{-1} NaCl."
	L-cysteine methyl ester	6.38 ± 0.02	U	+H			
		9.17 ± 0.02	U	+H			
	L-cysteine ethyl ester	6.54 ± 0.03	U	+H			
		9.36 ± 0.03	U	+H			
	L-tyrosine methyl ester	7.04 ± 0.02	U	+H			
		9.73 ± 0.03	U	+H			
	L-tyrosine ethyl ester	7.05 ± 0.02	U	+H			
		9.71 ± 0.02	U	+H			
	L-tryptophan methyl ester	7.10 ± 0.02	U	+H			
		10.06 ± 0.05	U	+H			
	L-tryptophan ethyl ester	7.10 ± 0.05	U	+H			
		10.79 ± 0.02	U	+H			
	L-lysine methyl ester	6.98 ± 0.04	U	+H			
		9.99 ± 0.05	U	+H			
	L-lysine ethyl ester	7.18 ± 0.04	U	+H			
		10.32 ± 0.04	U	+H			
	L-histidine methyl ester	4.96 ± 0.02	U	+H			
		7.10 ± 0.04	U	+H			

NB: See Water, below. For all compounds in this report, the reliability was assigned as "uncertain". This was because there was generally not good agreement with literature values for the unesterified amino acids themselves.

Cpd	30% EtOH (U)		50% EtOH (U)		70% EtOH (U)	
	pK_1	pK_2	pK_1	pK_2	pK_1	pK_2
Cys-Me	9.23 ± 0.05	6.09 ± 0.03	9.34 ± 0.01	5.88 ± 0.02	10.82 ± 0.02	4.24 ± 0.02
Cys-Et	9.37 ± 0.02	6.17 ± 0.02	9.48 ± 0.03	5.97 ± 0.02	11.13 ± 0.04	4.23 ± 0.02
Tyr-Me	10.10 ± 0.02	6.73 ± 0.01	10.49 ± 0.05	6.42 ± 0.01	10.62 ± 0.02	6.17 ± 0.03
Tyr-Et	10.13 ± 0.04	6.75 ± 0.02	10.52 ± 0.01	6.45 ± 0.01	10.63 ± 0.02	6.19 ± 0.04
Try-Me	11.27 ± 0.02	6.78 ± 0.05	11.56 ± 0.04	6.51 ± 0.05	11.39 ± 0.07	6.29 ± 0.02
Try-Et	11.28 ± 0.02	6.98 ± 0.02	11.70 ± 0.03	6.80 ± 0.01	11.77 ± 0.08	6.45 ± 0.05
Lys-Me	9.80 ± 0.03	6.90 ± 0.02	9.52 ± 0.01	6.63 ± 0.03	9.12 ± 0.02	6.46 ± 0.02
Lys-Et	9.70 ± 0.02	6.81 ± 0.02	9.45 ± 0.03	6.60 ± 0.04	9.00 ± 0.03	6.35 ± 0.04
His-Me	6.72 ± 0.02	4.78 ± 0.02	6.54 ± 0.03	4.61 ± 0.01	6.34 ± 0.02	4.39 ± 0.04

No.	Compound	pKa	Rel.	Type	Method	Conditions	Reference
35	4-Aminobenzoic acid ($C_7H_7NO_2$)	2.501	R	+H	Potentiometric	H_2O	Deviney ML, Anderson RC and Felsing WA, Application of the glass electrode to the determination of thermodynamic ionization constants of *p*-aminobenzoic acid, *JACS*, **79**, 2371–2373 (1957). Cited in Perrin Bases no. 3010, ref. D31. The study used a glass electrode without liquid junction potential, and simple linear extrapolation to zero ionic strength. Numerous other values cited as well.
	H_2N—(C₆H₄)—COOH	4.874	R	−H		$t = 25$; $I = 0.005$ to 0.03	
36	4-Aminobenzoic acid	2.42	A	+H	Potentiometric, Conductance	H_2O; $t = 25$	Bell PH and Roblin, RO, Jr., Studies in chemotherapy. VII. A theory of the relation of structure to activity of sulfanilamide-type compounds, *JACS*, **64**, 2905–2917 (1942).
		4.68	U	−H	Potentiometric	$I = 0.05$	"The acid constants were determined from the pK_a values obtained from 0.05N NaOH electrometric titration curves (hydrogen electrode). Experimentally it was impossible to measure all the compounds by this method, since some were extremely insoluble (*sic*) in water, and because of their high molecular weights, the pK_a values were not significant. For the very insoluble (*sic*) sulfanilamides it was possible to use 50% ethanol as the solvent.... It was found that compounds in the sulphanilamide series, which were measurable in water, when measured in 50% ethanol gave a smooth curve of p$K_{a(H20)}$ versus p$K_{a(50\%\ EtOH)}$. (NB: covering a pK_a range from ~2.8 to ~10.6). From this curve, it was possible to determine the acid constants for compounds in the same series.... Titration of acids weaker than ($K_a = 2 \times 10^{-11}$) did not give curves

(continued)

Appendix A *(continued)*

No.	Name	pK_a value(s)	Data quality	Ionization type	Method	Conditions t°C; I or c M	Comments and Reference(s)
							sufficiently different from a blank titration to be reliable. . . . The basic groups were all weak . . . and the results of 0.05-N HCl titrations were not significant unless the compounds were quite water soluble. Sulfanilamide, metanilamide and p-aminobenzoic acid were carefully studied by this method, using a hydrogen electrode. The base constants of sulphanilamide and p-aminobenzoic acid were also determined from conductance measurements on their hydrochlorides. These results agreed very well with the water titration values." NB: the basic groups for poorly water soluble compounds were obtained by a similar method to that used for the corresponding acid groups, except that titrations used perchloric acid in glacial acetic acid.
37	4-Aminobenzoic acid	2.41 ± 0.04 4.87 ± 0.02	A A	+H −H	Potentiometric, Spectro	H_2O $t = 25$ $I = 0.00$	Van de Graaf B, Hoefnagel AJ and Wepster BM, Substituent effects. 7. Microscopic dissociation constants of 4-amino- and 4-(dimethylamino)benzoic acid, *JOC*, **46**(4), 653–7 (1981). NB: The method gave a pK_a value for benzoic acid of 4.21 (comparable to the best in the literature) and hence the method was well validated. Results at low ionic strengths (0.04 to 0.06) were corrected to zero ionic strength.
38	4-Aminobenzoic acid	2.504	A	+H	Spectro	H_2O $t = 25$ $I = 0.00$	Klotz IM and Gruen DM, The isoelectric nature of sulfanilamide and p-aminobenzoic acid, *JACS*, **67**, 843–6 (1945). NB: Also reported pK_a values for dimethylsulfanilamide (2.058), diethylsulfanilamide (1.535), methyl *p*-aminobenzoate (2.404), ethyl *p*-aminobenzoate (2.366).
		2.45 4.85	A A	+H −H	Spectro	H_2O $t = 25$ $I = 0.00$	NB: Robinson RA, Biggs AI, Ionization constants of p-aminobenzoic acid in aqueous solution at 25°, *Aust. J. Chem.*, **10**, 128–134 (1957). Also reported benzoic acid (4.203); methyl p-aminobenzoate (2.465); ethyl p-aminobenzoate (2.508); n-propyl p-aminobenzoate (2.487); and n-butyl p-aminobenzoate (2.472).
39	4-Aminobenzoic acid	2.46	U	+H		H_2O $t = 50$ $I = 0.20$	Otomo M, Fukui K and Kodama K, Heterocyclic azomethine compounds and their reduction products as analytical reagents. II. Reaction of 1-picolylideneamino-2-naphthol with zinc(II), *Bull. Chem. Soc. Jpn.*, **47**(2), 455–457 (1974).

40	6-Aminopenicillanic acid ($C_8H_{12}N_2O_3S$) 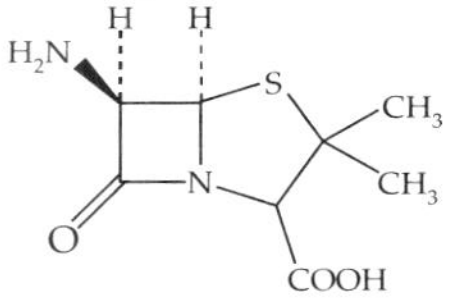	2.30 ± 0.05 4.90 ± 0.05	A A	−H +H	Potentiometric	H_2O $t = 25$ $c = 0.0081$	Rapson HDC and Bird AE, *J. Pharm. Pharmacol.*, **15**, 222T (1963). Cited in Perrin Bases Suppl. No. 7777 ref R6. NB: Potentiometric titrations used a glass electrode with an unsymmetrical cell and liquid junction potentials.
41	2-Amino-4-[4′-hydroxyphenyl]butane ($C_{10}H_{15}NO$) 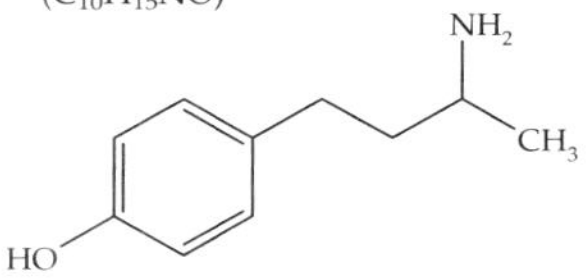	9.14 ± 0.02	U	−H	Potentiometric	H_2O $t = 25.0 \pm 0.2$ $I \leq 0.001$	Leffler EB, Spencer HM and Burger A, Dissociation constants of adrenergic amines, *JACS*, **73**, 2611–2613 (1951). NB: See Amphetamine for details.
42	2-Amino-5-[4′-hydroxyphenyl]pentane ($C_{11}H_{17}NO$) 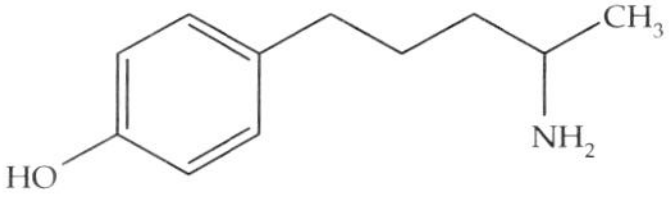	9.40 ± 0.05	U	−H	Potentiometric	H_2O $t = 25.0 \pm 0.2$ $I \leq 0.001$	Leffler EB, Spencer HM and Burger A, Dissociation constants of adrenergic amines, *JACS*, **73**, 2611–2613 (1951). NB: See Amphetamine for details.
43	2-Amino-2-methyl-3-hydroxyoctane ($C_8H_{21}NO$) 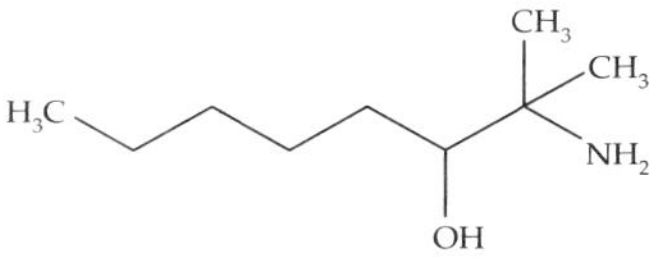	9.85 ± 0.02	U	−H	Potentiometric	H_2O $t = 25.0 \pm 0.2$ $I \leq 0.001$	Leffler EB, Spencer HM and Burger A, Dissociation constants of adrenergic amines, *JACS*, **73**, 2611–2613 (1951). NB: See Amphetamine for details.
44	2-Amino-4-phenylbutane ($C_{10}H_{15}N$) 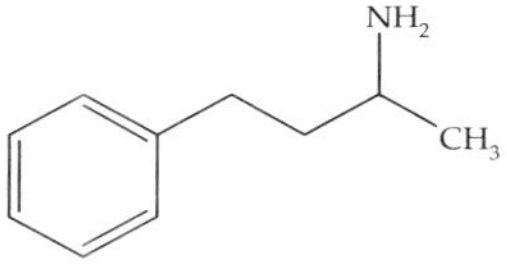	9.79 ± 0.02	U	−H	Potentiometric	H_2O $t = 25.0 \pm 0.2$ $I \leq 0.001$	Leffler EB, Spencer HM and Burger A, Dissociation constants of adrenergic amines, *JACS*, **73**, 2611–2613 (1951). NB: See Amphetamine for details.

(continued)

No.	Name	pK_a value(s)	Data quality	Ionization type	Method	Conditions t°C; I or c M	Comments and Reference(s)
45	2-Amino-5-phenylpentane ($C_{11}H_{17}N$) 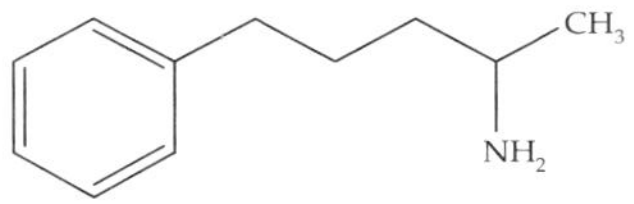	9.99 ± 0.05	U	−H	Potentiometric	H_2O $t = 25.0 \pm 0.2$ $I \leq 0.001$	Leffler EB, Spencer HM and Burger A, Dissociation constants of adrenergic amines, *JACS*, **73**, 2611–2613 (1951). NB: See Amphetamine for details.
46	Aminopyrine (aminophenazone) ($C_{13}H_{17}N_3O$) 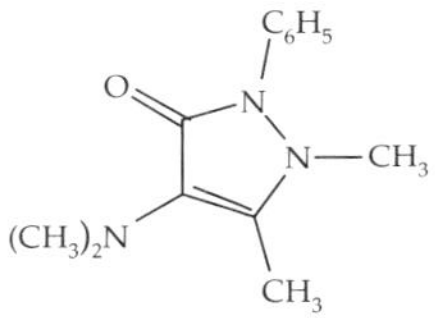	5.0	U	+H	partition	H_2O	Shore PA, Brodie BB and Hogben CAM, The gastric secretion of drugs: A pH-partition hypothesis, *JPET*, **119**, 361–369 (1957).
47	Aminopyrine (aminophenazone)	5.06 ± 0.01	A	+H	Potentiometric	H_2O $t = 25.0$ $I = 0.1$ (NaCl)	Takacs-Novak K and Avdeef A, Interlaboratory study of log P determination by shake-flask and potentiometric methods, *J. Pharm. Biomed. Anal.*, **14**, 1405–1413 (1996). NB: See Acetaminophen for further details.
48	2-Aminoquinoline ($C_9H_8N_2$) 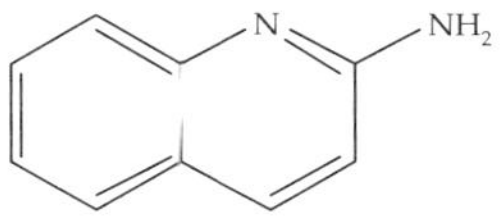	−5.7 12.7 −9.08 7.34	U U U U	+H −H +H −H	fluoro Spectro	H_2O t = RT I undefined	Kovi PJ, Capomacchia AC, Schulman SG, Electronic spectra of 2-aminoquinoline and 4-aminoquinaldine: evidence for the cyclic amidine structures of the singly protonated cations, *Anal. Chem.*, **44**, 1611–1615 (1972). "Electronic absorption, fluorescence, and IR spectroscopies have been employed to show that the singly protonated (at heterocyclic nitrogen) species derived from 2-aminoquinoline (I) and 4-aminoquinaldine (II) have the protonated amidine electronic structures in ground and lowest electronically excited singlet states. The neutral and doubly protonated compounds, however, appear to be well behaved arylamines and arylammonium ions, respectively, in ground and lowest excited singlet states. The anomalous pK_a values corresponding to ground and excited state prototropic equilibria of the I and II are attributed to the relative contributions of the basicity and acidity of the amidine species and those of the acidity and basicity of the arylamine and arylammonium ion species to the overall free energy of prototropic exchange."

No.	Compound	pK	Quality	Type	Method	Conditions	Comments and Reference
	4-aminoquinaldine ($C_9H_8N_2$) 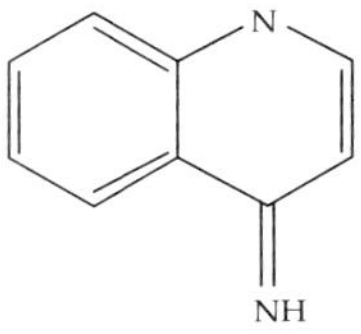	−3.4 12.6	U U	+H −H	fluoro		NB: Fluorescence spectra were obtained in sulphuric acid, borate buffer and sodium hydroxide solutions. See Brown EV and Plasz AC, Spectrophotometric determination of the second dissociation constants of the aminoquinolines, *J. Heterocycl. Chem.*, **7**, 335–338 (1970) for spectroscopic measurements on the substituted quinolines.
49	4-Aminosalicyclic acid ($C_7H_7NO_3$) 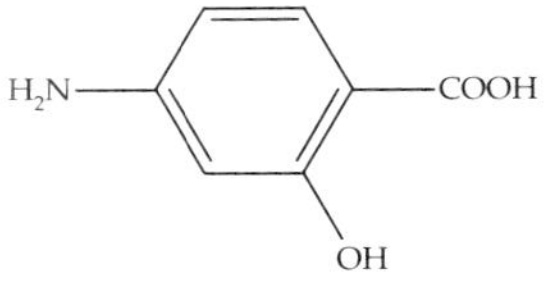	1.78 3.63	A A	+H −H	Potentiometric	H_2O $t = 20$ $I = 0.1$ (KCl)	Willi AV and Stocker JF, Kinetik der Decarboxylierung von *p*-Aminosalicylsaure, *Helv. Chim. Acta*, **37**, 1113–1121 (1954). Cited in Perrin Bases 3071 ref. W37. The study used a glass electrode in a cell without liquid junction potential.
		1.79	A	+H	Spectro	(HCl+KCl)	Perrin Bases no. 3071 reported several other values in the ranges $pK_{a1} = 1.7$ to 2.0, $pK_{a2} = 3.7$ to 3.9, and $pK_{a3} = 13.7$ to 14, without adequate validation information.
50	5-Aminosalicyclic acid ($C_7H_7NO_3$) 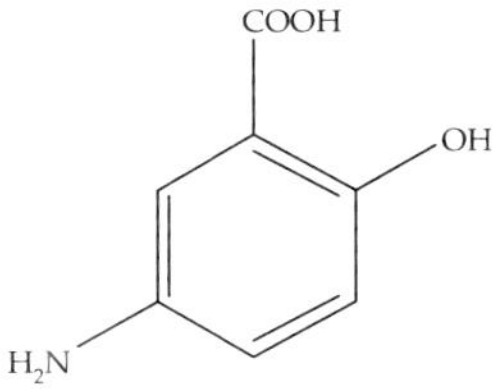	3.0 6.0 13.9	U U U	−H +H −H	^{13}C-NMR/pH	H_2O	Allgayer H, Sonnenbichler J, Kruis W and Paumgartner G, Determination of the pK_a values of 5-aminosalicylic acid and N-acetylaminosalicylic acid and comparison of the pH dependent lipid-water partition coefficients of sulphasalazine and its metabolites, *Arzneim.-Forsch.*, **35(9)**, 1457–1459 (1985). "Determinations of the pK_a values: . . . ^{13}C-NMR has been used to determine p*K* values. The compounds (5-ASA , AcASA) were dissolved in d_4 methanol/H_2O mixtures or in d_7 dimethyl-formamide/H_2O mixtures, titration was performed with HCl or NaOH. The p*K* values were determined form the turning points of the chemical shifts (ppm) obtained by a series of ^{13}CNMR spectra . . . at different pH values and extrapolated to water. The figures represent the mean of six experiments ±SD. . . ., p<0.01 was considered statistically significant. Results: By a series of ^{13}C-NMR spectra at different pH (from pH 1.0 to 14 in one unit steps) the chemical shifts were recorded. The turning points gave the p*K* values of the different substituents of 5-ASA and AcASA, respectively. . . .. Thus, the following p*K* values were obtained: 5-ASA -COOH: group: 3.0; -NH3+: 6.0; -OH: 13.9; AcASA -COOH: 2.7; -OH: 12.9."

(*continued*)

No.	Name	pK_a value(s)	Data quality	Ionization type	Method	Conditions t°C; I or c M	Comments and Reference(s)
51	Amiodarone ($C_{25}H_{29}I_2NO_3$) 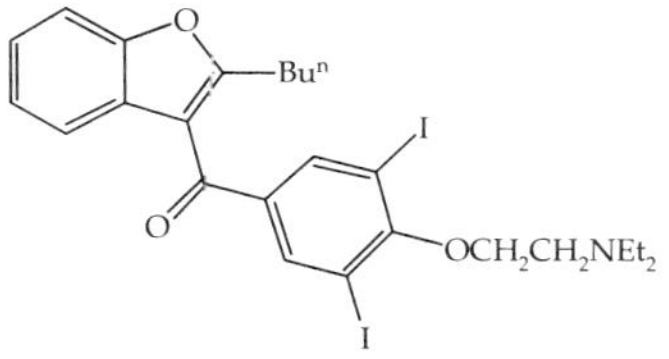	8.7 ± 0.2	U	+H	partition	H_2O $t = 25.0$ $I = 0.15$ (KCl)	Krämer SD, Gautier J-C and Saudemon P, Considerations on the potentiometric log P determination, *Pharm. Res.*, **15(8)**, 1310–1313 (1998). "To determine aqueous pK_a . . . values, titrations of 4 to 100 μmol compound in 20 ml 0.15M KCl, 20 ml methanol/ 0.15M KCl mixtures and in 20 to 90 ml of the biphasic system n-octanol/ 0.15 M KCl were performed on the PCA 101 titrator . . . at 25 °C for amiodarone, metoprolol and ramipril and at 37 °C for atenolol. The pK_a values of metoprolol, atenolol, and ramipril were determined in 0.15M KCl. For the extrapolation of the aqueous pK_a of amiodarone, we used the Yasuda-Shedlovsky plot (4). Apparent pK_a values (+ log [H_2O]) in the methanol/0.15M KCl mixtures at ratios between 47 and 71%(w/w) methanol were plotted against the inverse dielectric constant . . . and the pK_a of the molecule in 100% aqueous solution was extrapolated by linear regression."
52	Amiodarone 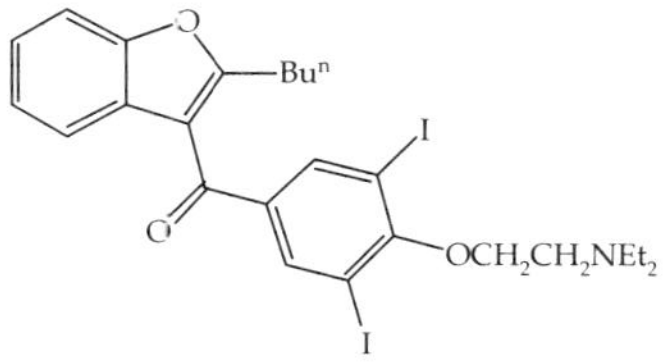	8.7 ± 0.5	U	+H	surface potential vs pH	H_2O $t = 20 \pm 0.5$ $I = 0.15$ (NaCl)	Ferreira J, Brasseur R, Chatelain P and Ruysschaest J, Properties of amiodarone monolayer spread at the air-water interface, *J. Pharm. Pharmacol.*, **38(2)**, 561–566 (1986). ". . . the extrapolated pK_i and pK_a values at $\alpha = 0$ are identical (8.7 ± 0.5).There are several reported pK_a values for amiodarone: 5.6, 7 (Andreason *et al.* 1981), 7.4 (Canada *et al.*, 1981) and 6.56 (Bonati *et al.*, 1984). The dispersion of the results and the low pK_a value obtained for a ternary amine might reflect the difficulties encountered in the determination of the aqueous dissociation constants of amiodarone, because of its low water solubility using classical potentiometric or UV spectrophotometric methods. This is also illustrated by the recent determination of a pK_a value of 9.12 (Gachon, 1981). The determination of the pK_i of amiodarone performed in the present work is not subject to the difficulties encountered when working in a solution. . .. It should thus be more reliable than previous determinations. It is also closer to the expected value for any ternary amine. . . . several characteristics of amiodarone were determined at the air-water interface using the unique property of the compound to form a stable monolayer. These characteristics are the area occupied per molecule in the close-packed state (0.44 nme^2/molecule) and the pK_i (8.7). This approach combined with a semi-empirical conformation analysis gave a molecular picture of the amiodarone conformation and orientation at the air-water interface."

Alpha	pK_i	pK_a
0.05	8.56	8.17
0.10	8.66	7.94
0.20	8.70	7.49
0.30	8.74	7.20
0.40	8.67	6.90
0.50	8.76	6.80

Andreason F, Agerback H, Bjerregaard P and Gotysch H, *Eur. J. Clin. Pharmacol.*, **19**, 293–299 (1981).
Canada AT, Lesko LJ and Hafferjee CI, *Curr. Ther. Res.*, **30**, 968–974 (1981).
Gachon R, Sanofi-Research. Internal Scientific and Technical Report. (1981).

53	Amiodarone	6.64 ± 0.28	VU	+H	Spectro (242 nm)	H_2O $t = 20.0$	Plomp TA, Amiodarone, *APDS*, **20**, 1–120 (1991).

Amiodarone structure: benzofuran ring with Bu^n, carbonyl to diiodophenyl ring bearing $OCH_2CH_2NEt_2$.

"The apparent dissociation constant (pK_a) was determined spectrophotometrically by Bonati *et al.* (4) and found to be 6.56 ± 0.06 at 25 °C. This . . . was substantially higher than the previous . . . pK_a of 5.6 (5,7). Our pK_a measurements of solutions of 4 batches of amiodarone hydrochloride reference substance using the spectrophotometric method of Clarke *et al.* (8) showed a mean pK_a value of 6.64 ± 0.28. . . . the pK_a value of solutions of 3 different reference batches of desethylamiodarone hydrochloride showed a mean value of 5.58 ± 0.35. The pK_a value of both compounds was assessed by UV titration at λ_{max} of aqueous solutions containing 10 µg/ml of the drugs respectively. . .. A range of pH 1.4 – 3.5 was obtained by addition of various volumes of 0.1M HCl, pH 5.5–8.8 by various volumes of phosphate/sodium hydroxide and borate/sodium hydroxide buffers, while values to pH 11.4 were reached by addition of various volumes of 0.1M sodium hydroxide. All solutions were prepared . . . by diluting methanolic stock solution of the drugs (10 mg/ml) . . . to concentrations of 10 µg/ml. The pH values were checked at 20 °C using a standard pH meter. . .. The dissociation constant was determined according to the (standard spectrophotometric) equation. . .. In the dissociation constant experiments we also found that the UV absorption curves of amiodarone and desethylamiodarone recorded at the different pHs showed characteristic bathochromic shifts of the maximum absorbance (λ_{max}) from pH 5.5 to 8.8. A shift of λ_{max} from 242 nm to 252 nm and from 242 nm to 250 nm were observed respectively for amiodarone and desethylamiodarone with in both cases a decrease of the specific extinction. These findings were in good agreement with previously reported observations (4,7).

(continued)

No.	Name	pK_a value(s)	Data quality	Ionization type	Method	Conditions t°C; I or c M	Comments and Reference(s)
	Amiodarone, desethyl (desethylamiodarone) ($C_{23}H_{25}I_2NO_3$)	5.58 ± 0.35	VU	+H			4. Bonati M, Gaspari F, D'Aranno V, Benfenati E, Neyroz P, Galletti F and Tognoni G, Physicochemical and analytical characteristics of amiodarone, *J. Pharm. Sci.*, **73**(**1**), 829–830 (1984). 5. Crispin S, Alexis J and Mouffe P, Methode de controle pour le chlorhydrate d'amiodarone - L3428 Qualité P. Report D.T. 8.029 S.A.Labaz 1980. 7. Moffat AC, Jackson JV, Moss MS and Widdop B, Clarke's Isolation and Identification of Drugs, 2nd ed., Pharmaceutical Press, London, pp. 344–345 (1986)."
54	Amiodarone	6.56 ± 0.06	U	+H	Spectro	H_2O t = 25.0 I controlled with NaCl Nonlinear fitting; triplicate	Bonati M, Gaspari F, D'Aranno V, Benfenati E, Neyroz P, Galletti F and Tognoni G, Physicochemical and analytical characteristics of amiodarone, *J. Pharm. Sci.*, **73**(**1**), 829–830 (1984). "According to a long-used method (Senstrom W and Goldsmith N, *J. Phys. Chem.*, 30, 1683, 1926) the pK_a values were determined by UV titration (at λ_{max}) of aqueous solutions containing known amounts of amiodarone. A range of pH 1–5 was achieved by addition of hydrochloric acid, pH 5.8–9.2, by 0.05M borate phosphate . . ., while higher values were obtained by addition of sodium hydroxide solutions. Sodium chloride was used to adjust the ionic strength. All solutions were prepared immediately before use and all pH values were checked at 25 °C using a standard pH-meter. In the temperature range of 20–30 °C there was no substantial effect on the true pH of the solutions, but only 3–4 h after preparation we found a 10% decrease in amiodarone concentration at pH values >7.0. The extinction coefficient was related to different hydrogen ion concentrations to check the λ_{max} shift. The dissociation constant (K_a) was determined according to the equation $E = (A_H[H+] + A_B \times K_a)/([H+] + K_a)$ where AH and AB are the extinction coefficients when amiodarone exists under acidic and basic conditions, respectively."

55	Amiodarone	8.73 ± 0.07	U	+H	Potentiometric	H_2O $t = 25$ $I = 0.15$ (KCl)	Sirius Technical Application Notes, vol. **2**, p. 114 (1995). Sirius Analytical Instruments Ltd., Forest Row and East Sussex, RH18 5DW, UK. NB: Extrapolated to 0% MeOH by Yasuda-Shedlovsky procedure from data in 49.7–75.5% aqueous MeOH. [Cited by Lombardo F, Obach RS, Shalaeva MY and Gao F, Prediction of human volume of distribution values for neutral and basic drugs. 2. Extended data set and leave-class-out statistics, *J. Med. Chem.*, **47**, 1242–1250 (2004); ref 278].
56	Amitriptyline ($C_{20}H_{23}N$) $CHCH_2CH_2N(CH_3)_2$	9.4	U	+H	soly	H_2O $t = 24 \pm 1$	Green AL, Ionization constants and water solubilities of some aminoalkylphenothiazine tranquillizers and related compounds, *J. Pharm. Pharmacol.*, **19**, 10–16 (1967). Cited in Blessel KW, Rudy BC, Senkowski BZ, Amitriptyline hydrochloride, *APDS*, **3**, 1974, 127–148; N&K. "The dissociation constant for amitriptyline was determined using a graphical method involving the pH dependence of the water solubility. The value for the pK_a determined by this method was 9.4." NB: Solubilities in 0.01M NaOH or 0.01M buffers were measured by a combination of spectrophotometric and turbidimetric methods.
57	Amitriptyline	9.48 ± 0.1	U	+H	Potentiometric	H_2O $t = 25$ I undefined Ar atmosphere	Seiler P, Simultaneous determination of partition coefficient and acidity constant of a substance, *Eur. J. Med. Chem.*, **9**, 663–665 (1974). Cited in: Lombardo F, Obach RS, Shalaeva MY and Gao F, Prediction of human volume of distribution values for neutral and basic drugs. 2. Extended data set and leave-class-out statistics, *J. Med. Chem.*, **47**, 1242–1250 (2004); (ref. 279). NB: Titrations were performed by autotitrator in the presence of high purity dodecane, to allow simultaneous measurement of the log P value. The pH meter was calibrated against three standard solutions.
58	Amitriptylline	9.49	U	+H	Potentiometric	H_2O $t = 25$	Bergström CAS, Strafford M, Lazorova L, Avdeef A, Luthman K and Artursson P, Absorption classification of oral drugs based on molecular surface properties, *J. Med. Chem.*, **46(4)**, 558–570 (2003). From extrapolation of aqueous-methanol mixtures to 0% methanol.
59	Amitriptyline	9.4	U	+H	Potentiometric	EtOH/H_2O	Thoma K and Albert K, Fast method for the potentiometric determination of pK_a values in solvent mixtures, *Arch. Pharm. Weinheim*, **314**, 1053–1055 (1981). "The potentiometric determination of the dissociation constants of amitriptyline HCl, doxepin HCl, imipramine HCl, and noxiptiline HCl in ethyl alcohol and water systems is described."
60	Ammonia NH_3	9.28	U	+H	Conductance ($K_b = 1.9 \times 10^{-5}$)	H_2O $t = 25$	Kendall J, Electrical conductivity and ionization constants of weak electrolytes in aqueous solution, *in* Washburn EW, Editor-in-Chief, *International Critical Tables*, Vol. **6**, McGraw-Hill, NY 259–304 (1929). NB: Other pK_a values: $t = 0$, 8.9; $t = 5$, 8.9; $t = 10$, 9.0; $t = 15$, 8.04; and $t = 20$, 8.08.

(continued)

No.	Name	pK_a value(s)	Data quality	Ionization type	Method	Conditions t°C; I or c M	Comments and Reference(s)
61	Amoxicillin ($C_{16}H_{19}N_3O_5S$)	2.4	U	−H		H_2O $t = 22$	Bird AE, Amoxicillin, *APDS*, **23**, 1994, 1–54.
		7.4	U	+H,−H			NB: Cited Marshall AC, personal communication.
		9.6	U	−H,+			
62	Amoxicillin	2.60	U	−H	Potentiometric	H_2O	Bergström CAS, Strafford M, Lazorova L, Avdeef A, Luthman K and Artursson P, Absorption classification of oral drugs based on molecular surface properties, *J. Med. Chem.*, **46(4)**, 558–570 (2003). NB: From extrapolation of aqueous-methanol mixtures to 0% methanol.
		7.31	U	+H,−H		$t = 25$	
		9.53	U	−H,+H			
63	Amoxicillin	2.87	U	−H	kinetic	H_2O	Zia H, Shalchian N and Borhanian F, Kinetics of amoxicillin in aqueous solutions, *Can. J. Pharm. Sci.*, **12**, 80–83 (1977). Cited in Bird AE, Amoxicillin, *APDS*, **23.**
		7.28	U	+H,−H		$t = 35$	
		9.65	U	−H,+H		$I = 0.5$ (KCl)	
64	Amoxicillin	2.67	U	−H	Potentiometric	H_2O	Tsuji A, Nakashima E, Hamano S and Yamana T, Physicochemical properties of amphoteric β-lactam antibiotics, *J. Pharm. Sci.*, **67**, 1059–1066 (1978). Cited in Bird AE, Amoxicillin, *APDS*, **23**, 1–54 (1994). NB: Also reported the following values:
		7.11	U	+H,−H	Potentiometric	$t = 37$	
		9.55	U	−H,+H	Spectro	$I = 0.5$ (KCl)	

	37 °C			35 °C		
Compound	pK_{a1}	pK_{a2}	pK_{a3}	pK_{a1}	pK_{a2}	pK_{a3}
Ampicillin	2.67	6.95	–	–	–	–
Cyclacillin	2.64	7.18	–	–	–	–
Amoxicillin	2.67	7.11	9.55	2.63	7.16	9.55
Epicillin	–	–	–	2.77	7.17	–

No.	Name	pK_a value(s)	Data quality	Ionization type	Method	Conditions t°C; I or c M	Comments and Reference(s)
65	Amoxicillin	2.63	U	−H	Potentio,	H_2O	Bundgaard H, Polymerization of penicillins. II. Kinetics and mechanism of dimerization and self-catalyzed hydrolysis of amoxicillin in aqueous solution. *Acta Pharm. Suecica*, **14**, 47–66 (1977); see also next entry. Cited in Bird AE, Amoxicillin, *APDS*, **23**, 1–54 (1994).
		7.55	U	+H,−H	Spectro	$t = 23$	
		9.64	U	−H,+H		$I = 1.0$ (KCl)	

No.	Compound	pKa	Data quality	Ionization	Method	Conditions	Reference/Comments
66	Amoxicillin	2.61 7.30 9.45	U U U	−H +H,−H −H,+H	Potentio, Spectro	H_2O $t = 35$ $I = 1.0$ (KCl)	Bundgaard H, Polymerization of penicillins. II. Kinetics and mechanism of dimerization and self-catalyzed hydrolysis of amoxicillin in aqueous solution. *Acta Pharm. Suecica*, **14**, 47–66 (1977). Cited in Bird AE, Amoxicillin, *APDS*, **23**, 1–54 (1994). NB: The four microdissociation constants in aqueous KCl ($I = 1.0$) for the four forms in which the carboxy group is ionised are:

T (°C)	pK_a NH_2 (OH)	pK_a NH_2 (O^-)	pK_a OH (NH_3^+)	pK_a OH (NH_2)
23	7.58	8.49	8.70	9.61
35	7.33	8.24	8.51	9.49

No.	Compound	pKa	Data quality	Ionization	Method	Conditions	Reference/Comments
67	Amphetamine ($C_9H_{13}N$) C_6H_5–$CH_2CH(CH_3)NH_2$	9.94	A	+H	Potentiometric	H_2O $t = 25.0 \pm 0.5$ $I = 0.01$	Warren RJ, Begosh PP and Zarembo JE, Identification of amphetamines and related sympathomimetic amines, *J. Assoc. Off. Anal. Chem.*, **54**, 1179–1191 (1971). "The apparent dissociation constants (pK_a') of the amphetamine-related compounds were determined by spectroscopic and potentiometric methods described by Albert and Serjeant (1). The pK_a' values for most of the compounds in this study are being reported for the first time; a few have been reported previously (9–12). These pK_a' values for the same compounds reported by different workers have not always been in very good agreement; procedural and/or conceptual errors can account for these differences (13). We have spared no effort to achieve a high degree of reliability in this study; every effort was made to eliminate or minimize experimental error. For example, the compounds were all of pharmaceutical grade purity, dissolved in carbon dioxide-free water, and titrated with carbonate-free potassium hydroxide solution, using a pH meter whose stability was checked before and after titrations with 2 different buffers (pH 8.48 and 10.00). . . . When available, both amine salt and free base were titrated and duplicate titrations, done on different days, were performed at 0.01 meq/ml at 25 ± 0.5 °C."
68	Amphetamine	9.77 ± 0.05	U	+H	Potentiometric	H_2O $t = 25.0 \pm 0.2$ $I \leq 0.001$	Leffler EB, Spencer HM and Burger A, Dissociation constants of adrenergic amines, *JACS*, **73**, 2611–2613 (1951). "The apparent dissociation constants of the amines were determined by measuring the pH of a solution containing equivalent concentrations of the amine and its salt. These solutions were obtained by adding to a solution of the salt the calculated amount of sodium hydroxide solution required for half-neutralization.

(continued)

No.	Name	pK_a value(s)	Data quality	Ionization type	Method	Conditions t°C; I or c M	Comments and Reference(s)
							The concentrations ranged from 0.0003 – 0.001 molal in salt and free base at the half-neutralization point. The pH values were measured with a hydrogen electrode or a pH meter. With the several compounds which were determined in both ways, good agreement was found . . ." NB: The hydrogen electrode was held at 25.0 ± 0.2° C; it was calibrated daily against saturated potassium hydrogen tartrate (pH = 3.57 ± 0.02 at 25 °C). The pH meter was a Beckman Model G. pH values were corrected for varying solution temperatures by a procedure of Hall and Sprinkle (*JACS*, **54**, 3469 (1932)), where amine pH values change by −0.02/deg, for amines with pK_b in the range 4–6. No exclusion of CO_2 reported for pH meter. Despite the efforts made, the data reported in this paper cannot be assessed as anything but U = uncertain, due to acceptance of the pK_a = pH of half-neutralization approximation.
69	Amphetamine, 4-hydroxy ($C_9H_{13}NO$) CH_3 NH_2 HO	9.70 10.53	A U	+H −H	Potentiometric Spectro	H_2O $t = 25.0 \pm 0.5$ $I = 0.01$	Warren RJ, Begosh PP and Zarembo JE, Identification of amphetamines and related sympathomimetic amines, *J. Assoc. Off. Anal. Chem.*, **54**, 1179–1191 (1971). "Our spectroscopically determined pK_a's (phenols) are in good agreement with those reported by others (12, 13). However, the potentiometric methods coupled with the Noyes method (16) of calculation for overlapping constants gave results for the phenolic pK_a' that were lower by 0.1 and 0.2 unit for (4-hydroxyamphetamine) and (4-hydroxymethamphetamine), respectively, and very imprecise results for (synephrine). Consequently, the pK_a' values for these compounds were determined by the same procedures used by Kappe and Armstrong (13) (spectroscopic value used to determine amine value from titration curve)." 12. Lewis GP, The importance of ionization in the activity of sympathomimetic amines, *Br. J. Pharmacol.*, **9**, 488–493 (1954). 13. Kappe T and Armstrong MD, Ultraviolet absorption spectra and apparent acidic dissociation constants of some phenolic amines, *J. Med. Chem.*, **8**, 368–374 (1965). 16. Britton HTS, *Hydrogen Ions*, D van Nostrand, Princeton, NJ (1956) 217." NB: See Amphetamine for further details. An alternative calculation method gave $pK_{a1} = 9.49$ and $pK_{a2} = 10.66$.

No.	Compound	pKa	Data quality	Ionization type	Method	Conditions	Reference
70	Amphetamine, 3-methoxy ($C_{10}H_{15}NO$) 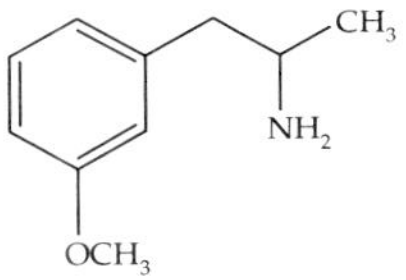	9.86	A	+H	Potentiometric	H_2O $t = 25.0 \pm 0.5$ $I = 0.01$	Warren RJ, Begosh PP and Zarembo JE, Identification of amphetamines and related sympathomimetic amines, *J. Assoc. Off. Anal. Chem.*, **54**, 1179–1191 (1971). NB: See Amphetamine for further details.
71	Amphetamine, 4-methoxy ($C_{10}H_{15}NO$) 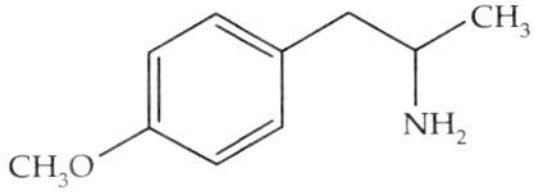	9.99	A	+H	Potentiometric	H_2O $t = 25.0 \pm 0.5$ $I = 0.01$	Warren RJ, Begosh PP and Zarembo JE, Identification of amphetamines and related sympathomimetic amines, *J. Assoc. Off. Anal. Chem.*, **54**, 1179–1191 (1971). NB: See Amphetamine for further details.
72	Ampicillin ($C_{16}H_{19}N_3O_4S$) 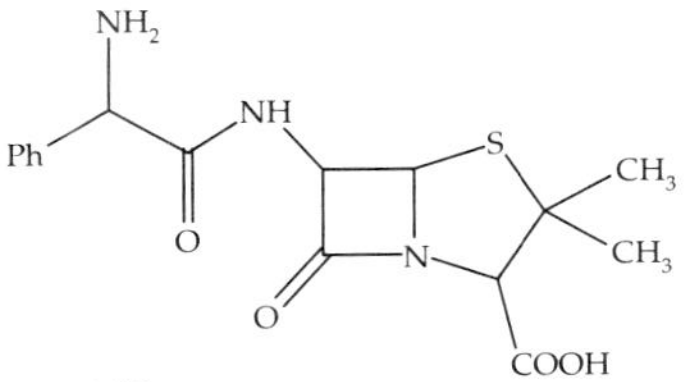	2.53 ± 0.04 7.24 ± 0.02	A A	−H +H		H_2O	Ivashkiv E, Ampicillin, *APDS*, **2**, 1–61 (1973). "Rapson and Bird reported ionization constants for ampicillin to be: $pK_1 = 2.53 \pm 0.004$ and $pK_2 = 7.24 \pm 0.02$. Jacobsen and Russo-Alesi calculated pK_2 for ampicillin trihydrate to be 7.24. Hou and Poole reported $pK_1 = 2.66 \pm 0.03$ and $pK_2 = 7.24 \pm 0.03$. Rapson HDC and Bird AE, *J. Pharm. Pharmacol.*, Suppl.15, 222T (1963). Jacobson H and Russo-Alesi F, The Squibb Institute for Medical Research, private communication (1969). Hou JP and Poole JW, *J. Pharm. Sci.*, **58**, 1510–1515 (1969)." NB: See next entry.
73	Ampicillin	2.65 ± 0.05 7.25 ± 0.03	A A	−H +H	Potentiometric	H_2O $t = 25.0 \pm 0.1$ $I = 0.00$ N_2 atmosphere	Hou JP and Poole JW, The aminoacid nature of ampicillin and related penicillins, *J. Pharm. Sci.*, **58**, 1510–1515 (1969). NB: Careful work, with N_2 atmosphere and KOH titrant prepared and stored carbonate-free according to Albert and Serjeant. Activity coefficients applied.
74	Ampicillin	2.67 6.95	U U	−H +H	Potentiometric Potentiometric	H_2O $t = 37$ $I = 0.5$ (KCl)	Tsuji A, Nakashima E, Hamano S and Yamana T, Physicochemical properties of amphoteric β-lactam antibiotics, *J. Pharm. Sci.*, **67**, 1059–1066 (1978). Cited in Bird AE, Amoxicillin, *APDS*, **23**, 1–54 (1994).
75	Ampicillin	2.53 ± 0.04 7.25 ± 0.03	A A	−H +H	Potentiometric	H_2O $t = 25$ $c = 0.079$	Rapson HDC, Bird AE, *J. Pharm. Pharmacol.*, Suppl.15, 222T (1963). Cited in Perrin Bases Suppl. No. 7778. Ref. R6. NB: The study used pH measurements with a glass electrode and junction potentials.
76	Ampicillin	2.5 7.04 2.41 6.94	U A U A	−H +H −H +H	CE/pH (+ve ion mode) CE/pH (-ve ion mode)	H_2O $t = 25$ $I = 0.025$	Wan H, Holmen AG, Wang Y, Lindberg W, Englund M, Nagard MB and Thompson RA, High-throughput screening of pK_a values of pharmaceuticals by pressure-assisted capillary electrophoresis and mass spectrometry, *Rapid Commun. Mass Spectrom.*, **17**, 2639–2648 (2003). NB: Reported predicted values (ACD Labs) of 2.44 and 6.76.

(continued)

No.	Name	pK_a value(s)	Data quality	Ionization type	Method	Conditions t°C; I or c M	Comments and Reference(s)
77	n-Amylpenilloic acid ($C_{13}H_{24}N_2O_3S$) 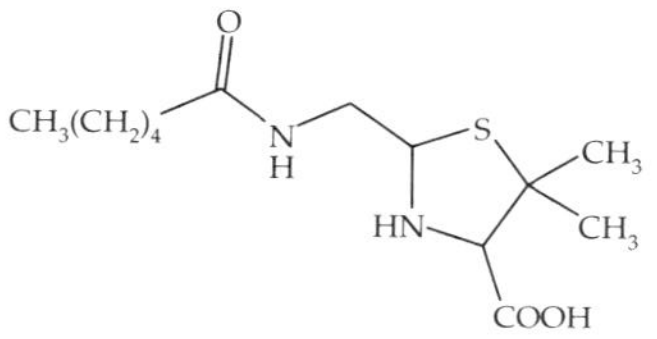	1.48 5.16	U U	−H +H		H_2O $t = 5$	Woodward RB, Neuberger A and Trenner NR, *in* Clarke H, Johnson JR and Robinson Sir R, (eds.), The Chemistry of Penicillin, Princeton University Press, Princeton, NJ 415–422 (1949) .
78	n-Amylpenilloic acid	1.44 4.85	U U	−H +H		H_2O $t = 25$	Woodward RB, Neuberger A and Trenner NR, *in* Clarke H, Johnson JR and Robinson Sir R (eds.), The Chemistry of Penicillin, Princeton University Press, Princeton, NJ (1949) 415–422.
79	Anagrelide ($C_{10}H_7Cl_2N_3O$) 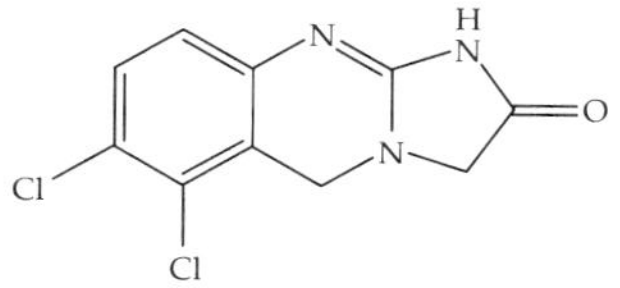	2.87 10	U U	+H −H	CE/pH	H_2O $t = 25$ $I = 0.025$	Wan H, Holmen AG, Wang Y, Lindberg W, Englund M, Nagard MB and Thompson RA, High-throughput screening of pK_a values of pharmaceuticals by pressure-assisted capillary electrophoresis and mass spectrometry, *Rapid Commun. Mass Spectrom.*, **17**, 2639–2648 (2003). NB: Reported predicted values (ACD Labs) of 2.43 and 11.79.
80	Anhydrochlortetracycline ($C_{22}H_{21}ClN_2O_7$) 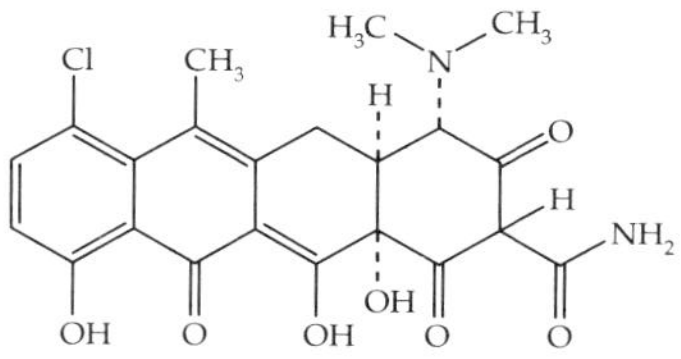	3.28 5.37 –	U U U	−H −H +H	Potentiometric	H_2O $t = 30.0 \pm 0.2$ $I = 0.01$ (KCl) N_2 atmosphere	Doluisio JT and Martin AN, Metal complexation of the tetracycline hydrochlorides, *J. Med. Chem.*, **6**, 16–20 (1963). NB: Metal-free solutions of the tetracycline were titrated with standard NaOH solution and the pH measured. No details were given of the pH meter calibration. Metal stability constants were determined from identical titrations in the presence of varying concentrations of nickel(II), zinc(II) or copper(II) ions.

No.	Compound	pKa	Method	Type	Assay	Conditions	Reference / Comments
81	Anisindione ($C_{16}H_{12}O_3$) 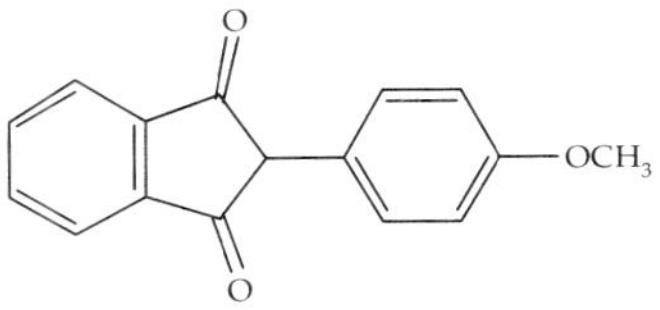	4.13	U	–H	Spectro	H_2O $t = 25.0 \pm 0.1$ $I = 0.1$ (NaCl)	Stella VJ and Gish R, Kinetics and mechanism of ionization of the carbon acids 4′-substituted 2-phenyl-1,3-indandiones, *J. Pharm. Sci.*, **68(8)**, 1047–1049 (1979). "The ionization kinetics of 1,3-diketone carbon acids are slow relative to those of classical acids and bases. The ionization kinetics of three 4′-substituted 2-phenyl-1,3-indandiones, 4′-chloro-4′-methoxy-, and 2-phenyl-1,3-indandione itself, were studied at 25° and ionic strength 0.1 using stopped–flow spectrophotometry and a pH jump technique."

Table I. Macroscopic Ionization Constants for Anisindione, Phenindione, and Clorindione determined spectrophotometrically at 25±0.1° and $\mu = 0.1$ with sodium chloride.

			Literature
Compound	λ (nm)	pK_a	pK_a values
Anisindione	330	4.13	4.09, 4.25, 5.6
Phenindione	326	4.09	4.10, 4.13, 5.4
Clorindione	284	3.59	3.54, 3.72, 4.8

NB: This data is in close agreement with data reported previously for 1 aqueous methanol solutions: Linabergs Y, Neiland, O, Veis A, Latu AN and Vanag G, Acidity and enolization of 2-aryl-1,3-indandiones. *Dokl. Akad. SSSR*, **154**, 1385–8 (1964); *Eng. Trans.*, **154**, 184 (1968). Numerous other indandione pK_a values were also reported.

No.	Compound	pKa	Method	Type	Assay	Conditions	Reference / Comments
82	Apomorphine ($C_{17}H_{17}NO_2$) 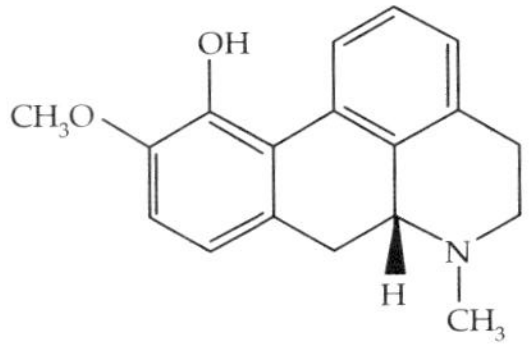	7.0, 7.2 8.9	U U	+H –H			Muhtadi FJ and Hifnawy MS, Apomorphine hydrochloride, *APDS*, **20**, 121–171 (1991). Kolthoff IM, The dissociation constants, solubility product and titration of alkaloids, *Biochem. Z.*, **162**, 289–353 (1925).
83	Apomorphine	7.20 8.92	U U	+H –H	Spectro	H_2O $t = 15$ $c = 0.0005$ to 0.002	Kolthoff IM, The dissociation constants, solubility product and titration of alkaloids, *Biochem. Z.*, **162**, 289–353 (1925). Cited in Perrin Bases 2860 ref. K47. NB: See Aconitine for details.

(continued)

No.	Name	pK_a value(s)	Data quality	Ionization type	Method	Conditions $t°C$; I or c M	Comments and Reference(s)
84	Apomorphine	7.0 8.9	U U	+H −H			Foye 1; also Clarke p. 357.
85	Aprindine ($C_{22}H_{30}N_2$) 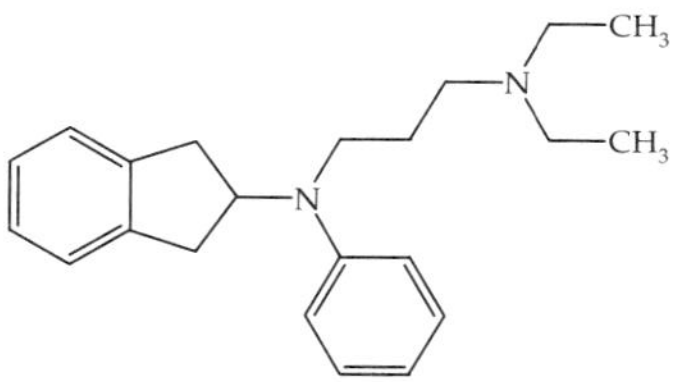	3.79 ± 0.02 9.43 ± 0.01 4.53 10.16	U U U U	+H +H +H +H	Potentiometric	40% EtOH $t = 25.0$ H_2O	Mannhold R, Rodenkirchen R and Bayer R and Haus W, The importance of drug ionization for the action of calcium antagonistsand related compounds, *Arzneim.-Forsch.*, **34**, 407–409 (1984). NB: pK_a values were determined by potentiometric microtitration. Ethanol-water (40:60) as solvent and drug concentrations of 0.6 mmol/l were used according to the low solubility and availability of the test compounds. pK_a taken as the apparent pH of half-neutralization. Mean of three determinations. These values were converted to apparent values in water by adding 0.74, based on the differences in the pK_a values for mebeverine and tiapamil, both of which were sufficiently soluble for direct titration in water.
86	Arecaidine ($C_7H_{11}NO_2$) 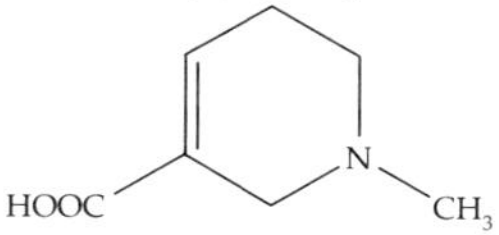	9.07	A	+H	Potentiometric	H_2O $t = 25$	Chilton J and Stenlake JB, Dissociation constants of some compounds related to lysergic acid: Beta-dimethylaminopropionic acid, dihydroarecaidine, ecgonine and their derivatives, *J. Pharm. Pharmacol.*, **7**, 1004–1011 (1955). Cited in Perrin 2862 ref. C27.
87	Arecaidine methyl ester ($C_8H_{13}NO_2$) 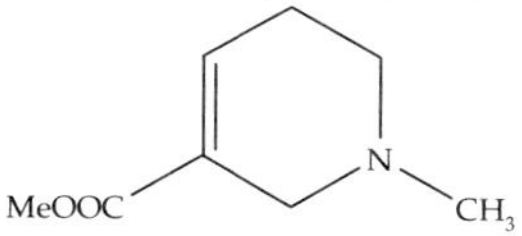	7.64	A	+H	Potentiometric	H_2O $t = 25$	Chilton J and Stenlake JB, Dissociation constants of some compounds related to lysergic acid: Beta-dimethylaminopropionic acid, dihydroarecaidine, ecgonine and their derivatives, *J. Pharm. Pharmacol.*, **7**, 1004–1011 (1955). Cited in Perrin 2862 ref C27.
88	Arecoline ($C_8H_{13}NO_2$) 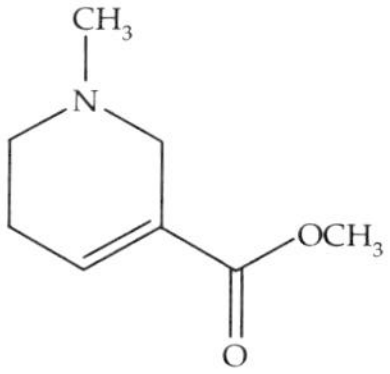	7.61	U	+H	Potentiometric	H_2O $t = 35$	Burgen ASV, Comparative activity of arecoline and arecoline *N*-metho salt, *J. Pharm. Pharmacol.*, **16**, 638 (1964).

89	Arecoline	7.41	U	+H	Potentiometric	H_2O $t = 17.5$	Muller F, *Z. Elektrochem.*, **30**, 587 (1924). Cited in Perrin Bases 2864 ref. M60. NB: Study used measurements of pH using hydrogen electrodes in an asymmetric cell with liquid junction potentials.
90	1-Arylpiperazine derivatives 1-Phenylpiperazine ($C_{10}H_{14}N_2$)	9.02 (20) 8.29 (37)	U U	+H +H	Spectro ($\lambda = 250$ to 280 nm)	H_2O $t = 20$ $t = 37$ *I* uncorrected	Caccia S, Fong MH and Urso R, Ionization constants and partition coefficients of 1-arylpiperazine derivatives, *J. Pharm. Pharmacol.*, **37**, 567–570 (1985). "The comparative ionization constant and lipophilicity, as determined by n-octanol aqueous buffer partition, of 14 1-arylpiperazines were investigated.... The ionization constant varied little across the entire series..."

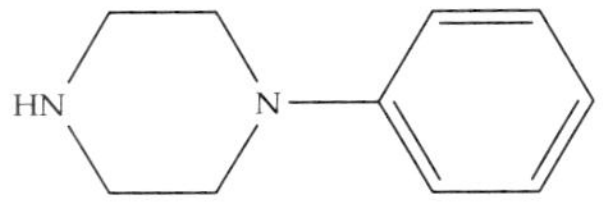

N-aryl substituent	pK_{a20}	pK_{a37}	$\Delta pK_a/\Delta t$	*N*-aryl substituent	pK_{a20}	pK_{a37}	$\Delta pK_a/\Delta t$
Phenyl	9.02	8.29	0.73	2-Methoxyphenyl	9.37	8.91	0.46
2-Pyrimidinyl	8.88	8.75	0.13	2-Methylphenyl	9.28	9.14	0.14
5-Fluoro-2-pyrimidinyl	8.91	8.50	0.41	2-Chlorophenyl	9.13	8.94	0.19
2-Thiazolyl	8.39	7.94	0.45	3-Chlorophenyl	8.85	8.64	0.21
2-Pyridyl	8.90	8.59	0.31	4-Chlorophenyl	8.90	8.49	0.41
2-Quinolinyl	8.82	8.76	0.07	3-Trifluoromethylphenyl	8.85	8.66	0.19
1,2-Benzisothiazol-3-yl	ND	8.68	ND	4-Fluorophenyl	8.98	8.55	0.43

NB:

(1) The range of pK_a values for each temperature is not really as small as the authors claimed. Further, the temperature dependences ($\Delta pK_a/\Delta t$) are interesting, in that the compounds fall into three distinct groups, based on the *N*-aryl substituent: high (phenyl); medium (5-fluoro-2-pyrimidinyl; 2-thiazolyl; 2-pyridyl; 2-methoxyphenyl; 4-chlorophenyl; 4-fluorophenyl); and low (2-pyrimidinyl; 2-quinolinyl; 2-methylphenyl; 2-chlorophenyl; 3-chlorophenyl; 3-trifluoromethylphenyl).

(2) Each compound should have another pK_a value (+H) in the region 1–3, due to protonation of the aromatic amine nitrogen.

(continued)

Appendix A *(continued)*

No.	Name	pK_a value(s)	Data quality	Ionization type	Method	Conditions t°C; I or c M	Comments and Reference(s)
91	Ascorbic acid ($C_6H_8O_6$)	4.05 ± 0.01 11.62 ± 0.04	A U	–H –H	Potentiometric	H_2O $t = 25.0$ $I = 0.1$ (NaCl)	Takacs-Novak K and Avdeef A, Interlaboratory study of log P determination by shake-flask and potentiometric methods, *J. Pharm. Biomed. Anal.*, **14**, 1405–1413 (1996). NB: See Acetaminophen for further details.
92	Ascorbic acid	GLpK_a: 4.03 ± 0.01 10.95 ± 0.02 A&S: 4.03 ± 0.02 11.2 ± 0.15	 A U U U	 –H –H –H –H	Spectro	H_2O $t = 25$ $I = 0.15$ (KCl) Ar atmosphere	Tam KY and Takacs-Novac K, Multi-wavelength spectrophotometric determination of acid dissociation constants, *Anal. Chim. Acta*, **434**, 157–167 (2001). NB: See Clioquinol for details.
93	Ascorbic acid	4.30 4.25 11.79	A U U	–H –H –H –H	Conductance method not given	H_2O $t = 25.00 \pm 0.01$ $I = 0.00$	Apelblat A, Manzurola E and Orekhova Z, Electrical conductance studies in aqueous solutions with ascorbate ions, *J. Solution Chem.*, **35**, 879–888 (2006). NB: Obtained from measurements on dilute solutions of the sodium, magnesium, calcium and ferrous salts of ascorbic acid, extrapolated to zero ionic strength and calculated from the reported K_1 value. Reported the following additional values: $t = 30.00$, 4.29; $t = 37.00$, 4.28 and $t = 40.00$, 4.27. Also cited numerous literature values, with the observation that the values were "not particularly close to each other." Lavrenov SN and Preobrazhenskaya MN, L-Ascorbic acid: Properties and ways of modification, *Pharmaceutical Chemistry Journal*, **39(5)** 251–264 (2005); Translated from: *Khimiko-Farmatsevticheskii Zhurnal*, Vol. **39(5)**, 26–39 (2005). NB: Cited Seib PA, Tolbert BM. Ascorbic acid: Chemistry, metabolism and uses, Adv. in Chemistry Series, ACS, Washington DC (1982).

No.	Compound	pKa	Type	Sign	Method	Conditions	Reference
94	Aspartame ($C_{14}H_{18}N_2O_5$) 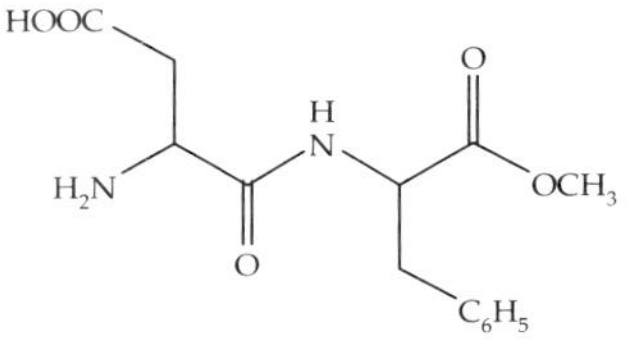	3.19 ± 0.01 7.87 ± 0.02	A A	–H +H	Potentiometric	H_2O $t = 25.0$ $I = 0.100$ (NaCl)	Skwierczynski RD and Connors KA, Demethylation kinetics of aspartame and L-phenylalanine methyl ester in aqueous solution, *Pharm. Res.*, **10(8)**, 1174–1180 (1993). NB: L-phenylalanine methyl ester; pK_a = 7.11 ± 0.02.
95	Aspirin ($C_9H_8O_4$) 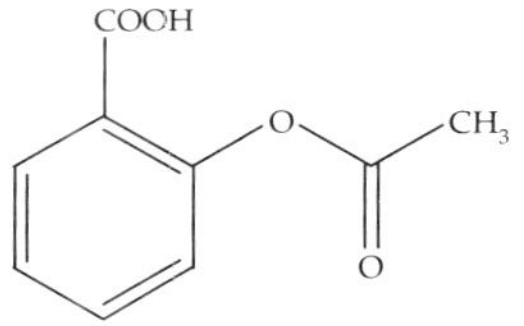	3.565	A	–H	Spectro	H_2O $t = 17$ $I = 0.00$	Edwards LJ, The hydrolysis of aspirin, *Trans. Farad. Soc.*, **46**, 723–735 (1950). NB: A very detailed account, using a spectrophotometric method based on the method of Flexser, Hammett and Dingwall (1935). pH values were calculated for acetic acid-acetate buffers using the electrometric data of Harned and Ehlers. Activity coefficient corrections used the full Debye-Huckel equation.
96	Aspirin	3.47 ± 0.01	A	–H	Potentiometric	H_2O $t = 25.0$ $I = 0.1$ (NaCl)	Takacs-Novak K and Avdeef A, Interlaboratory study of log P determination by shake-flask and potentiometric methods, *J. Pharm. Biomed. Anal.*, **14**, 1405–1413 (1996). NB: See Acetaminophen for further details. Also reported pK_a = 3.50 ± 0.01 at I = 0.15 M (KCl).
		3.41	A	–H	Potentiometric	H_2O $t = 37$ $I = 0.15$ (KCl)	Balon K, Riebesehl BU and Muller BW, Drug liposome partitioning as a tool for the prediction of human passive intestinal absorption, *Pharm. Res.*, **16**, 882–888 (1999). NB: Also reported the following values: Allopurinol, 9.00; Moxonidine, 7.36; Nizatidine; 2.44, 6.75; Olanzapine, 5.44, 7.80; Paromomycin, 5.99, 7.05, 7.57, 8.23, 8.90; Rifabutine, 6.90, 9.37 and Terbinafine, 7.05.
97	Aspirin Polymorph I Polymorph II	 8.99 9.19	 U U	 –H –H	Potentiometric	DMF	Kildsig DO, Denbo R and Peck GE, Structural differences in solutions derived from polymorphic modifications of aspirin, *J. Pharm. Pharmacol.*, **23**, 374–376 (1971). "Differences in the structure of solutions derived from 2 polymorphic modifications of aspirin were demonstrated through differences in apparent pK_a values. Polymorph I was prepared by slow crystallization at room temperature from a saturated solution of aspirin in 95% ethanol. Polymorph II was prepared by crystallization from a saturated solution of aspirin in n-hexane at room temperature. The apparent pK_a's were determined in dimethylformamide using tetrabutyl-ammonium hydroxide (in methanol-benzene solvent) as the titrant. The pK_a differences were ascribed to differences in intra- and intermolecular hydrogen bonding of the solute."

(continued)

No.	Name	pK_a value(s)	Data quality	Ionization type	Method	Conditions t°C; I or c M	Comments and Reference(s)
							NB: These data are very curious. It is difficult to imagine the hydrogen bonding interactions described for DMF solutions (and the consequent apparent pK_a differences) persisting in an aqueous environment. If these polymorphic differences persisted in DMF solution, then NMR and infrared solution spectra would be expected to confirm the data.
98	Astemizole ($C_{28}H_{31}FN_4O$) 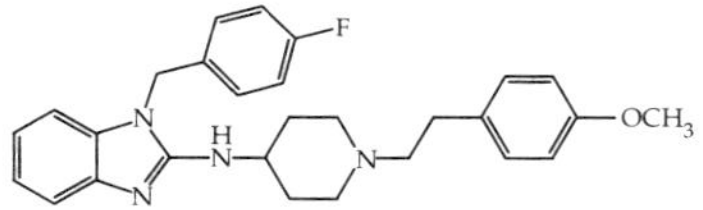	4.85 8.69	U U	−H +H	CE/pH (+ve ion mode)	H_2O $t = 25$ $I = 0.025$	Wan H, Holmen AG, Wang Y, Lindberg W, Englund M, Nagard MB and Thompson RA, High-throughput screening of pK_a values of pharmaceuticals by pressure-assisted capillary electrophoresis and mass spectrometry, *Rapid Commun. Mass Spectrom.*, **17**, 2639–2648 (2003). NB: Reported predicted values (ACD Labs) of 6.8 and 9.03.
99	Atenolol ($C_{14}H_{22}N_2O_3$) 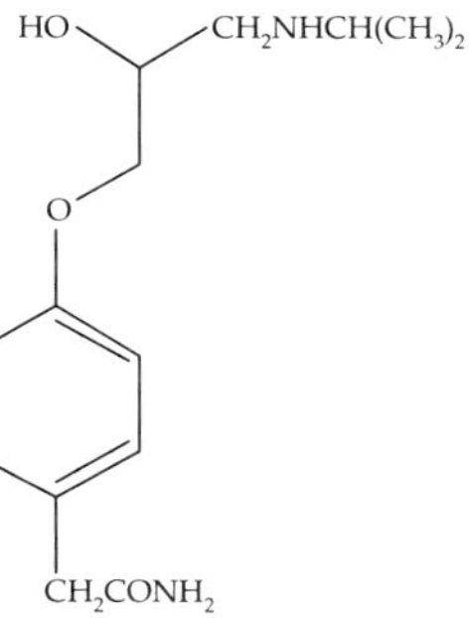 HO, $CH_2NHCH(CH_3)_2$, O, CH_2CONH_2	9.60	U	+H	Potentiometric	H_2O $t = 21–24$ (RT)	Schurmann W and Turner P, Membrane model of the human oral mucosa as derived from buccal absorption performance and physicochemical properties of the beta-blocking drugs atenolol and propranolol, *J. Pharm. Pharmacol.*, **30**, 137–147 (1978). "The pK_a was taken as the midpoint of the buffering plateau of the titration curve. The titration curves were established at room temperature from the measurement of the pH of a series of test tubes containing a constant volume of drug solution to which increasing volumes of titrant had been added. The pH-independent water solubility of the free base, So, was determined . . . from the pH and the final concentration of a drug solution that had been titrated with NaOH until precipitation became visible, according to $So = S/(1 + 10^{pKa - pHlim})$ where S is the drug concentration and the pHlim the pH at the titration point when solute crystallization is imminent. . . . The buccal absorption characteristics and physicochemical properties of the beta-adrenoceptor blocking agents propranolol hydrochloride (I) and atenolol (II) were investigated . . .. The dissociation constants, solubilities of free base, and n-heptane partition coefficients show that I in its unionized form is much more lipophilic than II, both drugs being bases with a similar pK_a. Buccal absorption was studied under conditions of varying drug concentration, contact time, and pH, and controlled through the use of a nonabsorbable marker. The absorption findings are in general agreement with the pH partition theory. A new compartmental diffusional model that

No.	Compound	pKa	Quality	Type	Method	Conditions	Reference / Notes
							includes membrane storage and a hypothetical aqueous pH buffering surface system allowed a more exhaustive interpretation to be made. . . . With human oral mucosa the intrinsic pH was near 6.7, and the buffering capacity of the system about 2.86. . . ."
100	Atenolol	9.58 ± 0.01	A	+H	Potentiometric	H_2O $t = 25.0$ $I = 0.15$ (NaCl)	Takacs-Novak K and Avdeef A, Interlaboratory study of log P determination by shake-flask and potentiometric methods, *J. Pharm. Biomed. Anal.*, **14**, 1405–1413 (1996). NB: See Acetaminophen for further details.
		9.25	U	+H	Potentiometric	H_2O $t = 37$ $I = 0.15$ (KCl)	Balon K, Riebesehl BU and Muller BW, Drug liposome partitioning as a tool for the prediction of human passive intestinal absorption, *Pharm. Res.*, **16**, 882–888 (1999).
101	Atenolol	9.56 ± 0.00	A	+H	partition	H_2O $t = 25.0$ $I = 0.15$ (KCl)	Krämer SD, Gautier J-C and Saudemon P, Considerations on the potentiometric log P determination, *Pharm. Res.*, **15(8)**, 1310–1313 (1998). NB: $pK_a(37\,°C) = 9.26 \pm 0.00$. See Amiodarone for details
		9.54 ± 0.01	A	+H	Potentiometric	H_2O $t = 25.0$ $I = 0.15$ (KCl)	Sirius Technical Application Notes, vol. **2**, pp. 67–68 (1995). Sirius Analytical Instruments Ltd., Forest Row, East Sussex, RH18 5DW, UK. NB: Concentration of analyte, 0.47–0.95 mM.
102	Atenolol	9.64	A	+H	CE/pH (+ve ion mode)	H_2O $t = 25$ $I = 0.025$	Wan H, Holmen AG, Wang Y, Lindberg W, Englund M, Nagard MB and Thompson RA, High-throughput screening of pK_a values of pharmaceuticals by pressure-assisted capillary electrophoresis and mass spectrometry, *Rapid Commun. Mass Spectrom.*, **17**, 2639–2648 (2003). NB: Reported a predicted value (ACD Labs) of 9.17.
103	Atorvastatin ($C_{33}H_{35}FN_2O_5$)	4.46	U	−H	soly	H_2O $t = 30$ I uncontrolled	Kearney AS, Crawford LF, Mehta SC, Radebaugh GW, Interconversion kinetics, equilibrium, and solubilities of the lactone and hydroxy-acid forms of the HMG-CoA reductase inhibitor, CI-981, *Pharm. Res.*, **10**, 1461–1465 (1993). "The pH dependence of the interconversion kinetics, equilibrium, and solubilities of the lactone and hydroxyacid forms of calcium (R-(R*,R*))-2-(4-fluorophenyl)-β,δ-dihydroxy-5-(1-methylethyl)-3-phenyl- 4-((phenylamino)-carbonyl)-1H-pyrrole-1-heptanoate (CI-981; atorvastatin calcium) are described. Over a pH range of 2.1–6.0 and at 30 °C, the apparent solubility of the sodium salt of CI-981 increases about 60-fold, and the profile yields a pK_a for the terminal carboxyl group of 4.46. In contrast, over a pH range of 2.3–7.7 at the same temperature, the apparent solubility of the lactone form of CI-981 varies little, and the mean solubility is 1.34 mcg/ml. The kinetics of interconversion and the equilibrium between the hydroxyacid and lactone forms have been studied as a function of pH, buffer concentration, and temperature at a fixed ionic strength. The rate constant for lactone formation is well described by specific acid-catalyzed and spontaneous lactonization pathways, whereas the rate constant for lactone hydrolysis is well described by specific acid-, water-, and specific base-catalyzed pathways."

(*continued*)

No.	Name	pK_a value(s)	Data quality	Ionization type	Method	Conditions t°C; I or c M	Comments and Reference(s)
104	Atrazine ($C_8H_{14}ClN_5$) $(CH_3)_2CHNH$ … Cl … $NHCH_2CH_3$	1.85 ± 0.04	U	+H	Spectro	H_2O *t* undefined *I* undefined	Weber JB, Spectrophotometrically determined ionization constants of 13 alkylamino-s-triazines and the relationships of molecular structure and basicity, *Spectrochim. Acta*, **23A**, 458–461 (1967) . NB: Calculations were performed according to Albert and Serjeant. No experimental details were given. Also reported the following pK_a values:
105	Atropine ($C_{17}H_{23}NO_3$) CH_3, N, O, C_6H_5, CH_2OH, H, O	9.85	U	+H	spectro	H_2O $t = 18$ $c = 0.005$ to 0.01	Kolthoff IM, The dissociation constants, solubility product and titration of alkaloids, *Biochem. Z.*, **162**, 289–353 (1925). Cited in Perrin DD, Dissociation constants of bases, Butterworths, London, 1965, no. 2866, ref. K47. NB: See aconitine for details.
		10.20	U	+H	Potentiometric	H_2O $t = 18$ $c = 0.005$ to 0.01	Potentiometric study used hydrogen electrode with liquid junction potentials Ref. M60 Muller F, *Z. Elektrochem.*, **30**, 587 (1924).

Sym-triazine derivative	pK_{a1}	pK_{a2}
2-methoxy-4-methyl-6-isopropylamino-	4.15	–
2-methoxy-4,6-bis(ethylamino)-	4.15	–
2-methoxy-4-ethylamino-6-isopropylamino-	4.20	–
2-methoxy-4,6-bis(isopropylamino)-	4.28	–
2-methoxy-4-ethylamino-6-diethylamino-	4.51	–
2-methoxy-4-isopropylamino-6-diethylamino-	4.54	–
2-methoxy-4,6-bis(diethylamino)-	4.54	–
2-methylthio-4,6-bis(isopropylamino)-	4.05	–
2-methylthio-4-isopropyl-6-diethylamino-	4.43	–
2-chloro-4,6-bis(isopropylamino)-	1.85	–
2-chloro-4-isopropylamino-6-diethylamino-	1.85	–(atrazine)
2-hydroxy-4,6-bis(isopropylamino)-	5.20	~11
2-hydroxy-4-isopropylamino-6-diethylamino-	5.32	~11

No.	Compound	pKa	Class	Type	Method	Conditions	Reference / Notes
106	Atropine	9.66	U	+H	Potentiometric	H_2O $t = 25$	Bergström CAS, Strafford M, Lazorova L, Avdeef A, Luthman K and Artursson P, Absorption classification of oral drugs based on molecular surface properties, *J. Med. Chem.*, **46(4)** 558–570 (2003). NB: From extrapolation of aqueous-methanol mixtures to 0% methanol.
107	Atropine	5.93	U	+H	Potentiometric	Acetic acid	Al-Badr AA and Muhtadi FJ, Atropine, *APDS*, **14**, 325–380 (1985). NB: No reference or supporting data given. NB: differing values in Merck Index (4.35), Foye (9.25), Connors *et al.* (9.8 at 18 °C; 10.2 at 16.5 °C). There is confusion here between pK_b and pK_a values, especially the Merck value. The value of 5.93 comes from Medwick T, Kaplan G, Weyer LG, Measurement of acidity and equilibria in glacial acetic acid with the glass calomel electrode system, *J. Pharm. Sci.*, **58**, 308–313 (1969); see also Bases (nonaqueous titrations). This value is mainly relevant to quantitative analysis by nonaqueous titration.
108	Azapropazone (apazone) ($C_{16}H_{20}N_4O_2$) 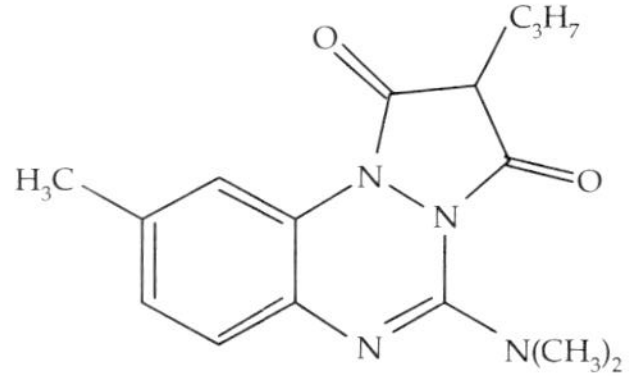	6.3	U	+H	soly	H_2O *t* undefined *I* undefined	Herzfeldt CD and Kümmel R, Dissociation constants, solubilities, and dissolution rates of some selected nonsteroidal antiinflammatories, *Drug Dev. Ind. Pharm.*, **9(5)**, 767–793 (1983). NB: See Ibuprofen for further details. "pK_a determination was only practicable by the solubility procedure, the spectrophotometric methods failed. . . . In some cases solubility data enable the determination of dissociation constants as described by Krebs and Speakman. The pK_a value results in the intercept with the abscissa when plotting log ([S/So] − 1) versus pH."
109	Azapropazone	6.58	U	+H	Potentiometric	50% EtOH *t* undefined *I* undefined	Jahn U and Wagner-Jauregg T, Wirkungsvergleich saurer Antiphlogistika im Bradykinin-, UV-Erythem- und Rattenpfotenödem-Test, *Arzneim.-Forsch.*, **24**, 494–499 (1974).
		6.4	U	+H		80% Me cellosolve	NB: Literature values were obtained from the pH of half-neutralization.
110	Azathioprine ($C_9H_7N_7O_2S$) 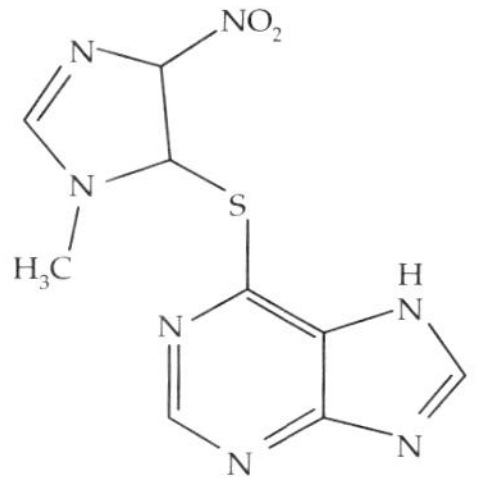	7.94 ± 0.04 7.87 ± 0.04	A A	−H −H	soly Spectro	H_2O $t = 25$ $I = 0.00$	Newton DW, Ratanamaueichatara S and Murray WJ, Dissociation, solubility and lipophilicity of azathioprine, *Int. J. Pharm.*, **11**, 209–213 (1982). "The pK_a of I (azathioprine) was determined by solubility and spectrophotometric methods. Triplicate samples of I in masses that were incompletely soluble were equilibrated at 25 °C for 48 h with 10 mL of 0.02N buffer at pH 4.00 and with 10 mL of 0.025M tromethamine buffers at six pH values from 7.00 to 8.60. The pH values of samples at 25 °C were recorded, then the filtrates were diluted and I concentrations were determined from absorbances at 280 nm by comparison to known I solutions at the same pH values. . . . Seven 2.89×10^{-5} M I solutions in 0.025 M tromethamine buffers wer prepared over the range of pH 7.30 – 8.40 at 25 °C. Absorbance

(continued)

Appendix A *(continued)*

No.	Name	pK_a value(s)	Data quality	Ionization type	Method	Conditions t°C; I or c M	Comments and Reference(s)
							values of samples were determined at 226 nm, the wavelength of maximum separation between spectra of the undissociated and anionic species. . . . Correction of the solubility/pH value to zero ionic strength (according to the method of Albert and Serjeant) gives $pK_a = 7.99$." NB: Accuracy was limited by a difference of only 0.24 abs units between the anion and neutral forms. Two pK_a values have been also recorded for 6-mercaptopurine itself (hydrolysis product of azathioprine) (see 6-Mercaptopurine, nos. 782–783): pK_{a1} 7.77 and 7.7 (Albert and Brown, *J. Chem. Soc*, 2060–2071 (1954)). pK_{a2} 10.84 and 11.17 (Fox *et al.*, *JACS*, **80** 1669–1675 (1958)).
111	Azelastine ($C_{22}H_{24}ClN_3O$) 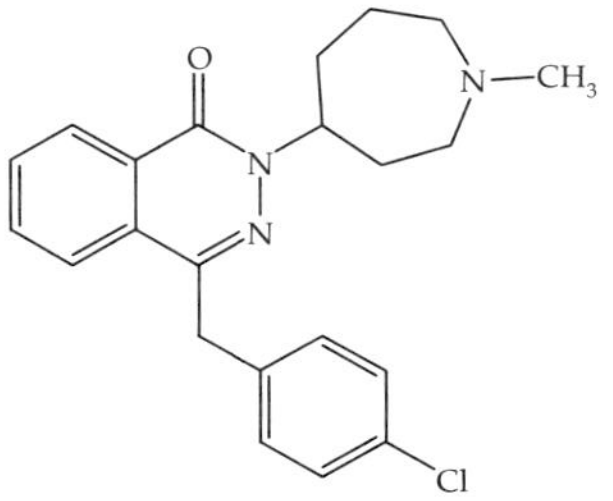	9.82	U	+H	CE/pH (+ve ion mode)	H_2O $t = 25$ $I = 0.025$	Wan H, Holmen AG, Wang Y, Lindberg W, Englund M, Nagard MB and Thompson RA, High-throughput screening of pK_a values of pharmaceuticals by pressure-assisted capillary electrophoresis and mass spectrometry, *Rapid Commun. Mass Spectrom.*, **17**, 2639–2648 (2003). NB: Reported a predicted value (ACD Labs) of 9.16.
112	Azelastine	9.54	U	+H	Potentiometric	H_2O $t = 25$ $I = 0.025$	Lombardo F, Obach RS, Shalaeva MY and Gao F, Prediction of human volume of distribution values for neutral and basic drugs. 2. Extended data set and leave-class-out statistics, *J. Med. Chem.*, **47**, 1242–1250 (2004); ref. not given: potentiometric titration.

113	Azithromycin ($C_{38}H_{72}N_2O_{12}$) 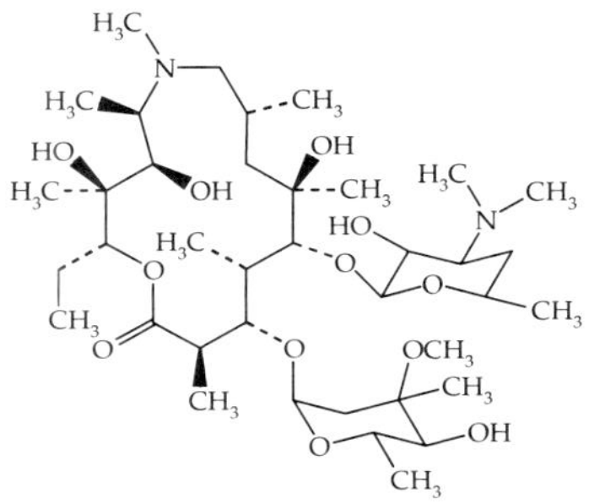	8.74 9.45	U U	+H +H	Potentiometric	H_2O $t = 25$ $I = 0.167$	McFarland JW, Berger CM, Froshauer SA, Hayashi SF, Hecker SJ, Jaynes BH, Jefson MR, Kamicker BJ, Lipinski CA, Lundy KM, Reese CP and Vu CB, Quantitative structure-activity relationships among macrolide antibacterial agents: *In vitro* and *in vivo* potency against Pasteurella multocida, *J. Med. Chem.*, **40**, 1340–1346 (1997). "Determinations of pK_a's and log P's were performed by the Sirius PCA 101 Potentiometric System. All macrolides were soluble in water at pH 6 or lower. An approximately 1 mM solution of each macrolide at a constant ionic strength of 0.167N NaCl was titrated from a low to a high pH. The acid used was 0.5N HCl , and the base, 0.5N NaOH. The acid and base were standardized to four decimal places using NIST standards. . . . The pK_a's and log P's were determined in triplicate under an argon atmosphere at . . . 25 °C. The values reported here have an average standard deviation of ± 0.07 for the pK_a . . . ."
114	Aztreonam ($C_{13}H_{17}N_5O_8S_2$) 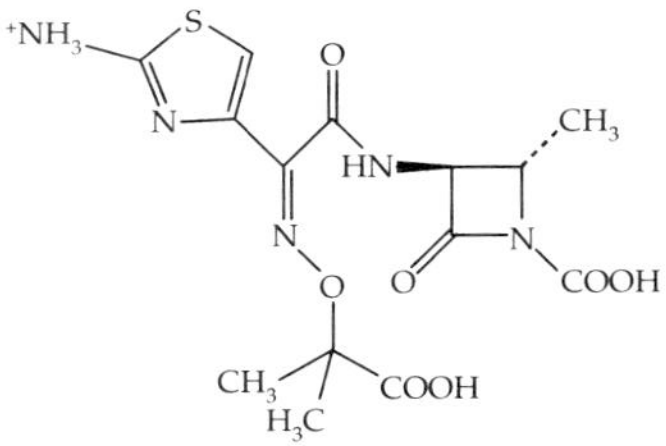	−0.7 2.75 3.91	U U U	−H +H −H	soly	H_2O t = RT I undefined	Florey K, Aztreonam, *APDS*, **17**, 1–40 (1988). "The apparent pK_a's of aztreonam at RT were determined by the method of Peck and Benet (Peck CC, Benet LZ, General method for determining macrodissociation constants of polyprotic amphoteric compounds from solubility measurements, *J. Pharm. Sci.*, 67, 12–16 (1978)) which uses the x-intercepts in a plot of log base 10 (S/So'') versus pH as the initial estimation of the apparent pK_a's. So'' is the lowest solubility measured. The pK_a values of the sulfonyl, amine and carboxyl groups are −0.7, 2.75, and 3.91 respectively. In the pH range of 0 to 10, aztreonam can exist as neutral zwitterion, monoanion, and dianion. The pK_a values are in good agreement with values obtained by potentiometric, spectrophotometric, and kinetic methods." Pipkin JD and Davidovich M, The Squibb Institute, 1983, personal communication. Pipkin JD, *in* Connors KA, Amidon GL and Stella VJ, eds., *Chemical Stability of Pharmaceuticals*, Wiley-Interscience, New York, (1986) 250–256.

(*continued*)

No.	Name	pK_a value(s)	Data quality	Ionization type	Method	Conditions t°C; I or c M	Comments and Reference(s)
115	Azuloic acid ($C_{11}H_8O_2$) 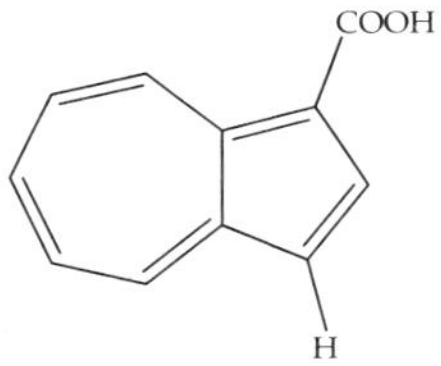	5.11 ± 0.03	U	−H	Spectro	H_2O $t = 20$ to 25 $I = 0.008$	Lichtenwalner MR and Speaker TJ, Substituent constants of azulene, *J. Pharm. Sci.*, **69**, 337–339 (1980). "The ionization constants in water for six 3-substituted azuloic acids were determined spectrophotometrically. Conversion of these physical constants to their pK_a values allowed a set of Hammett-type sigma values for the substituents on these acids to be calculated. Determination of partition coefficients for nine 1-substituted azulenes allowed Hansch-type values to be determined, using azulene as the model compound. The method employed for the determination of ionization constants were adapted from that of Albert and Serjeant. The acid of interest was dissolved in 0.01N HCl, 0.01N NaOH, or a graded series of buffers having pH values that bracketed the anticipated pK_a. The pH values of the resulting solutions were recorded to a thousandth of a pH unit, and the desired spectral characteristics were measured using a recording spectrophotometer. All measurements were made between 20 and 25° to eliminate making temperature corrections."
	3-Cl ($C_{11}H_7ClO_2$)	4.95 ± 0.05	U				
	3-Br ($C_{11}H_7BrO_2$)	4.87 ± 0.05	U				
	3-COCH$_3$ ($C_{13}H_{10}O_3$)	4.76 ± 0.07	U				
	3-CHO ($C_{12}H_8O_3$)	4.51 ± 0.04	U				
	3-NO$_2$ ($C_{11}H_7NO_4$)	4.32 ± 0.06	U				
116	Bamipine ($C_{19}H_{24}N_2$) 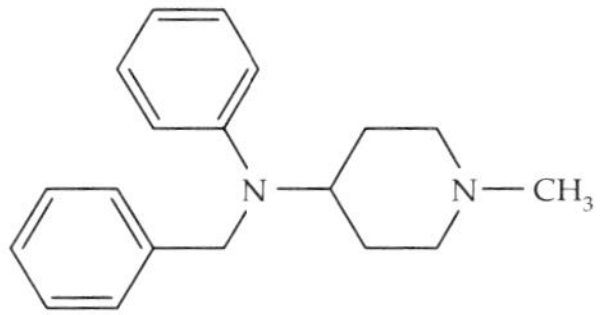	3.34 ± 0.04 8.04 ± 0.11	U U	+H +H	Potentiometric	H_2O t undefined $I = 0.30$ (NaCl)	Testa B and Murset-Rossetti L, The partition coefficient of protonated histamines, *Helv. Chim. Acta*, **61**, 2530–2537 (1978). NB: See Cycliramine for details.
117	Barbituric acid ($C_4H_4N_2O_3$) 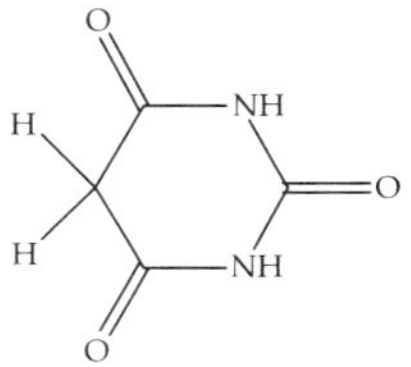	4.06	U	−H	Potentiometric	H_2O $t = 25 \pm 0.5$ N_2 atmosphere	Rubino JT, Electrostatic and non-electrostatic free energy contributions to acid dissociation constants in cosolvent-water mixtures, *Int. J. Pharm.*, **42**, 181–191 (1988). "Values of the molecular radii were estimated from molar volumes . . . Bondi (1968). . . . All pK_a values were adjusted for the solvent and concentration effects using the appropriate form of the Debye Huckel equation. . . . Thermodynamic values for dissociation constants of barbituric acid and several derivatives were determined in solvents containing 0–50% ethyl alcohol in water . . . . Results indicated that the type of hydrophilic functional group has a large influence on the non-electrostatic effect, necessitating a

No.	Compound	pKa	Data quality	Ionization	Method	Conditions	Comments / Reference
	Vol. fraction EtOH						model modification which divided lipophilic and hydrophilic components. Lipophilic nonelectrostatic effects correlated well with the hydrophobic surface area and log octanol-water partition coefficient of the solute. It was suggested that a linear free energy approach can be used to estimate dissociation constant changes for weak organic electrolytes in cosolvent-water mixtures."
	0.1	4.03	U				
	0.3	3.95	U				
	0.5	4.06	U				NB: Some standard acids, for example, acetic, were also titrated, with good agreement with the literature.
118	Barbituric acid, 5,5-dibromo ($C_4H_2Br_2N_2O_3$) 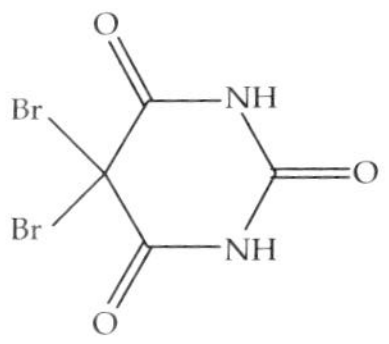	5.68 ± 0.02	A	–H	Potentiometric	H_2O $t = 25.0 \pm 0.02$ $I = 0.00$	McKeown RH, First thermodynamic dissociation constants of 5,5-disubstituted barbituric acids in water at 25 °C. Part 1, *JCS. Perkin II*, 504–514 (1980). NB: Thermodynamic value refined for activity coefficients, [H+], and [OH–]. Results validated through measured pK_a values for benzoic and 5,5-diethylbarbituric acids in agreement with best literature values.
119	Barbituric acid, 5,5-dichloro ($C_4H_2Cl_2N_2O_3$) 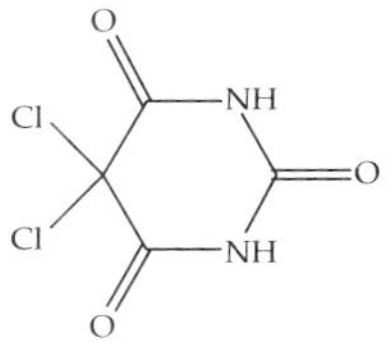	5.55 ± 0.02	A	–H	Potentiometric	H_2O $t = 25.0 \pm 0.02$ $I = 0.00$	McKeown RH, First thermodynamic dissociation constants of 5,5-disubstituted barbituric acids in water at 25 °C. Part 1, *JCS. Perkin II*, 504–514 (1980). NB: Thermodynamic value refined for activity coefficients, [H+], and [OH–]. Results validated through measured pK_a values for benzoic and 5,5-diethylbarbituric acids in agreement with best literature values.
120	Barbituric acid, 5,5-dimethyl ($C_6H_8N_2O_3$) 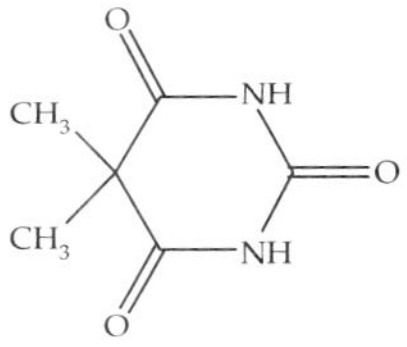	8.51 ± 0.02	A	–H	Potentiometric	H_2O $t = 25.0 \pm 0.02$ $I = 0.00$	McKeown RH, First thermodynamic dissociation constants of 5,5-disubstituted barbituric acids in water at 25 °C. Part 1, *JCS. Perkin II*, 504–514 (1980). NB: Thermodynamic value refined for activity coefficients, [H+], and [OH–]. Results validated through measured pK_a values for benzoic and 5,5-diethylbarbituric acids in agreement with best literature values.

(continued)

No.	Name	pK_a value(s)	Data quality	Ionization type	Method	Conditions t°C; I or c M	Comments and Reference(s)
121	Barbituric acid, 5,5-dimethyl	8.50 ± 0.03	A	–H	Potentiometric	H_2O $t = 25.0 \pm 0.02$ $I = 0.00$	Prankerd RJ, Physical Factors and Drug Activity, M. Pharm. thesis, University of Otago (1977). NB: Thermodynamic value refined for activity coefficients, [H+], and [OH−]. Results validated through measured pK_a values for benzoic and 5,5-diethylbarbituric acids in agreement with best literature values.

t (°C)	pK_a	t (°C)	pK_a
5	8.974 (0.01)	30	7.406 (0.02)
10	8.809 (0.01)	35	7.336 (0.02)
15	8.730 (0.01)	40	7.249 (0.02)
20	8.618 (0.02)	45	7.186 (0.03)
25	8.497 (0.01)		

Thermodynamic functions for ionization at 25 °C:
$\Delta G^\circ = 11.6$ kcal/mol; $\Delta H^\circ = 7.57$ kcal/mol; $\Delta S^\circ = 13.5$ cal/mol/K; $\Delta C^\circ_p = 73$ cal/mol/K.

No.	Name	pK_a value(s)	Data quality	Ionization type	Method	Conditions t°C; I or c M	Comments and Reference(s)
122	Barbituric acid, 5,5-dimethyl	7.1	VU	–H	Conductance	H_2O $t = 25.0 \pm 0.02$ $I = 0.00$	Kendall J, Electrical conductivity and ionization constants of weak electrolytes in aqueous solution, *in* Washburn EW, Editor-in-Chief, *International Critical Tables*, Vol. 6, McGraw-Hill, NY, 259–304 (1929). NB: From Wood JK, The acidic constants of some ureides and uric acid derivatives, *J. Chem. Soc*, **89**, 1831–1839 (1906). Of historical interest only.
123	Barbituric acid, 5-ethyl-5-methyl ($C_7H_{10}N_2O_3$) 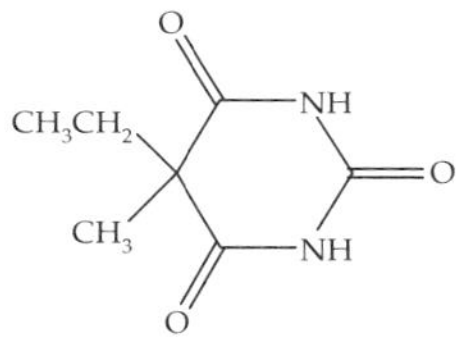	8.28 ± 0.02	A	–H	Potentiometric	H_2O $t = 25.0 \pm 0.02$ $I = 0.00$	McKeown RH, First thermodynamic dissociation constants of 5,5-disubstituted barbituric acids in water at 25 °C. Part 1, *JCS. Perkin II*, 504–514 (1980). NB: Thermodynamic value refined for activity coefficients, [H+], and [OH−]. Results validated through measured pK_a values for benzoic and 5,5-diethylbarbituric acids in agreement with best literature values.

No.	Compound	pKa	Data quality	Ionization	Method	Conditions	Reference / Comments
124	Barbituric acid, 5-methyl-5-isopropyl ($C_8H_{12}N_2O_3$) 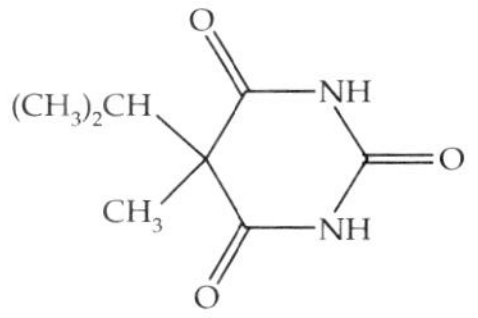	8.45 ± 0.02	A	–H	Potentiometric	H_2O $t = 25.0 \pm 0.02$ $I = 0.00$	McKeown RH, First thermodynamic dissociation constants of 5,5-disubstituted barbituric acids in water at 25 °C. Part 1, *JCS. Perkin II*, 504–514 (1980). NB: Thermodynamic value refined for activity coefficients, [H+], and [OH−]. Results validated through measured pK_a values for benzoic and 5,5-diethylbarbituric acids in agreement with best literature values.
125	Barbituric acid, 5-cyclohex-1′-enyl-1,5-dimethyl (hexobarbital) ($C_{12}H_{16}N_2O_3$) 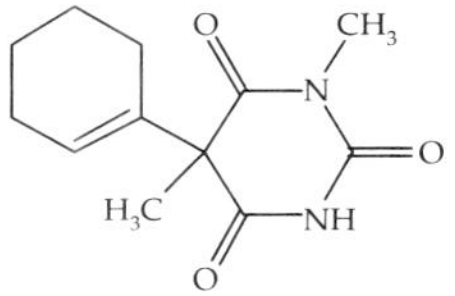	8.34	U	–H	Potentiometric	H_2O	Krahl ME, The effect of variation in ionic strength and temperature on the apparent dissociation constants of thirty substituted barbituric acids, *J. Phys. Chem.*, **44**, 449–463 (1940). NB: See Barbital, no. 129, for comment.
126	Barbituric acid, 5,5-diethyl (barbital) ($C_8H_{12}N_2O_3$)	7.9798	VR	–H	Electrometric	H_2O $t = 25.0$ $I = 0.00$	Manov GG, Schuette KE and Kirk FS, Ionization constant of 5,5-diethylbarbituric acid, *J. Res. Nat. Bur. Stand.*, **48**, 84–91 (1952). NB: Exceedingly careful work performed using an electrochemical (hydrogen) cell without liquid junction potentials.

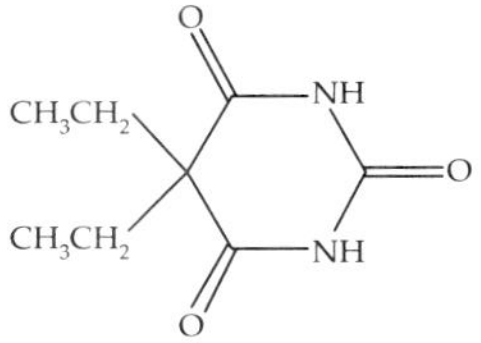

T(°C)	pK_a	T(°C)	pK_a	T(°C)	pK_a	T(°C)	pK_a
0	8.3971	20	8.0592	35	7.8471	50	7.6776
5	8.3040	25	7.9798	40	7.7858	55	7.6264
10	8.2171	30	7.9092	45	7.7290	60	7.5762
15	8.1367						

Thermodynamic functions for ionization at 25 °C:
$\Delta G^\circ = 10.89$ kcal/mol; $\Delta H^\circ = 5.81$ kcal/mol; $\Delta S^\circ = 17.0$ cal/deg/mol;

No.	Compound	pKa	Data quality	Ionization	Method	Conditions	Reference / Comments
127	Barbituric acid, 5,5-diethyl	7.98 ± 0.02 12.4	A U	–H –H	Potentiometric Spectro	H_2O $t = 25.0 \pm 0.02$ $I = 0.00$ (pK_1) $I = 0.3$ (pK_2)	McKeown RH, First thermodynamic dissociation constants of 5,5-disubstituted barbituric acids in water at 25 °C. Part 1, *JCS. Perkin II*, 504–514 (1980). NB: Thermodynamic pK_{a1} value refined for activity coefficients, [H+], and [OH−]. Results validated through measured pK_a value for benzoic acid in agreement with best literature values.

(continued)

Appendix A *(continued)*

No.	Name	pK_a value(s)	Data quality	Ionization type	Method	Conditions t°C; I or c M	Comments and Reference(s)
128	Barbituric acid, 5,5-diethyl Vol fraction EtOH	7.97 ± 0.02	A	–H	Potentiometric	H_2O $t = 25 \pm 0.5$ I undefined N_2 atmosphere	Rubino JT, Electrostatic and non-electrostatic free energy contributions to acid dissociation constants in cosolvent-water mixtures, *Int. J. Pharm.*, **42**, 181–191 (1988). NB: See Barbituric acid for details.
	0.1	8.03	U				
	0.3	8.43	U				
	0.5	8.80	U				
129	Barbituric acid, 5,5-diethyl	7.91	U	–H	Potentiometric	H_2O $t = 25$ $I = 0.00$	Krahl ME, The effect of variation in ionic strength and temperature on the apparent dissociation constants of thirty substituted barbituric acids, *J. Phys. Chem.*, **44**, 449–463 (1940). NB: Not reliable, not in agreement with several others (7.98 ± 0.02). The work of Krahl used careful potentiometric measurements, but was flawed by attempting to correct for ionic strengths that were too high for the version of the Debye-Hückel equation used.
130	Barbituric acid, 5,5-diethyl	7.86 ± 0.03	U	–H	Spectro	H_2O $t = 20 \pm 0.2$ $I = 0.1$	Mokrosz JL, Bojarski J and Welna W, The dissociation constants of some 1,5,5-trisubstituted barbituric acids, *Arch. Pharm. Weinheim*, **319**, 255–260 (1986). NB: The deviation (0.20 pK unit) between this value and the value given by Manov *et al.* (1952) (No. 126) for the same temperature, even when the ionic strength is taken into account (using the Davies' equation), raises doubts about the values reported in this paper for the 5,5-diethyl-1-substituted derivatives (see below), despite the good reproducibility of these results. The possibility of CO_2 interference cannot be discounted.
131	Barbituric acid, 5,5-diethyl	12.31 ± 0.05	U	–H	Spectro	H_2O $t = 38.0$ $I = 0.1$	Butler TC, Ruth JM and Tucker GF, The second ionization of 5, 5-disubstituted derivatives of barbituric acid, *JACS*, **77**, 1486–1488 (1955). NB: pK_2 value. Mean of nine values (over a pH range of 12.17–12.40)

132	Barbituric acid, 5,5-diethyl-1-methyl ($C_9H_{14}N_2O_3$)	8.30 ± 0.03	U	–H	Spectro	H_2O $t = 20 \pm 0.2$ $I = 0.1$	Mokrosz JL, Bojarski J and Welna W, The dissociation constants of some 1,5,5-trisubstituted barbituric acids, *Arch. Pharm. Weinheim*, **319**, 255–260 (1986). NB: See comment for the 5,5-diethylbarbituric acid value reported by the same authors. The pK_a values of 15 N-substituted 5,5-diethylbarbituric acids containing phenyl, benzyl, and benzoyl moieties were determined. The pK_a values for the remaining derivatives are as follows:

N-substituent	pK_a value	N-substituent	pK_a value
Phenyl	7.62±0.02	*p*-methylbenzoyl	6.64±0.02
p-nitrophenyl	7.32±0.03	*o*-methoxybenzoyl	6.54±0.03
Benzyl	8.12±0.03	*m*-methoxybenzoyl	6.51±0.02
p-chlorobenzyl	8.01±0.04	*p*-methoxybenzoyl	6.80±0.02
p-nitrobenzyl	7.48±0.03	*o*-bromobenzoyl	6.56±0.02
benzoyl	6.40±0.03	*m*-bromobenzoyl	6.39±0.02
o-methylbenzoyl	6.50±0.03	*p*-bromobenzoyl	6.55±0.03
m-methylbenzoyl	6.64±0.03		

133	Barbituric acids, *N*-methylated $R_1 = R_2 =$ n-Pr $R_1 =$ Me, $R_2 =$ i-Pr $R_1 =$ Me, $R_2 =$ Et $R_1 =$ Me, $R_2 =$ n-Bu $R_1 = R_2 =$ Allyl $R_1 =$ n-Pr, $R_2 =$ i-Pr		U	–H	Potentiometric	H_2O $t = 20.0$ $I = 0.10$	Doornbos DA and de Zeeuw RA, Determination of the acid dissociation constants of barbiturates by an accurate method of pH measurement. II. Correlations between normal and *N*-methylated barbiturates, *Pharm. Weekbl.*, **106**, 134–141 (1971). "The potentiometric determination of the "proton lost" dissociation constant of six *N*-methylated barbiturates was carried out. The constants are tabulated as association constants *K*H1 for an ionic strength *mu* = 0.10 and *t* = 20.0 °C. The influence of substituents at N1 and C5 on the acid strength is discussed. From the determination of the "proton gained" dissociation constants in strongly acid medium it could not be concluded how many protons can be bound. The 6 barbiturates studied were: 1-methyl-5,5-dipropyl-; 1,5-dimethyl-5-isopropyl-; 1,5-dimethyl-5-ethyl-; 1-methyl-5-ethyl-5-butyl-; 1-methyl-5,5-diallyl-; and 1-methyl-5-propyl-5-isopropylbarbituric acid."
134	Barbituric acid, 5-ethyl-5-isopropyl ($C_9H_{14}N_2O_3$)	8.14 ± 0.02	A	–H	Potentiometric	H_2O $t = 25.0 \pm 0.02$ $I = 0.00$	McKeown RH, First thermodynamic dissociation constants of 5,5-disubstituted barbituric acids in water at 25 °C. Part 1, *JCS. Perkin II*, 504–514 (1980). NB: Thermodynamic value refined for activity coefficients, [H+], and [OH−]. Results validated through measured pK_a values for benzoic and 5,5-diethylbarbituric acids in agreement with best literature values.

(continued)

No.	Name	pK_a value(s)	Data quality	Ionization type	Method	Conditions t°C; I or c M	Comments and Reference(s)
135	Barbituric acid, 5-ethyl-5-isopropyl (probarbital)	8.01	U	–H	Potentiometric	H_2O $t = 25$ $I = 0.00$	Krahl ME, The effect of variation in ionic strength and temperature on the apparent dissociation constants of thirty substituted barbituric acids, *J. Phys. Chem.*, **44**, 449–463 (1940). NB: See Barbital, no. 129, for comment.
136	Barbituric acid, 5-ethyl-5-isopropyl	12.59 ± 0.05	U	–H	Spectro	H_2O $t = 38.0$ $I = 0.1$	Butler TC, Ruth JM and Tucker GF, The second ionization of 5,5-disubstituted derivatives of barbituric acid, *JACS*, **77**, 1486–8 (1955). NB: pK_2 value. Mean of nine values.
137	Barbituric acid, 5-ethyl-5-butyl ($C_{10}H_{16}N_2O_3$) [structure: CH_3CH_2, n-C_4H_9, NH, NH, O, O, O]	7.98	U	–H	Potentiometric	H_2O $t = 25$	McKeown RH, personal communication (2006).
138	Barbituric acid, 5-ethyl-5-sec-butyl ($C_{10}H_{16}N_2O_3$) [structure: CH_3CH_2, NH, NH, O, O, O]	12.62 ± 0.05	U	–H	Spectro	H_2O $t = 38.0$ $I = 0.1$	Butler TC, Ruth JM and Tucker GF, The second ionization of 5,5-disubstituted derivatives of barbituric acid, *JACS*, **77**, 1486–1488 (1955). NB: pK_2 value. Mean of nine values.
139	Barbituric acid, 5-ethyl-5-iso-butyl (butabarbital) ($C_{10}H_{16}N_2O_3$)	7.95 ± 0.02	U	–H	Potentiometric	H_2O $t = 25 \pm 0.5$ N_2 atmosphere	Rubino JT, Electrostatic and non-electrostatic free energy contributions to acid dissociation constants in cosolvent-water mixtures, *Int. J. Pharm.*, **42**, 181–191 (1988). NB: See Barbituric acid for details.
	Vol fraction EtOH						
	0.1	8.22	U				
	0.3	8.59	U				
	0.5	9.04	U				

140	Barbituric acid, 5-ethyl-5-(1-methylbutyl) (pentobarbitone) ($C_{11}H_{18}N_2O_3$) 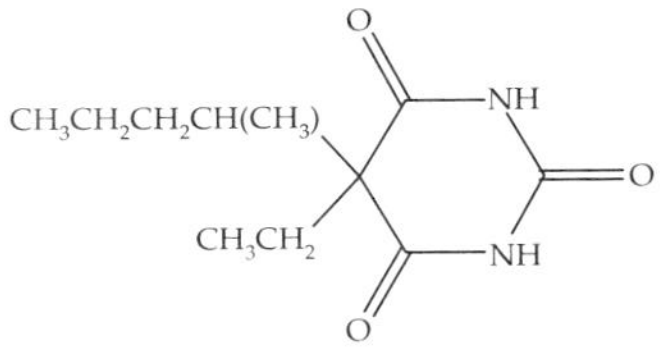	8.11	U	–H	Potentiometric	H_2O $t = 25$ $I = 0.00$	Krahl ME, The effect of variation in ionic strength and temperature on the apparent dissociation constants of thirty substituted barbituric acids, *J. Phys. Chem.*, **44**, 449–463 (1940). NB: See Barbital, no. 129, for comment. See also no. 155.
141	Barbituric acid, 5-ethyl-5-(1-methylbutyl) (pentobarbitone)	8.13 ± 0.02	A	–H	Spectro	H_2O $t = 25.0 \pm 0.02$ $I = 0.00$	Prankerd RJ, Physical Properties and Biological Activities of Barbituric Acids, Ph.D. thesis, University of Otago (1985). NB: Thermodynamic value refined for activity coefficients; corrected for effects of 2nd ionization on the spectrum of the monoanion. Results validated through measured pK_a values for benzoic and 5,5-diethylbarbituric acids in agreement with best literature values.
142	Barbituric acid, 5-ethyl-5-(1-methylbutyl) (pentobarbitone)	7.95	U	–H		H_2O $t = 37$	Ballard BE and Nelson E, Physicochemical properties of drugs that control absorption rate after subcutaneous implantation, *JPET*, **135**, 120–127 (1962).
143	Barbituric acid, 5-ethyl-5-(1-methylbutyl) (pentobarbitone)	12.67 ± 0.05	U	–H	Spectro	H_2O $t = 38.0$ $I = 0.1$	Butler TC, Ruth JM and Tucker GF, The second ionization of 5,5-disubstituted derivatives of barbituric acid, *JACS*, **77**, 1486–1488 (1955). NB: pK_2 value. Mean of nine values.
144	Barbituric acid, 5-ethyl-5-(1-methylbutyl)-2-thio (thiopentone)	7.6	U	–H	Spectro	H_2O $I = 0.1$	Shore PA, Brodie BB and Hogben CAM, The gastric secretion of drugs, *JPET*, **119**, 361–9 (1957). NB: The value was measured according to Flexser LA, Hammett LP and Dingwall A, The determination of ionization by ultraviolet spectrophotometry, *JACS*, **57**, 2103–2115 (1935).
145	Barbituric acid, 5-ethyl-5-(3-methylbutyl)- (amobarbital; amylobarbitone) ($C_{11}H_{18}N_2O_3$) 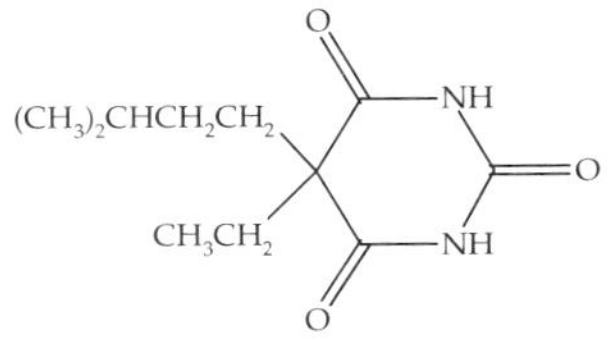	7.94	U	–H	Potentiometric	H_2O $t = 25$ $I = 0.00$	Krahl ME, The effect of variation in ionic strength and temperature on the apparent dissociation constants of thirty substituted barbituric acids, *J. Phys. Chem.*, **44**, 449–463 (1940). NB: See Barbital, no. 129, for comment.

(continued)

No.	Name	pK_a value(s)	Data quality	Ionization type	Method	Conditions t°C; I or c M	Comments and Reference(s)
146	Barbituric acid, 5-ethyl-5-(3-methylbutyl)- (amobarbital; amylobarbitone)	7.96 ± 0.04	A	–H	Spectro	H_2O $t = 25.0 \pm 0.01$ $I = 0.00$	Briggs AG, Sawbridge JE, Tickle P and Wilson JM, Thermodynamics of dissociation of some barbituric acids in aqueous solution, *J. Chem. Soc.* (*B*), 802–805 (1969). NB: All compounds were recrystallized to constant melting point and pH values were measured to ± 0.005. Ionic strengths were corrected to zero with Davies' equation. Results were discussed with reference to the IUPAC compilation data, primarily from AI Biggs, *JCS*, 2485–2488 (1956), which were obtained at relatively poorly controlled temperature.
147	Barbituric acid, 5-ethyl-5-(3-methylbutyl)- (amobarbital; amylobarbitone)	7.94 ± 0.02	A	–H	Potentiometric	H_2O $t = 25 \pm 0.5$ N_2 atmosphere	Rubino JT, Electrostatic and non-electrostatic free energy contributions to acid dissociation constants in cosolvent-water mixtures, *Int. J. Pharm.*, **42**, 181–191 (1988). NB: See Barbituric acid for details.
	Vol. fraction EtOH						
	0.1	8.06	U				
	0.3	8.39	U				
	0.5	8.89	U				
148	Barbituric acid, 5-ethyl-5-(3-methylbutyl)- (amobarbital; amylobarbitone)	12.42 ± 0.05	U	–H	Spectro	H_2O $t = 38.0$ $I = 0.1$	Butler TC, Ruth JM and Tucker GF, The second ionization of 5,5-disubstituted derivatives of barbituric acid, *JACS*, **77**, 1486–1488 (1955). NB: pK_2 value. Mean of nine values.
149	Barbituric acid, 5-ethyl-5-(1,3-dimethylbutyl) ($C_{12}H_{20}N_2O_3$) $(CH_3)_2CHCH_2CH(CH_3)$, CH_3CH_2, NH, NH, O, O, O	8.14 ± 0.02	A	–H	Spectro	H_2O $t = 25.0 \pm 0.02$ $I = 0.00$	Prankerd RJ, Physical Properties and Biological Activities of Barbituric Acids, Ph.D. thesis, University of Otago (1985). NB: Thermodynamic value refined for activity coefficients as well as for [H+], and [OH–]; corrected for effects of 2nd ionization on the spectrum of the monoanion. Results validated through measured pK_a values for benzoic and 5, 5-diethylbarbituric acids in agreement with best literature values.
150	Barbituric acid, 5-ethyl-5-cyclohex-1-enyl (Cyclobarbital) ($C_{12}H_{16}N_2O_3$) CH_3, NH, NH, O, O, O	7.50	U	–H	Potentiometric	H_2O $t = 25$ $I = 0.00$	Krahl ME, The effect of variation in ionic strength and temperature on the apparent dissociation constants of thirty substituted barbituric acids, *J. Phys. Chem.*, **44**, 449–463 (1940). NB: See Barbital, no. 129, for comment.

No.	Compound	pKa	Data quality	Ionization	Method	Conditions	Reference / Notes
151	Barbituric acid, 5-allyl-5-ethyl ($C_9H_{12}N_2O_3$) 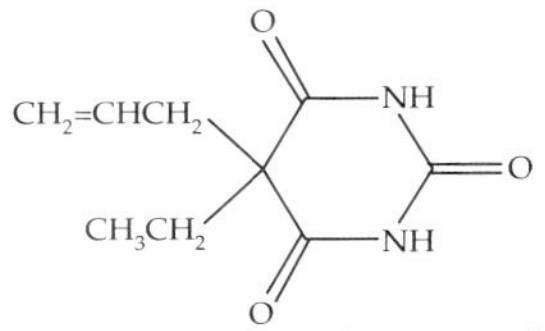	7.89 ± 0.02	A	–H	Potentiometric	H_2O $t = 25.0 \pm 0.02$ $I = 0.00$	Baird DR, Barbituric acids: Structure-reactivity, a quantitative examination, M. Pharm. thesis, University of Otago, 1979. NB: Thermodynamic value refined for activity coefficients, [H+], and [OH−]. Results were validated through measured pK_a values for benzoic and 5,5-diethylbarbituric acids in agreement with best literature values. Also reported values for 5-allyl-5-phenyl (pK_a = 7.40), 5-phenyl-5-isopropyl (pK_a = 7.76), 5-ethyl-5-phenyl (pK_a = 7.48) and -1,5′-spirocyclohexane (pK_a = 8.88).
152	Barbituric acid, 5-allyl-5-isopropyl ($C_{10}H_{14}N_2O_3$) 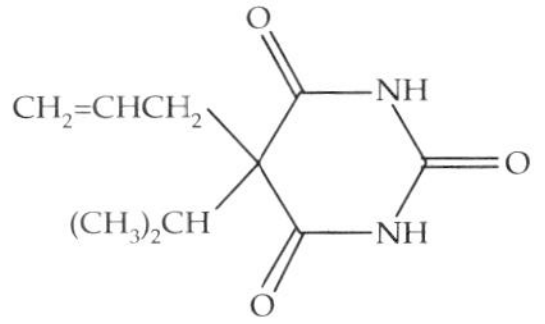	8.02 ± 0.02	A	–H	Potentiometric	H_2O $t = 25.0 \pm 0.02$ $I = 0.00$	McKeown RH, First thermodynamic dissociation constants of 5,5-disubstituted barbituric acids in water at 25 °C. Part 1, *JCS. Perkin II*, 504–514 (1980). NB: Thermodynamic value refined for activity coefficients, [H+], and [OH–]. Results validated through measured pK_a values for benzoic and 5,5-diethylbarbituric acids in agreement with best literature values. NB: See also no. 155.
153	Barbituric acid, 5-allyl-5-isopropyl (aprobarbital)	8.00 ± 0.02	A	–H	Potentiometric	H_2O $t = 25 \pm 0.5$ N_2 atmosphere	Rubino JT, Electrostatic and non-electrostatic free energy contributions to acid dissociation constants in cosolvent-water mixtures, *Int. J. Pharm.*, **42**, 181–191 (1988). NB: See Barbituric acid for details.
	Vol. fraction EtOH						
	0.1	8.11	U				
	0.3	8.56	U				
	0.5	8.99	U				
154	Barbituric acid, 5-allyl-5-isopropyl (aprobarbital)	12.52 ± 0.05	U	–H	Spectro	H_2O $t = 38.0$ $I = 0.1$	Butler TC, Ruth JM and Tucker GF, The second ionization of 5,5-disubstituted derivatives of barbituric acid, JACS, 77, 1486–1488 (1955). NB: pK_2 value. Mean of nine values.
155	Barbituric acid, 5-allyl-5-isobutyl ($C_{11}H_{16}N_2O_3$) 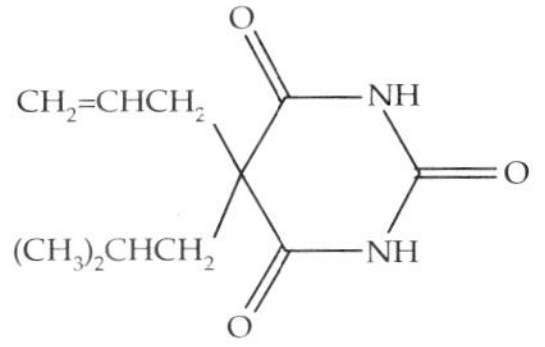	7.63 ± 0.1	U	–H	Potentiometric	H_2O $t = 24$	Maulding HV and Zoglio MA, pK_a determinations utilizing solutions of 7-(2-hydroxypropyl) theophylline, *J. Pharm. Sci.*, **60**, 309–311 (1971).
		7.86	U	–H			NB: By extrapolation of apparent values to 0% 7-(2-hydroxypropyl)-theophylline that was used as a complexing agent to increase solubility. TRIS was used as a pK_a reference (pK_a = 8.18 at 20 °C). See also Ritschel, who cited Fincher JH, Entrekin DN, Hartman CW, Surfactant-Base-Barbiturate Suppositories I Rectal absorption in rabbits, J. Pharm. Sci., 55, 23–28 (1966). Fincher in turn cited values from Shanker LS, Absorption of drugs from the rat colon, JPET, 126, 283–290 (1959) for 4 cpds: 5-allyl-5-isobutylbarbituric acid (7.86); 5-butyl-5-ethylbarbituric acid (8.10); 5-ethyl-5-(1-methylbutyl)barbituric acid (8.17) and 5-allyl-5-isopropylbarbituric acid (7.54).

(continued)

No.	Name	pK_a value(s)	Data quality	Ionization type	Method	Conditions t°C; I or c M	Comments and Reference(s)
156	Barbituric acid, 5-allyl-5-isobutyl	12.36 ± 0.05	U	−H	Spectro	H_2O $t = 38.0$ $I = 0.1$	Butler TC, Ruth JM and Tucker GF, The second ionization of 5,5-disubstituted derivatives of barbituric acid, *JACS*, **77**, 1486–1488 (1955). NB: pK_2 value. Mean of nine values.
157	Barbituric acid, 5-allyl-5-(1-methylbutyl) (secobarbital)	12.60 ± 0.05	U	−H	Spectro	H_2O $t = 38.0$ $I = 0.1$	Butler TC, Ruth JM, Tucker GF, The second ionization of 5, 5-disubstituted derivatives of barbituric acid, *JACS*, **77**, 1486–1488 (1955). NB: pK_2 value. Mean of nine values.
158	Barbituric acid, 5,5-diallyl (allobarbital) ($C_{10}H_{12}N_2O_3$)	7.81 ± 0.02	A	−H	Potentiometric	H_2O $t = 25.0 \pm 0.02$ $I = 0.00$	McKeown RH, First thermodynamic dissociation constants of 5,5-disubstituted barbituric acids in water at 25 °C. Part 1, *JCS.Perkin II*, 504–514 (1980). NB: Thermodynamic value refined for activity coefficients, [H+], and [OH−]. Results validated through measured pK_a values for benzoic and 5,5-diethylbarbituric acids in agreement with best literature values.
159	Barbituric acid, 5,5- diallyl (allobarbital)	7.73 ± 0.02	U	−H	Potentiometric	H_2O $t = 25 \pm 0.5$ N_2 atmosphere	Rubino JT, Electrostatic and non-electrostatic free energy contributions to acid dissociation constants in cosolvent-water mixtures, *Int. J. Pharm.*, **42**, 181–191 (1988). NB: See Barbituric acid for details.
	Vol fraction EtOH						
	0.1	7.88	U				
	0.3	8.34	U				
	0.5	8.75	U				
160	Barbituric acid, 5,5- diallyl (allobarbital)	7.79	U	−H	Potentiometric	H_2O $t = 25$ $I = 0.00$	Krahl ME, The effect of variation in ionic strength and temperature on the apparent dissociation constants of thirty substituted barbituric acids, *J. Phys. Chem.*, **44**, 449–463 (1940). NB: See Barbital, no. 129, for comment.

161	Barbituric acid, 5-methyl-5-(3-methyl-but-2-enyl) ($C_{10}H_{14}N_2O_3$) 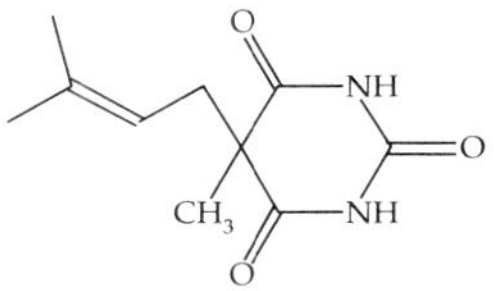	8.39 ± 0.02	A	–H	Spectro	H_2O $t = 25.0 \pm 0.02$ $I = 0.00$	Prankerd RJ, Physical Properties and Biological Activities of Barbituric Acids, Ph.D. thesis, University of Otago (1985). NB: Thermodynamic value refined for activity coefficients; corrected for effects of 2nd ionization on the spectrum of the monoanion. Results validated through measured pK_a values for benzoic and 5,5-diethylbarbituric acids in agreement with best literature values.
162	Barbituric acid, 5-ethyl-5-(3-methylbut-2-enyl) ($C_{11}H_{16}N_2O_3$) 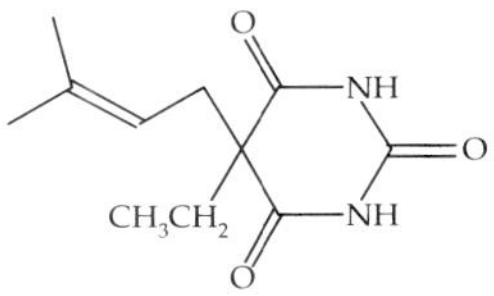	8.06 ± 0.02	A	–H	Spectro	H_2O $t = 25.0 \pm 0.02$ $I = 0.00$	Prankerd RJ, Physical Properties and Biological Activities of Barbituric Acids, Ph.D. thesis, University of Otago (1985). NB: Thermodynamic value refined for activity coefficients; corrected for effects of 2nd ionization on the spectrum of the monoanion. Results validated through measured pK_a values for benzoic and 5,5-diethylbarbituric acids in agreement with best literature values.
163	Barbituric acid, 5-isopropyl-5-(3-methylbut-2-enyl) ($C_{12}H_{18}N_2O_3$) 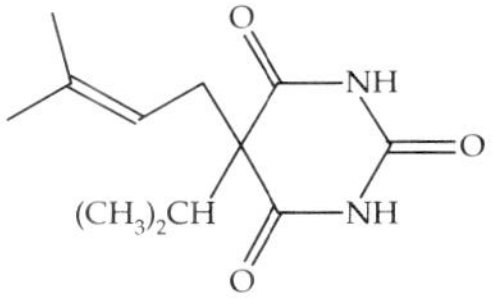	8.19 ± 0.02	A	–H	Spectro	H_2O $t = 25.0 \pm 0.02$ $I = 0.00$	Prankerd RJ, Physical Properties and Biological Activities of Barbituric Acids, Ph.D. thesis, University of Otago (1985). NB: Thermodynamic value refined for activity coefficients; corrected for effects of 2nd ionization on the spectrum of the monoanion. Results validated through measured pK_a values for benzoic and 5,5-diethylbarbituric acids in agreement with best literature values.
164	Barbituric acid, 5-t-butyl-5-(3-methylbut-2-enyl) ($C_{13}H_{20}N_2O_3$) 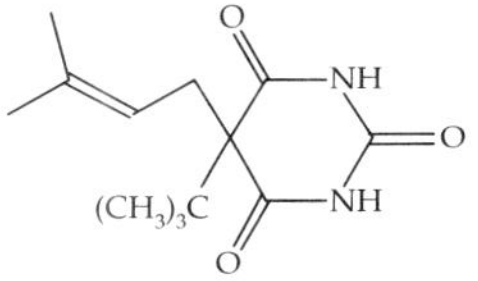	8.39 ± 0.02	A	–H	Spectro	H_2O $t = 25.0 \pm 0.02$ $I = 0.00$	Prankerd RJ, Physical Properties and Biological Activities of Barbituric Acids, Ph.D. thesis, University of Otago (1985). NB: Thermodynamic value refined for activity coefficients; corrected for effects of 2nd ionization on the spectrum of the monoanion. Results validated through measured pK_a values for benzoic and 5,5-diethylbarbituric acids in agreement with best literature values.

(continued)

No.	Name	pK_a value(s)	Data quality	Ionization type	Method	Conditions t°C; I or c M	Comments and Reference(s)
165	Barbituric acid, 5,5-di-(3-methylbut-2-enyl) ($C_{14}H_{20}N_2O_3$) 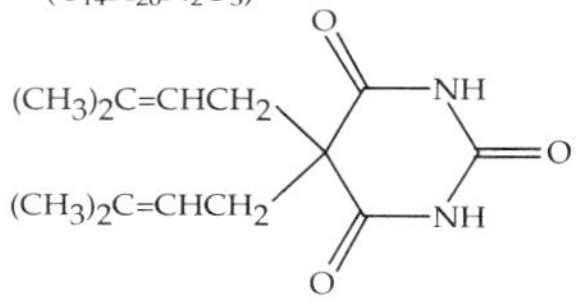	8.13 ± 0.02	A	–H	Spectro	H_2O t = 25.0 ± 0.02 I = 0.00	Prankerd RJ, Physical Properties and Biological Activities of Barbituric Acids, Ph.D. thesis, University of Otago (1985). NB: Thermodynamic value refined for activity coefficients; corrected for effects of 2nd ionization on the spectrum of the monoanion. Results validated through measured pK_a values for benzoic and 5,5-diethylbarbituric acids in agreement with best literature values.
166	Barbituric acid, 5-ethyl, 5-(1-methylbut-1-enyl ($C_{11}H_{16}N_2O_3$) 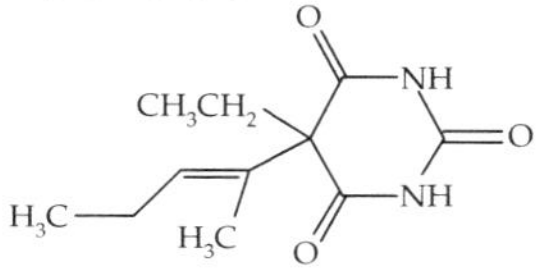	8.00	U	–H	Potentiometric	H_2O t = 25 I = 0.00	Krahl ME, The effect of variation in ionic strength and temperature on the apparent dissociation constants of thirty substituted barbituric acids, *J. Phys. Chem.*, **44**, 449–463 (1940). NB: Also called Vinbarbital. See Barbital, no. 129, for comment.
167	Barbituric acid, 5-ethyl-5-(1-methylbut-1-enyl) ($C_{11}H_{16}N_2O_3$) 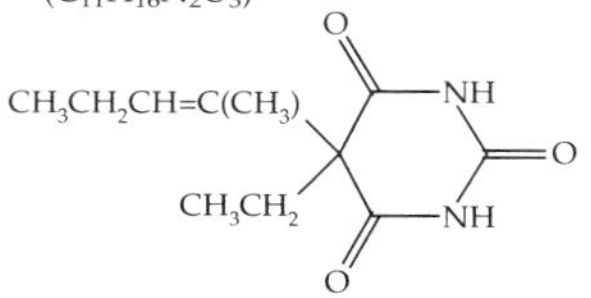	12.06 ± 0.05	U	–H	Spectro	H_2O t = 38.0 I = 0.1	Butler TC, Ruth JM and Tucker GF, The second ionization of 5,5-disubstituted derivatives of barbituric acid, *JACS*, **77**, 1486–1488 (1955). NB: pK_2 value. Mean of nine values.
168	Barbituric acid, 5-methyl-5-(4-methylpent-1-en-2-yl) ($C_{11}H_{16}N_2O_3$) 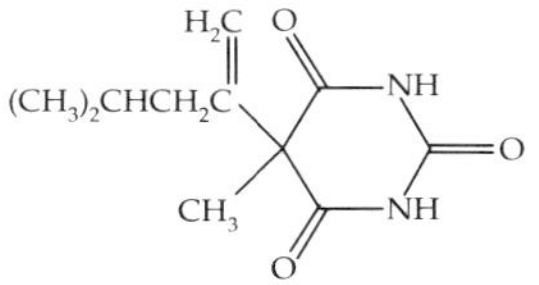	7.78 ± 0.02	A	–H	Spectro	H_2O t = 25.0 ± 0.02 I = 0.00	Prankerd RJ, Physical Properties and Biological Activities of Barbituric Acids, Ph.D. thesis, University of Otago (1985). NB: Thermodynamic value refined for activity coefficients; corrected for effects of 2nd ionization on the spectrum of the monoanion. Results validated through measured pK_a values for benzoic and 5,5-diethylbarbituric acids in agreement with best literature values.

169	Barbituric acid, 5-ethyl-5-(4-methyl-pent-1-en-2-yl) ($C_{12}H_{18}N_2O_3$) 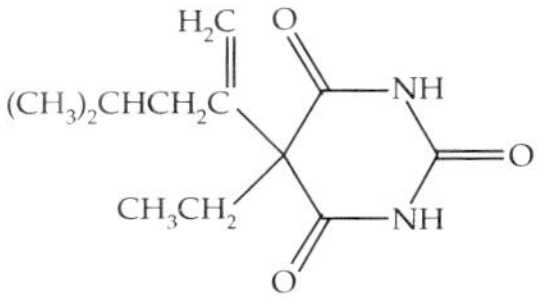	7.48 ± 0.02	A	–H	Spectro	H_2O $t = 25.0 \pm 0.02$ $I = 0.00$	Prankerd RJ, Physical Properties and Biological Activities of Barbituric Acids, Ph.D. thesis, University of Otago (1985). NB: Thermodynamic value refined for activity coefficients; corrected for effects of 2nd ionization on the spectrum of the monoanion. Results validated through measured pK_a values for benzoic and 5,5-diethylbarbituric acids in agreement with best literature values.
170	Barbituric acid, 5-methyl-5-phenyl ($C_{11}H_{10}N_2O_3$) 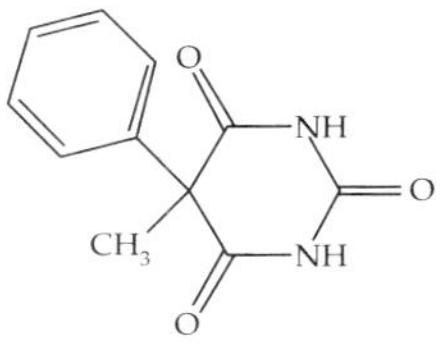	7.78 ± 0.02	A	–H	Potentiometric	H_2O $t = 25.0 \pm 0.02$ $I = 0.00$	McKeown RH, First thermodynamic dissociation constants of 5,5-disubstituted barbituric acids in water at 25 °C. Part 1, *JCS. Perkin II*, 504–514 (1980). NB: Thermodynamic value refined for activity coefficients, [H+], and [OH−]. Results validated through measured pK_a values for benzoic and 5,5-diethylbarbituric acids in agreement with best literature values.
171	Barbituric acid, 5-methyl-5-phenyl	7.63	U	–H	Spectro	H_2O $t = 20 \pm 0.2$ $I = 0.1$	Doornbos DA and de Zeeuw RA, Determination of the acid dissociation constants of barbiturates by an accurate method of pH measurement. II. Correlations between normal and N-methylated barbiturates, *Pharm. Weekbl.*, **106**, 134–141 (1971).
172	Barbituric acid, 5-methyl-5-phenyl-1-benzoyl ($C_{18}H_{14}N_2O_4$) 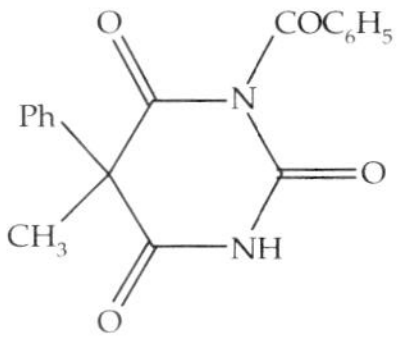	6.35 ± 0.06	U	–H	Spectro	H_2O $t = 20 \pm 0.2$ $I = 0.1$	Mokrosz JL, Bojarski J and Welna W, The dissociation constants of some 1,5,5-trisubstituted barbituric acids, *Arch. Pharm. Weinheim*, **319**, 255–260 (1986). NB: See comment for the 5,5-diethylbarbituric acid value reported by the same authors (above).
173	Barbituric acid, 5-ethyl-5-phenyl ($C_{12}H_{12}N_2O_3$) 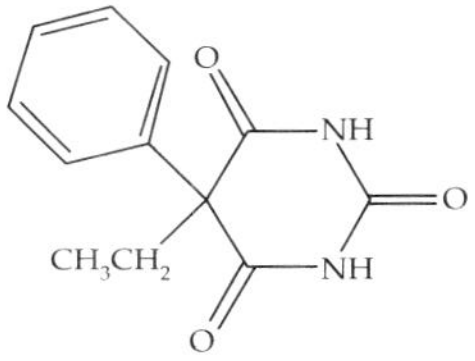	7.48 ± 0.02	A	–H	Potentiometric	H_2O $t = 25.0 \pm 0.02$ $I = 0.00$	McKeown RH, First thermodynamic dissociation constants of 5,5-disubstituted barbituric acids in water at 25 °C. Part 1, *JCS. Perkin II*, 504–514 (1980). NB: Thermodynamic value refined for activity coefficients, [H+], and [OH−]. Results validated through measured pK_a values for benzoic and 5,5-diethylbarbituric acids in agreement with best literature values.

(continued)

No.	Name	pK_a value(s)	Data quality	Ionization type	Method	Conditions t°C; I or c M	Comments and Reference(s)
174	Barbituric acid, 5-ethyl-5-phenyl	7.48 ± 0.02	A	−H	Spectro	H_2O $t = 25.0 \pm 0.02$ $I = 0.00$	Prankerd RJ, Physical Properties and Biological Activities of Barbituric Acids, Ph.D. thesis, University of Otago (1985). NB: Thermodynamic value refined for activity coefficients; corrected for effects of 2nd ionization on the spectrum of the monoanion. Results validated through measured pK_a values for benzoic and 5,5-diethylbarbituric acids in agreement with best literature values.
175	Barbituric acid, 5-ethyl-5-phenyl	7.48 ± 0.02	A	−H	Potentiometric	H_2O $t = 25 \pm 0.5$ N_2 atmosphere	Rubino JT, Electrostatic and non-electrostatic free energy contributions to acid dissociation constants in cosolvent-water mixtures, *Int. J. Pharm.*, **42**, 181–191 (1988). NB: See Barbituric acid for details.
	Vol. fraction EtOH						
	0.1	7.66	U				
	0.3	8.04	U				
	0.5	8.41	U				
176	Barbituric acid, 5-ethyl-5-phenyl	7.41	U	−H	Potentiometric	H_2O $t = 25$ $I = 0.00$	Krahl ME, The effect of variation in ionic strength and temperature on the apparent dissociation constants of thirty substituted barbituric acids, *J. Phys. Chem.*, **44**, 449–463 (1940). NB: See Barbital, no. 129, for comment.
177	Barbituric acid, 5-ethyl-5-phenyl	7.52 ± 0.1	U	−H	Potentiometric	H_2O $t = 24$	Maulding HV and Zoglio MA, pK_a determinations utilizing solutions of 7-(2-hydroxypropyl) theophylline, *J. Pharm. Sci.*, **60**, 309–311 (1971). NB: See Barbituric acid, 5-allyl-5-isobutyl for details.
178	Barbituric acid, 5-ethyl-5-phenyl	7.40 ± 0.02 12.2	U U	−H −H	Spectro Spectro	H_2O $t = 25.0 \pm 0.02$ $I = 0.00$ (pK_1) $I = 0.3$ (pK_2)	McKeown RH, First thermodynamic dissociation constants of 5,5-disubstituted barbituric acids in water at 25 °C. Part 1, *JCS. Perkin II*, 504–514 (1980). NB: Thermodynamic value refined for activity coefficients, [H+], and [OH−].
179	Barbituric acid, 5-ethyl-5-phenyl	11.77 ± 0.05	U	−H	Spectro	H_2O $t = 38.0$ $I = 0.1$	Butler TC, Ruth JM and Tucker GF, The second ionization of 5,5-disubstituted derivatives of barbituric acid, *JACS*, **77**, 1486–8 (1955). NB: pK_2 value. Mean of nine values.
180	Barbituric acid, 5-ethyl-5-phenyl	7.29	U	−H		H_2O $t = 37$	Ballard BE and Nelson E, Physicochemical properties of drugs that control absorption rate after subcutaneous implantation, *JPET*, **135**, 120–127 (1962).
181	Barbituric acid, 5-ethyl-5-phenyl	7.31 (0.16) 11.99 (0.16)	U U	−H −H	spectro (pK_1, 260 nm; pK_2, 257 nm)	H_2O $t = 20.0$	Wahbe AM, El-Yazbi FA, Barary MH and Sabri SM, Application of orthogonal functions to spectrophotometric analysis. Determination of dissociation constants, *Int. J. Pharm.*, **92(1)**, 15–22 (1993). NB: Alternative graphical method gave $pK_a = 7.3$ and 11.9. See Acetaminophen for further details.

182	Barbituric acid, 5-ethyl-5-phenyl	7.36	U	–H	Potentiometric	H_2O $t = 20 \pm 0.2$ $I = 0.1$	Doornbos DA and de Zeeuw RA, Determination of the acid dissociation constants of barbiturates by an accurate method of pH measurement. II. Correlations between normal and N-methylated barbiturates, *Pharm. Weekbl.*, **106**, 134–141 (1971).
183	Barbituric acid, 5-ethyl-5-phenyl	7.41 ± 0.03	U	–H	Potentiometric	H_2O $t = 25.0 \pm 0.1$ $I = 0.1$ (NaCl)	Takacs-Novak K, Box KJ and Avdeef A, Potentiometric pK_a determination of water-insoluble compounds: Validation study in methanol/water mixtures, *Int. J. Pharm.*, **151**, 235–248 (1997). NB: p$K_a = 7.41 \pm 0.05$ by extrapolation from 16.3 – 64.7 %w/w aqueous MeOH. See Acetaminophen for full details.
184	Barbituric acid, 5-ethyl-5-phenyl	7.43 ± 0.05	U	–H	Potentiometric	H_2O $t = 25.0$ $I = 0.1$ (NaCl)	Takacs-Novak K and Avdeef A, Interlaboratory study of log P determination by shake-flask and potentiometric methods, *J. Pharm. Biomed. Anal.*, **14**, 1405–1413 (1996). NB: See Acetaminophen for further details. Also reported p$K_a = 7.49 \pm 0.02$ at $I = 0.1$ (KNO_3).
185	Barbituric acid, 5-ethyl-5-phenyl-1-benzoyl ($C_{19}H_{16}N_2O_4$)	6.58	U	–H	Spectro	H_2O $t = 20 \pm 0.2$ $I = 0.1$	Paluchowska M, Ekiert L, Jochym K and Bojarski J, Hydrolysis of barbituric acid derivatives. Part V. Hydrolysis of 1-benzoyl-5-ethyl-5-phenylbarbituric acid, *Pol. J. Chem.*, **57**(7-8-9), 799–807 (1983); CA 102, 5311m, 1985.
186	Barbituric acid, 5-ethyl-5-(3-nitro-phenyl) ($C_{12}H_{11}N_3O_5$)	7.04 ± 0.02	A	–H	Spectro	H_2O $t = 25.0 \pm 0.02$ $I = 0.00$	McKeown RH, First thermodynamic dissociation constants of 5,5-disubstituted barbituric acids in water at 25 °C. Part 1, *JCS.Perkin II*, 504–514 (1980). NB: Thermodynamic value refined for activity coefficients. Results validated through measured pK_a values for benzoic and 5,5-diethylbarbituric acids in agreement with best literature values.
187	Barbituric acid, 5-ethyl-5-(3-nitro-phenyl)	6.99 ± 0.02	A	–H	Potentiometric	H_2O $t = 25 \pm 0.5$ N_2 atmosphere	Rubino JT, Electrostatic and non-electrostatic free energy contributions to acid dissociation constants in cosolvent-water mixtures, *Int. J. Pharm.*, **42**, 181–191 (1988). NB: See Barbituric acid for details.
	Vol. fraction EtOH						
	0.1	7.20	U				
	0.3	7.58	U				
	0.5	7.88	U				

(continued)

No.	Name	pK_a value(s)	Data quality	Ionization type	Method	Conditions t°C; I or c M	Comments and Reference(s)
188	Barbituric acid, 5-ethyl-5-(4-nitro-phenyl) ($C_{12}H_{11}N_3O_5$) 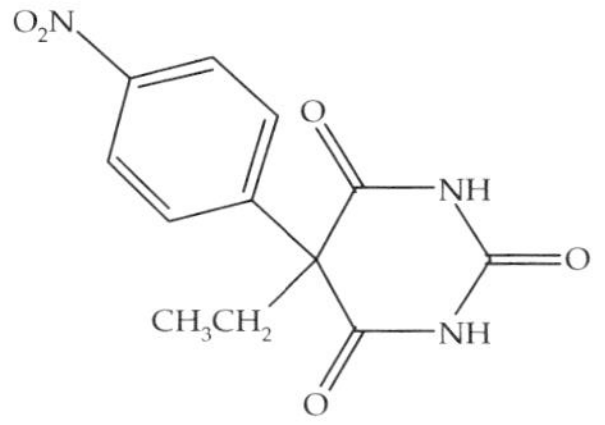	6.94 ± 0.02	A	–H	Spectro	H_2O $t = 25.0 \pm 0.02$ $I = 0.00$	McKeown RH, First thermodynamic dissociation constants of 5,5-disubstituted barbituric acids in water at 25 °C. Part 1, *JCS. Perkin II*, 504–514 (1980). NB: Thermodynamic value refined for activity coefficients. Results validated through measured pK_a values for benzoic and 5,5-diethylbarbituric acids in agreement with best literature values.
189	Barbituric acid, 5,5-diphenyl ($C_{16}H_{12}N_2O_3$) 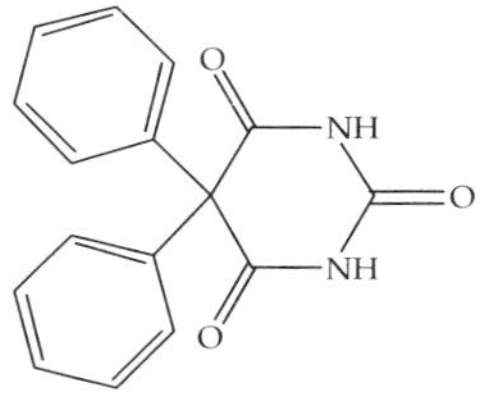	7.30 ± 0.02 11.9	A U	–H –H	spectro Spectro	H_2O $t = 25.0 \pm 0.02$ $I = 0.00$ (pK_1) $I = 0.3$ (pK_2)	McKeown RH, First thermodynamic dissociation constants of 5,5-disubstituted barbituric acids in water at 25 °C. Part 1, *JCS. Perkin II*, 504–514 (1980). NB: Thermodynamic value refined for activity coefficients. Results validated through measured pK_a values for benzoic and 5,5-diethylbarbituric acids in agreement with best literature values.
190	Barbituric acid, -1′,5-spiro(cyclopropane) ($C_6H_6N_2O_3$) 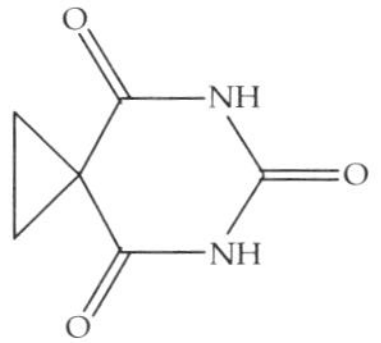	8.73 ± 0.03	A	–H	Potentiometric	H_2O $t = 25.0 \pm 0.02$ $I = 0.00$	McKeown RH and Prankerd RJ, First thermodynamic dissociation constants of 5,5-disubstituted barbituric acids in water at 25 °C. Part 3, *JCS. Perkin II*, 481–487 (1981). NB: Thermodynamic value refined for activity coefficients, [H+], and [OH–]. Results validated through measured pK_a values for benzoic and 5,5-diethylbarbituric acids in agreement with best literature values.

No.	Compound	pK	Type		Method	Conditions	Reference / Notes
191	Barbituric acid, -1′,5-spiro(cyclobutane) ($C_7H_8N_2O_3$) 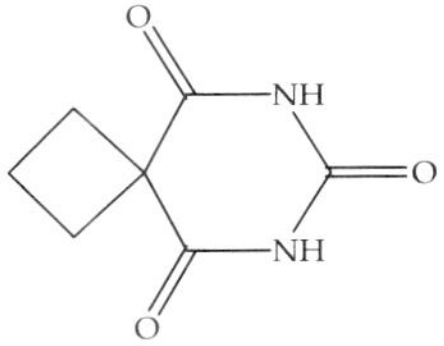	8.82 ± 0.02	A	−H	Potentiometric	H_2O $t = 25.0 \pm 0.02$ $I = 0.00$	McKeown RH and Prankerd RJ, First thermodynamic dissociation constants of 5,5-disubstituted barbituric acids in water at 25 °C. Part 3, *JCS. Perkin II*, 481–487 (1981). NB: Thermodynamic value refined for activity coefficients, [H+], and [OH−]. Results validated through measured pK_a values for benzoic and 5,5-diethylbarbituric acids in agreement with best literature values.
192	Barbituric acid, -1′,5-spiro(cyclopentane) ($C_8H_{10}N_2O_3$) 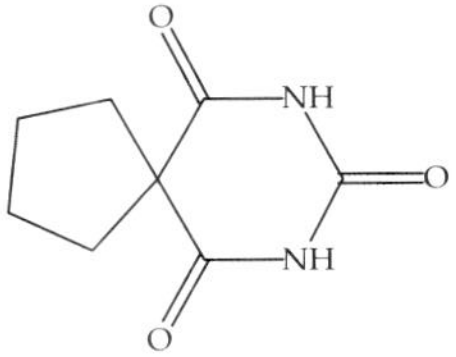	8.83 ± 0.03	A	−H	Potentiometric	H_2O $t = 25.0 \pm 0.02$ $I = 0.00$	McKeown RH and Prankerd RJ, First thermodynamic dissociation constants of 5,5-disubstituted barbituric acids in water at 25 °C. Part 3, *JCS. Perkin II*, 481–487 (1981). NB: Thermodynamic value refined for activity coefficients, [H+], and [OH−]. Results validated through measured pK_a values for benzoic and 5,5-diethylbarbituric acids in agreement with best literature values. ACD calculated value 9.30 ± 0.2.
193	Barbituric acid, -1′,5-spiro(cyclohexane) ($C_9H_{12}N_2O_3$) 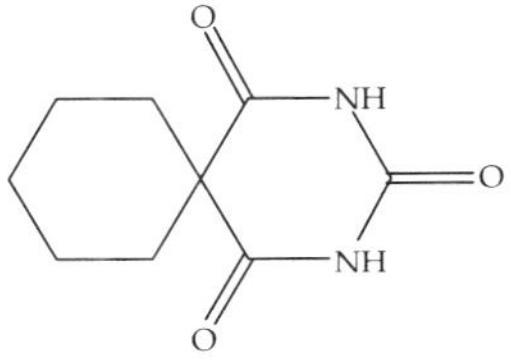	8.88 ± 0.03	A	−H	Potentiometric	H_2O $t = 25.0 \pm 0.02$ $I = 0.00$	McKeown RH and Prankerd RJ, First thermodynamic dissociation constants of 5,5-disubstituted barbituric acids in water at 25 °C. Part 3, *JCS. Perkin II*, 481–487 (1981). NB: Thermodynamic value refined for activity coefficients, [H+], and [OH−]. Results validated through measured pK_a values for benzoic and 5,5-diethylbarbituric acids in agreement with best literature values. This value was measured by DR. Baird, see Barbituric acid, 5-ethyl-5-butyl, for details.
194	Bases (nonaqueous titrations)	Acetous pK_b			Potentiometric	acetic acid	Castellano T, Medwick T, Shinkai JH and Bailey L, Differential titration of bases in glacial acetic acid, *J. Pharm. Sci.*, **70**, 104–105 (1981). "The overall basicity constants for 20 bases were measured in glacial acetic acid, and the differential titration of five binary mixtures of variable dissociation constant (pK_b) values was followed using a glass electrode-modified calomel electrode system. A leveling diagram was constructed that indicated that bases stronger than aqueous pK_b 10 are leveled to an acetous pK_b 5.69, whereas weaker
	Benzocaine	7.53	U	+H			
	Nicotinic acid	7.01	U	+H			
	Isonicotinic acid	6.86	U	+H			
	TRIS buffer	6.06	U	+H			
	Atropine	5.93	U	+H			

(*continued*)

No.	Name	pK_a value(s)	Data quality	Ionization type	Method	Conditions t°C; I or c M	Comments and Reference(s)
							bases are not leveled but instead exhibit their own intrinsic basicity, with the acetous pK_b to aqueous pK_b values being linearly related. A minimum acetous pK_b of 4 units is required for satisfactory differential titration of two bases in acetic acid.''
195	Bencyclane ($C_{19}H_{31}NO$) 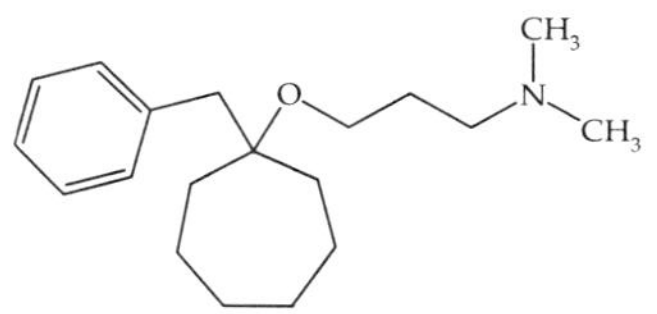	8.94 ± 0.04	U	+H	Potentiometric	40% EtOH $t = 25.0$	Mannhold R, Rodenkirchen R, Bayer R and Haus W, The importance of drug ionization for the action of calcium antagonistsand related compounds, *Arzneim.-Forsch.*, **34**, 407–409 (1984).
		9.68	U	+H		H_2O	NB: See Aprindine for details.
196	Bendroflumethiazide ($C_{15}H_{14}F_3N_3O_4S_2$) 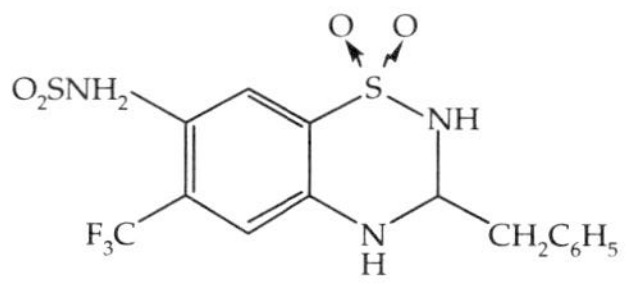	8.53 ± 0.05	U	−H	soly	H_2O $t = 25$ $I = 0.2$	Ågren A and Bäck T, Complex formation between macromolecules and drugs. VIII., *Acta Pharm. Suecica*, **10**, 223–228 (1973). ''The pK_a values of bendroflumethiazide . . . were therefore determined by using the solubility variation with pH according to Green [8].'' Also cited in Florey K and Russo-Alesi FM, Bendroflumethiazide, *APDS*, **5**, 1–19 (1976).
197	Bendroflumethiazide			−H	Potentiometric	H_2O	Hennig UG, Moskalyk RE, Chatten LG and Chan SF, Semiaqueous potentiometric determinations of apparent pK_{a1} values for benzothiadiazines and detection of decomposition during solubility variation with pH studies, *J. Pharm. Sci.*, **70**, 317–319 (1981). ''Semi-aqueous potentiometric determination of apparent pK_a values by extrapolation technique: The apparent pK_{a1} values of various benzothiadiazines were determined by the method of Chatten *et al.* (5) using acetone-water mixtures. Four concentrations of each benxothiadiazine (0.0005, 0.001, 0.0015, and 0.002 M) were prepared from 0.02M acetone stock solutions. The acetone-water ratios were 5:45, 10:40, 15:35, and 20:30, respectively. The solutions were titrated with 0.05N NaOH and the pH was measured after the addition of each 0.1 mL increment of titrant. The apparent pK_{a1} values were obtained by the usual extrapolation techniques (2, 5).''
	Flumethiazide	6.3	U				
	Hydrochlorothiazide	8.7	U				
	Hydroflumethiazide	8.5	U				
	Cyclothiazide	8.8	U				
	Trichloromethiazide	6.9	U				
	Methyclothiazide	9.5	U				
	Polythiazide	9.1	U				

							2. Chatten LG and Harris LE, Relationship between $pK_b(H_2O)$ of organic compounds and $E_{1/2}$ values in several nonaqueous solvents, *Anal. Chem.*, **34**, 1495–1501 (1962). 5. Chatten LG, Moskalyk RE, Locock RA and Scaefer FJ, Nonaqueous titration of methaqualone and its dosage forms, *J. Pharm. Sci.*, **63**, 1294–1296 (1974). NB: Hennig *et al.* claimed that the studies of Ågren and Bäck for bendroflumethiazide did not take into account degradation in solution during their solubility-pH measurements. However, they failed to report a value for the title compound, so comparisons with Ågren and Bäck could not be done.
198	Benidipine ($C_{28}H_{31}N_3O_6$) 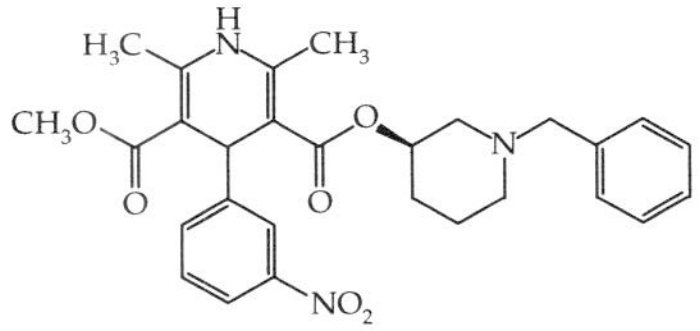	7.34	U	+H	Potentiometric	H_2O (long extrapolation)	Suzuki H, Ono E, Ueno H, Takemoto Y and Nakamizo N, Physicochemical properties and stabilities of the highly potent calcium antagonist benidipine hydrochloride, *Arzneim.-Forsch.*, **38**(11a), 1671–1676 (1988). "The pK_a value was measured by the titration method. About 250 mg of KW-3049 was dissolved in 100 ml aqueous methanol solution containing 90, 85, 80, and 75% methanol by volume, and 5 ml 0.1-N hydrochloric acid, respectively. These solutions were titrated with a 0.1-N sodium hydroxide. The pK_a in water (0% methanol) was obtained by extrapolating the pK_a values at the various methanol concentrations to methanol 0%."
199	2-Benzenesulfanilamidopyrimidine ($C_{10}H_9N_3O_2S$) 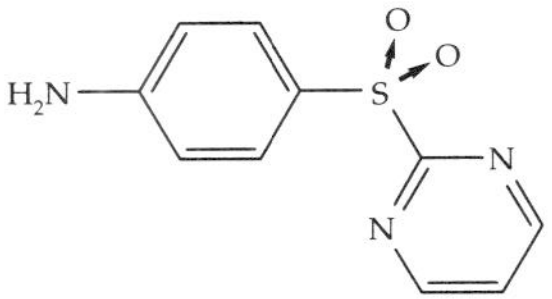	5.91	U	−H	Potentiometric	H_2O $t = 20$ $I = 0.1$ (KCl)	Willi AV and Meier W, Die Aciditatskonstanten von Benzolsulfonamiden mit heterocyclischer Amin-Komponente (The acidity constants for benzenesulfonamides with heterocyclic amine components), *Helv. Chim. Acta*, **39**, 54–56 (1956). NB: See Sulfapyridine.
200	2-Benzenesulfonamidopyridine ($C_{11}H_9NO_2S$) 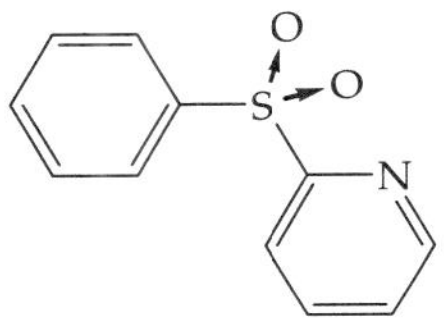	8.20	U	−H	Potentiometric	H_2O $t = 20$ $I = 0.1$ (KCl)	Willi AV and Meier W, Die Aciditatskonstanten von Benzolsulfonamiden mit heterocyclischer Amin-Komponente (The acidity constants for benzenesulfonamides with heterocyclic amine components), *Helv. Chim. Acta*, **39**, 54–56 (1956). NB: See Sulfapyridine.

(continued)

No.	Name	pK_a value(s)	Data quality	Ionization type	Method	Conditions t°C; I or c M	Comments and Reference(s)
201	Benzenesulfonamide and analogues ($C_6H_7NO_2S$) 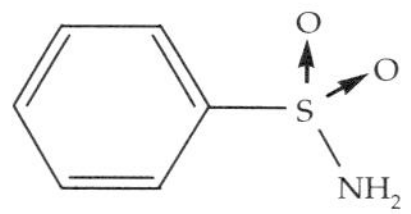	10.0	U	−H	Potentiometric	H_2O $t = 20$ $I = 0.1$ (KCl)	Willi AV 5. Die Aciditatskonstanten von Benzolsulfonamiden und ihre Beeinflussbarkeit durch Substitution (Effect of substitution on the acidity constants for benzenesulfonamides), *Helv. Chim. Acta*, **39**, 46–53 (1956). NB: This paper reported pK_a values for the following substituted benzenesulfonamides. The pK_a values were treated according to the Hammett equation:
202	Benzocaine ($C_9H_{11}NO_2$) 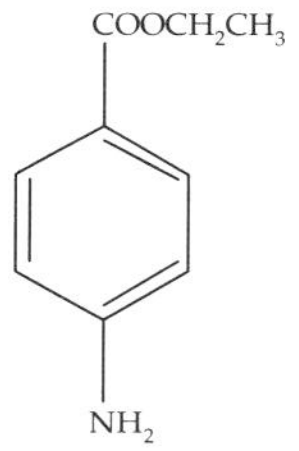	2.45	U	+H	CE/pH (+ve ion mode)	H_2O $t = 25$ $I = 0.025$	Wan H, Holmen AG, Wang Y, Lindberg W, Englund M, Nagard MB and Thompson RA, High-throughput screening of pK_a values of pharmaceuticals by pressure-assisted capillary electrophoresis and mass spectrometry, *Rapid Commun. Mass Spectrom.*, **17**, 2639–2648 (2003). NB: Reported a literature value (Tam KY and Takacs-Novac K, *Anal Chim Acta*, **434**, 157–167 (2001)) of 2.53 and a predicted value (ACD Labs) of 2.51.
		2.39 ± 0.01	U	+H	Potentiometric	H_2O $t = 25$ $I = 0.15$ (KCl)	Sirius Technical Application Notes, vol. **2**, p. 25 (1995). Sirius Analytical Instruments Ltd., Forest Row, East Sussex, RH18 5DW, UK.

Table for entry 201:

Substituent X	$X\text{-}C_6H_4SO_2NH_2$	$X\text{-}C_6H_4SO_2NH\text{-}C_6H_5$	$X\text{-}C_6H_4SO_2NH\text{-}C_6H_4\text{-}X$
H	10.00	8.31	8.31
p-CH_3	10.17	8.46	8.64
p-CH_3O	10.22	8.66	8.70
p-Cl	9.77	7.98	7.93
p-NO_2	9.14	7.415	6.20
m-NO_2	9.195	7.50	6.94
p-NH_2	10.575	8.89	9.05
p-CH_3CONH	10.02	–	–
p-$COCH_3$	–	–	6.94
p-COO-	–	–	7.75
m-OH, p-COO-	–	–	7.61

No.	Compound	pKa			Method	Conditions	Reference / Comments
203	Benzocaine	GLpK_a: 2.53 ± 0.02	A	+H	Spectro	H_2O $t = 25$	Tam KY and Takacs-Novac K, Multi-wavelength spectrophotometric determination of acid dissociation constants, *Anal. Chim. Acta*, **434**, 157–167 (2001).
		A&S: 2.50 ± 0.04	A	+H		$I = 0.15$ (KCl) Ar atmosphere	NB: See Clioquinol for details.
204	1,4-Benzodiazepines				Spectro	5% MeOH in H_2O	Barrett J, Smyth WF and Davidson IE. Examination of acid-base equilibria of 1,4-benzodiazepines by spectrophotometry, *J. Pharm. Pharmacol.*, **25**, 387–393 (1973).
	Chlordiazepoxide	4.6	U	+H		$t = 20$	"Changes of ultraviolet absorption spectra with pH in solution were used to determine pK_a values for six 1,4-benzodiazepines. Diazepam, chlordiazepoxide, and medazepam as a result of protonation of the molecule in acidic solutions, were found to each have one pK_a, while 2 pK_a values were observed for oxazepam, nitrazepam, and lorazepam, because of protonation in acid and deprotonation of the neutral molecule in alkaline media. The spectra are explained by considering them to be superimposed spectra of the 2 benzene rings, one monosubstituted and one trisubstituted, within the molecule. Sites of protonation (principally at nitrogen atoms in position 4 in the diazepine ring) and deprotonation (for oxazepam, nitrazepam, and lorazepam) are predicted and the differences in the observed pK_a values explained."
	Diazepam	3.3	U	+H		$I = 0.15$	
	Medazepam	4.4	U	+H			
	Lorazepam	1.3	U	+H			
	Nitrazepam	3.2	U	+H			
		10.8	U	−H			
	Oxazepam	1.6	U	+H			
		11.6	U	−H			
205	1,4-Benzodiazepine metabolites				Spectro	5% MeOH in H_2O	Barrett J, Smyth WF and Hart JP, Polarographic and spectral behavior of some 1,4-benzodiazepine metabolites: Application to differentiation of mixtures, *J. Pharm. Pharmacol.*, **26**, 9–17 (1974).
	7-Acetamidonitrazepam ($C_{17}H_{13}N_3O_4$)	3.2	U	+H		$t = 20$	"Changes of ultraviolet absorption spectra with solution pH were used to determine pK_a values for four 1,4-benzodiazepine metabolites. 7-Acetamidonitrazepam, desmethyldiazepam, and the chlordiazepoxide lactam all gave 2 pK_a values, corresponding to protonation in acid and deprotonation of the neutral molecule in alkaline media. 7-Aminonitrazepam gave 3 pK_a values, the third one being due to an additional protonation in acid media. The spectra are explained by considering them to be superimposed spectra of the 2 benzene rings, one monosubstituted and one trisubstituted, within the molecule. Sites of protonation and deprotonation are predicted and the differences in the observed pK_a values explained. Differences in the pK_a values or the polarographic behavior between the parent compounds and some of their metabolites are then used to effect novel separations after solvent extractions from aqueous buffered solutions."
		12.4	U	−H		$I = 0.15$	
	7-Aminonitrazepam ($C_{15}H_{12}N_4O_3$)	2.5	U	+H			
		4.6	U	+H			
		13.1	U	−H			
	Desmethyldiazepam ($C_{15}H_{11}ClN_2O$)	3.5	U	+H			
		12.0	U	−H			
	Chlordiazepoxide lactam ($C_{15}H_{11}ClN_2O_2$)	4.5	U	+H			
		11.5	U	−H			

(continued)

No.	Name	pK_a value(s)	Data quality	Ionization type	Method	Conditions t°C; I or c M	Comments and Reference(s)
206	Benzoic acid ($C_7H_6O_2$) COOH	4.190–4.203	R	–H	Conductance	H_2O $t = 25.0$ $I = 0.00$	Robinson RA and Stokes RH, Electrolyte Solutions, 2nd Revised Edn., Butterworths (Lond.), Appendix 12.1. Cited Ives DJG, Linstead RP and Riley HL, The olefinic acids. VIII. Dissociation constants, *J. Chem. Soc.*, 561–568 (1933) (4.190); Brockman FG and Kilpatrick M, The thermodynamic dissociation constant of benzoic acid from conductance measurements, *JACS*, **56**, 1483–1486 (1934) (4.199); Saxton B and Meier HF, The ionization constants of benzoic acid and the three monochlorbenzoic acids, at 25°, from conductance measurements, *JACS*, **56**, 1918–1921 (1934) (4.200); Dippy JFJ and Williams FR, Chemical constitution and the dissociation constants of monocarboxylic acids. II, *JCS*, 1888–1892 (1934) (4.203); Jeffery GH and Vogel AI, Thermodynamic dissociation constant of benzoic acid at 25° from conductivity measurements.*Phil. Mag.*, **18**, 901–909 (1934) (4.196).
207	Benzoic acid	4.212–4.218	R	–H	Electrometric	H_2O $t = 25.0$ $I = 0.00$	Robinson RA and Stokes RH, Electrolyte Solutions, 2nd Revised Edn., Butterworths (Lond.), Appendix 12.1. Cited Briscoe HT and Peake JS, Measurements of the ionization constant of benzoic acid using silver chloride electrodes, *J. Phys. Chem.*, **42**, 637–640 (1938) (4.218); Briegleb G and Bieber A, Dissociation constants of substituted benzoic acids at different temperatures and thermodynamic functions of acid dissociation. Potential measurements in cells without liquid junction potentials, *Z. Elektrochem.*, **55**, 250–259 (1951) (4.212).
208	Benzoic acid	4.203–4.216	R	–H	Spectro	H_2O $t = 25.0$ $I = 0.00$	Robinson RA and Stokes RH, Electrolyte Solutions, 2nd Revised Edn., Butterworths (Lond.), Appendix 12.1. Cited von Halban H and Brull J, The exact determination by means of indicators of the dissociation constants of moderately strong acids, *Helv. Chim. Acta*, **27**, 1719–1727 (1944) (4.216); Kilpatrick M and Arenberg CA, Effect of substituents on the protolytic constants of anilinium type protolypes, *JACS*, **75**, 3812–3821 (1953) (4.208); Robinson RA and Biggs AI, Ionization constants of p-aminobenzoic acid in aqueous solution at 25°, *Aust. J. Chem.*, **10**, 128–134 (1957) (4.203).
209	Benzoic acid	4.13 (0.11)	U	–H	Spectro (275 nm)	H_2O $t = 20.0$ I undefined	Wahbe AM, El-Yazbi FA, Barary MH and Sabri, SM, Application of orthogonal functions to spectrophotometric analysis. Determination of dissociation constants, *Int. J. Pharm.*, **92(1)** 15–22 (1993). NB: See Acetaminophen for further details. Alternative graphical method gave pK_a = 4.15.

210	Benzoic acid	4.05 ± 0.01	U	–H	Potentiometric	H_2O $t = 23.0$ I undefined Ar atmosphere	Clarke FH and Cahoon NM, Ionization constants by curve-fitting: Determination of partition and distribution coefficients of acids and bases and their ions, *J. Pharm. Sci.*, **76(8)** 611–620 (1987). "Automatic titrations were performed using (an) automatic titrator, an 80 mL beaker . . . and a combination glass electrode. . . Water was degassed by boiling for 1 hr under a stream of nitrogen, of HPLC grade water . . . was used. . . . Reagents and solutions were maintained at room temperature. . . . A solution of the salt of the compound . . . was prepared as follows. The salt (0.08 mmol if monoprotic, 0.04 mmol if diprotic) was placed in an 80 mL titration beaker and attached to the titrator. Air was displaced with a slow stream of argon and 25 mL of water was added, followed by 5 mL of 1.0 M NaCl solution. If the compound was the salt of a base, then 0.1 mL of 0.1 M HCl was added. If the compound was the salt of an acid, then 0.1 mL of 0.1 M NaOH was added. If the titrated compound was the free base or free acid, then the salt was prepared in situ by addition of 0.9 mL of 0.1 M HCl or NaOH, respectively. . . . The titrator was set to run under equilibrium conditions using 0.01 mL titrant increments. Usually the time to reach equilibrium was set at 5 s . . . ." [NB: difference titrations were performed to "correct" for the presence of dissolved carbon dioxide. The procedure was modified for simultaneous measurement of the O/W partition coefficient and pK_a. Authors acknowledged the importance of temperature control.]
211	Benzoic acid	3.98 ± 0.01	U	–H	Potentiometric	H_2O $t = 25.0$ $I = 0.1$ (NaCl)	Takacs-Novak K and Avdeef A, Interlaboratory study of log P determination by shake-flask and potentiometric methods, *J. Pharm. Biomed. Anal.*, **14**, 1405–1413 (1996). NB: See Acetaminophen for further details. Also reported $pK_a = 3.98 \pm 0.01$ at $I = 0.1$M (KNO_3). The same result was reported in Sirius Technical Application Notes, vol. **2**, p. 151 (1995). Sirius Analytical Instruments Ltd., Forest Row, East Sussex, RH18 5DW, UK. NB: From extrapolation to 0% DMSO from data in 10.7–58.3 wt% DMSO by the Yasuda-Shedlovsky procedure. Concentration of analyte, 0.64–1.16 mM;$I = 0.155$ M (KCl).
212	Benzoic acid, 4-R diethylaminoethyl esters R–C_6H_4–C(=O)–O–CH_2CH_2–$N(C_2H_5)_2$				Potentiometric	H_2O $t = 20.0$ $I = 0.1$	Buchi J, Bruhin HK and Perlia X, Relations between the physicochemical properties, the chemical reactivity and the local anesthetic activity of 4-substituted diethylaminoethyl esters of benzoic acid. Part 27. *Arzneim.-Forsch.*, **21**, 1003–1017 (1971).

(*continued*)

No.	Name	pK_a value(s)	Data quality	Ionization type	Method	Conditions t°C; I or c M	Comments and Reference(s)
	Sidechain (R)					N_2 atmosphere	"Some 4-substituted diethylaminoethyl esters of benzoic acid were prepared and their physico-chemical properties (pK_a, solubility, partition coefficient, surface activity), as well as IR spectra, rate constants, and binding to albumin were determined. A correlation between the physicochemical properties and the local anesthetic activity could not be found. There was a good correlation between the chemical reactivity (carbonyl frequency, rate constants, and sigma-values) and the activity. The compound with more electron releasing groups gave the most pronounced increase in activity. The binding to albumin and the activity also showed a satisfactory correlation. Measurement of the acetic acid pK_a was used as validation for the method."
	-NHC_2H_5	8.96	A	+H			
	-OC_2H_5	8.95	A	+H			
	-NH_2	9.00	A	+H			
	-H	8.80	A	+H			
	-CH_3	8.85	A	+H			
	-OH	9.09	A	+H			
	-Br	8.83	A	+H			
	-Cl	8.82	A	+H			
	-F	8.82	A	+H			
	-NO_2	8.67	A	+H			
	R—(benzene ring)—COOH						NB: Analogues of the local anaesthetic procaine. pK_a values were determined as described previously by Buchi J and Perlia X, Beziehungen zwischen de physikalisch-chemische Eigenschaften und der Wirkung von Lokalanasthetica, *Arzneim.-Forsch.*, **10**, 745–754 (1960). These were careful studies conducted by Dr. G Anderegg at the Analytical Chemistry laboratories of the ETH Zurich (Prof. G. Schwarzenbach). Potentiometric tirations with calibrated glass and calomel electrodes were performed under a nitrogen atmosphere. Solutions were of constant ionic strength 0.1 M (KCl) at 20 °C. Acetic acid was used as a test substance, with reference to the very accurate measurements of Harned. Some measurements were also performed spectrophotometrically. The experimental error was typically ± 0.03.
	Sidechain (R)						
	-NH_2	4.71	A	+H			
	-NHC_2H_5	4.77	A	+H			
	-OH	4.46	A	+H			
	-OC_2H_5	4.33	A	+H			
	-CH_3	4.30	A	+H			
	-H	4.13	A	+H			
	-F	4.08	A	+H			
	-Cl	3.98	A	+H			
	-Br	3.95	A	+H			
	-NO_2	3.50	U	+H			
213	Benzoylecgonine($C_{16}H_{19}NO_4$)	11.80	U	+H			Kolthoff IM, The dissociation constants, solubility product and titration of alkaloids, *Biochem. Z.*, **162**, 289–353 (1925). Cited in Perrin no. 2867 ref. K47. NB: See Aconitine for details.
	(structure: CH_3, N, COOH, O, O)						

214	Benzoylecgonine methyl ester ($C_{17}H_{21}NO_4$) 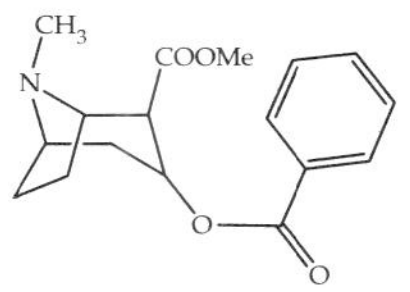	8.70 ± 0.05	A	+H	Potentiometric	H_2O $t = 25$	Chilton J and Stenlake JB, Dissociation constants of some compounds related to lysergic acid: Beta-dimethylaminopropionic acid, dihydroarecaidine, ecgonine and their derivatives, *J. Pharm. Pharmacol.*, **7**, 1004–1011 (1955). Cited in Perrin 2862 ref. C27. NB: The method used a glass electrode with liquid junction potentials.
215	Benzquinamide	5.9	U	+H		50% aqueous dioxane $t = 25$	Wiseman EH, Schreiber EC and Pinson R, The distribution, excretion and metabolism of benzquinamide, *Biochem. Pharmacol.*, **13**, 1421–1435, (1964). NB: Method not stated but probably potentiometric.
216	5-R-benzyl-2,4-diaminopyridines 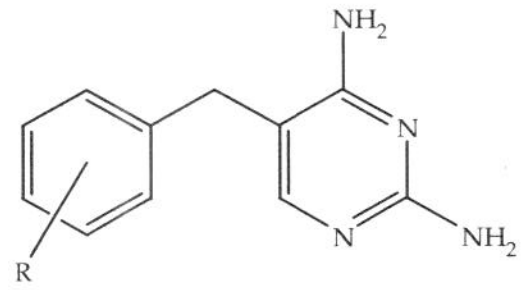	7.14 ± 0.08	U	+H	Potentiometric	H_2O $t = 25$	Seiler P, Bischuff O and Wagner R, Partition coefficients of 5-(substituted benzyl)-2,4-diaminopyridines, *Arzneim.-Forsch.*, **32(7)**, 711–714 (1982). "The average p*K* value of the 19 compounds measured is 7.14, the standard deviation within the series being 0.08 which is not greater than the standard deviation due to experimental error. Thus the small differences between our p*K*-values, though possibly really existing, are not judged significant, and in all further discussions we assume a mean p*K*-value of 7.14 for all compounds in Table 1." NB: The pK_a values were averaged arithmetically, corresponding to geometric averaging of the*K* values. The values were measured according to Albert and Serjeant (1971). Mean = 7.3966E−8; mean $pK_a = 7.13$ SD = 1.3691E−8 Sigma = 1.3326E−8

R_2	R_3	R_4	R_5	R_6	
	OEt	Pyrr	OEt		7.00
Cl	OMe		OMe	Cl	7.25
Br		O-CH_2-O			7.11
	OEt	$COHMe_2$	OEt		7.04
	OMe	OMe	OBz		7.15
	NMe_2	Me	NMe_2		7.16
OMe	Pyr	OMe			7.20
	OMe	Pyrr	OMe		7.02
	OMe	Pyr	OMe		7.05
	OMe	SMe	OMe		7.08
		O-CH_2-O			7.20
	OMe	O-CH_2-O			7.17
OMe		OMe	OMe		7.24
	NMe_2	OMe	NMeCOMe		7.07
	OMe	OMe	OMe		7.13
	OMe	NH_2	OMe		7.24
	OMe	OY	OMe		7.10
	OH	OMe	OH		7.18
	OMe	OH	OH		7.23

Y = $CH_2CH_2OCH_3$

(*continued*)

No.	Name	pK_a value(s)	Data quality	Ionization type	Method	Conditions t°C; I or c M	Comments and Reference(s)
217	Benzylpenicillin (Penicillin G) ($C_{16}H_{18}N_2O_4S$) 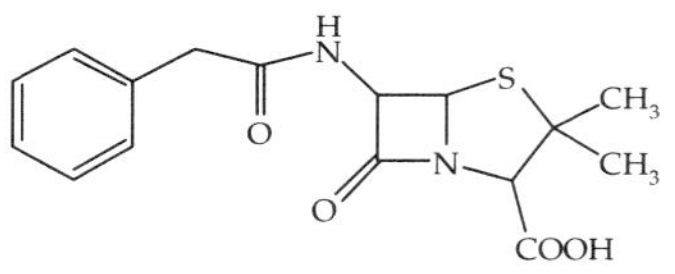	2.73 ± 0.03 or 2.71 ± 0.05	A A	–H –H	Potentiometric	H_2O $t = 25.0 \pm 0.1$ $c = 0.01$ N_2 atmos	Rapson HDC and Bird AE, Ionization constants of some penicillins and of their alkaline and penicillinase hydrolysis products, *J. Pharm. Pharmacol.*, Suppl.15, 222–231T (1963). NB: The two values were from slightly different initial concentrations. Results were rejected if pH calibration standards (4, 7, 9.15) shifted by > 0.02 over the course of a titration. "The p*K* of benzylpenicillin at 25° and in water at a concentration of 0.0099 M was 2.73 ± 0.03 and for 0.0093 M was 2.71 ± 0.05 using titration with 0.4-M HCl."
		2.78	U	–H	Potentiometric	H_2O $t = 60$	Finholt P, Jurgensen G and Kristiansen H, catalytic effect of buffers on degradation of penicillin G in aqueous solution, *J. Pharm. Sci.*, **54**, 387–393 (1965). NB: Kirschbaum incorrectly stated that this study used an iodimetric titration method.
		3.82 4.10	U U	–H –H	Potentiometric	20% MeOH 30% MeOH $t = 20$ $I = 0.15$	Salto F, Prieto JG and Alemany MT, Interactions of cephalosporins and penicillins with non-polar octadecylsilyl stationary phase, *J. Pharm. Sci.*, **69**, 501–506 (1980). NB: This study measured pK_a values in partially aqueous media that corresponded to chromatographic mobile phases, using the potentiometric method described by Albert and Serjeant.
218	Benzylpenicilloic acid ($C_{16}H_{20}N_2O_5S$) 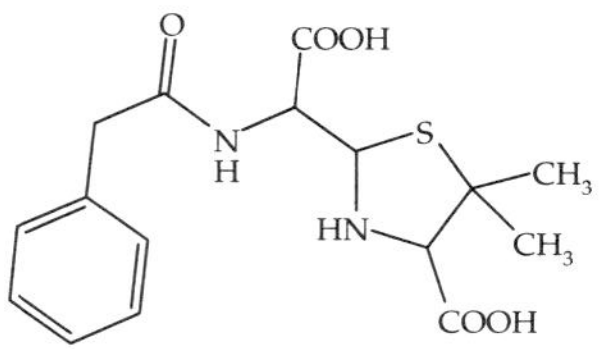	5.32 ± 0.05	A	+H	Potentiometric	H_2O $t = 25$ $c = 0.01M$	Rapson HDC and Bird AE, Ionization constants of some penicillins and of their alkaline and penicillinase hydrolysis products, *J. Pharm. Pharmacol.*, Suppl.15, 222–231T, (1963). Cited in Perrin Suppl. No. 7780 Ref. R6. Also cited in Kirschbaum J and Penicillin G, Potassium (Potassium Benzylpenicillin), *APDS*, **15**, 427–507 (1986). NB: The study used pH measurements with a glass electrode and liquid junction potentials.
219	Benzylpenicilloic acid	2.95 5.32	U U	–H +H	Method not given	H_2O $t = 23$	Woodward RB, Neuberger A and Trenner NR, *in* Clarke H, Johnson JR and Robinson Sir R (eds.), The Chemistry of Penicillin, Princeton University Press, Princeton, NJ, 415–422, 1949.

No.	Compound	pKa	Quality	Type	Method	Conditions	Reference
220	Benzylpenicilloic acid α-benzylamide ($C_{23}H_{25}N_3O_4S$) 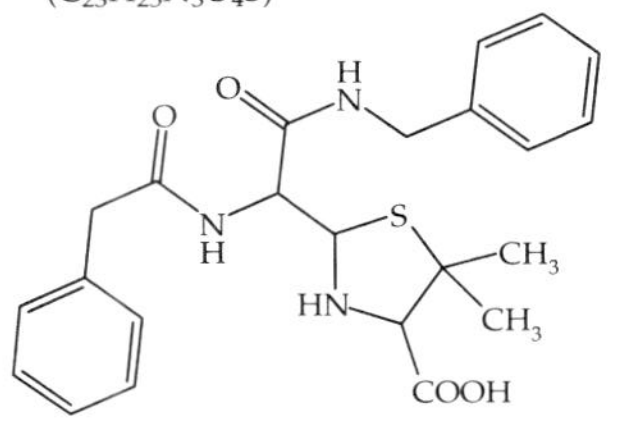	4.1	U	+H	Method not given	H_2O $t = 23$	Woodward RB, Neuberger A and Trenner NR, *in* Clarke H, Johnson JR and Robinson Sir R (eds.), The Chemistry of Penicillin, Princeton University Press, Princeton, NJ, 415–422, 1949.
221	Benzylpenicilloic acid α-benzylamide	3.96	U	+H	Method not given	H_2O $t = 23$	Woodward RB, Neuberger A and Trenner NR, *in* Clarke H, Johnson JR and Robinson Sir R (eds.), The Chemistry of Penicillin, Princeton University Press, Princeton, NJ, 415–422, 1949.
222	Benzylpenilloic acid ($C_{15}H_{20}N_2O_3S$) 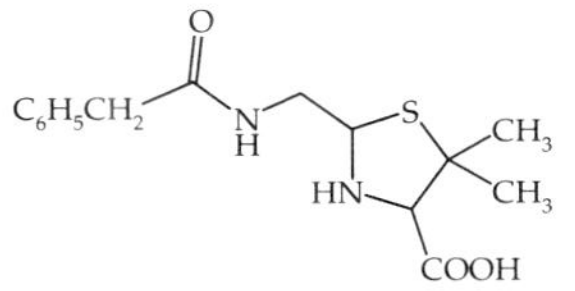	1.35 4.75	U U	−H +H	Method not given	H_2O $t = 23$	Woodward RB, Neuberger A and Trenner NR, *in* Clarke H, Johnson JR and Robinson Sir R (eds.), The Chemistry of Penicillin, Princeton University Press, Princeton, NJ, 415–422, 1949.
223	Bisoprolol ($C_{18}H_{31}NO_4$) 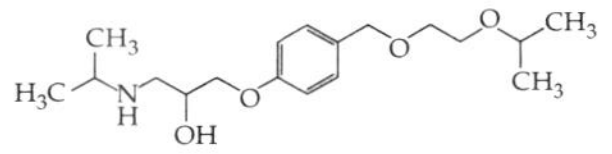	9.57	U	+H	Potentiometric	H_2O (extrap) $t = 25 \pm 1$ I undefined Ar atmosphere	Cheymol G, Poirier J-M, Carrupt PA, Testa B, Weissenburger J, Levron J-C and Snoeck E., Pharmacokinetics of β-adrenoceptor blockers in obese and normal volunteers, *Br. J. Clin. Pharmacol.*, **43**, 563–570 (1997).
224	Boric acid (H_3BO_3) H_3BO_3	9.234	VR	−H	Potentiometric	H_2O $t = 25.0$ $I = 0.00$	Manov GG, DeLollis NJ and Acree SF, Ionization constant of boric acid, J. *Res. Nat. Bur. Stand.*, **33**, 287–306 (1944). NB: Very careful work performed using an electrochemical cell without liquid junction potentials.

T (°C)	pK_a	T (°C)	pK_a	T (°C)	pK_a
0	9.5078	25	9.2340	45	9.1013
5	9.4374	30	9.1947	50	9.0766
10	9.3785	35	9.1605	55	9.0537
15	9.3255	40	9.1282	60	9.0310
20	9.2780				

(continued)

No.	Name	pK_a value(s)	Data quality	Ionization type	Method	Conditions t°C; I or c M	Comments and Reference(s)
225	Brequinar ($C_{23}H_{15}F_2NO_2$) 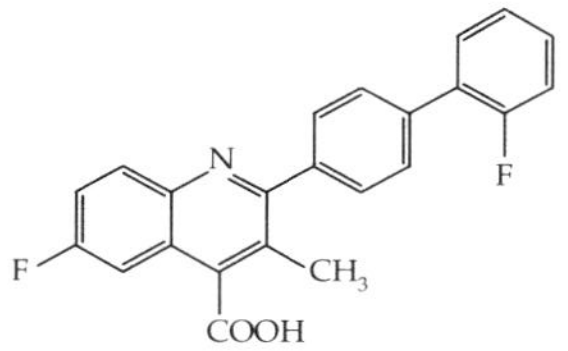	4.45	U	−H	Spectro (λ = 254 nm)	H_2O t = RT I = 0.10	S-YP King, AM Basista and G Torosian, Self-association and solubility behaviours of a novel anti-cancer agent, brequinar sodium, *J. Pharm. Sci.*, **78**, 95–100 (1989). "The apparent ionization constant was spectrophotometrically determined at RT in 0.1 M sodium acetate buffer at an ionic strength of 0.1. The analytical wavelength of 254 nm was chosen after examining pure spectra of both ionized (at pH 7.0) and molecular (at pH 2.85) species. These two species differed most in absorbance at this wavelength. Six pH values ranging from 3.75 to 5.04 were used. The ionization constant was calculated according to the method of Albert and Serjeant." NB: Only one pK_a value was found, presumably the –COOH. However, the quinoline nitrogen should also have partly ionized in the pH range of the study. Interference from partial ionization of this group may account for: (a) the wide range of results obtained (4.26–4.57); and (b) observed irregularities in the pH-solubility profile.
226	Brinzolamide ($C_{12}H_{21}N_3O_5S_3$) 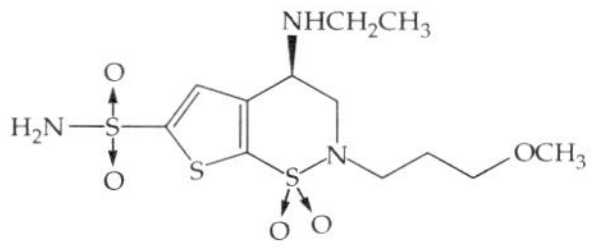	5.88 8.48	U U	+H −H	Potentiometric		Hall R, Havner G, Baker J, Stafford G, Schneider W, Lin W-Y, May J, Curtis M, Struble C, McCue B, Jasheway D and McGee D, Brinzolamide, *APDS*, **26**, 47–96 (1999). NB: No references were given, so data was presumably obtained in house (R&D, Alcon Laboratories, Ft Worth, TX, USA), but see Dorzolamide. The assignments would need to be confirmed by spectroscopy.
227	Bromocresol green ($C_{21}H_{14}Br_4O_5S$) 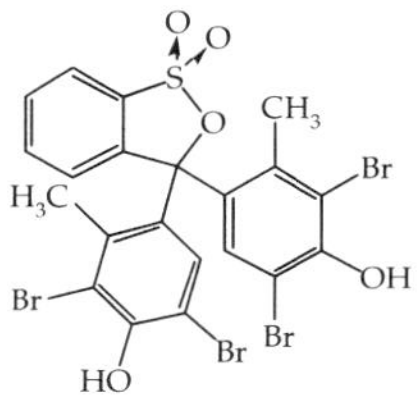	−0.85	U	−H	Spectro	H_2O t = RT	Das Gupta V and Reed JB Jr., First pK_a values of some acid-base indicators, *J. Pharm. Sci.*, **59**, 1683–1685 (1970).

Compound	pK_{a1}	Compound	pK_{a1}
Bromocresol green	−0.85	Cresol red	+1.05
Bromocresol purple	−0.75	Phenol red	+1.03
Bromophenol blue	−0.95		

No.	Compound	pKa	Data quality	Type	Method	Conditions	Comments and reference
228	Bromophenol blue ($C_{19}H_{10}Br_4O_5S$) 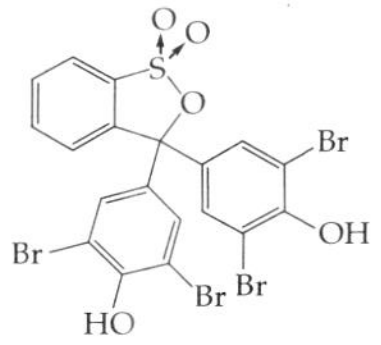	−0.95	U	−H	Spectro	H_2O t = RT	See Bromocresol green.
229	8-Bromotheophylline ($C_7H_7BrN_4O_2$) 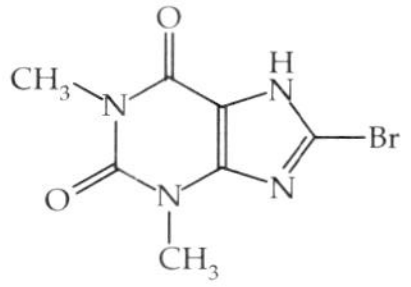	5.45 ± 0.1	U	−H	Potentiometric	H_2O t = 24	Maulding HV and Zoglio MA, pK_a determinations utilizing solutions of 7-(2-hydroxypropyl) theophylline, *J. Pharm. Sci.*, **60**, 309–311 (1971). NB: See Barbituric acid, 5-allyl-5-isobutyl for details.
230	Brompheniramine ($C_{16}H_{19}BrN_2$) 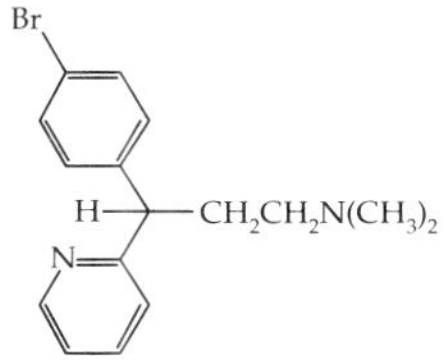	3.97 ± 0.03	U	+H	Potentiometric	H_2O t undefined I = 0.30 (NaCl)	Testa B and Murset-Rossetti L, The partition coefficient of protonated histamines, *Helv. Chim. Acta*, **61**, 2530–2537 (1978). NB: See Cycliramine for details.
		9.79 ± 0.12	U	+H			
231	Brucine ($C_{23}H_{26}N_2O_4$) 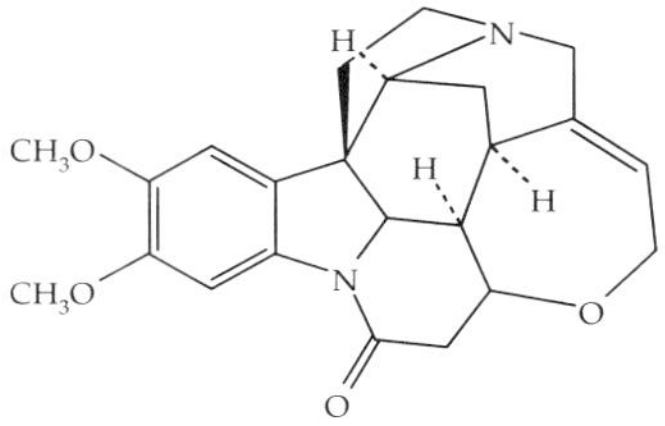	8.28	A	+H	Potentiometric	H_2O t = 25 N_2 atmos	Gage JC, The potentiometric titration of weak acids and bases in dilute aqueous solution, *Analyst*, **82**, 219–228 (1957). Cited in Perrin, IUPAC Bases, no. 2871 ref. G4. NB: This careful study used pH measurements with a glass electrode and junction potentials. Although no corrections for activities were made, the Henderson-Hasselbalch equation included corrections for hydronium and hydroxyl ion concentrations. Measured deviations from the known pK_{a2} for phthalic acid were used to further correct the results. Also included values for 11 mono-, di-, tri-, tetra-, and pentachlorophenols. It is not clear to which group the spectroscopic value of 2.50 should be assigned. Calculations using ACD/pK_a are in good agreement with a value of about 8.2 for the 3° amine, but the only other value appears to be for the extremely weakly basic properties of the aromatic amide group. A detailed spectroscopic study is needed. The spectrophotometric result is from Kolthoff; see Aconitine for details.
		2.50	U	+H	Spectro	H_2O t = 15	
		8.16	U	+H			

(continued)

No.	Name	pK_a value(s)	Data quality	Ionization type	Method	Conditions t°C; I or c M	Comments and Reference(s)
232	Bumetanide ($C_{17}H_{20}N_2O_5S$)	3.6 7.7	U U	−H −H	Potentiometric	H_2O $t = 20$	Orita Y, Ando A, Urakabe S and Abe H, A metal complexing property of furosemide and bumetanide: Determination of pK and stability constant, *Arzneim.-Forsch.*, **26**(1), 11–13 (1976). Cited in Tata PNV, Venkataramanan R and Sahota SK, Bumetanide, *APDS*, **22**, 107–144, 1993. "The pK_1 and pK_2 values of bumetanide are reported to be 3.6 and 7.7, respectively." NB: These pK_a values were obtained from measured pH values for the partially aqueous solutions (80% dioxane-water) that were claimed to be converted to aqueous solution pH values with an equation from the literature: pH in aq. = pH in dioxane-water + 1.8 (Hoeschle GK, Andelman JB, Gregor HP, *J. Phys. Chem.*, **62**, 1239–1244 (1958)).
233	Bumetanide	3.83	U	−H	CE/pH (−ve ion mode)	H_2O $t = 25$ $I = 0.025$	Wan H, Holmen AG, Wang Y, Lindberg W, Englund M, Nagard MB and Thompson RA, High-throughput screening of pK_a values of pharmaceuticals by pressure-assisted capillary electrophoresis and mass spectrometry, *Rapid Commun. Mass Spectrom.*, **17**, 2639–2648 (2003). NB: Reported literature values (Dollery C, Therapeutic Drugs, 1999) of 4 and 5.2 and a predicted value (ACD Labs) of 3.18.
234	Bupivacaine ($C_{18}H_{28}N_2O$)	8.09 (R, S, or racemate form)	U	+H	soly	H_2O $t = 23 \pm 1$ $I > 0.1$	Friberger P and Aberg G, Some physicochemical properties of the racemates and the optically active isomers of two local anaesthetic compounds, *Acta Pharm. Suecica*, **8**, 361–364 (1971).
235	Bupivacaine	8.17 (RS form)	U	+H	Potentiometric	H_2O $t = 25 \pm 0.02$ $I = 0.05$ (KCl) N_2 atmosphere	Kamaya H, Hayes JJ and Ueda I, Dissociation constants of local aesthetics and their temperature dependence, *Anesth. Analg.*, **62**, 1025–1030 (1983). Cited by Wilson TD, Bupivacaine, *APDS*, **19**, 59–94 (1990). NB: Other pK_a values: 8.495 (10 °C); 7.927 (38 °C); See Lidocaine for details.

236	Bupivacaine	8.10	U	+H	Potentiometric	H_2O $t = 25.0 \pm 0.2$ $I = 0.01$ (NaCl)	Johansson P-A, Liquid-liquid distribution of lidocaine and some structurally related anti-arrythmic drugs and local anaesthetics, *Acta Pharm. Suec.*, **19**, 137–142 (1982). NB: See Lidocaine for details.
237	Buprenorphine ($C_{29}H_{41}NO_4$)	8.24	U	+H	soly	H_2O $t = 23.0 \pm 0.3$	Garrett ER and Ravi Chandran V, Pharmacokinetics of morphine and its surrogates. Part 6. Bioanalysis, solvolysis kinetics, solubility, pK'_a values and protein binding of buprenorphine, *J. Pharm. Sci.*, **74**, 515–524 (1985).
		10.16	U	−H	spectro	25% EtOH	"Soly/pH method: Buffer solutions with pH values ranging between 6 and 13 were prepared. Powdered buprenorphine hydrochloride was added in excess of solubility to 3 ml of each buffer solution. The solutions were vortexed for 30 min and maintained at room temperature, 23.0 $\pm$ 0.3 °C overnight. The solutions were filtered through 100 um Millipore filters aided by reduced pressure. The clear filtrate was appropriately diluted and analysed by HPLC using fluorescence detection. Calibration curves were prepared simultaneously." NB: Assumptions in the treatment of the spectrophotometric measurements led to the expectation that the true pK_{a2} value (phenol) was higher than the extrapolated value of 9.39.
		9.85	U	−H	spectro	15% EtOH	
		9.7	U	−H	spectro	10% EtOH	
		9.39	U	−H	extrap	0% EtOH	
238	Buprenorphine	8.31 9.62	U U	+H −H	Potentiometric	$EtOH/H_2O$ $t = 25.0 \pm 0.1$ $I = 0.15$ (KCl) under Ar	Avdeef A, Barrett DA, Shaw PN, Knaggs RD and Davis SS, Octanol-, chloroform-, and propylene glycol dipelargonat-water partitioning of morphine-6-glucuronide and other related opiates, *J. Med. Chem.*, **39**, 4377–4381 (1996). NB: Extrapolated to 0% ethanol by Yasuda-Shedlovsky treatment. See Morphine for further details.
239	Buspirone ($C_{21}H_{31}N_5O_2$) 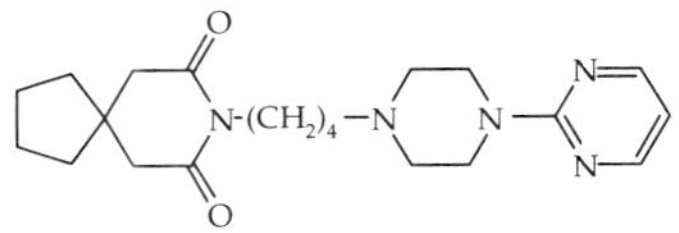	7.60 ± 0.01	U	+H	Potentiometric	H_2O $t = 25.0$ $I = 0.1$ (NaCl)	Takacs-Novak K and Avdeef A, Interlaboratory study of log P determination by shake-flask and potentiometric methods, *J. Pharm. Biomed. Anal.*, **14**, 1405–1413 (1996). NB: See Acetaminophen for further details.
240	Butanephrine ($C_{10}H_{15}NO_3$) 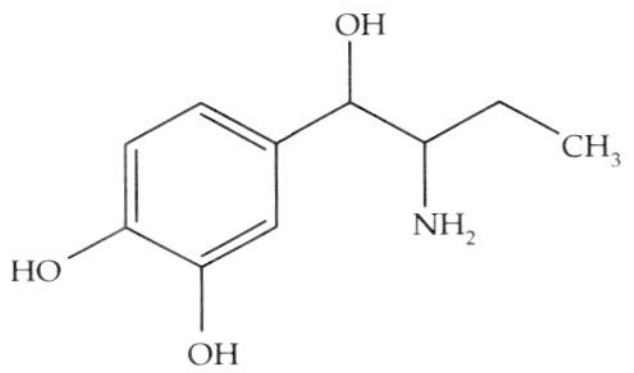	8.42	U	+H	Potentiometric	H_2O $t = 25.0 \pm 0.2$ $I \leq 0.001$	Leffler EB, Spencer HM and Burger A, Dissociation constants of adrenergic amines, *JACS*, **73**, 2611–2613 (1951). NB: See Amphetamine for details; from $pK_b = 5.58$; no attempt was made to unravel the microconstants.

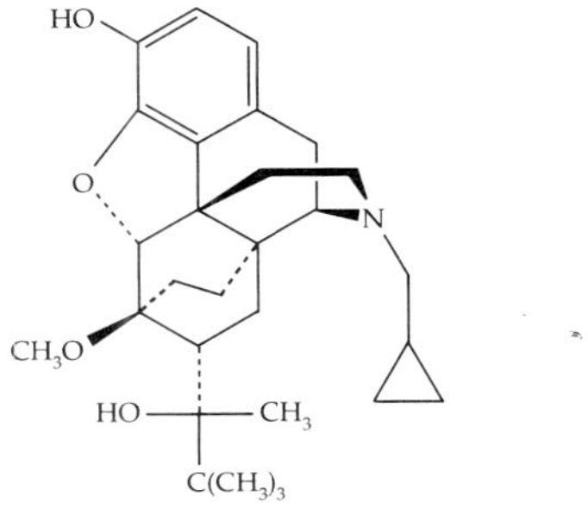

(continued)

No.	Name	pK_a value(s)	Data quality	Ionization type	Method	Conditions t°C; I or c M	Comments and Reference(s)
241	Caffeine ($C_8H_{10}N_4O_2$)	−0.13 ± 0.03	U	+H	^{13}CNMR/H_o	0.05 to 18-M H_2SO_4 t = 28.0	Benoit RL and Fréchette M, Protonation of hypoxanthine, guanine, xanthine and caffeine, *Can. J. Chem.*, **63**(11), 3053–3056 (1985).
	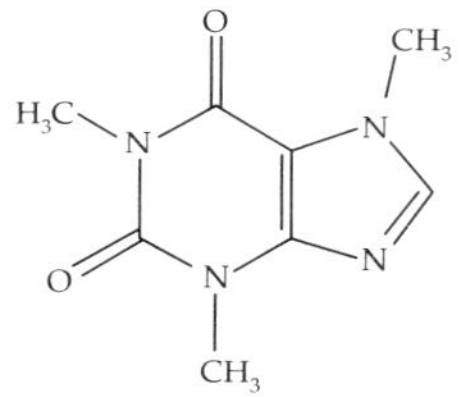	0.18 ± 0.01	U	+H	Spectro		(see table below)
242	Caffeine	−0.10	U	+H	kinetic	t = 40.1	Perrin Bases, Supplement, no. 7464. Ref. Wood JK, *J. Chem. Soc.*, **89**, 1831–1839 (1906).
243	Caffeine	1.22	VU	+H	kinetic	H_2O t = 55	Arnall F, The determination of the relative strengths of some nitrogen bases and alkaloids, *J. Chem. Soc.* **117**, 835–839 (1920). Cited in Perrin Bases 2873 Ref. A73. Relative values for a further 24 compounds were cited.
		<1	U	+H	Spectro	H_2O t = 25	NB: See also Ref T16: Turner A, Osol A, The spectrophotometric determination of the dissociation constants of theophylline, theobromine and caffeine, *J. Am. Phar. Assoc.*, **38**, 158–161 (1949) for spectrophotometric absorbance versus pH method.
244	Camptothecin ($C_{20}H_{16}N_2O_4$) 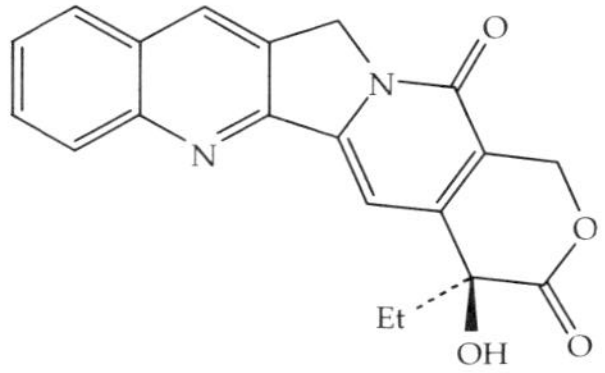	1.18	U	+H	Spectro	H_2O t = 25 I = 0.5	Fassberg J and Stella VJ, A kinetic and mechanistic study of the hydrolysis of camptothecin and some analogues, *J. Pharm. Sci.*, **81**, 676–684 (1992). Analogues: (see table below)

Caffeine structure labels: O, H_3C, N, CH_3, N, N, O, N, CH_3

Compound	pK_{NMR}	pK_{UV}	pK_{lit}
Guanine$_{H+}$			3.3
Guanine$_{2H+}$	−1.01 ± 0.03	−0.99 ± 0.01	−1.05
Hypoxanthine$_{H+}$			1.8, 2.0
Hypoxanthine$_{2H+}$	−3.12 ± 0.05	−3.65 ± 0.10	
Xanthine$_{H+}$		0.91 ± 0.01	0.8, 1.2
Xanthine$_{2H+}$		−10.0 ± 0.06	
Purine$_{H+}$			2.39
Purine$_{2H+}$	−1.66 ± 0.04		−1.5
Adenine$_{H+}$			4.19
Adenine$_{2H+}$	−0.43 ± 0.02	−0.47	−0.35
Adenine$_{3H+}$	−4.23 ± 0.2		

Camptothecin structure labels: N, N, O, O, O, Et, OH

Compound	pK_a (quinoline)	pK_a (phenol)	pK_a (benzyl-dimethylamine)
Camptothecin	1.18	–	–
10-hydroxycamptothecin	1.39	8.56	–
9-(methyl-N,N-dimethylamino) camptothecin	0.601	6.99	10.50

No.	Compound	pKa			Method	Conditions	Reference/Comments
245	Candesartan cilexetil ($C_{33}H_{34}N_6O_6$) 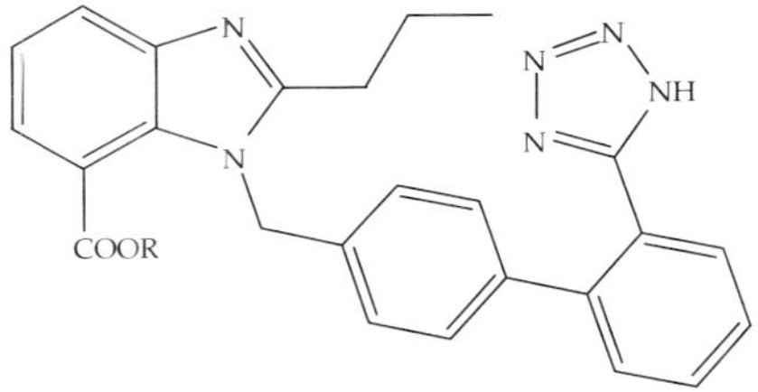 R = 1-[[(cyclohexyloxy)carbonyl]oxy]ethyl	3.55 5.91	U U	−H +H	CE/pH (−ve ion mode)	H_2O $t = 25$ $I = 0.025$	Wan H, Holmen AG, Wang Y, Lindberg W, Englund M, Nagard MB and Thompson RA, High-throughput screening of pK_a values of pharmaceuticals by pressure-assisted capillary electrophoresis and mass spectrometry, *Rapid Commun. Mass Spectrom.*, **17**, 2639–2648 (2003). NB: Reported literature values (Astra-Zeneca) of 4.5 and 6 and predicted values (ACD Labs) of 4.24 and 4.22.
246	Captopril ($C_9H_{15}NO_3S$) 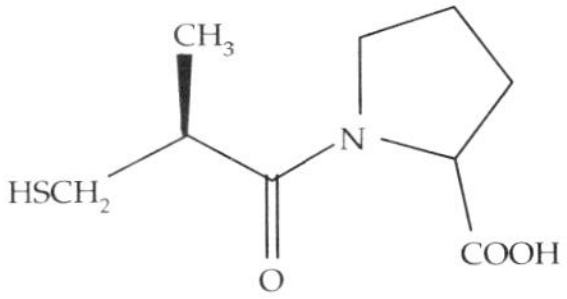	3.7 9.8	U U	−H −H	Potentiometric Spectro	H_2O	Kadin H and Captopril, *APDS*, **11**, 79–137 (1982). "The pK_a of the carboxyl of captopril (pK_1) is reported (29) to be 3.7. Whereas a carboxyl break was readily observed with alkali potentiometry, the sulfhydryl break could not be detected (17). Therefore, the pK_a of the sulfhydryl in captopril (pK_2) was not estimated by classical potentiometry. It was, however, estimated at 9.8 (pK_2) by Ondetti (19) and Weiss (30) using sulfhydryl u.v. shifts to higher wavelengths with increase in pH. The method utilized was adapted from Benesch and Benesch (31)." 17. Whigan DB, personal communication, April 1976 (presumably Squibb Institute for Medical Research). 19. Ondetti M, personal communication, February 1976 (presumably Squibb Institute for Medical Research). 29. Weiss AL, personal communication, March 1980 (presumably Squibb Institute for Medical Research). 30. Weiss AL, personal communication, May 1976 (presumably Squibb Institute for Medical Research). 31. Benesch R, The acid strength of the –SH group in cysteine and related compounds, *JACS*, **77**, 5877–5881 (1955).

(*continued*)

No.	Name	pK_a value(s)	Data quality	Ionization type	Method	Conditions t°C; I or c M	Comments and Reference(s)
247	Carbenoxolone ($C_{34}H_{50}O_7$) 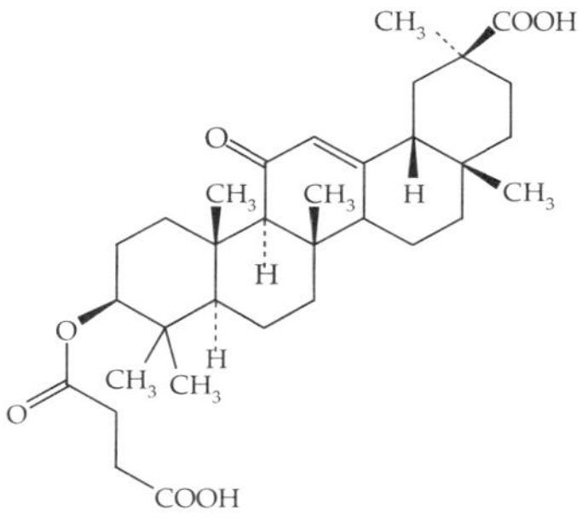	4.28 ± 0.1	U	–H	soly, partition	H_2O t = RT I = 0.5	Pindado S, Corrigan OI and O'Driscoll CM, Carbenoxolone sodium, *APDS*, **24**, 1–43 (1996). "Large differences in the values for the dissociation constants of carbenoxolone have been reported. These are summarized in the Table Table: Dissociation constants (pK_a) of carbenoxolone pK_{a1} | pK_{a2} | Method (reference) 6.7 | 7.1 | not stated (Downer *et al.*, 1970) 4.18 | 5.56 | Partition (Blanchard *et al.*, 1988) 4.38 | 5.11 | Solubility (no reference) (present) (Data is from Blanchard *et al.*)" Downer HD, Galloway RW, Horwich L, Parke DV, *J. Pharm. Pharmacol.*, **22**, 479–487 (1970). This reference gave no method or reference for the pK_a measurements, neither was there any discussion of the high values, compared to typical -COOH pK_a values. Blanchard J, Boyle JO, Van Wagenen S, *J. Pharm. Sci.*, **77**, 548–552 (1988). See no. 248.
		5.28 ± 0.3	U	–H			
248	Carbenoxolone	4.18	U	–H	partition	H_2O t = RT I = 0.5	Blanchard J, Boyle JO and Van Wagenen S, Determination of the partition coefficients, acid dissociation constants and intrinsic solubility of carbenoxolone, *J. Pharm. Sci.*, **77**, 548–552 (1988). NB: Partition and soly methods used ^{3}H-labeled compound, due to very low solubility in water. Partitioning method required fitting the apparent partition coefficient (APC) at each pH value to: $APC = \frac{TPC + \frac{C_1}{[H^+]} + \frac{C_2}{[H^+]^2}}{1 + \frac{K_1}{[H^+]} + \frac{K_1K_2}{[H^+]^2}}$ where TPC is the true partition coefficient, K_1 and K_2 are dissociation constants for the two carboxyl groups, $C_1 = K_{IP}[Na^+]K_1$ and $C_2 = K_{IP2}[Na^+]K_1K_2$, where K_{IP} and K_{IP2} are the equilibrium constants for ion-pair formation with the monoanion and dianion, respectively. The resulting pK_a values were compared favourably with model compounds, succinic acid, and 1-methylcyclohexanecarboxylic acid. A spectrophotometric method for enoxolone used buffers in the range $pH = 2.6$ to 7.6.
		5.56	U	–H			
		4.38	U	–H	soly		
		5.11	U	–H			
	Enoxolone ($C_{30}H_{46}O_4$) 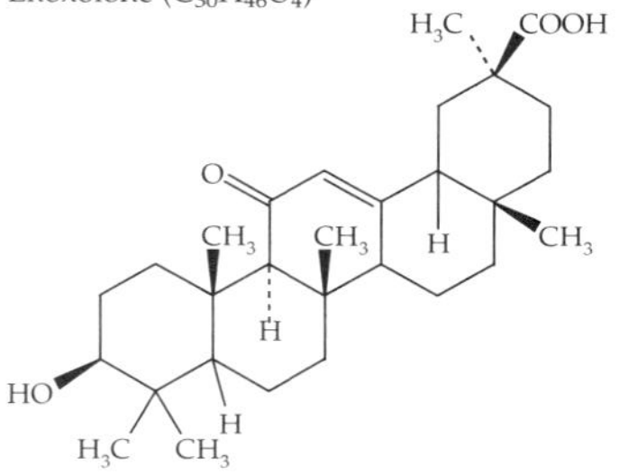	5.56 ± 0.1	U	–H	Spectro		

249	Carbinoxamine ($C_{16}H_{19}ClN_2O$)	3.77 ± 0.04 8.98 ± 0.04	U U	+H +H	Potentiometric	H_2O t undefined $I = 0.30$ (NaCl)	Testa B and Murset-Rossetti L, The partition coefficient of protonated histamines, *Helv. Chim. Acta*, **61**, 2530–2537 (1978). NB: See Cycliramine for details.
250	Carbomycin A (Magnamycin A) ($C_{42}H_{67}NO_{16}$)	7.61	U	+H	Potentiometric	H_2O $t = 25$ $I = 0.167$	McFarland JW, Berger CM, Froshauer SA, Hayashi SF, Hecker SJ, Jaynes BH, Jefson MR, Kamicker BJ, Lipinski CA, Lundy KM, Reese CP and Vu CB, Quantitative structure-activity relationships among macrolide antibacterial agents: *In vitro* and *in vivo* potency against Pasteurella multocida, *J. Med. Chem.*, **40**, 1340–1346 (1997). NB: See Azithromycin for details; average standard deviation of ± 0.07 for the pK_a.
	Carbomycin B ($C_{42}H_{67}NO_{15}$) Cf Carbomycin A	7.55	U	+H			
251	Carbonic acid H_2CO_3	6.352 10.329	R R	−H −H	Potentiometric	H_2O $t = 25$ $I = 0.000$	Harned HS and Scholes SR, The ionization constant of HCO_3^- from 0 to 50°, *JACS*, **63**, 1706–09 (1941); Harned HS and Davis R, The ionization constant of carbonic acid in water and the solubility of carbon dioxide in water and aqueous salt solutions from 0 to 50°, *JACS*, **65**, 2030–2037 (1943); Shedlovsky T, MacInnes DA, The first ionization constant of carbonic acid, 0 to 38°, from conductance measurements, *JACS*, **57**, 1705–1710 (1935) gave 6.583 ($t = 0$), 6.429 ($t = 15$), 6.366 ($t = 25$), and 7.317 ($t = 38$) by careful conductance work.

Results of Harned and Davis (pK_{a1}) and Harned and Scholes (pK_{a2})

T (°C)	pK_{a1}; pK_{a2}	T (°C)	pK_{a1}; pK_{a2}	T (°C)	pK_{a1}; pK_{a2}
0	6.5787; 10.625	20	6.3809; 10.377	40	6.2978; 10.220
5	6.5170; 10.557	25	6.3519; 10.329	45	6.2902; 10.195
10	6.4640; 10.490	30	6.3268; 10.290	50	6.2851; 10.172
15	6.4187; 10.430	35	6.3094; 10.250		

NB: The pK_{a1} measurements are for $CO_2 + H_2O \Leftrightarrow H^+ + HCO_3^-$. A further reaction of major interest is $H_2CO_3 \Leftrightarrow H^+ + HCO_3^-$, for which p$K_a$ has been shown by Roughton FJW, The kinetics and thermochemistry of carbonic acid, JACS, 63 2930–34 (1941) to be 3.62 at 15 °C and by Berg D, Patterson A, The high field conductance of aqueous solutions of carbon dioxide at 25°. The true ionization constant of carbonic acid, *ib.*, 75, 5197–5200 (1953) to be 3.88 at 25 °C. This difference arises because only about 0.3% of dissolved CO_2 (at 25 °C) is in the form of H_2CO_3 molecules.

		5.84–6.26	U	−H	Potentiometric	healthy human plasma $t = 37$	Flear CTG, Roberts SW, Hayes S, Stoddart JC, Covington AK, pK'_1 and bicarbonate concentration in plasma, *Clin. Chem.*, **33**, 13–20 (1987).

(continued)

No.	Name	pK_a value(s)	Data quality	Ionization type	Method	Conditions t°C; I or c M	Comments and Reference(s)
252	Carboxylic acids and phenols	Measured pK_a (lit)	U	–H	Potentiometric	H_2O $t = 25.0$ $I < 0.005$ N_2 atmos Presence of α-cyclodextrin	Connors KA and Lipari JM, Effect of cycloamyloses on apparent dissociation constants of carboxylic acids and phenols: equilibrium analytical selectivity induced by complex formation, *J. Pharm. Sci.*, **65**, 379–383 (1976). "Apparent dissociation constants of organic acids were determined by potentiometric titration in the presence of cyclohexaamylose (α-CD) or cycloheptaamylose (β-CD). This experimental approach has been elaborated in 3 systematic techniques for studying molecular complexes of acids and bases. The quantity $\Delta pK_a' = pK_a'$ (cycloamylose)- pK_a' (water) was positive or zero for all carboxylic acids studied and negative or zero for all phenols. The term $\Delta pK_a'$ can be related to the cyclo-amylose concentration, K_{11a}, and K_{11b}, where K_{11a} and K_{11b} are 1:1 stability constants for complexes of the acid and the anion, respectively. From the dependence of $\Delta pK_a'$ on cyclo-amylose concentration, estimates of K_{11a} and K_{11b} can be obtained. If $\Delta pK_a' \neq 0$, then $K_{11a} \neq K_{11b}$; for carboxylic acids, $K_{11a} > K_{11b}$; for phenols, $K_{11b} > K_{11a}$. Because of variable pK_a' shifts, it is possible to carry out differentiating titrations of some acid mixtures in cycloamylose solutions, whereas the same acids cannot be differentiated in water. If an acid is weakened by cycloamylose, its conjugate base is strengthened, and some carboxylate salts can be readily titrated in the presence of a cycloamylose."
	Effect of α-CD						
	Acetic acid	4.83 (4.76)					
	Propionic acid	5.16 (4.97)					
	Chloroacetic acid	3.30 (3.20)					
	Maleic acid	3.40 (3.32), 6.38 (6.23)					
	Malonic acid	3.69 (3.60), 5.92 (5.70)					
	Benzoic acid	5.20 (4.11)					
	Salicylic acid	3.63 (3.40)					
	m-OH-benzoic acid	4.95 (4.10)					
	p-OH-benzoic acid	5.83 (4.58)					
	o-MeO-benzoic acid	4.20 (4.15)					
	m-MeO-benzoic acid	5.13 (4.18)					
	o-nitrobenzoic acid	3.20 (3.19)					
	m-nitrobenzoic acid	3.83 (3.60)					
	p-nitrobenzoic acid	4.53 (3.74)					
	p-fluorobenzoic acid	4.98 (4.18)					
	o-phthalic acid	3.50 (3.4) 5.34 (5.20)					
	nicotinic acid	4.85 (4.85)					
	picolinic acid	5.33 (5.31)					
	gallic acid	4.53 (4.20)					
	cinnamic acid	5.80 (4.43)					
	o-hydroxycinnamic	5.82 (4.69)					
	m-hydroxycinnamic	5.64 (4.49)					
	p-hydroxycinnamic	5.88 (4.40)					
	o-methoxycinnamic	5.20 (4.70)					
	m-methoxycinnamic	5.76 (4.47)					
	p-methoxycinnamic	6.24 (4.90)					
	phenol	9.81 (9.81)					
	o-nitrophenol	7.21 (7.21)					
	m-nitophenol	8.00 (8.29)					
	p-nitrophenol	6.15 (7.09)					
	p-bromophenol	9.20 (9.20)					
	1-naphthol	9.20 (9.20)					
	2-naphthol	9.30 (9.40)					

Effect of β-CD on apparent pK_a values:

Compound	Measured	Lit	Compound	Measured	Lit
Benzoic acid	4.95	4.11	*o*-MeO-cinnamic acid	5.48	4.70
o-nitrobenzoic acid	3.33	3.19	*m*-MeO-cinnamic acid	4.88	4.47
p-nitrobenzoic acid	4.02	3.74	*p*-MeO-cinnamic acid	5.29	4.90
o-MeO-benzoic acid	4.48	4.15	*o*-phthalic acid	5.55	5.20
o-OH-benzoic acid	3.68	3.40	*p*-nitrophenol	6.70	7.09
Cinnamic acid	5.08	4.43	2-naphthol	9.05	9.40
o-OH-cinnamic acid	5.33	4.69			

No.	Compound	pKa			Method	Conditions	Reference / Notes
253	Carbutamide ($C_{11}H_{17}N_3O_3S$) 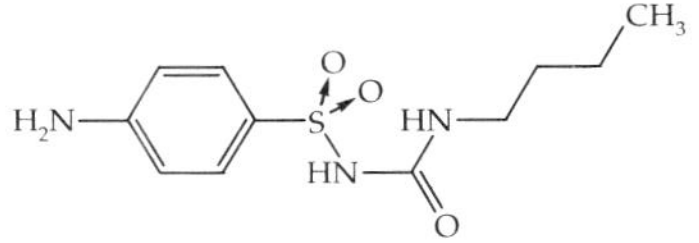	5.75	U	–H	Spectro	H_2O $t = 25.0 \pm 0.5$ $I = 0.2$	Elofsson R, Nilsson SO and Agren A, Complex formation between macromolecules and drugs. IV. *Acta Pharm. Suec.* **7**, 473–482 (1970). NB: See Sulphanilamide for details.
254	Cefadroxil ($C_{16}H_{17}N_3O_5S$) 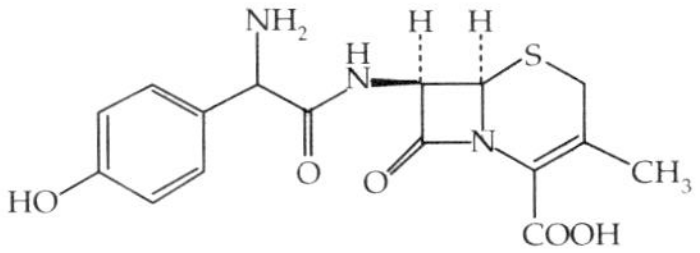	1.38 7.55 10.10	U U U	–H +H –H	Potentiometric	H_2O $t = 20 \pm 2$ N_2 atmosphere	Mariño EL and Dominquez-Gil A, Determination of the macro- and micro-ionization constants of a dipolar zwitterionic cephalosporin: cefadroxil, *Int. J. Pharm.*, **8(1)**, 25–33 (1981).
255	Cefazaflur ($C_{13}H_{13}F_3N_6O_4S_3$) 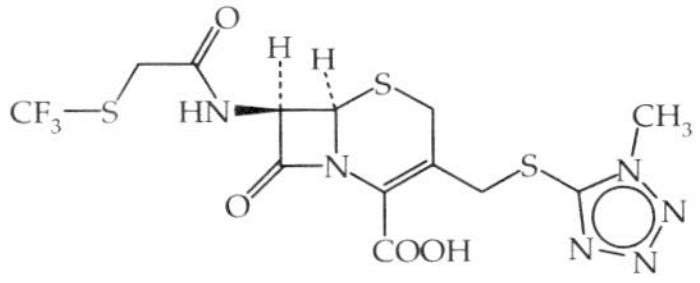	2.45	U	–H	Potentiometric	H_2O $t = 37$ $I = 0.2$	Irwin VP and Timoney RF, Cefazaflur: Kinetics of hydrolysis in aqueous solution, acid dissociation constant and alkaline decomposition to fluorescent products, *J. Pharm. Pharmacol.*, **41**, 360 (1989). NB: The potentiometric method followed one in the literature: Streng WH, Microionization constants of commercial cephalosporins, *J. Pharm. Sci.*, **67**, 666–669 (1978).
256	Cefazolin ($C_{14}H_{14}N_8O_4S_3$) 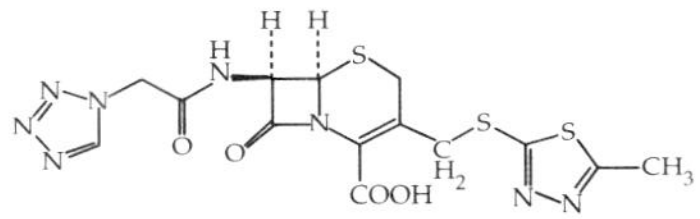	2.75	U	–H	Potentiometric	H_2O $t = 25.0 \pm 0.1$ $I = 0.1$	Streng WH, Microionization constants of commercial cephalosporins, *J. Pharm. Sci.*, **67**, 666–669 (1978). NB: See Cephalosporin derivatives for further details.

Macro constants

pK_1	pK_2	pK_3
1.4	7.5	10
1.38	7.55	10.10

Micro constants (spectrophotometric; $I = 0.05$ M)

	pK_a	pK_b	pK_c	pK_d
Edsall's Method	8.21	7.66	9.44	9.99
Linear Reg	8.35	7.63	10.37	11.09

NB: The similar magnitudes of the microconstants mean that all four species (–/–/+, –/+, –/– and – coexist in the pH range 6.8 to 11.8.

(continued)

No.	Name	pK_a value(s)	Data quality	Ionization type	Method	Conditions t°C; I or c M	Comments and Reference(s)
257	Cefotaxime ($C_{16}H_{17}N_5O_7S_2$) 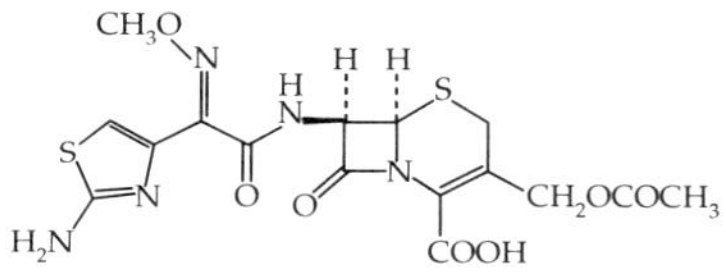	2.1 3.4 10.9	U U U	−H +H +H	Potentiometric	H_2O $t = 20$ $I < 0.005$	Fabre H, Hussam-Eddine N and Berge G, Degradation kinetics in aqueous solution of cefotaxime sodium, a third-generation cephalosporin, *J. Pharm. Sci.*, 73, 611–618 (1984). "The pK_a values of I were determined by the potentiometric titration of a 2.8×10^{-3} M cefotaxime aqueous solution with 0.1 M NaOH at 20 °C, under nitrogen. The two functional groups corresponding to the terminal amino group of the sidechain at C-7 and the carboxylic group at C-3 of the cephem ring were acidified with a stoichiometric amount of 0.1 M HCl before titration." NB: The values were determined using methods of Albert and Serjeant. No values were reported in Muhtadi FJ and Hassan MMA, Cefotaxime, *APDS*, **11**, 132–168 (1982).
258	Cefroxadine ($C_{16}H_{19}N_3O_5S$) 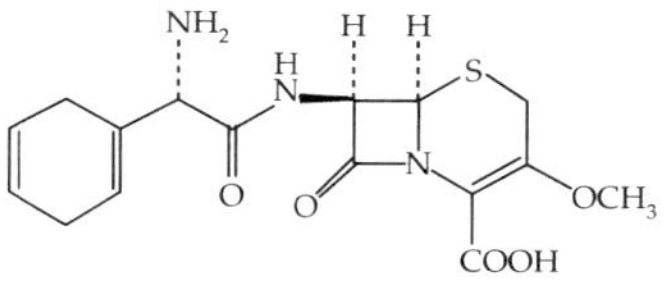	3.30 ± 0.02 7.00 ± 0.06	A U	−H +H	Potentiometric	H_2O $t = 35.0$ $I = 0.00$	Neito MJ, González JL, Dominguez-Gil A and Lanao JM, Determination of the thermodynamic ionization constants of cefroxadine, *J. Pharm. Sci.*, **76**, 228–237 (1987). "The values of thermodynamic ionization constants of the carboxylic (pK = 3.30 ± 0.02) and amine (pK = 7.00 ± 0.06) groups of cefroxadine were determined at 35.0 °C using potentiometric data. The apparent ionization constants of these groups were also determined at 35 °C, and at different values of ionic strength."
259	Cephalexin ($C_{16}H_{17}N_3O_4S$) 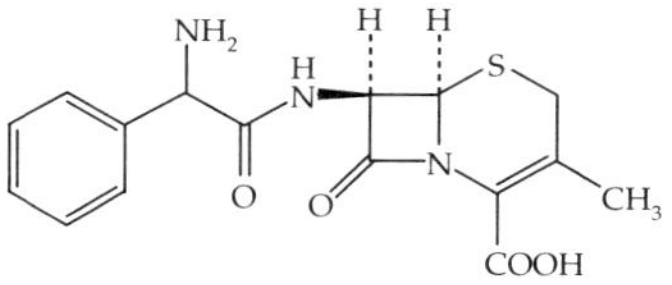	2.53 ± 0.02 7.14 ± 0.02	U U	−H +H	Potentiometric	H_2O $t = 25.0 \pm 0.1$ $I = 0.1$ (NaCl)	Takacs-Novak K, Box KJ and Avdeef A, Potentiometric pK_a determination of water-insoluble compounds: Validation study in methanol/water mixtures, *Int. J. Pharm.*, **151**, 235–248 (1997). "pK_{a1} = 2.72 ± 0.02 and pK_{a2} = 7.15 ± 0.02 by extrapolation from 14.8 – 55.7 %w/w aqueous MeOH." NB: See Acetaminophen for full details.
260	Cephalexin	5.2–5.3 7.3	U U	−H +H	Potentiometric	66% DMF	Marelli LP, Cephalexin, *APDS*, **4**, 21–46 (1975). Hargrove WW, Eli Lilly and Co., personal communication, 1967 Ryan CW, Simon RL and van Heyningen EM, Chemistry of cephalosporin antibiotics. XIII. Desacetoxycephalosporins. The synthesis of cephalexin and some analogs, *J. Med. Chem.*, **12**, 310–313 (1969). Flynn EH (ed.), *Cephalosporins and penicillins. Chemistry and Biology*, Academic Press, NY, 310 (1972). NB: Ryan *et al.* and Flynn gave values in 66% DMF. Hargrove also gave a value of 7.1 for the amino group in water.

No.	Compound	pKa			Method	Conditions	Reference
261	Cephalexin	2.45 7.62	U U	−H +H	Potentiometric, Spectro	H_2O $t = 25 \pm 2$ $I = 0.1$	Streng WH, Microionization constants of commercial cephalosporins, *J. Pharm. Sci.*, **67**, 666–669 (1978). NB: The two macroconstants were calculated from the four microionization constants, which were determined by a combination of spectrophotometry and potentiometry, giving the following values: pK_1: 2.48 (3.31e−3); pK_2: 7.59 (2.57e−8); pK_3: 6.46 (3.47e−7); pK_4: 3.60 (2.51e−4) where the microconstants K_1, K_2, K_3 and K_4 refer to the equilibria for $^+$HNRCOOH to $^+$HNRCOO$^-$, $^+$HNRCOO$^-$ to NRCOO$^-$, NRCOO$^-$ to NRCOOH and NRCOOH to $^+$HNRCOOH, respectively. Apparent values were corrected to zero ionic strength with Davies' equation.
262	Cephalexin	5.3 7.3	U U	−H +H	Potentiometric	66% DMF	Demarco PV, Nagarajan R, Physical-Chemical Properties *in* Flynn EH (ed.), *Cephalosporins and penicillins: Chemistry and Biology*, Academic Press, NY, 311–316 (1972).
263	Cephaloglycin ($C_{18}H_{19}N_3O_6S$) 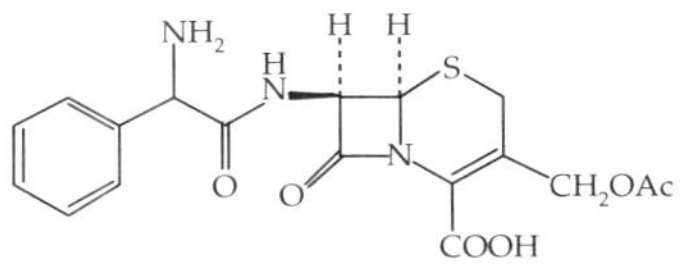	1.71 7.29	U U	−H +H	Potentiometric, Spectro	H_2O $t = 25 \pm 2$ $I = 0.1$	Streng WH, Microionization constants of commercial cephalosporins, *J. Pharm. Sci.*, **67**, 666–669 (1978). NB: The two macroconstants were calculated from the four microionization constants, which were determined by a combination of spectrophotometry and potentiometry, giving the following values: pK_1: 1.78 (1.66e−2); pK_2: 7.22 (6.03e−8); pK_3: 6.46 (3.47e−7); pK_4: 2.53 (2.95e−3) where the microconstants K_1, K_2, K_3 and K_4 refer to the equilibria for $^+$HNRCOOH to $^+$HNRCOO$^-$, $^+$HNRCOO$^-$ to NRCOO$^-$, NRCOO$^-$ to NRCOOH and NRCOOH to $^+$HNRCOOH, respectively. Apparent values were corrected to zero ionic strength with Davies' equation.
264	Cephaloglycin	4.7 7.4	U U	−H +H	Potentiometric	66% DMF	Demarco PV and Nagarajan R, Physical-Chemical Properties, in Flynn EH (ed.), *Cephalosporins and penicillins: Chemistry and Biology*, Academic Press, NY, 311–316 (1972).
265	L-Cephaloglycin	4.6 7.1	U U	−H +H	Potentiometric	66% DMF	Craig; N&K; Spencer JL, Flynn EH, Roeske RW and Siu FY, Chauvette RR, Chemistry of cephalosporin antibiotics. VII. Synthesis of cephaloglycin and some homologs, *J. Med. Chem.*, **9**, 746–750 (1966).

(*continued*)

No.	Name	pK_a value(s)	Data quality	Ionization type	Method	Conditions t°C; I or c M	Comments and Reference(s)
266	Cephaloridine ($C_{19}H_{17}N_3O_4S_2$) 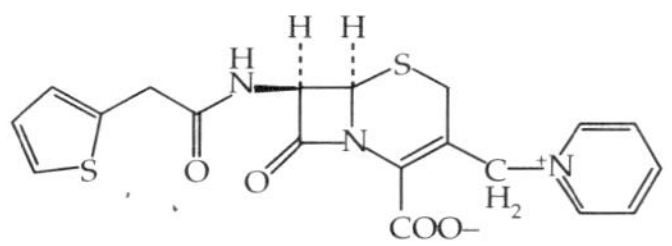	3.4	U	–H	Potentiometric	66% DMF	Craig; N&K; Flynn EH, ed., *Cephalosporins and penicillins Chemistry and Biology*, Academic Press, NY, 312–316 (1972).
267	Cephaloridine	3.4	U	–H	Potentiometric	66% DMF	Demarco PV and Nagarajan R, Physical-Chemical Properties, in Flynn EH (ed.), *Cephalosporins and penicillins: Chemistry and Biology*, Academic Press, NY, 311–316 (1972).
268	Cephalosporin derivative ($C_{16}H_{17}N_3O_7S$) 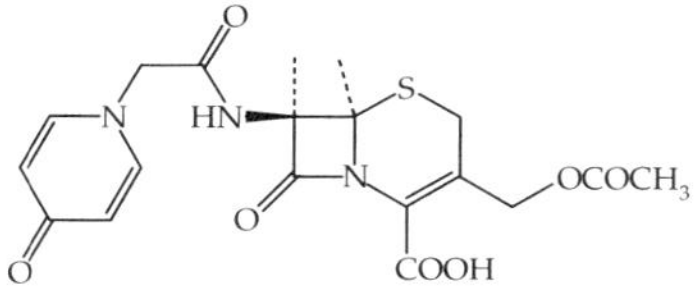				potentio, Spectro	H_2O $t = 25 \pm 2$ $I = 0.1$	Streng WH, Microionization constants of commercial cephalosporins, *J. Pharm. Sci.*, **67**, 666–669 (1978). NB: The four microionization constants were estimated with a combination of spectrophotometry and potentiometry: (see table below) where the microconstants K_1, K_2, K_3 and K_4 refer to the equilibria for $^+$HNRCOOH to $^+$HNRCOO$^-$, $^+$HNRCOO$^-$ to NRCOO$^-$, NRCOO$^-$ to NRCOOH and NRCOOH to $^+$HNRCOOH, respectively. Apparent values were corrected to zero ionic strength with Davies' equation.
269	Cephalosporin, 5-amino-5-carboxyvaleramido 7-substituted 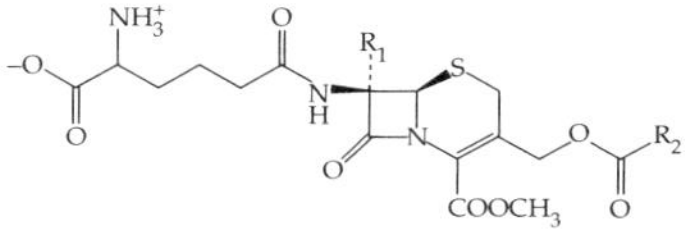				Potentiometric	66% DMF	Nagarajan R, β-Lactam antibiotics from Streptomyces, *in* Flynn EH (ed.), *Cephalosporins and penicillins: Chemistry and Biology*, Academic Press, NY, 636–647 (1972). Corresponding *N*-acyl derivatives:

pK_1	pK_2	pK_3	pK_4
1.70	2.80	1.97	2.53

No.	Compound	pK_a	Reliability	Ionization type	Method	Conditions	Comments and reference
	$R^1 = H$						
	(2) $R^2 = CH_3$ ($C_{17}H_{23}N_3O_8S$)	3.9	U	–H			
		5.3	U	–H			
		10.5	U	+H			
	(3) $R^2 = NH_2$ ($C_{16}H_{22}N_4O_8S$)	3.9	U	–H			
		5.3	U	–H			
		10.5	U	+H			
	$R^1 = OCH_3$						
	(4) $R^2 = CH_3$ ($C_{18}H_{25}N_3O_9S$)	4.0	U	–H			
		5.3	U	–H			
		10.5	U	+H			
	(5) $R^2 = NH_2$ ($C_{17}H_{24}N_4O_9S$)	4.2	U	–H			
		5.6	U	–H			
		10.4	U	+H			

Compound	pK_a values (all classified U)	
	—COOH	—COOH
N-chloroacetyl-(2)	4.8	6.2
N-phthalimido-(3)	5.1	6.5
N-chloroacetyl-(3)	5.0	6.5
N-phthalimido-(3)	5.2	6.5
N-chloroacetyl-(4)	4.8	6.1
N-phthalimido-(4)	5.5	6.7
N-chloroacetyl-(5)	5.4	6.8
N-phthalimido-(5)	5.2	6.5

No.	Compound	pK_a	Reliability	Ionization type	Method	Conditions	Comments and reference
270	Cephalosporin C ($C_{16}H_{21}N_3O_8S$)	3.9	U	–H	Potentiometric	66% DMF	Demarco PV and Nagarajan R, Physical-Chemical Properties *in* Flynn EH (ed.), *Cephalosporins and penicillins: Chemistry and Biology*, Academic Press, NY, 311–316 (1972).
		5.0	U	–H			
		10.4	U	+H			

Compound	pK_a	Reliability	Ionization type
Deacetoxycephalosporin C ($C_{14}H_{19}N_3O_7S$)	4.0	U	–H
	5.8	U	–H
	10.6	U	+H
N-Chloroacetylcephalosporin C ($C_{18}H_{22}ClN_3O_9S$)	4.8	U	–H
	6.2	U	–H
N-Chloracetyl deacetoxycephalo-sporin C ($C_{16}H_{20}ClN_3O_8S$)	5.4	U	–H
	6.7	U	–H

No.	Compound	pK_a	Reliability	Ionization type	Method	Conditions	Comments and reference
271	Cephalothin ($C_{16}H_{16}N_2O_6S_2$)	2.35	U	–H	Potentiometric	H_2O $t = 25 \pm 2$ $I = 0.1$	Streng WH, Microionization constants of commercial cephalosporins, *J. Pharm. Sci.*, **67**, 666–669 (1978). NB: See Cephalosporin derivative.
272	Cephalothin	5.0	U	–H	Potentiometric	66% DMF	Demarco PV and Nagarajan R, Physical-Chemical Properties *in* Flynn EH (ed.), *Cephalosporins and penicillins Chemistry and Biology*, Academic Press, NY, 311–316 (1972).

(continued)

No.	Name	pK_a value(s)	Data quality	Ionization type	Method	Conditions t°C; I or c M	Comments and Reference(s)
273	Cephapirin ($C_{17}H_{17}N_3O_6S_2$) 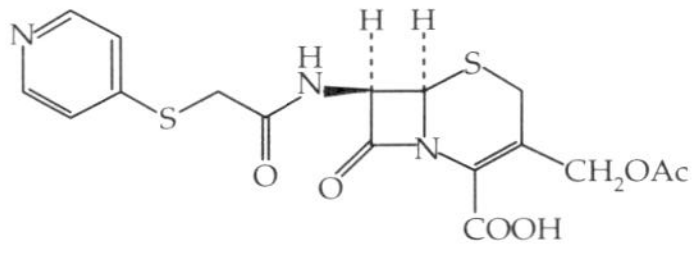	1.75 5.56	U U	−H +H	Potentiometric, Spectro	H_2O $t = 25 \pm 2$ $I = 0.1$	Streng WH, Microionization constants of commercial cephalosporins, *J. Pharm. Sci.*, **67**, 666–669, 1978. NB: The two macroconstants were calculated from the four microionization constants, which were determined by a combination of spectrophotometry and potentiometry. Two calculation methods gave the following values: (see table below) where the microconstants K_1, K_2, K_3 and K_4 refer to the equilibria for $^+$HNRCOOH to $^+$HNRCOO$^-$, $^+$HNRCOO$^-$ to NRCOO$^-$, NRCOO$^-$ to NRCOOH and NRCOOH to $^+$HNRCOOH, respectively. Apparent values were corrected to zero ionic strength with Davies' equation.
274	Cerivastatin ($C_{26}H_{34}FNO_5$) 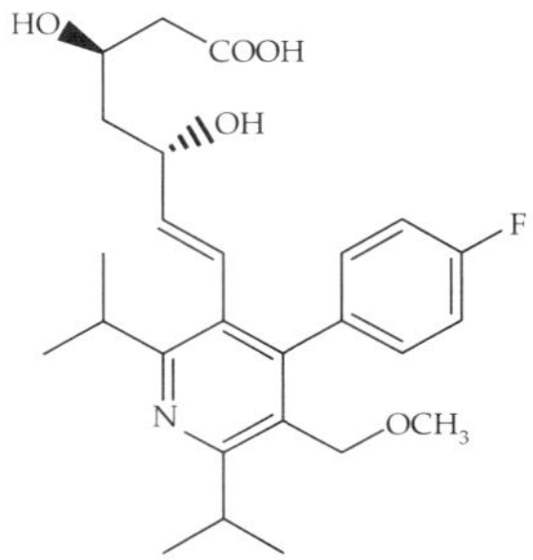	4.38 5.29	U U	+H +H	CE/pH (−ve ion mode)	H_2O $t = 25$ $I = 0.025$	Wan H, Holmen AG, Wang Y, Lindberg W, Englund M, Nagard MB and Thompson RA, High-throughput screening of pK_a values of pharmaceuticals by pressure-assisted capillary electrophoresis and mass spectrometry, *Rapid Commun. Mass Spectrom.*, **17**, 2639–2648 (2003). NB: Reported predicted values (ACD Labs) of 4.24 and 4.22. These data are identical to those given in the same source for Candesartan, so may be a typo.

pK_1	pK_2	pK_3	pK_4
1.83 (1.48e − 2)	5.48 (3.31e − 6)	4.78 (1.66e − 5)	2.53 (2.95e − 3)
1.85	5.44	4.47	2.81

No.	Compound	pKa	Data quality	Ionization	Method	Conditions	Reference and comments
275	Chenodeoxycholic acid ($C_{24}H_{40}O_4$)	5.03 ± 0.05	A	–H	Potentiometric	H_2O $t = 25.0 \pm 0.1$ $I = 0.00$	Fini A and Roda A, Chemical properties of bile acids. IV. Acidity constants of glycine-conjugated bile acids, *J. Lipid Res.*, **28(7)**, 755–759 (1987).

"The use of mixed aqueous-organic solvents was justified on the basis that the effects on pK_a of micelle formation were eliminated. . . . Bile acids were purified . . . by preparative thin layer chromatography. . . ."

NB: Apparent pK_a values were determined by titration in either aqueous DMSO (30–80 wt%) solutions or in aqueous methanol (10–50 wt%) solutions. Ionic strength effects were corrected with Davies' modification of the Debye-Huckel equation. Weight percent compositions were converted to mole fraction and plots (often exhibiting traces of curvature) of apparent pK_a were extrapolated to 100% water by linear regression analysis. The following values were reported for the pK_a value in water (pK_{aw}):

	pK_{aw}		
Compound	LR[a]	Extr[b]	Extr[c]
3α-hydroxy	5.03		
3α-hydroxy, gly conj	3.84		
3α,7β-dihydroxy	5.02	5.05	5.08
3α,7β-dihydroxy, gly conj	3.86		
3α,7α-dihydroxy	4.98	5.08	5.03
3α,7α-dihydroxy, gly conj	3.87		
3α,12α-dihydroxy	5.02		
3α,12α-dihydroxy, gly conj	3.88	3.81	3.93
3α-hydroxy 7-keto	5.00		
3α-hydroxy, 7-keto, gly conj	3.89		
3α,7α,12α-trihydroxy	5.00		
3α,7α,12α-trihydroxy, gly conj	3.88	3.81	3.93

[a] pK_{aw} calculated from a previously validated linear relationship between the apparent pK_a in 80% aqueous DMSO and the true pK_a

[b] pK_{aw} calculated by extrapolation of the apparent pK_a values in DMSO-water mixtures to 100% water

[c] pK_{aw} calculated by extrapolation of the apparent pK_a values in MeOH-water mixtures to 100% water

The pattern of hydroxyl group substitution does not significantly affect the pK_a values of either the bile acid –COOH or the corresponding glycine conjugate.

(*continued*)

No.	Name	pK_a value(s)	Data quality	Ionization type	Method	Conditions t°C; I or c M	Comments and Reference(s)
276	Chloral Hydrate ($C_2H_3Cl_3O_2$)	10.04	U	−H	Potentiometric	H_2O $t = 25.0 \pm 0.005$ $I = 0.00$	Bell RP and Onwood DP, Acid strengths of the hydrates of formaldehyde, acetaldehyde and chloral, *Farad. Soc. Trans.*, **58**, 1557–61 (1962). Cited by Fairbrother JE, Chloral Hydrate, *APDS*, **2**, 85–143 (1973). "The ionization constant, pK_a, was obtained by measuring the pH of buffered solutions of chloral hydrate (96). The mean of 25 experiments gave p$K_a = 10.04$, which differs considerably from the value of 11 obtained by Euler and Euler (97) and is closer to p$K_a = 9.77$, derived indirectly from kinetic measurements (98). 96. Bell RP and Onwood DP, *Farad. Soc. Trans.*, **58**, 1557–1561 (1962). 97. Euler H and Euler A, *Chem. Ber.*, **36**, 4246–4259 (1903). 98. Gustafsson C and Johanson M, Kinetic study of the decomposition of chloral hydrate by sodium hydroxide in aqueous solution, *Acta Chem. Scand.*, **2**, 42–48 (1948). This paper gave, from the kinetic results, $K = 1.1 \times 10^{-10}$ (15°); 1.4×10^{-10} (20°); 1.7×10^{-10} (25°); 2.0×10^{-10} (30°).
277	Chloral Hydrate	10.07 ± 0.01	U	−H	kinetic	H_2O $t = 25.0$ $I = 0.00$	Kucerova T and Mollin J, Effect of medium on the acid-catalyzed solvolysis of hydroxamic acids, *Collect. Czech. Chem Communs.*, **43**, 1571–1580 (1978).
		9.66 9.70	U U	−H −H	Potentiometric kinetic	H_2O $t = 25.0$? $I = 0.00$?	Gawron O and Draus F, Kinetic evidence for reaction of chloralate ion with p-nitrophenyl acetate in aqueous solution, *JACS*, **80**, 5392–5394 (1958). They state that Gustaffson and Johanssen reported 9.70, not 9.77 (see above, no. 276).
		10.54 ± 0.41	U	−H	comp		ACD/pK_a estimate
		13.96 ± 0.2	U	−H			Ref. 9.95/25.00/I = 0.10/Palm VA, 1975.
278	Chlorcyclizine ($C_{18}H_{21}ClN_2$)	2.12 ± 0.04 7.65 ± 0.04	A A	+H +H	Potentiometric	H_2O $t = 24.5 \pm 0.5$ $I = 0.00$	Newton DW, Murray WJ and Lovell MW, pK_a determination of benzhydrylpiperazine antihistamines in aqueous and aqueous methanol solutions, *J. Pharm. Sci.*, **71(12)**, 1363–1366 (1982). "The pK_{a1} and pK_{a2} values of three benzhydrylpiperazine antihistamines, cyclizine (I), chlorcyclizine (II), and hydroxyzine (III) were determined at 24.5 ± 0.5° by potentiometric titration in aqueous solution to be 2.16 ± 0.02 and 8.05 ± 0.03, 2.12 ± 0.04 and 7.65 ± 0.04, and 1.96 ± 0.05 and 7.40 ± 0.03, respectively. The pK_{a2} values were also determined by titration in seven aqueous methanol solutions in the range of 11.5–52.9% (w/w) methanol. The apparent dissociation constants of I - III in the aqueous methanol solutions, psK_{a2}, were plotted according to two linear regression equations from which the values in water, pwK_{a2}, were extrapolated. The plotted variables were psK_{a2} versus methanol concentration (%w/w) and psK_{a2} + log(water concentration, M) versus 1000/ϵ, where ϵ is the dielectric constant

							of the aqueous methanol solution. The maximum difference between pK_{a2} and pwK_{a2} was observed in the case of II where pwK_{a2} was 5.23% higher. . . . . The observed ratio of K_{a1}/K_{a2} in I–III, 2.75×10^5, was attributed to solvent and space-mediated field effects and electrostatic induction between nitrogen atoms in the piperazine ring.'' NB: This paper gave a good summary of the Yasuda-Shedlovsky procedure.
279	Chlordiazepoxide ($C_{16}H_{14}ClN_3O$) 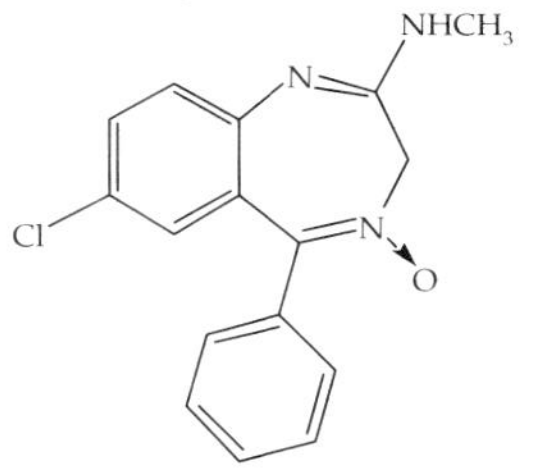	4.6	U	+H	Spectro	5% MeOH in H_2O $t = 20$ $I = 0.15$	Barrett J, Smyth WF and Davidson IE, Examination of acid-base equilibria of 1,4-benzodiazepines by spectrophotometry, *J. Pharm. Pharmacol.*, **25**, 387–393 (1973). NB: See 1,4-Benzodiazepines for details.
280	Chlorimipramine ($C_{19}H_{23}ClN_2$) 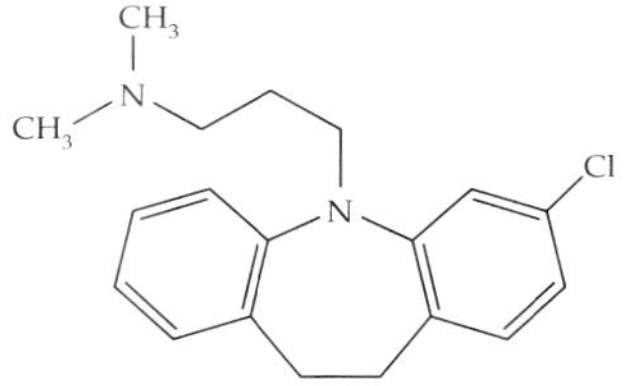	9.38	U	+H	Potentiometric	H_2O $t = 25$ I undefined Ar atmosphere	Seiler P, Simultaneous determination of partition coefficient and acidity constant of a substance, *Eur. J. Med. Chem.*, **9**, 663–665 (1974). NB: See Amitriptylline for details.

(*continued*)

No.	Name	pK_a value(s)	Data quality	Ionization type	Method	Conditions t°C; I or c M	Comments and Reference(s)
281	2-Chloro-2′,3′-dideoxy-adenosine (2-ClDDA) ($C_{10}H_{12}ClN_5O_2$) 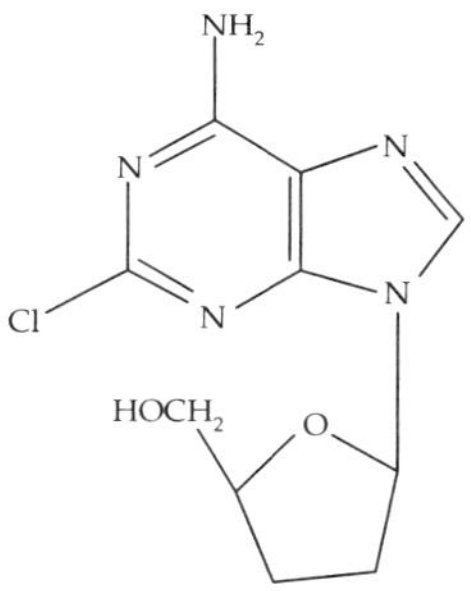	2.2	U	+H	Spectro	0.9% EtOH $t = 25.0 \pm 0.1$ $I = 0.15$	Al-Razzak LA and Stella VJ, Stability and solubility of 2-chloro-2′,3′-dideoxyadenosine, *Int. J. Pharm.*, **60**, 53–60 (1990). "The solubility in various solvents and the effects of pH, temperature, and buffer type and concentration on the stability of 2-chloro-2′,3′-dideoxyadenosine (2-CIDDA; I) were studied and used to develop some prototype parenteral dosage forms; pK_a and degradation rate constants were determined, and degradation products were identified using high pressure liquid chromatography. In the pH range 1.0–10.5, 2-chloroadenine was the principal degradation product detected. A mixture of propylene glycol, ethanol (ethyl alcohol), and water at several different ratios provided a clinically desirable solubility of >5.0 mg/ml. The pK_a of I was determined spectrophotometrically to be 2.2. Prototype solution formulations were developed with projected shelf-lives of >2 yr at room temperature; these shelf-lives were limited by the time for precipitation of 2-chloroadenine rather than the time for 10% degradation of I. It was concluded I degrades via the same mechanism and at comparable rates to 2′,3′-dideoxyadenosine." NB: Required extrapolation to zero time of absorbances in solutions of pH<3.0, due to rapid hydrolysis.
282	N^1-*p*-Chlorophenyl-N^5-alkylbiguanides 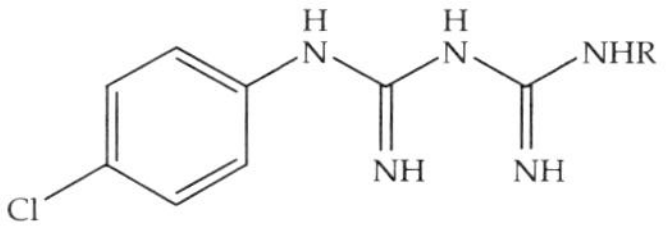				Potentiometric	H_2O $t = 25$ $I < 0.002$ N_2 atmos CO_2-free	Warner VD, Lynch DM and Ajemian RS, Synthesis, physicochemical parameters, and in vitro evaluation of N[1]-p-chlorophenyl-N5-alkylbiguanides, *J. Pharm. Sci.*, **65**, 1070–1072 (1976). "A series of N[1]-p-chlorophenyl-N[5]-alkylbiguanides were synthesized as potential inhibitors of dental plaque. Partition coefficients and pK_a values were determined by standard methods. Biological activity was evaluated against *Streptococcus mutans*, a pure strain of plaque forming bacteria. All compounds were compared to chlorhexidine acetate."
	R = CH_3 ($C_9H_{12}ClN_5$)	10.95 ± 0.04	A	+H			
	R = C_2H_5 ($C_{10}H_{14}ClN_5$)	11.01 ± 0.06	A	+H			
	R = n-C_3H_7 ($C_{11}H_{16}ClN_5$)	11.02 ± 0.06	A	+H			
	R = n-C_4H_9 ($C_{12}H_{18}ClN_5$)	10.83 ± 0.04	A	+H			

R	pK_a (25% EtOH)	Reliability	R	pK_a (25% EtOH)	Reliability
CH_3	10.60 ± 0.04	U	n-C_3H_7	10.78 ± 0.06	U
C_2H_5	10.82 ± 0.06	U	n-C_4H_9	10.78 ± 0.04	U
Chlorhexidine	10.78 ± 0.05	U			

283	3-(*p*-Chlorophenyl)-5,6-dihydro-2-ethyl imidazo [2,1-b]thiazole ($C_{13}H_{13}ClN_2S$) 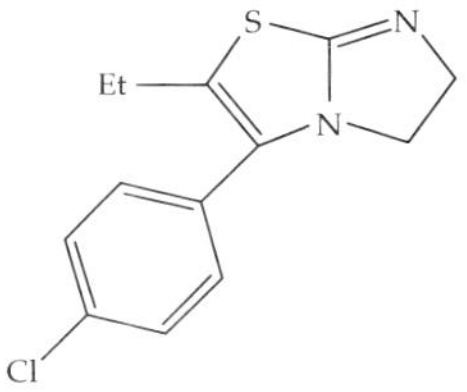	9.30 ± 0.1	U	−H	Potentiometric	H_2O $t = 24$	Maulding HV and Zoglio MA, pK_a determinations utilizing solutions of 7-(2-hydroxypropyl) theophylline, *J. Pharm. Sci.*, **60**, 309–311 (1971). NB: See Barbituric acid, 5-allyl-5-isobutyl for details.
284	3-(*p*-Chlorophenyl)-2-ethyl-2,3,5,6-tetrahydro-imidazo[2,1-b]thiazol-3-ol ($C_{13}H_{15}ClN_2OS$) 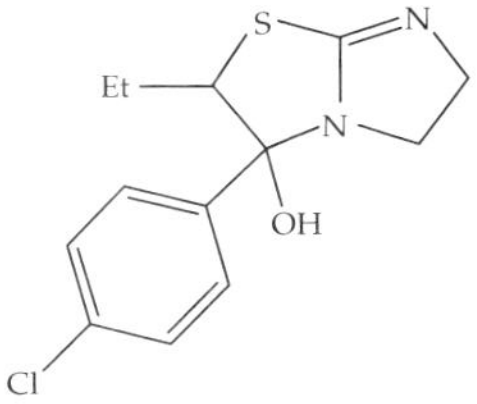	7.68 ± 0.1	U	−H	Potentiometric	H_2O $t = 24$	Maulding HV and Zoglio MA, pK_a determinations utilizing solutions of 7-(2-hydroxypropyl) theophylline, *J. Pharm. Sci.*, **60**, 309–311 (1971). NB: See Barbituric acid, 5-allyl-5-isobutyl for details.
285	Chloroquine ($C_{18}H_{26}ClN_3$) 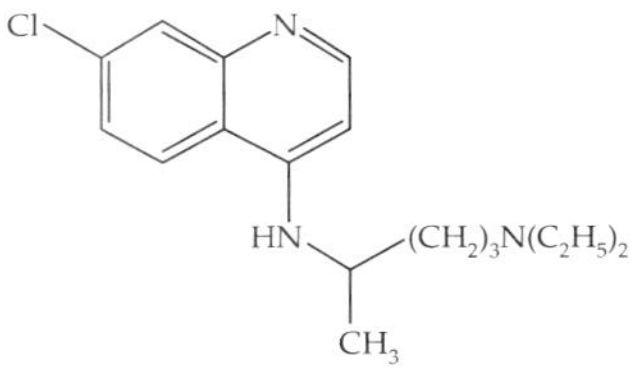	8.37	A	+H	spectro (λ = 343 nm)	H_2O $t = 20.0 \pm 0.1$ $I = 0.15$	Schill G, Photometric determination of amines and quaternary ammonium compounds with bromothymol blue. Part 5. Determination of dissociation constants of amines, *Acta Pharm. Suec.*, **2**, 99–108 (1965).
		10.76	A	+H	soly		"... In an aqueous solution, a change in absorbance was observed in the pH range 6–11. Beyond this range, up to pH 13 and down to pH 1, no further absorbance changes were observed.... The absorbance changes were assumed to be due to protolysis of the heterocyclic nitrogen, since the structurally related 4-aminoquinoline had pK_a = 9.17. ... When determining the constants, the possibility of overlapping values had to be taken into account." NB: Excellent agreement with Wahbe *et al.*, 1993, no. 286.
286	Chloroquine phosphate	8.34 (0.12)	A	+H	spectro (pK_1, λ = 334 nm; pK_2, λ = 346 nm)	H_2O $t = 20.0$	Wahbe AM, El-Yazbi FA, Barary MH and Sabri SM, Application of orthogonal functions to spectrophotometric analysis. Determination of dissociation constants, *Int. J. Pharm.*, **92(1)**, 15–22 (1993).
		10.75 (–)	A	+H			NB: Alternative graphical method gave pK_a = 8.3 and 10.75. See Acetaminophen for further details.

(continued)

No.	Name	pK_a value(s)	Data quality	Ionization type	Method	Conditions t°C; I or c M	Comments and Reference(s)
287	8-Chlorotheophylline ($C_7H_7ClN_4O_2$) 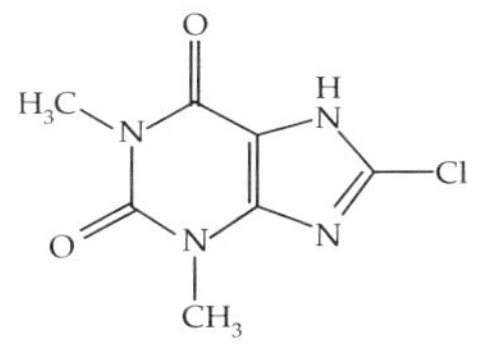	5.3	U	–H	Potentiometric	H_2O $t = 24$	Maulding HV and Zoglio MA, pK_a determinations utilizing solutions of 7-(2-hydroxypropyl) theophylline, *J. Pharm. Sci.*, **60**, 309–311 (1971). NB: See Barbituric acid, 5-allyl-5-isobutyl for details.
288	Chlorothiazide ($C_7H_6ClN_3O_4S_2$) 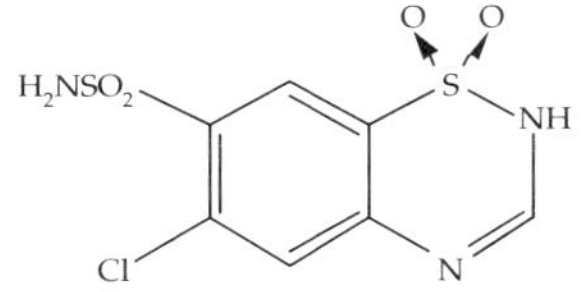	6.85 9.45	U U	–H –H	Potentiometric	H_2O	Brittain HG, Chlorothiazide, *APDS*, **18**, 33–56 (1989). "The relative insolubility of chlorothiazide in most common solvents has made the determination of ionization constants difficult. Several groups have used aqueous potentiometric titration to obtain pK_a data, with the general consensus being that p$K_{a1} = 6.85$ and p$K_{a2} = 9.45$." NB: See Whitehead CW, Traverso JJ, Sullivan HR and Morrison DE, Diuretics. IV. 6-Chloro-3-substituted-7-sulfamoyl-1,2,4-benzothiadiazine-1,1-dioxides, *JOC*, **26**, 2809–2813 (1961). Essig A, Competitive inhibition of renal transport of p-amino-hippurate by analogs of chlorothiazide, *Am. J. Physiol.*, **201**, 303–308 (1961); Charnicki WP, Bacher FA, Freeman SA and DeCesare DH, The pharmacy of chlorothiazide: A new orally effective diuretic agent, *J. Am. Pharm. Assoc., Sci Edn.*, **48**, 656–659 (1959).
289	Chlorothiazide	9.7	U	–H	Spectro	H_2O	Hennig UG, Chatten LG, Moskalyk RE and Ediss C, Benzothiadiazine dissociation constants. Part 1. Ultraviolet spectrophotometric pK_a determinations, *Analyst*, **106**, 557–564 (1981). NB: Thermodynamic values for the dissociation constants of a number of benzothiadiazines, including chlorothiazide, methyclothiazide, polythiazide and diazoxide, were determined by UV spectrophotometry. Resolution of the overlapping acidity constants was achieved by employing a computer program of successive approximations.

No.	Compound	pKa value(s)	Data quality	Ionization type	Method	Conditions	Comments and Reference(s)
290	Chlorpheniramine ($C_{16}H_{19}ClN_2$) 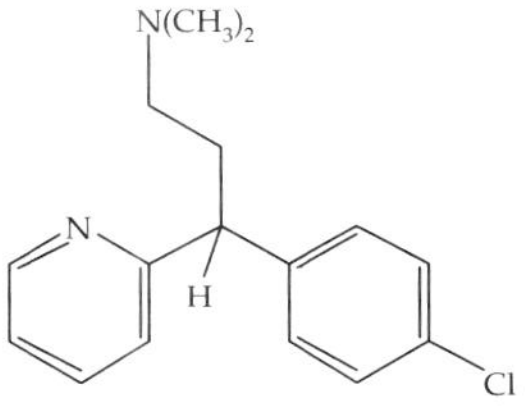 $N(CH_3)_2$ … H … N … Cl	3.99 ± 0.05 9.26 ± 0.02	U U	+H +H	Potentiometric	H_2O t undefined $I = 0.30$ (NaCl)	Testa B and Murset-Rossetti L, The partition coefficient of protonated histamines, *Helv. Chim. Acta*, **61**, 2530–2537 (1978). NB: See Cycliramine for details .
291	Chlorpromazine ($C_{17}H_{19}ClN_2S$) 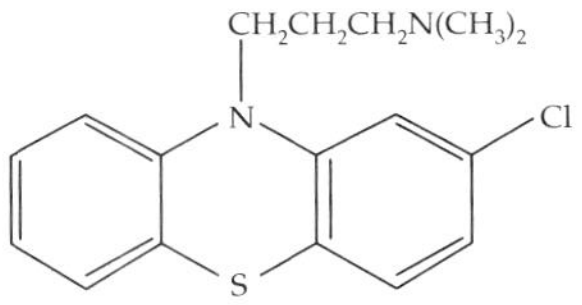 $CH_2CH_2CH_2N(CH_3)_2$ … N … Cl … S	9.32	U	+H	soly	H_2O $t = 20.0 \pm 0.1$ $I = 0.15$	Schill G, Photometric determination of amines and quaternary ammonium compounds with bromothymol blue, Part 5. Determination of dissociation constants of amines, *Acta Pharm. Suecica*, **2**, 99–108 (1965). NB: See Chlorquine for details.
292	Chlorpromazine	9.22	U	+H	Potentiometric	H_2O (extrap) $t = 24 \pm 1$ $I \sim 0.002$	Chatten LG and Harris LE, Relationship between $pK_b(H_2O)$ of organic compounds and $E_{1/2}$ values in several nonaqueous solvents, *Anal. Chem.*, **34**, 1495–1501 (1962). "Methanolic solutions of the free bases were prepared, and from these, accurately measured aliquots were taken. Known dilutions were prepared with distilled water, and titrations were carried to exactly half-neutralization with 0.1N HCl, so that the concentrations of total salt and free base did not exceed 0.001 to 0.002M. The pH of the solution was then measured on the Fisher Titrimeter. Three measurements were made for the same concentration of base but with varying percentages of methanol. These pH readings were plotted … and the straight line was extrapolated to 0% alcohol. Three such series were run giving a total of nine measurements for each … base. The … reading obtained … by extrapolation was plotted against milligrams of base. The straight line was extrapolated back to infinite dilution of the drug and the pH determined …. Because pH = pK_a at half-neutralization, this same reading gives the pK_a value of the base. The pK_b was then determined."

(continued)

No.	Name	pK_a value(s)	Data quality	Ionization type	Method	Conditions t°C; I or c M	Comments and Reference(s)
							NB: Cited Marshall PB, Some chemical and physical properties associated with histamine antagonism, *Br. J. Pharmacol.*, **10**, 270–278 (1955). Compounds were checked for purity (99%) by nonaqueous titration. They were stored over P_2O_5 and protected from light. All solutions were freshly prepared in light protected flasks and operations were performed in a semi-darked room. Results were reported as pK_b values. As no further corrections were performed, results must be regarded as uncertain.
293	Chlorpromazine	9.3	U	+H	soly	H_2O $t = 24 \pm 1$	Green AL, Ionization constants and water solubilities of some aminoalkylphenothiazine tranquilizers and related compounds, *J. Pharm. Pharmacol.*, **19**, 10–16 (1967). NB: See Amitriptylline for details. Value cited by Abadi A, Rafatullah S, Al-Badr AA, Chlorpromazine, *APDS*, **26**, 108 (1999), which cited Clarke as source.
294	Chlorpromazine	9.3	U	+H	partition/pH	H_2O $t = 25.0 \pm 0.1$	Persson BA, Extraction of amines as complexes with inorganic anions. Part 3, *Acta Pharm. Suec.*, **5**, 335–342 (1968). NB: Reference for methods, Persson BA, Schill G, Extraction of amines as complexes with inorganic anions. Part 2, *Acta Pharm. Suec.*, **3**, 281–302 (1966).
295	Chlorpromazine	9.24 ± 0.02	U	+H	Potentiometric	H_2O $t = 25.0 \pm 0.1$ $I = 0.1$ (NaCl)	Takacs-Novak K, Box KJ, Avdeef A, Potentiometric pK_a determination of water-insoluble compounds: Validation study in methanol/water mixtures, *Int. J. Pharm.*, **151**, 235–248 (1997). NB: By extrapolation from 34–50 %w/w aqueous MeOH. See Acetaminophen for full details.
296	Chlorpromazine	9.29 ± 0.13	U	+H	Potentiometric	H_2O $t = 25$ I undefined Ar atmosphere	Seiler P, Simultaneous determination of partition coefficient and acidity constant of a substance, *Eur. J. Med. Chem.*, **9**, 663–665 (1974). NB: See Amitriptylline for details.
297	Chlorpromazine	9.38	U	+H	CE/pH (+ve ion mode)	H_2O $t = 25$ $I = 0.025$	Wan H, Holmen AG, Wang Y, Lindberg W, Englund M, Nagard MB and Thompson RA, High-throughput screening of pK_a values of pharmaceuticals by pressure-assisted capillary electrophoresis and mass spectrometry, *Rapid Commun. Mass Spectrom.*, **17**, 2639–2648 (2003). NB: Reported a predicted value (ACD Labs) of 9.41.
298	Chlorpromazine	9.3	U	+H	Potentiometric	H_2O $t = 25$	Bergström CAS, Strafford M, Lazorova L, Avdeef A, Luthman K and Artursson P, Absorption classification of oral drugs based on molecular surface properties, *J. Med. Chem.*, **46(4)**, 558–570 (2003). NB: From extrapolation of aqueous-methanol mixtures to 0% methanol.

No.	Compound	pKa			Method	Conditions	Reference / Comments
299	Chlorpromazine	9.21	U	+H	Potentiometric	H_2O (extrap.) $t = 20$ N_2 atmosphere	Sorby DL, Plein EM and Benmaman D, Adsorption of phenothiazine derivatives by solid adsorbents, *J. Pharm. Sci.*, **55**, 785–794 (1966). NB: Followed the procedure of Marshall. Titrations were performed in hydroalcoholic solutions (10, 20, 30, 40, and 50%) and apparent values extrapolated to zero percent alcohol, using a linear relationship. Precipitation occurred and data post-precipitation were used, "assuming the concentration of the amine base actually in solution to be constant after the point where precipitation first occurred". This assumption seems to be unreasonable.
300	Chlorpromazine	9.40 ± 0.05	U	+H	partition/ pH	H_2O $t = 20 \pm 0.5$	Vezin WR and Florence AT, The determination of dissociation constants and partition coefficients of phenothiazine derivatives, *Int. J. Pharm.*, **3**, 231–237 (1979). NB: See also Promethazine for details. Compounds were purified before use and partitioned between buffers and cyclohexane. Only the aqueous phases were analysed, the cyclohexane phase concentrations being assessed by difference. The effect of ionic strength was tested over a fourfold range and found to cause no systematic deviations.
301	Chlortetracycline ($C_{22}H_{23}ClN_2O_8$) 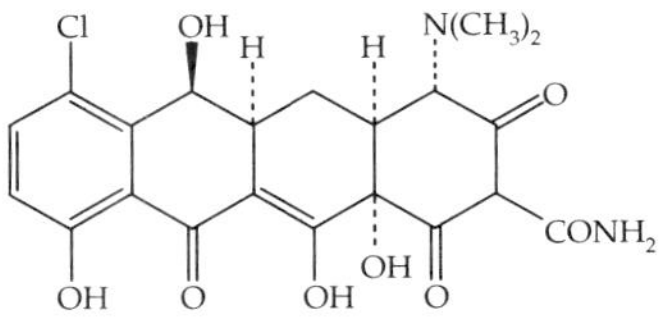	3.30 7.44 9.27	U U U	–H –H +H	Potentiometric	H_2O $I = 0.00$	Schwartzman G, Wayland L, Alexander T, Furnkranz K, Selzer G, and the US Antibiotics Standards Research Group, Chlortetracycline hydrochloride, *APDS*, **8**, (1979), 101–137 Leeson L, Krueger J and Nash R, *Tet. Lett.*, **18**, 1155–1160 (1963). Kalnins K and Belen'skii BG, A study of dissociation of tetracyclines by infrared spectroscopy, *Dokl. Akad. Nauk. SSSR*, **157(3)**, 619–621 (1964); CA 61:9381f. "CTC exhibits three acidic dissociation constants when titrated in aqueous solutions. Stephens *et al.* identified the three acidic groups, and reported thermodynamic pK_a values of 3.30, 7.44, and 9.27. Leeson *et al.* assigned pK_a values to the following acidic groups:

pK_a	Assignment
3.30	Tricarbonylmethane System (A)
7.44	Dimethylamino System (B)
9.27	Phenolic Diketone System (C)

Kalnins and Belen'skii verified the assignments by infrared (IR) spectroscopy of several tetracyclines."

(continued)

No.	Name	pK_a value(s)	Data quality	Ionization type	Method	Conditions t°C; I or c M	Comments and Reference(s)
302	Chlortetracycline 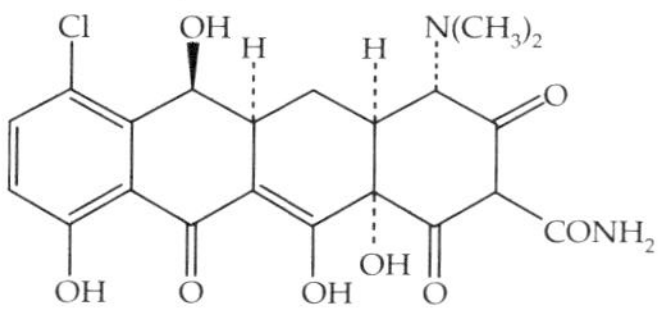	3.30 7.44 9.27	A A A	−H +H −H	Potentiometric	H_2O $t = 25$ $I = 0.00$	Stephens C, Murai K, Brunings K and Woodward RB, Acidity constants of the tetracycline antibiotics, *JACS*, **78**, 4155–4158 (1956). Cited in Perrin Bases 3324 ref. S73. NB: Used a glass electrode with liquid junction potentials. "The three observed dissociation constants of the antibiotics oxytetracycline, chlortetracycline and tetracycline have been assigned to specific acidic groupings. In each case, the first dissociation is due to the tricarbonyl system in ring A, the second to the dimethylammonium function and the third to the phenolic β-diketone system: pK_a — Assignment 3.30 — Tricarbonylmethane System (A) 7.44 — Dimethylamino System (B) 9.27 — Phenolic Diketone System (C) NB: A note added in proof indicated that these assignments were in error. The correct assignments interchange (B) and (C). Later papers, mainly Garrett (*J. Pharm. Sci.*, 1963) and Rigler *et al.*, (*Anal. Chem.*, 1965) provided strong evidence for the interchanged assignments.
303	Chlortetracycline	3.30 7.44 9.27	A A A	−H −H +H	Potentiometric	H_2O $t = 20$ $c = 0.001$	Albert A, Avidity of terramycin and aureomycin for metallic cations, *Nature*, **172**, 201 (1953). Cited in Perrin Bases 3324 ref A17. NB: The study used pH measurements with a glass electrode and junction potentials.
304	Chlortetracycline	3.66 7.40 9.06	U A U	−H −H +H	Potentiometric	H_2O $t = 30.0 \pm 0.2$ $I = 0.01$ (KCl) N_2 atmosphere	Doluisio JT and Martin AN, Metal complexation of the tetracycline hydrochlorides, *J. Med. Chem.*, **6**, 16–20 (1963). NB: Metal-free solutions of the tetracycline were titrated with standard NaOH solution and the pH measured. No details were given of the pH meter calibration. Metal stability constants were determined from identical titrations in the presence of varying concentrations of nickel(II), zinc(II), or copper(II) ions.

No.	Compound	pK_a			Method	Conditions	Reference / Notes
305	Chlortetracycline	3.27	A	−H	Potentiometric	H_2O	Benet LZ and Goyan JE, Determination of the stability constants of tetracycline complexes, *J. Pharm. Sci.*, **54**, 983–987 (1965). NB: Used carbonate-free KOH to titrate the hydrochloride salt and measured pH with a high quality pH meter (Beckman Research Model) that had been calibrated at pH 4.01. Activity corrections were applied.
		7.43	A	−H		$t = 25 \pm 0.05$	
		9.33	A	+H		$I = 0.01$	
306	Chlortetracycline	3.14	U	−H	Potentiometric	H_2O	Parke TV and Davis WW, Use of apparent dissociation constants in qualitative organic analysis, *Anal. Chem.*, **26**, 642–5 (1954). NB: No activity corrections were used, and the data therefore described as pK_a' values, that is, apparent pK_a.
		7.33	U	−H		t undefined	
		9.24	U	+H		I undefined	
307	Chlorthalidone ($C_{14}H_{11}ClN_2O_4S$) 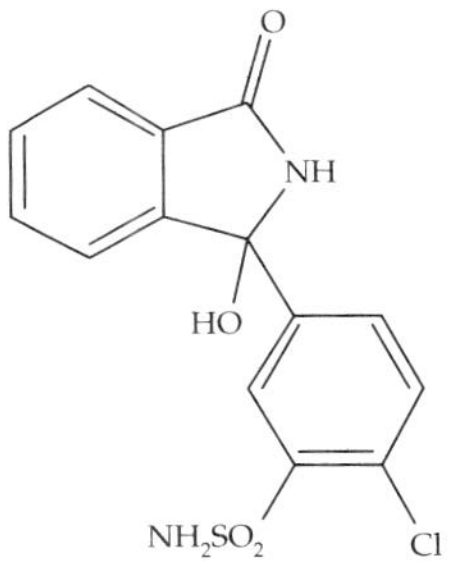	9.36	A	−H	soly	H_2O RT	Singer JM, O'Hare MJ, Rehm CR and Zarembo JE, Chlorthalidone, *APDS*, **14**, 1–34 (1985); Fleurren ALJ, van Ginneken CAM and van Rossum JM, Differential potentiometric method for determining dissociation constants of very slightly water soluble drugs applied to the sulfonamide diuretic chlorthalidone, *J. Pharm. Sci.*, **68(8)**, 1056–1058 (1979). NB: The sulfonamide function present in chlorthalidone, is considered to be responsible for the acid dissociation. The ionization constant of chlorthalidone was determined based on spectrophotometric measurements of the concentration [chlor] at various pH values:
		9.35 ± 0.02	A	−H	Potentiometric	H_2O	
		12.2 – 13.3	U	−H		$t = 25 \pm 1$; $I = 0.00$; N_2 atmosphere	

pH	[chlor] mg/ml	pH	[chlor] mg/ml	pH	[chlor] mg/ml
4.90	0.167	8.95	0.300	10.10	2.958
7.00	0.180	9.40	0.390	10.30	4.698
7.70	0.183	9.60	0.597	10.50	5.534
8.40	0.210	10.00	1.201	10.90	9.911
8.65	0.230				

"A single pK_a value of 9.36 in water (22° ± 1 °C) was obtained, which indicates that chlorthalidone is a weakly mono-acidic compound. ... Potentiometric difference titrations (Fleuren *et al.*) produced a value of 9.24 ± 0.02 in 0.1 M aqueous KCl which, when corrected for ionic strength, yields a thermodynamic constant of 9.35 (25 °C), which is in excellent agreement with the spectrophotometric (i.e., soly) determination."

(continued)

No.	Name	pK_a value(s)	Data quality	Ionization type	Method	Conditions t°C; I or c M	Comments and Reference(s)
308	Cimetidine ($C_{10}H_{16}N_6S$) 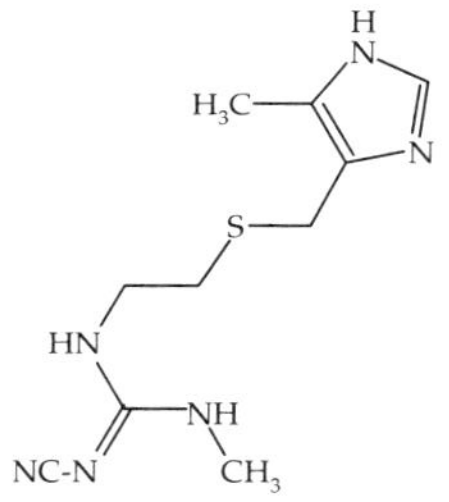	7.11 ± 0.04	U	+H	Potentiometric	H_2O $t = 25$ $I = 0.1$ (KCl)	Bavin PMG, Post A and Zarembo JE, Cimetidine, *APDS*, **13**, 127–182 (1984). Graham MJ, Smith Kline and French Ltd., unpublished results.
309	Cimetidine	6.89	U	+H	CE/pH (+ve ion mode)	H_2O $t = 25$ $I = 0.025$	Wan H, Holmen AG, Wang Y, Lindberg W, Englund M, Nagard MB and Thompson RA, High-throughput screening of pK_a values of pharmaceuticals by pressure-assisted capillary electrophoresis and mass spectrometry, *Rapid Commun. Mass Spectrom.*, **17**, 2639–2648 (2003). NB: Reported a predicted value (ACD Labs) of 6.72.
310	Cinchonidine ($C_{19}H_{22}N_2O$) 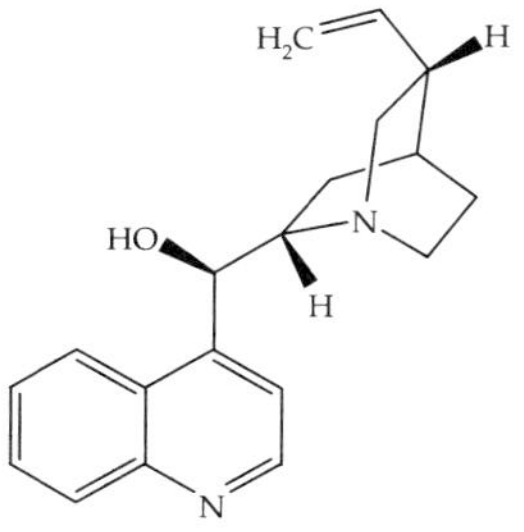	4.17 8.40	U U	+H +H	Spectro	H_2O $t = 15$ ($pK_w = 14.35$ at 15 °C)	Kolthoff IM, The dissociation constants, solubility product and titration of alkaloids, *Biochem. Z.*, **162**, 289–353 (1925). Cited in Perrin Bases 2876. NB: See Aconitine for details.

311	Cinchonine ($C_{19}H_{22}N_2O$)	4.28	U	+H	Spectro	H_2O	Kolthoff IM, The dissociation constants, solubility product and titration of alkaloids, *Biochem. Z.*, **162**, 289–353 (1925). Cited in Perrin Bases 2877. NB: See Aconitine for details.
		8.35	U	+H		$t = 15$ ($pK_w =$ 14.35 at 15°C)	
		5.21	U	+H	kinetic	H_2O $t = 55$	NB: Also by rate of inversion of sucrose, ref. A73, Arnall F, *JCS*, **117**, 835 (1920). See Caffeine (no. 243) for details.
		4.04	U	+H	calorimetric	H_2O	NB: Dragulescu C and Policec S, Thermometric titration of weak diacidic bases. *Studii Cercetari Chim.*, **9**, 33–40 (1962); CA 58:30272. Cited in Perrin Suppl. No. 7488. NB: From pK_b values 5.85 and 9.96 at unknown temperature (assumed 25 °C).
		8.15	U	+H			
312	Cinnamic acid ($C_9H_8O_2$)	4.45 ± 0.01	A	−H	Conductance	H_2O $t = 25$ $c = 0.004$–0.0004	Kendall J, Electrical conductivity and ionization constants of weak electrolytes in aqueous solution, *in* Washburn EW, Editor-in-Chief, *International Critical Tables*, Vol. 6, McGraw-Hill, NY, 259–304 (1929).
313	Cinnarizine ($C_{26}H_{28}N_2$)	7.47	U	+H	Potentiometric	H_2O $t = 25$	Peeters J, Determination of ionization constants in mixed aqueous solvents of varying composition by a single titration, *J. Pharm. Sci.*, **67**, 127–129 (1978). "A potentiometric titration method is proposed in which only one titration is necessary to obtain dissociation constants for different solvent compositions. The method allows the results to be extrapolated to the value for pure water. …. Titration data for 1-Me-1–1H-imidazole are given. The pK_a values determined by the proposed method …. are given for seperidol, cinnarizine, diphenoxylate, etomidate and miconazole." NB: The paper recognizes the problem of long extrapolations from aqueous-organic solvent mixtures to estimate pK_a values of poorly water-soluble substances. No activity corrections were applied.

(continued)

No.	Name	pK_a value(s)	Data quality	Ionization type	Method	Conditions t°C; I or c M	Comments and Reference(s)
314	Cinnopentazone ($C_{22}H_{22}N_2O_2$) 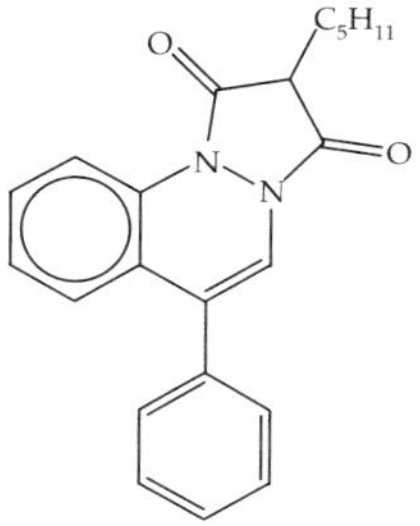	4.4	U	+H	Potentiometric	50% EtOH t undefined I undefined	Jahn U and Wagner-Jauregg T, Wirkungsvergleich saurer Antiphlogistika im Bradykinin-, UV-Erythem- und Rattenpfotenödem-Test, *Arzneim.-Forsch.*, **24**, 494–499 (1974). NB: Literature values obtained from the pH of half-neutralization. Also gave values (range 4.3–5.4) in 50% EtOH for 8 analogous structures.
		4.6	U	+H		80% Me cellosolve	
315	Cinoxacin ($C_{12}H_{10}N_2O_5$) 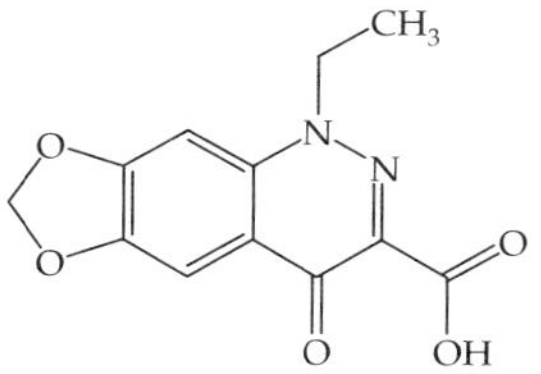	5.38	U	−H	CE/pH (−ve ion mode)	H_2O $t = 25$ $I = 0.025$	Wan H, Holmen AG, Wang Y, Lindberg W, Englund M, Nagard MB and Thompson RA, High-throughput screening of pK_a values of pharmaceuticals by pressure-assisted capillary electrophoresis and mass spectrometry, *Rapid Commun. Mass Spectrom.*, **17**, 2639–2648 (2003). NB: Reported a predicted value (ACD Labs) of 0.61 (typographical error?).
316	Ciprofloxacin ($C_{17}H_{18}FN_3O_3$) 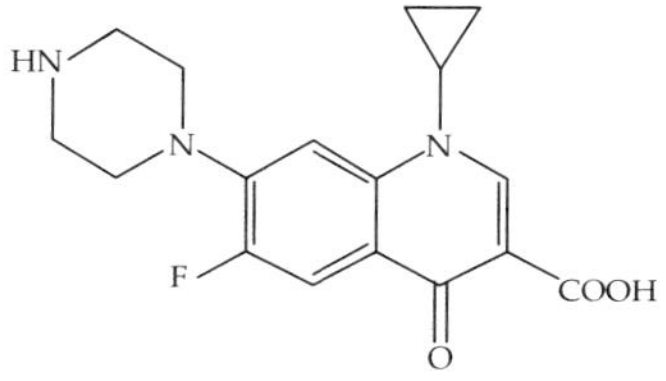	6.15 8.66	U U	−H +H	Potentiometric	H_2O $t = 25$	Bergström CAS, Strafford M, Lazorova L, Avdeef A, Luthman K and Artursson P, Absorption classification of oral drugs based on molecular surface properties, *J. Med. Chem.*, **46(4)** 558–570 (2003). NB: From extrapolation of aqueous-methanol mixtures to 0% methanol.

No.	Compound	pKa	Class	±H	Method	Conditions	Reference / Comments
317	Cisplatin degradation product (H_7ClN_2OPt) 	6.56 ± 0.01	U	+H	kinetic	H_2O $t = 37$ $I = 0.15$	Andersson A, Hedenmalm H, Elfsson B and Ehrsson H, Determination of the acid dissociation constant for *cis*-diammineaquachloroplatinum(II) ion: hydrolysis product of cisplatin, *J. Pharm. Sci.*, **83**, 859–862 (1994). "The determination of the pK_a of *cis*-diammineaquachloroplatinum(II) ion (monoaqua) at 37 °C by utilizing the fact that it rapidly reacts with chloride ions while the corresponding mono-hydroxy moiety remains unreactive is described.... The pK_a was determined to be 6.56 ± 0.01 (SEM). Thus, at physiological pH, the monoaqua complex is present mostly in its less reactive monohydroxy form." NB: U classification made for the following reasons: (1) kinetic results are often uncertain, as it is not easy to relate the equilibrium constant to the ground state or the transition state; and (2) the kinetic measurements were made at 37 °C, but the necessary pH measurements were made at 25 °C.
318	Citric acid ($C_6H_8O_7$) 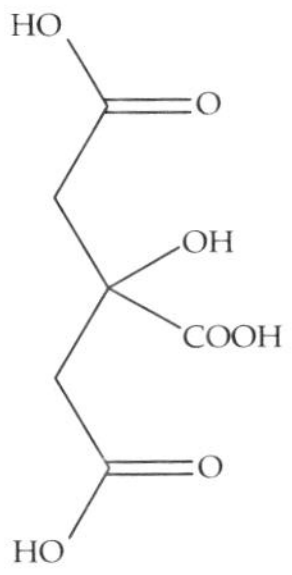	$3.128 \pm 0.001_8$ $4.761 \pm 0.000_9$ $6.396 \pm 0.001_1$	VR VR VR	−H −H −H	potentio (electro-chemical cells with the utmost technical refinement)	H_2O $t = 25$ I extrap to 0.000 (KCl)	Bates RG and Pinching GD, Resolution of the dissociation constants of citric acid at 0 to 50°, and determination of certain related thermodynamic functions, *JACS*, **71**, 1274–1283 (1949). NB: The maximum uncertainty in these data is ± 0.003. The three pK_a values for citric acid are temperature dependent, see Table. The pK_a values for carboxylic acids frequently pass through a minimum with temperature. The Table shows that for citric acid, the pK_{a2} value does this near 40 °C, the pK_{a3} minimum is near 12 °C, while the minimum for the pK_{a1} value is just above 50 °C.

Table: pK_a values for Citric Acid in water at various temperatures

Temperature °C (K)	pK_{a1}	pK_{a2}	pK_{a3}
0 (273.15)	3.220	4.837	6.393
5 (278.15	3.200	4.813	6.386
10 (283.15)	3.176	4.797	6.383
15 (288.15)	3.160	4.782	6.384
20 (293.15)	3.142	4.769	6.388
25 (298.15)	3.128	4.761	6.396
30 (303.15)	3.116	4.755	6.406
35 (308.15)	3.109	4.751	6.423
40 (313.15)	3.099	4.750	6.439
45 (318.15)	3.097	4.754	6.462
50 (323.15)	3.095	4.757	6.484

NB: See also Poerwono H, Higashiyama K, Kubo H, Poernomo AT, Suharjono, Sudiana IK, Indrayanto G and Brittain HG. Citric Acid, *APDS*, *28*, 13–14 (2001). Several references to the data are given, all of which are secondary sources.

(*continued*)

No.	Name	pK_a value(s)	Data quality	Ionization type	Method	Conditions t°C; I or c M	Comments and Reference(s)
319	Citric acid	–		−H	Potentiometric	H_2O $t = 23.0$	Clarke FH and Cahoon NM, Ionization constants by curve-fitting: Determination of partition and distribution coefficients of acids and bases and their ions, *J. Pharm. Sci.*, **76(8)**, 611–620 (1987). NB: See Benzoic acid for further details.
		4.81	U	−H			
		6.21	U	−H			
320	Clarithromycin ($C_{38}H_{69}NO_{13}$) 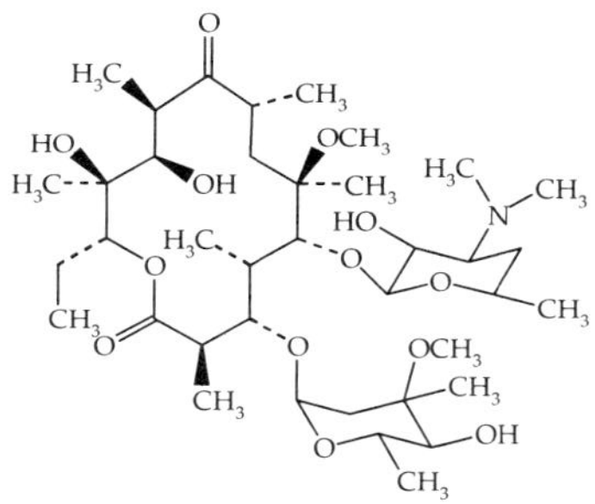	8.99	U	+H	Potentiometric	H_2O $t = 25$ $I = 0.167$	McFarland JW, Berger CM, Froshauer SA, Hayashi SF, Hecker SJ, Jaynes BH, Jefson MR, Kamicker BJ, Lipinski CA, Lundy KM, Reese CP and Vu CB, Quantitative Structure-activity relationships among macrolide antibacterial agents: In vitro and in vivo potency against Pasteurella multocida, *J. Med. Chem.*, **40**, 1340–1346 (1997). NB: See Azithromycin for details; average standard deviation of ± 0.07 for the pK_a.
321	Clarithromycin	8.76	U	+H	soly	H_2O $t = 37$	Nakagawa Y, Itai S, Yoshida T and Nagai T, Physicochemical properties and stability in the acidic solution of a new macrolide antibiotic, clarithromycin, in comparison with erythromycin, *Chem. Pharm. Bull.*, **40**, 725–728 (1992).
322	Clindamycin ($C_{18}H_{33}ClN_2O_5S$) 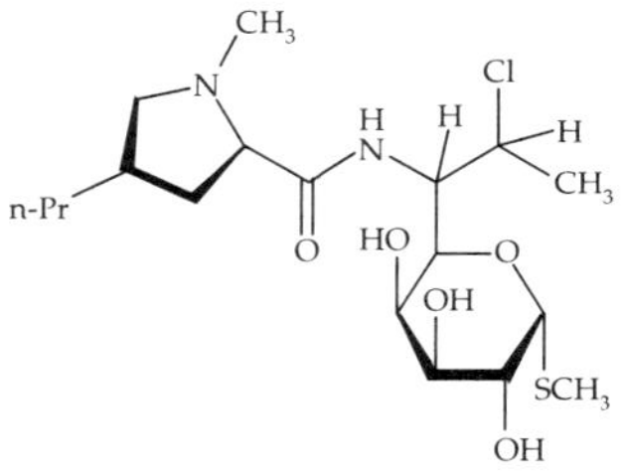	7.72 ± 0.04	A	+H	Potentiometric	H_2O $t = 25$ $I < 0.005$	Tarazska MJ, Absorption of clindamycin from the buccal cavity, *J. Pharm. Sci.*, **59**, 873–874 (1970).

323	Clindamycin	7.77	A	+H	CE/pH (+ve ion mode)	H_2O $t = 25$ $I = 0.025$	Wan H, Holmen AG, Wang Y, Lindberg W, Englund M, Nagard MB and Thompson RA, High-throughput screening of pK_a values of pharmaceuticals by pressure-assisted capillary electrophoresis and mass spectrometry, *Rapid Commun. Mass Spectrom.*, **17**, 2639–2648 (2003). NB: Reported a predicted value (ACD Labs) of 8.74.
324	Clindamycin	7.6	U	+H			Novak E, Wagner JG and Lamb DJ, Local and systemic tolerance, absorption and excretion of clindamycin hydrochloride after intramuscular administration, *Int. J. Clin. Pharm.*, **3(3)**, 201–208 (1970). Cited by Brown LW and Beyer WF, Clindamycin hydrochloride, *APDS*, **10**, 75–90 (1981).
325	Clindamycin	7.45 ± 0.02	U	+H		H_2O $t = 37$	Taraszka MJ, Transfer of clindamycin and 1′demethyl-4′-depropyl-4′-pentylclindamycin by the cannulated everted rat gut, *J. Pharm. Sci.*, **60**, 946–948 (1971). NB: Also reported pK_a = 7.96 ± 0.04 for 1′-demethyl-4′-depropyl-4′-pentylclindamycin.
326	Clindamycin 2-palmitate ($C_{34}H_{63}ClN_2O_6S$) 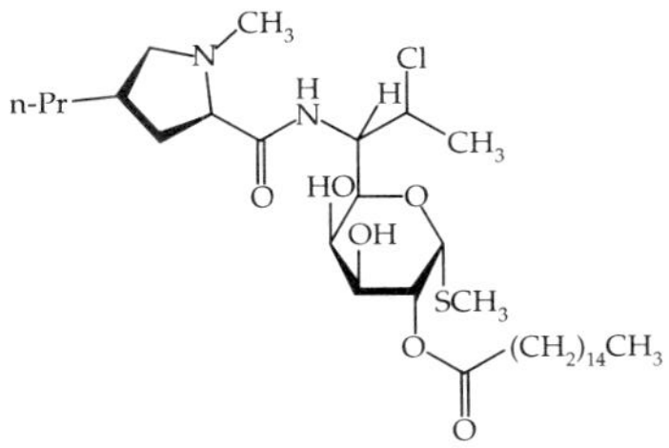	7.6	U	+H	soly	H_2O $t = 24.3 \pm 0.3$ I not reported but stated to be low	Rowe EL, Anomalous solution behavior of 2-palmitate esters of lincomycin and clindamycin, *J. Pharm. Sci.*, **68**, 1292–1296 (1979). NB: The low aqueous solubility of clindamycin-2-palmitate was recognized as a problem for accurate pK_a measurement, but it was rationalized that the pK_a value for the parent should not be greatly altered by the remote acyl substituent. This assumption appears justified.

(continued)

No.	Name	pK_a value(s)	Data quality	Ionization type	Method	Conditions t°C; I or c M	Comments and Reference(s)
327	Clindamycin-2-phosphate ($C_{18}H_{35}ClN_2O_8SP$) 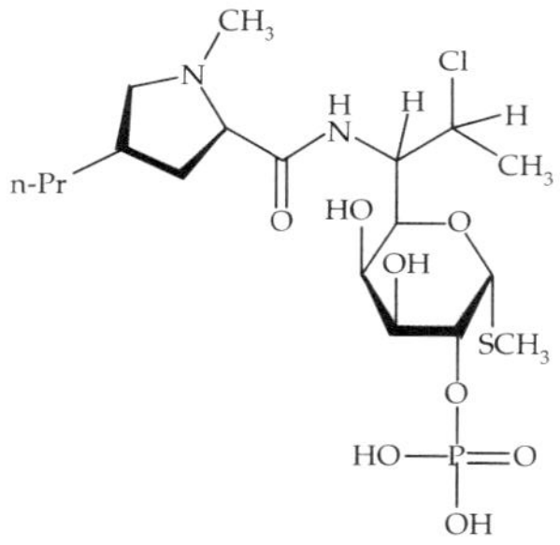	0.964 ± 0.06 (I = 0.11) 6.06 ± 0.06 (I = 0.008)	U U	–H –H	^{31}P-NMR	H_2O $t = 21$	Kipp JE, Smith III WJ and Myrdal PB, Determination of phosphate functional group acid dissociation constants of clindamycin 2-phosphate using ^{31}P Fourier transform NMR spectrometry, *Int. J. Pharm.*, **74**, 215–220 (1991). NB: 31P-NMR; values by fitting chemical shifts to double sigmoid curve. Fitting of coupling constants gave pK_1 = 1.31 ± 0.8; pK_2 = 6.47 ± 0.41. Errors calculated from 3(s.e.).
328	Clioquinol (C_9H_5ClINO) 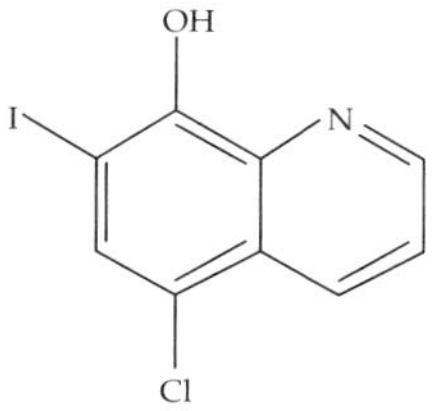	8.12	A	–H	Potentiometric	50% aq EtOH $t = 35.0 \pm 0.1$ $I = 0.00$ N_2 atmosphere	Agrawal YK and Patel DR, Thermodynamic proton-ligand and metal-ligand stability constants of some drugs, *J. Pharm. Sci.*, **75(2)**, 190–192 (1986). "The thermodynamic proton-ligand (pK_a) and metal-ligand stability constants of clioquinol, clofibrate, nitrofurazone, and tetracycline with Cu^{+2}, Zn^{+2}, Mn^{+2}, Mg^{+2} and Ca^{+2} have been determined at 35 °C in 50% ethanol-water media. An empirical pH correction for mixed aqueous media has been applied. . . . Determination of pK_a: The titration procedure was essentially that recommended by Agrawal *et al.*. A 0.5 M portion of a drug and 47.5 ml of solvent [50% (v/v) ethyl alcohol: water 50 ml of pure EtOH diluted to 100 ml with water] were placed in a three-necked titration vessel (thermostated at 35 °C ± 0.1 °C) carrying the combined electrode, microburette (5x 0.01ml), and inlet for nitrogen. Nitrogen gas, presaturated with carbonate-free water and the desired solvent mixture [50% (v/v) ethyl alcohol: water], was passed through the solutions. The solutions were titrated with potassium hydroxide (0.1M) which was also prepared to contain the required ethanol percentage by adding aliquots (0.5 ml) The highest appropriate drift-free reading on the meter was recorded."

329	Clioquinol	3.17 ± 0.11 8.08 ± 0.08	U U	+H −H	partition/ pH	H_2O $t = 25$	Tanaka H and Tamura Z, Determination of the partition coefficient and acid dissociation constants of iodochlorohydroxyquin by an improved partition method, *J. Pharm. Sci.*, **73(11)**, 1647–1649 (1984). ". . . . .The partition ratio at various pH values (1.2–9.8) was measured using five buffer systems. Figure 3 shows that the pH profile of the partition ratios in I was a bell-shaped curve with a maximum at pH 5.6. The constants, K_p, K_1', and K_2' were estimated from the data . . . as follows; $K_p = 2230 \pm 43$ (CV 1.9%), $n = 11$; $K_1' = 5.6 \times 10^{-4}$, $pK_1' = 3.25$; $K_2' = 1.29 \times 10^{-8}$, $pK_2' = 7.89$. Consequently Eq 2 can be expressed as Partition ratio $= 2230/10^{3.25\text{-pH}} + 1 + 10^{\text{pH-7.89}}$. . . . . .The values of pK_1' and pK_2', obtained from the spectrophotometric results at the same temperature and the same ionic strength as the partition method, were 3.17 ± 0.11 and 8.05 ± 0.08 at 25 °C, respectively, and agree well with those obtained by the partition method."
330	Clioquinol	2.42 8.16	U A	+H −H	CE/pH (−ve ion mode)	H_2O $t = 25$ $I = 0.025$	Wan H, Holmen AG, Wang Y, Lindberg W, Englund M, Nagard MB and Thompson RA, High-throughput screening of pK_a values of pharmaceuticals by pressure-assisted capillary electrophoresis and mass spectrometry, *Rapid Commun. Mass Spectrom.*, **17**, 2639–2648 (2003). NB: Reported predicted values (ACD Labs) of 2.32 and 7.23.
331	Clioquinol	GLpK_a: 2.96 ± 0.01 7.60 ± 0.01 A&S: No data	 U U	 +H −H	Spectro	H_2O $t = 25$ $I = 0.15$ (KCl) Ar atmosphere	Tam KY and Takacs-Novac K, Multi-wavelength spectrophotometric determination of acid dissociation constants, *Anal. Chim. Acta*, **434**, 157–167 (2001). NB: A new spectrophotometric method was validated by comparison with the standard method of Albert and Serjeant, using 10^{-5} M solutions in 0.15M KCl. No details of pH meter calibration. In the novel method, operational pH values were converted to p[H] by a multi-parametric equation (ref. 12, Avdeef A). Sample concentrations of 10–100 uM were titrated in a Sirius GLpK_a instrument and pH data were collected when drift was <0.02 pH/min. 15–25 wavelengths were used for each measurement. Note that errors are for reproducibility only, not accuracy.
332	Clioquinol	7.90	U	−H			Martell AE and Smith RM, *Critical Stability Constants*, Vol. 3, Plenum Press, NY (1977) . Not in M&S or in Foye, Ritschel, W&G, N&K.

(continued)

No.	Name	pK_a value(s)	Data quality	Ionization type	Method	Conditions t°C; I or c M	Comments and Reference(s)
333	Clofazimine ($C_{27}H_{22}Cl_2N_4$) 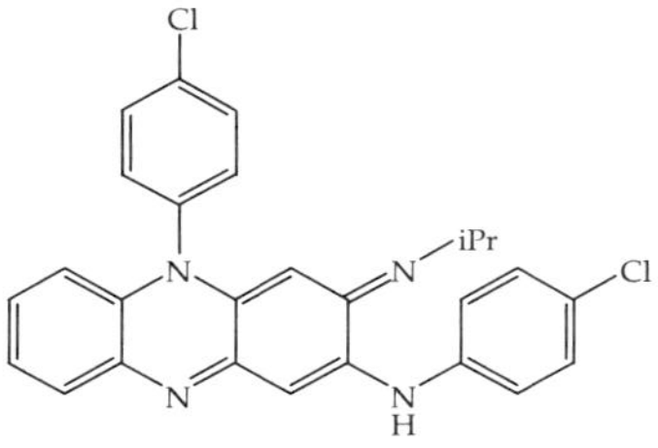	8.37	U	+H	Potentiometric	H_2O	O'Driscoll CM and Corrigan OI, Clofazimine, *APDS*, **21**, 75–108 (1992). Potentiometric method (Canavan B, Esmonde AG, Feely JP, Quigley JM, Timoney RF, The influence of lipophilic and steric properties on the transport of N2-substituted phenazines to spleen of mice following oral administration, *Eur. J. Med. Chem.*, **21**, 199–203 (1986)). NB: The paper gave no details of the potentiometric procedure beyond stating that the procedure followed that of Mizutani (1925). The Mizutani procedure used organic cosolvent mixtures to dissolve the compounds of interest, and ignored the effect of the solvent on pK_a when the percentage was low (<10–20%). No details of the cosolvent solution compositions was given. Also reported pK_a = 8.35, method unspecified (Morrison NE and Marley GM, Clofazimine binding studies with deoxyribonucleic acid, *Int. J. Lepr.*, **44**, 475–481 (1976)).
334	Clofazimine	8.511	U	+H	Spectro (λ = 482 nm)	H_2O t = 37 I unspecified	Quigley JM, Fahelelbom KMS, Timoney RF and Corrigan OI, Temperature dependence and thermodynamics of partitioning of clofazimine analogues in the n-octanol/water system, *Int. J. Pharm.*, **54**, 155–159 (1989); *ib.* **58**, 107–113 (1990).
335	Clofazimine	9.11	U	+H	CE/pH (+ve ion mode)	H_2O t = 25 I = 0.025	Wan H, Holmen AG, Wang Y, Lindberg W, Englund M, Nagard MB and Thompson RA, High-throughput screening of pK_a values of pharmaceuticals by pressure-assisted capillary electrophoresis and mass spectrometry, *Rapid Commun. Mass Spectrom.*, **17**, 2639–2648 (2003). NB: Reported a predicted value (ACD Labs) of 7.87.
336	Clofezone	4.5	U	−H	Spectro	H_2O t undefined I undefined	Herzfeldt CD and Kümmel R, Dissociation constants, solubilities, and dissolution rates of some selected nonsteroidal antiinflammatories, *Drug Dev. Ind. Pharm.*, **9(5)**, 767–793 (1983). NB: Used dA/dpH method. See Azapropazone and Ibuprofen for details.

Compound	pK_a	Compound	pK_a
H	8.480	CH(Me)$(CH_2)_3N(Et)_2$	8.662
$CHMe_2$ (clofazimine)	8.511	$(CH_2)_3$-N<>$(CH_2)_4$	8.596
$(CH_2)_2NEt_2$	8.850	$(CH_2)_3$-N<>$(CH_2)_5$	8.800
$(CH_2)_3NEt_2$	8.813	CH_2-CH<> [$(CH_2)_2NHCH_2CH_2$]	8.085

							NB: Martindale states that clofezone is a mixture of clofexamide (2-(4-chlorophenoxy)-N-2-(diethylaminoethyl)acetamide) and phenylbutazone. The value reported here appears to be for the phenylbutazone component, as the diethylaminoethyl group in clofexamide should have a value of about 7.5–7.8, and would not be expected to appear in a spectrophotometric study.
337	Clonidine ($C_9H_9Cl_2N_3$) 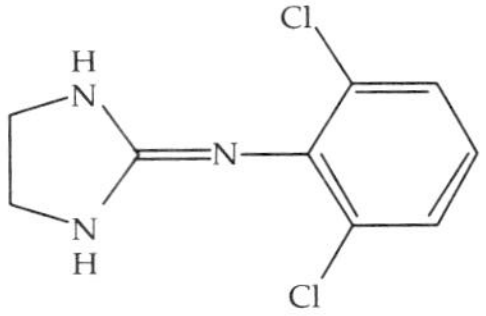	8.26	U	+H	CE/pH (+ve ion mode)	H_2O $t = 25$ $I = 0.025$	Wan H, Holmen AG, Wang Y, Lindberg W, Englund M, Nagard MB and Thompson RA, High-throughput screening of pK_a values of pharmaceuticals by pressure-assisted capillary electrophoresis and mass spectrometry, *Rapid Commun. Mass Spectrom.*, **17**, 2639–2648 (2003). NB: Reported a predicted value (ACD Labs) of 7.71.
338	Clonidine	GLpK_a: 8.12 ± 0.02	A	+H	Spectro	H_2O $t = 25$ $I = 0.15$ (KCl) Ar atmosphere	Tam KY and Takacs-Novac K, Multi-wavelength spectrophotometric determination of acid dissociation constants, *Anal. Chim. Acta*, **434**, 157–167 (2001).
		A&S: 8.04 ± 0.11	U	+H			NB: See Clioquinol for details.
339	Clopenthixol ($C_{22}H_{25}ClN_2OS$) 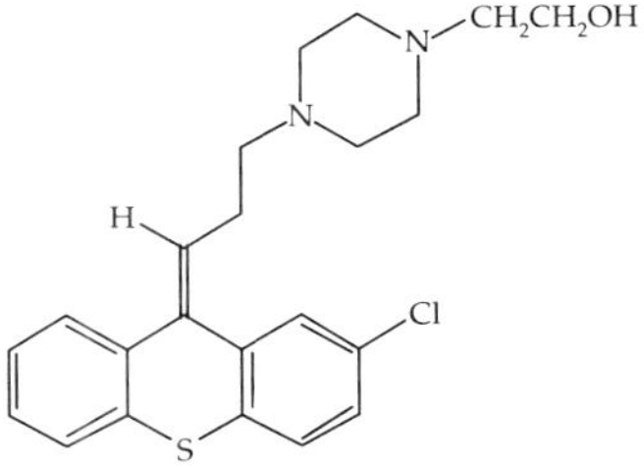	3.443	A	+H	Potentiometric	H_2O $I = 0.00$ $t = 25$	Lukkari S, Ionization of some thiaxanthene derivatives, clopenthixol and flupenthixol, in aqueous solutions, *Farm. Aikak.*, **80(4–5)**, 237–242 (1971).
		6.14	A	+H			"The acid ionization constants of clopenthixol HCl and flupenthixol HCl were determined potentiometrically in aqueous solutions at 25 °C. The values p$K_{a1,2}$ = 3.443, pK_{a1} = 6.14 are obtained for the acid ionization constant of clopenthixol and p$K_{a1,2}$ = 3.362, pK_{a1} = 6.18 for those of flupenthixol at zero ionic strength. The effect of the ionic strength on the ionization constants, as adjusted with sodium chloride, was determined."
340	Clorindione ($C_{15}H_9ClO_2$) 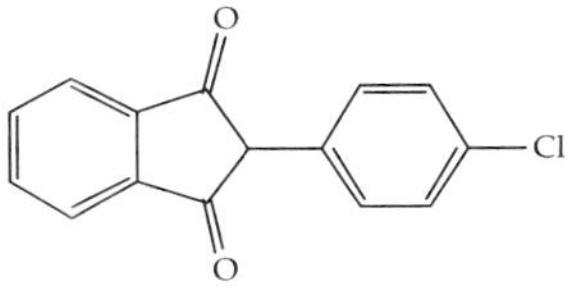	3.59	U	−H	Spectro	H_2O $I = 0.1$ $t = 25.0 \pm 0.1$	Stella VJ and Gish R, Kinetics and mechanism of ionization of the carbon acids 4′-substituted 2-phenyl-1,3-indandiones, *J. Pharm. Sci.*, **68(8)**, 1047–1049 (1979). NB: See Anisindione for details.

(continued)

No.	Name	pK_a value(s)	Data quality	Ionization type	Method	Conditions t°C; I or c M	Comments and Reference(s)
341	Clotrimazole ($C_{22}H_{17}ClN_2$) 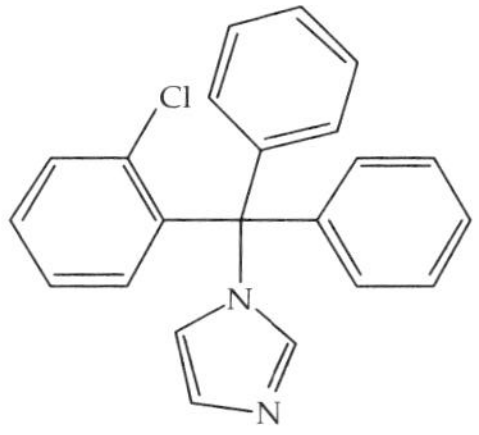	4.7	U	+H	Potentiometric	EtOH 50% aq.	Buechel KH, Draber W, Regel E and Plempel M, Synthesen und Eigenschaften von Clotrimazol und weiteren antimykotischen 1-Triphenylmethylimidazolen, *Arzneim.-Forsch.*, **22**, 1260–1272 (1972). NB: pH = pK_a at half-neutralization method. c = 5.3 mM titrated with 0.1 M HCl. Cited in Hoogerheide JG and Wyka BE, Clotrimazole, *APDS*, **11**, 225–255 (1982).
342	Cloxacillin ($C_{19}H_{18}ClN_3O_5S$) 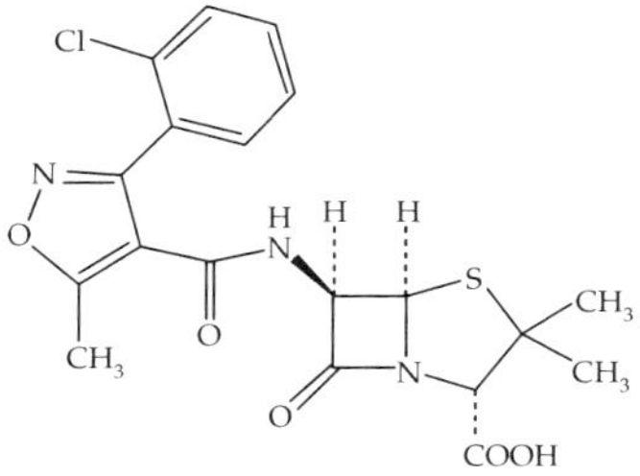	2.70 ± 0.07	U	−H	Potentiometric	H_2O t = 35.0 c = 0.0025	Mays DL, Cloxacillin sodium, *APDS*, **4**, 113–136 (1975). "Value calculated by averaging the data from Bundgaard and Ilver with that from Rapson and Bird. . . . Bundgaard and Ilver reported an apparent pK_a of 2.68 ± 0.05 at 35 °C, determined by measuring the pH of a partially neutralized 0.0025 M solution of sodium cloxacillin. Rapson and Bird obtained replicate apparent pK values of 2.73 ± 0.04 and 2.70 ± 0.03 at 25 °C by titrating 0.0025M sodium cloxacillin solutions." Bundgaard H and Ilver K, Kinetics of degradation of cloxacillin sodium in aqueous solution, *Dansk. Tidsskrift. Farm.*, **44**, 365–380 (1970). Rapson HDC and Bird AE, Ionization constants of some penicillins and of their alkaline and penicillinase hydrolysis products, *J. Pharm. Pharmacol.*, Suppl. 15, 222–231T (1963).
343	Clozapine ($C_{18}H_{19}ClN_4$) 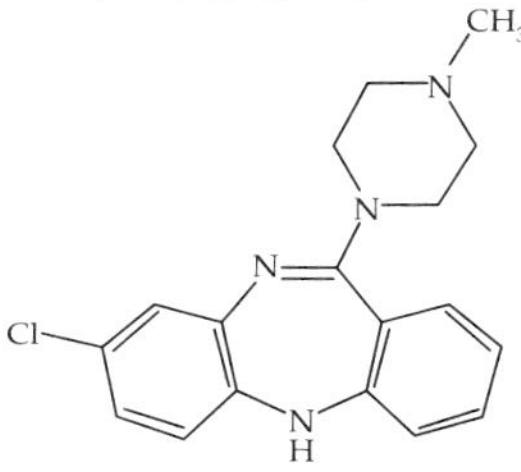	3.58 7.94	U U	+H +H	CE/pH (+ve ion mode)	H_2O t = 25 I = 0.025	Wan H, Holmen AG, Wang Y, Lindberg W, Englund M, Nagard MB and Thompson RA, High-throughput screening of pK_a values of pharmaceuticals by pressure-assisted capillary electrophoresis and mass spectrometry, *Rapid Commun. Mass Spectrom.*, **17**, 2639–2648 (2003). NB: Reported predicted values (ACD Labs) of 4.36 and 6.19.

No.	Compound	pKa	Data quality	Ionization	Method	Conditions	Comments and Reference(s)
344	Clozapine	7.50	U	+H		H_2O t undefined I undefined	El Tayar N, Kilpatrick GJ, van de Waterbeemd H, Testa B, Jenner P and Marsden CD, Interaction of neuroleptic drugs with rat striatal D-1 and D-2 dopamine receptors: a quantitative structure-affinity relationship study, *Eur. J. Med. Chem.*, **23**, 173–182 (1988). NB: Refered to El Tayar N, van de Waterbeemd H and Testa B, *J. Chromatogr.*, **320**, 305–312 (1985) for methodology.
345	Cobefrin (Nordefrin) ($C_9H_{13}NO_3$) 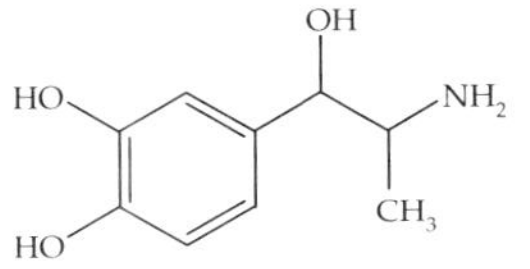	8.45	U	+H	Potentiometric	H_2O $t = 25.0 \pm 0.2$ $I \leq 0.001$	Leffler EB, Spencer HM and Burger A, Dissociation constants of adrenergic amines, *JACS*, **73**, 2611–2613 (1951). NB: See Amphetamine for details. Decomposes in aqueous solution.
346	Cocaine ($C_{17}H_{21}NO_4$) 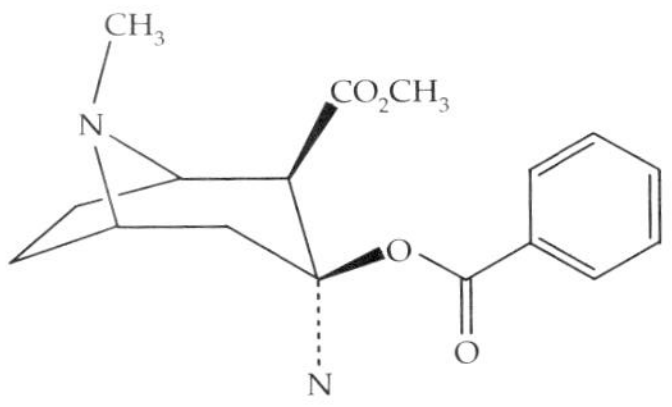	8.85 ± 0.03	A	+H	Potentiometric	H_2O $t = 20.0$ $I = 0.10$ (KCl) N_2 atmosphere	Buchi J and Perlia X, Beziehungen zwischen de physikalisch-chemische Eigenschaften und der Wirkung von Lokalanasthetica, *Arzneim.-Forsch.*, **10**, 745–754 (1960). NB: These were careful studies conducted by Dr. G. Anderegg at the Analytical Chemistry laboratories of the ETH Zurich (Prof. G. Schwarzenbach). Potentiometric titrations with calibrated glass and calomel electrodes were performed under a nitrogen atmosphere. Solutions were of constant ionic strength 0.1 M (KCl) at 20 °C. Acetic acid was used as a test substance, with reference to the very accurate measurements of Harned. Some measurements were also performed spectrophotometrically. The experimental error was typically ± 0.03
347	Cocaine	8.61	U	+H	Spectro	H_2O $t = 15$ ($pK_w = 14.35$ at 15 °C)	Kolthoff IM, The dissociation constants, solubility product and titration of alkaloids, *Biochem. Z.*, **162**, 289–353 (1925). Cited in Perrin Bases 2878. NB: See Aconitine for details.
		8.39	U	+H	Potentiometric	H_2O $t = 24$	NB: The potentiometric study (Muller F, *Z. Elektrochem*, **30**, 587, (1924)) used pH measurements in an asymmetric cell with a hydrogen electrode and liquid junction potentials.
348	Codeine ($C_{18}H_{21}NO_3$) 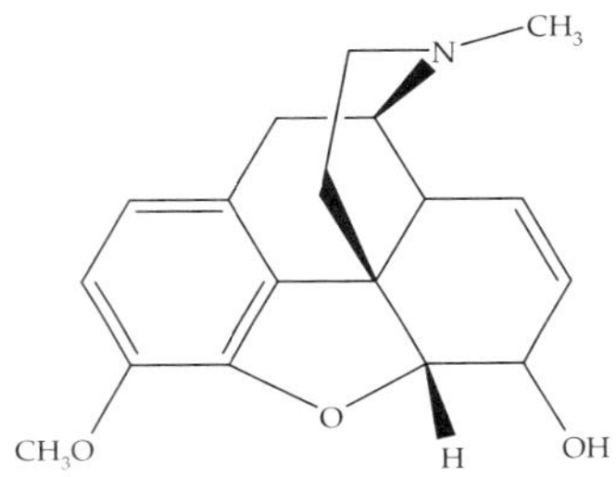	8.22 ± 0.01	A	+H	Potentiometric	H_2O $t = 25.0 \pm 0.1$ $I = 0.15$ (KCl) Ar atmosphere	Avdeef A, Barrett DA, Shaw PN, Knaggs RD and Davis SS, Octanol-, chloroform-, and propylene glycol dipelargonat-water partitioning of morphine-6-glucuronide and other related opiates, *J. Med. Chem.*, **39**, 4377–4381 (1996). NB: See Morphine for further details NB: The same value was reported in: Takacs-Novak K and Avdeef A, Interlaboratory study of log P determination by shake-flask and potentiometric methods, *J. Pharm. Biomed. Anal.*, **14**, 1405–1413 (1996).

(*continued*)

No.	Name	pK_a value(s)	Data quality	Ionization type	Method	Conditions t°C; I or c M	Comments and Reference(s)
349	Codeine	8.21	A	+H	Conductance	H_2O $t = 25$ $\kappa < 1.5$	Oberst FW and Andrews HL, The electrolytic dissociation of morphine derivatives and certain synthetic analgetic compounds, *JPET*, **71**, 38–41 (1941). Cited in Perrin Bases 2879 ref O1. NB: Results were reported as K_b values. For codeine, $K_b = 1.61 \times 10^{-6}$, giving $pK_b = 5.79$. The potentiometric study used an asymmetric cell with diffusion potentials.
		8.25	U	+H	Potentiometric	H_2O $t = 20$ $c = 0.01 - 0.03$M	NB: Also gave 8.04 (U, potentiometric with glass electrode and liquid junction potentials; H_2O); 8.15 (U, spectro; H_2O; $t = 15$; $c = 0.005$–0.01 M); 8.17 (U; quinhydrone electrode).
350	Codeine	8.25 ± 0.01	A	+H	Potentiometric	H_2O $t = 25.0$ $I = 0.1$ (NaCl)	Takacs-Novak K and Avdeef A, Interlaboratory study of log P determination by shake-flask and potentiometric methods, *J. Pharm. Biomed. Anal.*, **14**, 1405–1413 (1996). NB: See Acetaminophen for further details.
351	Codeine	8.21	A	+H	CE/pH (+ve ion mode)	H_2O $t = 25$ $I = 0.025$	Wan H, Holmen AG, Wang Y, Lindberg W, Englund M, Nagard MB and Thompson RA, High-throughput screening of pK_a values of pharmaceuticals by pressure-assisted capillary electrophoresis and mass spectrometry, *Rapid Commun. Mass Spectrom.*, **17**, 2639–2648 (2003). NB: Reported a predicted value (ACD Labs) of 8.25.
352	Codeine	8.18 ± 0.02	U	+H	Potentiometric	H_2O $t = 20$ $I < 0.01$	Kaufman JJ, Semo NM and Koski WS. Microelectrometric titration measurement of the pK_a's and partition and drug distribution coefficients of narcotics and narcotic antagonists and their pH and temperature dependence, *J. Med. Chem.*, **18**, 647–655 (1975).
		8.10 ± 0.02	U	+H		H_2O $t = 37$ $I < 0.01$ N_2 atmosphere	"... a small sample cell in which 0.04 mmol of the sample is dissolved in 5 ml of CO_2-free triply distilled water. A constant temperature circulator is employed to circulate water continuously through a jacket surrounding the ... cell. ... N_2 gas is maintained over the sample .... A pH meter type PH64 Radiometer utilizing calomel and glass electrodes is used .... A microburet attached to a microdelivery tip enables one to accurately add small quantities of titrant (2 N NaOH). A constant speed magnetic stirrer enables proper agitation of the sample solution and an externally mounted light source enables one to observe the first signs of any light scattering ... due to formation of a precipitate."
353	Codeine	9.27	U	+H		$t = 37$	Ballard BE and Nelson E, Physicochemical properties of drugs that control absorption rate after subcutaneous implantation, *JPET*, **135**, 120–127 (1962). NB: From $pK_b = 6.05$, $pK_w = 13.621$ at 37 °C.

No.	Compound	pKa	Data quality	Assignment	Method	Conditions	Reference / Notes
354	Colchicine ($C_{22}H_{25}NO_6$) 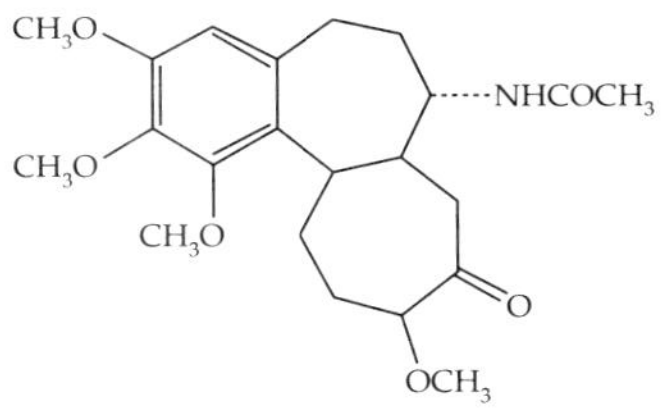	1.85	U	+H	Spectro	H_2O $t = 15$ $c = 0.03 - 0.05$	Kolthoff IM, The dissociation constants, solubility product and titration of alkaloids, *Biochem. Z.*, **162**, 289–353 (1925). Cited in Perrin Bases 2880. See Aconitine for details. NB: $pK_w = 14.35$ at 15 °C.
355	Coumermycin A_1 ($C_{55}H_{59}N_5O_{20}$) 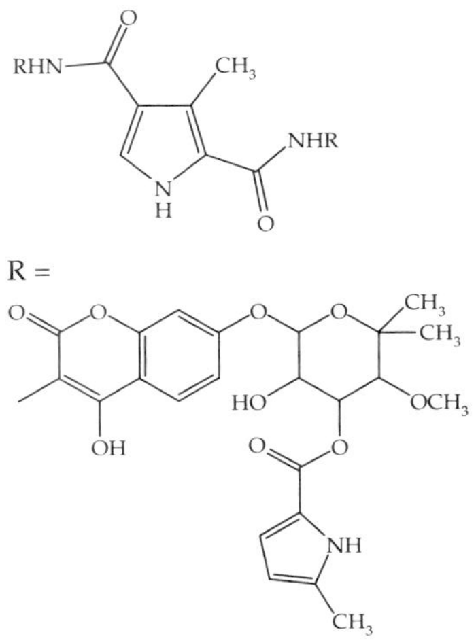	6.0 ± 0.2	U	–H (2x)		H_2O	Newmark HL and Berger J, Coumermycin A_1 - Biopharmaceutical studies, *J. Pharm. Sci.*, **59**, 1246–1248 (1970).
		6.35 7.76	U U	–H (2x) –H (2x)		75% aq DMF 75% aq dioxane	NB: Each value reported is a mean of two values. The values in partially-aqueous solvent mixtures are from: Kawaguchi H, Tsukiura H, Okanishi M, Miayaki T, Ohmori T, Fujisawa K and Koshiyama H, Coumermycin, a new antibiotic. I., *J. Antibiot. Ser. A*, **18**, 1–10 (1968); Kawaguchi H, Naito T and Tsukiura H, Studies on coumermycin, a new antibiotic. II., *ib.*, 11–25 (1968).

(*continued*)

No.	Name	pK_a value(s)	Data quality	Ionization type	Method	Conditions t°C; I or c M	Comments and Reference(s)
356	Creatine ($C_4H_9N_3O_2$)	2.79 ± 0.03 12.1 ± 0.09	U U	−H +H	Potentiometric	H_2O $t = 25.0$	Breccia A, Fini A, Girotti S and Stagni G, Correlation between physico-chemical parameters of phosphocreatine, creatine, and creatinine, and their reactivity with their potential diffusion in tissue, *Pharmatherapeutica*, **3**(**4**), 227–232 (1982). "The method utilized consisted of potentiometric titration with 0.1 N HCl or NaOH. The instrumentation used was a PHM 26 Radiometer potentiometer with glass and saturated calomel electrodes, standardized with phosphate buffer at pH 6.865, according to Robinson and Stoke's description, at 25 °C. The pK_a values obtained were not corrected for ionic strength. From the titrimetric curves, the values reported in Table 2 were obtained. The values in parenthesis were derived from the pK_a values of compounds having only one functional group in the molecule.
357	Creatinine ($C_4H_7N_3O$)	4.68 ± 0.02	U	+H	Potentiometric	H_2O $t = 25.0$	Breccia A, Fini A, Girotti S and Stagni G, Correlation between physico-chemical parameters of phosphocreatine, creatine, and creatinine, and their reactivity with their potential diffusion in tissue, *Pharmatherapeutica*, **3**(**4**), 227–232 (1982). NB: See Creatine for details.
358	Cresol red ($C_{21}H_{18}O_5S$)	1.05	U	−H	Spectro	H_2O RT	Das Gupta V and Reed JB Jr., First pK_a values of some acid-base indicators, *J. Pharm. Sci.*, **59**, 1683–1685 (1970). NB: See Bromocresol green.

Compound	pK_a		
Phosphocreatine	11.0 ± 0.05	4.7 ± 0.01	(-0.31)
Creatine	12.07 ± 0.09	2.79 ± 0.03	
Creatinine		4.86 ± 0.02"	

No.	Compound	pK	Rating	Type	Method	Conditions	Reference
359	Cyanocobalamin ($C_{63}H_{88}CoN_{14}O_{14}P$)	3.28 ± 0.04	U	–H	^{1}H-NMR	H_2O, D_2O $t = 23 \pm 0.5$ I unspecified	Brodie JC and Poe M, Proton magnetic resonance of vitamin B12 derivatives, *Biochemistry*, **11**, 2534–2542 (1972). Cited in: Kirshchbaum J and Cyanocobalamin, *APDS*, **10**, 183–235 (1981). "^{1}H-NMR; Using the dependence of the proton-NMR chemical shift on pH, base atom B-2 . . . gave a pK of 3.28 ± 0.04. This value is in excellent agreement with the previously-reported value of 3.3 (ref. 122). pK values for cobalamins and cobinamides have been discussed. The limiting conductance of the cyanocobalamin ion is 33 mhos. 122. Hill JA, Pratt JM and Williams RJP, The corphyrins, *J. Theor. Biol.*, **3**, 423–445 (1962)." NB: Calibration of the glass electrode was performed in such a way that it was claimed that pK results in H_2O were identical to those in D_2O. Results reported here are for measurements in D_2O.
360	2-Cyanoguanidinophenytoin ($C_{16}H_{12}N_4O$)	5.3	U	–H	Spectro (λ = 236 nm)	2.5% aq. MeOH	Lambert DM, Masereel B, Gallez B, Geurts M and Scriba GK, Bioavailability and anticonvulsant activity of 2-cyanoguanidinophenytoin, a structural analog of phenytoin, *J. Pharm. Sci.*, **85**, 1077–1081 (1996). NB: No details were given of temperature or ionic strength.
361	Cyclacillin ($C_{15}H_{23}N_3O_4S$)	2.64 7.18	A U	–H +H	Potentiometric Potentiometric	H_2O $t = 37$ $I = 0.5$ (KCl)	Tsuji A, Nakashima E, Hamano S and Yamana T, Physicochemical properties of amphoteric β-lactam antibiotics, *J. Pharm. Sci.*, **67**, 1059–1066 (1978). NB: See Amoxicillin.

(continued)

No.	Name	pK_a value(s)	Data quality	Ionization type	Method	Conditions t°C; I or c M	Comments and Reference(s)
362	Cyclacillin	2.68 ± 0.04 7.50 ± 0.02	A U	−H +H	Potentiometric	H_2O $t = 25$	Hou JP and Poole JW, The aminoacid nature of ampicillin and related penicillins, *J. Pharm. Sci.*, **58**, 1510–1515 (1969). NB: See Ampicillin for details.
363	Cyclamic acid ($C_6H_{13}NO_3S$) 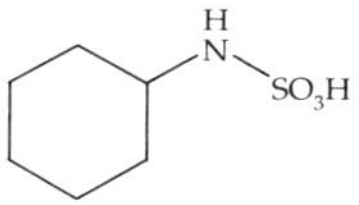	2.28	U	−H	Potentiometric	H_2O t = RT $c = 0.056$	Talmage JM, Chafetz L and Elefant M., Observation on the instability of cyclamate in hydro-alcoholic solution, *J. Pharm. Sci.*, **57**, 1073–1074 (1968). NB: The paper also gave the following apparent pK_a values in various %v/v ethanol-water mixtures: 20, 2.48; 40, 2.60; 60, 2.90; 80, 3.22; 100, 3.45.
364	Cycliramine ($C_{18}H_{15}ClN_2$) 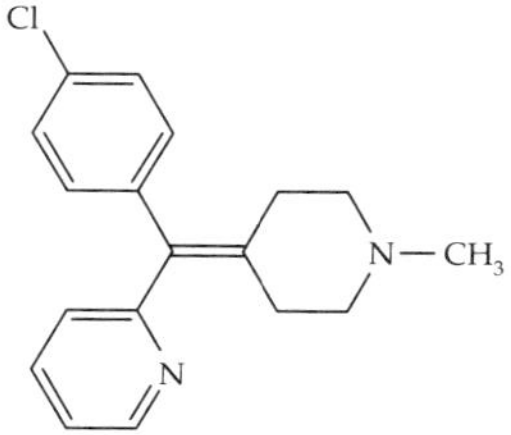	3.64 ± 0.12 8.78 ± 0.02	U U	+H +H	Potentiometric	H_2O t undefined I = 0.30 (NaCl)	Testa B and Murset-Rossetti L, The partition coefficient of protonated histamines, *Helv. Chim. Acta*, **61**, 2530–2537 (1978). "The amines (free base, 3 – 5 × 10^{-4} mol) were dissolved in 20 mL 0.05 N HCl + 100 mL 2% NaCl solution. The solutions (ionic strength 0.30) were titrated with 0.1N NaOH using a Metrohm E356 potentiograph . . . The pK_a values were calculated from the titration curves by the method of Benet and Goyan. . .. This method yields the stoichiometric dissociation constant."
365	Cyclizine ($C_{18}H_{22}N_2$) 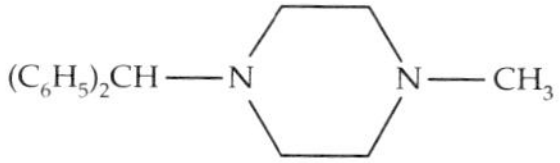	2.16 ± 0.02 8.05 ± 0.03	A A	+H +H	Potentiometric	H_2O $t = 24.5 \pm 0.5$	Newton DW, Murray WJ and Lovell MW, pK_a determination of benzhydrylpiperazine antihistamines in aqueous and aqueous methanol solutions, *J. Pharm. Sci.*, **71(12)**, 1363–1366 (1982). NB: See Chlorcyclizine for details.
366	Cyclizine	1.88 ± 0.12 8.32 ± 0.02	U U	+H +H	Potentiometric	H_2O t undefined I = 0.30 (NaCl)	Testa B and Murset-Rossetti L, The partition coefficient of protonated histamines, *Helv. Chim. Acta*, **61**, 2530–2537 (1978). NB: See Cycliramine for details.
367	Cyclohexaamylose (α-cyclodextrin) ($C_{36}H_{60}O_{30}$)	12.36	U	+H		H_2O $t = 30$	Gelb RI, Schwartz LM, Bradshaw JJ and Laufer DA, Acid dissociation of cyclohexamylose and cycloheptaamylose, *Bioorg. Chem.*, **9**, 299–304 (1980). NB: Cited in Larsen C, Macromolecular prodrugs. XIII. Determination of the ionization constant of dextran by potentiometric titration and from kinetic analysis of the hydrolysis of dextran indomethacin ester conjugates, *Int. J. Pharm.*, **52**, 55–61 (1989).

368	1-Cyclohexyl-2-aminopropane ($C_9H_{19}N$) 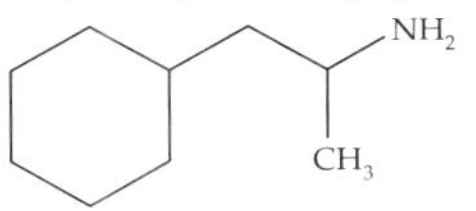	10.14 ± 0.1	U	+H	Potentiometric	H_2O $t = 25.0 \pm 0.2$ $I \leq 0.001$	Leffler EB, Spencer HM and Burger A, Dissociation constants of adrenergic amines, *JACS*, **73**, 2611–2613 (1951). NB: See Amphetamine for details.
369	Cyclopentamine ($C_9H_{19}N$) 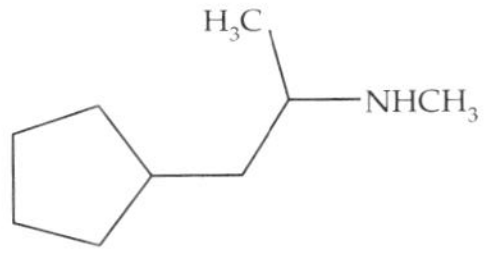	10.47	U	+H	Potentiometric	H_2O t undefined $I \sim 0.01$	Chatten LG and Harris LE, Relationship between $pK_b(H_2O)$ of organic compounds and $E_{1/2}$ values in several nonaqueous solvents, *Anal. Chem.*, **34**, 1495–1501 (1962). NB: From reported $pK_b = 3.53$ (assumed $t = 25\,^\circ C$).
370	Cyclopentolate ($C_{17}H_{25}NO_3$) 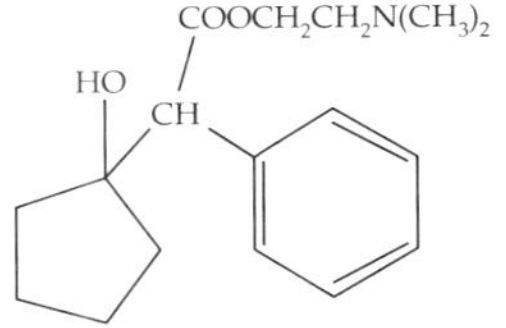	7.93	U	+H	Potentiometric	H_2O	Wang ESN and Hammarlund ER, Corneal absorption reinforcement of certain mydriatics, *J. Pharm. Sci.*, **59**, 1559–1563 (1970). "An aqueous solution of known concentration of cyclopentolate HCl was neutralized with stoichiometric quantity of standardized NaOH solution to liberate free cyclopentolate. A titration curve of the free cyclopentolate was obtained by titrating the neutralized solution with standardized dilute HCl, using a Beckman Zeromatic pH meter. The mid-point of the titration curve was ascertained, and the pK_a value of cyclopentolate HCl was determined by finding the pH value that corresponded to this midpoint."
371	D-Cycloserine ($C_3H_6N_2O_2$) 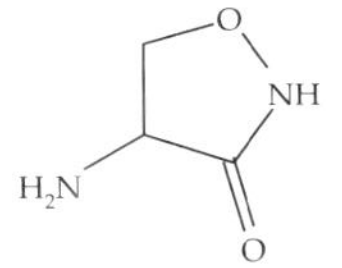	4.388 7.346		−H +H	Potentiometric	H_2O $t = 25.0$ $I = 0.1$	El-Obeid HA and Al-Badr AA, D-cycloserine, *APDS*, **18**, 567–597 (1989) . "Braibante *et al.* studied the equilibrium of D-cycloserine with protons . . . in aqueous solutions, the equivalent of D-cycloserine (HL) with the ions of H, Co, Ni, Cu, and Zn were studied potentiometrically at 25 °C and 0.1 mol/dm^3 KCl. The protonation constants are $\log K_1 = 7.346$ (5)(-NH3+) and $\log K_2 = 4.388$ (6)(-OH); the corresponding enthalpy changes are -32.25 (15) and -14.52 (15) KJ/mol respectively."

(continued)

No.	Name	pK_a value(s)	Data quality	Ionization type	Method	Conditions t°C; I or c M	Comments and Reference(s)
372	*iso*-Cyclosporin A ($C_{62}H_{111}N_{11}O_{12}$) (cyclosporin) 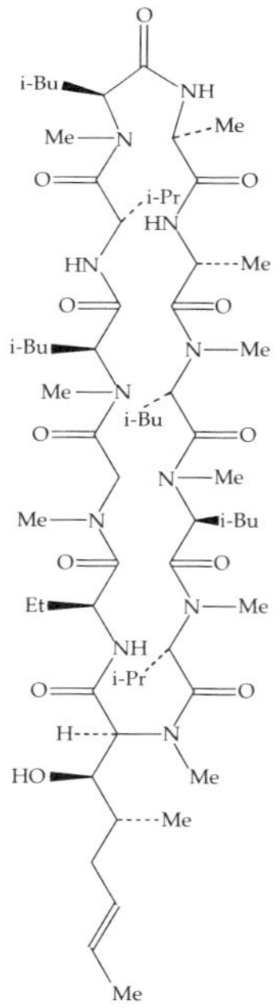	6.9	U	+H	kinetic	H_2O $t = 37.0 \pm 0\,0.2$ $I = 0.15$ (KCl)	Oliyai R and Stella VJ, Kinetics and mechanism of isomerization of cyclosporin A, *Pharm. Res.*, **9**, 617–622 (1992). "The kinetics of isomerization of cyclosporine (cyclosporin A) to isocyclosporin A were studied in various nonaqueous solvents as a function of temperature and added methanesulfonic acid; the rate of conversion of isocyclosporin to cyclosporine was also determined. The isomerization of cyclosporine was acid-catalyzed over the acid concentration studied. The choice of organic solvent markedly altered the isomerization rate. Isocyclosporin isomerization was extremely rapid compared to the forward reaction. Isocyclosporin showed a kinetically generated pK_a value of 6.9 for the secondary amine moiety. From pH 8–10, the pH-rate profile was linear, with a slope almost equal to unity, indicating hydroxide ion catalysis. The isomerization rate in plasma was comparable to that in pH 7.4 buffer solution. It was concluded that isomerization kinetics of cyclosporine in nonaqueous solution are acid-catalyzed and affected by the choice of organic solvent. . . . Due to the minimal aqueous solubility and lack of appropriate chromophoric properties of isoCsA, no attempt was made to determine the pK_a of isoCsA independently. . . . The kinetics of isoCsA to CsA coversion were studied in dilute aqueous solution as a function of pH, temperature and buffer concentration. . . . It was not possible to measure the apparent pK_a of isoCsA by other means . . . The kinetically generated pK_a value of 6.9 is reasonable for an aliphatic amine of the type seen . . . ."
373	Cyclothiazide ($C_{14}H_{16}ClN_3O_4S_2$) 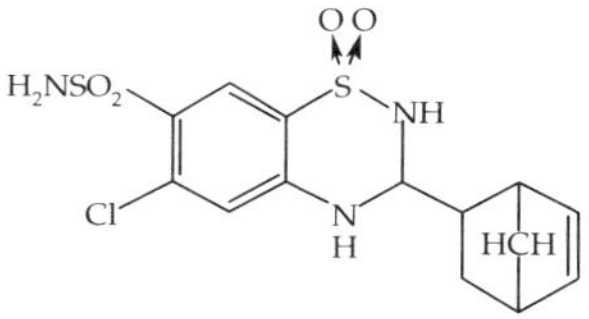	9.1 10.5	U U	−H −H	Potentiometric	30% EtOH in H_2O	Novello FC and Sprague JM, Structure-activity relations among the thiazide diuretics, *Ind. Chim. Belge.*, **32** (Special no.), 222–225 (1967). NB: Apparent pH at half-neutralization. NB: See also Whitehead CW and Traverso JJ, US Pat. 3,419,552 (1968), who reported 11.0 to 11.4 and 13.0 to 13.3 in 66% DMF.

No.	Compound	pKa	Grade	Type	Method	Conditions	Reference / Comments
374	Cyclothiazide	8.8	U	−H	Potentiometric	acetone/H_2O	Henning VG, Moskalyk RE, Chatten LG and Chan SF, Semi-aqueous potentiometric determination of apparent pK_{a1} values for benzothiadiazines and detection of decomposition during solubility variation with pH studies, *J. Pharm. Sci.*, **70(3)**, 317–319 (1981). NB: See Flumethiazide for details.
375	Cysteine ($C_3H_7NO_2$)	1.88 8.38 ± 0.02	U A	−H +H	Potentiometric	H_2O $t = 20.0 \pm 0.05$ $I = 0.15$ N_2 atmosphere	Zucconi TD, Janauer GE, Donahe S and Lewkowicz C, Acid dissociation and metal complex formation constants of penicillamine, cysteine, and antiarthritic gold complexes at simulated biological conditions, *J. Pharm. Sci.*, **68(4)**, 426–432 (1979). "The pH measurements were carried out in a 100-ml jacketed titration cell fitted with a magnetic stirrer and rubber stopper through which were inserted nitrogen and buret delivery tubes and glass and calomel electrodes. A research pH meter was used; the system was calibrated at pH 4 and 9.22. The ionic strength was 0.15 in all cases (KNO_3). The temperature was held to ± 0.05° of the desired value. . . . Each acid function was titrated with increments of 0.1000 M KOH, carbonate free." NB: The pK_a value for benzoic acid test substance was found in a validation experiment to be 4.25 ± 0.02 at 25 °C, slightly higher than the literature value (4.21).

SH, NH_2, COOH

Acid Dissociation Constants

	pK_a at 20°, μ = 0.15 (lit. value)			pK_a at 37°, μ = 0.15		
Compound	COOH	NH_3	SH	COOH	NH_3	SH
Penicillamine	2	8.11 ± 0.02 (8.03)	10.82 ± 0.04 (10.83)	–	7.83 ± 0.01	10.34 ± 0.04
Cysteine	1.88	8.38 ± 0.02 (8.32)	10.60 ± 0.03 (10.48)	–	8.04 ± 0.02	10.21 ± 0.06
Thiomalic Acid	4.68 ± 0.03 (4.68)	–	10.51 ± 0.03 (10.55)	4.68 ± 0.04	–	10.24 ± 0.02
Thioglucose	–	–	11.67 ± 0.05	–	–	11.51 ± 0.03

(continued)

No.	Name	pK_a value(s)	Data quality	Ionization type	Method	Conditions t°C; I or c M	Comments and Reference(s)
376	Daunorubicin ($C_{27}H_{29}NO_{10}$) 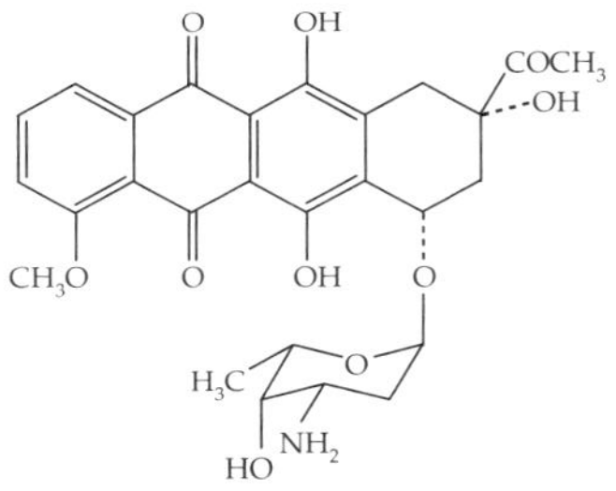	4.92 ± 0.16	U	+H	CZE/pH	H_2O $t = 25 \pm 0.5$ $I = 0.03$	Hu Q, Hu G, Zhou T and Fang Y, Determination of dissociation constants of anthracycline by capillary zone electrophoresis with amperometric detection, *J. Pharmaceut. Biomed. Anal.*, **31**, 679–684 (2003). " $$\mathrm{p}K_\mathrm{a} = \mathrm{pH} - \log\left(\frac{\mu_e}{\mu_a - \mu_e}\right) + \frac{0.5805z^2\sqrt{I}}{1 + 0.3281a\sqrt{I}} \quad (1)$$ $$\mathrm{p}K_\mathrm{a} = \mathrm{pH} + \log\left(\frac{\mu_e}{\mu_b - \mu_e}\right) - \frac{0.5805z^2\sqrt{I}}{1 + 0.3281a\sqrt{I}} \quad (2)$$ where μ_a is the electrophoretic mobility of the fully deprotonated acid, μ_b the electrophoretic mobility of the fully protonated base and μ_e is the electrophoretic mobility observed at the experimental pH. . . . Before each injection, the capillary was flushed with the . . . buffer solution for 5 min. Phenol . . . was chosen as a neutral marker . . .. The neutral marker was dissolved in deionized water (0.1 mol/l) and added to the sample solutions to measure electro-osmotic flow (EOF). The concentraton of neutral marker was adjusted to obtain a measurable reference peak. . . ., a separation potential of 14 kV was used. All samples were injected electro-kinetically, applying 14 kV for 10s. Data pairs of pH and μ_e were imported into MATHCAD . . . where μ_e (?μ_b) and pK_a were determined by performing a non-linear fit to Eqs. (1) and (2). Spectrophotometric pK_a values . . . were determined . . . using 1 cm cuvettes at 25 ± 0.5 °C. The experimental results showed that the three analytes were fully protonated and deprotonated at about pH 2.50 and 9.00, respectively. . . .. Absorbance measurements were taken at a wavelength which showed a significant difference as a function of pH for each solute, . . ." NB: Corrections for ionic strength effects were made with the Debye-Huckel equations as illustrated by Eqs (1) and (2). The meaning of these results must be considered very carefully. The authors have treated the ionizing group for daunorubicin (and its congeners, idarubicin, and pharmorubicin) as a base, for the raw results showed decreasing mobility with increasing pH, as expected where the more mobile form was protonated and the less mobile form was deprotonated. However, the results are in serious disagreement with all other anthracycline pK_a values. It is possible that a systematic error has been made in the calculations, and that the results are really pK_b values.
		4.99 ± 0.06	U	+H	Spectro ($\lambda = 495$ nm)	H_2O $t = 25 \pm 0.5$ $I = 0.03$	
	Idarubicin ($C_{26}H_{27}NO_9$) 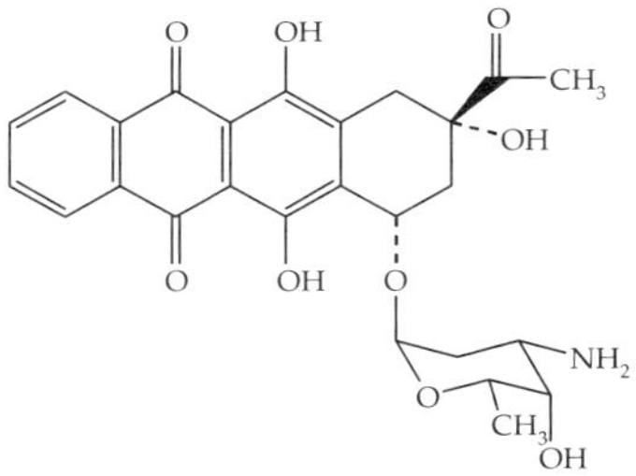	4.73 ± 0.21	U	+H	CZE/pH	H_2O $t = 25 \pm 0.5$ $I = 0.03$	
		4.64 ± 0.03	U	+H	Spectro ($\lambda = 495$ nm)	H_2O $t = 25 \pm 0.5$ $I = 0.03$	
	Pharmorubicin ($C_{27}H_{29}NO_{11}$) 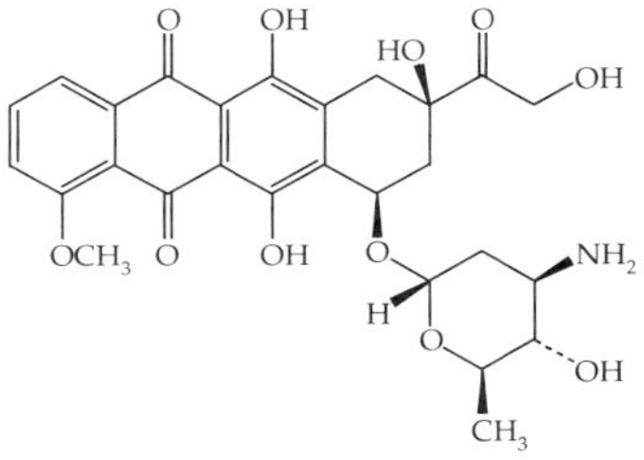	4.81 ± 0.13	U	+H	CZE/pH	H_2O $t = 25 \pm 0.5$ $I = 0.03$	
		4.92 ± 0.03	U	+H	Spectro ($\lambda = 495$ nm)	H_2O $t = 25 \pm 0.5$ $I = 0.03$	

No.	Compound	pKa	Rating	Type	Method	Conditions	Reference / Notes
377	Daunorubicin	8.2	U	+H			Couvreur P, Kante B, Roland M, Guiot P, Bauduin P and Speiser P, Polycyanoacrylate nanocapsules as potentional lysoomotropic carriers: preparation, morphological, and sorptive properties, *J. Pharm. Pharmacol.*, **31**, 331–332 (1979). NB: Value quoted without reference or experimental detail. There should be at least one –H value.
378	Decyl carnitine ($C_{17}H_{33}NO_4$) 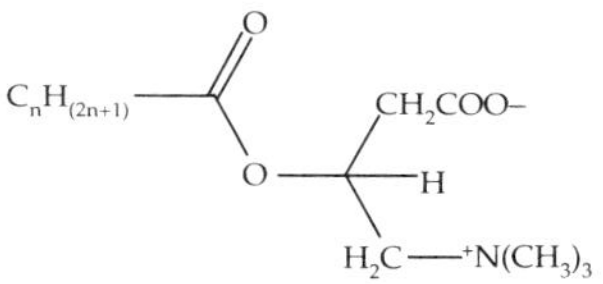	3.65 ($n = 9$) 3.60 ($n = 1$) 3.56 ($n = 3$) 3.60 ($n = 7$)	A	–H	Potentiometric	H_2O $t = 25$ $I < 0.02$ N_2 atmos CO_2-free H_2O	Yalkowsky SH and Zografi G, Potentiometric titration of monomeric and micellar acylcarnitines, *J. Pharm. Sci.*, **59**, 798–802 (1970). NB: Compound purity was checked by ensuring that surface tension *vs* concentration plots did not display minima. The paper has a detailed discussion on the effects of micellization on pK_a values. Related pK_a values:

Compound	pK_a	Compound	pK_a
Butyric acid ($C_4H_8O_2$)	4.83	Carnitine ($C_7H_{15}NO_3$) (as chloride)	3.80
γ-Butyrobetaine (GBB) ($C_7H_{15}NO_2$)	4.02	Norcarnitine ($C_6H_{13}NO_3$) (as chloride)	3.81
γ-Aminobutyric acid (GABA) ($C_4H_9NO_2$)	4.01	β-hydroxy-γ-aminobutyric acid ($C_4H_9NO_3$) (as cation)	3.80

No.	Compound	pKa	Rating	Type	Method	Conditions	Reference / Notes
379	Dehydroestrone, 6- ($C_{18}H_{20}O_2$)	10.17 ± 0.04	U	–H	Spectro	H_2O $t = 23 \pm 2$	Hurwitz AR, Liu ST, Determination of aqueous solubility and pK_a values of estrogens, *J. Pharm. Sci.*, **66**, 624–627 (1977). NB: See 17α-Estradiol for details.

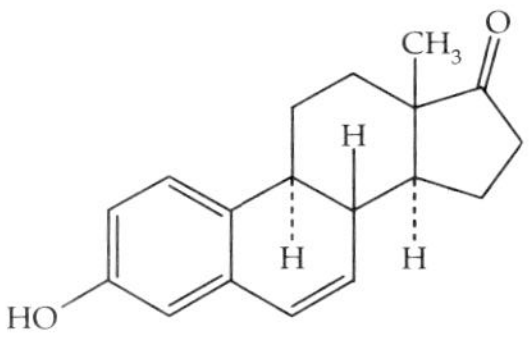

(*continued*)

No.	Name	pK_a value(s)	Data quality	Ionization type	Method	Conditions t°C; I or c M	Comments and Reference(s)
380	Demeclocycline (6-demethyl-7-chlorotetracycline) ($C_{21}H_{21}ClN_2O_8$) 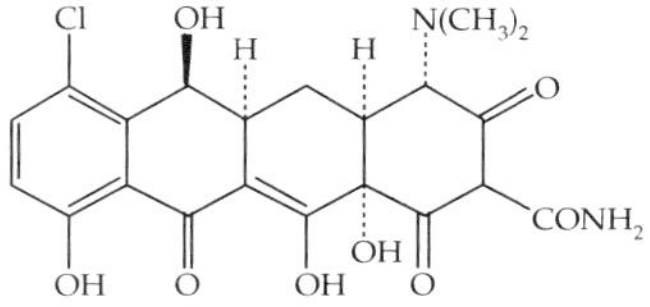	3.30 7.16 9.25	U U U	−H −H +H	Potentiometric	H_2O $t = 25 \pm 0.05$ $I = 0.01$	Benet LZ and Goyan JE, Determination of the stability constants of tetracycline complexes, *J. Pharm. Sci.*, **54**, 983–987 (1965). NB: See Chlortetracycline for details.
381	Demeclocycline (6-demethyl-7-chlorotetracycline)	3.85 7.31 9.23	U U U	−H −H +H	Potentiometric	H_2O $t = 30.0 \pm 0.2$ $I = 0.01$ (KCl) N_2 atmosphere	Doluisio JT and Martin AN, Metal complexation of the tetracycline hydrochlorides, *J. Med. Chem.*, **6**, 16–20 (1963). NB: Metal-free solutions of the tetracycline were titrated with standard NaOH solution and the pH measured. No details were given of the pH meter calibration. Metal stability constants were determined from identical titrations in the presence of varying concentrations of nickel(II), zinc(II) or copper(II) ions.
382	Deoxyepinephrine (Epinine) ($C_9H_{13}NO_2$) 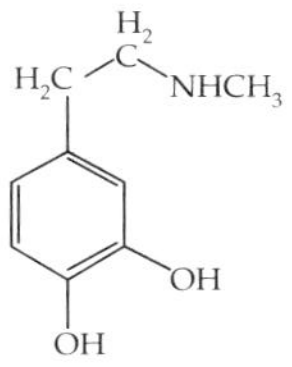	8.64 ± 0.04 9.70 ± 0.12	A U	−H, +H +H, −H	Potentiometric, Spectro	H_2O $t = 37$ $I = 1.0$	Schüsler-van Hees, MTIW, Reactivity of Catechol(amine)s, Ph.D. thesis, Ctr for Biopharm. Sci., Leiden University, 1–93 (1983). NB: Microscopic: 8.79 ± 0.01; 9.16 ± 0.11; 9.55 ± 0.16; 9.18 ± 0.04. Schüsler-van Hees, MTIW, Beijersbergen van Henegouwen GMJ and Driever MFJ, Ionization constants of catechols and catecholamines, *Pharm. Weekblad*, Sci. Edn., **5**, 102–108 (1983).
383	Deramciclane ($C_{19}H_{31}NO$) 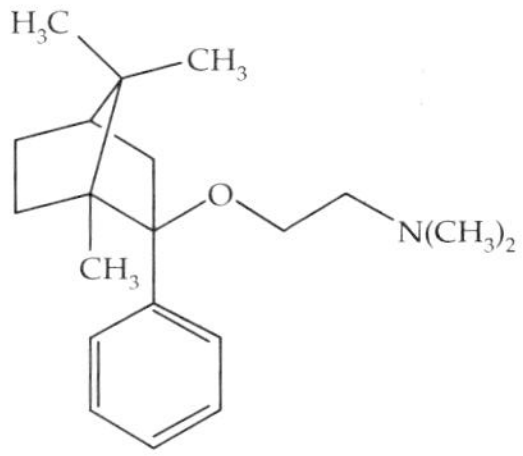	9.61 ± 0.03	U	+H	Potentiometric	H_2O $t = 25.0 \pm 0.1$ $I = 0.1$ (NaCl)	Takacs-Novak K, Box KJ and Avdeef A, Potentiometric pK_a determination of water-insoluble compounds: Validation study in methanol/water mixtures, *Int. J. Pharm.*, **151**, 235–248 (1997). NB: By extrapolation from 34–53 %w/w aqueous MeOH. See Acetaminophen for full details.

384	Desferrioxamine ($C_{25}H_{48}N_6O_8$) 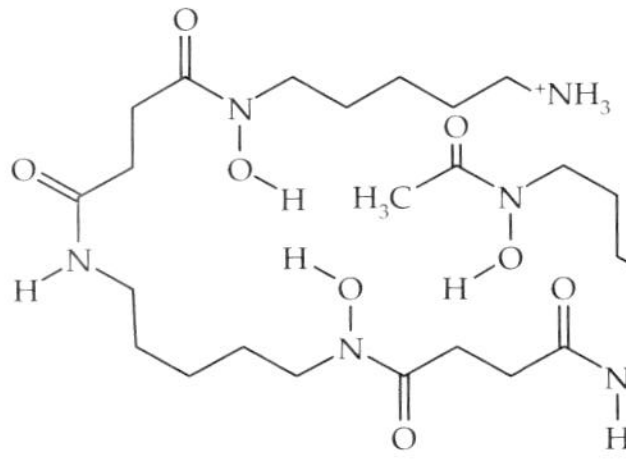	8.32 ± 0.07 9.16 ± 0.02 9.94 ± 0.05 11.44 ± 0.12	U A U U	−H −H −H +H	Potentiometric	H_2O $t = 25$ I not specified but low N_2 atmosphere	Ihnat P and Robinson DH, Potentiometric determination of the thermodynamic ionization constants of deferoxamine, *J. Pharm. Sci.*, **82**, 110–112 (1993). NB: Autotitrator used to titrate 100 mL of 0.001M deferoxamine mesylate (magnetically stirred) with 0.1M NaOH (standardized against potassium hydrogen phthalate). N_2 passed through the titrand (shown by others to be more effective in removing CO_2 than merely filling the headspace). The pH meter was standardized at $pH = 4$ and $pH = 10$. The iterative algorithm for fitting pH values to NaOH titre was expanded from earlier work by Lambert and Dalga (*Drug Dev. Ind. Pharm.*, **16**, 719–737 (1990)), and calculations were cycled until successive results for all four pK_a values differed by <0.01 log unit. The results were compared to literature values (Schwarzenbach G, Schwarzenbach K, Hydroxamatkomplexe I. Die stabilität der Eisen(III)-Komplexe einfacher Hydroxamsäuren und des Ferrioxamins B, *Helv. Chim. Acta*, **46**, 1390–1400 (1963)): 8.39, 9.03, 9.70, >11 (20 °C, $I = 0.1$M). The high pK_a value for the terminal amino group (normally 10.60 to 10.65) was attributed by the authors to hydrogen bonding, presumably to one or more ionized hydroxamic acid groups. Unpublished molecular models by the present author suggest that a stable arrangement is formed when the protonated amine is equidistant from all three ionized hydroxamic acid groups.
385	Desipramine ($C_{18}H_{22}N_2$) 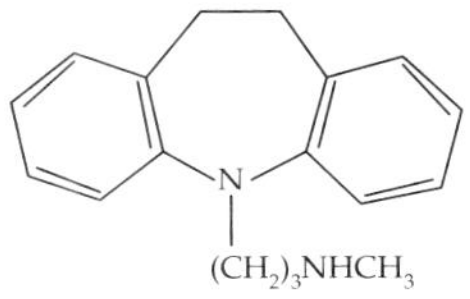	10.21 (−)	U	+H	Spectro (277 nm)	H_2O $t = 20.0$	Wahbe AM, El-Yazbi FA, Barary MH and Sabri SM, Application of orthogonal functions to spectrophotometric analysis. Determination of dissociation constants, *Int. J. Pharm.*, **92**(1), 15–22 (1993). NB: See Acetaminophen for further details. An alternative graphical method gave pK_a = 10.2.
386	Desipramine	10.2	U	+H	Soly	H_2O $t = 24 \pm 1$	Green AL, Ionization constants and water solublities of some aminoalkylphenothiazine tranquilizers and related compounds, *J. Pharm. Pharmacol.*, **19**, 10–16 (1967). NB: See Amitriptylline for details.
387	Desipramine	10.08	U	+H	Potentiometric	H_2O $t = 25$	Bergström CAS, Strafford M, Lazorova L, Avdeef A, Luthman K and Artursson P, Absorption classification of oral drugs based on molecular surface properties, *J. Med. Chem.*, **46**(4), 558–570 (2003). NB: From extrapolation of aqueous-methanol mixtures to 0% methanol.

(*continued*)

No.	Name	pK_a value(s)	Data quality	Ionization type	Method	Conditions t°C; I or c M	Comments and Reference(s)
388	Desmethyldoxepin ($C_{18}H_{19}NO$) 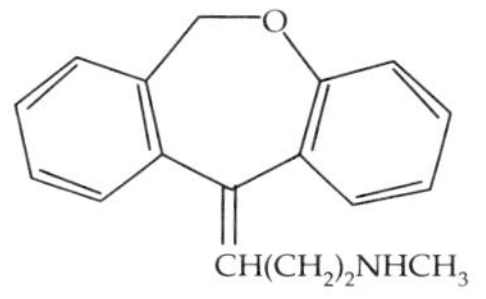	9.75	U	+H	Soly	$I = 0.00$ $t = 25.0$	Embil K and Torosian G, Solubility and ionization characteristics of doxepin and desmethyldoxepin, *J. Pharm. Sci.*, **71**(2), 191–193 (1982).
389	Desmethylpheniramine	4.02 ± 0.10 10.10 ± 0.06	U U	+H +H	Potentiometric	H_2O t undefined $I = 0.30$ (NaCl)	Testa B and Murset-Rossetti L, The partition coefficient of protonated histamines, *Helv. Chim. Acta*, **61**, 2530–2537 (1978). NB: See Cycliramine for details.
390	Desmycosin ($C_{39}H_{65}NO_{14}$) 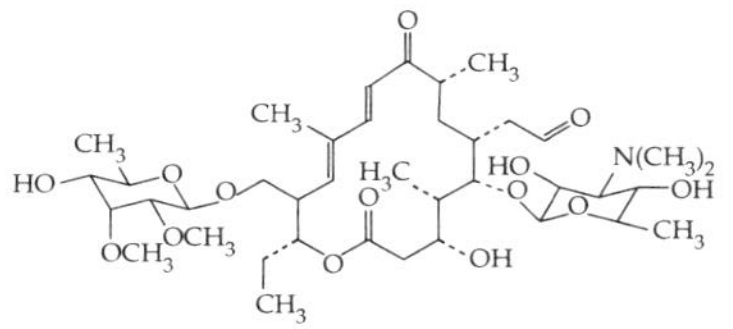	8.36	U	+H	Potentiometric	H_2O $t = 25$ $I = 0.167$	McFarland JW, Berger CM, Froshauer SA, Hayashi SF, Hecker SJ, Jaynes BH, Jefson MR, Kamicker BJ, Lipinski CA, Lundy KM, Reese CP and Vu CB, Quantitative Structure-activity relationships among macrolide antibacterial agents: In vitro and in vivo potency against Pasteurella multocida, *J. Med. Chem.*, **40**, 1340–1346 (1997). NB: See Azithromycin for details; average standard deviation of ± 0.07 for the pK_a. Paper also reported pK_a values for the following desmycosin analogues, where these were obtained by reducing the aldehyde (-CHO) group with an amine:
391	Desmycarosylcarbomycin A ($C_{38}H_{72}N_2O_{12}$) 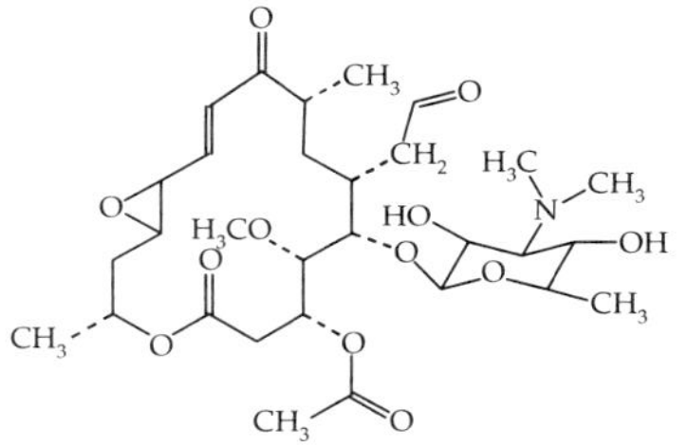	8.44	U	+H	Potentiometric	H_2O $t = 25$ $I = 0.167$	McFarland JW, Berger CM, Froshauer SA, Hayashi SF, Hecker SJ, Jaynes BH, Jefson MR, Kamicker BJ, Lipinski CA, Lundy KM, Reese CP and Vu CB, Quantitative Structure-activity relationships among macrolide antibacterial agents: *In vitro* and *in vivo* potency against Pasteurella multocida, *J. Med. Chem.*, **40**, 1340–1346 (1997). NB: See Azithromycin for details; average standard deviation of ± 0.07 for the pK_a.

Side chain:	pK_{a1}	pK_{a2}	pK_{a3}
$Me_2N(CH_2)_2NMe$	5.61	8.19	8.95
$Me_2N(CH_2)_3NMe$	7.45	8.43	9.77

No.	Compound	pKa	Data quality	Ionization type	Method	Conditions	Comments and Reference(s)
392	Dexamethasone-21-phosphate ($C_{22}H_{30}FO_8P$)	1.89	U	–H	partition	H_2O t unspecified I unspecified	Derendorf H, Rohdewald P, Hochhaus G, Mollmann H, HPLC determination of glucocorticoid alcohols, their phosphates and hydrocortisone in aqueous solutions and biological fluids, *J. Pharm. Biomed. Anal.*, **4**(2), 197–206 (1986).
		1.9	U	–H	Potentiometric		NB: Potentiometry was performed according to Albert and Serjeant (1971). pK_a values reported for triamcinolone acetonide phosphate were 1.70 (partition); corresponding values from potentiometry were 1.5 and 6.15.
		6.38	U	–H			
393	Dexamphetamine ($C_9H_{13}N$)	9.77	U	+H	Potentiometric	H_2O $t = 25.0 \pm 0.2$ $I \leq 0.001$	Leffler EB, Spencer HM and Burger A, Dissociation constants of adrenergic amines, *JACS*, **73**, 2611–2613 (1951).
394	Dextran				Potentiometric	H_2O $t = 37$	Larsen C, Macromolecular prodrugs. Part 8. Determination of the ionization constant of dextran by potentiometric titration and from kinetic analysis of the hydrolysis of dextran indomethacin ester conjugates, *Int. J. Pharm.*, **52**, 55–61 (1989). "The kinetics of hydrolysis of dextran indomethacin ester conjugates were studied in the pH range 6.81–9.13 at 37 °C, using high performance liquid chromatography. . . .. The kinetic experiments were in agreement with the ionization constant of dextran, determined by potentiometric titration to be 10–11.78."
	T-70 grade (Mw = 74 kDa)	11.78 ± 0.02	U	–H			
	T-500 grade (Mw = 488 kDa)	11.78 ± 0.02	U	–H			
395	Dextrose (glucose) ($C_6H_{12}O_6$) 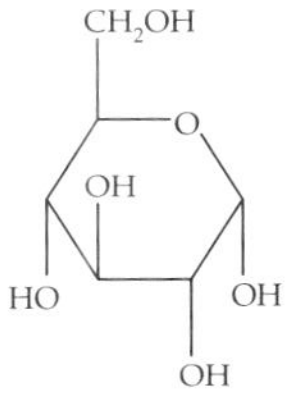	12.38 ± 0.1	U	–H	Potentiometric	H_2O $t = 25.0 \pm 0.1$ $I < 0.0005$	Woolley EM, Tomkins J and Hepler LG, Ionization constants for very weak organic acids in aqueous solution and apparent ionization constants for water in aqueous organic mixtures, *J. Solution Chem.*, **1**, 341–351 (1972). "Potential measurements were made as before (1) with a variety of wide-pH-range glass electrodes. The cells were maintained at 25.0 (± 0.1) °C, and potential readings were recorded when they became constant to ± 0.1 mV. Measurements were repeated at least three times with different combinations of electrodes, pH meters, and electrolyte concentrations ranging from 0.0005 to 0.01M. . . ." NB: Data cited by: Killion RB and Stella VJ, Nucleophilicity of dextrose, sucrose, sorbitol, and mannitol with p-nitrophenyl esters in aqueous solution, *Int. J. Pharm.*, **66**, 149–155 (1990). "In order to better understand the nucleophilicity of carbohydrates and polyhydric alcohols, the reactivities of dextrose, sucrose, sorbitol, and mannitol with a few simple p-nitrophenyl esters were studied at 25 °C and ionic strength 1.0 (KCl) in aqueous solution.

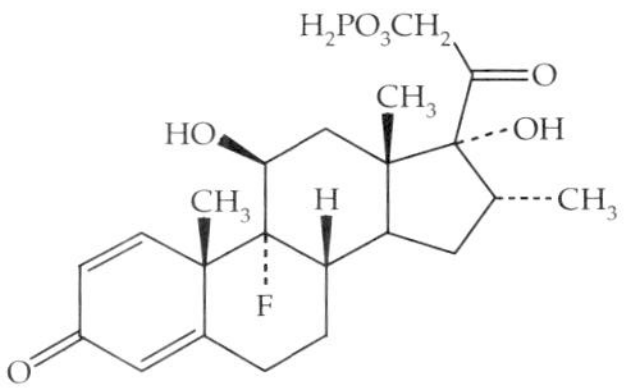

(continued)

No.	Name	pK_a value(s)	Data quality	Ionization type	Method	Conditions $t°C$; I or c M	Comments and Reference(s)
							Based on the pH dependency of the catalytic rate constants, the reactivity of the polyhydric alcohols was attributed to the anion derived from ionization of a hydroxy group in the polyhydric alcohol. The second-order rate constants representing the nucleophilic reactivity of the polyhydric alcohol anions were determined. Results showed that the nucleophilic reactivity of dextrose, sucrose, sorbitol and mannitol is similar to other alcohols of comparable pK_a. . . ."
396	Dextrose	11.81	U	−H	Potentiometric	H_2O $t = 37$	Bundgaard H and Larsen C, Kinetics and mechanism of reaction of benzylpenicillin and ampicillin with carbohydrates and polyhydric alcohols in aqueous solution, *Arch. Pharm. Chemi*, Sci. Edn., **8**, 184–200 (1978).
397	Diamorphine (heroin) ($C_{21}H_{23}NO_5$) 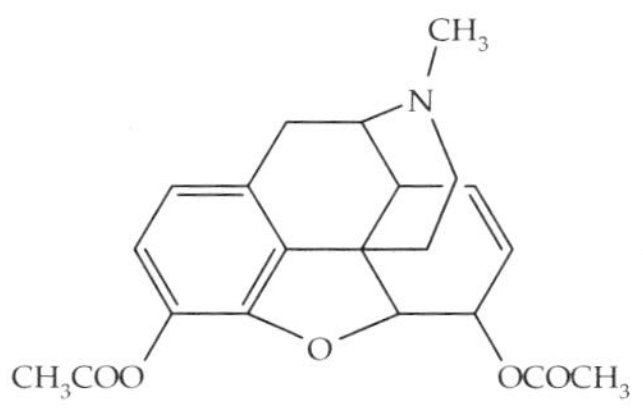 CH_3 N O CH_3COO $OCOCH_3$	7.95	U	+H	Potentiometric	H_2O $t = 25.0 \pm 0.1$ $I = 0.15$ (KCl) Ar atmosphere	Avdeef A, Barrett DA, Shaw PN, Knaggs RD and Davis SS, Octanol-, chloroform-, and propylene glycol dipelargonat-water partitioning of morphine-6-glucuronide and other related opiates, *J. Med. Chem.*, **39**, 4377–4381 (1996). NB: See Morphine for further details.
398	Diamorphine	7.6	U	+H	Spectro	H_2O $t = 23$ $c = 0.01$	Schoorl N, Dissociation constants and titration exponents of several less common alkaloids, *Pharm. Weekblad*, **76**, 1497–1501 (1939); CA 34:1900. Cited in Perrin Bases 2915 ref. S16. NB: The study used the indicator spectrophotometric (colorimetric) method as described by Kolthoff (1925). Reported also the pK_b values for the following monobasic alkaloids: Dicodide (hydrocodone) (6.05; $pK_a = 7.95$); Dilaudid (hydromorphone) (6.2; $pK_a = 7.8$); β-Eucaine (4.65; $pK_a = 9.35$); Eucodal (oxycodone) (5.4; $pK_a = 8.6$); Homatropine (4.3; $pK_a = 9.7$); Scopolamine (6.5; $pK_a = 7.95$); Stovaine (amylocaine) (6.1; $pK_a = 7.5$). The following dibasic alkaloids had the reported pK_b values: Alypine (amydricaine) (4.5, 10.2; $pK_a = 9.5, 3.8$); optochine (ethylhydrocupreine) (5.5, 9.95; $pK_a = 8.5, 4.05$); yohimbine (6.55, 11; $pK_a = 7.45, 3$).

No.	Compound	pKa	Class	Ionization	Method	Conditions	Reference
399	Diazepam ($C_{16}H_{13}ClN_2O$) 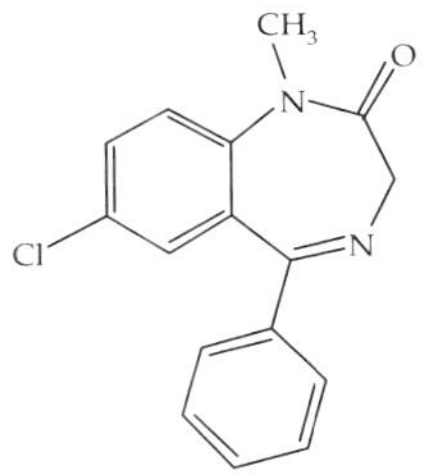	3.3	U	+H	CE/pH (+ve ion mode)	H_2O $t = 25$ $I = 0.025$	Wan H, Holmen AG, Wang Y, Lindberg W, Englund M, Nagard MB and Thompson RA, High-throughput screening of pK_a values of pharmaceuticals by pressure-assisted capillary electrophoresis and mass spectrometry, *Rapid Commun. Mass Spectrom.*, **17**, 2639–2648 (2003). NB: Reported a predicted value (ACD Labs) of 3.4.
400	Diazepam	3.42	U	+H	Spectro	H_2O $t = 25$ $I = 0.025$	Tam KY and Takacs-Novac K, Multi-wavelength spectrophotometric determination of acid dissociation constants, *Anal. Chim. Acta*, **434**, 157–167 (2001). NB: Reported a predicted value (ACD Labs) of 3.4.
401	Diazepam	3.3	U	+H	Spectro	5% MeOH in H_2O $t = 20$ $I = 0.15$	Barrett J, Smyth WF and Davidson IE. Examination of acid-base equilibria of 1,4-benzodiazepines by spectrophotometry, *J. Pharm. Pharmacol.*, **25**, 387–393 (1973). NB: See 1,4-Benzodiazepines for details.
402	Diazepam	3.3	U	+H			Konishi M, Hirai K and Mari Y, Kinetics and mechanism of the equilibrium reaction of triazolam in aqueous solution, *J. Pharm. Sci.*, **71**(12), 1328–1334 (1982). NB: See Triazolam.
403	Diazoxide ($C_8H_7ClN_2O_2S$)	8.75 ± 0.02	U	+H	Potentiometric	40% EtOH $t = 25.0$	Mannhold R, Rodenkirchen R, Bayer R and Haus W, The importance of drug ionization for the action of calcium antagonistsand related compounds, *Arzneim.-Forsch.*, **34**, 407–409 (1984).
	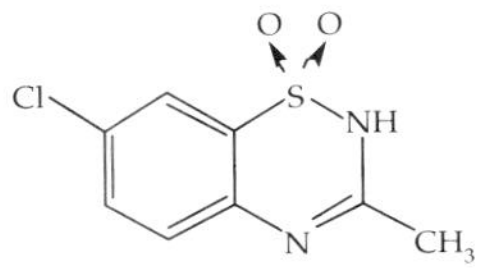	9.49	U	+H		H_2O	NB: See Aprindine for details.
404	Dibenzepine ($C_{18}H_{21}N_3O$) 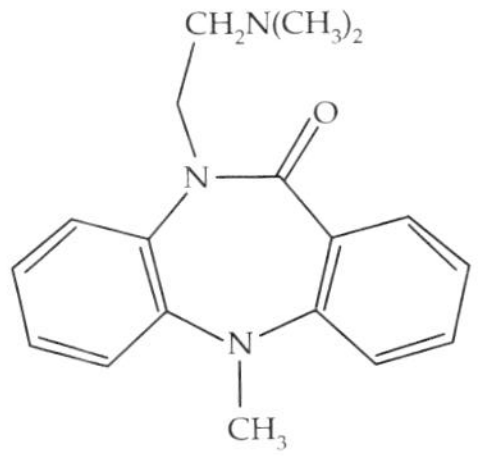	3.451 8.275	A U	+H +H	Potentiometric	H_2O $t = 25$ $I = 0.00$	Lukkari S, Ionization of some dibenzazepine derivatives, dibenzepine and opipramol, in aqueous solutions, *Farm. Aikak.*, **80** (**4-5**), 210–215 (1971). "The acid ionization constants of dibenzepine HCl and opipramol HCl were determined potentiometrically in aqueous solutions at 25 °C. The values p$K_1 = 8.275$ and p$K_{1,2} = 3.451$ respectively, were obtained for the acid ionization constants at zero ionic strength. The value p$K_1 = 8$ was obtained for opipramol at the ionic strength I = 0.01. The effect of ionic strength on the ionization constant, as adjusted with sodium chloride, was determined."

(*continued*)

Appendix A *(continued)*

No.	Name	pK_a value(s)	Data quality	Ionization type	Method	Conditions t°C; I or c M	Comments and Reference(s)
405	Dibenzepin	8.25 ± 0.05	U	+H	Potentiometric	H_2O $t = 21.0 \pm 1.0$	Egli A and Michaelis WR, Dibenzepin hydrochloride, *APDS*, **9**, 181–206 (1980). NB: No reference was given. "Titration of a 0.003 M solution in water at 20–22 °C yielded as pK_a 8.25 ± 0.05 for the 3′-nitrogen."
406	Dibucaine (cinchocaine) ($C_{20}H_{29}N_3O_2$) N, $O(CH_2)_3CH_3$, $CONHCH_2CH_2N(Et)_2$	8.83 ± 0.03	A	+H	Potentiometric	H_2O $t = 20.0$ $I = 0.10$ (KCl) N_2 atmosphere	Buchi J and Perlia X, Beziehungen zwischen de physikalisch-chemische Eigenschaften und der Wirkung von Lokalanasthetica, *Arzneim.-Forsch.*, **10**, 745–754 (1960). NB: See Cocaine for details.
407	Dibucaine (cinchocaine) homologues				Potentiometric, Spectro	H_2O $t = 20.0$ $I = 0.10$ (KCl) N_2 atmosphere	Buchi J and Perlia X, Beziehungen zwischen de physikalisch-chemische Eigenschaften und der Wirkung von Lokalanasthetica, *Arzneim.-Forsch.*, **10**, 745–754 (1960). NB: See Cocaine for details.
	O-R						
	CH_3	1.60; 8.82	A	+H; +H			
	C_2H_5	1.94; 8.89					
	C_3H_7	1.89; 8.69					
	C_4H_9	1.92; 8.87					
	C_5H_{11}	1.91; 8.83					
	C_6H_{12}	1.91; 8.83					
408	3,5-Dichlorophenol ($C_6H_4Cl_2O$) OH, Cl, Cl	8.48	U	−H	CE/pH (−ve ion mode)	H_2O $t = 25$ $I = 0.025$	Wan H, Holmen AG, Wang Y, Lindberg W, Englund M, Nagard MB and Thompson RA, High-throughput screening of pK_a values of pharmaceuticals by pressure-assisted capillary electrophoresis and mass spectrometry, *Rapid Commun. Mass Spectrom.*, **17**, 2639–2648 (2003). NB: Reported a predicted value (ACD Labs) of 8.04.

No.	Compound	pKa	Class		Method	Conditions	Reference / Comments
409	Diclofenac ($C_{14}H_{11}Cl_2NO_2$) 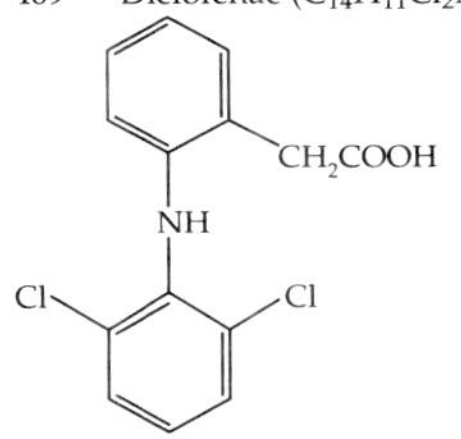	3.99 ± 0.01	A	–H	Potentiometric	H_2O $t = 25 \pm 0.5$ $I = 0.15$ (KCl)	pH-metric log P 10. Determination of liposomal membrane-water partition coefficients of ionizable drugs, Avdeef A, Box KJ, Comer JEA, Hibbert C and Tam KY, *Pharm. Res.*, **15(2)** 209–215 (1998). NB: Used a Sirius PCA101 autotitrator. Also gave log P (octanol-water) and log P (dioleylphosphatidyl-choline unilamellar vesicles). The same result was reported in Sirius Technical Application Notes, vol. 2, p. 146–147 (1995). Sirius Analytical Instruments Ltd., Forest Row, East. Sussex, RH18 5DW, UK. NB: Concentration of analyte, 0.46–0.70 mM.
		4.01	U	–H	Potentiometric	H_2O $t = 37$ $I = 0.15$ (KCl)	Balon K, Riebesehl BU and Muller BW, Drug liposome partitioning as a tool for the prediction of human passive intestinal absorption, *Pharm. Res.*, **16**, 882–888 (1999).
410	Diclofenac	4.2	U	–H	Potentiometric	H_2O $t = 20$ I undefined but very low	Rainer VG, Krüger U and Klemm K. Syntheses und physicalisch-chemische Eigenschaften von Lonazolac-Ca einem neuen Antiphlogistikum/Antirheumatikum, *Arzneim.-Forsch.*, **31(4)**, 649–655 (1981). NB: See Indomethacin for details.
411	Diclofenac	4.7	U	–H	Spectro	H_2O t undefined I undefined	Herzfeldt CD and Kümmel R, Dissociation constants, solubilities, and dissolution rates of some selected nonsteroidal antiinflammatories, *Drug Dev. Ind. Pharm.*, **9(5)**, 767–793 (1983). NB: Used dA/dpH method. See Azapropazone and Ibuprofen for details. The exact pK_a determination should be optimized by another method because of the irregularity in the dA/dpH verses mean pH plot at ~pH 4.
412	Diclofenac	4.0	U	–H	Potentiometric	H_2O $t = 25$	Maitani Y, Nakagaki M and Nagai T, Determination of the acid dissociation constants in ethanol-water mixtures and partition coefficients for diclofenac, *Int. J. Pharm.*, **74**(2,3), 105–114 (1991). "Plot given of pK_a vs EtOH concentration (almost linear from 90% down to 18%), however precipitation of diclofenac at EtOH% <50% needed approximate treatment of the titration curves; value from extrapolation to 0% EtOH. Also, there was an approximately linear plot of apparent pK_a and the reciprocal of the dielectric constant, which led to a similar value. The dissociation constants of diclofenac sodium in ethyl alcohol . . . were determined using the titration method. The acid dissociation constant of the drug was decreased by the increase in the concentration of ethanol in the aqueous solution. . . .. It is suggested that ethanol, which is used as an enhancer for percutaneous absorption, assumes another role, increasing the proportion of unionized form of the drug and forming ion pairs in low dielectric media. . . . The distribution behavior of diclofenac is . . . affected in the presence of added cations. Above pH 7, ion pair formation promotes the distribution of the drug into lipophilic environment."

(*continued*)

No.	Name	pK_a value(s)	Data quality	Ionization type	Method	Conditions t°C; I or c M	Comments and Reference(s)
413	Diclofenac	4.21	U	−H	CE/pH (−;ve ion mode)	H_2O $t = 25$ $I = 0.025$	Wan H, Holmen AG, Wang Y, Lindberg W, Englund M, Nagard MB and Thompson RA, High-throughput screening of pK_a values of pharmaceuticals by pressure-assisted capillary electrophoresis and mass spectrometry, *Rapid Commun. Mass Spectrom.*, **17**, 2639–2648 (2003). NB: Reported a predicted value (ACD Labs) of 4.18.
414	Dicloxacillin ($C_{19}H_{17}Cl_2N_3O_5S$) 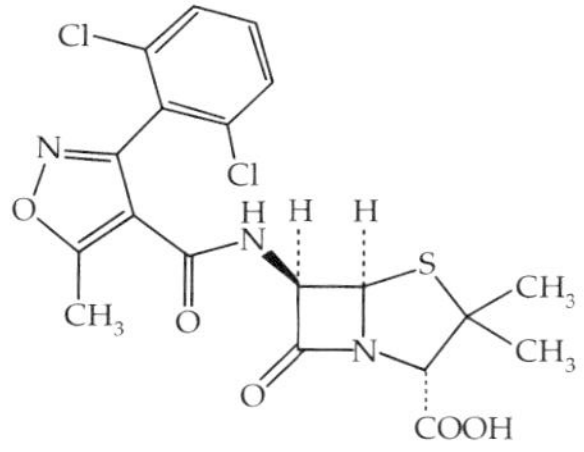	2.67	U	−H	Potentiometric	H_2O $t = 25.0 \pm 0.1$ $c = 0.006$	Hou JP and Poole JW, The aminoacid nature of ampicillin and related penicillins, *J. Pharm. Sci.*, **58**, 1510–1515 (1969). NB: Apparent pK_a values were extrapolated to 0 % alcohol.
415	Dicyclomine ($C_{19}H_{35}NO_2$) 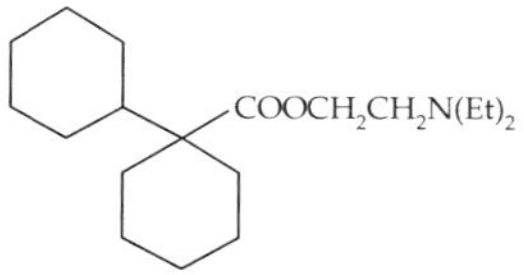	9.0	U	+H			Foye 3rd; see Azatadine; from McEvoy.
416	Didanosine ($C_{10}H_{12}N_4O_3$) 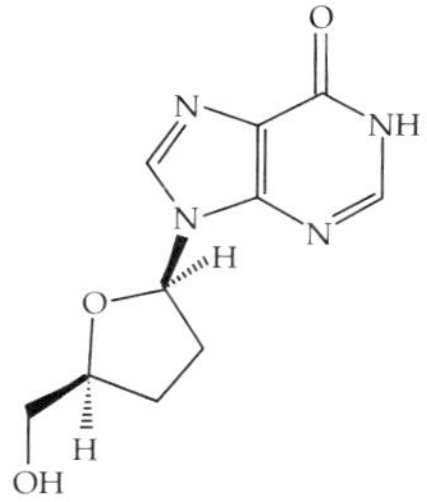	9.12 ± 0.02	U	+H	Potentiometric	H_2O RT	Nassar MN, Chen T, Reff MJ and Agharkar SN, Didanosine, *APDS*, **22**, 185–227 (1993). "The apparent pK_a value of didanosine, uncorrected for activity coefficients, was obtained by titration of a 0.01 M solution of didanosine in water with standardised solution of 0.1N NaOH at room temperature. The apparent pK_a of didanosine was found to be 9.12 ± 0.02 (7) 7. Anderson BD, Wygant MB, Xiang T-X, Waugh WA and Stella VJ, Preformulation solubility and kinetic studies of 2′,3′-dideoxypurine nucleosides: potential anti-AIDS agents, *Int. J. Pharm.*, **45**, 27–37 (1988)."

No.	Compound	pKa	Class	Ionization	Method	Conditions	Reference / comments
417	2′,3′-Dideoxyadenosine ($C_{10}H_{13}N_5O_2$) 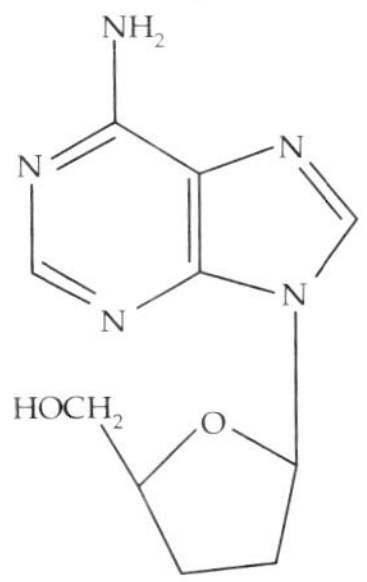	3.84	U	+H	Spectro	H_2O $t = 25.0 \pm 0.1$ $I = 0.01$	Darrington RT, Xiang T and Anderson BD, Inclusion complexes of purine nucleosides with cyclodextrins. Part 1. Complexation and stabilization of a dideoxypurine nucleoside with 2-hydroxypropyl-beta-cyclodextrin, *Int. J. Pharm.*, **59**, 35–44 (1990). "The effects of complexation and stabilization of 2′,3′-dideoxy-adenosine (I) with 2-hydroxypropyl-beta-cyclodextrin (II) were studied, and the results of kinetic studies and pK_a determinations are reported. Although hydrolysis is 100% suppressed in both the protonated and neutral complexes, due to the small binding constants, the maximum stabilization attainable in a 0.1 M solution of II at 25 °C was approximately 5-fold at pH 5 and 2-fold at pH 2. Possible inclusion geometries are considered in an attempt to account for the kinetic data.

T °C	pK_a	T °C	pK_a
4	3.95	37	3.69–3.75
25	3.92–3.94	50	3.61

No.	Compound	pKa	Class	Ionization	Method	Conditions	Reference / comments
418	Didesmethylpheniramine ($C_{14}H_{16}N_2$) 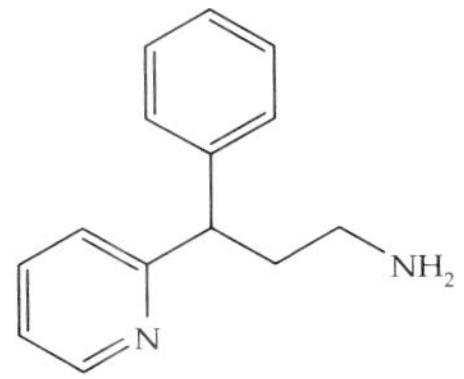	4.22 ± 0.08 9.88 ± 0.06	U U	+H +H	Potentiometric	H_2O t undefined $I = 0.30$ (NaCl)	Testa B and Murset-Rossetti L, The partition coefficient of protonated histamines, *Helv. Chim. Acta*, **61**, 2530–2537 (1978). NB: See Cycliramine for details. NB: Monodesmethylpheniramine ($C_{15}H_{18}N_2$) under the same conditions gave pK_a values of 4.02 ± 0.10 and 10.10 ± 0.06.
419	Diethylstilbestrol ($C_{18}H_{20}O_2$) 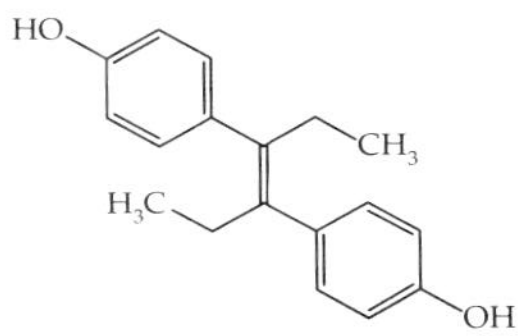	9.02	U	−H	CE/pH (−ve ion mode)	H_2O $t = 25$ $I = 0.025$	Wan H, Holmen AG, Wang Y, Lindberg W, Englund M, Nagard MB and Thompson RA, High-throughput screening of pK_a values of pharmaceuticals by pressure-assisted capillary electrophoresis and mass spectrometry, *Rapid Commun. Mass Spectrom.*, **17**, 2639–2648 (2003). NB: Reported a predicted value (ACD Labs) of 10.06.

(continued)

No.	Name	pK_a value(s)	Data quality	Ionization type	Method	Conditions t°C; I or c M	Comments and Reference(s)
420	Diflunisal ($C_{13}H_8F_2O_3$) 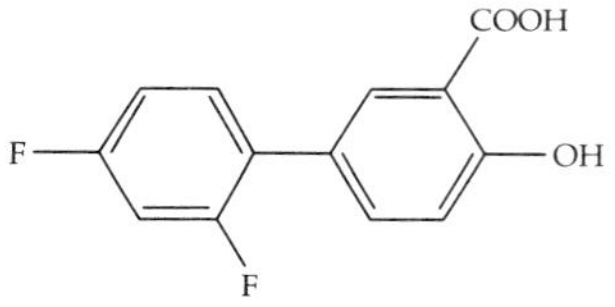	3.3	U	–H	soly	H_2O $I = 0.1$	Cotton ML and Hux RA, Diflunisal, *APDS*, **14**, 491–524 (1985). "From the solubility data in acetate buffers and the limiting solubility of the free acid in hydrochloric acid solution, a pK_a value of 3.3 was estimated for solutions of 0.1 ionic strength. The pK_a of the phenolic group has been estimated by a spectrophotometric method to be 14. These values correlate reasonably well with pK_a values for salicylic acid of 3.0 and 13.9." DeMarco JD, Merck Sharp and Dohme Laboratories, West Point, PA, unpublished data.
		14	U	–H	Spectro		
421	Diflunisal	4.91	U	–H	soly	H_2O $t = 37$ $I = 0.5$ (KCl)	Najib NM and Suleiman MS, Kinetics of dissolution of diflunisal and diflunisal-polyethylene glycol solid dispersion, *Int. J. Pharm.*, **57**, 197–203 (1989). "... solid dispersion of diflunisal (I) in polyethylene glycol (4000) (II) was prepared and studied to determine the effects of II on the dissociation constant, dissolution kinetics, diffusion coefficient and hydrodynamic layer thickness of I; the influence of pH was also evaluated. Dispersing I in II reduced its (apparent) dissociation constant. ..."
	1:1 dispersion in PEG	5.07	U	–H			
	1:3 dispersion in PEG	5.15	U	–H			
	1:5 dispersion in PEG	5.19	U	–H			
	1:7 dispersion in PEG	5.21	U	–H			
422	α-(2,4-Difluorophenyl)-α-[1-(2-(2-pyridyl)-phenylethenyl)]-1H-1,2,4-triazole-1-ethanol (XD405, bis-mesylate salt) ($C_{13}H_8F_2O_3$) 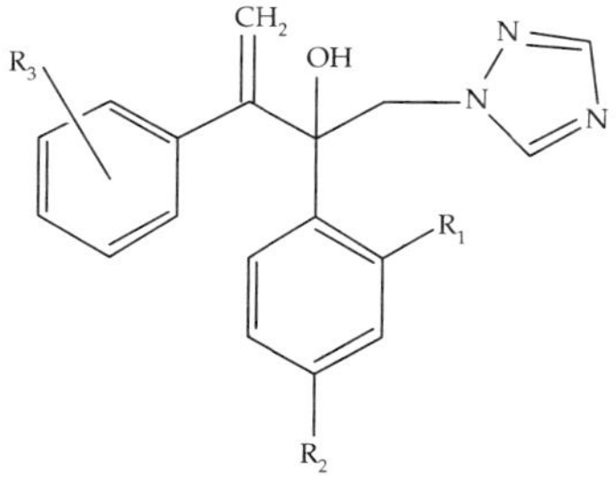 $R_1 = R_2 = F$ $R_3 = 4$ 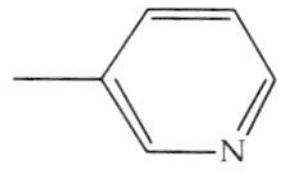 —	1.68 ± 0.14	U	+H	soly	H_2O $t = 22$ I undefined	Maurin MB, Vickery RD, Gerard CA and Hussain M, Solubility of ionization behavior of the antifungal α-(2,4-difluorophenyl)-α-[1-(2-(2-pyridyl)phenylethenyl)]-1H-1,2,4-triazole-1-ethanol bismesylate (XD405), *Int. J. Pharm.* **94**, 11–14 (1993). "Solubility studies were carried out by placing excess XD405 into a suitable container with deionized water, adding various amounts of either hydrochloric acid or sodium hydroxide to adjust the pH and rotating end-to-end for 24 h at room temperature (22 °C). .... Preliminary experiments indicated that 24 h provided sufficient time to reach equilibrium. The suspension was passed through a 0.45 μm filter (Acrodisc® LC13 PVDF, Gelman Sciences) with the first portion discarded to ensure saturation of the filter. An aliquot of the filtrate was diluted and analyzed by HPLC and the remainder of the filtrate was employed for pH determination. .... The (solubility-pH) data were analyzed by regression analysis with a zero intercept second order polynomial that was based on eqn. (5) and employed an intrinsic solubility of 3 μg/mL. The resulting polynomial had a correlation coefficient of 0.997 and provided ionization constants ± standard error of $pK_{a1} = 1.68 \pm 0.14$ and $pK_{a2} = 4.93 \pm 0.096$.
		4.93 ± 0.096	U	+H			

							$(S_T/[A]) - 1 = [H^+]^2/(K_{a1})(K_{a2}) + [H^+]/K_{a2}$ where S_T is the total solubility and [A] is the intrinsic solubility of the free base (3 μg/mL). For 2 analogues: DuP 860 $R_1 = H$, $R_2 = Cl$, $R_3 = 2\text{-}Cl$; $pK_a = 1.79$ DuP 991 $R_1 = R_2 = F$, $R_3 = 2\text{-}Cl$; $pK_a = 2.15$ (Maurin MB, Addicks WJ, Rowe SM and Hogan R, Physicochemical properties of α-styryl carbinol antifungal agents, *Pharm. Res.*, **10**, 309–312 (1993)."
423	Dihydroarecaidine 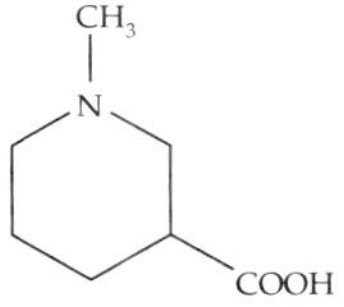	9.70	A	+H	Potentiometric	H_2O $t = 25$ $c = 0.02$	Chilton J and Stenlake JB, Dissociation constants of some compounds related to lysergic acid: Beta-dimethylaminopropionic acid, dihydroarecaidine, ecgonine and their derivatives, *J. Pharm. Pharmacol.*, **7**, 1004–1011 (1955). Cited in Perrin 2862 ref. C27. NB: The study used measurements of pH with a glass electrode.
424	Dihydroarecaidine methyl ester 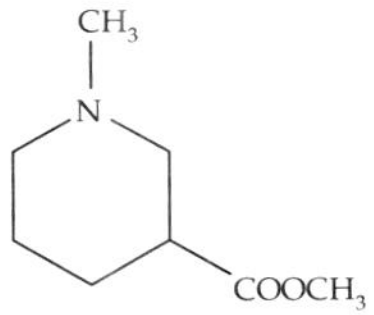	8.39	A	+H	Potentiometric	H_2O $t = 25$ $c = 0.02$	Chilton J and Stenlake JB, Dissociation constants of some compounds related to lysergic acid: Beta-dimethylaminopropionic acid, dihydroarecaidine, ecgonine and their derivatives, *J. Pharm. Pharmacol.*, **7**, 1004–1011 (1955). Cited in Perrin 2862 ref. C27. NB: The study used measurements of pH with a glass electrode.
425	Dihydrocodeine ($C_{18}H_{23}NO_3$) 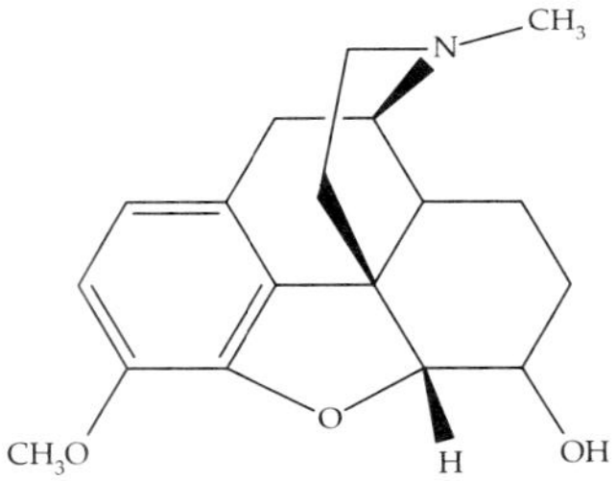	8.75	A	+H	Conductance	H_2O $t = 25$ $\kappa < 1.5$	Oberst FW and Andrews HL, The electrolytic dissociation of morphine derivatives and certain synthetic analgetic compounds, *JPET*, **71**, 38–41 (1941). Cited in Perrin Bases 2890 ref. O1. NB: Results were reported as K_b values. For dihydrocodeine, $K_b = 5.65 \times 10^{-6}$, giving $pK_b = 5.25$. See Codeine for details.

(*continued*)

No.	Name	pK_a value(s)	Data quality	Ionization type	Method	Conditions t°C; I or c M	Comments and Reference(s)
426	Dihydrodesoxycodeine ($C_{18}H_{23}NO_2$) 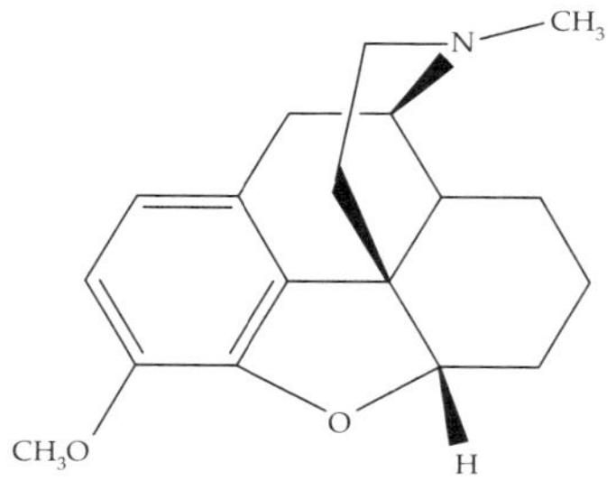	8.83	U	+H	Potentiometric	H_2O t undefined I undefined	Rapaport H and Masamune S, The stereochemistry of 10-hydroxycodeine derivatives, *JACS*, **77**, 4330–4335 (1955). Cited in Perrin Bases Supplement 7466 ref. R3. NB: The study used measurements of pH with a glass electrode and liquid junction potentials.
427	Dihydrodesoxynorcodeine ($C_{17}H_{21}NO_2$) 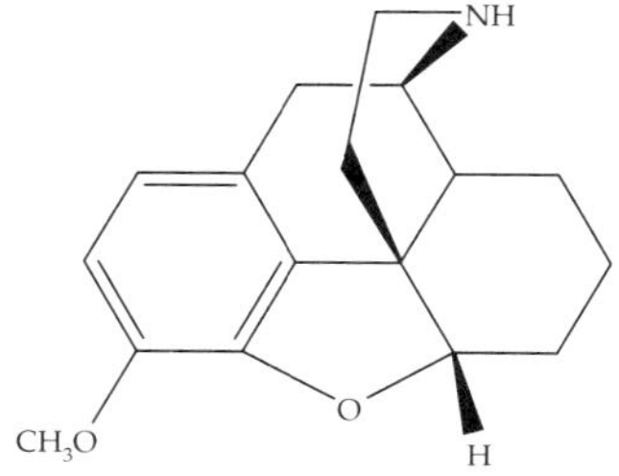	9.62	U	+H	Potentiometric	H_2O	Rapaport H and Masamune S, The stereochemistry of 10-hydroxycodeine derivatives, *JACS*, **77**, 4330–4335 (1955). Cited in Perrin Bases Supplement 7467 ref. R3. NB: The study used measurements of pH with a glass electrode and liquid junction potentials.
428	Dihydroequilin, 17α- ($C_{18}H_{22}O_2$) 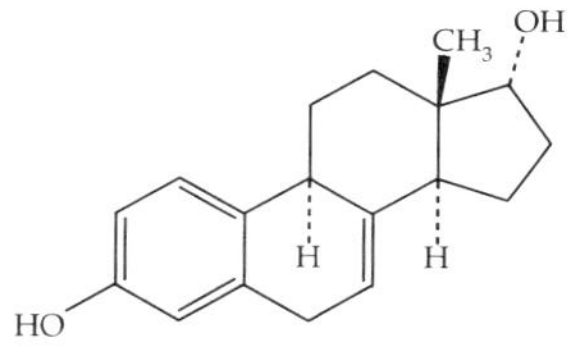	10.29 ± 0.02	A	−H	Spectro	H_2O $t = 23 \pm 2$	Hurwitz AR and Liu ST, Determination of aqueous solubility and pK_a values of estrogens, *J. Pharm. Sci.*, **66**, 624–627 (1977). NB: See 17α-Estradiol for details.

No.	Compound	pKa	Data quality	Ionization	Method	Conditions	Reference and comments
429	Dihydroequilinen, 17β- ($C_{18}H_{22}O_2$) 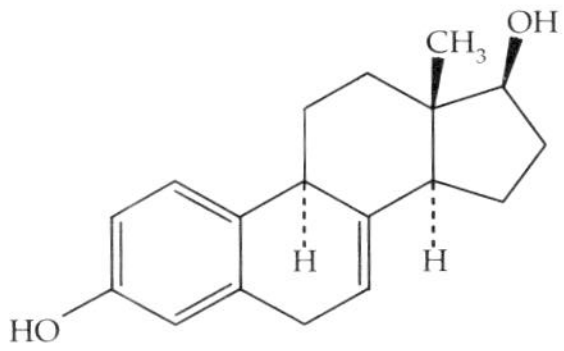	9.77 ± 0.04	A	−H	Spectro	H_2O $t = 23 \pm 2$	Hurwitz AR and Liu ST, Determination of aqueous solubility and pK_a values of estrogens, J. Pharm. Sci., 66, 624–627 (1977). NB: See 17α-Estradiol for details.
430	Dihydroergocornine ($C_{31}H_{41}N_5O_5$) 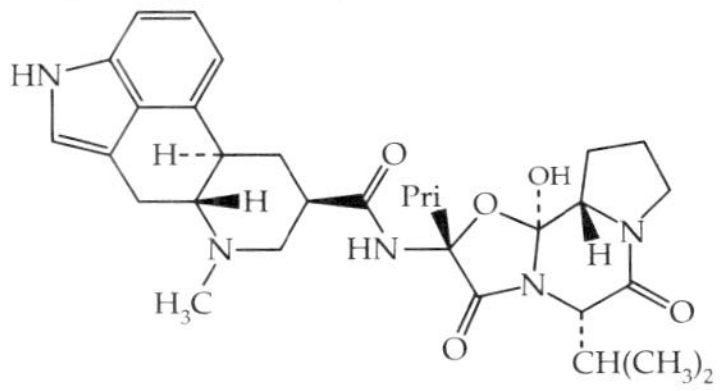	6.91 ± 0.07	U	+H	Potentiometric	H_2O $t = 24$ $I < 0.01$	Maulding HV and Zoglio MA, Physical chemistry of ergot alkaloids and derivatives. I. Ionization constants of several medicinally active bases, *J. Pharm. Sci.*, **59**, 700–701 (1970). NB: Solubility increased by complexation with 7-β-hydroxypropyltheophylline (7HPT) and the apparent pK_a values extrapolated to [7HPT] = 0. Solutions were titrated with carbonate-free KOH solution. See methysergide for further details. Ritschel gave 6.76, also cited Maulding and Zoglio; this is the apparent value in 10% 7HPT.
431	Dihydroergocriptine ($C_{32}H_{43}N_5O_5$) 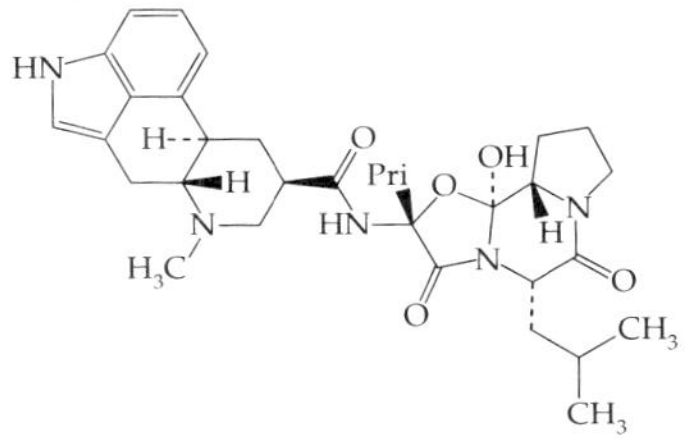	6.89 ± 0.07	U	+H	Potentiometric	H_2O $t = 24$ $I < 0.01$	Maulding HV and Zoglio MA, Physical chemistry of ergot alkaloids and derivatives. I. Ionization constants of several medicinally active bases, *J. Pharm. Sci.*, **59**, 700–701 (1970). NB: See Dihydroergocornine and methysergide for details. Ritschel gave 6.74, also cited Maulding and Zoglio; this is the apparent value in 10% 7HPT.
432	Dihydroergocristine ($C_{35}H_{41}N_5O_5$) 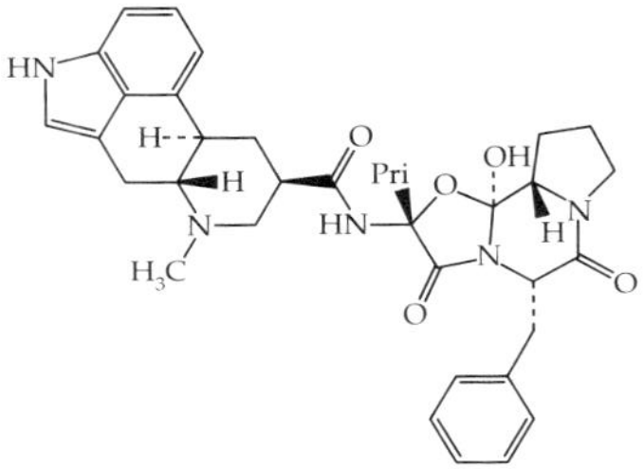	6.89 ± 0.07	U	+H	Potentiometric	H_2O $t = 24$ $I < 0.01$	Maulding HV and Zoglio MA, Physical chemistry of ergot alkaloids and derivatives. I. Ionization constants of several medicinally active bases, *J. Pharm. Sci.*, **59**, 700–701 (1970). NB: See Dihydroergocornine and methysergide for details. Ritschel gave 6.74, also cited Maulding and Zoglio; this is the apparent value in 10% 7HPT.

(continued)

No.	Name	pK_a value(s)	Data quality	Ionization type	Method	Conditions t°C; I or c M	Comments and Reference(s)
433	Dihydroergonovine ($C_{19}H_{25}N_3O_2$) 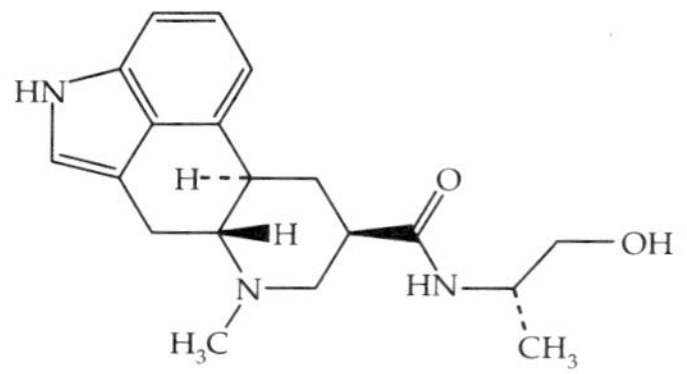	7.38	U	+H	Potentiometric	H_2O $t = 24$ $c = 0.002$	Craig LC, Shedlovsky T, Gould RG and Jacobs WA, The ergot alkaloids. XIV. The positions of the double bond and the carboxyl group in lysergic acid and its isomer. *J. Biol. Chem.*, **125**, 289–298 (1938). Cited in Perrin Bases 2894 ref. C54. NB: The study used an asymmetric cell with a glass electrode and liquid junction potential. The pK_a value is the pH reported at half-neutralisation. Values are also reported for several other lysergic acid derivatives.
434	Dihydroergotamine ($C_{33}H_{37}N_5O_5$) 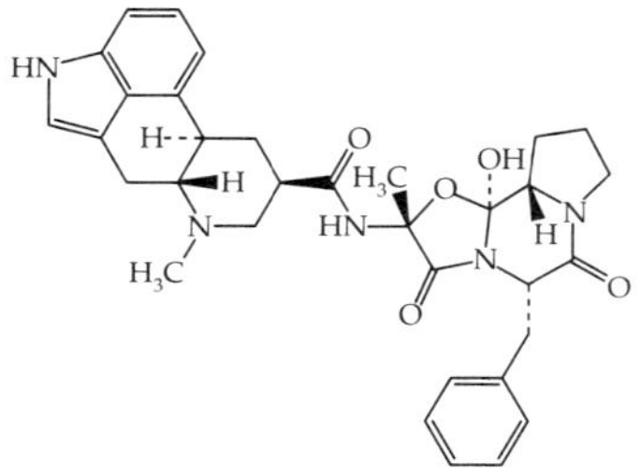	6.90 ± 0.07	U	+H	Potentiometric	H_2O $t = 24$ $I < 0.01$	Maulding HV and Zoglio MA, Physical chemistry of ergot alkaloids and derivatives. I. Ionization constants of several medicinally active bases, *J. Pharm. Sci.*, **59**, 700–701 (1970). NB: See Dihydroergocornine and Methysergide for details. Ritschel gave 6.75, also cited Maulding and Zoglio; this is the apparent value in 10% 7HPT.
435	Dihydroergotoxine A 1:1:1 mixture of dihydroergocornine, dihydroergocristine and dihydroergocryptine	6.9 ± 0.07		+H	Potentiometric	H_2O $t = 24.0$ I undefined	Schoenleber WD, Jacobs AL and Brewer GA, Jr., Dihydroergotoxine methanesulfonate, *APDS*, **7**, 81–147 (1978) (Wehrli A, Sandoz Ltd., personal communication). "Maulding and Zoglio have reported the ionization constants for the dihydroergotoxine components at 24° in water as: dihydroergocornine 6.91 ± 0.07 dihydroergocristine 6.89 ± 0.07 dihydroergokryptine 6.89 ± 0.07

No.	Compound	pKa	Data quality	Assignment	Method	Conditions	Comments / Reference
							The pK_{aMCS} values (MCS = methyl cellosolve) according to the method of Simon have been determined as: dihydroergocornine 5.84 dihydroergocristine 5.83 dihydro-α-ergokryptine 5.81 dihydro-ß-ergocryptine 5.84"
436	Dihydrofolic acid ($C_{19}H_{21}N_7O_6$)	1.38 3.84 9.54 0.28	VU VU VU VU	+H +H −H +H	Spectro	H_2O $t = 25$ $I = 0.10$ $I = 0.1$ to 4	Poe M, Acidic dissociation constants of folic acid, dihydrofolic acid and methotrexate, *J. Biol. Chem.*, **252(11)**, 3724–3728 (1977). NB: The compound aqueous solubility was insufficient to measure the pK_a values for the two carboxyl groups. Furthermore, dihydrofolic acid is relatively unstable in acidic solutions. See methotrexate for additional details.
437	Dihydromorphine ($C_{17}H_{21}NO_3$) 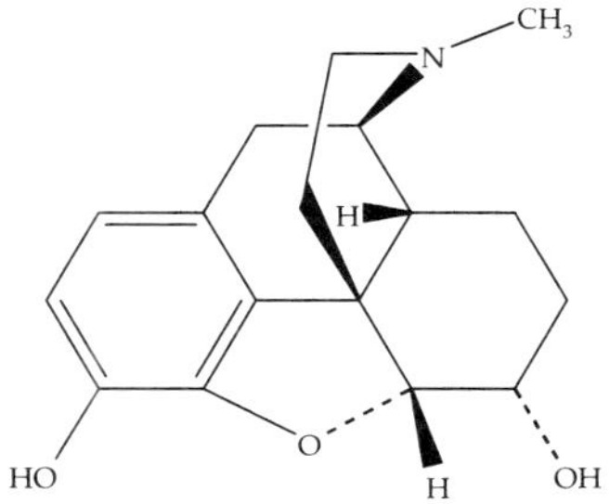	9.35	A	+H	Conductance	H_2O $t = 25$ $\kappa < 1.5$	Oberst FW and Andrews HL, The electrolytic dissociation of morphine derivatives and certain synthetic analgetic compounds, *JPET*, **71**, 38–41 (1941). Cited in Perrin Bases 2893 ref. O1. NB: Results were reported as K_b values. For dihydromorphine, $K_b = 2.26 \times 10^{-6}$, giving $pK_b = 5.65$. See Codeine for details.
438	2,8-Dihydroxyadenine ($C_5H_5N_5O_2$)	2.45 ± 0.06 8.12 ± 0.05 11.4 ± 0.2 2.49 8.09 11.5	U U U U U U	+H −H −H +H −H −H	soly Spectro	H_2O $t = 37.0$ $I = 0.1$	Peck CC and Benet LZ, General method for determining macrodissociation constants of polyprotic, amphoteric compounds from solubility measurements, *J. Pharm. Sci.*, **67**, 12–16 (1978). "A general method for estimating dissociation values, given a set of solubility and pH measurements for tyrosine and 2,8-dihydroxyadenine, is presented. Equations are derived extending solubility, pH and dissociation constant relationships from weak acids and bases to polyprotic, amphoteric compounds. Included in the estimation procedure is a subroutine for approximating thermodynamic dissociation constants."

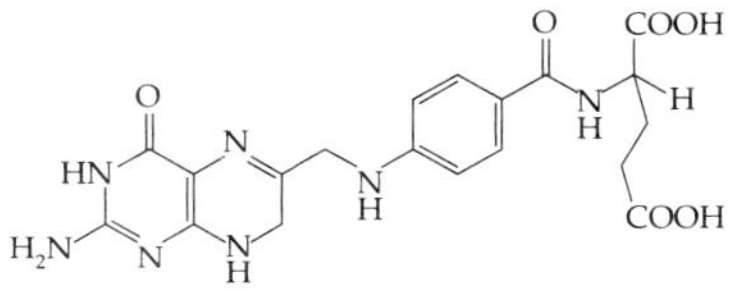

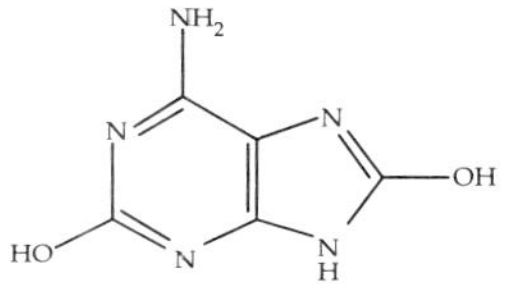

(continued)

No.	Name	pK_a value(s)	Data quality	Ionization type	Method	Conditions t°C; I or c M	Comments and Reference(s)
439	3,5-Di-iodo-L-tyrosine ($C_9H_9I_2NO_3$) 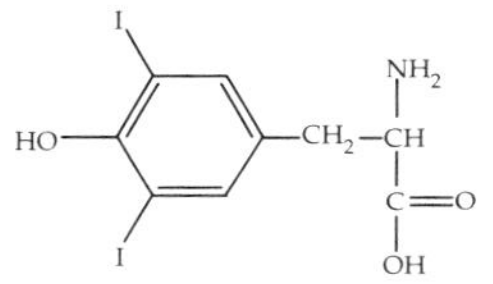	2.10 6.46 7.52	U U U	−H +H −H		H_2O $t = 37.0$	Ballard BE and Nelson E, Physicochemical properties of drugs that control absorption rate after subcutaneous implantation, *JPET*, **135**, 120–127 (1962). NB: pK_a = 2.10 from pK_b = 11.52 where pK_w = 13.621 at 37 °C; secondary source W&G assumed the result was at 25 and used pK_w = 14.00.
440	Dilazep ($C_{31}H_{44}N_2O_{10}$) 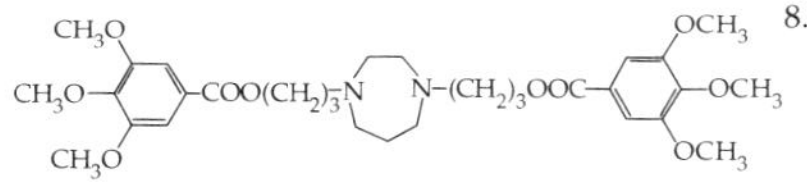	5.14 ± 0.11 8.25 ± 0.08	U U	+H +H	Potentiometric	H_2O $t = 25.0 \pm 0.1$ $I = 0.1$ N_2 atmosphere	IJzerman AP, Limiting solubilities and ionization constants of sparingly soluble compounds: Determination from aqueous potentiometric data only, *Pharm. Res.*, **5**(12), 772–775 (1988). NB: Nitrogen atmosphere; glass electrode; KOH titrant; S_o = 320 ± 3 μM.
441	Dilazep	4.46 ± 0.03 8.14 ± 0.02 5.19 8.88	U U U U	+H +H +H +H	Potentiometric	40% EtOH $>t = 25.0$ H_2O	Mannhold R, Rodenkirchen R, Bayer R and Haus W, The importance of drug ionization for the action of calcium antagonists and related compounds, *Arzneim.-Forsch.*, **34**, 407–409 (1984). NB: See Aprindine for details.
442	Diltiazem ($C_{22}H_{26}N_2O_4S$) 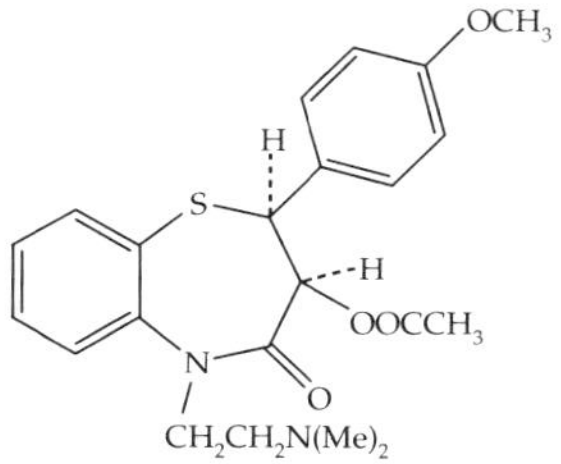	7.32 ± 0.02 8.06	U U	+H +H	Potentiometric	40% EtOH $t = 25.0$ H_2O	Mannhold R, Rodenkirchen R and Bayer R, Haus W, The importance of drug ionization for the action of calcium antagonistsand related compounds, *Arzneim.-Forsch.*, **34**, 407–409 (1984). NB: See Aprindine for details.
443	Dimethoxyamphetamine ($C_{11}H_{17}NO_2$) 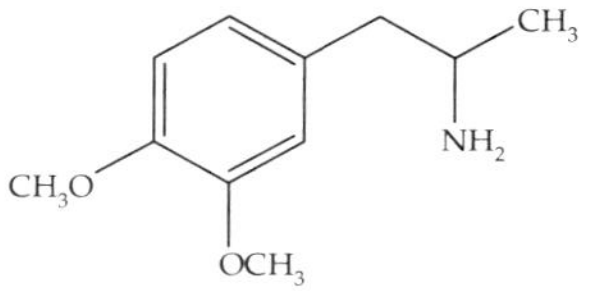	9.60 ± 0.05	U	+H	Potentiometric	H_2O $t = 25.0 \pm 0.2$ $I \leq 0.001$	Leffler EB, Spencer HM and Burger A, Dissociation constants of adrenergic amines, *JACS*, **73**, 2611–2613 (1951). NB: See Amphetamine for details. From pK_b = 4.40.

444	1-(3,4-Dimethoxyphenyl)-2-*N*-methylaminopropane ($C_{12}H_{19}NO_2$) 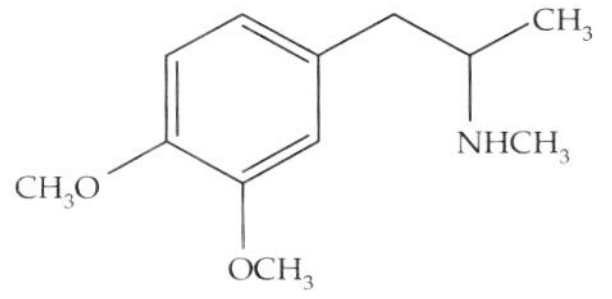	9.81 ± 0.02	U	+H	Potentiometric	H_2O $t = 25.0 \pm 0.2$ $I \leq 0.001$	Leffler EB, Spencer HM and Burger A, Dissociation constants of adrenergic amines, *JACS*, **73**, 2611–2613 (1951). NB: see Amphetamine for details. From $pK_b = 4.19$. This compound was insufficiently soluble in water; pH measurements were performed in a series of ethanol-water solutions, which were then extrapolated back to 0 % ethanol.
445	(3,4-Dimethyl-5-isoxazolyl)-4-amino-1, 2-naphthoquinone ($C_{15}H_{12}N_2O_3$) 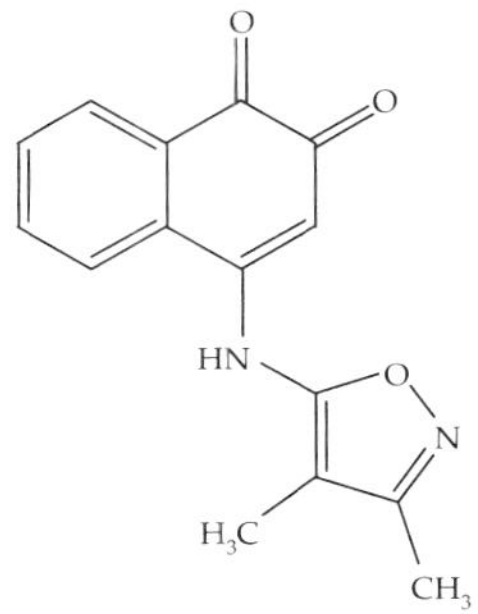	1.10	U	+H	kinetic	H_2O $t = 70$ $I = 0.5$	Longhi MR and De Bertorello MM, Isoxazoles. Part 6. Aspects of the chemical stability of a new naphthoquinone-amine in acidic aqueous solution, *J. Pharm. Sci.*, **79**, 754–757 (1990). "The chemical degradation of N-(3,4-dimethyl-5-isoxazolyl)-4-amino-1,2-naphthoquinone was studied as a function of pH and temperature. In acid and neutral pH, 4 main degradation products were identified. . . .. No significant buffer effects were observed for the buffer species used. The pH rate profile exhibited a specific acid catalysis which was important at pH values under 3.5 and an inflection point at pH 1.1 corresponding to a pK_a value. From Arrhenius plots, the activation energy was found to be 17.8+−0.3 kcal/mol. It was concluded that the compound undergoes irreversible acid hydrolysis on the carbon-4 of the naphthoquinone ring, with the aminoisoxazolyl substituent as the leaving group; specific acid catalysis with a pH-independent plateau occurs from pH 3.5 to 7.3." NB: Further details in Longhi MR, Ph.D. Thesis, Universidad Nacional de Cordoba (1989).
446	5,5-Dimethyl-3-(α,α,α, 4-tetrafluoro-*m*-tolyl) hydantoin ($C_{12}H_{10}F_4N_2O_2$) 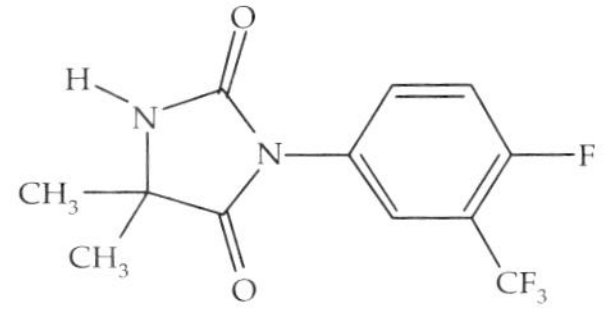	12.14 ± 0.04	U	−H	dissolution rate/ pH	H_2O $t = 25$ $I = 0.1$	Hansen JB and Hafliger O, Determination of the dissociation constant of a weak acid using a dissolution rate method, *J. Pharm. Sci.*, **72**, 429–431 (1983).
		12.11 ± 0.03	U	−H	Spectro		"A method for the determination of dissociation constants of a weak monoprotic acid, based on theories of diffusion controlled mass transport for dissolution processes, was developed and applied to the study of 5,5-dimethyl-3-(α,α,α,4-tetrafluoro-m-tolyl)hydantoin. The method includes measurements of the initial dissolution rate as a function of pH, using the rotating disk technique and determination of the intrinsic solubility. Comparison of results with an established spectrophotometric technique showed good agreement."

(continued)

No.	Name	pK_a value(s)	Data quality	Ionization type	Method	Conditions t°C; I or c M	Comments and Reference(s)
447	Dimethylamphetamine ($C_{11}H_{17}N$) 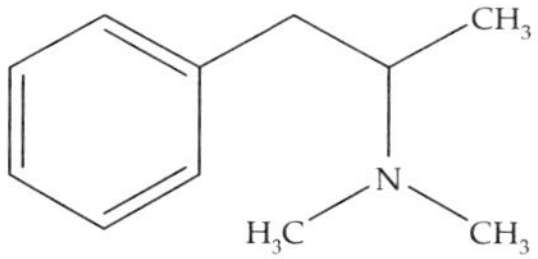	9.40 ± 0.05	U	+H	Potentiometric	H_2O $t = 25.0 \pm 0.2$ $I \leq 0.001$	Leffler EB, Spencer HM and Burger A, Dissociation constants of adrenergic amines, *JACS*, **73**, 2611–2613 (1951). NB: See Amphetamine for details. From pK_b = 4.60. NB: This is the actual value reported by Leffler *et al*. The values cited in the next two entries are claimed to come from Leffler *et al*., but it is clear that they didn't come from this source.
448	Dimethylamphetamine	8.16	U	+H			Leffler EB, Spencer HM and Burger A, Dissociation constants of adrenergic amines, *JACS*, **73**, 2611–13 (1951). Cited in: Ritschel; Beckett *et al*., *JPP*, **20**, 92–97 (1968).
449	Dimethylamphetamine	9.80	U	+H			Leffler EB, Spencer HM and Burger A, Dissociation constants of adrenergic amines, *JACS*, **73**, 2611–13 (1951). Cited in: Vree, Muskens and van Rossum, *JPP*, **21**, 774–775 (1969).
450	Dimethyloxytetracycline ($C_{24}H_{28}N_2O_9$) 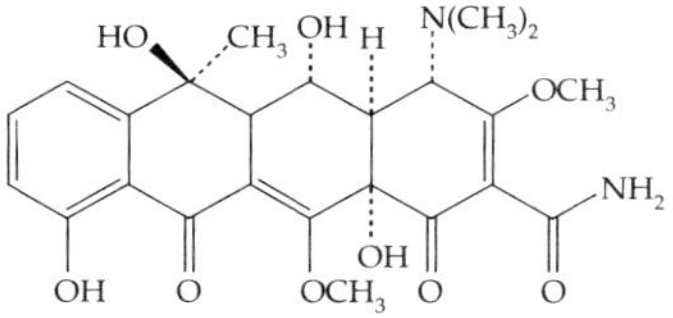	7.5 9.4	U U	−H +H	Potentiometric	H_2O $t = 25$ $c = 0.005$	Stephens C, Murai K, Brunings K and Woodward RB, Acidity constants of the tetracycline antibiotics, *JACS*, **78**, 4155–4158 (1956). Cited in Perrin 3326 ref S73. NB: The study used pH measurements with a glass electrode and liquid junction potentials.
451	Dinoprost (prostaglandin $F_{2\alpha}$) ($C_{20}H_{34}O_5$) 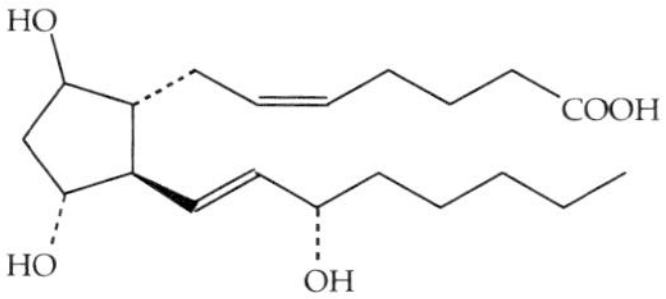	4.90	U	−H	Potentiometric	H_2O $t = 25$ $c <$ CMC ($\sim 10^{-2}$ M) N_2 atmos Activity corrections	Roseman TJ and Yalkowsky SW, Physicochemical properties of prostaglandin F2α (tromethamine salt). Solubility behaviour, surface properties and ionization constants, *J. Pharm. Sci.*, **62**, 1680–1685 (1973). NB: pK_a value increases to 5.53 ± 0.04 at $c = 2 \times 10^{-1}$ M. Solubility *versus* pH at $t = 25$ °C gave pK_a = 4.99.

No.	Compound	pKa	Reliability	Ionization	Method	Conditions	Reference
452	1,2-Dioleoylphosphatidylethanolamine (DOPE) ($C_{41}H_{78}NO_8P$)	7.55 ± 0.01	A	+H	potentio (buffer capacity *vs* pH)	H_2O $t = 25.0 \pm 0.1$ $I = 0.01$	Lobo BA, Koe GS, Koe JG and Middaugh CR, Thermodynamic analysis of binding and protonation in DOTAP/DOPE (1:1): DNA complexes using isothermal titration calorimetry, *Biophys. Chem.*, **104**, 67–78 (2003).
		7.7 ± 0.1	U (− DNA)	+H	ITC *versus* pH	$I = 0.15$	NB: ITC = isothermal titration microcalorimetry. The enthalpy of ionization was plotted as a function of solution pH; − DNA = titration performed in the absence of DNA; + DNA = titration performed in the presence of DNA complexed with DOPE.
		8.8 ± 0.1	U (+ DNA)	+H	ITC *versus* pH		
453	Dipeptides				Potentiometric	H_2O $t = 25 \pm 1$ $I = 0.1$ (KCl) Ar atmosphere	Vallat P, Gaillard G, Carrupt P-A, Tsai R-S and Testa B, Structure-lipophlicity and structure-polarity relationships of amino acids and peptides, *Helv. Chim. Acta*, **78**, 471–485 (1995). NB: pK_a values were measured with a Sirius PCA101 autotitrator as described elsewhere (Avdeef A, Kearney DL, Brown JA and Chemotti AR, Jr., Bjerrum plots for the determination of systematic concentration errors in titration data, *Anal. Chem.*, **54**, 2322–26 (1982); Avdeef A, pH-metric log P. Part 1. Difference plots for determining ion-pair octanol-water partition coefficients of multiprotic substances, *Quant-Struct. Act. Relat.*, **11**, 510–517 (1992)). The dipeptide Leu-His displayed a further value for the histidinyl residues of 6.69. All values are assigned a reliability of U, as no experimental errors were reported for the measured pK_a values.
	Ala-Phe	3.08; 7.91		−H; +H			
	Ala-Ile	3.34; 8.01		−H; +H			
	Ala-Leu	3.35; 8.02		−H; +H			
	Phe-Phe	2.98; 7.17		−H; +H			
	Phe-Gly	3.60; 7.38		−H; +H			
	Phe-Leu	3.41; 7.20		−H; +H			
	Phe-Ser	3.02; 7.48		−H; +H			
	Phe-Tyr	3.19; 7.14		−H; +H			
	Gly-Phe	2.93; 8.12		−H; +H			
	Gly-Gly	3.10; 8.08		−H; +H			
	Gly-Trp	3.14; 8.06		−H; +H			
	Phe-Phe	3.20 ± 0.01	A	−H		H_2O $t = 25.0$ $I = 0.15$ (KCl)	Sirius Technical Application Notes, vol. **2**, p. 6 (1995). Sirius Analytical Instruments Ltd., Forest Row, East Sussex, RH18 5DW, UK.
		7.18 ± 0.01	A	+H			
	Phe-Phe-Phe	3.37 ± 0.01	A	−H			
		7.04 ± 0.01	A	+H			

Dipeptide	pK_{a1}; pK_{a2}	Ionization	Dipeptide	pK_{a1}; pK_{a2}	Ionization
Leu-Phe	3.25; 7.70	−H; +H	Ser-Leu	3.35; 7.30	−H; +H
Leu-His	2.76; 7.78	−H; +H	Val-Gly	3.21; 8.00	−H; +H
Leu-Tyr	3.32; 7.69	−H; +H	Val-Tyr	3.23; 7.78	−H; +H
Met-Leu	3.39; 7.31	−H; +H	Trp-Phe	3.20; 7.30	−H; +H
Ser-Phe	2.93; 7.23	−H; +H	Trp-Gly	3.12; 7.76	−H; +H

(continued)

No.	Name	pK_a value(s)	Data quality	Ionization type	Method	Conditions t°C; I or c M	Comments and Reference(s)
454	Diphenhydramine ($C_{17}H_{21}NO$) $(C_6H_5)_2CHOCH_2CH_2N(Me)_2$	9.12	U	+H		H_2O $t = 25.0$	Andrews AC, Lyons TD, O'Brien TD and *JCS*, 1776–80 (1962). NB: Cited in Holcomb IJ, Fusari SA, Diphenhydramine hydrochloride, *APDS*, **3**, 173–232 (1974).
		9.67	U	+H		H_2O $t = 0.0$	"Andrews determined the ionization constant of diphenhydramine at 0°, $pK'_a = 9.67$, and 25°, $pK'_a = 9.12$ in water. These values compare well with those obtained by Lordi (NB: Lordi NG and Christian JE, Physical properties and pharmacological activity: antihistaminics, *J. Am. Pharm. Assn.*, Sci. Edn., **45**, 300–305 (1956)) of $pK'_a = 9.00$ in water. DeRoos (NB: deRoos AM, Rekker RF and Nauta WT, Arzneim.-Forsch., 20, 1763–65 (1970)) has determined the pK'_a at 20° to be 9.06 in water. The pK_a of diphenhydramine USP has been determined in a water:methanol (1:1) system to be 8.4 (Spurlock CH and Parke Davis & Co., personal communication). Since the pK'_a varies slightly with the alcohol content, the value obtained is acceptable."
455	Diphenhydramine	9.00	U	+H	Potentiometric	H_2O $t = 25$	Lordi NG and Christian JE, Physical properties and pharmacological activity: antihistaminics, *J. Am. Pharm. Assn., Sci. Edn.*, **45**, 300–305 (1956). NB: $pK'_a = 9.00$ in water. See Chlorpheniramine (no. 1704) for details.
456	Diphenhydramine	9.02 ± 0.09	U	+H	Potentiometric	H_2O $t = 20.0$	deRoos AM, Rekker RF and Nauta WT, The base strength of substituted 2-(diphenylmethoxy)-N,N-dimethylethylamines, *Arzneim.-Forsch.*, **20**, 1763–1765 (1970).
	Analogues						"The dissociation constants were established potentiometrically. As on titration in an aqueous medium the base liberated from the salt would precipitate, the pK_a was detemined in several alcohol-water mixtures (80/20, 70/30, etc.). The straight line through the measuring (*sic*) points was obtained by the method of least squares using equidistant X values. Next the pK_a value in water was determined by extrapolation. In the same way the experimental error was calculated."
	2-methyl	8.91 ± 0.11	U				NB: As is well-known, this method of extrapolation often leads to errors due to the non-linearity of these plots, especially for amines at low alcohol percentages. It is a pity that the raw data (apparent pK_a vs alcohol-water composition) was not available, so that more recent procedures, such as the Yasuda-Shedlovsky equation, could be used in an attempt to extract more reliable pK_a values.
	3-methyl	8.95 ± 0.10	U				NB: The substitutions indicated in the first column that are primed numbers refer to one phenyl ring, while unprimed numbers are the corresponding positions on the other ring.
	4-methyl	9.03 ± 0.11	U				
	2,2′-dimethyl	8.99 ± 0.04	U				
	4,4′-dimethyl	9.18 ± 0.09	U				
	2,6-dimethyl	9.02 ± 0.12	U				
	3,5-dimethyl	8.88 ± 0.15	U				
	2,2′,6-trimethyl	8.77 ± 0.11	U				
	2,2′,6,6′-tetramethyl	8.55 ± 0.10	U				
	3,3′,5,5′-tetramethyl	9.11 ± 0.16	U				
	2,2′-diethyl	8.52 ± 0.21	U				
	2,6-diethyl	8.53 ± 0.15	U				
	2,2′,6,6′-tetraethyl	8.33 ± 0.02	U				
	2-t-butyl	9.01 ± 0.09	U				
	4-methyl-3′-bromo	8.84 ± 0.13	U				
	4-methyl-3′-trifluoromethyl	8.70 ± 0.15	U				
	4-methyl-4′-chloro	8.84 ± 0.06	U				

	4-methyl-4′- trifluoromethyl	8.77 ± 0.13	U				
	4-methyl-4′-nitro	8.72 ± 0.18	U				
	4-methoxy	9.04 ± 0.07	U				
	3,3′-dimethoxy	8.86 ± 0.13	U				
	4,4′-dimethoxy	9.19 ± 0.09	U				
	2-trifluoromethyl	8.98 ± 0.08	U				
	3-bromo	8.80 ± 0.10	U				
	4-fluoro	8.84 ± 0.05	U				
	4-nitro	8.73 ± 0.08	U				
	4-methoxy-4′-chloro	9.03 ± 0.01	U				
457	Diphenic acid ($C_{14}H_{10}O_4$) 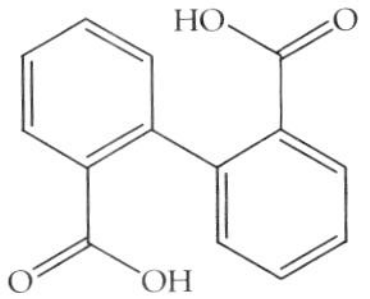	3.20 5.06	U U	−H −H	Potentiometric	H_2O $t = 23.0$	Clarke FH and Cahoon NM, Ionization constants by curve-fitting: Determination of partition and distribution coefficients of acids and bases and their ions, *J. Pharm. Sci.*, **76**, 611–620 (1987). NB: See Benzoic acid for further details.
458	Diphenoxylate ($C_{30}H_{32}N_2O_2$) 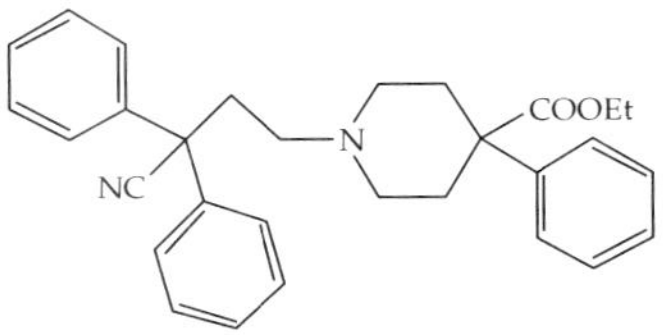	7.07	U	+H	Potentiometric	H_2O $t = 25$	Peters JJ, Determination of ionization constants in mixed aqueous solvents of varying composition by a single titration, *J. Pharm. Sci.*, **67**, 127–129 (1978). NB: The paper recognized the problem of long extrapolations from aqueous-organic solvent mixtures to estimate pK_a values of poorly water-soluble substances. No activity corrections were applied. See Cinnarizine for further details.
459	Diphenylpyraline ($C_{19}H_{23}NO$) 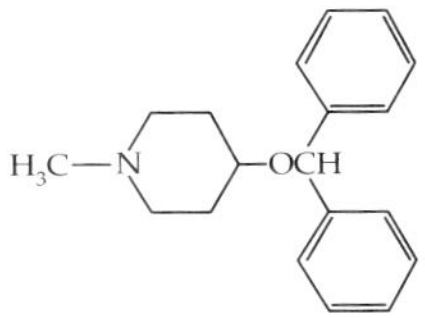	8.90 ± 0.06	U	+H	Potentiometric	H_2O t undefined $I = 0.30$ (NaCl)	Testa B and Murset-Rossetti L, The partition coefficient of protonated histamines, *Helv. Chim. Acta*, **61**, 2530–2537 (1978). NB: See Cycliramine for details.

(continued)

No.	Name	pK_a value(s)	Data quality	Ionization type	Method	Conditions t°C; I or c M	Comments and Reference(s)
460	Disopyramide ($C_{21}H_{29}N_3O$) 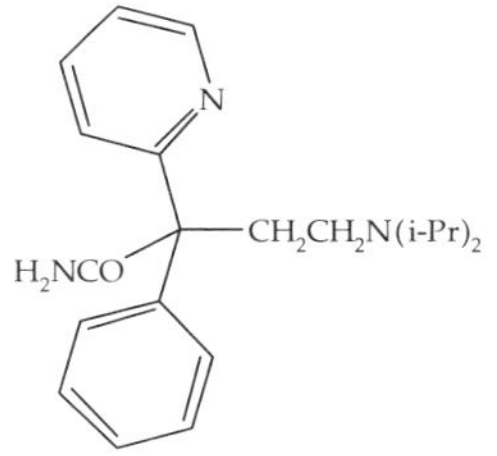	< 2.0 10.2	U U	+H +H	Potentiometric	H_2O $t = 20–23$	Czeisler JL and El-Rashidy RM, Pharmacologically active conformation of disopyramide: Evidence from apparent pK_a measurements, *J. Pharm. Sci.*, **74**, 750–754 (1985). "Autotitrator (Parke and Davis method). . . . Evidence for an intramolecular hydrogen bond between the amido and pyridine groups of disopyramide is presented, and effects of the resultant constraint on the rotation of the pyridine ring on anti-arrhythmic activity and suppression of anticholinergic activity are discussed . . . ."
		10.45	U	+H	Potentiometric	H_2O *t* undefined	Hinderling PH, Bres J and Garrett ER, Protein binding and erythrocyte partitioning of disopyramide and its monodealkylated metabolite, *J. Pharm. Sci.*, **63,** 1684 (1974). NB: This paper reported pK_2 = 10.45, also from the Parke and Davis differential method.
461	Disopyramide	8.36	U	+H			Anon: Norpace (disopyramide phosphate) Prescribing information brochure 7N17, Searle Laboratories, Chicago, IL, 1977.
462	Disopyramide	7.76	U	+H	soly	H_2O $t = 27$	Calculated by this author from the following solubility-pH data in Wickham A and Finnegan P, Disopyramide phosphate, *APDS*, **13**, 183–209 (1984):

pH	solubility (mg/mL)	pH	solubility (mg/mL)
3.93	62.0	7.98	4.32
5.85	32.0	9.86	3.78
7.45	8.71	12.7	1.29

No.	Name	pK_a value(s)	Data quality	Ionization type	Method	Conditions t°C; I or c M	Comments and Reference(s)
463	DMP-777 ($C_{31}H_{40}N_4O_6$) 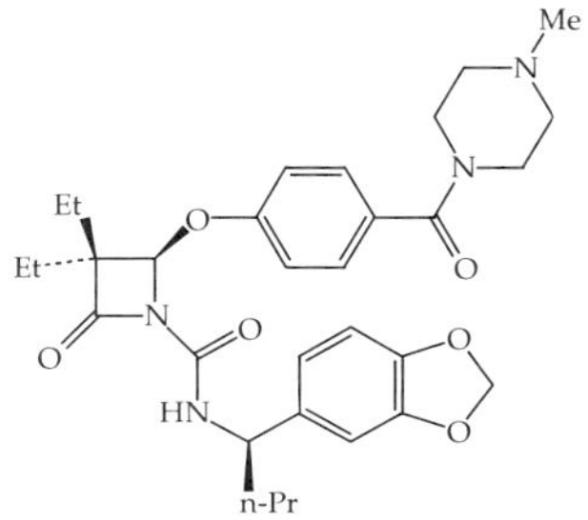	7.04	U	+H	Potentiometric	H_2O t undefined *I* undefined	Raghavan KS, Gray DB, Scholz TH, Nemeth GA and Hussain MA, Degradation kinetics of DMP-777, an elastase inhibitor, *Pharm. Res.*, **13**, 1815–1820 (1996). "The apparent dissociation constant of the protonated piperazine nitrogen in DMP 777 was determined at 22 °C by potentiometric titration in a mixture of water and methanol, and extrapolating the pK_a values so obtained to 100% water. The pK_a of the protonated piprazine nitrogen in DMP 777 was estimated to be 7.04."

No.	Compound	pKa	Rating	Type	Method	Conditions	Reference
464	DOPA, L- (Levodopa) ($C_9H_{11}NO_4$) 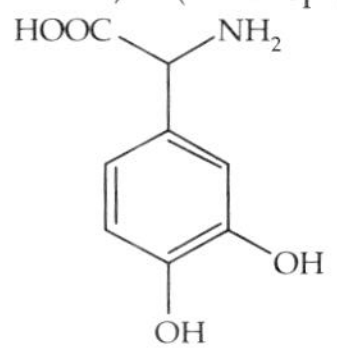 HOOC, NH_2, OH, OH	8.81 ± 0.03 9.94 ± 0.07	A U	−H, +H +H, −H	Potentiometric, Spectro	H_2O $t = 37$ $I = 1.0$	Schüsler-van Hees, MTIW, Reactivity of Catechol(amine)s, Ph.D. thesis, Ctr for Biopharm. Sci., Leiden University, 1–93 (1983). NB: Microscopic values: 9.01 ± 0.01; 9.24 ± 0.06; 9.73 ± 0.10; and 9.50 ± 0.03. Schüsler-van Hees, MTIW, Beijersbergen van Henegouwen GMJ, Driever MFJ, Ionization constants of catechols and catecholamines, *Pharm. Weekblad*, Sci. Edn., **5**, 102–108 (1983).
465	Dopamine ($C_8H_{11}NO_2$) 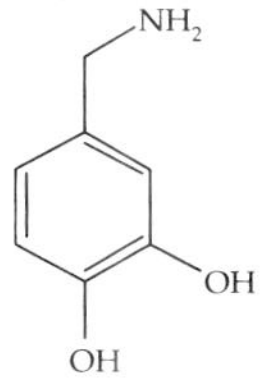 NH_2, OH, OH	8.57 ± 0.03 10.08 ± 0.10	A U	+H,−H −H,+H	Potentiometric, Spectro	H_2O $t = 37$ $I = 1.0$	Schüsler-van Hees, MTIW, Reactivity of Catechol(amine)s, Ph.D. thesis, Ctr for Biopharm. Sci., Leiden University, 1–93 (1983). NB: Microscopic values: 8.79 ± 0.01; 8.98 ± 0.05; 9.87 ± 0.13; and 9.68 ± 0.07. Schüsler-van Hees, MTIW, Beijersbergen van Henegouwen GMJ, Driever MFJ, Ionization constants of catechols and catecholamines, *Pharm. Weekblad*, Sci. Edn. **5**, 102–108 (1983).
466	Dothiepin (dosulepine) ($C_{19}H_{21}NS$) 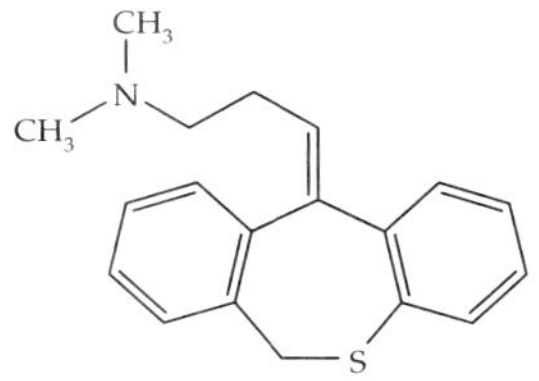 CH_3, N, CH_3, S	9.25	U	+H	Potentiometric	H_2O $t = 25$ I undefined Ar atmosphere	Seiler P, Simultaneous determination of partition coefficient and acidity constant of a substance, *Eur. J. Med. Chem.*, **9**, 663–665 (1974). NB: See Amitriptylline for details.
467	Doxepin ($C_{19}H_{21}NO$) 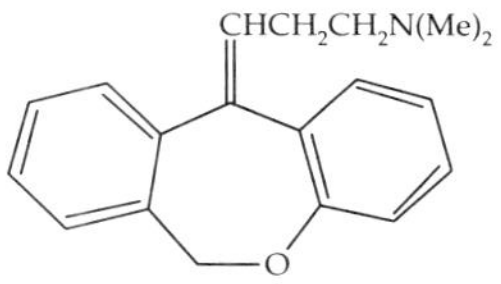 $CHCH_2CH_2N(Me)_2$, O	8.96	U	+H	soly	H_2O $I = 0.00$ $t = 25.0$	Embil K and Torosian G, Solubility and ionization characteristics of doxepin and desmethyldoxepin, *J. Pharm. Sci.*, **71**(2), 191–193 (1982).

(continued)

No.	Name	pK_a value(s)	Data quality	Ionization type	Method	Conditions t°C; *I* or c M	Comments and Reference(s)
468	Doxepin	9.4	U	+H	Potentiometric	mixed solvs	Thoma K and Albert K, Kuramitteilungen Schnell-methode zur potentiometrischen Bestimmung von pK_a-Werten in gemischten Lösungsmittelsystemen, *Arch. Pharm.*, **314**(12), 1053–1055 (1981).''Doxepin-HCl, Amitriptylin-HCl, und Imipramin-HCl errechmen sich Dissoziationskanstanten von 9,4, 9,5, bzw. 9,7.''
469	Doxorubicin (Adriamycin) $C_{27}H_{29}NO_{11}$) 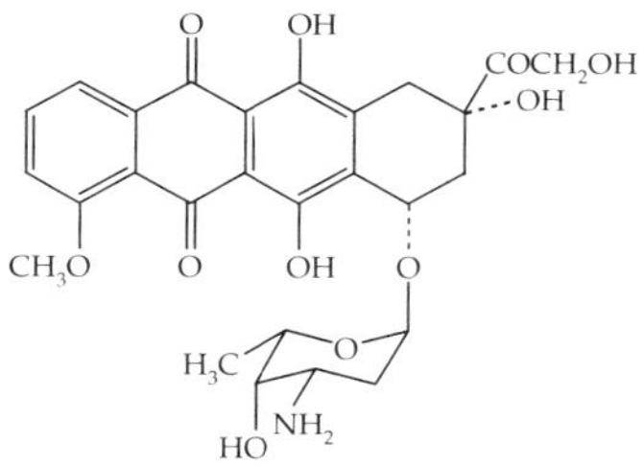	−5.9 ± 0.05 8.15 ± 0.07 10.16 ± 0.09 13.2 ± 0.2	U U U U	+H −H +H −H	Spectro	H_2O	Vigevani A and Williamson MJ, Doxorubicin, *APDS*, **9**, 245–274 (1980). ''Solutions of doxorubicin show indicator-like properties, turning from orange-red to blue-violet about pH = 9 (ref 13). Values of −5.9, 8.2, 10.2, and 13.2 for pK_1, pK_2, pK_3, and pK_4, determined by spectrophotometric methods, have been reported (Ref 19). A pK_a of 8.22 was determined for the hydrochloride with N/20 sodium hydroxide. 13. Arcamone F, Cassinelli G, Franceschi G, *et al.*, Intl. Symposium on Adriamycin, Carter SK, DiMarco A, Ghione M, *et al.* (eds.), Springer-Verlag, Berlin, 1–22 (1972). 19. Sturgeon R and Schulman SG, *J. Pharm. Sci.*, **66**, 958–961 (1977).'' NB: See below.
		8.22	U	−H	Potentiometric	H_2O	
470	Doxorubicin	8.22 10.2	U U	−H +H	Spectro	H_2O	Sturgeon R and Schulman SG, Electronic absorption spectra and protolytic equilibria of doxorubicin: Direct spectrophotometric determination of microconstants, *J. Pharm. Sci.*, **66**, 958–961 (1977).
471	Doxycycline ($C_{22}H_{24}N_2O_8$) 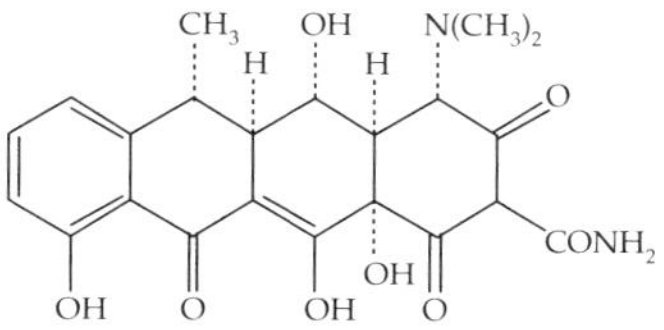	3.21 7.56 8.85 11.54	U U U U	−H −H +H −H	Potentiometric	H_2O *t* = 25	Bergström CAS, Strafford M, Lazorova L, Avdeef A, Luthman K and Artursson P, Absorption classification of oral drugs based on molecular surface properties, *J. Med. Chem.*, **46**(4), 558–570 (2003). NB: From extrapolation of aqueous-methanol mixtures to 0% methanol.
472	Ebifuramin ($C_{19}H_{22}N_4O_7$) 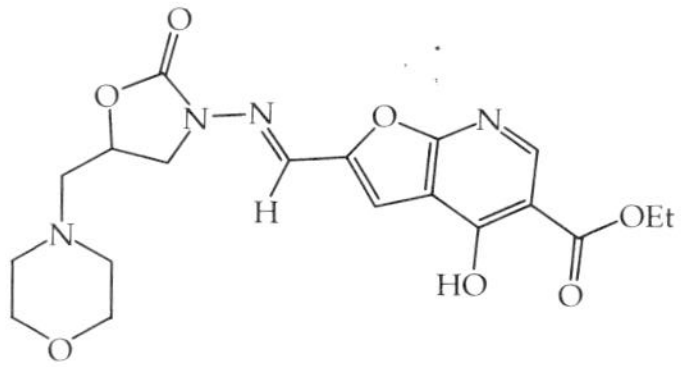	5.24 ± 0.04 (25) 5.16 ± 0.03 (t = 37)	A A	−H −H	Spectro	H_2O *t* = 25 *t* = 37 *I* = 0.1 (NaCl or KCl)	Prankerd RJ and Stella VJ, Equilibria and kinetics of hydrolysis of ebifuramin (NSC-201047), an azomethine-containing structure exhibiting a reversible degradation step in acidic solutions, *Int. J. Pharm.*, **52**, 71–78 (1989). ''Spectrophotrometric and potentiometric pK_a values were obtained by standard methods (Albert and Serjeant, 1971). UV-vis spectra (200–500 nm) were obtained with HP8451A Diode Array or Shimadzu UV-260 spectrophotometers. pH values were measured with Fisher Accumet 610A or Brinkman Metrohm 632 pH meters. The following buffer systems (μ = 0.1M ; NaCl or KCl) were employed for pK_a determinations: HCl (0.1M)and 0.01M), sodium

No.	Compound	pKa	Data quality	Ionization	Method	Conditions	Reference / Notes
							acetate (pH 4.0–5.4), potassium succinate (pH 5.9), sodium phosphate (pH 6.9, 7.5) and Tris-HCl (pH 8.0). Aliquots (25 uL) of stock solutions (2.4×10^{-2} M) of I and III (in dilute methanesulfonic acid) and II (in 95% ethanol) were diluted to 25.0 mL immediately before spectra were determined. Time-dependent observations showed that no significant degradation occurred during measurements. IV was titrated with standard HCl (Fisher Certified) at an ionic strength of 0.1M.''
473	Ecgonine ($C_9H_{15}NO_3$) 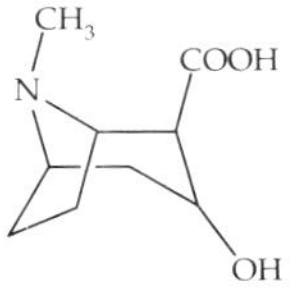	10.21	U	+H	Potentiometric	H_2O $t = 25$ $c = 0.02$	Chilton J and Stenlake JB, Dissociation constants of some compounds related to lysergic acid: Beta-dimethylaminopropionic acid, dihydroarecaidine, ecgonine and their derivatives, *J. Pharm. Pharmacol.*, **7**, 1004–1011 (1955). Cited in Perrin 2862 ref. C27. NB: The potentiometric study used pH measurements with a glass electrode and junction potentials. See Aconitine for details of the spectrophotometric study.
		11.15	U	+H	Spectro		
474	Ecgonine	11.1	U	+H	optical rotation versus pH	H_2O	Avico U, Optical rotation of ecgonine and anhydroecgonine. Rend 1st Super. Sanita 26, 1024–30 (1963); CA 61:11361b. Cited in Perrin Bases Supplement 7472. NB: Reported $pK_b = 11.2$ for ecgonine. For anhydroecgonine, reported $pK_a = 9.8$ and $pK_b = 10.2$.
475	Ecgonine methyl ester ($C_{10}H_{17}NO_3$) 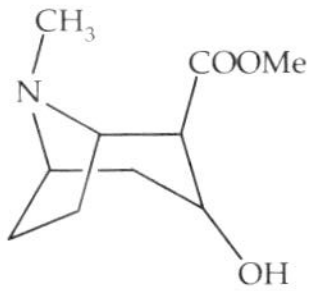	9.16	U	+H	Potentiometric	H_2O $t = 25$ $c = 0.02$	Chilton J and Stenlake JB, Dissociation constants of some compounds related to lysergic acid: Beta-dimethylaminopropionic acid, dihydroarecaidine, ecgonine and their derivatives, *J. Pharm. Pharmacol.*, **7**, 1004–1011 (1955). Cited in Perrin 2862 ref. C27. The study used pH measurements with a glass electrode and liquid junction potentials.
476	Edetic acid (EDTA) ($C_{10}H_{16}N_2O_8$) 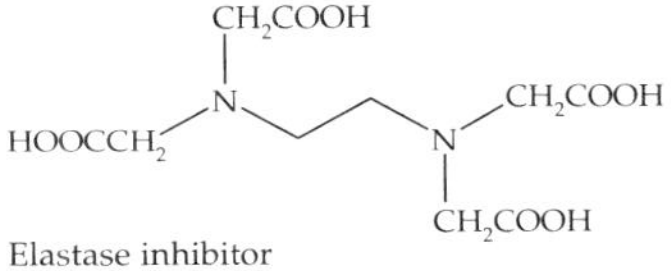	2	U	−H			Belal F and Al-Badr AA, Edetic Acid, *APDS*, **29**, 63, 2002. NB: See also Connors KA, *A Textbook of Pharmaceutical Analysis*, 3rd Edn., Wiley-Interscience, New York, p. 78 (1982). NB: Also called ethylenediaminetetraacetic acid.
		2.67	U	−H			
		6.16	U	−H			
		10.26	U	−H			
477	Elastase inhibitor See DMP-777	7.04	U	+H	Spectro	H_2O	Raghavan KS, Gray DB, Scholz TH, Nemeth GA and Hussain MA, Degradation kinetics of DMP-777, an elastase inhibitor, *Pharm. Res.*, **13**, 1815–1820 (1996). NB: The pK_a for the protonated piperazine nitrogen was estimated to be 7.04.

(*continued*)

No.	Name	pK_a value(s)	Data quality	Ionization type	Method	Conditions t°C; I or c M	Comments and Reference(s)
478	Emetine ($C_{29}H_{40}N_2O_4$) 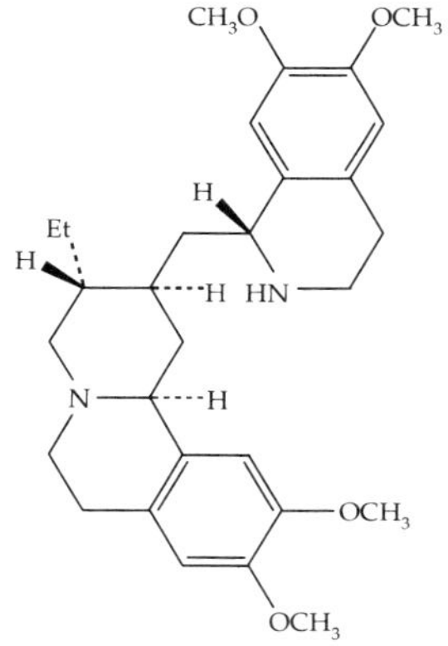	7.20 8.07	U U	+H +H		H_2O $t = 40.0$ ($K_w = 13.535$) I undefined	Feyns LV and Grady LT, Emetine hydrochloride, *APDS*, **10**, 289–322 (1981). NB: The following pK_b values have been reported for water: at 15 ° 5.77 and 6.64 (refs 22, 60); at 40° 5.47 and 6.34 (ref 61) Other (pK_b) values reported: 5.73 and 6.74 (no indication on the temperature) (ref. 59). 22. Beilstein's Handbuch der Organischen Chemie, 4 Aufl., 2. Erg.-Werk, Bd. XXIII, Springer-Verlag, Berlin, 1954, p. 449 59. Das Gupta V and Herman HB, Selection of best pH range for extraction of amine-bromthymol blue complexes, *J. Pharm. Sci.*, **62(2)**, 311–313 (1973); this reference merely cited Parrott's Pharmaceutical Technology or Martin's Physical Pharmacy. 60. Auterhoff H and Moll F, The general importance of the reaction of alkaloids of the secondary amine type with formaldehyde, *Arch. Pharm.*, **293**, 132–141 (1960) (see below, no. 479). 61. Volpi A, Relation between physical constants therapeutic doses of some organic bases, *Boll. Chim. Farm.*, **115**, 466–474 (1976).
479	Emetine	7.71 8.58	U U	+H +H		H_2O $t = 15.0$ ($K_w = 14.346$) I undefined	Auterhoff H and Moll F, The general importance of the reaction of alkaloids of the secondary amine type with formaldehyde, *Arch. Pharm.*, **293**, 132–141 (1960). NB: This paper also reported K_b values for the following bases: piperidine (1.6×10^{-3} at 25 °C); coniine (1.3×10^{-3} at 25 °C); conhydrine (2×10^{-4} at 18 °C); theophylline (1.9×10^{-14} at 25 °C); theobromine (1.3×10^{-14} at 25 °C); yohimbine (1×10^{-11} and 2.8×10^{-7} at 23 °C).
480	Emetine	7.56 8.43	U U	+H +H	Spectro	H_2O $t = 15.0$ $c = 0.003$ to 0.005	Kolthoff IM, The dissociation constants, solubility product and titration of alkaloids, *Biochem. Z.*, **162**, 289–353 (1925). Cited in Perrin Bases 2901. NB: See Aconitine for details of the spectrophotometric method. The data were not corrected for overlapping pK_a values. The potentiometric study used hydrogen electrodes in an asymmetric cell with liquid junction potentials.
		8.22	U	+H	Potentiometric	H_2O $t = 16.0$	NB: pK_w = 14.346 at 15 °C; pK_w = 14.382 at 16 °C
481	Enalapril ($C_{20}H_{28}N_2O_5$) 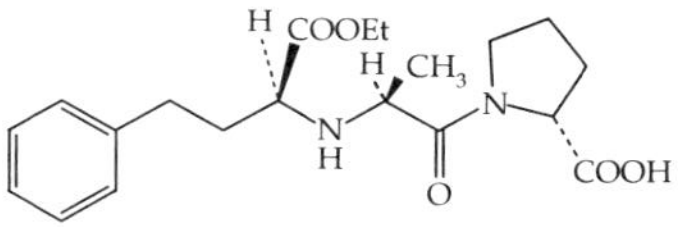	2.97 5.35	A A	−H +H	Potentiometric	H_2O $t = 25.0$	Ip DP and Brenner GS, Enalapril Maleate, *APDS*, **16**, 207–243 (1987). "Aqueous acidic/basic potentiometric titration yielded pK_a values of 2.97 and 5.35 at 25 °C for enalapril." McCauley JA, Merck, Sharp and Dohme Research Laboratories, personal communication.

No.	Compound	pKa			Method	Conditions	Reference
482	Enalapril	2.98	A	−H	CE/pH	H_2O	Wan H, Holmen AG, Wang Y, Lindberg W, Englund M, Nagard MB and Thompson RA, High-throughput screening of pK_a values of pharmaceuticals by pressure-assisted capillary electrophoresis and mass spectrometry, *Rapid Commun. Mass Spectrom.*, **17**, 2639–2648 (2003). NB: Reported predicted values (ACD Labs) of 3.17 and 5.42.
		5.37	A	+H	(+ve ion mode)	$t = 25$	
						$I = 0.025$	
483	Enkephalin (met-enkephalin) ($C_{27}H_{35}N_5O_7S$)	3.68 ± 0.05	A	−H	Potentiometric	H_2O	Ishimitsu T and Sakurai H, Stucture-ionization relationships of enkephalin and related fragments in aqueous solution, *Int. J. Pharm.*, **12(3)**, 271–274 (1982).
		7.77 ± 0.03	A	+H		$t = 25.0$	
		9.89 ± 0.03	A	−H		$I = 0.1$	
		9.96 ± 0.08	U	−H	CTS		

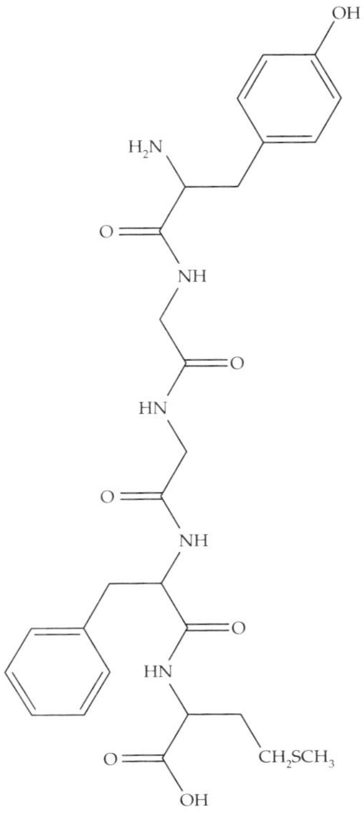

"The micro-constants of enkephalin and tyrosine peptides... were determined according to the modified method of complementary tri-stimulus colorimetry (CTS method) (Flaschka H, Applications of complementary tri-stimulus colorimetry – I. Analysis of binary and ternary colorant systems, *Talanta*, **7**, 90–106 (1960). The acid dissociation and micro-constants thus determined are summarized in Table 1 and Table 2.

Table 1: Acid Dissociation constants of Tyrosine - Containing Peptides[a]

Cpd	Titration			CTS
	pK_{COOH}	pK_1	pK_2	pK_2[b]
Enkephalin	3.68±0.05	7.77±0.03	9.89±0.03	9.96±0.08
Tyr-gly-gly	3.21±0.03	7.75±0.02	9.78±0.06	9.92±0.09
Gly-tyr-gly	3.10±0.04	8.06±0.03	9.78±0.05	9.97±0.08
Gly-gly-tyr	3.18±0.03	8.17±0.07	9.75±0.09	9.94±0.07
Tyr-gly	3.51±0.02	7.77±0.04	10.04±0.02	9.97±0.07
Gly-tyr	3.24±0.06	8.23±0.04	10.55±0.06	10.39±0.04

[a] μ=0.1 ($NaClO_4$) 25 °C.
[b] The pK_2 value, which corresponds to the proton dissociation of aromatic hydroxyl group, determined by CTS method.

Table 2: Microscopic Acid Dissociation Constants and Tautomeric Constants of Tyrosine Peptides

Compound	pK_1	pK_2	pK_{12}	pK_{21}	Kt[a] (k_2/k_1)
Enkephalin	8.61 ±0.03	7.99 ±0.04	9.63 ±0.04	10.03	4.2

[a] Kt-value was calculated from the ratio k_2/k_1.

(continued)

No.	Name	pK_a value(s)	Data quality	Ionization type	Method	Conditions t°C; I or c M	Comments and Reference(s)
484	Ephedrine ($C_{10}H_{15}NO$) 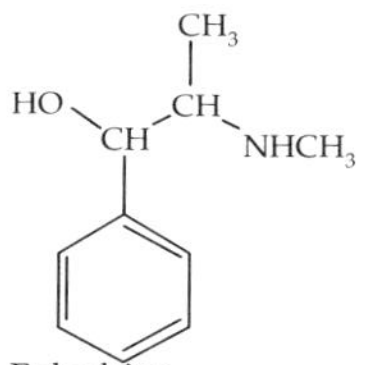	9.68 ± 0.01	A	+H	Potentiometric	H_2O $t = 20.0$ $I = 0.00$	Everett DH and Hyne JB, Dissociation constants of the isomeric (-)-ephedrinium and (+)-ψ-ephedrinium ions in water from 0° to 60°, *J. Chem. Soc.*, 1636–1642 (1958). Cited in Ali, S.L., Ephedrine hydrochloride, *APDS*, **15**, 233–283 (1986). NB: High quality work done with electrometric cells with liquid junction potentials. Thermodynamic ionization functions calculated.
485	Ephedrine	9.54 ± 0.01	A	+H	Potentiometric	H_2O $t = 25.0$ $I = 0.00$	Everett DH and Hyne JB, Dissociation constants of the isomeric (-)-ephedrinium and (+)-ψ-ephedrinium ions in water from 0° to 60°, *JCS*, 1636–1642 (1958).
486	Ephedrine	9.64 ± 0.03	A	+H	Potentiometric	H_2O $t = 25.0 \pm 0.1$ $I = 0.1$ (NaCl)	Takacs-Novak K, Box KJ and Avdeef A, Potentiometric pK_a determination of water-insoluble compounds: Validation study in methanol/water mixtures, *Int. J. Pharm.*, **151**, 235–248 (1997). NB: pK_a = 9.60 ± 0.02 by extrapolation from 15.7–64.6 %w/w aqueous MeOH. See Acetaminophen for full details.
487	Ephedrine	9.63	A	+H	Potentiometric	H_2O $t = 25.0 \pm 0.5$ $I = 0.01$	Warren RJ, Begosh PP and Zarembo JE, Identification of amphetamines and related sympathomimetic amines, *J. Assoc. Off. Anal. Chem.*, **54**, 1179–1191 (1971). NB: See Amphetamine for further details.
488	Ephedrine	9.64 ± 0.03	A	+H	Potentiometric	H_2O $t = 25.0$ $I = 0.1$ (NaCl)	Takacs-Novak K and Avdeef A, Interlaboratory study of log P determination by shake-flask and potentiometric methods, *J. Pharm. Biomed. Anal.*, **14**, 1405–1413 (1996). NB: See Acetaminophen for further details. Also reported pK_a = 9.65 ± 0.01 at I = 0.15 (KCl). The same result was reported in Sirius Technical Application Notes, vol. **2**, pp. 131–132 (1995). Sirius Analytical Instruments Ltd., Forest Row, East Sussex, RH18 5DW, UK NB: Concentration of analyte, 2.0 – 3.3 mM.
489	Ephedrine	9.58 ± 0.02	A	+H	Potentiometric	H_2O $t = 25.0 \pm 0.2$ $I \leq 0.001$	Leffler EB, Spencer HM and Burger A, Dissociation constants of adrenergic amines, *JACS*, **73**, 2611–2613 (1951). Cited in: Chatten LG and Harris LE, Relationship between pK_b(H_2O) of organic compounds and $E_{1/2}$ values in several nonaqueous solvents, *Anal. Chem.*, **34**, 1495–1501 (1962). NB: See Amphetamine for details. From pK_b = 4.42.
490	Ephedrine	GLpK_a: 9.66 ± 0.01 A&S: 9.65 ± 0.07	 A U	 +H +H	Spectro	H_2O $t = 25$ $I = 0.15$ (KCl) Ar atmosphere	Tam KY and Takacs-Novac K, Multi-wavelength spectrophotometric determination of acid dissociation constants, *Anal. Chim. Acta*, **434**, 157–167 (2001). NB: See Clioquinol for details.

No.	Compound	pKa	Type	Assignment	Method	Conditions	Reference
491	Epianhydrotetracycline 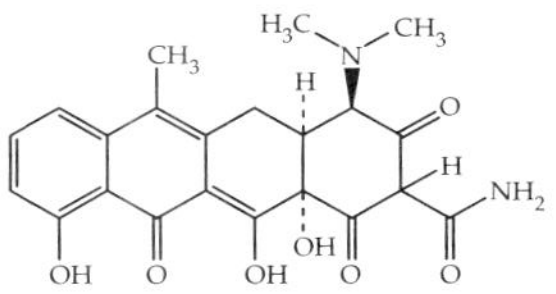	3.48 5.87 8.86	U U U	−H −H +H	Potentiometric	H_2O $t = 25 \pm 0.05$ $I = 0.01$	Benet LZ and Goyan JE, Determination of the stability constants of tetracycline complexes, *J. Pharm. Sci.*, **54**, 983–987 (1965). NB: See Chlortetracycline for details.
492	Epianhydrotetracycline (anhydro-4-epitetracycline)	4.38 – 8.95	U U U	−H −H +H	Potentiometric	H_2O $t = 30.0 \pm 0.2$ $I = 0.01$ (KCl) N_2 atmosphere	Doluisio JT and Martin AN, Metal complexation of the tetracycline hydrochlorides, *J. Med. Chem.*, **6**, 16–20 (1963). NB: Metal-free solutions of the tetracycline were titrated with standard NaOH solution and the pH measured. No details were given of the pH meter calibration. Metal stability constants were determined from identical titrations in the presence of varying concentrations of nickel(II), zinc(II), or copper(II) ions.
493	Epichlortetracycline 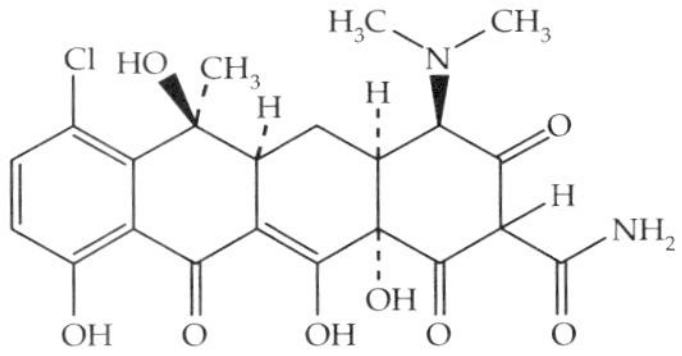	3.65 7.65 9.2	U U U	−H −H +H	Potentiometric	H_2O $t = 25 \pm 0.05$ $I = 0.01$	Benet LZ and Goyan JE, Determination of the stability constants of tetracycline complexes, *J. Pharm. Sci.*, **54**, 983–987 (1965). NB: See Chlortetracycline for details.
494	Epichlortetracycline (7-chloro-4-epitetracycline)	4.07 7.56 9.26	U U U	−H −H +H	Potentiometric	H_2O $t = 30.0 \pm 0.2$ $I = 0.01$ (KCl) N_2 atmosphere	Doluisio JT and Martin AN, Metal complexation of the tetracycline hydrochlorides, *J. Med. Chem.*, **6**, 16–20 (1963). NB: Metal-free solutions of the tetracycline were titrated with standard NaOH solution and the pH measured. No details were given of the pH meter calibration. Metal stability constants were determined from identical titrations in the presence of varying concentrations of nickel(II), zinc(II), or copper(II) ions.
495	Epicillin ($C_{16}H_{21}N_3O_4S$) 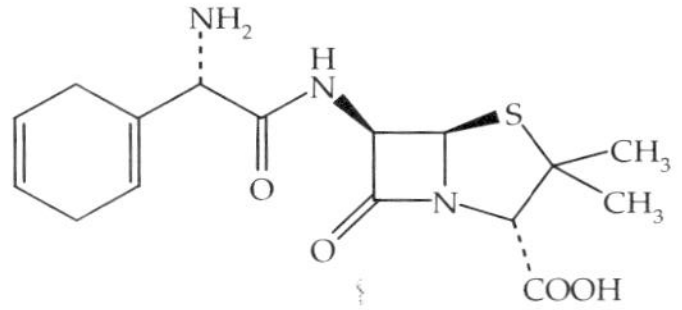	2.77 7.17	U U	−H +H	Potentiometric Potentiometric	H_2O $t = 35$ $I = 0.5$ (KCl)	Tsuji A, Nakashima E, Hamano S and Yamana T, Physicochemical properties of amphoteric β-lactam antibiotics, *J. Pharm. Sci.*, **67**, 1059–1066 (1978). NB: See Amoxicillin for details.

(continued)

No.	Name	pK_a value(s)	Data quality	Ionization type	Method	Conditions t°C; I or c M	Comments and Reference(s)
496	Epinastine ($C_{16}H_{15}N_3$) 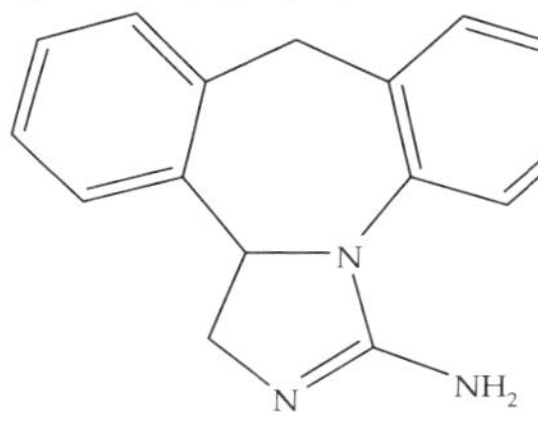	11.2	U	+H	Potentiometric	H_2O $t = 20$	Walther G, Daniel H, Bechtel WD and Brandt K, New tetracyclic guanidine derivatives with H_1-antihistaminic properties: chemistry of epinastine, *Arzneim.-Forsch.*, **40**, 440–446 (1990). "The pK_a values were determined potentiometrically at 20 °C using the method Ebel *et al.* (Ebel S, Binder A, Mohr H, Metrohm Monographien, p. 24, Metrohm AG, Herisau, 1977)." NB: The paper also reported a pK_a value of 7.5 for Mianserin.
497	Epinephrine (adrenaline) ($C_9H_{13}NO_3$) 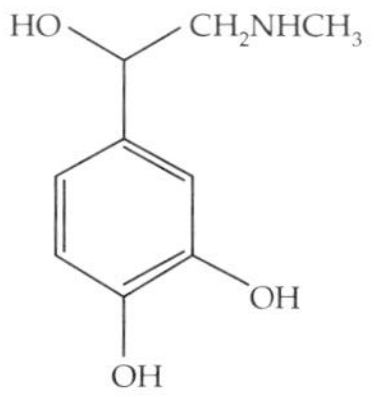	8.69 9.90	U U	−H, +H +H, −H	Spectro	H_2O $t = 25$ $I = 0.1$	Martin RB, Zwitterion formation upon deprotonation in L-3,4-dihydroxyphenylalanine and other phenolic amines, *J. Phys. Chem.*, **75**, 2657–2661 (1971). Cited in Szulczewski DH, Hong W-H and Epinephrine, *APDS*, **7**, 193–229 (1978); other references were also given. microscopic: 8.79; 10.10; 8.88; 9.51 $t = 25$, $I \sim 0$ microscopic: 8.66; 9.95; 8.72; 9.57 $t = 25$, $I = 0.1$ microscopic: 8.71; 9.90; 8.81; 9.39 $t = 20$, $I = 0.1$ NB: Also gave corresponding data for p-tyramine, tyrosine ethyl ester, tyrosine, DOPA, dopamine, norepinephrine, and isopropylnorepinephrine.
498	Epinephrine (adrenaline)	8.42 ± 0.02 9.66 ± 0.10	U U	−H, +H +H, −H	Potentiometric, Spectro	H_2O $t = 37$ $I = 1.0$	Schüsler-van Hees MTIW, Reactivity of Catechol(amine)s, Ph.D. thesis, Ctr for *Biopharm. Sci.*, Leiden University, 1–93 (1983). NB: Gave also several literature refs. Microscopic: 8.56 ± 0.01; 8.99 ± 0.06; 9.52 ± 0.10; 9.09 ± 0.03; macroscopic: 8.73 and 10.14.
		8.73 10.14	U U				Schüsler-van Hees, MTIW, Beijersbergen van Henegouwen GMJ and Driever MFJ, Ionization constants of catechols and catecholamines, *Pharm. Weekblad*, Sci. Edn., **5**, 102–108 (1983).
499	Epinephrine (adrenaline)	8.73 10.14	U U	−H, +H +H, −H	Spectro	H_2O $t = 25.0 \pm 0.05$ $I = 0.10$	Ijzerman AP, BultsmaT, Timmerman H and Zaagsma J, The ionization of β-adrenoceptor agonists: A method for unravelling ionization schemes, *J. Pharm. Pharmacol.*, **36(1)**, 11–15 (1984). NB: Microscopic: 8.69 and 10.14; macroscopic: 8.73 and 10.14. See Isoprenalin.
500	Epinephrine (adrenaline)	8.75 9.89	U U	−H, +H +H, −H	Spectro Potentiometric	pK_{a1}: H_2O $t = 25$ I undefined pK_{a2}: H_2O $t = 35$ I undefined	Kappe T and Armstrong MD, Ultraviolet absorption spectra and apparent acidic dissociation constants of some phenolic amines, *J. Med. Chem.*, **8**, 368–374 (1965). Cited in: Schwender CF, Levarterenol bitartrate, *APDS*, **1**, 339–361 (1972). NB: See Levarterenol (Norepinephrine) for details.

501	Epinephrine	8.50	U	+H	Potentiometric	H_2O $t = 25.0 \pm 0.2$ $I \leq 0.001$	Leffler EB, Spencer HM and Burger A, Dissociation constants of adrenergic amines, *JACS*, **73**, 2611–2613 (1951). NB: See Amphetamine for details; from $pK_b = 5.50$; no attempt made to unravel microconstants.
502	Epinephrine	8.55	U	+H	Potentiometric	H_2O t undefined I undefined	Tuckerman MM, Mayer JR and Nachod FC, Anomalous pK_a values of some substituted phenylethylamines, *JACS*, **81**, 92–94 (1959). NB: Method as described by Parke and Davis, 1945.
503	4-Epitetracycline	4.8 8.0 9.3	U U U	−H −H +H	NMR, Potentiometric	$MeOH/H_2O$ (1:1) $t = 30 \pm 2$ (NMR) $t = 26 \pm 1$ (potentio)	Rigler NE, Bag SP, Leyden DE, Sudmeier JL and Reilley CN, Determination of a protonation scheme of tetracycline using nuclear magnetic resonance, *Anal. Chem.*, **37**, 872–875 (1965). NB: See Tetracycline for details.
504	Equilenin ($C_{18}H_{18}O_2$)	9.72	U	−H	Soly	H_2O $t = 25.0 \pm 0.02$	Hurwitz AR and Liu ST, Determination of aqueous solubility and pK_a values of estrogens, *J. Pharm. Sci.*, **66**, 624–627 (1977).
		9.75 ± 0.06	U	−H	Spectro	H_2O $t = 23 \pm 2$	NB: Spectroscopic study used 2 cm cells for improved sensitivity. See 17α-Estradiol for further details.
505	Equilin ($C_{18}H_{20}O_2$)	10.26 ± 0.04	U	−H	Spectro	H_2O $t = 23 \pm 2$	Hurwitz AR and Liu ST, Determination of aqueous solubility and pK_a values of estrogens, *J. Pharm. Sci.*, **66**, 624–627 (1977). See 17α-Estradiol for details.
506	Ergonovine ($C_{19}H_{23}N_3O_2$)	6.73	U	+H	Potentiometric	H_2O $t = 24$ $c = 0.002$	Craig LC, Shedlovsky T, Gould RG and Jacobs WA, The ergot alkaloids. XIV. The positions of the double bond and the carboxyl group in lysergic acid and its isomer, *J. Biol. Chem.*, **125**, 289–298 (1938). Cited in Perrin Bases 2905 ref. C54. The studies used glass electrode measurements of the pH of solutions containing equal concentrations of the base and salt.
		6.80	U	+H	Potentiometric	$t = 22$	NB: Several other ergot alkaloids were reported in Perrin Bases - nos. 2902 to 2905; see also Perrin Bases Suppl. 7473 to 7475.
507	Ergonovine	6.91	U	+H	Potentiometric	H_2O $t = 25$	Bergström CAS, Strafford M, Lazorova L, Avdeef A, Luthman K and Artursson P, Absorption classification of oral drugs based on molecular surface properties, *J. Med. Chem.*, **46(4)** 558–570 (2003). NB: From extrapolation of aqueous-methanol mixtures to 0% methanol.

(*continued*)

No.	Name	pK_a value(s)	Data quality	Ionization type	Method	Conditions t°C; I or c M	Comments and Reference(s)
508	Ergostine ($C_{35}H_{39}N_5O_5$) 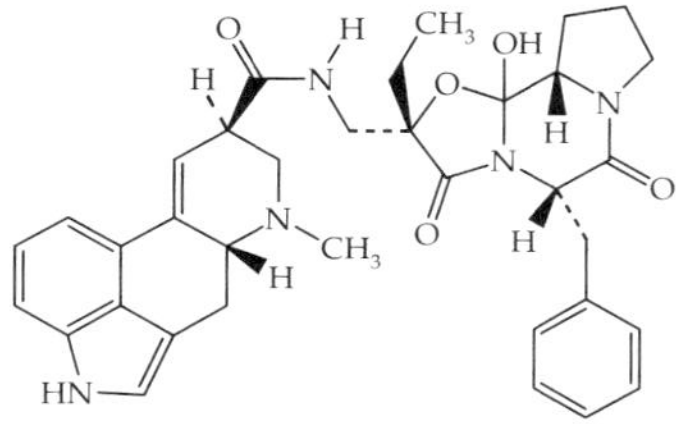	6.30 ± 0.04	U	+H	Potentiometric	H_2O $t = 24$ $I < 0.01$	Maulding HV and Zoglio MA, Physical chemistry of ergot alkaloids and derivatives. I. Ionization constants of several medicinally active bases, *J. Pharm. Sci.*, **59**, 700–701 (1970). NB: See Methysergide for details. This is the apparent value found in 15–20% 7HPT solution.
509	Ergostine	6.45 ± 0.09	U	+H	Potentiometric	H_2O $t = 24$ $I < 0.01$	Maulding HV and Zoglio MA, Physical chemistry of ergot alkaloids and derivatives. I. Ionization constants of several medicinally active bases, *J. Pharm. Sci.*, **59**, 700–701 (1970). NB: See Methysergide for details. This is the value found after extrapolation to 0% 7HPT. Also cited in Perrin Bases Suppl. no. 7473.
510	Ergotamine ($C_{33}H_{35}N_5O_5$) 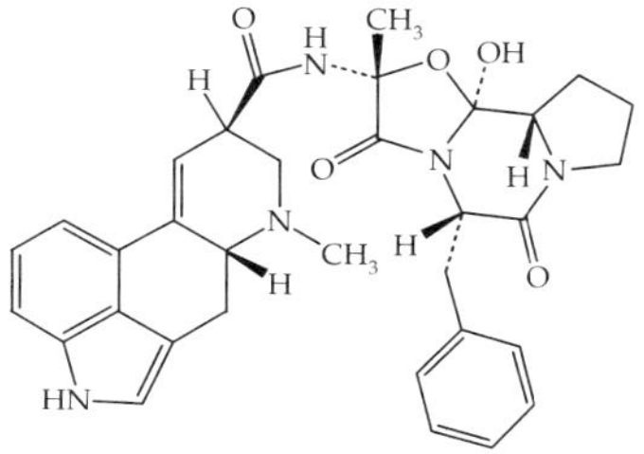	6.25 ± 0.04	U	+H	Potentiometric	H_2O $t = 24$ $I < 0.01$	Maulding HV and Zoglio MA, Physical chemistry of ergot alkaloids and derivatives. I. Ionization constants of several medicinally active bases, *J. Pharm. Sci.*, **59**, 700–701 (1970). NB: See Methysergide for details. This is the apparent value found in 15–20% 7HPT solution.
511	Ergotamine	6.40 ± 0.09	U	+H	Potentiometric	H_2O $t = 24$ $I < 0.01$	Maulding HV and Zoglio MA, Physical chemistry of ergot alkaloids and derivatives. I. Ionization constants of several medicinally active bases, *J. Pharm. Sci.*, **59**, 700–701 (1970). NB: See Methysergide for details. This is the value found after extrapolation to 0% 7HPT. Also cited in Perrin Suppl. no. 7474.
512	Ergotaminine ($C_{33}H_{35}N_5O_5$)	6.72 ± 0.04	U	+H	Potentiometric	H_2O $t = 24$ $I < 0.01$	Maulding HV and Zoglio MA, Physical chemistry of ergot alkaloids and derivatives. I. Ionization constants of several medicinally active bases, *J. Pharm. Sci.*, **59**, 700–701 (1970). NB: See Methysergide for details. This is the apparent value found in 15–20% 7HPT solution. Structure: Isomer of ergotamine.

No.	Compound	pKa			Method	Conditions	Reference
513	Ergotaminine	6.87 ± 0.09	U	+H	Potentiometric	H_2O $t = 24$ $I < 0.01$	Maulding HV and Zoglio MA, Physical chemistry of ergot alkaloids and derivatives. I. Ionization constants of several medicinally active bases, *J. Pharm. Sci.*, **59**, 700–701 (1970). NB: See Methysergide for details. This is the value found after extrapolation to 0% 7HPT. Also cited in Perrin Suppl. no. 7475.
514	Erythromycin ($C_{37}H_{67}NO_{13}$) 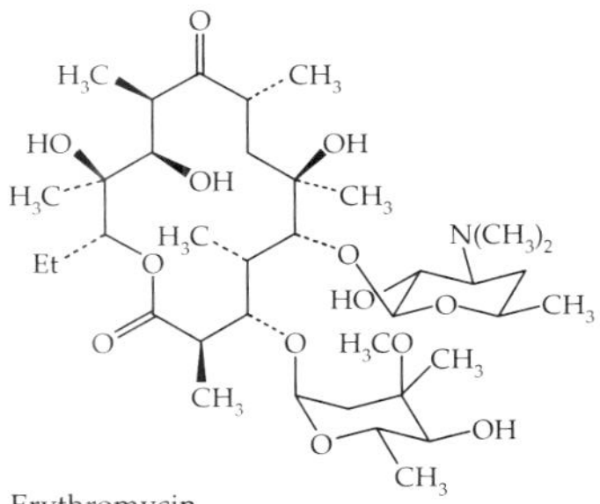	8.6	U	+H		66% aqueous DMF	Kobrehel G, Tamburasev Z and Djokic S, Erythromycin series. IV. Thin layer chromatography of erythromycin, erythromycin oxime, erythromycyclamine and their acyl derivatives, *J. Chromatogr.*, **133**, 415–419 (1977).
515	Erythromycin	9.1	U	+H	NMR	10% D_2O in H_2O $t = 30$ I undefined	Goldman RC, Fesik SW and Doran CC, Role of protonated and neutral forms of macrolides in binding to ribosomes from Gram-positive and Gram-negative bacteria, *Antimicrob. Agents Chemother.*, **34**, 426–431 (1990). NB: Also reported $pK_a = 8.6–8.9$ from aqueous titration.
516	Erythromycin	8.8	U	+H	NMR	D_2O $t = 20$ I undefined	Gharbi-Benarous J, Delaforge M, Jankowski CK and Girault J-P, A comparative NMR study between the macrolide antibiotic roxithromycin and erythromycin A with different biological properties, *J. Med. Chem.*, **34**, 1117–1125 (1991).
517	Erythromycin ($C_{37}H_{67}NO_{13}$)	8.88	U	+H	Potentiometric	H_2O $t = 25$ $I = 0.167$	McFarland JW, Berger CM, Froshauer SA, Hayashi SF, Hecker SJ, Jaynes BH, Jefson MR, Kamicker BJ, Lipinski CA, Lundy KM, Reese CP and Vu CB, Quantitative Structure-activity relationships among macrolide antibacterial agents: *In vitro* and *in vivo* potency against Pasteurella multocida, *J. Med. Chem.*, **40**, 1340–1346 (1997). NB: See Azithromycin for details; average standard deviation of ± 0.07 for the pK_a value.
518	Erythromycin	8.36	U	+H	soly	H_2O $t = 37$	Nakagawa Y, Itai S, Yoshida T and Nagai T, Physicochemical properties and stability in the acidic solution of a new macrolide antibiotic, clarithromycin, in comparison with erythromycin, *Chem. Pharm. Bull.*, **40**, 725–728 (1992).
519	Erythromycin	8.80	U	+H	Potentiometric	H_2O $t = 25$	Bergström CAS, Strafford M, Lazorova L, Avdeef A, Luthman K and Artursson P, Absorption classification of oral drugs based on molecular surface properties, *J. Med. Chem.* **46(4)** 558–570 (2003). NB: From extrapolation of aqueous-methanol mixtures to 0% methanol.

(*continued*)

No.	Name	pK_a value(s)	Data quality	Ionization type	Method	Conditions t°C; I or c M	Comments and Reference(s)
520	Erythromycyclamine ($C_{37}H_{70}N_2O_{12}$) 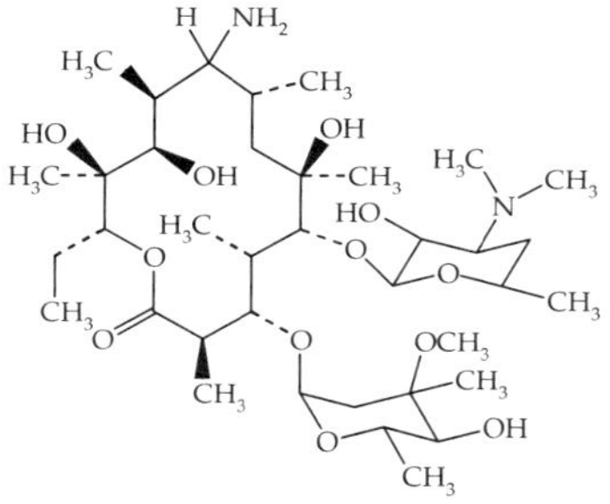	8.96 9.95	U U	+H +H	Potentiometric	H_2O $t = 25$ $I = 0.167$	McFarland JW, Berger CM, Froshauer SA, Hayashi SF, Hecker SJ, Jaynes BH, Jefson MR, Kamicker BJ, Lipinski CA, Lundy KM, Reese CP and Vu CB, Quantitative structure-activity relationships among macrolide antibacterial agents: *In vitro* and *in vivo* potency against Pasteurella multocida, *J. Med. Chem.*, **40**, 1340–1346 (1997). NB: See Azithromycin for details; average standard deviation of ± 0.07 for the pK_a value. The paper also cited Massey EH, Kitchell BS, Martin LD and Gerzon K, Antibacterial activity of 9(S)-erythromycyclamine-aldehyde condensation products, *J. Med. Chem.*, **17**, 105–107 (1974), who reported pK_{a1} = 8.8, pK_{a2} = 9.8 in 66% aqueous DMF (titration), but no other details. A second paper (Kobrehel G, Tamburasev Z and Djokic S, Erythromycin series. IV. Thin layer chromatography of erythromycin, erythromycin oxime, erythromycyclamine and their acyl derivatives, *J. Chromatogr.*, **133**, 415–419 (1977), which reported pK_{a1} = 8.4, but no other details.
521	Erythromycyclamine-11,12-carbonate ($C_{38}H_{68}N_2O_{13}$) 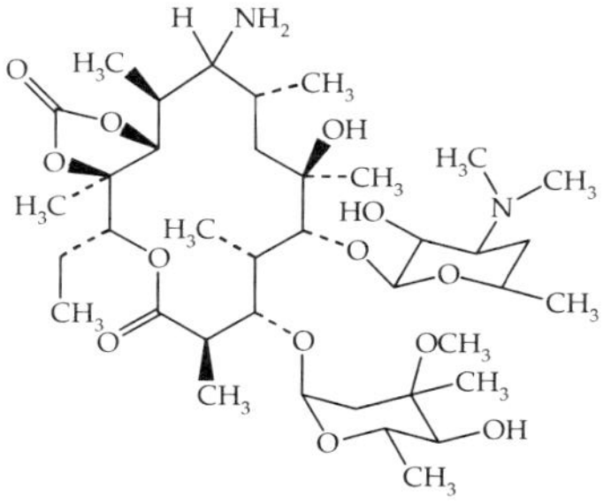	8.31 9.21	U U	+H +H	Potentiometric	H_2O $t = 25$ $I = 0.167$	McFarland JW, Berger CM, Froshauer SA, Hayashi SF, Hecker SJ, Jaynes BH, Jefson MR, Kamicker BJ, Lipinski CA, Lundy KM, Reese CP and Vu CB, Quantitative structure-activity relationships among macrolide antibacterial agents: *In vitro* and *in vivo* potency against Pasteurella multocida, *J. Med. Chem.*, **40**, 1340–1346 (1997). NB: See Azithromycin for details; average standard deviation of ± 0.07 for the pK_a value.
522	Estazolam ($C_{16}H_{11}ClN_4$) 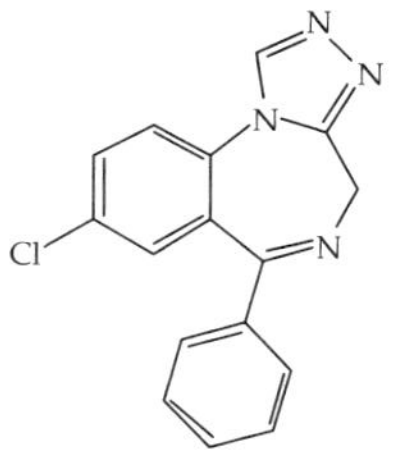	2.84	U	+H		H_2O	Koyama H, Yamada M and Matsuzawa T, Physicochemical studies on a new potent central nervous system depressant, 8-chloro-6-phenyl-4H-s-triazolo[4,3-a][1,4]benzodiazepine (D-40TA), *J. Takeda Res. Lab.*, **32**, 77–90 (1973). Cited in Konishi M, Hirai K and Mari Y, Kinetics and mechanism of the equilibrium reaction of triazolam in aqueous solution, *J. Pharm. Sci.*, **71(12)**, 1328–1334 (1982). NB: See also Inotsume N and Nakano M, Reversible ring-opening reactions of triazolobenzo- and triazolothienodiazepines in acidic media at around body temperature, *Chem. Pharm. Bull.*, **28**, 2536–2540 (1980).

No.	Compound	pKa	Rating	Group	Method	Conditions	Reference / Comments
523	Estradiol, 17α- ($C_{18}H_{24}O_2$) 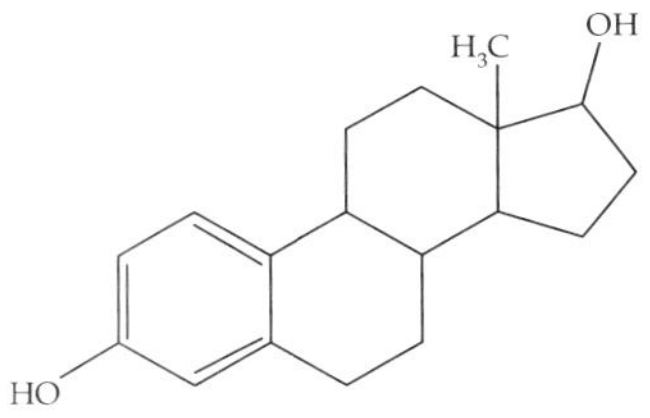	10.46 ± 0.03	U	–H	Spectro (λ = 240, 248, 295, 300 nm; 5 cm cell)	H_2O (0.12 - 0.5% EtOH) t = 23 ± 2 I unspecified but low	Hurwitz AR and Liu ST, Determination of aqueous solubility and pK_a values of estrogens, *J. Pharm. Sci.*, **66**, 624–627 (1977). "The thermodynamic ionization constants of 9 phenolic steroids were studied by UV spectrophotometry. A solubility method was employed to determine the ionization constant of equilenin. Aqueous solubility values of estrone, estradiol, equilin, and equilenin were also determined by either UV or GLC methods. No evidence was obtained for long range D to A ring electronic transmission affecting pK_a. Significant differences in pK_a values resulted only when conjugated unsaturation was added into the B ring of estrone or estradiol. . . . Stock solutions of estrogens, 3.0 x 10^{-3}, M were prepared by dissolving weighed amounts in 100 mL of pure ethanol. Aliquots of 0.12 – 0.5 mL were then diluted to 100 mL with distilled water . . . The solution pH was adjusted with 0.1 N HCl or 0.1 N NaOH. The UV spectrum of the solution was recorded immediately after a stable pH reading was obtained. Spectrophotometric determinations were carried out at . . . 23 ± 2°C" NB: This paper reported otherwise good work that was compromised by poor temperature and ionic strength control. The remark regarding lack of evidence for long range D to A ring electronic transmission related to: Legrand M, Delaroff V and Mathieu J, Distance effects in the steroid series, *Bull. Soc. Chim. Fr.*, 1346–1348 (1961), in which such effects were claimed. This claim appeared to be weakly supported, but was subsequently shown to be based on flawed data (see Lewis and Archer, below).
	Estradiol, 17β-	10.71 ± 0.02	A	–H	spectro (λ = 297 nm; 1 cm cell)	H_2O (0.1% p-dioxane) t = 25 ± 0.1 I = 0.03 (KCl)	Lewis KM and Archer RD, pK_a values of estrone, 17β-estradiol and 2-methoxyestrone. *Steroids*, **34**(5), 485–499 (1979). NB: Care was taken to exclude both carbon dioxide and oxygen. The pH meter was calibrated with multiple pH standards. The spectrophotometer was checked for wavelength and absorbance accuracy (Haupt J, *J. Res. Nat. Bur. Stand.*, **48**, 414 (1952)). Also reported estrone, 10.77 ± 0.02 (see below) and 2-methoxyestrone (λ = 300), 10.81 ± 0.03 (A).
524	Estriol ($C_{18}H_{24}O_3$) 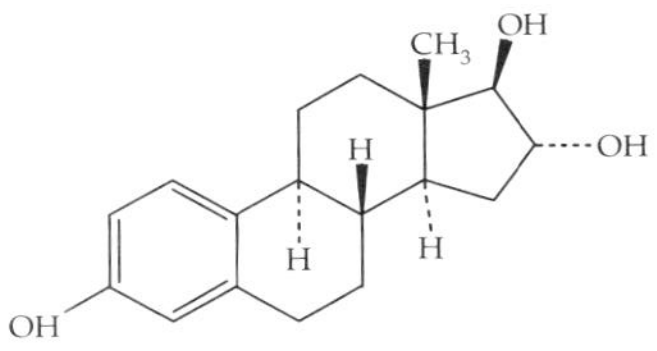	10.38 ± 0.02	U	–H	Spectro (λ = 240, 248, 295, 300 nm) (5 cm cell)	H_2O t = 23 ± 2 I unspecified	Hurwitz AR and Liu ST, Determination of aqueous solubility and pK_a values of estrogens, *J. Pharm. Sci.*, **66**, 624–627 (1977). NB: See 17α-Estradiol for details.

(continued)

No.	Name	pK_a value(s)	Data quality	Ionization type	Method	Conditions t°C; I or c M	Comments and Reference(s)
525	Estrone ($C_{18}H_{22}O_2$) H_3C O HO	10.77 ± 0.02	A	–H	Spectro (λ = 297 nm; 1 cm cell)	H_2O	Lewis KM and Archer RD, pK_a values of estrone, 17β-estradiol and 2-methoxyestrone. Steroids, 34(5), 485–499 (1979). Cited in Both D, Estrone, *APDS*, **12**, 135–174 (1983)
		10.34 ± 0.05	A	–H	Spectro (λ = 240, 248, 295, 300 nm) (5 cm cell)	H_2O t = 23 ± 2 I unspecified	"The reported acid ionization constant (pK_a) of estrone shows great variation ranging from 9.36 to 11.0 (132, 133). Previous methods of measurement included: back titration, conductimetric and most recently U.V. spectrophotometric. Recent work (99, 134) places the pK_a between 10.34 and 10.914. The most recent spectrophotometric determination (135) reported the pK_a to be 10.77 ± 0.02 with seven determinations." 99. Hurwitz AR and Liu ST, Determination of aqueous solubility and pK_a values of estrogens, *J. Pharm. Sci.*, **66**, 624–627 (1977). NB: This paper reported otherwise good work that was compromised by poor temperature and ionic strength control. 132. Marrian GF, Chemistry of estrin. IV. The chemical nature of crystalline preparations, *Biochem. J.*, **24**, 1021–1030 (1930). NB: Historical value only. 133. Butenandt A, The female sexual hormone. XI. Constitution of the follicular hormone. 2. The degree of saturation and the aromatic character of the follicular hormone, *Z. Physiol. Chem.*, **223**, 147–168 (1934). NB: Historical value only. Gave $K = 0.44 \times 10^{-9}$, i.e., pK_a = 9.36 (U). 134. Kirdani RY and Burgett M, The pK of five estrogens in aqueous solutions, *Arch. Biochem. Biophys.*, **118(1)**, 33–36 (1967). NB: From spectroscopic work, reported estrone (pK_a = 10.91) , 17α-estradiol (pK_a = 10.098) , 17β-estradiol (pK_a = 10.078), 6-hydroxyestradiol (pK_a = 9.744) and 6-ketoestradiol (pK_a = 9.079). These values appear to be systematically low (U). 135. Lewis KM and Archer RD, pK_a values of estrone, 17β-estradiol and 2-methoxyestrone, *Steroids*, **34(5)**, 485–499 (1979). NB: Reported estrone (pK_a = 10.77), 17β-estradiol (pK_a = 10.71) and 2-methoxyestrone (pK_a = 10.81) (A).
526	Ethambutol ($C_{10}H_{24}N_2O_2$) $HOCH_2$ NH CH_3CH_2 CH_2OH HN CH_2CH_3	6.6 9.5	U U	+H +H		H_2O	Shepherd RG, Baughn C, Cantrall ML, Goodstein B and Wilkinson RG, Structure-activity studies leading to ethambutol, a new type of antituberculous compound. *Ann. NY Acad. Sci.* **135**, 686–710 (1966). NB: See also Lee C and Benet, LZ, Ethambutol hydrochloride, *APDS*, **7**, 231–249(1978); C. Lee, Ph.D. thesis, Univ. Calif. San Francisco.

527	Ethinyl estradiol ($C_{20}H_{24}O_2$) 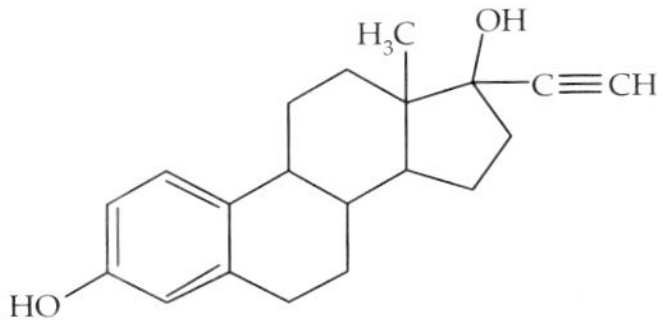	10.40 ± 0.01	A	−H	Spectro	H_2O $t = 23 \pm 2$ I unspecified	Hurwitz AR and Liu ST, Determination of aqueous solubility and pK_a values of estrogens, *J. Pharm. Sci.*, **66**, 624–627 (1977). NB: See 17α-Estradiol for details.
528	Ethinyl estradiol	10.41	U	−H	Potentiometric	H_2O $t = 25$	Bergström CAS, Strafford M, Lazorova L, Avdeef A, Luthman K and Artursson P, Absorption classification of oral drugs based on molecular surface properties, *J. Med. Chem.*, **46(4)**, 558–570 (2003). NB: From extrapolation of aqueous-methanol mixtures to 0% methanol.
529	Ethionamide ($C_8H_{10}N_2S$) 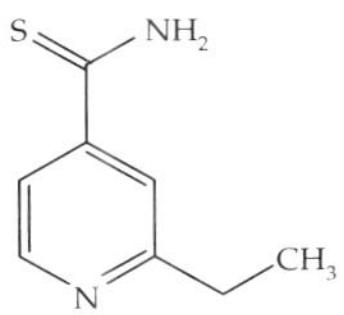	4.37	U	+H	CE/pH (+ve ion mode)	H_2O $t = 25$ $I = 0.025$	Wan H, Holmen AG, Wang Y, Lindberg W, Englund M, Nagard MB and Thompson RA, High-throughput screening of pK_a values of pharmaceuticals by pressure-assisted capillary electrophoresis and mass spectrometry, *Rapid Commun. Mass Spectrom.*, **17**, 2639–2648 (2003). NB: Reported a predicted value (ACD Labs) of 4.34.
530	Ethopropazine ($C_{19}H_{24}N_2S$) 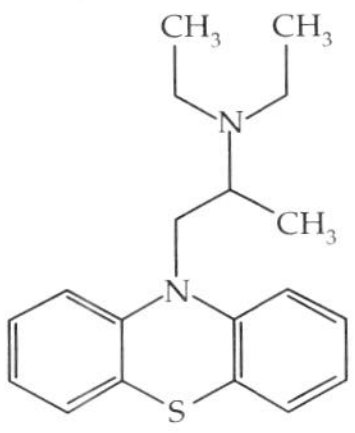	9.50	U	+H	Potentiometric	H_2O (extrap) $t = 20$ N_2 atmosphere	Sorby DL, Plein EM and Benmaman D, Adsorption of phenothiazine derivatives by solid adsorbents, *J. Pharm. Sci.*, **55**, 785–794 (1966). NB: See Chlorpromazine for details.

(continued)

No.	Name	pK_a value(s)	Data quality	Ionization type	Method	Conditions t°C; I or c M	Comments and Reference(s)
531	Ethox(a)zolamide ($C_9H_{10}N_2O_3S_2$) 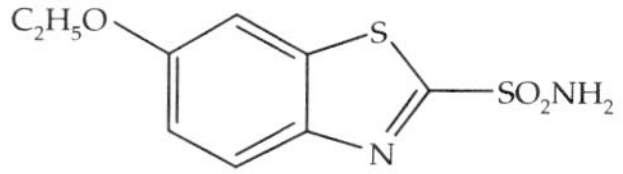	8.12	U	−H	soly; Potentiometric	H_2O $t = 25.0$	Eller MG, Schoenwald RD, Dixson JA, Segarra T and Barfknecht CF, Topical carbonic anhydrase inhibitors III: Optimization model for corneal penetration of ethoxzolamide analogues, *J. Pharm. Sci.*, **74(1)**, 155–160 (1985).
532	α-Ethyl n-amylpenilloate	1.25 4.12	U U	−H +H	Method not given	H_2O $t = 5$	Woodward RB, Neuberger A and Trenner NR, *in* Clarke H, Johnson JR, Robinson Sir R (eds.), *The Chemistry of Penicillin*, Princeton University Press, Princeton, NJ, 415–422, 1949.
533	α-Ethyl n-amylpenilloate	1.31 3.95	U U	−H +H	Method not given	H_2O $t = 25$	Woodward RB, Neuberger A, Trenner NR, *in* Clarke H, Johnson JR, Robinson Sir R (eds.), *The Chemistry of Penicillin*, Princeton University Press, Princeton, NJ, 415–422, 1949.
534	α-Ethylphenylpenilloate	1.34 4.11	U U	−H +H	Method not given	H_2O $t = 5$	Woodward RB, Neuberger A and Trenner NR, *in* Clarke H, Johnson JR, Robinson Sir R (eds.), *The Chemistry of Penicillin*, Princeton University Press, Princeton, NJ, 415–422, 1949.
535	α-Ethylphenylpenilloate	1.35 3.92	U U	−H +H	Method not given	H_2O $t = 25$	Woodward RB, Neuberger A and Trenner NR, *in* Clarke H, Johnson JR, Robinson Sir R (eds.), *The Chemistry of Penicillin*, Princeton University Press, Princeton, NJ, 415–422, 1949.
536	Ethylmorphine ($C_{19}H_{23}NO_3$) 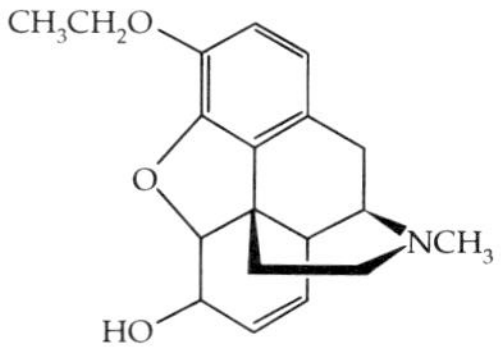	8.08	U	+H	Spectro	H_2O $t = 15$ $c = 0.005$ to 0.01	Kolthoff IM, The dissociation constants, solubility product and titration of alkaloids, *Biochem. Z.*, **162**, 289–353 (1925). NB: Cited in Perrin Bases no. 2906. The spectrophotometric method used colorimetric determination of the extent of ionization by use of an indicator of known pK_a value.
		8.2	U	+H			NB: pK_w = 14.35 at 15 °C.

Compound	pK_a	Compound	pK_a
6-Hydrogen	7.91	6-Nitro	7.38
6-Hydroxy	pK_1 = 7.81; pK_2 = 9.26	6-Ethoxy (ethoxzolamide)	8.12
6-Chloro	7.69	6-Hydroethoxy	7.88
4,6-Dichloro	7.64	6-Benzyloxy	8.14
6-Amino	8.03	6-Acetamido	7.93

No.	Compound	pKa			Method	Conditions	Reference / Comments
537	Etidocaine ($C_{17}H_{28}N_2O$) 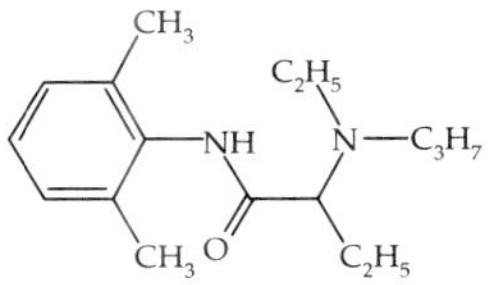	7.86	U	+H	Potentiometric	H_2O $t = 25.0 \pm 0.2$ $I = 0.01$ (NaCl)	Johansson P-A, Liquid-liquid distribution of lidocaine and some structurally related anti-arrythmic drugs and local anaesthetics, *Acta Pharm. Suec.*, **19**, 137–142 (1982). NB: See Lidocaine for details.
538	Etilefrine (ethylphenylephrine) ($C_{10}H_{15}NO_2$) 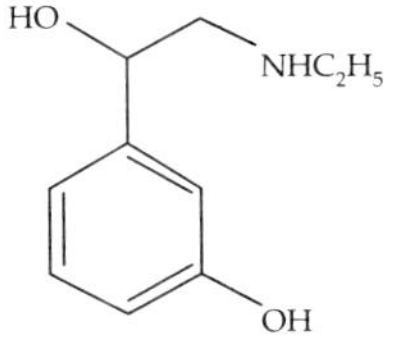	9.0 8.9	U U	+H +H	Potentiometric	H_2O 90.7% MeOH	Wagner J, Grill H and Henschler D, Prodrugs of etilefrine: Synthesis and evaluation of 3′-(O-acyl) derivatives, *J. Pharm. Sci.*, **69**(**12**), 1423–1427 (1980). "(3-hydroxy)phenylethanolamines in solution represent a mixture of the uncharged form and ionic species (cation, anion, and zwitterion). At half-neutralization, these compounds are in a specific equilibrium, which is represented by an apparent average pK_a value. The pK_a value is dependent on the substituent at the amino nitrogen and is increased from 8.67 (norfenefrine) to 8.9 (phenylephrine) or 9.0 (etilefrine) for the 3′-hydroxyphenylethanol-amines when introducing a methyl or ethyl radical into the amino group. The same value also was obtained for etilefrine in $1.5x10^{-3}$ M aqueous solution." NB: See also 3′-(*O*-Pivaloyl)-etilefrine.
539	Etomidate ($C_{14}H_{16}N_2O_2$) 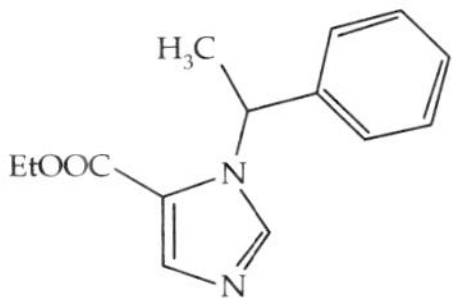	4.24	U	+H	Potentiometric, Spectro	H_2O $t = 25$	Peters JJ, Determination of ionization constants in mixed aqueous solvents of varying composition by a single titration, *J. Pharm. Sci.*, **67**, 127–129 (1978). NB: The paper recognizes the problem of long extrapolations from aqueous-organic solvent mixtures to estimate pK_a values of poorly water-soluble substances. No activity corrections were applied. See Cinnarizine for further details.

(*continued*)

No.	Name	pK_a value(s)	Data quality	Ionization type	Method	Conditions t°C; I or c M	Comments and Reference(s)
540	Etoposide ($C_{29}H_{32}O_{13}$)	9.8	U	−H	Spectro	1% MeOH $t = 25$ $I = 0.1$	Beijnen JH, Holthuis JJM, Kerkdijk HG, van der Houwen OAGJ, Paalman ACA, Bult A and Underberg WJM, Degradation kinetics of etoposide in aqueous solution, *Int. J. Pharm.*, **41**, 169–178 (1988). Cited in Holthius JJM, Ketlenes-van-den Bosch JJ, Bult A, Etoposide, *APDS*, **18**, 121–151 (1989).
		10.1	U	−H	kinetic	1.5% MeOH $t = 25$ $I = 0.1$	"Absorbances of 0.1mM solutions of etoposide in 0.05 M sodium borate buffers containing 4% methanol were recorded at 264 nm with a Shimadzu UV-140 double beam spectrometer." NB: Also reported pK_a = 10.8 for *cis*-etoposide from kinetic measurements.
541	β-Eucaine ($C_{15}H_{21}NO_2$)	9.35	U	+H	Spectro	H_2O $t = 23$ $c = 0.005$	Schoorl N, Dissociation constants and titration exponents of several less common alkaloids, *Pharm. Weekblad*, **76**, 1497–1501 (1939); CA 34:1900. Cited in Perrin Bases no. 2908 ref. S16. NB: The method used colorimetric determination of the extent of ionization by use of an indicator of approximately known pK_a value. See Diamorphine for details.

	Name	pK_a value(s)	Data quality	Ionization type	Method	Conditions t°C; I or c M	Comments and Reference(s)
542	Famotidine ($C_8H_{15}N_7O_2S_3$)	6.76	U	–H	Spectro	H_2O $t = 23$	Islam MS and Narurkar MM, Solubility, stability and ionization behavior of famotidine, *J. Pharm. Pharmacol.*, **45**, 682–686 (1993).
		6.98 6.89	U U	–H –H	Soly Partition		"Famotidine was studied *in vitro* to determine the solubility, stability, and ionization behavior. Using spectrophotometric, solubility, and partitioning methods, the pK_a at 23 °C was 6.76, 6.98, and 6.89, respectively. The pH-solubility profile indicated an intrinsic solubility of 2.7 mM at 23 °C. At pH 1–11, drug degradation followed pseudo-first-order kinetics at 37 °C and ionic strength of 0.5. The pH-rate profile was accounted for by specific acid and base catalyzed reactions and water-catalyzed decomposition of protonated and free drug. A pK_a of 6.6, determined by potentiometry at 37 °C, was used in kinetic calculations. Maximum stability occurred at pH 6.3. Studies indicated a partition coefficient of 0.23 for free drug at 23 °C. It was concluded that famotidine shows poor lipophilicity, poor aqueous solubility, and susceptibility to gastric degradation."
		6.6	U	–H	Potentiometric	H_2O $t = 37$	
543	Famotidine	6.45	U	–H	Soly	H_2O $t = 37$ $I = 0.5$ (KCl)	Najib NM and Suleiman MS, Determination of some parameters influencing the dissolution rate of famotidine, *Int. J. Pharm.*, **61**, 173–178 (1990). "The influence of pH, temperature and stirring rates on the dissolution of famotidine (I) was studied. The dissociation constant, intrinsic solubility, diffusion coefficient, hydrodynamic layer thickness and activation energy of dissolution of I were determined. . . .intrinsic solubility is 0.278 mg.ml^{-1}."
		6.56 11.02	U U	+H –H	Potentiometric	H_2O $t = 37$ $I = 0.15$ (KCl)	Balon K, Riebesehl BU and Muller BW, Drug liposome partitioning as a tool for the prediction of human passive intestinal absorption, *Pharm. Res.*, **16**, 882–888 (1999).
544	Fenbufen ($C_{16}H_{14}O_3$)	4.3	U	–H	Spectro	H_2O t undefined I undefined	Herzfeldt CD and Kümmel R, Dissociation constants, solubilities, and dissolution rates of some selected nonsteroidal antiinflammatories, *Drug Dev. Ind. Pharm.*, **9(5)**, 767–793 (1983). NB: Used dA/dpH method. See Azapropazone and Ibuprofen for details.

(continued)

	Name	pK_a value(s)	Data quality	Ionization type	Method	Conditions t°C; I or c M	Comments and Reference(s)
545	Fendiline ($C_{23}H_{25}N$) 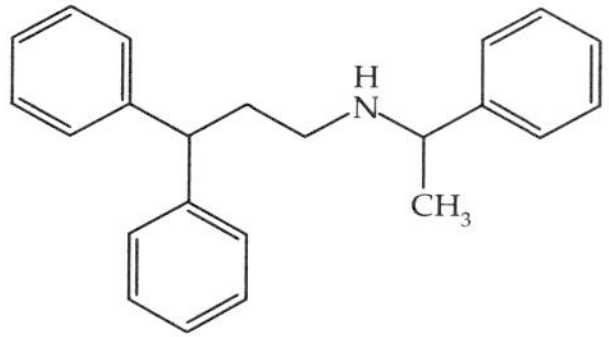	8.59 ± 0.03	U	+H	Potentiometric	40% EtOH $t = 25.0$	Mannhold R, Rodenkirchen R, Bayer R and Haus W, The importance of drug ionization for the action of calcium antagonistsand related compounds, *Arzneim.-Forsch.*, **34**, 407–409 (1984).
		9.33	U	+H		H_2O	NB: See Aprindine for details.
546	Fenoprofen ($C_{15}H_{14}O_3$) 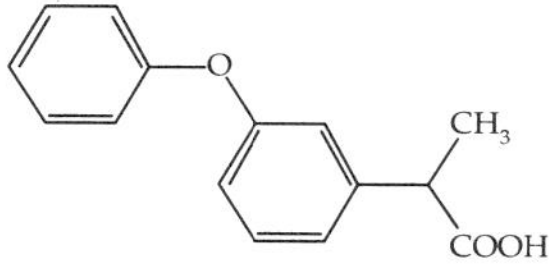	4.5	U	−H	Spectro	H_2O t undefined I undefined	Herzfeldt CD and Kümmel R, Dissociation constants, solubilities, and dissolution rates of some selected nonsteroidal antiinflammatories, *Drug Dev. Ind. Pharm.*, **9(5)**, 767–793 (1983). NB: Used dA/dpH method. See Azapropazone and Ibuprofen for details.
547	Fentanyl ($C_{22}H_{28}N_2O$) 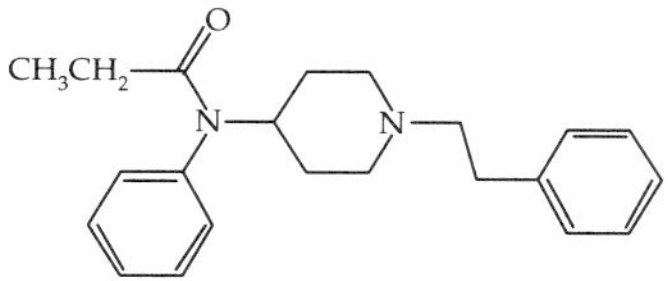	8.99	U	+H	Soly	H_2O $t = 35.0$	Roy SD and Flynn GL, Solubility behavior of narcotic analgesics in aqueous media: solubilities and dissociation constants of morphine, fentanyl, and sufentanil, *Pharm. Res.*, **6(2)**, 147–151 (1989). "Dissociation constants and corresponding pK' values of the drugs were obtained from measured free-base solubilities (determined at high pH's) and the concentrations of saturated solutions at intermediate pH's. Morphine, fentanyl, and sufentanil exhibited pK_a' values of 8.08, 8.99, and 8.51, respectively. Over the pH range of 5 to 12.5 the apparent solubilities are determined by the intrinsic solubility of the free base plus the concentration of ionized drug necessary to satisfy the dissociation equilibrium at a given pH."

No.	Compound	pKa	Data quality	Ionization type	Method	Conditions	Reference
548	Flavoxate ($C_{24}H_{25}NO_4$) 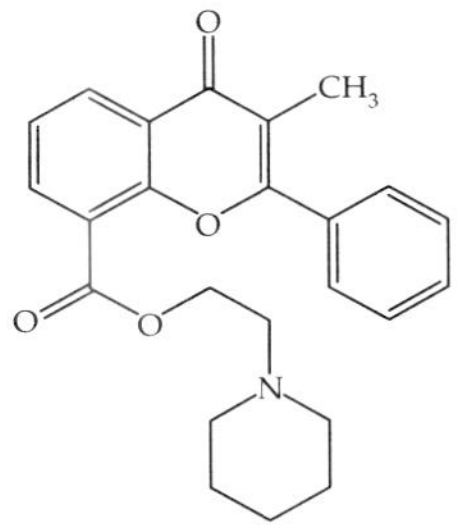	7.3	U	+H	Potentiometric	H_2O $t = 37$	Fontani F and Setnikar I, Chemical and physical properties of flavoxate, *Pharmazeutische Industrie*, **36**, 802–805 (1974). Cited in Goldberg Y, Flavoxate hydrochloride, *APDS*, **28**, 82 (2001).
549	Fluconazole ($C_{13}H_{12}F_2N_6O$) 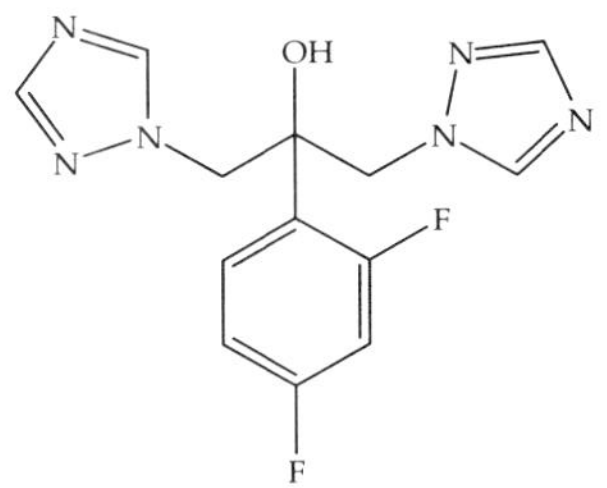	1.76 ± 0.10	U	−H	Soly	H_2O $t = 24$ $I = 0.1$ (NaCl)	Dash AK and Elmquist WF, Fluconazole *APDS* **27**, 85–87 (2000). "The ionization constant (pK_a) of fluconazole was determined by solubility measurements, and found at 24 °C in 0.1M NaCl solution to be 1.76 ± 0.10 [8]. Fluconazole is a weak base, and undergoes protonation at the N-4 nitrogen. The predominate protonation of the N-4 nitrogen of the 1H-1,2,4-triazole moiety has also been confirmed by ^{15}N NMR spectroscopy. . . . [8] Unpublished data from Pfizer, Inc."
550	Fludiazepam ($C_{16}H_{12}ClFN_2O$) 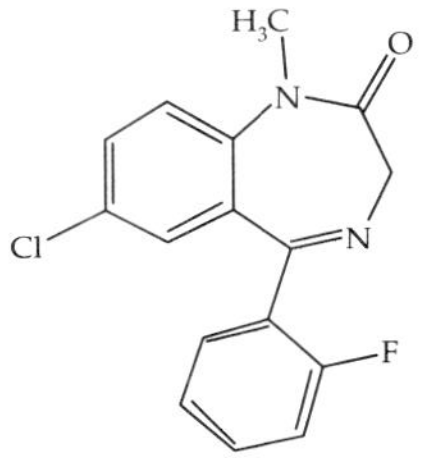	2.29 ± 0.009	U	+H	Spectro	H_2O	Inotsume N and Nakano M, Reversible azomethinine bond cleavage of 2′-fluoro derivatives of benzodiazepines in acidic solutions at body temperature, *Int. J. Pharm.*, **6**, 147–154 (1980). NB: See Konishi M, Hirai K and Mari Y, Kinetics and mechanism of the equilibrium reaction of triazolam in aqueous solution, *J. Pharm. Sci.*, **71(12)**, 1328–1334 (1982). See also Triazolam and Flunitrazepam.

(continued)

	Name	pK_a value(s)	Data quality	Ionization type	Method	Conditions t°C; I or c M	Comments and Reference(s)
551	Flufenamic acid ($C_{14}H_{10}F_3NO_2$)	5.0	VU	−H	Spectro	H_2O t undefined I undefined	Herzfeldt CD and Kümmel R, Dissociation constants, solubilities, and dissolution rates of some selected nonsteroidal anti-inflammatories, *Drug Dev. Ind. Pharm.*, **9(5)**, 767–793 (1983). NB: Used dA/dpH method. See Azapropazone and Ibuprofen for details. The pK_a determination supports only the region of the possible value.
552	Flufenamic acid	5.84	U	−H	Potentiometric	5–10% aq. acetone $t = 25 \pm 0.1$ $I = 0.11$	Terada H and Muraoka S, Physicochemical propertiesand uncoupling activity of 3′-substituted analogues of N-phenylanthranilic acid, *Mol. Pharmacol.*, **8**, 95–103 (1972).
	Substituted fenamic acids:						"For determinations of pK_a values, 2–3 mg of the compound to be tested were dissolved in water or in dilute NaOH solution containing 5–10% acetone. The solution was titrated with 1 N HCl using a pH-stat. The influence of organic solvents on pK values was regarded as negligible, since this effect has been reported to be significant only when the solution contains at least 20% (v/v) organic solvent (Mizutani M, *Z. Phys. Chem.*, **118**, 318–326 (1925); *ib.* 327–341; Cavill GWK, Gibson NA and Nyholm RS, Dissociation constants of some p-alkoxybenzoic acids, *JCS*, 2466–2470 (1949))."
	3′-H	5.28	U				
	3′-Me	5.84	U				
	3′-NH_2	4.72	U				
	3′-OMe	5.17	U				
	3′-acetyl	5.05	U				
553	Flufenamic acid	3.85	U	−H	Soly	H_2O $t = 25 \pm 0.1$ $I = 0.11$	Terada H, Muraoka S and Fujita T, Structure-activity relationships of fenamic acids, *J. Med. Chem.*, **17**, 330–334 (1974). NB: Methodology used a three hour equilibration time, which is possibly not long enough. These values are very different to those found in an earlier paper, where titration in dilute acetone-water mixtures was used. Could postulate internal H-bond giving either stabilization or destabilization of the anion. Further computed values were suggested, based on a structure-reactivity relationship (LFER) using sigma values: 3′-Cl, 3.86; 3′-NO_2, 3.59; 3′-OH, 4.05; 3′-$N(CH_3)_2$, 4.31; 3′-COC_6H_5, 3.87; 3′-C_6H_5, 4.10; 2′,3′-$(CH)_4$-, 4.11.
	Substituted fenamic acids:						
	3′-Me	4.15	U				
	3′-NH_2	4.35	U				
	3′-OMe	3.95	U				
	3′-acetyl	3.90	U				
554	Flufenamic acid	5.94	VU	+H	Potentiometric	50% EtOH t undefined I undefined	Jahn U and Wagner-Jauregg T, Wirkungsvergleich saurer Antiphlogistika im Bradykinin-, UV-Erythem- und Rattenpfotenödem-Test, *Arzneim.-Forsch.*, **24**, 494–499 (1974).
		6.0	VU	+H		80% Me cellosolve	NB: Literature values obtained from the pH of half-neutralization (Parke-Davis). Also gave a value of 7.5 in water. This last value is clearly in error.

No.	Compound	pKa	Assessment	Ionization	Method	Conditions	Reference / Notes
555	Flumequine ($C_{14}H_{12}FNO_3$)	6.38 ± 0.04	A	–H	Potentiometric	H_2O $t = 25.0 \pm 0.1$ $I = 0.1$ (NaCl)	Takacs-Novak K, Box KJ and Avdeef A, Potentiometric pK_a determination of water-insoluble compounds: Validation study in methanol/water mixtures, *Int. J. Pharm.*, **151**, 235–248 (1997).
		6.35	U	–H	Spectro		NB: By extrapolation from 3–44%w/w aqueous MeOH. See Acetaminophen for full details. Takacs-Novak K and Avdeef A, Interlaboratory study of log P determination by shake-flask and potentiometric methods, *J. Pharm. Biomed. Anal.*, **14**, 1405–1413 (1996) also reported pK_a = 6.27 ± 0.01 at I = 0.15 M (NaCl) (potentio).
556	Flumequine	6.42	A	–H	CE/pH (–ve ion mode)	H_2O $t = 25$ $I = 0.025$	Wan H, Holmen AG, Wang Y, Lindberg W, Englund M, Nagard MB and Thompson RA, High-throughput screening of pK_a values of pharmaceuticals by pressure-assisted capillary electrophoresis and mass spectrometry, *Rapid Commun. Mass Spectrom.*, **17**, 2639–2648 (2003). NB: Reported a predicted value (ACD Labs) of 3.62.
557	Flumequine	GLpK_a: 6.31 ± 0.02 A&S: 6.35 ± 0.05	 U A	 –H –H	Spectro	H_2O $t = 25$ $I = 0.15$ (KCl) Ar atmosphere	Tam KY and Takacs-Novac K, Multi-wavelength spectrophotometric determination of acid dissociation constants, *Anal. Chim. Acta*, **434**, 157–167 (2001). NB: See Clioquinol for details.
558	Flumethiazide ($C_8H_6F_3N_3O_4S_2$)	6.3	U	–H	Potentiometric	acetone /H_2O	Henning VG, Moskalyk R.E, Chatten LG and Chan SF, Semiaqueous potentiometric determination of apparent pK_{a1} values for benzothiadiazines and detection of decomposition during solubility variation with pH studies, *J. Pharm. Sci.*, **70**, 317–319 (1981). "Four concentrations of each benzothiadiazine (0.0005, 0.001, 0.0015, and 0.002M) were prepared from 0.02 M acetone stock solutions. The acetone-water ratios were 5:45, 10:40, 15:35, and 20:30 ml, respectively. The solutions were titrated with 0.5 N NaOH and the pH was measured after the addition of each 0.1 ml increment of titrant. The apparent pK_{a1} values then were obtained by the usual extrapolation techniques

	Literature Aq. Poten. Tit.	Semiaq Poten. Tit.
Benzothiadiazine	6.44	6.3
Flumethiazide	7.0, 7.9, 8.6, 8.8	8.7
Hydrochlorothiazide	8.9, 8.45	8.5
Cyclothiazide	–	8.8
Trichloromethiazide	–	6.9
Methyclothiazide(II)	–	9.5
Polythiazide	–	9.1

(continued)

	Name	pK_a value(s)	Data quality	Ionization type	Method	Conditions t°C; I or c M	Comments and Reference(s)
559	Flunitrazepam ($C_{16}H_{12}FN_3O_3$) 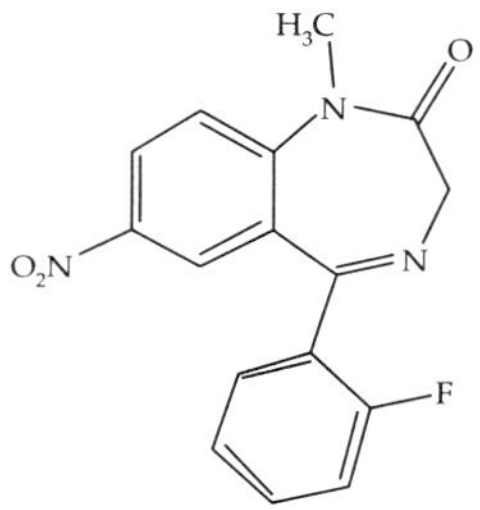	1.71 ± 0.003	U	+H	Spectro	H_2O	Inotsume N and Nakano M, Reversible azomethinine bond cleavage of 2′-fluoro derivatives of benzodiazepines in acidic solutions at body temperature, *Int. J. Pharm.*, **6**, 147–154 (1980). NB: Methodology referred to Albert and Serjeant 1971. Absorbances were read immediately following solution preparation to minimize the effect of reaction at the azomethine bond.
		1.82	U	+H	Spectro	H_2O t undefined $I = 0.01$	Boxenbaum H, data on file, Hoffman La Roche, cited in Boxenbaum HG, Posmanter HN, Macasieb T and Geitner KA, *et al.*, Pharmacokinetics of flunitrazepam following single- and multiple-dose oral administration to healthy human subjects, *J. Pharmacokin. Biopharm.*, **6**, 283–293 (1978).
560	Flunitrazepam	1.71	U	+H			Konishi M, Hirai K and Mari Y, Kinetics and mechanism of the equilibrium reaction of triazolam in aqueous solution, *J. Pharm. Sci.*, **71(12)**, 1328–1334 (1982). NB: See Triazolam.
561	Fluorescein ($C_{20}H_{12}O_5$) 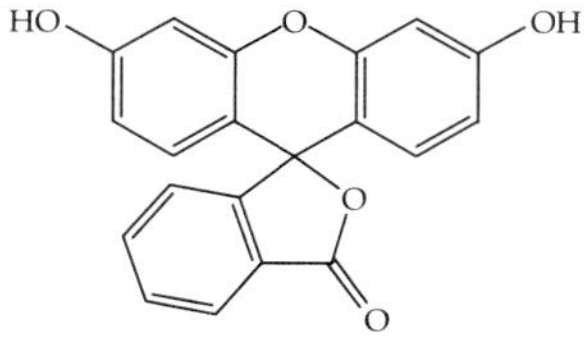	2.24 ± 0.06 4.36 ± 0.06	U U	−H −H	Soly	H_2O $t = 20.0$ $I = 0.10$	Asuero AG, Evaluation of acidity constants of two-step overlapping equilibria of amphoteric substances from solubility measurements, *J. Pharmaceut. Biomed. Anal.*, **6(3)**, 313–316 (1988). NB: Intrinsic solubility (S_o) = (3.38 ± 0.30) × 10^{-4} M. Also called fluorescein yellow.
562	5-Fluorouracil ($C_4H_3FN_2O_2$) 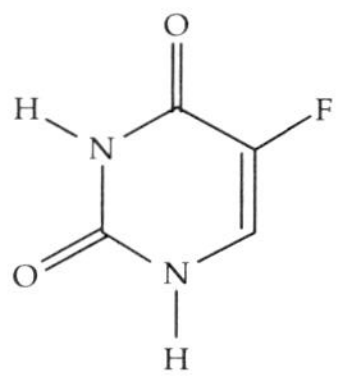	8.0 ± 0.1 13.0 ± 0.1	U U	−H −H	Spectro	H_2O	Rudy BC and Senkowski BZ, Fluorouracil, *APDS*, **2**, 221–244 (1973). "When the pK_a's for uracil were determined in a similar manner they were found to be 9.4 ± 0.1 and >13.5. These latter values are in good agreement with the pK_a's reported by Shugar and Fox for uracil. The above values for 5-FU are from Berens and Shugar, *Acta Biochem. Pol.*, **10**, 25 (1963)." Motchane A, Hoffman-La Roche Inc., unpublished data.

563	Fluorouracil, 1-methyloxycarbonyl ($C_6H_5FN_2O_4$) 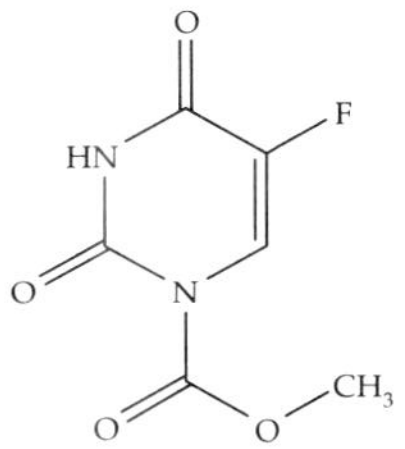	6.8	U	–H	Spectro ($\lambda = 235$ nm)	H_2O $t = 37$ $I = 0.5$	Buur A and Bundgaard H, Prodrugs of 5-fluorouracil V. 1-Alkoxycarbonyl derivatives as potential prodrug forms for improved rectal or oral delivery of 5-fluorouracil, *J. Pharm. Sci.*, **75**, 522–527 (1986). NB: pK_a values for several alkyloxycarbonyl derivatives of 5-FU were reported, in the range 6.7 to 6.9.
564	Fluorouracil, 3-ethyloxycarbonyl (R = Et; $C_7H_7FN_2O_4$) 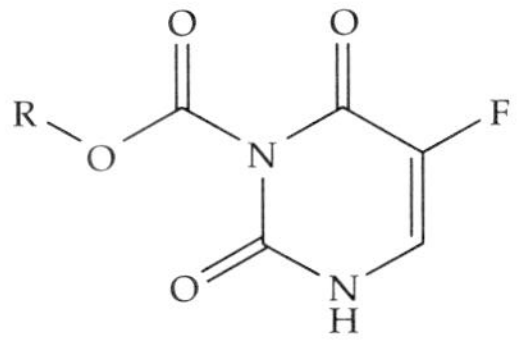	8.6	U	–H	Spectro ($\lambda = 300$ nm)	H_2O $t = 22$ $I = 0.5$	Buur A and Bundgaard H, Prodrugs of 5-fluorouracil. II. Hydrolysis kinetics, bioactivation, solubility and lipophilicity on N-alkoxycarbonyl derivatives of 5-fluorouracil, *Arch. Pharm. Chem., Sci. Edn.*, **12(2)**, 37–44 (1984). NB: Reported pK_a value for: 3-Phenyloxycarbonyl-5-FU, 6.6.
565	Fluorouracil, 1-ethyl-oxycarbonyloxymethyl ($C_8H_9FN_2O_5$) 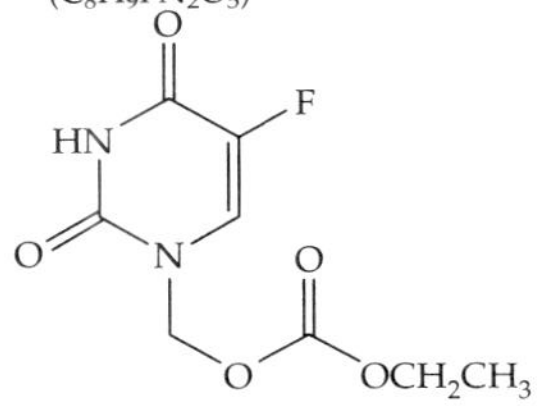	7.3	U	–H	Spectro ($\lambda = 230$ nm)	H_2O $t = 22$ $I = 0.5$	Buur A, Bundgaard H and Falch E, Prodrugs of 5-fluorouracil, *Acta Pharm. Suecica*, **23**, 205–216 (1986). "... The ionization constants for the derivatives were determined at 22 °C and $\mu = 0.5$ by spectrophotometry according to Albert and Serjeant..." NB: A value of 7.2 was derived from kinetic measurements.

(continued)

	Name	pK_a value(s)	Data quality	Ionization type	Method	Conditions t°C; I or c M	Comments and Reference(s)
566	Fluorouracil, 3-ethyl-oxycarbonyloxymethyl	7.9	U	–H	Spectro ($\lambda = 306$ nm)	H_2O $t = 22$ $I = 0.5$	Buur A, Bundgaard H and Falch E, Prodrugs of 5-fluorouracil, *Acta Pharm. Suecica*, **23**, 205–216 (1986). NB: See Fluorouracil, 1-ethyl-oxycarbonyloxymethyl, for details.
567	Fluorouracil, 1-phenyl-oxycarbonyloxymethyl ($C_{12}H_9FN_2O_5$) 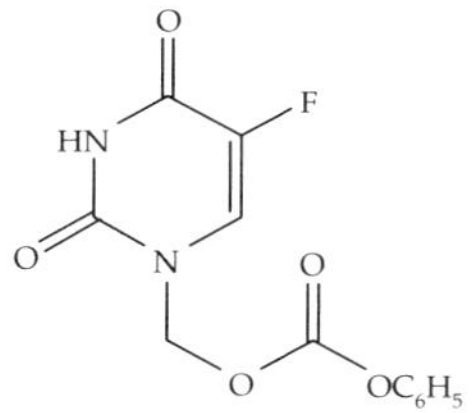	7.3	U	–H	Spectro ($\lambda = 306$ nm)	H_2O $t = 22$ $I = 0.5$	Buur A, Bundgaard H and Falch E, Prodrugs of 5-fluorouracil, *Acta Pharm. Suecica*, **23**, 205–216 (1986). NB: See Fluorouracil , 1-ethyl-oxycarbonyloxymethyl, for details. A value of 7.2 was derived from kinetic measurements.
568	Fluorouracil, 3-phenyl-oxycarbonyloxymethyl ($C_{12}H_9FN_2O_5$)	7.9	U	–H	Spectro ($\lambda = 306$ nm)	H_2O $t = 22$ $I = 0.5$	Buur A, Bundgaard H and Falch E, Prodrugs of 5-fluorouracil, *Acta Pharm. Suecica*, **23**, 205–216 (1986). NB: See Fluorouracil , 1-ethyl-oxycarbonyloxymethyl, for details.
569	Fluorouracil, 1-acetyl-oxymethyl ($C_7H_7FN_2O_4$) 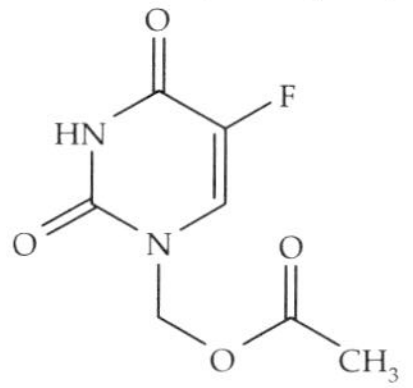	7.3	U	–H	Spectro ($\lambda = 236$ nm)	H_2O $t = 22$ $I = 0.5$	Buur A, Bundgaard H and Falch E, Prodrugs of 5-fluorouracil IV, *Int. J. Pharm.*, **24**, 43–60 (1985). "The ionization constants for the derivatives I-IV and VII were determined at 22 °C and $\mu = 0.5$ by spectrophotometry, according to Albert and Serjeant (1971)." NB: Also reported the same result for 1-propionyloxymethyl-5-FU; 1-butyryloxymethyl-5-FU; and 1-pivaloyloxymethyl-5-FU.
570	Fluorouracil, 3-acetyl-oxymethyl ($C_7H_7FN_2O_4$)	8.0	U	–H	Spectro ($\lambda = 300$ nm)	H_2O $t = 22$ $I = 0.5$	Buur A, Bundgaard H and Falch E. Prodrugs of 5-fluorouracil IV, *Int. J. Pharm.*, **24**, 43–60 (1985). "The ionization constants for the derivatives I-IV and VII were determined at 22 °C and $\mu = 0.5$ by spectrophotometry, according to Albert and Serjeant (1971)."

571	Flupenthixol ($C_{23}H_{25}F_3N_2OS$) 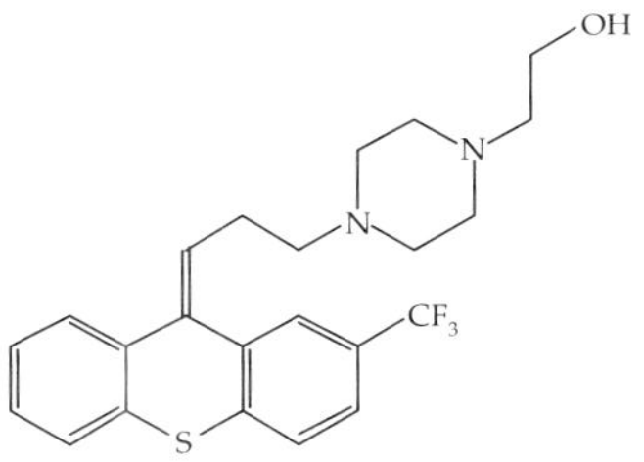	3.362 6.18	U U	+H +H	Potentiometric	H_2O $I = 0.00$ $t = 25$	Lukkari S, Ionization of some thiaxanthene derivatives, clopenthixol and flupenthixol, in aqueous solutions, *Farm. Aikak.*, **80(4-5)**, 237–242 (1971). NB: See Clopenthixol for details.
572	Fluphenazine ($C_{22}H_{26}F_3N_3OS$) 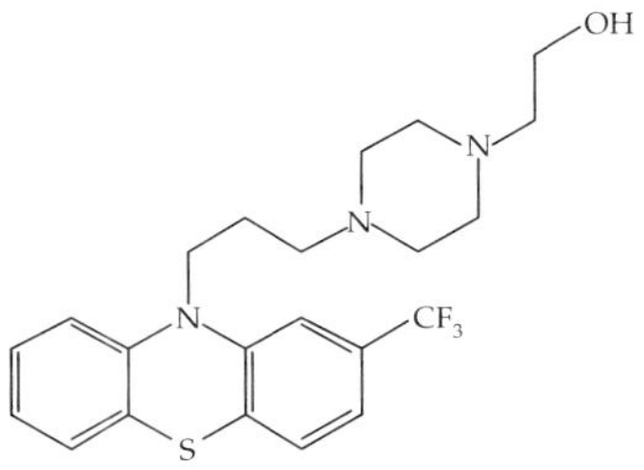	7.98 ± 0.03	U	+H	partition/pH	H_2O $t = 20 \pm 0.5$	Vezin WR and Florence AT, The determination of dissociation constants and partition coefficients of phenothiazine derivatives, *Int. J. Pharm.*, **3**, 231–237 (1979). NB: I was not reported but pK_a was stated to be independent of I. See Promethazine for details.
573	Fluphenazine	3.90 8.05	U U	+H +H	Potentiometric	H_2O (extrap.) $t = 20$ N_2 atmosphere	Sorby DL, Plein EM and Benmaman D, Adsorption of phenothiazine derivatives by solid adsorbents, *J. Pharm. Sci.*, **55**, 785–794 (1966). NB: See Chlorpromazine for details.
574	Flu(o)promazine ($C_{18}H_{19}F_3N_2S$)	9.2	U	+H	Soly	H_2O $t = 24 \pm 1$	Green AL, Ionization constants and water solublities of some aminoalkylphenothiazine tranquilizers and related compounds, *J. Pharm. Pharmacol.*, **19**, 10–16 (1967). NB: See Amitriptylline for details. Same structure as trifluopromazine.
575	Flurbiprofen ($C_{15}H_{13}FO_2$) 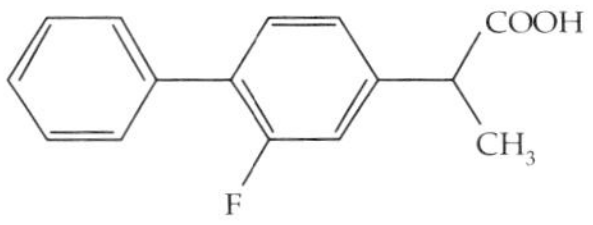	4.9	U	−H	Spectro	H_2O t undefined I undefined	Herzfeldt CD and Kümmel R, Dissociation constants, solubilities, and dissolution rates of some selected nonsteroidal antiinflammatories, *Drug Dev. Ind. Pharm.*, **9(5)**, 767–793 (1983). NB: Used dA/dpH method. See Azapropazone and Ibuprofen for details.

(continued)

No.	Name	pK_a value(s)	Data quality	Ionization type	Method	Conditions t°C; I or c M	Comments and Reference(s)
576	Folic acid ($C_{19}H_{19}N_7O_6$) 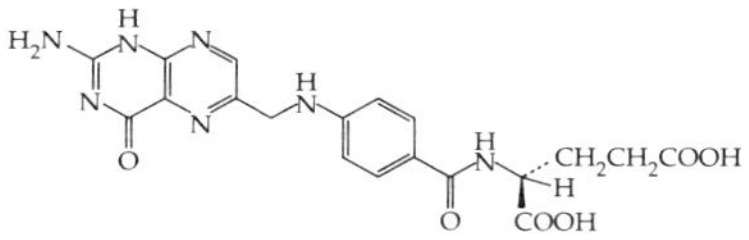	2.35 ± 0.1 8.38	U U	+H −H	Spectro	H_2O $t = 25$ $I = 0.10$	Poe M, Acidic dissociation constants of folic acid, dihydrofolic acid and methotrexate, *J. Biol. Chem.*, **252(11)**, 3724–3728 (1977). NB: The compound aqueous solubility was insufficient to measure the pK_a values for the two carboxyl groups. See Methotrexate for additional details.
		0.20 <−1.5	U U	+H +H		$I = 0.1$ to 4	
577	Folic acid analogues						See Pteridines
578	Folinic acid (Leucovorin) ($C_{20}H_{23}N_7O_7$) 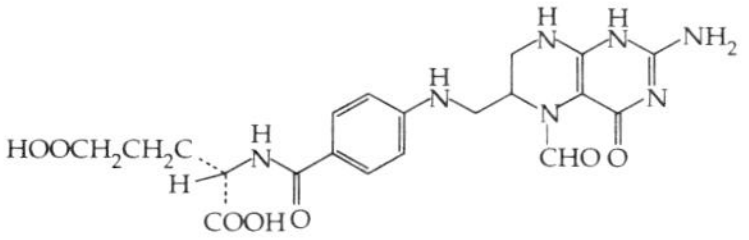	3.1 4.8 10.4	U U U	−H −H −H	Potentiometric	H_2O t undefined I undefined	Flynn EH, Bond TJ, Bardos TJ and Shive W, *JACS*, **73**, 1979–1982 (1951); May M, Bardos TJ, Barger FL, Lansford M, Ravel JM, Sutherland GL and Shive W, Synthetic and degradative investigations of the structure of folinic acid-SF, *ib.*, 3067–75. Cited in Pont LO, Cheung APK and Lim P, Leucovorin Calcium, *APDS*, **8**, 315–350 (1979). "Three pK_a values have been reported for leucovorin (free acid); they are 3.1, 4.8, and 10.4, as determined by electrometric titration. The first two values are attributed to the glutamyl carboxyls, and 10.4 is assigned to the hydroxyl group at the 4 position, by comparison to model compounds."
579	Folinic acid	3.10 4.56 10.15	U U U	−H −H +H	Potentiometric	H_2O $t = 25$	Bergström CAS, Strafford M, Lazorova L, Avdeef A, Luthman K and Artursson P, Absorption classification of oral drugs based on molecular surface properties, *J. Med. Chem.* **46(4)**, 558–570 (2003). NB: From extrapolation of aqueous-methanol mixtures to 0% methanol.
580	Formic acid (CH_2O_2) HCOOH	3.742	R	−H	Potentiometric	H_2O $t = 25.0$ $I = 0.00$	Bates RG, Siegel GL and Acree SF, Dissociation constants and titration curves at constant ionic strength from electrometric titrations in cells without liquid junction, *J. Res. Nat. Bur. Stand.*, **30**, 347–359 (1943). NB: Very careful work using hydrogen and silver-silver chloride half cells. Also reported acetic acid $pK_a = 4.754$, cf. Acetic acid, no. 12 (above), $pK_a = 4.756$ by conductance and potentiometry.
		3.772	A	−H	Conductance	H_2O $t = 25.0$ $I = 0.00$	Bell RP and Miller WBT, Dissociation constants of formic acid and d-formic acid, *Trans. Farad. Soc.*, **59**, 1147–1148 (1963). NB: Reported values for the title compounds of 3.772 ± 0.001 and 3.737 ± 0.001, respectively.

581	Furosemide (frusemide) ($C_{12}H_{11}ClN_2O_5S$) 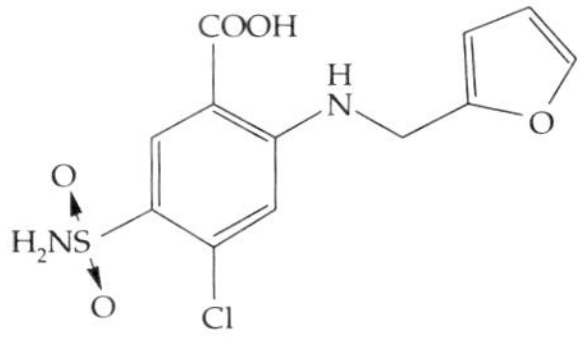	3.9 ± 0.1	U	−H	Potentiometric	H_2O	Hadju P and Haussler A, Untersuchungen mit dem Salidiureticum 4-chlor-N-(2-furylmethyl)-5-sulfamyl-Anthranilsaure, *Arzneim.-Forsch.*, **14**, 709–710 (1964). NB: Potentiometric titrations of 0.5–2.0 mM solutions in one equivalent of NaOH with 0.01N HCl, as previously described for Rastinon (Haussler A and Hadju P, Dissociation constants and solubility of *N*-butyl-*N*-(*p*-tolylsulphonyl)urea, *Arch. Pharm. Weinheim*, **291**, 531–535 (1958)).
582	Furosemide (frusemide)	3.65 ± 0.15	U	−H	partition/pH	H_2O	Hadju P and Haussler A, Untersuchungen mit dem Salidiureticum 4-chlor-N-(2-furylmethyl)-5-sulfamyl-Anthranilsaure, *Arzneim.-Forsch.*, **14**, 709–710 (1964). NB: Partitioned between ether and citrate or phosphate buffers.
583	Furosemide (frusemide)	3.64 10.6	U U	−H +H	CE/pH (−ve ion mode)	H_2O $t = 25$ $I = 0.025$	Wan H, Holmen AG, Wang Y, Lindberg W, Englund M, Nagard MB and Thompson RA, High-throughput screening of pK_a values of pharmaceuticals by pressure-assisted capillary electrophoresis and mass spectrometry, *Rapid Commun. Mass Spectrom.*, **17**, 2639–2648 (2003). NB: Reported predicted values (ACD Labs) of 3.04 and 9.79. The assignment in this paper of the second ionization as a base seems erroneous, as this value is more likely to be ascribed to the sulfonamido group.
584	Furosemide (frusemide)	GLpK_a: 3.65 ± 0.02 10.24 ± 0.04 A&S: 3.74 ± 0.12 10.37 ± 0.09	 A U U U	 −H −H −H −H	Spectro	H_2O $t = 25$ $I = 0.15$ (KCl) Ar atmosphere	Tam KY and Takacs-Novac K, Multi-wavelength spectrophotometric determination of acid dissociation constants, *Anal. Chim. Acta*, **434**, 157–167 (2001). NB: See Clioquinol for details.
585	Furosemide (frusemide)	3.6 ± 0.15	U	−H	Soly	H_2O $t = 27$	Hadju P and Haussler A, Untersuchungen mit dem Salidiureticum 4-chlor-N-(2-furylmethyl)-5-sulfamyl-Anthranilsaure, *Arzneim.-Forsch.*, **14**, 709–710 (1964). NB: Solubilities were determined in Britton-Robinson buffers. Although the solubilities increased in buffer solutions containing 2–5% w/v urea, the pK_a value remained constant. See Glibenclamide for further details.

(*continued*)

	Name	pK_a value(s)	Data quality	Ionization type	Method	Conditions t°C; I or c M	Comments and Reference(s)
586	Furosemide (frusemide)	3.8 7.5	U U	–H –H	Potentiometric	H_2O $t = 20$	Orita Y, Ando A, Urakabe S and Abe H, A metal complexing property of furosemide and bumetanide: Determination of pK and stability constant, *Arzneim.-Forsch.*, **26**(**1**), 11–13 (1976). NB: See Bumetanide for details.
		3.61 10.24	U U	–H –H	Potentiometric	H_2O $t = 37$ $I = 0.15$ (KCl)	Balon K, Riebesehl BU and Muller BW, Drug liposome partitioning as a tool for the prediction of human passive intestinal absorption, *Pharm. Res.*, **16**, 882–888 (1999).
587	Gallic acid	GLpK_a: 4.21 ± 0.01 8.54 ± 0.04 A&S: 4.20 ± 0.15	 A U U	 –H –H –H	Spectro	H_2O $t = 25$ $I = 0.15$ (KCl) Ar atmosphere	Tam KY and Takacs-Novac K, Multi-wavelength spectrophotometric determination of acid dissociation constants, *Anal. Chim. Acta*, **434**, 157–167 (2001). NB: See Clioquinol for details. Cited Palm VA, Table of Rate and Equilibrium Constants of Heterolytic Organic Reactions, Vol I, Viniti, Moscow, 1975, with values of 4.4, 8.7, 11.4 and 13.0.
588	Gelatin	3.7 to 4.5	U	–H	Potentiometric	H_2O $t = 25$ $I = 0.00$ N_2 atmosphere	Ofner CM and Schott H, Shifts in the apparent ionization constant of the carboxylic acid groups of gelatin, *J. Pharm. Sci.*, **74**, 1317–1321 (1985).
		3.8 to 4.4	U			$I = 0.1$	"The shifts in the apparent dissociation constants of carboxylic acid groups of gelatin as a function of the degree of ionization and the presence of an added electrolyte are discussed. The shifts were ascribed to changes in the electrostatic interference of neighboring ionic groups with the ionization process of the carboxylic acid groups. Cationic groups enhanced the ionization while carboxylate ions hampered it. The added electrolyte reduced the magnitude of both types of interference."
589	Glafenine ($C_{19}H_{17}ClN_2O_4$) HO, OH, O, O, H N, Cl, N	7.2	U	+H	Spectro	H_2O $t = 20$	Badwan AA, Zughul MB and Al Omari M, Glafenine, *APDS*, **21**, 197–232 (1992). "The pK_a was determined spectrophotometrically at 20 °C in accordance with an earlier reported method (5). Stock solution of glafenine in 10^{-3} N HCl was prepared and diluted with suitable buffer solutions ranging from pH 6–10 to obtain a final glafenine concentration of 10μg/mL. The absorbance of theses solutions was measured at the maximum of 342.5 nm. This method yielded 7.2 as the pK_a of glafenine at 20 °C (4)."

							4. Badwan AA and Al-Omari MM, Unpublished data, The Jordanian Pharmaceutical Manufacturing Co., Jordan 5. Takla PG and Dakas CJ, A study of interactions of chloroquine with ethanol, sugars and glycerol using ultraviolet spectrophotometry, *Int. J. Pharm.*, **43**, 225–232 (1988).
590	Glibenclamide (glyburide) ($C_{23}H_{28}ClN_3O_5S$) 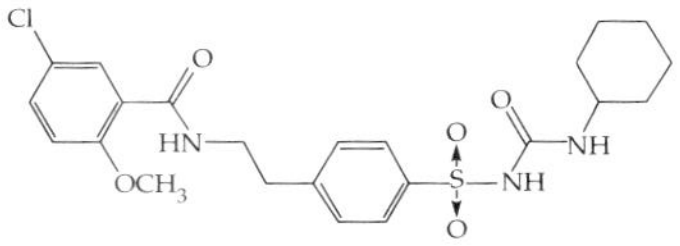	5.3 ± 0.1	U	−H	Potentiometric	H_2O	Hadju P, Kohler KF, Schmidt FH and Spingler H, Physicalisch-chemische und analytische untersuchungen an HB 419, *Arzneim.-Forsch.*, **19**, 1381–1386 (1969). Cited in Takla, P.G., Glibenclamide, *APDS*, **10**, 337–352 (1981). "Glibenclamide is a weak acid. It has been concluded that it has the same dissociation constant as tolbutamide (5.3 ± 0.1), since both compounds show the same dissociation at half-neutralization in solvent mixtures such as methylcellosolve and water or methanol and water. The direct determination of its pK_a in water is impossible owing to its low solubility." NB: The poor water solubility of glibenclamide indicates that it is a good candidate for pK_a determination by the solubility-pH method (see no. 591)
591	Glibenclamide	5.3 ± 0.1	U	−H	Potentiometric	H_2O	Hadju P, Kohler KF, Schmidt FH and Spingler H, Physicalisch-chemische und analytische untersuchungen an HB 419, *Arzneim.-Forsch.*, **19**, 1381–1386 (1969).
		6.3	U	−H	Partition/pH	H_2O $t = 27$	NB: Solubility-pH dependence in Britton-Robinson buffer solutions was used to estimate the pK_a value, after equilibration, filtration and appropriate dilution with 0.01N NaOH, followed by UV absorption measurements (λ_{max} not specified). Partition-pH dependence measurements were between ether and Britton-Robinson buffers in the pH range 2.4–10.4. The substantially different values from each type of measurement (see also no. 592) deserves further investigation.
		6.8 ± 0.15	U	−H	Soly		
592	Glibenclamide (glyburide)	5.22	U	+H	CE/pH (+ve ion mode)	H_2O $t = 25$ $I = 0.025$	Wan H, Holmen AG, Wang Y, Lindberg W, Englund M, Nagard MB and Thompson RA, High-throughput screening of pK_a values of pharmaceuticals by pressure-assisted capillary electrophoresis and mass spectrometry, *Rapid Commun. Mass Spectrom.*, **17**, 2639–2648 (2003). NB: Reported a predicted value (ACD Labs) of 4.99.

(continued)

	Name	pK_a value(s)	Data quality	Ionization type	Method	Conditions t°C; I or c M	Comments and Reference(s)
593	Gluconic acid ($C_6H_{12}O_7$) 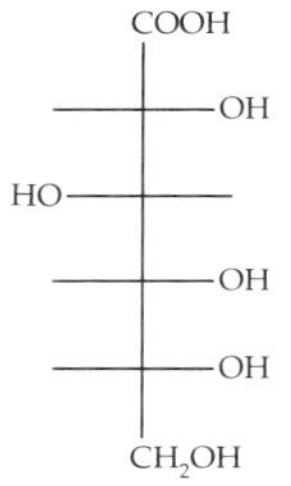	3.62 ± 0.02	A	−H	Potentiometric	H_2O $t = 25$	Skibsted LH and Kilde G, Strength and lactone formation of gluconic acid, *Dansk. Tidsskr. Farm.*, **45(9)**, 320–324 (1971). "The strength of gluconic acid in water at 25 °C has been determined by the measurement of hydrogen ion activity using a glass electrode in gluconic acid/sodium gluconate buffer solutions. The hydrogen ion activity immediately after mixing the buffer solutions indicated that the acid ionization constant of gluconic acid is $pK_a = 3.62$ (estimated standard deviation 0.02). However, the activity changes with time and when equilibrium is reached the apparent acid ionization constant of gluconic acid is $pK_{app} = 3.84$ (0.04). The constants have been extrapolated to zero ionic strength. The difference between the 2 acid ionization constants is attributed to lactone formation, and for the equilibrium 'lactone ⇔ acid' a value of $K = 0.60$ (0.05) has been determined."
594	D-Glucosamine ($C_6H_{13}NO_5$) 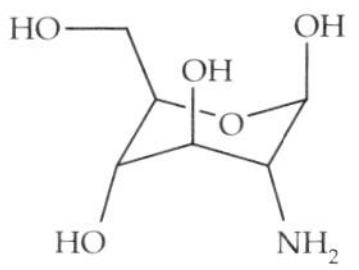	8.04	A	+H	Potentiometric	H_2O $t = 15.5$ $I = 0.00$ N_2 atmosphere	Zimmerman HK, Studies on glucosammonium chloride. II. Protolysis constants and the stability of glucosamine in weakly alkaline aqueous solution, *J. Phys. Chem.*, **62**, 963–965 (1958). Cited in Perrin Bases 3407 ref. Z3. NB: Used pH measurements with a glass electrode and liquid junction potentials. Results at various ionic strengths were extrapolated to zero. For the same conditions, also gave 7.75 ($t = 23.9$); 7.71 ($t = 25.5$); 7.50 ($t = 29.0$); 7.35 ($t = 39.0$); 7.19 ($t = 51.0$) (all A).
595	D(+)-Glucosamine-6-phosphate ($C_6H_{14}NO_8P$) 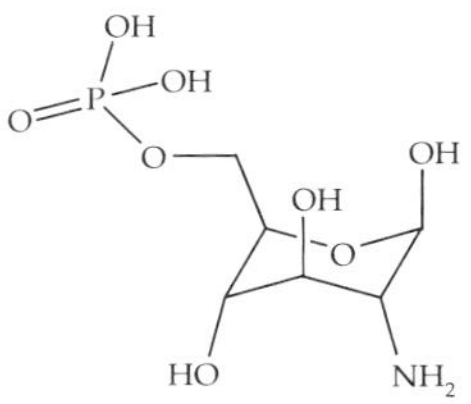	6.10 8.14	A A	−H +H	Potentiometric	H_2O $t = 26.3$ $c = 0.001$ N_2 atmosphere	Brown DH, Action of phosphoglucomutase on D-glucosamine-6-phosphate, *J. Biol. Chem.*, **204**, 877–889 (1953). Cited in Perrin Bases 3408 ref. B111. NB: Used pH measurements with a glass electrode and liquid junction potentials. Correction of the reported apparent pK_a' values of 6.08 and 8.10 led to the given pK_a values (Perrin). The first dissociation constant of the phosphate group could not be determined at the very low concentrations used.

No.	Compound	pKa	Data quality	Ionization	Method	Conditions	Reference / Notes
596	D-Glucose ($C_6H_{12}O_6$)	12.43 12.92	A A	−H −H	Potentiometric	H_2O $t = 18$ $t = 0$	Thamsen J, The acidic dissociation constants of glucose, mannitol and sorbitol, as measured by means of the hydrogen electrode and the glass electrode at 0° and 18 °C, *Acta Chem. Scand.*, **6**, 270–284 (1952). NB: See also Dextrose
597	Glutethimide ($C_{13}H_{15}NO_2$) 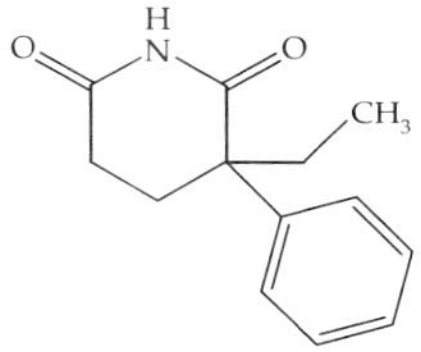	11.8	U	−H			DeLuca PP, Lachman L and Schroeder HG, Physical chemical properties of substituted amides in aqueous solution and evaluation of their potential use as solubilizing agents, *J. Pharm. Sci.*, **62**, 1320–1327 (1973). NB: No details reported for measurement, or references given.
598	Glycerol ($C_3H_8O_3$) 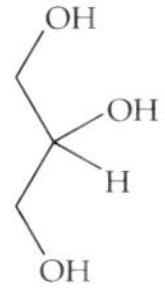	14.07 ± 0.2	U	−H	Potentiometric	H_2O $t = 25.0 \pm 0.1$ $I < 0.0005$	Woolley EM, Tomkins J and Hepler LG, Ionization constants for very weak organic acids in aqueous solution and apparent ionization constants for water in aqueous organic mixtures, *J. Solution Chem.*, **1**, 341–351 (1972). NB: See Dextrose for details.
599	Glycine xylidide ($C_{10}H_{14}N_2O$) 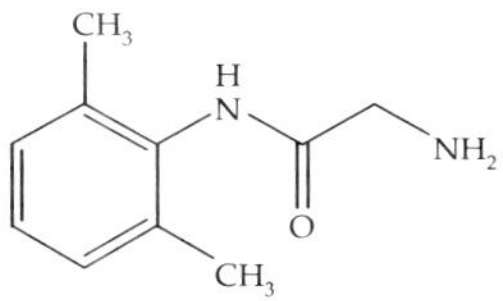	7.68	U	+H	Potentiometric	H_2O $t = 25.0 \pm 0.2$ $I = 0.01$ (NaCl)	Johansson P-A, Liquid-liquid distribution of lidocaine and some structurally related anti-arrythmic drugs and local anaesthetics, *Acta Pharm. Suec.*, **19**, 137–142 (1982). NB: See Lidocaine for details.
600	Glycocholic acid ($C_{26}H_{43}NO_6$) 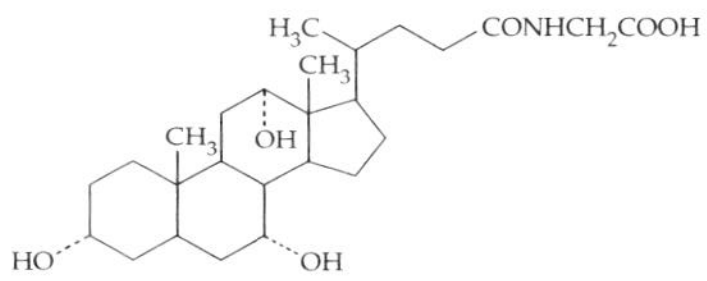	3.87 ± 0.06	U	−H	Potentiometric	H_2O (extrap.) $t = 25.0 \pm 0.1$ $I = 0.00$	Fini A and Roda A, Chemical properties of bile acids. IV. Acidity constants of glycine-conjugated bile acids, *J. Lipid Res.*, **28(7)**, 755–759 (1987). NB: See Chenodeoxycholic acid for more detail.

(continued)

	Name	pK_a value(s)	Data quality	Ionization type	Method	Conditions t°C; I or c M	Comments and Reference(s)
601	Glycodeoxycholic acid ($C_{26}H_{43}NO_5$) 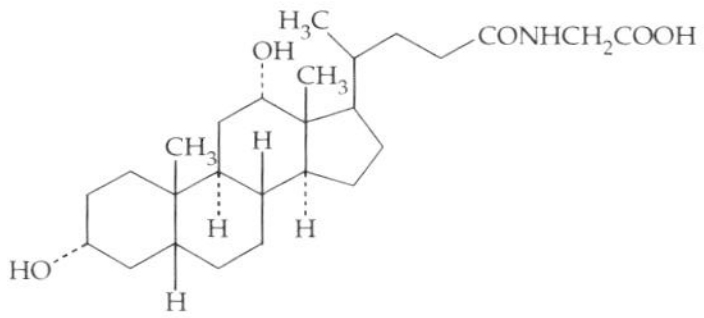	3.87 ± 0.06	U	–H	Potentiometric	H_2O (extrap.) $t = 25.0 \pm 0.1$ $I = 0.00$	Fini A and Roda A, Chemical properties of bile acids. IV. Acidity constants of glycine-conjugated bile acids, *J. Lipid Res.*, **28**(**7**), 755–759 (1987). NB: See Chenodeoxycholic acid for more detail.
602	Guanazole prodrugs 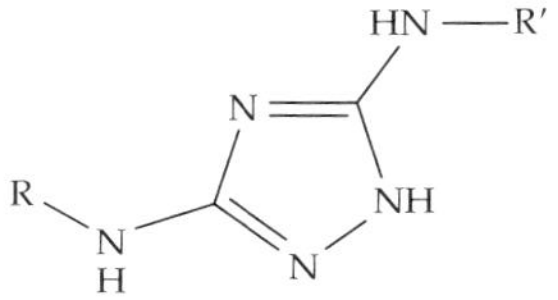				Spectro	H_2O $t = 25.0$ $I = < 0.1$	Alhaider AA, Selassie CD, Chua S and Lien EJ, Measurements of ionization constants and partition coefficients of guanazole prodrugs, *J. Pharm. Sci.*, **71**(**1**), 89–94 (1982). "A previously described procedure (Albert and Serjeant, 1972) was used for the determination of ionization constants. Its accuracy was checked with an authentic sample of salicylic acid (at 25°), and the result (2.94) was in close agreement with that reported (2.97). . . ."
	3,5-diacetamido-1,2,4-triazole ($C_4H_9N_5O_2$)	9.35	U	–H			
	3,5-dipropionamido-1,2,4-triazole ($C_6H_{13}N_5O_2$)	9.34	U	–H			Further guanazole prodrug pK_a values:

Compound	R	R′	pK_a
3,5-di(trifluoroacetamido)-1,2,4-triazole	$COCF_3$	$COCF_3$	5.90
3-amino-5-heptafluorobutyramido-1,2,4-triazole	H	$COCF_2CF_2CF_3$	6.10
3,5-dibenzamido-1,2,4-triazole	COC_6H_5	COC_6H_5	10.21
3,5-di(*p*-nitrobenzamido)-1,2,4-triazole	$COC_6H_4NO_2$	$COC_6H_4NO_2$	8.64

No.	Compound	pKa	Data quality	Ionization	Method	Conditions	Reference / Notes
603	Haloperidol ($C_{21}H_{23}ClFNO_2$) 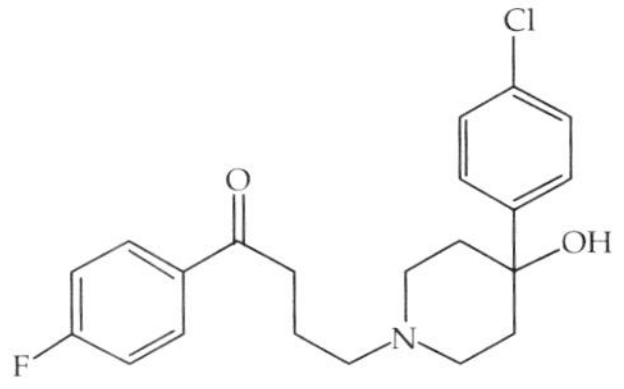	8.96	U	+H	CE/pH (+ve ion mode)	H_2O $t = 25$ $I = 0.025$	Wan H, Holmen AG, Wang Y, Lindberg W, Englund M, Nagard MB and Thompson RA, High-throughput screening of pK_a values of pharmaceuticals by pressure-assisted capillary electrophoresis and mass spectrometry, *Rapid Commun. Mass Spectrom.*, **17**, 2639–2648 (2003). NB: Reported a predicted value (ACD Labs) of 8.25.
604	Haloperidol	8.65 ± 0.05	U	+H	Potentiometric	H_2O $t = 25.0 \pm 0.1$ $I = 0.1$ (NaCl)	Takacs-Novak K, Box KJ and Avdeef A, Potentiometric pK_a determination of water-insoluble compounds: Validation study in methanol/water mixtures, *Int. J. Pharm.*, **151**, 235–248 (1997). NB: By extrapolation from 40–60%w/w aqueous MeOH. See Acetaminophen for full details.
605	Heliotridane ($C_8H_{15}N$) 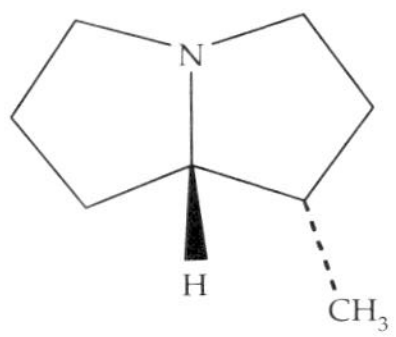	11.40	A	+H	Potentiometric	H_2O $t = 27$ $c = 0.02$	Adams R, Carmack M and Mahan JE, *JACS*, 64, 2593 (1942). Cited in Perrin 2913 ref. A5. NB: Used pH measurements on equimolar solutions of the base and salt with a glass electrode and liquid junction potentials. Use of a temperature coefficient for amine pK_a values (Hall and Sprinkle, *JACS*, **54**, 3469 (1932)) gave the value at t = 25 of 11.48.
606	Heliotridene ($C_8H_{13}N$) 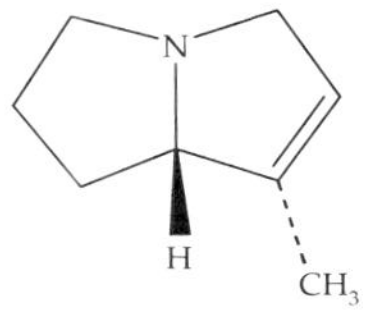	10.55	A	+H	Potentiometric	H_2O $t = 25.5$ $c = 0.02$	Adams R, Carmack M and Mahan JE, *JACS*, **64**, 2593 (1942). Cited in Perrin 2913 ref. A5. Used pH measurements on equimolar solutions of the base and salt with a glass electrode and liquid junction potentials. Use of a temperature coefficient for amine pK_a values (Hall and Sprinkle, *JACS*, **54**, 3469 (1932)) gave the value at $t = 25$ of 10.60. pK_a values for several related compounds were also reported.
607	Hexachlorophene ($C_{13}H_6Cl_6O_2$) 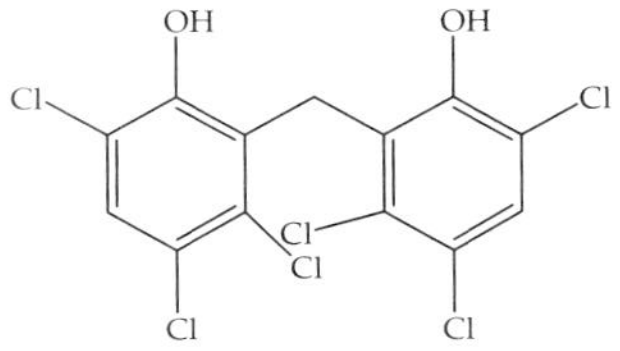	4.89 ± 0.02 10.83 ± 0.02	U U	−H −H	Potentiometric	H_2O $t = 25.0 \pm 0.1$ $I = 0.1$ (NaCl)	Takacs-Novak K, Box KJ and Avdeef A, Potentiometric pK_a determination of water-insoluble compounds: Validation study in methanol/water mixtures, *Int. J. Pharm.*, **151**, 235–248 (1997). NB: By extrapolation from 42–60 %w/w aqueous MeOH. See Acetaminophen for full details.

(continued)

Appendix A *(continued)*

	Name	pK_a value(s)	Data quality	Ionization type	Method	Conditions t°C; I or c M	Comments and Reference(s)
608	Hexachlorophene	GLpK_a: 5.21 ± 0.01 10.27 ± 0.09	 U U	Spectro +H +H	H_2O $t = 25$ $I = 0.15$ (KCl)	Ar atmosphere	Tam KY and Takacs-Novac K, Multi-wavelength spectrophotometric determination of acid dissociation constants, *Anal. Chim. Acta*, **434**, 157–167 (2001). NB: See Clioquinol for details; extrapolated from partially aqueous mixtures, using the Yasuda-Shedlovsky approximation.
609	Hexachlorophene	5.65	U	−H	Soly	H_2O $t = 25 \pm 0.1$ $I = 0.11$	Terada H, Muraoka S and Fujita T, Structure-activity relationships of fenamic acids, *J. Med. Chem.*, **17**, 330–334 (1974). NB: See Fluphenamic acid for details.
610	Hexetidine ($C_{21}H_{45}N_3$) 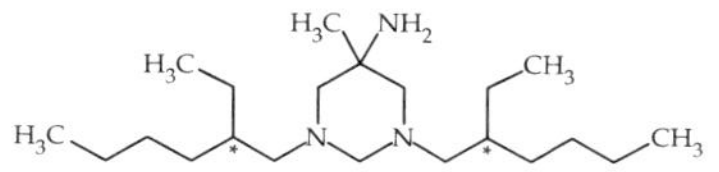	8.3	U	+H	Potentiometric	50% aqueous EtOH	Satzinger G, Herrmann W and Zimmermann F, Hexetidine, *APDS*, **7**, 277–295 (1978). NB: Apparent pH of half-neutralization on titration with 0.1 M HCl.
611	Hexetidine	7.9	U	+H	CE/pH (+ve ion mode)	H_2O $t = 25$ $I = 0.025$	Wan H, Holmen AG, Wang Y, Lindberg W, Englund M, Nagard MB and Thompson RA, High-throughput screening of pK_a values of pharmaceuticals by pressure-assisted capillary electrophoresis and mass spectrometry, *Rapid Commun. Mass Spectrom.*, **17**, 2639–2648 (2003). NB: Reported a predicted value (ACD Labs) of 8.74.
612	Hexobendine ($C_{30}H_{44}N_2O_{10}$) 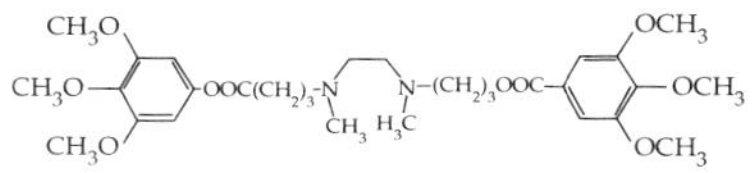	4.52 ± 0.01 8.47 ± 0.01	A A	+H +H	Potentiometric	H_2O $I = 0.1$ $t = 25.0 \pm 0.1$ N_2 atmosphere	IJzerman AP, Limiting solubilities and ionization constants of sparingly soluble compounds: Determination from aqueous potentiometric data only, *Pharm. Res.*, **5(12)**, 772–775 (1988). NB: Glass electrode; KOH titrant. So = 567 ± 8 μM.
613	Hexobendine	4.04 ± 0.02 7.94 ± 0.02	U U	+H +H	Potentiometric	40% EtOH $t = 25.0$	Mannhold R, Rodenkirchen R, Bayer R and Haus W, The importance of drug ionization for the action of calcium antagonistsand related compounds, *Arzneim.-Forsch.*, **34**, 407–409 (1984).
		5.09 8.68	U U	+H +H		H_2O	NB: See Aprindine for details.

No.	Compound	pK	Quality	Assignment	Method	Conditions	Reference
614	Histamine ($C_5H_9N_3$)	6.04 9.75	A A	+H +H	Potentiometric	H_2O $t = 25.0 \pm 0.1$ $I = 0.00$ N_2 atmosphere	Paiva TB, Tominaga M and Pavia ACM, Ionization of histamine, *N*-acetylhistamine and their iodinated derivatives, *J. Med. Chem.*, **13**, 689–692 (1970). NB: Cited by Al-Badr AA and El-Subbagh HI, Histamine Phosphate, *APDS*, **27**, 168 (2000).
615	Histamine	5.784 9.756	R R	+H +H	Potentiometric	H_2O $t = 25$ $I = 0.01$ to 0.5	Von Schalien SNR, Potentiometric studies on histamine and its metal chelates. II. Cd and Ni chelates of histamine in aqueous solutions, *Suomen Kemistilehti*, **32B**, 148–153 (1959); CA 54:27393. Cited in Perrin Bases no. 1396 ref. V18. NB: Data from titrations using a glass electrode in an unsymmetrical cell with liquid junction potentials. Graphical extrapolation to zero ionic strength was used. Also reported $t = 37$, 5.595 & 9.386; $t = 50$, 5.358 & 9.047. Numerous other approximate (A) and uncertain (U) values were also reported by Perrin.
616	Histamine, monoiodo ($C_5H_8IN_3$)	4.06 9.20 11.88	A A A	+H +H +H	Potentiometric	H_2O $t = 25.0 \pm 0.1$ $I = 0.00$ N_2 atmosphere	Paiva TB, Tominaga M and Pavia ACM, Ionization of histamine, N-acetylhistamine and their iodinated derivatives, *J. Med. Chem.*, **13**, 689–692 (1970).

Structure labels (614): HN, N, $CH_2CH_2NH_2$
Structure labels (616): HN, N, I, $CH_2CH_2NH_2$

Table (614):

Ionization	$t = 15°C$	$t = 20°C$	$t = 25°C$	$t = 30°C$	$t = 35°C$
pK_1	6.23	6.14	6.04	5.94	5.84
pK_2	10.03	9.93	9.75	9.61	9.47
pK_3	–	–	14.36[a]	–	–

[a] spectrophotometric.

Table (616):

Ionization	$t = 12°C$	$t = 15°C$	$t = 20°C$	$t = 25°C$	$t = 30°C$	$t = 35°C$	$t = 38°C$
pK_1	4.21	4.17	4.11	4.06	3.99	-	3.89
pK_2	9.62	9.52	9.37	9.20	9.08	-	8.86
pK_3	–	12.32	12.11	11.88	11.72	11.60	11.38

(continued)

	Name	pK_a value(s)	Data quality	Ionization type	Method	Conditions t°C; I or c M	Comments and Reference(s)
617	Histamine, diiodo ($C_5H_7I_2N_3$)	2.31 8.20 10.11	A A A	+H +H +H	Potentiometric	H_2O $t = 25.0 \pm 0.1$ $I = 0.00$ N_2 atmosphere	Paiva TB, Tominaga M and Pavia ACM, Ionization of histamine, N-acetylhistamine and their iodinated derivatives, *J. Med. Chem.* **13**, 689–692 (1970).
618	Histapyrrodine ($C_{19}H_{24}N_2$)	2.99 ± 0.12 7.95 ± 0.02	U U	+H +H	Potentiometric	H_2O t undefined $I = 0.30$ (NaCl)	Testa B and Murset-Rossetti L, The partition coefficient of protonated histamines, *Helv. Chim. Acta*, **61**, 2530–2537 (1978). NB: See Cycliramine for details.
619	HNB-1 ($C_{27}H_{26}NO_4$)	8.49 ± 0.02 9.20 ± 0.05	U U	+H −H	Potentiometric	H_2O $t = 25.0 \pm 0.1$ $I = 0.1$ (NaCl)	Takacs-Novak K, Box KJ and Avdeef A, Potentiometric pK_a determination of water-insoluble compounds: Validation study in methanol/water mixtures, *Int. J. Pharm.*, **151**, 235–248 (1997). NB: By extrapolation from 33–60 %w/w aqueous MeOH. See Acetaminophen for full details.

Ionization	$t = 12$°C	$t = 15$°C	$t = 20$°C	$t = 25$°C	$t = 30$°C	$t = 38$°C
pK_1	2.51	2.44	2.38	2.31	2.25	2.12
pK_2	8.52	8.45	8.29	8.20	8.08	7.93
pK_3	10.58	10.47	10.28	10.11	9.97	9.72

No.	Compound	pKa	Data quality	Ionization type	Method	Conditions	Reference / Comments
620	HNB-5 ($C_{26}H_{25}N_2O_4$)	4.31 ± 0.05 8.64 ± 0.05 9.42 ± 0.05	U U U	+H +H −H	Potentiometric	H_2O $t = 25.0 \pm 0.1$ $I = 0.1$ (NaCl)	Takacs-Novak K, Box KJ and Avdeef A, Potentiometric pK_a determination of water-insoluble compounds: Validation study in methanol/water mixtures, *Int. J. Pharm.*, **151**, 235–248 (1997). NB: By extrapolation from 24–44 %w/w aqueous MeOH. See Acetaminophen for more details.
621	Hydralazine ($C_8H_8N_4$)	7.1 6.820 ± 0.005	U U	+H +H	Potentiometric Conductance	H_2O H_2O $t = 25$ I undefined	Evstratova KI and Ivanova AI, *Farmatsiya* (Moscow) **17(2)**, 41–45, (1968); CA 69:46128a; Evstratova KI, Goncharova NA and Solomko VY, *Farmatsiya* (Moscow) **17(4)**, 33–36, (1968); CA 69:99338a (1968). "Evstratova and Ivanova reported a pK_a of 7.1. Evstratova *et al.* reported a pK_a of 7.1 in water, a pK_a of 4.7 in aqueous 90 percent acetone, and a pK_b of 15.6 in the latter solvent. Naik *et al.* reported a pK_a of 6.9 for the dissociation of the monohydrochloride and 0.5 for dissociation of the dihydrochloride, determined absorptiometrically. Artamanov *et al.* reported a pK_a of 7.2." Artamonov BP, Konenkova TY and Maiofis SL, Conductimetric analysis in the manufacture and control of drugs. III. Conductimetric titration and some physicochemical characteristics of Apressin, *Med. Prom. SSSR*, **19(9)**, 57–9 (1965); CA 63:17800f. Cited in: Orzech CE, Nash NG and Daley RD, Hydralazine hydrochloride, *APDS*, **8**, 283–314 (1979).
622	Hydralazine	0.5 6.9	U U	+H +H		H_2O	W&G: Naik DV, Davis BR, Minnet KM and Schulman SG, Fluorescence of hydralazine in concentrated sulfuric acid, *J. Pharm. Sci.*, **65**, 274–6 (1976).
623	Hydrochlorothiazide ($C_7H_8ClN_3O_4S_2$)	8.6 9.9	U U	−H −H	Potentiometric	H_2O $t = 60.0 \pm 0.05$ I undefined	Mollica JA, Rehm, CR, Smith JB and Govan HK, Hydrolysis of benzothiadiazines, *J. Pharm. Sci.*, **60**, 1380–1384 (1971). Ref. 26 in Deppeler HP, Hydrochlorothiazide, *APDS*, **10**, 405–436 (1981).
		8.7 ± 0.05	U	−H	Spectro	H_2O $t = 25.0 \pm 0.05$ I undefined	NB: Potentiometric titration data was analysed according to the method of Noyes for substances with two ionizing groups with pK_a values less than 2.7 units apart. Also reported pK_a values under the same conditions for *N*1-ethylhydrochlorothiazide, 9.5 ± 0.05; C2-ethylhydrochlorothiazide, 8.8 ± 0.05.
		8.81 ± 0.05 10.4 ± 0.1	U U	−H −H	Spectro	H_2O t undefined I undefined	Stahl PH, Ciba-Geigy Ltd., Basle, data on file (1978).

(continued)

	Name	pK_a value(s)	Data quality	Ionization type	Method	Conditions t°C; I or c M	Comments and Reference(s)
624	Hydrochlorothiazide	8.5	U U	−H −H	Potentiometric	mixed aq.	Henning VG, Moskalyk RE, Chatten LG and Chan SF, Semiaqueous potentiometric determination of apparent pK_{a1} values for benzothiadiazines and detection of decomposition during solubility variation with pH studies, *J. Pharm. Sci.*, **70(3)**, 317–319 (1981). NB: See Flumethiazide for details. The second pK_a value was not reported.
625	Hydrochlorothiazide	8.57 ± 0.04 10.22 ± 0.02	U U	−H (ring) −H (side)	Potentiometric	H_2O $t = 25.0 \pm 0.1$ $I = 0.1$ (NaCl)	Takacs-Novak K, Box KJ and Avdeef A, Potentiometric pK_a determination of water-insoluble compounds: Validation study in methanol/water mixtures, *Int. J. Pharm.*, **151**, 235–248 (1997). NB: By extrapolation from 23–37 %w/w aqueous MeOH. See Acetaminophen for full details.
626	Hydrochlorothiazide	GLpK_a: 8.78 ± 0.01 9.96 ± 0.02 A&S: 8.90 ± 0.08 10.20 ± 0.07	 A U U U	 +H +H +H +H	Spectro	H_2O $t = 25$ $I = 0.15$ (KCl) Ar atmosphere	Tam KY and Takacs-Novac K, Multi-wavelength spectrophotometric determination of acid dissociation constants, *Anal. Chim. Acta*, **434**, 157–167 (2001). NB: See Clioquinol for details.
627	Hydrocortisone hydrogen succinate ($C_{25}H_{34}O_8$) 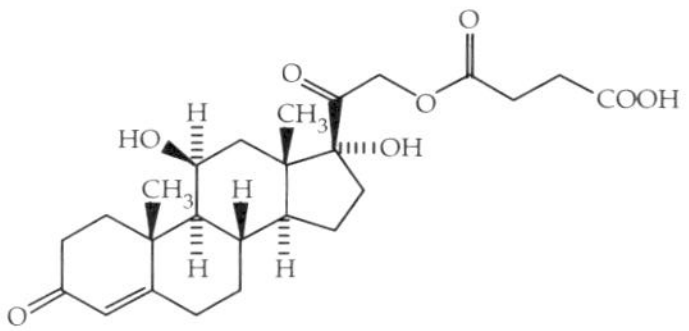	5.10 5.64	U	−H	Potentiometric	20% aq EtOH 50% aq EtOH	Garrett ER, Prediction of stability in pharmaceutical preparations. X. Alkaline hydrolysis of hydrocortisone hemisuccinate, *J. Pharm. Sci.*, **51**, 445–450 (1962).
628	Hydrocortisone, imidazole-1-carboxylic acid prodrug ($C_{25}H_{32}N_2O_6$) 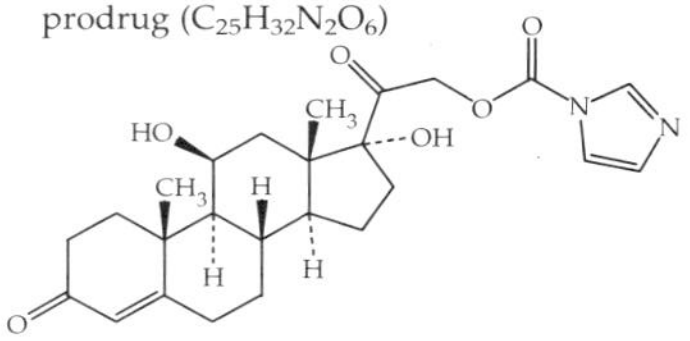	3.3	U	+H	kinetic	H_2O $t = 25.0$ $I = 0.5$ (KCl)	Klixbull U and Bundgaard H, Prodrugs as drug delivery systems. Part 29. Imidazole-1-carboxylic acid esters of hydrocortisone and testosterone, *Arch. Pharm. Chemi., Sci. Edn.*, **11(4)**, 101–110 (1983). "The kinetics of hydrolysis of imidazole-1-carboxylic acid esters of hydrocortisone and testosterone were studied to assess their suitability as prodrug forms. The pH-rate profiles of the 2 derivatives were derived in the range pH 1–12 and were accounted for by assuming spontaneous hydrolysis of the protonated forms (pK_a 3.3–3.5) and hydroxide ion-catalyzed hydrolysis of the free base forms. . . ."

No.	Compound	pKa			Method	Conditions	Reference
629	Hydroflumethiazide ($C_8H_8F_3N_3O_4S_2$) 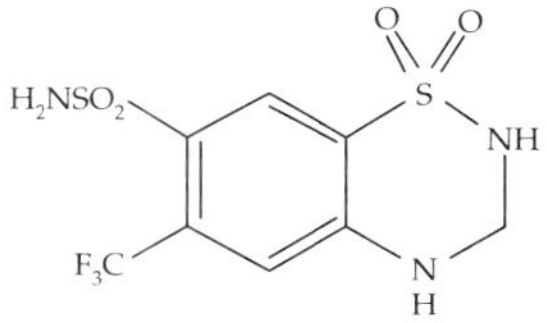	8.5	U U	–H –H	Potentiometric	mixed aq.	Henning VG, Moskalyk R.E, Chatten LG and Chan SF, Semiaqueous potentiometric determination of apparent pK_{a1} values for benzothiadiazines and detection of decomposition during solubility variation with pH studies, *J. Pharm. Sci.*, **70**, 317–319 (1981). NB: See Flumethiazide.
630	Hydroflumethiazide	8.9 10.7	U U	–H –H		RT	Kobinger W and Lund FJ, Investigations into a new oral diuretic, rontyl (6-trifluoromethyl-7-sulfamyl-3, 4-dihydro-1,2,4-benzothiadiazine-1,1-dioxide), *Acta Pharmacol. Toxicol.*, **15**, 265–274 (1959); CA 53:14332e. Cited in Orzech CE, Nash NG and Daley RD, Hydroflumethiazide, *APDS*, **7**, 297–317 (1978). "Kobinger and Lund report a pK_1 of 8.9 and a pK_2 of 10.7 at room temperature."
631	Hydroflumethiazide	8.73 8.79	U U	–H –H	Spectro ($\lambda = 273$ nm) fluoro ($\lambda_{ex} = 333$ nm; $\lambda_{em} = 393$ nm)	H_2O	Smith RB, Smith RV and Yakatan GJ, Spectrofluorometric determination of hydroflumethiazide in plasma and urine, *J. Pharm. Sci.*, **65**, 1208–1211 (1976). NB: A value of 10.5 is reported in Sunshine I, Handbook of analytical toxicology, CRC Press, Boca Raton, FL, USA. Cited by Orzech CE, Nash NG and Daley RD, Hydroflumethiazide, *APDS*, **7**, 297–317 (1978).
632	Hydrogen peroxide (H_2O_2) H-O-O-H	>11.3	U	–H	Conductance	H_2O $t = 25$	Kendall J, Electrical conductivity and ionization constants of weak electrolytes in aqueous solution, *in* Washburn EW, Editor-in-Chief, *International Critical Tables*, vol **6**, McGraw-Hill, NY 259–304 (1929).
633	Hydroquinone ($C_6H_6O_2$) 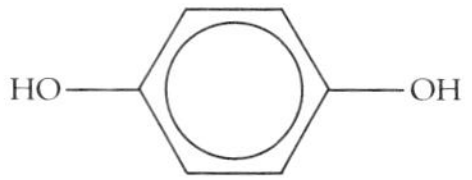	8.87	U	–H	Spectro	H_2O $t = 15$ $c = 0.0025$ to 0.005	Kolthoff IM, The dissociation constants, solubility product and titration of alkaloids, *Biochem. Z.*, **162**, 289–353 (1925). Cited in Perrin Bases 2919, ref. K47. NB: The method used colorimetric determination of the extent of ionization by use of an indicator of known pK_a value.
634	*N*-hydroxyamphetamine ($C_9H_{13}NO$) 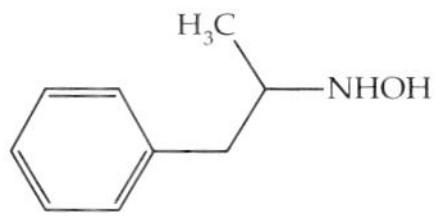	5.77	U	+H		H_2O t =25 $I = 1.0$ (KCl)	Lindeke B, Anderson E, Lundkvist G, Jonsson U and Eriksson SO, Autoxidation of *N*-hydroxyamphetamine and *N*-hydroxyphentermine: the formation of 2-nitroso-1-phenyl-propanes and 1-phenyl-2-propanone oxime, *Acta Pharm. Suec.*, **12(2)**, 183–198 (1975). NB: See *N*-hydroxyphentermine for full details.

(continued)

	Name	pK_a value(s)	Data quality	Ionization type	Method	Conditions t°C; I or c M	Comments and Reference(s)
635	4-Hydroxyamphetamine (Paredrine) ($C_9H_{13}NO$) 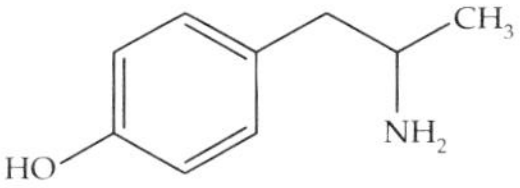	9.31	U	+H	Potentiometric	H_2O $t = 25.0 \pm 0.2$ $I \leq 0.001$	Leffler EB, Spencer HM and Burger A, Dissociation constants of adrenergic amines, *JACS*, **73**, 2611–2613 (1951). NB: See Amphetamine for details; from pK_b = 4.69.
636	10-Hydroxycodeine ($C_{18}H_{21}NO_4$) 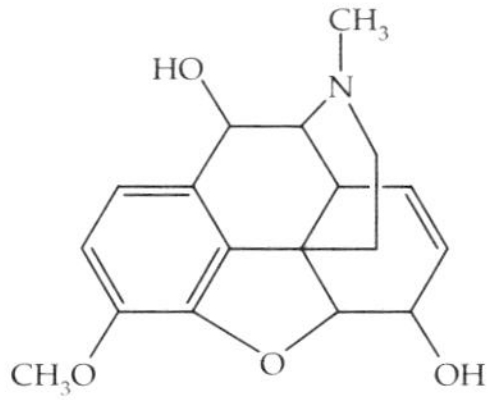	7.12	U	+H	Potentiometric	H_2O	Rapaport H and Stevenson GW, 10-Hydroxycodeine, *JACS*, **76**, 1796–99 (1954). Cited in Perrin Bases 2920 ref. R3. NB: The study used measurements of pH with a glass electrode and liquid junction potentials.
637	4-Hydroxycoumarin derivatives 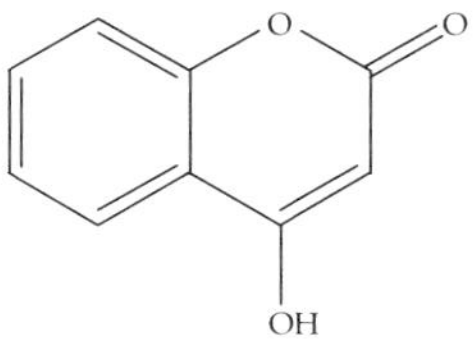	4.13	U	−H	Potentiometric spectro partition	H_2O $t = 25$ $I = 0.1$	van der Glesen WF and Janssen LHM, Influence of ionization and ion-pair formation on lipophilicity of some 4-hydroxycoumarin derivatives in the octanol-water system, *Int. J. Pharm.*, **12(3)**, 231–249 (1982). "The pH dependence of six 4-hydroxycoumarins was studied in octanol-water systems, with partition coefficients being calculated in systems of pH 1–11. At high pH values, the apparent partition coefficient is independent of pH but dependent on the concentration of counterions, indicating ion pair extraction. . . . It was found that partition values obtained under conditions where the compounds are ionized in the aqueous phase depend on both the partition coefficient and the dissociation constant of the acid or salt."

Compound	Partition/pH	Potentio	Spectro
4-Hydroxycoumarin	4.13 (A)	4.10 (A)	–
Warfarin	5.02 (A)	5.00 (A)	–
Acenocoumarin	4.98 (A)	5.05 (A)	–
Phenprocoumon	4.02 (U)	4.20 (U)	–
Ethyl biscoumacetate	–	–	1.25 (U)
	7.50 (A)	7.50 (A)	–
Coumetarol	–	–	0.65 (U)
	–	9.00 (U)	–

No.	Compound	pKa	Quality	Type	Method	Solvent	Reference
638	14-*cis*-Hydroxydihydro-codeinone ($C_{18}H_{21}NO_4$) 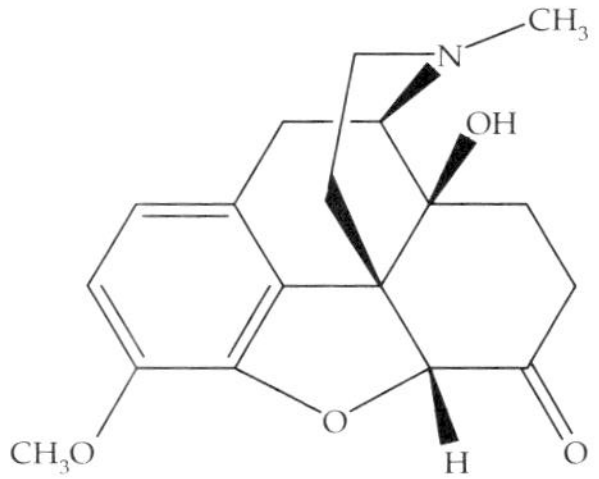	8.53	U	+H	Potentiometric	H_2O	Rapaport H and Masamune S, The stereochemistry of 10-hydroxycodeine derivatives, *JACS*, **77**, 4330–4335 (1955). Cited in Perrin Bases Supplement 7476 ref. R3. NB: The study used measurements of pH with a glass electrode and liquid junction potentials.
639	10-*cis*-Hydroxydihydrodesoxycodeine ($C_{18}H_{23}NO_3$) 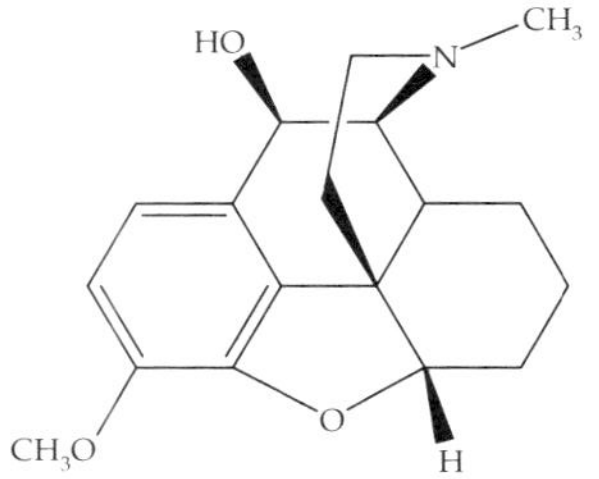	9.41	U	+H	Potentiometric	H_2O	Rapaport H and Masamune S, The stereochemistry of 10-hydroxycodeine derivatives, *JACS*, **77**, 4330–4335 (1955). Cited in Perrin Bases Supplement 7477 ref. R3. NB: The study used measurements of pH with a glass electrode and liquid junction potentials.
640	10-*trans*-Hydroxydihydrodesoxycodeine ($C_{18}H_{23}NO_3$) 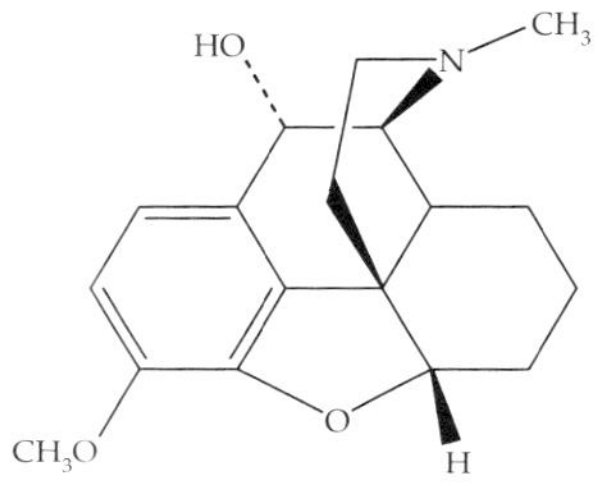	7.71	U	+H	Potentiometric	H_2O	Rapaport H and Masamune S, The stereochemistry of 10-hydroxycodeine derivatives, *JACS*, **77**, 4330–4335 (1955). Cited in Perrin Bases Supplement 7478 ref. R3. NB: The study used measurements of pH with a glass electrode and liquid junction potentials.

(continued)

	Name	pK_a value(s)	Data quality	Ionization type	Method	Conditions t°C; I or c M	Comments and Reference(s)
641	10-*cis*-Hydroxydihydronorcodeine ($C_{17}H_{21}NO_4$) 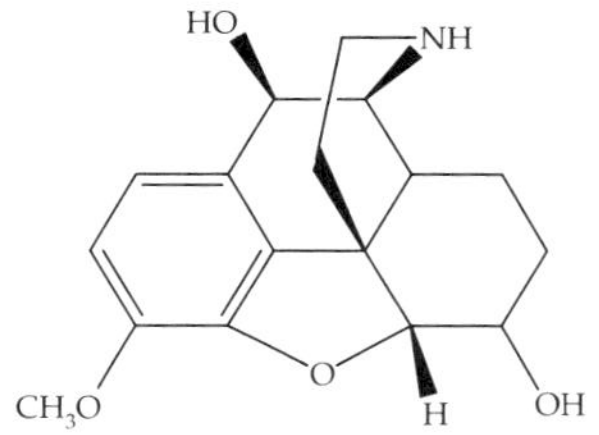	9.17	U	+H	Potentiometric	H_2O	Rapaport H and Masamune S, The stereochemistry of 10-hydroxycodeine derivatives, *JACS*, **77**, 4330–4335 (1955). Cited in Perrin Bases Supplement 7479 ref. R3. NB: The study used measurements of pH with a glass electrode and liquid junction potentials.
642	10-*trans*-Hydroxydihydronorcodeine ($C_{17}H_{21}NO_4$) 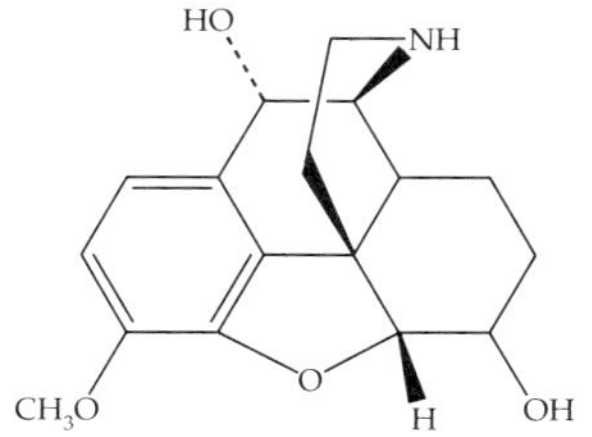	8.72	U	+H	Potentiometric	H_2O	Rapaport H and Masamune S, The stereochemistry of 10-hydroxycodeine derivatives, *JACS*, **77**, 4330–4335 (1955). Cited in Perrin Bases Supplement 7480 ref. R3. NB: The study used measurements of pH with a glass electrode and liquid junction potentials.
643	4-(2-Hydroxy-3-isopropylamino-propoxy) indole ($C_{14}H_{20}N_2O_2$) 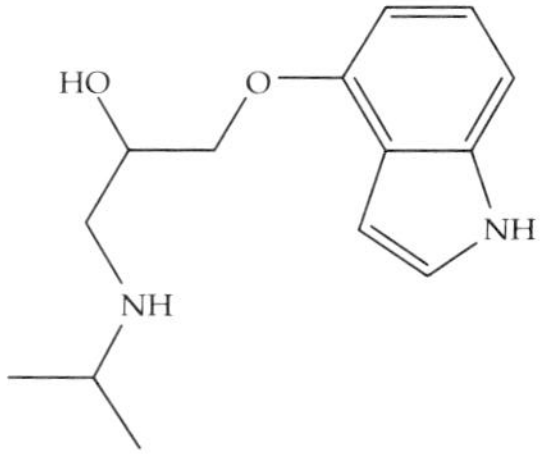	9.65 ± 0.1	U	−H	Potentiometric	H_2O $t = 24$	Maulding HV and Zoglio MA, pK_a determinations utilizing solutions of 7-(2-hydroxypropyl) theophylline, *J. Pharm. Sci.*, **60**, 309–311 (1971). NB: See Barbituric acid, 5-allyl-5-isobutyl for details.

No.	Compound	pKa	Data quality	Ionization type	Method	Conditions	Comments and reference(s)
644	3-Hydroxy-α-(methyl-amino)methylbenzene-methanol ($C_9H_{13}NO_2$) 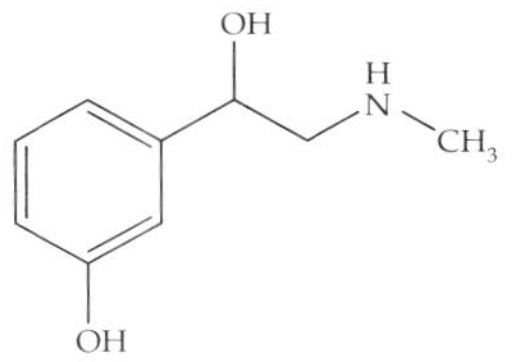	9.48 9.71	U U	+H, −H −H, +H	Spectro	H_2O $t = 20.0 \pm 0.1$ $I = 0.16$	Quintero B, Baena E, Cabeza MC, Thomas J and Alvarez JM, Determination of microscopic dissociation constants of 3-hydroxy-α-(methylamino) methylbenzenemethanol by a spectral deconvolution method, *J. Pharm. Pharmacol.*, **41(7)**, 485–488 (1989). "Microscopic dissociation constants of 3-hydroxy-α-(methylamino) methylbenzene-methanol have been calculated from the titration spectrophotometer data (c = 3.8×10^{-4} M. Ionic strength = 0.16; buffer system: H_3BO_3/KOH) by application of a spectral deconvolution method. The results found ($pK_a = 9.48$; $pK_b = 9.71$; $pK_c = 10.12$; $pK_d = 9.88$) are in good concordance with those obtained from the conventional regression linear method ($pK_a = 9.45$; $pK_b = 9.77$; $pK_c = 10.14$; $pK_d = 9.81$)."
645	5-[1-Hydroxy-2-[(1-methylethyl)amino]-ethyl]-1,3-benzenediol (metaproterenol) ($C_{11}H_{17}NO_3$) 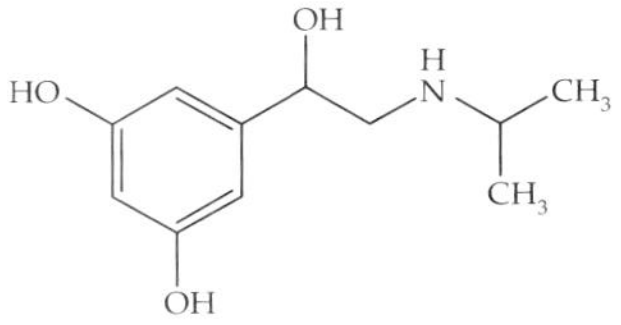	8.84 10.28	U U	+H, −H −H, +H	Spectro	H_2O $I = 0.02$	Quintero B, Lopez J and Thomas J, Determination of dissociation constants of 5-[1-hydroxy-2-[(1-methylethyl)-amino]ethyl]-1,3-benzenediol, *J. Pharm. Sci.*, **74(1)**, 72–75 (1985).
		8.88 10.28 11.70	U U U	+H, −H −H, +H −H			Hernandez Gaina A, Ph.D. Thesis, University of Granada, 1982.
646	*N*-Hydroxyphentermine ($C_{10}H_{15}NO$) 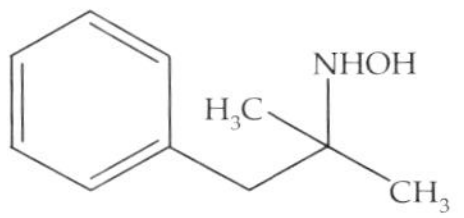	6.03	U	+H	Potentiometric	H_2O $t = 25$ $I = 1.0$ (KCl)	Lindeke B, Anderson E, Lundkvist G, Jonsson U and Eriksson SO, Autoxidation of *N*-hydroxyamphetamine and *N*-hydroxyphentermine: the formation of 2-nitroso-1-phenyl-propanes and 1-phenyl-2-propanone oxime, *Acta Pharm. Suec.*, **12(2)**, 183–198 (1975).
		5.80	U	+H		29% EtOH $t = 25$ $I = 0.1$ (KCl)	"The autoxidation of *N*-hydroxyamphetamine (I) and *N*-hydroxyphentermine (II) was investigated. The oxidation rate, in aqueous solution, is pH dependent with a considerable oxidation taking place around pK_a. . . . The pK_a values were determined for *N*-hydroxyamphetamine and *N*-hydroxyphentermine. The reported constants are the stoichiometric ones, and were determined at 25 °C and fixed ionic strength, using methods earlier described [15]. . . . *N*-hydroxyamphetamine (0.2 meq) was titrated with HCl (0.5 M) and *N*-hydroxyphentermine-HCl (0.2 meq) was titrated with KOH (0.5 M). The stoichiometric pK_a values are calculated from about 25 pH readings in the range −log [H+] = $pK_a \pm 0.7$ and are expected to be correct within ± 0.02 units. . . . Beckett *et al.* have reported a pK_a value of 8.25 for

(continued)

	Name	pK_a value(s)	Data quality	Ionization type	Method	Conditions t°C; I or c M	Comments and Reference(s)
							N-hydroxyamphetamine [5]. (Note added in proof: Since this paper was submitted Beckett *et al.* . . . report a pK_a value of 8.1 for N-hydroxyphentermine. The pK_a values reported by Beckett *et al.* are in good agreement with what we consider to be the pK_b values.)" 5. Beckett AH and Al-Sarraj S, *J. Pharm. Pharmacol.*, **25**, 328–334 (1973). NB: See no. 1922. 15. Eriksson S-O and Holst C, Hydrolysis of anilides. II. Hydrolysis of trifluoro- and trichloroacetanilide by hydroxyl ions and by some bifunctional catalysts. *Acta Chem. Scand.* **20**, 1892–1906 (1966). NB: MO calculations of the electron density at nitrogen (R. Prankerd, unpublished results) support the Beckett value as the pK_a and Lindeke's values as pK_b values.
647	*p*-Hydroxy-norephedrine ($C_9H_{13}NO_2$) 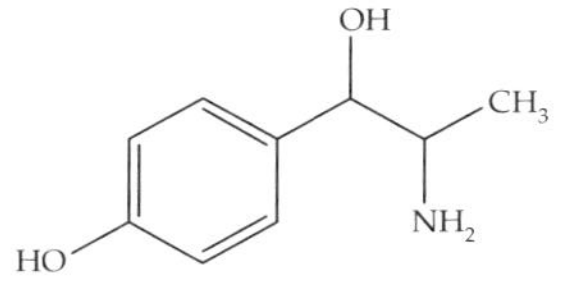	8.70 ± 0.02	U	+H	Potentiometric	H_2O $t = 25.0 \pm 0.2$ $I \leq 0.001$	Leffler EB, Spencer HM and Burger A, Dissociation constants of adrenergic amines, *JACS*, **73**, 2611–13 (1951). NB: See Amphetamine for details; from $pK_b = 5.30$.
648	7-(2-Hydroxypropyl)theophylline ($C_{10}H_{14}N_4O_3$) 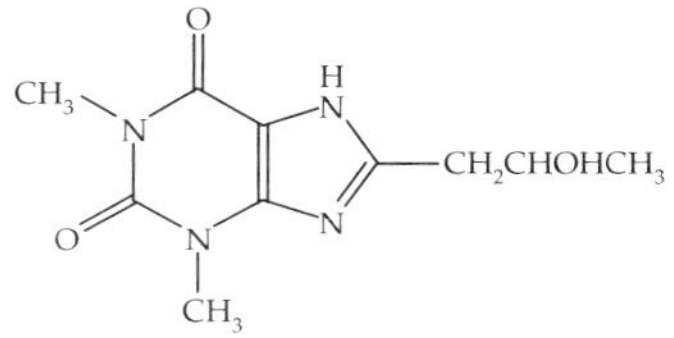						Maulding HV and Zoglio MA, pK_a determinations utilizing solutions of 7-(2-hydroxypropyl)-theophylline, *J. Pharm. Sci.*, **60**, 309–311 (1971). "A method exhibiting applicability in the evaluation of ionization constants of certain difficultly soluble compounds by potentiometric titration is discussed. The neutral xanthine derivative, 7-(2-hydroxypropyl)theophylline, is employed as an aid in dissolution of poorly soluble molecules, allowing their subsequent titration with acid or base in the usual manner. Examples of the utility of this treatment are given and pertinent data included as they apply to several structural prototypes. . . ."

No.	Compound	pKa	Type	Ionization	Method	Conditions	Reference / Notes
649	4-Hydroxysalicylic acid ($C_7H_6O_4$)	3.23	U	−H	Potentiometric	H_2O $t = 25 \pm 0.1$	Pothisiri P and Carstensen JT, Solid state decomposition of para-substituted salicyclic acids, *J. Pharm. Sci.*, **64**, 1931–1935 (1975).

Substituent (X)	pK_a for p-XC_6H_4COOH	pK_a for p-XC_6H_3(*o*-OH)COOH
−H	4.21	2.90[a]
$-OC_2H_5$	4.45	3.16
−OH	4.59	3.23
$-N(Me)_2$	5.05 (6.01)	3.76
$-N(Et)_2$	(6.19)	3.84

[a] Cited from Kortum, Vogel, Andrussow, *Dissociation Constants of Organic Acids in Water*, IUPAC (1961). NB: Serjeant and Dempsey, *Ionization Constants of Organic Acids in Water* (Supplement), IUPAC (1979) reported three pK_{a1} values for 4-Hydroxysalicylic acid in the range 3.10–3.30 at 25°C and $I = 0.1$ M. Reported pK_{a2} values were 9.12 at 25°C and 8.91 at 30°C.

No.	Compound	pKa	Type	Ionization	Method	Conditions	Reference / Notes
650	Hydroxyzine ($C_{21}H_{27}ClN_2O_2$)	1.96 ± 0.05 7.40 ± 0.03	A U	+H +H	Potentiometric	H_2O $t = 24.5 \pm 0.5$	Newton DW, Murray WJ and Lovell MW, pK_a determination of benzhydrylpiperazine antihistamines in aqueous and aqueous methanol solutions, *J. Pharm. Sci.*, **71(12)**, 1363–1366 (1982). NB: See Chlorcyclizine for details.
651	Hydroxyzine	1.99 ?	A	+H +H	Potentiometric	H_2O $t = 25$ $I = 0.00$ (NaCl)	Lukkari S, Potentiometric studies on the ionization of some psychotropic diphenylmethane derivatives, hydroxyzine and meclizine., *Farm. Aikak.* **80,** 161–165 (1971); CA 75:41145t. Cited in Tsau J and DeAngelis N, Hydroxyzine dihydrochloride, *APDS*, **7,** 319–341 (1978). NB: Activity effects were estimated with the Guggenheim equation (constant term = 0.160). Also found pK_a = 2.0 (spectro).

(*continued*)

	Name	pK_a value(s)	Data quality	Ionization type	Method	Conditions t°C; I or c M	Comments and Reference(s)
652	Hyoscyamine ($C_{17}H_{23}NO_3$)	9.68	U	+H	Partition	H_2O $t = 21 \pm 2$ $c = 0.028$	Bottomley W, Mortimer PI, Partition separation of tropane alkaloids, *Aust. J. Chem.*, **7**, 189–196 (1954). Cited in Perrin Bases 2921 ref. B86.
		9.65	U	+H			NB: Used a glass electrode in an unsymmetrical cell with liquid junction potentials to measure the pH at which 50% extraction of the alkaloid is complete with equal volumes of aqueous and chloroform phases. Also reported studies on hyoscine, norhyoscyamine, tigloidine, valeroidine and apoatropine.
653	Hypoxanthine ($C_5H_4N_4O$) 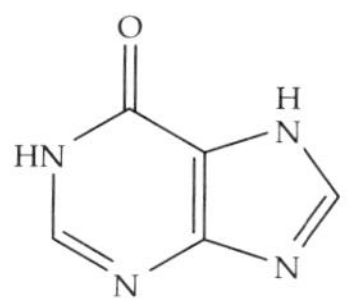	8.50 ± 0.06	U	−H	Spectro	H_2O $t = 25$ $I = 0.3$	Cohen JL and Connors KA, Stability and structure of some organic molecular complexes in aqueous solution, *J. Pharm. Sci.*, **59**, 1271–1276 (1970).

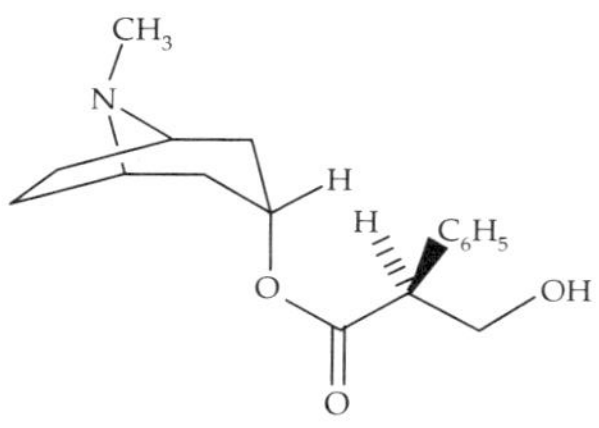

Compound	pK_a
Xanthine	9.95 ± 0.05
8-nitrotheophylline	3.55 ± 0.05
Benzimidazole-2-acetonitrile	4.20 ± 0.02; 11.76 ± 0.02
6-nitrobenzimidazole	2.89 ± 0.03; 10.69 ± 0.05

No.	Compound Name	pK_a value(s)	Data quality	Ionization type	Method	Conditions	Comments and Reference(s)
			See comments	(+H or −H)			Data reliability cut-off points: R: Reliable = ±<0.02; A: Approx = ±0.02 to ±0.06; U: Uncertain = ±>0.06 These overall data reliability cut-offs apply when all other aspects of the pK_a values have been considered. Where key variables (such as temperature, ionic strength, solvent composition) have not been reported, or the value was obtained from a computer program, a value is automatically assessed as U: Uncertain. A few results have been classified as VU: Very Uncertain.
654	Ibuprofen ($C_{13}H_{18}O_2$)	4.45 ± 0.04	A	−H	Potentiometric	H_2O t = 25 ± 0.5 I = 0.15 (KCl)	Avdeef A, Box KJ, Comer JEA, Hibbert C and Tam KY, pH-metric log P 10. Determination of liposomal membrane-water partition coefficients of ionizable drugs, *Pharm. Res.*, **15**(2), 209–215 (1998). NB: Used a Sirius PCA101 autotitrator. Also reported log P (octanol-water) and log P (dioleylphosphatidylcholine unilamellar vesicles-water). The same result was reported in Sirius Technical Application Notes, vol. **2**, pp. 26–27 (1995). Sirius Analytical Instruments Ltd., Forest Row, East Sussex, RH18 5DW, UK. NB: from extrapolation to 0% MeOH from data in 16–74 wt% MeOH by the Yasuda-Shedlovsky procedure. Concentration of analyte, 1.3–2.5 mM.
655	Ibuprofen	4.43	A	+H	CE/pH (−ve ion mode)	H_2O t = 25 I = 0.025	Wan H, Holmen AG, Wang Y, Lindberg W, Englund M, Nagard MB and Thompson RA, High-throughput screening of pK_a values of pharmaceuticals by pressure-assisted capillary electrophoresis and mass spectrometry, *Rapid Commun. Mass Spectrom.*, **17**, 2639–2648 (2003). NB: Reported a predicted value (ACD Labs) of 4.41.
656	Ibuprofen	GLpK_a: 4.24 ± 0.03 A&S: 4.45 ± 0.05	U A	−H −H	Spectro	H_2O t = 25 I = 0.15 (KCl) Ar atmosphere	Tam KY and Takacs-Novac K, Multi-wavelength spectrophotometric determination of acid dissociation constants, *Anal. Chim. Acta*, **434**, 157–167 (2001). NB: See Clioquinol for details.

(continued)

No.	Compound Name	pK_a value(s)	Data quality	Ionization type	Method	Conditions	Comments and Reference(s)
657	Ibuprofen	4.51 ± 0.07	U	–H	Potentiometric	H_2O $t = 25.0 \pm 0.1$ $I = 0.1$ (NaCl)	Takacs-Novak K, Box KJ and Avdeef A, Potentiometric pK_a determination of water-insoluble compounds: validation study in methanol/water mixtures, *Int. J. Pharm.*, **151**, 235–248 (1997). NB: By extrapolation from 16–51%w/w aqueous MeOH. See Acetaminophen for full details.
658	Ibuprofen	4.31 ± 0.05	U	–H	Spectro	H_2O $t = 22.0$ $I = 0.2$ (NB: compare with other values taking into account effects of I and t)	Ueda H, Pereira-Rosario R, Riley CM and Perrin JH, Some pharmaceutical applications of diode array spectrophotometers, *Drug Dev. Ind. Pharm.*, **11**(4), 833–843 (1985). "An ibuprofen concentration of 8×10^{-5} M was used for the measurements in phosphate buffers of ionic strength 0.2. The unionised spectrum was obtained in 0.1 N HCl and the ionised spectrum in 0.01-N NaOH. The spectra were obtained between 210 and 280 nm. The analysis was performed . . . assuming a two component mixture. . . . Ibuprofen has been reported to have pK_as varying from 4.13 to 5.2 (Davis, 1975; Whitlam *et al.*, 1979). The measurements . . . were obtained of the unionised and ionised species and then of solutions of accurately known pH at a constant ionic strength of 0.2. Table 1 shows the apparent pK_as determined by a) single wavelength determinations, b) by analysis of a two component mixture over a wavelength range of ten nanometers and c) by analysis of a two component mixture using the first derivative technique over the same wavelength range:

Table 1: pK_a determination . . . at 22 °C and ionic strength 0.2

pK_a	nm	deriv.	n
4.25±0.03	226	0	12
4.31±0.05	224–234	0	11
4.41±0.02	224–234	0	12

The pK_a value of 4.31 ± 0.05 . . . seems to be the most reliable because the data is averaged over the 10 nm wavelength range of maximum difference between the spectra of the ionised and unionised forms. The derivative method gave a slightly higher result, which may be due to the increased signal to noise ratio associated with obtaining the derivative spectra, although the statistics associated with the measurements are excellent."

No.	Compound	pKa			Method	Conditions	Reference / Comments
659	Ibuprofen	4.13 ± 0.05	U	–H	soly-pH	H_2O t undefined I undefined	Whitlam JB, Crooks MJ, Brown KF and Pedersen PV, Binding of nonsteroidal antiinflammatory agents to proteins I. Ibuprofen-serum albumin interaction, *Biochem. Pharmacol.*, **28**, 675–678 (1979). NB: The pK_a was determined by the solubility technique of Albert and Serjeant (1971) to be 4.13 ± 0.05.
660	Ibuprofen	4.45	U	–H	Potentiometric	H_2O $t = 37$ $I = 0.15$ (KCl)	Balon K, Riebesehl BU and Muller BW, Drug liposome partitioning as a tool for the prediction of human passive intestinal absorption, *Pharm. Res.*, **16**, 882–888 (1999).
661	Ibuprofen	5.2	U	–H	partition	H_2O $t = 25.0 \pm 0.1$	Persson B-A, Extraction of amines as complexes with inorganic anions. Part 3, *Acta Pharm. Suec.*, **5**, 335–342 (1968). NB: See Chlorpromazine for details.
		5.2	U	–H	not stated	60% EtOH	Davis LJ, Ibuprofen, *Drug Intell. Clin. Pharm.*, **9**, 501–503 (1975).
662	Ibuprofen	5.3	U	–H	Spectro	H_2O t undefined I undefined	Herzfeldt CD and Kümmel R, Dissociation constants, solubilities, and dissolution rates of some selected nonsteroidal antiinflammatories, *Drug Dev. Ind. Pharm.*, **9**(5), 767–793 (1983). "Drug solutions were prepared by dissolving an amount of 0.5 to 2.0 mg (of the drug) in 100 mL of both solution A and solution B. Different amounts dissolved depend on the minimum solubility in solution A (acidic) and on an optimum UV-absorption within 0.1 and 0.9 absorption units. Both solutions have been mixed to selected and controlled pH-values according to the mixing diagram. The solutions were measured spectrophotometrically in a 1-cm cell between 200 and 400 nm.

Buffer solutions

Solution A:		Solution B:	
HCl 1 M	94.0 mL	$Na_2HPO_4 \cdot 2H_2O$	20.5 g
Glycocoll	0.5 g	KH_2PO_4	2.8 g
NaCl	3.68 g	NaCl	0.15 g
H_2O	to 1000.0 mL	H_2O	to 1000.0 mL

Method A for substances having a change in the wavelength of the absorption maximum with pH. . . . The wavelength of maximum absorption λ has been plotted *vs.* pH as shown in the tables. The dissociation constant has been either extrapolated graphically and calculated by determining the maximum change with pH.

(continued)

No.	Compound Name	pK_a value(s)	Data quality	Ionization type	Method	Conditions	Comments and Reference(s)
							Method B for substances showing a change in absorption height at the same wavelength with pH. . . .. The change in absorption height with pH (dA/dpH) has been differentially plotted as shown in the tables. The dissociation constant has been either extrapolated graphically or calculated by estimation of the maximum change of height with pH."
663	Ibuprofen	4.5–4.6	U	−H	Potentiometric		Higgins JD, Gilmor TP, Martellucci SA, Bruce RD and Brittain HG, Ibuprofen, *APDS*, **27** (2000). Cited: Takacs-Novak K, Box KJ and Avdeef A, Potentiometric pK_a determination of water-insoluble compounds: Validation study in methanol/water mixtures, *Int. J. Pharm.*, **151**, 235–248 (1997).
		4.52	U	−H	Potentiometric	H_2O $t = 25 \pm 0.5$ $I = 0.00$ (KCl)	Rafols C, Roses M and Bosch E, Dissociation constants of several non-steroidal anti-inflammatory drugs in isopropyl alcohol /water mixtures, *Anal. Chim. Acta*, **350**, 249–255 (1997). NB: Potentiometric method using i-PrOH/water mixtures in the range 10/90 to 90/10 v/v. Metrohm autotitrator; careful electrode calibration. Activity corrections applied and extrapolated to zero [i-PrOH]. Alternative data handling procedure gave $pK_a = 4.64 \pm 0.04$.
664	Ibuprofen	4.30–5.16	U	−H	Potentiometric	H_2O $t = 25 \pm 0.5$ $I = 0.15$ (KCl)	Avdeef A, Box KJ, Comer JEA, Gilges M, Hadley M, Hibbert C, Patterson W and Tam KY, PH-metric log P 11. pK_a determination of water-insoluble drugs in organic solvent-water mixtures, *J. Pharm. Biomed. Anal.*, **20**, 631–641 (1999). NB: Used a Sirius PCA101 autotitrator. Titrations were performed in a range of aqueous organic cosolvent mixtures, as follows:

Cosolvent	pK_a (SD)	Cosolvent	pK_a (SD)
Acetonitrile	4.31 (0.04)	Ethanol	4.33 (0.01)
Dimethyl formamide	4.30 (0.05)	Ethylene glycol	4.34 (0.06)
Dimethyl sulfoxide	4.35 (0.03)	Methanol	4.45 (0.04)
1,4-Dioxane	4.46 (0.10)	Tetrahydrofuran	5.16 (0.07)

No.	Compound Name	pK_a value(s)	Data quality	Ionization type	Method	Conditions	Comments and Reference(s)
665	Idarubicin ($C_{26}H_{27}NO_9$)	4.73 ± 0.21	U	+H			NB: See Daunorubicin for experimental details.
666	Idoxuridine	8.25	U	−H			Prusoff WH, Recent advances in chemotherapy of viral diseases, *Pharmacol. Rev.*, **19**, 209–250 (1967). NB: No reference given.

667	Imidazole ($C_3H_4N_2$) 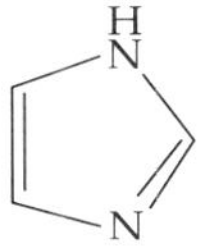	6.953	U	+H	soly (auto-titrator)	H_2O $t = 25 \pm 0.2$ I undefined	Cho MJ, Kurtz RR, Lewis C, Machkovech SM and Houser DJ, Metronidazole phosphate-a water-soluble prodrug for parenteral solutions of metronidazole, *J. Pharm. Sci.*, **71**, 410–414 (1982). NB: See Metronidazole for details. The paper gave pK_a values of the following substituted imidazoles:
668	Imidazole	7.0	U	+H	Potentiometric	H_2O $t = 25$ $I = 0.0$	Bates RG, Amine buffers for pH control, *Ann. NY Acad. Sci.*, **92**, 341–356 (1961).
669	Imidazole, *N*-acetyl ($C_5H_6N_2O$) 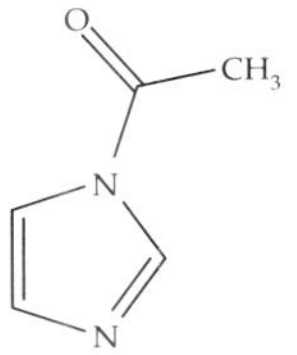	3.6	U	+H	kinetic	H_2O $t = 25$ $I = 0.2$	Jencks WP and Carriulo J, Imidazole catalysis. II. Acyl transfer and the reactions of acetyl imidazole with water and oxygen anions, *J. Biol. Chem.*, **234**, 1272–1279 (1959). Cited in: Klixbull U and Bundgaard H, Prodrugs as drug delivery systems XXIX. Imidazole-1-carboxylic acid esters of hydrocortisone and testosterone, *Arch. Pharm. Chem., Sci. Edn.*, **11**, 101–110 (1983).
670	Imipenem ($C_{12}H_{17}N_3O_4S$) 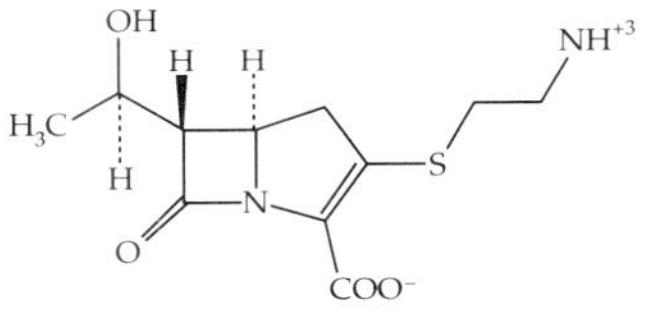	~3.2 ~9.9	U U	−H +H	Potentiometric	H_2O $t = 25.0$	Oberholtzer, E.R., Imipenem, *APDS*, **17**, 73–114 (1988). McCauley J, Merck Sharp and Dohme Research Laboratories, Rahway, NJ, internal communication. "The pK_a values for imipenem determined by aqueous acidic/basic potentiometric titration at 25 °C are pK_{a1} ~3.2 and pK_{a2} ~9.9."

Table for entry 667:

Compound	pK_a	Compound	pK_a
Imidazole	6.953	1-CH_3-4-NO_2-	−0.530
1-C_2H_5-	7.300	2-CH_3-	7.851
1-CH_3-5-NO_2-	2.130	Metronidazole	2.62

(continued)

No.	Compound Name	pK_a value(s)	Data quality	Ionization type	Method	Conditions	Comments and Reference(s)
671	Imipramine ($C_{19}H_{24}N_2$) $(CH_2)_3N(CH_3)_2$	9.62	U	+H	Spectro	H_2O $t = 22.0$	Kender DN and Schiesswohl RE, Imipramine hydrochloride, *APDS*, **14**, 37–72 (1985). Temperature (°C) (ref.) / Method/Conditions / pK_a: 22 (9) / Photometric titration / 9.62 24 (9, 15) / Solubility / 9.5 –(9) / Extrapolated from water/methylcellosolve mixture; Potentiometric titration / 9.5 9. Analytical Dept., Ciba-Geigy Ltd., personal communication (Nov. 1979). 15. Green AL, Ionization constants and water solubilities of some aminoalkylphenothiazine tranquillizers and related compounds, *J. Pharm. Pharmacol.*, **19**, 10–16 (1967).
672	Imipramine	9.5	U	+H	soly	H_2O $t = 24 \pm 1$	Green AL, Ionization constants and water solublities of some aminoalkylphenothiazine tranquilizers and related compounds, *J. Pharm. Pharmacol.*, **19**, 10–16 (1967). NB: See Amitriptylline for details.
673	Imipramine	9.66	U	+H	CE/pH (+ve ion mode)	H_2O $t = 25$ $I = 0.025$	Wan H, Holmen AG, Wang Y, Lindberg W, Englund M, Nagard MB and Thompson RA, High-throughput screening of pK_a values of pharmaceuticals by pressure-assisted capillary electrophoresis and mass spectrometry, *Rapid Commun. Mass Spectrom.*, **17**, 2639–2648 (2003). NB: Reported a predicted value (ACD Labs) of 9.49.
674	Imipramine	9.40	U	+H	Potentiometric	H_2O $t = 25$ I undefined Ar atmosphere	Seiler P, Simultaneous determination of partition coefficient and acidity constant of a substance. *Eur. J. Med. Chem.*, **9**, 663–665 (1974). NB: See Amitriptylline for details.

No.	Compound	pKa			Method	Conditions	Reference / Comments
675	Indapamide ($C_{16}H_{16}ClN_3O_3S$)	8.8 ± 0.2	U	–H	soly	H_2O $t = 25$ I undefined	Difeo TJ and Shuster JE, Indapamide, *APDS*, **23**, 229–268 (1994). Cited Rosoff M and Serajuddin A, USV Pharmaceutical Corporation, Tuckahoe, NY, USA, personal communication. NB: Soly-pH data in aqueous buffers (Moody JE, O'Hare MJ and Sapio JP, USV Pharmaceutical Corporation, Tuckahoe, NY, USA, personal communication):

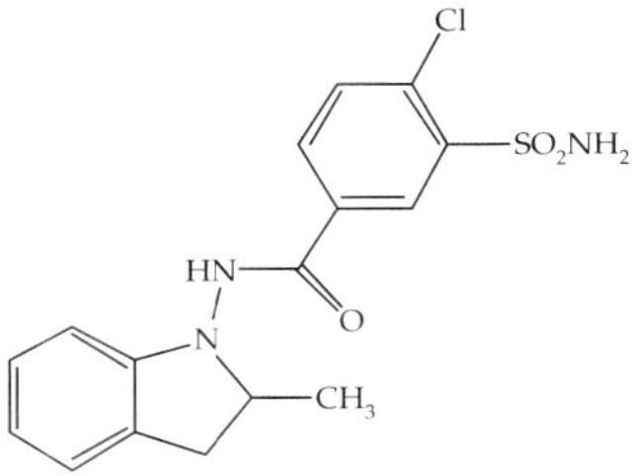

pH	soly (25 °C) (mg/mL)	soly (37 °C) (mg/mL)	pH	soly (25 °C) (mg/mL)	soly (37 °C) (mg/mL)
1.0	0.063	0.095	7.8	0.069	–
2.0	0.061	0.094	8.0	0.076	0.103
3.5	0.061	0.093	8.4	0.088	0.130
4.5	0.059	0.091	8.7	0.111	0.178
5.5	0.061	0.094	9.3	0.231	0.370
6.5	0.059	0.097	9.9	0.573	0.860
7.5	0.063	0.098			

The very small apparent increase in solubility below pH 4.5 suggests that protonation of the indole nitrogen is occurring to a slight extent.

No.	Compound	pKa			Method	Conditions	Reference / Comments
676	Indicators				Spectro	H_2O	Das Gupta V and Reed JB, Jr., First pK_a values of some acid-base indicators, *J. Pharm. Sci.*, **59**, 1683–1685 (1970).
	Bromocresol green	–0.85	U	+H		t undefined	NB: Used a method described by Reilley CN, Sawyer DT, *Experiments for Instrumental Methods*, McGraw-Hill, New York, NY, 153–155 (1961).
	Bromocresol purple	–0.75	U	+H		I undefined	
	Bromophenol blue	–0.95	U	+H		but >0.1	
	Cresol red	1.05	U	+H			
	Phenol red	1.03	U	+H			
677	Indinavir ($C_{36}H_{47}N_5O_4$)	3.8	U	+H	soly	H_2O	Johnson BD, Howard A, Varsolona R, McCauley J and Ellison DK, Indinavir Sulfate, *APDS*, **26**, 319–357 (1999).
		6.2	U	+H	Spectro	t undefined I undefined	

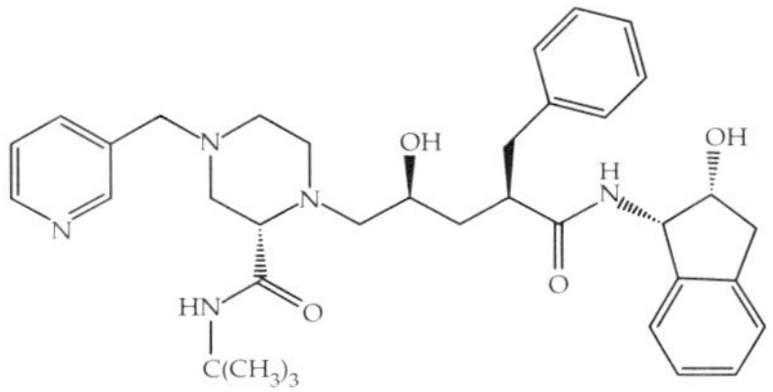

Table: pH Solubility profile

pH	Solubility (mg/mL)	pH	Solubility (mg/mL)	pH	Solubility (mg/mL)
3.4	61.2	4.0	5.13	6.9	0.019
3.5	31.7	4.9	0.312	7.0	0.023
3.6	20.3	5.9	0.037	8.0	0.017

NB: The spectroscopic pK_a value was stated to be the result of in-house measurements. A pK_a value of about 6 can also be estimated from the dependence of log P (octanol-aqueous buffer) on pH.

(continued)

No.	Compound Name	pK_a value(s)	Data quality	Ionization type	Method	Conditions	Comments and Reference(s)
678	Indomethacin ($C_{19}H_{16}ClNO_4$) O, Cl, N, CH_3, CH_3O, CH_2COOH	4.3	U	–H	Potentiometric	H_2O $t = 20$ I undefined but very low	Rainer VG, Krüger U and Klemm K, Syntheses und physicalisch-chemische Eigenschaften von Lonazolac-Ca einem neuen Antiphlogistikum/Antirheumatikum, *Arzneim.-Forsch.*, **31**(4), 649–655 (1981). NB: The pK_a values were measured potentiometrically in 10^{-5}–10^{-4} molar solutions of the agent in dioxan-water mixtures (3–20 vol% dioxan), with 0.1N NaOH delivered from a microburette with "Folgepotentiometer" and a pH meter at 20 °C under nitrogen. The analysis of the measurements used a non-linear regression program on the CDC 6500 computer . . . . "Aus den über dem Molenbruch des Wassers aufgetragenen gemessenen pK_a–Werten (NB: 8–16 measurements per compound) wurden die auf reines Wasser bezogenen pK_a–Werte durch graphische Extrapolation ermittelt."
679	Indomethacin	4.55	U	–H	Spectro	H_2O t undefined I undefined	Herzfeldt CD and Kümmel R, Dissociation constants, solubilities, and dissolution rates of some selected nonsteroidal antiinflammatories, *Drug Dev. Ind. Pharm.*, **9**(5), 767–793 (1983). NB: Used dA/dpH method. NB: See Azapropazone and Ibuprofen for details.
680	Indomethacin	4.5	U	–H	soly	H_2O t undefined I undefined	O'Brien M, McCauley J and Cohen E, Indomethacin, *APDS*, **13**, 211–235 (1984). From: Rodgers DH, Merck Sharp and Dohme Research Laboratories, West Point, PA, personal communication. NB: Also reported pK_a = 4.5 (potentiometric data, extrapolated from 50% MeOH–water) (Mulligan RE, Merck Sharp and Dohme Res. Lab., personal communication).
681	Indomethacin	4.15	U	–H	CE/pH (+ve ion mode)	H_2O $t = 25$ $I = 0.025$	Wan H, Holmen AG, Wang Y, Lindberg W, Englund M, Nagard MB, Thompson RA, High-throughput screening of pK_a values of pharmaceuticals by pressure-assisted capillary electrophoresis and mass spectrometry, *Rapid Commun. Mass Spectrom.*, **17**, 2639–2648 (2003).
		4.06	U	–H	CE/pH (–ve ion mode)		NB: Reported a predicted value (ACD Labs) of 4.18.
682	Indomethacin	4.14	U	–H	Potentiometric	H_2O $t = 25$ I unspecified	Bergström CAS, Strafford M, Lazorova L, Avdeef A, Luthman K and Artursson P, Absorption classification of oral drugs based on molecular surface properties, *J. Med. Chem.*, **46**(4), 558–570 (2003). NB: From extrapolation of aqueous-methanol mixtures to 0% methanol.

No.	Compound	pKa			Method	Conditions	Comments and Reference(s)
683	Indomethacin	5.3	U	+H	Potentiometric	50% EtOH t undefined I undefined	Jahn U and Wagner-Jauregg T, Wirkungsvergleich saurer Antiphlogistika im Bradykinin-, UV-Erythem- und Rattenpfotenödem-Test, *Arzneim.-Forsch.*, **24**, 494–499 (1974).
		6.4	U	+H		80% Me cellosolve	NB: Literature values obtained from the pH of half-neutralization. Also gave a value of 4.6 in water (Merck and Co.). This last value is in good agreement with other values.
684	Insulin, A21-aspartyl —— Asn —Tyr— Cys —Asp-COOH —— Val — Cys Gly	4.06 (rsd = 1.23%)	U	–H	kinetic	H_2O $t = 35$ $I = 0.1$ (NaCl)	Darrington RT and Anderson BD, Role of intramolecular nucleophilic catalysis and the effects of self-association on the deamidation of human insulin at low pH, *Pharm. Res.*, **11**, 784–793 (1994). NB: A21-aspartylinsulin is a hydrolysis product of native insulin.
685	Insulin, A21-aspartyl	Apparent pK_a			kinetic	Solid state (adsorbed surface water) $t = 35$ I undefined	Strickley RG and Anderson BD, Solid-state stability of human insulin. Part 1. Mechanism and the effect of water on the kinetics of degradation in lyophiles from pH 2–5 solutions, *Pharm. Res.*, **13**, 1142–1153 (1996). "The mechanisms and kinetics of degradation of human insulin lyophilized powders at pH 2–5 were studied as a function of water content. . . . The degradation rate increased with decreasing apparent pH ('pH'), yielding, at any given water content, solid state 'pH' rate profiles parallel to the solution pH rate profile. The 'pH' dependence was accounted for in terms of the fraction of insulin A21 carboxyl in its neutral form, with an apparent pK_a of ~4, independent of water content."
	% adsorbed water 3	3.85 ± 0.16	U	–H			
	% adsorbed water 10	4.04 ± 0.14	U	–H			
	% adsorbed water 14	4.07 ± 0.14	U	–H			
	% adsorbed water 52	4.13 ± 0.10	U	–H			
	Solution	3.92 ± 0.09	U	–H		H_2O $t = 35$ $I = 0.1$ (NaCl)	
686	Iodamide ($C_{12}H_{11}I_3N_2O_4$) COOH, I, I, CH_3CONH, $CH_2NHCOCH_3$, I	1.88	U	–H	Potentiometric	H_2O $t = 25$	Pitré D, Iodamide, *APDS*, **15**, 337–365 (1986). "The apparent ionization constant, determined in methylcellosolve/water (w/w) was pK_{MCS} (25) = 4.15; extrapolation to aqueous solution gave a pK_{H_2O}(25) = 1.88 (7). Other authors (6) reported a pK_a = 3.7, determined by titration. 6. Engelen AJM, Rodrigues de Miranda JF and Ariens EJ, Pharmacokinetics of renal contrast media. I. Renal excretion processes studied in dogs by the stop-flow technique, *Invest. Rad.*, **8**, 210–218 (1973). 7. Felder E, Pitre D and Grandi M, Radioopaque contrast media. XXIII. Apparent acidity constants of radioopaque acids in the system methylcellosolve-water. *Il Pharmaco, Sci. Edn.*, **28**, 485–493 (1973)."
		3.7	U	–H	Potentiometric	H_2O	

(*continued*)

No.	Compound Name	pK_a value(s)	Data quality	Ionization type	Method	Conditions	Comments and Reference(s)
687	Iodamide	1.34	U	–H	soly	H_2O $t = 25$ I undefined	Calculated by this author from data on solubility as a function of pH (Felder E, Pitre D and Grandi M, Radiopaque contrast media. XLIII. Physicochemical properties of iodamide, an intravenous uro-angiographic contrast agent, *Il Farmaco, Sci. Edn.*, **32**, 755–766 (1977)):
688	Iodipamide ($C_{20}H_{14}I_6N_2O_6$) 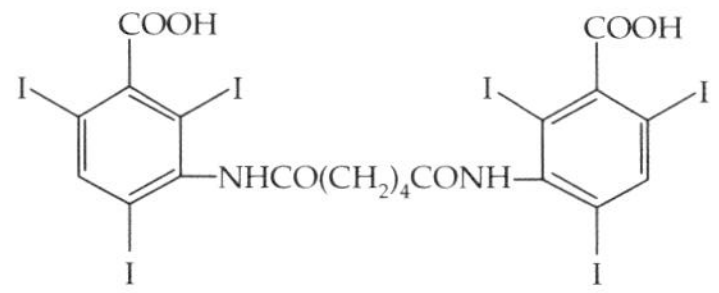	3.5	U	–H	Spectro	H_2O	Neudert W and Röpke H, The physico-chemical properties of the disodium salt of adipic acid bis(2,4,6-triiodo-3-carboxyanilide) and other triiodobenzene derivatives, *Chem. Ber.*, **87**, 659–667 (1954). Cited in Lerner HH, Iodipamide, *APDS*, **3**, 333–363 (1974). NB: This pK_a value may be a statistically modified composite of pK_{a1} and pK_{a2}, since both dissociation constants can be expected to be very similar in value.
689	Iodoxamic acid ($C_{26}H_{26}I_6N_2O_{10}$) 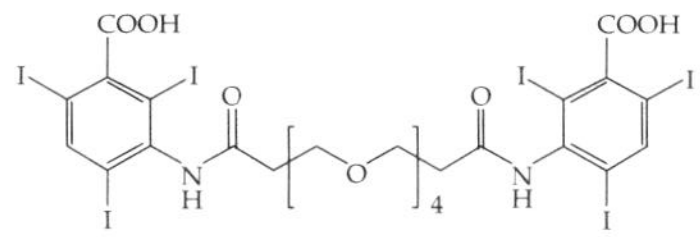	1.8 2.8	U U	–H –H	Potentiometric	H_2O (extrap from 80% MCS-H_2O) $t = 25$	Pitré D, Davies A and Grandi M, Iodoxamic acid, *APDS*, **20**, 303–335 (1991). "The ionization constants of iodoxamic acid were determined by potentiometric titration in the methylcellosolve (MCS)–water (3) and ethanol-water systems (2). In both cases 2 equivalents of alkali were required per mole of acid and the resulting curves showed only one inflection point. Data are reported in Table 9.

pH ± 0.05	Solubility (g/100 mL)	pH ± 0.05	Solubility (g/100 mL)	pH ± 0.05	Solubility (g/100 mL)
1.0	0.03	3.6	5.41	4.5	52.9
2.0	0.14	3.8	11.1	5.0	73.2
3.0	1.16				

Table 9: Apparent ionization constants of iodoxamic acid

solvent	t (°C)	pK_1	pK_2	pK_{av}
80% MCS	25	4.06	5.07	–
30% EtOH	21	–	–	2.96
50% EtOH	21	–	–	3.22
70% EtOH	21	–	–	3.67

							For the 80% MCS system, extrapolation to water solvent yields values of $pK_1 = 1.8$ and $pK_2 = 2.8$ (8,9). 2. Miyake Y and Asahi Y, Physicochemical properties, stabilities, and analysis of iodoxamic acid, *J. Takeda Res. Lab.*, **33**, 73–86 (1974). 3. Felder E, Pitre D and Grandi M, Radioopaque contrast media. XXV. Physicochemical properties of iodoxamic acid, a new intravenous cholecystographic agent, *Il Farmaco, Sci. Edn.*, **28**, 925–936 (1973). 8. Simon W, Structure and acidity of organic compounds [1], *Angew. Chem.*, **76**, 772–780 (1964). 9. Sommer P, Dissertation, ETH, Zurich, Juris-Verlag, 1961."
690	Iopamidol ($C_{17}H_{22}I_3N_3O_8$)	10.70	U	–H	Potentiometric	H_2O $t = 25.0$	Felder E, Grandi M, Pitré D and Vittadini G, Iopamidol, *APDS*, **17**, 115–154 (1988). NB: No reference given to the original data source. This value probably relates to the deprotonation of the anilido proton. Semiempirical MO calculations (MNDO) indicate that this proton is the most electropositive, i.e., that the N–H bond involved is the most polarised (R. Prankerd, unpublished results, 1998).
691	Iopanoic acid ($C_{11}H_{12}I_3NO_2$)	5.06	U	–H	Potentiometric	H_2O (extrap)	Felder E, Pitre D and Grandi M, Radioopaque contrast media. XXIII. Apparent acidity constants of radioopaque acids in the system methylcellosolve-water, *Il Farmaco, Sci. Edn.*, **28**, 485–493 (1973). Cited in Pitre D, Iopanoic Acid, *APDS*, **14**, 181–203 (1985). NB: Extrapolated from results in 80% aqueous methylcellosolve. Other authors reported $pK_a = 5.9$ [McChesney EW and Hoppe JO, The metabolism of iodopanoic acid, *Arch. Int. Pharmacodyn.*, **99**, 127–140 (1954). CA 49:33374].
692	Iophenoxic acid ($C_{11}H_{11}I_3O_3$)	7.5	U	–H	Potentiometric	H_2O (extrap)	Anton AH, A drug-induced change in the distribution and renal excretion of sulfonamides, *JPET*, **134**, 291–303 (1961). "The pK_a was obtained . . . by personal communication . . . (from) Schering through the courtesy of Dr. Kabasakalian."

(continued)

No.	Compound Name	pK_a value(s)	Data quality	Ionization type	Method	Conditions	Comments and Reference(s)
693	Isocarboxazid ($C_{12}H_{13}N_3O_2$) 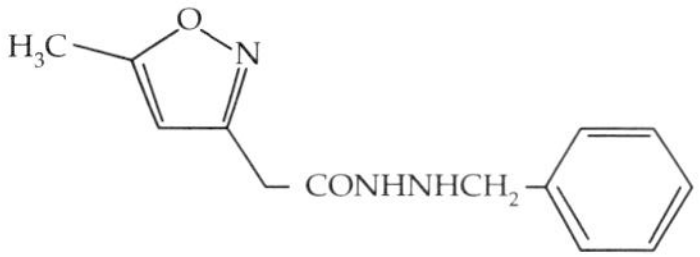	10.4	U	+H	Spectro		Rudy BC and Senkowski BZ, Isocarboxazid, *APDS*, **2**, 295–314 (1973). Lau E, Hoffmann-La Roche, unpublished data.
694	Isochlorotetracycline ($C_{22}H_{23}ClN_2O_8$) 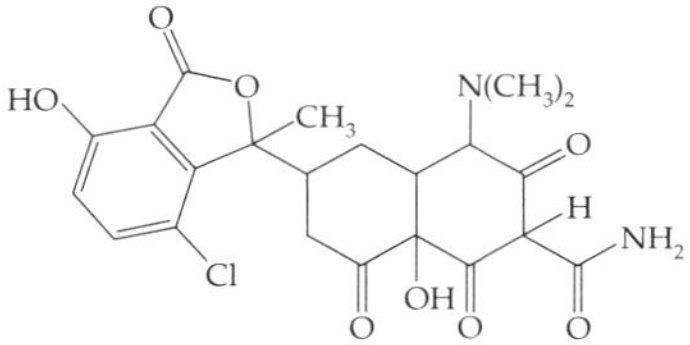	3.1 6.7 8.3	U U U	–H –H +H	Potentiometric	H_2O $t = 25$ $c = 0.005$	Stephens C, Murai K, Brunings K and Woodward RB, Acidity constants of the tetracycline antibiotics, *JACS*, **78**, 4155–4158 (1956). Cited in Perrin Bases 3327, Ref. S73. NB: The data were obtained in the presence of calcium ions, with a glass electrode in an asymmetric cell with liquid junction potentials. Stephens *et al.* cited Waller CW, Hutchings BL, Wolf CF, Goldman AA, Broschard RW and Williams JH, *JACS*, **74**, 4981 (1952) as the source of earlier values for comparison. However, this latter reference reported only two values, i.e., $pK_{a2} = 6.8$; $pK_{a3} = 8.1$.
695	Isochlortetracycline	3.96 6.73 7.93	U U U	–H –H +H	Potentiometric	H_2O $t = 30.0 \pm 0.2$ $I = 0.01$ (KCl) N_2 atmosphere	Doluisio JT and Martin AN, Metal complexation of the tetracycline hydrochlorides, *J. Med. Chem.*, **6**, 16–20 (1963). NB: Metal-free solutions of the tetracycline titrated with standard NaOH solution and the pH measured. No details were given of the pH meter calibration. Metal stability constants were determined from identical titrations in the presence of varying concentrations of nickel(II), zinc(II) or copper(II) ions.
696	Isolysergic acid ($C_{16}H_{16}N_2O_2$) 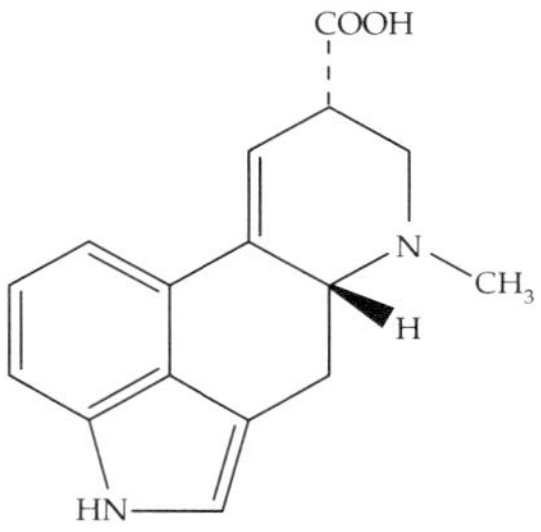	3.33 8.46 3.50 8.40	U U U U	–H +H –H +H	Potentiometric	H_2O $t = 24$ $c = 0.002$ H_2O $t = 38$ $c = 0.002$	Craig LC, Shedlovsky T, Gould RG and Jacobs WA, The ergot alkaloids. XIV. The positions of the double bond and the carboxyl group in lysergic acid and its isomer, *J. Biol. Chem.*, **125**, 289–298 (1938). Cited in Perrin Bases 2922 ref. C54. NB: The study used an asymmetric cell with a glass electrode and liquid junction potential. The pK_a value is the pH reported at half-neutralisation.

No.	Compound	pKa			Method	Solvent	Reference
697	DL-Isomethadone ($C_{21}H_{27}NO$) 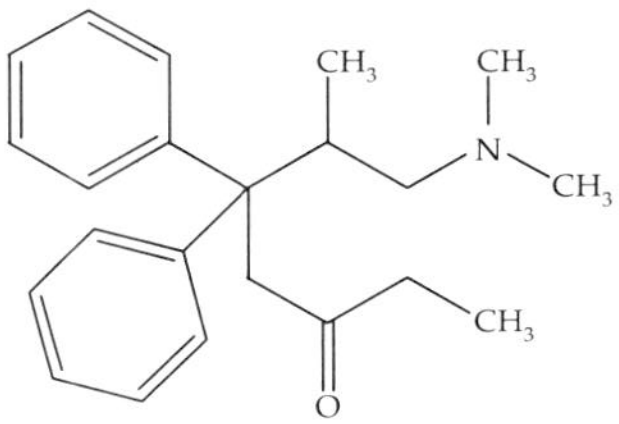	8.07	U	+H		50% EtOH	Farmilo CG, Oestreicher PM and Levi L, Physical methods for the identification of narcotics. I B Common physical constants for identification of ninety-five narcotics and related compounds, *Bull. Narcotics*, UN Dept. Social Affairs, vol. **6**, pp. 7–19 (1954). CA 48:69490. NB: Cited in Beckett AH, Analgesics and their antagonists. I., *J. Pharm. Pharmacol.*, **8**, 848–859 (1956); also cited in Clouet DH (ed.), *Narcotic Drugs Biochemical Pharmacology*, Plenum Press, New York, 52–53 (1971). NB: See Alphaprodine for details.
698	DL-Isomethadone	8.21	U	+H			Taylor JF, Ph.D. Thesis, Univ London (1968).
699	Isoniazid ($C_6H_7N_3O$)	2.00	U	+H	Spectro	H_2O	Brewer, G.A., Isoniazid, *APDS*, **6**, 183–258 (1977).
		3.60	U	+H			
		10.8	U	−H			

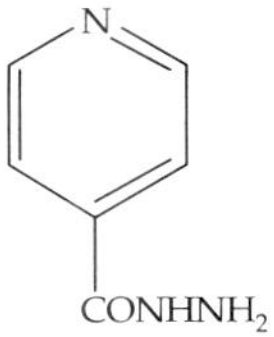

CONHNH$_2$

"There is a discrepancy in the literature on the dissociation constants of isoniazid. . . . in part due to the different methods of measurement employed. Fallab (34) determined the basic dissociation constant as 3e−11 measured conductometrically. Canic and Djordjevic (35) established that the 1st basic constant should be ascribed to the pyridine nitrogen and the 2nd to the hydrazine group. This is contrary to previous work by Cinglani and Gaudiano (36). Nagano and coworkers (37) determine the dissociation constants potentiometrically as pK_1 = 2.13, pK_2 = 3.81, pK_3 = 11.03. Salvesen and Glendrange (38) determined the dissociation constants in 1.0 M sodium chloride solution as K_1 = 9.80e−3 and K_2 = 1.42e−4. Zommer and Szuszkiewicz (11) have established pK_1 = 10.75 and pK_2 = 11.15 and protonation constants of 3.57 for the pyridine N and 1.75 for the hydrazine N. Rekker and Nauta (65) found that solutions of isoniazid became yellow at pH 10 and 2.7. The color is reversible on changing the pH. They explained this behavior on the basis of the existance of two positive ions, a monovalent yellow positive ion, a monovalent yellow positive ion and a divalent colorless positive ion. The pK values are pK' = 2.00, pK'' = 3.6, and pK''' = 10.8."

(11) Zommer S and Szuszkiewicz J, *Chem. Anal.*, **14**, 1075–1083 (1969); CA 72:89616n (1970).
(34) Fallab S, *Helv. Chim. Acta*, **36**, 3–5 (1953).
(35) Canic VD and Djordjevic Rd, *Beograd*, **21**, 193–199 (1956); CA 5:16026f (1958).
(36) Cingolani and Gaudiano, *Rend. ist. super. sanita*, **17**, 601 (1954).
(37) Nagano K, Tsukahara H, Kinoshita H and Tamura Z, *Chem. Pharm. Bull.*, **11**, 797–805 (1963); CA 59:4867h (1963).

(*continued*)

No.	Compound Name	pK_a value(s)	Data quality	Ionization type	Method	Conditions	Comments and Reference(s)
							(38) Salvesen B and Glendrange JH, *Medd. Norsk. Farm. Selskap.*, **27**, 135–144 (1965); CA 64:9507f (1966).
							(65) Rekker RF and Nauta WT, *Pharm.Weekblad*, **99**, 1157–1165 (1964); CA 63:435c (1965).
							NB: See following for further details from Refs. 11, 37, and 65.
		1.75	U	+H	Spectro	H_2O	Ref. 11: Zommer S and Szuszkiewicz J. Spectrophotometric investigations of acid-base properties of hydrazides of isonicotinic and cyanoacetic acids. *Chemia Analityczna* (Warsaw, Poland), **14**(5), 1075–1083 (1969); CA 72:89616n.
		3.57	U	+H			
		10.75	U	−H			
		11.15	U	−H			
							"Isonicotinic acid hydrazide (I) in an acid medium exhibits two absorption max. (molar absorptivity. . .): 265 mμ, 5400 and 212 mμ, 5040; in alk. media the former disappears and a new broad max. at approx. 300 mμ appears. The uv spectrum of $CNCH_2CONHNH_2$ (II) shows only one max. λ_{max} 200 mμ at pH 1.6; the band is shifted towards longer wavelengths with increasing pH. The values of the dissocn. consts. (pK) are: 10.75 and 11.15 and those of the protonation consts.: pK_2 (pyridine N) 3.57, and pK_1 (hydrazide N) 1.75 and 2.3 for I and II, resp."
		2.13	U	+H	Potentiometric		Ref. 37: Nagano K, Tsukahara H, Kinoshita H and Tamura Z. Metal complexes of isonicotinoylhydrazine and related compounds. II. Acid dissociation constants and ultraviolet absorption spectra of isonicotinoylhydrazine and related compounds, *Chem. Pharm. Bull.*, **II(6)**, 797–805 (1963); CA 59:4867h.
		3.81	U	+H			
		11.03	U	−H			
							"Successive acid dissocn. consts. for a series of related compds. were detd. under N atm. by pH titrn. The solns. were M KNO_3, 4×10^{-2}N HNO_3, and 2×10^{-2}M in the compd. studied. Values detd. were: isonicotinoylhydrazine, pK_1 = 2.13, pK_2 = 3.81, pK_3 = 11.03; nicotinoylhydrazine, 2.26, 3.63, 11.49; picolinoylhydrazine, 1.26, 3.07, 12.25; benzoylhydrazine, 3.27; 12.53; p-itrobenzoylhydrazine, 2.90, 11.28; isonicotinamide, 3.82; 1-isonicotinoyl-L-methylhydrazine, 1.03, 4.17; 1-isonicotinoyl-2-methylhydrazine, 2.46, 4.04, 10.96."
		2.00	U	+H	Spectro		Ref. 65: Rekker RF and Nauta WT. Spectrophotometric investigation of isonicotinoyl hydrazide (INH). A contribution to the knowledge of the yellow isonicotinoyl hydrazide-p-aminosalicylic acid combination (INH-PAS), *Pharm. Weekblad*, **99**(42), 1157–1165 (1964); CA 63:435c.
		3.60	U	+H			
		10.8	U	−H			

No.	Compound	pKa			Method	Conditions	Comments/Reference
							"The combination p-aminosalicylic acid (I) and isonicotinoyl hydrazide (II) in the proportion 1:1 has been described variously as a salt, an addn. compd., or a mol. complex (Reinstein and Higuchi, CA 53:1639g). I did not show any yellowing regardless of pH, whereas II was colored at pH 10 and 2.70, the yellow color being stronger in alk. soln. but both reversible. The colors can be explained by the existence of 2 pos. II ions, one monovalent yellow pos. ion and a divalent colorless pos. ion. The pK values were $pK' = 2.00$, $pK'' = 3.60$, and $pK''' = 10.80$."
700	Isonicotinic acid ($C_6H_5NO_2$)	1.78 ± 0.10 4.85 ± 0.08	U U	+H −H	Spectro	H_2O $t = 20 \pm 1$ $I = 0$	Asuero AG, Navas MJ, Herrador MA and Recamales AF, Spectrophotometric evaluation of acidity constants of isonicotinic acid, *Int. J. Pharm.*, **34**, 81–92 (1986). "All photometric measurements were made with a Bausch and Lomb Spectronic 2000 instrument. Matched silica cuvettes were used in all measurements and the light path was 10 mm. The pH was measured with a Crison Model 501 pH meter fitted with a combined glass electrode. The pH meter was standardised against 0.05 M potassium hydrogen phthalate (pH = 4.01 at 25 °C). . . . Solutions for absorbance and pH measurements were prepared by mixing 2 mL of 2×10^{-3} M stock solution of isonicotinic acid in water, 2.5 mL of sodium perchlorate and a few drops of sodium hydroxide or perchloric acid at different concentrations. The solutions were then diluted to 25 mL with distilled water. Absorbance was measured against a solvent blank and the pH checked after the absorbance measurements. The temperature was kept at 20 ± 1 °C." NB: Several computational approaches were used to deconvolute the overlapping pK_a values. Mean ionic activity coefficients were estimated with a Güntelberg correction. Another paper by the same authors reported an almost identical pK_{a2} value of 4.88 for isonicotinic acid (Asuero AG, Herrador MA and Camean AM, Spectrophotometric evaluation of acidity constants of diprotic acids: Errors involved as a consequence of an erroneous choice of the limit absorbances, *Analytical Letters*, **19**, 1867–1880 (1986))
701	Isopilocarpine ($C_{11}H_{16}N_2O_2$)	7.18	U	+H	Potentiometric	H_2O $t = 15$	Muller F, *Z. Elektrochem.*, **30**, 587 (1924). Cited in Perrin Bases 2923 ref. M60. NB: The study used potentiometric titration with hydrogen electrodes in an unsymmetric cell with liquid junction potentials.

(continued)

No.	Compound Name	pK_a value(s)	Data quality	Ionization type	Method	Conditions	Comments and Reference(s)
702	Isoproterenol ($C_{11}H_{17}NO_3$) HO, HO– (3,4-dihydroxyphenyl)–CH(OH)–$CH_2NHCH(CH_3)_2$	8.60 10.1 12.0	U U U	+H, −H −H, +H −H		H_2O $t = 20.0$	Tariq M and Al-Badr AA, Isoproterenol, *APDS*, **14**, 391–420 (1985). *BPC*, 11th Edn., Pharmaceutical Press, London 472 (1979).
703	Isoproterenol (DL-isoprenaline) HO, HO– (3,4-dihydroxyphenyl)–CH(OH)–$CH_2NHCH(CH_3)_2$	8.65 ± 0.02 10.07 ± 0.01	A A	−H, +H +H, −H	Spectro, Potentiometric	H_2O $t = 25.0 \pm 0.05$ $I = 0.10$ (KCl) N_2 atmosphere pH calibration standards: 7.00 ± 0.02, 9.94	Ijzerman AP, BultsmaT, Timmerman H and Zaagsma J, The ionization of β-adrenoceptor agonists: A method for unravelling ionization schemes, *J. Pharm. Pharmacol.*, **36**(1), 11–15 (1984). NB: Experimental data was fitted with the Marquardt algorithm. "The macroscopic ionisation constants for noradrenaline, adrenaline and isoprenaline are in good agreement with the values obtained by others ... overlap in macroscopic ionisation constants has often confused the interpretation of results. ... Jameson and Neillie (1965) and Rajan *et al.* (1972) (NB: See no. 731) related pK_1 with the dissociation of the protonated amino function, whereas Antikainen and Witkainen (1973) and Granot (1976) assigned the phenolic function as the more acidic group in accordance with some previous interpretations (Lewis, 1954; Kappe and Armstrong, 1965). (The present data) substantiates the latter view; the red shift in the uv-spectra of all amines, caused by the transition phenol-phenolate, yields pK_{1z}, which value is in all cases very close to pK_1. .... Ultraviolet-spectrophotometric measurements: determination of microscopic ionization constant K_{1z}. ... The compounds ... were analysed at 25.00 ° ± 0.05 °C ... catecholamines also with exclusion of light. ...

Compound	pK_{1z}	pK_{2z}	Compound	pK_{1z}	pK_{2z}
Noradrenaline	8.68	9.68	Salbutamol	9.22	10.22
Adrenaline	8.69	10.14	Orciprenaline	8.67	9.92
Isoprenalin	8.68	10.03	Terbutaline	8.73	10.06
Th 1206	8.68	10.31	Cp22352-1	8.02	10.25
AH 3021	9.23	9.93	Pirbuterol	7.95	10.64

	pK1	pK2		pK1	pK2
Noradrenaline	8.63	9.73	Salbutamol	9.07	10.37
Adrenaline	8.73	10.14	Orciprenaline	8.70	9.92
Isoprenalin	8.65	10.07	Terbutaline	8.70	10.09
Th 1206	8.72	10.31	Cp 22352-1	8.08	10.25
AH 3021	9.01	10.15	Pirbuterol	8.01	10.64

Electrochemical titrations: Determination of the macroscopic ionization constants, K_1 and K_2....The compounds (0.6 mM KCl) were ... titrated with 0.1 M KOH from a calibrated Mettler DV10 micropipetter at 25.00 ± 0.05 °C ... (catecholamines also with exclusion of light), using a digital Philips PW 9414 ion activity meter. The volume ... was 25–30 ml, and for the spectroscopic measurements, the ionic strength was assumed to be that of 0.1 M KCl, the activity coefficient was taken as 0.775. The low concentrations ... were chosen to obtain results comparable to the UV-spectrophotometric ones...."

704	Isoproterenol	8.57	U	+H	Potentiometric	H_2O $t = 25.0 \pm 0.2$ $I \leq 0.001$	Leffler EB, Spencer HM and Burger A, Dissociation constants of adrenergic amines *JACS*, **73**, 2611–2613 (1951). NB: See Amphetamine for details. Decomposes in aqueous solution. Result claimed to be accurate to ± 0.02 log unit, but recognised (oxidative) discoloration of solution even when pH was measured as rapidly as possible.
705	Isoproterenol	8.7 9.9	U U	+H −H	Potentio-metric, Spectro	H_2O t undefined $I = 0.1$ (glycine)	Lewis GP, The importance of ionization in the activity of sympathomimetic amines, *Br. J. Pharmacol.*, **9**, 488–493 (1954). NB: Method similar to Kappe and Armstrong, see Levarterenol (no. 729).
706	Isoproterenol (DL-isoprenaline)	8.58 ± 0.03 9.93 ± 0.12	A U	−H, +H +H, −H	Potentio-metric, Spectro	H_2O $t = 37$ $I = 1.0$	Schüsler-van Hees, MTIW, Reactivity of Catechol(amine)s, Ph.D. thesis, Ctr for Biopharm. Sci., Leiden University, 1–93 (1983). NB: Microscopic: 8.69 ± 0.01; 9.22 ± 0.08; 9.83 ± 0.14; 9.30 ± 0.05. Schüsler-van Hees, MTIW, Beijersbergen van Henegouwen GMJ, Driever MFJ, Ionization constants of catechols and catecholamines, *Pharm. Weekblad Sci. Edn.* **5**, 102–108 (1983).

(continued)

No.	Compound Name	pK_a value(s)	Data quality	Ionization type	Method	Conditions	Comments and Reference(s)
707	Isopyridoxal ($C_8H_9NO_3$) 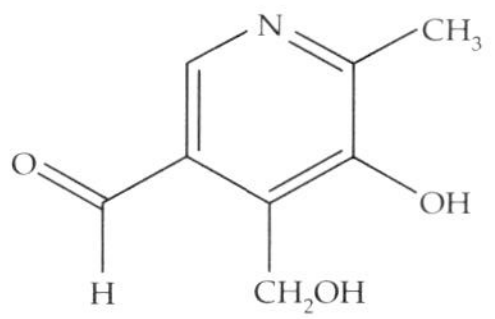	4.00	U	+H	Spectro	H_2O $t = 25$ $I > 0.1$	Pocker A and Fischer EH, Synthesis of analogs of pyridoxal-5′-phosphate. *Biochem.*, **8**, 5181–5188 (1969). Cited in Perrin Bases suppl. no. 7785 ref. P41a. NB: The study used spectrophotometric and electrometric measurements (following the procedure of Metzler DE and Snell EE, Spectra and ionization constants of the vitamin B_6 group and related 3-hydroxypyridine derivatives, *JACS*, **77**, 2431–2437 (1955)) in citrate-phosphate buffers. Although the work was quite careful, U classification has been applied as presented figures showed some aberrations in the isosbestic points, suggesting that some decomposition had occurred. The paper also reported the corresponding value for isopyridoxal phosphate, $pK_{a1} = 4.25$, $pK_{a2} = 8.80$, and numerous other values.
708	Itanoxone ($C_{17}H_{13}ClO_3$) 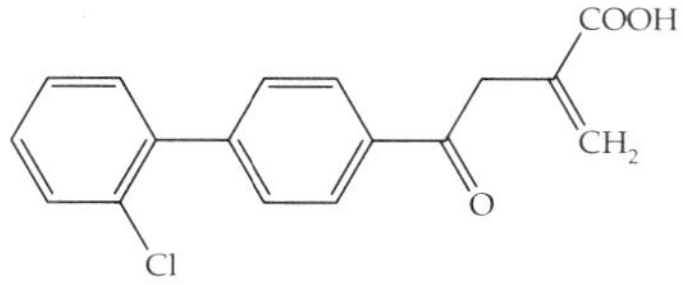	5.30	U	−H	Potentiometric	H_2O (extrap) $t = 25.0$ N_2 atmosphere	Cousse H, Mouzin G, Ribet J and Vezin J, Physicochemical and analytical characteristics of itanoxone, *J. Pharm. Sci.*, **70**(11), 1245–1248 (1981). "…maintained the temperature of the titrating vessel at 25°. Stirring was … by introducing a slow nitrogen stream under the surface of the solution to be titrated. The titrant of carbonate-free potassium hydroxide (0.100N KOH) was delivered from a micrometer syringe; for each addition of titrant, the pH was measured and the pK_a was determined from the half neutralization point … The values of g(s) for different percentage volumes of dioxane and the pK_a values of salicylic acid (the reference compound) are shown in Table 1. The difference between the two pK_a values is a linear function of g (s). The determination of the pK_a of I in water by extrapolation of measurements made on water-dioxane mixtures is plotted versus g(s) determined for each water-dioxane system. The pK_a of salicylic acid determined in water is 3.0. The pK_a of I determined in water-dioxane with extrapolation of the results to zero dioxane concentration is 3.0 + 2.3 = 5.30."
709	Kanamycin A ($C_{18}H_{36}N_4O_{11}$)	6.40 7.55 8.40 9.40	U U U U	+H +H +H +H	Potentiometric	H_2O	Claes PJ, Dubost M and Vanderhaeghe H, Kanamycin Sulfate, *APDS*, **6**, 259–296 (1977). Vanderhaeghe H and Claes PJ, unpublished results. NB: Derived from a pH-titration (0.5N HCl titrant). Avery lists a single value of 7.2, which appears to be a composite. However, see Amikacin (no. 1557) and Gentamicin (no. 1888).

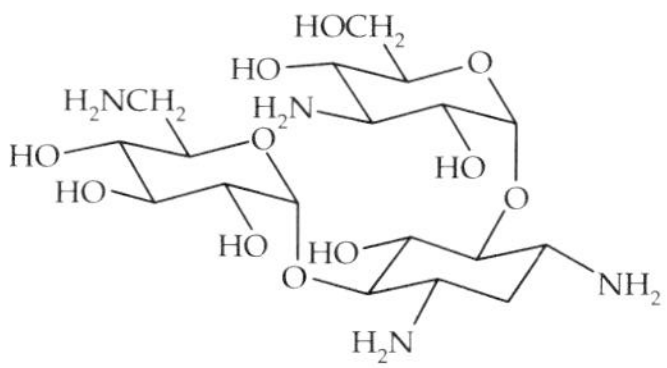

710	Ketamine ($C_{13}H_{16}ClNO$) 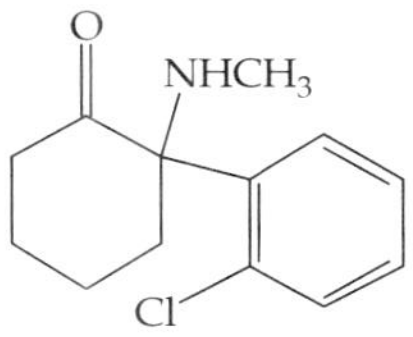	7.5	U	+H			Sass WC and Fusari SA, Ketamine, *APDS*, **6**, 297–322 (1977). McCarthy DA, Parke-Davis and Co., personal communication. "N-dealkylated ketamine, $pK_a = 6.65$ . . . pH values for 10, 50 and 100 mg/mL solutions of the hydrochloride are 4.63, 4.16 and 3.92, respectively."
711	Ketamine	7.5	U	+H	Potentiometric	H_2O	Cohen ML and Trevor AJ, Cerebral accumulation of ketamine and the relation between metabolism of the drug and its pharmacological effects, *JPET*, **189**, 351–358 (1974).
712	Ketobemidone ($C_{15}H_{21}NO_2$) 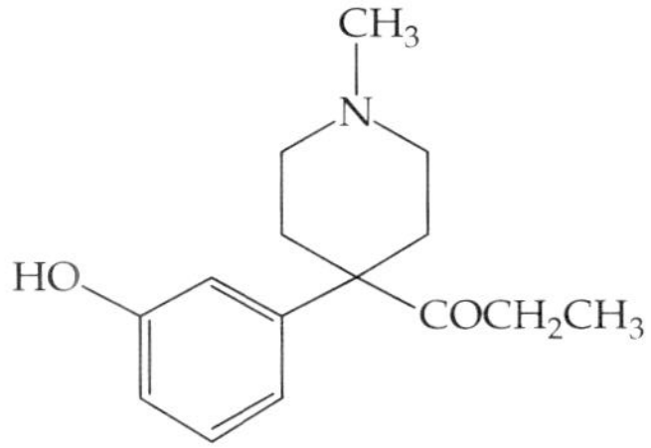	8.52 9.65	U U	+H, −H −H, +H	Spectro ($\lambda = 300$ nm)	H_2O	Ahnfelt N, Abrahamsson L, Bondesson V and Hartvig P, Reaction of the aminophenols ketobemidone and norketobemidone in buffered aqueous solution with ethyl chloroformate, *Acta Pharm. Suec.*, **19**, 355–366 (1982). "The acid dissociation constants were determined according to the method outlined in [6] using a a concentration of ketobemidone and norketobemidone of 2×10^{-4}M. A portion of 10.00 ml of the aminophenol solution is mixed with 5.00 ml of buffer. The pH and the absorbance at 300 nm were recorded using as reference solution a mixture of 5.00 ml of 0.1 M HCl and 10.00 ml of the aminophenol solution with the absorbance, A_{acid}. The maximum absorbance, A_{max}, was obtained in a solution of 5.00 ml of 0.3 M NaOH and 10.00 ml of the aminophenol solution. Macroconstants: $pK_1 = 8.52$, $pK_2 = 9.65$, Microconstants: $pk_1 = 9.10$, $pk_2 = 8.65$, $pk_3 = 9.07$, and $pk_4 = 9.52$."
713	Ketobemidone	8.67	U	+H	Potentiometric	50% aq EtOH	Farmilo CG, Oestreicher PM and Levi L, Physical methods for the identification of narcotics. I B Common physical constants for identification of ninety-five narcotics and related compounds, *Bull. Narcotics*, UN Dept. Social Affairs vol. **6**, pp. 7–19 (1954). CA 48:69490. NB: Cited in Clouet DH (ed.), *Narcotic Drugs Biochemical Pharmacology*, Plenum Press, New York, 52–53 (1971).

(continued)

No.	Compound Name	pK_a value(s)	Data quality	Ionization type	Method	Conditions	Comments and Reference(s)
714	Ketoconazole ($C_{26}H_{28}Cl_2N_4O_4$) 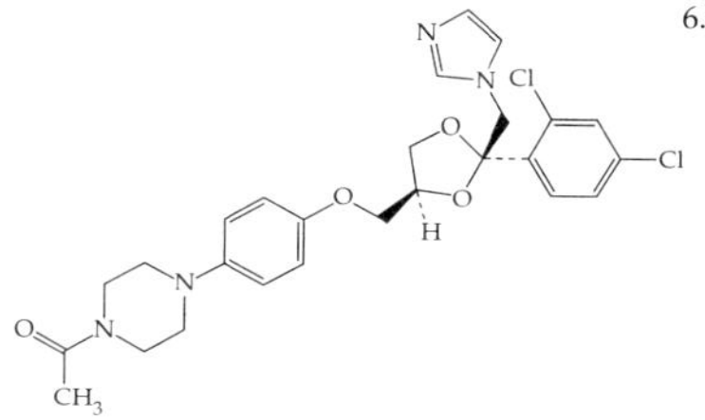	3.25 6.22	U U	+H +H	CE/pH (+ve ion mode)	H_2O $t = 25$ $I = 0.025$	Wan H, Holmen AG, Wang Y, Lindberg W, Englund M, Nagard MB and Thompson RA, High-throughput screening of pK_a values of pharmaceuticals by pressure-assisted capillary electrophoresis and mass spectrometry, *Rapid Commun. Mass Spectrom.*, **17**, 2639–2648 (2003). NB: Reported predicted values (ACD Labs) of 2.92 and 6.54.
715	Ketoprofen ($C_{16}H_{14}O_3$) 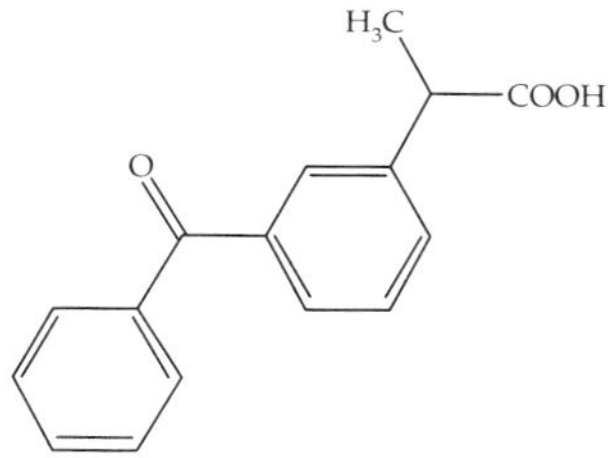	5.94 (apparent)	U	−H		MeOH/H_2O	Liversidge GG, Ketoprofen, *APDS*, **10**, 443–469 (1981). App pK_a = 7.2 (dioxan:water (2:1)) (20) App pK_a = 5.02 (acetonitrile:water (3:1)) (5) App pK_a = 5.937 (methanol:water (3:1)) (14) 5. Blazevic N, Zinic M, Kovac T, Sunjic V and Kajfez F, *Acta Pharm. Jugoslav.*, **25**(3), 155–64 (1975). 14. Unterhalt B, *Pharm. Zeitung.*, **123**(41), 1801–1803 (1978). 20. Brunet JP and Cometti A, *Ger. Pat.*, **2**, 538,985.
716	Ketoprofen	3.7	U	−H	soly	H_2O t unspecified *I* unspecified	Herzfeldt CD and Kümmel R, Dissociation constants, solubilities, and dissolution rates of some selected nonsteroidal antiinflammatories, *Drug Dev. Ind. Pharm.*, **9**(5), 767–793 (1983). NB: See Azapropazone and Ibuprofen for details.
717	KHL 8430 ($C_{25}H_{29}NO_2$) 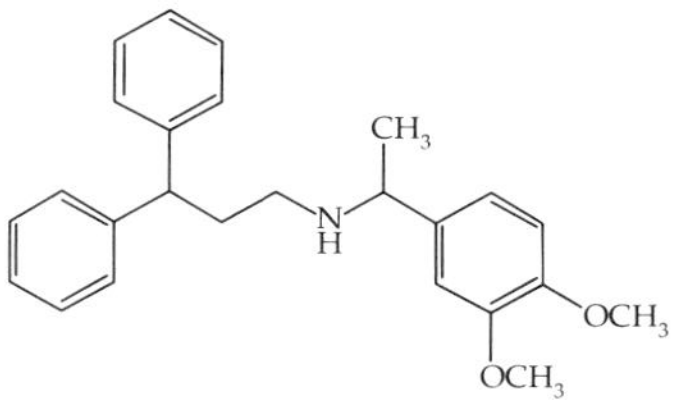	10.60 ± 0.05	U	+H	Potentiometric	H_2O $t = 25.0 \pm 0.1$ $I = 0.1$ (NaCl)	Takacs-Novak K, Box KJ and Avdeef A, Potentiometric pK_a determination of water-insoluble compounds: Validation study in methanol/water mixtures, *Int. J. Pharm.*, **151**, 235–248 (1997). NB: By extrapolation from 42–60%w/w aqueous MeOH. See Acetaminophen for full details.

No.	Compound	pKa value	Data quality	Ionization type	Method	Conditions	Reference and comments
718	Labetalol ($C_{19}H_{24}N_2O_3$) 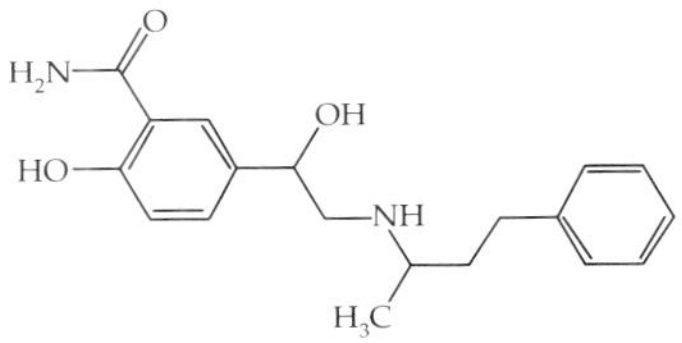	7.44 9.38	U U	+H −H	Potentiometric	H_2O (extrap) $t = 25 \pm 1$ I undefined Ar atmosphere	Cheymol G, Poirier J-M, Carrupt PA, Testa B, Weissenburger J, Levron J-C and Snoeck E., Pharmacokinetics of β-adrenoceptor blockers in obese and normal volunteers, *Br. J. Clin. Pharmacol.*, **43**, 563–570 (1997). NB: Determined by extrapolation from MeOH-water solutions, using the Yasuda-Shedlovsky procedure.
719	Labetalol	7.4	U	+H		H_2O	Barbato F, Caliendo G, LaRotonda MI, Morrica P, Silipo C and Vittoria A. Relationships between octanol-water partition data, chromatographic indices and their dependence on pH in a set of beta-adrenoceptor blocking agents, *Farmaco*, **45**, 647–663 (1990). Cited in: Lombardo F, Obach RS, Shalaeva MY and Gao F, Prediction of human volume of distribution values for neutral and basic drugs. 2. Extended data set and leave-class-out statistics, *J. Med. Chem.*, **47**, 1242–1250 (2004) (ref. 275).
720	Labetalol	GLpK_a: 7.41 ± 0.01 9.36 ± 0.01 A&S: 7.49 ± 0.10 9.27 ± 0.03	 A A U A	 +H +H	Spectro	H_2O $t = 25$ $I = 0.15$ (KCl) Ar atmosphere	Tam KY and Takacs-Novac K, Multi-wavelength spectrophotometric determination of acid dissociation constants, *Anal. Chim. Acta*, **434**, 157–167 (2001). NB: See Clioquinol for details. Cited Pagliara A, Carrupt PA, Caron G, Gaillard P and Testa B, Lipophilicity profiles of ampholytes, *Chem. Rev.*, **97**, 3385–3400 (1997), with values of 7.48 and 9.39.
721	Lactic acid, D(−) ($C_3H_6O_3$) 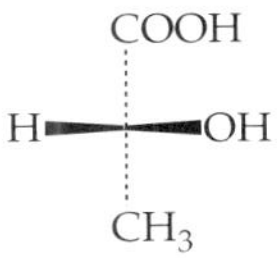	3.83		−H			Al-Shammary FJ, Mian NAA and Mian MS, Lactic Acid, *APDS*, **22**, 263–316 (1993). NB: Martindale 29th Edn. See also Crutchfeld CA, McNabb WM and Hazel JF, Complexes of uranyl ion with some simple organic acids, *J. Inorg. Nucl. Chem.*, **24**, 291–298 (1962).
722	Lactic acid, L(+) ($C_3H_6O_3$) 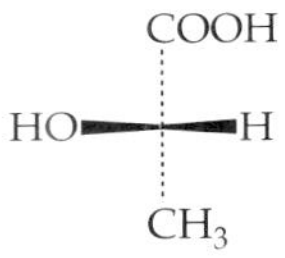	3.79		−H			Al-Shammary FJ, Mian NAA and Mian MS, Lactic Acid, *APDS*, **22**, 263–316 (1993). Martindale 29th Edn.

(continued)

No.	Compound Name	pK_a value(s)	Data quality	Ionization type	Method	Conditions	Comments and Reference(s)
723	Lactic acid, (±) ($C_3H_6O_3$)	3.861 ± 0.002	R	–H	Conductance	H_2O $t = 25.0 \pm 0.01$ $I = 0.00$ (NaCl, KCl)	Partanen JI, Juusola PM and Minkkinen PO, Determination of stoichiometric dissociation constants of lactic acid in aqueous salt solutions at 291.15 and 298.15 K. *Fluid Phase Equilibria*, **204**, 245–266 (2003). NB: Recalculated from the raw conductance data of Martin AW and Tartar HV, *JACS*, **59**, 2672–2675 (1937).
724	Lactic acid (±)	3.863 ± 0.02	A	–H	Potentiometric	H_2O $t = 25.0 \pm 0.01$ $I = 0.00$ (NaCl, KCl)	Partanen JI Juusola PM and Minkkinen PO, Determination of stoichiometric dissociation constants of lactic acid in aqueous salt solutions at 291.15 and 298.15 K, *Fluid Phase Equilibria*, **204**, 245–266 (2003). "Potentiometric LacOH titrations were carried out in aqueous NaCl and KCl solutions at 298.15 K. To avoid . . . dimerisation of LacOH by ester formation, lactic acid was prepared in situ in the titration vessel at the beginning of each titration by adding equivalent amounts of sodium lactate (NaLacO) and hydrochloric acid. . . . The solutions titrated were prepared by mixing a volume of 10.00 cm^3 of the NaLacO solution, 10.00 cm^3 of . . . the HCl solutions, 100.0 cm^3 of a salt solution (NaCl or KCl) and 25.00 cm^3 of water. . .. The NaCl solutions were titrated by using the (0.200 M) NaOH reagent and the KCl solutions by using the (0.200 M) KOH reagent. During the titrations the cell potential difference was measured by means of an N62 combination electrode and a CG841 pH-meter . . .. The resolution of the meter was 0.1 mV. The titrant was added by using increments of 0.050 cm^3 by a Dosimat (Metrohm). . . . Standard buffer solutions of pH = 4.005 and 6.865 were used to check the stability of the measuring system between titrations. The pH-meter usually reproduced the same reading within 0.2 mV(+/– 0.003 pH) in these buffer solution tests. To check further the stability of (the) measuring system, a titration of formic acid was carried out . . .." [NB: the formic acid titration gave pK_a = 3.740; this is acceptably close to the literature value, pK_a = 3.752 (Robinson and Stokes; Eberson)]
725	Lauric acid ($C_{12}H_{24}O_2$) $CH_3(CH_2)_{10}COOH$	4.92	U	–H	Potentiometric	H_2O RT	Johns WH and Bates TR, Quantification of the binding tendencies of cholestyramine II. Mechanism of interaction with bile salts and fatty acid salt anions, *J. Pharm. Sci.*, **59**, 329–333 (1970).
		5.3	U	–H	Potentiometric	H_2O $t = 20$	Nyren V and Back E, The ionization constant, solubility product, and solubility of lauric and myristic acids, *Acta Chem. Scand.*, **12**, 1305–1311 (1958). Also reported pK_a = 6.3 for myristic acid.

726	Levallorphan tartrate ($C_{19}H_{25}NO.C_4H_6O_6$) 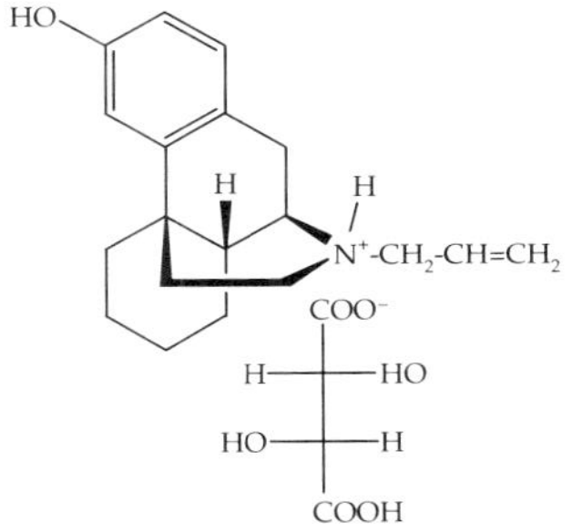	4.5 6.9	U U	−H +H	Potentiometric	H_2O RT	Rudy BC and Senkowski BZ, Levallorphan tartrate, *APDS*, **2**, 339–361 (1973). Cited Lau E and Yao C, Hoffmann-La Roche, unpublished data. NB: The pK_{a1} value (4.5) refers to the hydrogen tartrate counterion.
727	Levallorphan ($C_{19}H_{25}NO$)	8.73	U	+H	Potentiometric	H_2O $t = 20$ $I < 0.01$	Kaufman JJ, Semo NM and Koski WS, Microelectrometric titration measurement of the pK_a's and partition and drug distribution coefficients of narcotics and narcotic antagonists and their pH and temperature dependence, *J. Med. Chem.*, **18**, 647–655 (1975). NB: See Codeine for further details.
		8.43	U	+H		H_2O $t = 37$ $I < 0.01$	
728	Levallorphan ($C_{19}H_{25}NO$)	8.30	U	+H	Potentiometric	50% EtOH	Farmilo CG, Oestreicher PM and Levi L, Physical methods for the identification of narcotics. I B Common physical constants for identification of ninety-five narcotics and related compounds, *Bull. Narcotics*, UN Dept. Social Affairs, vol. **6**, pp. 7–19 (1954). CA 48:69490. NB: Cited in Clouet DH (ed.), *Narcotic Drugs Biochemical Pharmacology*, Plenum Press, New York, 52–53, 1971.
729	Levarterenol (L-noradrenaline; L-norepinephrine) ($C_8H_{11}NO_3$) 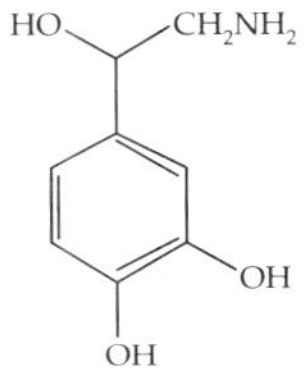	8.72	U	−H,+H	Spectro	pK_{a1}: H_2O $t = 25$ I undefined	Kappe T and Armstrong MD, Ultraviolet absorption spectra and apparent acidic dissociation constants of some phenolic amines. *J. Med. Chem.*, **8**, 368–374 (1965). Cited in Schwender CF, Levarterenol bitartrate, *APDS*, **1**, 339–361 (1972). "... spectrophotometric methods were used for determining the phenolic pK_{a1} values. These values were then used to correct potentiometric titration curves to determine the amine pK_{a2} value, as the two steps were overlapping. pH measurements were made with a titrator that had been standardised with commercial pH standards. Potentiometric titrations were performed in an argon atmosphere using standard methods, including correction for sodium ion errors (<0.02 pH at worst). No corrections were made for activity coefficients and ionic strengths were not reported."
		9.72	U	+H,−H	Potentiometric	pK_{a2}: H_2O $t = 35$ I undefined	

(*continued*)

No.	Compound Name	pK_a value(s)	Data quality	Ionization type	Method	Conditions	Comments and Reference(s)
							NB: These studies do not address the separation of microconstants. Also, the use of different temperatures for the two series of measurements is a deficiency, as amine pK_a values are temperature-dependent. However, the paper is ambiguous on this point, as all results are reported in a Table . . . for 25 °C (below). Hence, values are apparent and approximate.

Apparent dissociation constants at 25 °C

Compound	pK_{a1} (OH)	pK_{a2} (NH)
phenethylamine	–	9.88
2-MeO-phenethylamine	–	10.20
3-MeO-phenethylamine	–	9.89
3-HO-phenethylamine	9.58	10.50
N-methyl-4-HO-phenethylamine	9.76	10.71
N,*N*-dimethyl-4-HO-phenethylamine	9.78	10.02
2,4-di-HO-phenethylamine	8.91 (11.7)	10.8
β-(2-hydroxyphenyl)ethanolamine	9.42	9.90
β-(4-hydroxyphenyl)ethanolamine	9.57	9.66
N-methyl-β-(4-hydroxyphenyl)ethanolamine	9.57	9.66
N,*N*-dimethyl-β-(4-hydroxyphenyl) ethanolamine	9.58	9.50
β-(4-hydroxy-3-methoxyphenyl)ethanolamine	9.54	9.56
N-methyl-β-(4-hydroxy-3-methoxyphenyl) ethanolamine	9.54	9.56
N,*N*-dimethyl-β-(4-hydroxy-3-methoxyphenyl) ethanolamine	9.54	9.52

No.	Compound Name	pK_a value(s)	Data quality	Ionization type	Method	Conditions	Comments and Reference(s)
730	Levarterenol (L-noradrenaline; L-norepinephrine)	8.90 ± 0.06 9.78 ± 0.09	U U	+H,−H −H,+H	Potentiometric, Spectro	H_2O *t* unspecified $I = 0.1$ (glycine)	Lewis GP, The importance of ionization in the activity of sympathomimetic amines, *Br. J. Pharmacol.*, **9**, 488–493 (1954). Cited in Schwender CF, Levarterenol bitartrate, *APDS*, **1**, 339–361 (1972). NB: Method similar to Kappe and Armstrong, see Levarterenol (no. 729).

731	Levarterenol (L-noradrenaline; L-norepinephrine)	8.57 9.73 11.13	U U U	+H,−H −H,+H −H			Wilson TD, Levarterenol bitartrate, *APDS*, **11**, 555–586 (1982).

Table: Dissociation constants

pK_1	pK_2	pK_3	References
8.72	9.72	>12	Foye 3rd
9.3	10.3	13	Granot J, *FEBS Letts*, **67**, 271–275 (1976). NB: These studies used NMR chemical shifts as a function of pD.
8.57	9.73	11.13	Rajan KS, Davis JM, Colburn RW and Jarke FH, *J. Neurochem.*, **19**, 1099–1116 (1972).

NB: Cited Ganellin C, *J. Med. Chem.*, **20**, 579–581 (1977), who gave a short summary of the catecholamine pK_a literature and calculated the proportions of neutral and zwitterionic species from the microconstant data of Sinistri C, Villa L, *Farmaco, Sci. Edn.*, **17**, 949–966 (1962).

732	Levodopa ($C_9H_{11}NO_4$)	2.31 8.71 9.74 13.40	U U U U	−H +H −H −H			N&K; APDS 5. NB: Appears to be the Gorton data (no. 733).
733	Levodopa	2.3 8.7 9.7 13.4	U U U U	−H +H −H −H	Potentiometric	H_2O $t = 25 \pm 0.02$ $I = 1$ (KNO_3)	Gorton JE and Jameson RF, Complexes of doubly chelating ligands. I. Proton and copper(II) complexes of L-β-(3,4-dihydroxyphenyl) alanine (DOPA), *J. Chem. Soc. (A)*, 2615–2618 (1968).
734	Levorphanol ($C_{17}H_{23}NO$)	9.79	U	+H	Potentiometric	H_2O $t = 20$ $I < 0.01$	Kaufman JJ, Semo NM and Koski WS, Microelectrometric titration measurement of the pK_a's and partition and drug distribution coefficients of narcotics and narcotic antagonists and their pH and temperature dependence, *J. Med. Chem.*, **18**, 647–655 (1975). NB: See Codeine for further details.
		9.37	U	+H		H_2O $t = 37$ $I < 0.01$	

(continued)

No.	Compound Name	pK_a value(s)	Data quality	Ionization type	Method	Conditions	Comments and Reference(s)
735	Levorphanol	9.2	U	+H	partition	H_2O t undefined $I = 0.1$	Shore PA, Brodie BB and Hogben CAM, The gastric secretion of drugs, *JPET*, **119**, 361–369 (1957). NB: Levorphan; L-3-hydroxy-*N*-methylmorphinan; the method followed the partition-pH method of Butler TC, Quantitative studies of the demethylation of *N*-methylbarbital, *JPET*, **108**, 474–480 (1953).
736	Levorphan(ol)	8.18	U	+H	Potentiometric	50% aq EtOH	Farmilo CG, Oestreicher PM, Levi L, Physical methods for the identification of narcotics. I B Common physical constants for identification of ninety-five narcotics and related compounds, *Bull. Narcotics*, UN Dept. Social Affairs vol. **6**, pp. 7–19 (1954). CA 48:69490. NB: Cited in Beckett AH, Analgesics and their antagonists. I., *J. Pharm. Pharmacol.*, **8**, 848–859 (1956). NB: See Alphaprodine for details.
737	Levulinic acid ($C_5H_8O_3$) CH_3 O OH O	4.65	U	−H	Conductance	H_2O $t = 25$ $c = 0.03–0.001$	Kendall J, Electrical conductivity and ionization constants of weak electrolytes in aqueous solution, *in* Washburn EW, Editor-in-Chief, International Critical Tables, vol. **6**, McGraw-Hill, NY, 259–304 (1929). NB: Other values: $t = 0$, 4.66; $t = 35.5$, 4.65.
738	Lidocaine (lignocaine) ($C_{14}H_{22}N_2O$) CH_3 NH N(Et)$_2$ O CH_3	7.96 ± 0.02	A	+H	Potentiometric	H_2O $t = 25 \pm 0.5$ $I = 0.15$ (KCl)	Avdeef A, Box KJ, Comer JEA, Hibbert C and Tam KY, pH-metric log P 10. Determination of liposomal membrane-water partition coefficients of ionizable drugs, *Pharm. Res.*, **15**(2), 209–215 (1998). NB: Used a Sirius PCA101 autotitrator. Also gave log P (octanol-water) and log P (dioleylphosphatidyl-choline unilamellar vesicles). ACD calculated value = 8.53 ± 0.25.
739	Lidocaine (lignocaine)	7.84	U	+H	Potentiometric	H_2O $t = 25.0 \pm 0.2$ $I = 0.01$ (NaCl)	Johansson P-A, Liquid-liquid distribution of lidocaine and some structurally related anti-arrythmic drugs and local anaesthetics, *Acta Pharm. Suec.*, **19**, 137–142 (1982). NB: See also Lofgren, Dissertation, Univ. Stockholm, 1948, potentiometric, 7.855.

No.	Compound	pKa	Quality	Ionization	Method	Conditions	Reference / Comments
							"All constants were determined ... by potentiometric titrations according to ... [19]. ... 1×10^{-3} or 2×10^{-3} M solutions of the amine salts were titrated. The ionic strength was adjusted to 0.01 with sodium chloride. Groningsson K, Lindgren J-E, Lundberg E, Sandberg R, Wahlen A, Lidocaine base and hydrochloride, *APDS*, **14**, 207–235 (1985)."
740	Lidocaine (lignocaine)	7.86	U	+H			Narahashi TI, Frazier DT and Yamada M, The site of action and active form of local anesthetics, *JPET*, **171**, 32–44 (1970). NB: No experimental details given. Values also given for two lidocaine analogues: *N*-methyl-*N*-β-methoxyethyl analogue, $pK_a = 6.3$; analogue with $-(CH_2)_3-$ bridge instead of methylene bridge, $pK_a = 9.8$.
741	Lidocaine (lignocaine)			+H	Potentiometric	H_2O $I = 0.05$ (KCl) N_2 atmosphere	Powell MF, Lidocaine and lidocaine hydrochloride, *APDS*, **15**, 761–779 (1986).
		8.24	A			$t = 10.0$	"The pK_a of Lidocaine has been measured by potentiometric titration at ionic strength 0.05 M (KCl) (Kamaya H, Hayes JJ and Ueda I, Dissociation constants of local anesthetics and their temperature dependence, *Anesth. Analg.*, **62**, 1025–1030 (1983)) ..."
		7.92	A			$t = 25.0$	
		7.57	A			$t = 38.0$	NB: Very careful work with calibrated microsyringes and glass-calomel electrodes. Activity coefficient corrections. Also reported pK_a-temperature dependence for the following additional local anesthetics (the last significant figure is uncertain):

Compound	pK_a (10 °C)	pK_a (25 °C)	pK_a (38 °C)
Tetracaine	8.662	8.463	8.267
Dibucaine	9.046	8.729	8.412
Mepivacaine	8.021	7.768	7.556
Chloroprocaine	9.372	8.974	8.774
Procaine	9.380	9.052	8.662

No.	Compound	pKa	Quality	Ionization	Method	Conditions	Reference / Comments
742	Lidocaine	7.96	A	+H	CE/pH (+ve ion mode)	H_2O $t = 25$ $I = 0.025$	Wan H, Holmen AG, Wang Y, Lindberg W, Englund M, Nagard MB and Thompson RA, High-throughput screening of pK_a values of pharmaceuticals by pressure-assisted capillary electrophoresis and mass spectrometry, *Rapid Commun. Mass Spectrom.*, **17**, 2639–2648 (2003). NB: Reported a predicted value (ACD Labs) of 8.84.

(*continued*)

No.	Compound Name	pK_a value(s)	Data quality	Ionization type	Method	Conditions	Comments and Reference(s)
743	Lidocaine (lignocaine)	7.86 (0.03)	A	+H	Potentiometric	H_2O $t = 24.0 \pm 0.1$ $I = 0.0050–0.0775$ (KCl)	Powell MF, Lidocaine and lidocaine hydrochloride, *APDS*, **15**, 761–779 (1986). I / pK_a / Stats: 0.0050 7.850 mean K_a = 1.38318E−08; 0.0100 7.825 mean pK_a = 7.859; 0.0240 7.907 std dev = 0.0327; 0.0300 7.835; 0.0775 7.884. "The ionic strength, when adjusted with KCl from 0.005 to 0.077 M, showed little effect on the acidity constant (9). 9. Levy RH and Rowland M, *J. Pharm. Pharmacol.*, **24**, 841–7 (1972)."
744	Lidocaine (lignocaine)	7.94 ± 0.01	A	+H	Potentiometric	H_2O $t = 25.0$ $I = 0.15$ (KCl)	Sirius Technical Application Notes, 1994, vol. **1**, pp. 122–123. Sirius Analytical Instruments Ltd., Forest Row, East Sussex, RH18 5DW, UK. [Cited in Lombardo F, Obach RS, Shalaeva MY and Gao F, Prediction of human volume of distribution values for neutral and basic drugs. 2. Extended data set and leave-class-out statistics, *J. Med. Chem.*, **47**, 1242–1250 (2004); ref. 282]
745	Lidocaine (lignocaine)	7.18	U	+H	Conductance	H_2O $t = 25.0 \pm 0.02$ $c = 0.0003–0.00657$	Sjoberg H, Karami K, Beronius P and Sundelof L-O, Ionization constants... A revised pK_a of lidocaine hydrochloride, *Int. J. Pharm.*, **141**, 63–70 (1996). NB: Corrected to infinite frequency. Limiting conductance 17.70–17.87. NB: The large difference between this value and several very similar values in the literature (7.84, 7.855, 7.92, 7.96; all at 25.0 °C) demands some explanation. The conductance method assumes that no ion-pairing occurs. It seems likely that the deviation of the conductance result from the others is due to ion-pairing. EJ King, Acid-Base Equilibria (p. 36, 1965) stated that the conductance method is not suitable for determination of pK_a values for weak acids >5.
746	Lidocaine homologues -NH-CO-R		A	+H	Potentiometric	H_2O	Buchi J and Perlia X, Beziehungen zwischen de physikalisch-chemische Eigenschaften und der Wirkung von Lokalanasthetica. *Arzneim.-Forsch.*, **10**, 745–754 (1960). NB: These values were cited from Lofgren N, *Studies on local anesthetics, Xylocaine, a new synthetic drug*, J. Hoeggstrom, Stockholm, 1948.
	CH_2-NH-CH_3	7.99					
	CH_2-CH_2-N$(C_2H_5)_2$	8.96					
	CH_2-NH-CH_2CH_3	8.08					
	CH_2-N$(CH_3)_2$	7.36					
	CH_2-NH-C_3H_7	8.01					
	CH_2-NH-C_4H_9	8.07					
	CH_2-NH-*i*-C_4H_9	7.83					
	CH(CH_3)-N$(C_2H_5)_2$	8.11					
	CH_2-N$(C_2H_5)_2$	7.88					
	CH_2-N$(C_3H_7)_2$	7.68					

747	Lincomycin ($C_{18}H_{34}N_2O_6S$) CH3 HO CH3 N H N HO O CH3CH2CH2 O OH SCH3 OH	7.5	U	+H		H_2O	Hoeksma H, Bannister B, Birkenmeyer RD, Kagan F, Magerlein BJ, MacKellar FA, Schroeder W, Slomp G and Herr RR, Chemical studies on lincomycin, *JACS*, **86**, 4223–4224 (1964). No experimental details given. Morozowich W, Lamb DJ, Karnes HA, Mackellar FA, Stern KF and Rowe EL, Synthesis and bioactivity of lincomycin-2-phosphate, *J. Pharm. Sci.*, **58**, 1485–1489 (1969). NB: No experimental details given. Assumed that the amine function of lincomycin-2-phosphate had the same pK_a value as lincomycin, which was reported as 7.5, without supporting experimental details or references.
748	Lincomycin	7.6	U	+H			Muti HY and Al-Hajjar FH, Lincomycin Hydrochloride, *APDS*, **23**, 269–319 (1994). Eble TE, Lincomycin related antibiotics, *J. Chromatogr. Libr.*, **15**, 231–271 (1978).
749	Linoleic acid ($C_{18}H_{32}O_2$) CH3 (CH2)6COOH	5.10	U	+H	Potentiometric	H_2O RT	Johns WH and Bates TR, Quantification of the binding tendencies of cholestyramine II. Mechanism of interaction with bile salts and fatty acid salt anions, *J. Pharm. Sci.*, **59**, 329–333 (1970).
750	Lisinopril ($C_{21}H_{31}N_3O_5$) NH2 H COOH H CH2 N N H O COOH	2.5 4.0 6.7 10.1	U U U U	−H −H +H +H	Potentiometric	H_2O $t = 25$	Ip DP, DeMarco JD and Brooks MA, Lisinopril, *APDS*, **21**, 233–276, 1992. Cited McCauley JA, Merck Sharp and Dohme Research Laboratories, Rahway, NJ. NB: Foye 1 gave values: 1.7, 3.3, 7.0 and 11.1.
751	Lisinopril	3.13 7.2	U U	+H +H	CE/pH (+ve ion mode)	H_2O $t = 25$ $I = 0.025$	Wan H, Holmen AG, Wang Y, Lindberg W, Englund M, Nagard MB and Thompson RA, High-throughput screening of pK_a values of pharmaceuticals by pressure-assisted capillary electrophoresis and mass spectrometry, *Rapid Commun. Mass Spectrom.*, **17**, 2639–2648 (2003). NB: Reported predicted values (ACD Labs) of 3.18 and 7.53.
		3.07 7.25	U U	+H +H	CE/pH (−ve ion mode)		

(continued)

No.	Compound Name	pK_a value(s)	Data quality	Ionization type	Method	Conditions	Comments and Reference(s)
752	Lisuride ($C_{20}H_{26}N_4O$) 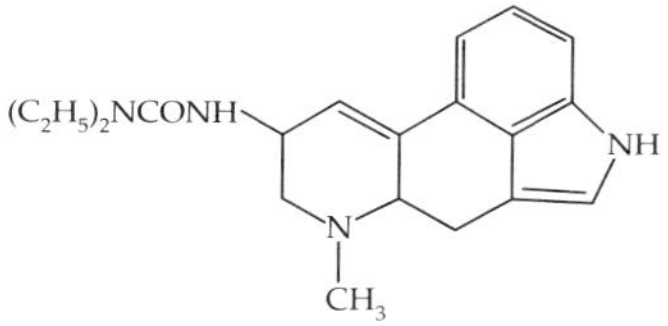	7.29	U	+H	soly	H_2O $t = 37$ $I = 0.15$	Zimmerman I, Determination of pK_a values from solubility data, *Int. J. Pharm.*, **13**(1), 57–65 (1983). NB: See Pyrazolic acid for details.
753	Lonazolac ($C_{17}H_{13}ClN_2O_2$) 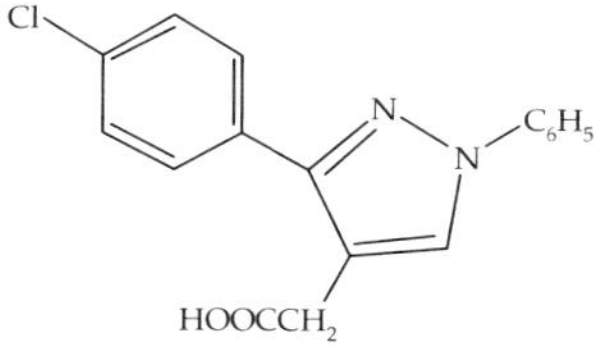	4.3	U	–H	Potentiometric	H_2O $t = 20$ I undefined	Rainer VG, Krüger U and Klemm K Syntheses und physicalisch-chemische Eigenschaften von Lonazolac-Ca einem neuen Antiphlogistikum/Antirheumatikum, *Arzneim.-Forsch.*, **31**(4), 649–655 (1981). NB: See Indomethacin for details.
754	Loperamide ($C_{29}H_{33}ClN_2O_2$) 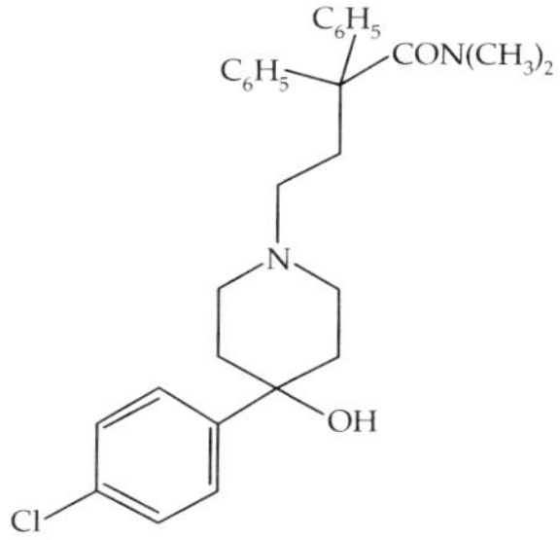	8.66	U	+H	Potentiometric	H_2O	van Rompay J and Carter JE, Loperamide hydrochloride, *APDS*, **19**, 341–365 (1990). NB: Cited R. Caruwels, Janssen Research Foundation, personal communication, as the source of the pK_a value, which had been obtained using a procedure reported by Peeters (Peeters J, Determination of ionization constants in mixed aqueous solvents of varying composition by a single titration, *J. Pharm. Sci.*, **67**, 127–129 (1978)). This involved extrapolation of the pK_a values in methanol-water mixtures to zero methanol content.

No.	Compound	pKa	Class	±H	Method	Conditions	Reference / Comments
755	Lorazepam ($C_{15}H_{10}Cl_2N_2O_2$) 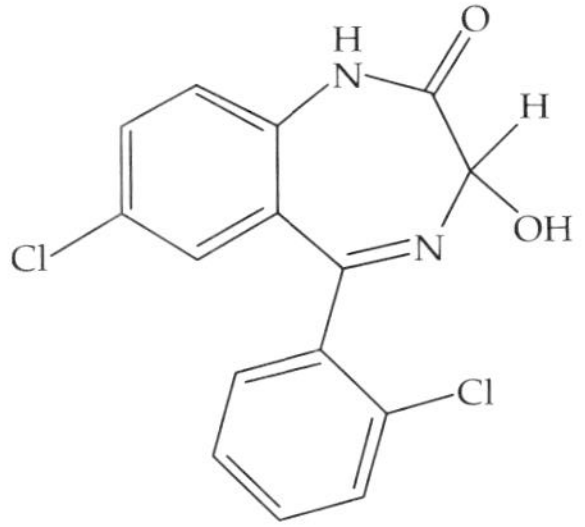	1.3 11.5 1.8	U U U	+H −H +H	Spectro polarography	H_2O	Rutgers JG and Shearer CM, Lorazepam, *APDS*, **9**, 398–424 (1980). "Two pK_a's are observed for lorazepam (15, 5). ... pK_a values, determined spectrophotometrically in aqueous buffers, are 1.3 and 11.5. pK_{a1} = 1.8 (polarography) (16). Barrett *et al.* (5) have proposed that three species, a protonated, a neutral and a deprotonated form are involved in the equilibria. Protonation at low pH occurs at the nitrogen in the 4 position. Deprotonation occurs at high pH with the loss of the hydrogen atom from the 3-hydroxyl group. 5. Barrett J, Smyth WF and Davidson IE, An examination of acid-base equilibria of 1,4-benzodiazepines by spectrophotometry, *J. Pharm. Pharmacol.*, **25**, 387–393 (1973); 15. Davidson IE and Smyth WF, *Proc. Soc. Anal. Chem.*, **9**, 209–211 (1972); 16. Goldsmith JA, Jenkins HA, Grant J and Smyth WF, The polarographic behaviour of the 1,4-benzodiazepines..., *Anal. Chem. Acta*, **66**, 427–434 (1973)."
756	Lorazepam	1.3	U	+H	Spectro	5% MeOH in H_2O $t = 20$ $I = 0.15$	Barrett J, Smyth WF, Davidson IE. Examination of acid-base equilibria of 1,4-benzodiazepines by spectrophotometry, *J. Pharm. Pharmacol.*, **25**, 387–393 (1973). NB: See 1,4-Benzodiazepines for details. The same value is repeated by Konishi *et al.* (Konishi M, Hirai K and Mari Y, Kinetics and mechanism of the equilibrium reaction of triazolam in aqueous solution, *J. Pharm. Sci.*, **71**(12), 1328–1334 (1982); see Triazolam).
757	Lysergic acid ($C_{16}H_{16}N_2O_2$) 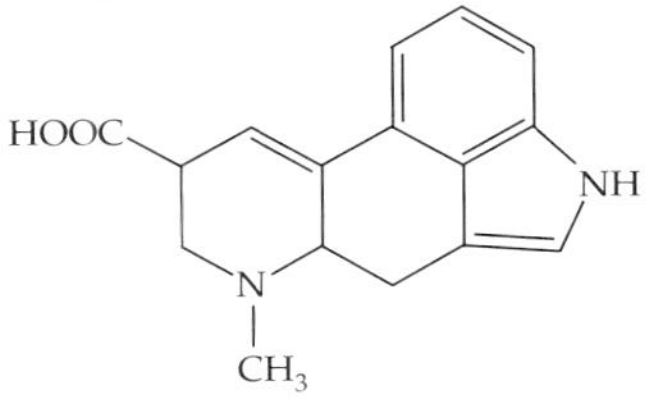	3.32 7.82 3.33 7.73	U U	−H +H	Potentiometric	H_2O $t = 24$ $I < 0.005$ H_2O $t = 38$ $I < 0.005$	Craig LC, Shedlovsky T, Gould RG and Jacobs WA, The ergot alkaloids. XIV. The positions of the double bond and the carboxyl group in lysergic acid and its isomer, *J. Biol. Chem.*, **125**, 289–298 (1938). Cited in Perrin Bases 2925 ref. C54. NB: The study used an asymmetric cell with a glass electrode and liquid junction potential. The pK_a values reported here are corrected values reported by Perrin from apparent pK_a' values (pH at half-neutralisation) in the original source.
758	Maleic acid ($C_4H_4O_4$) 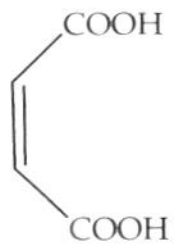	1.94	A	−H	Conductance	H_2O $t = 25$ $c = 0.06–0.001$	Kendall J, Electrical conductivity and ionization constants of weak electrolytes in aqueous solution, *in* Washburn EW, Editor-in-Chief, International Critical Tables, vol. **6**, McGraw-Hill, NY, 259–304 (1929). NB: Other value: $t = 18$, p$K_a = 1.86$.
		1.921 6.225	A A	−H −H	Potentiometric	H_2O $t = 25$	German WL, Vogel AI and Jeffery GH, Thermodynamic primary and secondary dissociation constants of fumaric and maleic acids, *Phil. Mag.*, **22**, 790–800 (1936). NB: See Fumaric acid for details.

(continued)

No.	Compound Name	pK_a value(s)	Data quality	Ionization type	Method	Conditions	Comments and Reference(s)
759	Malic acid ($C_4H_6O_5$)	3.459 5.097	A A	–H –H		H_2O $t = 25$ $I = 0.00$	Brittain HG, Malic Acid, *APDS*, **28**, 158 (2001). NB: All of these data are also cited in Martell AE and Smith RM, *Critical Stability Constants*, vol. **3**, Plenum Press, NY (1977). Kortum also cited a value at 25 °C for pK_{a2} = 5.14 (U).
		3.24 ± 0.03 4.71 ± 0.01	A A	–H –H		$I = 0.1$	
		3.11 4.45 ± 0.01	A A	–H –H		$I = 1.0$	
760	Malonic acid ($C_3H_4O_4$) 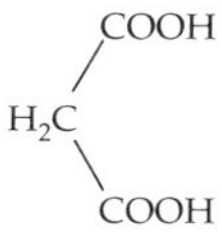	2.80	U	–H	Conductance	H_2O $t = 25$ $c = 0.06–0.001$	Kendall J, Electrical conductivity and ionization constants of weak electrolytes in aqueous solution, *in* Washburn EW, Editor-in-Chief, International Critical Tables, vol. **6**, McGraw-Hill, NY, 259–304 (1929). NB: Other value: $t = 0$, 2.84.
		2.85 5.66	A A	–H –H	Potentiometric	H_2O $t = 25.0 \pm 0.01$ N_2 atmosphere	German WL and Vogel AI, The primary and secondary dissociation constants of malonic, succinic and glutaric acids by potentiometric titration, *JACS*, **58**, 1546–1549 (1936). NB: Used quinhydrone electrode in cells with liquid junction potential.
761	Mandelic acid ($C_8H_8O_3$)	3.40 ± 0.01	U	–H		H_2O $t = 25$ $I = 0.0$	Brittain HG, Mandelic Acid, *APDS*, **29**, 185–186 (2002). NB: These data are cited from Martell AE and Smith RM, *Critical Stability Constants*, vol. **3**, Plenum Press, NY (1977). Computed value from ACD PhysChem is 3.41 ± 0.2. The hydroxyl group was predicted to have a pK_a value of 15.7 ± 0.25.
		3.19 ± 0.01 3.17 ± 0.04	U U	–H –H		$I = 0.1$ $I = 1.0$	
762	D-(−)-Mandelic acid and analogues	3.35	U	–H	Potentiometric	H_2O $t = 25.0$ I undefined	Randinitis EJ, Barr M, Wormser HC and Nagwekar JB, Kinetics of urinary excretion of D-(−)-mandelic acid and its homologs. I. Mutual inhibitory effect of D-(−)-mandelic acid and its certain homologs on their renal tubular secretion in rats, *J. Pharm. Sci.*, **59**, 806–812 (1970). NB: The acid (20 mg) was dissolved in water (10 ml) and titrated with NaOH (0.01 N) solution. The pH of half-neutralization was recorded as the pK_a value. No details were given of conditions or pH meter calibration. Also reported the following pK_a values for mandelic acid analogues: D-(−)-benzyllactic acid, 3.85; D-(−)-4-hydroxy-4-phenylbutanoic acid, 4.70.

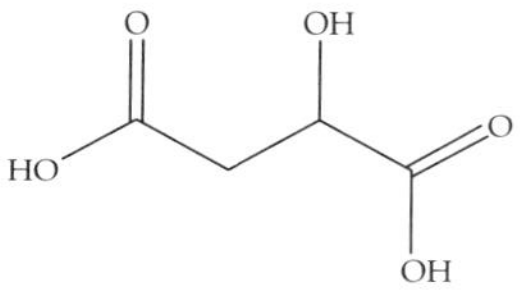

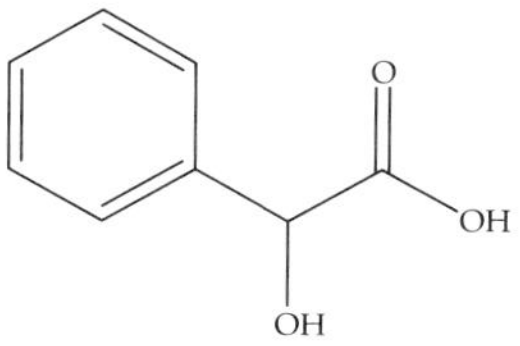

No.	Compound	pKa	Data quality	Ionization type	Method	Conditions	Reference
763	Mannitol ($C_6H_{14}O_6$) 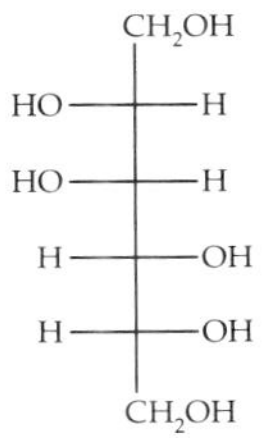	13.50 14.09	A A	−H −H	Potentiometric	H_2O $t = 18$ $t = 0$	Thamsen J, The acidic dissociation constants of glucose, mannitol and sorbitol, as measured by means of the hydrogen electrode and the glass electrode at 0 and 18 °C. *Acta Chem. Scand.*, **6**, 270–284 (1952).
764	Mebendazole ($C_{16}H_{13}ClN_3O_3$) 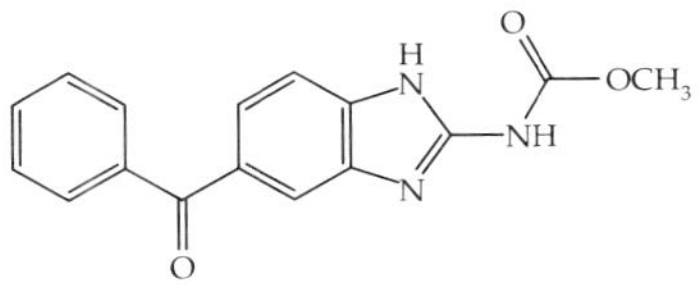	3.43 9.93	U U	+H +H	CE/pH (+ve ion mode)	H_2O $t = 25$ $I = 0.025$	Wan H, Holmen AG, Wang Y, Lindberg W, Englund M, Nagard MB and Thompson RA, High-throughput screening of pK_a values of pharmaceuticals by pressure-assisted capillary electrophoresis and mass spectrometry, *Rapid Commun. Mass Spectrom.*, **17**, 2639–2648 (2003). NB: Reported predicted values (ACD Labs) of 5.02 and 8.52.
765	Mebeverine ($C_{25}H_{35}NO_5$) 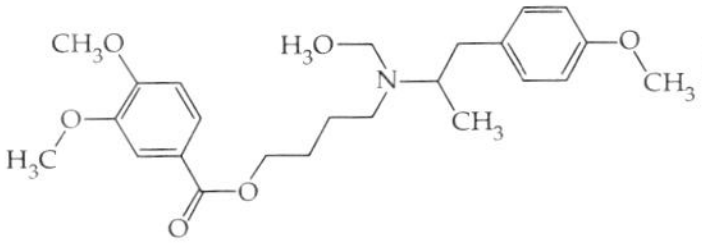	9.07 ± 0.02 9.8	U U	+H +H	Potentiometric	40% EtOH $t = 25.0$ H_2O	Mannhold R, Rodenkirchen R, Bayer R and Haus W, The importance of drug ionization for the action of calcium antagonistsand related compounds, *Arzneim.-Forsch.*, **34**, 407–409 (1984). NB: See Aprindine for details.
766	Mebhydroline ($C_{19}H_{20}N_2$) 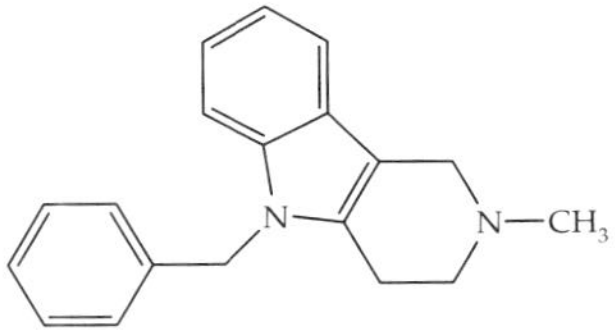	2.55 ± 0.18 7.63 ± 0.04	U U	+H +H	Potentiometric	H_2O t undefined $I = 0.30$ (NaCl)	Testa B and Murset-Rossetti L, The partition coefficient of protonated histamines, *Helv. Chim. Acta*, **61**, 2530–2537 (1978). NB: See Cycliramine for details.

(continued)

No.	Compound Name	pK_a value(s)	Data quality	Ionization type	Method	Conditions	Comments and Reference(s)
767	Mecamylamine ($C_{11}H_{21}N$) 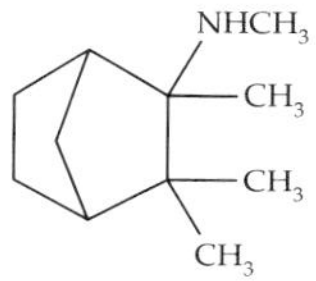	11.2	U	+H	partition	H_2O t undefined $I = 0.1$	Schanker LS, Shore PA, Brodie BB and Hogben CAM, Absorption of drugs from the stomach. I. The rat, *JPET*, **120**, 528–539 (1957). NB: The pK_a value was determined by the partitioning method of Butler TC, Quantitative studies of the demethylation of *N*-methyl barbital (metharbital, gemonil), *JPET*, **108**, 474–480 (1953).
768	Meclizine ($C_{25}H_{27}ClN_2$) 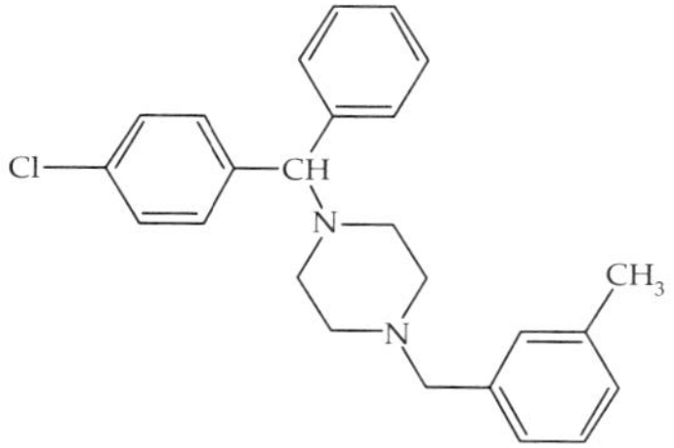	3.1 6.2	U U	+H +H	partition	H_2O	Persson BA and Schill G, Extraction of amines as complexes with inorganic anions, *Acta Pharm. Suec.*, **3**, 291–302 (1966). NB: No details were given for the measurement of what are described as "approximate acid dissociation constants." Newton DW, Murray WJ and Lovell MW, pK_a determination of benzhydrylpiperazine antihistamines in aqueous and aqueous methanol solutions, *J. Pharm. Sci.*, **71**(12), 1363–1366 (1982). NB: Attempted to measure pK_a for meclizine by extrapolation of apparent pK_a values in cosolvent-water mixtures to 0% cosolvent, but found that meclizine could only be titrated in >47.8% methanol solutions, which was too high.
769	Meclizine	2.05	U	+H	Potentiometric	H_2O $t = 25$ $I = 0.0051$ (NaCl)	Lukkari S, Potentiometric studies on the ionization of some psychotropic diphenylmethane derivatives, hydroxyzine and meclizine., *Farm. Aikak.*, **80**, 161–165 (1971); CA 75:41145t. NB: Activity effects estimated with the Guggenheim equation (constant term = 0.160). Also found pK_a = 2.0 (spectro).
770	Medazepam ($C_{16}H_{15}ClN_2$) 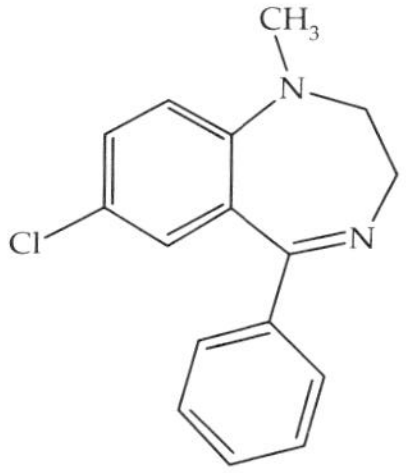	6.2	U	+H	Spectro soly	H_2O	Le Petit GF, Medazepam pK_a determined by spectrophotometric and solubility methods, *J. Pharm. Sci.*, **65**, 1094–1095 (1976). "The pK_a of medazepam was determined by a UV spectrophotometric method and compared with data obtained from the solubility method. Good correlation was found between the results of the 2 methods."

No.	Compound	pKa	Class	Type	Method	Conditions	Reference / Comments
771	Medazepam	4.4	U	+H	Spectro	5% MeOH in H_2O $t = 20$ $I = 0.15$	Barrett J, Smyth WF and Davidson IE. Examination of acid-base equilibria of 1,4-benzodiazepines by spectrophotometry, *J. Pharm. Pharmacol.*, **25**, 387–393 (1973). NB: Significant discrepancy with the other two values, but this one looks more likely, compared to other benzodiazepines. See 1,4-Benzodiazepines for details.
772	Mefenamic acid ($C_{15}H_{15}NO_2$) COOH NH CH_3 CH_3	4.55 ± 0.06	U	−H	Spectro	H_2O $t = 20$ I extr. to 0.00 ($NaClO_4$)	Zommer-Urbanska S, Bojarowicz H, Spectrophotometric investigations of protolytic equilibria of mefenamic acid and determination by means of Fe(III) in methanol-aqueous media, *J. Pharm. Biomed. Anal.*, **4**(4), 475–481 (1986). "... studies were carried out on the effect of pH on solutions of mefenamic acid (10^{-4} M) at various ionic equilibria ($\mu = 0.100$, 0.075, 0.040 and 0.025 M) in methanol-water (50:50, v/v). Constant ionic strength was maintained by ... sodium perchlorate. Changes in pH were achieved by adding perchloric acid and sodium hydroxide. ... after pH measurements had been made the spectra were recorded. ... To determine the thermodynamic constant, measurements were carried out at various ionic strengths; the values for the pK_a were extrapolated to ... ionic strength of zero. ... the value of pK_a = 5.54 ± 0.06 was obtained at 20 °C. After correction for the ... methanol, the pK_a for aqueous solutions was 4.55 ± 0.06."
773	Mefenamic acid	7.0	U	+H	Potentiometric	50% EtOH t unspecified I unspecified	Jahn U and Wagner-Jauregg T, Wirkungsvergleich saurer Antiphlogistika im Bradykinin-, UV-Erythem- und Rattenpfotenödem-Test, *Arzneim.-Forsch.*, **24**, 494–499 (1974).
		6.2	U	+H		80% Me cellosolve	NB: Literature pK_a values identified with the pH of half-neutralization.
774	Mefenamic acid	4.33	U	−H	comp	H_2O $t = 25 \pm 0.1$ $I = 0.11$	Terada H, Muraoka S and Fujita T, Structure-activity relationships of fenamic acids, *J. Med. Chem.* **17**, 330–334 (1974). NB: This result was computed from a structure-reactivity relationship.
775	Melphalan ($C_{13}H_{18}Cl_2N_2O_2$) HOOC H_2N CH_2CH_2Cl CH_2CH_2Cl	1.42 2.75 9.17	U U U	+H −H +H	kinetic	H_2O $t = 37.0$ $I = 0.5$	Stout SA and Riley CM, The hydrolysis of L-phenylalanine mustard (melphalan), *Int. J. Pharm.*, **24**, 193–208 (1985). NB: Estimated from hydrolytic rate constants in the pH range 1–13. Taken from: Stout SA, *Chemical stability and in vitro cytotoxicity of melphalan*, Ph.D. Thesis, University of Florida, 1987, 36–42.

(continued)

No.	Compound Name	pK_a value(s)	Data quality	Ionization type	Method	Conditions	Comments and Reference(s)
776	Mepazine (pecazine) ($C_{19}H_{22}N_2S$) 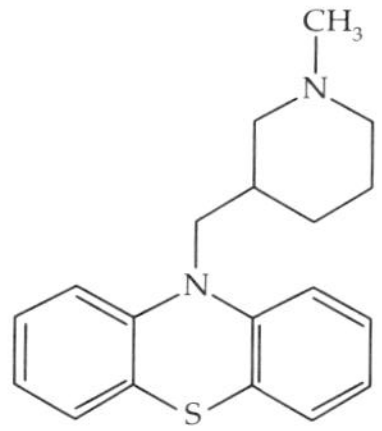	9.25	U	+H	Potentiometric	H_2O (extrap) $t = 20$ I undefined N_2 atmosphere	Sorby DL, Plein EM and Benmaman D, Adsorption of phenothiazine derivatives by solid adsorbents, *J. Pharm. Sci.*, **55**, 785–794 (1966). NB: There should be a weaker value as well for the thiazine ring nitrogen. See Chlorpromazine for details.
777	Meperidine (pethidine) ($C_{15}H_{21}NO_2$) 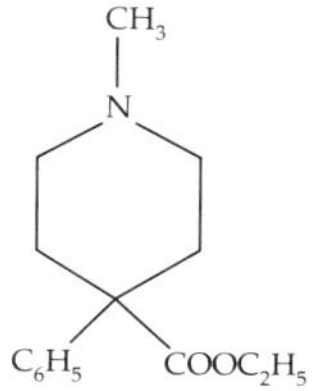	8.68	U	+H	Potentiometric	H_2O $t = 20$ $I < 0.01$	Kaufman JJ, Semo NM and Koski WS. Microelectrometric titration measurement of the pK_a's and partition and drug distribution coefficients of narcotics and narcotic antagonists and their pH and temperature dependence, *J. Med. Chem.*, **18**, 647–655 (1975). NB: See Codeine for further details. Cited in: Lombardo F, Obach RS, Shalaeva MY and Gao F, Prediction of human volume of distribution values for neutral and basic drugs. 2. Extended data set and leave-class-out statistics, *J. Med. Chem.*, **47**, 1242–1250 (2004) (ref. 285).
		8.50	U	+H		H_2O $t = 37$ $I < 0.01$	
778	Mephentermine ($C_{11}H_{17}N$) 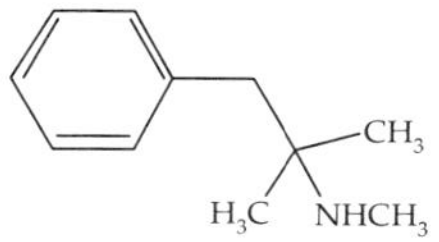	10.37	A	+H	Potentiometric	H_2O $t = 25.0 \pm 0.5$ $I = 0.01$	Warren RJ, Begosh PP and Zarembo JE, Identification of amphetamines and related sympathomimetic amines, *J. Assoc. Off. Anal. Chem.*, **54**, 1179–1191 (1971). NB: See Amphetamine for further details.
779	Mephentermine	10.38	U	+H	Potentiometric	H_2O (extrap) $t = 24 \pm 1$ $I \sim 0.002$	Chatten LG and Harris LE, Relationship between pK_b(H_2O) of organic compounds and $E_{1/2}$ values in several nonaqueous solvents, *Anal. Chem.*, **34**, 1495–1501 (1962). NB: See Chlorpromazine for details.

780	Mepivacaine ($C_{15}H_{22}N_2O$)	7.78	U	+H	soly	H_2O $t = 23 \pm 1$ $I > 0.1$	Friberger P and Aberg G, Some physicochemical properties of the racemates and the optically active isomers of two local anaesthetic compounds, *Acta Pharm. Suec.*, **8**, 361–364 (1971).
781	Mepivacaine	7.73	U	+H	Potentiometric	H_2O $t = 25.0 \pm 0.2$ $I = 0.01$ (NaCl)	Johansson P-A, Liquid-liquid distribution of lidocaine and some structurally related anti-arrythmic drugs and local anaesthetics, *Acta Pharm. Suec.*, **19**, 137–142 (1982). NB: See Lidocaine for details.
782	6-Mercaptopurine ($C_5H_4N_4S$)	7.77 ± 0.02 10.88	A U	−H −H	Potentiometric	H_2O $t = 20$	Albert A and Brown DJ, Purine studies. I. Stability to acid and alkali. Solubility. Ionization, *J. Chem. Soc.*, 2060–2071 (1954). Cited in Newton DW, Ratanamaueichatara S and Murray WJ, Dissociation, solubility and lipophilicity of azathioprine, *Int. J. Pharm.*, **11**, 209–213 (1982). NB: See also Brown DJ and Mason SF, Purine Studies. Part III. The structure of the monohydroxymonomercaptopurines: Some thiazolo[5:4-D] pyrimidines, *J. Chem. Soc.*, 682–689 (1957).
783	6-Mercaptopurine	7.7 ± 0.1 11.17 ± 0.06	U U	−H −H	Spectro, Potentiometric	H_2O $t = 23$ $c = 0.01$ N_2 atmosphere	Fox JJ, Wempen I, Hampton A and Doerr I, *JACS*, **80**, 1669–1675 (1958). Cited in Benezra SA and Foss PRB, 6-Mercaptopurine, *APDS*, **7**, 343–357 (1978). NB: This paper also reported the pK_a values several other purines and purine nucleosides. The pK_{a2} value was determined potentiometrically by titration of carbonate-free NaOH into a nitrogen stirred solution of the purine made in pre-boiled water. Spectrophotometric studies followed previously described procedures: Fox JJ, Cavalieri LF and Chang N, *JACS*, **75**, 4315–4317 (1953); Fox JJ, Chang N, Davoll J and Brown GB, *ib.* **78**, 2117–2122 (1956); Shugar D and Fox JJ, *Bull. Soc. Chim. Belges*, **61**, 293–309 (1952); Fox JJ and Shugar D, *Biochim. Biophys. Acta*, **9**, 369–384 (1952).

(*continued*)

No.	Compound Name	pK_a value(s)	Data quality	Ionization type	Method	Conditions	Comments and Reference(s)
784	Metaclazepam ($C_{18}H_{18}BrClN_2O$) 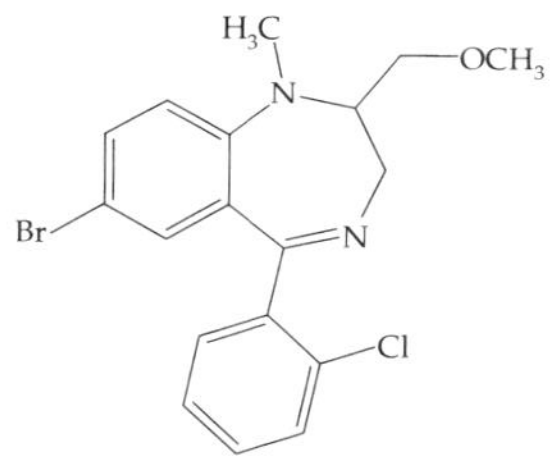	5.50	U	+H	Spectro	H_2O	Fernandez-Arciniega MA and Hernandez L, Analytical properties of metaclazepam a new 1,4-benzodiazepine, *Farmaco Edn. Prat.*, **40**, 81–86 (1985). CA 102:191007. "The physicochemical properties of metaclazepam hydrochloride (Talis) were studied using UV, IR, and NMR spectrometry and TLC. The pK_a value was 5.50."
785	Metaproterenol (orciprenaline) ($C_{11}H_{17}NO_3$) 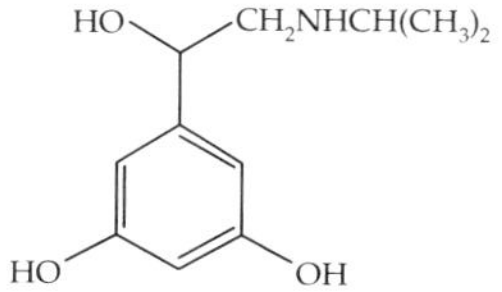	8.84 10.28	U U	−H, +H +H, −H	Spectro	H_2O $I = 0.02$	Quintero B, López J and Thomas J, Determination of dissociation constants of 5-[1-hydroxy-2-[(1-methyl-ethyl)amino]ethyl]-1-3-benzenediol, *J. Pharm. Sci.*, **74**(1), 72–75 (1985). NB: Microconstants: pK_a = 8.90; pK_b = 10.0; pK_c = 10.25; pK_d = 9.15; K_z = 12.5 (K_z = equilibrium constant between the zwitterion and the neutral form).
786	Metaproterenolol (orciprenaline)	8.70 9.92	U U	+H, −H −H, +H	Spectro, Potentiometric	H_2O $t = 25.0 \pm 0.05$ $I = 0.10$ (KCl) N_2 atmosphere	Ijzerman AP, BultsmaT, Timmerman H and Zaagsma J, The ionization of β-adrenoceptor agonists: a method for unravelling ionization schemes, *J. Pharm. Pharmacol.*, **36**(1), 11–15 (1984). NB: pH calibration standards: 7.00 ± 0.02, 9.94. Microscopic: 8.67 and 9.92; macroscopic: 8.70 and 9.92. See Isoprenalin.
787	Metformin ($C_4H_{11}N_5$) 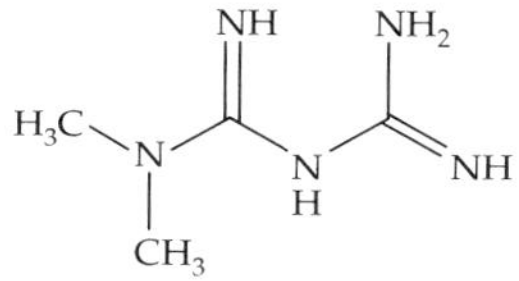	2.8 11.5	U U	+H +H		H_2O $t = 32$	Moffat AC, *Clarke's Isolation and Identification of Drugs*, 2nd ed., Pharmaceutical Press, London 740–741 (1986). Cited in Bretnall AE and Clarke GS, Metformin Hydrochloride, *APDS*, **25**, 243–293 (1997).

788	Methadone analogues		U	+H	Potentiometric	H_2O $t = 20.0$ $I = 0.013$	Beckett AH, Analgesics and their antagonists. I., *J. Pharm. Pharmacol.*, **8**, 848–859 (1956). NB: Further analogues were also reported:
	R = piperidino						
	R′ = H	8.96					
	R′ = $COCH_2CH_3$	8.86					
	R′ = CN	8.07					
	R = morpholino						
	R′ = H	7.25					
	R′ = $COCH_2CH_3$	7.00					
	R′ = CN	6.09					
	R = dimethylamino						
	R′ = H	9.23					
	R′ = $COCH_2CH_3$	9.40					
	R′ = CN	8.31					

	pK_a		pK_a
R = piperidino		R = piperidino	
R′ = H	8.80	R′ = CN	7.54
R′ = $COCH_2CH_3$	8.58	R = dimethylamino	
R′ = CN	7.73	R′ = CN	7.90
R = morpholino			
R′ = H	6.90		
R′ = $COCH_2CH_3$	6.73		
R′ = CN	6.10		
R = dimethylamino			
R′ = H	9.48		
R′ = $COCH_2CH_3$	8.99		
R′ = CN	8.31		

789	Methamphetamine (methylamphetamine) ($C_{10}H_{15}N$)	9.87 ± 0.02	U	+H	Potentiometric	H_2O $t = 25.0 \pm 0.2$ $I \leq 0.001$	Leffler EB, Spencer HM and Burger A and Dissociation constants of adrenergic amines *JACS*, **73**, 2611–2613 (1951). NB: See Amphetamine for details. From $pK_b = 4.13$. Cited as $pK_a = 10.11$ in Vree TB, Muskens ATJM and van Rossum JM, Some physicochemical properties of amphetamine and related drugs, *J. Pharm. Pharmacol.*, **21**, 774–775 (1969). The source of this value is not known.
790	Methamphetamine	9.99	A	+H	Potentiometric	H_2O $t = 25.0 \pm 0.5$ $I = 0.01$	Warren RJ, Begosh PP and Zarembo JE, Identification of amphetamines and related sympathomimetic amines, *J. Assoc. Off. Anal. Chem.*, **54**, 1179–1191 (1971). NB: See Amphetamine for further details.

(continued)

Appendix A *(continued)*

No.	Compound Name	pK_a value(s)	Data quality	Ionization type	Method	Conditions	Comments and Reference(s)
791	Methamphetamine, 4-hydroxy ($C_{10}H_{15}NO$) 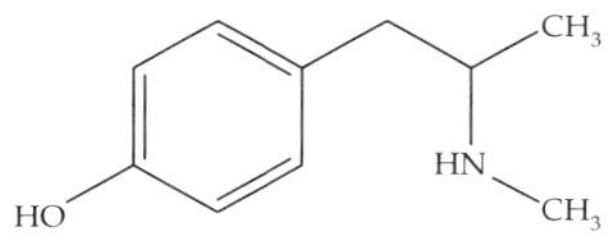	9.62 10.75	A U	+H −H	Potentiometric Spectro	H_2O $t = 25.0 \pm 0.5$ $I = 0.01$	Warren RJ, Begosh PP and Zarembo JE, Identification of amphetamines and related sympathomimetic amines, *J. Assoc. Off. Anal. Chem.*, **54**, 1179–1191 (1971). NB: See Amphetamine and Amphetamine, 4-hydroxy, for further details. Alternative calculation method gave p$K_{a1} = 9.49$ and p$K_{a2} = 10.78$.
792	Methaphenilene ($C_{15}H_{20}N_2S$) 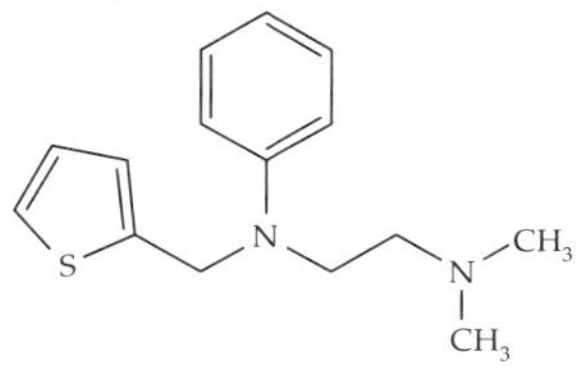	3.02 ± 0.05 8.24 ± 0.11	U U	+H +H	Potentiometric	H_2O t unspecified $I = 0.30$ (NaCl)	Testa B and Murset-Rossetti L, The partition coefficient of protonated histamines. *Helv. Chim. Acta*, **61**, 2530–2537 (1978). NB: See Cycliramine for details.
793	Methaqualone ($C_{16}H_{14}N_2O$) 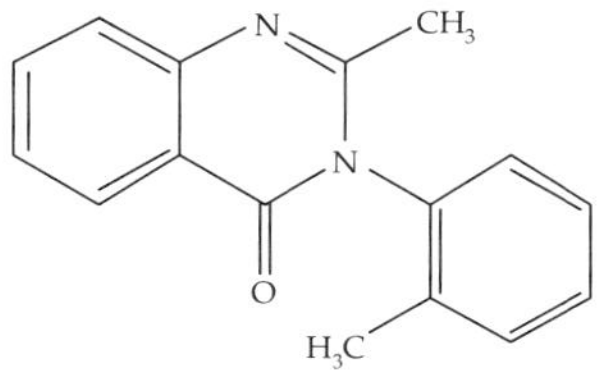	2.54	U	+H	Spectro	H_2O t unspecified $I = 0.1$	Zalipsky JJ, Patel DM, Darnowski RJ and Reavey-Cantwell NH, pK_a Determination of methaqualone, *J. Pharm. Sci.*, **65**, 460–461 (1976). Cited in Patel DP, Visalli AJ, Zalipsky JJ and Reavey-Cantwell NH, Methaqualone, *APDS*, **4**, 245–267 (1975). "The pK_a value for methaqualone (I) was determined using a UV . . . method at 2 . . . analytical wavelengths, 286 and 315 nm. A pK_a value of 2.54 at 316 nm was found. This value differs from those determined using potentiometric titration of the drug in varying compositions of an acetone-water solvent system."
		3.56	U	+H	Potentiometric	acetone-water	Chatten LG, Moskalyk RE, Locock RA and Schaefer FJ, Non-aqueous titration of methaqualone and its dosage forms, *J. Pharm. Sci.*, **63**, 1294–1296 (1974). NB: From the pH of half-neutralization.

794	Methicillin ($C_{17}H_{20}N_2O_6S$) 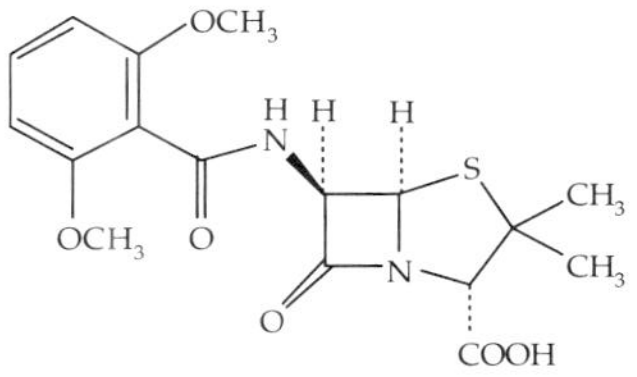	2.77 ± 0.04	U	−H	Potentiometric	H_2O $t = 25$ $c = 0.0097$	Rapson HDC and Bird AE, Ionization constants of some penicillins and of their alkaline and penicillinase hydrolysis products, *J. Pharm. Pharmacol.*, Suppl. **15**, 222–231T (1963).
		2.77 ± 0.03	U	−H	Potentiometric	H_2O $t = 37$ $I = 0.15$	Tsuji A, Kubo O, Miyamoto E and Yamana T, Physicochemical properties of β-lactam antibiotics: oil-water distribution, *J. Pharm. Sci.*, **66**, 1675–1679 (1977).
795	Methotrexate ($C_{20}H_{22}N_8O_5$) 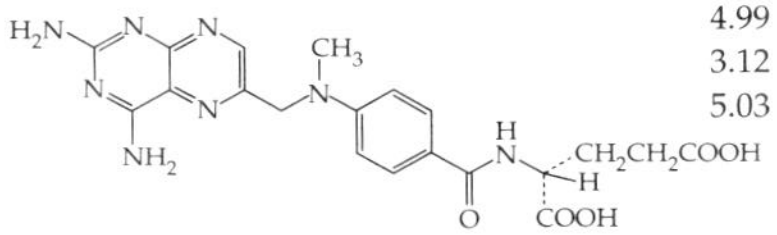	3.04 4.99 3.12 5.03	U U U U	−H +H −H +H	CE/pH (+ve ion mode)	H_2O $t = 25$ $I = 0.025$	Wan H, Holmen AG, Wang Y, Lindberg W, Englund M, Nagard MB and Thompson RA, High-throughput screening of pK_a values of pharmaceuticals by pressure-assisted capillary electrophoresis and mass spectrometry, *Rapid Commun. Mass Spectrom.*, **17**, 2639–2648 (2003). NB: Reported predicted values (ACD Labs) of 3.54 and 5.09.
796	Methotrexate	5.71	U	+H	Spectro	H_2O $t = 25$ $I = 0.10$ (NaCl)	Poe M, Acidic dissociation constants of folic acid, dihydrofolic acid and methotrexate, *J. Biol. Chem.*, **252**(11), 3724–3728 (1977). NB: Spectrophotometric measurements were made (under nitrogen) as a function of either pH or the acidity function H_o. Measured values correspond to the structure as follows: <−1.5 (N5); 0.5 (N10); 3.36 (α-carboxyl); 4.70 (γ-carboxyl); 5.71 (N1).
		3.36 4.70	U U	−H −H	Potentiometric	N_2 atmosphere	
		0.50 ± 0.05 <−1.5	U U	+H +H	Spectro	$I = 0.1$ to 4	
797	Methotrexate	3.31 4.00 5.39	U U U	−H −H +H	Potentiometric	H_2O $t = 25$	Bergström CAS, Strafford M, Lazorova L, Avdeef A, Luthman K and Artursson P, Absorption classification of oral drugs based on molecular surface properties, *J. Med. Chem.*, **46**(4), 558–570 (2003). NB: From extrapolation of aqueous-methanol mixtures to 0% methanol.
798	Methotrexate alkyl esters	5.71	U	+H	Spectro	H_2O $t = 25$ $I = 0.10$ (NaCl) N_2 atmosphere	Fort JJ and Mitra AK, Solubility and stability characteristics of a series of methotrexate dialkyl esters, *Int. J. Pharm.*, **59**, 271–279 (1990); Fort JJ, Mitra AK, Physicochemical properties and chromatographic behavior of a homologous series of methotrexate-α,γ-dialkyl esters, *Int. J. Pharm.*, **36**, 7–16 (1987). NB: Made the reasonable assumption that the esters would have the same pK_a value for the pteridine ring as in unchanged methotrexate. Cited the value from Poe (no. 796).

(continued)

No.	Compound Name	pK_a value(s)	Data quality	Ionization type	Method	Conditions	Comments and Reference(s)
799	Methotrimeprazine (Levomepromazine) ($C_{19}H_{24}N_2OS$) 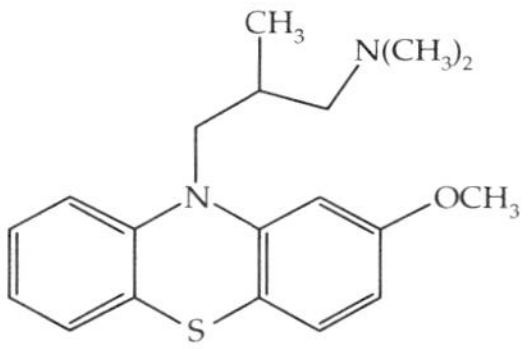	9.19	U	+H		H_2O	Mannhold R, Dross KP and Reffer RF, Drug lipophilicity in QSAR practice: I. A comparison of experimental with calculative approaches. *Quant. Struct. Act. Relat.*, **9**, 21–28 (1990). NB: Cited in Lombardo F, Obach RS, Shalaeva MY and Gao F, Prediction of human volume of distribution values for neutral and basic drugs. 2. Extended data set and leave-class-out statistics, *J. Med. Chem.*, **47**, 1242–1250 (2004); ref. 277.
800	Methotrimeprazine (levomepromazine)	9.15	U	+H	Potentiometric	H_2O (extrap) $t = 24 \pm 1$ $I \sim 0.002$	Chatten LG and Harris LE, Relationship between $pK_b(H_2O)$ of organic compounds and $E_{1/2}$ values in several nonaqueous solvents, *Anal. Chem.*, **34**, 1495–1501 (1962). NB: See Chlorpromazine for details.
801	Methoxamine ($C_{11}H_{17}NO_3$) 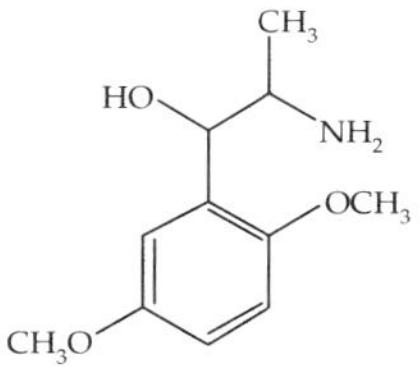	9.32	A	+H	Potentiometric	H_2O $t = 25.0 \pm 0.5$ $I = 0.01$	Warren RJ, Begosh PP and Zarembo JE, Identification of amphetamines and related sympathomimetic amines, *J. Assoc. Off. Anal. Chem.*, **54**, 1179–1191 (1971). NB: See Amphetamine for further details.
802	Methoxamine	9.18	U	+H	Potentiometric	H_2O (extrap) $t = 24 \pm 1$ $I \sim 0.002$	Chatten LG and Harris LE, Relationship between $pK_b(H_2O)$ of organic compounds and $E_{1/2}$ values in several nonaqueous solvents. *Anal. Chem.*, **34**, 1495–1501 (1962). NB: See Chlorpromazine for details.
803	4-Methoxyamphetamine ($C_{10}H_{15}NO$) 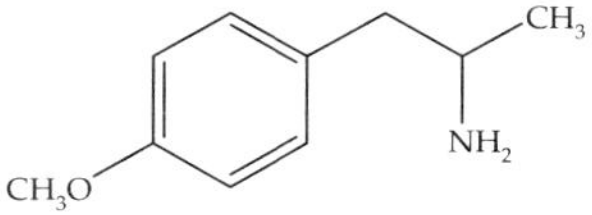	9.53 ± 0.05	U	+H	Potentiometric	H_2O $t = 25.0 \pm 0.2$ $I \leq 0.001$	Leffler EB, Spencer HM and Burger A, Dissociation constants of adrenergic amines, *JACS*, **73**, 2611–2613 (1951). NB: See Amphetamine for details. From $pK_b = 4.47$.

No.	Compound	pKa	Class	Type	Method	Conditions	Reference / Notes
804	Methoxyphenamine ($C_{11}H_{17}NO$) 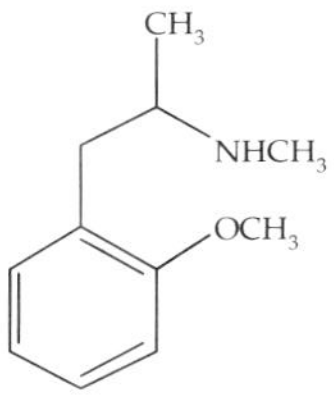	10.45	A	+H	Potentiometric	H_2O $t = 25.0 \pm 0.5$ $I = 0.01$	Warren RJ, Begosh PP and Zarembo JE, Identification of amphetamines and related sympathomimetic amines, *J. Assoc. Off. Anal. Chem.*, **54**, 1179–1191 (1971). NB: See Amphetamine for further details.
805	Methyclomethiazide ($C_9H_{11}Cl_2N_3O_4S_2$) 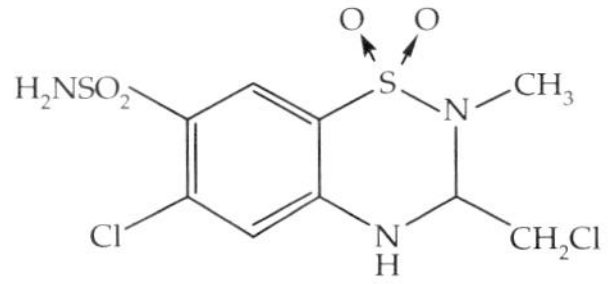	9.5	U	−H	Potentiometric	acetone/H_2O	Henning VG, Moskalyk R.E, Chatten LG and Chan SF, Semiaqueous potentiometric determination of apparent pK_{a1} values for benzothiadiazines and detection of decomposition during solubility variation with pH studies, *J. Pharm. Sci.*, **70**(3), 317–319 (1981). NB: See Flumethiazide for details.
806	Methyclothiazide	9.4	U	−H	Potentiometric	H_2O (extrap)	Raihle JA, Methyclothiazide, *APDS*, **5**, 307–326 (1976). NB: Extrapolated from acetone-water mixtures. No range of values given for the compositions. No reference, so presumably work done by this author.
807	DL-*N*-Methyladrenaline ($C_{10}H_{15}NO_3$) 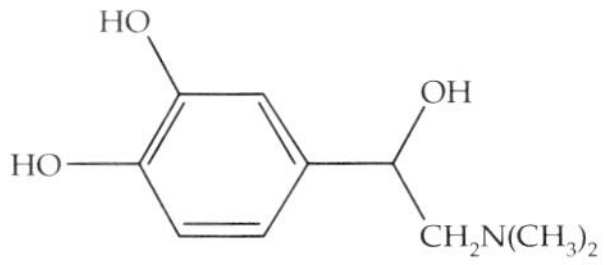	8.48 ± 0.01 9.89 ± 0.06	A U	+H, −H −H, +H	Potentiometric Spectro	H_2O $t = 37$ $I = 1.0$	Schüsler-van Hees, MTIW, *Reactivity of Catechol(amine)s*, Ph.D. thesis, Ctr for Biopharm. Sci., Leiden University, 1983, 1–93. NB: Microscopic: 8.60 ± 0.01; 9.09 ± 0.04; 9.77 ± 0.08; 9.28 ± 0.03. Schüsler-van Hees, MTIW, Beijersbergen van Henegouwen GMJ and Driever MFJ, Ionization constants of catechols and catecholamines, *Pharm. Weekblad*, Sci. Edn. **5**, 102–108 (1983).
808	2-Methylamino-5-chlorobenzophenone ($C_{14}H_{12}ClNO$) 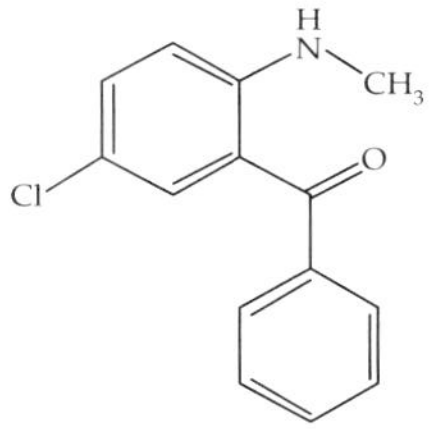	1.45 ± 0.04	U	+H	Spectro ($\lambda = 415$ nm)	7% EtOH in H_2O $t = 25$ 0.02 to 0.40 M in HCl	Newton DW, pK_a of 2-methylamino-5-chlorobenzophenone, a diazepam hydrolysis product, *J. Pharm. Sci.*, **68**, 937–938 (1979). "Estimation of the dissociation constant (pK_a) of 2-methylamino-5-chlorobenzophenone, using spectrophotometric methods, is presented. The pK_a was calculated as 1.45 ± 0.04. Comparison of this value with other aniline derivatives is also presented." NB: No activity corrections were performed.

(continued)

No.	Compound Name	pK_a value(s)	Data quality	Ionization type	Method	Conditions	Comments and Reference(s)
809	α-Methyl benzyl penilloate ($C_{16}H_{22}N_2O_3S$) 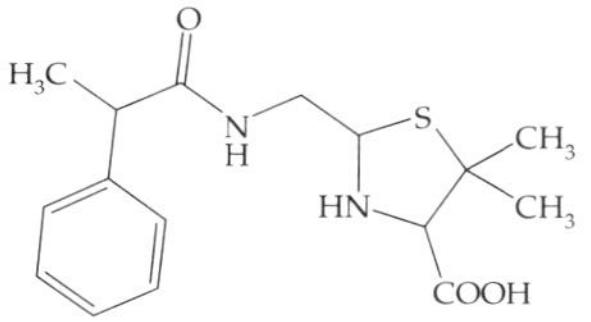	3.65	U	+H		H_2O $t = 23$	Woodward RB, Neuberger A, Trenner NR, *in* Clarke H, Johnson JR and Robinson Sir R. (eds.), *The Chemistry of Penicillin*, Princeton University Press, Princeton, NJ, 415–422, 1949.
810	L-α-Methyldopa ($C_{10}H_{13}NO_4$) 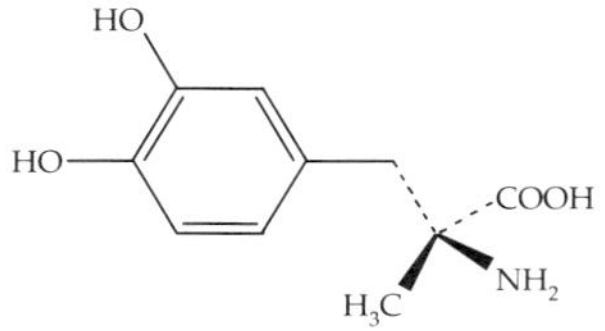	2.218 9.157 10.629 12.00	A U U U	−H −H, +H +H, −H −H	Potentiometric Spectro	H_2O $t = 25$ $I = 0.00$ (KCl)	Halmekoski J and Lukkari S, Ionization constants of pharmaceuticals. II. α-Methyl-3,4-dihydroxyphenylalanine (Aldomet), *Farm. Aikak.*, **74**, 173–180 (1965). NB: The effect of ionic strength was corrected with the Debye-Huckel limiting law. The pK_a value for the more acidic phenolic group was also determined spectrophotometrically.
811	L-α-Methyldopa	GLpK_a: 2.31 ± 0.02 8.89 ± 0.02 10.06 ± 0.06 13.77 ± 0.15 A&S: 2.29 ± 0.02	 A A U U A	 −H −H, +H +H, −H −H −H	Spectro	H_2O $t = 25$ $I = 0.15$ (KCl) Ar atmosphere	Tam KY and Takacs-Novac K, Multi-wavelength spectrophotometric determination of acid dissociation constants, *Anal. Chim. Acta*, **434**, 157–167 (2001). NB: See Clioquinol for details.
812	L-α-Methyldopa	? 8.87 ± 0.02 9.92 ± 0.07 ?	 A U 	−H +H, −H −H, +H −H	Potentiometric Spectro	H_2O $t = 37$ $I = 1.0$	Schüsler-van Hees, MTIW, *Reactivity of Catechol(amine)s*, Ph.D. thesis, Ctr for Biopharm. Sci., Leiden University, 1983, 1–93. NB: Microscopic: 9.03 ± 0.01; 9.38 ± 0.07; 9.75 ± 0.10; 9.40 ± 0.02. Schüsler-van Hees, MTIW, Beijersbergen van Henegouwen GMJ and Driever MFJ, Ionization constants of catechols and catecholamines, *Pharm. Weekblad, Sci. Edn.*, **5**, 102–108 (1983).

No.	Compound	pKa	Quality	Type	Method	Conditions	Reference
813	D-α-Methyldopamine ($C_9H_{13}NO_2$) 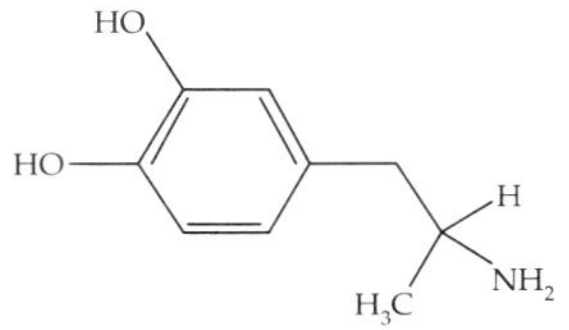	8.68 ± 0.01 10.17 ± 0.07	A U	+H, −H −H, +H	Potentiometric Spectro	H_2O $t = 37$ $I = 1.0$	Schüsler-van Hees, MTIW, *Reactivity of Catechol(amine)s*, Ph.D. thesis, Ctr for Biopharm. Sci., Leiden University, 1983, 1–93. NB: Microscopic: 8.85 ± 0.01; 9.18 ± 0.04; 10.00 ± 0.09; 9.67 ± 0.04. Schüsler-van Hees, MTIW, Beijersbergen van Henegouwen GMJ, Driever MFJ, Ionization constants of catechols and catecholamines, *Pharm. Weekblad Sci. Edn.* **5**, 102–108 (1983).
814	Methylenedioxyamphet-amine (MDA) ($C_{12}H_{19}NO_2$) 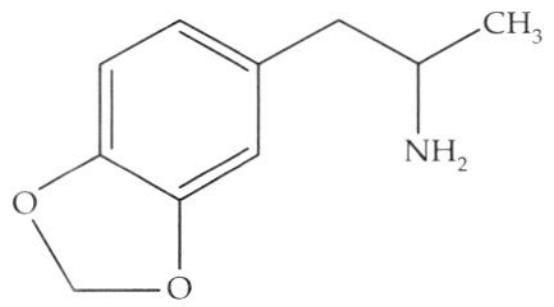	9.67 ± 0.02	U	+H	Potentiometric	H_2O $t = 25.0 \pm 0.2$ $I \leq 0.001$	Leffler EB, Spencer HM and Burger A, Dissociation constants of adrenergic amines *JACS*, **73**, 2611–2613 (1951). NB: See Amphetamine for details. From $pK_b = 4.33$.
815	*N*-Methylephedrine ($C_{11}H_{17}NO$) 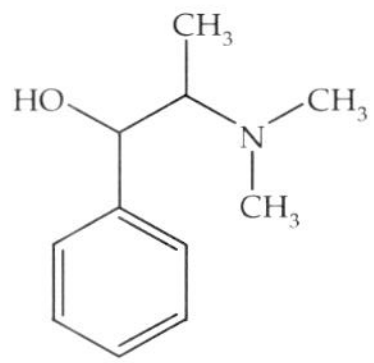	9.257	U	+H	Potentiometric	H_2O $t = 25$ $I = 0.0$ (by extra-polation)	Halmekoski J and Hannikainen H, Ionization constants of pharmaceuticals. Part 4. *Acta Pharm. Suec.*, **3**, 145–152 (1966). "For calibrating the pH meter, the following buffer solutions were used . . . for pH 4.008, a 0.05 M potassium biphthalate solution and for pH 9.180 a 0.01 M borax solution . . . The potentiometric measurements were carried out with a Radiometer pH meter PHM4c, equipped with a Radiometer Type G 202 C NH glass electrode and a Radiometer K 401 saturated calomel electrode. The titrations were made in a closed glass vessel of about 100 ml. During titration, the titration vessel was kept in a waterbath at 25 °C, and purified nitrogen was led trough the vessel when titrated. The weighed amount of the amine in the salt form was dissolved in 50 ml of distilled water in the titration vessel. 0.1-M sodium hydroxide was used as a titrant. When titrated at higher ionic strengths, potassium chloride was also added to the titrant to give a concentration of 0.5, 1.0 and 2.0 moles/litre. . . . The titrant was added to the titration vessel by means of a 10 ml piston burette . . . having a dial division of 0.01 ml. The titration solution was mixed by a magnetic stirrer. The apparent pK_a values were determined at zero, 0.5, 1.0 and 2.0 M concentrations of potassium chloride. . . ."
	Vasculat (DL-1-(p-hydroxyphenyl)-2-butylaminoethanol) ($C_{12}H_{19}NO_2$)	9.034 10.16	U U	+H −H			
	Effortil (DL-1-(m-hydroxyphenyl)-2-ethylaminoethanol) ($C_{10}H_{15}NO_2$)	8.983 10.16	U U	+H −H			

(*continued*)

No.	Compound Name	pK_a value(s)	Data quality	Ionization type	Method	Conditions	Comments and Reference(s)
816	Methylergonovine ($C_{20}H_{25}N_3O_2$) 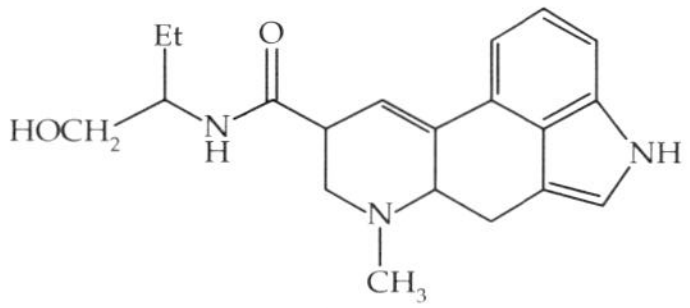	6.65 ± 0.03	A	+H	Potentiometric	H_2O $t = 24$ $I < 0.01$	Maulding HV, Zoglio MA, Physical chemistry of ergot alkaloids and derivatives. I. Ionization constants of several medicinally active bases, *J. Pharm. Sci.*, **59**, 700–701 (1970). Cited in Perrin Bases suppl. No. 7482. NB: See Methysergide for details.
817	*N*-Methylglucamine ($C_7H_{17}NO_5$) 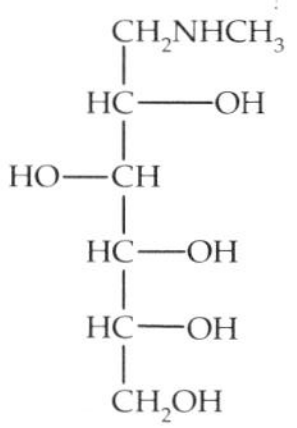	9.39	A	+H	Potentiometric	H_2O $t = 30.0$ $I = 0.05$ N_2 atmosphere	Juvet RS, The *N*-methylglucamine complexes. I. The lead *N*-methylglucamine system, *JACS*, **81**, 1796–1801 (1959). Cited in Perrin Bases 3409 ref. J23. NB: The pK_a value is calculated from pK_b = 4.35; pK_w at 30 °C = 13.833.
818	Methylhexaneamine ($C_7H_{17}N$) 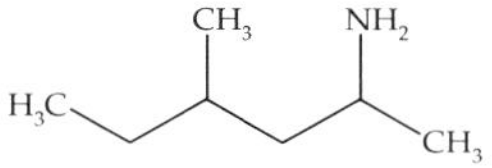	10.54	U	+H	Potentiometric	H_2O (extrap) $t = 24 \pm 1$ $I \sim 0.002$	Chatten LG, Harris LE, Relationship between pK_b(H_2O) of organic compounds and $E_{1/2}$ values in several nonaqueous solvents, *Anal. Chem.*, **34**, 1495–1501 (1962). NB: See Chlorpromazine for details.
819	1-Methyl-1H-imidazole ($C_4H_6N_2$) 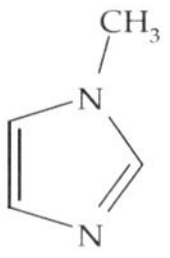	7.24	U	+H	Potentiometric	H_2O $t = 25$	Peeters J, Determination of ionization constants in mixed aqueous solvents of varying composition by a single titration, *J. Pharm. Sci.*, **67**, 127–129 (1978). NB: See Cinnarizine for details. For 1-methyl-1H-imidazole, multiple titrations, each at a different percentage of methanol, gave pK_a = 7.16 on extrapolation to 0% MeOH.

No.	Compound	pKa	Data quality	Ionization type	Method	Conditions	Reference
820	Methylparaben ($C_8H_8O_3$)	8.15 (0.05)	U	−H	Spectro ($\lambda = 297$ nm)	H_2O $t = 20.0$	Wahbe AM, El-Yazbi FA, Barary MH and Sabri SM, Application of orthogonal functions to spectrophotometric analysis. Determination of dissociation constants, *Int. J. Pharm.*, **92**(1), 15–22 (1993). NB: See Acetaminophen for further details. An alternative graphical method gave pK_a = 8.15.
821	*N*-Methylphenethylamine ($C_9H_{13}N$)	10.31	U	+H	Potentiometric	H_2O *t* undefined *I* undefined	Tuckerman MM, Mayer JR and Nachod FC, Anomalous pK_a values of some substituted phenylethylamines, *JACS*, **81**, 92–94 (1959). NB: Method as described by Parke and Davis, 1945.
822	Methylphenidate ($C_{14}H_{19}NO_2$)	8.9	U	+H	Potentiometric	H_2O $t = 25$ $I = 0.05$ (KCl) N_2 atmosphere	Smith J, Piskorik H, Ciba-Geigy, personal communication, using the method of Benet LZ, Goyan JE, *J. Pharm. Sci.*, **54**, 1179–1182 (1965). NB: Calibration pH values, 4.010, 9.180. Titration with carbonate-free KOH used a 1 ml microburette.
		9.0	U	+H	Potentiometric	H_2O	Padmanabham, GR., Methylphenidate hydrochloride, *APDS*, **10**, 473–497 (1981).
823	Methylphenidate	8.8	U	+H			Siegel S, Lachman L and Malspeis L, A kinetic study of the hydrolysis of methyl-DL-α-phenyl-2-piperidyl acetate, *J. Am. Pharm. Assoc., Sci. Edn.*, **48**, 431–439 (1959).
824	Methylprednisolone-21-phosphate ($C_{22}H_{31}O_8P$)	2.55 6.04	U U	−H −H	Potentiometric	H_2O $t = 70$ $I = 0.15$ (NaCl) $c = 1.33$ mg/mL	Flynn GL and Lamb DJ, Factors influencing solvolysis of corticosteroid 21-phosphate esters, *J. Pharm. Sci.*, **59**, 1433–1438 (1970). NB: Apparent pK_a values reported from the pH of half-neutralization.

(continued)

No.	Compound Name	pK_a value(s)	Data quality	Ionization type	Method	Conditions	Comments and Reference(s)
825	Methysergide ($C_{21}H_{27}N_3O_2$) 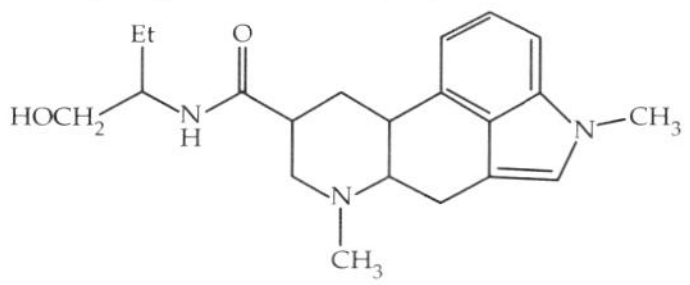	6.62 ± 0.02	A	+H	Potentiometric	H_2O $t = 24$ $I < 0.01$	Maulding HV and Zoglio MA, Physical chemistry of ergot alkaloids and derivatives. I. Ionization constants of several medicinally active bases, *J. Pharm. Sci.*, **59**, 700–701 (1970). NB: Aqueous solubility was increased by complexation with 7-β-hydroxypropyltheophylline (7HPT) and the apparent pK_a values extrapolated to [7HPT] = 0. Solutions were titrated with carbonate-free KOH solution. This is the value found in the absence of 7HPT, compared to 6.50 when extrapolated to 0% 7HPT (from 2–30% 7HPT). Measurements under these two conditions on methylergonovine gave a similar discrepancy for the pK_a value. Extrapolated values for all other ergot alkaloids were corrected for this discrepancy.
826	Methysergide	6.62	A	+H	Potentiometric	H_2O $t = 24$	Maulding HV and Zoglio MA, Physical chemistry of ergot alkaloids and derivatives. I. Ionization constants of several medicinally active bases, *J. Pharm. Sci.*, **59**, 700–701 (1970). Cited in Perrin Bases suppl. no. 7483. NB: Used a glass electrode in a cell with liquid junction potentials.
827	Metoclopramide ($C_{14}H_{22}ClN_3O_2$) 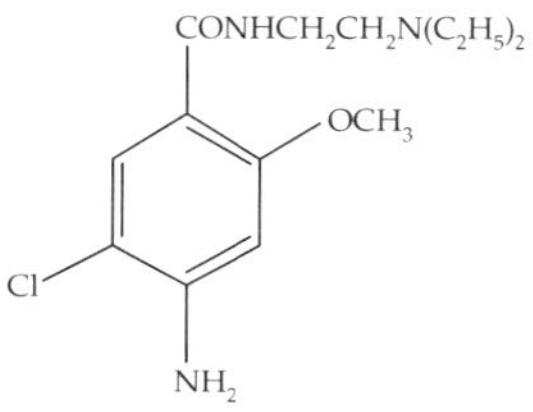	0.42 ± 0.03 9.71 ± 0.02	A A	+H +H	Spectro	H_2O	Hanocq M, Topart J, van Damme M and Molle L, Determination of ionization constants for some substituted *N*-(2-diethylaminoethyl) benzamides, *J. Pharm. Belge*, **28**, 649–662 (1975). Cited in Pitre D, Stradi R, Metoclopramide hydrochloride, *APDS*, **16**, 327–360 (1987). "Metoclopramide . . . shows two ionisation constants: pK_1 = 9.71 and pK_2 = 0.42. The determination was carried out spectrophotometrically in aqueous solution . . . values are a mean of 16 determinations with a standard deviation of 0.03 and 0.02." NB: Spectrophotometric pK_a values were similar to potentiometric values.
828	Metoclopramide	9.51	U	−H	Potentiometric		Mannhold R, Dross KP and Reffer RF, Drug lipophilicity in QSAR practice: I. A comparison of experimental with calculative approaches. *Quant. Struct. Act. Relat.*, **9**, 21–28 (1990). Cited in Lombardo F, Obach RS, Shalaeva MY and Gao F, Prediction of human volume of distribution values for neutral and basic drugs. 2. Extended data set and leave-class-out statistics, *J. Med. Chem.*, **47**, 1242–1250 (2004); ref. 277.

No.	Compound	pKa	Reliability	Type	Method	Conditions	Comments
829	Metoprolol ($C_{15}H_{25}NO_3$)	8.9 ± 0.2	U	+H	Potentiometric	H_2O $t = 25.0$ $I = 0.1$	Luch JR, Metoprolol Tartrate, *APDS*, **12**, 325–354 (1983). 8.9 ± 0.2 (Stahl PH, Ciba-Geigy, Ltd., personal communication).
		9.68 ± 0.02	U				9.68 ± 0.02 (Research Laboratories AB Hässle, personal communication).
		9.5 ± 0.2	U				9.5 ± 0.2 (Jäkel K, Moser P, Ciba-Geigy Ltd., personal communication). NB: It is not clear from the *APDS* source which of the given conditions is applicable in each case and why there are significant discrepancies between them. See no. 830.
830	Metoprolol	9.55 ± 0.00	A	+H	partition	H_2O $t = 25.0$ $I = 0.15$ (KCl)	Krämer SD, Gautier J-C, Saudemon P, Considerations on the potentiometric log P determination, *Pharm. Res.*, **15**(8), 1310–1313 (1998). NB: See Amiodarone for details.
		9.56 ± 0.01	A	+H	Potentiometric	H_2O $t = 25.0$ $I = 0.15$ (KCl)	Sirius Technical Application Notes, 1995, vol. **2**, pp. 75–76. Sirius Analytical Instruments, Ltd., Forest Row, East Sussex, RH18 5DW, UK. NB: Concentration of analyte, 0.39–1.05 mM.
831	Metronidazole ($C_6H_9N_3O_3$)	2.62	U	+H	soly (auto-titrator)	H_2O $t = 25 \pm 0.2$ I undefined	Cho MJ, Kurtz RR, Lewis C, Machkovech SM, Houser DJ, Metronidazole phosphate - A water-soluble prodrug for parenteral solutions of metronidazole, *J. Pharm. Sci.*, **71**(4), 410–414 (1982). "…. The aqueous dissociation constant for (metronidazole phosphate) is given. In acidic medium, the compound behaved as a zwitterion, with minimum solubility at pH 2. At pH 7, the solubility was ~50 times that of metronidazole."

pK_a values of substituted imidazoles (a)

Compound	pK_a	Reliability
Imidazole	6.953	R
Imidazole, 1-C_2H_5-	7.300	U
Imidazole, 1-CH_3-5-NO_2-	2.130	A
Imidazole, 1-CH_3-4-NO_2-	−0.530	U
Imidazole, 2-CH_3-	7.851	R
Metronidazole	2.62 (b)	U

(a) Perrin DD, *Dissociation constants of organic bases in aqueous solution*, Butterworth, Washington, DC, nos. 1393, 1405, 1411, 1413, 1414.
(b) reported work.

No.	Compound	pKa	Reliability	Type	Method	Conditions	Comments
832	Metronidazole	2.55	U	+H	Spectro	H_2O t undefined I undefined	Gallo GG, Pasqualucci CR, Radaeill P, Lancini GC, The ionization constants of some imidazoles, *JOC*, **29**, 862–865 (1964). Cited by Schwartz DE, Jeunet F, Comparative pharmacokinetic studies of ornidazole and metronidazole in man, *Chemotherapy*, **22**, 19–29 (1976). NB: The pK_a values for imidazole and an additional 15 substituted imidazoles were also reported.

(continued)

No.	Compound Name	pK_a value(s)	Data quality	Ionization type	Method	Conditions	Comments and Reference(s)
833	Mexiletine ($C_{11}H_{17}NO$) 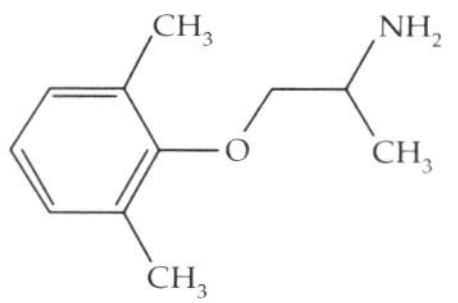	9.00	U	+H	Potentiometric	H_2O $t = 25.0 \pm 0.2$ $I = 0.01$ (NaCl)	Johansson P-A, Liquid-liquid distribution of lidocaine and some structurally related anti-arrythmic drugs and local anaesthetics, *Acta Pharm. Suec.*, **19**, 137–142 (1982). NB: See Lidocaine for details.
834	Mexiletine	9.14 ± 0.01	A	+H	Potentiometric	H_2O $t = 25.0 \pm 0.1$ $I = 0.1$ (NaCl)	Takacs-Novak K, Box KJ and Avdeef A, Potentiometric pK_a determination of water-insoluble compounds: validation study in methanol/water mixtures, *Int. J. Pharm.*, **151**, 235–248 (1997). NB: $pK_a = 9.07 \pm 0.03$ by extrapolation from 16.2–63.9% w/w aqueous MeOH. See Acetaminophen for full details.
835	Mezlocillin ($C_{21}H_{25}N_5O_8S_2$) 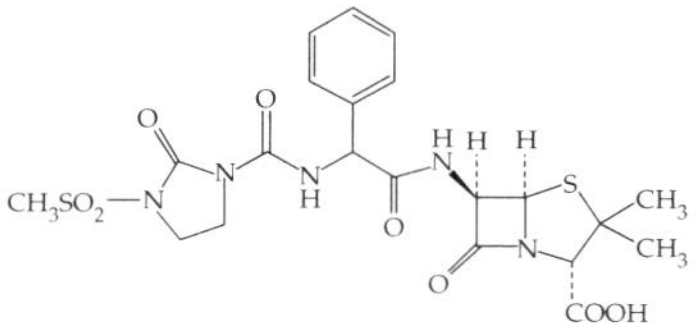	2.7	U	−H			Foye 3rd; see Azatadine from McEvoy.
836	Mianserin ($C_{18}H_{20}N_2$) 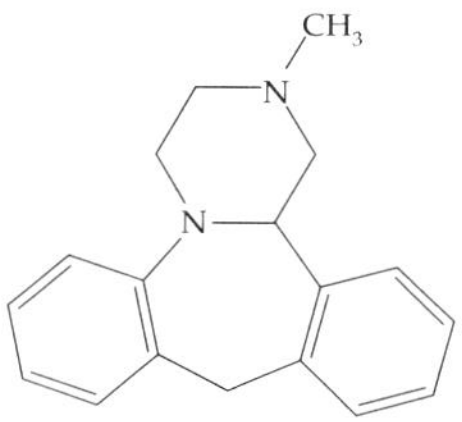	7.5	U	+H	Potentiometric	H_2O $t = 20$	Walther G, Daniel H, Bechtel WD and Brandt K, New tetracyclic guanidine derivatives with H_1-antihistaminic properties: chemistry of epinastine, *Arzneim.-Forsch.*, **40**, 440–446 (1990).
		7.4	U	+H	Potentiometric	47% MeOH in H_2O $t = 25$ $I = 0.15$ (KCl)	Kelder J, Funcke C, De Boer T, Delbressine L, Leysen D and Nickolson V, A comparison of the physicochemical and biological properties of mirtazapine and mianserin, *J. Pharm. Pharmacol.*, **49**, 403–411 (1997). NB: Used a Sirius PCA101 autotitrator.

837	Miconazole ($C_{18}H_{14}Cl_4N_2O$)	6.91	U	+H	Potentiometric	H_2O $t = 25$	Peeters JJ, Determination of ionization constants in mixed aqueous solvents of varying composition by a single titration, *J. Pharm. Sci.*, **67**, 127–129 (1978). NB: See Cinnarizine for details. Reported literature value pK_a = 6.65 after extrapolation of potentiometric data to 100% water.
		6.12	U	+H	Potentiometric	H_2O $t = 37$ $I = 0.15$ (KCl)	Balon K, Riebesehl BU and Muller BW, Drug liposome partitioning as a tool for the prediction of human passive intestinal absorption, *Pharm. Res.*, **16**, 882–888 (1999).
838	Midazolam ($C_{18}H_{13}ClFN_3$)	1.7 ± 0.1 6.15 ± 0.1	U	+H			Walser A, Benjamin LE, Flynn T, Mason C, Schwartz R and Fryer RI, Quinazolines and 1,4-benzodiazepines. Synthesis and reactions of imidazo [1,5-a][1,4]benzodiazepines, *J. Org. Chem.*, **43**, 936–944 (1978). NB: No experimental details reported. Acknowledgement to "Dr V Toome for . . . p*K* determinations. . . ." Cited by Loftsson T, Gudmundsdottir H, Sigurjonsdottir JF, Sigurdsson JH, Sigfusson SD, Masson M and Steffanson E, Cyclodextrin solubilization of benzodiazepines: Formulation of midazolam nasal spray, *Int. J. Pharm.*, **212**, 29–40 (2001).
839	Minoxidil ($C_9H_{15}N_5O$)	4.61	U	+H	Spectro Potentiometric	H_2O t undefined $I = 0.01$	Cowden WB and Jacobsen NW, *Aust. J. Chem.*, **37**, 1195 (1984). Cited in: Gorecki DKJ, Minoxidil, *APDS*, **17**, 185–219 (1988). Brown MA, The Upjohn Company of Canada, personal communication. Confirmed the result of Cowden and Jacobsen.
840	Mirtazepine ($C_{17}H_{19}N_3$)	7.1	U	+H	Potentiometric	47% MeOH in H_2O $t = 25$ $I = 0.15$ (KCl)	Kelder J, Funcke C, De Boer T, Delbressine L, Leysen D and Nickolson V, A comparison of the physicochemical and biological properties of mirtazapine and mianserin, *J. Pharm. Pharmacol.*, **49**, 403–411 (1997). NB: Used Sirius autotitrator. Cited in Lombardo F, Obach RS, Shalaeva MY and Gao F, Prediction of human volume of distribution values for neutral and basic drugs. 2. Extended data set and leave-class-out statistics, *J. Med. Chem.*, **47**, 1242–1250 (2004). Ref. 292 = Dallet P, Labat L, Richard M, Langlois MH and Dubost JP, A reversed phase HPLC method development for the separation of new antidepressants, *J. Liq. Chrom. Rel. Technol.*, **25**, 101–111 (2002).

(continued)

No.	Compound Name	pK_a value(s)	Data quality	Ionization type	Method	Conditions	Comments and Reference(s)
841	Mitomycin C ($C_{15}H_{18}N_4O_5$)	2.8 12.44	U U	+H −H	kinetic Spectro	H_2O RT	Beijnen JH, Bult A, Underberg WJM and Mitomycin C, *APDS*, **16**, 361–401 (1987).

"Mitomycin C contains several prototropic functions. Basic groups are the 7-amino group, the N-4 nitrogen and the aziridine nitrogen. Due to rapid degradation, the pK_a values of the conjugated acid concerning the 7-amino group and the N-4 nitrogen can not be deduced from titration experiments with intact mitomycin C. Therefore the acid dissociation constants for these functions were derived from titrations with stable analogs (37). Results are listed in Table VIII. The pK_a of the aziridine nitrogen has been determined titrimetically (11) and kinetically from the pH-rate relationships in degradation studies (49–51) although the titrimetric determination also has been influenced by degradation.

TABLE VIII Prototropic functions of mitomycin C

Function	Method	pK_a	Ref.	Function	Method	pK_a	Ref.
N-4	spectro	−1.2	37	aziridine	kinetic	2	50
7-amino	spectro	−1.3	37	aziridine	kinetic	2.74	51
aziridine	potentio	3.2	11	aziridine	kinetic	2.50 (49.5 °C)	49
aziridine	kinetic	2.8	49	7-amino	spectro	12.44	37

11. Stevens CL, Taylor KG, Munk ME, Marshall WS, Noll K, Shah GD, Shah LG and Uzu K, *J. Med. Chem.*, **8**, 1–10 (1965) (See no. 843).
37. Underberg WJM and Lingeman H, *J. Pharm. Sci.*, **72**, 553–556 (1983) (See no. 842).
49. McClelland RA and Lam K, *JACS*, **107**, 5182–5186 (1985).
50. Den Hartigh J, *Analysis, electrochemistry and pharmacokinetics of mitomycin C*, Thesis, Utrecht, The Netherlands (1986).
51. Beijnen JH, van der Houwen OAGJ, Rosing H and Underberg WJM, A systematic study on the chemical stability of mitomycin A and mitomycin B, *Chem. Pharm. Bull.*, **34**, 2900–2913 (1986)."

842	Mitomycin	−1.3 −1.2 ~1.5 12.44	U U U U	+H +H +H −H	Potentiometric Spectro ($\lambda = 363$ nm)	H_2O t = RT	Underberg WJM and Lingeman H, Determination of pK_a values of some prototropic functions in mitomycin and porfiromycin, *J. Pharm. Sci.*, **72**, 553–556 (1983). NB: Rates of ionization for pK_{a4} were measured by stopped-flow spectrophotometry, with half-lives found to range from 7 to 25 seconds.
843	Mitomycin	3.20	U	+H	Potentiometric	50% MeOH t undefined I undefined	Stevens CL, Taylor KG, Munk ME, Marshall WS, Noll K, Shah GD, Shah LG and Uzu K, Chemistry and structure of Mitomycin C, *J. Med. Chem.*, **8**, 1–10 (1965). NB: No further experimental details on Mitomycin C pK_a measurements. Other titration measurements were made on various degradation products.
844	Mitomycin	2.8	U	+H	kinetic	H_2O $t = 25$ $I = 0.1$	McClelland RA and Lam K, Kinetics and mechanism of the acid hydrolysis of mitomycins, *JACS*, **107**, 5182–5186 (1985). NB: First order behaviour observed. Other observed kinetic pK_a values: mitomycin C, 38 °C, 2.7; mitomycin C, 49.5 °C, 2.50; mitomycin C, 25 °C in D_2O, 3.3; porfiromycin, 25 °C, 2.4; mitomycin A, 25 °C, 2.7.
845	Molindone ($C_{16}H_{24}N_2O_2$) 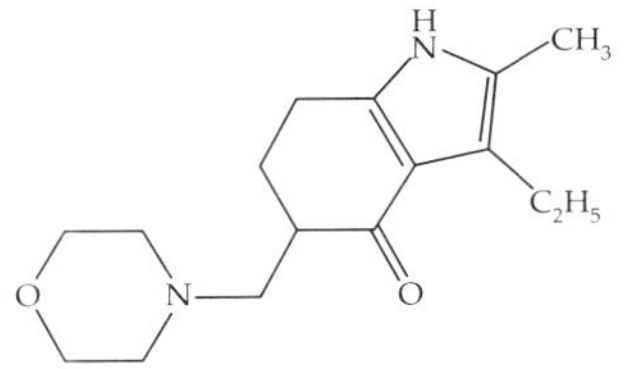	6.9	U	+H	Potentiometric	H_2O $t = 25$ $I = 0.2$	Dudzinski J, Lachman L, Shami E and Tingstad J, Preformulation studies. I. Molindone hydrochloride, *J. Pharm. Sci.*, **62**, 622–624 (1973). NB: Reported the pH of half-neutralization in carbonate-free solutions as the apparent pK_a value.
846	Monoethylglycine xylidide ($C_{12}H_{18}N_2O$) 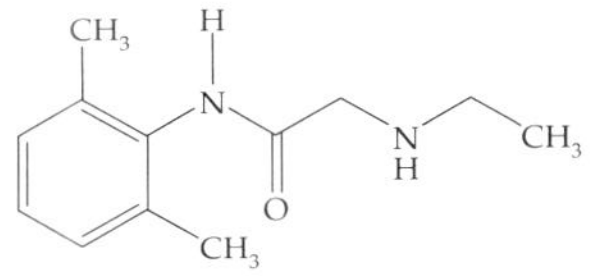	8.04	U	+H	Potentiometric	H_2O $t = 25.0 \pm 0.2$ $I = 0.01$ (NaCl)	Johansson P-A, Liquid-liquid distribution of lidocaine and some structurally related anti-arrythmic drugs and local anaesthetics, *Acta Pharm. Suec.*, **19**, 137–142 (1982). NB: See Lidocaine for details.

(*continued*)

No.	Compound Name	pK_a value(s)	Data quality	Ionization type	Method	Conditions	Comments and Reference(s)
847	Morizicine ($C_{22}H_{24}N_3O_4S$) 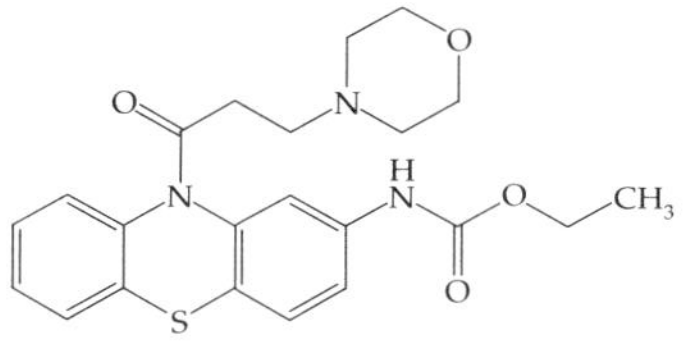	6.13	U	+H	CE/pH (+ve ion mode)	H_2O $t = 25$ $I = 0.025$	Wan H, Holmen AG, Wang Y, Lindberg W, Englund M, Nagard MB and Thompson RA, High-throughput screening of pK_a values of pharmaceuticals by pressure-assisted capillary electrophoresis and mass spectrometry, *Rapid Commun. Mass Spectrom.*, **17**, 2639–2648 (2003). NB: Reported a predicted value (ACD Labs) of 6.77.
848	Morphine ($C_{17}H_{19}NO_3$) 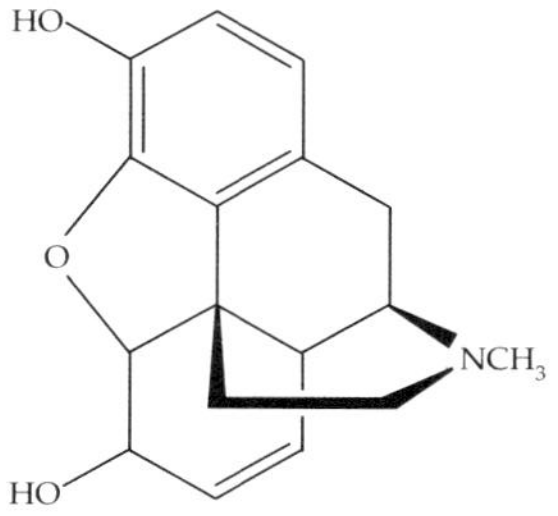	8.21	A	+H	Conductance	H_2O $t = 25$ $\kappa < 1.5$	Oberst FW and Andrews HL, The electrolytic dissociation of morphine derivatives and certain synthetic analgetic compounds, *JPET*, **71**, 38–41 (1941). Cited in Perrin Bases 2929 ref. O1. NB: Results reported as K_b values. For morphine, $K_b = 1.64 \times 10^{-6}$ giving pK_b = 5.79.
		8.03	U	+H	Potentiometric	H_2O $t = 20$	Perel'man Y, *Sb. Nauchn. Tr. Leningr. Khim-Farmatsevt. Inst.* **2**, 38 (1957); CA 54: 11382b. Cited in Perrin Bases 2929 ref. P16. NB: Unsymmetrical cell with diffusion potentials.
849	Morphine	8.07 9.85	U U	+H −H	Spectro	H_2O $t = 15$	Kolthoff IM, The dissociation constants, solubility product and titration of alkaloids, *Biochem. Z.*, **162**, 289–353 (1925). NB: See Aconitine for details. Muhtadi, *APDS*, 17, 259–366 (1988): Has other values (8.0, 9.9 at 20°), but these are citations from Merck **10**, Dictionary of Organic Compounds or BPC 11.

850	Morphine	8.31 9.51	U U	+H –H	Spectro ($\lambda = 300$ nm)	H_2O $t = 20.0$ I undefined	Schill G and Gustavii K, Acid dissociation constants of morphine, *Acta Pharm. Suec.*, **1**, 24–35 (1964). Cited in Perrin Bases Suppl no. 7484. NB: Glass electrode was only calibrated at pH = 9.20 at 20 °C. The values were validated by comparisons with pK_a values for monofunctional morphine derivatives and checked by potentiometric and partitioning (against dichloromethane) experiments. Reported also codeine (8.26); ethylmorphine (8.33); morphine *N*-methyl bromide (8.83).
		8.29 9.49	U U	+H –H	Potentiometric	H_2O $t = 20.0$ I undefined N_2 atmosphere	Microconstants were determined spectrophotometrically and with reference to data for corresponding monofunctional compounds, using calculations described by Edsall JT, Martin RB and Hollingworth BR, *Proc. Natl. Acad. Sci. USA*, **44**, 505–518 (1958): pK_1, 8.87; pK_2, 8.45; pK_{12}, 8.95; pK_{21}, 9.37.
851	Morphine	8.02	U	+H	Potentiometric	H_2O $t = 20$ $I < 0.01$	Kaufman JJ, Semo NM and Koski WS, Microelectrometric titration measurement of the pK_as and partition and drug distribution coefficients of narcotics and narcotic antagonists and their pH and temperature dependence, *J. Med. Chem.*, **18**, 647–655 (1975).
		7.93	U	+H		H_2O $t = 37$ $I < 0.01$	NB: See Codeine for further details.
852	Morphine	8.18 9.26	A U	+H –H	Potentiometric	H_2O $t = 25.0 \pm 0.1$ $I = 0.15$ (KCl) Ar atmosphere	Avdeef A, Barrett DA, Shaw PN, Knaggs RD and Davis SS, Octanol-, chloroform-, and propylene glycol dipelargonat-water partitioning of morphine-6-glucuronide and other related opiates, *J. Med. Chem.*, **39**, 4377–4381 (1996). NB: Used a Sirius autotitrator. Also gave $I = 0.15$ M (NaCl): pK_a (amine), 8.17; pK_a (phenolic), 9.26; $I = 0.001$ M (NaCl): pK_a (amine), 8.13; pK_a (phenolic), 9.46.
853	Morphine	8.18 ± 0.01 9.34 ± 0.01	A A	+H –H	Potentiometric	H_2O $t = 25.0 \pm 0.1$ $I = 0.1$ (NaCl)	Takacs-Novak K, Box KJ and Avdeef A, Potentiometric pK_a determination of water-insoluble compounds: Validation study in methanol/water mixtures, *Int. J. Pharm.*, **151**, 235–248 (1997). NB: p$K_{a1} = 8.24 \pm 0.02$, p$K_{a2} = 9.46 \pm 0.03$ by extrapolation from 15.6–59.7% w/w aqueous MeOH. See Acetaminophen for full details.
854	Morphine	8.81 ± 0.01* 9.34 ± 0.01	A A	+H –H	Potentiometric	H_2O $t = 25.0$ $I = 0.1$ (NaCl)	Takacs-Novak K and Avdeef A and Interlaboratory study of log P determination by shake-flask and potentiometric methods, *J. Pharm. Biomed. Anal.*, **14**, 1405–1413 (1996). NB: See Acetaminophen for further details. Also reported p$K_{a1} = 8.18 \pm 0.01$, p$K_{a2} = 9.26 \pm 0.01$ at $I = 0.15$ (KCl). *The reported value 8.81 seems to be a typo for 8.18.

(continued)

Appendix A *(continued)*

No.	Compound Name	pK_a value(s)	Data quality	Ionization type	Method	Conditions	Comments and Reference(s)
855	Morphine	8.13 ± 0.01 9.46 ± 0.01	A A	+H −H	Potentiometric	H_2O $t = 25.0$ $I = 0.001$ (NaCl)	Sirius Technical Application Notes, 1994, vol. **1**, pp. 84–86. Sirius Analytical Instruments Ltd., Forest Row, East Sussex, RH18 5DW, UK. NB: Also reported 8.17 ± 0.01(A) and 9.26 ± 0.01 (A), both at $t = 25.0$, $I = 0.15$ (NaCl).
856	Morphine	8.08	U	+H	soly	H_2O $t = 35.0$ I undefined	Roy SD and Flynn GL, Solubility behavior of narcotic analgesics in aqueous media: Solubilities and dissociation constants of morphine, fentanyl, and sufentanil, *Pharm. Res.*, **6**(2), 147–151 (1989). NB: See Fentanyl for details.
857	Morphine	7.6 9.6	U U	+H −H		H_2O $t = 37$	Ballard BE and Nelson E, Physicochemical properties of drugs that control absorption rate after subcutaneous implantation, *JPET*, **135**, 120–127 (1962). NB: pK_a = 7.6 from pK_b = 6.0 where pK_w = 13.621 at 37 °C; secondary source W&G assumed the result was at 25 °C and used pK_w = 14.00.
858	Morphine-3-glucuronide ($C_{23}H_{27}NO_9$)	2.83 ± 0.05	A	−H	Potentiometric	H_2O t undefined $I = 0.1$	Carrupt PA, Testa B, Bechalany A, El Tayar N, Descas P and Perrissoud D, Morphine-6-glucuronide and morphine-3-glucuronide as molecular chameleons with unexpected lipophilicity, *J. Med. Chem.*, **34**, 1272–1275 (1991). NB: See Walther B, Carrupt PA, El Tayar N and Testa B, 8-Substituted xanthines as phosphodiesterase inhibitors: Conformation-dependent lipophilicity and structure-activity relationships, *Helv. Chim. Acta*, **72**, 507–517 (1989).
		2.88 ± 0.01 8.21 ± 0.01	A A	−H +H	Potentiometric	H_2O $t = 25.0$ $I = 0.154$ (KCl)	Sirius Technical Application Notes, 1994, vol. **1**, pp. 91–93. Sirius Analytical Instruments, Ltd. Forest Row, East Sussex, RH18 5DW, UK.

No.	Compound	pKa			Method	Conditions	Reference
859	Morphine-3-glucuronide	2.86 8.21	A U	−H +H	Potentiometric	H_2O $t = 25.0 \pm 0.1$ $I = 0.15$ (KCl) Ar atmosphere	Avdeef A, Barrett DA, Shaw PN, Knaggs RD and Davis SS, Octanol-, chloroform-, and propylene glycol dipelargonat-water partitioning of morphine-6-glucuronide and other related opiates, *J. Med. Chem.*, **39**, 4377–4381 (1996). NB: See Morphine for further details.
860	Morphine-6-glucuronide ($C_{23}H_{27}NO_9$)	3.23 ± 0.05	U	−H	Potentiometric	H_2O t undefined $I = 0.1$	Carrupt PA, Testa B, Bechalany A, El Tayar N, Descas P and Perrissoud D, Morphine-6-glucuronide and morphine-3-glucuronide as molecular chameleons with unexpected lipophilicity, *J. Med. Chem.*, **34**, 1272–1275 (1991). NB: See Walther B, Carrupt PA, El Tayar N and Testa B, 8-Substituted xanthines as phosphodiesterase inhibitors: Conformation-dependent lipophilicity and structure-activity relationships, *Helv. Chim. Acta*, **72**, 507–517 (1989).
		2.89 ± 0.01 8.14 ± 0.01 9.36 ± 0.01	A A A	−H −H +H	Potentiometric	H_2O $t = 25.0$ $I = 0.156$ (KCl)	Sirius Technical Application Notes, 1994, vol. **1**, pp. 87–89. Sirius Analytical Instruments Ltd., Forest Row, East Sussex, RH18 5DW, UK
861	Morphine-6-glucuronide	2.77 8.22 9.42	U U U	−H +H −H	Potentiometric	H_2O $t = 25.0 \pm 0.1$ $I = 0.15$ (KCl) Ar atmosphere	Avdeef A, Barrett DA, Shaw PN, Knaggs RD and Davis SS, Octanol-, chloroform-, and propylene glycol dipelargonat-water partitioning of morphine-6-glucuronide and other related opiates, *J. Med. Chem.*, **39**, 4377–4381 (1996). NB: See Morphine for further details.
862	Muroctasin (Romurtide) ($C_{43}H_{78}N_6O_{13}$)	5.70 ± 0.05	U	−H	Potentiometric	90% MeOH $t = 35$	Moroi R, Yamazaki K, Hirota T, Watanabe S, Kataoka K and Ichinose M, Physico-chemical properties of muroctasin, *Arzneim.-Forsch.*, **38**, 953–959 (1988). "The dissociation constant of MDP-Lys(L18) (muroctasin; I) was measured by the titration method. MDP-Lys(L18) was dissolved in mixed solvents of MeOH-H_2O (9:1)." NB: Solubility of this compound is poor in most solvents (e.g., H_2O, <1 g/10,000 mL).

(continued)

No.	Compound Name	pK_a value(s)	Data quality	Ionization type	Method	Conditions	Comments and Reference(s)
863	5-O-Mycaminosyltylo-nolide (OMT) ($C_{31}H_{51}NO_{10}$) 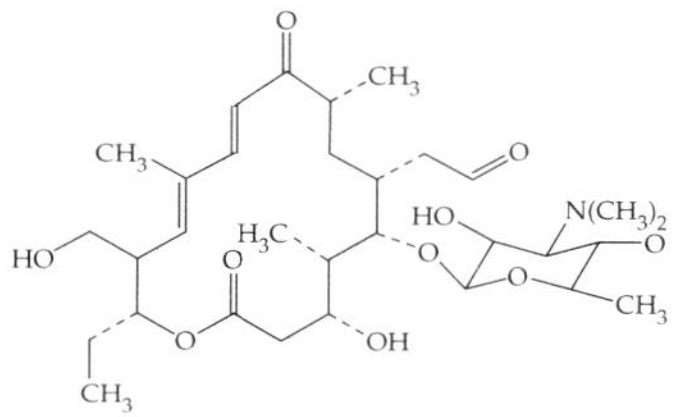	8.40	U	+H	Potentiometric	H_2O $t = 25$ $I = 0.167$	McFarland JW, Berger CM, Froshauer SA, Hayashi SF, Hecker SJ, Jaynes BH, Jefson MR, Kamicker BJ, Lipinski CA, Lundy KM, Reese CP and Vu CB, Quantitative Structure-activity relationships among macrolide antibacterial agents: *In vitro* and *in vivo* potency against Pasteurella multocida, *J. Med. Chem.*, **40**, 1340–1346 (1997). NB: See Azithromycin for details; average standard deviation of $\pm$ 0.07 for the pK_a value.
864	Nafcillin ($C_{21}H_{22}N_2O_5S$) 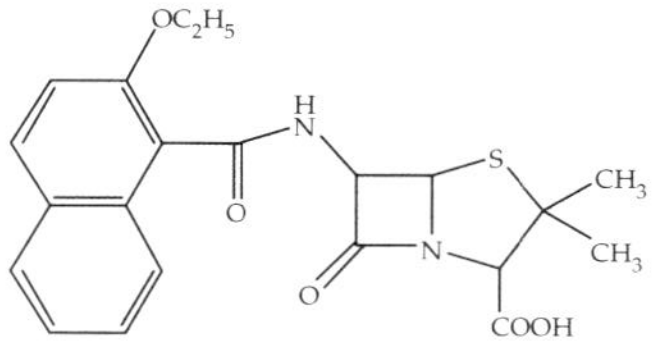	2.65	U	−H	Potentiometric	H_2O $t = 25.0 \pm 0.1$ $c = 0.006$	Hou JP and Poole JW, The aminoacid nature of ampicillin and related penicillins, *J. Pharm. Sci.*, **58**, 1510–1515 (1969). NB: Apparent pK_a values extrapolated to zero percent alcohol. Cited in Foye 1 (2 refs.); See Idoxuridine.
865	Nalidixic acid ($C_{12}H_{12}N_2O_3$) 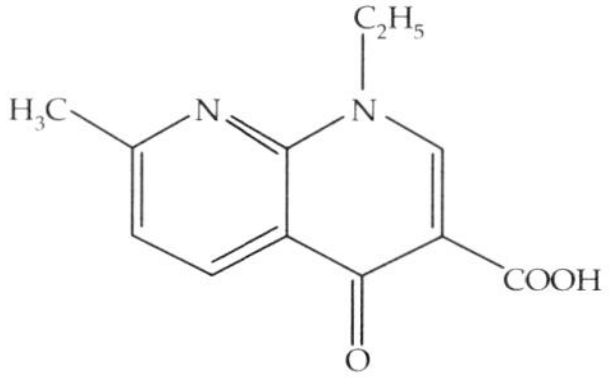	−0.94 6.02 5.99 $\pm$ 0.03	U U A	+H −H −H	Spectro Potentiometric	H_2O	Staroscik R and Sulkowska J, Acid-base equilibria of nalidixic acid, *Acta Pol. Pharm.*, **28**(6), 601–606 (1971); CA 76:158322k. Cited in Grubb PE, Nalidixic Acid, *APDS*, **8**, 371–397 (1979). "The pK_a of the protonation of the nitrogen in position 8 has been reported as 6.02 and the pK_a for the carboxylate anion formulation has been reported as −0.94. These were determined by . . . a spectrophotometric method. Further study by the same workers on the partition equilibria of nalidixic acid between water and various organic solvents led to . . . pK_a values of 5.99 $\pm$ 0.03 for *N*-protonation and for carboxylate . . . formation. Takasugi and coworkers reported the apparent pK_a of nalidixic acid to be 5.9 at 28° by a spectrophotometric method."

							NB: See also Sulkowska J and Staroscik R, *Pharmazie*, **30**, 405–406 (1975).
866	Nalidixic acid	6.11 ± 0.02	A	−H	Spectro	H_2O	Vincent WR, Schulman SG, Midgely JM, van Oort WJ and Sorel RHA, Prototropic and metal complexation equilibria of nalidixic acid in the physiological pH region, *Int. J. Pharm.*, **9**, 191–198 (1981). "The change in the absorption of the nalidixic acid as a function of pH corresponds to a calculated pK_a of 6.11 ± 0.02. As shown . . .., this dissociation may be assigned to the carboxylic acid group (K (CZ) or K(NA)), the nitrogen in position 8 (K(CN) or K(NA)) or both in combination considering the possibility of two closely overlapping equilibria."
867	Nalorphine ($C_{19}H_{21}NO_3$) 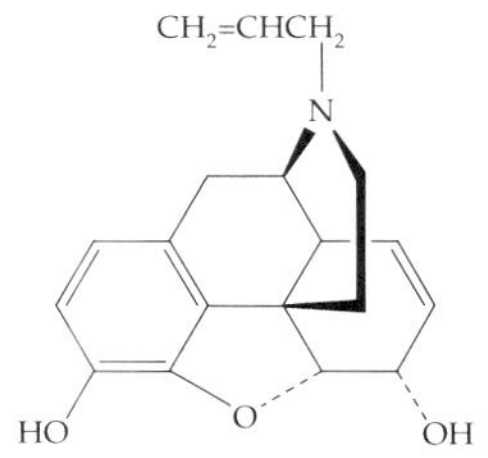	7.73	U	+H	Potentiometric	H_2O $t = 20$ $I < 0.01$	Kaufman JJ, Semo NM and Koski WS. Microelectrometric titration measurement of the pK_as and partition and drug distribution coefficients of narcotics and narcotic antagonists and their pH and temperature dependence, *J. Med. Chem.*, **18**, 647–655 (1975). NB: See Codeine for further details.
		7.59	U	+H		H_2O $t = 37$ $I < 0.01$	
868	Naloxone ($C_{19}H_{21}NO_4$) 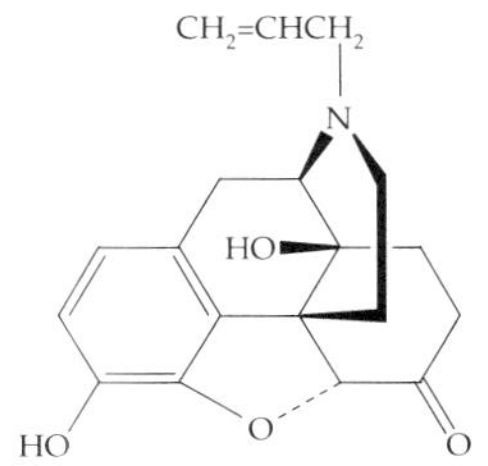	7.94	U	+H	Potentiometric	H_2O $t = 20$ $I < 0.01$	Kaufman JJ, Semo NM and Koski WS. Microelectrometric titration measurement of the pK_as and partition and drug distribution coefficients of narcotics and narcotic antagonists and their pH and temperature dependence, *J. Med. Chem.*, **18**, 647–655 (1975). NB: See Codeine for further details. Cited in: Lombardo F, Obach RS, Shalaeva MY and Gao F, Prediction of human volume of distribution values for neutral and basic drugs. 2. Extended data set and leave-class-out statistics, *J. Med. Chem.*, **47**, 1242–1250 (2004) (ref. 285).
		7.82	U	+H		H_2O $t = 37$ $I < 0.01$	

(continued)

No.	Compound Name	pK_a value(s)	Data quality	Ionization type	Method	Conditions	Comments and Reference(s)
869	Naltrexone ($C_{20}H_{23}NO_4$) 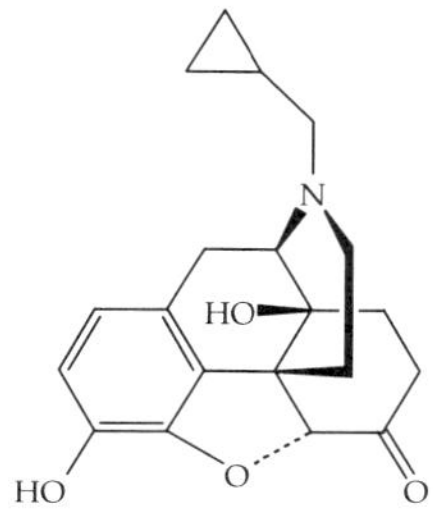	8.38	U	+H	Potentiometric	H_2O $t = 20$ $I < 0.01$	Kaufman JJ, Semo NM and Koski WS, Microelectrometric titration measurement of the pK_as and partition and drug distribution coefficients of narcotics and narcotic antagonists and their pH and temperature dependence, *J. Med. Chem.*, **18**, 647–655 (1975). NB: See Codeine for further details.
		8.13	U	+H		H_2O $t = 37$ $I < 0.01$	
870	Naphazoline ($C_{14}H_{14}N_2$) 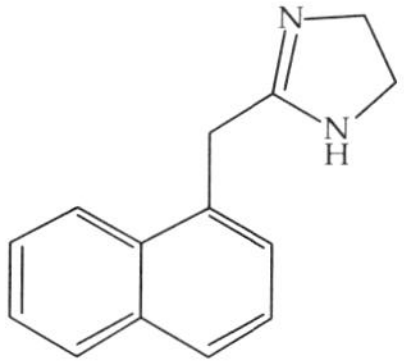	10.35 ± 0.02	U	+H	kinetic	H_2O $t = 25$ I undefined	Wall GM, Naphazoline Hydrochloride, *APDS*, **21**, 307–344, (1992). NB: pK_a = 10.9 at 20 °C (4); 10.13 ± 0.02 at 35 °C and 9.92 ± 0.03 at 45 °C (16). 4. Reynolds JEF (ed.), *Martindale's Extra Pharmacopoeia*, 29th Edn., Pharmaceutical Press, London 1470 (1989). 16. Stern MJ, King LD and Marcus AD, Kinetics of the specific base-catalyzed hydrolysis of naphazoline, *J. Am. Pharm. Assn.*, **48**, 641–647 (1959).
871	Naphazoline	10.14	U	+H	Potentiometric	H_2O (extrap) $t = 24 \pm 1$ $I \sim 0.002$	Chatten LG and Harris LE, Relationship between pK_b(H_2O) of organic compounds and $E_{1/2}$ values in several nonaqueous solvents, *Anal. Chem.*, **34**, 1495–1501 (1962). NB: See Chlorpromazine for details.
872	Naproxen ($C_{14}H_{14}O_3$) 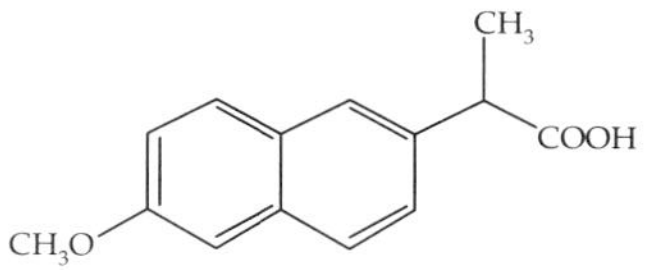	4.15	U	−H	soly	H_2O $t = 25$ I undefined	Chowhan ZT, pH solubility profiles of organic carboxylic acids and their salts, *J. Pharm. Sci.*, **67**, 1257–1260 (1978). "The solubilities of naproxen (I), 7-methylthio-2-xanthonecarboxylic acid (tixanox), 7-methylsulfinyl-2-xanthonecarboxylic acid . . . were determined as a function of pH. The results on the solubility of I and its salts were in excellent agreement with the theoretical profiles describing the relationship between pH values of the solutions and the dissociation constant of the acid. . .."

No.	Compound	pKa value	Quality	Ionization	Method	Conditions	Comments and Reference(s)
873	Naproxen	4.1	U	−H	Spectro	H_2O t unspecified I unspecified	Herzfeldt CD and Kümmel R, Dissociation constants, solubilities, and dissolution rates of some selected nonsteroidal antiinflammatories, *Drug Dev. Ind. Pharm.*, **9**(5), 767–793 (1983). NB: See Azapropazone and Ibuprofen for details. Al-Shammary FJ, Mian NAA and Mian MS, Naproxen, *APDS*, **21**, 345–373 (1992) gave $pK_a = 4.2$, citing Moffat AC, *Clarke's Isolation and Identification of Drugs*, 2nd Edn. (1986).
874	Naproxen	4.28 ± 0.02	U	−H	HPLC/pH	H_2O $t = 25$ $I = 0.01$	Unger SH, Cook JR and Hollenberg JS, Simple procedure for determining octanol-aqueous partition, distribution, and ionization coefficients by reversed phase high pressure liquid chromatography, *J. Pharm. Sci.*, **67**, 1364–1367 (1978).
		4.21 ± 0.02	U			$I = 0.1$	
	Benzoic acid	4.33 ± 0.02	U			H_2O $t = 25$ $I = 0.01$	"A liquid chromatographic method for the determination of partition, distribution and ionization coefficients of various compounds in octanol-aqueous solutions is described. . . . . A stable baseline is obtained rapidly, and the log relative retention times are highly correlated with unit slope to log distribution or partition coefficients obtained from the classical shake-flask procedures. In addition, if the apparent dissociation constant of an ionizable compound lies within the pH operating range of the column support, the apparent dissociation constant usually can be determined simultaneously with the log of the partition coefficient by measuring the log distribution coefficient at several pH values."
		4.38 ± 0.02	U			$I = 0.1$	NB: The poor agreement between the experimental values reported here for benzoic acid and the best literature values (see Benzoic acid, nos. 206–208) suggest that the remainder of the values in this paper must be classified as Uncertain.
875	Nebivolol ($C_{22}H_{25}F_2NO_4$) 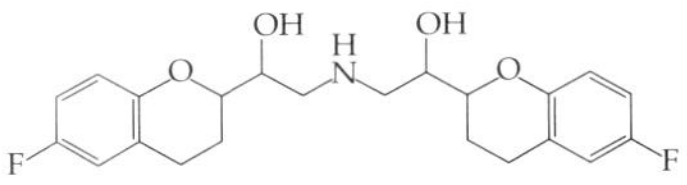	8.22	U	+H	Potentiometric	H_2O (extrap) $t = 25 \pm 1$ I undefined Ar atmosphere	Cheymol G, Poirier J-M, Carrupt PA, Testa B, Weissenburger J, Levron J-C and Snoeck E., Pharmacokinetics of β-adrenoceptor blockers in obese and normal volunteers, *Br. J. Clin. Pharmacol.*, **43**, 563–570 (1997). NB: Determined by extrapolation from MeOH-water solutions, using the Yasuda-Shedlovsky procedure. Cited in: Lombardo F, Obach RS, Shalaeva MY and Gao F, Prediction of human volume of distribution values for neutral and basic drugs. 2. Extended data set and leave-class-out statistics, *J. Med. Chem.*, **47**, 1242–1250 (2004); ref. 14.
876	Neurophysin I (histidine residue)	6.87	U	+H	NMR/pH	D_2O $t = 22 \pm 1$ $I = 0.1$ (NaCl)	Griffin JH, Cohen JS, Cohen P and Camier M, Drug-biomolecule interactions: Proton magnetic resonance studies of complex formation between bovine neurophysins and oxytocin at molecular level, *J. Pharm. Sci.*, **64**, 507–511 (1975).
	Neurophysin I in a 1:1 complex with oxytocin	6.67	U	+H			"Proton magnetic resonance spectroscopy was used to monitor individual amino acid residues in bovine neurophysin, in the nonapeptide hormone oxytocin, and in the complex formed between them. For neurophysin I alone, a normal titration curve

(continued)

No.	Compound Name	pK_a value(s)	Data quality	Ionization type	Method	Conditions	Comments and Reference(s)
							for the C-2 proton resonance of the lone histidine residue was obtained with an apparent ionization constant of 6.97. Addition of oxytocin to a solution of neurophysin I at pH 6.5 resulted in several changes in the spectrum. The effect on the histidine C-2 proton resonance signal indicated a slow exchange process between 2 states, probably representing a conformational change in the protein. The apparent pK of the histidine residue in the hormonal complex was shifted to 6.67, indicating a slightly more positive (less electron dense) environment for the histidine residue. . .''
877	Nicotinamide (niacinamide) ($C_6H_6N_2O$) N, $CONH_2$	3.328 ± 0.010	R	+H	Spectro	H_2O $t = 20.0 \pm 0.5$ $I < 0.01$ to 0.1 (KCl)	Willi AV, Die Ionizationskonstante von Nicotinsaureamid in wasseriger Losung, *Helv. Chim. Acta*, **37**, 602–606 (1954). NB: Careful spectrophotometric work performed with CO_2-free solutions.
		3.35	A	+H	Spectro ($\lambda = 261.5$ nm)	H_2O $t = 20 \pm 2$ $I = 0.01$	Jellinek HHG and Wayne MG, Nicotinamide: Ultraviolet spectra and dissociation constants, *J. Phys. Chem.*, **55**, 173–179 (1951). NB: Used glass electrode to measure pH values.
		3.20	A	+H	Potentiometric	H_2O $t = 20$	No reference cited by Perrin. Used glass electrode in an unsymmetrical cell with liquid junction potentials.
		also: 0.50	U	+H	Spectro	$c = 0.01$	NB: All values cited in Perrin Bases no. 1073.
878	Nicotine ($C_{10}H_{14}N_2$) N, CH_3, N	3.22 8.11	A A	+H +H	Potentiometric	H_2O $t = 20$ $I < 0.05$	Vickery HB and Pucher GW, The determination of ''free nicotine'' in tobacco: The apparent dissociation constants of nicotine, *J. Biol. Chem.*, **84**, 233–241 (1929).
		3.55 8.13	U A	+H +H		H_2O $t = 0$ $I < 0.2$	Fowler RT, Redetermination of ionization constants of nicotine, *J. Appl. Chem. (Lond.)*, **4**, 449–452 (1954). NB: Results were reported as K_b values. Gave also p$K_{a1} = 3.42$ and p$K_{a2} = 8.02$ at 25 °C.
		3.22 7.65	U U	+H +H		H_2O $t = 40$ I undefined	Kolthoff IM, The dissociation constants, solubility product and titration of alkaloids, *Biochem. Z.*, **162**, 289–353 (1925). NB: All values cited in Perrin Bases no. 2933. The Vickery and Pucher data at 20 °C were corrected to values at 25 °C (3.12, 8.02) by Perrin.
879	Nicotine	3.13 8.24	A U	+H +H	CE/pH (+ve ion mode)	H_2O $t = 25$ $I = 0.025$	Wan H, Holmen AG, Wang Y, Lindberg W, Englund M, Nagard MB and Thompson RA, High-throughput screening of pK_a values of pharmaceuticals by pressure-assisted capillary electrophoresis and mass spectrometry, *Rapid Commun. Mass Spectrom.*, **17**, 2639–2648 (2003). NB: Reported predicted values (ACD Labs) of 3.21 and 8.

No.	Compound	pKa	Code	Assignment	Method	Conditions	Reference
880	Nicotine analogues				Potentiometric	H_2O	Rondahl L and Ingman F, Synthetic analogues of nicotine VII. Acid dissociation constants of some nicotine analogues, *Acta Pharm. Suec.*, **16**(1), 56–63 (1979).
	5-H	3.41	U	+H		$t = 25.0 \pm 0.05$	"The experimental set-up used for determining the dissociation constants has been described elsewhere [Ingman F, Johansson A, Johansson S and Karlsson R, Titration of mixtures of acids of nearly equal strengths, *Anal. Chim. Acta*, **64**, 113–120 (1973)]. A known amount of the acid, or in some cases, the salt to which a known amount of hydrochloric acid had been added, was titrated in a 0.5 M ionic medium (KCl) with sodium hydroxide solution of the same ionic strength. . . .. The dissociation constants were evaluated from the titration data by the ETITR version of the general error minimising program LETAGROP [Brauner P, Sillen L and Whiteker R, *Ark. Kemi.*, **31**, 365–376 (1969)].
		8.21	U	+H		$I = 0.5$ (KCl)	
	5-fluoro	1.07	U	+H		corrections were made	
		7.72	U	+H		for $[H^+]$	
	5-chloro	1.15	U	+H		and $[OH^-]$	
		7.68	U	+H		from K_w	
	5-bromo	1.42	U	+H			
		7.77	U	+H			
	5-iodo	1.70	U	+H			
		7.67	U	+H			
	5-hydroxy	3.30	U	+H			
		8.20	U	+H			
	γ-nicotine	3.62	U	+H			
		7.87	U	+H			

Other compounds:

Substituted pyridines	pK_a (pyr N)	pK_a (alicyclic or aliphatic N)
N-Me 3-(2-aminoethyl)	4.69	8.83
N-Me 3-(3-aminopropyl)	5.30	9.66
N,N-diMe 3-(aminomethyl)	3.40	8.04
N,N-diMe 3-(2-aminoethyl)	4.36	8.90
N,N-diMe 3-(3-aminopropyl)	5.31	9.70
N,N-diMe 3-(4-aminobutyl)	5.57	10.00
3-(pyrrolidinomethyl)	3.44	8.82
3-(pyrrolidino-2-ethyl)	4.34	9.32
3-(pyrrolidino-3-propyl)	5.16	10.08
3-(pyrrolidino-4-butyl)	5.34	10.26
Substituted tetrahydroquinolines:		
5-amino	4.65	8.72
5-dimethylamino	4.25	8.20

No.	Compound	pKa	Code	Assignment	Method	Conditions	Reference
881	Nicotinic acid	2.03	A	+H,−H	Potentiometric	H_2O	Jaffe HH and Doak GO, The basicities of substituted pyridines and their 1-oxides, *JACS*, **77**, 4441–4444 (1955). Used a glass electrode in unsymmetrical cell with junction potentials.
	(structure: pyridine ring, N, with COOH)	4.83	A	−H,+H		$t = 24 \pm 1$	
						$c = 0.04$	
		2.00	A	+H,−H	Spectro	H_2O	Evans RF, Herington EFG and Kynaston W, Determination of dissociation constants of the pyridine-monocarboxylic acids by ultraviolet photoelectric spectrophotometry. *Trans. Farad. Soc.*, **49**, 1284–1292 (1953).
		4.82	A	−H,+H		$t = 25$	
						$I = 0.03$	

(*continued*)

No.	Compound Name	pK_a value(s)	Data quality	Ionization type	Method	Conditions	Comments and Reference(s)
		2.07 4.81	A A	+H,−H −H,+H	Potentiometric	H_2O $t = 22$ $c = 0.05$ to 0.1 N_2 atmosphere	Green RW and Tong HK, The constitution of the pyridine monocarboxylic acids in their isoelectric forms, *JACS*, **78**, 4896–4900 (1956). Used a glass electrode standardized with phthalate solution (pH = 4.00) and the Guntelberg equation to correct for *I*. Estimated the microconstants from spectrophotometric data on the acid and its methyl ester: pK_A, 2.11; pK_B, 3.13; pK_C, 4.77; pK_D, 3.75; where the subscripts represent the following equilibria: A, diprotonated to zwitterion; B, diprotonated to neutral; C, zwitterion to fully deprotonated; D, neutral to fully deprotonated. Also reported the corresponding data for picolinic and isonicotinic acids.
		4.76	A	−H,+H	Spectro	H_2O $t = 25.0$ $I = 0.005$–0.025	Fischer A, Galloway WJ and Vaughan J, Structure and reactivity in the pyridine series. I. Acid dissociation constants of pyridinium ions, *J. Chem. Soc. B*, 3591–3596 (1964). All the above data were cited in Perrin Bases no. 1076; Perrin Bases suppl. no. 5080.
882	Nicotinic acid	4.75	A	−H,+H	CE/pH (+ve ion mode)	H_2O $t = 25$ $I = 0.025$	Wan H, Holmen AG, Wang Y, Lindberg W, Englund M, Nagard MB and Thompson RA, High-throughput screening of pK_a values of pharmaceuticals by pressure-assisted capillary electrophoresis and mass spectrometry, *Rapid Commun. Mass Spectrom.*, **17**, 2639–2648 (2003). NB: Reported a predicted value (ACD Labs) of 4.8.
		4.92	A	−H,+H	Conductance	H_2O $t = 25.00 \pm 0.01$ $I = 0.00$	Orekhova Z, Ben-Hamo M, Manzurola E and Apelblat A, Electrical conductance and volumetric studies in aqueous solutions of nicotinic acid, *J. Sol. Chem.*, **34**, 687–700 (2005). NB: Obtained from measurements on dilute solutions of the sodium salt of nicotinic acid, extrapolated to zero ionic strength and calculated from the reported K_1 value. Also reported the following values: $t = 15.00$, 4.99; 20.00, 4.95; 30.00, 4.89; $t = 35.00$, 4.85; $t = 40.00$, 4.82; $t = 45.00$, 4.79; $t = 50.00$, 4.76.
883	Nicotinic acid	GLpK_a: 2.10 ± 0.01 4.63 ± 0.01 A&S: 2.14 ± 0.02 4.82 ± 0.05	 A U A A	 +H,−H −H,+H +H,−H −H,+H	Spectro	H_2O $t = 25$ $I = 0.15$ (KCl) Ar atmosphere	Tam KY and Takacs-Novac K, Multi-wavelength spectrophotometric determination of acid dissociation constants, *Anal. Chim. Acta*, **434**, 157–167 (2001). NB: See Clioquinol for details. Cited Tam KY and Takacs-Novac K, Multiwavelength spectrophotometric determination of acid dissociation constants. Part II. First derivative versus target function analysis, *Pharm. Res.*, **16**, 374–381 (1999); values of 2.00 ± 0.01 and 4.63 ± 0.01.

No.	Compound	pKa	Data quality	Ionization	Method	Conditions	Reference / Comments
884	iso-Nicotinic acid ($C_6H_5NO_2$)	1.74 4.96	A A	+H,−H −H,+H	Potentiometric	H_2O $t = 24 \pm 1$ $c = 0.04$	Jaffe HH and Doak GO, The basicities of substituted pyridines and their 1-oxides, *JACS*, **77**, 4441–4444 (1955). NB: Used a glass electrode in an unsymmetrical cell with junction potentials.
		1.77 4.84	A A	+H,−H −H,+H	Spectro	H_2O $t = 25$ $I = 0.03$	Evans RF, Herington EFG and Kynaston W, Determination of dissociation constants of the pyridine-monocarboxylic acids by ultraviolet photoelectric spectrophotometry, *Trans. Farad. Soc.*, **49**, 1284–1292 (1953).
		1.84 4.86	A A	+H,−H −H,+H	Potentiometric	H_2O $t = 22$ $c = 0.05–0.1$ N_2 atmosphere	Green RW and Tong HK, The constitution of the pyridine monocarboxylic acids in their isoelectric forms, *JACS*, **78**, 4896–4900 (1956). NB: Used a glass electrode standardized with phthalate solution (pH = 4.00) and the Guntelberg equation to correct for I. Estimated the microconstants from spectrophotometric data on the acid and its methyl ester: pK_A, 2.11; pK_B, 3.13; pK_C, 4.77; pK_D, 3.75; where the subscripts represent the following equilibria: A, diprotonated to zwitterion; B, diprotonated to neutral; C, zwitterion to fully deprotonated; D, neutral to fully deprotonated. Also reported the corresponding data for picolinic and isonicotinic acids.
		1.70 4.95	A A	+H,−H −H,+H	Spectro	H_2O $t = 20 \pm 2$ $I = 0.01$; $c < 5 \times 10^{-4}$	Jellinek HHG and Urwin JR, Ultraviolet absorption spectra and dissociation constants of picolinic, isonicotinic acids and their amides, *J. Phys. Chem.*, **58**, 548–550 (1954). NB: Used glass electrode to measure pH values and recorded spectra with a Hilger Uvispek spectrophotometer. Ionic strength was extrapolated to zero with the Debye-Huckel equation.
		4.90	A	−H,+H	Spectro	H_2O $t = 25.0$ $I = 0.005–0.025$	Fischer A, Galloway WJ and Vaughan J, Structure and reactivity in the pyridine series. I. Acid dissociation constants of pyridinium ions, *J. Chem. Soc. B*, 3591–3596 (1964). All data were cited in Perrin Bases no. 1076–77; suppl. no. **5081**.
885	Nifedipine ($C_{17}H_{18}N_2O_6$)	−0.9 >13	U U	+H −H	Potentiometric	DMF t undefined	Mannhold R, Rodenkirchen R and Bayer R, Haus W, The importance of drug ionization for the action of calcium antagonistsand related compounds, *Arzneim.-Forsch.*, **34**, 407–409 (1984). NB: Tetrabutylammonium hydroxide used as base. Cited in Nifedipine, Ali SL, *APDS*, **18**, 231–288 (1989). NB: The information given does not correspond exactly to the cited paper (below).

(continued)

Appendix A *(continued)*

No.	Compound Name	pK_a value(s)	Data quality	Ionization type	Method	Conditions	Comments and Reference(s)
886	Nifedipine	–	U	+H	Potentiometric	40% EtOH	Mannhold R, Rodenkirchen R, Bayer R and Haus W, The importance of drug ionization for the action of calcium antagonistsand related compounds, *Arzneim.-Forsch.*, **34**, 407–409 (1984). NB: Ionization constants stated to be too extreme to measure. See Aprindine for details.
887	Niflumic acid ($C_{13}H_9F_3N_2O_2$) 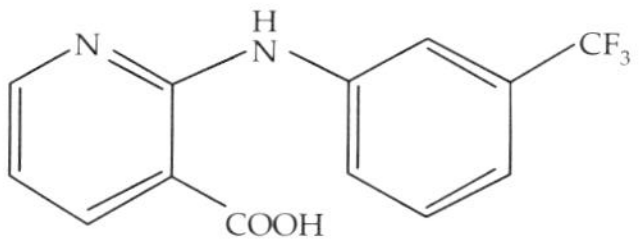	2.13 ± 0.04 5.07 ± 0.04	U/A U/A	+H –H	soly	H_2O $t = 20.0$	Asuero AG, Evaluation of acidity constants of two-step overlapping equilibria of amphoteric substances from solubility measurements, *J. Pharmaceut. Biomed. Anal.*, **6**(3), 313–316 (1988). Intrinsic solubility (So) = 22.57 ± 1.65 mg/L. Previously reported values: pK_1 = 2.15; pK_2 = 5.05 (Bres et al., *Trav. Soc. Pharm.* Montpellier, **36**, 331–364 (1976)). See also Asuero AG, *Int. J. Pharm.*, **89**, 103–110 (1993); Asuero AG, *Int. J. Pharm.*, **88**, 15–22 (1992); Asuero AG, *Int. J. Pharm.*, **52**, 129–137 (1989).
888	Niflumic acid	2.26 ± 0.08 4.44 ± 0.03	U A	+H –H	Potentiometric	H_2O $t = 25.0 \pm 0.1$ $I = 0.1$ (NaCl)	Takacs-Novak K, Box KJ and Avdeef A, Potentiometric pK_a determination of water-insoluble compounds: validation study in methanol/water mixtures, *Int. J. Pharm.*, **151**, 235–248 (1997). NB: By extrapolation from 30–55% w/w aqueous MeOH. See Acetaminophen for full details.
889	Niflumic acid	4.5	U	–H	Spectro	H_2O t unspecified I unspecified	Herzfeldt CD and Kümmel R, Dissociation constants, solubilities, and dissolution rates of some selected nonsteroidal antiinflammatories, *Drug Dev. Ind. Pharm.*, **9**(5), 767–793 (1983). NB: Used dλ/dpH method. See Azapropazone and Ibuprofen for details.
890	Nikethamide ($C_{10}H_{14}N_2O$) 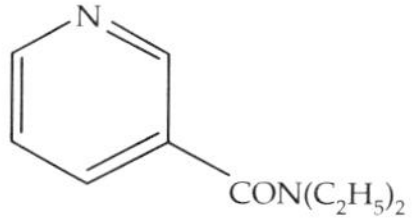	3.46 (0.72)	U	+H	Spectro ($\lambda = 261$ nm)	H_2O $t = 20.0$	Wahbe AM, El-Yazbi FA, Barary MH and Sabri SM, Application of orthogonal functions to spectrophotometric analysis. Determination of dissociation constants, *Int. J. Pharm.*, **92**(1), 15–22 (1993). NB: See Acetaminophen for further details. An alternative graphical method gave pK_a = 3.4.

891	Nimesulide ($C_{13}H_{12}N_2O_5S$) 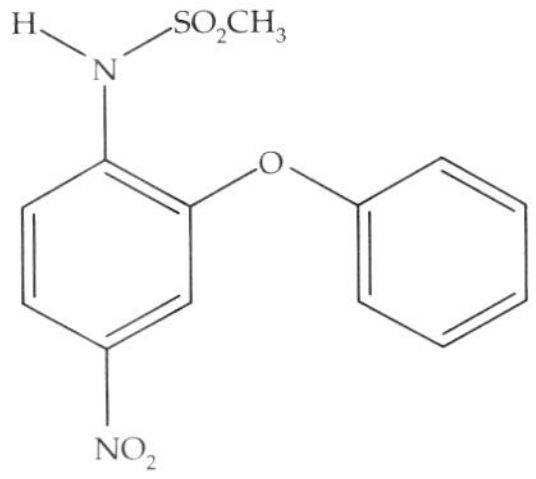	6.46	A	–H	Potentiometric	H_2O $t = 25.0$ $I < 0.01$ activity corrections	Fallavena PRB and Schapoval EES, pK_a determination of nimesulide in methanol-water mixtures by potentiometric titrations, *Int. J. Pharm.*, **158**(1), 109–112 (1997). NB: The reported pK_a value has been obtained by extrapolation of apparent values in water-methanol mixtures (34.47–60.00 wt%) to 0% methanol, using Yasuda-Shedlovsky plots.
892	Nimesulide	6.56 ± 0.03	A	–H	Spectro	H_2O $t = 25$ $I = 0.02$	Singh S, Sharda N and Mahajan L, Spectrophotometric determination of pK_a of nimesulide, *Int. J. Pharm.*, **176**, 261–264 (1999). "The pH were determined on a pH meter . . . equipped with a combined glass electrode that was standardised at 25° using standard buffers (Bates, 1962). The absorbance readings and spectra were recorded on two spectrophotometers, model DU640i and diode array model 700 (both Beckman, USA). A thermostatic bath equipped with a precision controller . . . was used for control of temperature. . . . The procedure for determination and calculation of pK_a was essentially the same as described by Albert and Serjeant (1962). The buffers were prepared by mixing predetermined volumes of stock 0.2M NaH_2PO_4, 0.1M K_2HPO_4, 0.2M NaCl solutions and water to give final buffer molarity 0.01M and ionic strength 0.02. Aliquots of 5 ml of each buffer were distributed into five tubes each. Each tube 50 μl drug stock solution in methanol (2 mg ml^{-1}) added to give a final drug strength of 20 μg ml^{-1}. For a validation study in which the effect of drug concentration was looked for, the drug strength was doubled to 40 μg ml^{-1}. The solutions were mixed and the tubes were placed in the thermostatic bath set at 25 °C. The absorbance was determined for each solution at 393 nm. Full scans were also taken for selected samples. . . . The absorbance of neutral and ionic species of the drug were determined in a similar manner employing 0.01N HCl and 0.01N NaOH, respectively. The ionic strength of these solutions was also preadjusted to 0.02 by addition of suitable quantity of sodium chloride."

(continued)

No.	Compound Name	pK_a value(s)	Data quality	Ionization type	Method	Conditions	Comments and Reference(s)
							NB: Four replicate values were reported (no correction for activity effects): 6.5579 ± 0.0139 (Instr. 1; person 1, day 1, [drug] = 20 μg ml^{-1}) 6.5593 ± 0.0102 (Instr. 1; person 1, day 2, [drug] = 20 μg ml^{-1}) 6.5584 ± 0.0317 (Instr. 1; person 1, day 3, [drug] = 40 μg ml^{-1}) 6.5451 ± 0.0251 (Instr. 2; person 2, day 4, [drug] = 20 μg ml^{-1})
893	Nimesulide	6.51 ± 0.05	A	−H			Singh A, Singh P and Kapoor VK, Nimesulide, *APDS*, **28**, 203 (2001). "Various pK_a values have been reported: 5.9 [38]; 6.46 [39]; 6.50 [40], 6.56 [41]. [38] Hansch C, Sammes PG and Taylor JB, *Compr. Med. Chem.*, **6**, Pergamon, Oxford, p. 711 (1990). [39] Fallavena PRB and Schapoval EES, pK_a determination of nimesulide in methanol-water mixtures by potentiometric titrations, *Int. J.Pharm.*, **158**(1), 109–112 (1997). NB: See separate entry. [40] Magni E, *Drug Invest* (Suppl.) **3**, 1 (1991). [41] Singh S, Sharda N and Mahajan L, *Int. J. Pharm.*, **176**, 261–264 (1999)." NB: See separate entry.
894	Nimetazepam ($C_{16}H_{13}N_3O_3$)	2.53	U	+H	Spectro (λ = 280 nm)	EtOH/H_2O t = 37 I undefined	Inotsume N and Nakano M, Reversible ring opening reactions of nimetazepam and nitrazepam in acidic media at body temperature, *J. Pharm. Sci.*, **69**, 1331–1334 (1980). NB: The percentage of ethanol in the experimental systems (from dilution of ethanolic stock solutions with aqueous buffers) was not stated, but was presumably small. Used the spectrophotometric procedure of Albert and Serjeant (1971).
895	Nitrazepam ($C_{15}H_{11}N_3O_3$)	2.84 ± 0.2 10.51 ± 0.05	U U	+H −H	Potentiometric	H_2O t = 25.0 ± 0.1 I = 0.1 (NaCl)	Takacs-Novak K, Box KJ and Avdeef A, Potentiometric pK_a determination of water-insoluble compounds: Validation study in methanol/water mixtures, *Int. J. Pharm.*, **151**, 235–248 (1997). NB: By extrapolation from 10–64%w/w aqueous MeOH. See Acetaminophen for full details.

896	Nitrazepam	GLpK_a: 2.90 ± 0.05 10.39 ± 0.04 A&S: 2.94 ± 0.05 10.66 ± 0.04	 A A A A	 +H +H +H +H	Spectro	H_2O $t = 25$ $I = 0.15$ (KCl) Ar atmos-phere	Tam KY and Takacs-Novac K, Multi-wavelength spectrophotometric determination of acid dissociation constants, *Anal. Chim. Acta*, **434**, 157–167 (2001). NB: See Clioquinol for details.
897	Nitrazepam	3.2 10.8	U U	+H −H	Spectro	5% MeOH in H_2O $t = 20$ $I = 0.15$	Barrett J, Smyth WF and Davidson IE. Examination of acid-base equilibria of 1,4-benzodiazepines by spectrophotometry, *J. Pharm. Pharmacol.*, **25**, 387–393 (1973). NB: See 1,4-Benzodiazepines for details.
898	Nitrazepam	2.77	U	+H	Spectro ($\lambda = 280$ nm)	$EtOH/H_2O$ $t = 37$ I undefined	Inotsume N and Nakano M, Reversible ring opening reactions of nimetazepam and nitrazepam in acidic media at body temperature, *J. Pharm. Sci.*, **69**, 1331–1334 (1980). NB: See Nimetazepam for details.
899	Nitrazepam	3.2	U	+H		H_2O	Konishi M, Hirai K and Mari Y, Kinetics and mechanism of the equilibrium reaction of triazolam in aqueous solution, *J. Pharm. Sci.*, **71**(12), 1328–1334 (1982). NB: See Triazolam.
900	Nitrofurazone ($C_6H_6N_4O_4$) 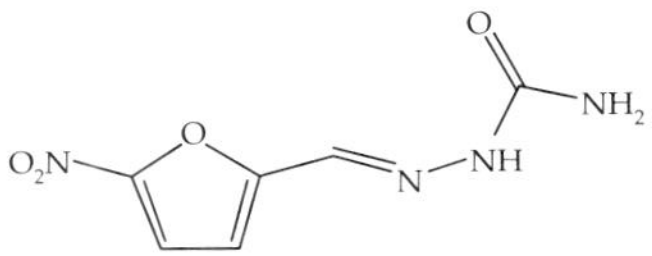	9.28 ± 0.03	U	−H	Potentiometric	EtOH $t = 35.0 \pm 0.1$ $I = 0.00$	Agrawal YK and Patel DR, Thermodynamic proton-ligand and metal-ligand stability constants of some drugs, *J. Pharm. Sci.*, **75**(2), 190–192 (1986). NB: See Clioquinol for details.
901	8-Nitrotheophylline ($C_7H_7N_5O_4$) 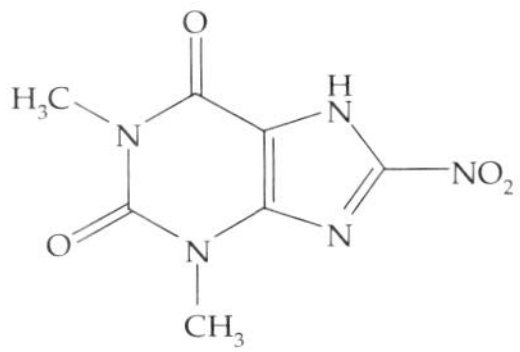	3.55 ± 0.05	U	−H	Spectro	H_2O $t = 25$ $I = 0.3$	Cohen JL and Connors KA, Stability and structure of some organic molecular complexes in aqueous solution, *J. Pharm. Sci.*, **59**, 1271–1276 (1970). NB: See Hypoxanthine.

(*continued*)

No.	Compound Name	pK_a value(s)	Data quality	Ionization type	Method	Conditions	Comments and Reference(s)
902	8-Nitrotheophylline	2.07	U	−H			Eichman ML, Guttman DE, van Winkle Q and Guth EP, Interactions of xanthine molecules with bovine serum albumin. I, *J. Pharm. Sci.* **51**, 66–71 (1962).
		3.6	U	−H			Guttman DE and Gadzala AE, Interactions of xanthine molecules with bovine serum albumin. II, *J. Pharm. Sci.*, **54**, 742–746 (1965). NB: In both cases, the pK_a value was quoted without experimental details. The lower value appears to be incorrect. Cited in Mayer MC and Guttman DE, Interactions of xanthine derivatives with bovine serum albumin III. Inhibition of binding, *J. Pharm. Sci.*, **57**, 245–249 (1968).
903	Norcodeine ($C_{17}H_{19}NO_3$) 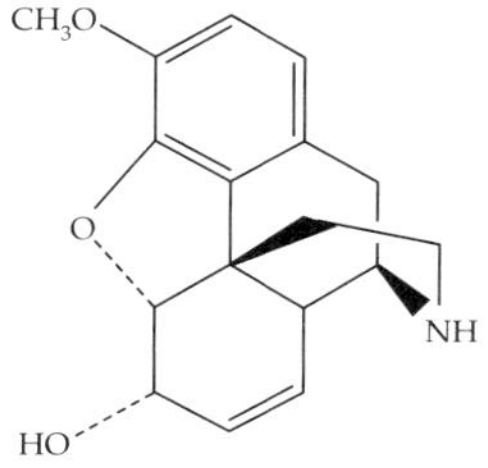	9.23 ± 0.01	A	+H	Potentiometric	H_2O $t = 25.0 \pm 0.1$ $I = 0.15$ (KCl) under Ar	Avdeef A, Barrett DA, Shaw PN, Knaggs RD and Davis SS, Octanol-, chloroform-, and propylene glycol dipelargonat-water partitioning of morphine-6-glucuronide and other related opiates, *J. Med. Chem.*, **39**, 4377–4381 (1996). NB: See Morphine for further details. The same result was reported in Sirius Technical Application Notes, vol. **2**, p. 151, 1995. Sirius Analytical Instruments Ltd., Forest Row, East Sussex, RH18 5DW, UK. NB: Concentration of analyte, 0.7–0.8 mM.
904	Norcodeine	9.10	U	+H	Potentiometric	H_2O t undefined I undefined	Rapaport H and Masamune S, The stereochemistry of 10-hydroxycodeine derivatives, *JACS*, **77**, 4330–4335 (1955). Cited in Perrin Bases Supplement 7485 ref. R3. NB: The study used measurements of pH with a glass electrode and liquid junction potentials.
905	Norcodeine, *N*-(2-cyano)ethyl ($C_{20}H_{22}N_2O_3$) 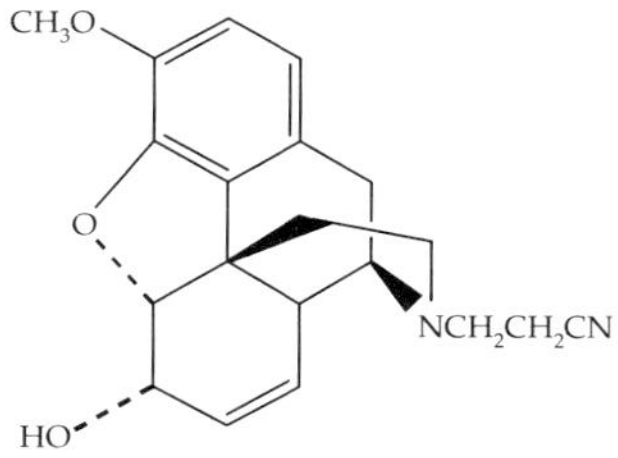	5.68	A	+H	Potentiometric	H_2O $t = 25 \pm 0.1$ $I < 0.02$	Stephenson GW, Williamson D, Base strengths of cyanoamines, *JACS*, **80**, 5943–5947 (1958). Cited in Perrin Bases. No. 2937 ref. S75. NB: Study used a glass electrode with liquid junction potential. Careful calibration of the pH meter was reported and activity effects taken into account. Also reported a pK_a value for methamphetamine, *N*-(2-cyano)ethyl = 6.95. The mean difference in pK_a caused by the presence of the 2-cyanoethyl group was about −3.28 log unit.

No.	Compound	pKa			Method	Conditions	Reference
906	Nordefrin (cobefrin) ($C_9H_{13}NO_3$) 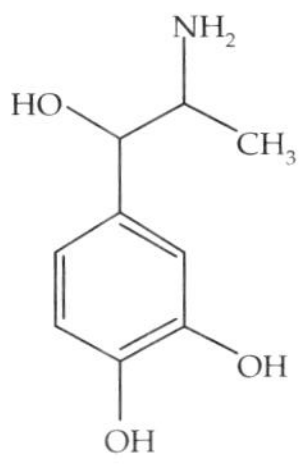	8.5	U	+H	Potentiometric	H_2O $t = 25.0 \pm 0.2$ $I \leq 0.001$	Leffler EB, Spencer HM and Burger A, Dissociation constants of adrenergic amines *JACS*, **73**, 2611–2613 (1951). NB: See Amphetamine for details.
907	Norephedrine ($C_9H_{13}NO$) 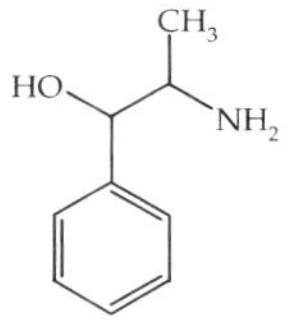	9.05	A	+H	Potentiometric	H_2O $t = 25.0 \pm 0.5$ $I = 0.01$	Warren RJ, Begosh PP and Zarembo JE, Identification of amphetamines and related sympathomimetic amines, *J. Assoc. Off. Anal. Chem.*, **54**, 1179–1191 (1971). NB: See Amphetamine for further details.
908	Norepinephrine (DL-noradrenaline) ($C_8H_{11}NO_3$) 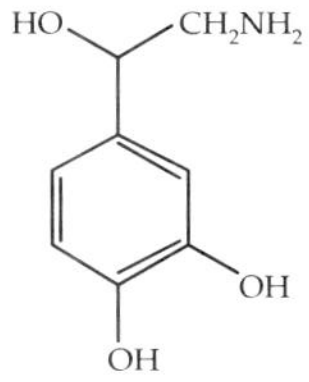	8.55 ± 0.02 9.82 ± 0.10	A U	+H, −H −H, +H	Potentio-metric, Spectro	H_2O $t = 37$ $I = 1.0$	van Hees, MTIW, Reactivity of Catechol(amine)s, Ph.D. thesis, Ctr for *Biopharm.* Sci., Leiden University, 1983, 1–93. NB: Microscopic: 8.72 ± 0.01; 9.03 ± 0.05; 9.65 ± 0.09; 9.34 ± 0.03. Schüsler-van Hees, MTIW, Beijersbergen van Henegouwen GMJ and Driever MFJ, Ionization constants of catechols and catecholamines, *Pharm. Weekblad Sci. Edn.* 5, 102–108 (1983).
909	Norepinephrine (noradrenaline)	8.63 9.73	A U	+H, −H −H, +H	Spectro	H_2O $t = 25.0 \pm 0.05$ $I = 0.10$	Ijzerman AP, BultsmaT, Timmerman H and Zaagsma J, The ionization of β-adrenoceptor agonists: A method for unravelling ionization schemes, *J. Pharm. Pharmacol.*, **36**(1), 11–15 (1984). NB: Microscopic: 8.68 and 9.68; macroscopic: 8.63 and 9.73. See Isoprenalin.
910	Norepinephrine	8.58	U	+H	Potentiometric	H_2O t undefined I undefined	Tuckerman MM, Mayer JR and Nachod FC, Anomalous pK_a values of some substituted phenylethylamines, *JACS*, **81**, 92–94 (1959). NB: Method as described by Parke and Davis, 1945.

(continued)

No.	Compound Name	pK_a value(s)	Data quality	Ionization type	Method	Conditions	Comments and Reference(s)
911	Norfenefrine ($C_8H_{11}NO_2$)	8.67	U	+H, −H	Potentiometric	H_2O t undefined I undefined	Wagner J, Grill H and Henschler D, Prodrugs of etilefrine: Synthesis and evaluation of 3′-(O-Acyl) derivatives, *J. Pharm. Sci.*, **69**(12), 1423–1427 (1980). "Phenolethanolamines in solution represent a mixture of the uncharged form and ionic species (cation, anion, and zwitterion). At half-neutralization, these compounds are in apparent average pK_a value. The pK_a value . . . increased from 8.67 (norfenefrine) to 8.9 (phenylephrine) or 9.0 (etilefrine) for the 3′-hydroxyphenylethanolamines when introducing a methyl or ethyl . . . into the amino group. The same value also was obtained for etilefrine in 1.5×10^{-3} M aqueous solution."
912	Norhyoscyamine ($C_{16}H_{21}NO_3$)	10.28	U	+H	Potentiometric	H_2O $t = 21 \pm 2$ I undefined	Bottomley W and Mortimer PI, Partition separation of tropane alkaloids. *Aust. J. Chem.*, **7**, 189–196 (1954). Cited in Perrin Bases 2939 ref. B86. NB: Used a glass electrode in an unsymmetrical cell with liquid junction potentials. See Hyoscyamine for details.
913	Norketobemidone ($C_{14}H_{19}NO_2$)	9.00 9.84	U U	+H, −H −H, +H	Spectro ($\lambda = 300$ nm)	H_2O	Ahnfelt N, Abrahamsson L, Bondesson V and Hartvig P, Reaction of the aminophenols ketobemidone and norketobemidone in buffered aqueous solution with ethyl chloroformate, *Acta Pharm. Suec.*, **19**, 355–366 (1982). Macroconstants: pK_1 = 9.00, pK_2 = 9.84. Microconstants: pk_1 = 9.20, pk_2 = 9.44, pk_3 = 9.64, and pk_4 = 9.40.

No.	Compound	pKa			Method	Conditions	Reference
914	Normorphine ($C_{16}H_{17}NO_3$) 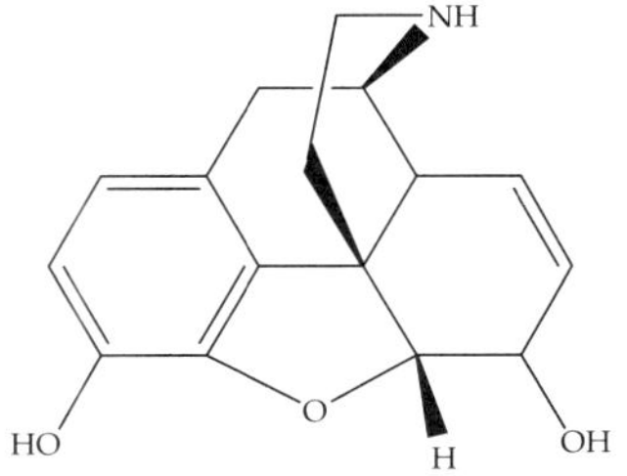	8.66 ± 0.01 9.80 ± 0.01	A A	+H −H	Potentiometric	H_2O $t = 25.0 \pm 0.1$ $I = 0.15$ (KCl) Ar atmosphere	Avdeef A, Barrett DA, Shaw PN, Knaggs RD and Davis SS, Octanol-, chloroform-, and propylene glycol dipelargonat-water partitioning of morphine-6-glucuronide and other related opiates, *J. Med. Chem.*, **39**, 4377–4381 (1996). NB: See Morphine for further details. The same result was reported in Sirius Technical Application Notes, vol. **2**, p. 151, (1995). Sirius Analytical Instruments Ltd., Forest Row, East Sussex, RH18 5DW, UK. NB: Concentration of analyte, 1.1–1.3 mM.
915	Normorphine	9.76	U	−H	Potentiometric	H_2O *t* undefined *I* undefined	Rapaport H and Masamune S, The stereochemistry of 10-hydroxycodeine derivatives, *JACS*, **77**, 4330–4335 (1955). Cited in Perrin Bases Supplement 7486 ref. R3. NB: The study used measurements of pH with a glass electrode and liquid junction potentials.
916	Norparamethadione ($C_6H_9NO_3$) 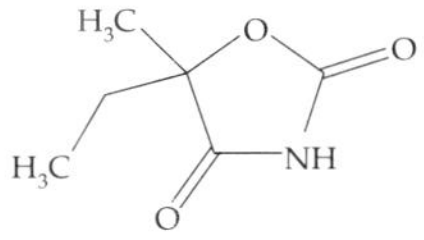	6.1	U	−H	Spectro	H_2O RT *I* undefined	Butler TC, The effects of *N*-methylation in 5,5-disubstituted derivatives of barbituric acid, hydantoin and 2,4-oxazolidinedione, *J. Am. Ph. Assoc.*, **44**, 367–370 (1955). NB: Other reported values: barbital (7.8); metharbital (8.2); phenobarbital (7.3); mephobarbital (7.7); *N*-norhexobarbital (7.9); hexobarbital (8.3). As the values for barbital and phenobarbital are about 0.2 units less than the best values in the literature, the remaining values should be regarded as low by the same amount.
917	Norpseudoephedrine ($C_9H_{13}NO$) 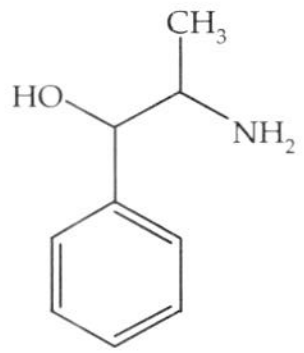	9.19	A	+H	Potentiometric	H_2O $t = 25.0 \pm 0.5$ $I = 0.01$	Warren RJ, Begosh PP and Zarembo JE, Identification of amphetamines and related sympathomimetic amines, *J. Assoc. Off. Anal. Chem.*, **54**, 1179–1191 (1971). NB: See Amphetamine for further details.

(*continued*)

No.	Compound Name	pK_a value(s)	Data quality	Ionization type	Method	Conditions	Comments and Reference(s)
918	Nortrimethadione ($C_5H_7NO_3$) 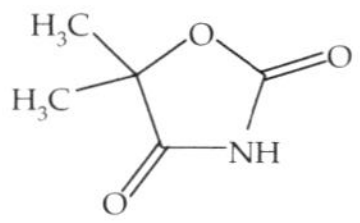	6.2	U	−H	Spectro	H_2O RT *I* undefined	Butler TC, The effects of *N*-methylation in 5,5-disubstituted derivatives of barbituric acid, hydantoin and 2,4-oxazolidinedione, *J. Am. Ph. Assoc.*, **44**, 367–370 (1955). NB: See Norparamethadione for additional details.
919	Nortriptyline ($C_{19}H_{21}N$) 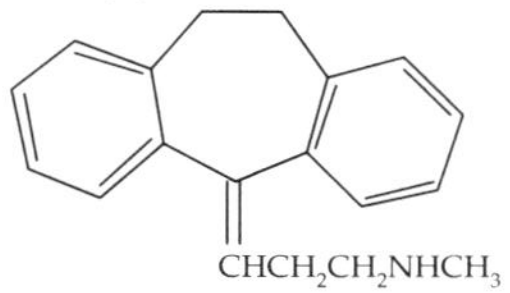	9.7	U	+H			Craig; N&K; Avery.
920	Noscapine (narcotine) ($C_{22}H_{23}NO_7$) 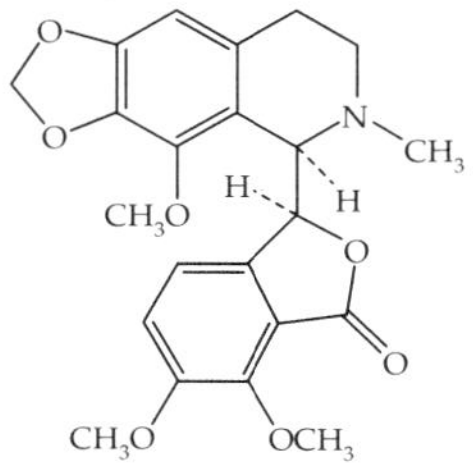	6.37	U	+H	Spectro	H_2O $t = 15$ $c = 0.0005$ to 0.002	Kolthoff IM, The dissociation constants, solubility product and titration of alkaloids, *Biochem. Z.*, **162**, 289–353 (1925). Cited in Perrin Bases no. 2932 ref. K47. NB: See Aconitine for details.
921	Noscapine (narcotine)	5.86	U	+H	Potentiometric	H_2O $t = 24$	Muller F, *Z. Elektrochem.*, **30**, 587 (1924). Cited in Perrin Bases 2932 ref. M60. NB: Study used measurements of pH using hydrogen electrodes in an asymmetric cell with liquid junction potentials.
922	Noscapine (narcotine)	4.38	VU	+H	kinetic	H_2O $t = 55$	Arnall F, The determination of the relative strengths of some nitrogen bases and alkaloids, *J. Chem. Soc.*, **117**, 835–839 (1920). Cited in Perrin Bases no. 2932 ref. A73.

No.	Compound	pKa value(s)	Data quality	Ionization type	Method	Conditions	Comments and Reference(s)
923	Octodrine ($C_8H_{19}N$) 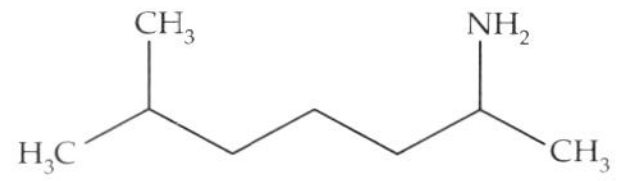	10.28 ± 0.1	U	+H	Potentiometric	H_2O $t = 25.0 \pm 0.2$ $I \leq 0.001$	Leffler EB, Spencer HM and Burger A, Dissociation constants of adrenergic amines *JACS*, **73**, 2611–2613 (1951). NB: See Amphetamine for details.
924	Oleandomycin ($C_{35}H_{61}NO_{12}$) 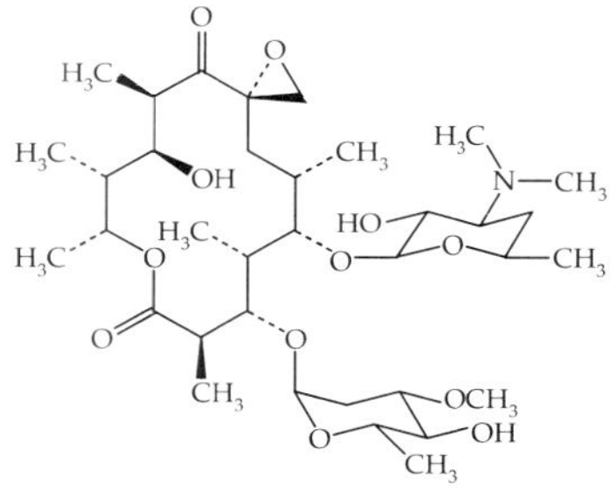	8.5	U	+H		50% aq EtOH t unspecified *I* unspecified	Els H, Celmer WD and Murai K, Oleandomycin (PA-105). II. Chemical characterization (I), *JACS*, **80**, 3777–3782 (1958). Cited in: Celmer WO, Els H and Murai K, *Antibiotics Ann.*, 476–483, 1957–58; NB: See also Winningham DG, Nemoy NJ, Stamey TA. Diffusion of antibiotics from plasma into prostatic fluid, *Nature* **219**, 139–143 (1968). Method not given but probably potentiometric.
925	Oleandomycin	8.84	U	+H	Potentiometric	H_2O $t = 25$ $I = 0.167$	McFarland JW, Berger CM, Froshauer SA, Hayashi SF, Hecker SJ, Jaynes BH, Jefson MR, Kamicker BJ, Lipinski CA, Lundy KM, Reese CP and Vu CB, Quantitative Structure-activity relationships among macrolide antibacterial agents: *In vitro* and *in vivo* potency against Pasteurella multocida, *J. Med. Chem.*, **40**, 1340–1346 (1997). NB: See Azithromycin for details; average standard deviation of ± 0.07 for the pK_a.
926	Oleic acid ($C_{18}H_{34}O_2$)  $CH_3(CH_2)_6$–CH=CH–$(CH_2)_6COOH$	5.35	U	−H	Potentiometric	H_2O RT	Johns WH and Bates TR, Quantification of the binding tendencies of cholestyramine II. Mechanism of interaction with bile salts and fatty acid salt anions, *J. Pharm. Sci.*, **59**, 329–333 (1970).
927	Omeprazole ($C_{17}H_{19}N_3O_3S$) 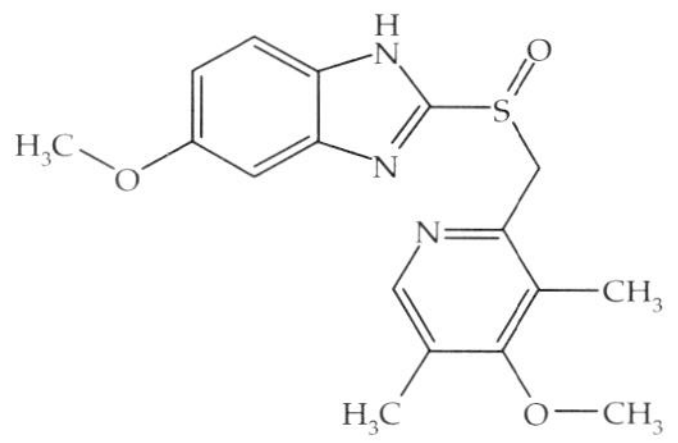	4.14 8.9	U U	+H −H	CE/pH (+ve ion mode)	H_2O $t = 25$ $I = 0.025$	Wan H, Holmen AG, Wang Y, Lindberg W, Englund M, Nagard MB and Thompson RA, High-throughput screening of pK_a values of pharmaceuticals by pressure-assisted capillary electrophoresis and mass spectrometry, *Rapid Commun. Mass Spectrom.*, **17**, 2639–2648 (2003). NB: Reported predicted values (ACD Labs) of 4.5 and 7.34.

(continued)

No.	Compound Name	pK_a value(s)	Data quality	Ionization type	Method	Conditions	Comments and Reference(s)
928	Opipramol ($C_{23}H_{29}N_3O$) 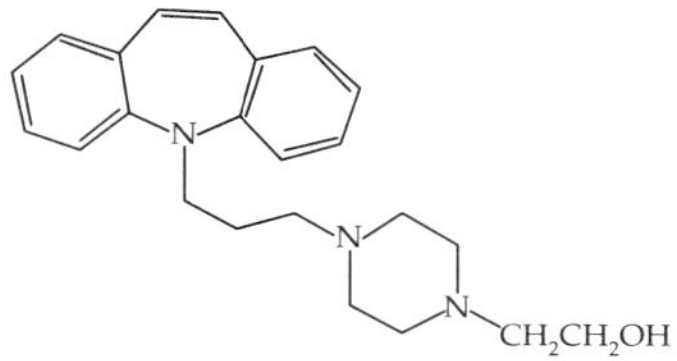	8	U	+H	Potentiometric	H_2O $t = 25$ $I = 0.01$ (NaCl)	Lukkari S, Ionization of some dibenzazepine derivatives, dibenzepine and opipramol, in aqueous solutions, *Farm. Aikak.*, **80**(4–5), 210–215 (1971). NB: See Dibenzepine for details.
929	Orciprenaline (metaproterenol) ($C_{11}H_{17}NO_3$) 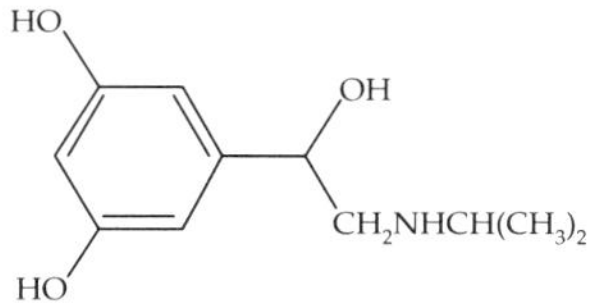	8.70 9.92	U U	−H, +H +H, −H	Spectro	H_2O $t = 25.0 \pm 0.05$ $I = 0.10$	Ijzerman AP, BultsmaT, Timmerman H and Zaagsma J, The ionization of β-adrenoceptor agonists: a method for unravelling ionization schemes, *J. Pharm. Pharmacol.*, **36**(1), 11–15 (1984). NB: Microscopic: 8.67 and 9.92. Negligible formation of the neutral form. See Isoprenalin.
930	Oxamniquine ($C_{14}H_{21}N_3O_3$) 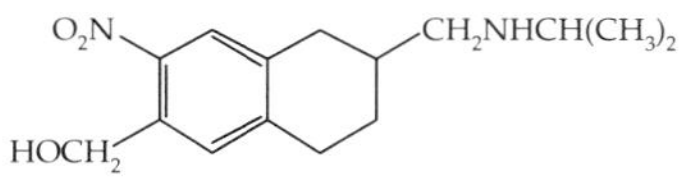	3.28 ± 0.07 9.53	U U	+H +H	log P/pH soly	H_2O $t = 25.0$	Kofitsekpo WM, An experimental evaluation of the log P of oxamniquine—a new schistosomicide, *Drugs Exptl. Clin. Res.*, **6**(5), 421–426 (1980). NB: Reported intrinsic solubility $S_o = 7.8853 \times 10^{-5}$ M.
931	Oxazepam ($C_{15}H_{11}ClN_2O_2$) 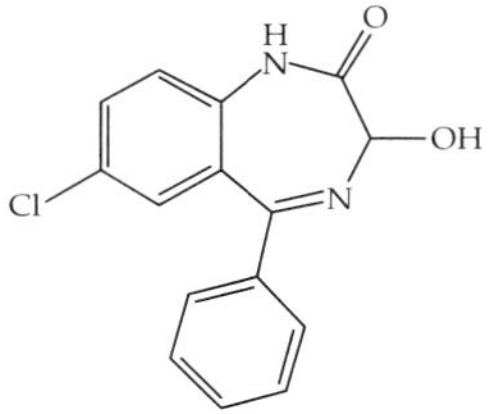	1.6 11.6	U U	+H −H	Spectro	5% MeOH in H_2O $t = 20$ $I = 0.15$	Barrett J, Smyth WF and Davidson IE. Examination of acid-base equilibria of 1,4-benzodiazepines by spectrophotometry, *J. Pharm. Pharmacol.*, **25**, 387–393 (1973). NB: See 1,4-Benzodiazepines for details.

No.	Compound	pKa	Type	Sign	Method	Conditions	Reference
932	Oxazepam	1.8 11.1	U U	+H −H	Spectro	H_2O	Shearer CM and Pilla CR, Oxazepam, *APDS*, **3**, 441–464 (1974). NB: No reference given.
933	Oxazepam	1.7	U	+H		H_2O	Konishi M, Hirai K and Mari Y, Kinetics and mechanism of the equilibrium reaction of triazolam in aqueous solution, *J. Pharm. Sci.*, **71**, 1328–1334 (1982). NB: See Triazolam.
934	Oxycodone	8.53	U	+H	soly	H_2O $t = 37$	Kuo PC, Liu JC, Chang SF and Chien YW, *In vitro* transdermal permeation of oxycodone. I. Effect of pH, delipidation and skin stripping, *Drug Dev. Ind. Pharm.*, **15**, 1199–1215 (1989). NB: No further details given. Value was repeated in Tien J-H., Transdermal-controlled administration of oxycodone, *J. Pharm. Sci.*, **80**, 741–743 (1991); Lombardo F, Obach RS, Shalaeva MY and Gao F, Prediction of human volume of distribution values for neutral and basic drugs. 2. Extended data set and leave-class-out statistics, *J. Med. Chem.*, **47**, 1242–1250 (2004).
935	Oxyphenbutazone ($C_{19}H_{20}N_2O_3$) OH N—N O O H $CH_2CH_2CH_2CH_3$	4.60	U	−H	Potentiometric	80% Me cellosolve	Girod E, Delley R and Hafliger F, Uber derivate des Phenylbutazons III. Die Struktur der Reduktionsprodukte des γ-keto-phenylbutazons, *Helv. Chim. Acta*, **40**, 408–428 (1957). NB: Values obtained by titration in 80% methylcellosolve with tetramethylammonium hydroxide. Apparent pK_{aMCS} values for 29 other analogues also reported.
936	Oxyphenbutazone	4.5	U	−H	Spectro	H_2O *t* undefined *I* undefined	Herzfeldt CD and Kümmel R, Dissociation constants, solubilities, and dissolution rates of some selected nonsteroidal antiinflammatories, *Drug Dev. Ind. Pharm.*, **9**(5), 767–793 (1983). NB: Used dλ/dpH method. See Azapropazone and Ibuprofen for details.
937	Oxyphenbutazone	5.85	U	+H	Potentiometric	80% Me cellosolve	Jahn U and Wagner-Jauregg T, Wirkungsvergleich saurer Antiphlogistika im Bradykinin-, UV-Erythem- und Rattenpfotenödem-Test. *Arzneim.-Forsch.*, **24**, 494–9 (1974). NB: Literature values obtained from the pH of half-neutralization. Also gave a value of 4.7 in water. This last value is in good agreement with other values.

(continued)

No.	Compound Name	pK_a value(s)	Data quality	Ionization type	Method	Conditions	Comments and Reference(s)
938	Oxyphenbutazone	4.7 10.0 ± 0.2	U U	−H −H	Spectro	H_2O t = RT I undefined	Perel JM, Snell MM, Chen W and Dayton PG, *Biochem. Pharmacol.*, **13**, 1305–1317 (1964). NB: performed in "dilute buffer solutions." Quoted in W&G incorrectly as pK_{a1} = 4.5. See: Phenylbutazone analogs.
939	Oxytetracycline ($C_{22}H_{24}N_2O_9$) 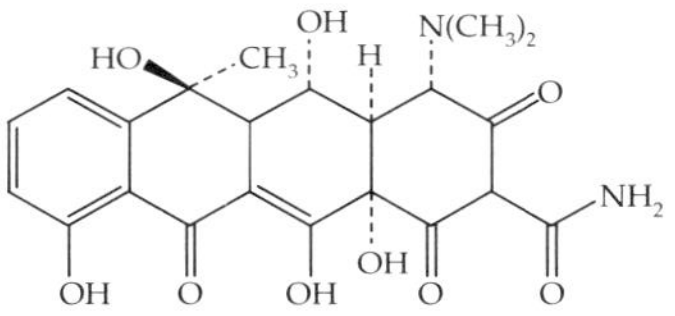	3.27 7.32 9.11	A A A	−H −H +H	Potentiometric	H_2O t = 25 I < 0.01	Stephens C, Murai K, Brunings K and Woodward RB, Acidity constants of the tetracycline antibiotics, *JACS*, **78**, 4155–4158 (1956). Cited in Perrin Bases 3329 ref. S73. NB: Used a glass electrode with liquid junction potentials.
940	Oxytetracycline	3.47 7.57 9.27	A A A	−H −H +H	Potentiometric	H_2O t = 28 I < 0.005	Regna PP, Solomons IA, Murai K, Timreck AE, Brunings KJ and Lazier WA, *JACS*, **73**, 4211–4215 (1951). Cited in Perrin Bases 3329 ref. R9. NB: The study used pH measurements with a glass electrode and junction potentials. Minor corrections were made, based on ionic strengths.
941	Oxytetracycline	3.09 7.28 9.14	U A A	−H −H +H	Potentiometric	H_2O t = 20 I < 0.002	Albert A, Avidity of terramycin and aureomycin for metallic cations. *Nature*, **172**, 201 (1953). Cited in Perrin Bases 3329 ref. A17. NB: The study used pH measurements with a glass electrode and liquid junction potentials.
942	Oxytetracycline	3.04 8	U U	+H −H	CE/pH (+ve ion mode)	H_2O t = 25 I = 0.025	Wan H, Holmen AG, Wang Y, Lindberg W, Englund M, Nagard MB and Thompson RA, High-throughput screening of pK_a values of pharmaceuticals by pressure-assisted capillary electrophoresis and mass spectrometry, *Rapid Commun. Mass Spectrom.*, **17**, 2639–2648 (2003). NB: Reported predicted values (ACD Labs) of 4.72 and 8.53. The assignment of the first pK_a value to a basic group disagrees with the consensus of literature, which assigns this value to the tricarbonyl carbon acid system.
943	Oxytetracycline	GLpK_a: 3.23 ± 0.01 7.22 ± 0.02 8.82 ± 0.02 A&S: 3.25 ± 0.05	 A A A U	 −H −H +H −H	Spectro	H_2O t = 25 I = 0.15 (KCl) Ar atmosphere	Tam KY and Takacs-Novac K, Multi-wavelength spectrophotometric determination of acid dissociation constants, *Anal. Chim. Acta*, **434**, 157–167 (2001). NB: See Clioquinol for details.

944	Oxytetracycline	3.60 7.42 9.11	U U U	−H −H +H	Potentiometric	H_2O $t = 30.0 \pm 0.2$ $I = 0.01$ (KCl) N_2 atmosphere	Doluisio JT and Martin AN, Metal complexation of the tetracycline hydrochlorides, *J. Med. Chem.*, **6**, 16–20 (1963). NB: Metal-free solutions of the tetracycline were titrated with standard NaOH solution and the pH measured. No details given of the pH meter calibration. Metal stability constants determined from identical titrations in the presence of varying concentrations of nickel(II), zinc(II) or copper(II) ions.
945	Oxytetracycline	3.3 7.3 9.1	U U U	−H −H +H			Foye 1 N&K; Avery; W&G
946	Papaverine ($C_{20}H_{21}NO_4$) 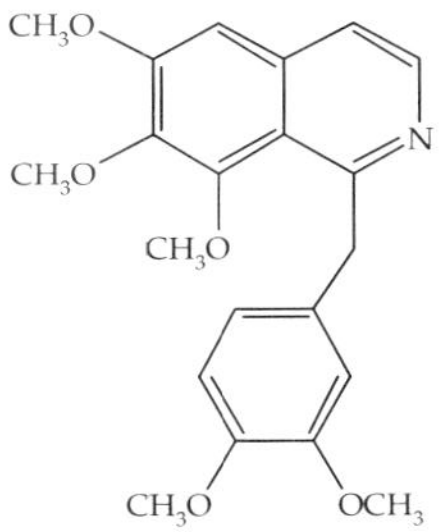	6.40	A	+H	Spectro	H_2O $t = 25 \pm 1$ $c = 1.5 \times 10^{-5}$	Biggs AI, *Trans. Farad. Soc.*, **50**, 800–802 (1954). Cited in Perrin Bases no. 2942, ref. B64. NB: Used measurements of absorption combined with pH measurements and reported as the pK_b.
		6.13	U	+H	Spectro	H_2O $t = 15$ $c = 0.0006–0.003$	Kolthoff IM, The dissociation constants, solubility product and titration of alkaloids, *Biochem. Z.*, **162**, 289–353 (1925). Cited in Perrin Bases no. 2942, ref. K47. See Aconitine for details.
947	Papaverine	6.4 7.60 (pK_b)	U U	+H +H	Spectro	$t = 25$	Hifnawy, MS and Muhtadi, FJ, Papaverine hydrochloride, *APDS*, **17**, 367–447 (1988). Ref.: Briggs AI, *Trans Farad. Soc.*, **50**, 800, 1954. NB. Error for Biggs AI (no. 946). NB: The pK_a and pK_b values could easily become confused for papaverine.
948	Papaverine	6.25	U	+H	CE/pH (+ve ion mode)	H_2O $t = 25$ $I = 0.025$	Wan H, Holmen AG, Wang Y, Lindberg W, Englund M, Nagard MB and Thompson RA, High-throughput screening of pK_a values of pharmaceuticals by pressure-assisted capillary electrophoresis and mass spectrometry, *Rapid Commun. Mass Spectrom.*, **17**, 2639–2648 (2003). NB: Reported a predicted value (ACD Labs) of 6.32.
949	Papaverine	6.18 ± 0.03	U	+H	Potentiometric	H_2O $t = 25.0 \pm 0.1$ $I = 0.1$ (NaCl)	Takacs-Novak K, Box KJ, Avdeef A, Potentiometric pK_a determination of water-insoluble compounds: validation study in methanol/water mixtures, *Int. J. Pharm.*, **151**, 235–248 (1997). NB: $pK_a = 6.18 \pm 0.03$ by extrapolation from 16.4–64.7% w/w aqueous MeOH. See Acetaminophen for full details.

(continued)

No.	Compound Name	pK_a value(s)	Data quality	Ionization type	Method	Conditions	Comments and Reference(s)
950	Papaverine	6.38 ± 0.03 6.39 ± 0.01	A A	+H +H	Potentiometric	H_2O $t = 25.0$ $I = 0.1$ (NaCl) $I = 0.15$ (KCl)	Takacs-Novak K and Avdeef A, Interlaboratory study of log P determination by shake-flask and potentiometric methods, *J. Pharm. Biomed. Anal.*, **14**, 1405–1413 (1996). NB: See Acetaminophen for more details. Also reported pK_a = 6.39 ± 0.01 ($I = 0.15$; KCl). The same result was reported in Sirius Technical Application Notes, 1995, vol. **2**, p. 151. Sirius Analytical Instruments Ltd., Forest Row, East Sussex, RH18 5DW, UK. NB: Concentration of analyte, 1.1–1.3 mM.
951	Papaverine	GLpK_a: 6.47 ± 0.01 A&S: 6.51 ± 0.04	A A	+H +H	Spectro	H_2O $t = 25$ $I = 0.15$ (KCl) Ar atmosphere	Tam KY and Takacs-Novac K, Multi-wavelength spectrophotometric determination of acid dissociation constants, *Anal. Chim. Acta*, **434**, 157–167 (2001). NB: See Clioquinol for details.
952	Pecazine(mepazine) ($C_{19}H_{22}N_2S$)	9.48 ± 0.03	U	+H	partition/pH	H_2O $t = 20 \pm 0.5$	Vezin WR and Florence AT, The determination of dissociation constants and partition coefficients of phenothiazine derivatives, *Int. J. Pharm.*, **3**, 231–237 (1979). NB: I not reported but pK_a was stated to be independent of I. See Chlorpromazine and Promethazine for additional details.
953	Pecazine(mepazine)	9.7	U	+H	soly	H_2O $t = 24 \pm 1$	Green AL, Ionization constants and water solublities of some aminoalkylphenothiazine tranquilizers and related compounds, *J. Pharm. Pharmacol.*, **19**, 10–16 (1967). NB: See Amitriptylline for details.
954	Pelargonic acid ($C_9H_{18}O_2$) $CH_3(CH_2)_7COOH$	4.96	U	−H			Ritschel: Johns and Bates (1970).

No.	Compound	pKa	Data quality	Ionization	Method	Conditions	Reference / Comments
955	Penicillamine ($C_5H_{11}NO_2S$) HS NH$_2$ | | $(CH_3)_2C$—CHCOOH	1.8	U	−H	Potentiometric	H_2O	Chiu CC and Grady LT, Penicillamine, *APDS*, **10**, 601–637 (1981).
		7.9	U	+H		$t = 25.0$	"Recently, the ionization constants for acid functions of D-penicillamine were verified by pH titration at 37 °C and 0.15 M ionic strength (17,18). These results correspond to that previously obtained by other workers (19–23).
		10.5	U	−H		I undefined	
		7.88	A	+H	Potentiometric (Lenz & Martell)	H_2O t=25.05±0.05 $I = 0.10$ (KNO_3) N_2 atmosphere	17. Laurie SH, Prime DH and Sarkar B, Analytical potentiometric and spectroscopic study of the equilibriums in the aqueous nickel(II)-triethylenetetramine and nickel(II)-D-penicillamine systems, *Can. J. Chem.*, **57**, 1411–1417 (1979).
		10.43	U	−H			18. Zucconi TD, Janauer GE, Donahe S and Lewkowicz C, *J. Pharm. Sci.*, **68**, 426–432 (1979) (see no. 956).
		>2	U	−H	Potentiometric (Doornbos & Faber)		19. Doornbos DA and Faber J, Metal complexes of drugs. D-penicillamine and *N*-acetyl-D-penicillamine, *Pharm. Weekbl.*, **99**, 289–309 (1964).
		8.01	U	+H			20. Kuchinskas E and Rosen Y, Metal chelates of DL-penicillamine, *Arch. Biochem. Biophys.*, **97**, 370–372 (1962).
		10.67	U	−H			21. Lenz GR and Martell AE, Metal chelates of some sulphur-containing amino acids, *Biochem.*, **3**, 745–750 (1964). NB: Report also the pK_a values for DL-methionine (9.04); S-methyl-L-cysteine (8.73); DL-ethionine (9.02) and L-cysteine (8.13, 10.11) under the same conditions. 22. Perrin DD and Sayce IG, Complex formation by nickel and zinc with penicillamine and cysteine, *J. Chem. Soc, A*, 53–57 (1968). 23. Ritsma JH and Jellinek F, Stereoselectivity in the complex formation of penicillamine with nickel(II), *Recueil. Trav. Chim.* **91**, 923–928 (1972)."
956	Penicillamine	7.83 ± 0.01	A	+H	Potentiometric	H_2O t=37.0 ± 0.05 $I = 0.15$	Zucconi TD, Janauer GE, Donahe S and Lewkowicz C, Acid dissociation and metal complex formation constants of penicillamine, cysteine, and antiarthritic gold complexes at simulated biological conditions, *J. Pharm. Sci.*, **68**(4), 426–432 (1979). NB: See Cysteine for details.
957	Penicilloic acid, benzyl ($C_{16}H_{20}N_2O_5S$) [structure: O, COOH, N H, S, CH_3, CH_3, HN, OH, O]	5.31 ± 0.05	U	+H	Potentiometric	H_2O $t = 25$ $c = 0.01$	Rapson HDC and Bird AE, Ionization constants of some penicillins and of their alkaline and penicillinase hydrolysis products, *J. Pharm. Pharmacol.*, Suppl. **15**, 222–231T (1963).

(continued)

No.	Compound Name	pK_a value(s)	Data quality	Ionization type	Method	Conditions	Comments and Reference(s)
958	Penicillin G (benzylpenicillin) ($C_{16}H_{18}N_2O_4S$)	 2.73 ± 0.03 2.71 ± 0.05 2.75 ± 0.03	 A A A	 –H –H –H	 Potentiometric Potentiometric	H_2O $t = 25$ $I = 0.0099$ $I = 0.0093$ H_2O $t = 37$ $I = 0.15$ (KCl)	Rapson HDC and Bird AE, Ionization constants of some penicillins and of their alkaline and penicillinase hydrolysis products, *J. Pharm. Pharmacol.*, Suppl. **15**, 222–231T (1963). Cited in Kirschbaum J, Penicillin G, Potassium, *APDS*, **15**, 427–507 (1986). NB: For $t = 20\,°C$ and $I = 0.15$ M, p$K_a = 3.82$ (20% aq. MeOH) and 4.10 (30% aq. MeOH) (Salto F, Prieto JG and Alemany MT, Interactions of cephalosporins and penicillins with non-polar octadecylsilyl stationary phase, *J. Pharm. Sci.*, **69**, 501–6 (1980)). Tsuji A, Kubo O, Miyamoto E and Yamana T, Physicochemical properties of β-lactam antibiotics: Oil-water distribution, *J. Pharm. Sci.*, **66**, 1675–1679 (1977).
959	Penicillin G (benzylpenicillin)	2.76	U	–H		H_2O $t = 25$	Woodward, RB, Neuberger A and Trenner NR, Other Physical Methods *in* Clarke HT, Johnson JR and Robinson Sir R (eds.), *The Chemistry of Penicillin*, Princeton University Press, Princeton NJ, p. 419 (1949). NB: p$K_a = 2.70$ in water at 23 °C; p$K_a = 2.74$ in water at 5 °C. Also reported a pK_a value of 4.84 in 80% EtOH. Other values in water at 23 °C in this reference: *p*-Hydroxybenzylpenicillin, 2.62; *n*-Heptylpenicillin, 2.66. Benzyldesthiopenicillin, p$K_a = 3.48$ in 8% aqueous EtOH at 23 °C.
960	Penicillin V (phenoxymethylpenicillin) ($C_{16}H_{18}N_2O_5S$) 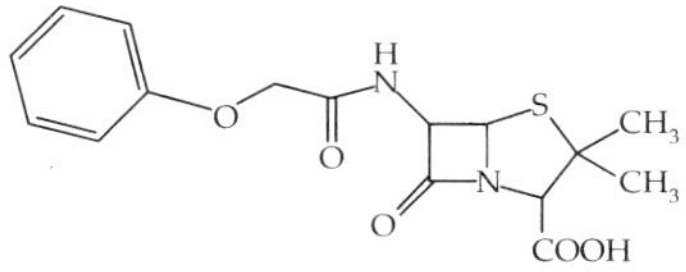	2.735 ± 0.05	A	–H	Potentiometric	H_2O $t = 25$ $c < 0.01$	Rapson HDC and Bird AE, Ionization constants of some penicillins and of their alkaline and penicillinase hydrolysis products, *J. Pharm. Pharmacol.*, Suppl. **15**, 222–231T (1963). Cited in Dunham JM, Potassium Phenoxymethyl Penicillin, *APDS*, **1**, 260 (1972).
961	Penicillin V (phenoxymethylpenicillin)	2.79 ± 0.03	A	–H	Potentiometric	$H_2Ot = 37$ $I = 0.15$ (KCl) $c = 0.01$	Tsuji A, Kubo O, Miyamoto E and Yamana T, Physicochemical properties of β-lactam antibiotics: Oil-water distribution, *J. Pharm. Sci.*, **66**, 1675–9 (1977). Cited in Sieh DH, Potassium penicillin V, *APDS*, **17** Supplement, 677–748 (1988). NB: Also reported the following values: propicillin, 2.76; phenethicillin, 2.80; benzylpenicillin, 2.75; methicillin, 2.77.

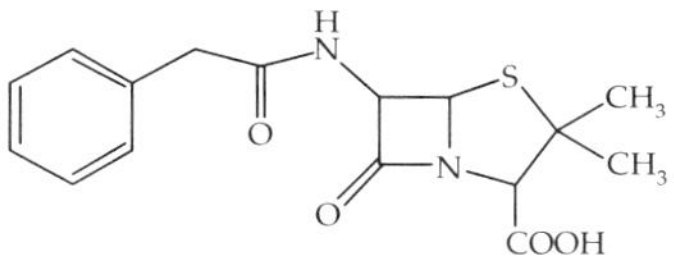

962	Penicillin V (phenoxymethylpenicillin)	2.78	A	−H	Potentiometric	H_2O $t = 37$ $I = 0.15$ (KCl) $c = 0.01$	Tsuji A, Kubo O, Miyamoto E and Yamana T, Physicochemical properties of β-lactam antibiotics: oil-water distribution, *J. Pharm. Sci.*, **66**, 1675–1679 (1977). Cited in Sieh DH, Potassium penicillin V, *APDS*, **17** Supplement, 677–748 (1988). NB: This result was obtained by extrapolation to zero % ethanol from the following apparent pK_a' values: 16.0, 3.15; 32.8, 3.61; 41.9, 3.85; 51.4, 4.06. The close agreement in this case (and also for oxacillin) between the extrapolated value and the value measured directly in water was used as validation for extrapolation to zero % ethanol for the following compounds: indanylcarbenicillin, 2.94; phenylcarbenicillin, 2.91; dicloxacillin, 2.76; floxacillin, 2.76; cloxacillin, 2.78.
963	Pentazocine ($C_{19}H_{27}NO$) 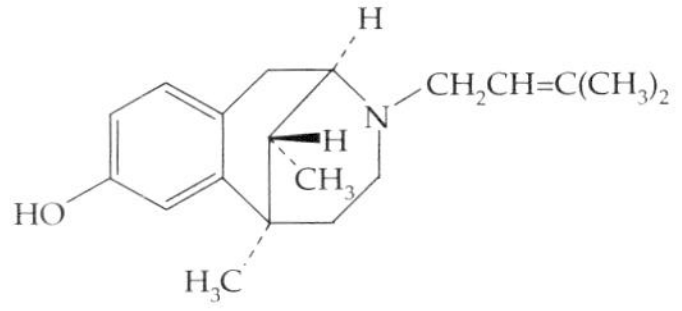	9.68 ± 0.05 11.23 ± 0.05	U U	+H,−H −H,+H	Spectro	H_2O $t = 20$ $I = 0.1$	Borg K and Mikaelsson A, Fluorometric determination of pentazocine in biological samples by partition chromatography as ion-pair in micro-columns. *Acta Pharm. Suecica*, **7**, 673–680 (1970). Cited in Wilson TD, Pentazocine, *APDS*, **13**, 361–411 (1984). "The four microscopic acid dissociation constants of pentazocine were determined according to [4,5] at 20 °C using sodium carbonate-bicarbonate buffer solutions of ionic strength 0.1. Apparent microconstants: $pK_1 = 9.74$, $pK_2 = 10.56$, $pK_{12} = 11.17$ and $pK_{21} = 10.35$. Macroscopic dissociation constants were determined from the relations: $K_1 = (a_{H+})([II] + [III])/[I]$; $K_2 = (a_{H+})[IV]/([II] + [III])$. where [I] is the concentration of the cation, [II] is the concentration of the zwitterion, [III] is the concentration of the neutral molecule and [IV] is the concentration of the anion."
964	Pentostatin ($C_{11}H_{16}N_4O_4$) 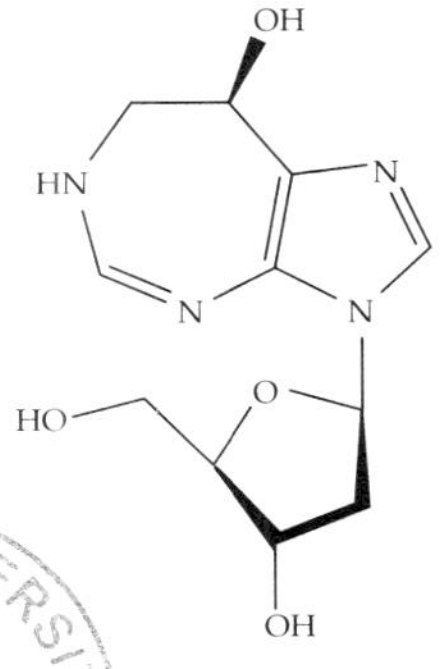	2.03 ± 0.03 5.57 ± 0.02 5.50 ± 0.02 1.67 ± 0.03	A A A A	+H +H +H +H	Spectro Potentiometric Spectro	H_2O $t = 25.0 \pm 0.1$ $I = 0.15$ (NaCl) H_2O $t = 25.0 \pm 0.1$ $I = 0.00$	Al-Razzak LA, Benedetti AE, Waugh WN and Stella VJ, Chemical stability of pentostatin (NSC-218321), a cytotoxic and immunosuppressive agent, *Pharm. Res.*, **7**, 452–460 (1990). "The pK_a values of pentostatin were determined using uv spectroscopy and by potentiometric titration. For the spectroscopic method a 100 μL aliquot of a 1.5×10^{-3} M solution of pentostatin in water was diluted with 1.0 ml of 0.05 M buffer solution . . . at various pH values (pH range, 1.0–8.0). Samples of the resulting solutions were placed in a cuvette thermostated at 25 °C. The changes in the absorbance were monitored at 279 and 300 nm for the determination of pK_{a1} and pK_{a2}, respectively. . . . At low pH . . . (pH < 3.0) it was necessary to extrapolate the absorbance measurement to zero time since pentostatin readily undergoes degradation under acidic conditions. The lower pK_a was also deterimined at zero buffer concentration with no ionic strength adjustment by measuring the pK_a at three different buffer concentrations (0.01, 0.1 and 0.5 M) . . . For potentiometric titration a pentostatin solution (0.1 M, initial pH ~8) containing 0.15 M NaCl was prepared using glass-distilled water and thermostated at 25 °C. The resulting solution was titrated with standardized 0.1 N HCl solution. After each addition, 1 min was allowed prior to pH reading."

(continued)

No.	Compound Name	pK_a value(s)	Data quality	Ionization type	Method	Conditions	Comments and Reference(s)
965	Perhexilene ($C_{19}H_{35}N$) 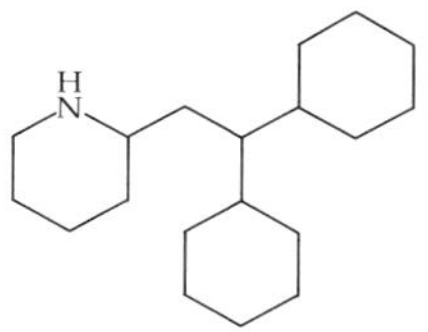	10.36 ± 0.06	U	+H	Potentiometric	40% EtOH $t = 25.0$	Mannhold R, Rodenkirchen R, Bayer R and Haus W, The importance of drug ionization for the action of calcium antagonistsand related compounds, *Arzneim.-Forsch.*, **34**, 407–409 (1984).
		11.0	U	+H		H_2O	NB: See Aprindine for details.
966	Perphenazine ($C_{21}H_{26}ClN_3OS$) 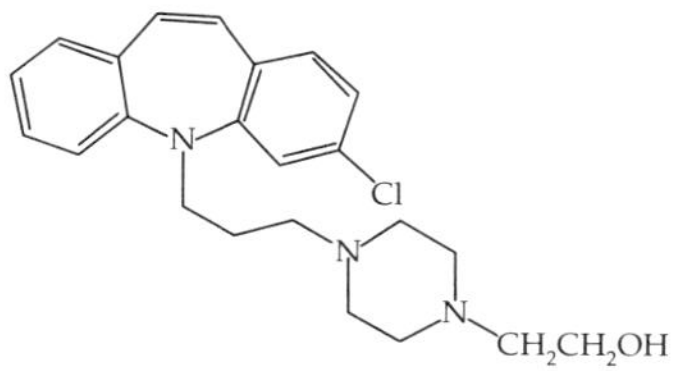	7.8	U	+H	soly	H_2O $t = 24 \pm 1$	Green AL, Ionization constants and water solublities of some aminoalkylphenothiazine tranquilizers and related compounds, *J. Pharm. Pharmacol.*, **19**, 10–16 (1967). NB: See Amitriptylline for details.
967	Perphenazine	8.11	U	+H	CZE/pH		Mannhold R, Dross KP and Reffer RF, Drug lipophilicity in QSAR practice: I. A comparison of experimental with calculative approaches, *Quant-Struct.-Act. Relat.*, **9**, 21–28 (1990). NB: Cited in Lombardo F, Obach RS, Shalaeva MY and Gao F, Prediction of human volume of distribution values for neutral and basic drugs. 2. Extended data set and leave-class-out statistics, *J. Med. Chem.*, **47**, 1242–1250 (2004). Ref. 277.
968	Pethidine	8.7	U	+H			NB: See Meperidine.
969	Pharmorubicin	4.81 ± 0.13	U	+H			NB: See Daunorubicin for experimental details.
970	Phenazocine ($C_{22}H_{27}NO$) 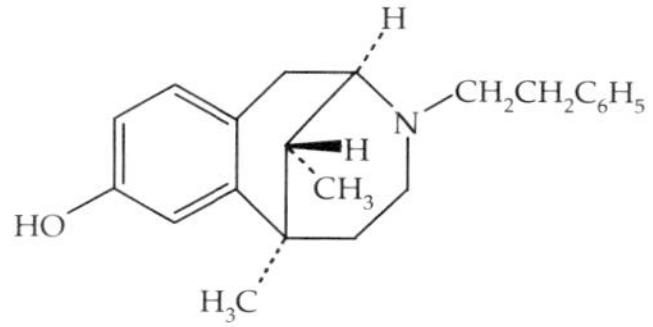	8.50	U	+H	Potentiometric	50% aq EtOH	Clouet DH (ed.), *Narcotic Drugs Biochemical Pharmacology*, Plenum Press, New York, 52–53 (1971); cited from Farmilo CG, Oestreicher PM and Levi L, Physical methods for the identification of narcotics. IB Common physical constants for identification of ninety-five narcotics and related compounds, *Bull. Narcotics*, UN Dept. Social Affairs, vol. **6**, pp. 7–19 (1954). NB: See Alphaprodine for details.

No.	Compound	pKa	Data quality	Ionization	Method	Conditions	Reference / Notes
971	Phenazopyridine ($C_{11}H_{11}N_5$) 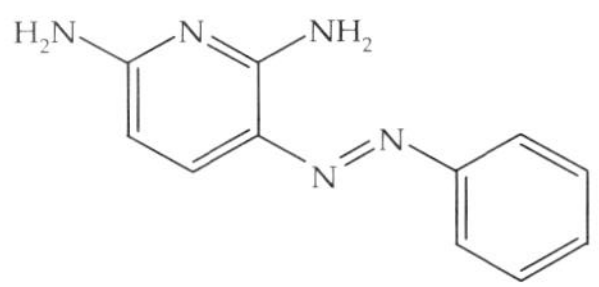	5.15	U	+H	Potentiometric	H_2O $t = 25$	Bergström CAS, Strafford M, Lazorova L, Avdeef A, Luthman K and Artursson P, Absorption classification of oral drugs based on molecular surface properties, *J. Med. Chem.* **46**(4), 558–570 (2003). NB: From extrapolation of aqueous-methanol mixtures to 0% methanol.
972	Phenethicillin ($C_{17}H_{20}N_2O_5S$) 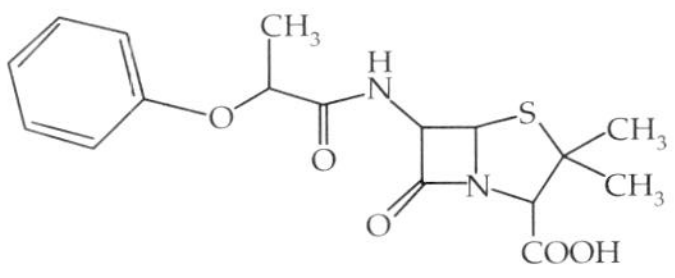	2.73 ± 0.04	A	–H	Potentiometric	H_2O $t = 25$ $c = 0.0091$	Rapson HDC and Bird AE, Ionization constants of some penicillins and of their alkaline and penicillinase hydrolysis products, *J. Pharm. Pharmacol.*, Suppl. **15**, 222–231T (1963).
		2.78	U	–H			Rollo IM, Physicochemical properties of some semisynthetic penicillins, *Can. J. Physiol. Pharmacol.*, **50**, 976–985 (1972); cited in N&K.
973	α-Phenethylamine (1-amino-1-phenylethane) ($C_8H_{11}N$) 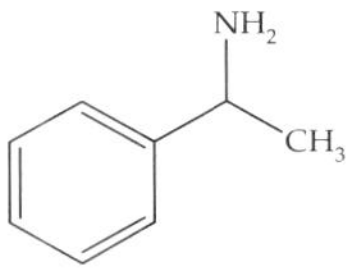	9.08 ± 0.02	U	+H	Potentiometric	H_2O $t = 25.0 \pm 0.2$ $I \leq 0.001$	Leffler EB, Spencer HM and Burger A, Dissociation constants of adrenergic amines, *JACS*, **73**, 2611–2613 (1951). NB: See Amphetamine for details. From $pK_b = 4.92$.
974	Phenformin ($C_{10}H_{15}N_5$) 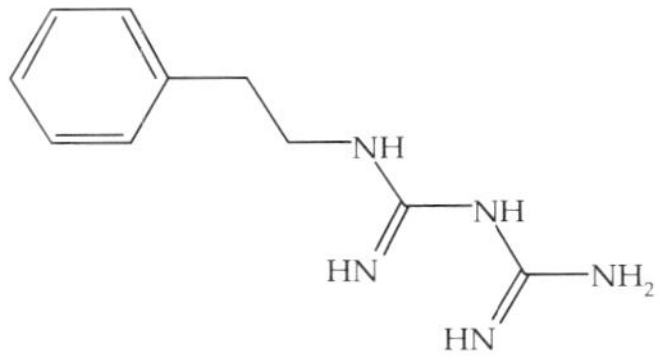	2.7 11.8	U U	+H +H	Potentiometric	H_2O $t = 32$ I undefined	Moody JE, Phenformin hydrochloride, *APDS*, **4**, 319–332 (1975). "Phenformin is a strongly basic substance and consequently exists in di and mono ionic forms. Ray has reported the ionization constants, $pK_a' = 11.8$ and $pK_a'' = 2.7$ at 32°. In our laboratory (Moody JE, USV Pharmaceutical Corp, Tuckahoe, NY), the apparent $pK_a'' = 3.1$ was measured at 25° by potentiometric titration. The second ionization constant, pK_a', could not be reliably measured in aqueous medium by our method. However, Garrett has calculated an approximate pK_a' of 13.0 from plots of the reciprocals of the apparent partition coefficients against the hydrogen ion concentration." Ray P, *Chem. Rev.*, **61**, 313–359 (1972).
		3.1	U	+H		H_2O $t = 25$	
		12.9 ± 0.1	U	+H	partition	H_2O $t = 25$ $I > 0.1$	Garrett ER, Tsau J and Hinderling PH, Application of ion-pair methods to drug extraction from biological fluids. II, *J. Pharm. Sci.*, **61**, 1411–1418 (1972). NB: From the average of results using CH_2Cl_2/water and 1:1 $CHCl_3$ - tert amyl alcohol/water partitioning systems. Activity coefficients from the literature were used to correct the high NaOH concentrations to activities, hence this value is a thermodynamic value.

(continued)

Appendix A *(continued)*

No.	Compound Name	pK_a value(s)	Data quality	Ionization type	Method	Conditions	Comments and Reference(s)
975	Phenindamine ($C_{19}H_{19}N$) 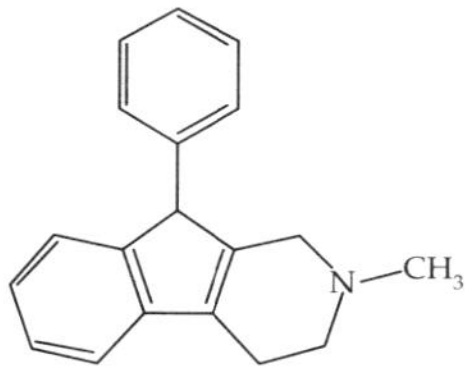	8.11 ± 0.08	U	+H	Potentiometric	H_2O t undefined $I = 0.30$ (NaCl)	Testa B and Murset-Rossetti L, The partition coefficient of protonated histamines. *Helv. Chim. Acta*, **61**, 2530–2537 (1978). NB: See Cycliramine for details.
976	Phenindione ($C_{10}H_{15}O_2$) 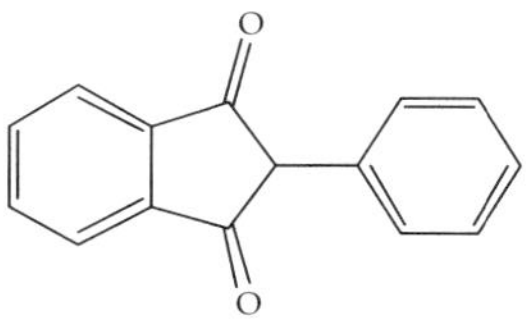	4.09	U	−H	Spectro	H_2O $I = 0.1$ $t = 25.0$	Stella VJ and Gish R, Kinetics and mechanism of ionization of the carbon acids 4′-substituted 2-phenyl-1,3-indandiones, *J. Pharm. Sci.*, **68**(8), 1047–1049 (1979). NB: See Anisindione for details.
977	Pheniramine ($C_{16}H_{20}N_2$) 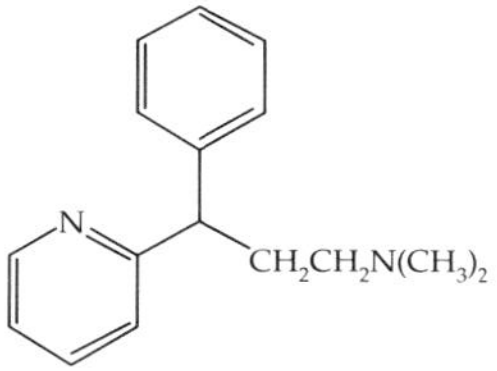	4.03 ± 0.08 9.32 ± 0.06	U U	+H +H	Potentiometric	H_2O t undefined $I = 0.30$ (NaCl)	Testa B and Murset-Rossetti L, The partition coefficient of protonated histamines, *Helv. Chim. Acta*, **61**, 2530–2537 (1978). NB: See Cycliramine for details.

No.	Name	pKa	Data quality	Ionization	Method	Conditions	Reference/Comments
978	Phenolphthalein ($C_{20}H_{14}O_4$) 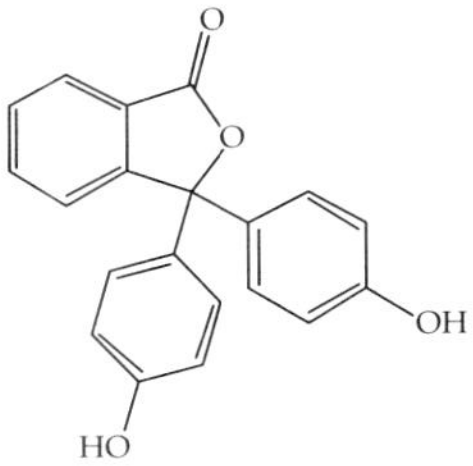	9.70	U	–H		$t = 25$	Al-Shammary FJ, Mian MS and Mian NAA, Phenolphthalein, *APDS*, **20**, 627–664 (1991). Clarke Drug Information **90**, ASHP, p. 1639. N&K; Ritschel gave 9.70; ref. Parrott EL, 1970, p. 217.
979	Phenolphthalein		U	–H			Lalanne JR, Determination of pK_a of phenolphthalein and its discoloration rate, *J. Chem. Educ.*, **48**, 266–268 (1971).
980	Phenolsulphonphthalein (phenol red) ($C_{20}H_{32}O_5S$) 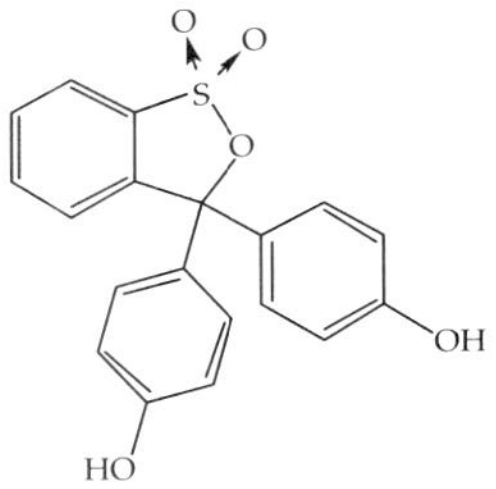	1.03	U	–H	Spectro	H_2O RT	NB: See Bromocresol green.
981	Phenothiazine $R_1 = R_2 = H$ ($C_{12}H_9NS$) 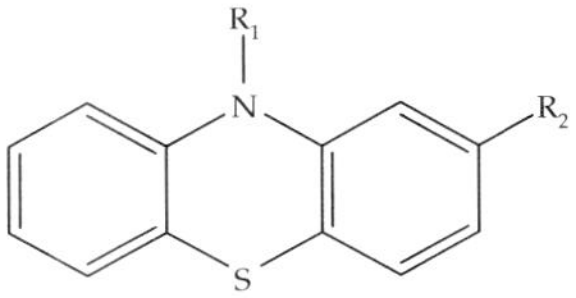	2.52	U	+H			Foye 1

(*continued*)

No.	Compound Name	pK_a value(s)	Data quality	Ionization type	Method	Conditions	Comments and Reference(s)
982	Phenothiazines						Zografi G and Munshi MV, Effect of chemical modification on the surface activity of some phenothiazine derivatives, *J. Pharm. Sci.*, **59**, 819–822 (1970). "The relative surface activity of various phenothiazine derivatives has been measured under solution conditions which for the first time allow meaningful comparison of structural effects. . . . A primary factor to consider also is how substitution influences the dissociation constant(s) of the amino group(s) and hence the degree of ionization at the pH being utilized for comparison. A table listing data on 12 phenothiazine derivatives and numerous graphs plotting surface pressure *vs.* molar concentration are included."
983	Phenothiazine, 10-[*N*-(4-carbamoyl) piperidinyl]-propyl-2-chloro (pipamazine) ($C_{20}H_{32}ClN_3OS$) R_1 = (structure: propyl chain to piperidine N, 4-$CONH_2$) R_2 = Cl	8.60	U	+H	Potentiometric	H_2O *t* undefined $c < 0.002$	Chatten LG and Harris LE, Relationship between pK_b (H_2O) of organic compounds and $E_{1/2}$ values in several non-aqueous solvents, *Anal. Chem.*, **34**, 1495–1501 (1962). Cited in Perrin Bases suppl. no. 7428 ref. C10. NB: Study used glass electrode measurements of the pH value of solutions containing equimolar proportions of the free base and the salt. From reported pK_b = 5.40 at unknown temperature (assumed 25 °C). See parent structure above. See also Pipamazine, no. 1119.
984	Phenothiazine, 2-chloro-10-(3-dimethylamino-propyl)-(chlorpromazine) ($C_{17}H_{19}ClN_2S$) $R_1 = (CH_2)_3N(CH_3)_2$ R_2 = Cl	9.3	U	+H	soly	H_2O $t = 24 \pm 1$	Green AL, Ionization constants and water solubilities of some aminoalkylphenothiazine tranquillizers and related compounds, *J. Pharm. Pharmacol.*, **19**, 10–16 (1967). Cited in Perrin suppl. no. 7429 ref. G29. NB: See Phenothiazine parent structure above. See Amitriptylline for details.
985	Phenothiazine, 2-chloro-10-(3-dimethylamino-propyl)-(chlorpromazine)	9.22	U	+H	Potentiometric	H_2O *t* undefined	Chatten LG and Harris LE, Relationship between pK_b (H_2O) of organic compounds and $E_{1/2}$ values in several non-aqueous solvents, *Anal. Chem.*, **34**, 1495–1501 (1962). Cited in Perrin Bases suppl. no. 7429 ref. C10. NB: Study used glass electrode measurements of the pH value of solutions containing equimolar proportions of the free base and the salt. From reported pK_b = 4.78 at unknown temperature.

986	Phenothiazine, 2-chloro-10-[*N*-(2-hydroxy)ethyl]-piperazinylpropyl- ($C_{21}H_{22}ClN_3OS$) R_1 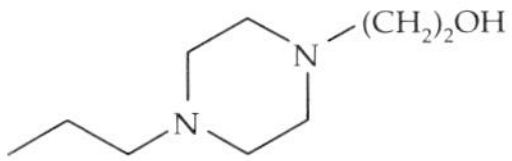 = propyl–N-piperazinyl–N'–$(CH_2)_2OH$ R_2 = Cl	7.8	U	+H	soly	H_2O $t = 24 \pm 1$	Cited in Perrin suppl. no. 7430 ref. G29. See no. 984 for details. NB: See Phenothiazine parent structure above.
987	Phenothiazine, 2-chloro-10-[*N*-methyl] piperazinyl-propyl- (prochlorperazine) ($C_{20}H_{24}ClN_3S$) R_1 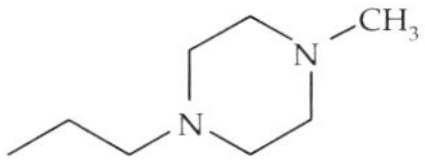 = propyl–N-piperazinyl–N'–CH_3 R_2 = Cl	8.1	U	+H	soly	H_2O $t = 24 \pm 1$	Cited in Perrin suppl. no. 7431 ref. G29. See no. 984 for details. NB: See Phenothiazine parent structure above. See also separate entries for Prochlorperazine.
988	Phenothiazine, 2-chloro-10-[*N*-methyl] piperazinyl-propyl- (prochlorperazine)	3.60 7.54	U U	+H +H	Potentiometric	H_2O *t* undefined	Chatten LG and Harris LE, Relationship between pK_b (H_2O) of organic compounds and $E_{1/2}$ values in several non-aqueous solvents, *Anal. Chem.*, **34**, 1495–1501 (1962). Cited in Perrin Bases suppl. no. 7431 ref. C10. NB: Study used glass electrode measurements of the pH value of solutions containing equimolar proportions of the free base and the salt. From reported pK_b = 6.46, 10.40 at unknown temperature (assumed 25 °C).
989	Phenothiazine, 2-chloro-10-[*N*-(2-propionyloxy)-ethyl] piperazinylpropyl- ($C_{24}H_{26}ClN_3O_2S$) R_1 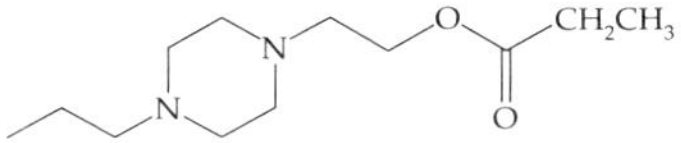 = propyl–N-piperazinyl–N'–CH_2CH_2–O–C(=O)–CH_2CH_3 R_2 = Cl	7.3	U	+H	soly	H_2O $t = 24 \pm 1$	Cited in Perrin suppl. no. 7432 ref. G29. See no. 984 for details. NB: See Phenothiazine parent structure above.

(*continued*)

No.	Compound Name	pK_a value(s)	Data quality	Ionization type	Method	Conditions	Comments and Reference(s)
990	Phenothiazine, 2-chloro-10-[*N*-(2-propionyloxy)-ethyl] piperazinylpropyl-	3.20 7.15	U U	+H +H	Potentiometric	H_2O *t* undefined	Chatten LG and Harris LE, Relationship between pK_b (H_2O) of organic compounds and $E_{1/2}$ values in several non-aqueous solvents, *Anal. Chem.*, **34**, 1495–1501 (1962). Cited in Perrin Bases suppl. no. 7432 ref. C10. NB: Study used glass electrode measurements of the pH value of solutions containing equimolar proportions of the free base and the salt. From reported $pK_b = 6.85$, 10.80 at unknown temperature (assumed 25 °C).
991	Phenothiazine, 10-(2-diethylaminoethyl)- ($C_{18}H_{22}N_2S$) $R_1 = (CH_2)_2N(CH_2CH_3)_2$ $R_2 = H$	7.0	U	+H	Potentiometric	H_2O *t* undefined	Chatten LG and Harris LE, Relationship between pK_b (H_2O) of organic compounds and $E_{1/2}$ values in several non-aqueous solvents, *Anal. Chem.*, **34**, 1495–1501 (1962). Cited in Perrin Bases suppl. no. 7433 ref. C10. NB: Study used glass electrode measurements of the pH value of solutions containing equimolar proportions of the free base and the salt. From reported $pK_b = 7.0$ (assumed $t = 25$ °C). See parent structure above.
992	Phenothiazine, 10-(2-di-methylaminomethyl)-propyl-2-methoxy- (methotrimeprazine) ($C_{19}H_{24}N_2OS$) $R_1 = CH_2CH(CH_3)CH_2N(CH_3)_2$ $R_2 = OCH_3$	9.15	U	+H	Potentiometric	H_2O *t* undefined	Chatten LG and Harris LE, Relationship between pK_b (H_2O) of organic compounds and $E_{1/2}$ values in several non-aqueous solvents, *Anal. Chem.*, **34**, 1495–1501 (1962). Cited in Perrin Bases suppl. no. 7434 ref. C10. NB: Study used glass electrode measurements of the pH value of solutions containing equimolar proportions of the free base and the salt. From reported $pK_b = 4.85$ at unknown temperature (assumed 25 °C). See parent structure above. See Methotrimeprazine separate entry—appears to be from the Chatten and Harris 1962 result.
993	Phenothiazine, 10-(2-dimethylaminopropyl)- ($C_{17}H_{20}N_2S$) (promethazine) $R_1 = CH_2CH(CH_3)N(CH_3)_2$ $R_2 = H$	9.1	U	+H	soly	H_2O $t = 24 \pm 1$	Cited in Perrin suppl. no. 7435 ref. G29. See no. 984 for details. NB: See Phenothiazine parent structure above.
994	Phenothiazine, 10-(2-dimethylaminopropyl)-	9.1	U	+H	Potentiometric	H_2O *t* undefined	Chatten LG and Harris LE, Relationship between pK_b (H_2O) of organic compounds and $E_{1/2}$ values in several non-aqueous solvents, *Anal. Chem.*, **34**, 1495–1501 (1962). Cited in Perrin Bases suppl. no. 7435 ref. C10. NB: Study used glass electrode measurements of the pH value of solutions containing equimolar amounts of the free base and the salt. From reported $pK_b = 4.90$ at unknown temperature (assumed 25 °C).

995	Phenothiazine, 10-(3-dimethylaminopropyl)- ($C_{17}H_{20}N_2S$) (promazine) $R_1 = (CH_2)_3N(CH_3)_2$ $R_2 = H$	9.4	U	+H	soly	H_2O $t = 24 \pm 1$	Perrin suppl. no. 7436 ref. G29. See no. 984 for details. NB: See Phenothiazine parent structure above.
996	Phenothiazine, 10-(3-dimethylaminopropyl)-2-trifluoromethyl- (trifluopromazine) ($C_{18}H_{19}F_3N_2S$) $R_1 = (CH_2)_3N(CH_3)_2$ $R_2 = CF_3$	9.2	U	+H	soly	H_2O $t = 24 \pm 1$	Perrin suppl. no. 7437 ref. G29. See no. 984 for details. NB: See Phenothiazine parent structure above. See also Trifluopromazine separate entry.
997	Phenothiazine, 10-(3-dimethylaminopropyl)-2-trifluoromethyl-	9.41	U	+H	Potentiometric	H_2O t undefined	Chatten LG and Harris LE, Relationship between pK_b (H_2O) of organic compounds and $E_{1/2}$ values in several non-aqueous solvents, *Anal. Chem.*, **34**, 1495–1501 (1962). Cited in Perrin Bases suppl. no. 7437 ref. C10. NB: Study used glass electrode measurements of the pH value of solutions containing equimolar proportions of the free base and the salt. From reported p*K* (assumed 25 °C) = 4.59 at unknown temperature (assumed 25 °C).
998	Phenothiazine, 10-(*N*-methyl) piperazinylpropyl-2-trifluoromethyl- (trifluoperazine) ($C_{21}H_{24}F_3N_3S$) $R_1 =$ $R_1 =$ [3-(4-methylpiperazin-1-yl)propyl structure] $R_2 = CF_3$	8.1	U	+H	soly	H_2O $t = 24 \pm 1$	Perrin suppl. no. 7438 ref. G29. See no. 984 for details. NB: See Phenothiazine parent structure above and Trifluoperazine separate entry.
999	Phenothiazine, 10-(*N*-methyl) piperazinylpropyl-2-trifluoromethyl- (trifluoperazine)	3.90 8.40	U	+H	Potentiometric	H_2O t undefined	Chatten LG and Harris LE, Relationship between pK_b (H_2O) of organic compounds and $E_{1/2}$ values in several non-aqueous solvents, *Anal. Chem.*, **34**, 1495–1501 (1962). Cited in Perrin Bases suppl. no. 7438 ref. C10. NB: Study used glass electrode measurements of the pH value of solutions containing equimolar proportions of the free base and the salt. From reported values for pK_b = 5.60, 10.10 at unknown temperature (assumed 25 °C).

(*continued*)

No.	Compound Name	pK_a value(s)	Data quality	Ionization type	Method	Conditions	Comments and Reference(s)
1000	Phenothiazine, 10-[2-(*N*-methyl) piperidinyl]ethyl-2-methylthio- ($C_{21}H_{22}N_2S_2$) H_3C — N R_1 = R_2 = SCH_3	9.5	U	+H	soly	H_2O $t = 24 \pm 1$	Perrin suppl. no. 7439 ref. G29. See no. 984 for details. NB: See Phenothiazine parent structure above.
1001	Phenothiazine, 10-[2-(*N*-methyl) piperidinyl]ethyl-2-methylthio-	9.16	U	+H	Potentiometric	H_2O t undefined	Chatten LG and Harris LE, Relationship between pK_b (H_2O) of organic compounds and $E_{1/2}$ values in several non-aqueous solvents, *Anal. Chem.*, **34**, 1495–1501 (1962). Cited in Perrin Bases suppl. no. 7439 ref. C10. NB: Study used glass electrode measurements of the pH value of solutions containing equimolar proportions of the free base and the salt. From reported pK_b = 4.84 at unknown temperature (assumed 25 °C).
1002	Phenothiazine, 10-(*N*-methyl)piperidinyl] methyl- (mepazine) ($C_{19}H_{22}N_2S$) H_3C — N R_1 = R_2 = H	9.7	U	+H	soly	H_2O $t = 24 \pm 1$	Perrin suppl. no. 7440 ref. G29. See no. 984 for details. NB: See Phenothiazine parent structure above and Mepazine separate entry.
1003	Phenothiazine, 10-(*N*-pyrrolidinyl)ethyl- (pyrathiazine) ($C_{18}H_{20}N_2S$) N R_1 = R_2 = H	8.91	U	+H	Potentiometric	H_2O t undefined	Chatten LG and Harris LE, Relationship between pK_b (H_2O) of organic compounds and $E_{1/2}$ values in several non-aqueous solvents, *Anal. Chem.*, **34**, 1495–1501 (1962). Cited in Perrin Bases suppl. no. 7441 ref. C10. NB: Study used glass electrode measurements of the pH value of solutions containing equimolar proportions of the free base and the salt. From reported pK_b = 5.09 at unknown temperature (assumed 25 °C). See parent structure above. See separate entry for Pyrathiazine.

1004	Phenoxyacetic acid ($C_8H_8O_3$) 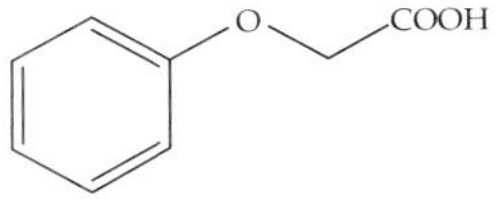	3.13	U	–H	Conductance	H_2O $t = 25$ $c = 0.03$–0.001	Kendall J, Electrical conductivity and ionization constants of weak electrolytes in aqueous solution, *in* Washburn EW, Editor-in-Chief, International Critical Tables, vol. **6**, McGraw-Hill, NY, 259–304 (1929).
1005	Phenoxyethylpenicilloic acid ($C_{17}H_{22}N_2O_6S$) 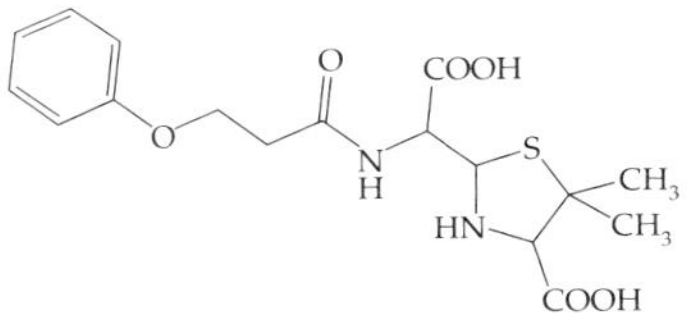	5.29 ± 0.05	A	+H	Potentiometric	H_2O $t = 25$ $c = 0.01$	Rapson HDC and Bird AE, Ionization constants of some penicillins and of their alkaline and penicillinase hydrolysis products, *J. Pharm. Pharmacol.*, Suppl. **15**, 222–231T (1963). Cited in Perrin Bases suppl. no. 7787 ref. R6. NB: Potentiometric titrations used a glass electrode with an unsymmetrical cell and liquid junction potentials.
1006	Phenoxymethylpenicilloic acid ($C_{16}H_{20}N_2O_6S$) 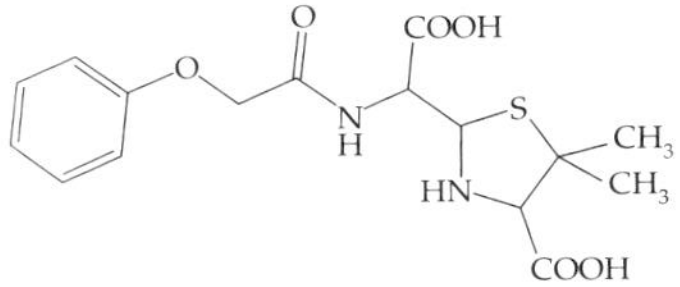	5.15 ± 0.07	U	+H	Potentiometric	H_2O $t = 25$ $c = 0.01$	Rapson HDC and Bird AE, Ionization constants of some penicillins and of their alkaline and penicillinase hydrolysis products, *J. Pharm. Pharmacol.*, Suppl. **15**, 222–231T (1963). Cited in Perrin Bases suppl. no. 7788 ref. R6. NB: Potentiometric titrations used a glass electrode with an unsymmetrical cell and liquid junction potentials.
1007	Phenoxypropylpenicilloic acid ($C_{18}H_{24}N_2O_6S$) 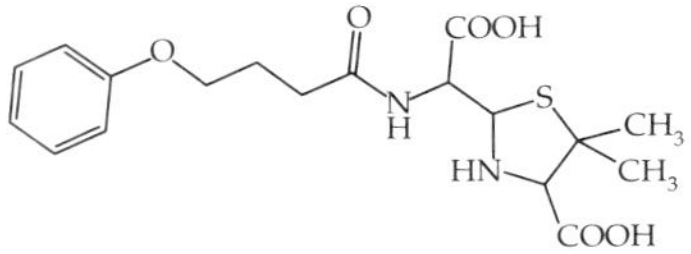	5.30 ± 0.05	A	+H	Potentiometric	H_2O $t = 25$ $c = 0.01$	Rapson HDC and Bird AE, Ionization constants of some penicillins and of their alkaline and penicillinase hydrolysis products, *J. Pharm. Pharmacol.*, Suppl. **15**, 222–231T (1963). Cited in Perrin Bases suppl. no. 7789 ref. R6. NB: Potentiometric titrations used a glass electrode with an unsymmetrical cell and liquid junction potentials. Also reported $pK_a = 2.72 \pm 0.05$ for the corresponding penicillin.

(continued)

No.	Compound Name	pK_a value(s)	Data quality	Ionization type	Method	Conditions	Comments and Reference(s)
1008	*N*-Phenylanthranillic acid ($C_{13}H_{11}NO_2$) 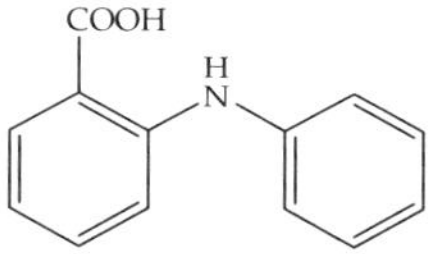	4.15	U	–H	soly	H_2O $t = 25 \pm 0.1$ $I = 0.11$	Terada H, Muraoka S and Fujita T, Structure-activity relationships of fenamic acids, *J. Med. Chem.*, **17**, 330–334 (1974). NB: See Flufenamic acid for details.
1009	Phenylbutazone ($C_{19}H_{20}N_2O_2$) 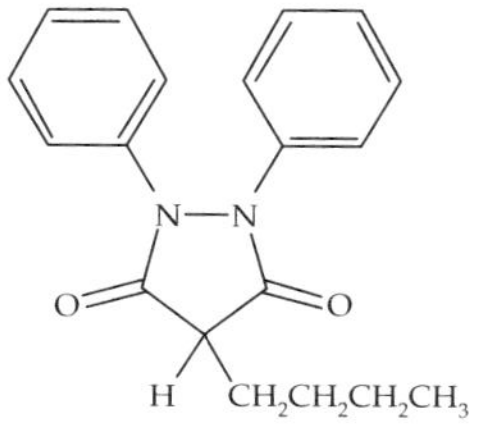	4.52	A	–H	Spectro	H_2O $t = 23$ $I = 0.025$	Girod E, Delley R, Hafliger F, Uber derivate des Phenylbutazons III. Die Struktur der Reduktionsprodukte des γ-keto-phenylbutazons, *Helv. Chim. Acta*, **40**, 408–428 (1957).
		5.25	U	–H	Potentiometric	80%MCS	NB: Also reported apparent pK_a values from titrimetry with tetramethylammonium hydroxide in 80% methocellosolve for 29 other related compounds.
1010	Phenylbutazone	4.53 ± 0.06	A	–H	Spectro	H_2O $t = 25.0 \pm 0.05$ $I = 0.00$	Prankerd RJ, Determination of the pK_a of phenylbutazone, 4th Year Undergraduate Research Project, Department of Pharmacy, University of Otago, Dunedin, New Zealand, 1973. NB: The Guggenheim modification of the Debye-Hückel equation was used to account for ionic strength effects by extrapolating to $I = 0.00$.
1011	Phenylbutazone	4.53	A	+H	CE/pH (+ve ion mode)	H_2O $t = 25$ $I = 0.025$	Wan H, Holmen AG, Wang Y, Lindberg W, Englund M, Nagard MB and Thompson RA, High-throughput screening of pK_a values of pharmaceuticals by pressure-assisted capillary electrophoresis and mass spectrometry, *Rapid Commun. Mass Spectrom.*, **17**, 2639–2648 (2003). NB: Reported a predicted value (ACD Labs) of 6.39. The assignment of the pK_a value to a basic group is in disagreement with the consensus of literature, which assigns this value to the carbon acid system.
1012	Phenylbutazone	GLpK_a: 4.30 ± 0.01 A&S: 4.34 ± 0.04	 A A	 –H –H	Spectro	H_2O $t = 25$ $I = 0.15$ (KCl) Ar atmosphere	Tam KY and Takacs-Novac K, Multi-wavelength spectrophotometric determination of acid dissociation constants, *Anal. Chim. Acta* **434**, 157–167 (2001). NB: See clioquinol for details.

No.	Compound	pKa	Type	Ionization	Method	Conditions	Reference / Comments
1013	Phenylbutazone	4.43	U	–H	Spectro	H_2O $I = 0.1$ $t = 25.0 \pm 0.2$	Stella VJ and Pipkin JD, Phenylbutazone ionization kinetics, *J. Pharm. Sci.*, **65**(8), 1161–1165 (1976). NB: $pK_a = 4.33$ by stopped flow pH jump spectrophotometry.
1014	Phenylbutazone	4.5	U	–H	Potentiometric	H_2O $t = 20$ I undefined	Rainer VG, Krüger U and Klemm K, Syntheses und physicalisch-chemische Eigenschaften von Lonazolac-Ca einem neuen Antiphlogistikum/Antirheumatikum, *Arzneim.-Forsch.*, **31**(4), 649–655 (1981). NB: See Indomethacin for details.
1015	Phenylbutazone	4.70 ± 0.2	U	–H	Potentiometric	H_2O $t = 24$	Maulding HV and Zoglio MA, pK_a determinations utilizing solutions of 7-(2-hydroxypropyl) theophylline, *J. Pharm. Sci.*, **60**, 309–311 (1971). NB: See Barbituric acid, 5-allyl-5-isobutyl for details.
1016	Phenylbutazone	4.6	U	–H	Spectro	H_2O t undefined I undefined	Herzfeldt CD and Kümmel R, Dissociation constants, solubilities, and dissolution rates of some selected nonsteroidal antiinflammatories, *Drug Dev. Ind. Pharm.*, **9**(5), 767–793 (1983). NB: Used dλ/dpH method. NB: See Azapropazone and Ibuprofen for details.
1017	Phenylbutazone	5.47	U	–H	Potentiometric	50% EtOH t undefined I undefined	Jahn U and Wagner-Jauregg T, Wirkungsvergleich saurer Antiphlogistika im Bradykinin-, UV-Erythem- und Rattenpfotenödem-Test, *Arzneim.-Forsch.*, **24**, 494–499 (1974).
		5.45	U	–H		80% Me cellosolve	NB: Literature values obtained from the pH of half-neutralization. Also gave a value of 4.5 in water. This last value is in good agreement with other values.
1018	Phenylbutazone analogs		U	–H	Spectro	H_2O t = RT I undefined	Perel JM, Snell MM, Chen W and Dayton PG, A study of structure-activity relationships in regards to species difference in the phenylbutazone series, *Biochem. Pharmacol.*, **13**, 1305–1317 (1964). NB: All compounds where R_1 = OH have a pK_{a2} value of 10.0 ± 0.2.

Compound	R_1	R_2	R_3	pK_{a1}
G-34208	OH	H	*t*-Bu	7.1
G-35716	OH	H	*i*-Pr	5.8
G-13838	H	H	*i*-Pr	5.5
Oxyphenbutazone	OH	H	*n*-Bu	4.7
Phenylbutazone	H	H	*n*-Bu	4.5
G-32170	F	F	*n*-Bu	4.5
G-29665	OH	H	$-(CH_2)_2CH(Me)OH$	4.3
G-25592	H	H	$-(CH_2)_2$-O-Ph	4.2
G-33378	OH	H	$-(CH_2)_2$-S-Ph	4.1
G-28231	H	H	$-(CH_2)_2CH(Me)OH$	4.0
G-25671	H	H	$-(CH_2)_2$-S-Ph	3.9
G-28234	NO_2	H	*n*-Bu	3.2
G-32642	OH	H	$-(CH_2)_2$-SO-Ph	3.1
Sulfinpyrazone	H	H	$-(CH_2)_2$-SO-Ph	2.8
G-32567	CH_3SO_2-	CH_3SO_2-	*n*-Bu	2.6
G-29701	OH	H	$-C(=O)CH_2CH_2CH_3$	2.3

(continued)

No.	Compound Name	pK_a value(s)	Data quality	Ionization type	Method	Conditions	Comments and Reference(s)
1019	5-Phenyl-1,3-dihydro-1,4-benzodiazepin-2-one ($C_{15}H_{12}N_2O$) 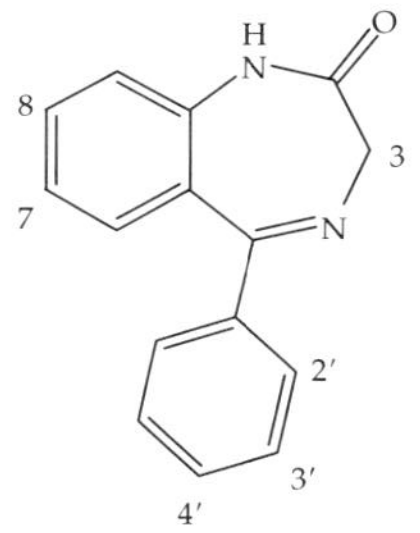	4.17 12.49	U U	+H −H	Spectro	2% aq. EtOH $t = 25$	Seiler P and Zimmermann I, 5-Phenyl-1,3-dihydro-1,4-benzodiazepin-2-ones. Experimental verification of substituent constants, *Arzneim.-Forsch.*, **33**, 1519–1522 (1983). "The acidity constants (p*K*-values) in the aqueous phase with 2% ethanol were determined spectrophotometrically at 25 °C according to published methods (Albert and Serjeant, 1971). Each p*K*-value was detemined from the absorbance at three or more pH-values." NB: Partition coefficients were also detemined. The data were used as in linear free energy relationships to obtain substituent constants.
1020	5-Phenyl-1,3-dihydro-1,4-benzodiazepin-2-one, 2′-fluoro-7,8-dichloro ($C_{15}H_9Cl_2FN_2O$)	2.26 11.20	U U	+H −H	Spectro	2% aq. EtOH $t = 25$	Seiler P and Zimmermann I, 5-Phenyl-1,3-dihydro-1,4-benzodiazepin-2-ones. Experimental verification of substituent constants, *Arzneim.-Forsch.*, **33**, 1519–1522 (1983). NB: See parent compound for details.
1021	5-Phenyl-1,3-dihydro-1,4-benzodiazepin-2-one, 1,3-dimethyl-2′,7-dichloro ($C_{17}H_{14}Cl_2N_2O$)	2.07	U	+H	Spectro	2% aq. EtOH $t = 25$	Seiler P and Zimmermann I, 5-Phenyl-1,3-dihydro-1,4-benzodiazepin-2-ones. Experimental verification of substituent constants, *Arzneim.-Forsch.*, **33**, 1519–1522 (1983). NB: See parent compound for details.
1022	5-Phenyl-1,3-dihydro-1,4-benzodiazepin-2-one, 4′-trifluoromethyl ($C_{16}H_{11}F_3N_2O$)	3.06 12.30	U U	+H −H	Spectro	2% aq. EtOH $t = 25$	Seiler P and Zimmermann I, 5-Phenyl-1,3-dihydro-1,4-benzodiazepin-2-ones. Experimental verification of substituent constants, *Arzneim.-Forsch.*, **33**, 1519–1522 (1983). NB: See parent compound for details.
1023	5-Phenyl-1,3-dihydro-1,4-benzodiazepin-2-one, 3-methyl-7-chloro ($C_{16}H_{13}ClN_2O$)	3.57 12.29	U U	+H −H	Spectro	2% aq. EtOH $t = 25$	Seiler P and Zimmermann I, 5-Phenyl-1,3-dihydro-1,4-benzodiazepin-2-ones. Experimental verification of substituent constants, *Arzneim.-Forsch.*, **33**, 1519–1522 (1983). NB: See parent compound for details.
1024	5-Phenyl-1,3-dihydro-1,4-benzodiazepin-2-one, 1-ethyl-7-chloro ($C_{17}H_{15}ClN_2O$)	3.18	U	+H	Spectro	2% aq. EtOH $t = 25$	Seiler P and Zimmermann I, 5-Phenyl-1,3-dihydro-1,4-benzodiazepin-2-ones. Experimental verification of substituent constants, *Arzneim.-Forsch.*, **33**, 1519–1522 (1983). NB: See parent compound for details.
1025	5-Phenyl-1,3-dihydro-1,4-benzodiazepin-2-one, 3′-trifluoromethyl ($C_{16}H_{11}F_3N_2O$)	3.22 12.26	U U	+H −H	Spectro	2% aq. EtOH $t = 25$	Seiler P and Zimmermann I, 5-Phenyl-1,3-dihydro-1,4-benzodiazepin-2-ones. Experimental verification of substituent constants, *Arzneim.-Forsch.*, **33**, 1519–1522 (1983). NB: See parent compound for details.
1026	5-Phenyl-1,3-dihydro-1,4-benzodiazepin-2-one, 2′-trifluoro-methyl-7-chloro ($C_{16}H_{10}ClF_3N_2O$)	1.73 11.95	U U	+H −H	Spectro	2% aq. EtOH $t = 25$	Seiler P and Zimmermann I, 5-Phenyl-1,3-dihydro-1,4-benzodiazepin-2-ones. Experimental verification of substituent constants, *Arzneim.-Forsch.*, **33**, 1519–1522 (1983). NB: See parent compound for details.

1027	5-Phenyl-1,3-dihydro-1,4-benzodiazepin-2-one, 2′-bromo-7-chloro ($C_{15}H_{10}BrClN_2O$)	2.09 11.84	U U	+H −H	Spectro	2% aq. EtOH $t = 25$	Seiler P and Zimmermann I, 5-Phenyl-1,3-dihydro-1,4-benzodiazepin-2-ones. Experimental verification of substituent constants, *Arzneim.-Forsch.*, **33**, 1519–1522 (1983). NB: See parent compound for details.
1028	5-Phenyl-1,3-dihydro-1,4-benzodiazepin-2-one, 2′,7-dichloro ($C_{15}H_{10}Cl_2N_2O$)	2.17 11.80	U U	+H −H	Spectro	2% aq. EtOH $t = 25$	Seiler P and Zimmermann I, 5-Phenyl-1,3-dihydro-1,4-benzodiazepin-2-ones. Experimental verification of substituent constants, *Arzneim.-Forsch.*, **33**, 1519–1522 (1983). NB: See parent compound for details.
1029	5-Phenyl-1,3-dihydro-1,4-benzodiazepin-2-one, 1-methyl-4′-methoxy-7-chloro ($C_{17}H_{15}ClN_2O_2$)	4.00	U	+H	Spectro	2% aq. EtOH $t = 25$	Seiler P and Zimmermann I, 5-Phenyl-1,3-dihydro-1,4-benzodiazepin-2-ones. Experimental verification of substituent constants, *Arzneim.-Forsch.*, **33**, 1519–1522 (1983). NB: See parent compound for details.
1030	5-Phenyl-1,3-dihydro-1,4-benzodiazepin-2-one, 3′-methoxy-7-chloro ($C_{16}H_{13}ClN_2O_2$)	3.29 11.93	U U	+H −H	Spectro	2% aq. EtOH $t = 25$	Seiler P and Zimmermann I, 5-Phenyl-1,3-dihydro-1,4-benzodiazepin-2-ones. Experimental verification of substituent constants, *Arzneim.-Forsch.*, **33**, 1519–1522 (1983). NB: See parent compound for details.
1031	5-Phenyl-1,3-dihydro-1,4-benzodiazepin-2-one, 1-methyl-2′-fluoro-7-iodo ($C_{16}H_{12}FIN_2O$)	2.45	U	+H	Spectro	2% aq. EtOH $t = 25$	Seiler P and Zimmermann I, 5-Phenyl-1,3-dihydro-1,4-benzodiazepin-2-ones. Experimental verification of substituent constants, *Arzneim.-Forsch.*, **33**, 1519–1522 (1983). NB: See parent compound for details.
1032	5-Phenyl-1,3-dihydro-1,4-benzodiazepin-2-one, 1-methyl-2′,7-dichloro ($C_{16}H_{12}Cl_2N_2O$)	1.92	U	+H	Spectro	2% aq. EtOH $t = 25$	Seiler P and Zimmermann I, 5-Phenyl-1,3-dihydro-1,4-benzodiazepin-2-ones. Experimental verification of substituent constants, *Arzneim.-Forsch.*, **33**, 1519–1522 (1983). NB: See parent compound for details.
1033	5-Phenyl-1,3-dihydro-1,4-benzodiazepin-2-one, 2′-methyl-7-chloro ($C_{16}H_{13}ClN_2O$)	3.14 11.96	U U	+H −H	Spectro	2% aq. EtOH $t = 25$	Seiler P and Zimmermann I, 5-Phenyl-1,3-dihydro-1,4-benzodiazepin-2-ones. Experimental verification of substituent constants, *Arzneim.-Forsch.*, **33**, 1519–1522 (1983). NB: See parent compound for details.
1034	5-Phenyl-1,3-dihydro-1,4-benzodiazepin-2-one, 7-trifluoromethyl ($C_{16}H_{11}F_3N_2O$)	3.25 11.68	U U	+H −H	Spectro	2% aq. EtOH $t = 25$	Seiler P and Zimmermann I, 5-Phenyl-1,3-dihydro-1,4-benzodiazepin-2-ones. Experimental verification of substituent constants, *Arzneim.-Forsch.*, **33**, 1519–1522 (1983). NB: See parent compound for details.
1035	5-Phenyl-1,3-dihydro-1,4-benzodiazepin-2-one, 1-methyl-4′-chloro-7-fluoro ($C_{16}H_{12}ClF_2O$)	1.92	U	+H	Spectro	2% aq. EtOH $t = 25$	Seiler P and Zimmermann I, 5-Phenyl-1,3-dihydro-1,4-benzodiazepin-2-ones. Experimental verification of substituent constants, *Arzneim.-Forsch.*, **33**, 1519–1522 (1983). NB: See parent compound for details.
1036	5-Phenyl-1,3-dihydro-1,4-benzodiazepin-2-one, 1-methyl-2′,6′,7-trichloro ($C_{16}H_{11}Cl_3N_2O$)	0.80	U	+H	Spectro	2% aq. EtOH $t = 25$	Seiler P and Zimmermann I, 5-Phenyl-1,3-dihydro-1,4-benzodiazepin-2-ones. Experimental verification of substituent constants, *Arzneim.-Forsch.*, **33**, 1519–1522 (1983). NB: See parent compound for details.

(continued)

No.	Compound Name	pK_a value(s)	Data quality	Ionization type	Method	Conditions	Comments and Reference(s)
1037	5-Phenyl-1,3-dihydro-1,4-benzodiazepin-2-one, 1,2′-dimethyl-7-chloro ($C_{17}H_{15}ClN_2O$)	2.93	U	+H	Spectro	2% aq. EtOH $t = 25$	Seiler P and Zimmermann I, 5-Phenyl-1,3-dihydro-1,4-benzodiazepin-2-ones. Experimental verification of substituent constants, *Arzneim.-Forsch.*, **33**, 1519–1522 (1983). NB: See parent compound for details.
1038	5-Phenyl-1,3-dihydro-1,4-benzodiazepin-2-one, 2′-thiomethyl-7-chloro ($C_{16}H_{13}ClN_2OS$)	2.65 11.95	U U	+H −H	Spectro	2% aq. EtOH $t = 25$	Seiler P and Zimmermann I, 5-Phenyl-1,3-dihydro-1,4-benzodiazepin-2-ones. Experimental verification of substituent constants, *Arzneim.-Forsch.*, **33**, 1519–1522 (1983). NB: See parent compound for details.
1039	5-Phenyl-1,3-dihydro-1,4-benzodiazepin-2-one, 1-methyl-7-thiomethyl ($C_{17}H_{16}N_2OS$)	3.70	U	+H	Spectro	2% aq. EtOH $t = 25$	Seiler P and Zimmermann I, 5-Phenyl-1,3-dihydro-1,4-benzodiazepin-2-ones. Experimental verification of substituent constants, *Arzneim.-Forsch.*, **33**, 1519–1522 (1983). NB: See parent compound for details.
1040	5-Phenyl-1,3-dihydro-1,4-benzodiazepin-2-one, 7-chloro ($C_{15}H_{11}ClN_2O$)	3.48 11.82	U U	+H −H	Spectro	2% aq. EtOH $t = 25$	Seiler P and Zimmermann I, 5-Phenyl-1,3-dihydro-1,4-benzodiazepin-2-ones. Experimental verification of substituent constants, *Arzneim.-Forsch.*, **33**, 1519–1522 (1983). NB: See parent compound for details.
1041	5-Phenyl-1,3-dihydro-1,4-benzodiazepin-2-one, 7-thiomethyl ($C_{16}H_{14}N_2OS$)	3.96 12.27	U U	+H −H	Spectro	2% aq. EtOH $t = 25$	Seiler P and Zimmermann I, 5-Phenyl-1,3-dihydro-1,4-benzodiazepin-2-ones. Experimental verification of substituent constants, *Arzneim.-Forsch.*, **33**, 1519–1522 (1983). NB: See parent compound for details.
1042	5-Phenyl-1,3-dihydro-1,4-benzodiazepin-2-one, 2′-fluoro-7-ethyl ($C_{17}H_{15}FN_2O$)	3.41 12.25	U U	+H −H	Spectro	2% aq. EtOH $t = 25$	Seiler P and Zimmermann I, 5-Phenyl-1,3-dihydro-1,4-benzodiazepin-2-ones. Experimental verification of substituent constants, *Arzneim.-Forsch.*, **33**, 1519–1522 (1983). NB: See parent compound for details.
1043	5-Phenyl-1,3-dihydro-1,4-benzodiazepin-2-one, 4′-fluoro-7-chloro ($C_{15}H_{10}ClF_2O$)	3.21 12.07	U U	+H −H	Spectro	2% aq. EtOH $t = 25$	Seiler P and Zimmermann I, 5-Phenyl-1,3-dihydro-1,4-benzodiazepin-2-ones. Experimental verification of substituent constants, *Arzneim.-Forsch.*, **33**, 1519–1522 (1983). NB: See parent compound for details.
1044	5-Phenyl-1,3-dihydro-1,4-benzodiazepin-2-one, 2′,6′-difluoro-8-chloro ($C_{15}H_{10}ClF_2N_2O$)	1.95 11.36	U U	+H −H	Spectro	2% aq. EtOH $t = 25$	Seiler P and Zimmermann I, 5-Phenyl-1,3-dihydro-1,4-benzodiazepin-2-ones. Experimental verification of substituent constants, *Arzneim.-Forsch.*, **33**, 1519–1522 (1983). NB: See parent compound for details.
1045	5-Phenyl-1,3-dihydro-1,4-benzodiazepin-2-one, 1-methoxymethyl-7-chloro ($C_{17}H_{15}ClN_2O_2$)	2.74	U	+H	Spectro	2% aq. EtOH $t = 25$	Seiler P and Zimmermann I, 5-Phenyl-1,3-dihydro-1,4-benzodiazepin-2-ones. Experimental verification of substituent constants, *Arzneim.-Forsch.*, **33**, 1519–1522 (1983). NB: See parent compound for details.

1046	5-Phenyl-1,3-dihydro-1,4-benzodiazepin-2-one, 1-methyl-7-chloro ($C_{16}H_{15}ClN_2O$)	3.17	U	+H	Spectro	2% aq. EtOH $t = 25$	Seiler P and Zimmermann I, 5-Phenyl-1,3-dihydro-1,4-benzodiazepin-2-ones. Experimental verification of substituent constants, *Arzneim.-Forsch.*, **33**, 1519–1522 (1983). NB: See parent compound for details.
1047	5-Phenyl-1,3-dihydro-1,4-benzodiazepin-2-one, 1-methyl-2′-fluoro-7-chloro ($C_{16}H_{12}ClFN_2O$)	2.22	U	+H	Spectro	2% aq. EtOH $t = 25$	Seiler P and Zimmermann I, 5-Phenyl-1,3-dihydro-1,4-benzodiazepin-2-ones. Experimental verification of substituent constants, *Arzneim.-Forsch.*, **33**, 1519–1522 (1983). NB: See parent compound for details.
1048	5-Phenyl-1,3-dihydro-1,4-benzodiazepin-2-one, 3-methyl-2′-chloro-7-nitro ($C_{16}H_{13}ClN_3O_3$)	1.61 10.79	U U	+H −H	Spectro	2% aq. EtOH $t = 25$	Seiler P and Zimmermann I, 5-Phenyl-1,3-dihydro-1,4-benzodiazepin-2-ones. Experimental verification of substituent constants, *Arzneim.-Forsch.*, **33**, 1519–1522 (1983). NB: See parent compound for details.
1049	5-Phenyl-1,3-dihydro-1,4-benzodiazepin-2-one, 2′-fluoro-7-chloro ($C_{15}H_{10}ClFN_2O$)	2.57 11.77	U U	+H −H	Spectro	2% aq. EtOH $t = 25$	Seiler P and Zimmermann I, 5-Phenyl-1,3-dihydro-1,4-benzodiazepin-2-ones. Experimental verification of substituent constants, *Arzneim.-Forsch.*, **33**, 1519–1522 (1983). NB: See parent compound for details.
1050	5-Phenyl-1,3-dihydro-1,4-benzodiazepin-2-one, 2′,6′-difluoro-7-chloro ($C_{15}H_9ClF_2N_2O$)	1.63 11.64	U U	+H −H	Spectro	2% aq. EtOH $t = 25$	Seiler P and Zimmermann I, 5-Phenyl-1,3-dihydro-1,4-benzodiazepin-2-ones. Experimental verification of substituent constants, *Arzneim.-Forsch.*, **33**, 1519–1522 (1983). NB: See parent compound for details.
1051	5-Phenyl-1,3-dihydro-1,4-benzodiazepin-2-one, 2′-methoxy-7- chloro ($C_{16}H_{13}ClN_2O_2$)	3.66 12.05	U U	+H −H	Spectro	2% aq. EtOH $t = 25$	Seiler P and Zimmermann I, 5-Phenyl-1,3-dihydro-1,4-benzodiazepin-2-ones. Experimental verification of substituent constants, *Arzneim.-Forsch.*, **33**, 1519–1522 (1983). NB: See parent compound for details.
1052	5-Phenyl-1,3-dihydro-1,4-benzodiazepin-2-one, 7-methyl ($C_{16}H_{14}N_2O$)	4.39 12.44	U U	+H −H	Spectro	2% aq. EtOH $t = 25$	Seiler P and Zimmermann I, 5-Phenyl-1,3-dihydro-1,4-benzodiazepin-2-ones. Experimental verification of substituent constants, *Arzneim.-Forsch.*, **33**, 1519–1522 (1983). NB: See parent compound for details.
1053	5-Phenyl-1,3-dihydro-1,4-benzodiazepin-2-one, 2′-trifluoromethyl ($C_{16}H_{11}F_3N_2O$)	2.39 12.24	U U	+H −H	Spectro	2% aq. EtOH $t = 25$	Seiler P and Zimmermann I, 5-Phenyl-1,3-dihydro-1,4-benzodiazepin-2-ones. Experimental verification of substituent constants, *Arzneim.-Forsch.*, **33**, 1519–1522 (1983). NB: See parent compound for details.
1054	5-Phenyl-1,3-dihydro-1,4-benzodiazepin-2-one, 7-dimethylamino ($C_{17}H_{17}N_3O$)	4.60 12.76	U U	+H −H	Spectro	2% aq. EtOH $t = 25$	Seiler P and Zimmermann I, 5-Phenyl-1,3-dihydro-1,4-benzodiazepin-2-ones. Experimental verification of substituent constants, *Arzneim.-Forsch.*, **33**, 1519–1522 (1983). NB: See parent compound for details.
1055	5-Phenyl-1,3-dihydro-1,4-benzodiazepin-2-one, 2′-chloro-7-nitro ($C_{15}H_{10}ClN_3O_3$)	1.57 10.50	U U	+H −H	Spectro	2% aq. EtOH $t = 25$	Seiler P and Zimmermann I, 5-Phenyl-1,3-dihydro-1,4-benzodiazepin-2-ones. Experimental verification of substituent constants, *Arzneim.-Forsch.*, **33**, 1519–1522 (1983). NB: See parent compound for details.

(continued)

No.	Compound Name	pK_a value(s)	Data quality	Ionization type	Method	Conditions	Comments and Reference(s)
1056	5-Phenyl-1,3-dihydro-1,4-benzodiazepin-2-one, 3-hydroxy-2′,7-dichloro ($C_{15}H_{10}Cl_2N_2O_2$)	0.39 11.03	U U	+H −H	Spectro	2% aq. EtOH $t = 25$	Seiler P and Zimmermann I, 5-Phenyl-1,3-dihydro-1,4-benzodiazepin-2-ones. Experimental verification of substituent constants, *Arzneim.-Forsch.*, **33**, 1519–1522 (1983). NB: See parent compound for details.
1057	5-Phenyl-1,3-dihydro-1,4-benzodiazepin-2-one, 4′-fluoro ($C_{15}H_{11}FN_2O$)	4.01 12.31	U U	+H −H	Spectro	2% aq. EtOH $t = 25$	Seiler P and Zimmermann I, 5-Phenyl-1,3-dihydro-1,4-benzodiazepin-2-ones. Experimental verification of substituent constants, *Arzneim.-Forsch.*, **33**, 1519–1522 (1983). NB: See parent compound for details.
1058	5-Phenyl-1,3-dihydro-1,4-benzodiazepin-2-one, 7-fluoro ($C_{15}H_{11}FN_2O$)	3.52 12.14	U U	+H −H	Spectro	2% aq. EtOH $t = 25$	Seiler P and Zimmermann I, 5-Phenyl-1,3-dihydro-1,4-benzodiazepin-2-ones. Experimental verification of substituent constants, *Arzneim.-Forsch.*, **33**, 1519–1522 (1983). NB: See parent compound for details.
1059	5-Phenyl-1,3-dihydro-1,4-benzodiazepin-2-one, 7-methoxy ($C_{16}H_{14}N_2O_2$)	4.04 12.46	U U	+H −H	Spectro	2% aq. EtOH $t = 25$	Seiler P and Zimmermann I, 5-Phenyl-1,3-dihydro-1,4-benzodiazepin-2-ones. Experimental verification of substituent constants, *Arzneim.-Forsch.*, **33**, 1519–1522 (1983). NB: See parent compound for details.
1060	5-Phenyl-1,3-dihydro-1,4-benzodiazepin-2-one, 3-hydroxy-7-chloro ($C_{15}H_{11}ClN_2O_2$)	1.62 11.24	U U	+H −H	Spectro	2% aq. EtOH $t = 25$	Seiler P and Zimmermann I, 5-Phenyl-1,3-dihydro-1,4-benzodiazepin-2-ones. Experimental verification of substituent constants, *Arzneim.-Forsch.*, **33**, 1519–1522 (1983). NB: See parent compound for details.
1061	5-Phenyl-1,3-dihydro-1,4-benzodiazepin-2-one, 1-methyl-7-nitro ($C_{16}H_{13}N_3O_3$)	2.63	U	+H	Spectro	2% aq. EtOH $t = 25$	Seiler P and Zimmermann I, 5-Phenyl-1,3-dihydro-1,4-benzodiazepin-2-ones. Experimental verification of substituent constants, *Arzneim.-Forsch.*, **33**, 1519–1522 (1983). NB: See parent compound for details.
1062	5-Phenyl-1,3-dihydro-1,4-benzodiazepin-2-one, 7-nitro ($C_{15}H_{12}N_3O_3$)	2.88 11.88	U U	+H −H	Spectro	2% aq. EtOH $t = 25$	Seiler P and Zimmermann I, 5-Phenyl-1,3-dihydro-1,4-benzodiazepin-2-ones. Experimental verification of substituent constants, *Arzneim.-Forsch.*, **33**, 1519–1522 (1983). NB: See parent compound for details.
1063	5-Phenyl-1,3-dihydro-1,4-benzodiazepin-2-one, 1-methyl-2′-fluoro-7-nitro ($C_{16}H_{13}FN_3O_3$)	1.87	U	+H	Spectro	2% aq. EtOH $t = 25$	Seiler P and Zimmermann I, 5-Phenyl-1,3-dihydro-1,4-benzodiazepin-2-ones. Experimental verification of substituent constants, *Arzneim.-Forsch.*, **33**, 1519–1522 (1983). NB: See parent compound for details.
1064	5-Phenyl-1,3-dihydro-1,4-benzodiazepin-2-one, 1-methoxymethyl-7-nitro ($C_{17}H_{16}N_3O_4$)	2.09	U	+H	Spectro	2% aq. EtOH $t = 25$	Seiler P and Zimmermann I, 5-Phenyl-1,3-dihydro-1,4-benzodiazepin-2-ones. Experimental verification of substituent constants, *Arzneim.-Forsch.*, **33**, 1519–1522 (1983). NB: See parent compound for details.

1065	5-Phenyl-1,3-dihydro-1,4-benzodiazepin-2-one, 7-cyano ($C_{16}H_{11}N_3O$)	2.88 11.17	U U	+H −H	Spectro	2% aq. EtOH $t = 25$	Seiler P and Zimmermann I, 5-Phenyl-1,3-dihydro-1,4-benzodiazepin-2-ones. Experimental verification of substituent constants, *Arzneim.-Forsch.*, **33**, 1519–1522 (1983). NB: See parent compound for details.
1066	5-Phenyl-1,3-dihydro-1,4-benzodiazepin-2-one, 1-[butane-2,4-diol]-2′-fluoro-7-iodo ($C_{19}H_{18}FIN_2O_3$)	2.31	U	+H	Spectro	2% aq. EtOH $t = 25$	Seiler P and Zimmermann I, 5-Phenyl-1,3-dihydro-1,4-benzodiazepin-2-ones. Experimental verification of substituent constants, *Arzneim.-Forsch.*, **33**, 1519–1522 (1983). NB: See parent compound for details.
1067	5-Phenyl-1,3-dihydro-1,4-benzodiazepin-2-one, 1-ethyl-7-amino ($C_{17}H_{17}N_3O$)	4.07	U	+H	Spectro	2% aq. EtOH $t = 25$	Seiler P and Zimmermann I, 5-Phenyl-1,3-dihydro-1,4-benzodiazepin-2-ones. Experimental verification of substituent constants, *Arzneim.-Forsch.*, **33**, 1519–1522 (1983). NB: See parent compound for details.
1068	5-Phenyl-1,3-dihydro-1,4-benzodiazepin-2-one, 2′-fluoro-7-acetyl ($C_{17}H_{13}FN_2O_2$)	2.67 11.22	U U	+H −H	Spectro	2% aq. EtOH $t = 25$	Seiler P and Zimmermann I, 5-Phenyl-1,3-dihydro-1,4-benzodiazepin-2-ones. Experimental verification of substituent constants, *Arzneim.-Forsch.*, **33**, 1519–1522 (1983). NB: See parent compound for details.
1069	5-Phenyl-1,3-dihydro-1,4-benzodiazepin-2-one, 2′,7-dinitro ($C_{15}H_{10}N_4O_5$)	0.92 10.45	U U	+H −H	Spectro	2% aq. EtOH $t = 25$	Seiler P and Zimmermann I, 5-Phenyl-1,3-dihydro-1,4-benzodiazepin-2-ones. Experimental verification of substituent constants, *Arzneim.-Forsch.*, **33**, 1519–1522 (1983). NB: See parent compound for details.
1070	5-Phenyl-1,3-dihydro-1,4-benzodiazepin-2-one, 1-[butane-2,4-diol]-2′-fluoro-7-chloro ($C_{19}H_{18}ClFN_2O_3$)	2.26	U	+H	Spectro	2% aq. EtOH $t = 25$	Seiler P and Zimmermann I, 5-Phenyl-1,3-dihydro-1,4-benzodiazepin-2-ones. Experimental verification of substituent constants, *Arzneim.-Forsch.*, **33**, 1519–1522 (1983). NB: See parent compound for details.
1071	5-Phenyl-1,3-dihydro-1,4-benzodiazepin-2-one, 1-methyl-7-amino ($C_{16}H_{15}N_3O$)	4.24	U	+H	Spectro	2% aq. EtOH $t = 25$	Seiler P and Zimmermann I, 5-Phenyl-1,3-dihydro-1,4-benzodiazepin-2-ones. Experimental verification of substituent constants, *Arzneim.-Forsch.*, **33**, 1519–1522 (1983). NB: See parent compound for details.
1072	5-Phenyl-1,3-dihydro-1,4-benzodiazepin-2-one, 1-methoxymethyl-7-amino ($C_{17}H_{17}N_3O_2$)	3.81	U	+H	Spectro	2% aq. EtOH $t = 25$	Seiler P and Zimmermann I, 5-Phenyl-1,3-dihydro-1,4-benzodiazepin-2-ones. Experimental verification of substituent constants, *Arzneim.-Forsch.*, **33**, 1519–1522 (1983). NB: See parent compound for details.
1073	5-Phenyl-1,3-dihydro-1,4-benzodiazepin-2-one, 1- methyl-2′-fluoro-7-amino ($C_{16}H_{14}FN_3O$)	3.55	U	+H	Spectro	2% aq. EtOH $t = 25$	Seiler P and Zimmermann I, 5-Phenyl-1,3-dihydro-1,4-benzodiazepin-2-ones. Experimental verification of substituent constants, *Arzneim.-Forsch.*, **33**, 1519–1522 (1983). NB: See parent compound for details.
1074	5-Phenyl-1,3-dihydro-1,4-benzodiazepin-2-one, 1-methoxymethyl-2′-fluoro-7-amino ($C_{12}H_{16}FN_3O_2$)	3.34	U	+H	Spectro	2% aq. EtOH $t = 25$	Seiler P and Zimmermann I, 5-Phenyl-1,3-dihydro-1,4-benzodiazepin-2-ones. Experimental verification of substituent constants, *Arzneim.-Forsch.*, **33**, 1519–1522 (1983). NB: See parent compound for details.

(continued)

No.	Compound Name	pK_a value(s)	Data quality	Ionization type	Method	Conditions	Comments and Reference(s)
1075	5-Phenyl-1,3-dihydro-1,4-benzodiazepin-2-one, 7-amino ($C_{15}H_{13}N_3O$)	4.51 12.98	U U	+H −H	Spectro	2% aq. EtOH $t = 25$	Seiler P and Zimmermann I, 5-Phenyl-1,3-dihydro-1,4-benzodiazepin-2-ones. Experimental verification of substituent constants, *Arzneim.-Forsch.*, **33**, 1519–1522 (1983). NB: See parent compound for details.
1076	5-Phenyl-1,3-dihydro-1,4-benzodiazepin-2-one, 1-[butane-2,4-diol]-7-nitro ($C_{19}H_{19}N_3O_5$)	2.36	U	+H	Spectro	2% aq. EtOH $t = 25$	Seiler P and Zimmermann I, 5-Phenyl-1,3-dihydro-1,4-benzodiazepin-2-ones. Experimental verification of substituent constants, *Arzneim.-Forsch.*, **33**, 1519–1522 (1983). NB: See parent compound for details.
1077	Phenylephrine (Neo-synephrine) ($C_9H_{13}NO_2$) HO, CH_2NHCH_3, OH	8.86	U	+H, −H	Potentiometric	H_2O (extrap) $t = 25.0 \pm 0.2$ I ~0.01	Leffler EB, Spencer HM and Burger A, Dissociation constants of adrenergic amines, *JACS*, **73**, 2611–13 (1951). NB: See Amphetamine for details. From pK_b = 5.14.
1078	Phenylethylamine ($C_8H_{11}N$) NH_2	9.83	U	+H	Potentiometric	H_2O $t = 25.0 \pm 0.2$ $I \leq 0.001$	Leffler EB, Spencer HM and Burger A, Dissociation constants of adrenergic amines, *JACS*, **73**, 2611–2613 (1951). NB: See Amphetamine for details.
1079	Phenylethylamine	9.78	U	+H	Potentiometric	H_2O t undefined I undefined	Tuckerman MM, Mayer JR and Nachod FC, Anomalous pK_a values of some substituted phenylethylamines, *JACS*, **81**, 92–94 (1959). NB: Method as described by Parke and Davis (1945).
		9.88 ± 0.1	U	+H	Potentiometric	H_2O $t = 25$	Kappe T, Armstrong MD, Ultraviolet absorption spectra and apparent acidic dissociation constants of some phenolic amines, *J. Med. Chem.*, **8**, 368–374 (1965). See Levarterenol (noradrenaline) (no. 729) and *m*-Hydroxyphenylethylamine, 2-hydroxy (no. 1081) for details.
1080	Phenylethylamine, 2-hydroxy ($C_8H_{11}NO$)	8.90	U	+H	Potentiometric	H_2O t undefined I undefined	Tuckerman MM, Mayer JR and Nachod FC, Anomalous pK_a values of some substituted phenylethylamines, *JACS*, **81**, 92–94 (1959). NB: Method as described by Parke and Davis (1945).

No.	Compound	pKa	Class	Type	Method	Conditions	Reference
1081	*m*-Hydroxyphenylethylamine, 2-hydroxy ($C_8H_{11}NO_2$)	8.67	U	+H	Potentiometric	H_2O *t* undefined *I* undefined	Tuckerman MM, Mayer JR and Nachod FC, Anomalous pK_a values of some substituted phenylethylamines, *JACS*, **81**, 92–94 (1959). NB: Method as described by Parke and Davis (1945).
		9.56 ± 0.05	U	+H	Spectro	H_2O	Kappe T and Armstrong MD, Ultraviolet absorption spectra and apparent acidic dissociation constants of some phenolic amines, *J. Med. Chem.*, **8**, 368–374 (1965). NB: The pK_a for the phenolic group was determined spectrophotometrically and the value then used to correct the potentiometric titration curve for the ionization of this group. The resulting difference titration curve was then use to estimate the pK_a value for the amine group. As the amine group pK_a value was estimated by this approach, the experimental error is necessarily larger than for the phenolic group.
		9.63 ± 0.1	U	−H	Potentiometric	*t* = 25	
1082	*m*-Hydroxyphenylethylamine, 2-hydroxy, *N*-methyl ($C_9H_{13}NO_2$)	8.89	U	+H	Potentiometric	H_2O *t* undefined *I* undefined	Tuckerman MM, Mayer JR and Nachod FC, Anomalous pK_a values of some substituted phenylethylamines, *JACS*, **81**, 92–94 (1959). NB: Method as described by Parke and Davis (1945).
1083	*p*-Hydroxyphenylethylamine ($C_8H_{11}NO$)	9.22	U	+H	Potentiometric	H_2O *t* undefined *I* undefined	Tuckerman MM, Mayer JR and Nachod FC, Anomalous pK_a values of some substituted phenylethylamines, *JACS*, **81**, 92–94 (1959). NB: Method as described by Parke and Davis (1945).
		9.74 ± 0.05	U	+H	Spectro	H_2O	Kappe T and Armstrong MD, Ultraviolet absorption spectra and apparent acidic dissociation constants of some phenolic amines, *J. Med. Chem.*, **8**, 368–374 (1965). NB: See Levarterenol (noradrenaline (no. 729) and *m*-Hydroxyphenylethylamine, 2-hydroxy (no. 1081) for details).
		10.52 ± 0.1	U	−H	Potentiometric	*t* = 25	
1084	*p*-Hydroxyphenylethylamine, *N*-methyl ($C_9H_{13}NO$)	9.36	U	+H	Potentiometric	H_2O *t* undefined *I* undefined	Tuckerman MM and Mayer JR, Nachod FC, Anomalous pK_a values of some substituted phenylethylamines, *JACS*, **81**, 92–94 (1959). NB: Method as described by Parke and Davis (1945).
		9.76 ± 0.05	U	+H	Spectro	H_2O	Kappe T and Armstrong MD, Ultraviolet absorption spectra and apparent acidic dissociation constants of some phenolic amines, *J. Med. Chem.*, **8**, 368–374 (1965). NB: See Levarterenol (noradrenaline (no. 729) and *m*-Hydroxyphenylethylamine, 2-hydroxy (no. 1081) for details).
		10.71 ± 0.1	U	−H	Potentiometric	*t* = 25	
1085	*m*,*p*-Dihydroxyphenylethylamine ($C_8H_{11}NO_2$)	8.93	U	+H	Potentiometric	H_2O *t* undefined *I* undefined	Tuckerman MM, Mayer JR and Nachod FC, Anomalous pK_a values of some substituted phenylethylamines, *JACS*, **81**, 92–94 (1959). NB: Method as described by Parke and Davis (1945).
1086	*m*,*p*-Dihydroxyphenylethylamine, *N*-methyl ($C_9H_{13}NO_2$)	8.78	U	+H	Potentiometric	H_2O *t* undefined *I* undefined	Tuckerman MM, Mayer JR and Nachod FC, Anomalous pK_a values of some substituted phenylethylamines, *JACS*, **81**, 92–94 (1959). NB: Method as described by Parke and Davis (1945).

(continued)

No.	Compound Name	pK_a value(s)	Data quality	Ionization type	Method	Conditions	Comments and Reference(s)
1087	*p*-Hydroxyphenylethylamine, 2-hydroxy ($C_8H_{11}NO_2$)	8.81	U	+H	Potentiometric	H_2O *t* undefined *I* undefined	Tuckerman MM, Mayer JR and Nachod FC, Anomalous pK_a values of some substituted phenylethylamines, *JACS*, **81**, 92–94 (1959). NB: Method as described by Parke and Davis (1945).
		9.57 ± 0.05 9.66 ± 0.1	U U	+H −H	Spectro Potentiometric	H_2O $t = 25$	Kappe T and Armstrong MD, Ultraviolet absorption spectra and apparent acidic dissociation constants of some phenolic amines, *J. Med. Chem.*, **8**, 368–374 (1965). NB: See Levarterenol (noradrenaline (no. 729) and *m*-Hydroxyphenylethylamine, 2-hydroxy (no. 1081) for details).
1088	*p*-Hydroxyphenylethylamine, 2-hydroxy, *N*-methyl ($C_9H_{13}NO_2$)	8.62	U	+H	Potentiometric	H_2O *t* undefined *I* undefined	Tuckerman MM, Mayer JR and Nachod FC, Anomalous pK_a values of some substituted phenylethylamines, *JACS*, **81**, 92–94 (1959). NB: Method as described by Parke and Davis (1945).
1089	Phenylethylamine, 2-hydroxy, *N*-methyl ($C_9H_{13}NO$)	9.31	U	+H	Potentiometric	H_2O *t* undefined *I* undefined	Tuckerman MM, Mayer JR and Nachod FC, Anomalous pK_a values of some substituted phenylethylamines, *JACS*, **81**, 92–94 (1959). NB: Method as described by Parke and Davis (1945).
1090	Phenylpenilloic acid ($C_{14}H_{17}N_2O_3S$) 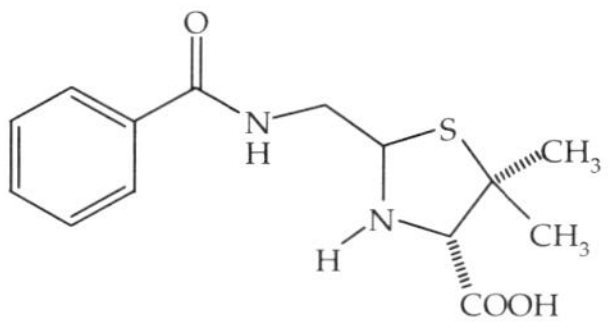	1.51 5.18	U U	−H +H		H_2O $t = 5$	Woodward RB, Neuberger A and Trenner NR, *in* Clarke H, Johnson JR and Robinson Sir R. (eds.), *The Chemistry of Penicillin*, Princeton University Press, Princeton, NJ, 415–422 (1949). NB: Method was not described in this paper.
1091	Phenylpenilloic acid	1.50 4.90	U U	−H +H		H_2O $t = 25$	Woodward RB, Neuberger A and Trenner NR, *in* Clarke H, Johnson JR and Robinson Sir R. (eds.), *The Chemistry of Penicillin*, Princeton University Press, Princeton, NJ, 415–422 (1949). NB: Method was not described in this paper.
1092	Phenylpropanolamine ($C_9H_{13}NO$) 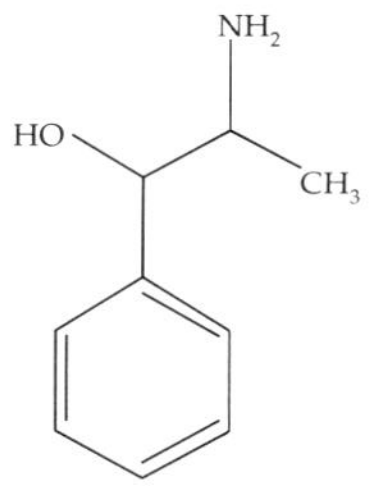	9.014	A	+H	Potentiometric	H_2O $t = 25$ $I = 0.00$	Lukkari S, Electrolyte effect on the ionization constant of pharmaceuticals, alpha-((2-pyridylamino)methyl)-benzyl alcohol (phenyramidol) and alpha-(1-aminoethyl)benzyl alcohol (norephedrine; phenylpropanolamine), *Farm. Aikak.*, **79**, 95–99 (1970). "The acid ionization constants of phenyramidol and phenylpropanolamine in aqueous solutions at 25 °C. were determined potentiometrically. The values pK^o = 6.488 and pK^o = 9.014, respectively, were obtained for the acid ionization constants at zero ionic strength. The effect of ionic strength on the ionization constant, as adjusted with sodium perchlorate, was determined."

No.	Compound	pKa	Rating	Type	Method	Conditions	Reference / Comments
1093	Phenylpropanolamine	9.05	U	+H	Potentiometric	H_2O $t = 25.0 \pm 0.5$ $I = 0.01$	Warren RJ, Begosh PP and Zarembo JE, Identification of amphetamines and related sympathomimetic amines, *J. Assoc. Off. Anal. Chem.*, **54**, 1179–1191 (1971). NB: See amphetamine for further details.
1094	Phenylpropanolamine	9.44 ± 0.04	A	+H	Potentiometric	H_2O $t = 20.0$ $I = 0.1$ (glycine)	Lewis GP, The importance of ionization in the activity of sympathomimetic amines, *Br. J. Pharmacol.*, **9**, 488–493 (1954). NB: Reported pK_a values for a further 23 sympathomimetic amines. Where compounds contained phenolic group as well as the amine, both potentiometric and spectrophotometric methods were used. Methods were similar to Kappe and Armstrong, see Levarterenol (no. 729). Cited in Kanfer I, Haigh JM and Dowse R, Phenylpropanolamine hydrochloride, *APDS*, **12**, 357–380 (1983).
1095	Phenylpropylmethylamine (1-Methylamino-2-phenylpropane) ($C_{10}H_{15}N$) 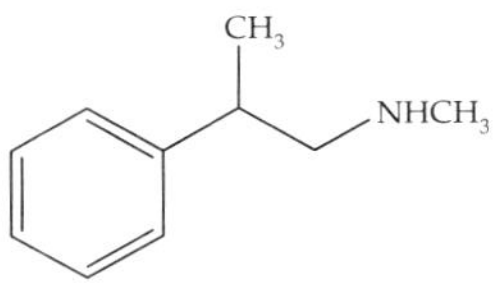	10.07	A	+H	Potentiometric	H_2O $t = 25.0 \pm 0.5$ $I = 0.01$	Warren RJ, Begosh PP and Zarembo JE, Identification of amphetamines and related sympathomimetic amines, *J. Assoc. Off. Anal. Chem.*, **54**, 1179–1191 (1971). NB: See Amphetamine for further details.
1096	Phenylpropylmethyl-amine (Vonedrine)	9.88 ± 0.02	U	+H	Potentiometric	H_2O (extrap) $t = 25.0 \pm 0.2$ $I \sim 0.01$	Leffler EB, Spencer HM and Burger A, Dissociation constants of adrenergic amines, *JACS*, **73**, 2611–13 (1951). NB: See Amphetamine for details. From pK_b = 4.12. Cited in: Chatten LG and Harris LE, Relationship between pK_b(H_2O) of organic compounds and $E_{1/2}$ values in several nonaqueous solvents, *Anal. Chem.*, **34**, 1495–1501 (1962).
1097	5-Phenylvaleric acid ($C_{10}H_{14}O_2$) 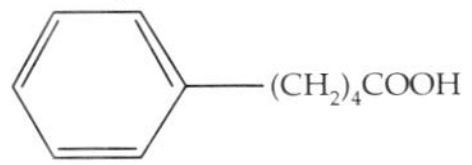	4.59 ± 0.02	U	−H	Potentiometric	H_2O $t = 25 \pm 0.5$ $I = 0.15$ (KCl)	Avdeef A, Box KJ, Comer JEA, Hibbert C and Tam KY, pH-metric log P 10. Determination of liposomal membrane-water partition coefficients of ionizable drugs, *Pharm. Res.*, **15**(2), 209–215 (1998). NB: Used a Sirius PCA101 autotitrator. Also gave log P (octanol-water) and log P (dioleylphosphatidylcholine unilamellar vesicles).
1098	Phenyramidol ($C_{13}H_{14}N_2O$) 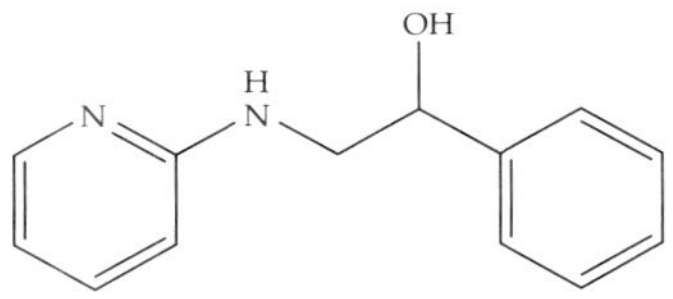	6.488	A	+H	Potentiometric	H_2O $I = 0.00$ $t = 25$	Lukkari S, Electrolyte effect on the ionization constant of pharmaceuticals, α-((2-pyridylamino)methyl)benzyl alcohol (phenyramidol) and α-(1-aminoethyl)benzyl alcohol (norephedrine; phenylpropanolamine), *Farm. Aikak.*, **79**, 95–99 (1970). "The acid ionization constants of phenyramidol and phenylpropanolamine in aqueous solutions at 25 °C. were determined potentiometrically. The values pK^o = 6.488 and pK^o = 9.014, respectively, were obtained for the acid ionization constants at zero ionic strength. The effect of ionic strength on the ionization constant, as adjusted with sodium perchlorate, was determined."

(continued)

No.	Compound Name	pK_a value(s)	Data quality	Ionization type	Method	Conditions	Comments and Reference(s)
1099	Phenyramidol	5.85	U	+H	Potentiometric	60:40 DMF-H_2O $t = 25$ $I = \sim 0.01$	Gray AP, Heitmeier DE and Spinner EE, *JACS*, **81**, 4351–4355 (1959) reported the following data:

Name	pK_a'
2-mandelamidopyridine	2.94
2-mandelamido-4-picoline	3.25
2-(β-hydroxyphenethylamino)-pyridine	5.85
4-(β-hydroxyphenethylamino)-pyridine	8.49
2-(β-hydroxyphenethylamino)-4-picoline	6.50
6-(β-hydroxyphenethylamino)-3-picoline	6.30
2-(β-hydroxyphenethylamino)-5-chloropyridine	3.70
2-(2-hydroxypropylamino)-pyridine	6.10
2-(β-hydroxy-β?β-diphenylethylamino)-pyridine	5.50
2-(β-acetoxyphenethylamino)-pyridine	5.85 (approx)
1-(β-hydroxyphenethyl)-2-imino-1,2-dihydropyridine	11.6 (approx)
1-(β-acetoxyphenethyl)-2-acetylimino-1,2-dihydropyridine	5.83
2-(γ-hydroxy-γ-phenylpropylamino)-pyridine	6.15
2-phenethylaminopyridine	6.10

"Apparent pK_a's were measured by titration of ca. 0.01 M solutions of the hydrochloride salts in 60:40 dimethylformamide—carbon dioxide-free water with 0.1 N NaOH at 25°. A Beckman model G pH meter was used. The listed values are the averages of at least two determinations."

NB: All values assessed as U = uncertain. See also Gray AP and Heitmeier DE, Aminopyridines. I. β-hydroxyalkylaminopyridines via glycolamidopyridines, *JACS*, **81**, 4347–4350 (1959); cited pK_a values for 2-aminopyridine (6.86) and 4-aminopyridine (9.17), from Albert A, Goldacre R and Phillips J, *J. Chem. Soc.* 2240–2249 (1948).

No.	Compound	pKa			Method	Conditions	Reference / Comments
1100	Phenytoin ($C_{15}H_{12}N_2O_2$)	8.31 ± 0.04	U	−H	Spectro (λ = 236 nm)	1% EtOH in H_2O *t* undefined *I* undefined	Agarwal SP and Blake MJ, Determination of the p$K_a{}'$ value for 5,5-diphenylhydantoin *J. Pharm. Sci.*, **57**, 1434–1435 (1968). NB: Cited by Philip J, Holcomb IJ and Fusari SA, Phenytoin, *APDS*, **13**, 417–440 (1984). Spectrophotometric method according to Albert and Serjeant (1962). Potentiometric results from linear extrapolation to 0% of apparent values for titration of sodium phenytoin with 1-N HCl in 20%, 30%, 40%, and 50% v/v aqueous ethanol.
		8.33	U	−H	Potentiometric	H_2O *t* undefined $I = 0.01$	
1101	Phenytoin	8.06	U	−H	soly	H_2O/MeOH (0–4%) pH = 4.8–8.4	Schwartz PA, Rhodes CT and Cooper JW, Solubility and ionization characteristics of phenytoin, *J. Pharm. Sci.*, **66**, 994–997 (1977). "The solubility of phenytoin (I) was determined in pH 7.4 and 5.4 phosphate buffers at 5 temperatures; in methanol 0–4%; and in pH 4.8–8.4 buffer solutions. The data obtained from the buffer solutions were used to calculate the apparent dissociation constant of I as 8.06. . . ."
1102	Phenytoin	8.32 ± 0.01	U	−H	Spectro, Potentiometric	H_2O	Schwartz PA, Rhodes CT and Cooper JW, Solubility and ionization characteristics of phenytoin, *J. Pharm. Sci.*, **66**, 994–997 (1977).
1103	Phenytoin	8.43	U	+H	CE/pH (−ve ion mode)	H_2O $t = 25$ $I = 0.025$	Wan H, Holmen AG, Wang Y, Lindberg W, Englund M, Nagard MB and Thompson RA, High-throughput screening of pK_a values of pharmaceuticals by pressure-assisted capillary electrophoresis and mass spectrometry, *Rapid Commun. Mass Spectrom.*, **17**, 2639–2648 (2003). NB: Reported a predicted value (ACD Labs) of 8.33. Assignment of the pK_a value to a basic group is in disagreement with the consensus of literature, which assigns this value to the imide acid system.
1104	Phenytoin	8.21 (0.14)	U	−H	Spectro (253 nm)	H_2O $t = 20.0$	Wahbe AM, El-Yazbi FA, Barary MH and Sabri SM, Application of orthogonal functions to spectrophotometric analysis. Determination of dissociation constants, *Int. J. Pharm.*, **92**(1), 15–22 (1993). NB: See Acetaminophen for further details. An alternative graphical method gave pK_a = 8.2.

(continued)

No.	Compound Name	pK_a value(s)	Data quality	Ionization type	Method	Conditions	Comments and Reference(s)
1105	Phosphocreatine ($C_4H_{10}N_3O_5P$) 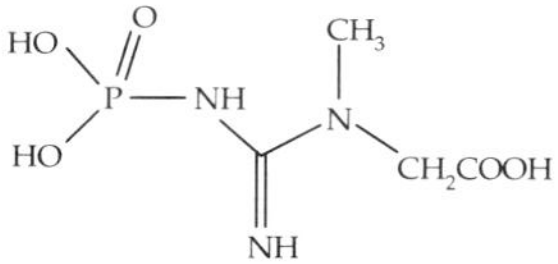	4.7 ± 0.01 11.0 ± 0.05	A U	−H +H	Potentiometric	H_2O $t = 25.0$	Breccia A, Fini A, Girotti S and Stagni G, Correlation between physico-chemical parameters of phosphocreatine, creatine, and creatinine, and their reactivity with their potential diffusion in tissue, *Pharmatherapeutica*, **3**(4), 227–232 (1982). NB: See Creatine for details.
1106	Phosphocreatine	2.7 4.5 12	U U U	−H −H +H	Potentiometric	H_2O $t = 37$	Meyerhof O and Lohmann K, The natural guanidinophosphoric acids (phosphagens) in striated muscle. II. Physico-chemical properties of the guanidinophosphoric acids, *Biochem. Z.*, **196**, 49–72 (1928). Cited in Perrin Bases 3410 ref. M41. CA 23:7362.
1107	Phthalic acid ($C_8H_6O_4$) 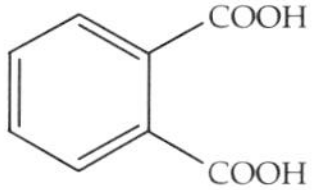	2.950 5.408	R R	−H −H	Potentiometric	H_2O $t = 25.0 \pm 0.01$ $I = 0.000$ (KCl)	Hamer WJ, Pinching GD and Acree SF, First dissociation constant of *o*-phthalic acid and related pH values of phthalate buffers from 0° to 60°, *J. Res. Nat. Bur. Stand.*, **35**, 539–564 (1945); Hamer WJ and Acree SF, *J. Res. Nat. Bur. Stand.*, **35**, 381–416 (1945). NB: Used electrochemical cell without liquid junction potentials; detailed procedures were required to extrapolate ionic strength effects to zero.
1108	Physostigmine salicylate ($C_{22}H_{27}N_3O_5$) 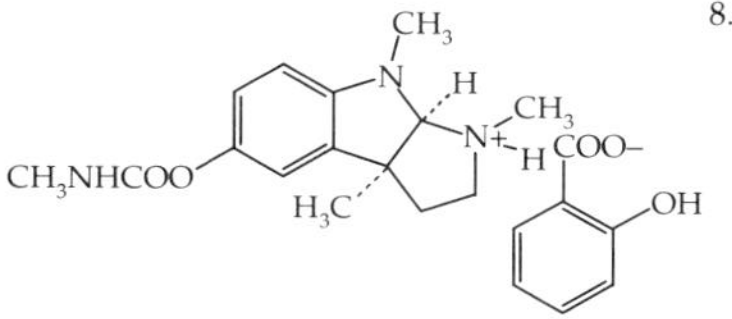	1.96 8.08	U U	+H +H	Spectro	H_2O $t = 15$	Kolthoff IM, The dissociation constants, solubility product and titration of alkaloids, *Biochem. Z.*, **162**, 289–353 (1925). Cited in Perrin Bases 2947 ref. K47. NB: See Aconitine for details.
1109	Physostigmine	GLpK_a: 8.17 ± 0.02 A&S: 8.10 ± 0.09	 A U	 +H +H	Spectro	H_2O $t = 25$ $I = 0.15$ (KCl) Ar atmosphere	Tam KY and Takacs-Novac K, Multi-wavelength spectrophotometric determination of acid dissociation constants, *Anal. Chim. Acta*, **434**, 157–167 (2001). NB: See Clioquinol for details.

No.	Compound	pKa	Data quality	Ionization	Method	Conditions	Reference / Comments
1110	Picolinic acid ($C_6H_5NO_2$) 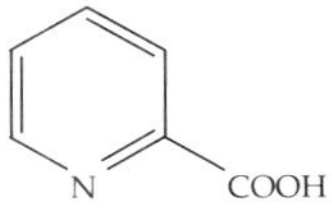	1.5 5.49	U A	+H −H	Potentiometric	H_2O $t = 18 \pm 2$ $c = 0.04$	Holmes F and Crimmin WRC, The stabilities of metal chelate compounds formed by some heterocyclic acids. I. Studies in aqueous solution, *J. Chem. Soc.*, 1175–1180 (1955).
		0.99 5.39	A A	+H −H	Spectro	H_2O $t = 25$ $I = 0.03$	Evans RF, Herington EFG and Kynaston W, Determination of dissociation constants of the pyridine-monocarboxylic acids by ultraviolet photoelectric spectrophotometry. *Trans. Farad. Soc.*, **49**, 1284–1292 (1953). NB: Data for pK_{a1} was obtained with $I > 0.03$.
		1.01 5.32	A A	+H −H	Potentiometric	H_2O $t = 22$ $c = 0.05$ to 0.1 N_2 atmosphere	Green RW and Tong HK, The constitution of the pyridine monocarboxylic acids in their isoelectric forms, *JACS*, **78**, 4896–4900 (1956). Used a glass electrode standardized with phthalate solution (pH = 4.00) and the Guntelberg equation to correct for I. Estimated the microconstants from spectrophotometric data on the acid and its methyl ester: pK_A, 1.04; pK_B, 2.21; pK_C, 5.29; pK_D, 4.12; where the subscripts represent the following equilibria: A, diprotonated to zwitterion; B, diprotonated to neutral; C, zwitterion to fully deprotonated; D, neutral to fully deprotonated. Also reported the corresponding data for picolinic and isonicotinic acids.
		1.60 5.40	U A	+H −H	Spectro	H_2O $t = 20 \pm 2$ $I = 0.01$; $c < 5 \times 10^{-4}$	Jellinek HHG and Urwin JR, Ultraviolet absorption spectra and dissociation constants of picolinic, isonicotinic acids and their amides, *J. Phys. Chem.*, **58**, 548–550 (1954). Used glass electrode to measure pH values and recorded spectra with a Hilger Uvispek spectrophotometer. Ionic strength was extrapolated to zero with the Debye-Huckel equation.
		5.3	A	−H	Spectro	H_2O $t = 25.0$ $I = 0.005$–0.025	Fischer A, Galloway WJ and Vaughan J, Structure and reactivity in the pyridine series. I. Acid dissociation constant of pyridinium ions, *J. Chem. Soc.* B, 3591–3596 (1964). All data were cited in Perrin Bases no. 1076–77; Perrin Bases suppl. no. 5081.
1111	Pilocarpine ($C_{11}H_{16}N_2O_2$) 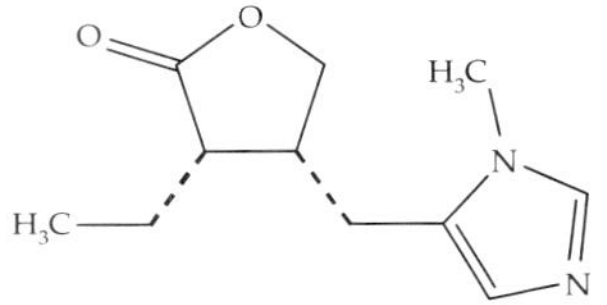	7.00	A	+H	Potentiometric	H_2O $t = 25$ $I = 0.00$ (NaCl, $NaClO_4$)	Lukkari S and Palonen M, Potentiometric and spectrophotometric studies on the ionization of pilocarpine in aqueous solution, *Suomen Kemistilehti B*, **41**, 225–8 (1968). Cited in Perrin Bases suppl. no. 7487.
1112	Pilocarpine	7.08 ± 0.02	A	+H	Potentiometric	H_2O $t = 25.0$ $I = 0.1$ (NaCl)	Takacs-Novak K and Avdeef A, Interlaboratory study of log P determination by shake-flask and potentiometric methods, *J. Pharm. Biomed. Anal.*, **14**, 1405–1413 (1996). NB: See Acetaminophen for further details.

(continued)

No.	Compound Name	pK_a value(s)	Data quality	Ionization type	Method	Conditions	Comments and Reference(s)
1113	Pilocarpine	6.98	A	+H	CE/pH (+ve ion mode)	H_2O $t = 25$ $I = 0.025$	Wan H, Holmen AG, Wang Y, Lindberg W, Englund M, Nagard MB and Thompson RA, High-throughput screening of pK_a values of pharmaceuticals by pressure-assisted capillary electrophoresis and mass spectrometry, *Rapid Commun. Mass Spectrom.*, **17**, 2639–2648 (2003). NB: Reported a predicted value (ACD Labs) of 7.02.
1114	Pilocarpine	GLpK_a: 7.06 ± 0.02 A&S: 7.12 ± 0.14	 A U	 +H +H	Spectro	H_2O $t = 25$ $I = 0.15$ (KCl) Ar atmosphere	Tam KY and Takacs-Novac K, Multi-wavelength spectrophotometric determination of acid dissociation constants, *Anal. Chim. Acta*, **434**, 157–167 (2001). NB: See Clioquinol for details.
1115	Pilocarpine	7.07	U	+H			Longwell A, Birss S, Keller N and Moore D, Effect of topically applied pilocarpine on tear film pH, *J. Pharm. Sci.*, **65**, 1654–1657 (1976). "The reduction in tear film pH produced by I eyedrops or spray solution is attributable to the acid pH and buffer capacity of these solutions. Delivery of I base without pH change was achieved with ocular therapeutic systems, because the drug (pK_a = 7.07) was delivered free, or virtually so, of excipients."
1116	Pilocarpine	1.63 7.05	U U	+H +H	Spectro	H_2O $t = 15$	Kolthoff IM, The dissociation constants, solubility product and titration of alkaloids, *Biochem. Z.*, **162**, 289–353 (1925). Cited in Perrin Bases 2949 ref. K47. NB: See Aconitine for details.
1117	Pilocarpine	7.15 12.57	U U	+H +H		$t = 20.0$	Al-Badr AA and Aboul-Enein HY, Pilocarpine, *APDS*, **12**, 385–427 (1983). NB: These are clearly pK_b values.
1118	Pilocarpine	6.97	A	+H	Potentiometric	H_2O $t = 30$ $I = 0.07$	Cowgill RW and Clark WM, Coordination of imidazoles with ferrimesoporphyrin, *J. Biol. Chem.*, **198**, 33–61 (1952). Cited in Perrin Bases 2949 ref. C51 (pK_a value was misquoted as 6.87). Study used a glass electrode (carefully calibrated) and solutions were titrated with carbonate-free KOH.
1119	Pipamazine ($C_{20}H_{32}ClN_3OS$)	8.60	U	+H	Potentiometric	H_2O (extrap) $t = 24 \pm 1$ $I \sim 0.002$	Chatten LG and Harris LE, Relationship between pK_b (H_2O) of organic compounds and $E_{1/2}$ values in several non-aqueous solvents, *Anal. Chem.*, **34**, 1495–1501 (1962). Cited in Perrin Bases suppl. no. 7438 ref. C10. NB: Study used glass electrode measurements of the pH value of solutions containing equimolar proportions of the free base and the salt. Results for several different concentrations in methanol were extrapolated back to 0% methanol.

O NH2 N N Cl S

1120	Piperazine ($C_4H_{10}N_2$)	5.333 9.731	R R	+H +H	Potentiometric	H_2O $t = 25$ $I = 0.01$ to 0.10	Hetzer HA, Robinson RA, Bates RG, Dissociation constants of piperazinium ion and related thermodynamic quantities from 0 to 50°, *J. Phys. Chem.*, **72**, 2081–2086 (1968). Cited in Perrin Bases suppl. no. 5572 ref. H36. NB: Very careful work using an electrochemical cell without liquid junction potentials. Activity coefficients were calculated for measurements made on carbonate-free solutions at the following temperatures:
1121	Piperazine	5.55 9.81	U U	+H +H	Potentiometric	H_2O $t = 25.0$ I undefined	Hall HK, Field and inductive effects on the base strength of amines, *JACS*, **78**, 2570–2572 (1956). Cited in Perrin Bases no. 1465 ref. H5. NB: Used glass electrode in cell with liquid junction potentials. No activity corrections. See also Piperazine, Newton and Kluza. Foye 1 gave 5.7; 9.8 (cited Merck Index).
		5.57 9.81	U U	+H +H	calorimetry	H_2O	Dragulescu C, Policec S, Thermometric titration of weak diacidic bases. *Studii Cercetari Chim.*, **9**, 33–40 (1962). CA 58:30272. Cited in Perrin Bases suppl. no. 7488. NB: From pK_b values 4.19 and 8.43 at unknown temperature (assumed 25 °C).
1122	Piperazine estrone sulfate ($C_{22}H_{32}N_2O_5S$)	3.6 9.7	U U	+H +H	Potentiometric	acetonitrile, 80% v/v	Chang ZL, Piperazine Estrone Sulfate, *APDS*, **5**, 375–402 (1976). Cited Wimer DC, Abbott Laboratories, personal communication. "The apparent pK_a value of the (free base) piperazine nitrogen was found to be 3.6, by titration in acetonitril-water (80/20, v/v) with aqueous soldium hydroxide. Attempts to find systems to extrapolate the pK_a to 100% water were unsuccessful. . . . The pK_a value of the protonated piperazine nitrogen (proton lost) was found to be 9.7 by titration in pyridine-water mixtures with methanolic KOH, and extrapolation to 100% water."

T (°C)	pK_{a1}	pK_{a2}	T (°C)	pK_{a1}	pK_{a2}	T (°C)	pK_{a1}	pK_{a2}
0	5.816	10.407	20	5.424	9.864	40	5.066	9.367
5	5.712	10.259	25	5.333	9.731	45	4.981	9.252
10	5.614	10.126	30	5.246	9.609	50	4.896	9.142
15	5.518	9.996	35	5.153	9.485			

(continued)

No.	Compound Name	pK_a value(s)	Data quality	Ionization type	Method	Conditions	Comments and Reference(s)
1123	Piperidine ($C_5H_{11}N$)	11.20	U	+H	Potentiometric	H_2O $t = 25.0 \pm 0.2$ $I = 0.2$ (NaCl) N_2 atmosphere	Rosenblatt DH, Hull LA, DeLuca DC, Davis GT, Weglein RC and Williams HKR, Oxidations of amines. II. Substituent effects in chlorine dioxide oxidations, *JACS*, 89 1158–1163 (1967). Cited in Perrin Bases suppl. no. 4961, ref. R22a. Used a glass electrode in a cell with liquid junction potentials. Solutions were carbonate-free. No activity corrections were made. Calibrated pH meter at pH = 7.00 and 10.00.
1124	Piperidine	11.123	R	+H	Potentiometric	H_2O $t = 25.0 \pm 0.2$ $I = 0.01$ to 0.10 (NaCl) N_2 atmosphere	Bates RG and Bower VE, Dissociation constant of piperidinium ion from 0° to 50° and related thermodynamic quantities, *J. Res. Natl. Bur. Stand.*, **57**, 153–157 (1956) Perrin Bases, no. 967 ref. B28. NB: Very careful work using an electrochemical cell without liquid junction potentials. Activity coefficients were calculated for measurements made on carbonate-free solutions at the following temperatures: T (°C) / pK_a: 0 / 11.963; 5 / 11.786; 10 / 11.613; 15 / 11.443; 20 / 11.280; 25 / 11.123; 30 / 10.974; 35 / 10.818; 40 / 10.670; 45 / 10.526; 50 / 10.384 See also Hong W-H and Connors KA, *J. Pharm. Sci.*, **57**, 1789–90 (1968), who reported $pK_a = 11.22$, but without measurement details or references.
1125	Piperine ($C_{17}H_{19}NO_3$)	1.98	U	+H	Spectro	H_2O $t = 15.0$	Kolthoff IM, The dissociation constants, solubility product and titration of alkaloids, *Biochem. Z.*, **162**, 289–353 (1925). Cited in Perrin Bases 2950 ref. K47. NB: See Aconitine for details; Arnall F, The determination of the relative strengths of some nitrogen bases and alkaloids, *J. Chem. Soc.* **117**, 835–839 (1920). Reported $pK_a = 1.42$ at $t = 55$ °C from kinetic measurements. Cited in Perrin Bases 2950 ref. A73.

T (°C)	pK_a	T (°C)	pK_a	T (°C)	pK_a
0	11.963	20	11.280	40	10.670
5	11.786	25	11.123	45	10.526
10	11.613	30	10.974	50	10.384
15	11.443	35	10.818		

No.	Compound	pKa	Quality	Type	Method	Conditions	Reference
1126	DL-Pipidone ($C_{24}H_{31}NO$)	6.80	U	+H	Potentiometric	50% aq EtOH	Farmilo CG, Oestreicher PM and Levi L, Physical methods for the identification of narcotics. IB Common physical constants for identification of ninety-five narcotics and related compounds, *Bull. Narcotics*, UN Dept. Social Affairs, vol. **6**, pp. 7–19 (1954); cited in Clouet DH (ed.), *Narcotic Drugs Biochemical Pharmacology*, Plenum Press, New York, 52–53 (1971). NB: See Alphaprodine for details.
1127	Pirbuterol ($C_{12}H_{20}N_2O_3$)	8.01 10.64	U U	−H +H	Spectro	H_2O $t = 25.0 \pm 0.05$ $I = 0.10$	Ijzerman AP, BultsmaT, Timmerman H and Zaagsma J, The ionization of β-adrenoceptor agonists: A method for unravelling ionization schemes, *J. Pharm. Pharmacol.*, **36**(1), 11–15 (1984). NB: Microscopic: 7.95 and 10.64. See Isoprenalin.
1128	Piribedil (1-(3,4-methylenedioxybenzyl)-4-(2-pyrimidinyl)piperazine) ($C_{16}H_{18}N_4O_2$)	6.94 ± 0.01	U	+H	Potentiometric	H_2O $I = 0.1$ $t = 25.0 \pm 0.2$	Tsai R-S, El Tayar N, Carrupt P-A and Testa B, Physicochemical properties and transport behaviour of piribedil: Considerations on its membrane-crossing potential, *Int. J. Pharm.*, **80**(1) 39–49 (1992). NB: Spectrophotometric value, 6.92 ± 0.22. pK_{a2} reported to be 1.3 (Caccia S, Fong MH and Guiso G, Disposition of the psychotropic drugs buspirone, MJ-13805 and piribedil, and of their common active metabolite 1-(2-pyrimidinyl)-piperazine in the rat, *Xenobiotica*, **15** 835–844, (1985)).

(continued)

No.	Compound Name	pK_a value(s)	Data quality	Ionization type	Method	Conditions	Comments and Reference(s)
1129	Piroxicam ($C_{15}H_{13}N_3O_4S$) 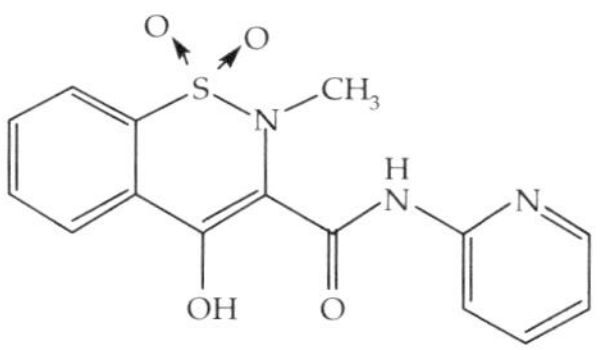	5.3	U	–H	Spectro	H_2O *t* undefined *I* undefined	Herzfeldt CD and Kümmel R, Dissociation constants, solubilities, and dissolution rates of some selected nonsteroidal antiinflammatories, *Drug Dev. Ind. Pharm.*, **9**(5), 767–793 (1983). NB: Used dλ/dpH method. See Azapropazone and Ibuprofen for details. Craig gave 4.6 (no ref.).
1130	Piroxicam	6.3	U	–H	Potentiometric	H_2O-dioxan (2:1)	Mihalic M, Hofman H, Kuftinec J, Krile B, Capler V, Kajfez F and Blazevic N, Piroxicam, *APDS*, **15**, 509–531 (1986). NB: Cited refs 4, 7 and 18: 4. Wiseman EH, Chang Y-H and Lombardino JG, Piroxicam, a novel anti-inflammatory agent, *Arzneim.-Forsch.*, **26**, 1300–1303 (1976). NB: "An acidity constant (pK_a) was determined by potentiometric titration of piroxicam in 2:1 dioxane-water . . . at 25 °C. The apparent pK_a value was estimated from the half-neutralization point . . . ." 7. Lombardino JG, Wiseman EH and McLamore WM, Synthesis and antiinflammatory activity of some 3-carboxamides of 2-alkyl-4-hydroxy-2*H*-1,2-benzothiazine-1,1-dioxide, *J. Med. Chem.*, **14**, 1171–1175 (1971). NB: This reference also reported apparent pK_a values for numerous analogues of piroxicam. 18. Lombardino JG, Wiseman EH and Chiaini J, Potent antiinflammatory *N*-heterocyclic 3-carboxamides of 4-hydroxy-2-methyl-2*H*-1,2-benzothiazine-1,1-dioxide, *J. Med. Chem.*, **16**, 493–496 (1973). NB: This reference also reported apparent pK_a values for a further 14 heterocyclic analogues of piroxicam.
1131	Piroxicam	2.33 ± 0.18 5.07 ± 0.03	U A	+H –H	Potentiometric	H_2O $t = 25.0$ $I = 0.15$ (KCl)	Sirius Technical Application Notes, vol. **2**, pp. 110–111 (1995). Sirius Analytical Instruments Ltd., Forest Row, East Sussex, RH18 5DW, UK. NB: Analyte concentration, 0.2–0.7 mM. Extrapolated to 0 wt% MeOH from data obtained in 27–67 wt% MeOH by the Yasuda-Shedlovsky procedure.

1132	(O-Pivaloyl)-etilefrine ($C_{15}H_{23}NO_3$) 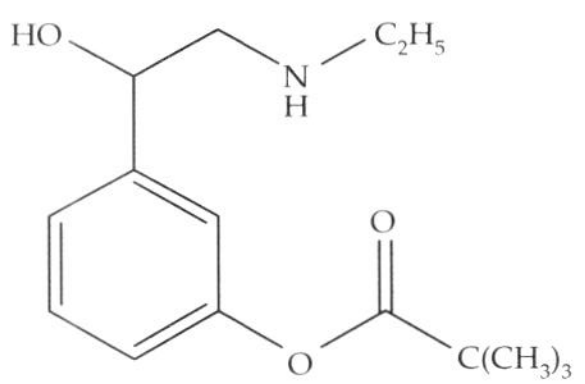	9.4	U	+H	Potentiometric	H_2O	Wagner J, Grill H and Henschler D, Prodrugs of etilefrine: synthesis and evaluation of 3′-(O-acyl) derivatives, *J. Pharm. Sci.*, **69**, 1423–1427 (1980).
		8.7	U	+H		90.7% MeOH	"A series of nineteen 3′-(O-acyl) derivatives of etilefrine HCl was synthesized and their physical characteristics were evaluated. Correlations between structure and solubility, dissociation constant values, lipophilicity and esterase catalyzed hydrolysis were demonstrated. 3′-(O-Pivaloyl)-etilefrine showed favorable characteristics of solubility, lipophilicity and stability against enzymatic cleavage in blood." NB: The same apparent pK_a values were measured for four other substituted alkanoic acid esters: propionic; butanoic, 2-methylpropanoic; and pentanoic. Thirteen other compounds, all alkanoic or aromatic acid esters were too poorly water-soluble to be measured in aqueous solution. They all were reported to have apparent $pK_a = 8.7$ in 90.7% aqueous methanol.
1133	Polythiazide ($C_{11}H_{13}ClF_3N_3O_4S_3$) 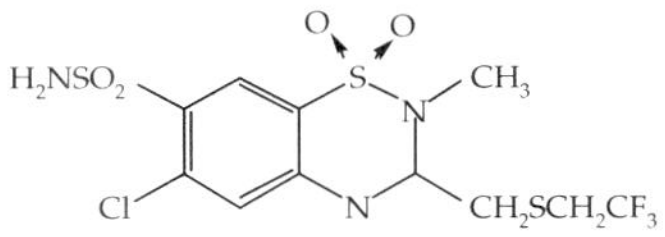	9.10	U	−H	Potentiometric	acetone/H_2O	Henning VG, Moskalyk RE, Chatten LG and Chan SF, Semiaqueous potentiometric determination of apparent pK_{a1} values for benzothiadiazines and detection of decomposition during solubility variation with pH studies, *J. Pharm. Sci.*, **70**(3), 317–319 (1981).
		9.58	U	−H	Spectro	H_2O $t = 25$	
		9.82 ± 0.04	U	−H	partition	H_2O	NB: See Flumethiazide. Cited in Negendra Vara Prasad T, Rao, V, Sastry, S and Surya Prakasa Sastry C, Polythiazide, *APDS*, **20**, 665–692 (1991), which reported the following values:

Value	Ref.	Source
9.10	18	Hennig et al. (this entry).
	20	Hennig UG, MSc Thesis Univ. Alberta, Canada.
9.82 ± 0.04	19	Khan AS and Cantwell FF, Measurement of acidity constants of benzothiadiazines by solvent extraction with use of a membrane phase separator, Talanta, **33**, 119–123 (1986).
11.0	21	Scriabine A, Schreiber EC, Yu M, Wiseman EH, Renal clearance of polythiazide, *Proc. Soc. Exptl. Biol. Med.*, **110**, 872–875 (1962).
9.58	22	Yamazaki M, Suzuka T, Ito Y, Oith S, Kitamura M, et al., Biopharmaceutical studies of thiazide diuretics. I., *Chem. Pharm. Bull.*, **32**, 2380–2386 (1984).

(continued)

No.	Compound Name	pK_a value(s)	Data quality	Ionization type	Method	Conditions	Comments and Reference(s)
1134	Porfiromycin ($C_{16}H_{20}N_4O_5$) 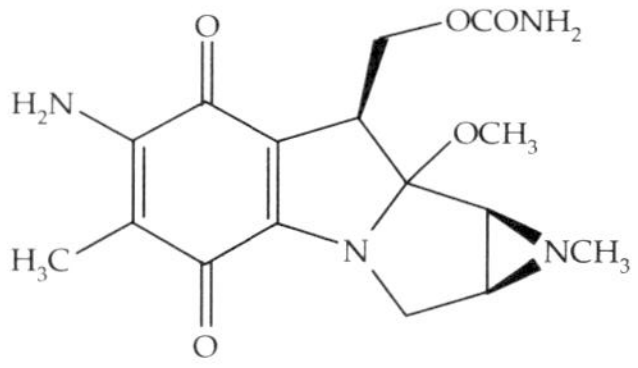	−1.3 −1.2 ~1.5 12.44 2.4	U U U U U	+H +H +H −H +H	Potentiometric Spectro ($\lambda = 363$ nm) kinetic	H_2O t = RT H_2O t = 25	Underberg WJM and Lingeman H, Determination of pK_a values of some prototropic functions in mitomycin and porfiromycin, *J. Pharm. Sci.*, **72**, 553–556 (1983). NB: See also Mitomycin C. The prototropic properties of mitomycin and porfiromycin were studied, and the dissociation constants for 2 potentially basic groups and one acidic function were established by titration. The kinetics of the tautomerization preceding the prototropic reaction in an alkaline medium are also discussed. McClelland RA and Lam K, Kinetics and mechanism of the acid hydrolysis of mitomycins, *JACS*, **107**, 5182–5186 (1985). NB: First order behaviour observed. Other observed kinetic pK_a values: Porfiromycin, 25 °C, 2.4.
1135	Pralidoxime chloride ($C_7H_9ClN_2O$) 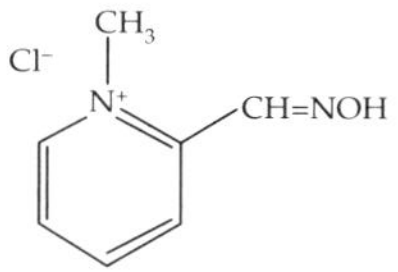	7.8–8	U	+H			Banakar UV and Patel UN, Pralidoxime chloride, *APDS*, **17**, 533–569 (1988). - Oxime group (Connors, Amidon & Stella, Chemical Stability of Pharmaceuticals, 1986). ~8-Pralidoxime (AHFS, 1987).
1136	Prenylamine ($C_{24}H_{27}N$) 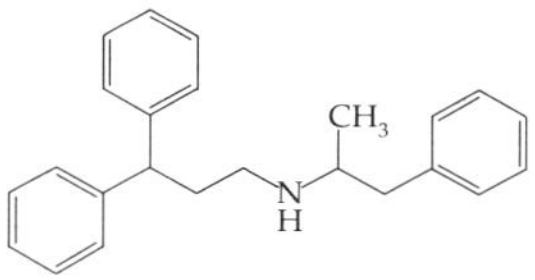	8.74 ± 0.02 9.47	U U	+H +H	Potentiometric	40% EtOH t = 25.0 H_2O	Mannhold R, Rodenkirchen R, Bayer R and Haus W, The importance of drug ionization for the action of calcium antagonistsand related compounds, *Arzneim.-Forsch.*, **34**, 407–409 (1984). NB: See Aprindine for details.
1137	Primaquine ($C_{15}H_{21}N_3O$)	3.74 9.99	U U	+H +H	Potentiometric	H_2O t = 25	Bergström CAS, Strafford M, Lazorova L, Avdeef A, Luthman K and Artursson P, Absorption classification of oral drugs based on molecular surface properties, *J. Med. Chem.*, **46**(4), 558–570 (2003). NB: From extrapolation of aqueous-methanol mixtures to 0% methanol.

No.	Compound	pKa	Data quality	Type	Method	Conditions	Reference and notes
1138	Pristinamycin I_A; $R_1 = H$; $R_2 = -N(CH_3)_2$ (vernamycin Bα) 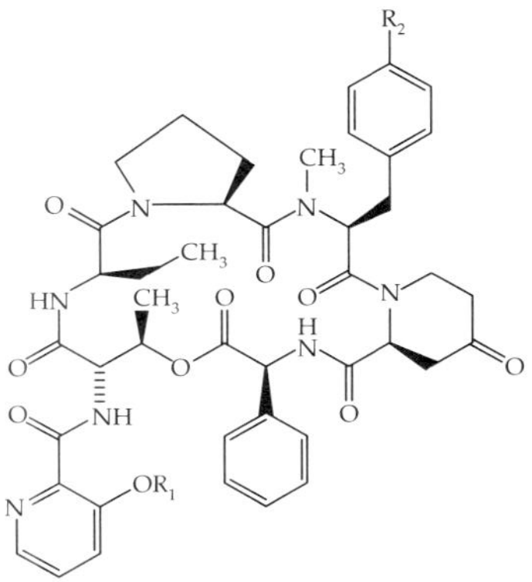	1.00 ± 0.05 7.20 ± 0.05 2.40 ± 0.05 7.85 ± 0.05	U U U U	+H −H +H −H	Potentiometric Spectro ($\lambda = 280$–400 nm)	10% MeOH in H_2O $t = 25$ $I = 0.2$ ($LiClO_4$)	Largeron M and Fleury MB, Acid-base properties of pristinamycin IA and related compounds, *J. Pharm. Sci.*, **81**, 565–568 (1992). "The pK_a values of acid-base equilibria involved in the protonation of pristinamycin IA were determined by UV-visible absorption spectrometry and potentiometric titration. The equilibrium between dipolar ionic and uncharged neutral forms was investigated spectrometrically in methanol-water solutions. With pristinamycin, the dipolar ionic form predominated in aqueous solutions buffered to the isoelectric pH. The effects of structure on the ionization constants are briefly discussed." NB: A model compound restricted to the essential features of the 3-hydroxypyridine portion (giving the second set of pK_a values) was used to simplify the data analysis. A third pK_a value, corresponding to the dimethylamino group, has not been reported. The microconstants for the 3-hydroxypyridine system were reported from a combination of potentiometric and spectrophotometric results.
1139	Probenecid ($C_{13}H_{19}NO_4S$) 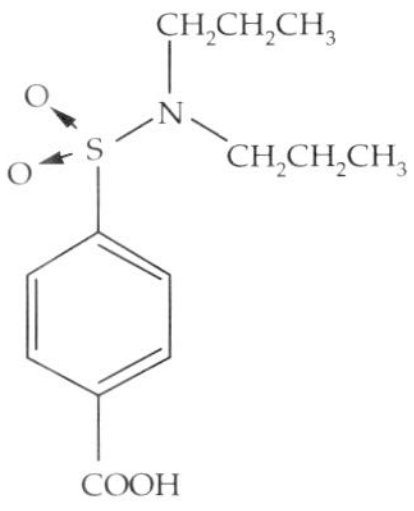	3.4	U	−H	partition	H_2O t undefined $I = 0.1$	Shore PA, Brodie BB and Hogben CAM, The gastric sectretion of drugs: A pH-partition hypothesis, *JPET*, **119**, 361–369 (1957).

(*continued*)

No.	Compound Name	pK_a value(s)	Data quality	Ionization type	Method	Conditions	Comments and Reference(s)
1140	Procaine ($C_{13}H_{20}N_2O_2$) 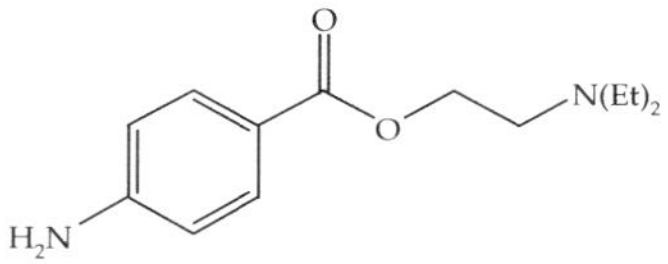	2.29 ± 0.01 9.04 ± 0.01	A A	+H +H	Potentiometric	H_2O $t = 25 \pm 0.5$ $I = 0.15$ (KCl)	Avdeef A, Box KJ, Comer JEA, Hibbert C and Tam KY, pH-metric log P 10. Determination of liposomal membrane-water partition coefficients of ionizable drugs, *Pharm. Res.*, **15**(2), 209–215 (1998). NB: Used a Sirius PCA101 autotitrator. Also gave log P (octanol-water) and log P (dioleylphosphatidylcholine unilamellar vesicles).
1141	Procaine	9.04 ± 0.01	A	+H	Potentiometric	H_2O $t = 25.0$ $I = 0.1$ (NaCl)	Takacs-Novak K and Avdeef A, Interlaboratory study of log P determination by shake-flask and potentiometric methods, *J. Pharm. Biomed. Anal.*, **14**, 1405–1413 (1996). NB: See Acetaminophen for further details.
1142	Procaine	9.09 ± 0.03	A	+H	Potentiometric	H_2O $t = 20.0$ $I = 0.10$ (KCl) N_2 atmosphere	Buchi J and Perlia X, Beziehungen zwischen de physikalisch-chemische Eigenschaften und der Wirkung von Lokalanasthetica, *Arzneim. -Forsch.*, **10**, 745–754 (1960). NB: See Cocaine for details.
1143	Procarbazine ($C_{12}H_{19}N_3O$) 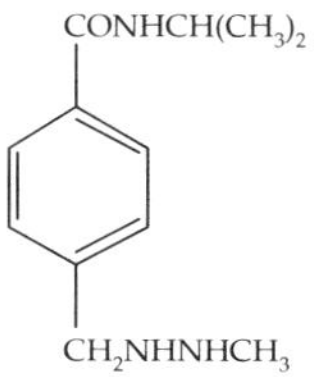	6.8	U	+H	Potentiometric	H_2O $I = 0.1$	Rucki RJ, Procarbazine hydrochloride, *APDS*, **5**, 403–427 (1976). NB: From Philip C, Hoffmann-La Roche Inc., personal communication.
1144	Prochlorperazine 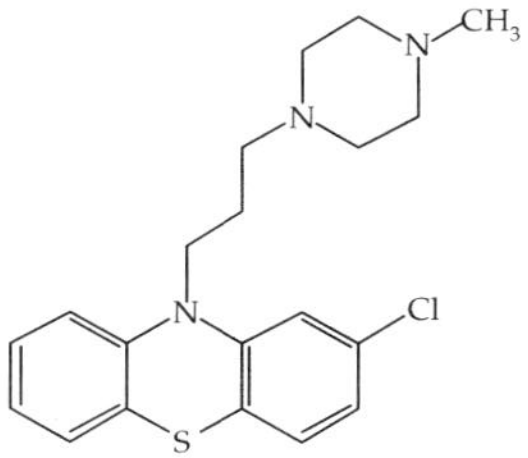	8.14 ± 0.06	U	+H	partition/pH	H_2O $t = 20 \pm 0.5$ I not reported but pK_a was stated to be independent of I.	Vezin WR and Florence AT, The determination of dissociation constants and partition coefficients of phenothiazine derivatives, *Int. J. Pharm.*, **3**, 231–237 (1979). NB: See Chlorpromazine and Promethazine for additional details. See separate entries for Phenothiazine, 2-chloro-10-(*N*-methylpiperazinyl)-3-propyl.

No.	Compound	pKa	Quality	Type	Method	Conditions	Reference
1145	Prochlorperazine	8.1	U	+H	soly	H_2O $t = 24 \pm 1$ I undefined	Green AL, Ionization constants and water solublities of some aminoalkylphenothiazine tranquilizers and related compounds, *J. Pharm. Pharmacol.*, **19**, 10–16 (1967). NB: See Amitriptylline for details.
1146	Prochlorperazine	3.60 7.54	U U	+H +H	Potentiometric	H_2O (extrap) $t = 24 \pm 1$ $I \sim 0.002$	Chatten LG and Harris LE, Relationship between $pK_b(H_2O)$ of organic compounds and $E_{1/2}$ values in several nonaqueous solvents, *Anal. Chem.*, **34**, 1495–1501 (1962).
1147	Prochlorperazine	7.5 8.1	U U	+H +H			Zografi G, Munshi MV, Effect of chemical modification on the surface activity of some phenothiazines, *J. Pharm. Sci.*, **59**, 819–822 (1970). NB: Cited the values of Green (1967) and Chatten and Harris (1962).
1148	Progabide ($C_{17}H_{16}ClFN_2O_2$) 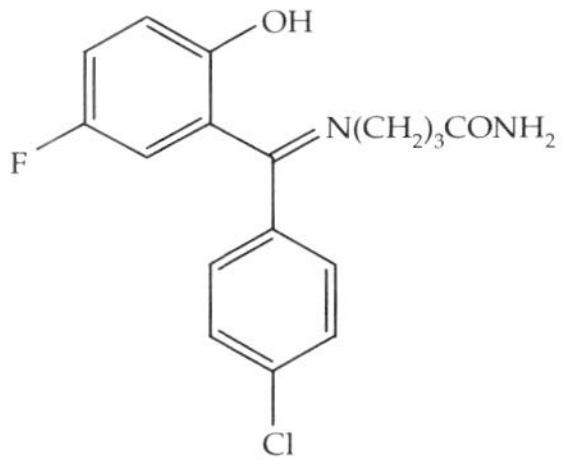 OH F $N(CH_2)_3CONH_2$ Cl	3.41 ± 0.04 12.95 ± 0.07	A U	+H −H	Spectro	0.4% MeOH $t = 37.0$ I undefined	Farraj NF, Davis SS, Parr GD and Stevens HNE, Dissociation and partitioning of progabide and its degradation product, *Int. J. Pharm.*, **46**, 231–239 (1988). "The pK_{a1} value . . . was determined to be 3.41 with a SD of 0.02. . . . the pK_{a2} value was found to be 12.95. . . . the pK_a (of degradation product (3-fluoro-6-hydroxy-4′-chlorobenzophenone, SL79.182)) was found to be 8.94 ± 0.18 (phenolic hydroxyl)."
1149	Promazine ($C_{17}H_{20}N_2S$) 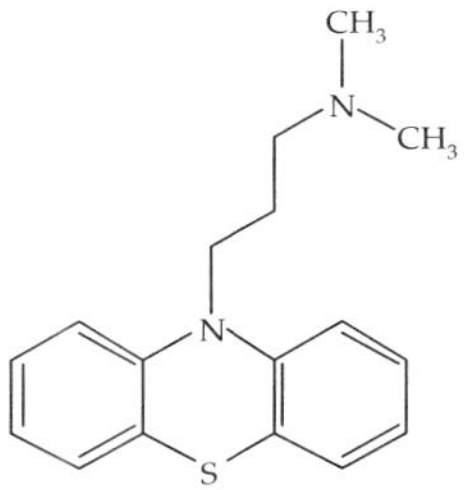 CH_3 N CH_3 N S	9.40	U	+H	soly	H_2O $t = 20.0 \pm 0.1$ $I = 0.15$	Schill G, Photometric determination of amines and quaternary ammonium compounds with bromothymol blue, Part 5. Determination of dissociation constants of amines, *Acta Pharm. Suec.* **2**, 99–108 (1965). NB: See Chloroquine for details. See separate entries under Phenothiazine.
1150	Promazine	9.4	U	+H	soly	H_2O $t = 24 \pm 1$	Green AL, Ionization constants and water solublities of some aminoalkylphenothiazine tranquilizers and related compounds, *J. Pharm. Pharmacol.*, **19**, 10–16 (1967). NB: See Amitriptylline for details.

(continued)

No.	Compound Name	pK_a value(s)	Data quality	Ionization type	Method	Conditions	Comments and Reference(s)
1151	Promazine	9.34	U	+H	Potentiometric	H_2O $t = 25$ I undefined Ar atmosphere	Seiler P, Simultaneous determination of partition coefficient and acidity constant of a substance, *Eur. J. Med. Chem.*, **9**, 663–665 (1974). NB: See Amitriptylline for details.
1152	Promazine	9.39	U	+H	Potentiometric	H_2O (extrap) $t = 20$ N_2 atmosphere	Sorby DL, Plein EM and Benmaman D, Adsorption of phenothiazine derivatives by solid adsorbents, *J. Pharm. Sci.*, **55**, 785–794 (1966). NB: See Chlorpromazine for details.
1153	Promethazine ($C_{17}H_{20}N_2S$)	8.99 ± 0.03	A	+H	partition	H_2O $t = 20 \pm 0.5$ I not reported but pK_a was stated to be independent of I.	Vezin WR and Florence AT, The determination of dissociation constants and partition coefficients of phenothiazine derivatives, *Int. J. Pharm.*, **3**, 231–237 (1979). "The solubility dependence on pH has been used for phenothiazines and tricyclic drugs by Green, but there are discrepancies of up to 0.9 pK units in some of the results from these two methods. . . . Table 1 gives the pK_a value determined by the method described and by titration (Chatten and Harris, *Anal. Chem.*, **34**, 1495–1501 (1962) unless otherwise stated) and by solubility (Green AL, Ionization constants and water solubilities of some aminoalkylphenothiazine tranquillizers and related compounds, *J. Pharm. Pharmacol.*, **19**, 10–16 (1967)) . . . .

H$_3$C, N, CH$_3$, CH$_3$, N, S

Base	Titrn	Soly	This work	Base	Titrn	Soly	This work
Promethazine	9.09	9.1	8.99	Thioridazine	9.16	9.5	9.62
Chlorpromazine	9.26	9.3	9.40	Prochlorperazine	7.54	8.1	8.14
Triflupromazine	9.41	9.2	9.29	Trifluperazine	8.40	8.1	8.05
Pecazine	-	9.7	9.48	Fluphenazine	8.05	-	7.98

. . . the measurement of the pH dependency of partition coefficients was . . . applied to 8 phenothiazine derivatives . . .. Precise spectrophotometry was essential in this method, but this coupled with a large number of data points can lead to a precision of <±0.03 in pK_a."

1154	Promethazine	9.00	U	+H	Potentiometric	H_2O $t = 25$	Bergström CAS, Strafford M, Lazorova L, Avdeef A, Luthman K and Artursson P, Absorption classification of oral drugs based on molecular surface properties, *J. Med. Chem.*, **46**(4), 558–570 (2003). NB: From extrapolation of aqueous-methanol mixtures to 0% methanol.
1155	Promethazine	9.10	U	+H	Potentiometric	H_2O (extrap) $t = 24 \pm 1$ $I \sim 0.002$	Chatten LG and Harris LE, Relationship between $pK_b(H_2O)$ of organic compounds and $E_{1/2}$ values in several nonaqueous solvents, *Anal. Chem.*, **34**, 1495–1501 (1962). NB: See Chlorpromazine for details.
1156	Promethazine	9.1	U	+H	soly	H_2O $t = 24 \pm 1$	Green AL, Ionization constants and water solublities of some aminoalkylphenothiazine tranquilizers and related compounds, *J. Pharm. Pharmacol.*, **19**, 10–16 (1967). NB: See Amitriptylline for details.
1157	Proparacaine ($C_{16}H_{26}N_2O_3$) $COOCH_2CH_2N(Et)_2$ NH_2 $OCH_2CH_2CH_3$	3.22	U	+H	Spectro	H_2O	Hefferren J, Klessig R and Dietz C, Ultraviolet absorption of local anaesthetics with an aromatic amino group as a function of pH, *J. Dental Res.*, **42**, 793–802 (1963). NB: Cited in Whigan DB, Proparacaine hydrochloride, *APDS*, **6**, 423–456 (1977). Also reported values for procaine, butethamine, tetracaine, chlorprocaine, propoxycaine, butylaminobenzoate, benzocaine, benoxinate, metabutethamine and metabutoxycaine.
1158	Propionic acid ($C_3H_6O_2$) CH_3CH_2COOH	4.874 ± 0.003	R	–H	Potentiometric	H_2O $t = 25.0 \pm 0.01$ $I = 0.00$ (KCl)	Harned HS and Ehlers RW, The dissociation constant of propionic acid from 0 to 60°, *JACS*, **55**, 2379–83 (1933). NB: Used hydrogen electrodes in cells without liquid junction potentials. Solutions were prepared with carbonate-free sodium hydroxide. The effects of ionic strength from added KCl extrapolated to $I = 0.00$.
1159	Propofol ($C_{12}H_{18}O$) OH $(CH_3)_2CH$ $CH(CH_3)_2$	11.10	A	–H	Spectro	H_2O $t = 20$	Serjeant E and Dempsey B, *Ionisation constants of organic acids in aqueous solution*, IUPAC Chemical Data Series No. **23**, Pergamon Press, no. 5583, p. 621 (1979). Demerseman P, Lechartier JP, Reynaud R, Cheutin A, Royer R and Rumpf P, *Bull. Soc. Chim. Fr.*, 2559–2563 (1963).

(continued)

Appendix A *(continued)*

No.	Compound Name	pK_a value(s)	Data quality	Ionization type	Method	Conditions	Comments and Reference(s)
1160	Propranolol ($C_{16}H_{21}NO_2$)	9.53 ± 0.01	A	+H	Potentiometric	H_2O $t = 25 \pm 0.5$ $I = 0.15$ (KCl)	Avdeef A, Box KJ, Comer JEA, Hibbert C and Tam KY, pH-metric log P 10. Determination of liposomal membrane-water partition coefficients of ionizable drugs, *Pharm. Res.*, **15**(2), 209–215 (1998). NB: Used a Sirius PCA101 autotitrator. Also gave log P (octanol-water) and log P (dioleylphosphatidylcholine unilamellar vesicles).
	HO, $CH_2NHCH(CH_3)_2$, O	9.40 ± 0.01	A	+H	Potentiometric	H_2O $t = 25 \pm 0.5$ $I = 0.001$ (KCl)	Sirius Technical Application Notes, vol. **2**, pp. 81–82 (1995). Sirius Analytical Instruments, Ltd., Forest Row, East Sussex, RH18 5DW, UK.
		9.14	U	+H	Potentiometric	H_2O $t = 37$ $I = 0.15$ (KCl)	Balon K, Riebesehl BU and Muller BW, Drug liposome partitioning as a tool for the prediction of human passive intestinal absorption, *Pharm. Res.*, **16**, 882–888 (1999).
1161	Propranolol	9.57	A	+H	CE/pH (+ve ion mode)	H_2O $t = 25$ $I = 0.025$	Wan H, Holmen AG, Wang Y, Lindberg W, Englund M, Nagard MB and Thompson RA, High-throughput screening of pK_a values of pharmaceuticals by pressure-assisted capillary electrophoresis and mass spectrometry, *Rapid Commun. Mass Spectrom.*, **17**, 2639–2648 (2003). NB: Reported a predicted value (ACD Labs) of 9.15.
1162	Propranolol	9.51	A	+H	Potentiometric	MeOH/H_2O $t = 25$	Irwin WJ and Belaid KA, Drug delivery by ion exchange. Part 2. Physicochemical properties of ester prodrugs of propranolol, *Drug Dev. Ind. Pharm.*, **13**(9–11), 2033–2045 (1987). "The pK_a values, solubilities and partition coefficients of a series of O-n-acylpropranolol prodrugs . . . were determined. Titrations in a range of aqueous methanolic solutions were used to estimate the pK_a, while titration under non-logarithmic conditions (when excess undissolved base was present in the system) was used to determine solubilities."

1163	Propranolol	9.32 (0.03)	U	+H	Potentiometric	H_2O $t = 20.0 \pm 0.5$	Zaagsma J and Nauta WT, β-Adrenoceptor studies. 1. *In vitro* β-adrenoceptor blocking antiarrhythmic and local anaesthetic activities of a new series of aromatic bis(2-hydroxy-3-isopropylaminopropyl) ethers, *J. Med. Chem.*, **17**, 507–513 (1974). "pK_a values were determined by potentiometric titration of the bases[21], which were dissolved in 30, 40, 50, 60, 70 and 80% aqueous EtOH by volume, with ethanolic HCl at 20° ± 0.5. From the pK_a values obtained, the intercept at 100% H_2O with its S.E. was calculated by the method of least squares." Ref. 21: Roos AM, Rekker RF and Nauta WT, Base strength of substituted dimethyl[2-(diphenyl-methoxy)ethyl]amines. *Arzneim.-Forsch.*, **20**, 1763–1765 (1970); Roos AM, Rekker RF, Nauta WT, *Pharm. Acta Helv.*, **38**, 569–576 (1963). NB: Lists propranolol and 11 related compounds with pK_a = 9.22–9.45. NB: This example illustrates the pitfalls of using titration in partly aqueous mixtures followed by extrapolation to 100% water content. Compare with the examples above, even allowing for the I and temperature differences. Studies on lidocaine suggest that I has little influence on amine pK_a values (in the range $I = 0.005$ to 0.077), while lower temperature should give a stronger base, not weaker, as here.
1164	Propranolol	9.72	U	+H	Potentiometric	H_2O $t = 23.0$	Clarke FH and Cahoon NM, Ionization constants by curve-fitting: Determination of partition and distribution coefficients of acids and bases and their ions, *J. Pharm. Sci.*, **76**(8), 611–620 (1987). NB: See Benzoic acid for further details.
1165	Propranolol	9.43	U	+H	Potentiometric	H_2O $t = 22$ $I = 0.5$	Quigley JM, Jordan CGM and Timoney RF, The synthesis, hydrolysis kinetics and lipophilicity of O-acyl esters of propranolol, *Int. J. Pharm.*, **101**, 145–163 (1994). NB: Potentiometric titrations were also done in the presence of octanol; the difference gave the partition coefficient. Also reported values for a series of propranolol esters.
1166	Propranolol	9.45	U	+H	Potentiometric	H_2O $t = 21–24$ (RT)	Schurmann W and Turner P, Membrane model of the human oral mucosa as derived from buccal absorption performance and physicochemical properties of the beta-blocking drugs atenolol and propranolol, *J. Pharm. Pharmacol.*, **30**, 137–147 (1978). NB: See Atenolol for details.

(*continued*)

No.	Compound Name	pK_a value(s)	Data quality	Ionization type	Method	Conditions	Comments and Reference(s)
1167	Propranolol	9.23	U	+H	Potentiometric	H_2O $t = 35$ $c = 0.001$ N_2 atmosphere	Schoenwald RD and Huang HS, Corneal penetration behaviour of β-blocking agents. I. Physicochemical factors, *J. Pharm. Sci.*, **72**, 1266–1272 (1983). NB: Used 1-mL syringe burette and Metrohm autotitrator with glass electrode. The pK_a value was determined from a modified Gran plot. Also reported the following values: (see table below)
1168	Propranolol	9.40	U	+H	Potentiometric	H_2O $t = 20.0$	Mannhold R, Dross KP and Reffer RF, Drug lipophilicity in QSAR practice: I. A comparison of experimental with calculative approaches, *Quant-Struct.-Act. Relat.*, **9**, 21–28 (1990).
1169	Propranolol	9.50	U	+H	Potentiometric	H_2O (extrap) $t = 25 \pm 1$ I undefined Ar atmosphere	Cheymol G, Poirier J-M, Carrupt PA, Testa B, Weissenburger J, Levron J-C and Snoeck E., Pharmacokinetics of β-adrenoceptor blockers in obese and normal volunteers, *Br. J. Clin. Pharmacol.*, **43**, 563–570 (1997).
1170	Propranolol	9.7	U	+H	partition/pH	H_2O $t = 37$	Kramer SD and Wunderli-Allenspach H, pH-dependence in the partitioning behavior of (RS)-(^{3}H)propranolol between MDCK cell lipid vesicles and buffer, *Pharm. Res.*, **13**, 1851–1855 (1996). "The pH dependent partitioning of (RS)-propranolol hydrochloride (DL-propranolol hydrochloride) between kidney epithelial (MDCK) cell lipids and buffer was studied using equilibrium dialysis at 37 C and pH 7–11. . . . The highest apparent partition coefficient was 1797 at pH 9.7, the lowest was 805 at pH 6.9. Curve fitting with a combination of Henderson-Hasselbach equations revealed an inflection point at 9.7, the apparent pK_a of propranolol, and 2 additional pK_a values, 7.7 and 10. These values corresponded to the pK_a of free fatty acids in lipid bilayers and the pK_a of phosphatidylethanolamine, respectively. True partition coefficients (P) of the neutral and ionized solute were fitted for each ionization status of the membrane."

Values reported for No. 1167:

Compound	pK_a	Compound	pK_a	Compound	pK_a
acebutolol	9.20	levobunolol	9.32	penbutolol	9.26
atenolol	9.32	metoprolol	9.24	sotalol	8.15, 9.65
bevantolol	8.38	nadolol	9.39	timolol	9.21
bufuralol	8.97	oxprenolol	9.32		

No.	Compound	pKa	Data quality	Ionization	Method	Conditions	Reference / Comments
1171	Propylhexedrine ($C_{10}H_{21}N$) 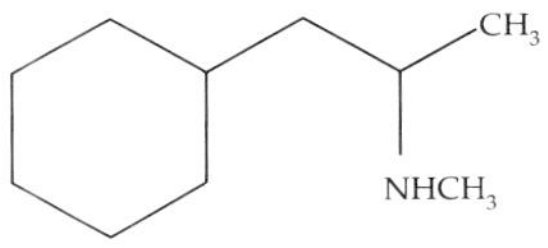	10.52	U	+H	Potentiometric	H_2O $t = 25.0 \pm 0.2$ $I \leq 0.001$	Leffler EB, Spencer HM and Burger A, Dissociation constants of adrenergic amines *JACS*, **73**, 2611–2613 (1951). NB: See Amphetamine for details. This compound was insufficiently soluble in water; pH measurements were performed in a series of ethanol-water solutions, which were then extrapolated back to 0% ethanol.
1172	Prostaglandin E_1	4.85 ± 0.07	U	–H	Potentiometric	H_2O $t = 25.0 \pm 0.1$ $I = 0.1$ (NaCl)	Takacs-Novak K, Box KJ and Avdeef A, Potentiometric pK_a determination of water-insoluble compounds: Validation study in methanol/water mixtures, *Int. J. Pharm.*, **151**, 235–248 (1997). NB: By extrapolation from 13–34% w/w aqueous MeOH. See Acetaminophen for full details.
1173	Prostaglandin E_2 ($C_{20}H_{32}O_5$) 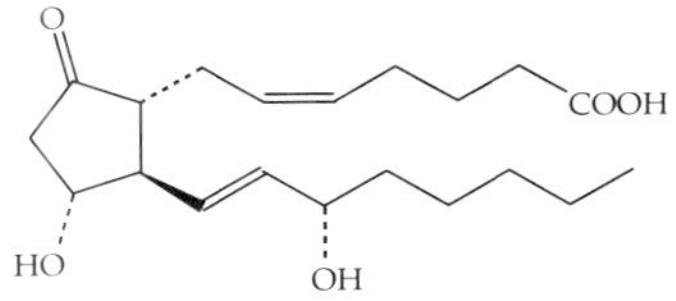	4.77 ± 0.09	U	–H	Potentiometric	H_2O $t = 25.0 \pm 0.1$ $I = 0.1$ (NaCl)	Takacs-Novak K, Box KJ and Avdeef A, Potentiometric pK_a determination of water-insoluble compounds: Validation study in methanol/water mixtures, *Int. J. Pharm.*, **151**, 235–248 (1997). NB: By extrapolation from 9–25%w/w aqueous MeOH. See Acetaminophen for full details. See separate entry for Dinoprostone.
1174	Prostaglandin $F_{2\alpha}$ ($C_{20}H_{34}O_5$) 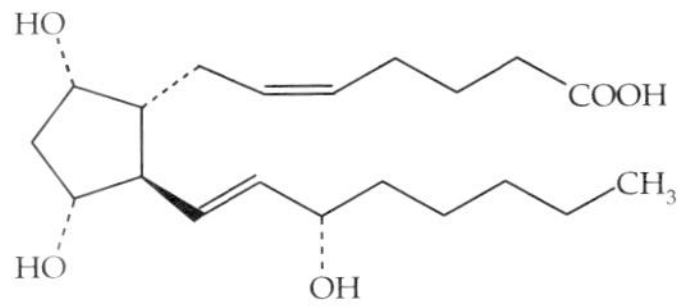	4.90	U	–H	Potentiometric	H_2O $t = 25 \pm 2$ $I = 0.0$ $c < 0.01$ N_2 atmosphere	Roseman TJ and Yalkowsky SW, Physicochemical properties of prostaglandin F2α (tromethamine salt). Solubility behaviour, surface properties and ionization constants, *J. Pharm. Sci.*, **62**, 1680–1685 (1973). NB: Used carbonate-free solutions and corrected for effects of ionic strength with modified Debye-Huckel equation. The value given is for concentrations below the CMC. At concentrations >CMC, the apparent pK_a value increased to 5.6, due to micelle formation.
1175	Pseudoecgonine ($C_9H_{15}NO_3$) 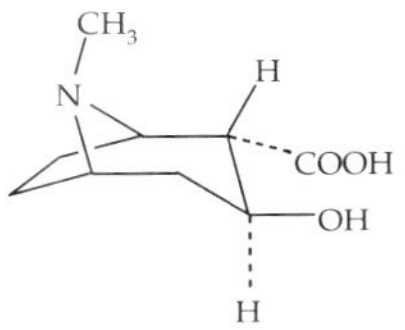	9.70	A	+H	Potentiometric	H_2O $t = 25$ $I < 0.02$	Chilton J and Stenlake JB, Dissociation constants of some compounds related to lysergic acid: Beta-dimethylaminopropionic acid, dihydroarecaidine, ecgonine and their derivatives, *J. Pharm. Pharmacol.*, **7**, 1004–1011 (1955). Cited in Perrin 2862 ref. C27. NB: The study used measurements of pH with a glass electrode. There should be another pK_a value for the –COOH group.

(continued)

Appendix A *(continued)*

No.	Compound Name	pK_a value(s)	Data quality	Ionization type	Method	Conditions	Comments and Reference(s)
1176	Pseudoecgonine, methyl ester ($C_{10}H_{17}NO_3$)	8.15	A	+H	Potentiometric	H_2O $t = 25$ $I < 0.02$	Chilton J and Stenlake JB, Dissociation constants of some compounds related to lysergic acid: Beta-dimethylaminopropionic acid, dihydroarecaidine, ecgonine and their derivatives, *J. Pharm. Pharmacol.*, **7**, 1004–1011 (1955). Cited in Perrin 2862 ref. C27. NB: The study used measurements of pH with a glass electrode. See also Jensen HH, Lyngbye L, Jensen A and Bols M, Stereoelectronic substituent effects in polyhydroxylated piperidines and hexahydropyridazines, *Chem. Eur. J.*, **8**(5), 1218–1226 (2002). This paper reported pK_a values for methyl pseudoecgonine (8.2) and methyl ecgonine (9.1), along with more than 60 other piperidines or closely related compounds.
1177	Pseudoephedrine ($C_{10}H_{15}NO$)	9.73	A	+H	Potentiometric	H_2O $t = 25.0 \pm 0.5$ $I = 0.01$	Warren RJ, Begosh PP and Zarembo JE, Identification of amphetamines and related sympathomimetic amines, *J. Assoc. Off. Anal. Chem.*, **54**, 1179–1191 (1971). NB: See Amphetamine for further details.
1178	Pseudoephedrine	9.22	U	+H	Potentiometric	80% aqueous MCS	Prelog V and Häfliger O, Cinchona alkaloids. IX. The influence of configuration on basicity and the relative cofiguration at carbons 8 and 9, *Helv. Chim. Acta*, **33**, 2021–2029 (1950). CA 45:29705. MCS = methylcellosolve. Cited in Benezra SA and McRae JW, Pseudoephedrine hydrochloride, *APDS*, **8**, 489–507 (1979). NB: Shown to be a stronger base than ephedrine, p$K_a = 9.14$. Similarly, *N*-methylephedrine (p$K_a = 8.50$) was found to be a weaker base than *N*-methylpseudoephedrine (p$K_a = 8.81$).

No.	Compound	pKa	Data quality	Type	Method	Conditions	Reference / Notes
1179	Pseudotropine ($C_8H_{15}NO$) 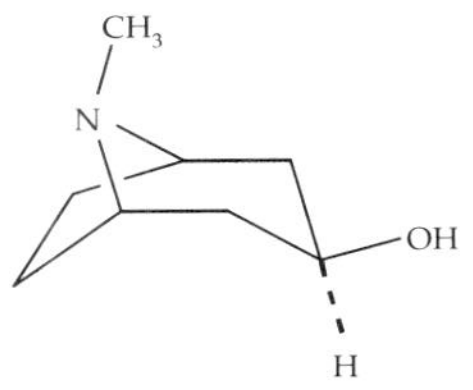	9.86 ± 0.05	A	+H	Potentiometric	H_2O $t = 25$ $I = 0.0006$ N_2 atmosphere	Geissman TA, Wilson BD and Medz RB, The base strengths of cis- and trans-1,2-aminoalcohols, *JACS*, **76**, 4182–4183 (1954). Cited in Perrin Bases No. 2956 ref. G7. NB: Used pH measurements on solutions containing equimolar amounts of the base and salt.
		10.26	A	+H	Potentiometric	H_2O $t = 25$ $I < 0.1$	Smith PF and Hartung WH, Cis- and trans-tropine (tropanol), *JACS*, **75**, 3859–60 (1953). Cited in Perrin Bases No. 2956 ref. S56. NB: Used pH measurements on solutions containing equimolar amounts of the base and salt.
		10.40	U	+H	Spectro	H_2O $t = 15$ $c = 0.03–0.04$	Kolthoff IM, The dissociation constants, solubility product and titration of alkaloids, *Biochem. Z.*, **162**, 289–353 (1925). Cited in Perrin Bases No. 2956 ref. K47. NB: See Aconitine for details.
1180	Pteridine, 2,4-diamino-6-methyl ($C_7H_8N_6$) 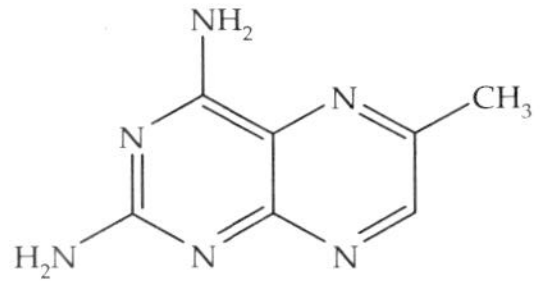	5.5	U	+H	Spectro	H_2O t unspecified $I \sim 0.1$	Zakrewski SF, Relationship between basicity of certain folate analogues and their affinities for folate reductase, *J. Biol. Chem.*, **238**(12), 4002–4004 (1963). NB: Stock solutions of each compound (100 ug/mL) were diluted (1:9) with a buffer consisting of KCl, glycine, monobasic potassium phosphate and citric acid (each at 25 mM) that had been adjusted to the desired pH with 5 M NaOH or HCl. UV spectra recorded with a Beckman DU. No record of pH meter or calibration procedure. The pK_a values appear to have been read directly from plots of optical density versus pH.
1181	Pteridine, 2,4-diamino-6-formyl ($C_7H_6N_6O$) 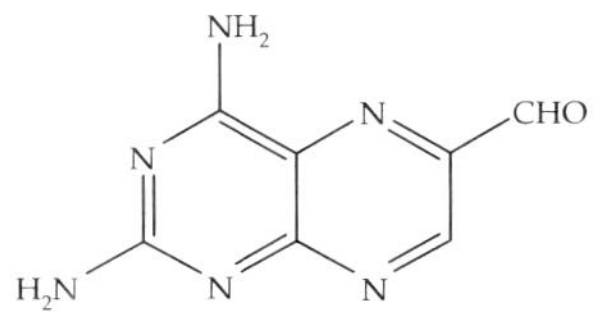	5.6	U	+H	Spectro	H_2O t unspecified $I \sim 0.1$	Zakrewski SF, *J. Biol. Chem.*, **238**(12), 4002–4004 (1963). NB: See Pteridine, 2,4-diamino-6-methyl, for details. No indication of stability problems that can be anticipated with such an easily oxidisable compound as this. This paper also reported the following data: 1. 2,6-diaminopurine ($C_5N_6N_6$); pK_a = 5.1 2. 2,4-diamino-6-methylpyrimidine ($C_5N_8N_4$); pK_a = 7.7 3. 2,4-diamino-6-hydroxypyrimidine ($C_4N_6N_4O$); pK_a = 3.5
1182	Pteridine, 2,4-diamino-6-hydroxy ($C_6H_6N_6O$) 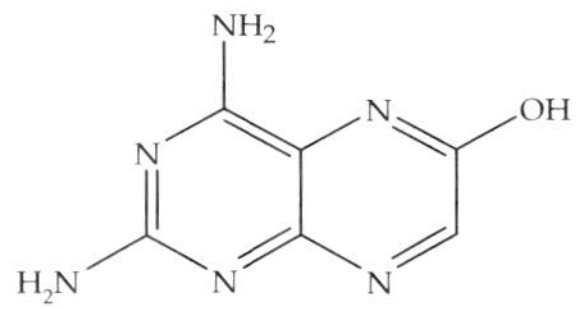	4.3	U	+H	Spectro	H_2O t unspecified $I \sim 0.1$	Zakrewski SF, *J. Biol. Chem.*, **238**(12), 4002–4004 (1963). NB: See Pteridine, 2,4-diamino-6-methyl, for details.

(continued)

No.	Compound Name	pK_a value(s)	Data quality	Ionization type	Method	Conditions	Comments and Reference(s)
1183	Pteridine, 2-amino-4-hydroxy-6-formyl ($C_7H_5N_5O_2$) 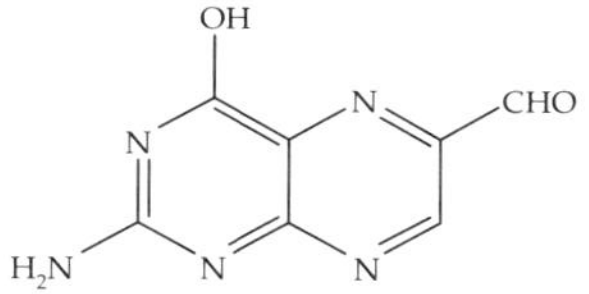	2.0	U	+H	Spectro	H_2O t unspecified $I \sim 0.1$	Zakrewski SF, *J. Biol. Chem.*, **238**(12), 4002–4004 (1963). NB: See Pteridine, 2,4-diamino-6-methyl, for details.
1184	Pteridine, 2-amino-4-hydroxy-6-methyl ($C_7H_7N_5O$) 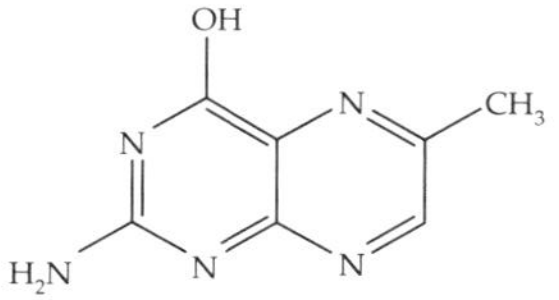	2.5	U	+H	Spectro	H_2O t unspecified $I \sim 0.1$	Zakrewski SF, *J. Biol. Chem.*, **238**(12), 4002–4004 (1963). NB: See Pteridine, 2,4-diamino-6-methyl, for details.
1185	Pyrathiazine ($C_{18}H_{20}N_2S$) 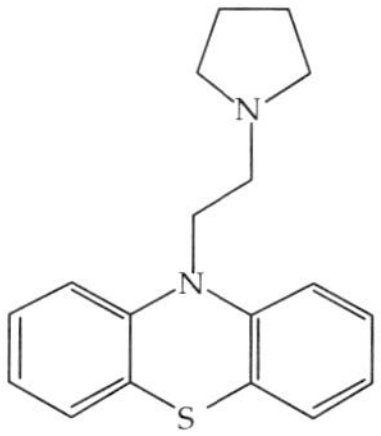	8.91	U	+H	Potentiometric	H_2O (extrap) $t = 24 \pm 1$ $I \sim 0.002$	Chatten LG and Harris LE, Relationship between pK_b(H_2O) of organic compounds and $E_{1/2}$ values in several nonaqueous solvents, *Anal. Chem.*, **34**, 1495–1501 (1962). Cited in: Foye 1 (2 refs), see Idoxuridine; N&K; Chatten LG (ed.), *Pharmaceutical Chemistry*, vol. **1**, Dekker, New York, 1966, pp. 85–87. NB: See Methdilazine and separate entry for Phenothiazine, 10-(*N*-pyrrolidinyl-2-ethyl).
1186	Pyrathiazine	9.36	U	+H	Potentiometric	H_2O (extrap) $t = 20$ N_2 atmosphere	Sorby DL, Plein EM and Benmaman D, Adsorption of phenothiazine derivatives by solid adsorbents, *J. Pharm. Sci.*, **55**, 785–794 (1966). NB: See Chlorpromazine for details.

No.	Compound	pKa	Data quality	Ionization	Method	Conditions	Reference and comments
1187	Pyrazolic acid ($C_{17}H_{13}ClN_2O_2$) 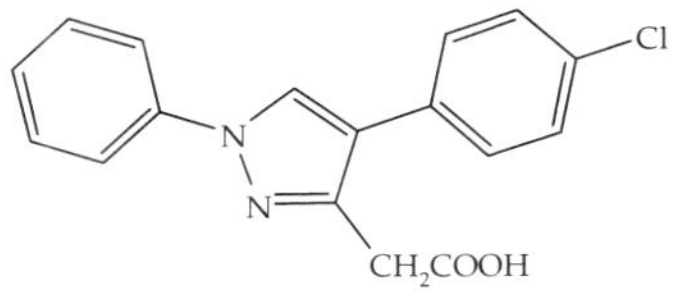	4.90	U	−H	soly	H_2O $t = 37$ $I = 0.15$	Zimmerman I, Determination of pK_a values from solubility data, *Int. J. Pharm.*, **13**(1), 57–65 (1983).
1188	Pyridoxal ($C_8H_9NO_3$) 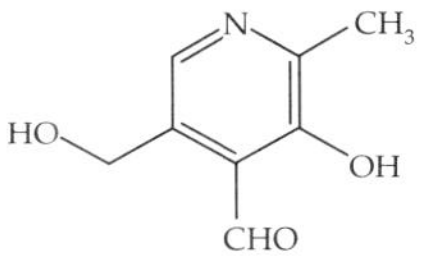	4.20 8.66 13	A A U	+H −H −H	Spectro	H_2O $t = 25$ $I = 0.1$	Metzler DE and Snell EE, Spectra and ionization constants of the vitamin B_6 group and related 3-hydroxypyridine derivatives, *JACS*, **77**, 2431–2437 (1955). Cited in Perrin Bases. No. 3330 ref. M40. NB: See Isopyridoxal for details.
1189	Pyridoxal	4.23 8.70	A A	+H −H	Potentiometric	H_2O $t = 25$ $I = 0.15$	Williams VR and Neilands JB, Apparent ionization constants, spectral properties, and metal chelation of the cotransaminases and related compounds, *Arch. Biochem. Biophys.*, **53**, 56–70 (1954). CA 49:57032. Cited in Perrin Bases. No. 3330 ref. W38.
1190	Pyridoxal	4.13 8.37 13.04	A A A	+H −H −H	Spectro	H_2O $t = 50$	Nagano K and Metzler DE, Machine computation of equilibrium constants and plotting of spectra of individual ionic species in the pyridoxal-alanine system, *JACS*, **89**, 2891–2900 (1967). Cited in Perrin Bases suppl. no. 7791 ref. N2. NB: Used spectrophotometric measurements combined with pH measurements (pK_{a1}, pK_{a2}) or solutions of known high alkali concentration (pK_{a3}).

Entry 1187 (continued):

Compound	pK_{a1}	pK_{a2}	So (mg/L)
Sulphadiazine	6.42	2.07	58.6
Pyrazolic Acid	5.35		48.9
Lisuride Hydrogen Maleate	7.24		20.6

Refined calculations using weighted least squares regression (Lewis, *Int. J. Pharm.*, **18**, 272–212 (1984)) gave (sd):

Compound	pK_{a1}	pK_{a2}	So (mg/L)
Sulphadiazine	6.44 (0.01)	2.06 (0.09)	61.0 (1.6)
Pyrazolic Acid	4.90 (0.09)		18.6 (3.6)
Lisuride Hydrogen Maleate	7.29 (0.02)		18.4 (0.6)

NB: This paper gave a detailed discussion of the solubility-pH dependence equation. Values for the correlation coefficient ($r = 0.99$ to 1.00) indicated very good fit of the experimental data to the solubility-pH equation for a single ionizing group.

(continued)

No.	Compound Name	pK_a value(s)	Data quality	Ionization type	Method	Conditions	Comments and Reference(s)
1191	Pyridoxal, 3-methoxy ($C_9H_{11}NO_3$)	4.55	U	+H	Spectro	H_2O $t = 25$	Pocker A, Fischer EH, Synthesis of analogs of pyridoxal-5′-phosphate, *Biochem.*, **8**, 5181–5188 (1969). Cited in Perrin Bases suppl. no. 7792 ref. P41a. NB: See Isopyridoxal for details.
1192	Pyridoxal, *N*-methyl ($C_9H_{12}NO_3$)	3.90	U	−H	Spectro	H_2O $t = 25$	Pocker A and Fischer EH, Synthesis of analogs of pyridoxal-5′-phosphate, *Biochem.*, **8**, 5181–5188 (1969). Cited in Perrin Bases suppl. no. 7793 ref. P41a. NB: See Isopyridoxal for details.
1193	Pyridoxal-5-phosphate ($C_8H_{10}NO_6P$)	4.14 6.20 8.69 <2.5	A A A U	+H −H −H −H	Potentiometric	H_2O $t = 25$ $I = 0.15$	Williams VR, Neilands JB, Apparent ionization constants, spectral properties, and metal chelation of the cotransaminases and related compounds, *Arch. Biochem. Biophys.*, **53**, 56–70 (1954). CA 49:57032. Cited in Perrin Bases. No. 3331 ref. W38. NB: See Pyridoxal for details.
1194	Pyridoxamine ($C_8H_{12}N_2O_2$)	3.31 7.90 10.4	A U U	−H +H +H	Spectro Potentiometric Potentiometric	H_2O $t = 25$ $I = 0.1$	Metzler DE and Snell EE, Spectra and ionization constants of the vitamin B_6 group and related 3-hydroxypyridine derivatives, *JACS*, **77**, 2431–37 (1955). Cited in Perrin Bases No. 3332 ref. M40. NB: See Isopyridoxal for details.
1195	Pyridoxamine	3.54 8.21 10.63	A A A	−H +H +H	Potentiometric	H_2O $t = 25$ $I = 0.15$ (NaCl)	Williams VR and Neilands JB, Apparent ionization constants, spectral properties, and metal chelation of the cotransaminases and related compounds, *Arch. Biochem. Biophys.*, **53**, 56–70 (1954). CA 49:57032. Cited in Perrin Bases No. 3332 ref. W38. NB: See Pyridoxal for details.

1196	Pyridoxamine	3.37 8.01 10.13	A A A	−H +H +H	Potentiometric	H_2O $t = 25$ $I = 0.1$ (KNO_3)	Gustafson RL and Martell AE, Stabilities of metal chelates of pyridoxamine, *Arch. Biochem. Biophys.*, **68**, 485–498 (1957). CA 51:85010. Cited in Perrin Bases No. 3332 ref. G62.
1197	Pyridoxamine	3.14 8.19 10.28	U U U	−H +H +H	Spectro	H_2O	Morozov YV, Bazhulina NP, Karpeiskii MY, Ivanov VI and Kuklin AI, Optical and luminescent properties of vitamin B6 and its derivatives. III. Pyridoxannne and pyridoxamine-5′-phosphate, *Biofizica*, **11**, 228–36 (1966); CA 65:8935. Cited in Perrin Bases suppl. no. 7794 ref. M21. NB: Also includes fluorescence data.
1198	Pyridoxamine, 3-methoxy ($C_9H_{14}N_2O_2$) 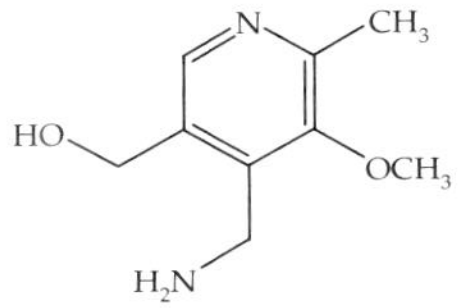	3.40	U	+H	Spectro	H_2O $t = 25$	Pocker A and Fischer EH, Synthesis of analogs of pyridoxal-5′-phosphate, *Biochem.*, **8**, 5181–5188 (1969). Cited in Perrin Bases suppl. no. 7795 ref. P41a. NB: See Isopyridoxal for details.
1199	Pyridoxamine, *N*-methyl ($C_9H_{15}N_2O_2$) 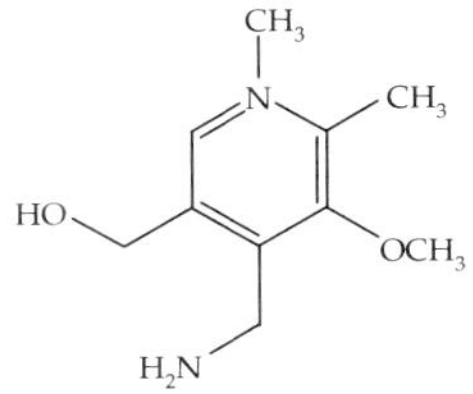	2.95	U	+H	Spectro	H_2O $t = 25$	Pocker A and Fischer EH, Synthesis of analogs of pyridoxal-5′-phosphate, *Biochem.*, **8**, 5181–5188 (1969). Cited in Perrin Bases suppl. no. 7796 ref. P41a. NB: See Isopyridoxal for details.
1200	Pyridoxamine-5-phosphate ($C_8H_{13}N_2O_6P$) 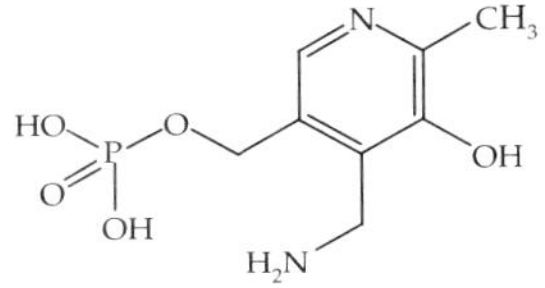	2.5 3.69 5.76 8.61 10.92	U A A A A	−H +H −H +H −H	Potentiometric	H_2O $t = 25$ $I = 0.15$	Williams VR and Neilands JB, Apparent ionization constants, spectral properties, and metal chelation of the cotransaminases and related compounds, *Arch. Biochem. Biophys.*, **53**, 56–70 (1954). CA 49:57032. Cited in Perrin Bases No. 3333 ref. W38. NB: See Pyridoxal for details.

(continued)

Appendix A *(continued)*

No.	Compound Name	pK_a value(s)	Data quality	Ionization type	Method	Conditions	Comments and Reference(s)
1201	Pyridoxamine-5-phosphate	3.12 8.33 10.39	U U U	+H +H −H	Spectro	H_2O t = ??	Morozov YV, Bazhulina NP, Karpeiskii MY, Ivanov VI and Kuklin AI, *Biofizica* **11**, 228–36 (1966); CA 65:1620f. Cited in Perrin Bases suppl. no. 7797 ref. M21. NB: See Pyridoxamine for details.
1202	Pyridoxine ($C_8H_{11}NO_3$) H_3C, N, HO, CH_2OH, CH_2OH	5.00 8.96	U A	+H −H	Potentiometric	H_2O $t = 25.0$ $I = 0.15$ (mixed)	Williams VR and Neilands JB, Apparent ionization constants, spectral properties, and metal chelation of the cotransaminases and related compounds, *Arch. Biochem. Biophys.*, **53**, 56–70 (1954). CA 49:57032. Cited in Perrin Bases No. 3334 ref. W38. NB: See Pyridoxal for details.
1203	Pyridoxine	4.67 9.02	U A	+H −H	CE/pH (+ve ion mode)	H_2O $t = 25$ $I = 0.025$	Wan H, Holmen AG, Wang Y, Lindberg W, Englund M, Nagard MB and Thompson RA, High-throughput screening of pK_a values of pharmaceuticals by pressure-assisted capillary electrophoresis and mass spectrometry, *Rapid Commun. Mass Spectrom.*, **17**, 2639–2648 (2003). NB: Reported predicted values (ACD Labs) of 5.06 and 8.37.
1204	Pyridoxine	4.87 ± 0.01 8.89 ± 0.01	U A	+H −H	Potentiometric	H_2O $t = 25.0$ $I = 0.1$ (NaCl)	Takacs-Novak K and Avdeef A, Interlaboratory study of log P determination by shake-flask and potentiometric methods, *J. Pharm. Biomed. Anal.*, **14**, 1405–1413 (1996). NB: See Acetaminophen for further details. Also reported pK_{a1} = 4.84 ± 0.01 pK_{a2} = 8.87 ± 0.01 at $I = 0.1$ (KNO_3).
1205	Pyridoxine	5.00 8.97	U U	+H −H	Spectro	H_2O $t = 16.5 \pm 0.5$ $c = 0.0001$	Lunn AK and Morton RA, Ultra-violet absorption spectra of pyridoxine and related compounds, *Analyst*, **77**, 718–731 (1952). Cited in Perrin Bases no. 3334 ref. L67. NB: Used a combination of spectrophotometric and pH measurements. See also Aboul-Enein HY and Loutfy MA, Pyridoxine hydrochloride, *APDS*, **13**, 447–478 (1984).
1206	Pyridoxine	2.7 5.00 9.0	U U U	+H? +H −H		H_2O t undefined	Snell EE, Chemical structure in relation to biological activities of vitamin B6, *Vitam. Horm.*, **16**, 77–125 (1958); CA 53:2789.

No.	Compound	pKa value(s)	Data quality	Ionization type	Method	Conditions	Comments and Reference(s)
1207	1,3-bis[(2-Pyridyl)-methyleneamino]-guanidine ($C_{13}H_{13}N_7$) 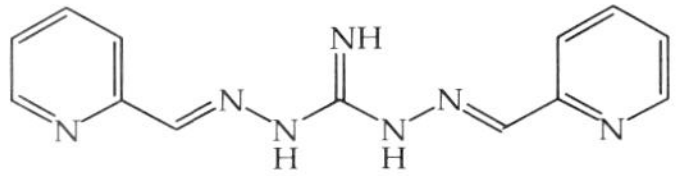	2.91 ± 0.05 6.22 ± 0.08	U U	+H +H	Spectro	H_2O $t = 20 \pm 1$ $I = 0.1$	Asuero AG, Herrador MA and Camean AM, Spectrophotometric evaluation of acidity constants of diprotic acids: Errors involved as a consequence of an erroneous choice of the limit absorbances, *Analytical Letters*, **19**, 1867–1880 (1986).
1208	Pyridylmethyl phosphate esters ($C_6H_8NO_4P$) 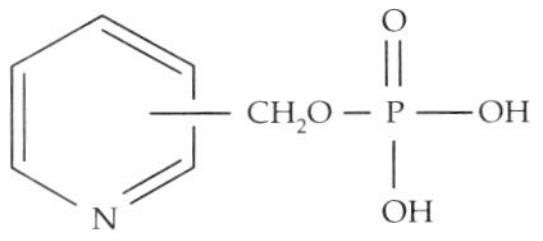	2-subst. ~1 4.15 6.54 3-subst. ~1 4.43 6.48 4-subst. ~1 4.73 6.42	 U U U U U U U U U	 −H +H −H −H +H −H −H +H −H	Potentiometric	H_2O $t = 80$ I ~0.002	Murakami Y and Takagi M, Solvolysis of organic phosphates. I. Pyridylmethylphosphate. "The acid dissociation constants of pyridylmethyl phosphates were evaluated at 80° by titrating 2×10^{-3} M solutions with standard base or acid. Titrations were carried out as quickly as possible to minimize any possible errors caused by the partial hydrolysis of the phosphates and checked by duplicate runs." NB: these values were measured for use with data on the kinetics for hydrolysis of these esters. Details given for pH meter calibration in the kinetics section and it is assumed that the same procedures were used for the pK_a measurements. An experimental method was used to estimate activity coefficients at $I = 0.1$M.
1209	Pyrrolo[2,3-d]pyrimidine derivatives 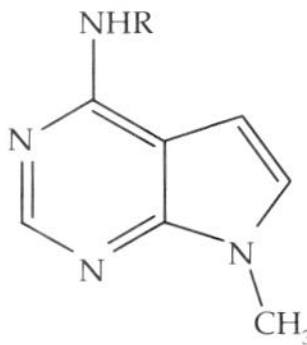	5.64 (R = ethyl) 5.52 (R = n-propyl) 5.66 (R = n-pentyl)	U U U	+H +H +H	Potentiometric	H_2O $t = 25$ I undefined	Hammer RH, Pyrrolo (2,3-d) pyrimidines, *J. Pharm. Sci.*, **57**, 1616–1619 (1968). NB: Used glass electrode to measure pH in titrations. Also reported R = 1-piperidyl, pK_a = 5.28 (U).
1210	Quinacrine ($C_{23}H_{30}ClN_3O$) 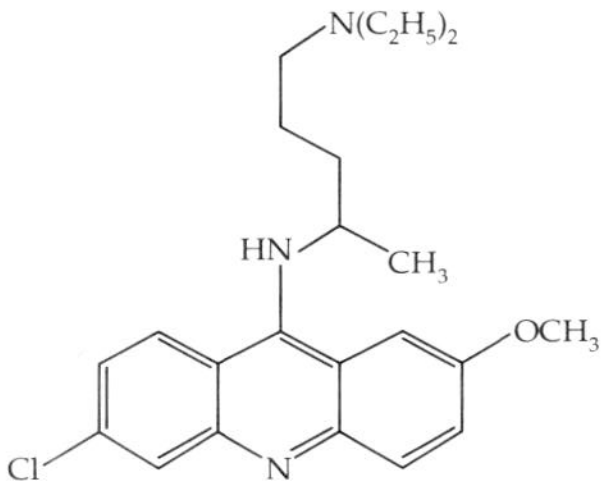	-6.3 8.2 10.2	U U U	+H +H +H	Spectro Spectro Potentiometric	H_2O t undefined I not reported but low	Capomacchia AC and Schulman SG, Electronic absorption and fluorescence spectrometry of quinacrine, *Anal. Chim. Acta*, **77**, 79–85 (1975). NB: Irvin JL, McQuaid and Irvin EM, *JACS*, **72**, 2750–2752 (1950) reported a value of −6.49 by a spectroscopic method at 30 °C in strongly acidic solutions.
		< 3 7.53 10.12	U U U	+H +H +H	Potentiometric	H_2O $t = 20$	Christophers SR, Dissociation constants and solubilities of bases of anti-malarial compounds. I. Quinine. II. Atebrin, *Ann. Trop. Med. Parasitol.*, **31**, 43–69 (1937). CA 31:58602. This paper reported pK_b values of >11, 6.47 and 3.88.

(continued)

No.	Compound Name	pK_a value(s)	Data quality	Ionization type	Method	Conditions	Comments and Reference(s)
1211	Quinidine ($C_{20}H_{24}N_2O_2$) 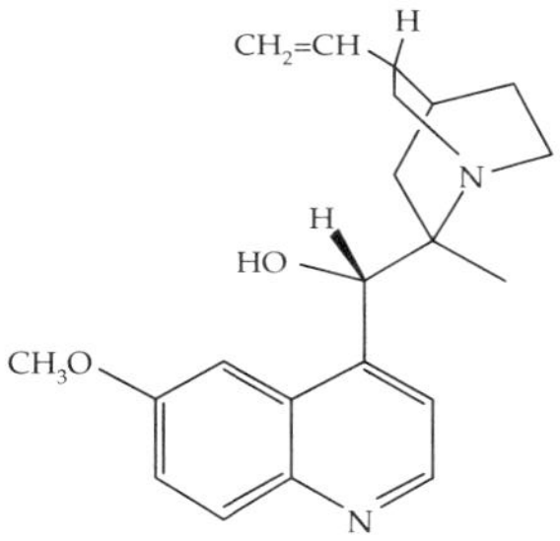	4.2 8.77	U U	+H +H	Spectro	H_2O $t = 15.0$ $c = 0.006$ to 0.01	Kolthoff IM, The dissociation constants, solubility product and titration of alkaloids, *Biochem. Z.*, **162**, 289–353 (1925). Cited in Perrin no. 2957 ref. K47. See Aconitine for details. See also Quinine no. 1220.
		4.12	U	+H	kinetic	$t = 55$	Perrin also cited a pK_a of 4.12 at 55 °C from the rate of inversion of sucrose (ref. A73, Arnall F, *JCS*, **117**, 835–839 (1920)).
1212	Quinidine	4.00 8.54	U U	+H +H	calorimetric	H_2O t undefined	Dragulescu C and Policec S, Thermometric titration of weak diacidic bases. *Studii Cercetari Chim.*, **9**, 33–40 (1962). CA 58:30272. Cited in Perrin suppl. no. 7488. NB: From pK_b values 5.46 and 10.0 at unknown temperature (assumed 25 °C).
1213	Quinidine	4.2 8.8	U U	+H +H		H_2O $t = 25.0$	Loutfy MA, Hassan MMA and Muhtadi FJ, Quinidine Sulfate, *APDS*, **12**, 483–536 (1983). Cited *The Pharmaceutical Codex* 11th Edn., The Pharmaceutical Press, London (1979). NB: also states pK_a values at 20° are 5.4 and 10–these look like pK_b values.
1214	Quinine ($C_{20}H_{24}N_2O_2$) 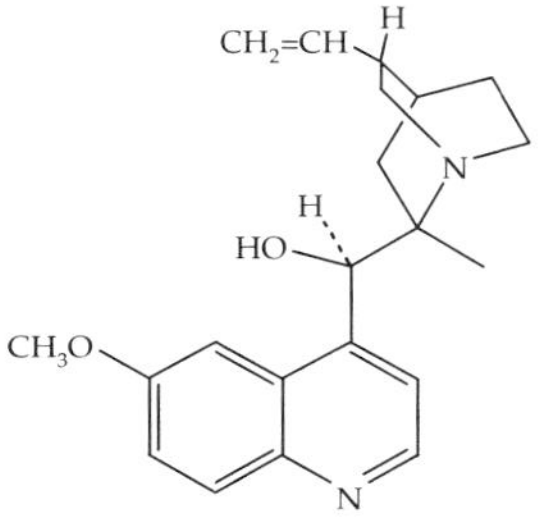	8.34	U	+H	Potentiometric	H_2O $t = 20$ $I = 0.02$	Christophers SR, Dissociation constants and solubilities of bases of anti-malarial compounds. I. Quinine. II. Atebrin, *Ann. Trop. Med. Parasitol.* **31**, 43–69 (1937). CA 31:58602. Cited in Perrin Bases no. 2958 ref. C30. NB: This paper reported pK_b values of 5.70 (equivalent to $pK_a = 8.47$) and 9.85 (equivalent to $pK_a = 4.32$), using $pK_w = 14.167$ at 20 °C. These do not correspond exactly with the values quoted by Perrin. Also reported the solubility-temperature dependence.
		4.21	U"mixed constant"	+H	Potentiometric	H_2O $t = 20$ $I = 0.002$	

1215	Quinine	8.52 4.13	A	+H +H	Potentiometric	H_2O $t = 25$ $c = 0.001$	These data are attributed in Perrin no. 2958 to ref. B130 (Brown HC and Mihm XR, Steric effects in displacement reactions, *JACS*, **77**, 1723 (1955)), but are actually from Gage JC, *Analyst*, **82** (1957). See Brucine for full details.
1216	Quinine	3.95 8.6	U U	+H +H	CE/pH (+ve ion mode)	H_2O $t = 25$ $I = 0.025$	Wan H, Holmen AG, Wang Y, Lindberg W, Englund M, Nagard MB and Thompson RA, High-throughput screening of pK_a values of pharmaceuticals by pressure-assisted capillary electrophoresis and mass spectrometry, *Rapid Commun. Mass Spectrom.*, **17**, 2639–2648 (2003). NB: Reported predicted values (ACD Labs) of 4.77 and 9.04.
1217	Quinine	GLpK_a: 4.33 ± 0.01 8.59 ± 0.01 A&S: 4.30 ± 0.04 8.56 ± 0.05	 A A U A	 +H +H +H +H	Spectro	H_2O $t = 25$ $I = 0.15$ (KCl) Ar atmosphere	Tam KY and Takacs-Novac K, Multi-wavelength spectrophotometric determination of acid dissociation constants, *Anal. Chim. Acta*, **434**, 157–167 (2001). NB: See Clioquinol for details.
1218	Quinine	4.6	U	+H	fluoro	H_2O t undefined $c = 0.0003$	Eisenbrand J, The determination of dissociation constants of fluorescing materials by quantitative measurements of fluorescence, *Z. Physik. Chem.*, **144**, 441–462 (1929). CA 24:6898. Cited in Perrin Bases no. 2958 ref. E19. NB: This result is slightly higher than those reported by methods that measure ground state ionization constant.
1219	Quinine	4.32 8.4	A U "mixed constant"	+H	Potentiometric	H_2O $t = 30$ $I = 0.1$	Irvin JL and Irvin EM, Apparent dissociation exponents of quinine, pamaquine, and a quinolylpiperidylcarbinol: Application of an extended pH scale, *J. Biol. Chem.*, **174**, 577–587 (1939). Cited in Perrin Bases no. 2958 ref. I6. NB: Also reported pK_2 of 4.33 by a spectrophotometric method, same conditions (U).
1220	Quinine	4.5 8.23	U U	+H	Spectro	H_2O $t = 15$ $c = 0.003$–0.01	Kolthoff IM, The dissociation constants, solubility product and titration of alkaloids, *Biochem. Z.*, **162**, 289–353 (1925). Cited in Perrin Bases no. 2958 ref. K47. See Aconitine for details.
1221	Quinine	7.73	U	+H	Potentiometric	80% aqueous MCS	Prelog V and Häfliger O, Cinchona alkaloids. IX. The influence of configuration on basicity and the relative configuration at carbons 8 and 9, *Helv. Chim. Acta*, **33**, 2021–2029 (1950). CA 45:29705. NB: Paper also reported the following: Epiquinine, pK_a = 8.44; quinidine, pK_a = 7.95; 9-epiquinidine, pK_a = 8.32. These data were useful in determining the relative configurations of the cinchona alkaloids.
1222	Quinine	4.24 ± 0.09 8.55 ± 0.04	U A	+H +H	Potentiometric	H_2O $t = 25.0 \pm 0.1$ $I = 0.1$ (NaCl)	Takacs-Novak K, Box KJ and Avdeef A, Potentiometric pK_a determination of water-insoluble compounds: Validation study in methanol/water mixtures, *Int. J. Pharm.*, **151**, 235–248 (1997). NB: By extrapolation from 15–69% w/w aqueous MeOH. See Acetaminophen for full details.

(continued)

No.	Compound Name	pK_a value(s)	Data quality	Ionization type	Method	Conditions	Comments and Reference(s)
1223	Quinine	3.85 to 4.32 8.15 to 8.58	U U	+H +H	Potentiometric	H_2O $t = 25.0 \pm 0.1$ $I = 0.15$ (KCl)	Avdeef A, Box KJ, Comer JEA, Gilges M, Hadley M, Hibbert C, Patterson W and Tam KY, PH-metric log P 11. pK_a determination of water-insoluble drugs in organic solvent-water mixtures, *J. Pharm. Biomed. Anal.*, **20**, 631–641 (1999). NB: Used a Sirius PCA101 autotitrator. Titration results were extrapolated from a range of aqueous organic cosolvent mixtures, as follows:
1224	Quinolone—*N*-acetylnorfloxacin ($C_{18}H_{20}FN_3O_4$) 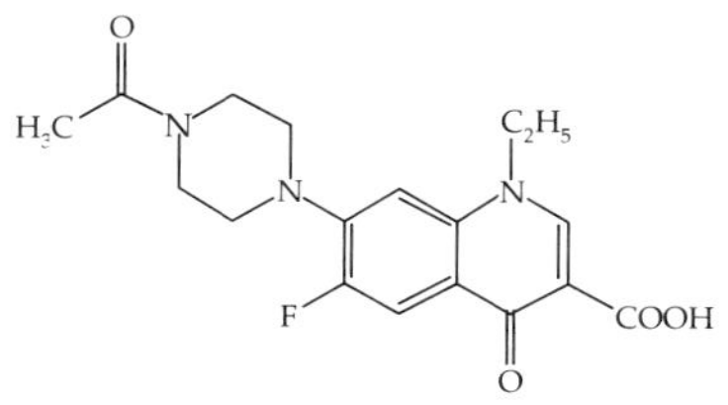	6.53 ± 0.05	U	−H	Potentiometric Spectro	H_2O $t = 25.0 \pm 0.1$ $I = 0.2$ (NaCl) $c = 0.005$ N_2 atmosphere	Takàcs-Novák K, Noszal B, Hermecz I, Kereszturi G, Podanyi B and Szasz G, Protonation equilibria of quinolone antibacterials, *J. Pharm. Sci.*, **79**, 1023–1028 (1990). NB: Used autoburette with accuracy of ± 0.005 cm^3. Data for *N*-acetylnorfloxacin and norfloxacin ethyl ester was used in in conjunction with spectrophotometric data to estimate the microconstants for quinolones that had two overlapping pK_a values.

Cosolvent	pK_1 (SD)	pK_2(SD)
Acetonitrile	4.13 (0.01)	8.52 (0.03)
Dimethyl formamide	3.85 (0.07)	8.15 (0.06)
Dimethylsulfoxide	4.32 (0.03)	8.51 (0.01)
1,4-Dioxane	4.25 (0.01)	8.57 (0.01)
Ethanol	4.24 (0.11)	8.55 (0.06)
Ethylene glycol	4.21 (0.06)	8.54 (0.04)
Methanol	4.24 (0.09)	8.55 (0.04)
Tetrahydrofuran	4.07 (0.09)	8.58 (0.04)

No.	Compound	pKa			Method	Conditions	Reference / Comments
1225	Quinolone—amifloxacin ($C_{16}H_{19}FN_4O_3$) 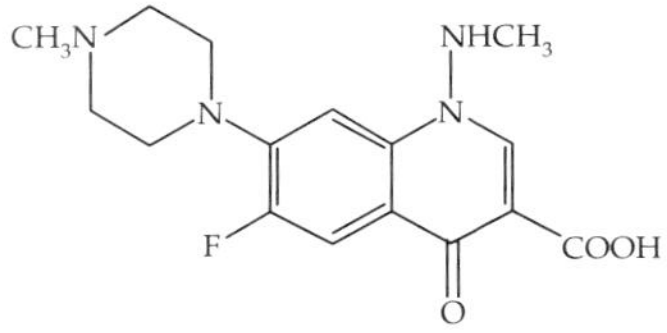	6.28 7.39	U U	−H +H	soly	H_2O $t = 25.0 \pm 0.1$ $I = 0.15$	Ross DL and Riley CM, Aqueous solubilities of some variously substituted quinolone antimicrobials, *Int. J. Pharm.*, **63**, 237–250 (1990). "The pK_a values of amifloxacin, ciprofloxacin HCl and difloxacin HCl were determined by the solubility method. In these experiments, excess drug was added to a series of buffer solutions from pH 4 to 9 (pH 4–6, 0.15M acetate buffer; pH 6.5–8, 0.05M phosphate buffer; pH 8.5–9.0, 0.15M borate buffer; $\mu = 0.15$ with NaCl). Solutions were protected from light and allowed to equilibrate for at least 48 h in a shaking water bath at 25(± 0.1) °C. The sample was filtered (5 µm), diluted with mobile phase and assayed by LC. All solubility experiments were conducted in triplicate. $St = So\{([H^+]^2 + K_1[H^+] + K_1K_2)/K_1[H^+]\}$ where St is the total solubility, So the intrinsic solubility of the neutral species (or zwitterion) and K_1 and K_2 represent apparent dissociation constants. . . .. Intrinsic solubilities, K_1 and K_2 values for amifloxacin, ciprofloxacin, and difloxacin were determined by fitting Eqn 4 to experimental solubility data with an RS1 curve-fitting algorithm."
1226	Quinolone—amifloxacin	5.42 ± 0.05 7.57 ± 0.05	U U	−H +H	Potentiometric	H_2O $t = 25.0 \pm 0.1$ $I = 0.2$ (NaCl) $c = 0.005$ N_2 atmosphere	Takàcs-Novák K, Noszal B, Hermecz I, Kereszturi G, Podanyi B, Szasz G, Protonation equilibria of quinolone antibacterials, *J. Pharm. Sci.*, **79**, 1023–1028 (1990). NB: See Quinolone—*N*-acetylnorfloxacin for details. Microconstants: C(ation) → Z(witterion), 5.7; C(ation) → N(eutral), 5.8; A(nion) → Z(witterion), 7.3; A(nion) → N(eutral), 7.2.
1227	Quinolone—ciprofloxacin ($C_{17}H_{18}FN_3O_3$) 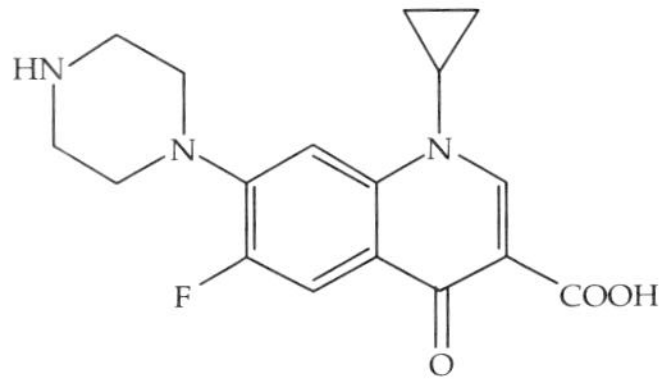	6.09 8.74	U U	−H +H	Soly	H_2O $t = 25.0 \pm 0.1$ $I = 0.15$	Ross DL and Riley CM, Aqueous solubilities of some variously substituted quinolone antimicrobials, *Int. J. Pharm.*, **63**, 237–250 (1990). NB: See Amifloxacin (no. 1225).
1228	Quinolone—ciprofloxacin	6.12 8.83	U	−H +H	Spectro	H_2O	Yu XQ, Britten NJ and Davidson GWR, Solubilities and apparent macroscopic dissociation constants of ciprofloxacin and norfloxacin, *Pharm. Res.*, **9**, **101**, S-219 (1992). NB: From alternative solubility method, $pK_{a1} = 6.24$; $pK_{a2} = 8.73$.

(continued)

No.	Compound Name	pK_a value(s)	Data quality	Ionization type	Method	Conditions	Comments and Reference(s)
1229	Quinolone—8-desfluorolomefloxacin ($C_{17}H_{20}FN_3O_3$) 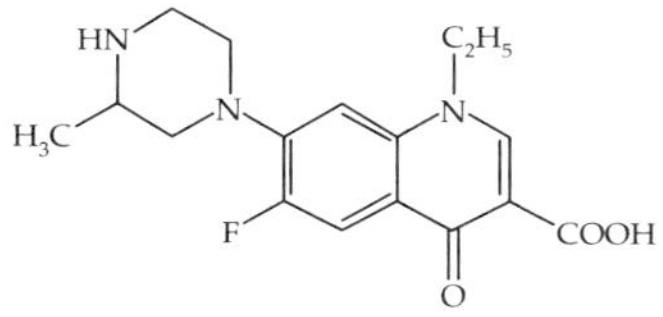	5.98 ± 0.05 8.39 ± 0.05	U U	−H +H	Potentiometric	H_2O $t = 25.0 \pm 0.1$ $I = 0.2$ (NaCl) $c = 0.005$ N_2 atmosphere	Takàcs-Novák K, Noszal B, Hermecz I, Kereszturi G, Podanyi B and Szasz G, Protonation equilibria of quinolone antibacterials, *J. Pharm. Sci.*, **79**, 1023–1028 (1990). NB: See Quinolone—*N*-acetylnorfloxacin for details. See Quinolone—amifloxacin for microconstant symbols. Microconstants: C→Z, 6.0; C→N, 7.0; A→Z, 8.4; A→N, 7.4.
1230	Quinolone—difloxacin ($C_{21}H_{19}F_2N_3O_3$) 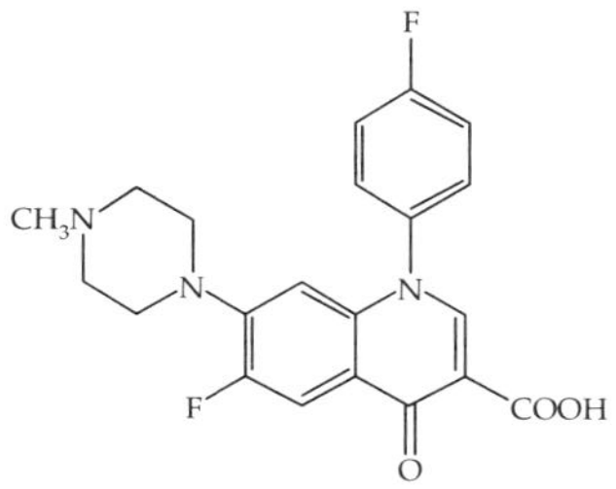	6.06 7.63	U U	−H +H	soly	H_2O $t = 25.0 \pm 0.1$ $I = 0.15$	Ross DL and Riley CM, Aqueous solubilities of some variously substituted quinolone antimicrobials, *Int. J. Pharm.*, **63**, 237–250 (1990). NB: See Amifloxacin (no. 1225).
1231	Quinolone—enoxacin ($C_{15}H_{17}FN_4O_3$) 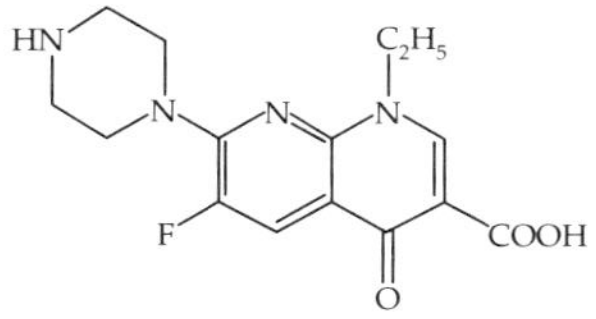	6.06 7.63	U U	−H +H	Spectro ($\lambda = 272$)	H_2O $t = 22 \pm 1$ $I = 0.15$ (NaCl)	Ross DL and Riley CM, Aqueous solubilities of some variously substituted quinolone antimicrobials, *Int. J. Pharm.*, **63**, 237–250 (1990). "The pK_a values of all the compounds, except amifloxacin, ciprofloxacin HCl, and difloxacin HCl, were determined spectrophotometrically at ambient temperature (22 ± 1 °C). A wavelength was chosen ... where the absorbance of the three species (cation, zwitterion, and anion) varied the greatest. The change of absorbance, at the selected wavelength, with pH was also monitored. The same total concentration of drug (between 2 × 10^{-4} and 2 × 10^{-5} M...) was used for all measurements. Ionic strength was held ... at 0.15 with NaCl. Linearity of absorbance with respect to concentration, ..., was established for all compounds in their cationic and anionic forms. All spectrophotometric measurements were made in triplicate.

							$AT = (AM[H^+] + AAK_1)/([H^+] + K_1)$ (1) $AT = (AC[H^+]^2 + AMK_1[H^+] + AAK_1K_2)/([H^+]^2 + K_1[H^+] + K_1K_2)$ (2) Spectrophotometrically determined apparent K_1 and K_2 values . . . were calculated by fitting Eqn 2 to the experimental (absorbance) data with an RS1 curve-fitting algorithm. For nalidixic acid, Eqn 1 was used. The value for r^2 was greater than 0.999 in all cases.''
1232	Quinolone—fleroxacin ($C_{17}H_{18}F_3N_3O_3$) 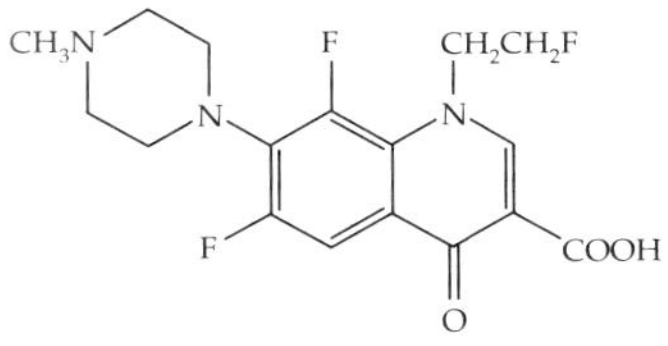	5.46 8.10	U U	−H +H	Spectro ($\lambda = 272$)	H_2O $t = 22 \pm 1$ $I = 0.15$	Ross DL and Riley CM, Aqueous solubilities of some variously substituted quinolone antimicrobials, *Int. J. Pharm.*, **63**, 237–250 (1990). NB: See Enoxacin (no. 1231).
1233	Quinolone—8-fluoronorfloxacin ($C_{16}H_{17}F_2N_3O_3$) 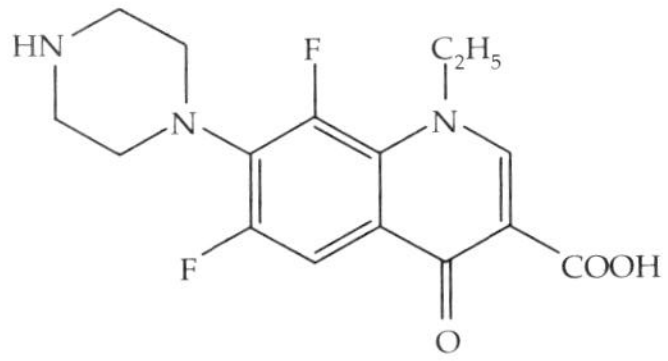	5.55 ± 0.05 9.33 ± 0.05	U U	−H +H	Potentiometric	H_2O $t = 25.0 \pm 0.1$ $I = 0.2$ (NaCl) $c = 0.005$ N_2 atmosphere	Takàcs-Novák K, Noszal B, Hermecz I, Kereszturi G, Podanyi B and Szasz G, Protonation equilibria of quinolone antibacterials, *J. Pharm. Sci.*, **79**, 1023–1028 (1990). NB: See Quinolone—*N*-acetylnorfloxacin for details. See Quinolone—amifloxacin for microconstant symbols. Microconstants: C→Z, 5.7; C→N, 6.1; A→Z, 9.2; A→N, 8.8.
1234	Quinolone—8-fluoropefloxacin ($C_{17}H_{19}F_2N_3O_3$) 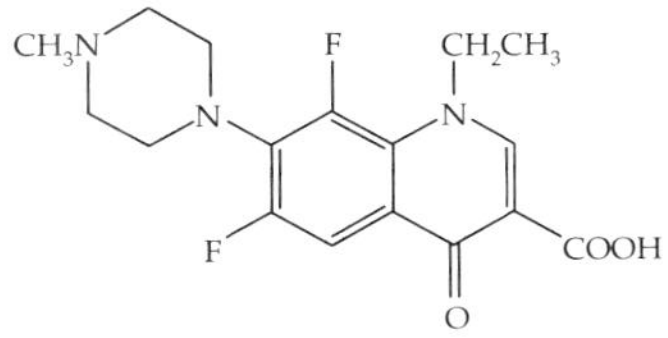	5.33 ± 0.05 8.13 ± 0.05	U U	−H +H	Potentiometric	H_2O $t = 25.0 \pm 0.1$ $I = 0.2$ (NaCl) $c = 0.005$ N_2 atmosphere	Takàcs-Novák K, Noszal B, Hermecz I, Kereszturi G, Podanyi B and Szasz G, Protonation equilibria of quinolone antibacterials, *J. Pharm. Sci.*, **79**, 1023–1028 (1990). NB: See Quinolone—*N*-acetylnorfloxacin for details. See Quinolone—amifloxacin for microconstant symbols. Microconstants: C→Z, 5.6; C→N, 5.7; A→Z, 7.9; A→N, 7.7.

(*continued*)

No.	Compound Name	pK_a value(s)	Data quality	Ionization type	Method	Conditions	Comments and Reference(s)
1235	Quinolone—lomefloxacin ($C_{17}H_{19}F_2N_3O_3$) 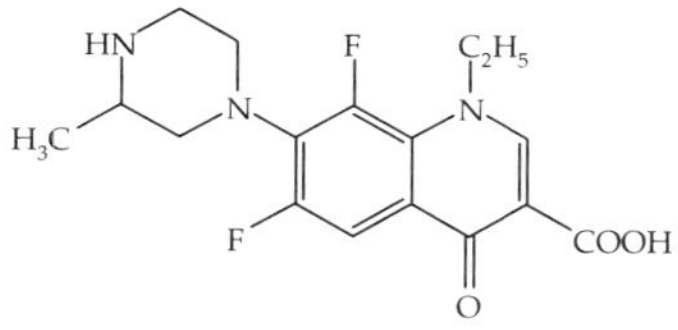	5.82 9.30	U U	–H +H	Spectro ($\lambda = 266$)	H_2O $t = 22 \pm 1$ $I = 0.15$	Ross DL and Riley CM, Aqueous solubilities of some variously substituted quinolone antimicrobials, *Int. J. Pharm.*, **63**, 237–250 (1990). NB: See Enoxacin (no. 1231). Also quoted in Sanzgiri YD, Knaub SR and Riley CM, Lomefloxacin, *APDS*, **23**, 321–369 (1994).
1236	Quinolone—lomefloxacin	5.49 ± 0.05 8.78 ± 0.05	U U	–H +H	Potentiometric	H_2O $t = 25.0 \pm 0.1$ $I = 0.2$ (NaCl) $c = 0.005$ N_2 atmosphere	Takàcs-Novák K, Noszal B, Hermecz I, Keresztúri G, Podanyi B and Szasz G, Protonation equilibria of quinolone antibacterials, *J. Pharm. Sci.*, **79**, 1023–1028 (1990). NB: See Quinolone—*N*-acetylnorfloxacin for details. See Quinolone—amifloxacin for microconstant symbols. Also quoted in Sanzgiri YD, Knaub SR and Riley CM, Lomefloxacin, *APDS*, **23**, 321–369 (1994). Microconstants: C→Z, 5.7; C→N,6.0; A→Z, 8.6; A→N, 8.3.
1237	Quinolone—nalidixic acid ($C_{12}H_{12}N_2O_3$) 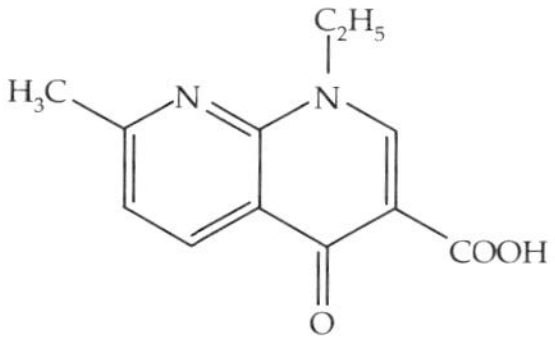	5.95	U	–H	Spectro ($\lambda = 310$)	H_2O $t = 22 \pm 1$ $I = 0.15$	Ross DL and Riley CM, Aqueous solubilities of some variously substituted quinolone antimicrobials, *Int. J. Pharm.*, **63**, 237–250 (1990). NB: See Enoxacin (no. 1231).
1238	Quinolone—nalidixic acid	6.13 ± 0.05	U	–H	Spectro	H_2O $t = 25.0 \pm 0.1$ $I = 0.2$ (NaCl) $c = 0.005$ N_2 atmosphere	Takàcs-Novák K, Noszal B, Hermecz I, Keresztúri G, Podanyi B and Szasz G, Protonation equilibria of quinolone antibacterials, *J. Pharm. Sci.*, **79**, 1023–1028 (1990). NB: See Quinolone—*N*-acetylnorfloxacin for details.

1239	Quinolone—norfloxacin ($C_{16}H_{18}FN_3O_3$) 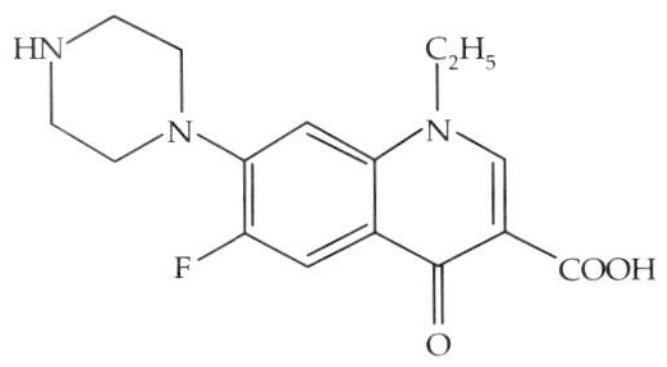	6.30 8.38	U U	−H +H	Spectro ($\lambda = 274$)	H_2O $t = 22 \pm 1$ $I = 0.15$	Ross DL and Riley CM, Aqueous solubilities of some variously substituted quinolone antimicrobials, *Int. J. Pharm.*, **63**, 237–250 (1990). NB: See Enoxacin (no. 1231).
1240	Quinolone—norfloxacin	6.22 ± 0.05 8.51 ± 0.05	U U	−H +H	Potentiometric	H_2O $t = 25.0 \pm 0.1$ $I = 0.2$ (NaCl) $c = 0.005$ N_2 atmosphere	Takàcs-Novák K, Noszal B, Hermecz I, Keresztúri G, Podanyi B and Szasz G, Protonation equilibria of quinolone antibacterials, *J. Pharm. Sci.*, **79**, 1023–1028 (1990). NB: See Quinolone—*N*-acetylnorfloxacin for details. See Quinolone—amifloxacin for microconstant symbols. Microconstants: C→Z, 6.3; C→N, 7.2; A→Z, 8.5; A→N, 7.6.
1241	Quinolone—norfloxacin	6.34 ± 0.06 8.75 ± 0.07	U U	−H +H	Potentiometric	H_2O $t = 25$	Mazuel C, Norfloxacin, *APDS*, **20**, 557–600 (1991). "The pK_a values for norfloxacin were determined by dissolution of the compound in dilute aqueous sodium hydroxide or hydrochloric acid solution and potentiometric titrations of the solutions at 25 °C with 0.1N HCl or 0.1N sodium hydroxide. The pK_{a1} (carboxylic acid) and pK_{a2} (protonated piperazine nitrogen) are respectively 6.34 ± 0.06 and 8.75 ± 0.07 (21). 21. Kyorin Pharmaceutical Co. Ltd., Tokyo, Japan, unpublished data."
1242	Quinolone—norfloxacin	6.30 8.69	U U	−H +H	Spectro	H_2O	Yu XQ, Britten NJ and Davidson GWR, Solubilities and apparent macroscopic dissociation constants of ciprofloxacin and norfloxacin, *Pharm. Res.*, **9**, **101**, S-219 (1992). NB: From alternative solubility method, p$K_{a1} = 6.45$; p$K_{a2} = 8.69$.
1243	Quinolone—norfloxacin ethyl ester ($C_{18}H_{22}FN_3O_3$) 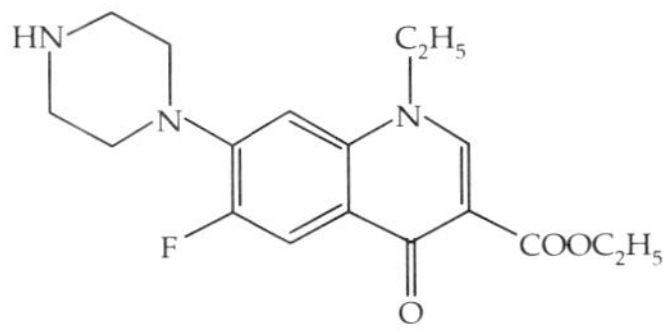	8.48 ± 0.05	U	+H	Spectro	H_2O $t = 25.0 \pm 0.1$ $I = 0.2$ (NaCl) $c = 0.005$ N_2 atmosphere	Takàcs-Novák K, Noszal B, Hermecz I, Keresztúri G, Podanyi B and Szasz G, Protonation equilibria of quinolone antibacterials, *J. Pharm. Sci.*, **79**, 1023–1028 (1990). NB: See Quinolone—*N*-acetylnorfloxacin for details.

(continued)

No.	Compound Name	pK_a value(s)	Data quality	Ionization type	Method	Conditions	Comments and Reference(s)
1244	Quinolone—ofloxacin ($C_{18}H_{20}FN_3O_4$) 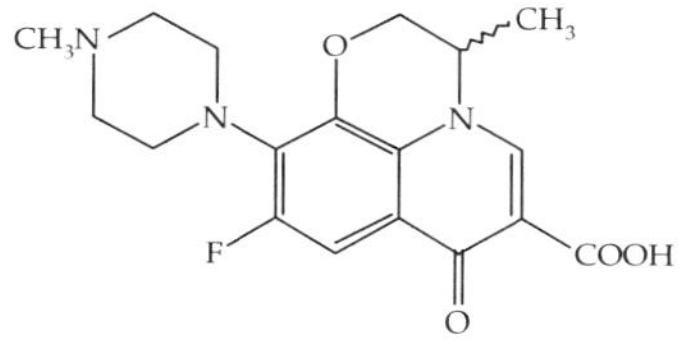	6.05 8.22	U U	−H +H	Spectro ($\lambda = 258$)	H_2O $t = 22 \pm 1$ $I = 0.15$	Ross DL and Riley CM, Aqueous solubilities of some variously substituted quinolone antimicrobials, *Int. J. Pharm.*, **63**, 237–250 (1990). NB: See Enoxacin (no. 1231).
1245	Quinolone—ofloxacin	6.08 ± 0.01 8.31 ± 0.01	U U	−H +H	Potentiometric	H_2O $t = 25.0$ $I = 0.1$ (NaCl)	Takacs-Novak K and Avdeef A, Interlaboratory study of log P determination by shake-flask and potentiometric methods, *J. Pharm. Biomed. Anal.*, **14**, 1405–1413 (1996). NB: See Acetaminophen for further details. Also reported pK_a1 = 6.09 ± 0.01; pK_a2 = 8.31 ± 0.01 at $I = 0.15$ (NaCl).
1246	Quinolone—orbifloxacin ($C_{19}H_{20}F_3N_3O_3$) 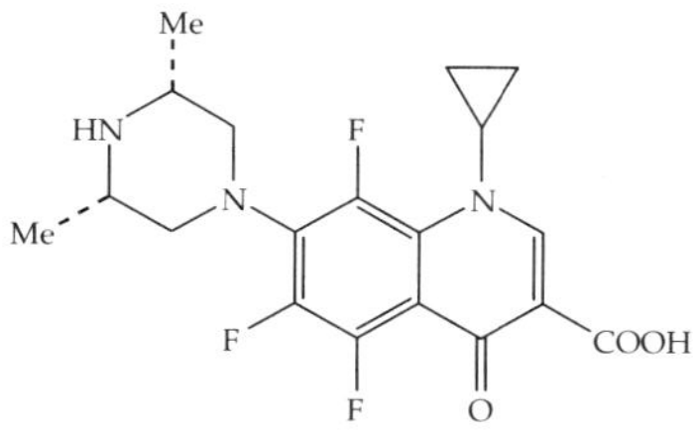	~6 ~9	U U	−H +H	kinetic	H_2O $t = 100$–120 $c = 0.05$ to 0.2	Morimura T, Ohno T, Matsukura H and Nobuhara Y, Degradation kinetics of the new antibacterial fluoroquinolone derivative, orbifloxacin, in aqueous solution, *Chem. Pharm. Bull.*, **43**, 1052–1054 (1995). "The degradation of fluoroquinolone, orbifloxacin, was studied as a function of pH (1.5–10.5), temperature (100–120 °C), and buffer concentration (0.05–0.2 M) using HPLC. Degradation followed apparent first-order kinetics under all experimental conditions. . . .. The log k-pH profiles indicated specific-acid and specific-base catalyses and there were inflection points near pH 6 and 9 corresponding to the p$K_{a\text{-}1}$ and p$K_{a\text{-}2}$ values. Arrhenius data showed that the degradation at room temperature was negligible at all pH values studied."
1247	Quinolone—pefloxacin ($C_{17}H_{20}FN_3O_3$) 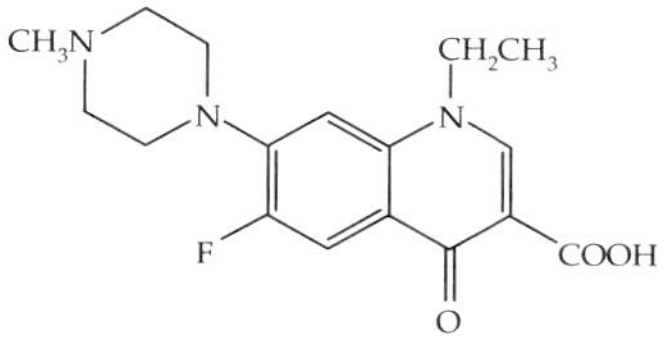	6.02 ± 0.05 7.80 ± 0.05	U U	−H +H	Potentiometric	H_2O $t = 25.0 \pm 0.1$ $I = 0.2$ (NaCl) $c = 0.005$ N_2 atmosphere	Takàcs-Novák K, Noszal B, Hermecz I, Kereszturi G, Podanyi B and Szasz G, Protonation equilibria of quinolone antibacterials, *J. Pharm. Sci.*, **79**, 1023–1028 (1990). NB: See Quinolone—*N*-acetylnorfloxacin for details. See Quinolone—amifloxacin for microconstant symbols. Microconstants: C→Z, 6.1; C→N, 6.7; A→Z, 7.7; A→N, 7.1.

1248	Quinolone—quinolone carboxylic acid ($C_{12}H_{11}NO_3$) 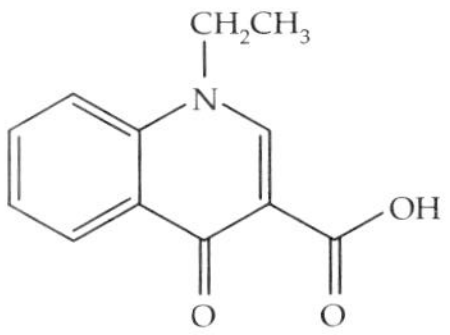	6.51 ± 0.05	U	–H	Spectro	H_2O $t = 25.0 \pm 0.1$ $I = 0.2$ (NaCl) $c = 0.005$ N_2 atmosphere	Takàcs-Novák K, Noszal B, Hermecz I, Kereszturi G, Podanyi B and Szasz G, Protonation equilibria of quinolone antibacterials, *J. Pharm. Sci.*, **79**, 1023–1028 (1990). NB: See Quinolone—*N*-acetylnorfloxacin for details.
1249	Quinolone—temafloxacin ($C_{21}H_{18}F_3N_3O_3$) 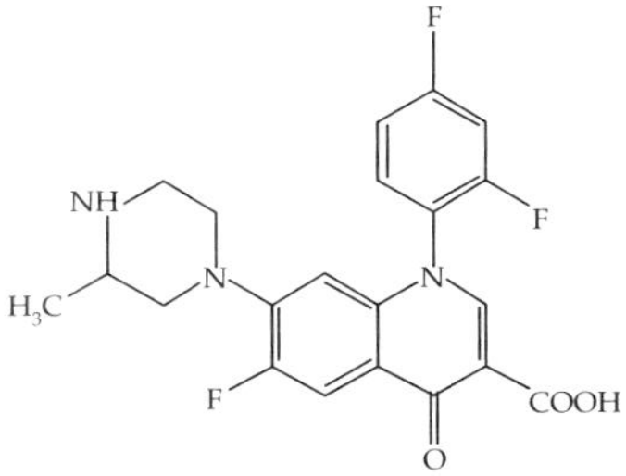	5.61 8.75	U U	–H +H	Spectro ($\lambda = 264$)	H_2O $t = 22 \pm 1$ $I = 0.15$	Ross DL and Riley CM, Aqueous solubilities of some variously substituted quinolone antimicrobials, *Int. J. Pharm.*, **63**, 237–250 (1990). NB: See Enoxacin (no. 1231).
1250	Ramipril ($C_{23}H_{32}N_2O_5$) 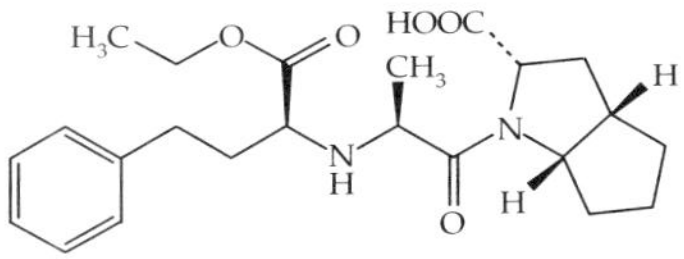	3.30 ± 0.01 5.75 ± 0.00	U U	–H +H	partition-pH	H_2O $t = 25.0$ $I = 0.15$ (KCl)	Krämer SD, Gautier J-C and Saudemon P, Considerations on the potentiometric log P determination, *Pharm. Res.*, **15**(8), 1310–1313 (1998). NB: See Amiodarone for details.

(continued)

No.	Compound Name	pK_a value(s)	Data quality	Ionization type	Method	Conditions	Comments and Reference(s)
1251	Remoxipride ($C_{16}H_{23}BrN_2O_3$) 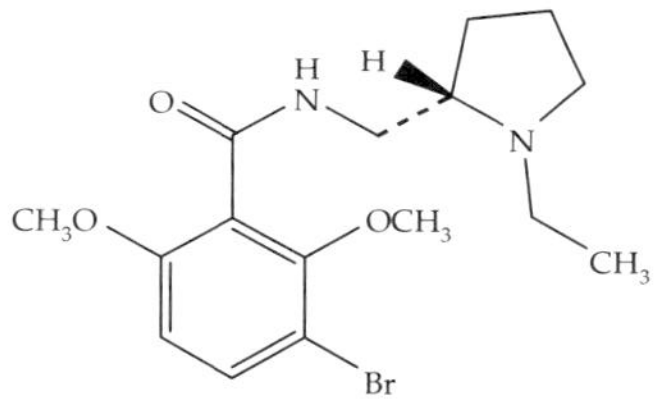	8.9	U	+H			Lombardo F, Obach RS, Shalaeva MY and Gao F, Prediction of human volume of distribution values for neutral and basic drugs. 2. Extended data set and leave-class-out statistics, *J. Med. Chem.*, **47**, 1242–50 (2004). Ref. 299 = Hogberg T, Ramsby S, de Paulis T, Stensland B, Csorregh I and Wagner A, Solid state conformations and antidopaminergic effects of remoxipride hydrochloride and a closely related salicylamide, FLA797, in relation to dopamine receptor models, *Mol. Pharmacol.*, **30**, 345–351 (1986).
1252	Repromicin ($C_{38}H_{72}N_2O_{12}$) 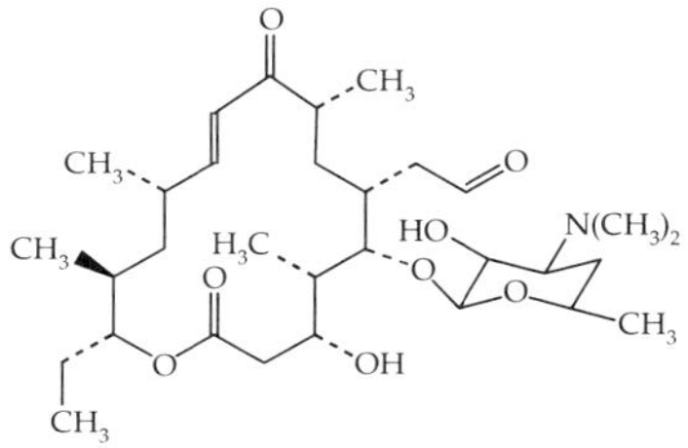	8.83	U	+H	Potentiometric	H_2O $t = 25$ $I = 0.167$	McFarland JW, Berger CM, Froshauer SA, Hayashi SF, Hecker SJ, Jaynes BH, Jefson MR, Kamicker BJ, Lipinski CA, Lundy KM, Reese CP and Vu CB, Quantitative Structure-activity relationships among macrolide antibacterial agents: *In vitro* and *in vivo* potency against Pasteurella multocida, *J. Med. Chem.*, **40**, 1340–1346 (1997). NB: See Azithromycin for details; average standard deviation of ± 0.07 for the pK_a. Paper also reported pK_a values for the following repromycin analogues, where these were obtained by reducing the aldehyde (-CHO) group with an amine:

Side chain:	pK_{a1}	pK_{a2}	pK_{a3}
Azetidin-1-yl	8.49	9.54	
$HOCH_2CH_2NH$	8.59	9.62	
$HOCH_2CH_2NCH_3$	8.56	9.18	
$HOCH_2CMe_2CH_2NH$	8.53	9.72	
$HOCH_2CH(OH)CH_2NH$[m]	8.55	9.02	
(2,6-dihydroxycylohexyl)amino[n]	8.56	9.31	
$Me_2N(CH_2)_3NH$	7.97	8.88	10.13
Gly-$NMe(CH_2)_2NMe$	6.31	8.10	8.95
L-Ala-$NMe(CH_2)_2NMe$	6.81	8.13	8.90
$Me_2N(CH_2)_3N$(Gly)	7.85	8.74	9.81
$Me_2N(CH_2)_3N$(L-Ala)	7.81	8.78	9.60
L-Ala-$NMe(CH_2)_3NMe$	7.03	8.53	9.16

[m] mixture of epimers; [n] hydroxy groups cis to each other and trans to the amino group.

No.	Compound	pK			Method	Conditions	Reference
1253	Riboflavine ($C_{17}H_{20}N_4O_6$) 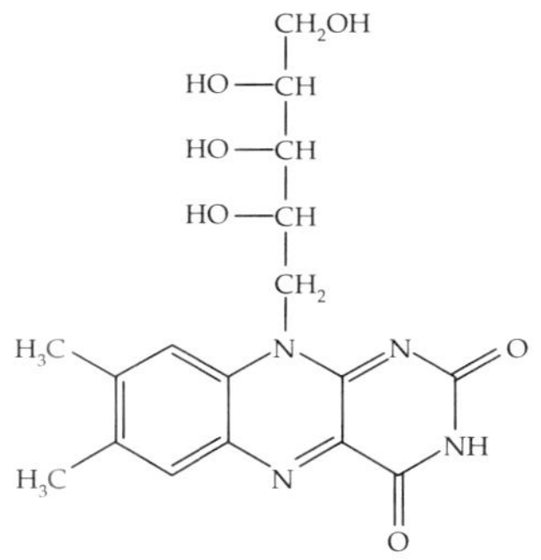	10.02	A	–H	Potentiometric	H_2O	Harkins TR and Freiser H, The chelating tendency of riboflavin, *J. Phys. Chem.*, **63**, 309–310 (1959). Cited in Perrin Bases no. 3335 ref. H30. NB: Also reported free energy change for ionization ($\Delta G = -13.2$ kcal/mol); enthalpy change for ionization ($\Delta H = -8.3$ kcal/mol) and entropy change for ionization ($\Delta S = 16$ cal/mol/K).
		9.69	A	–H		$t = 10$	
		9.40	A	–H		$t = 25$	
						$t = 40$; I low but not stated	
		9.93	A	–H	Potentiometric	$t = 20$; $c = 0.001$	Albert AA, Quantitative studies of the avidity of naturally occurring substances for trace metals, *Biochem. J.*, **54**, 646–654 (1953). Cited in Perrin Bases no. 3335 Ref. A15.
1254	Riboflavine	–0.2	U	+H	Spectro	H_2O	Michaelis, L, Schubert MP and Smythe CV, Potentiometric study of the flavins, *J. Biol. Chem.* **116**, 587–607 (1936). Cited in Perrin Bases no. 3335 ref. M44. NB: Study used spectrophotometric measurements in strongly acid solutions (pK_{a1}) and a potentiometric method (pK_{a2}).
		9.8	VU	–H	Potentiometric		
1255	Riboflavine	1.7	VU	+H	fluoro	H_2O	Kuhn R and Moruzzi G, Dissociation constants of the flavins; dependence of the fluorescence on the pH. *Ber.*, **67B**, 888–891 (1934). CA 28:42085. Cited in Perrin Bases no. 3335 Ref. K65.
		10.2	VU	–H			
1256	Rifampin ($C_{43}H_{58}N_4O_{12}$) 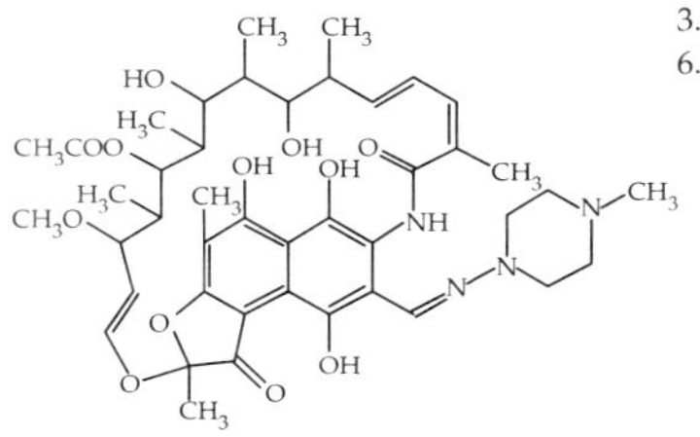	1.7	U	–H	Spectro	H_2O	Maggi N, Pasqualucci CR, Ballotta R and Sensi P, Rifampicin: A new orally active rifamycin, *Chemotherapia*, **11**, 285–292 (1966). Cited in Gallo GG, Radaelli P, Rifampin, *APDS*, **5**, 467–513 (1976).
		7.9	U	+H	Potentiometric		"The ionization properties of rifamycins have been used in conjunction with UV-VIS spectrophotometry to obtain information on the chromophoric part of the molecule and as a quantitative method of analysis. The pK values for rifampin have been determined spectrophotometrically and potentiometrically in solution in water and are reported in the table. Rifampin exists in water solution as the zwitterion with isoelectric point equal to 4.8."
		3.6	U	–H	Potentiometric	80% MCS	
		6.7	U	+H	Potentiometric		

	pK_a	pK_{MCS}	Attribution
proton lost	1.7	3.6	hydroxyl at C-8
proton gained	7.9	6.7	piperazine N-4

(continued)

No.	Compound Name	pK_a value(s)	Data quality	Ionization type	Method	Conditions	Comments and Reference(s)
							NB: The reported assignments have been questioned (see Prankerd RJ, Walters JM and Parnes JH, Kinetics for degradation of rifamipicin, an azomethine-containing drug which exhibits reversible hydrolysis in acidic solutions, *Int. J. Pharm.* **78**, 59–69 (1992)), chiefly due to the low pK_a value (1.7) claimed for the trihydroxyaromatic system. This low value was not supported by calculations using ACD/pK_a (Ver. 9), which suggested a value for the most acidic aromatic –OH of about 5 (author's calculations; see also no. 1257). The value of 7.9 appears more likely to be due to the piperazine nitrogen, as the computed value was found to be about 7.3. No pK_a value has been reported for the azomethine nitrogen. This value is expected to be in the range 1–4, as found for the benzodiazepines. See also: Gallo GG, Sensi P and Radaelli P, *Farmaco (Pavia)*, Edn. Prat., **15**, 283 (1960); Gallo GG, Pasqualucci CR, Radaelli P, *Farmaco (Pavia)*, Edn. Prat., **18**, 78 (1963); Pasqualucci CR, Vigevani A, Radaelli P, Maggi N, *Farmaco (Pavia)*, Edn. Prat., **24**, 46 (1969); Oppolzer W and Prelog V, Uber die Konstitution und die Konfiguration der Rifamycine B, O, S und SV, *Helv. Chim. Acta*, **56**, 2287–2313 (1973).
1257	Rifampicin	1.52 7.42	U U	–H +H	CE/pH (+ve ion mode)	H_2O $t = 25$ $I = 0.025$	Wan H, Holmen AG, Wang Y, Lindberg W, Englund M, Nagard MB and Thompson RA, High-throughput screening of pK_a values of pharmaceuticals by pressure-assisted capillary electrophoresis and mass spectrometry, *Rapid Commun. Mass Spectrom.*, **17**, 2639–2648 (2003). NB: Reported predicted values (ACD Labs) of 4.92 and 6.42. Chem Abs database reported predicted values (ACD Labs) of 4.96 and 7.30.
1258	Rosaramicin (juvenimicin) ($C_{31}H_{51}NO_9$) 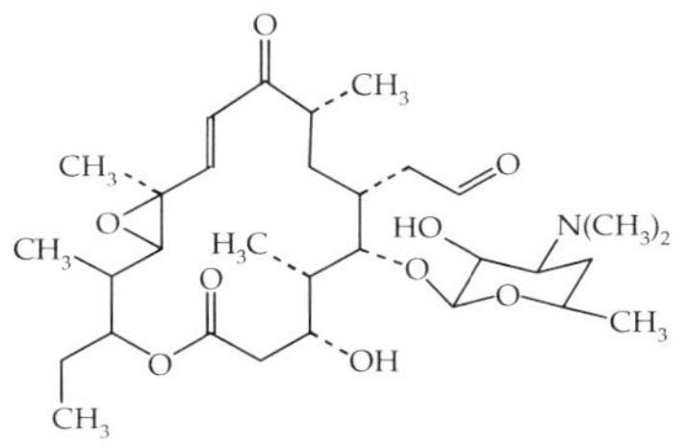 O, CH$_3$, CH$_3$, O, O, CH$_3$, H$_3$C, HO, N(CH$_3$)$_2$, O, O, O, CH$_3$, O, OH, CH$_3$	8.79	U	+H	Potentiometric	H_2O $t = 25$ $I = 0.167$	McFarland JW, Berger CM, Froshauer SA, Hayashi SF, Hecker SJ, Jaynes BH, Jefson MR, Kamicker BJ, Lipinski CA, Lundy KM, Reese CP and Vu CB, Quantitative Structure-activity relationships among macrolide antibacterial agents: *In vitro* and *in vivo* potency against Pasteurella multocida, *J. Med. Chem.*, **40**, 1340–1346 (1997). NB: See Azithromycin for details; average standard deviation of ± 0.07 for the pK_a. Also cited pK_a = 8.4 in 66% DMF, potentio (Kishi T, Harada S, Yamana H, Miyake A, Studies on Juvenimicin, a new antibiotic, *J. Antibiot.*, **29**, 1171–1181 (1976)).

No.	Compound	pKa	Type	Ionization	Method	Conditions	Reference / Notes
1259	Roxithromycin ($C_{41}H_{76}N_2O_{15}$) 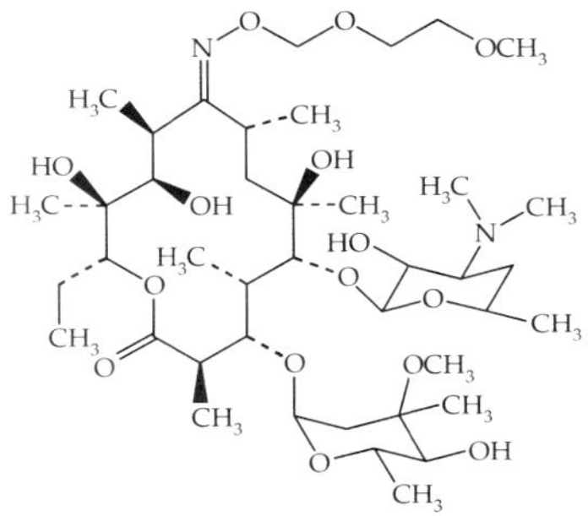	9.2	U	+H	NMR	D_2O $t = 20$ I undefined	Gharbi-Benarous J, Delaforge M, Jankowski CK and Girault J-P, A comparative NMR study between the macrolide antibiotic roxithromycin and erythromycin A with different biological properties, *J. Med. Chem.*, **34**, 1117–1125 (1991).
		9.27 ± 0.01	U	+H	Potentiometric	H_2O $t = 25.0$ $I = 0.15$ (KCl)	*Sirius Technical Application Notes*, vol. **2**, pp. 107–108, 165–166 (1995). Sirius Analytical Instruments Ltd., Forest Row, East Sussex, RH18 5DW, UK. NB: An alternative method gave 9.12 ± 0.01 under the same conditions.
1260	Salbutamol (albuterol) ($C_{13}H_{21}NO_3$) 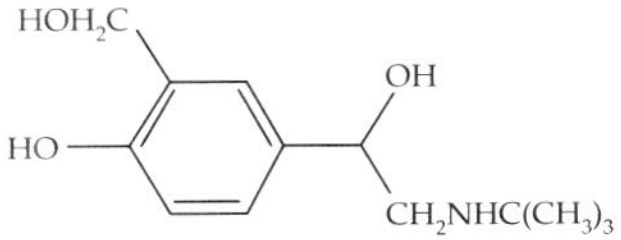	9.07 10.37	U U	−H, +H +H, −H	Spectro	H_2O $t = 25.0 \pm 0.05$ $I = 0.10$	Ijzerman AP, BultsmaT, Timmerman H and Zaagsma J, The ionization of β-adrenoceptor agonists: a method for unravelling ionization schemes, *J. Pharm. Pharmacol.*, **36**(1), 11–15 (1984). NB: Microscopic: 9.22, 10.22, 9.60 and 9.84. See Isoprenaline.
1261	Salicylamide ($C_7H_7NO_2$) 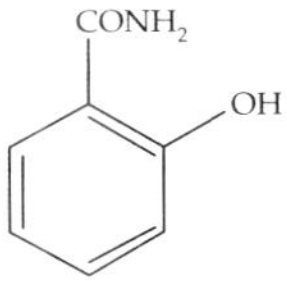	8.13	U	−H		H_2O $t = 37$	Ballard BE and Nelson E, Physicochemical properties of drugs that control absorption rate after subcutaneous implantation, *JPET*, **135**, 120–127 (1962). Babhair SA, Al-Badr AA and Aboul-Enien HY, Salicylamide, *APDS*, **13**, 521–548 (1984).
1262	Salicylamide	8.47 (0.13)	U	−H	Spectro (328 nm)	H_2O $t = 20.0$	Wahbe AM, El-Yazbi FA, Barary MH and Sabri SM, Application of orthogonal functions to spectrophotometric analysis. Determination of dissociation constants, *Int. J. Pharm.*, **92**(1), 15–22 (1993). NB: The orthogonal method is intended to correct for the effects of spectra which overlap for the protonated and deprotonated forms of the ionising species. pH values were measured at 20 °C but it was not clear if the spectral data were obtained at this temperature. Alternative graphical method gave $pK_a = 8.5$. See Acetaminophen for further details.

(continued)

No.	Compound Name	pK_a value(s)	Data quality	Ionization type	Method	Conditions	Comments and Reference(s)
1263	Salicylamide	8.2	U	–H			Bates TR, Lambert DA and Johns WH, Correlation between the rate of dissolution and absorption of salicylamide from tablet and suspension dosage forms, *J. Pharm. Sci.*, **58**, 1468–1470 (1969).
1264	Salicylic acid ($C_7H_6O_3$) COOH OH	2.95	U	–H	Potentiometric	H_2O $t = 23.0$	Clarke FH and Cahoon NM, Ionization constants by curve-fitting: Determination of partition and distribution coefficients of acids and bases and their ions, *J. Pharm. Sci.*, **76**, 611–620 (1987). NB: See Benzoic acid for further details. The phenolic group, due to proximity and Coulombic repulsion effects, has a very high value for pK_{a2}, e.g., Foye gave 13.4.
1265	Salicylic acid	2.99	U	–H	CE/pH (–ve ion mode)	H_2O $t = 25$ $I = 0.025$	Wan H, Holmen AG, Wang Y, Lindberg W, Englund M, Nagard MB, Thompson RA, High-throughput screening of pK_a values of pharmaceuticals by pressure-assisted capillary electrophoresis and mass spectrometry, *Rapid Commun. Mass Spectrom.*, **17**, 2639–2648 (2003). NB: Reported a predicted value (ACD Labs) of 3.01.
1266	Salicylic acid	2.83 ± 0.03	A	–H	Potentiometric	H_2O $t = 25.0$ $I = 0.1$ (NaCl)	Takacs-Novak K and Avdeef A, Interlaboratory study of log P determination by shake-flask and potentiometric methods, *J. Pharm. Biomed. Anal.*, **14**, 1405–1413 (1996). NB: See Acetaminophen for further details. Also reported $pK_a = 2.88 \pm 0.01$ at $I = 0.15$ (KCl).
1267	Salicylic acid	2.75 ± 0.01	A	–H	Potentiometric	H_2O $t = 25.0 \pm 0.1$ $I = 0.1$ (NaCl)	Takacs-Novak K, Box KJ, Avdeef A, Potentiometric pK_a determination of water-insoluble compounds: validation study in methanol/water mixtures, *Int. J. Pharm.*, **151**, 235–248 (1997). NB: $pK_a = 2.73 \pm 0.03$ by extrapolation from 15.4–58.1% w/w aqueous MeOH. See Acetaminophen for full details.
		2.78 ± 0.01 13.77 ± 0.19	A U	–H –H	Potentiometric	H_2O $t = 25.0$ $I = 0.167$ (KCl)	Sirius Technical Application Notes, vol. **2**, p. 6 (1995). Sirius Analytical Instruments, Ltd., Forest Row, East Sussex, RH18 5DW, UK. NB: Analyte concentration, 15.0–15.8 mM.
1268	Salicylic acid	3.52 ± 0.03 3.29 ± 0.03	U U	–H –H	HPLC retention/pH	H_2O $t = 25$ $I = 0.01$ $I = 0.1$	Unger SH, Cook JR and Hollenberg JS, Simple procedure for determining octanol-aqueous partition, distribution, and ionization coefficients by reversed phase high pressure liquid chromatography, *J. Pharm. Sci.*, 67; 1364–1367 (1978). NB: See Naproxen for details.

No.	Compound	pKa	Type	Ionization	Method	Conditions	Reference
1269	Salicylic acid derivatives				Potentiometric	H_2O	Hanninen K, Kaukonen AM, Kankkunen T and Hirvonen J, Rate and extent of ion-exchange process—the effect of physico-chemical properties of salicylates, *J. Controlled Release*, **91**, 449–463 (2003). NB: Extrapolated to aqueous solutions using the Yasuda-Shedlovsky procedure from mixtures in 30%, 40% and 50% aqueous MeOH. 5-Aminosalicylic acid was titrated in 4% aqueous DMSO (without further refinement), rather than MeOH-water mixtures, where it was poorly soluble. Additional values for the –COOH group (except where noted) (all U, –H):
	Salicylic acid	2.90 ± 0.02	U	–H		$t = 25 \pm 0.5$	
	5-Aminosalicylic acid	2.41 ± 0.02	U	–H		$I = 0.150$	
	5-NH_2	5.54 ± 0.01	U	+H		(KCl)	
	Acetylsalicylic acid	3.27 ± 0.01	U	–H			
	5-Hydroxysalicylic acid	2.72 ± 0.03	U	–H			
	5-OH	10.07 ± 0.04	U	–H			

Compound	pK_a	Compound	pK_a
5-Fluorosalicylic acid	2.76 ± 0.02	5-Bromosalicylic acid	2.68 ± 0.11
5-Methylsalicylic acid	2.78 ± 0.05	3-i-Propylsalicylic acid	2.89 ± 0.01
5-Chlorosalicylic acid	2.64 ± 0.03	5-Carboxysalicylic acid	3.44 ± 0.03
		5-COOH	4.16 ± 0.03

No.	Compound	pKa	Type	Ionization	Method	Conditions	Reference
1270	Scopolamine ($C_{17}H_{21}NO_4$)	7.55	U	+H	Spectro	H_2O $t = 23$	Schoorl N, Dissociation constants and titration exponents of several less common alkaloids, *Pharm. Weekblad*, **76**, 1497–1501 (1939); CA 34:1900. Cited in Perrin Bases no 2961 ref. S16. See Diamorphine for details.
		8.15	U	+H	Potentiometric	H_2O $t = 20$	Perel'man Y, Sb. Nauchn and Tr. Leningr, *Khim-Farmatsevt. Inst*, **2**, 38 (1957); CA 54: 11382b. Ref. P16 NB: Used a glass electrode in an unsymmetrical cell with liquid junction potentials.
		7.62	U	+H	Potentiometric	H_2O $t = 21 \pm 2$	Bottomley W and Mortimer PI, Partition separation of tropane alkaloids, *Aust. J. Chem.*, **7**, 189–196 (1954). Cited in Perrin Bases 2921 ref. B86. NB: See Hyoscyamine for details.

(*continued*)

No.	Compound Name	pK_a value(s)	Data quality	Ionization type	Method	Conditions	Comments and Reference(s)
1271	Seglitide ($C_{44}H_{56}N_8O_7$) 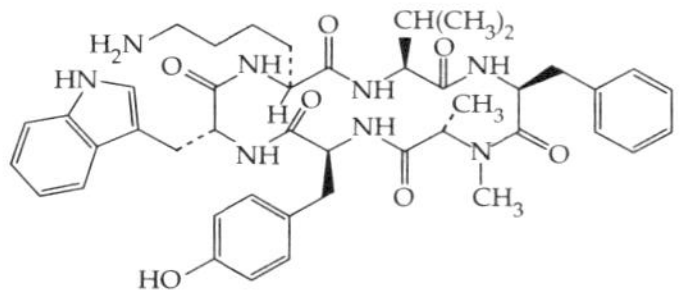 Cyclic (*N*-methyl-L-alanyl-L-tyrosyl-D-tryptophyl-L-lysyl-L-valyl-L-phenyl-alanyl)	9.7	U	+H	kinetic	H_2O	Krishnamoorthy R and Mitra AK, Kinetics and mechanism of degradation of a cyclic hexapeptide (somatostatin analog) in aqueous solution, *Pharm. Res.*, **9**, 1314–1320 (1992). "The mechanism and kinetics of degradation of seglitide (L-363586) were studied in aqueous solution. . . .. The pH rate profile exhibited specific acid catalysis at a pH less than 3 and base catalysis above pH 10.5. The kinetic pK_a was 9.7. This pK_a was attributed to the tyrosine residue. . . .. It was concluded that pH and temperature have a significant effect on seglitide degradation in aqueous solution, with different mechanisms of degradation under acidic and alkaline conditions."
1272	Selegiline (Deprenyl) ($C_{13}H_{17}N$) 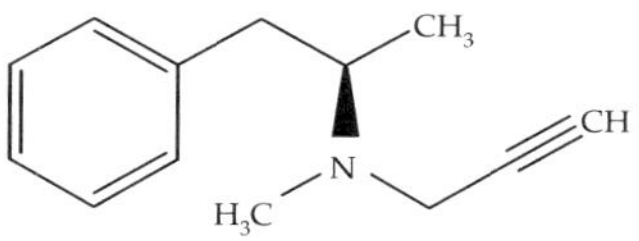	7.38 ± 0.01	U	+H	Potentiometric	H_2O $t = 25.0$ $I = 0.1$ (NaCl)	Takacs-Novak K and Avdeef A, Interlaboratory study of log P determination by shake-flask and potentiometric methods, *J. Pharm. Biomed. Anal.*, **14**, 1405–1413 (1996). NB: See Acetaminophen for further details. Also reported pK_a = 7.42 ± 0.01 at $I = 0.15$ (KCl).
		7.48 ± 0.01	U	+H	Potentiometric	H_2O $t = 25.0$ $I = 0.15$ (KCl)	Sirius Technical Application Notes, vol. **2**, pp. 26–27 (1995). Sirius Analytical Instruments Ltd., Forest Row, East Sussex, RH18 5DW, UK. NB: Concentration of analyte, 0.46–0.70 mM.
1273	Seperidol (clofluperol) ($C_{22}H_{22}ClF_4NO_2$) 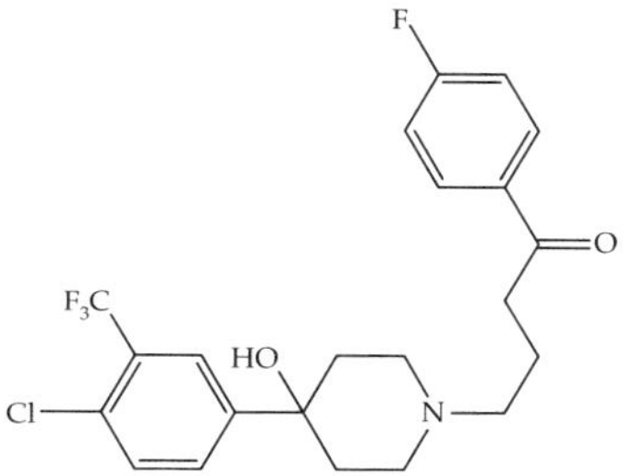	8.43	U	+H	Potentiometric	H_2O $t = 25$	Peeters JJ, Determination of ionization constants in mixed aqueous solvents of varying composition by a single titration, *J. Pharm. Sci.*, **67**, 127–129 (1978). NB: See Cinnarizine for details. Reported literature value pK_a = 8.44 (potentiometric data extrapolated to 100% water).

1274	Sertraline ($C_{17}H_{17}Cl_2N$) 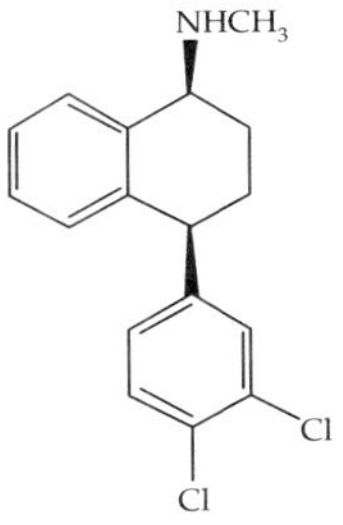	9.48 ± 0.04	A	+H	Potentiometric	H_2O $I > 0$	Johnson BM and Chang P-TL, Sertraline Hydrochloride, *APDS*, **24**, 443–486 (1996). "The pK_a of sertraline hydrochloride as determined by potentiometric titration in ethanol:water (1:1, v/v) was found to be 8.5. The pK_a determined in methanol:water (40:60, v/v) was 8.6. Titration of sertraline hydrochloride in water was carried out in the presence of sodium chloride and a measured excess of HCl. Titration with NaOH provided a curve that was evaluated by the method of Clarke and Cahoon (Clarke FH, Cahoon NM, *J. Pharm. Sci.*, **8**, 611–620 (1987)). The pK_a in water calculated by this method was 9.48 ± 0.04."
1275	Silybin ($C_{25}H_{22}O_{10}$) 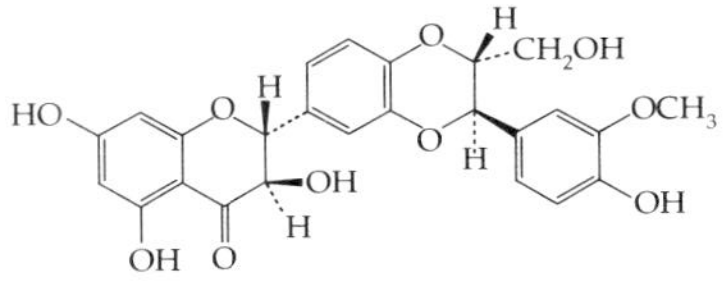	6.42 ± 0.04	U	−H	Spectro		Koch H, Demeter T and Zinsberger G, pK_a und Ionisationsprofil von Silybin, Silydianin und Silychristin, *Arch. Pharm.* (Weinheim), **313**, 565–571 (1980). NB: Reported also Silycristin, 6.41 ± 0.05; Silydianin, 6.37 ± 0.06.
1276	Soluflazine ($C_{28}H_{30}Cl_2FN_5O_2$) 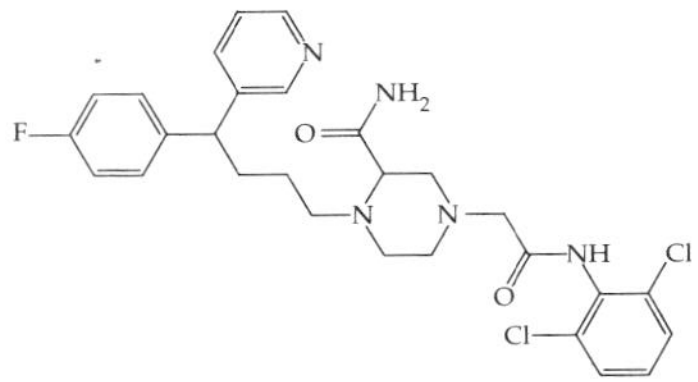	5.09 ± 0.09 5.74 ± 0.08	U U	+H +H	Potentiometric	H_2O $t = 25.0 \pm 0.1$ $I = 0.1$ N_2 atmosphere	IJzerman AP, Limiting solubilities and ionization constants of sparingly soluble compounds: Determination from aqueous potentiometric data only, *Pharm. Res.*, **5**(12), 772–775 (1988). NB: Glass electrode; KOH titrant; So = 343 ± 7 μM.

(continued)

No.	Compound Name	pK_a value(s)	Data quality	Ionization type	Method	Conditions	Comments and Reference(s)
1277	Sorbic acid ($C_6H_8O_2$)	4.51 ± 0.05 (trans-trans)	A	–H	Potentiometric	H_2O $t = 25$	Mansfield GH and Whiting MC, Investigations on acetylenic compounds. LIII. The relative strengths of some unsaturated carboxylic acids, *JCS*, 4761–4764 (1956). Cited in Serjeant, Dempsey, Ionization Constants of Organic Acids in Water, IUPAC, Butterworths, London (1979). Values for numerous other unsaturated carboxylic acids were also given.
		4.49 ± 0.05 (cis-cis)	A	–H		$t = 0.1$ (NaCl)	
		4.8	U	–H			Kendall J, Electrical conductivity and ionization constants of weak electrolytes in aqueous solution, *in* Washburn EW, Editor-in-Chief, *International Critical Tables*, vol. **6**, McGraw-Hill, NY, 259–304 (1929).
1278	Sorbitol ($C_6H_{14}O_6$) 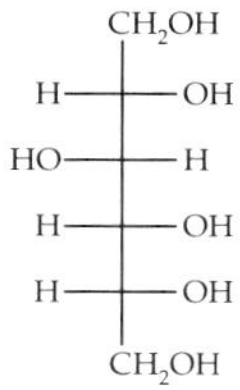	13.57	A	–H	Potentiometric	H_2O $t = 18$	Thamsen J, The acidic dissociation constants of glucose, mannitol and sorbitol, as measured by means of the hydrogen electrode and the glass electrode at 0° and 18 °C, *Acta Chem. Scand.*, **6**, 270–284 (1952).
		14.14	A	–H		$t = 0$	
1279	Sotalol ($C_{12}H_{20}N_2O_3S$) 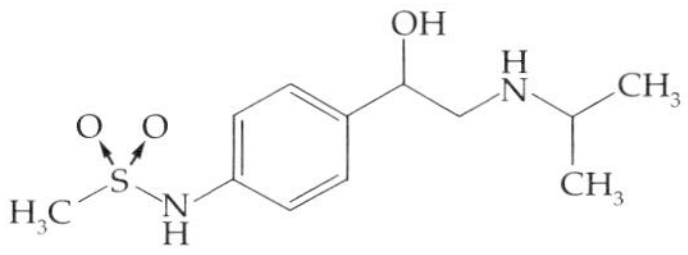	8.28 ± 0.01	A	–H	Potentiometric	H_2O (extrap) $t = 25 \pm 1$ I undefined Ar atmosphere	Sirius Technical Application Notes, vol. **2**, pp. 83–84, 167–168 (1995). Sirius Analytical Instruments, Ltd., Forest Row, East Sussex, RH18 5DW, UK.The same results were given by Cheymol G, Poirier J-M, Carrupt PA, Testa B, Weissenburger J, Levron J-C, Snoeck E, Pharmacokinetics of β-adrenoceptor blockers in obese and normal volunteers, *Br. J. Clin. Pharmacol.*, **43**, 563–570 (1997).
		9.72 ± 0.01	A	+H			
		8.30	A	–H	Potentiometric	H_2O (extrap) $t = 25 \pm 1$ $I = 0.20$ (KCl)	Garrett ER and Schnelle K, Separation and spectrofluorometric assay of the β-adrenergic blocker sotalol from blood and urine, *J. Pharm Sci.*, **60**, 833–9 (1971). NB: Also reported p$K_{a1} = 8.35$; p$K_{a2} = 9.80$ from spectrophotometric measurements ($\lambda = 248$ nm) under the same conditions.
		9.80	U	+H			
1280	Stearic acid ($C_{18}H_{36}O_2$) $CH_3(CH_2)_{16}COOH$	5.75	U	–H	Potentiometric	H_2O $t = 35$	Johns WH and Bates TR, Quantification of the binding tendencies of cholestyramine II. Mechanism of interaction with bile salts and fatty acid salt anions, *J. Pharm. Sci.*, **59**, 329–333 (1970).

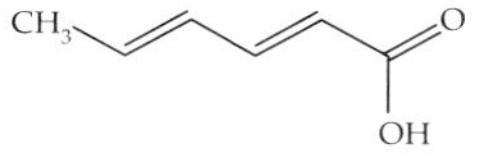

No.	Compound	pKa	Data quality	Assignment	Method	Conditions	Reference
1281	Streptomycin A ($C_{21}H_{39}N_7O_{12}$) 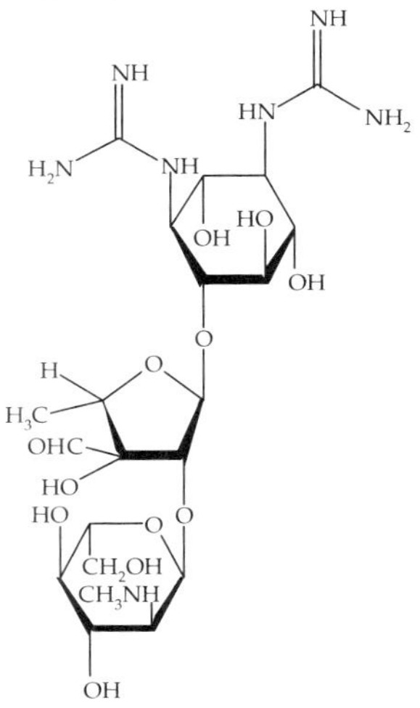	7.84 11.54 >12	U U U	+H +H +H	Potentiometric	H_2O $t = 25$ $I = 0.1$	Barbosa CM, Lima JL and Roque Da Silva MM, Determination of the acidity constants of streptomycin, *Rev. Port. Farm.*, **45**, 130–136 (1995). "The acidity constants of streptomycin were determined at 25 °C by potentiometry with an automatic system in aqueous medium at a constant ionic strength of 0.1 M. The acidity constants were calculated by algorithms based on material balance equations. Experimental data were analyzed using Gran's method. Values of 7.84 (pK_{a1}), 11.54 (pK_{a2}), and >12 (pK_{a3}) were found."
1282	Streptomycin derivatives	8.90	A	+H	Potentiometric	H_2O $t = 23 \pm 1$ $I = 0.01$	Inouye S, Prediction of the pK_a values of amino sugars, *Chem. Pharm. Bull.* Jpn., **16**, 1134–1137 (1968). CA 69:106993. Cited in Perrin suppl. nos. 7773–7776 Ref. I13. NB: Used a glass electrode in a cell with liquid junction potentials to measure pH of solutions at half-neutralization.
1283	Streptovitacin A dehydration product	10.8	U	+H	Spectro ($\lambda = 242$, 297 nm)	H_2O $t = 25$	Ritschel: Notari RE and Caiola SM, Catalysis of streptovitacin A dehydration: Kinetics and mechanism, *J. Pharm. Sci.*, **58**, 1203–1208 (1969). NB: The structure is a substituted 2,4,6-trialkylphenol.
1284	Strychnine ($C_{21}H_{22}N_2O_2$) 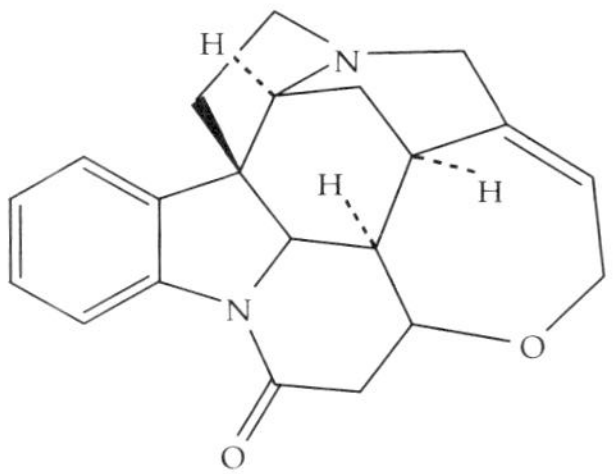	8.26	A	+H	Potentiometric	H_2O $t = 25$ $c = 10^{-4}$–10^{-5}	Everett AJ, Openshaw HT, Smith GF, Constitution of aspidospermine. III., *J. Chem. Soc.*, 1120–1123 (1957), CA 51:56814. Cited in Perrin Bases 2967 ref. E35. NB: Used glass electrode in cells with liquid junction potential.
		2.50 8.20	U U	+H +H	Spectro	H_2O $t = 15$ $c = 0.002$ to 0.02	Kolthoff IM, The dissociation constants, solubility product and titration of alkaloids, *Biochem. Z.*, **162**, 289–353 (1925). Cited in Perrin Bases no. 2967 K47. See Aconitine for details.

(continued)

No.	Compound Name	pK_a value(s)	Data quality	Ionization type	Method	Conditions	Comments and Reference(s)
1285	Succinic acid ($C_4H_6O_4$) 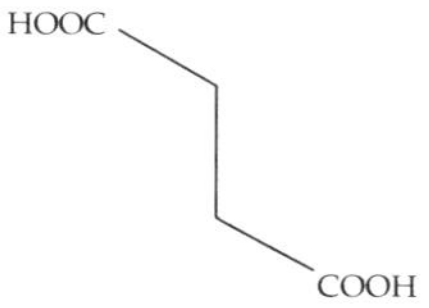	4.207 5.638 4.19 5.57	R R U U	–H –H –H –H	Potentiometric	H_2O $t = 25.0$ $I = 0.00$ H_2O $t = 25$	Pinching GD and Bates RG, First dissociation constant of succinic acid from 0° to 50° and related thermodynamic quantities, *J. Res. Nat. Bur. Stand.*, **45**, 444–449 (1950); Pinching GD and Bates RG, Second dissociation constant of succinic acid from 0° to 50° and related thermodynamic quantities, *J. Res. Nat. Bur. Stand.*, **45**, 322–328 (1950). NB: Parrott EL, *Pharmaceutical Technology: Fundamental Pharmaceutics*, Burgess Publishing Co., Minneapolis MN, p. 218 (1970).
1286	Succinic acid	4.21 5.64	U U	–H –H		H_2O	W&G: Pitman IH, Paulssen RB and Higuchi T, Interaction of acetic anhydride with di- and tricarboxylic acids in aqueous solution, *J. Pharm. Sci.*, **57**, 239–245 (1968). NB: Appears to be the Pinching and Bates data after rounding.
1287	Succinic acid	4.12 5.45	U U	–H –H	Potentiometric	H_2O $t = 23.0$	Clarke FH and Cahoon NM, Ionization constants by curve-fitting: Determination of partition and distribution coefficients of acids and bases and their ions, *J. Pharm. Sci.*, **76**(8), 611–620 (1987). NB: See Benzoic acid for further details.
1288	Succinimide ($C_4H_5NO_2$) 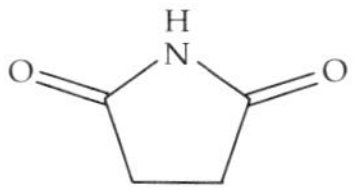	9.6	U	–H			W&G: Connors KA, *Textbook of Pharmaceutical Analysis*, 1st Edn., Wiley, NY, p. 475 (1967). NB: The value was given without references or experimental detail.
1289	Sucrose ($C_{12}H_{22}O_{11}$) 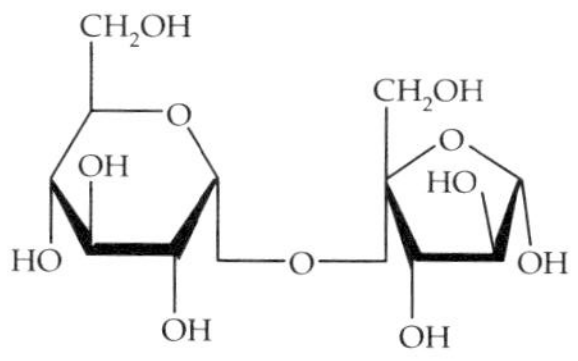	12.75–12.80	U	–H	Potentiometric	H_2O $t = 25.0 \pm 0.1$ $I < 0.0005$	Woolley EM, Tomkins J and Hepler LG, Ionization constants for very weak organic acids in aqueous solution and apparent ionization constants for water in aqueous organic mixtures, *J. Solution Chem.*, **1**, 341–351 (1972). NB: See Dextrose for details.

No.	Compound	pK_a	Quality	Ionization	Method	Conditions	Reference / Comments
1290	Sulfabenz ($C_{12}H_{12}N_2O_2S$) 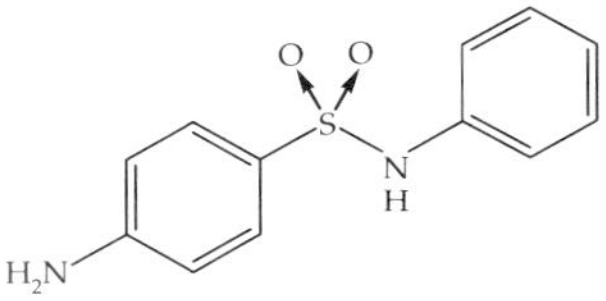	10.94	U				Ritschel: Merck 8, p. 994.
1291	Sulfacetamide ($C_8H_{10}N_2O_3S$)	GLpK_a:			Spectro	H_2O	Tam KY and Takacs-Novac K, Multi-wavelength spectrophotometric determination of acid dissociation constants, *Anal. Chim. Acta*, **434**, 157–167 (2001).
		1.76 ± 0.04	A	−H		$t = 25$	NB: See Clioquinol for details.
		5.22 ± 0.01	A	+H		$I = 0.15$ (KCl)	
		A&S:				Ar atmosphere	
		1.95 ± 0.11	U	−H			
		5.30 ± 0.05	A	+H			
1292	Sulfacetamide	5.27	U	−H	CE/pH (+ve ion mode)	H_2O $t = 25$ $I = 0.025$	Wan H, Holmen AG, Wang Y, Lindberg W, Englund M, Nagard MB and Thompson RA, High-throughput screening of pK_a values of pharmaceuticals by pressure-assisted capillary electrophoresis and mass spectrometry, *Rapid Commun. Mass Spectrom.*, **17**, 2639–2648 (2003). NB: Reported a predicted value (ACD Labs) of 5.62.
1293	Sulfadiazine ($C_{10}H_{10}N_4O_2S$)	2.21	U	+H	Spectro	H_2O $t = 25$ $I = 0.5$ (NaCl)	Stober H and DeWitte W, Sulfadiazine, *APDS*, **11**, 523–546 (1982).
		2.00	U	+H			
		6.48	U	−H			
		6.28	U	−H	soly	H_2O $I = 0.1$	
		6.35	U	−H	Potentiometric	H_2O $t = 20$ $I = 0.1$ (KCl)	

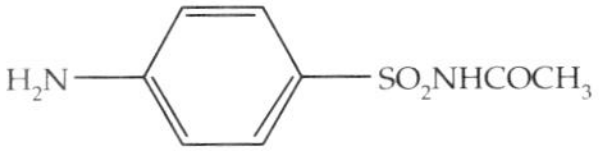

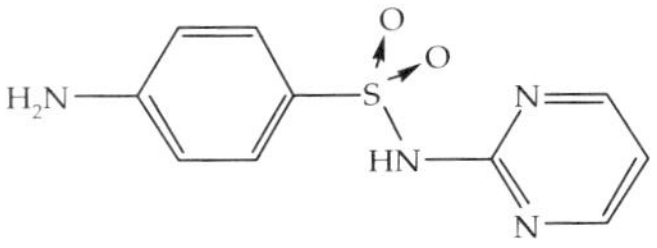

"Sulfadiazine is an ampholyte, and in aqueous solutions can exist in the protonated, neutral and anionic forms. Salvesen and Schroder-Neilsen (24) . . . reported a pK_a value of 2.21 for the anilinium ion associated with sulfadiazine. These authors also stated that the pK_a of the pyrimidinium ion of 2-sulfanilamidopyrimidines is less than zero. . . ..

Reported pK_a values of sulfadiazine

Group		method
-NH_3	SO_2-N-H	
2.21	-	spectro 0.5M NaCl. $t = 25$ °C (24)
2.00	6.48	not stated (27)
-	6.28	Solubility, $\mu = 0.1$ M (25)
-	6.35	Titrimetry, $\mu = 0.1$ M (KCl), $t = 20$ °C (26)

24. Salvesen B and Schroder-Nielson M, *Medd. Norsk. Farm. Selskap.*, **32**, 87–96 (1971).
25. Krebs HA and Speakman JC, *BMJ*, **1**, 47–50 (1946).
26. Willi AV and Meier W, *Helv. Chim. Acta*, **39**, 54–56 (1956). (see separate entry).
27. Koizumi T, Arita T and Kakemi K, *Chem. Pharm. Bull.*, **12**, 413–420 (1964)."

(continued)

No.	Compound Name	pK_a value(s)	Data quality	Ionization type	Method	Conditions	Comments and Reference(s)
1294	Sulfadiazine	3.93 6.42	U U	+H −H	soly	H_2O $t = 37$ $I = 0.15$	Zimmerman I, Determination of pK_a values from solubility data, *Int. J. Pharm.*, **13**, 57–65 (1983). NB: See Pyrazolic acid for details.
1295	Sulfadiazine	6.40 ± 0.06	U	−H	Potentiometric	H_2O $I = 0.10$	Krebs HA and Speakman JC, Dissolution constant, solubility and the pH value of the solvent, *J. Chem. Soc.*, 593–595 (1945).
1296	Sulfadiazine (2-sulfanilamido-pyrimidine)	6.35	U	−H	Potentiometric	H_2O $t = 20$ $I = 0.1$ (KCl)	Willi AV and Meier W, 6. Die Aciditatskonstanten von Benzolsulfonamiden mit heterocyclischer Amin-Komponente (The acidity constants for benzenesulfonamides with heterocyclic amine components), *Helv. Chim. Acta*, **39**, 54–56 (1956). NB: See Sulfapyridine.
1297	Sulfadiazine	2.00	U	+H		H_2O (extrap) $t = 24 \pm 1$ $I \sim 0.002$	Chatten LG and Harris LE, Relationship between $pK_b(H_2O)$ of organic compounds and $E_{1/2}$ values in several nonaqueous solvents, *Anal. Chem.*, **34**, 1495–1501 (1962).
1298	Sulfadiazine	2.00 6.48	U U	+H −H	Potentiometric	H_2O $t = 25$ $I = 0.05$	Bell PH and Roblin, RO, Jr., Studies in chemotherapy. VII. A theory of the relation of structure to activity of sulfanilamide-type compounds, *JACS*, **64**, 2905–2917 (1942). NB: See *p*-Aminobenzoic acid for details.
1299	Sulfadiazine	6.28	U	−H		H_2O $t = 37$	Ballard BE and Nelson E, Physicochemical properties of drugs that control absorption rate after subcutaneous implantation, *JPET*, **135**, 120–127 (1962). NB: Secondary source W&G quoted $pK_a = 6.5$, not 6.28, in citing this ref.
1300	Sulfadiazine	6.34	U	−H	Spectro	H_2O $t = 25.0 \pm 0.5$ $I = 0.2$	Elofsson R, Nilsson SO and Agren A, Complex formation between macromolecules and drugs. IV, *Acta Pharm. Suec.*, **7**, 473–482 (1970). NB: See Sulphanilamide for details.
1301	Sulfadiazine, N^4-acetyl ($C_{12}H_{12}N_4O_3S$) 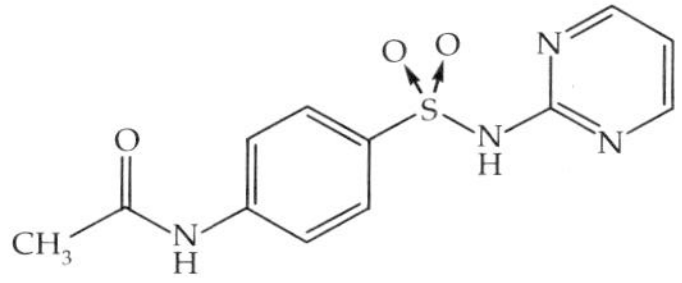	6.1	U	−H	Potentiometric	H_2O $t = 25$	Scudi JW and Plekss OJ, Chemotherapeutic activity of some sulfapyridine-1-oxides, *Proc. Soc. Exptl. Biol. Med.*, **97**, 639–641 (1958). NB: See Sulfapyridine-1-oxide.

1302	Sulfadimethoxine ($C_{12}H_{14}N_4O_4S$) 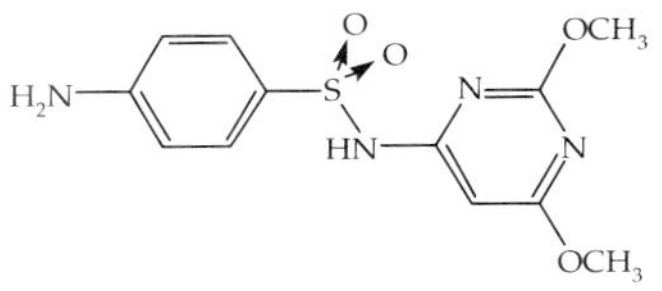	5.94	U	–H	Spectro	H_2O $t = 25.0 \pm 0.5$ $I = 0.2$	Elofsson R, Nilsson SO and Agren A, Complex formation between macromolecules and drugs. IV, *Acta Pharm. Suec.*, **7**, 473–482 (1970). NB: See Sulphanilamide for details.
1303	Sulfadimethoxytriazine ($C_{11}H_{13}N_5O_4S$) 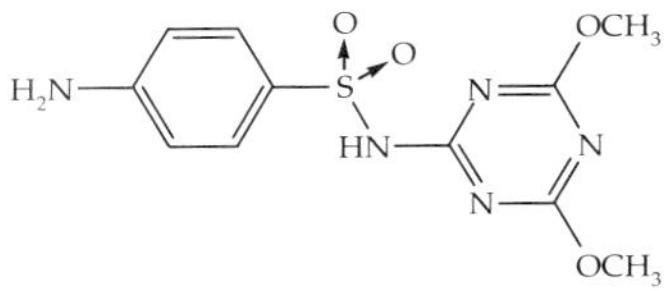	5.0	VU	–H	Potentiometric	H_2O $t = 25$	Scudi JW and Plekss OJ, Chemotherapeutic activity of some sulfapyridine-1-oxides, *Proc. Soc. Exptl. Biol. Med.*, **97**, 639–641 (1958). NB: See Sulfapyridine-1-oxide.
1304	Sulfadimidine ($C_{12}H_{14}N_4O_2S$) 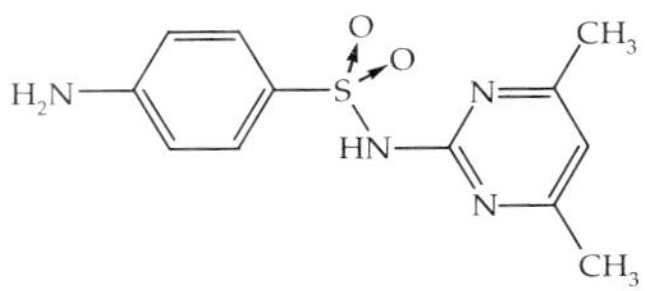	7.59	U	–H	Spectro	H_2O $t = 25.0 \pm 0.5$ $I = 0.2$	Elofsson R, Nilsson SO and Agren A, Complex formation between macromolecules and drugs. IV, *Acta Pharm. Suec.*, **7**, 473–482 (1970). NB: See Sulphanilamide for details.
1305	Sulfadimidine (2-sulfanilamido-4,6-dimethylpyrimidine)	2.36 7.37	U U	+H –H	Potentiometric	H_2O $t = 25$ $I = 0.05$	Bell PH and Roblin, RO, Jr., Studies in chemotherapy. VII. A theory of the relation of structure to activity of sulfanilamide-type compounds, *JACS*, **64**, 2905–2917 (1942). NB: See *p*-Aminobenzoic acid for details.
1306	Sulfadimidine (2-sulfanilamido-4,6-dimethylpyrimidine)	7.51	U	–H	Potentiometric	H_2O $t = 20$ $I = 0.1$ (KCl)	Willi AV and Meier W, 6. Die Aciditatskonstanten von Benzolsulfonamiden mit heterocyclischer Amin-Komponente (The acidity constants for benzenesulfonamides with heterocyclic amine components), *Helv. Chim. Acta*, **39**, 54–56 (1956). NB: See Sulfapyridine.

(continued)

No.	Compound Name	pK_a value(s)	Data quality	Ionization type	Method	Conditions	Comments and Reference(s)
1307	Sulfadimidine (sulfamethazine) ($C_{12}H_{14}N_4O_2S$) 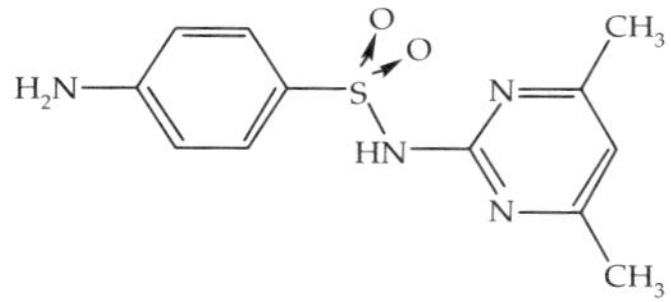	2.65 ± 0.2 7.4 ± 0.2	U U	+H −H		H_2O *t* undefined *I* undefined	Papastephanou C and Frantz M, Sulfamethazine, *APDS*, **7**, 401–422 (1978). Perlman S, The Squibb Institute, personal communication.
1308	Sulfadimidine (sulfamethazine)	7.38	U	−H		H_2O $t = 37$	Ballard BE and Nelson E, Physicochemical properties of drugs that control absorption rate after subcutaneous implantation, *JPET*, **135**, 120–127 (1962).
1309	Sulfaethidole ($C_{10}H_{12}N_4O_2S$) 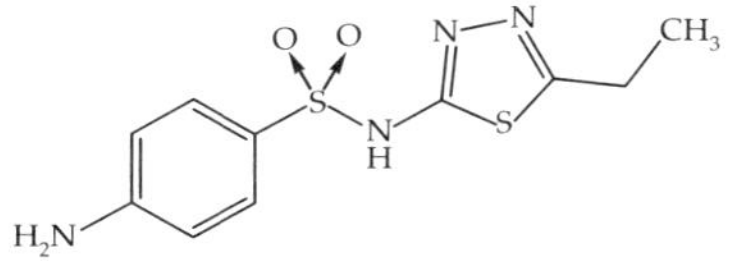	5.36	U	−H		H_2O $t = 37$	Ballard BE and Nelson E, Physicochemical properties of drugs that control absorption rate after subcutaneous implantation, *JPET*, **135**, 120–127 (1962).
1310	Sulfafurazole ($C_{11}H_{13}N_3O_3S$) 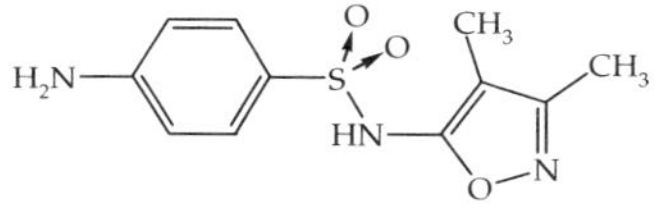	4.90	U	−H	Spectro	H_2O $t = 25.0 \pm 0.5$ $I = 0.2$	Elofsson R, Nilsson SO and Agren A, Complex formation between macromolecules and drugs. IV, *Acta Pharm. Suec.*, **7**, 473–482 (1970). NB: See Sulphanilamide for details.
1311	Sulfaguanidine ($C_7H_{10}N_4O_2S$) 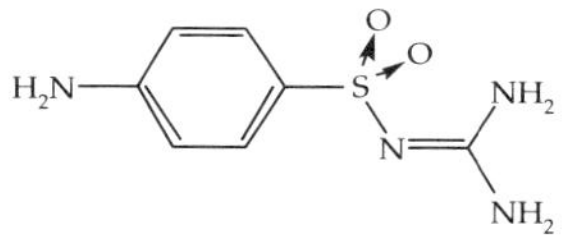	2.37	U	+H		H_2O $t = 37$	Ballard BE and Nelson E, Physicochemical properties of drugs that control absorption rate after subcutaneous implantation, *JPET*, **135**, 120–127 (1962). NB: From pK_b = 11.25 at 37 °C, where pK_w = 13.621. Secondary source W&G gave pK_a = 2.8 in citing this ref.

1312	Sulfalene ($C_{11}H_{12}N_4O_3S$) 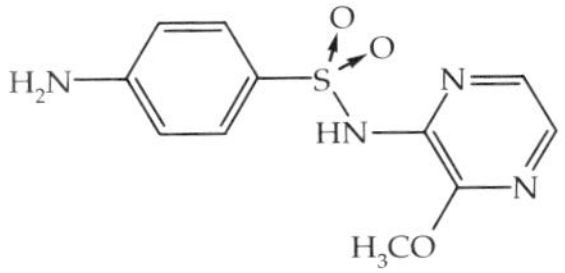	6.20	U	–H	Spectro	H_2O $t = 25.0 \pm 0.5$ $I = 0.2$	Elofsson R, Nilsson SO and Agren A, Complex formation between macromolecules and drugs. IV, *Acta Pharm. Suec.*, **7**, 473–482 (1970). NB: See Sulphanilamide for details.
1313	Sulfamerazine ($C_{11}H_{12}N_4O_2S$) 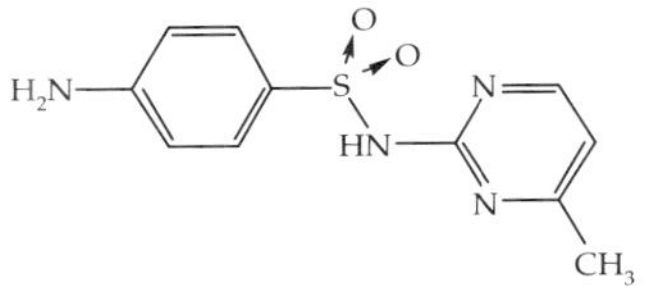	2.29	U	+H	Spectro, Potentiometric	H_2O $t = 24.0$ $I = 0.5$ (NaCl)	Woolfenden RDG, Sulphamerazine, *APDS*, **6**, 515–577 (1977). Koizumi T, Arita T and Kakemi K, *Chem. Pharm. Bull.*, **12**, 413–420 (1964). NB: reported $pK_a = 2.26$.
		7.00 ± 0.05	U	–H	soly	H_2O $t = 38$ $I = 0.1$ (NaCl)	Krebs HA and Speakman JC, *Br. Med. J.*, **1**, 47 (1946). NB: $S_o = 41$ mg/100 ml). Sjogren B, Ortenblad B, *Acta Chem. Scand.*, **1**, 605–618 (1947).
1314	Sulfamerazine (2-sulfanilamido-4-methylpyrimidine)	6.84	U	–H	Potentiometric	H_2O $t = 20$ $I = 0.1$ (KCl)	Willi AV and Meier W, 6. Die Aciditatskonstanten von Benzolsulfonamiden mit heterocyclischer Amin-Komponente (The acidity constants for benzenesulfonamides with heterocyclic amine components), *Helv. Chim. Acta*, **39**, 54–56 (1956). NB: See Sulfapyridine.
1315	Sulfamerazine	6.95	U	–H		H_2O $t = 37$	Ballard BE and Nelson E, Physicochemical properties of drugs that control absorption rate after subcutaneous implantation, *JPET*, **135**, 120–127 (1962).
1316	Sulfamerazine	6.90	U	–H	Spectro	H_2O $t = 25.0 \pm 0.5$ $I = 0.2$	Elofsson R, Nilsson SO and Agren A, Complex formation between macromolecules and drugs. IV, *Acta Pharm. Suec.*, **7**, 473–482 (1970). NB: See Sulphanilamide for details.
1317	Sulfamethine ($C_{11}H_{12}N_4O_2S$) 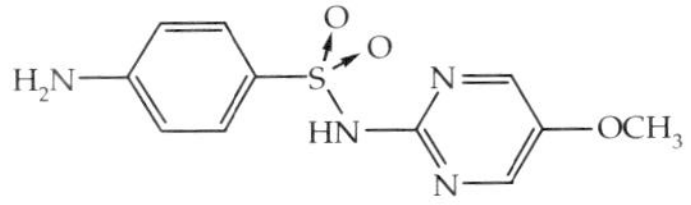	6.66	U	–H	Spectro	H_2O $t = 25.0 \pm 0.5$ $I = 0.2$	Elofsson R, Nilsson SO and Agren A, Complex formation between macromolecules and drugs. IV, *Acta Pharm. Suec.*, **7**, 473–482 (1970). NB: See Sulphanilamide for details.

(continued)

No.	Compound Name	pK_a value(s)	Data quality	Ionization type	Method	Conditions	Comments and Reference(s)
1318	Sulfamethizole (sulfamethylthiadiazole) ($C_9H_{10}N_4O_2S_2$) 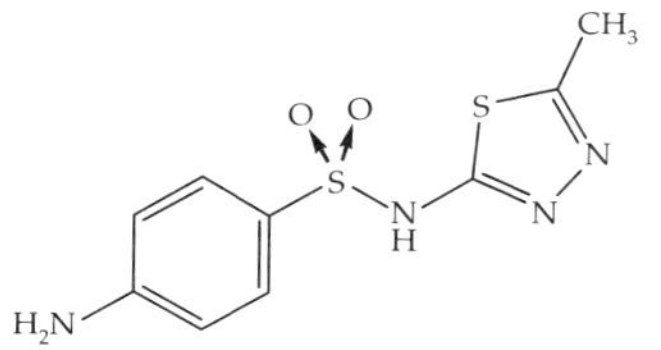	5.4	VU	–H	Potentiometric	H_2O $t = 25$	Scudi JW and Plekss OJ, Chemotherapeutic activity of some sulfapyridine-1-oxides, *Proc. Soc. Exptl. Biol. Med.*, **97**, 639–641 (1958). NB: See Sulfapyridine-1-oxide.
1319	Sulfamethizole	5.42	U	–H		H_2O $t = 37$	Ballard BE and Nelson E, Physicochemical properties of drugs that control absorption rate after subcutaneous implantation, *JPET*, **135**, 120–127 (1962).
1320	Sulfamethizole	5.28	U	–H	Spectro	H_2O $t = 25.0 \pm 0.5$ $I = 0.2$	Elofsson R, Nilsson SO and Agren A, Complex formation between macromolecules and drugs. IV, *Acta Pharm. Suec.*, **7**, 473–482 (1970). NB: See Sulphanilamide for details.
1321	Sulfamethizole, N^4-acetyl ($C_{11}H_{12}N_4O_3S_2$) 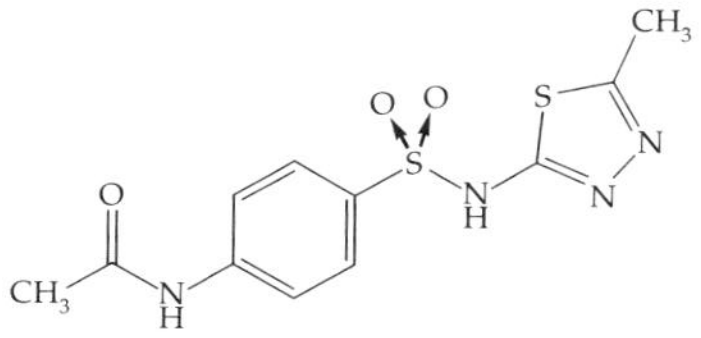	5.2	VU	–H	Potentiometric	H_2O $t = 25$	Scudi JW and Plekss OJ, Chemotherapeutic activity of some sulfapyridine-1-oxides, *Proc. Soc. Exptl. Biol. Med.*, **97**, 639–641 (1958). NB: See Sulfapyridine-1-oxide.
1322	Sulfamethomidine ($C_{12}H_{14}N_4O_3S$) 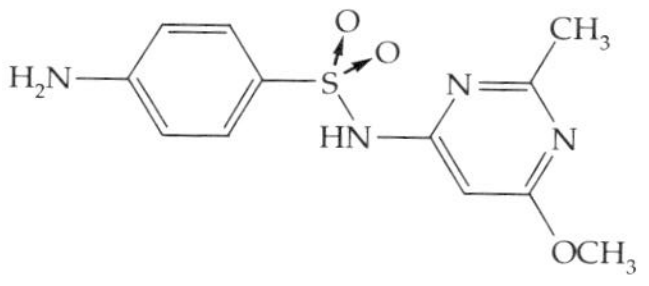	7.06	U	–H	Spectro	H_2O $t = 25.0 \pm 0.5$ $I = 0.2$	Elofsson R, Nilsson SO and Agren A, Complex formation between macromolecules and drugs. IV, *Acta Pharm. Suec.*, **7**, 473–482 (1970). NB: See Sulphanilamide for details.

No.	Compound	pKa	Data quality	Ionization	Method	Conditions	Reference / Notes
1323	Sulfamethoxydiazine (also called sulfameter)	6.8	U	–H			N&K; Merck 9.
1324	Sulfamethoxypyridazine ($C_{11}H_{12}N_4O_3S$) 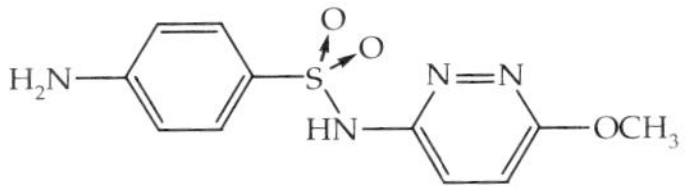	7.19	U	–H	Potentiometric	H_2O $t = 25.0 \pm 0.5$ $I = 0.2$	Elofsson R, Nilsson SO and Agren A, Complex formation between macromolecules and drugs. IV, *Acta Pharm. Suec.*, **7**, 473–482 (1970). NB: See Sulphanilamide for details.
1325	Sulfamethoxypyridazine	7.2	U	–H	Potentiometric	H_2O $t = 25$	Scudi JW and Plekss OJ, Chemotherapeutic activity of some sulfapyridine-1-oxides, *Proc. Soc. Exptl. Biol. Med.*, **97**, 639–641 (1958). NB: See Sulfapyridine-1-oxide.
1326	Sulfamethoxypyridazine, N^4-acetyl ($C_{13}H_{14}N_4O_3S$) 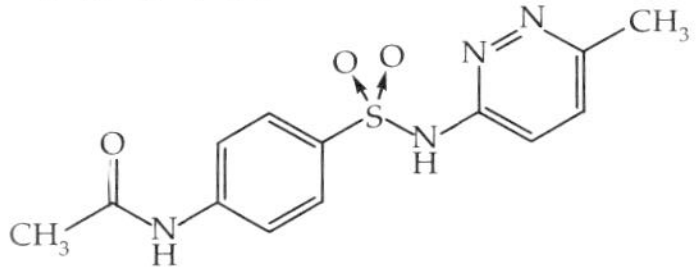	6.9	U	–H	Potentiometric	H_2O $t = 25$	Scudi JW and Plekss OJ, Chemotherapeutic activity of some sulfapyridine-1-oxides, *Proc. Soc. Exptl. Biol. Med.*, **97**, 639–641 (1958). NB: See Sulfapyridine-1-oxide.
1327	Sulfamilyl-3,4-xylamide ($C_{15}H_{16}N_2O_3S$) 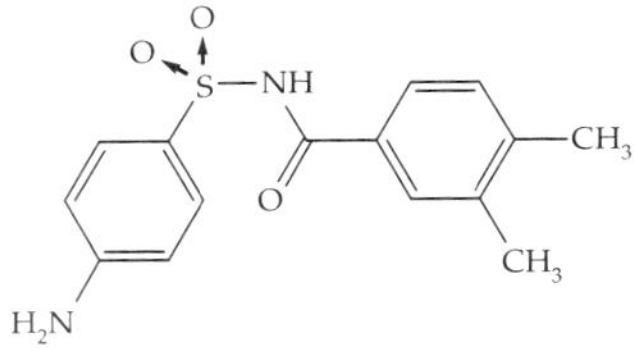	4.37	U	–H	Spectro	H_2O $t = 27 \pm 1$ $I = 0.2$	Yoshioka M, Hamamoto K and Kubota T, Acid dissociation constants of sulfanilamides and substituent effects on the constants, *Yakugaku Zasshi*, **84**, 90–93 (1964). NB: Note significant acid-strengthening from adjacent carbonyl group. Reported pK_a values for 10 further analogues. Cited in Ritschel.
1328	Sulfanilamide ($C_6H_8N_2O_2S$) 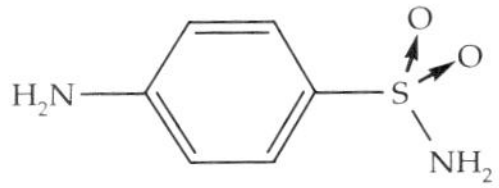	10.65	U	–H	Spectro	H_2O $t = 25.0 \pm 0.5$ $I = 0.2$	Elofsson R, Nilsson SO and Agren A, Complex formation between macromolecules and drugs. IV, *Acta Pharm. Suec.*, **7**, 473–482 (1970). "The spectrophotometric method proposed by Bates and Schwarzenbach [10, Bates RG and Schwarzenbach G, Die Bestimmung thermodynamischer Aciditatskonstanten, *Helv. Chim. Acta*, **37**, 1069–1079 (1954)] was used . . . ."

(continued)

No.	Compound Name	pK_a value(s)	Data quality	Ionization type	Method	Conditions	Comments and Reference(s)
1329	Sulfanilamide	2.36 10.43	U U	+H −H	−H by Potentio-metric +H by potentio and Condu-ctance	H_2O $t = 25$ $I = 0.05$	Bell PH and Roblin, RO, Jr., Studies in chemotherapy. VII. A theory of the relation of structure to activity of sulfanilamide-type compounds, *JACS*, **64**, 2905–2917 (1942). NB: See *p*-Aminobenzoic acid for details.
1330	Sulfanilamide	10.58	U	−H		H_2O $t = 37$	Ballard BE and Nelson E, Physicochemical properties of drugs that control absorption rate after subcutaneous implantation, *JPET*, **135**, 120–127 (1962).
1331	Sulfanilamide, N^1-acetyl ($C_8H_{10}N_2O_3S$) 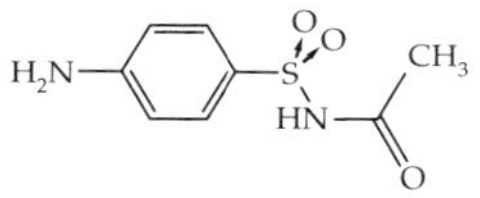	1.78 5.38	U U	+H −H	Potentiometric	H_2O $t = 25$ $I = 0.05$	Bell PH and Roblin, RO, Jr., Studies in chemotherapy. VII. A theory of the relation of structure to activity of sulfanilamide-type compounds, *JACS*, **64**, 2905–2917 (1942). NB: See *p*-Aminobenzoic acid for details.
1332	Sulfanilamide, N^1-*p*-aminobenzoyl ($C_{13}H_{13}N_3O_3S$) 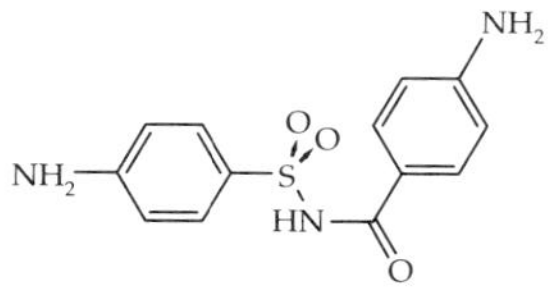	1.48 2.43 5.20	U U U	+H +H −H	Potentiometric	H_2O $t = 25$ $I = 0.05$	Bell PH and Roblin, RO, Jr., Studies in chemotherapy. VII. A theory of the relation of structure to activity of sulfanilamide-type compounds, *JACS*, **64**, 2905–2917 (1942). NB: See *p*-Aminobenzoic acid for details.
1333	Sulfanilamide, N^1-benzoyl ($C_{13}H_{12}N_2O_3S$) 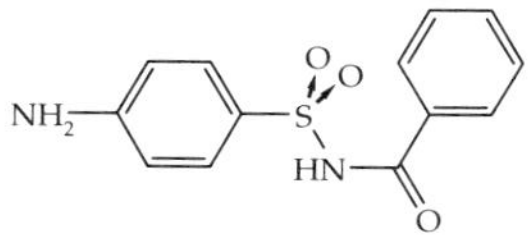	1.78 4.57	U U	+H −H	Potentiometric	H_2O $t = 25$ $I = 0.05$	Bell PH and Roblin, RO, Jr., Studies in chemotherapy. VII. A theory of the relation of structure to activity of sulfanilamide-type compounds, *JACS*, **64**, 2905–2917 (1942). NB: See *p*-Aminobenzoic acid for details.
1334	Sulfanilamide, N^1-ethylsulphonyl ($C_8H_{12}N_2O_4S_2$)	1.48 3.10	U U	+H −H	Potentiometric	H_2O $t = 25$ $I = 0.05$	Bell PH and Roblin, RO, Jr., Studies in chemotherapy. VII. A theory of the relation of structure to activity of sulfanilamide-type compounds, *JACS*, **64**, 2905–2917 (1942). NB: See *p*-Aminobenzoic acid for details.

1335	Sulfanilamide, N[1]-sulfanilyl ($C_{12}H_{13}N_3O_4S_2$)	— 2.89	 U	 −H	Potentiometric	H_2O $t = 25$ $I = 0.05$	Bell PH and Roblin, RO, Jr., Studies in chemotherapy. VII. A theory of the relation of structure to activity of sulfanilamide-type compounds, *JACS*, **64**, 2905–2917 (1942). NB: See *p*-Aminobenzoic acid for details; not sufficiently soluble in acetic acid for determination of the basic group.
1336	Sulfanilamide, N[1]-*p*-aminophenyl ($C_{13}H_{15}N_3O_2S$)	1.85 >5.0 10.22	U U U	+H +H −H	Potentiometric	H_2O $t = 25$ $I = 0.05$	Bell PH and Roblin, RO, Jr., Studies in chemotherapy. VII. A theory of the relation of structure to activity of sulfanilamide-type compounds, *JACS*, **64**, 2905–2917 (1942). NB: See *p*-Aminobenzoic acid for details.
1337	Sulfanilamide, N[1]-chloroacetyl ($C_8H_9ClN_2O_3S$)	1.60 3.79	U U	+H −H	Potentiometric	H_2O $t = 25$ $I = 0.05$	Bell PH and Roblin, RO, Jr., Studies in chemotherapy. VII. A theory of the relation of structure to activity of sulfanilamide-type compounds, *JACS*, **64**, 2905–2917 (1942). NB: See *p*-Aminobenzoic acid for details.
1338	Sulfanilamide, N[1]-furfuryl ($C_{11}H_{12}N_2O_3S$)	2.26 10.88	U U	+H −H	Potentiometric	H_2O $t = 25$ $I = 0.05$	Bell PH and Roblin, RO, Jr., Studies in chemotherapy. VII. A theory of the relation of structure to activity of sulfanilamide-type compounds, *JACS*, **64**, 2905–2917 (1942). NB: See *p*-Aminobenzoic acid for details.
1339	Sulfanilamide, N[1]-methyl ($C_7H_{10}N_2O_2S$)	2.20 10.77	U U	+H −H	Potentiometric	H_2O $t = 25$ $I = 0.05$	Bell PH and Roblin, RO, Jr., Studies in chemotherapy. VII. A theory of the relation of structure to activity of sulfanilamide-type compounds, *JACS*, **64**, 2905–2917 (1942). NB: See *p*-Aminobenzoic acid for details.
1340	Sulfanilamide, N[1]-phenyl ($C_{12}H_{12}N_2O_2S$)	2.15 9.60	U U	+H −H	Potentiometric	H_2O $t = 25$ $I = 0.05$	Bell PH and Roblin, RO, Jr., Studies in chemotherapy. VII. A theory of the relation of structure to activity of sulfanilamide-type compounds, *JACS*, **64**, 2905–2917 (1942). NB: See *p*-Aminobenzoic acid for details.
1341	Sulfanilamide, N[1]-o-tolyl ($C_{13}H_{14}N_2O_2S$)	2.04 9.96	U U	+H −H	Potentiometric	H_2O $t = 25$ $I = 0.05$	Bell PH and Roblin, RO, Jr., Studies in chemotherapy. VII. A theory of the relation of structure to activity of sulfanilamide-type compounds, *JACS*, **64**, 2905–2917 (1942). NB: See *p*-Aminobenzoic acid for details.
1342	Sulfanilamide, N[1]-m-tolyl ($C_{13}H_{14}N_2O_2S$)	2.11 9.74	U U	+H −H	Potentiometric	H_2O $t = 25$ $I = 0.05$	Bell PH and Roblin, RO, Jr., Studies in chemotherapy. VII. A theory of the relation of structure to activity of sulfanilamide-type compounds, *JACS*, **64**, 2905–2917 (1942). NB: See *p*-Aminobenzoic acid for details.
1343	Sulfanilamide, N[1]-*p*-tolyl ($C_{13}H_{14}N_2O_2S$)	2.15 9.82	U U	+H −H	Potentiometric	H_2O $t = 25$ $I = 0.05$	Bell PH and Roblin, RO, Jr., Studies in chemotherapy. VII. A theory of the relation of structure to activity of sulfanilamide-type compounds, *JACS*, **64**, 2905–2917 (1942). NB: See *p*-Aminobenzoic acid for details.
1344	Sulfanilamide, N[1],N[1]-dimethyl ($C_8H_{12}N_2O_2S$)	2.11	U	+H	Potentiometric	H_2O $t = 25$ $I = 0.05$	Bell PH and Roblin, RO, Jr., Studies in chemotherapy. VII. A theory of the relation of structure to activity of sulfanilamide-type compounds, *JACS*, **64**, 2905–2917 (1942). NB: See *p*-Aminobenzoic acid for details.

(continued)

No.	Compound Name	pK_a value(s)	Data quality	Ionization type	Method	Conditions	Comments and Reference(s)
1345	Sulfanilamide, N^1-hydroxyethyl ($C_8H_{12}N_2O_3S$)	2.30 10.92	U U	+H −H	Potentiometric	H_2O $t = 25$ $I = 0.05$	Bell PH and Roblin, RO, Jr., Studies in chemotherapy. VII. A theory of the relation of structure to activity of sulfanilamide-type compounds, *JACS*, **64**, 2905–2917 (1942). NB: See *p*-Aminobenzoic acid for details.
1346	2-Sulfanilamido-5-aminopyridine ($C_{11}H_{12}N_4O_2S$)	1.48 3.00 8.47	U U U	+H +H −H	Potentiometric	H_2O $t = 25$ $I = 0.05$	Bell PH and Roblin, RO, Jr., Studies in chemotherapy. VII. A theory of the relation of structure to activity of sulfanilamide-type compounds, *JACS*, **64**, 2905–2917 (1942). NB: See *p*-Aminobenzoic acid for details.
1347	5-Sulfanilamido-2-aminopyridine ($C_{11}H_{12}N_4O_2S$)	1.90 4.20 8.82	U U U	+H +H −H	Potentiometric	H_2O $t = 25$ $I = 0.05$	Bell PH and Roblin, RO, Jr., Studies in chemotherapy. VII. A theory of the relation of structure to activity of sulfanilamide-type compounds, *JACS*, **64**, 2905–2917 (1942). NB: See *p*-Aminobenzoic acid for details.
1348	2-Sulfanilamido-4-aminopyrimidine ($C_{10}H_{11}N_5O_2S$)	3.13 9.44	U U	+H −H	Potentiometric	H_2O $t = 25$ $I = 0.05$	Bell PH and Roblin, RO, Jr., Studies in chemotherapy. VII. A theory of the relation of structure to activity of sulfanilamide-type compounds, *JACS*, **64**, 2905–2917 (1942). NB: See *p*-Aminobenzoic acid for details.
1349	2-Sulfanilamido-5-bromopyridine ($C_{11}H_{10}BrN_3O_2S$)	1.90 7.15	U U	+H −H	Potentiometric	H_2O $t = 25$ $I = 0.05$	Bell PH and Roblin, RO, Jr., Studies in chemotherapy. VII. A theory of the relation of structure to activity of sulfanilamide-type compounds, *JACS*, **64**, 2905–2917 (1942). NB: See *p*-Aminobenzoic acid for details.
1350	5-Sulfanilamido-2-bromopyridine ($C_{11}H_{10}BrN_3O_2S$)	2.00 7.12	U U	+H −H	Potentiometric	H_2O $t = 25$ $I = 0.05$	Bell PH and Roblin, RO, Jr., Studies in chemotherapy. VII. A theory of the relation of structure to activity of sulfanilamide-type compounds, *JACS*, **64**, 2905–2917 (1942). NB: See *p*-Aminobenzoic acid for details.
1351	5-Sulfanilamido-2-chloropyrimidine ($C_{10}H_9ClN_4O_2S$)	— 5.80	 U	+H −H	Potentiometric	H_2O $t = 25$ $I = 0.05$	Bell PH and Roblin, RO, Jr., Studies in chemotherapy. VII. A theory of the relation of structure to activity of sulfanilamide-type compounds, *JACS*, **64**, 2905–2917 (1942). NB: See *p*-Aminobenzoic acid for details.
1352	2-Sulfanilamidoimidazole ($C_9H_{10}N_4O_2S$)	— 9.72	 U	+H −H	Potentiometric	H_2O $t = 25$ $I = 0.05$	Bell PH and Roblin, RO, Jr., Studies in chemotherapy. VII. A theory of the relation of structure to activity of sulfanilamide-type compounds, *JACS*, **64**, 2905–2917 (1942). NB: See *p*-Aminobenzoic acid for details.
1353	3-Sulfanilamido-4-methylfurazan ($C_9H_{10}N_4O_3S$)	1.90 4.40	U U	+H −H	Potentiometric	H_2O $t = 25$ $I = 0.05$	Bell PH and Roblin, RO, Jr., Studies in chemotherapy. VII. A theory of the relation of structure to activity of sulfanilamide-type compounds, *JACS*, **64**, 2905–2917 (1942). NB: See *p*-Aminobenzoic acid for details.

1354	5-Sulfanilamido-3-methylisoxazole ($C_{10}H_{11}N_3O_3S$)	— 4.2	 U	+H −H	Potentiometric	H_2O $t = 25$ $I = 0.05$	Bell PH and Roblin, RO, Jr., Studies in chemotherapy. VII. A theory of the relation of structure to activity of sulfanilamide-type compounds, *JACS*, **64**, 2905–2917 (1942). NB: See *p*-Aminobenzoic acid for details.
1355	2-Sulfanilamido-5-methyloxadiazole ($C_9H_{10}N_4O_3S$)	1.70 4.40	U U	+H −H	Potentiometric	H_2O $t = 25$ $I = 0.05$	Bell PH and Roblin, RO, Jr., Studies in chemotherapy. VII. A theory of the relation of structure to activity of sulfanilamide-type compounds, *JACS*, **64**, 2905–2917 (1942). NB: See *p*-Aminobenzoic acid for details.
1356	2-Sulfanilamido-5-methylthiadiazole ($C_9H_{10}N_4O_2S_2$)	2.20 5.45	U U	+H −H	Potentiometric	H_2O $t = 25$ $I = 0.05$	Bell PH and Roblin, RO, Jr., Studies in chemotherapy. VII. A theory of the relation of structure to activity of sulfanilamide-type compounds, *JACS*, **64**, 2905–2917 (1942). NB: See *p*-Aminobenzoic acid for details.
1357	2-Sulfanilamido-4-methylthiazole ($C_{10}H_{11}N_3O_2S_2$)	2.36 7.79	U U	+H −H	Potentiometric	H_2O $t = 25$ $I = 0.05$	Bell PH and Roblin, RO, Jr., Studies in chemotherapy. VII. A theory of the relation of structure to activity of sulfanilamide-type compounds, *JACS*, **64**, 2905–2917 (1942). NB: See *p*-Aminobenzoic acid for details.
1358	2-Sulfanilamidooxazole ($C_9H_9N_3O_3S$)	— 6.5	 U	+H −H	Potentiometric	H_2O $t = 25$ $I = 0.05$	Bell PH and Roblin, RO, Jr., Studies in chemotherapy. VII. A theory of the relation of structure to activity of sulfanilamide-type compounds, *JACS*, **64**, 2905–2917 (1942). NB: See *p*-Aminobenzoic acid for details.
1359	2-Sulfanilamidopyrazine ($C_{10}H_{10}N_4O_2S$)	1.78 6.04	U U	+H −H	Potentiometric	H_2O $t = 25$ $I = 0.05$	Bell PH and Roblin, RO, Jr., Studies in chemotherapy. VII. A theory of the relation of structure to activity of sulfanilamide-type compounds, *JACS*, **64**, 2905–2917 (1942). NB: See *p*-Aminobenzoic acid for details.
1360	3-Sulfanilamidopyridazine ($C_{10}H_{10}N_4O_2S$)	1.30 2.48 7.06	U U U	+H +H −H	Potentiometric	H_2O $t = 25$ $I = 0.05$	Bell PH and Roblin, RO, Jr., Studies in chemotherapy. VII. A theory of the relation of structure to activity of sulfanilamide-type compounds, *JACS*, **64**, 2905–2917 (1942). NB: See *p*-Aminobenzoic acid for details.
1361	3-Sulfanilamidopyridine ($C_{11}H_{11}N_3O_2S$)	1.60 3.00 7.89	U U U	+H +H −H	Potentiometric	H_2O $t = 25$ $I = 0.05$	Bell PH and Roblin, RO, Jr., Studies in chemotherapy. VII. A theory of the relation of structure to activity of sulfanilamide-type compounds, *JACS*, **64**, 2905–2917 (1942). NB: See *p*-Aminobenzoic acid for details.
1362	4-Sulfanilamidopyrimidine ($C_{10}H_{10}N_4O_2S$)	1.30 3.34 6.17	U U U	+H +H −H	Potentiometric	H_2O $t = 25$ $I = 0.05$	Bell PH and Roblin, RO, Jr., Studies in chemotherapy. VII. A theory of the relation of structure to activity of sulfanilamide-type compounds, *JACS*, **64**, 2905–2917 (1942). NB: See *p*-Aminobenzoic acid for details.
1363	5-Sulfanilamidopyrimidine ($C_{10}H_{10}N_4O_2S$)	1.90 6.62	U U	+H −H	Potentiometric	H_2O $t = 25$ $I = 0.05$	Bell PH and Roblin, RO, Jr., Studies in chemotherapy. VII. A theory of the relation of structure to activity of sulfanilamide-type compounds, *JACS*, **64**, 2905–2917 (1942). NB: See *p*-Aminobenzoic acid for details.

(continued)

No.	Compound Name	pK_a value(s)	Data quality	Ionization type	Method	Conditions	Comments and Reference(s)
1364	2-Sulfanilamido-1,3,4-thiadiazole ($C_8H_8N_4O_2S_2$)	2.15 4.77	U U	+H −H	Potentiometric	H_2O $t = 25$ $I = 0.05$	Bell PH and Roblin, RO, Jr., Studies in chemotherapy. VII. A theory of the relation of structure to activity of sulfanilamide-type compounds, *JACS*, **64**, 2905–2917 (1942). NB: See *p*-Aminobenzoic acid for details.
1365	4-Sulfanilamido-1,2,4-triazole ($C_8H_9N_5O_2S$)	1.85 4.66	U U	+H −H	Potentiometric	H_2O $t = 25$ $I = 0.05$	Bell PH and Roblin, RO, Jr., Studies in chemotherapy. VII. A theory of the relation of structure to activity of sulfanilamide-type compounds, *JACS*, **64**, 2905–2917 (1942). NB: See *p*-Aminobenzoic acid for details.
1366	Sulfanilylaminoguanidine ($C_7H_{11}N_5O_2S$)	1.30 2.48 —	U U	+H +H −H	Potentiometric	H_2O $t = 25$ $I = 0.05$	Bell PH and Roblin, RO, Jr., Studies in chemotherapy. VII. A theory of the relation of structure to activity of sulfanilamide-type compounds, *JACS*, **64**, 2905–2917 (1942). NB: See *p*-Aminobenzoic acid for details; not sufficiently soluble in acetic acid for determination of the basic group.
1367	Sulfanilylcyanamide ($C_7H_7N_3O_2S$)	— 2.92	 U	 −H	Potentiometric	H_2O $t = 25$ $I = 0.05$	Bell PH and Roblin, RO, Jr., Studies in chemotherapy. VII. A theory of the relation of structure to activity of sulfanilamide-type compounds, *JACS*, **64**, 2905–2917 (1942). NB: See *p*-Aminobenzoic acid for details; not sufficiently soluble in acetic acid for determination of the basic group.
1368	Sulfanilylglycine ($C_7H_{10}N_2O_2S$)	3.52	U	−H	Potentiometric	H_2O $t = 25$ $I = 0.05$	Bell PH and Roblin, RO, Jr., Studies in chemotherapy. VII. A theory of the relation of structure to activity of sulfanilamide-type compounds, *JACS*, **64**, 2905–2917 (1942). NB: See *p*-Aminobenzoic acid for details. Not sufficiently soluble in glacial acetic acid for measurement of the basic group.
1369	Sulfanilylguanidine ($C_7H_{10}N_4O_2S$)	2.75 4.77 —	U U	+H +H −H	Potentiometric	H_2O $t = 25$ $I = 0.05$	Bell PH and Roblin, RO, Jr., Studies in chemotherapy. VII. A theory of the relation of structure to activity of sulfanilamide-type compounds, *JACS*, **64**, 2905–2917 (1942). NB: See *p*-Aminobenzoic acid for details; not sufficiently soluble in acetic acid for determination of the basic group.
1370	N^3-Sulfanilylmetanilamide ($C_{12}H_{13}N_3O_4S_2$)	1.90 7.85	U U	+H −H	Potentiometric	H_2O $t = 25$ $I = 0.05$	Bell PH and Roblin, RO, Jr., Studies in chemotherapy. VII. A theory of the relation of structure to activity of sulfanilamide-type compounds, *JACS*, **64**, 2905–2917 (1942). NB: See *p*-Aminobenzoic acid for details.
1371	N^4-Sulfanilylsulfanilamide ($C_{12}H_{13}N_3O_4S_2$)	3.20 8.23	U U	+H −H	Potentiometric	H_2O $t = 25$ $I = 0.05$	Bell PH and Roblin, RO, Jr., Studies in chemotherapy. VII. A theory of the relation of structure to activity of sulfanilamide-type compounds, *JACS*, **64**, 2905–2917 (1942). NB: See *p*-Aminobenzoic acid for details.

1372	Sulfanilylurea ($C_7H_9N_3O_3S$)	1.78 5.42	U U	+H −H	Potentiometric	H_2O $t = 25$ $I = 0.05$	Bell PH and Roblin, RO, Jr., Studies in chemotherapy. VII. A theory of the relation of structure to activity of sulfanilamide-type compounds, *JACS*, **64**, 2905–2917 (1942). NB: See *p*-Aminobenzoic acid for details; not sufficiently soluble in acetic acid for determination of the basic group.
1373	Sulfaperine ($C_{11}H_{12}N_4O_2S$) 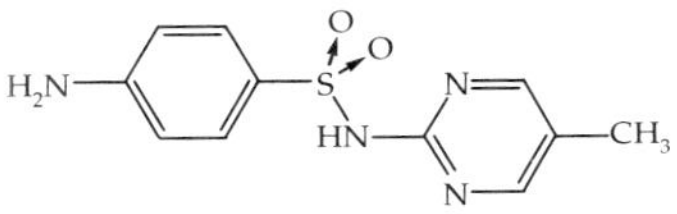	6.65	U	−H	Spectro	H_2O $t = 25.0 \pm 0.5$ $I = 0.2$	Elofsson R, Nilsson SO and Agren A, Complex formation between macromolecules and drugs. IV, *Acta Pharm. Suec.*, **7**, 473–482 (1970). NB: See Sulphanilamide for details.
1374	Sulfaperine (also called sulfamethyldiazine; 4-methylsulfadiazine; 2-sulfanilamido-4-methylpyrimidine)	2.08 7.06	U U	+H −H	Potentiometric	H_2O $t = 25$ $I = 0.05$	Bell PH and Roblin, RO, Jr., Studies in chemotherapy. VII. A theory of the relation of structure to activity of sulfanilamide-type compounds, *JACS*, **64**, 2905–2917 (1942). NB: See *p*-Aminobenzoic acid for details.
1375	Sulfaphenazole ($C_{15}H_{14}N_4O_2S$) 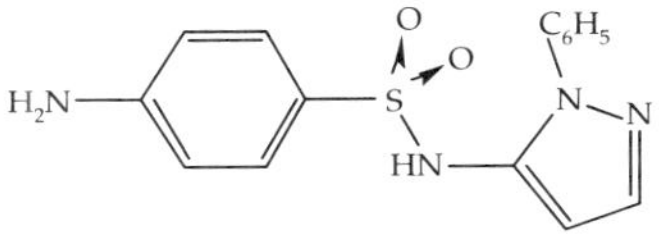	5.71	U	−H		H_2O $t = 28$	Chatten LG (ed.), *Pharmaceutical Chemistry*, Vol. 1, Dekker, New York, 1966, pp. 85–87; Nakagaki M, Koga N and Terada H, Physicochemical studies on the binding of chemicals with proteins. I. The binding of several sulphonamides with serum albumin, *J. Pharm. Soc. Jpn.*, **83**, 586–590 (1963).
1376	Sulfapyridine ($C_{11}H_{11}N_3O_2S$) 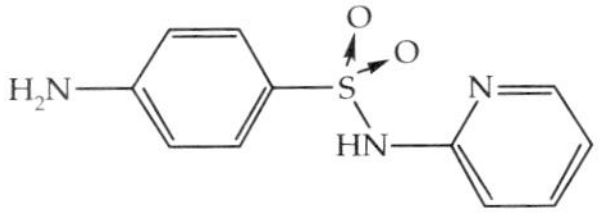	8.48	U	−H	Potentiometric	H_2O $t = 20$ $I = 0.1$ (KCl)	Willi AV and Meier W, Die Aciditatskonstanten von Benzolsulfonamiden mit heterocyclischer Amin-Komponente (The acidity constants for benzenesulfonamides with heterocyclic amine components), *Helv. Chim. Acta*, **39**, 54–56 (1956). NB: used previously described titrimetric method (Schwarzenbach G, Willi A and Bach RO, Komplexone IV. Die Aciditat und die Erdalkalikomplexe der Anilin-diessigsaure und ihrer Substitutionsprodukte, *Helv. Chim. Acta*, **30**, 1303–1320 (1947)) with glass electrode. Solutions made in carbonate-free sodium hydroxide and then back-titrated with perchloric acid titrant. Sodium ion corrections were made at pH values >9.8.
1377	Sulfapyridine (2-sulfanilamidopyridine)	1.00 2.58 8.43	U U U	+H +H −H	Potentiometric	H_2O $t = 25$ $I = 0.05$	Bell PH and Roblin, RO, Jr., Studies in chemotherapy. VII. A theory of the relation of structure to activity of sulfanilamide-type compounds, *JACS*, **64**, 2905–2917 (1942). NB: See *p*-Aminobenzoic acid for details.

(continued)

No.	Compound Name	pK_a value(s)	Data quality	Ionization type	Method	Conditions	Comments and Reference(s)
1378	Sulfapyridine	1.00 2.58	U U	+H +H	Potentiometric	H_2O (extrap) $t = 24 \pm 1$ $I \sim 0.002$	Chatten LG and Harris LE, Relationship between pK_b(H_2O) of organic compounds and $E_{1/2}$ values in several nonaqueous solvents. *Anal. Chem.*, **34**, 1495–1501 (1962).
1379	Sulfapyridine	8.45	U	–H	Spectro	H_2O $t = 25.0 \pm 0.5$ $I = 0.2$	Elofsson R, Nilsson SO and Agren A, Complex formation between macromolecules and drugs. IV, *Acta Pharm. Suec.*, **7**, 473–482 (1970). NB: See Sulphanilamide for details.
1380	Sulfapyridine	8.32	U	–H		H_2O $t = 37$	Ballard BE and Nelson E, Physicochemical properties of drugs that control absorption rate after subcutaneous implantation, *JPET*, **135**, 120–127 (1962).
1381	Sulfapyridine-1-oxide ($C_{11}H_{10}N_3O_3S$) 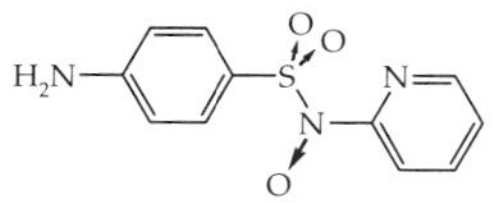	5.2	VU	–H	Potentiometric	H_2O $t = 25$ I undefined	Scudi JW and Plekss OJ, Chemotherapeutic activity of some sulfapyridine-1-oxides, *Proc. Soc. Exptl. Biol. Med.*, **97**, 639–641 (1958). "The pK_as of the sulfapyridine-1-oxides and reference drugs as well as their acetylated derivatives were approximated by titrating 0.1% solutions of the drugs in water at 25 °C with one-half of an equivalent of 0.1 N sodium hydroxide. . . ." Further values: 6′-methylsulfapyridine-1-oxide ($C_{12}H_{12}N_3O_3S$), pK_a = 5.9, 6′-ethylsulfapyridine-1-oxide ($C_{13}H_{14}N_3O_3S$), pK_a = 6.1, 4′,6′-dimethylsulfapyridine-1-oxide ($C_{13}H_{14}N_3O_3S$), pK_a = 6.2.
	3′-methylsulfapyridine-1-oxide ($C_{12}H_{12}N_3O_3S$)	6.1	VU	–H			
	4′-methylsulfapyridine-1-oxide ($C_{12}H_{12}N_3O_3S$)	5.5	VU	–H			
	5′-methylsulfapyridine-1-oxide ($C_{12}H_{12}N_3O_3S$)	5.7	VU	–H			
1382	Sulfapyridine-1-oxide, N[4]-acetyl, 6′-methyl ($C_{14}H_{14}N_3O_4S$) 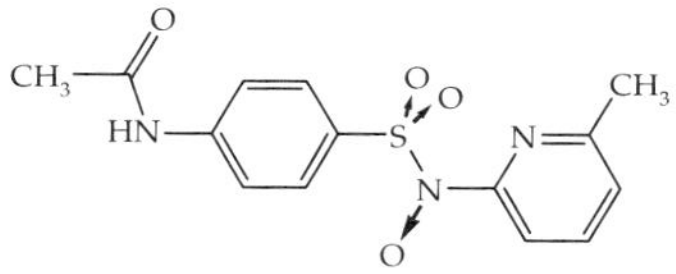	5.4	VU	–H	Potentiometric	H_2O $t = 25$	Scudi JW and Plekss OJ, Chemotherapeutic activity of some sulfapyridine-1-oxides, *Proc. Soc. Exptl. Biol. Med.*, **97**, 639–641 (1958). NB: See Sulfapyridine-1-oxide.

1383	Sulfapyridine, N^4-acetyl ($C_{13}H_{13}N_3O_3S$) 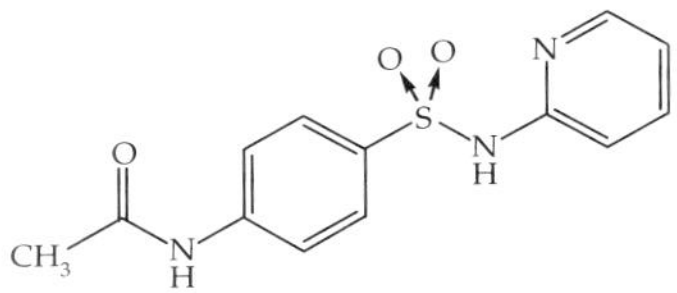	8.2	U	–H	Potentiometric	H_2O $t = 25$	Scudi JW and Plekss OJ, Chemotherapeutic activity of some sulfapyridine-1-oxides, *Proc. Soc. Exptl. Biol. Med.*, **97**, 639–641 (1958). NB: See Sulfapyridine-1-oxide.
1384	Sulfasalazine ($C_{18}H_{14}N_4O_5S$) 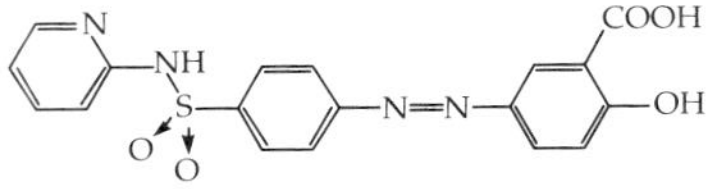	0.6 2.4 9.7 11.8	U U U U	+H –H –H –H	Spectro	H_2O $t = 20$ $I < 0.001$	Nygard B, Olofsson J and Sandberg M, Some physicochemical properties of salicylazosulphapyridine, including its solubility, protolytic constants and general spectrochemical and polarographic behaviour, *Acta Pharm. Suec.*, **3**, 313–342 (1966). Cited in McDonnell JP, Sulfasalazine, *APDS*, **5**, 515–532 (1976).
1385	Sulfathiazole ($C_9H_9N_3O_2S_2$) 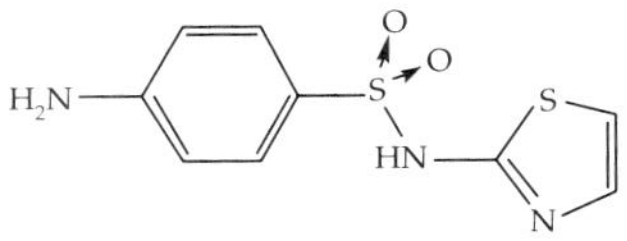	2.36 7.12	U U	+H –H	Potentiometric	H_2O $t = 25$ $I = 0.05$	Bell PH and Roblin, RO, Jr., Studies in chemotherapy. VII. A theory of the relation of structure to activity of sulfanilamide-type compounds, *JACS*, **64**, 2905–2917 (1942). NB: See *p*-Aminobenzoic acid for details.
1386	Sulfathiazole (2-sulfanilamidothiazole)	7.49	U	–H	Potentiometric	H_2O $t = 20$ $I = 0.1$ (KCl)	Willi AV and Meier W, 6. Die Aciditatskonstanten von Benzolsulfonamiden mit heterocyclischer Amin-Komponente (The acidity constants for benzenesulfonamides with heterocyclic amine components), *Helv. Chim. Acta*, **39**, 54–56 (1956). NB: See Sulfapyridine.
1387	Sulfathiazole	7.10	U	–H		H_2O $t = 37$	Ballard BE and Nelson E, Physicochemical properties of drugs that control absorption rate after subcutaneous implantation, *JPET*, **135**, 120–127 (1962).
1388	Sulfinpyrazone ($C_{23}H_{20}N_2O_3S$) 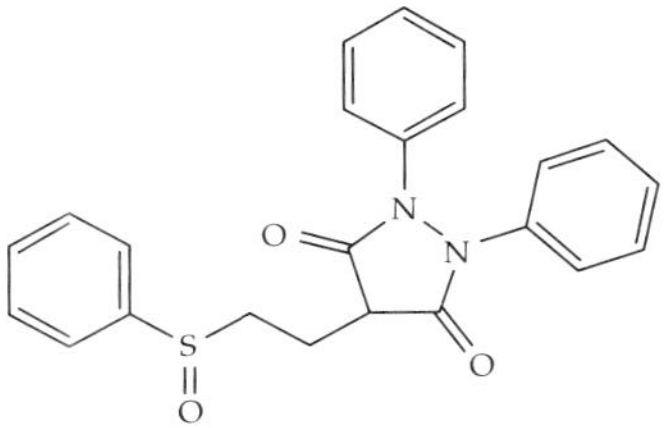	2.8	U	–H		H_2O $t =$ RT I undefined	Perel JM, Snell MM, Chen W and Dayton PG, *Biochem. Pharmacol.*, **13**, 1305–1317 (1964). NB: performed in ''dilute buffer solutions.'' See: Phenylbutazone analogs.

(continued)

No.	Compound Name	pK_a value(s)	Data quality	Ionization type	Method	Conditions	Comments and Reference(s)
1389	Sulfisomidine (2-sulfanilamido-2,4-dimethylpyrimidine) ($C_{12}H_{14}N_4O_2S$) 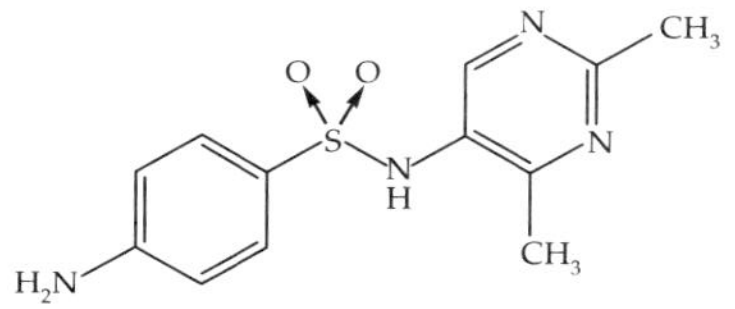	7.55	U	–H	Spectro	H_2O $t = 25.0 \pm 0.5$ $I = 0.2$	Elofsson R, Nilsson SO and Agren A, Complex formation between macromolecules and drugs. IV, *Acta Pharm. Suec.*, **7**, 473–482 (1970). NB: See Sulphanilamide for details.
1390	Sulfisomidine (4-sulfanilamido-2,6-dimethylpyrimidine)	7.49	U	–H	Potentiometric	H_2O $t = 20$ $I = 0.1$ (KCl)	Willi AV and Meier W, 6. Die Aciditatskonstanten von Benzolsulfonamiden mit heterocyclischer Amin-Komponente (The acidity constants for benzenesulfonamides with heterocyclic amine components), *Helv. Chim. Acta*, **39**, 54–56 (1956). NB: See: Sulfapyridine.
1391	Sulfisoxazole	4.96	U	–H	Potentiometric	H_2O $t = 37$	Ballard BE and Nelson E, Physicochemical properties of drugs that control absorption rate after subcutaneous implantation, *JPET*, **135**, 120–127 (1962).
1392	Sulfisoxazole 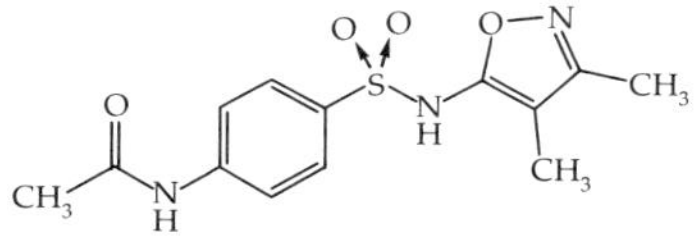	5.0	VU	–H	Potentiometric	H_2O $t = 25$	Scudi JW and Plekss OJ, Chemotherapeutic activity of some sulfapyridine-1-oxides, *Proc. Soc. Exptl. Biol. Med.*, **97**, 639–641 (1958). NB: See Sulfapyridine-1-oxide.
1393	Sulfisoxazole, N[4]-acetyl ($C_{13}H_{15}N_3O_4S$) 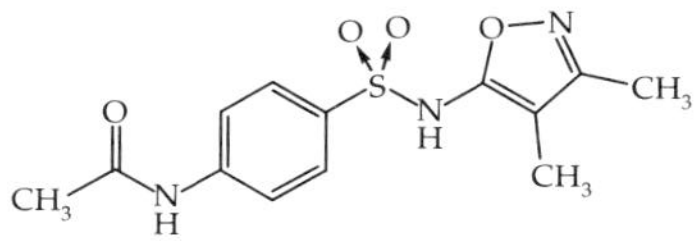	4.4	VU	–H	Potentiometric	H_2O $t = 25$	Scudi JW and Plekss OJ, Chemotherapeutic activity of some sulfapyridine-1-oxides, *Proc. Soc. Exptl. Biol. Med.*, **97**, 639–641 (1958). NB: See Sulfapyridine-1-oxide.

No.	Compound	pKa	Data quality	Ionization	Method	Conditions	Reference / Comments
1394	Sulfonazole ($C_9H_9N_3O_2S_2$) 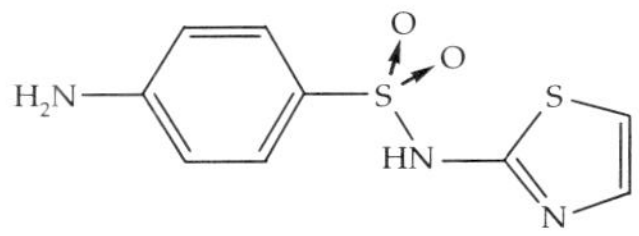	7.27	U	−H	Spectro	H_2O $t = 25.0 \pm 0.5$ $I = 0.2$	Elofsson R, Nilsson SO and Agren A, Complex formation between macromolecules and drugs. IV, *Acta Pharm. Suec.*, **7**, 473–482 (1970). NB: See Sulphanilamide for details.
1395	Sulfone, bis(4-aminophenyl) (4,4-diaminodiphenylsulphone; dapsone)	1.30 2.49 —	U U	+H +H −H	Potentiometric	H_2O $t = 25$ $I = 0.05$	Bell PH and Roblin, RO, Jr., Studies in chemotherapy. VII. A theory of the relation of structure to activity of sulfanilamide-type compounds, *JACS*, **64**, 2905–2917 (1942). NB: See *p*-Aminobenzoic acid for details.
1396	Sulindac ($C_{20}H_{17}FO_3S$) 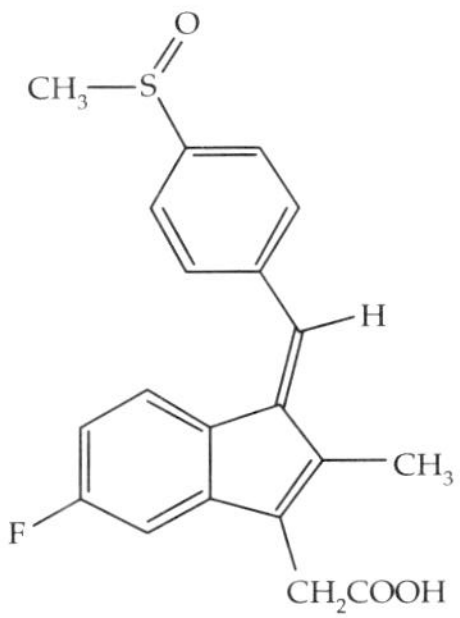	4.7	U	−H		$t = 25.0$	Plakogiannis FM and McCauley JA, Sulindac, *APDS*, **13**, 573–595 (1984). no reference — Foye 1 says 4.5
1397	Sulpiride ($C_{15}H_{23}N_3O_4S$) 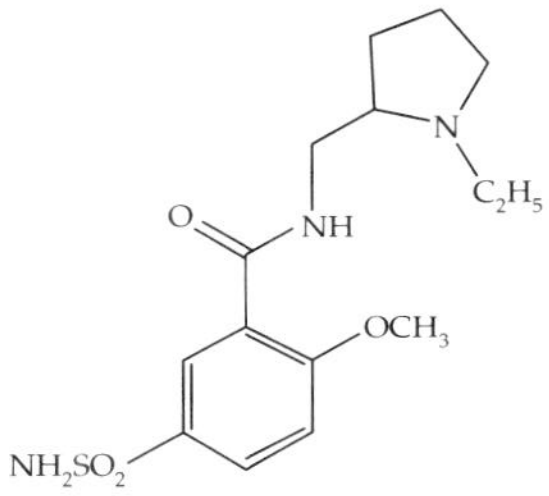	8.99 ± 0.01 10.2 ± 0.3	A U	+H −H	Potentio-metric, Spectro	H_2O	Pitré D, Stradi R and Nathansohn G, Sulpiride, *APDS*, **17**, 607–641 (1988). NB: Summarized the ionization constants of sulpiride:

Method	Solvent	pK_{a1}	pK_{a2}	
spectro	H_2O	8.99	–	
potentio	H_2O	9.00	10.19	Mean value of K_2 from all three results = 6.2268×10^{-11}, which gives a mean $pK_{a2} = 10.21 \pm 0.27$
potentio	7.7% MeOH	8.98	10.05	
potentio	42.4% MeOH	8.36	10.48	

(continued)

No.	Compound Name	pK_a value(s)	Data quality	Ionization type	Method	Conditions	Comments and Reference(s)
							Van Damme M, Hanocq M, Topart J and Molle L, Determination of ionization constants of *N*-(1-ethylpyrrolidin-2-ylmethyl)-2-methoxy-5-sulphamoylbenzamide (sulpiride), *Analysis*, **4**, 299–307 (1976). Van Damme M, Hanocq M and Molle L, Critical study of potentiometric methods in aqueous and hydroalcoholic media for the determination of acid dissociation constants of certain benzamides, *Analysis*, **7**, 499–504 (1979).
1398	Synephrine ($C_9H_{13}NO_2$) 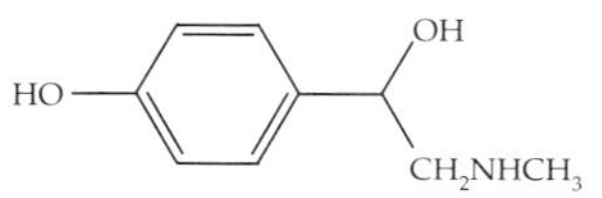 HO, OH, CH_2NHCH_3	9.60 ± 0.02	A	+H	Potentiometric	H_2O $t = 25.0 \pm 0.5$ $I = 0.01$	Warren RJ, Begosh PP and Zarembo JE, Identification of amphetamines and related sympathomimetic amines, *J. Assoc. Off. Anal. Chem.*, **54**, 1179–1191 (1971). NB: See Amphetamine and Amphetamine, 4-hydroxy, for further details. Lewis (see no. 1094, Phenylpropanolamine) reported p$K_{a1} = 9.59$ and p$K_{a2} = 9.71$.
1399	Synephrine (Sympatol) ($C_9H_{13}NO_2$)	8.90 ± 0.02	U	+H	Potentiometric	H_2O $t = 25.0 \pm 0.2$ $I \leq 0.001$	Leffler EB, Spencer HM and Burger A, Dissociation constants of adrenergic amines, *JACS*, **73**, 2611–2613 (1951). NB: See Amphetamine for details. From p$K_b = 5.10$. Careful reading of the paper suggests that the compound was investigated in the form of its tartrate salt. This would inevitably cause a problem with the overlapping of the pK_b value with the pK_{a2} of tartaric acid (4.3, below), as shown by the substantial discrepancy with the value found by Warren *et al.* (no. 1398).
1400	Tamoxifen ($C_{26}H_{29}NO$) 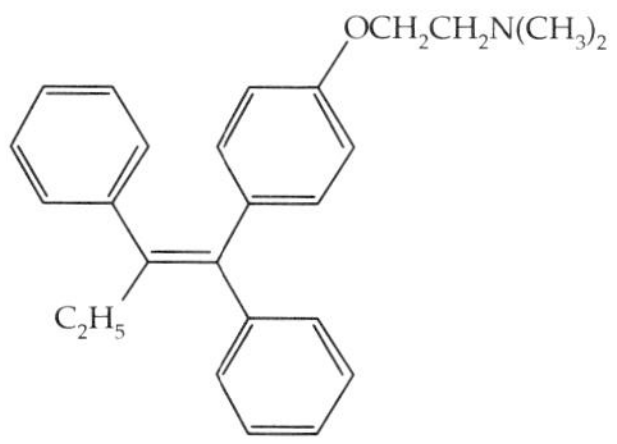 $OCH_2CH_2N(CH_3)_2$, C_2H_5	8.71	U	+H	CE/pH (+ve ion mode)	H_2O $t = 25$ $I = 0.025$	Wan H, Holmen AG, Wang Y, Lindberg W, Englund M, Nagard MB and Thompson RA, High-throughput screening of pK_a values of pharmaceuticals by pressure-assisted capillary electrophoresis and mass spectrometry, *Rapid Commun. Mass Spectrom.*, **17**, 2639–2648 (2003). NB: Reported a predicted value (ACD Labs) of 8.69.

No.	Compound	pKa	Rating	Ionization	Method	Conditions	Reference
1401	Tartaric acid ($C_4H_6O_6$) HOOCCH(OH)CH(OH)COOH	3.033 4.366	R R	−H −H	Potentiometric	H_2O $t = 25.0$ $I = 0.000$ (NaCl or KCl)	Bates RG and Canham RG, Resolution of the dissociation constants of D-tartaric acid from 0° to 50°, *J. Res. Nat. Bur. Stand.*, **47**, 343–348 (1951). NB: Used an electrochemical cell without liquid junction potentials. Extrapolated ionic strength to zero to give thermodynamic pK_a values, which were then used to estimate the thermodynamic functions for ionization.
1402	Taurocholic acid ($C_{26}H_{45}NO_7S$) 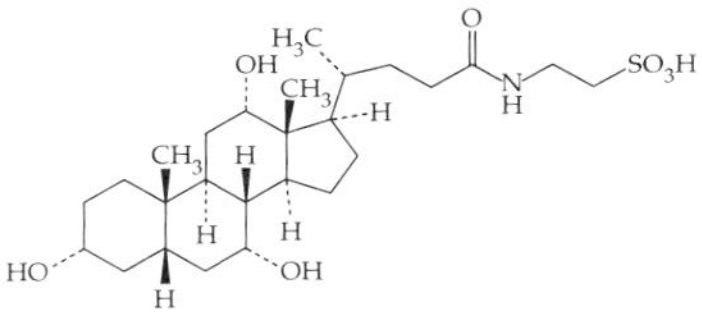	1.56	U	−H		H_2O $c < CMC$	Johns WH and Bates TR., Quantification of the binding tendencies of cholestyramine I: effect of structure and added electrolytes on the binding of unconjugated and conjugated bile salt anions, *J. Pharm. Sci.*, **58**, 179–183 (1969). NB: These values were quoted from Ekwall P, Rosendahl T and Lofman N, Bile salt solutions. I. The dissociation constants of the cholic and deoxycholic acids, *Acta Chem. Scand.*, **11**, 590–598 (1957). They were measured at concentrations both above and below the critical micellar concentration range.
		3.33	U	−H		H_2O $c > CMC$	
1403	Tecomine (tecomanine) ($C_{11}H_{17}NO$) 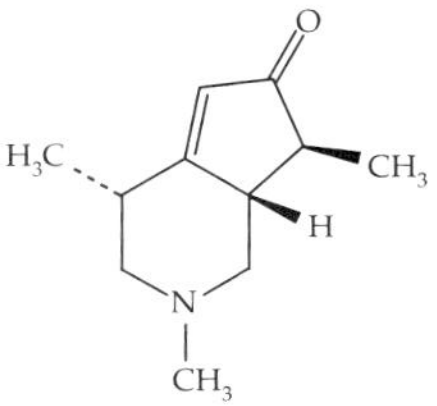	7.2	U	+H	Potentiometric	H_2O t undefined I undefined	Hammouda Y and Khalafallah N, Stability of tecomine, the major antidiabetic factor of *Tecoma stans* (Juss.) f. Bignoniaceae, *J. Pharm. Sci.*, **60**, 1142–1145 (1971). NB: The free base was titrated with standard NaOH to the pH of half-neutralization, which was taken as the pK_a value, and reported as pK_b = 6.8, thus assuming that the result was valid at 25 °C.
1404	Teniposide ($C_{32}H_{32}O_{13}S$) 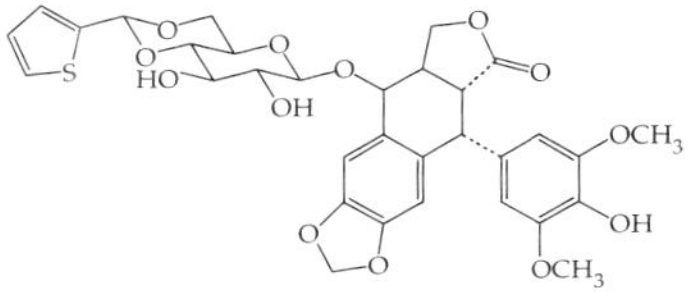	10.13	U	−H	Spectro	5% aq. DMSO t undefined $I = 0.15$	Holthuis JJM, Vendrig DEMM, van Oort WJ and Zuman P, Electrochemistry of podophyllotoxin derivatives. Part II. Oxidation of teniposide and some epipodophyllotoxin derivatives, *J. Electroanal. Chem. Interfacial Electrochem.*, **220**, 101–124 (1987). Cited in Kettenes-van den Bosch JJ, Holthuis JJM and Bult A, Teniposide, *APDS*, **19**, 575–600, 1990.

(continued)

No.	Compound Name	pK_a value(s)	Data quality	Ionization type	Method	Conditions	Comments and Reference(s)
1405	Terazosin ($C_{19}H_{25}N_5O_4$) 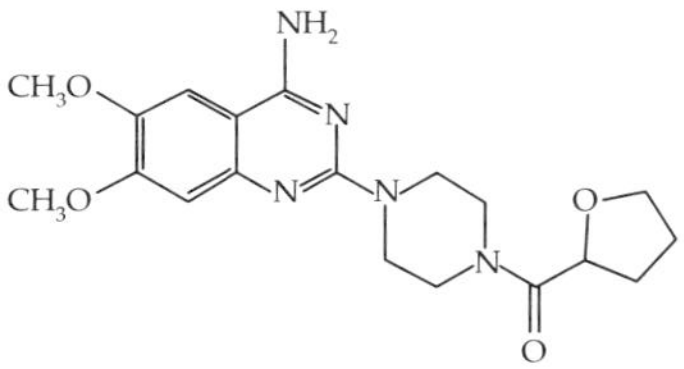	7.1	U	+H	Potentiometric	H_2O	Chang ZL and Bauer JF, Terazosin, *APDS*, **20**, 693–727 (1991). NB: Titration of terazosin (presumably the HCl or some other salt) with 0.1-N aqueous NaOH using water as the sample solvent gave a pK_a value of 7.1.
1406	Terbutaline ($C_{12}H_{19}NO_3$) 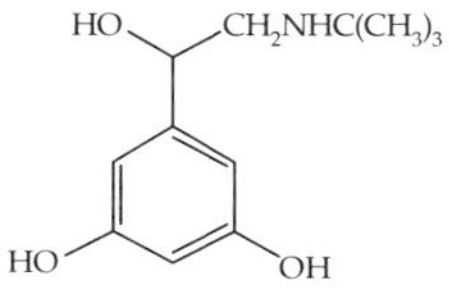	8.70 10.09	A U	+H −H	Spectro	H_2O $t = 25.0 \pm 0.05$ $I = 0.10$	Ijzerman AP, Bultsma T, Timmerman H and Zaagsma J, The ionization of β-adrenoceptor agonists: a method for unravelling ionization schemes, *J. Pharm. Pharmacol.*, **36**(1), 11–15 (1984). NB: microscopic: 8.73 and 10.06; Negligible formation of the neutral form. See Isoproterenol.
1407	Terbutaline	8.62	A	+H	CE/pH (+ve ion mode)	H_2O $t = 25$ $I = 0.025$	Wan H, Holmen AG, Wang Y, Lindberg W, Englund M, Nagard MB and Thompson RA, High-throughput screening of pK_a values of pharmaceuticals by pressure-assisted capillary electrophoresis and mass spectrometry, *Rapid Commun. Mass Spectrom.*, **17**, 2639–2648 (2003). NB: Reported a predicted value (ACD Labs) of 9.11.
1408	Terbutaline	GLpK_a: 8.64 ± 0.06 10.76 ± 0.03	 U U	 +H −H	Spectro	H_2O $t = 25$ $I = 0.15$ (KCl) Ar atmosphere	Tam KY and Takacs-Novac K, Multi-wavelength spectrophotometric determination of acid dissociation constants, *Anal. Chim. Acta*, **434**, 157–167 (2001). NB: See Clioquinol for details. Cited Takacs-Novac K, Noszal B, Tokeskovesdi M, Szasz G, *J. Pharm. Pharmacol.*, **47**, 431 (1995) with values of 8.57, 9.89, 11.0.

1409	Terfenadine ($C_{32}H_{41}NO_2$) 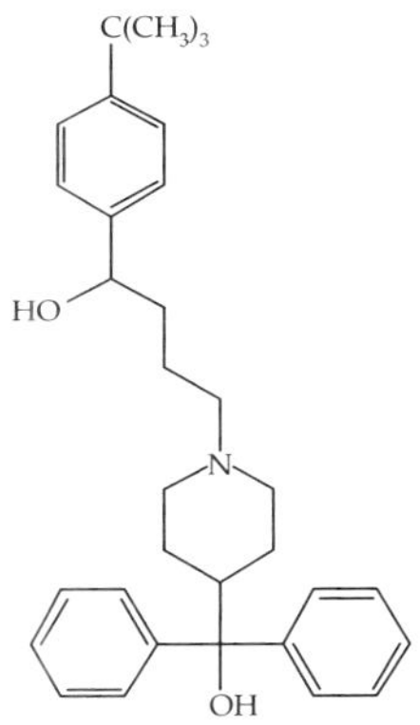	9.21	U	+H	CE/pH (+ve ion mode)	H_2O $t = 25$ $I = 0.025$	Wan H, Holmen AG, Wang Y, Lindberg W, Englund M, Nagard MB and Thompson RA, High-throughput screening of pK_a values of pharmaceuticals by pressure-assisted capillary electrophoresis and mass spectrometry, *Rapid Commun. Mass Spectrom.*, **17**, 2639–2648 (2003). NB: Reported a predicted value (ACD Labs) of 9.57.
1410	Testosterone, imidazole-1-carboxylic acid prodrug ($C_{23}H_{30}N_2O_3$) 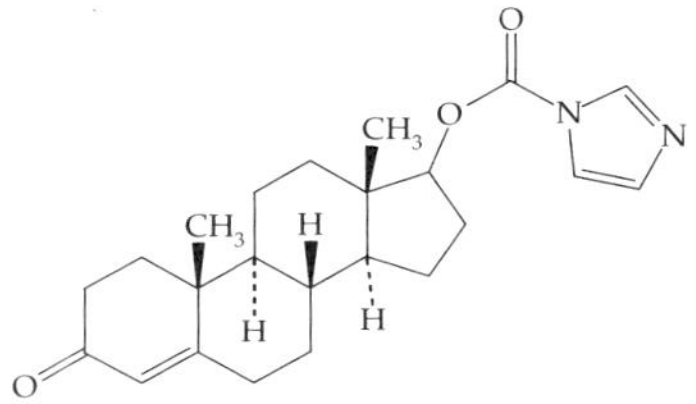	3.5	U	+H	kinetic	H_2O $t = 25.0$ $I = 0.5$ (KCl)	Klixbull U and Bundgaard H, Prodrugs as drug delivery systems. Part 29. Imidazole-1-carboxylic acid esters of hydrocortisone and testosterone, *Arch. Pharm. Chemi., Sci. Edn.*, **11**, 101–110 (1983). "The kinetics of hydrolysis of imidazole-1-carboxylic acid esters of hydrocortisone and testosterone were studied to assess their suitability as prodrug forms. The pH-rate profiles of the 2 derivatives were derived in the range pH 1–12 and were accounted for by assuming spontaneous hydrolysis of the protonated forms (pK_a 3.3–3.5) and hydroxide ion-catalyzed hydrolysis of the free base forms. . . ."

(continued)

No.	Compound Name	pK_a value(s)	Data quality	Ionization type	Method	Conditions	Comments and Reference(s)
1411	Tetracaine ($C_{15}H_{24}N_2O_2$)	8.33 ± 0.03	A	+H	Potentiometric	H_2O $t = 20.0$ $I = 0.10$ (KCl) under N_2	Buchi J and Perlia X, Beziehungen zwischen de physikalisch-chemische Eigenschaften und der Wirkung von Lokalanasthetica, *Arzneim. -Forsch.*, **10**, 745–754 (1960). NB: See Cocaine for details.
1412	Tetracaine	2.39 ± 0.02 8.49 ± 0.01	U U	+H +H	Potentiometric	H_2O $t = 25.0$ $I = 0.15$ (KCl)	Avdeef A, Box KJ, Comer JEA, Hibbert C and Tam KY, pH-metric log P 10. Determination of liposomal membrane-water partition coefficients of ionizable drugs, *Pharm. Res.*, **15**(2), 209–215 (1998). NB: Used a Sirius PCA101 autotitrator. Also gave log P (octanol-water) and log P (dioleylphosphatidylcholine unilamellar vesicles). See also Lidocaine (no. 741). The same results were also given in Sirius Technical Application Notes, vol. **2**, p. 94–95 (1995). Sirius Analytical Instruments Ltd., Forest Row, East Sussex, RH18 5DW, UK.
1413	Tetracycline ($C_{22}H_{24}N_2O_8$)	3.30 7.68 9.69	A A A	−H −H +H	Potentiometric	H_2O $t = 25$	Stephens C, Murai K, Brunings K and Woodward RB, Acidity constants of the tetracycline antibiotics, *JACS*, **78**, 4155–4158 (1956). Cited in Perrin Bases 3336 ref. S73. NB: Used a glass electrode with liquid junction potentials.
		3.33 7.84 9.59	A A A	−H −H +H	Potentiometric	H_2O $t = 20$	
1414	Tetracycline	3.33 7.70 9.50	U U U	−H −H +H	Potentiometric	H_2O $t = 25 \pm 0.05$ $I = 0.01$	Benet LZ and Goyan JE, Determination of the stability constants of tetracycline complexes, *J. Pharm. Sci.*, **54**, 983–987 (1965). NB: See Chlortetracycline for details. Other literature values:

pK_1	pK_2	pK_3	T	I	Ref.
3.35	7.82	9.57	20	0.01	Albert (1956)

Albert A and Rees, CW, Avidity of the tetracyclines for the cations of metals, *Nature*, **177**, 433–434 (1956).

1415	Tetracycline	4.4 7.8 9.4 10.67	U U U U	−H −H +H −H	NMR, Potentiometric Potentiometric	$MeOH/H_2O$ (1:1) $t = 30 \pm 2$ (NMR) $t = 26 \pm 1$ (potentio)	Rigler NE, Bag SP, Leyden DE, Sudmeier JL and Reilley CN, Determination of a protonation scheme of tetracycline using nuclear magnetic resonance, *Anal. Chem.*, **37**, 872–875 (1965). NB: The value assigned to $pK_{a4} = \sim 10.7$ was obtained by potentiometric titration of tetracycline methiodide, that is, where the dimethylamino group had been quaternized. For three compounds (tetracycline, epitetracycline, and tetracycline methiodide), the paper also reported (from NMR measurements) the following microconstants for the groups corresponding to pK_{a1}, pK_{a2}, and pK_{a3}, where the three functional groups, triketomethine (A), conjugated enolic-phenolic (B) and dimethylamino (C) are described as positive (+), negative (−) double negative (=) or neutral (o) charged:

		Value for compound:		
Transition	Micro-constant	Tetra-cycline	Epitetra-cycline	Tetra methiodide
$A^{\circ}B^{+}C^{\circ}$ to $A^{-}B^{+}C^{\circ}$	pk_1	4.49	4.91	3.98
$A^{\circ}B^{+}C^{\circ}$ to $A^{\circ}B^{\circ}C^{\circ}$	pk_2	5.40	negligible	NA
$A^{\circ}B^{+}C^{\circ}$ to $A^{\circ}B^{+}C^{-}$	pk_3	5.45	5.44	4.67
$A^{-}B^{+}C^{\circ}$ to $A^{-}B^{\circ}C^{\circ}$	pk_{12}	8.00	negligible	NA
$A^{-}B^{+}C^{\circ}$ to $A^{-}B^{+}C^{-}$	pk_{13}	8.51	7.93	7.72
$A^{\circ}B^{\circ}C^{\circ}$ to $A^{-}B^{\circ}C^{\circ}$	pk_{21}	7.09	negligible	NA
$A^{\circ}B^{\circ}C^{\circ}$ to $A^{\circ}B^{\circ}C^{-}$	pk_{23}	7.29	negligible	NA
$A^{\circ}B^{+}C^{-}$ to $A^{-}B^{+}C^{-}$	pk_{31}	7.55	7.41	7.03
$A^{\circ}B^{+}C^{-}$ to $A^{\circ}B^{\circ}C^{-}$	pk_{32}	7.24	8.36	NA
$A^{-}B^{\circ}C^{\circ}$ to $A^{-}B^{\circ}C^{-}$	pk_{123}	9.11	negligible	NA
$A^{-}B^{+}C^{-}$ to $A^{-}B^{\circ}C^{-}$	pk_{132}	8.60	9.45	NA
$A^{-}B^{\circ}C^{-}$ to $A^{-}B^{\circ}C^{=}$	pk_{134}	NA	NA	10.80
$A^{\circ}B^{\circ}C^{-}$ to $A^{-}B^{\circ}C^{-}$	pk_{321}	8.92	8.50	NA

1416	Tetracycline	3.31 (0.14) 7.13 (0.21) 9.64 (–)	U U U	−H −H +H	Spectro	H_2O $t = 20.0$	Wahbe AM, El-Yazbi FA, Barary MH and Sabri SM, Application of orthogonal functions to spectrophotometric analysis. Determination of dissociation constants, *Int. J. Pharm.*, **92**(1), 15–22 (1993). NB: See Acetaminophen for further details. Alternative graphical method gave $pK_a = 3.3, 7.15, 9.65$.
1417	Tetracycline	7.68	U	−H		H_2O $t = 37$	Ballard BE and Nelson E, Physicochemical properties of drugs that control absorption rate after subcutaneous implantation, *JPET*, **135**, 120–127 (1962).

(continued)

No.	Compound Name	pK_a value(s)	Data quality	Ionization type	Method	Conditions	Comments and Reference(s)
1418	Tetracycline	3.69	U	−H	Potentiometric	H_2O $t = 30.0 \pm 0.2$ $I = 0.01$ (KCl) N_2 atmosphere	Doluisio JT and Martin AN, Metal complexation of the tetracycline hydrochlorides, *J. Med. Chem.*, **6**, 16–20 (1963). NB: Metal-free solutions of the tetracycline titrated with standard NaOH solution and the pH measured. No details were given of the pH meter calibration. Metal stability constants were determined from identical titrations in the presence of varying concentrations of nickel(II), zinc(II), or copper(II) ions.
		7.63	U	−H			
		9.24	U	+H			
1419	Tetracycline methiodide ($C_{23}H_{27}IN_2O_8$)	3.9	U	−H	NMR, Potentiometric	MeOH/H_2O (1:1) $t = 30 \pm 2$ (NMR) $t = 26 \pm 1$ (potentio)	Rigler NE, Bag SP, Leyden DE, Sudmeier JL and Reilley CN, Determination of a protonation scheme of tetracycline using nuclear magnetic resonance, *Anal. Chem.*, **37**, 872–875 (1965). NB: See Tetracycline for details.
		7.8	U	−H			
		10.67	U	−H	Potentiometric		
1420	Tetrahydro-α-morphimethine ($C_{18}H_{25}NO_3$) 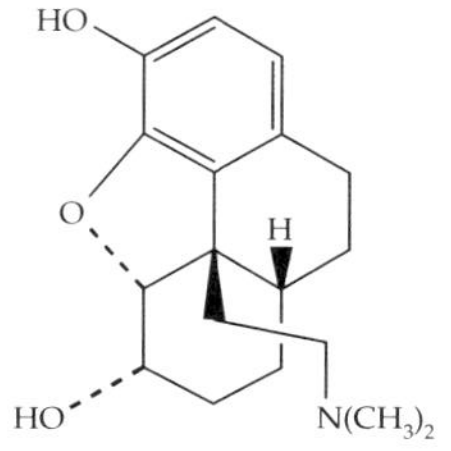	8.65	A	+H	Conductance	H_2O $t = 25$ $\kappa < 1.5$	Oberst FW and Andrews HL, The electrolytic dissociation of morphine derivatives and certain synthetic analgetic compounds, *JPET*, **71**, 38–41 (1941). Cited in Perrin Bases 2969 ref. O1. NB: Results were reported as K_b values. For tetrahydro-α-morphimethine, $K_b = 4.50 \times 10^{-6}$, giving p$K_b$ = 5.35. See Codeine for details.
1421	Tetrahydrozoline ($C_{13}H_{16}N_2$) 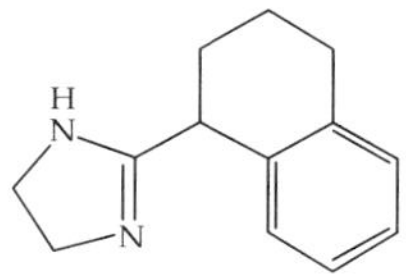	10.51	U	+H			Foye 1

No.	Compound	pKa			Method	Conditions	Reference
1422	Th 1206 ($C_{12}H_{19}NO_3$) 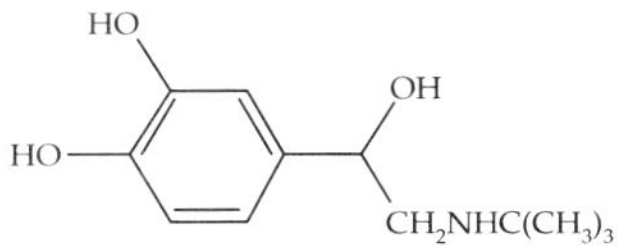	8.72 10.31	U U	+H +H	Spectro	H_2O $t = 25.0 \pm 0.05$ $I = 0.10$	Ijzerman AP, BultsmaT, Timmerman H and Zaagsma J, The ionization of β-adrenoceptor agonists: a method for unravelling ionization schemes, *J. Pharm. Pharmacol.*, **36**(1), 11–15 (1984). NB: Microscopic: 8.68 and 10.31; Negligible formation of the neutral form. See Isoprenalin.
1423	Thebaine ($C_{19}H_{21}NO_3$) 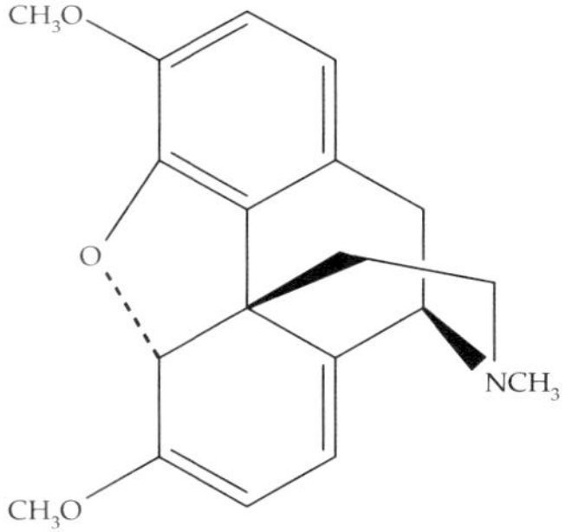	8.15	U	+H	Spectro	H_2O $t = 15$ $I < 0.010$	Kolthoff IM, The dissociation constants, solubility product and titration of alkaloids, *Biochem. Z.*, **162**, 289–353 (1925). Cited in Perrin Bases no. 2971 K47. NB: See Aconitine for details.
1424	Thenalidine ($C_{17}H_{22}N_2S$) 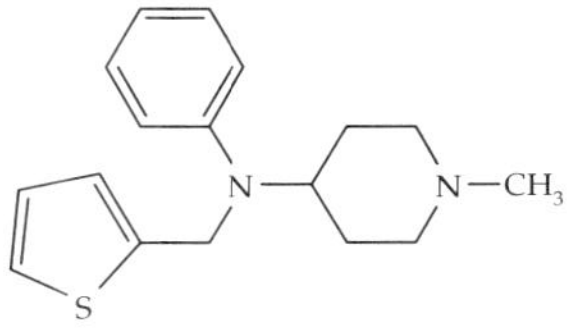	3.36 ± 0.06 8.39 ± 0.07	U U	+H +H	Potentiometric	H_2O t undefined $I = 0.30$ (NaCl)	Testa B and Murset-Rossetti L, The partition coefficient of protonated histamines, *Helv. Chim. Acta*, **61**, 2530–2537 (1978). NB: See Cycliramine for details.

HO, HO, OH, $CH_2NHC(CH_3)_3$

CH_3O, O, NCH_3, CH_3O

N, $N-CH_3$, S

(continued)

No.	Compound Name	pK_a value(s)	Data quality	Ionization type	Method	Conditions	Comments and Reference(s)
1425	Theobromine ($C_7H_8N_4O_2$)	1.08	U	+H	kinetic	H_2O $t = 55$	Arnall F, The determination of the relative strengths of some nitrogen bases and alkaloids, *JCS*, **117**, 835–839 (1920). Cited in Perrin Bases 2972 ref. A73. NB: Gave relative values for a further 24 compounds.
		<1	U	+H	Spectro	H_2O $t = 25$	Ref. T16: Turner A and Osol A, The spectrophotometric determination of the dissociation constants of theophylline, theobromine and caffeine, *J. Am. Phar. Assoc.*, **38**, 158–161 (1949) for the spectrophotometric absorbance versus pH method.
		10.0	U	−H			
		9.9	U	−H	Potentiometric	H_2O $t = 18 \pm 0.5$	Ogston AG, Constitution of the purine nucleosides. III. Potentiometric determination of the dissociation constants of methylated xanthines, *JCS*, 1376–1379 (1935). CA 30:667. NB: Used hydrogen electrodes to measure pH changes during titration. Gave the following values in H_2O and 90% EtOH:
		11.3	U	−H		90% ethanol $t = 18 \pm 0.5$	

Xanthine	pK_a	pK_a (90% EtOH)	Xanthine	pK_a	pK_a (90% EtOH)
Xanthine	7.7	9.3	1,3-dimethyl	8.6	8.7
1-methyl	7.7	9.2	1,7-dimethyl	8.5	8.7
3-methyl	8.5	8.6	1,9-dimethyl	6.3	6.6
7-methyl	8.5	8.8	3,7-dimethyl	9.9	11.3
9-methyl	6.3	6.8	xanthosine	6.0	6.6

No.	Compound Name	pK_a value(s)	Data quality	Ionization type	Method	Conditions	Comments and Reference(s)
1426	Theobromine	0.29	U	+H		H_2O $t = 37$	Ballard BE and Nelson E, Physicochemical properties of drugs that control absorption rate after subcutaneous implantation, *JPET*, **135**, 120–127 (1962). NB: pK_a = 0.29 from pK_b = 13.33 at 37 °C (pK_w = 13.621). Secondary source W&G reported pK_a = 0.7 when citing this ref.
		8.8	U	−H			
1427	Theobromine	−0.16	U	+H	soly	H_2O $t = 40.1$	Wood JK, The affinity constants of xanthine and its methyl derivatives, *JCS*, **89**, 1839–1847 (1906). Cited in Perrin Bases suppl. no. 7489.
1428	Theobromine	9.96	U	−H	kinetic	H_2O $t = 25$	Wood JK, The affinity constants of xanthine and its methyl derivatives, *JCS*, **89**, 1839–1847 (1906). Cited in Perrin Bases suppl. no. 7489. NB: pK_a value estimated from catalytic effect of the base on the rate of hydrolysis of methylacetate.

No.	Compound	pKa	Data quality	Ionization	Method	Conditions	Reference
1429	Theophylline ($C_7H_8N_4O_2$) 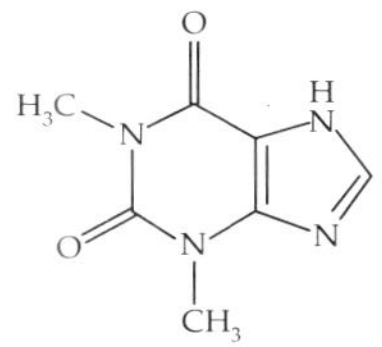	(0.5–2.5) 8.6	U U	+H −H	Spectro Potentiometric	H_2O $t = 25$ H_2O $I \sim 0.5$	Miyamoto H, Dissociation constant of 1,3-dimethylxanthine and thermodynamic quantities, *Sci. Rep. Niigata Univ.*, Ser. C, 23–30 (1969); CA 72:83560t (1969). Cited in Cohen JL, Theophylline, *APDS*, **4**, 466–493 (1975); Cohen JL, Ph.D. Dissertation, Univ. Wisconsin, 1969. NB: See also: Evstratova KJ and Ivanova AI, Dissociation constants and methods for analysis of some organic bases, *Farmatsiya*, **17**, 41–45, CA 69:46128a (1968); Linek K and Peciai C, *Chem. Zvesti.*, **16**, 692 (1962); CA., 58:8409h (1963).
1430	Theophylline	<1 8.6 8.6	U U U	+H −H −H	Spectro	H_2O $t = 25$ H_2O $t = 18 \pm 0.5$	Turner A and Osol A, The spectrophotometric determination of the dissociation constants of theophylline, theobromine and caffeine, *J. Am. Phar. Assoc.*, **38**, 158–61 (1949). Cited in Perrin Bases no. 2973 ref. T16. Ogston AG, *JCS*, 1376–39 (1935). CA 30:667. Cited in Perrin Bases no. 2973 ref. O3. NB: Used hydrogen electrodes to measure pH changes during titration.
1431	Theophylline	8.55 ± 0.01	A	−H	Potentiometric	H_2O $t = 25$ $I = 0.151$ (KCl)	Sirius Technical Application Notes, vol. **2**, p. 6 (1995). Sirius Analytical Instruments Ltd., Forest Row, East Sussex, RH18 5DW, UK. NB: From replicated titrations in aqueous solutions. The same result was reported in Bergström CAS, Strafford M, Lazorova L, Avdeef A, Luthman K and Artursson P, Absorption classification of oral drugs based on molecular surface properties, *J. Med. Chem.*, **46**(4) 558–570 (2003). NB: From extrapolation of aqueous-methanol mixtures to 0% methanol.
1432	Theophylline	8.75 ± 0.1	U	−H	Potentiometric	H_2O $t = 24$	Maulding HV and Zoglio MA, pK_a determinations utilizing solutions of 7-(2-hydroxypropyl) theophylline, *J. Pharm. Sci.*, **60**, 309–311 (1971). NB: See Barbituric acid, 5-allyl-5-isobutyl for details.
1433	Theophylline	0.36 8.77	U U	+H −H		H_2O $t = 37$	Ballard BE and Nelson E, Physicochemical properties of drugs that control absorption rate after subcutaneous implantation, *JPET*, **135**, 120–127 (1962). NB: pK_a = 0.36 from pK_b = 13.26 at 37 °C where pK_w = 13.621. Secondary source W&G reported pK_a = 0.7 when citing this ref.
1434	Theophylline	−0.24 8.79	U U	+H −H	kinetic	H_2O $t = 40.1$	Wood JK, The affinity constants of xanthine and its methyl derivatives, *JCS*, **89**, 1839–47 (1906). Cited in Perrin Bases suppl no. 7490. NB: The pK_a value was estimated from a catalytic effect on the rate of hydrolysis of methylacetate.

(continued)

No.	Compound Name	pK_a value(s)	Data quality	Ionization type	Method	Conditions	Comments and Reference(s)
1435	Thiomalic acid ($C_4H_6O_4S$) 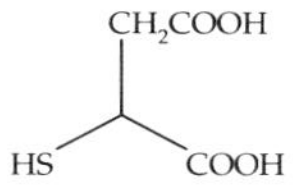	4.68 ± 0.04	U	−H	Potentiometric	H_2O $t = 20.0 \pm 0.05$ $I = 0.15$	Zucconi TD, Janauer GE, Donahe S and Lewkowicz C, Acid dissociation and metal complex formation constants of penicillamine, cysteine, and antiarthritic gold complexes at simulated biological conditions, *J. Pharm. Sci.*, **68**(4), 426–432 (1979). NB: See Cysteine for details.
1436	Thiopentone						See Barbituric acid, 5-ethyl-5-(1-methylbutyl)-2-thio.
1437	Thiopropazate ($C_{23}H_{28}ClN_3O_2S$) 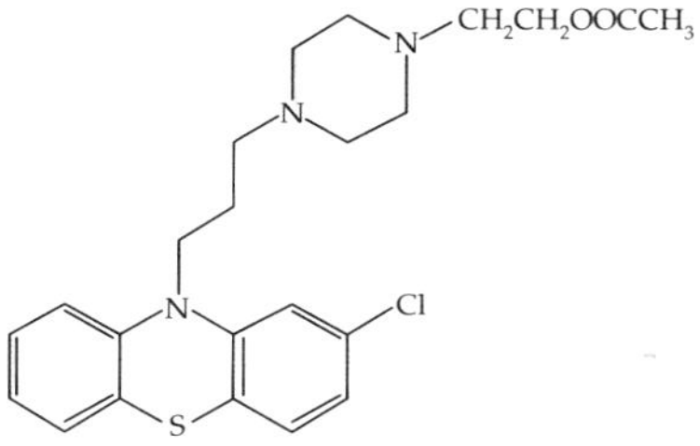	3.20 7.15	U U	+H +H	Potentiometric	H_2O (extrap) $t = 24 \pm 1$ $I \sim 0.002$	Chatten LG and Harris LE, Relationship between pK_b(H_2O) of organic compounds and $E_{1/2}$ values in several nonaqueous solvents, *Anal. Chem.*, **34**, 1495–1501 (1962). Cited in Foye 1 (2 refs); N&K; Chatten LG (ed.), *Pharmaceutical Chemistry*, vol. **1**, Dekker, NY, 1966, pp. 85–87.
1438	Thiopropazate	7.3	U	+H	soly	H_2O $t = 24 \pm 1$	Green AL, Ionization constants and water solublities of some aminoalkylphenothiazine tranquilizers and related compounds, *J. Pharm. Pharmacol.*, **19**, 10–16 (1967). NB: See Amitriptylline for details.
1439	Thioridazine ($C_{21}H_{26}N_2S_2$) 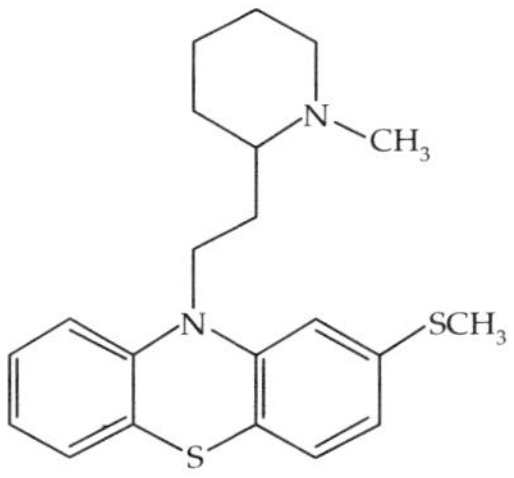	9.16	U	+H	Potentiometric	H_2O (extrap) $t = 24 \pm 1$ $I \sim 0.002$	Chatten LG and Harris LE, Relationship between pK_b(H_2O) of organic compounds and $E_{1/2}$ values in several nonaqueous solvents, *Anal. Chem.*, **34**, 1495–1501 (1962). NB: See Chlorpromazine for details.

1440	Thioridazine	9.62 ± 0.04	A	+H	partition/ pH	H_2O $t = 20 \pm 0.5$	Vezin WR and Florence AT, The determination of dissociation constants and partition coefficients of phenothiazine derivatives, *Int. J. Pharm.*, **3**, 231–237 (1979). NB: I was not reported but pK_a was stated to be independent of I. See Chlorpromazine and Promethazine for additional details.
1441	Thioridazine	9.5	U	+H	soly	H_2O $t = 24 \pm 1$	Green AL, Ionization constants and water solublities of some aminoalkylphenothiazine tranquilizers and related compounds, *J. Pharm. Pharmacol.*, **19**, 10–16 (1967). NB: See Amitriptylline for details.
1442	Thioridazine	9.45	U	+H	Potentiometric	H_2O (extrap) $t = 20$ N_2 atmos-phere	Sorby DL, Plein EM and Benmaman D, Adsorption of phenothiazine derivatives by solid adsorbents, *J. Pharm. Sci.*, **55**, 785–794 (1966). NB: See Chlorpromazine for details.
1443	Thyroxine, L- (Levothyroxine) ($C_{15}H_{11}I_4NO_4$)	2.2 6.7 10.1	U U U	−H −H +H	Potentiometric	H_2O (ref. 1)	Post A and Warren RJ, Levothyroxine sodium, *APDS*, **5**, 225–281 (1976).
		6.73 3.83 8.09 9.14	A A A A	−H −H −H +H	Spectro Potentiometric	H_2O (ref. 2) 75% DMSO (ref. 3) $t = 25$ $I = 0.1$ (KNO_3)	
1444	Tiapamil ($C_{26}H_{37}NO_8S_2$)	7.74 ± 0.01 8.48	U U	+H +H	Potentiometric	40% EtOH $t = 25.0$ H_2O	Mannhold R, Rodenkirchen R, Bayer R and Haus W, The importance of drug ionization for the action of calcium antagonistsand related compounds, *Arzneim.-Forsch.*, **34**, 407–409 (1984). NB: See Aprindine for details.

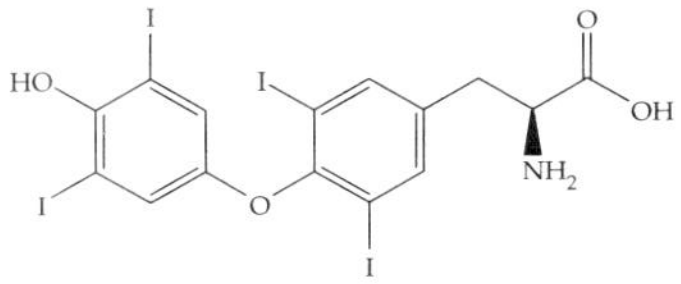

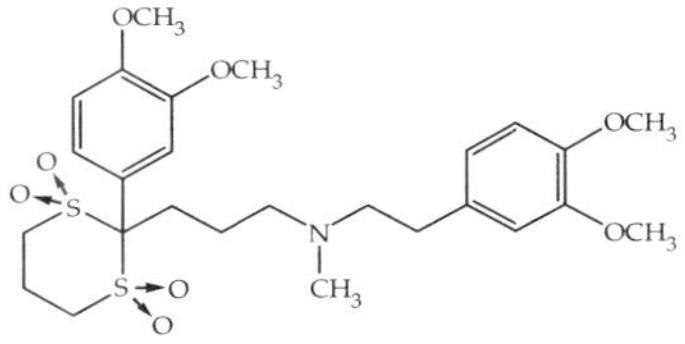

Notes to entry 1443:

"Function	pKa	pKa*
carboxyl	2.2 (1)	3.832 (3)
phenolic hydroxyl	6.7 (2)	8.085 (3)
amino	10.1 (1)	9.141 (3)"

(1) Dawson RM *et al.* (eds.), *Data for Biomedical Research*, 2nd Edn., OUP, NY (1969).
(2) Gemmill CL, The apparent ionization constants of the phenolic hydroxyl groups of thyroxine and related compounds, *Arch. Bichem. Biophys.*, **54**, 359–367 (1955). NB: Also reported corresponding values for: diiodotyrosine, 6.36; triiodothyronine, 8.45; diiodothyronine, 9.29; thyronine, 9.55.
(3) Wilson MF, Ionization of L-thyroxine in a 75% dimethylsulfoxide-water mixture. *Suomen Kemistilehto*, **45**, 53–55 (1972). NB: From K_a values of 1.471×10^{-4}, 8.226×10^{-9}, 7.236×10^{-10}.

(continued)

No.	Compound Name	pK_a value(s)	Data quality	Ionization type	Method	Conditions	Comments and Reference(s)
1445	Tienoxolol ($C_{21}H_{28}N_2O_5S$) 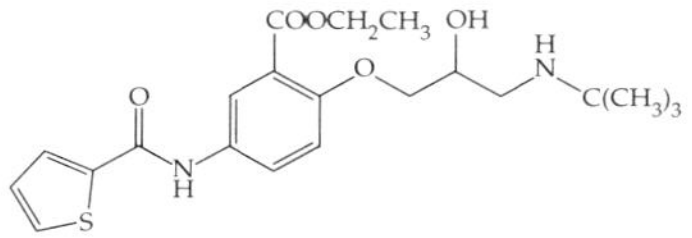	9.80 ± 0.1	U	+H	Potentiometric	H_2O	Guechot C, Bertrand A, Cramaille P and Teulon JM, Analytical profile of the new diuretic β-blocking agent tienoxolol hydrochloride, *Arzneim.-Forsch.*, **38**, 655–660 (1988). "The pK_a of the tienoxolol base was determined by titrimetry according to the OCDE's guideline [Ligne directice de l'OCDE pour les essais des produits chimiques N° **112**, OCDE, May 1981]. The pK_a was found to be 9.80 ± 0.1."
1446	Tilmicosin ($C_{46}H_{80}N_2O_{13}$) 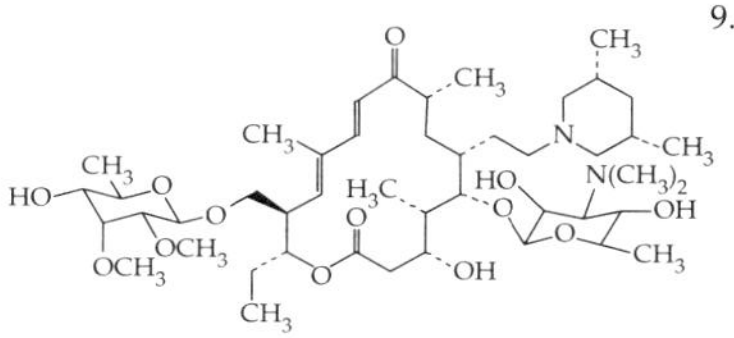	8.18 9.56	U U	+H +H	Potentiometric	H_2O $t = 25$ $I = 0.167$	McFarland JW, Berger CM, Froshauer SA, Hayashi SF, Hecker SJ, Jaynes BH, Jefson MR, Kamicker BJ, Lipinski CA, Lundy KM, Reese CP and Vu CB. Quantitative Structure-activity relationships among macrolide antibacterial agents: *In vitro* and *in vivo* potency against Pasteurella multocida, *J. Med. Chem.*, **40**, 1340–1346 (1997). NB: See Azithromycin for details; average standard deviation of ± 0.07 for the pK_a.
1447	Timolol ($C_{13}H_{24}N_4O_3S$) 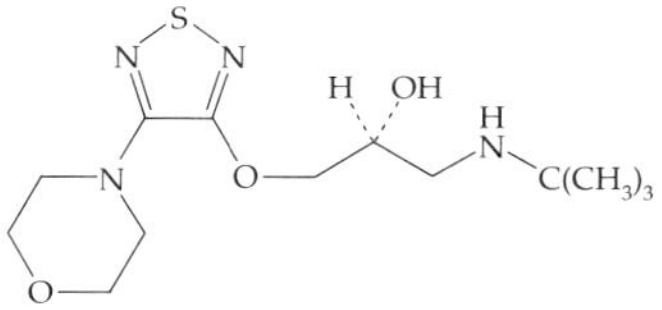	~9.2	U	+H	Potentiometric	H_2O $t = 25.0$	Mazzo DJ and Loper AE, Timolol Maleate, *APDS*, **16**, 641–692 (1987). Cited Oberholtzer E and Bondi JV, Dept. Pharmaceutical Research and Development, Merck Sharp and Dohme Research Laboratories, West Point PA, personal communication.
1448	Tixanox (7-methylsulfinyl-2-xanthone-carboxylic acid) ($C_{15}H_{10}O_5S$) 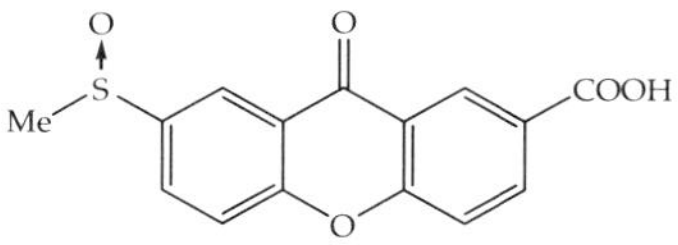	3.8	U	−H	soly	H_2O $t = 25$ I undefined	Chowhan ZT, pH solubility profiles of organic carboxylic acids and their salts, *J. Pharm. Sci.*, **67**, 1257–1260 (1978). NB: See Naproxen for details. Also reported an apparent pK_a value of 3.8 for 7-methylthio-2-xanthonecarboxylic acid.

1449	Tocainide ($C_{11}H_{16}N_2O$) 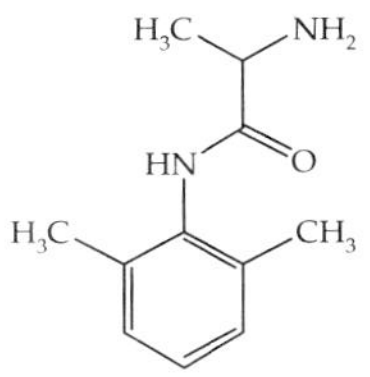	7.75	U	+H	Potentiometric	H_2O $t = 25.0 \pm 0.2$ $I = 0.01$ (NaCl)	Johansson P-A, Liquid-liquid distribution of lidocaine and some structurally related anti-arrythmic drugs and local anaesthetics, *Acta Pharm. Suec.*, **19**, 137–142 (1982).
1450	Tolazamide ($C_{14}H_{21}N_3O_3S$) 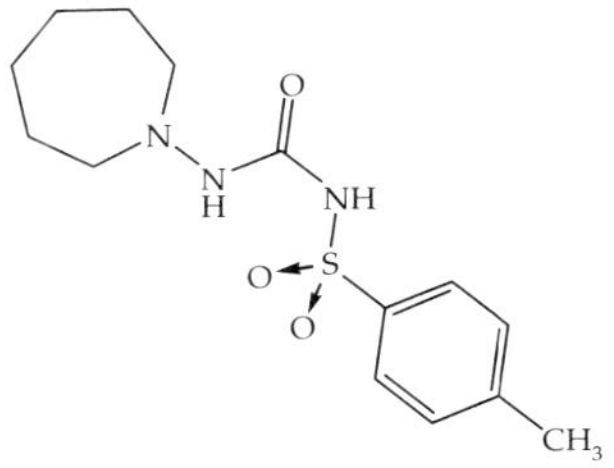	3.6 5.68	U U	−H +H		$t = 25$ $t = 37.5$	Lee JK, Chrzan K and Witt RH, Tolazamide, *APDS*, **22**, 489–516, 1993. NB: Appear to ascribe both values to the sulphonamide only. No reference for the 25 °C value. Reference for the 37.5 °C value cited as Remington 17th Edn.
1451	Tolazoline ($C_{10}H_{12}N_2$) 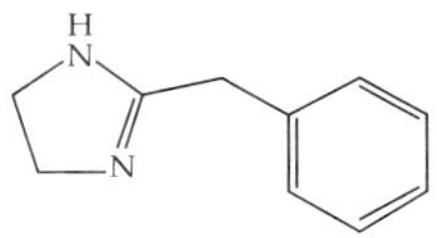	10.3	U	+H	Potentiometric	H_2O t undefined $I = 0.1$	Shore PA, Brodie BB and Hogben CAM, The gastric secretion of drugs, *JPET*, **119**, 361–369 (1957).
1452	Tolbutamide ($C_{12}H_{18}N_2O_3S$) 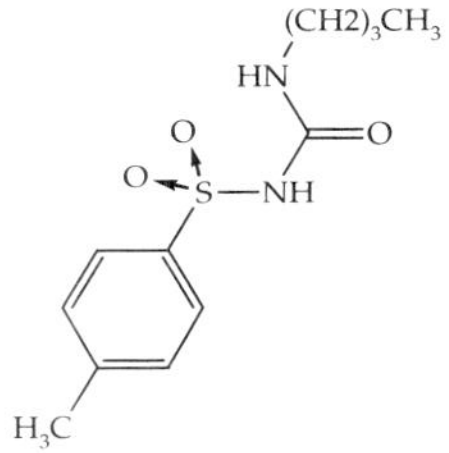	5.3 ± 0.1	U	−H	Potentiometric	H_2O t undefined I undefined	Haussler A and Hadju P, Dissociation constants and solubility of *N*-butyl-*N*-(*p*-tolylsulphonyl)urea, *Arch. Pharm. Weinheim*, **291**, 531–535 (1958). NB: No activity corrections.
		5.32	U	−H		H_2O $t = 37$	Ballard BE and Nelson E, Physicochemical properties of drugs that control absorption rate after subcutaneous implantation, *JPET*, **135**, 120–127, 1962.

(continued)

Appendix A *(continued)*

No.	Compound Name	pK_a value(s)	Data quality	Ionization type	Method	Conditions	Comments and Reference(s)
1453	Tolbutamide	5.3 ± 0.04	U	–H	Potentiometric	H_2O *t* undefined *I* undefined	Crooks MJ and Brown KF, The binding of sulphonylureas to serum albumin, *J. Pharm. Pharmacol.*, **26**, 305–311 (1974).
1454	Tolmetin ($C_{15}H_{15}NO_3$) 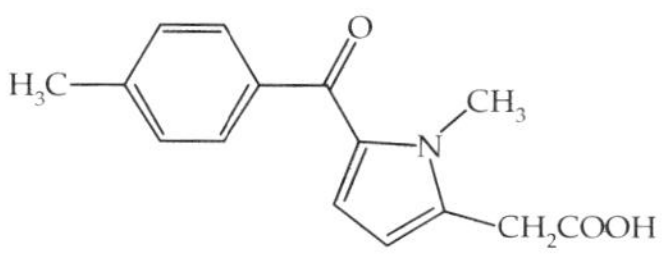	3.5	U	–H	Spectro	H_2O *t* undefined *I* undefined	Herzfeldt CD and Kümmel R, Dissociation constants, solubilities, and dissolution rates of some selected nonsteroidal antiinflammatories, *Drug Dev. Ind. Pharm.*, **9**(5), 767–793 (1983). NB: Used dA/dpH method. NB: See Azapropazone and Ibuprofen for details.
1455	Tolpropamide ($C_{18}H_{23}N$) 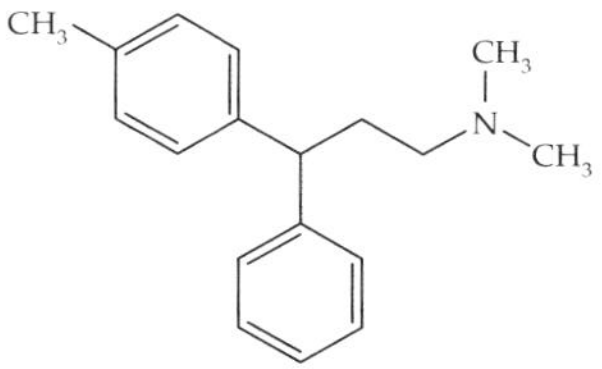	8.57 ± 0.08	U	+H	Potentiometric	H_2O *t* undefined $I = 0.30$ (NaCl)	Testa B and Murset-Rossetti L, The partition coefficient of protonated histamines, *Helv. Chim. Acta*, **61**, 2530–2537 (1978). NB: See Cycliramine for details.
1456	*p*-Toluic acid ($C_8H_8O_2$) 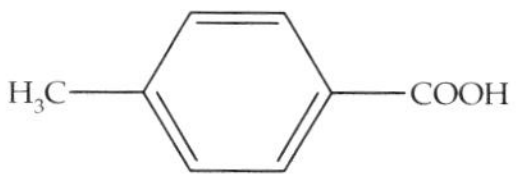	4.30 ± 0.09 4.41 ± 0.01	U U	–H	HPLC retention/ pH	H_2O $t = 25$ $I = 0.01$ $I = 0.1$	Unger SH, Cook JR and Hollenberg JS, Simple procedure for determining octanol-aqueous partition, distribution, and ionization coefficients by reversed phase high pressure liquid chromatography, *J. Pharm. Sci.*, 67; 1364–1367 (1978). NB: See Naproxen for details.

1457	Trazodone ($C_{19}H_{22}ClN_5O$) 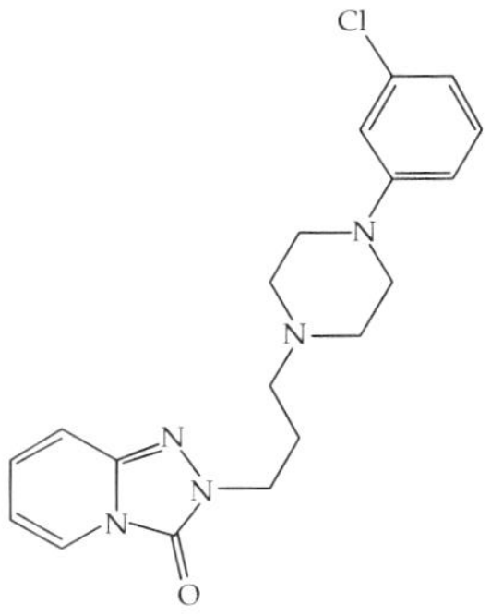	6.71 ± 0.02	U	+H	Potentiometric	H_2O $t = 20$ $c = 0.0001$	Suzuki H, Akimoto K, Nakagawa H and Sugimoto I, Quantitative analysis of trazodone hydrochloride in tablets by an ion-selective electrode, *J. Pharm. Sci.*, **78**, 62–65 (1989). NB: Glass/calomel electrode was used.
1458	Trazodone	6.14	U	+H	Potentiometric	50% EtOH	Baiocchi L, Chiari A, Frigerio A and Ridolfi P, Analytical profile of trazodone, *Arzneim.-Forsch.*, **23**, 400–406 (1973). NB: Glass/calomel electrode was used. No other experimental details given.
1459	Trazodone	6.14	U	+H	Potentiometric	EtOH	Gorecki, D.K.J. and Verbeeck, R.K., Trazodone hydrochloride, *APDS*, **16**, 693–730 (1987).
1460	Triazolam ($C_{17}H_{12}Cl_2N_4$) 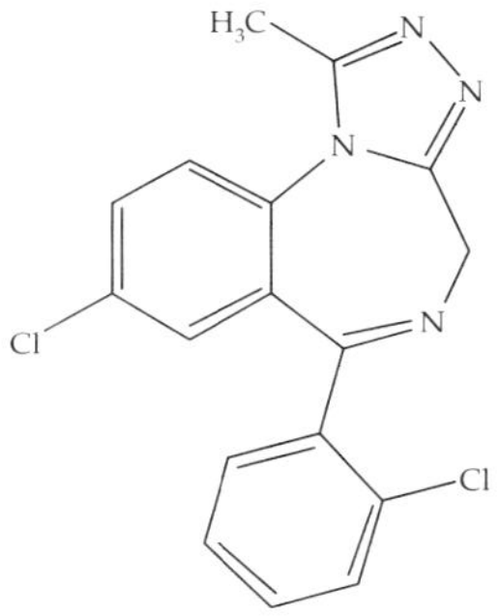	1.52 6.50	U U	+H +H	kinetic	H_2O	Konishi M, Hirai K, Mari Y, Kinetics and mechanism of the equilibrium reaction of triazolam in aqueous solution, *J. Pharm. Sci.*, **71**(12), 1328–1334 (1982). "The pK_a for the conjugate acid of Compounds I (NB: triazolam ring-opened hydrolysis product) and II (NB: triazolam) was estimated (NB: from the pH-rate profile) to be 6.50 and 1.52, respectively. No information has been reported about the pK_as for these compounds because the reaction in aqueous solution is too fast for measurement of the dissociation constant spectrometrically. . . . the pK_a for the conjugate acid of 1,4-benzodiazepines which possess a 2′-halogen substituent in the 5-phenyl group exhibit a pK_a value lower than that for corresponding nonsubstituent compounds. Namely, the pK_as for fludiazepam, flunitrazepam, and lorazepam were reported to be 2.29, 1.71, and 1.3, respectively, which were significantly lower than those of the corresponding compounds without halogen in the 2′-position, diazepam (pK_a = 3.3), nitrazepam (pK_a = 3.2), and oxazepam (pK_a = 1.7). The pK_a value for 8-chloro-6-phenyl-4H-s-triazolol[4,3-a][1,4] benzodiazepine (estazolam) was reported to be 2.84 from the UV absorption spectral change (NB: Koyama H, Yamada M and Matsuzawa T, *J. Takeda Res. Lab.*, **32**, 77–90 (1973)). Considering the structural difference mentioned, the estimated pK_a value for triazolam, 1.52, is reasonable."

(continued)

No.	Compound Name	pK_a value(s)	Data quality	Ionization type	Method	Conditions	Comments and Reference(s)
1461	Trichloromethiazide ($C_8H_8Cl_3N_3O_4S_2$) 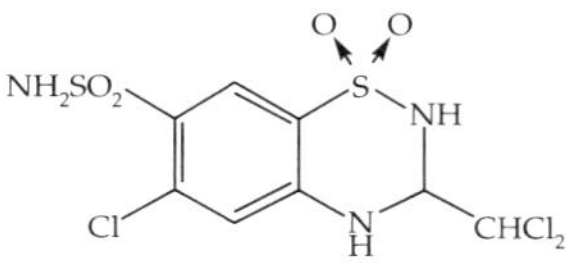	6.9	U	−H	Potentiometric	acetone/H_2O	Henning VG, Moskalyk RE, Chatten LG and Chan SF, Semiaqueous potentiometric determination of apparent pK_{a1} values for benzothiadiazines and detection of decomposition during solubility variation with pH studies, *J. Pharm. Sci.*, **70**, 317–319 (1981). NB: See Flumethiazide.
1462	Trifluoperazine ($C_{21}H_{24}F_3N_3S$) 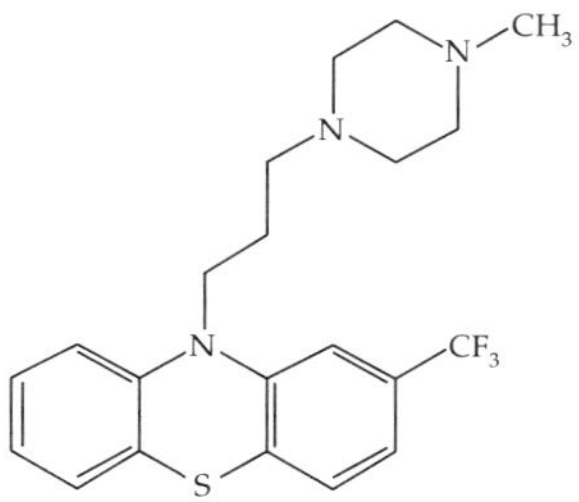	8.05 ± 0.04	A	+H	partition/ pH	H_2O $t = 20 \pm 0.5$ I not reported but pK_a was stated to be independent of I.	Vezin WR, Florence AT, The determination of dissociation constants and partition coefficients of phenothiazine derivatives, *Int. J. Pharm.*, **3**, 231–237 (1979). NB: See Chlorpromazine and Promethazine for additional details. See separate entries for Phenothiazine, 2-chloro-10-(*N*-methylpiperazinyl)-3-propyl-.
1463	Trifluoperazine	3.9 8.4	U U	+H +H	Potentiometric	H_2O $I = 0.00$	Attwood D and Natarajan, R., Effect of pH on the micellar properties of amphiphilic drugs in aqueous solution, *J. Pharm. Pharmacol.*, **33**(2), 136–140 (1981). "...pK_a determination. pK_a values were determined by potentiometric titration (Albert & Serjeant 1971) using a Dye Model 290 pH meter with a combined glass-silver chloride electrode ... An indication of the possible magnitude of such error was ascertained by determinations on chlorpromazine hydrochloride, which has a single ionizable group, at pH values well below the pK_a of 9.2 (Sorby et al., 1966). .... Trifluoperazine: The pK_{a1} and pK_{a2} values of this drug are 3.8 and 8.4 respectively ... The extrapolated value of 3.9 for pK_{a1} at zero concentration is identical with that quoted by Chatten LG, Harris LE, Relationship between p$K_b(H_2O)$ of organic compounds and $E_{1/2}$ values in several nonaqueous solvents, *Anal. Chem.*, **34**, 1495–1501 (1962)."

No.	Compound	pKa			Method	Conditions	Reference
1464	Trifluoperazine	8.1	U	+H	soly	H_2O $t = 24 \pm 1$	Green AL, Ionization constants and water solublities of some aminoalkylphenothiazine tranquilizers and related compounds, *J. Pharm. Pharmacol.*, **19**, 10–16 (1967). NB: See Amitriptylline for details.
1465	Trifluoperazine	4.10 8.36	U U	+H +H	Potentiometric	H_2O (extrap) $t = 20$ N_2 atmosphere	Sorby DL, Plein EM and Benmaman D, Adsorption of phenothiazine derivatives by solid adsorbents, *J. Pharm. Sci.*, **55**, 785–794 (1966). NB: See Chlorpromazine for details.
1466	Trifluoperazine	3.90 8.40	U U	+H +H	Potentiometric	H_2O (extrap) $t = 24 \pm 1$ $I \sim 0.002$	Chatten LG and Harris LE, Relationship between $pK_b(H_2O)$ of organic compounds and $E_{1/2}$ values in several nonaqueous solvents. *Anal. Chem.*, **34**, 1495–1501 (1962). NB: See Chlorpromazine for details.
1467	Trifluoperazine	4.04 8.08	U U	+H +H	Potentiometric	H_2O $t = 23.0$	Clarke FH and Cahoon NM, Ionization constants by curve-fitting: Determination of partition and distribution coefficients of acids and bases and their ions, *J. Pharm. Sci.*, **76**, 611–620 (1987). NB: See Benzoic acid for further details.
1468	Triflupromazine ($C_{18}H_{19}F_3N_3S$) 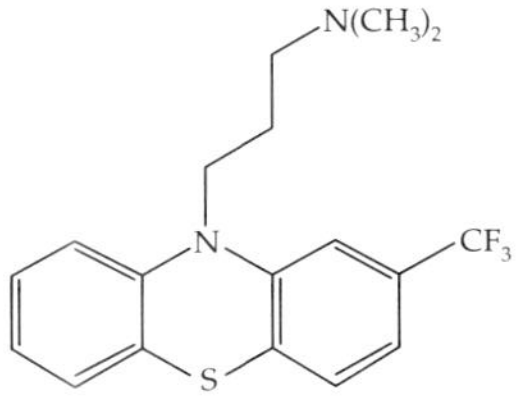	9.41	U	+H	Potentiometric	H_2O (extrap) $t = 24 \pm 1$ $I \sim 0.002$	Chatten LG and Harris LE, Relationship between $pK_b(H_2O)$ of organic compounds and $E_{1/2}$ values in several nonaqueous solvents, *Anal. Chem.*, **34**, 1495–1501 (1962).
1469	Triflupromazine	9.29 ± 0.04	U	+H	partition/ pH	H_2O $t = 20 \pm 0.5$	Vezin WR and Florence AT, The determination of dissociation constants and partition coefficients of phenothiazine derivatives, *Int. J. Pharm.*, **3**, 231–237 (1979). NB: I was not reported but the pK_a was stated to be independent of I. See Chlorpromazine and Promethazine for additional details.
1470	Trimethoprim ($C_{14}H_{18}N_4O_3$) 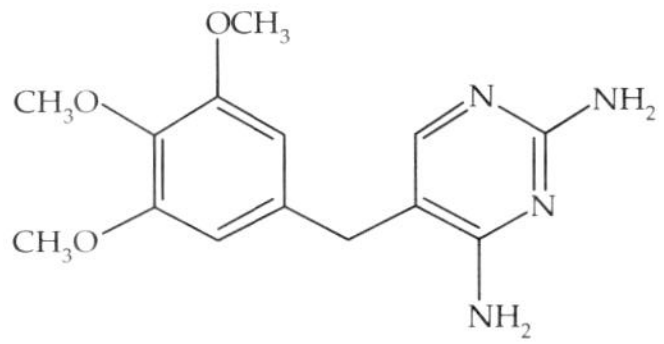	6.6	U	+H	Potentiometric	H_2O (extrap) *t* undefined *I* undefined	Manius GJ, Trimethoprim, *APDS*, **7**, 445–475 (1978). Cited Piccio E, Lau E, Senkowski BZ, Hoffmann-La Roche, unpublished data. NB: Value extrapolated from linear plot of apparent values in EtOH-water mixtures. A good example (when compared with no. 1471) of the risks of linear extrapolation of results from cosolvent-water mixtures.

(continued)

No.	Compound Name	pK_a value(s)	Data quality	Ionization type	Method	Conditions	Comments and Reference(s)
1471	Trimethoprim	7.12 ± 0.03	A	+H	Spectro	H_2O $t = 20 \pm 0.2$ I = see text	Roth B and Strelitz JZ, The protonation of 2,4-diaminopyrimidines. I. Dissociation constants and substituent effects, *J. Org. Chem.*, **34**(4), 821–836 (1969). NB: Solutions were prepared from dilutions of stock solution in water or EtOH; 1–2% of EtOH was found not to have observable effects on the spectra. Buffers were prepared by dilution of stock solutions with freshly boiled double-distilled water. Solutions in acetate (0.01 or 0.002 M) or phosphate (0.0067 or 0.0022 M) buffers gave pK_a values that were slightly lower in the more dilute buffers; these were taken without further correction to be the thermodynamic pK_a values. Most of the values were from measurements with a Beckman Model G pH meter; later measurements also used a Beckman Research Model pH meter, which gave values that differed by not more than 0.01 log unit from the values determined simultaneously with the Model G. Spectral readings from at least 12 wavelengths were used for each pK_a value. Data were rejected if the spectra did not show a precise isosbestic point. Recorded deviations are the range of values, not the standard deviation. pK_a values for a total of about 70 substituted 2,4-diaminopyrimidines were reported.
	Related 2,4-diaminopyrimidines:						
	5-methyl	7.69 ± 0.03	A	+H			
	5-benzyl	7.27 ± 0.03	A	+H			
	5-(*p*-chloro)benzyl	7.17 ± 0.05	A	+H			
	Trimethoprim	7.19 ± 0.03	A	+H		H_2O $t = 20 \pm 0.2$ $I = 0.1$ (NaCl)	
	Related 2,4-diaminopyrimidines:						
	5-methyl	7.76 ± 0.03	A	+H			
	5-(*p*-chloro)benzy	7.25 ± 0.03	A	+H			
1472	Tripelennamine ($C_{16}H_{21}N_3$) 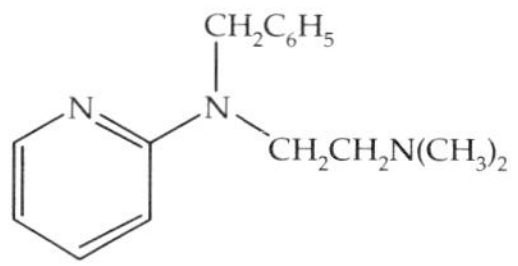	3.90 ± 0.08 8.68 ± 0.06	U U	+H +H	Potentiometric	H_2O t undefined $I = 0.30$ (NaCl)	Piskorik HG, Tripelennamine hydrochloride, *APDS*, **14**, 107–131 (1985). Testa B and Murset-Rossetti L, The partition coefficient of protonated histamines, *Helv. Chim. Acta*, **61**, 2530–2537 (1978). NB: See Cycliramine for details. Martindale 28th Ed gave 3.9 and 9.0 at 25 °C.
1473	Triprolidine ($C_{19}H_{22}N_2$) 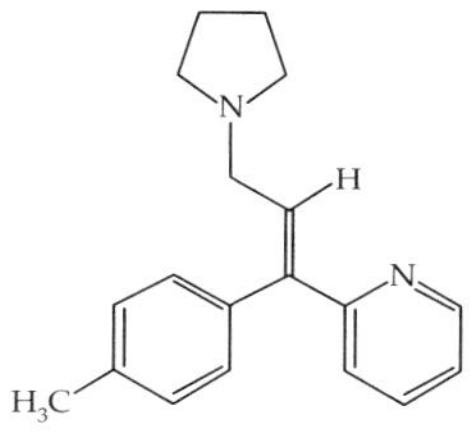	4.01 9.69	U U	+H +H	Potentiometric	H_2O $t = 23.0$	Clarke FH and Cahoon NM, Ionization constants by curve-fitting: Determination of partition and distribution coefficients of acids and bases and their ions, *J. Pharm. Sci.*, **76**, 611–620 (1987). NB: See Benzoic acid for further details.

No.	Compound	pKa	Rating	Type	Method	Conditions	Reference / Notes
1474	Tromethamine ($C_4H_{11}NO_3$) 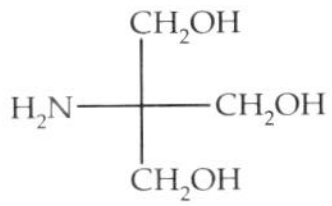	8.08 ± 0.005	R	+H	Potentiometric	H_2O $t = 25$ $I = 0.00$	Bates RG and Pinching GD, Dissociation constants of weak bases from electromotive-force measurements of solutions of partially hydrolyzed salts, *J. Res. Nat. Bur. Stand.*, **43**, 519–526 (1949). Cited in Perrin Bases, ref. B31. NB: Used glass electrodes in a highly refined electrochemical cell. Other reliable values reported for temperatures between 0 and 50 °C.
1475	Tropacocaine ($C_{15}H_{19}NO_2$) 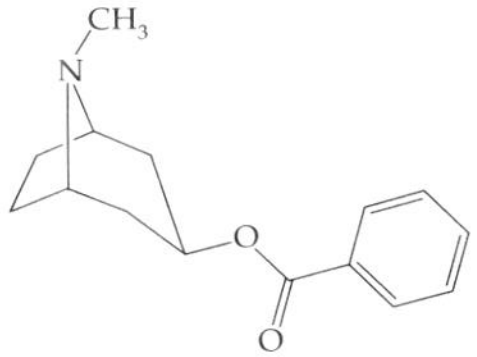	9.51 ± 0.03	A	+H	Potentiometric	H_2O $t = 20.0$ $I = 0.10$ (KCl) under N_2	Buchi J and Perlia X, Beziehungen zwischen de physikalisch-chemische Eigenschaften und der Wirkung von Lokalanasthetica, *Arzneim.-Forsch.*, **10**, 745–754 (1960). NB: See Cocaine for details.
1476	Tropacocaine	9.88	U	+H	Spectro	H_2O $t = 15$ $I < 0.02$	Kolthoff IM, The dissociation constants, solubility product and titration of alkaloids, *Biochem. Z.*, **162**, 289–353 (1925). Cited in Perrin Bases no. 2975 ref. K47. NB: See Aconitine for details.
1477	DL-Tropic acid ($C_9H_{10}O_3$) 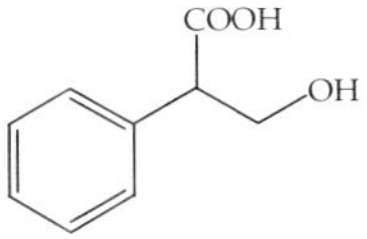	3.38	U	−H	Conductance	H_2O $t = 25$ $c = 0.03–0.001$	Kendall J, Electrical conductivity and ionization constants of weak electrolytes in aqueous solution, *in* Washburn EW, Editor-in-Chief, *International Critical Tables*, Vol. 6, McGraw-Hill, NY, 259–304 (1929).
		4.20	U	−H	Potentiometric	H_2O $t = 25$	Randinitis EJ, Barr M, Wormser HC and Nagwekar JB, Kinetics of urinary excretion of D-(−)-mandelic acid and its homologs. I. Mutual inhibitory effect of D-(−)-mandelic acid and its certain homologs on their renal tubular secretion in rats, *J. Pharm. Sci.*, **59**, 806–812 (1970). NB: See Mandelic acid for details.
1478	Tropine ($C_8H_{15}NO$) 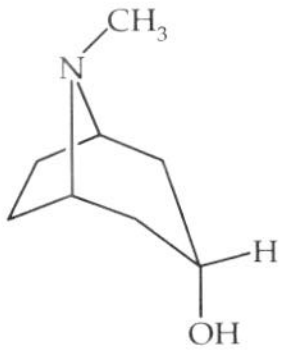	10.33	A	+H	Potentiometric	H_2O $t = 25$ $I = 0.0007$	Geissman TA, Wilson BD and Medz RB, The base strengths of cis- and trans-1,2-aminoalcohols, *JACS*, **76**, 4182–4183 (1954). Cited in Perrin Bases no. 2976 ref. G7. NB: See Pseudotropine for details.
		11.02	U	+H		$I = 0.05$	Smith PF and Hartung WH, Cis- and trans-tropine (tropanol), *JACS*, **75**, 3859–3860 (1953). Cited in Perrin Bases no. 2976 ref. S56. NB: See Pseudotropine for details.

(continued)

No.	Compound Name	pK_a value(s)	Data quality	Ionization type	Method	Conditions	Comments and Reference(s)
1479	Tropine	10.44	U	+H	Conductance	H_2O $t = 25$ $c = 0.06–0.01$	Kendall J, Electrical conductivity and ionization constants of weak electrolytes in aqueous solution, *in* Washburn EW, Editor-in-Chief, *International Critical Tables*, vol. **6**, McGraw-Hill, NY, 259–304 (1929). NB: Other values: $t = 10$, $pK_a = 10.27$; $t = 50$, $pK_a = 10.59$.
1480	Tryptophan ($C_{11}H_{12}N_2O_2$)	2.34 9.51	A A	–H +H	CE/pH (+ve ion mode)	H_2O $t = 25$ $I = 0.025$	Wan H, Holmen AG, Wang Y, Lindberg W, Englund M, Nagard MB and Thompson RA, High-throughput screening of pK_a values of pharmaceuticals by pressure-assisted capillary electrophoresis and mass spectrometry, *Rapid Commun. Mass Spectrom.*, **17**, 2639–2648 (2003). NB: Reported predicted values (ACD Labs) of 2.3 and 9.51.
		2.30 9.30	A A	–H +H	Potentiometric	H_2O $t = 25.0$ $I = 0.15$ (KCl)	Sirius Technical Application Notes, vol. **2**, p. 10 (1995). Sirius Analytical Instruments Ltd., Forest Row, East Sussex, RH18 5DW, UK.
1481	Tryptophan	2.43 9.44	A A	–H +H	Potentiometric	H_2O $t = 25$ $c = 0.025$	Schmidt CLA, Appleman WK and Kirk PL, The apparent dissociation constants of tryptophane and of histidine, *J. Biol. Chem.*, **85**, 137–140 (1929). Cited in Perrin Bases no. 3301 r,ef. S13. NB: Used hydrogen electrodes to measure pH of titrations. From a K_a value of 4.05×10^{-10} and a K_b value of 2.4×10^{-12}. Also reported for histidine a K_a value of 6.7×10^{-10} and K_b values of 1.15×10^{-8} and 2.90×10^{-13}, corresponding to the following pK_a values: 1.47, 6.07, 9.17.
		2.18 9.57	A A	–H +H	Potentiometric	H_2O $t = 20$ $c = 0.005$ N_2 atmosphere	Albert A, Quantitative studies of the avidity of naturally occurring substances for trace metals. *Biochem. J.* **47**, 531–538 (1950). Cited in Perrin Bases no. 3301 ref. A12. NB: Used glass electrodes for measurement of pH during titrations in carbonate-free conditions. All pK_a values were reliable to ± 0.05. NB: Other less reliable values were also cited by Perrin.
1482	Tuaminoheptane ($C_7H_{17}N$)	10.48	U	+H	Potentiometric	H_2O (extrap) $t = 24 \pm 1$ $I \sim 0.002$	Chatten LG and Harris LE, Relationship between $pK_b(H_2O)$ of organic compounds and $E_{1/2}$ values in several nonaqueous solvents. *Anal. Chem.*, **34**, 1495–1501 (1962). Cited in: Foye 1 (1 ref), see idoxuridine; N&K; Chatten LG (ed.), *Pharmaceutical Chemistry*, vol. **1**, Dekker, NY, 1966, pp. 85–87.

1483	Tylosin ($C_{46}H_{77}NO_{17}$) 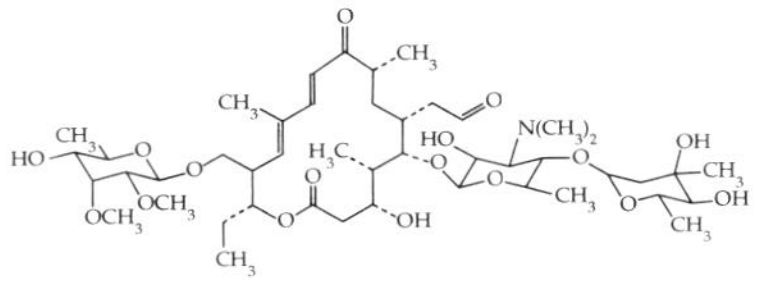	7.73	U	+H	Potentiometric	H_2O $t = 25$ $I = 0.167$	McFarland JW, Berger CM, Froshauer SA, Hayashi SF, Hecker SJ, Jaynes BH, Jefson MR, Kamicker BJ, Lipinski CA, Lundy KM, Reese CP and Vu CB, Quantitative Structure-activity relationships among macrolide antibacterial agents: *In vitro* and *in vivo* potency against Pasteurella multocida, *J. Med. Chem.*, **40**, 1340–1346 (1997). NB: See Azithromycin for details; average standard deviation of ± 0.07 for the pK_a.
1484	Tyrosine ($C_9H_{11}NO_3$) 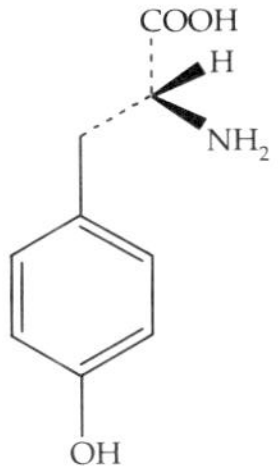	2.18 9.21 10.47 9.11 ± 0.01 10.13 ± 0.01	U A A A A	−H +H −H +H −H	Potentiometric Potentiometric	H_2O $t = 20$ $c = 0.005$ H_2O $t = 25$ $I = 0.16$ (KCl)	Albert A, Quantitative studies of the avidity of naturally occurring substances for trace metals, *Biochem. J.*, **50**, 690–7 (1952). Cited in Perrin Bases 1961 no. 3304 ref. A14. NB: Used glass electrodes for measurement of pH during titrations in carbonate-free conditions. All pK_a values were reliable to ± 0.05. Other less reliable values were also given: $pK_1 = 2.6$; $pK_2 = 9.5$; $pK_3 = 10.3$ (spectro, no temperature reported). Edsall JT, Martin RB and Hollingworth BR, Ionizatin of individual groups in dibasic acids, with application to the amino and hydroxyl groups of tyrosine, *Proc. Natl. Acad. Sci. USA*, **44**, 505–518 (1958). Cited in Perrin Bases 1961 no. 3304 ref. E7. NB: Used a glass electrode in cells with liquid junction potentials. These results were from direct titrations. pK_a values calculated from the measured microconstants were very similar.
1485	Tyrosine	2.06 ± 0.05 9.18 ± 0.06 10.40 ± 0.09	A A U	−H +H −H	soly	H_2O $t = 25.0$ $I = 0.00$	Peck CC and Benet LZ, General method for determining macrodissociation constants of polyprotic amphoteric compounds from solubility measurements, *J. Pharm. Sci.*, **67**, 12–16 (1978). NB: Ionic strengths were NMT 0.07M and then extrapolated to zero. Intrinsic solubility, $S_o = 2.89$ µM. See Dihydroxyadenine for further details.
1486	Ursodeoxycholic acid (ursodiol) ($C_{24}H_{40}O_4$) 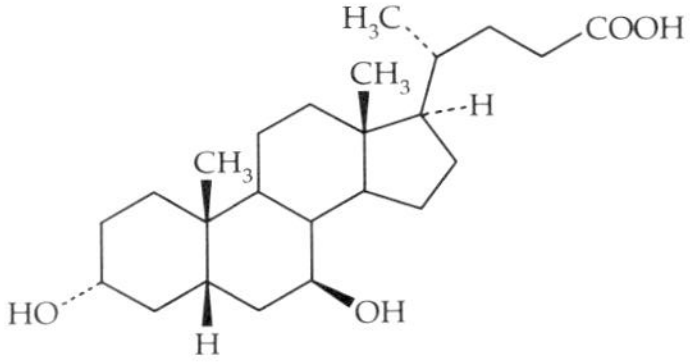	5.04 ± 0.04	A	−H	Potentiometric	H_2O $t = 25.0 \pm 0.1$ $I = 0.00$	Fini A and Roda A, Chemical properties of bile acids. IV. Acidity constants of glycine-conjugated bile acids, *J. Lipid Res.*, **28**(7), 755–759 (1987). NB: See Chenodeoxycholic acid for full details.

(continued)

No.	Compound Name	pK_a value(s)	Data quality	Ionization type	Method	Conditions	Comments and Reference(s)
1487	Valproic acid ($C_8H_{16}O_2$) 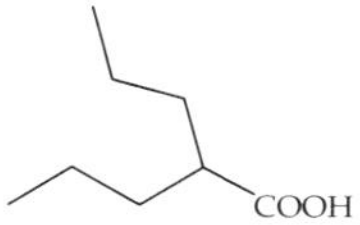	4.6	U	–H	Potentiometric	H_2O (extrap)	Chang ZL, Sodium Valproate and Valproic Acid, *APDS*, **8**, 529–556 (1979). NB: These results from titration of valproic acid with NaOH and extrapolated to 100% water from acetone–water mixtures. Titration of aqueous sodium valproate solution with aqueous HCl gave pK_a = 4.8.
1488	Vancomycin ($C_{66}H_{75}Cl_2N_9O_{24}$) 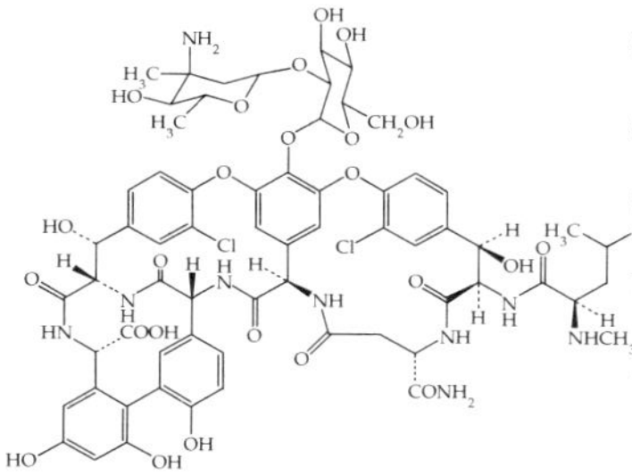	2.18 ± 0.08	U	–H (COOH)	Potentiometric	H_2O t = 25.0 ± 0.1 I = 0.2 (NaCl) N_2 atmosphere	Takacs-Novak K, Noszal B, Tokes-Kovesdi M and Szasz G, Acid-base properties and proton speciation of vancomycin, *Int. J. Pharm.*, **89**, 261–263 (1993). NB: The functional group in parentheses is the group mainly responsible for each macroconstant.
		7.75 ± 0.02	U	+H ($NHCH_3$)			
		8.89 ± 0.01	U	+H (NH_2)			
		9.59 ± 0.01	U	–H (phenol)			
		10.40 ± 0.02	U	–H (phenol)			
		12.00 ± 0.08	U	–H (phenol)			
1489	Vancomycin	2.64 ± 0.01	U	–H (COOH)	Potentiometric	H_2O t = 25.0 ± 0.1 I = 0.1 (NaCl)	Takacs-Novak K, Box KJ and Avdeef A, Potentiometric pK_a determination of water-insoluble compounds: Validation study in methanol/water mixtures, *Int. J. Pharm.*, **151**, 235–248 (1997). NB: See Acetaminophen for full details. The functional group in parentheses is the group mainly responsible for each macroconstant. pK_{a1} = 2.68 ± 0.05, pK_{a2} = 7.46 ± 0.03, pK_{a3} = 8.71 ± 0.04, pK_{a4} = 9.27 ± 0.03, pK_{a5} = 10.33 ± 0.07, pK_{a6} = 11.77 ± 0.08, by extrapolation from 6–37% w/w aqueous MeOH. Very similar results (± 0.03 at worst) for I = 0.166 M (KCl) were reported in Sirius Technical Application Notes, vol. **2**, pp. 32–35 (1995). Sirius Analytical Instruments Ltd., Forest Row, East Sussex, RH18 5DW, UK. Results were also obtained in 0.5 M KCl (2.63, 7.56, 8.63, 9.21, 9.94, 11.46).
		7.48 ± 0.01	U	+H ($NHCH_3$)			
		8.62 ± 0.01	U	+H (NH_2)			
		9.28 ± 0.01	U	–H (phenol)			
		10.15 ± 0.01	U	–H (phenol)			
		11.88 ± 0.01	U	–H (phenol)			

1490	Vancomycin	6.84 ± 0.05 7.82 ± 0.05	U U	+H	NMR	H_2O $t = 50$ I unspecified	Antipas AS, Vander Velde D and Stella VJ, Factors affecting the deamidation of vancomycin in aqueous solutions, *Int. J. Pharm.*, **109**, 261–269 (1994). "The degradation kinetics of vancomycin hydrochloride were investigated in 2 mM aqueous solutions at 50 °C at pH 1–9.8, and amine pK_a values were titrated by ^{1}H-NMR at 50 °C. The deamidation of vancomycin between pH 3 and 9.8 followed pseudo-first order kinetics. In addition, the pK_a values were lower than those reported at 25 °C. Rate constants obtained from curve fitting the pH-rate profile to the observed data indicate that the reactivity of vancomycin toward deamidation in the region of pH 6–9 is influenced by its ionic state. . . . the ionic state of vancomycin may influence its degradation rate."
1491	Vanillic acid ($C_8H_8O_4$) HO– –COOH H_3CO	4.53	U	–H	Conductance	H_2O $t = 25$ $c = 0.016$–0.001	Kendall J, Electrical conductivity and ionization constants of weak electrolytes in aqueous solution, *in* Washburn EW, Editor-in-Chief, *International Critical Tables*, vol. **6**, McGraw-Hill, NY, 259–304 (1929).
1492	Vanillin ($C_8H_8O_3$) CHO OCH_3 OH	7.396 ± 0.004	R	–H	Spectro	H_2O $I = 0.00$ $t = 25.0 \pm 1.0$	Robinson RA and Kiang AK, The ionization constants of vanillin and two of its isomers, *Tr. Farad. Soc.*, **51**, 1398–1402 (1955). NB: Compounds were recrystallized to constant melting point. pH values were defined by carefully prepared phosphate solutions for which pH values were known from the electrometric measurements of Bates at the US National Bureau of Standards. Activities were estimated from Davies' equation.
1493	Vanillin, *iso* ($C_8H_8O_3$) H O OCH_3 OH	$8.88_9 \pm 0.01_9$	A	–H	Spectro	H_2O $I = 0.00$ $t = 25.0 \pm 1.0$	Robinson RA and Kiang AK, The ionization constants of vanillin and two of its isomers, *Tr. Farad. Soc.*, **51**, 1398–1402 (1955). NB: See Vanillin for further details.

(continued)

No.	Compound Name	pK_a value(s)	Data quality	Ionization type	Method	Conditions	Comments and Reference(s)
1494	Vanillin, *ortho* ($C_8H_8O_3$) 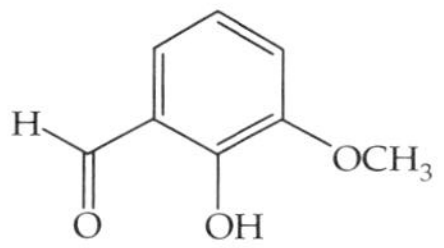	$7.91_2 \pm 0.03_6$	A	−H	Spectro	H_2O $I = 0.00$ $t = 25.0 \pm 1.0$	Robinson RA and Kiang AK, The ionization constants of vanillin and two of its isomers, *Tr. Farad. Soc.*, **51**, 1398–1402 (1955). NB: See Vanillin for further details.
1495	Verapamil (dexverapamil) ($C_{27}H_{38}N_2O_4$) 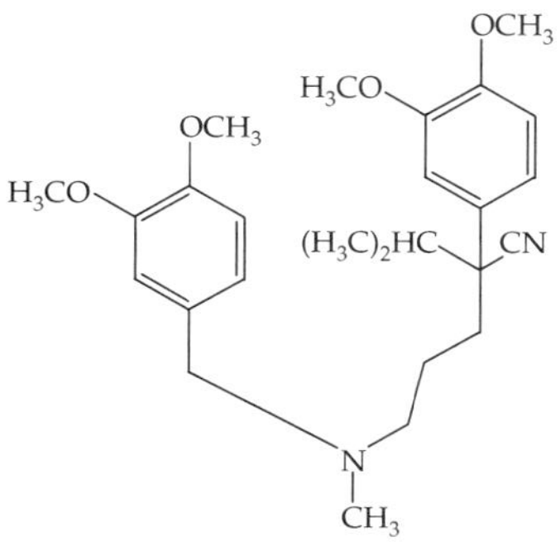	8.90	U	+H	soly	H_2O	Surakitbanharn Y, McCandless R, Krzyzaniak JF, Dannenfelser RM and Yalkowsky SH, Self-association of dexverapamil in aqueous solution, *J. Pharm. Sci.*, **84**, 720–723 (1995). "The self-association of dexverapamil hydrochloride and its effect on the solubility-pH profile of the drug were investigated. The pK_a and intrinsic solubility of monomeric dexverapamil were determined from its pH-solubility profile to be 8.90 and 6.6×10^{-5} M, respectively. . . .. The apparent pK_a of the self-associated drug was estimated to be 7.99. . . .. The dependence of the drug solubility on pH and the solubilization of naphthalene and anthracene as a function of ionized drug concentration suggest that the self-associated dexverapamil is a cationic dimer."
1496	Verapamil	8.75	U	+H	partition	H_2O	Hasegawa J, Fujita T, Hayashi Y, Iwamoto K and Watanabe J, pK_a determination of verapamil by liquid-liquid partition, *J. Pharm. Sci.*, **73**, 442–445 (1984). "The pK_a of verapamil hydrochloride (I) was determined by measuring the partition coefficient of I between n-heptane and aqueous buffer solution at various pH values. The effect of ionic strength and temperature on the pK_a was also measured. The estimated pK_a in human plasma was 8.75."
1497	Verapamil	9.07	U	+H	Potentiometric	H_2O $t = 25$	Bergström CAS, Strafford M, Lazorova L, Avdeef A, Luthman K and Artursson P, Absorption classification of oral drugs based on molecular surface properties, *J. Med. Chem.*, **46**(4), 558–570 (2003). NB: from extrapolation of aqueous-methanol mixtures to 0% methanol.

1498	Verapamil	7.99 ± 0.02	U	+H	Potentiometric	40% EtOH $t = 25.0$	Mannhold R, Rodenkirchen R, Bayer R and Haus W, The importance of drug ionization for the action of calcium antagonistsand related compounds, *Arzneim.-Forsch.*, **34**, 407–409 (1984). NB: See Aprindine for details.
		8.73	U	+H		H_2O	
1499	Vinblastine (VLB; vincaleukoblastine) ($C_{46}H_{58}N_4O_9$) 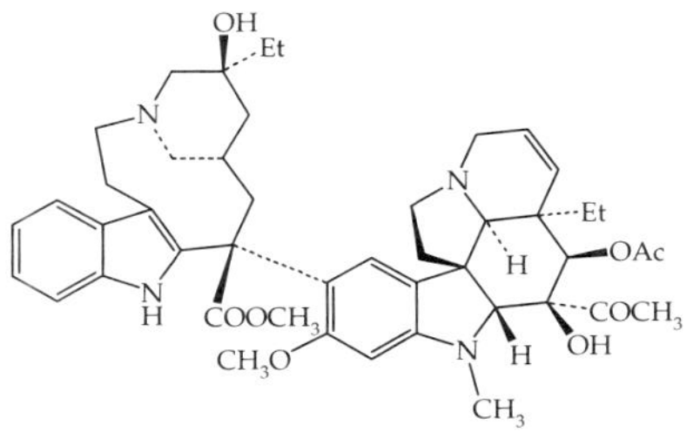	5.4 7.4	U U	+H +H	Potentiometric	H_2O t unspecified *I* unspecified	Neuss N, Gorman M, Svoboda GH, Maciak G and Beer T, Vinca alkaloids. III. Characterization of leurosine and vincaleukoblastine, new alkaloids from *Vinca rosea* Linn., *JACS*, **81**, 4754–4755 (1959). Cited in Burns JH, Vinblastine sulfate, *APDS*, **1**, 450 (1972). NB: Two pK_a' values were also reported for the closely related alkaloid leurosine, $pK_a' = 5.5, 7.5$ (water).
1500	Vincristine ($C_{46}H_{56}N_4O_{10}$) 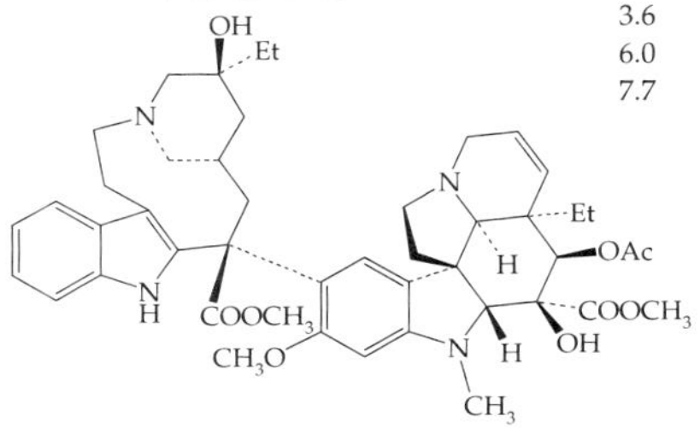	−0.8 3.6 6.0 7.7	U U U U	+H +H +H +H	kinetic	H_2O $t = 80$ $I = 0.00$	Vendrig DE, Beijnen JH, van der Houwen OA and Holthuis JJ, Degradation kinetics of vincristine sulfate and vindesine sulfate in aqueous solutions, *Int. J. Pharm.*, **50**, 189–196 (1989). "The degradation kinetics of vincristine sulfate (I) and vindesine sulfate (II) in the pH range from −2 to 11 at 80 °C were studied, and the effects of pH, buffer concentrations, ionic strength, and temperature were considered. The relationship between pH and log kobs was modeled by using a non-linear least squares curve fitting computer program. From this plot pK_a values of II were calculated."
1501	Vincristine	5.0 7.4	U U	+H +H		33% DMF/ H_2O	Burns JH, Vincristine sulfate, *APDS*, **1**, 470 (1972). Muhtadi FJ, Afify AFAA, Vincristine sulfate, *APDS*, **22**, 517–553 (1993). Svoboda GH, *Lloydia*, **24**, 173–178 (1961).

(*continued*)

No.	Compound Name	pK_a value(s)	Data quality	Ionization type	Method	Conditions	Comments and Reference(s)
1502	Vindesine ($C_{43}H_{55}N_5O_7$)	−0.8 3.5 5.8 8.5	U U U U	+H +H +H +H	kinetic	H_2O $t = 80$ $I = 0.00$	Vendrig DE, Beijnen JH, van der Houwen OA and Holthuis JJ, Degradation kinetics of vincristine sulfate and vindesine sulfate in aqueous solutions, *Int. J. Pharm.*, **50**, 189–196 (1989). NB: See Vincristine for details.
1503	Vindesine	6.04 7.67 5.39 7.36	U U U U	+H +H +H +H	Potentiometric	H_2 O *t* undefined *I* undefined 66% DMF t undefined I undefined	Barnett CJ, Cullinan GJ, Gerzon K, Hoying RC, Jones WE, Newton WM, Poore GA, Robison RL, Sweeney MJ, Todd GC, Dyke RW and Nelson RL, Structure-activity relationships of dimeric *Catharanthus* alkaloids. 1. Deacetylvinblastine amide (Vindesine) sulfate, *J. Med. Chem.*, **21**, 88–96 (1978). NB: Apparent pK_a' values were reported for several vindesine derivatives, for example, the monohydrazide and the acid azide.
1504	Warfarin ($C_{19}H_{16}O_4$)	4.90 ± 0.01	A	−H	Potentiometric	H_2O $t = 25 \pm 0.5$ $I = 0.15$ (KCl)	Avdeef A, Box KJ, Comer JEA, Hibbert C and Tam KY, pH-metric log P 10. Determination of liposomal membrane-water partition coefficients of ionizable drugs, *Pharm. Res.*, **15**(2), 209–215 (1998). NB: Used a Sirius PCA101 autotitrator. Also gave log P (octanol-water) and log P (dioleylphosphatidylcholine unilamellar vesicles-water).
1505	Warfarin	5.08	A	−H	CE/pH (−ve ion mode)	H_2O $t = 25$ $I = 0.025$	Wan H, Holmen AG, Wang Y, Lindberg W, Englund M, Nagard MB and Thompson RA, High-throughput screening of pK_a values of pharmaceuticals by pressure-assisted capillary electrophoresis and mass spectrometry, *Rapid Commun. Mass Spectrom.*, **17**, 2639–2648 (2003). NB: Reported a predicted value (ACD Labs) of 4.5.

No.	Compound	pKa	Rating	Type	Method	Conditions	Reference / Comments
1506	Warfarin	5.03 ± 0.01	A	–H	Spectro (λ = 272–284 nm)	H_2O $t = 25 \pm 0.2$ $I = 0.1$ (KCl)	Stella VJ, Mooney KG and Pipkin JD, Dissolution and ionization of warfarin, *J. Pharm. Sci.*, **73**, 946–948 (1984). "A study is described which investigated the ionization and ionization kinetics of warfarin (I), to confirm the probable existence of a cyclic hemiketal in aqueous I solution and to determine the possible consequences of the cyclic hemiketal to acyclic enol equilibrium and ionization kinetics on the dissolution rate of I. The equilibrium aqueous solubility of unionized I at 25 °C and 0.5 ionic strength was 1.28×10^{-5} M, with an observed dissociation constant of 5.03–5.06. Hemiketal-acyclic enol ratio was estimated to be diff 20:1. Ionization rates were also calculated. Solving equations for instantaneous ionizing acids demonstrated that dissolution of I was not affected by hemiketal formation." NB: Also reported a pK_a value for phenprocoumon, 3.77 ± 0.01 at $I = 0.1$ M.
		5.03 ± 0.02	A	–H		$I = 0.5$ (KCl)	
		5.06	A	–H	soly	$I = 0.1$ (KCl)	
1507	Warfarin	4.82	U	–H	Potentiometric	H_2O $t = 25$	Bergström CAS, Strafford M, Lazorova L, Avdeef A, Luthman K and Artursson P, Absorption classification of oral drugs based on molecular surface properties, *J. Med. Chem.*, **46**(4), 558–570 (2003). NB: From extrapolation of aqueous-methanol mixtures to 0% methanol.
1508	Warfarin	5.05 ± 0.1	U	–H	Spectro (λ = 274 nm)	H_2O t undefined $I = 0.1$	Hiskey CF, Bullock E and Whitman G, Spectrophotometric study of aqueous solutions of warfarin sodium, *J. Pharm. Sci.*, **51**, 43–46 (1962). NB: No details on calibration of pH meter or other experimental factors.
1509	Water (H_2O) H_2O $2H_2O$ <=> $H_3O^+ + OH^-$	13.996 ± 0.001	R	autoprotolysis	Potentiometric	H_2O $t = 25.00$ $I = 0.000$	Harned HS and Hamer WJ, The ionization constant of water and the dissociation of water in potassium chloride solutions from electromotive forces of cells without liquid junction, *JACS*, **55**, 2194–2205 (1933). NB: See also Harned HS and Robinson RA, Temperature variation of the ionization constants of weak electrolytes, *Trans. Farad. Soc.*, **36**, 973–978 (1940); Ramette RW, On deducing the pK-temperature equation, *J. Chem. Edn.*, **54**(5), 280–283 (1977). For pK_w values as a function of temperature, see Introductory text, Section 2.2.3.
1510	Water (H_2O) H_2O	13.78 ± 0.01	A	autoprotolysis	Potentiometric	H_2O $t = 25.00$ $I = 0.1$ (NaCl)	Canel E and Gultepe A, Dogan A, Kihc E, The determination of protonation constants of some amino acids and their esters by potentiometry in different media, *J. Solution Chem.*, **35**(1), 5–19 (2006). NB: See Amino acid esters for details. Other values for pK_w were reported in cosolvents at the same ionic strength, and compared with closely similar literature values:

Solvent	30% EtOH	50% EtOH	70% EtOH
pK_w	14.17±0.03	14.40±0.02	14.67±0.01

(continued)

No.	Compound Name	pK_a value(s)	Data quality	Ionization type	Method	Conditions	Comments and Reference(s)
1511	WY- 7953 (1′-amino-cyclopentanecarboxamidopenicillin) ($C_{14}H_{21}N_3O_4S$) 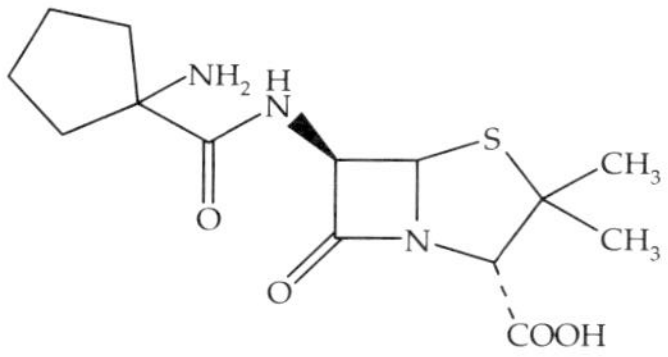	2.62 ± 0.05 7.61 ± 0.05	A U	−H +H	Potentiometric	H_2O $t = 25.0 \pm 0.1$ $I = 0.00$	Hou JP and Poole JW, The aminoacid nature of ampicillin and related penicillins, *J. Pharm. Sci.*, **58**, 1510–1515 (1969). NB: See Ampicillin for details. Also reported WY-4508, the corresponding cyclohexane homologue, which had p$K_{a1} = 2.68 \pm 0.04$ and p$K_{a2} = 7.50 \pm 0.02$ under the same conditions.
1512	WY- 8542 (1′-amino-3′-methylcyclopentanecarboxamidopenicillin) ($C_{15}H_{23}N_3O_4S$) 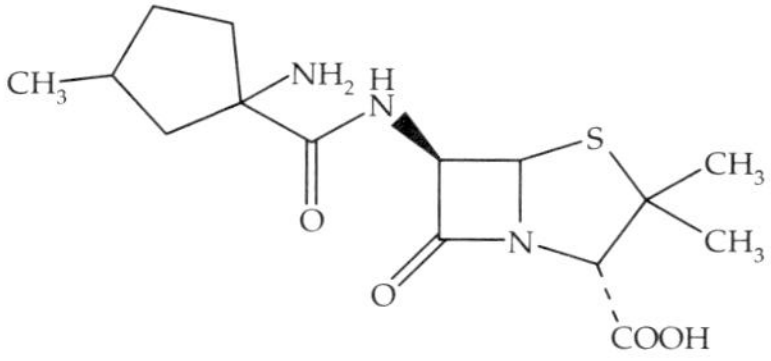	- 7.65 ± 0.06	 U	−H +H	Potentiometric	H_2O $t = 25.0 \pm 0.1$ $I = 0.00$	Hou JP and Poole JW, The aminoacid nature of ampicillin and related penicillins, *J. Pharm. Sci.*, **58**, 1510–1515 (1969). NB: See Ampicillin for details.
1513	Xanthine Derivatives 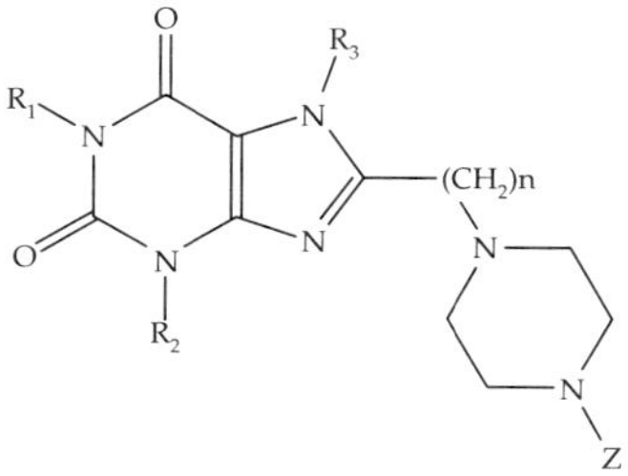					H_2O $t = 21 \pm$ $I = 0.1$ N_2 atmosphere	Walther B, Carrupt P-A, El Tayar N and Testa B, 8-Substituted xanthines as phosphodiesterase inhibitors: Conformation-dependent lipophilicity and structure-activity relationships, *Helv. Chim. Acta*, **72**, 507–517 (1989). Potentiometric Method: Solutions (final concentration 7.5×10^{-4} M) were prepared in distilled water which had been boiled to remove O_2 and CO_2 and saturated with N_2. The ionic strength was fixed at 0.1M using KCl. An excess of HCl was added, and the solutions were back-titrated with 0.01N NaOH using a Metrohm 670 titroprocessor. Titration curves were determined in triplicate for each compound and the pK_a calculated using a nonlogarithmic linearization of the titration curve proposed by Benet and Goyan [9] and modified by Leeson and Brown [10] to overcome the problem of dilution during titration. The temp was 21 ± 1.

No.	Compound	pKa			Method	Conditions	Reference / comments
	R1 R2 R3 n Z						Solubility method:
	Me i-Bu H 2 Ph_2CH	6.0	U	+H	soly		... pK_a values were determined from solubility data at 21 ± 1. For each compound, 5–7 buffer solutions (1/15M phosphate)were prepared in the pH range 3.5–7.5. These solns were saturated by addition of an excess of the compound and vigorously shaken for 3 h. The pH was then measured, the suspension filtered, and the solute concentration measured by RP-HPLC ... The pK_a values were then calculated according to Zimmermann [11].
	Me i-Bu H 2[a] Ph_2CH	5.6	U	+H	soly		NB: No details provided of pH meter calibration for either method. No demonstration that saturation had been reached in the solubility-pH method. Use was not made of the improved data handling procedure for the solubility method proposed by Lewis (1984) based on Zimmermann (see introductory text, refs. 52 and 53).
	H i-Bu H 2 Ph_2CH	5.2	U	+H	soly		
	Me i-Bu Me 2 Ph_2CH	6.4	U	+H	Potentiometric		
	Me Me H 2 Ph_2CH	5.6	U	+H	Potentiometric		
	Me i-Bu H 2 H	7.9	U	+H	Potentiometric		
	Me i-Bu H 2 4-F-Ph_2CH	5.2	U	+H	Potentiometric		
	Me i-Bu Me 2 4-F-Ph_2CH	6.1	U	+H	Potentiometric		
	Me i-Bu H 1 Ph_2CH	5.4	U	+H	soly		
	Me Me H 1 Ph_2CH	5.3	U	+H	soly		
	Me Me H 3 Ph_2CH	6.6	U	+H	Potentiometric		
	Me Me Me 3 Ph_2CH	6.1	U	+H	soly		
	Me i-Bu Me 3 4-F-Ph_2CH	5.6	U	+H	Potentiometric		
	Me Ph Me 3 Ph_2CH	6.8	U	+H	Potentiometric		
	Me Pr Me 3 Ph_2CH	6.4	U	+H	Potentiometric		
	Et i-Bu H 3 Ph_2CH	6.9	U	+H	Potentiometric		
	Et i-Bu Me 3 Ph_2CH	6.2	U	+H	Potentiometric		
	Me i-Bu H 3 Ph_2CH	5.4	U	+H	Potentiometric		
	Me i-Bu Me 3 Ph_2CH	6.3	U	+H	soly		
	[a] -CH2CH(OH)-						
1514	Xanthine derivative ($C_{30}H_{37}N_5O_2$) 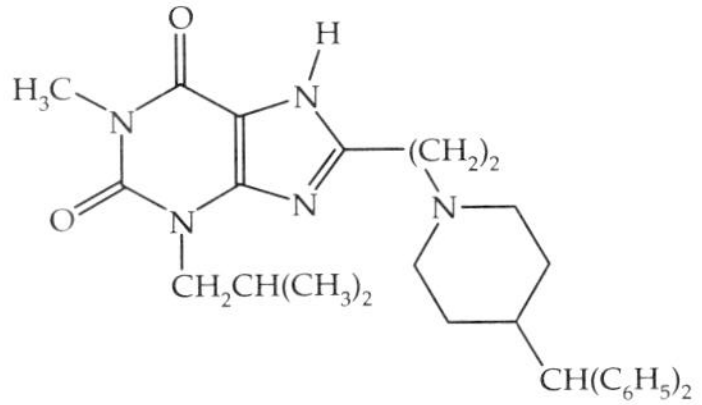	6.5	U	+H	Potentiometric	H_2O $t = 21 \pm 1$ $I = 0.1$ N_2 atmosphere	Walther B, Carrupt P-A, El Tayar N and Testa B, 8-Substituted xanthines as phosphodiesterase inhibitors: Conformation-dependent lipophilicity and structure-activity relationships, *Helv. Chim. Acta*, **72**, 507–517 (1989). NB: See Xanthine Derivatives (above) for details.
1515	Xanthine derivative ($C_{30}H_{40}F_2N_6O_2$) 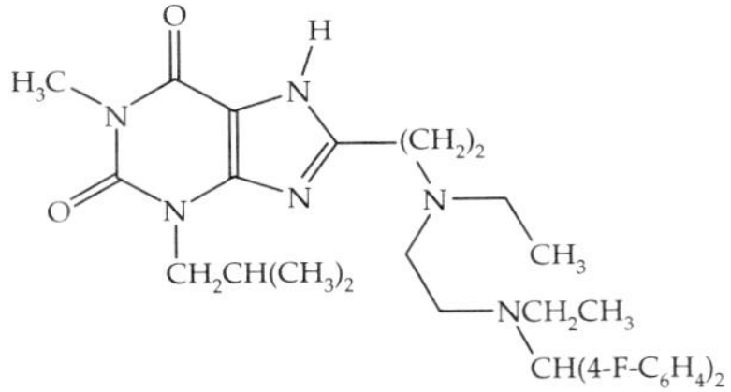	6.7	U	+H	Potentiometric	H_2O $t = 21 \pm 1$ $I = 0.1$ N_2 atmosphere	Walther B, Carrupt P-A, El Tayar N and Testa B, 8-Substituted xanthines as phosphodiesterase inhibitors: Conformation-dependent lipophilicity and structure-activity relationships. *Helv. Chim. Acta*, **72**, 507–517 (1989). NB: See Xanthine Derivatives (above) for details.

(continued)

No.	Compound Name	pK_a value(s)	Data quality	Ionization type	Method	Conditions	Comments and Reference(s)
1516	Xipamide ($C_{15}H_{15}ClN_2O_4S$) 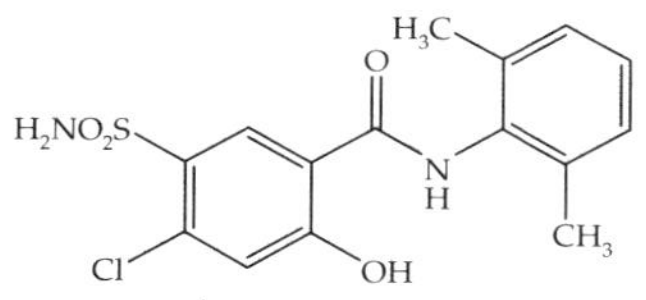	4.75 ± 0.04 10.0	U U	−H −H	Spectro	0.4% MeOH in H_2O	Hempelmann FW, Untersuchungen mit Xipamid (4-Chlor-5-sulfamoyl-2′,6′-salicyloxylidid). Teil I. Physicalisch-chemische und chemische Eigenschaften, *Arzneim.-Forsch.*, **27**, 2140–2143 (1977). NB: Spectrophotometric measurements used 0.127 mM solutions of xipamid in Britton-Robinson buffer at various pH values. Temperature was not stated. The paper assigned the 4.8 value to the phenol and 10.0 to the sulphonamide group, based on the absence of the lower value in measurements on the corresponding des-sulphamoyl model compound. ACD/pK_a calculations confirmed these assignments.
		4.58 10.47	U U	−H −H	Potentiometric	H_2O $t = 37$ $I = 0.15$ (KCl)	Balon K, Riebesehl BU and Muller BW, Drug liposome partitioning as a tool for the prediction of human passive intestinal absorption, *Pharm. Res.*, **16**, 882–888 (1999).
1517	Xylometazoline ($C_{16}H_{24}N_2$) 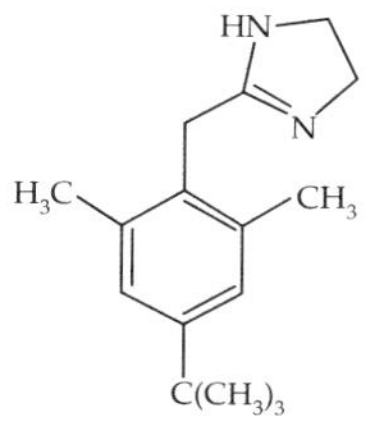	10.6 ± 0.1	U	+H	Spectro	H_2O $t = 22.0$	Golander Y and De Witte WJ, Xylometazoline hydrochloride, *APDS*, **14**, 135–155 (1985). Cited Stott AF, Ciba-Geigy Ltd., private communication.
1518	Zidovudine ($C_{10}H_{13}N_5O_4$) 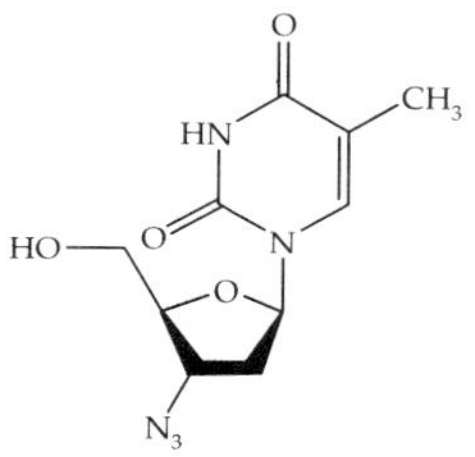	9.53	U	+H	Potentiometric	H_2O $t = 25$	Bergström CAS, Strafford M, Lazorova L, Avdeef A, Luthman K and Artursson P, Absorption classification of oral drugs based on molecular surface properties, *J. Med. Chem.* **46**(4), 558–570 (2003). NB: From extrapolation of aqueous-methanol mixtures to 0% methanol.
		9.45	U	+H	Potentiometric	H_2O $t = 37$ $I = 0.15$ (KCl)	Balon K, Riebesehl BU and Muller BW, Drug liposome partitioning as a tool for the prediction of human passive intestinal absorption, *Pharm. Res.*, **16**, 882–888 (1999).

No.	Compound	pKa			Method	Conditions	Reference / Comments
1519	Zileuton ($C_{11}H_{12}N_2O_2S$)	10.3	U	–H	Potentiometric	H_2O *t* undefined *I* undefined	Chang ZL, Zileuton, *APDS*, **25**, 535–575 (1997). "The pK_a of Zileuton was determined using the Calvin-Bjerrum titration technique (Irving HM and Rossoti HS, The calculation of formation curves of metal complexes from pH titration curves in mixed solvents, *J. Chem. Soc.*, 2904–2910 (1954)). pH measurements were made during the acid titration of aqueous methanolic solutions of sodium hydroxide, both in the presence and in the absence of zileuton. Extrapolation to pure water gave a pK_a value of 10.3.
		10.5	U	–H	soly		The pH dependence of zileuton solubility (Table) is consistent with a pK_a value of approximately 10.5."

Table: Solubility of zileuton in water at various pH values

pH	Solubility (mg/mL)	pH	Solubility (mg/mL)
1.87	0.18	10.1	0.28
3.80	0.10	10.5	1.50
7.50	0.12	11.6	73.7

No.	Compound	pKa			Method	Conditions	Reference / Comments
1520	Zileuton	10.51 (0.05)	U	–H	Spectro	H_2O t = RT I = 0.2–0.26 (KCl, NaCl)	Alvarez FJ and Slade RT, Kinetics and mechanism of degradation of zileuton, a potent 5-lipoxygenase inhibitor, *Pharm. Res.*, **9**, 1465–1473 (1992). "Fifty microliters of a stock solution of zileuton in methanol (4.23 mM) was mixed with 2450 uL of buffer and the absorbance at 260, 270 and 280 nm was recorded with a Hewlett-Packard Model 8451-A spectrophotometer. The pK_a was determined by nonlinear least-squares fit of the data to a model for the dissociation of a monoprotic acid as a function of pH. Similar values were recorded at the three wavelengths."
1521	Zomepirac ($C_{15}H_{14}ClNO_3$)	4.75	U	–H	Potentiometric	H_2O-EtOH (1:1 v/v)	Zinic M, Kuftinec J, Hofman H, Kajfez F and Meic Z, Zomepirac sodium, *APDS*, **15**, 673–698 (1986). "The pK_a value of zomepirac was determined by potentiometric titration using an automatic burette. Model ABU 13, coupled with the recording unit Titrigraph SBR-2C of a Titrator TTT2 (all equipment from Radiometer-Copenhagen). The glass electrode G-202C, was used against a calomel k 401 reference electrode. Zomepirac accurately weighed and dissolved in an ethanol/ water mixture 1:1 (v/v), was titrated with 0.1 M NaOH solution. Potentiometric curves were recorded between pH 3.45–12.0 and the pK_a value of 4.75 was obtained by calculation according to ref. 10."

(continued)

Appendix A *(continued)*

No.	Compound Name	pK_a value(s)	Data quality	Ionization type	Method	Conditions	Comments and Reference(s)
							10. Tenford C and Wawzanek S, *Physical Methods of Organic Chemistry*, Weissberger A (ed.), Intersc. Publ. Co., vol. I Bd. **4**, p. 2915.
1522	Zopiclone ($C_{17}H_{17}ClN_6O_3$)	−1.5 ± 0.1	U	+H	Spectro ($\lambda = 305$ nm)	10% ACN in aq. H_2SO_4 $t = 25.0$	Largeron M and Fleury MB, Acid catalyzed hydrolysis of a series of zopiclone analogues, *J. Pharm. Sci.*, **78**(8), 627–631 (1989). NB: The pyrazine nitrogens were shown by spectrophotometric measurements not to protonate.
		6.8	U	+H	Potentiometric	25% aq. ACN $t = 25.0$ $I = 0.075$ ($LiClO_4$) $c = 0.002$	
		6.76	U	+H	Potentiometric	H_2O $t = 37$ $I = 0.15$ (KCl)	Balon K, Riebesehl BU and Muller BW, Drug liposome partitioning as a tool for the prediction of human passive intestinal absorption, *Pharm. Res.*, **16**, 882–888 (1999).

APPENDIX B

Supplementary List—pK_a values found with little or no data quality information—mainly secondary literature

Reliability data has recently been found for a very small number of compounds in this listing. Efforts are continuing in the search for reliability data to support the remaining values in the primary literature. Structures not given in this list will be found in the Main List (Appendix A). Common secondary sources are given in this Appendix in abbreviated form:

Anon	Anon, American Hospital Formulary Service, ASHP, Washington DC (1977).
Avery	Speight TM, *Avery's Drug Treatment*, 3rd Edn., Publishing Sciences Group Inc., Littleton, MA pp. 1352–1380 (1987).
BPC	*British Pharmaceutical Codex*, 11th Edn., Pharmaceutical Press, London (1979).
Clarke	Moffat A (ed.), *Clarke's Isolation and Identification of Drugs*, 2nd Edn., Pharmaceutical Press London (1986).
Craig	Craig PN, Drug Compendium, *in* Hansch C, Sammes PG, Taylor JB (eds.), *Comprehensive Medicinal Chemistry*, vol. 6 pp. 236–965 (1990). The bulk (516) of the pK_a values are from the database of the Medicinal Chemistry Project (A. J. Leo, Director, Medicinal Chemistry Project, Chemistry Dept., Pomona College, Claremont, CA 91711 USA) at Pomona College. These were augmented by several values obtained from "*Clarke's Isolation and Identification of Drugs*" and "*The Merck Index*", 10th Edn.
Foye	Foye WO, Lemke TL and Williams DA, *Principles of Medicinal Chemistry*, 4th Edn., Williams and Wilkins, Philadelphia PA (1995).
Foye 3rd	Foye WO, *Principles of Medicinal Chemistry*, 3rd Edn., Lea and Febiger, Philadelphia PA (1989).
Hoover	Hoover JE (ed.), *Dispensing of Medication*, 8th Edn., Mack Publishing Co., Easton, PA, USA pp. 230, 247, 418–426, 468–634 (1976).
Kortum	Kortum G, Vogel W and Andrussow K, *Dissociation constants of organic acids in water*, IUPAC, Butterworths, London (1961).
Martin	Martin AN, Swarbrick J and Cammarata A, *Physical Pharmacy*, 3rd Edn., Lea and Febiger, Philadelphia, PA (1983) Tables 9–2 and 9–3; See also the corresponding Tables in later editions.
McEvoy	McEvoy GK (ed.), American Hospital Formulary Service, Drug Information, Am. Soc. Hosp. Pharm., Washington, DC (1994).

Merck 9	M. Windholz, *Merck Index*, 9th Edn., Merck and Co., Inc., Rahway, NJ (1976).
Merck 10	S. Budavari, *Merck Index*, 10th Edn., Merck and Co., Inc., Rahway, NJ (1983).
Merck 11	S. Budavari, *Merck Index*, 11th Edn., Merck and Co., Inc., Rahway, NJ (1989).
N&K	Newton DW and Kluza RB, pK_a values of medicinal compounds in pharmacy practice, *Drug Intelligence and Clinical Pharmacy*, **12**, 546–554 (1978).
Perrin Bases	Perrin DD, *Dissociation Contants of Organic Bases in Aqueous Solution*, IUPAC, Butterworths, London (1965).
Perrin Bases Suppl	Perrin DD, *Dissociation Constants of Organic Bases in Aqueous Solution: Supplement*, IUPAC, Butterworths, London (1972).
Ritschel	Ritschel WA, *in* Francke DE and Whitney HAK, Jr. (eds.), *Perspectives in Clinical Pharmacy*, 1st Edn., Drug Intelligence Publications, Hamilton, IL, USA pp. 325–367 (1972).
S&R	Smith SE and Rawlins MD, *Variability in Human Drug Response*, Butterworths, London pp. 154–165 (1973).
W&G	Delgado JN and Remers WA (eds.), *Wilson and Gisvold's Textbook of Organic Medicinal and Pharmaceutical Chemistry*, 10th Edn., Lippincott-Raven, Philadelphia and New York 915–921 (1998).

No.	Name	pK_a value(s)	Data quality	Ionization type	Method	Conditions t °C; I or c M	Comments and Reference(s)
			See comments	Type (+H or −H)		Solvent t (°C) Ionic strength (I) or analyte concentration (c) in molar (M) units	Date reliability cutoff points R: Reliable = ±<0.02 A: Approx = ±0.02 to ±0.06 U: Uncertain = ±>0.06 These overall data reliability cutoffs apply when all other aspects of the pK_a values have been considered. Where key variables (temperature, ionic strength and solvent composition) have not been reported, or the value was obtained from a computer program, a value is automatically assessed as U: Uncertain. A few results have been classified as VU: Very Uncertain.
1523	Acebutolol ($C_{18}H_{28}N_2O_4$) [structure: benzene ring bearing $OCH_2CH(OH)CH_2NHCH(CH_3)_2$, $COCH_3$ and $NHCOCH_2CH_2CH_3$]	9.40	U	+H			Kaye CM and Long AD, The influence of pH on the buccal absorption and plasma and renal elimination of acebutolol, *Br. J. Clin. Pharmacol.*, **3**(1–3), 196–7 (1976). "Acebutolol has a pK_a of 9.40 and a water solubility of 0.31%."
1524	Acebutolol	9.67	U	+H			Hinderling PH, Schmidlin O and Seydel JK, Quantitative relationships between structure and pharmacokinetics of beta-adrenoceptor blocking agents in man, *J. Pharmacokin. Biopharm.*, **12**, 263–287 (1984); Ref. Tempelton R, May & Baker Ltd., Dagenham, England. Personal communication.
1525	Acebutolol	9.67	U	+H			Barbato F, Caliendo G, LaRotonda MI, Morrica P, Silipo C and Vittoria A, Relationships between octanol-water

(*continued*)

No.	Name	pK_a value(s)	Data quality	Ionization type	Method	Conditions t °C; I or c M	Comments and Reference(s)
							partition data, chromatographic indices and their dependence on pH in a set of beta-adrenoceptor blocking agents, *Farmaco*, **45**, 647–663 (1990). Cited in Lombardo F, Obach RS, Shalaeva MY and Gao F, Prediction of human volume of distribution values for neutral and basic drugs. 2. Extended data set and leave-class-out statistics, *J. Med. Chem.*, **47**, 1242–1250 (2004); ref. 275.
1526	Acebutolol	9.4	U	+H			Clarke, p. 309. Cited in Foster RT and Carr RA, Acebutolol, *APDS*, **19**, 1–26 (1990).
1527	Acenocoumarol ($C_{19}H_{15}NO_6$)	4.7	U	−H			Anton AH, A drug-induced change in the distribution and renal excretion of sulfonamides, *JPET*, **134**, 291–303 (1961). "… The value was obtained from Dayton PG, personal communication."
1528	Acetaminophen (paracetamol)	10.15	U	−H			Chow YP and Repta A, *J. Pharm. Sci.*, **61**(9), 1454–58 (1972). Cited in Fairbrother JE, Acetaminophen, *APDS*, **3**, 27(1974). NB: No experimental data was given in the Chow and Repta reference, which simply stated: Abstracted in part from a M.S. thesis by Y.P. Chow, U. Kansas, 1972. Also presented to the Basic Pharmaceutics Section, APHA Academy Meeting, April 1972.
1529	Acetaminophen (paracetamol)	9.9	U	+H			Rhodes HJ, DeNardo JJ, Bode DW and Blake MI, Differentiating non-aqueous titration of aspirin, acetaminophen and salicylamide

							mixtures, *J. Pharm. Sci.*, **64**, 1386–1388 (1975). NB: Value quoted without reference or conditions.
1530	Acetaminophen (paracetamol)	9.86±0.13	U	–H	comp		ACD/pKa estimate
1531	Acetanilide	0.61	U	+H		H_2O $t = 40$	Chatten LG (ed.), Pharmaceutical Chemistry, vol. 1, Dekker, New York, pp. 85–87 (1966); Martin.
1532	Acetazolamide	7.2	U	–H			Maren TH, Mayer E and Wadsworth BC, Carbonic anhydrase inhibition I. The pharmacology of Diamox®, 2-acetylamino-1,3,4-thiadiazole-5-sulfonamide, *Bull. Johns Hopkins Hosp.*, **95**, 199–243 (1954). Cited by: Anton AH, A drug-induced change in the distribution and renal excretion of sulfonamides, *JPET*, **134**, 291–303 (1961).
1533	Acetazolamide	7.4	U	–H			Kunka RL and Mattocks AM, Relationship of pharmacokinetics to pharmacological response for acetazolamide, *J. Pharm. Sci.*, **68**(3), 347–349 (1979). NB: Value was quoted without references or experimental details.
1534	Acetazolamide	7.2	U	–H			Parasrampuria J, Acetazolamide, *APDS*, **22**, 1–32 (1993). NB: No references or experimental details.
1535	Acetazolamide	7.4	U	–H		$t = 37$	Wallace SM and Riegelman S, Uptake of acetazolamide by human erythrocytes *in vitro*, *J. Pharm. Sci.*, **66**, 729–731 (1977).
1536	Acetohydroxamic acid ($C_2H_5NO_2$) $CH_3CONHOH$	9.4	U	–H			Craig
1537	α-Acetylmethadol	8.6	U	+H			Craig
1538	Acyclovir	2.27	U	+H			Laskin OL, Clinical pharmacokinetics of acyclovir, *Clin. Pharmacokin.*, **8**, 187–201. (1983).
		9.25	U	–H			"The rise in serum creatine in the absence of renal dysfunction suggests that acyclovir, which is both a weak acid and base (pK_a 2.27 and 9.25) may compete with renal organic base transport of creatinine."

(continued)

No.	Name	pK_a value(s)	Data quality	Ionization type	Method	Conditions t °C; I or c M	Comments and Reference(s)
1539	Adenosine	3.5	U	+H			Cheung AP and Kenney D, *J. Chromatogr.*, **506**, 119–131 (1990). Cited in Mahler GS, Adenosine, *APDS*, **25**, 1–37 (1997).
1540	Ajmaline ($C_{20}H_{26}N_2O_2$) H HO N OH N H_3C H CH_2CH_3	8.2	U	+H			Craig
1541	Albuterol	9.3 10.3	U U	+H −H			Avery
1542	Alclofenac ($C_{11}H_{11}ClO_3$) Cl $CH_2{=}CHCH_2O$ CH_2COOH			−H	potentio	DMSO/H_2O	Chiarini A, Tartarini A and Fini A, pH-Solubility relationship and partition coefficients for some anti-inflammatory arylaliphatic acids, *Arch. Pharm. Weinheim*, **317**, 268–273 (1984). "The pK_a values of some anti-inflammatory arylaliphatic acids, including alclofenac and ibuprofen, were measured potentiometrically in dimethyl sulfoxide/water. The values were confirmed by assessing the solubility of the acids as a function of the pH of the aqueous solutions. In this way, the intrinsic partition coefficient in octanol/water was defined."

No.	Compound	pKa	Method	Type	Reference
1543	Alclofenac	4.3	U	−H	Craig
1544	Alfentanil ($C_{21}H_{32}N_6O_3$) 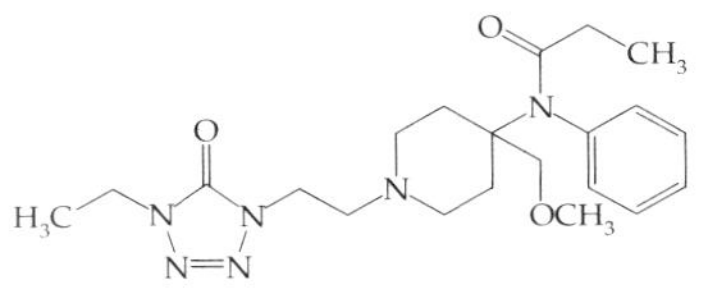	6.5	U	+H	Larijani GE and Goldberg ME, Alfentanil Hydrochloride: A new short-acting narcotic analgesic for surgical procedures, *Clinical Pharmacy*, **6**, 275–282 (1987). "... Alfentanil is a tertiary amine with an ionization constant of 6.5 resulting in approximately 10% ionization at physiologic pH. When compared with fentanyl (II), I has ... a much greater unionized fraction at physiologic pH."
1545	Alfentanil	6.5	U	+H	Meuldermans WEG, Hurkmans RMA and Heykants JJP, Plasma protein binding and distribution of fentanyl, sulfentanil, alfentanil and lofentanil in blood, *Arch. Int. Pharmacodyn.*, **257**, 4–19 (1982). Cited in Lombardo F, Obach RS, Shalaeva MY and Gao F, Prediction of human volume of distribution values for neutral and basic drugs. 2. Extended data set and leave-class-out statistics, *J. Med. Chem.*, **47**, 1242–1250 (2004); ref. 276.
1546	Alginic acids				Selleri R, Orzalesi G, Mari F, Bertol E, Influence of certain types of alginic acids on the activity of drugs used in gastric disorders, *Boll. Chim. Farm.*, **119**, 41–51 (1980).

(continued)

No.	Name	pK_a value(s)	Data quality	Ionization type	Method	Conditions t °C; I or c M	Comments and Reference(s)
	Alginate building blocks: 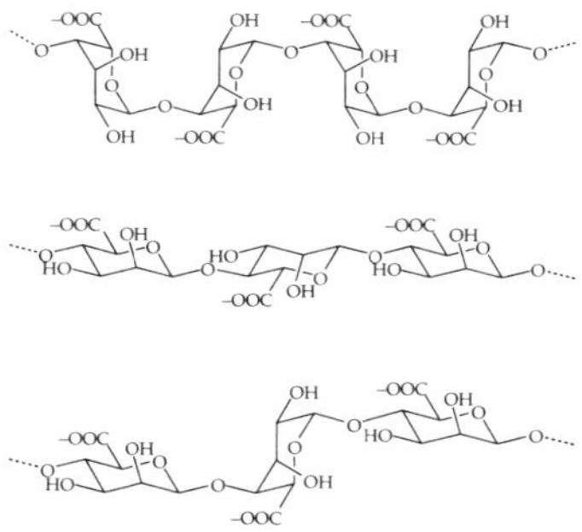 Monomer: $C_6H_{10}O_7$	Mannuronic acid: 3.38 Guluronic acid: 3.65	U U	–H –H			Onsøyen E, Hydration induced swelling of alginate based matrix tablets at GI-tract pH conditions, *in* Karsa DR and Stephenson RA (eds.), Excipients and delivery systems for pharmaceutical formulations, *Roy. Soc. Chem.*, London pp. 108–122 (1995). NB: Linear glycuronan polymer consisting of a mixture of β-(1-4)-D-mannosyluronic acid and α-(1-4)-L-gulosyluronic acid residues (RCOOH). "The organoleptic characteristics, level of impurities, exchange capacities, pK_a, pH, affinity for inorganic ions, intrinsic viscosities, molecular weights, buffer capacity in artificial gastric fluids and other properties of 3 different types of alginic acids used in antacid preparations are discussed."
1547	Allopurinol ($C_5H_4N_4O$) 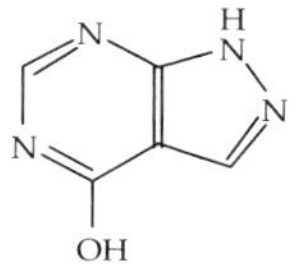	10.2	U	–H	spectro		Benezra SA and Bennett TR, Allopurinol, *APDS*, **7**, 1–17 (1978). Spence J and Jones A, Burroughs Wellcome, personal communication. NB: See also Foye, gave 9.4.
1548	Allopurinol	9.00 9.4	U U	–H –H	potentio	H_2O $t = 37$ $I = 0.15$ (KCl)	McEvoy NB: See Aspirin, no. 96.

1549	Alphaprodine ($C_{16}H_{23}NO_2$) 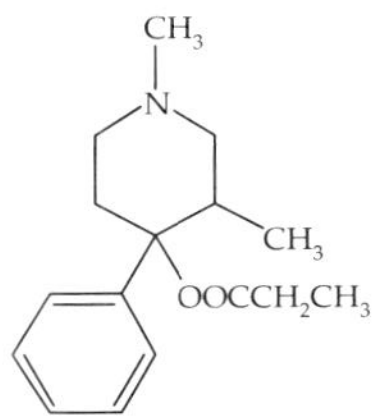	8.73	U	+H	potentio	50% aq EtOH	Farmilo CG, Oestreicher PM and Levi L, Physical methods for the identification of narcotics. IB. Common physical constants for identification of ninety-five narcotics and related compounds, *Bull. Narcotics*, UN Dept. Social Affairs, vol. **6**, pp. 7–19 (1954). CA 48:69490; updated by Martin L, Genest K, Cloutier JAK and Farmilo CG, *Bull. Narcotics*, UN Dept. Social Affairs, **15** (3–4), 17–38 (1963); NB: Cited in Beckett AH, Analgesics and their antagonists. I., *J. Pharm. Pharmacol.*, **8**, 848–859 (1956); also cited by Taylor JF, Methods of Chemical Analysis, Ch. 2, in Clouet DH (ed.), *Narcotic Drugs Biochemical Pharmacology*, Plenum Press NY (1971). "The pK_a values . . . are from the data of Farmilo and others, who used aqueous ethanol as solvent in many of the determinations."
1550	Alprenolol	9.63	U	+H			Avery; Heel and Avery, Drug Treatment, *in* Avery GS, 2nd Edn., ADIS Press, Sydney, 1212–1222 (1980). NB: See Acebutolol, no. 1524; ref: Bodin NO, Borg KO, Johannson R, Obianwu H and Svensson R, Absorption, distribution, and excretion of alprenolol in man, dog, and rat, *Acta Pharmacol. Toxicol.*, **35** (4), 261–269 (1974); gave pK_a = 9.7.
1551	Altretamine (hexamethylmelamine) ($C_9H_{18}N_6$) 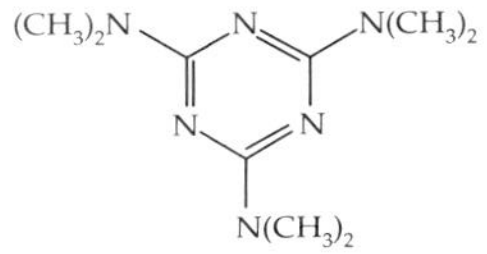	10.3	U	+H			Craig

(*continued*)

No.	Name	pK_a value(s)	Data quality	Ionization type	Method	Conditions t °C; I or c M	Comments and Reference(s)
1552	Amantadine	10.40	U	+H			Hoskey CF, personal communication, Endo Laboratories, Garden City, NY (1975).
1553	Amantadine	10.1	U	+H			Aoki FY and Long AD, Clinical pharmacokinetics of amantadine hydrochloride, *Clin. Pharmacokin.*, **14** (1), 35–41 (1988). "Amantadine (1-adamantanamine) is an aliphatic primary amine with a pK_a of 10.1. . . ."
1554	Amantadine	10.68	U	+H	potentio		Lombardo F, Obach RS, Shalaeva MY and Gao F, Prediction of human volume of distribution values for neutral and basic drugs. 2. Extended data set and leave-class-out statistics, *J. Med. Chem.*, **47**, 1242–1250 (2004). Ref. not given: Potentiometric titration.
1555	Amantadine	10.8	U	+H	comp		Albert A, *Selective Toxicity*, 4th Edn., London, Methuen, p. 281 (1968). NB: Stated that the pK_a value was calculated.
1556	Amdinocillin	3.4	U	−H			Craig
		8.9	U	+H			
1557	Amikacin ($C_{22}H_{43}N_5O_{13}$) 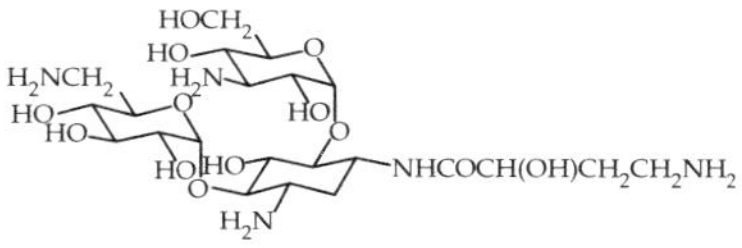	8.1	U	+H	potentio	H_2O	Monteleone PM, Muhammed N, Brown RD, McGrory JP and Hanna SA, Amikacin sulfate, *APDS*, **12**, 37–71 (1983). NB: Ref. 17 illustrates with Gentamicin the same observation that was made here for amikacin, i.e., the pK_a values are all very closely overlapping. "Amikacin (0.1 mmole) was dissolved in water, and 0.5 mmole of potassium hydroxide was added. The solution was titrated with 0.5-N hydrochloric acid using SCE/glass

						electrodes. Instrumentation included a Radiometer recording titration system with PM-64 pH meter TTT60 titrator, REC-61 servograph and an ABU-13 autoburet (16). If the four amine groups in amikacin are considered to be equivalent, then an apparent pK_a value of 8.1 may be estimated from the half neutralization point (17). 16. Hull DA, Bristol Labs, *Syracuse*, NY, data on file. 17. Rosenkrantz BE, Greco JR, Hoogerheide JG and Oden EM, *APDS*, **9**, 310 (1980).''
1558	Amiloride	8.7	U	+H		Avery
1559	Aminacrine (9-aminoacridine) ($C_{13}H_{10}N_2$)	10.0	U	+H		Craig
1560	Amino acids					See Perrin DD, *Dissociation Constants of Organic Bases in Aqueous Solution*, 1965, Butterworths, Lond. (1965), for detailed and extensive coverage of pK_a values for amino acids and small peptides.
1561	4-Aminobenzoic acid	2.40 4.90	U A	+H −H	H_2O $t = 25$	El-Obeid HA and Al-Badr AA, Aminobenzoic acid, *APDS*, **22**, 33–106 (1993). NB: Merck 11, no. 434: 4.65, 4.80; see also Clarke; N&K; Kortum (gives several earlier values).
1562	4-Aminobenzoic acid	4.83	U	−H	H_2O $t = 25$ $I = 0.00$	Palm VA, ed., *Tables of rate and equilibrium constants of heterolytic organic reactions*, Moscow, 1975. NB: Mentre I, *Ann. Chim. Fr.*, **7**, 333–341 (1972) gave 2.39 at 20.0 °C and $I = 0.00$.

(*continued*)

No.	Name	pK_a value(s)	Data quality	Ionization type	Method	Conditions t °C; I or c M	Comments and Reference(s)
1563	ε-Aminocaproic acid ($C_6H_{13}NO_2$)	4.4	U	−H			Craig
	$H_2N(CH_2)_5COOH$	10.8	U	+H			N&K: S&R.
1564	4-Aminohippuric acid 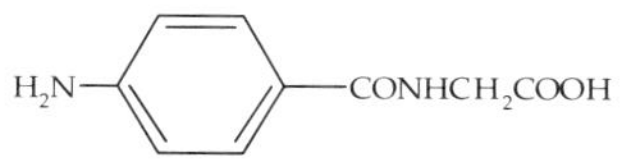	3.8	U	−H			Craig
	$H_2N-C_6H_4-CONHCH_2COOH$						
1565	4-Aminohippuric acid	3.64	U	−H			Hoover
1566	6-Aminopenicillanic acid	2.3	U	−H			W&G: Rapson HDC and Bird AE, Ionization constants of some penicillins and of their alkaline and penicillinase hydrolysis products, *J. Pharm. Pharmacol.*, Suppl., **15**, 222–231T (1963).
		4.9	U	+H			
1567	Aminophylline ($C_{16}H_{24}N_{10}O_4$) 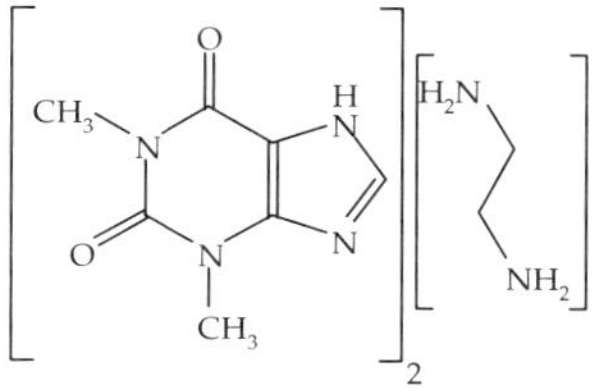	5.0	U	+H			Anon; N&K.
1568	Aminopterin ($C_{19}H_{20}N_8O_5$) 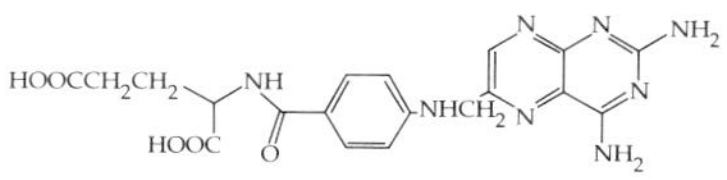	5.5	U	−H	spectro		Craig; N&K. Baker BR and Jordaan JH, Analogues of tetrahydrofolic acid XXVIII, *J. Pharm. Sci.*, **54**, 1740–1745 (1965); referred to: Zakrzewski and Sigmund F, Relation between basicity of certain folate analogs and their affinities for folate reductase, *J. Biol. Chem.*, **238**, 4002–4004 (1963). This paper only gave pK_a data for the pterin nucleus.

No.	Compound	pKa	Method	Ionization	Reference
1569	Aminopyrine	5.0	U	+H	W&G: Kolthoff IM and Stenger VA, Volumetric Analysis, vol. **1**, 2nd Edn., NY, Interscience (1942).
1570	4-Aminosalicyclic acid	1.7	U	+H	Kortum. NB: N&K.
		3.9	U	−H	
1571	Aminothiadiazole ($C_2H_3N_3S$) 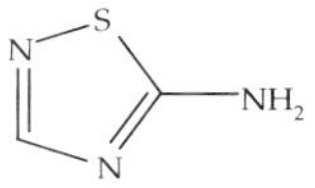	3.2	U	+H	Craig
1572	Amlodipine ($C_{20}H_{25}ClN_2O_5$) 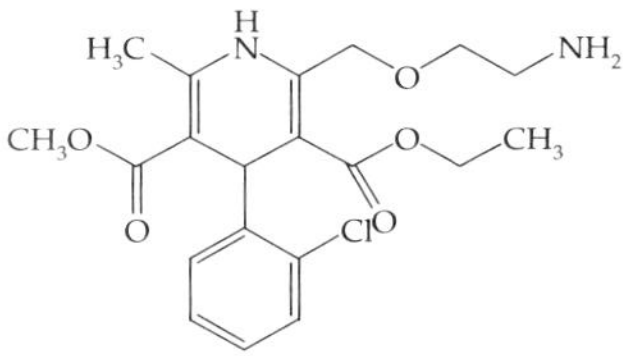	8.6	U	+H	Kass RS and Arena JP, Influence of pHo on calcium channel block by amlodipine, a charged dihydropyridine compound. Implications for location of the dihydropyridine receptor, *J. Gen. Physiol.*, **93**(6), 1109–1127 (1989). NB: Although the pK_a is repeatedly stated as 8.6, no experimental detail is given, nor any reference. The paper also claims an estimated pK_a value of <3.5 for nisoldipine (ref: Hugenholtz PG and Meyer J (eds.), Nisoldipine, Springer-Verlag, Berlin, 3–348 (1987).
1573	Amoxapine ($C_{17}H_{16}ClN_3O$) 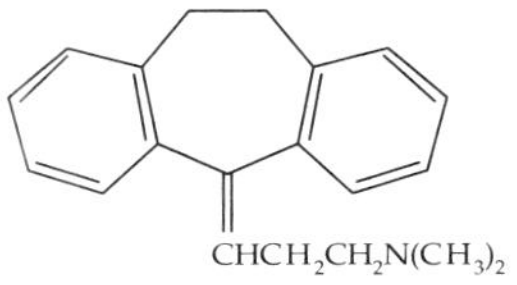	7.6	U	+H	Craig; McEvoy
1574	Amoxicillin	2.4	U	−H	Rolinson GN, Laboratory evaluation of amoxicillin, *J. Infect. Dis.*, **129**, S139–S145 (1974); N&K.
		9.6	U	−H,+H	
1575	Amphetamine	10.0	U	+H	Craig

(continued)

No.	Name	pK_a value(s)	Data quality	Ionization type	Method	Conditions t °C; I or c M	Comments and Reference(s)
1576	Amphotericin B ($C_{47}H_{73}NO_{17}$) 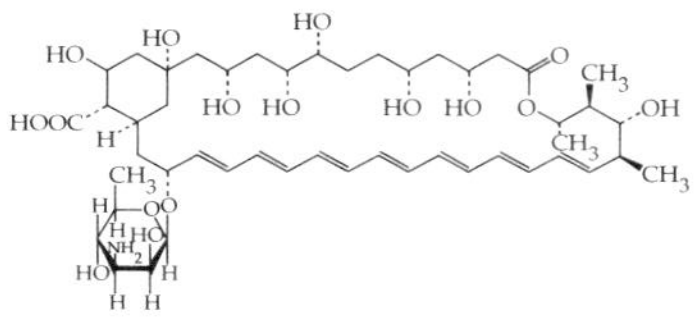	~5.7 10.0	U U	−H +H	potentio	DMF/H_2O	Asher IM, Schwartzman G and the USASRG, Amphotericin B, *APDS*, **6**, 1–42 (1977). Etingov ED and Kholodova GV, Kul'bakh VO and Karnatushkina AI, Acid-base properties of amphotericin B, *Antibiotiki*, **17**, 301–305 (1972). "Titration of 66% aqueous dimethylformamide solutions of Amphotericin B with methanolic HCl and KOH yields pK's near 5.7 and 10.0. Comparison with N-acetyl-Amphotericin B (pK=6.5) and Amphotericin B-methyl ester (pK=8.8) assigns the two pK's to carboxyl and amino groups respectively. Amphotericin B is found to be almost completely zwitterionic in this solution (tautomeric equilibrium constant K_t = 1000 with respect to the neutral molecule)."
1577	Amphotericin B	5.5 10.0	U U	−H +H			S&R
1578	Anileridine ($C_{22}H_{28}N_2O_2$) 	3.7 7.5	U U	+H +H			Craig; N&K
1579	Antazoline ($C_{17}H_{19}N_3$) 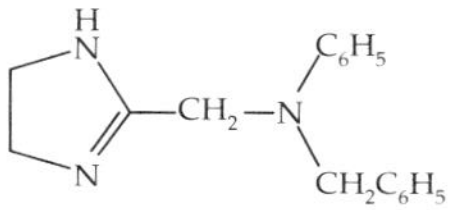	2.5 10.1	U U	+H +H			Craig
1580	Antazoline	7.2	U	+H			N&K; Ritschel; ref. Robson JM and Stacey RS, *Recent advances in pharmacology*, 4th Edn., Little Brown and Co., Boston, p. 108 (1968).

No.	Compound	pKa			Method	Solvent		Reference
1581	Antazoline	10.0	U	+H				Marshall PB, Some chemical and physical properties associated with histamine antagonism, *Br. J. Pharmacol.*, **10**, 270–278 (1955).
1582	Antipyrine ($C_{11}H_{12}N_2O$)	1.4	U	+H	spectro	H_2O		Mayer S, Maickel RP and Brodie BB, Kinetics of penetration of drugs and other foreign compounds into cerebrospinal fluid and brain, *JPET*, **127**, 205–211 (1959). NB: Followed the method of Flexser *et al.* (1935). This value cited by Anton AH, A drug-induced change in the distribution and renal excretion of sulfonamides, *JPET*, **134**, 291–303 (1961). Mayer *et al.* also reported values for thiopental (7.6), aniline (4.6), aminopyrine (5.1), 4-aminoantipyrine (4.1), acetanilide (1.0), barbital (7.8), N-acetyl-4-aminoantipyrine (NAAP; 0.5), salicylic acid (3.0). With the exceptions of acetanilide and NAAP, all additional p*K*a values were obtained by the potentiometric method. The barbital value is lower than the best value and suggests that carbon dioxide absorption interfered, however, the remaining values should not be affected by this factor.
1583	Antipyrine	1.5	U	+H				Craig; S&R; N&K report 1.4.
1584	Antipyrine	2.2	U	+H				W&G: Evstratova KI, Goncharova NA and Solomko VY, *Farmatsiya*, **17**(4), 33–36 (1968).
1585	Ascorbic Acid	4.17	U	−H				Al-Meshal IA and Hassan MMA, Ascorbic acid, *APDS*, **11**, 45–76. NB: Ref. Merck 9.
		11.57	U	−H				

(*continued*)

No.	Name	pK_a value(s)	Data quality	Ionization type	Method	Conditions t °C; I or c M	Comments and Reference(s)
1586	Aspirin	3.55	U	−H		H_2O $t = 25$	Florey K, Aspirin, *APDS*, **8**, 1–46 (1979); quoted Merck value of 3.49. NB: See Springer A and Jones HC, Study of the conductivity and dissociation of certain organic acids at different temperatures, *Am. Chem. J.*, **48**, 411–453 (1913).
1587	Aspirin	3.49	U	−H		H_2O $t = 25$	Chatten LG (ed.), Pharmaceutical Chemistry, vol. **1**, Dekker, New York, pp. 85–87 (1966); Martin.
1588	Atenolol	9.55	A	+H			NB: See Acebutolol, no. 1524. Ref. Adam HK, Imperial Chemical Industries Ltd., Macclesfield, England. Personal communication.
1589	Atenolol	9.6	U	+H			Lombardo F, Obach RS, Shalaeva MY and Gao F, Prediction of human volume of distribution values for neutral and basic drugs. 2. Extended data set and leave-class-out statistics, *J. Med. Chem.*, **47**, 1242–1250 (2004); ref. 275, 277.
1590	Atenolol	9.6	U	+H			Marmo E, On the pharmacology of atenolol, *Drugs Exptl. Clin. Res.*, **6**, 639–663 (1980); Cited in Caplar V, Mikotic-Mihun Z, Hofman H, Kuftinec J, Karijfez F, Nagl A and Blazevic N, Atenolol, *APDS*, **13**, 1–23 (1984).
1591	Atomoxetine 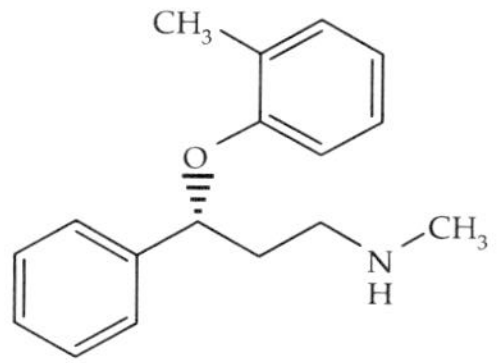 CH_3, O, N, H, CH_3	10.1	U	+H	capillary electro-phoresis		Lombardo F, Obach RS, Shalaeva MY and Gao F, Prediction of human volume of distribution values for neutral and basic drugs. 2. Extended data set and leave-class-out statistics, *J. Med. Chem.*, **47**, 1242–1250 (2004); ref. not given: capillary electrophoresis.

1592	Atrazine ($C_8H_{14}ClN_5$) $(CH_3)_2CHNH$, Cl, N, N, N, $NHCH_2CH_3$	1.64	U	+H	spectro (λ = 228 nm)	H_2O	Browne JE, Feldkamp JR, White JG and Hem SL, Potential of organic cation-saturated montmorillonite as treatment for poisoning by weak bases, *J. Pharm. Sci.*, **69**, 1393–1395 (1980). "Atrazine (2-chloro-4-ethylamino-6-isopropylamino-s-triazine) was chosen as a model weak base because it was expected to be neutral in the pH range of the GI tract due to its pK of 1.64 [Weber JB, *Spectrochim. Acta*, **23A**, 458–461 (1967)]." NB: The paper actually reported a value of 1.85.
1593	Atropine	9.84	U	+H	potentio		Lombardo F, Obach RS, Shalaeva MY and Gao F, Prediction of human volume of distribution values for neutral and basic drugs. 2. Extended data set and leave-class-out statistics, *J. Med. Chem.*, **47**, 1242–1250 (2004); ref. not given: potentiometric titration.
1594	Atropine	9.25	U	+H			N&K; cited Perrin DD, *Dissociation constants of bases*, Butterworths, London, Suppl. 1972.
1595	Atropine	9.7	U	+H			W&G: Kolthoff IM, The dissociation constants, solubility product and titration of alkaloids, *Biochem. Z.*, **162**, 289–353 (1925); cited in Christophers SR, *Trans. Farad. Soc.*, **39**, 333–338 (1943).
1596	Atropine	9.27	U	+H		$t = 37$	Ballard BE and Nelson E, Physicochemical properties of drugs that control absorption rate after subcutaneous implantation, *JPET*, **135**, 120–127 (1962). NB: From $pK_b = 4.35$, $pK_w = 13.621$ at 37°C.

(continued)

No.	Name	pK_a value(s)	Data quality	Ionization type	Method	Conditions t °C; I or c M	Comments and Reference(s)
1597	Azatadine ($C_{20}H_{22}N_2$) 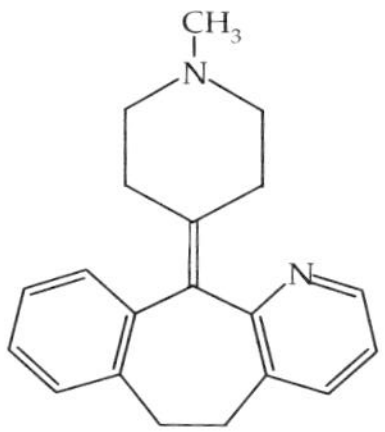	9.3	U	+H			Foye 3rd; cited McEvoy
1598	Azathioprine	8.2	U	–H		$t = 25$	Wilson WP and Benezra SA, Azathioprine, *APDS*, **10**, 29–53 (1981); cited Griffith GR, Wellcome Foundation Ltd., personal communication, 1980.
1599	Azlocillin ($C_{20}H_{23}N_5O_6S$) 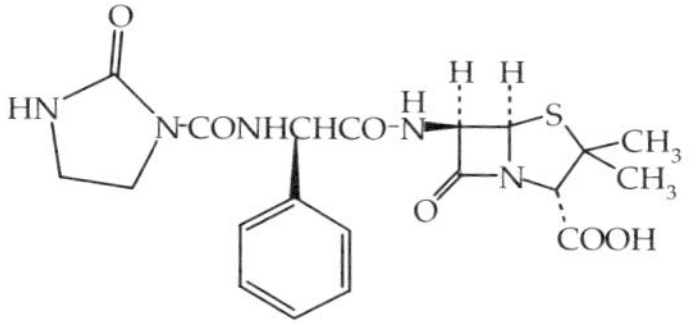	2.8	U	–H			Craig
1600	Bacampicillin ($C_{21}H_{27}N_3O_7S$) 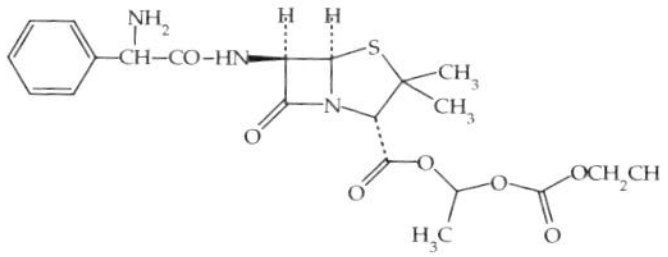	6.8	U	+H			Craig
1601	Baclofen ($C_{10}H_{12}ClNO_2$)	3.87 ± 0.1	U	–H		H_2O	Ahuja S, Baclofen, *APDS*, 14, 527–549 (1985).
		9.62 ± 0.1	U	+H		$t = 20.0$	NB: No reference to method, but probably potentiometric.

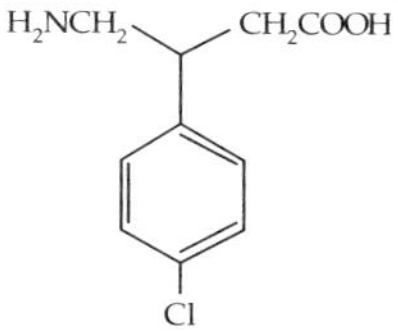

1602	Barbituric acid, 5,5-diethyl (barbital)	7.82	U	+H		$t = 37$	Ballard BE and Nelson E, Physicochemical properties of drugs that control absorption rate after subcutaneous implantation, *JPET*, **135**, 120–127 (1962).
1603	Barbituric acid, 5,5-diethyl-1-methyl (metharbital)	8.2	U	−H			Craig; this value is from Butler TC, the effects of N-methylation in 5,5-disubstituted derivatives of barbituric acid, hydantoin and 2,4-oxazolidinedione, *J. Am. Pharm. Assoc.*, **44**, 367–370 (1955).
1604	Barbituric acid, 5,5-diethyl-1-methyl (metharbital)	8.17	U	−H			Suzuki A, Higuchi WI and Ho NFH, Theoretical model studies of drug absorption and transport in the gastrointestinal tract, *J. Pharm. Sci.*, **59**, 651–659 (1970). N&K.
1605	Barbituric acid, 5-ethyl-5-iso-butyl (butabarbital)	7.9	U	−H			Avery
1606	Barbituric acid, 5-ethyl-5-(1-methylbutyl)-2-thio (thiopentone) ($C_{11}H_{18}N_2O_2S$) 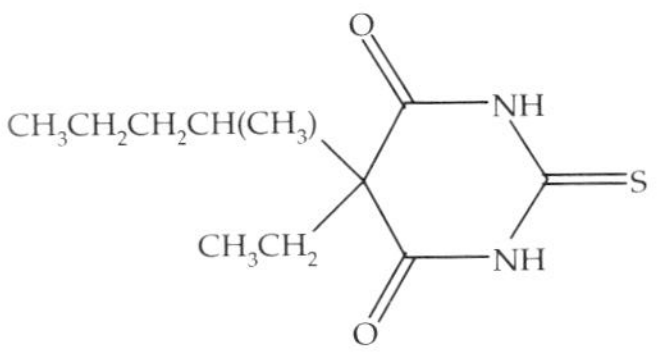	7.45	U	−H			Newton DW and Kluza RB, pK_a values of medicinal compounds in pharmacy practice, Drug Intelligence and Clinical Pharmacy, **12**, 546–554 (1978). Suzuki A, Higuchi WI and Ho NFH, Theoretical model studies of drug absorption and transport in the gastrointestinal tract. II, *J. Pharm. Sci.*, **59**, 651–659 (1970). Other values: 7.6 (Remington 17th ed. p. 1047; BPC 11th ed., 1979, p. 941).
1607	Barbituric acid, 5-ethyl-5-(1-methylbutyl)-2-thio (thiopentone)	7.3	U	−H			Anton AH, A drug-induced change in the distribution and renal excretion of sulfonamides, *JPET*, **134**, 291–303 (1961). NB: The value was obtained from Waddell WJ and Butler TC, Distribution and excretion of phenobarbital, *J. Clin. Invest.*, **36**, 1217–1226 (1957).

(continued)

No.	Name	pK_a value(s)	Data quality	Ionization type	Method	Conditions t °C; I or c M	Comments and Reference(s)
1608	Barbituric acid, 5-allyl-5-(1-methylbutyl) (secobarbital) ($C_{12}H_{18}N_2O_3$) 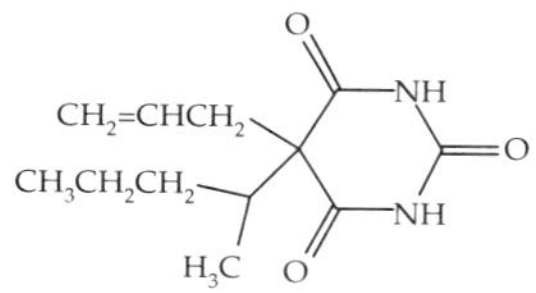	7.92	U	–H	potentio	H_2O $t = 20$ $I = 0.7$	Krahl ME, The effect of variation in ionic strength and temperature on the apparent dissociation constants of thirty substituted barbituric acids, *J. Phys. Chem.*, **44**, 449–463 (1940). NB: See Barbital, no. 129, for comment. Clowes GHA, Keltch AK and Krahl ME, Extracellular and intracellular hydrogen ion concentration in relation to anesthetic effects of barbituric acid derivatives, *JPET*, **68**, 312–329 (1940).
		12.60	U	–H	spectro	H_2O $t = 38$	Butler TC, Ruth JM and Tucker GF, The second ionization of 5,5-disubstituted derivatives of barbituric acid, *JACS*, 1486–1488 (1955). NB: pK_2 value. Mean of nine values.
1609	Barbituric acid, 5-allyl-5-(1-methylpent-2-ynyl)-N-methyl (methohexital) ($C_{14}H_{18}N_2O_3$) 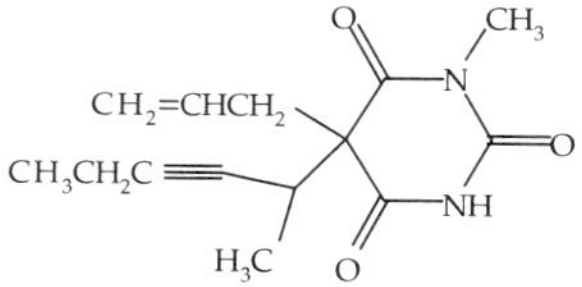	8.3	U	–H			Craig; N&K; S&R
1610	Barbituric acid, 5-allyl-5-(1-methylbutyl)-2-thio, (Thiamylal) ($C_{12}H_{18}N_2O_2S$) 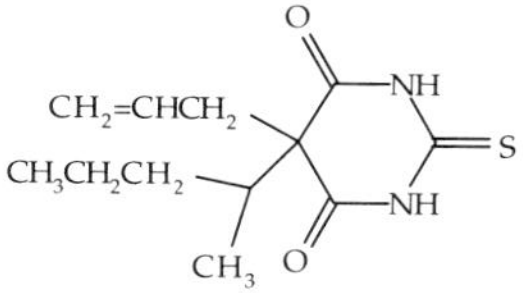	7.48	U	–H			Craig NB: See also Barbituric acid, 5-allyl-5-(1-methylbutyl)-2-thio.

1611	Barbituric acid, 5-allyl-5-(1-methylpropyl), (Talbutal) ($C_{11}H_{16}N_2O_3$) 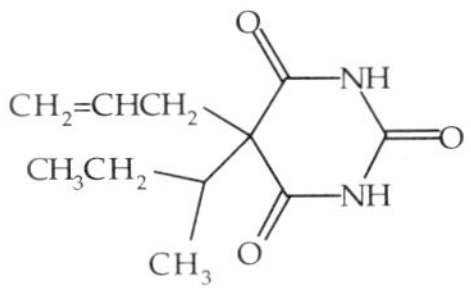	7.91	U	–H			Craig NB: See also Barbituric acid, 5-allyl-5-(1-methylpropyl).
1612	Barbituric acid, 5-ethyl-5-phenyl	7.3 11.8	U U	–H –H	spectro spectro		Chao MKC, Albert KS and Fusari SA, Phenobarbital, *APDS*, **7**, 359–399 (1978).
1613	Barbituric acid, 5-ethyl-5-phenyl	7.52	U	–H	spectro	H_2O	Maulding HV and Zoglio MA, pK_a determinations utilizing solutions of 7-(2-hydroxypropyl) theophylline, *J. Pharm. Sci.*, **60**, 309–311 (1971).
1614	Barbituric acid, 1-methyl-5-ethyl-5-phenyl ($C_{13}H_{14}N_2O_3$) 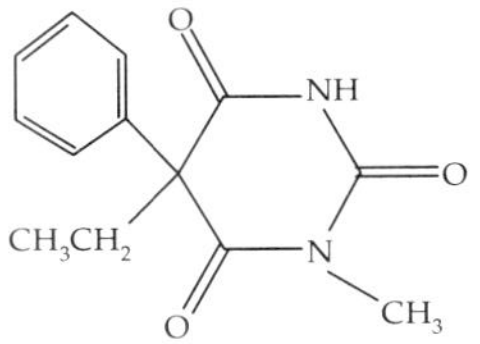	7.8	U	–H			Craig NB: Also called mephobarbital; N-methylphenobarbital.
1615	Bemegride ($C_8H_{13}NO_2$) 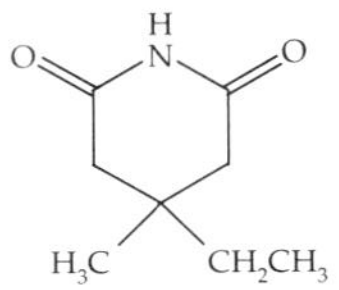	11.7	U	–H	potentio		Peinhardt G, Acidity and acidimetric titration of medically used glutaric acid imide, *Pharmazie*, **32**, 726–727 (1977) cited in CA 88:141587; also W&G.

(*continued*)

No.	Name	pK_a value(s)	Data quality	Ionization type	Method	Conditions t °C; I or c M	Comments and Reference(s)
1616	Ben(e)peridol ($C_{22}H_{24}FN_3O_2$) 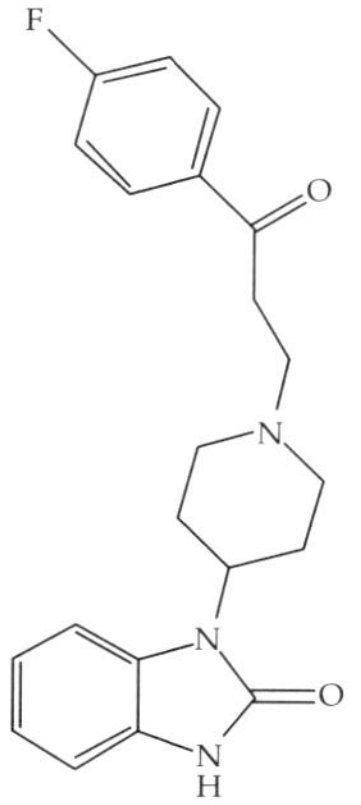	7.90	U	+H			Janssen Pharmaceutical, Internal reports. Cited in Takla PG, James KC and Gassim AEH, Beneperidol, *APDS*, **14**, 245–270 (1985). "Beneperidol has a pK_a of 7.90, and is a sufficiently weak base to be extracted by organic solvents from aqueous acid solution. It has been found in the present work that, due to the presence of the benzimidazolone ring system, beneperidol can also act as a weak acid when dissolved in strong alkali (pH11). Evidence of salt formation is seen in the bathochromic shift, from 279nm to 288nm, which occurs in the ultraviolet absorption spectrum of beneperidol when its aqueous solution is made alkaline."
1617	Benzocaine	2.5	U	+H	spectro		Ali SL, Benzocaine, *APDS*, **12**, 73–104 (1983). "The acid dissociation constant of benzocaine has been given as pK_a 2.5 (12). Estimation of microequilibrium constant of ethyl p-aminobenzoate was done spectrophotometrically and a *K* value of 0.00417 (NB: $pK_a = 2.38$) was obtained. Spectrophotometric method applied for estimating microequilibrium constants is simpler, faster, and more accurate than the conventional method employing the dissociation constant of the alkylated derivative (13)."
		2.38	U	+H	spectro		

					12. Europäisches Arzneibuch, Kommentar, p. 570, Wissenschaftliche Verlagsgesellschaft, Stuttgart (1976). 13. Schulman SG, Rosenberg LS and Sturgeon RJ, *J. Pharm. Sci.*, **67**, 334–337 (1978). NB: This reference actually took the data from Johnson J and Cummings AC, *Z. Phys. Chem.*, **57**, 557, 574 (1907). NB: The APDS reviewer did not seem to realize that only a single macro-equilibrium constant is relevant for benzocaine. Ref. 13 deals with determination of microconstants for aminobenzoic acids, which required the pK_a values for the corresponding esters. As noted above, Ref. 13 did not measure the benzocaine pK_a, as the reviewer appeared to suggest.
1618	Benzocaine	2.78	U	+H	N&K; Chatten LG (ed.), Pharmaceutical Chemistry, vol. **1**, Dekker, New York pp. 85–87 (1966); Martin; Craig.
1619	Benzocaine	2.8	U	+H	W&G: Kolthoff IM, The dissociation constants, solubility product and titration of alkaloids, *Biochem. Z.*, **162**, 289–353 (1925).
1620	Benzphetamine ($C_{17}H_{21}N$)	6.55	U	+H	Vree TB, Muskens AT and van Rossum JM, Some physicochemical properties of amphetamine and related drugs, *J. Pharm. Pharmacol.*, **21**, 774–775 (1969).

(*continued*)

No.	Name	pK_a value(s)	Data quality	Ionization type	Method	Conditions t °C; I or c M	Comments and Reference(s)
1621	Benzquinamide ($C_{22}H_{32}N_2O_5$) 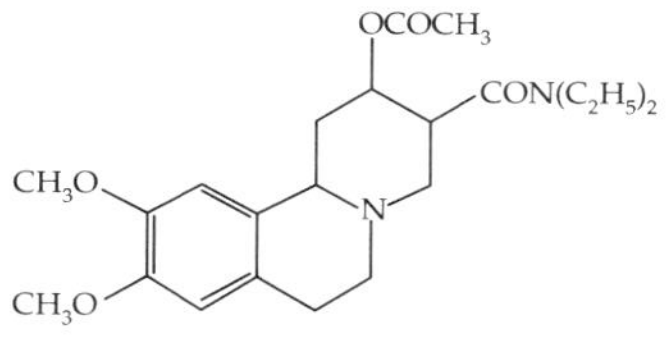	5.9	U	+H			Foye; Craig; N&K: Anon. Craig cited Koe BK and Pinson R, Isolation and characterization of urinary metabolites of benzquinamide and benzquinamide alcohol, *J. Med. Chem.*, **7**, 635–640 (1964). This paper has no pK_a data.
1622	Benztropine ($C_{21}H_{25}NO$) 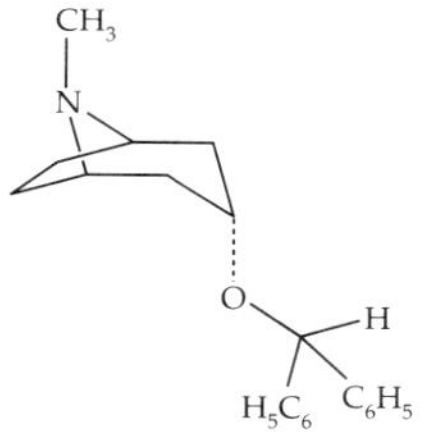	10.0	U	+H			Foye 3rd; see Azatadine. NB: From Benet LZ, Massoud N and Gambertoglio JG, *Pharmacokinetic basis for drug treatment*, New York, Raven Press, pp. 12–13 (1984).
1623	Benzylamphetamine ($C_{16}H_{19}N$) 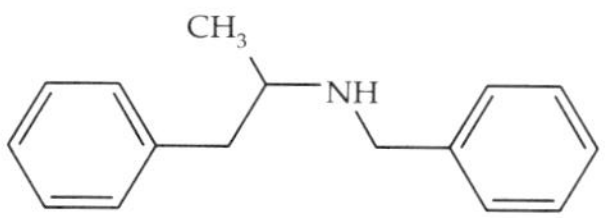	7.50	U	+H			Vree TB, Muskens ATJM and van Rossum JM, Some physicochemical properties of amphetamine and related drugs, *J. Pharm. Pharmacol.*, **21**, 774–775 (1969).
1624	Betahistine ($C_8H_{12}N_2$) 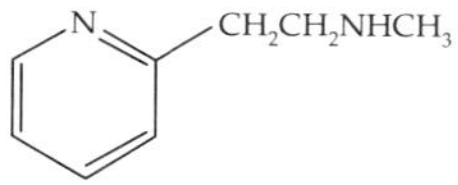	3.46 9.78	U U	+H +H			Craig; Walter LA, Hunt WH and Fosbinder RJ, β-(2- and 4-pyridylalkyl)-amines, *JACS*, **63**, 2771–2773 (1941). NB: This paper has only synthetic and structural details. N&K; S&R.
1625	Betaprodine ($C_{16}H_{23}NO_2$) 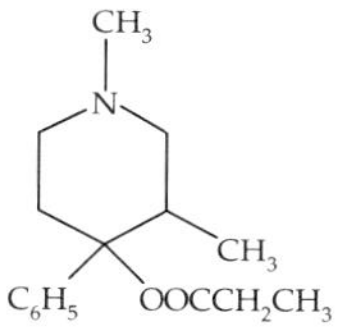	8.7	U	+H			Beckett AH, Casy AF, Kirk G and Walker J, α- and β-Prodine type compounds. Configurational studies, *JPP*, **9**, 939–948 (1957). NB: This paper has no pK_a data, only synthesis. See Craig.

No.	Compound	pKa			Method	Solvent	Reference
1626	Betaxolol ($C_{18}H_{29}NO_3$) 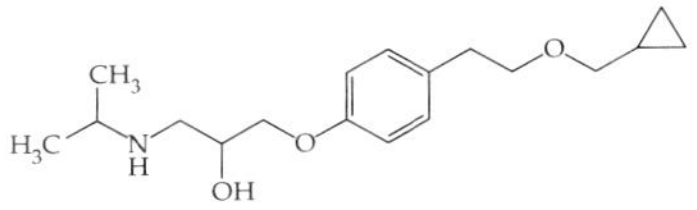	9.21	U	+H			Bree F, El Tayar N, van de Waterbeemd H, Testa B and Tillement JP, The binding of agonists and antagonists to rat lung beta-adrenergic receptors as investigated by thermodynamics and structure-activity relationships, *J. Recept. Res.*, **6**, 381–409 (1986). NB: *in* Lombardo F, Obach RS, Shalaeva MY and Gao F, Prediction of human volume of distribution values for neutral and basic drugs. 2. Extended data set and leave-class-out statistics, *J. Med. Chem.*, **47**, 1242–1250 (2004); ref. 281.
1627	Bethanidine ($C_{10}H_{15}N_3$) 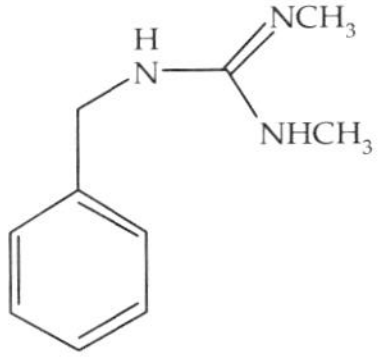	10.6	U	+H	partition	H_2O	Hengstmann JH, Falkner FC, Watson JT and Oates J, Quantitative determination of guanethidine and other guanido-containing drugs in biological fluids by gas chromatography with flame ionization detection and multiple ion detection, *Anal. Chem.*, **46**, 34–39 (1974). NB: Bethanidine was partitioned between aqueous buffers and ethylene chloride, then the aqueous phase pH increased to >13.5 and extracted again with organic solvent. The dried extracts were analysed by GLC and the pK_a determined by inspection of the plot of peak height versus pH. No further details were presented; cited in N&K.
1628	Biscoumacetic acid ($C_{20}H_{12}O_8$) 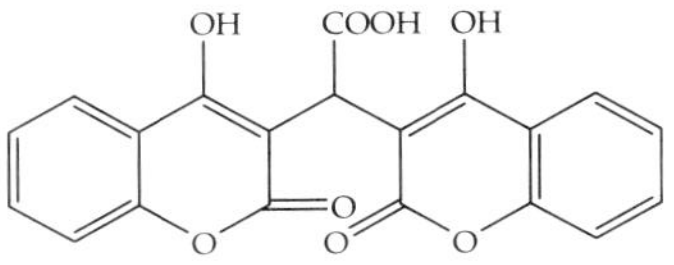	3.1 7.8	U U	−H −H	potentio	90% EtOH	W&G; Anton AH, A drug-induced change in the distribution and renal excretion of sulfonamides, *JPET*, **134**, 291–303 (1961). NB: The value was obtained from Dayton PG, personal communication. Originally measured by Burns JJ, Wexler S and Brodie BB, The isolation and characterization of a metabolic product of 3,3′-carboxymethylene-bis-(4-hydroxycoumarin) ethyl ester (Tromexan) from human urine, *JACS*, **75**, 2345–2346 (1953).

(continued)

No.	Name	pK_a value(s)	Data quality	Ionization type	Method	Conditions t °C; I or c M	Comments and Reference(s)
1629	Bromazepam ($C_{14}H_{10}BrN_3O$) 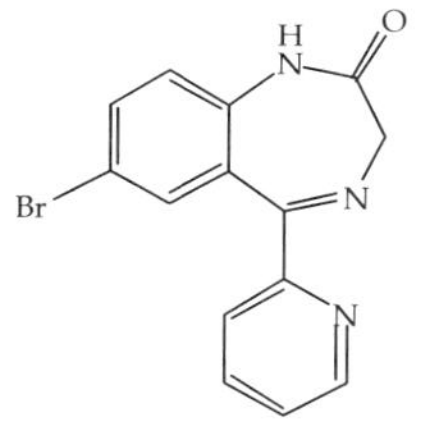	2.5 5.2 11.8	U U U	+H +H −H		spectro, polarographic	Smyth MR, Beng TS and Smyth WF, A spectral and polarographic study of the acid-base and complexing behavior of bromazepam, *Anal. Chem. Acta*, **92**, 129–138 (1977). Cited in Hassan MMA and Abounassif MA, Bromazepam, *APDS*, **16**, 1–51 (1987). "Bromazepam has three pK_a values of 2.5, 5.2, and 11.8, corresponding to protonation at the azomethine and pyridine nitrogen atoms and deprotonation at the nitrogen in position 1, respectively. Three ... values were determined by spectral and polarographic analysis."
1630	Bromazepam	2.9 11.0	U U	+H −H			N&K; Kaplan SA, Jack ML, Weinfeld RE, Glover W, Weissman L and Cotler S, Biopharmaceutical and clinical pharmacokinetic profile of bromazepam, *J. Pharmacokinet. Biopharm.*, **4**, 1–16 (1976).
1631	Bromocriptine ($C_{32}H_{40}BrN_5O_5$) 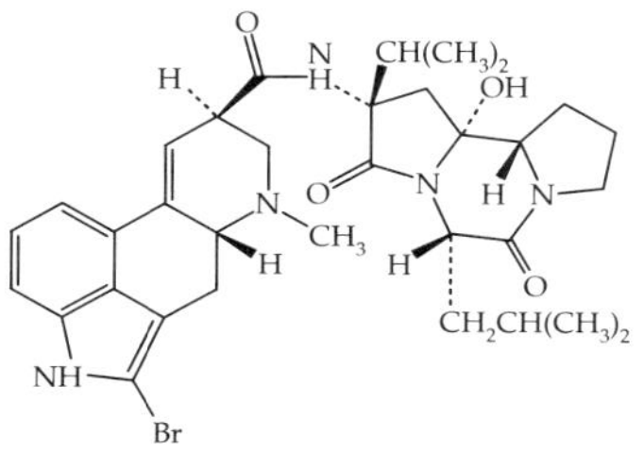	4.90±0.05	U	+H	potentio	80% MCS t = RT	Giron-Forest DA and Schönleber WD, Bromocriptine, *APDS*, **8**, 47–81 (1979). "Due to the low solubility of bromocriptine mesylate in water, the pK_a value had to be determined in methyl cellosolve/water 8:2 (w/w). Titration at ambient temperature yielded pK_a as 4.90 ± 0.05 for a 0.0078 M solution." NB: No references were given.

1632	Bromocriptine	9.8	U	+H	potentio	80% MCS	Foye. NB: This value is close to the pK_b value from no. 1631
1633	Bromodiphenhydramine ($C_{17}H_{20}BrNO$) 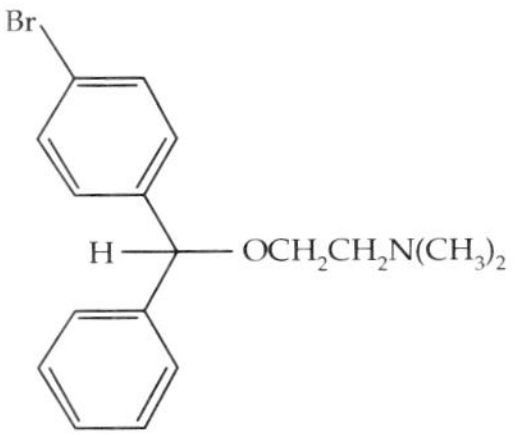	8.64	U	+H	potentio	H_2O $t = 25$ $c = 0.002$ to 0.01	Lordi NG and Christian JE, Physical properties and pharmacological activity: Antihistamines, *J. Am. Pharm. Assoc., Sci. Edn.*, **45**, 300–305 (1956). See Chlorpheniramine (no. 1704) for details. Also N&K.
1634	Bromothen ($C_{14}H_{18}BrN_3S$) 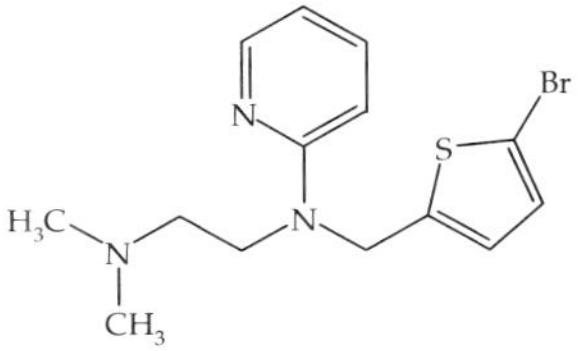	8.63	U	+H	potentio	H_2O $t = 20$ $c = 0.0025$	Marshall PB, Some chemical and physical properties associated with histamine antagonism, *Br. J. Pharmacol.*, **10**, 270–278 (1955). Cited in Perrin Bases 1068 ref. M14. Used glass electrode in cell with liquid junction potentials. NB: See also W&G.
1635	Brompheniramine	3.59	U	+H			Foye 3rd; see Azatadine; from McEvoy.
		9.12	U	+H			
1636	D-Brompheniramine	9.3	U	+H			Foye
1637	Brucine	2.5	U	+H			Craig; Merck 12 reported $pK_1 = 6.04$, $pK_2 = 11.7$. These values are approximately the Foye values subtracted from 14. ACD/pK_a calculated values are -1.09 ± 0.2 (protonation of amide); 8.27 ± 0.2 (protonation of alicyclic 3° amine). This confirms that the Merck values are pK_b values.
		8.2	U	+H			

(continued)

No.	Name	pK_a value(s)	Data quality	Ionization type	Method	Conditions t °C; I or c M	Comments and Reference(s)
1638	Bufuralol ($C_{16}H_{23}NO_2$) 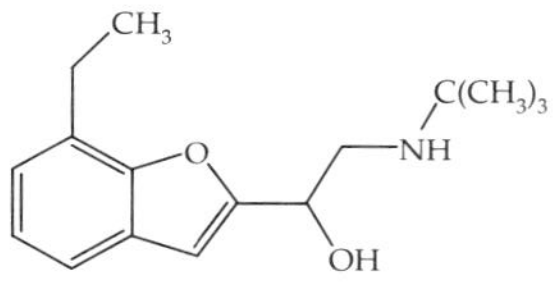	8.97	U	+H			Craig; Magometschnigg D, Bonelli J, Hitzenberger G, Kaik G and Korn A, Decrease of peripheral resistance after acute intravenous application of a new beta-receptor blocking agent, bufuralol hydrochloride, *Int. J. Clin. Pharm. Biopharm.*, **16**, 54–58 (1978).
1639	Bufuralol	9.2	U	+H			Hinderling PH, Schmidlin O and Seydel JK, Quantitative relationships between structure and pharmacokinetics of beta-adrenoceptor blocking agents in man, *J. Pharmacokin. Biopharm.*, **12**, 263–287 (1984): ref. Eckert M, Hoffman-LaRoche Inc., Basle, Switzerland. Personal communication.
1640	Bumetanide	5.2	U	–H			Craig
		10.0	U	–H			
1641	Bunolol (levobunolol) ($C_{17}H_{25}NO_3$) 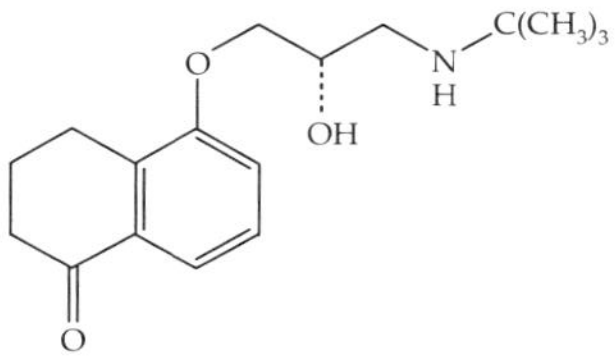	9.32	U	+H			Craig
1642	Bunolol	9.32	U	+H			Craig
1643	Bupropion ($C_{13}H_{18}ClNO$) 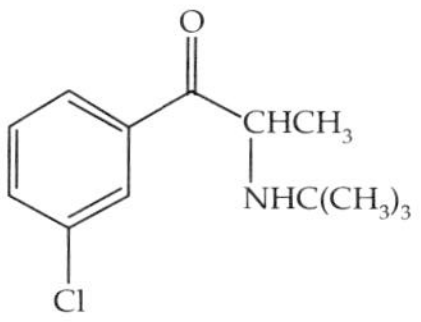	7.0	U	+H			Foye; cited Florey K, *APDS*, **9**, (1980). NB: This reference appears to be incorrect. ACD/pK_a predicted pK_a = 7.16.

No.	Compound	pKa			Method	Reference
1644	Burimamide ($C_9H_{16}N_4S$)	7.5	U	+H		N&K; S&R
		7.25	U	+H	potentio	Shankley NP, Black JW, Ganellin CR and Mitchell RC, *Br. J. Pharmacol.*, **94**, 264–274 (1988). NB: See Ranitidine for details.
1645	Butacaine ($C_{18}H_{30}N_2O_2$) 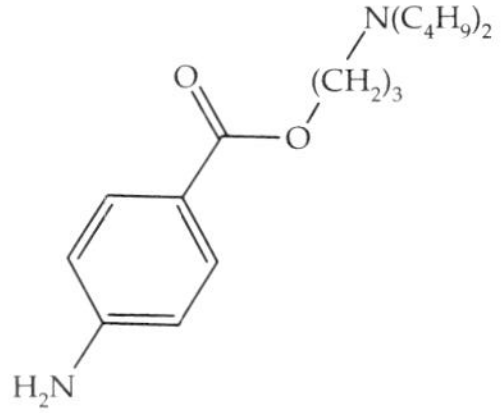	8.96	U	+H		Craig
1646	Butaclamol ($C_{25}H_{31}NO$) 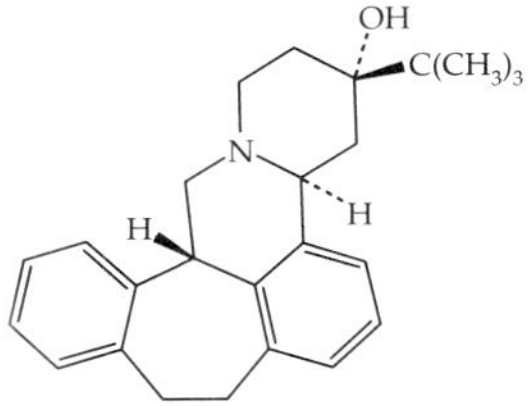	7.20	U	+H		Craig Also called Ayerst AY-23, 028.
1647	Butamben ($C_{11}H_{15}NO_2$) 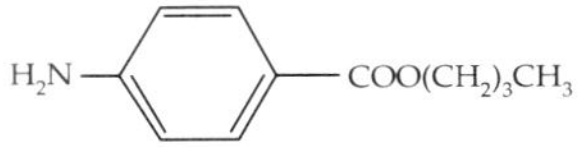	5.38	U	+H		Craig
1648	Butorphanol ($C_{21}H_{29}NO_2$) 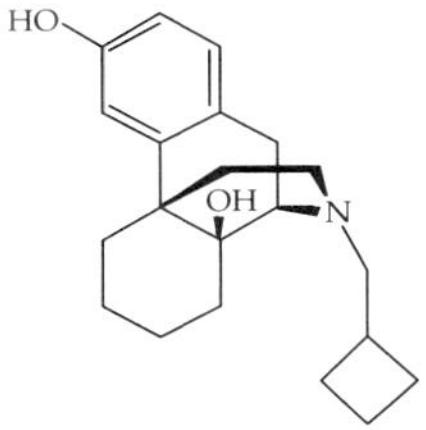	8.6	U	+H		Foye 3rd; see Azatadine; from McEvoy.

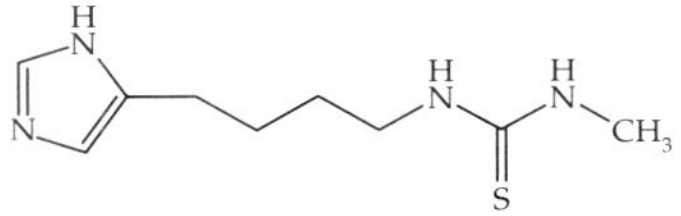

(continued)

No.	Name	pK_a value(s)	Data quality	Ionization type	Method	Conditions t °C; I or c M	Comments and Reference(s)
1649	Butorphanol	8.19	VU	+H			Lombardo F, Obach RS, Shalaeva MY and Gao F, Prediction of human volume of distribution values for neutral and basic drugs. 2. Extended data set and leave-class-out statistics, *J. Med. Chem.*, **47**, 1242–1250 (2004); NB: ref. not given: "estimated to be similar to codeine and morphine."
1650	Butylated hydroxytoluene ($C_{15}H_{24}O$)	17.5	VU	−H			Craig. NB: ACD/pK_a estimate is 12.76 ± 0.4.
	OH; $(CH_3)_3C$; $C(CH_3)_3$; CH_3	14	VU	−H	potentio	H_2O $t = 25$	Steigman J and Sussman D, Acid-base reactions in concentrated aqueous quaternary ammonium salt solutions. II., *JACS*, **89**, 6406–6410 (1967). NB: This study had to dissolve the analyte in 10 molal tetrabutylammonium bromide, and resulted in data with an indistinct endpoint.
1651	Butylparaben ($C_{11}H_{14}O_3$)	8.5	U	−H			Merck 11
	$CO_2(CH_2)_3CH_3$; OH	8.47	U	−H			Craig; cited Merck 10
1652	Butylparaben	8.4	U	−H			W&G: Tammilehto S and Buchi J, p-Hydroxybenzoic acid esters (Nipagins). I. Physicochemical properties, *Pharm. Acta Helv.*, **43**, 726–738 (1968).
1653	Caffeine	0.6	U	+H			N&K; Martin
		3.6	U	+H			
		14.0	U	−H			

No.	Compound	pKa			Method	Conditions	Comments/Reference
1654	Camptothecin	10.83	U	−H			Craig
1655	Cannabidiol ($C_{21}H_{30}O_2$) 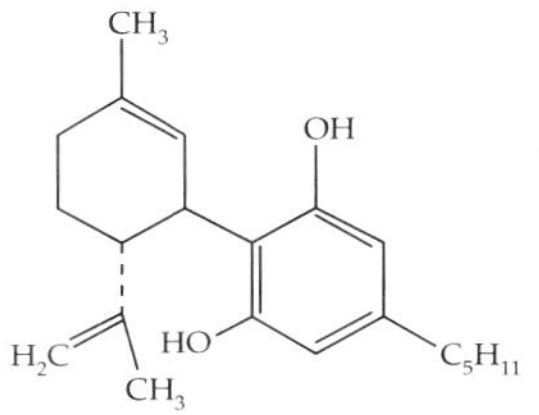		U	−H	spectro		El-Darawy ZI, Abu-Eitah R and Mobarak ZM, Studies on hashish. Part 5. Identification of cannabidiol and cannabidiolic acid by UV spectrophotometry, *Pharmazie*, **28**, 129–133 (1973); CA 78:144062. "The identification of cannabidiol (I) and cannabidiolic acid (II) by UV spectrometry was investigated. The absorption spectra of II in different solvents and at different pH values was investigated and the value of pK_a for I was calculated. The effect of heat on the spectra of I and II was noted."
1656	Capreomycin	6.2	U	+H	potentio	66% DMF	N&K; Merck 9
		8.2	U	+H			NB: Mixture including Capreomycin IA, OH/H = OH (25%) ($C_{25}H_{44}N_{14}O_8$); Capreomycin IB, OH/H = H (67%) ($C_{25}H_{44}N_{14}O_7$).
		10.1	U	+H			
		13.3	U	+H			
1657	Carbachol ($C_6H_{15}ClN_2O_2$) $[NH_2COOCH_2CH_2N(CH_3)_3]^+Cl^-$	4.8	U	+H			Craig; N&K; Ritschel gave 4.8; ref. Robson JM and Stacey RS, *Recent advances in pharmacology*, 4th Edn., Little Brown and Co., Boston, p. 108 (1968).
1658	Carbenicillin ($C_{17}H_{18}N_2O_6S$) 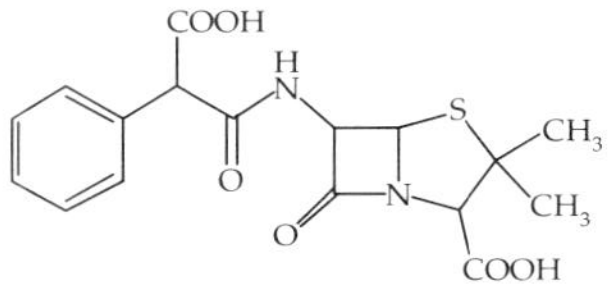	2.22 ± 0.05	U	−H	potentio	H_2O $t = 25.0$ $I = 0.15$ (KCl)	Sirius Technical Application Notes, vol. **2**, p. 109 (1995). Sirius Analytical Instruments Ltd., Forest Row East Sussex, RH18 5DW, UK. NB: Concentration of analyte, 0.54 mM.
		3.25 ± 0.02	U	−H			
		2.7	U	−H			Craig; N&K cited a 2nd pK_a value of 2.6, as well as 2.7; S&R.

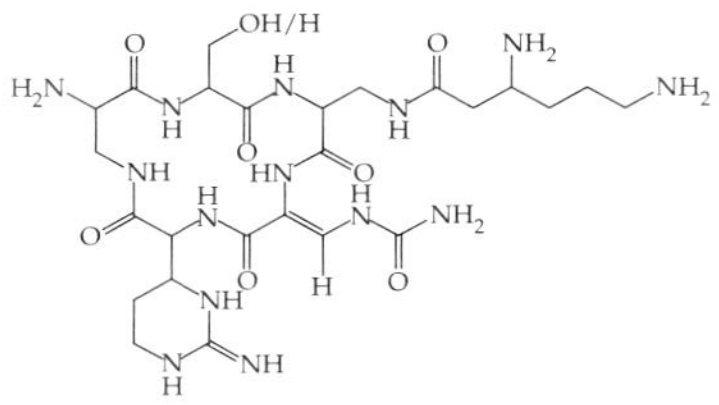

(continued)

No.	Name	pK_a value(s)	Data quality	Ionization type	Method	Conditions t °C; I or c M	Comments and Reference(s)
1659	Carbenoxolone	6.7 7.1	U U	−H −H			N&K; Foye. NB: These values are high for carboxylic acid groups, and are from Downer HD, Galloway RW, Horwich L and Parke DV, *J. Pharm. Pharmacol.*, **22**, 479–487 (1970).
1660	Carbinoxamine	8.1	U	+H	potentio		Craig; N&K; Borodkin S and Yunker MH, Interaction of amine drugs with a polycarboxylic acid ion-exchange resin, *J. Pharm. Sci.*, **59**, 481–486 (1970). NB: See also Rotoxamine (levo-isomer).
1661	Carbonic acid H_2CO_3	6.37 10.25	U U	−H −H		H_2O $t = 18$	W&G: Kolthoff IM and Bosch W, The influence of neutral salts on acid-salt equilibria. III. The second dissociation constant of carbonic acid and the influence of salts on the activity of the hydrogen ions in a bicarbonate-carbonate mixture, *Rec. Trav. Chim. Pays-Bas Belg.*, **47**, 819–825 (1928).
1662	Carisoprodol ($C_{12}H_{24}N_2O_4$) $H_2NCOOCH_2$–C($CH_2CH_2CH_3$)(CH_3)–$CH_2OOCNHCH(CH_3)_2$	4.2	U	+H			McEvoy
1663	Carpindolol ($C_{19}H_{28}N_2O_4$) [structure: O, O-iPr, NH, $(CH_3)_3C$, NH, O, HO]	8.75	U	+H			Craig

1664	Cefaclor ($C_{15}H_{14}ClN_3O_4S$) 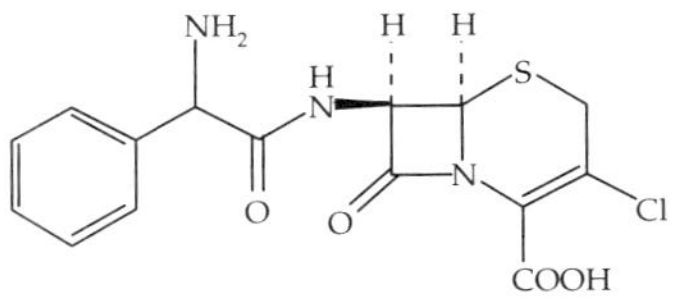	1.5±0.2	U	−H	potentio	H_2O	Lorenz LJ, Cefaclor, *APDS*, **9**, 107–123 (1980). NB: No references supplied. Solvent / pK$_a$ Carboxyl / pK$_a$ Amino: H_2O: 1.5 ± 0.2 / 7.17 DMF: 4.33 / 7.34
		7.17	U	+H			
1665	Cefamandole ($C_{18}H_{18}N_6O_5S_2$) 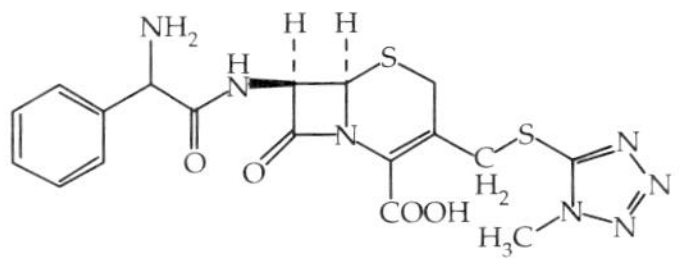	2.6–2.9	U	−H	potentio		Rickard EC and Cooke GG, Electrochemical analysis of the cephalosporin cefamandole nafate, *J. Pharm. Sci.*, **66**, 379–384 (1977). Cited in Bishara RH and Rickard EC, Cefamandole nafate, *APDS*, **9**, 125–154 (1980). NB: No supporting references, experimental data or conditions were presented to verify these data.
		3.0	U	−H	spectro		
1666	Cefazolin	2.15	U	−H	spectro		Fujisawa Pharmaceutical Co., Ltd., Japan, personal communication. Cited in Zappala AF, Walter WH and Post A, Cefazolin, *APDS*, **4**, 1–20 (1975). "The pK_a is 2.15 determined spectrophotometrically. A pK_a of 2.05 determined titrimetrically has been reported."
1667	Cefazolin	2.10	U	−H			N&K; Zappala AF, Walter WH and Post A, Cefazolin, *APDS*, **4**, 1–20 (1975).
1668	Cefazolin	2.3	U	−H			Nightingale CH, Greene DS and Quintiliani R, Pharmacokinetics and clinical use of cephalosporin antibiotics, *J. Pharm. Sci.*, **64**, 1899–1927 (1975). NB: The only pK_a reference is to an Eli Lilly package insert.

(continued)

No.	Name	pK_a value(s)	Data quality	Ionization type	Method	Conditions t °C; I or c M	Comments and Reference(s)
1669	Cefazolin						See Cephazolin
1670	Cefixime ($C_{16}H_{15}N_5O_7S_2$)	2.10 2.69 3.73	U U U	−H +H −H			Okeke CC, Srinavasan VS and Brittain HG, Cefixime, *APDS*, **25**, 39–83 (1997). NB: No reference given. "pK_{a1}: 2.10 (cephem -COOH); pK_{a2}: 2.69 (-NH_2); pK_{a3}: 3.73 (sidechain -COOH)."
1671	Cefoperazone ($C_{25}H_{27}N_9O_8S_2$) 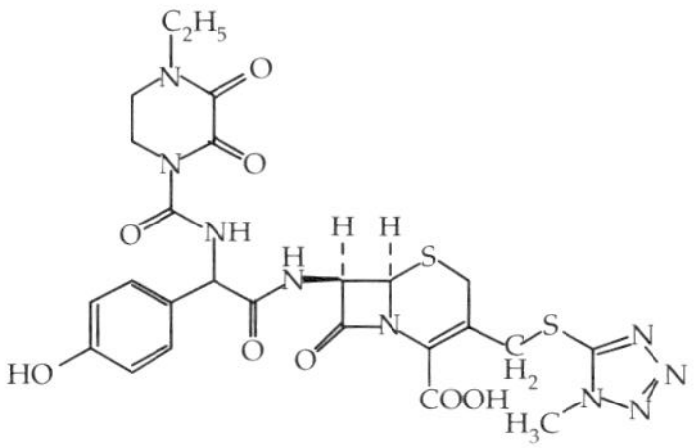	2.6	U	−H			McEvoy
1672	Cefoxitin ($C_{16}H_{17}N_3O_7S_2$) 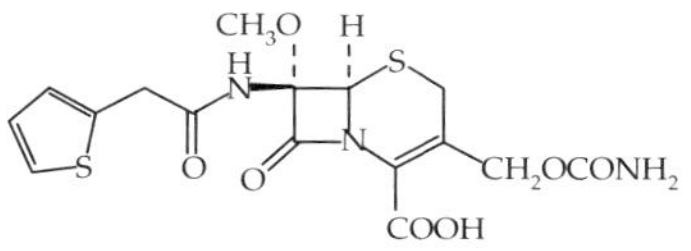	2.2	U	−H	potentio		Brenner GS, Cefoxitin sodium, *APDS*, **11**, 162–195 (1982). Cited Bicker G, Merck Sharp and Dohme Research Laboratories, personal communication; McCauley JA, Shah A, Merck Sharp and Dohme Research Laboratories, personal communication. NB: Also stated a similar value attributed to solubility data.

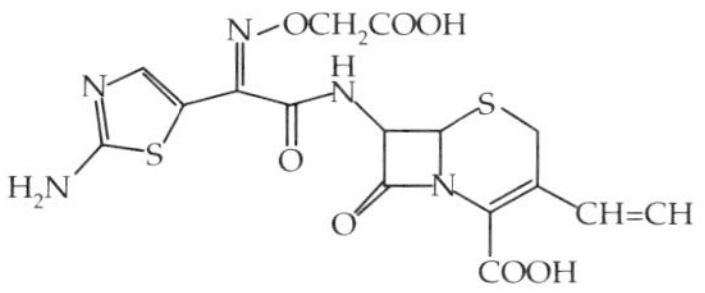

No.	Compound	pKa			Method	Conditions	Reference
1673	Ceftazidime ($C_{22}H_{22}N_6O_7S_2$)	1.8	U	–H			Abounassif M, Mian NAA, Mian MS, Ceftazidime, *APDS*, **19**, 95–121 (1990).
		2.7	U	–H			NB: See also McEvoy.
		4.1	U	+H			
1674	Ceftizoxime ($C_{13}H_{13}N_5O_5S_2$)	2.1	U	+H			McEvoy
		2.7	U	–H			
1675	Ceftriaxone ($C_{18}H_{18}N_8O_7S_3$)	–3.2 ± 0.6			comp		ACD/pK_a estimate
		–1.8 ± 0.2					
		2.57 ± 0.50	U	+H			
		2.90 ± 0.50	U	–H			
		8.03 ± 0.40	U	–H			
		10.0 ± 0.20					
1676	Ceftriaxone	3.2	U	+H			Craig
		3.2	U	–H			
		4.1	U	–H			
1677	Cefuroxime ($C_{16}H_{16}N_4O_8S$)	2.5	U	–H		H_2O	Wozniak TJ, Hicks JR, Cefuroxime Sodium, *APDS*, **20**, 209–236 (1991).
		5.1	U			DMF	NB: Gave no references. See also McEvoy.

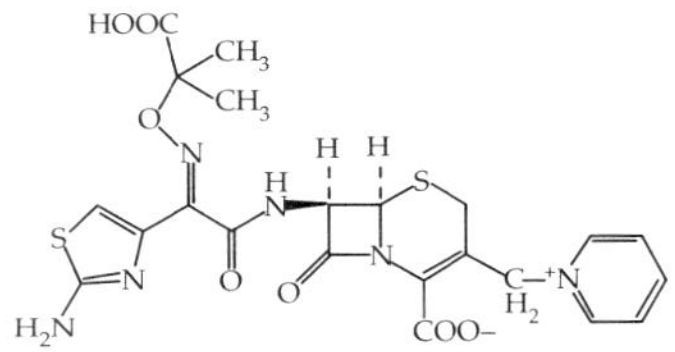

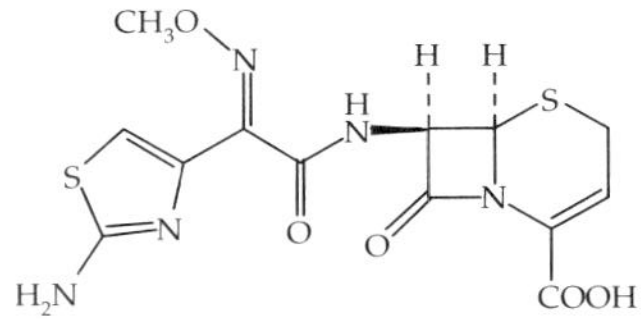

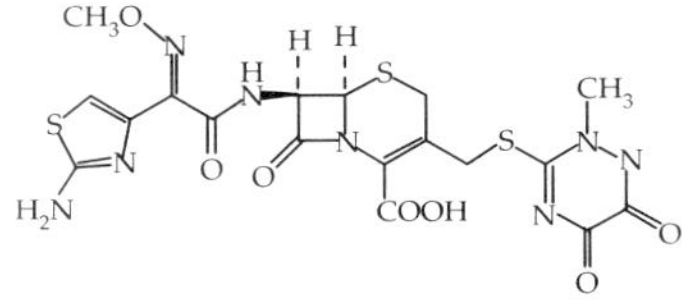

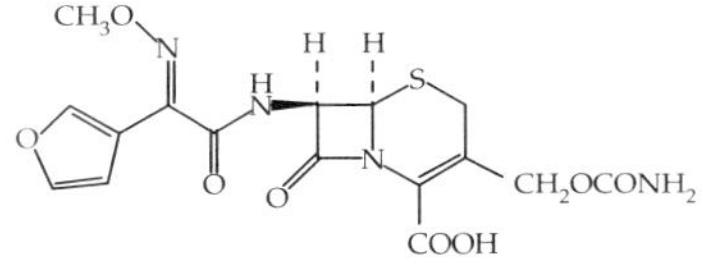

(continued)

No.	Name	pK_a value(s)	Data quality	Ionization type	Method	Conditions t °C; I or c M	Comments and Reference(s)
1678	Celiprolol ($C_{20}H_{33}N_3O_4$) 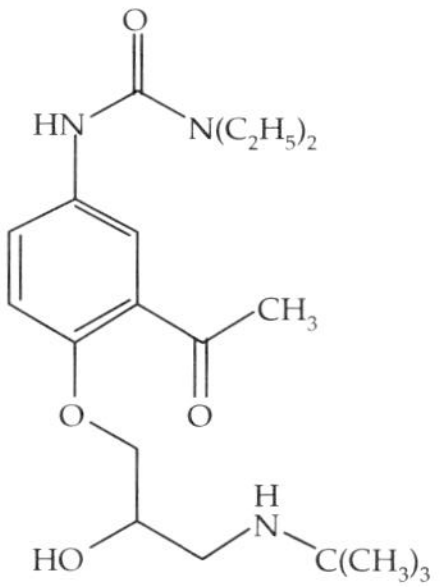	~9.7	U	+H	potentio	H_2O $t = 25$	Mazzo DJ, Obetz CL and Shuster JE, Celiprolol hydrochloride, *APDS*, **20**, 237–301 (1991). "The pK_a of celiprolol base obtained by potentiometric titration in water at 25 °C is approximately 9.7 (36)." 36. O'Hare MJ and Zarembo JE, Revlon Health Care, Tuckahoe NY, personal communication.
1679	Cellulose acetate phthalate 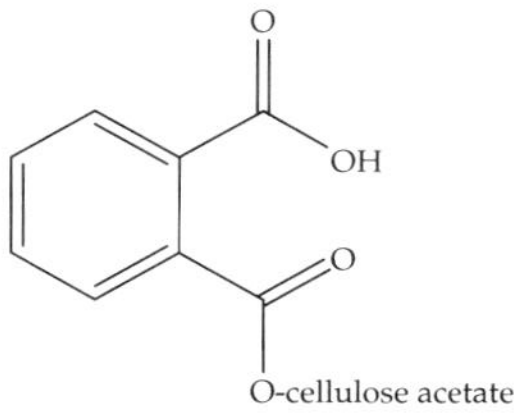						Spitael J, Kinget R and Naessens K, Dissolution rate of cellulose acetate phthalate and the Bronsted catalysis law, *Pharm. Ind.*, **42**(8), 846–849 (1980). "The rate constants for the dissolution of cellulose acetate phthalate in solutions of different basic salts were determined using an automatic pH-stat method. A linear relationship between the logarithm of the rate constants and the pK_a of the basic salts was observed. Evidence was given that the rate of proton transfer, which leads to the dissociation and, consequently, to the dissolution of the polymer, is the rate determining step and is governed by the Bronsted catalysis law."

No.	Compound	pKa	Data quality	Type	Method	Conditions	Reference
1680	Cephacetrile ($C_{13}H_{13}N_3O_6S$)	1.97	U	−H			Merck 10; Merck 11
1681	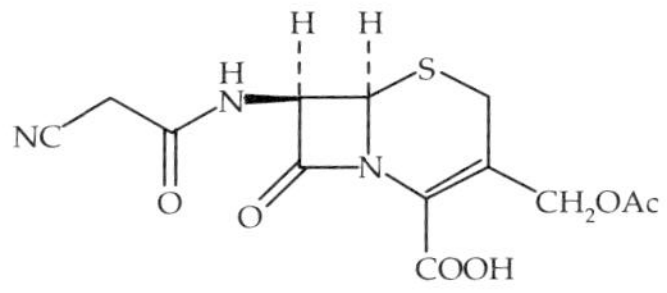 Cephalexin	5.25	U	−H		66% DMF	N&K; APDS 4
		7.1	U	+H		H_2O	
1682	Cephalexin	7.1	U	+H		H_2O	Marelli LP, Cephalexin, *APDS*, **4**, 21–46 (1975).
1683	Cephalothin	2.5	U	−H			Avery; NB: See *APDS*, **1**, 319.
1684	Cephapirin	2.15					Craig; McEvoy; N&K; Hoover
1685	Cephradine ($C_{16}H_{15}N_5O_7S_2$)	2.63	U	−H	potentio		Florey K, Cephradine, *APDS*, **5**, 21–59 (1976).
		7.27	U	+H			H. Jacobson, The Squibb Institute, personal communication.
1686	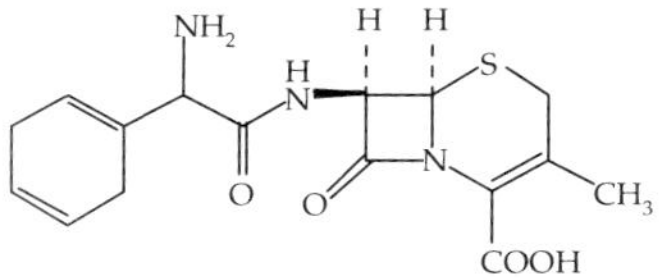 Chenodiol	4.34	U	−H			Craig; McEvoy. NB: Same compound as chenodeoxycholic acid.
1687	Chlorambucil ($C_{14}H_{19}Cl_2NO_2$)	~1.3	U	+H	spectro	H_2O	Linford JH, The influence of pH on the reactivity of chlorambucil, *Biochem. Pharmacol.*, **12**, 317–324 (1963). Cited in Tariq M and Abdullah AA, Chlorambucil, *APDS*, **16**, 85–118 (1987).
	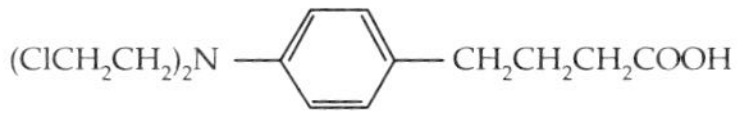 $(ClCH_2CH_2)_2N$–C_6H_4–$CH_2CH_2CH_2COOH$	5.8	U	−H	potentio	$I \sim 0.05$	
1688	Chlorambucil	5.8	U	−H			W&G; cited Linford JH, *Biochem. Pharmacol.*, **12**, 317–324 (1963).
1689	Chlorcyclizine	2.43	A	+H	potentio	H_2O	W&G; Lordi NG and Christian JE, Physical properties and pharmacological activity: antihistaminics, *J. Am. Pharm. Assn., Sci.* Edn., **45**, 300–305 (1956).
		7.81	U	+H		$t = 25$; $c = 0.002$ to 0.01	See Chlorpheniramine (no. 1704) for details.

(*continued*)

No.	Name	pK_a value(s)	Data quality	Ionization type	Method	Conditions t °C; I or c M	Comments and Reference(s)
1690	Chlordiazepoxide	4.76±0.05	U	+H	spectro		MacDonald A, Michaelis AF and Senkowski BZ, Chlordiazepoxide hydrochloride, *APDS*, **1**, p. 26 (1972). Toome V, Hoffman-La Roche Inc., personal communication.
		4.9	U	+H	potentio		Yao C, Lau E, Hoffman-La Roche Inc., personal communication.
1691	Chlordiazepoxide	4.8	U	+H			Van der Kleijn E, Protein binding and lipophilic nature of ataractics of the meprobamate and diazepine group, *Arch. Int. Pharmacodyn.*, **179**, 225–250 (1969). NB: No references or experimental details given; cited in: Van der Kleijn E, Kinetics of distribution and metabolism of diazepam and chlordiazepoxide in mice, *Arch. Int. Pharmacodyn.*, **178**, 193–215 (1969).
1692	Chlorhexidine ($C_{22}H_{30}Cl_2N_{10}$) 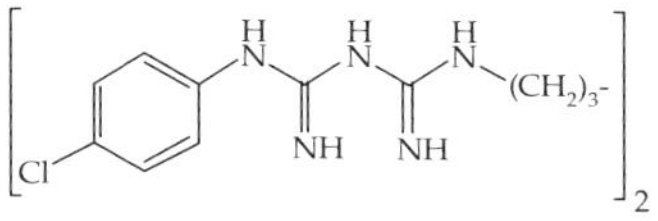	10.78	U	+H			Craig
1693	4-Chloroamphetamine ($C_9H_{12}ClN$) 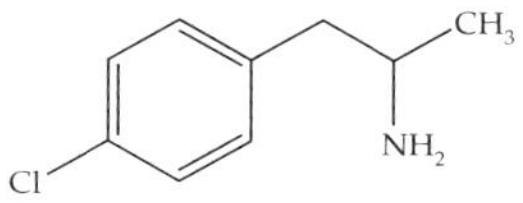	9.80	U	+H			Vree TB, Muskens ATJM and van Rossum JM, Some physicochemical properties of amphetamine and related drugs, *J. Pharm. Pharmacol.*, **21**, 774–775 (1969).

1694	Chlorocresol (C_7H_7ClO) 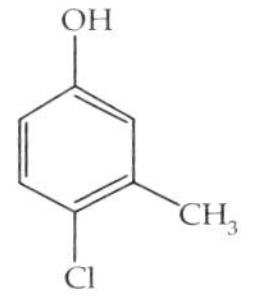	9.55	U	–H		$t = 25$	Craig
1695	(S)-6-Chloro-4-(cyclo-propylethynyl)-1,4-dihydro-4-(trifluoro-methyl)-2H-3,1-benz-oxazin-2-one (Efavirenz; DMP-266) ($C_{14}H_9ClF_3NO_2$) 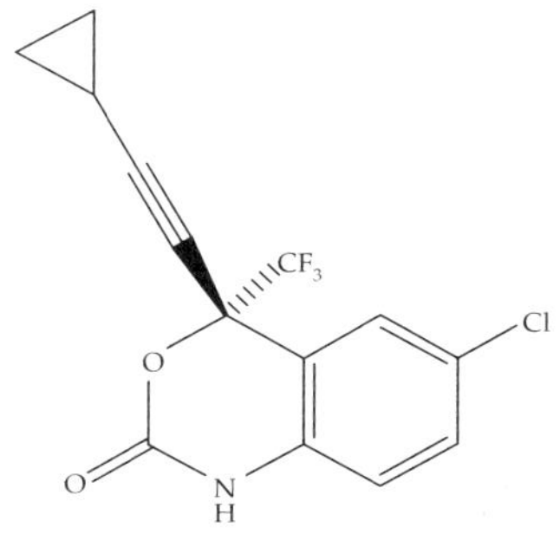	10.1–10.2	U	–H	spectro	H_2O	Rabel SR, Maurin MB, Rowe SM and Hussain M, Determination of the pK_a and pH-solubility behavior of an ionizable cyclic carbamate, (S)-6-chloro-4-(cyclopropylethynyl)-1, 4-dihydro-4-(trifluoromethyl)-2H-3,1-benzoxazin-2-one (DMP-266), *Pharm. Dev. Technol.*, **1**(1), 91–95 (1996). "The solubility of a nonnucleoside reverse transcriptase inhibitor, L-743726 ((S)-6-chloro-4-(cyclopropylethynyl)-1,4-dihydro-4-(trifluoromethyl)-2H-3,1-benzoxazin-2-one; DMP-266) was investigated as a function of pH. A dramatic increase in the aqueous solubility was observed at pH >= 10, which was consistent with going from a neutral to a charged species. The ionization of the proton positioned on the carbamate functionality was confirmed spectrophotometrically (pK_a = 10.1). The spectrophotometric result was in excellent agreement with that obtained from the solubility studies (pK_a = 10.2). The ionization behavior of L-743726 represents a unique case in which the pK_a for a carbamate functional group is quite low. It was concluded that the anomalous pK_a value may be attributed to stabilization of the negatively charged species through inductive effects, which originate

(*continued*)

No.	Name	pK_a value(s)	Data quality	Ionization type	Method	Conditions t °C; I or c M	Comments and Reference(s)
							from the surrounding substituents and delocalization of the negative charge via resonance effects."
1696	Chloroquine	8.4 10.8	U U	+H +H		H_2O $t = 20$	Tariq M and Al-Badr AA, Chloroquine, *APDS*, **13**, 95–123 (1984). NB: Values appear to be rounded from Schill, 1965. NB: See also BPC, p. 176.
1697	Chloroquine phosphate	8.10 9.94	U U	+H +H	potentio		Hong DD, Chloroquine phosphate, *APDS*, **5**, 61–85 (1976). Cited Dederick P, Sterling-Winthrop Research Institute, unpublished data.
1698	Chlorothen ($C_{14}H_{18}ClN_3S$)	8.4	U	+H	potentio	H_2O $t = 20$ $c = 0.0025$	Marshall PB, Some chemical and physical properties associated with histamine antagonism, *Br. J. Pharmacol.*, **10**, 270–278 (1955). Cited in Perrin Bases 1088 ref. M14. Used glass electrode in cell with liquid junction potentials.
		8.42	U	+H		H_2O $t = 25$ $c = 0.002$ to 0.01	Lordi NG and Christian JE, Physical properties and pharmacological activity: antihistaminics, *J. Am. Pharm. Assn., Sci. Edn.*, **45**, 300–305 (1956). See Chlorpheniramine (no. 1704) for details.
1699	8-Chlorotheophylline 8-Nitrotheophylline ($C_7H_7N_5O_4$)	5.28 2.07	U	−H			Mayer MC and Guttman DE, Interactions of xanthine derivatives with bovine serum albumin III. Inhibition of binding, *J. Pharm. Sci.*, **57**, 245–249 (1968). NB: The pK_a value was quoted without experimental details. Refs.: Eichman ML, Guttman DE, van Winkle Q and Guth EP, Interactions of xanthine

							molecules with bovine serum albumin. I., *J. Pharm. Sci.*, **51**, 66–71 (1962); Guttman DE and Gadzala AE, Interactions of xanthine molecules with bovine serum albumin. II., *J. Pharm. Sci.*, **54**, 742–746 (1965).
1700	8-Chlorotheophylline	4.6 ± 0.7	U	–H	comp		ACD/pK_a estimate
1701	8-Chlorotheophylline	8.2	U	–H			N&L; Charles RL, Searle Laboratories, Chicago, IL. Personal communication. NB: Based on the computed pK_a value above and the other measured values, this value appears more likely to be a pK_b for the conjugate base.
1702	Chlorothiazide	6.7 9.5	U U	–H –H			W&G: Resetarits DE and Bates TR, Errors in chlorothiazide bioavailability estimates based on a Bratton-Marshall colorimetric method for chlorothiazide in urine, *J. Pharm. Sci.*, **68**, 126–127 (1979). NB: Values given with no reference or experimental detail.
1703	Chlorothiazide	6.0 ± 0.4 9.7 ± 0.2	U U	–H –H	comp		ACD/p*K*a estimate
1704	Chlorpheniramine	9.16 ± 0.02	A	+H	potentio	H_2O $t = 25$ $c = 0.002$ to 0.01	Lordi NG and Christian JE, Physical properties and pharmacological activity: antihistaminics, *J. Am. Pharm. Ass., Sci. Edn.*, **45**, 300–305 (1956). Cited in Perrin Bases 1083 ref. L54. NB: Used glass electrode in cell with liquid junction potentials; electrode calibrated at pH = 3.75 ± 0.02 and 7.00 before every run. Titrant added with microsyringe until cloudiness from precipitated free base occurred. Where comparisons with Tolstoouhov (1955) (see Pheniramine, no. 2081) were possible, there were discrepancies of up to 2 log units.

(continued)

No.	Name	pK_a value(s)	Data quality	Ionization type	Method	Conditions t °C; I or c M	Comments and Reference(s)
1705	Chlorpheniramine	9.2	U	+H			Eckhart CG and McCorkle T, Chlorpheniramine maleate, *APDS*, **7**, 53 (1978). NB: See Perrin DD, Dissociation constants of organic bases in aqueous solutions, Butterworths, London (1965).
1706	Chlorpheniramine	9.16	U	+H	potentio	H_2O $t = 20$ $c = 0.0025$	Marshall PB, Some chemical and physical properties associated with histamine antagonism, *Br. J. Pharmacol.*, **10**, 270–278 (1955). Cited in Perrin Bases 1083 ref. M14. NB: Used glass electrode in cell with liquid junction potentials. See also Craig.
1707	Chlorpheniramine	8.99	U	+H			Chatten LG (ed.), *Pharmaceutical Chemistry*, vol. **1**, Dekker, New York, pp. 85–87 (1966).
1708	Chlorpheniramine	9.26	U	+H	potentio		Lombardo F, Obach RS, Shalaeva MY and Gao F, Prediction of human volume of distribution values for neutral and basic drugs. 2. Extended data set and leave-class-out statistics, *J. Med. Chem.*, **47**, 1242–1250 (2004); ref. not given: potentiometric titration.
1709	Chlorphentermine ($C_{10}H_{14}ClN$) Cl–C$_6$H$_4$–CH$_2$–C(CH$_3$)$_2$–NH$_2$	9.60	U	+H			Vree TB, Muskens ATJM and van Rossum JM, Some physicochemical properties of amphetamine and related drugs, *J. Pharm. Pharmacol.*, **21**, 774–775 (1969).
1710	Chlorpromazine	9.24±0.02	U	+H	potentio	H_2O $t = 25 \pm 0.5$ $I = 0.15$ (KCl)	Sirius Technical Application Notes, 1994, vol. **1**, Sirius Analytical Instruments Ltd., Forest Row, East Sussex, RH18 5DW, UK. NB: First result extrapolated to 0% MeOH by Yasuda-Shedlovsky procedure from data obtained in 35–50 wt% MeOH. Second result extrapolated to 0%
		9.02±0.04	U	+H	potentio	H_2O $t = 25 \pm 0.5$ $I = 0.15$ (KCl)	

No.	Compound	pKa	Data quality	Ionization type	Method	Conditions	Reference / Comments
							dioxane by Y-S method from data obtained in 23–52 wt% dioxane.
1711	Chlorpromazine sulfoxide ($C_{17}H_{19}ClN_2OS$) 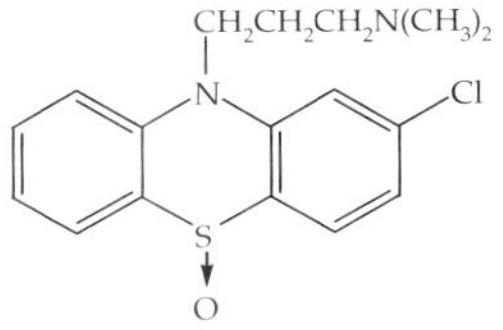	9.0	U	+H			W&G. NB: The reference given in this source (Nightingale CH, *et al.*, *J. Pharm. Sci.*, **64**, 1907 (1975)) is incorrect. The source of this data is not known.
1712	Chlorpropamide ($C_{10}H_{13}ClN_2O_3S$) 	4.8	U	–H			Avery
1713	Chlorpropamide	4.95±0.04	U	–H	potentio	H_2O t undefined I undefined	Crooks MJ and Brown KF, The binding of sulphonylureas to serum albumin, *J. Pharm. Pharmacol.*, **26**, 305–311 (1974). NB: Chlorpropamide was recrystallized from 95% EtOH, m.p. 127–129 °C.
1714	Chlorprothixene ($C_{18}H_{18}ClNS$) 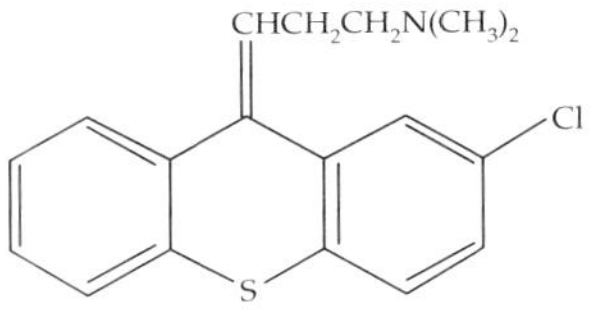	8.4	U	+H	spectro	H_2O	Rudy BC and Senkowski BZ, Chloroprothixene, *APDS*, **2**, 63–84 (1973).
		7.5	U	+H	potentio	50% iPrOH	"The apparent pK_a for chloroprothixene has been determined spectrometrically to be 8.4 (Lau). The apparent pK_a has also been determined from the titration curve in an isopropanol: water (1:1) mixture and found to be 7.5 (Toome and Raymond). In water, the trialkylamino type compounds are stronger bases, on the average, by 0.9 pK_a units (Lau, Toome and

Structure labels (1711): $CH_2CH_2CH_2N(CH_3)_2$, N, S, O, Cl

Structure labels (1712): Cl, $SO_2NHCONHCH_2CH_2CH_3$

Structure labels (1714): $CHCH_2CH_2N(CH_3)_2$, S, Cl

(continued)

No.	Name	pK_a value(s)	Data quality	Ionization type	Method	Conditions t °C; I or c M	Comments and Reference(s)
							Raymond). Therefore, the estimated pK_a in water is 8.4 which is in good agreement with that found spectrophotometrically. Lau E, Hoffmann-La Roche Inc., unpublished data. Toome V, Raymond G, Hoffmann-La Roche Inc., unpublished data."
1715	Chlortetracycline	3.3 7.4 9.3	U U U	−H −H +H			Avery NB: See Demeclocycline for details.
1716	Chlorthalidone	9.4	U	−H			Anon
1717	Chlorzoxazone ($C_7H_4ClNO_2$) 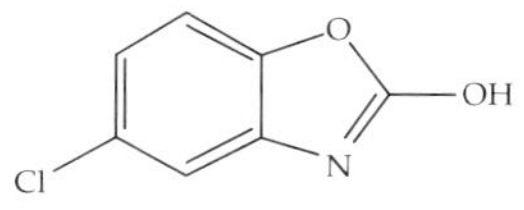	8.3	U	−H			Stewart JT and Janicki CA, Chlorzoxazone, *APDS*, **16**, 119–144 (1987); NB: No reference. N&K; Hoover
1718	Cholic acid ($C_{24}H_{40}O_5$) 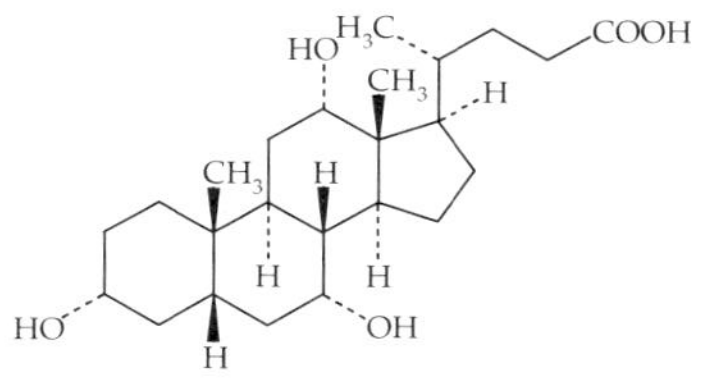	4.98±0.05	U	+H	potentio	H_2O $t = 20$ $c < 0.014$ (CMC)	Johns WH and Bates TR, Quantification of the binding tendencies of cholestyramine I: effect of structure and added electrolytes on the binding of unconjugated and conjugated bile salt anions, *J. Pharm. Sci.*, **58**, 179–183 (1969). NB: These values were quoted from Ekwall P, Rosendahl T and Lofman N, Bile salt solutions. I. The dissociation constants of the cholic and deoxycholic acids. *Acta Chem. Scand.*, **11**, 590–598 (1957). They were measured at concentrations below the critical micellar concentration range. At concentrations above 0.05 M, the pK_a reaches a stable value.

1719	Chromonar ($C_{20}H_{27}NO_5$) 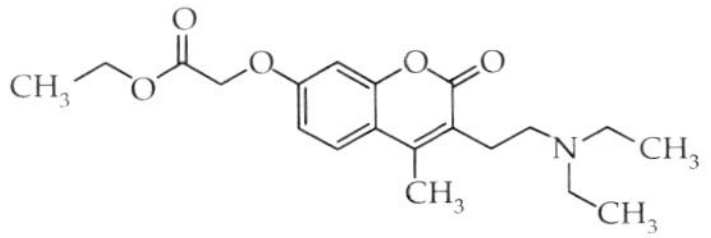	8.3	U	+H	potentio	Borodkin S and Yunker MH, Interaction of amine drugs with a polycarboxylic acid ion-exchange resin, *J. Pharm. Sci.*, **59**, 481–486 (1970).
1720	Cimetidine	6.80	U	+H		Brimblecombe RW, Duncan WAM, Durant GJ *et al.*, Cimetidine – a non-thiourea H2-receptor antagonist, *J. Int. Med. Res.*, **3**, 86–92 (1975). Also see Shankley NP, Black JW, Ganellin CR and Mitchell RC, *Br. J. Pharmacol.*, **94**, 264–274 (1988). NB: See Ranitidine for details.
1721	Cimetidine	6.97	U	+H	potentio	Lombardo F, Obach RS, Shalaeva MY and Gao F, Prediction of human volume of distribution values for neutral and basic drugs. 2. Extended data set and leave-class-out statistics, *J. Med. Chem.*, **47**, 1242–1250 (2004); ref. not given: potentiometic titration.
1722	Cinchonine	4.0 8.2	U U	+H +H		W&G: Kolthoff IM, The dissociation constants, solubility product and titration of alkaloids, *Biochem. Z.*, **162,** 289–353 (1925); Christophers SR, *Trans. Farad. Soc.*, **39**, 333–338 (1943).
1723	Cinchonine	4.04 8.15	U U	+H +H		Dragulescu C and Policec S, Acad Rep Populare Romine, Baza Cercetari Stiint Timisoara, *Studii Cercetari Chim.* **9**, 33–40, (1962); CA 58:5085f. Cited in Perrin Bases supplement no. 7465. NB: From $pK_{b2} = 5.85$ and $pK_{b1} = 9.96$.
1724	Ciprofloxacin	6.0 8.8	U U	−H +H		Foye

(*continued*)

No.	Name	pK_a value(s)	Data quality	Ionization type	Method	Conditions t °C; I or c M	Comments and Reference(s)
1725	Citalopram ($C_{20}H_{21}FN_2O$) 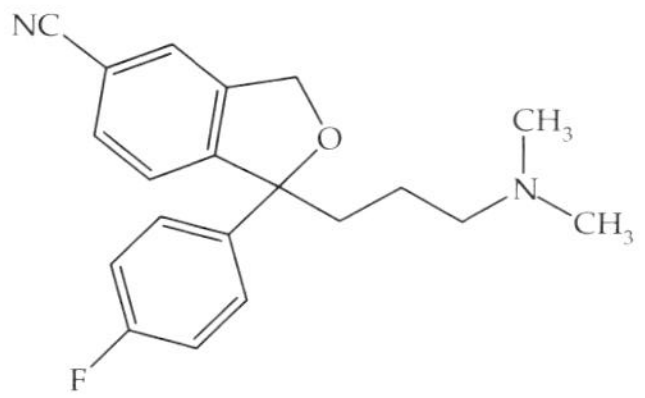	9.38	U	+H	CZE/pH	H_2O	Lombardo F, Obach RS, Shalaeva MY and Gao F, Prediction of human volume of distribution values for neutral and basic drugs. 2. Extended data set and leave-class-out statistics, *J. Med. Chem.*, **47**, 1242–1250 (2004); ref. not given: capillary electrophoresis.
1726	Clomethiazole (chlormethiazole) (C_6H_8ClNS) 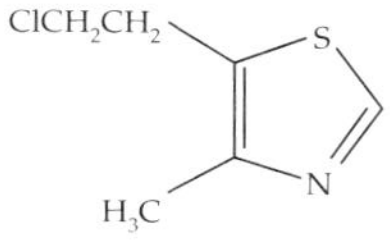	3.4	U	+H			Upton RN, Runciman WB and Mather LE, Relationship between some physico-chemical properties of ionizable drugs and their sorption into medical plastics, Aust. *J. Hosp. Pharm.*, **17**, 267–270 (1987). NB: no references or experimental details were given to support the stated value. "The sorption of drugs into medication plastics encountered in drug administration systems was examined for ... ionizable drugs .... Only I was subject to sorption, the greatest total decrease in concentration occurring with polyvinyl chloride (85%), then rubber (44%), polyethylene (21%) and polypropylene (6%). When sorption was compared with pK_a and lipophilicity, I was not the most lipophilic drug, but by virtue of its low pK_a, was completely un-ionized in solution unlike any of the other drugs. It was concluded that a high degree of un-ionization, even of

						relatively unlipophilic drugs, is a situation which is likely to result in the sorptive loss of the drug into medical plastics.''
1727	Clomethiazole	3.2	U	+H		Avery
1728	Clomethiazole	3.2	U	+H		N&K; Avery
1729	Clomipramine ($C_{19}H_{23}ClN_2$) 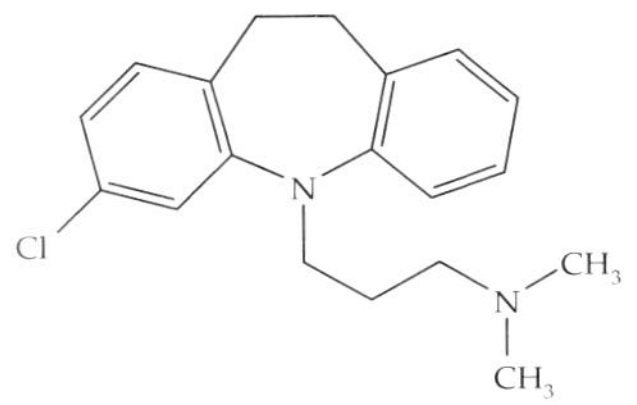	9.38	U	+H	H_2O	Lombardo F, Obach RS, Shalaeva MY and Gao F, Prediction of human volume of distribution values for neutral and basic drugs. 2. Extended data set and leave-class-out statistics, *J. Med. Chem.*, **47**, 1242–1250 (2004); ref. 279: Seiler P, Simultaneous determination of partition coefficient and acidity constant of a substance, *Eur. J. Med. Chem.*, **9**, 663–665 (1974).
1730	Clonazepam ($C_{15}H_{10}ClN_3O_3$) 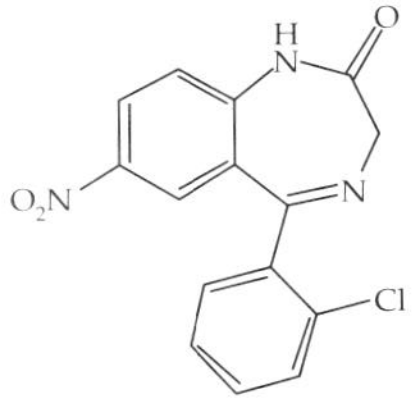	1.5 10.5	U U	+H −H		Kaplan SA, Alexander K, Jack ML, Puglisi CV, DeSilva JAF, Lee TL and Wenfeld RE, Pharmacokinetic profiles of clonazepam in dog and human and of flunitrazepam in dog, *J. Pharm. Sci.*, **63**, 527–532 (1974). NB: See Winslow WC, Clonazepam, *APDS*, **6**, 61–81 (1977). NB: no supporting references or experimental details
1731	Clonidine	8.2	U	+H		Abounassif MA, Mian MS and Mian NAA, Clonidine Hydrochloride, *APDS*, **21**, 109–147. NB: Clarke, pp. 481–482. See also Gennaro A, *et al.*, (eds.), *Remington's Pharmaceutical Sciences*, 16th edn., Mack Publishing Co., Easton PA 785 (1985).
1732	Clonidine	8.25	U	+H		Avery
1733	Clonidine	8.0	U	+H		W&G: Timmermans PBMWM and van Zweiten PA, Dissociation constants of clonidine and structurally related imidazolidines, *Arzneim.-Forsch.*, **28**, 1676–1681 (1978).

(continued)

No.	Name	pK_a value(s)	Data quality	Ionization type	Method	Conditions t °C; I or c M	Comments and Reference(s)
1734	Clonidine	8.05	U	+H		H_2O	Lombardo F, Obach RS, Shalaeva MY and Gao F, Prediction of human volume of distribution values for neutral and basic drugs. 2. Extended data set and leave-class-out statistics, *J. Med. Chem.*, **47**, 1242–1250 (2004); ref. 283: Timmermans PBMWM, Brands A and van Zwieten PA, Lipophilicity and brain disposition of clonidine and structurally related imidazolidines, *Naunyn-Schiedeberg's Arch. Pharmacol.*, **300**, 217–226 (1977).
1735	Clopenthixol	6.69 7.60	U U	+H +H			Craig
1736	Clorazepate ($C_{16}H_{11}ClN_2O_3$) 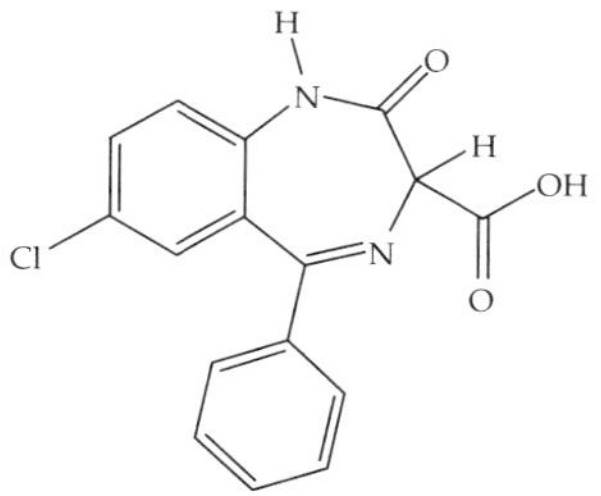	–	U	–H	potentio	H_2O	Raihle JA and Papendick VE, Clorazepate dipotassium, *APDS*, **4**, 91–112 (1975). "Attempts to measure the pK_a of the carboxyl group by titration (of chlorazepate dipotassium) in water with hydrochloric acid were unsuccessful. Only the KOH which is liberated on dissolving chlorazepate dipotassium in H_2O is titrated (Wimer DC, Abbott Laboratories, personal communication)." NB: The description of chlorazepate dipotassium as a monopotassium salt with a KOH adduct is odd. A solid isolated from a solution of chlorazepate in potassium hydroxide solution could also be the dipotassium salt, formed through ionization of both the –COOH and the cyclic amide groups. The latter of

							these would be expected to have a very high pK_a value, similar to KOH in solution. Further study is needed, preferably spectroscopic.
1737	Clozapine	3.70 7.60	U U	+H +H			McLeish MJ, Capuano B and Lloyd EJ, Clozapine, *APDS*, **22**, 145–184 (1993). $pK_1 = 3.70$ (3); $pK_2 = 7.60$ (4) 3. Schmutz J, Eichenberger E, *in* Bindra JS, Lednicer D, (eds.), *Chronicles of Drug Discovery*, vol. **1**, Wiley, NY 39–59 (1982). 4. Richter K, J. Chromatogr., **434**, 465–468 (1988).
1738	Clozapine	7.63	U	+H	potentio	H_2O	Lombardo F, Obach RS, Shalaeva MY and Gao F, Prediction of human volume of distribution values for neutral and basic drugs. 2. Extended data set and leave-class-out statistics, *J. Med. Chem.*, **47**, 1242–1250 (2004) ; ref. not given: potentiometric titration. NB: Craig, 8.0.
1739	Cocaine	8.76	U	+H		H_2O $t = 15.0$	Muhtadi FJ and Al-Badr AA, Cocaine hydrochloride, *APDS*, **15**, 151–231 (1986). NB: No reference to data source. This reference gave only the $pK_b = 5.59$, that is $pK_a = 8.76$ (pK_w=14.35 at 15 °C). Craig gave 8.7.
1740	Cocaine	8.5	U	+H			N&K
1741	Cocaine	8.7	U	+H			Tencheva J, Velinov, G and Budevsky O, New approach of the extrapolation procedure in the determination of acid-base constants of poorly soluble pharmaceuticals. *Arzneim.-Forsch.*, **29**, 1331–1334 (1979). Cited in Lombardo F, Obach RS, Shalaeva MY and Gao F, Prediction of human volume of distribution values for neutral and basic drugs. 2. Extended data set and leave-class-out statistics, *J. Med. Chem.*, **47**, 1242–1250 (2004); ref. 284.

(continued)

No.	Name	pK_a value(s)	Data quality	Ionization type	Method	Conditions t °C; I or c M	Comments and Reference(s)
1742	Codeine	8.22	U	+H	potentio	50% aq. EtOH	Beckett AH, Analgesics and their antagonists. I., *J. Pharm. Pharmacol.*, **8**, 848–859 (1956); cited from Farmilo CG, Oestreicher PM and Levi L, Physical methods for the identification of narcotics. IB Common physical constants for identification of ninety-five narcotics and related compounds, *Bull. Narcotics*, UN Dept. Social Affairs (1954) vol. **6**, pp. 7–19; also cited in Clouet DH (ed.), *Narcotic Drugs Biochemical Pharmacology*, Plenum Press, New York, 52–53 (1971). NB: See Alphaprodine for details.
1743	Codeine	8.2	U	+H		$t = 20$	Muhtadi FJ and Hassan MMA, Codeine phosphate, *APDS*, **10**, 93–134 (1982). BPC 11th Edn., The Pharmaceutical Press, London (1979).
1744	Codeine	8.22±0.01	A	+H	potentio	H_2O $t = 25.0$ $I = 0.15$ (KCl)	Sirius Technical Application Notes, vol. **1**, pp. 99b–99c (1994). Sirius Analytical Instruments Ltd., Forest Row, East Sussex, RH18 5DW, UK. NB: Concentration of analyte, 1.0–1.7 mM.
1745	Colchicine	12.35 (pK_b)	U	–H		$t = 20.0$	Wyatt DK, Grady LT and Sun S, Colchicine, *APDS*, **10**, 139–177 (1981). NB: Merck 9, p. 318 Gennaro *et al.*, Remington's Pharmaceutical Sciences p. 1049 (1975).
1746	Colchicine	1.65	U	+H			N&K; Merck 9, p. 318

No.	Name	pKa			Reference
1747	Cromolyn ($C_{23}H_{16}O_{11}$) 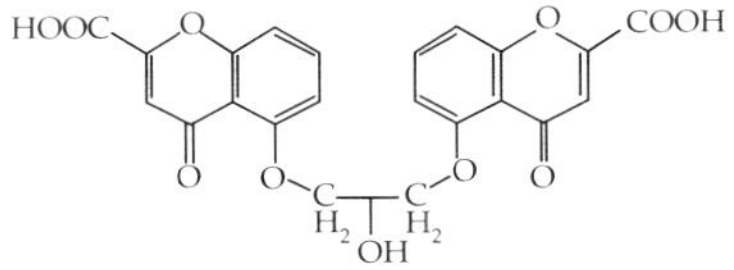	2.0	U	–H	Avery. NB: Probably an average value for the two groups.
1748	Cyanopromazine ($C_{18}H_{19}N_3S$) 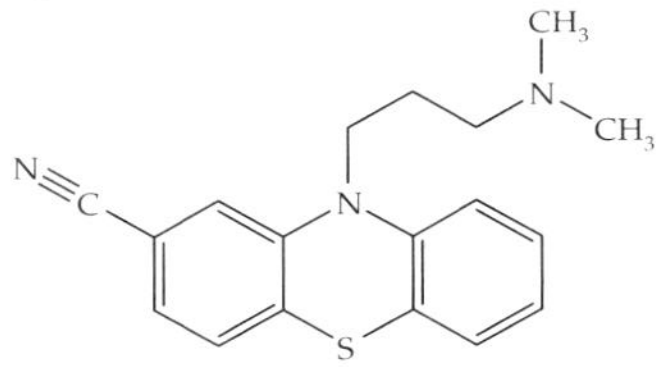	9.3	U	+H	W&G: Hulshoff A and Perrin J, Dissociation constants in water and methanol-water mixtures of some 10-(3-dimethylaminopropyl) phenothiazines, *Pharm. Acta Helv.*, **51**, 65–71 (1976).
1749	Cyclacillin	2.7	U	–H	McEvoy
		7.5	U	+H	
1750	Cyclazocine ($C_{18}H_{25}NO$) 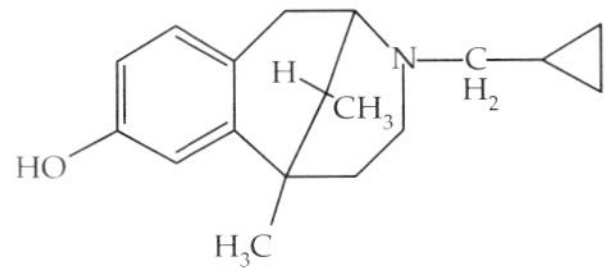	9.38	U	+H	Craig
1751	Cyclizine	8.16	U	+H	W&G: Barlow RB, *Introduction to Chemical Pharmacology*, 2nd Edn., Wiley, New York, pp. 124, 357 (1964).
1752	Cyclobenzaprine ($C_{20}H_{21}N$) 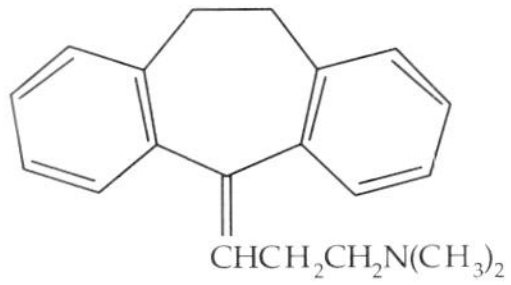	8.47	U	–H	Cotton ML and Down GRB, Cyclobenzaprine hydrochloride, *APDS*, **17**, 41–72 (1988). Downing GV, Merck Sharp and Dohme Research Labs, Rahway, NJ, unpublished data.

(continued)

No.	Name	pK_a value(s)	Data quality	Ionization type	Method	Conditions t °C; I or c M	Comments and Reference(s)
1753	Cyclopentamine	11.47	U	+H			Craig; also cited in: N&K; Chatten LG (ed.), *Pharmaceutical Chemistry*, vol. **1**, Dekker, New York pp. 85–87 (1966).
1754	D-Cycloserine	4.5	U	−H			N&K; Ritschel gave 4.5 and 7.4; ref. Brunner R and Machek G, Die Antibiotica, Band II, *Die Mittleren Antibiotica*, p. 132 (1965).
		7.4	U	+H			
1755	Cyclothiazide	9.1	U	−H		30% EtOH	N&K; APDS 1
		10.5	U	−H			
1756	Cyproheptadine ($C_{21}H_{21}N$) 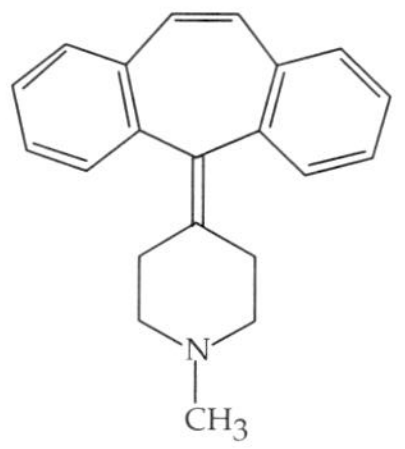	8.87	U	+H			McEvoy; Craig
1757	Cytarabine ($C_9H_{13}ClN_3O_5$) 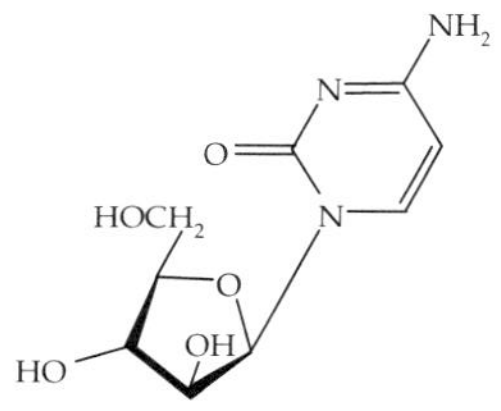	4.3	U	+H			Avery

No.	Compound	pKa	Class	Type	Method	Conditions	Reference
1758	Dacarbazine ($C_6H_{10}N_6O$) 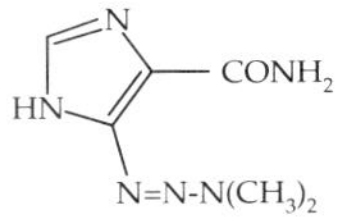	4.42	U	+H			Craig N&K; Anon
1759	Dantrolene ($C_{14}H_{10}N_4O_5$) 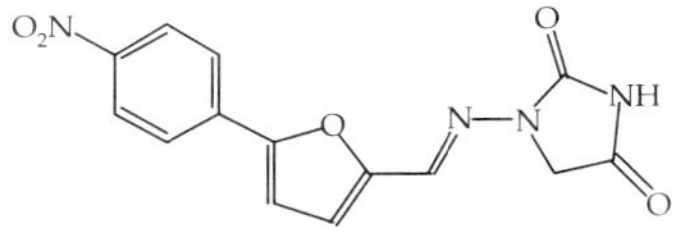	7.5	U	−H			Craig N&K; Anon
1760	Dantrolene	7.5	U	−H			Vallner JJ, Sternson LA and Parsons DL, Interaction of dantrolene sodium with human serum albumin, *J. Pharm. Sci.*, **65**, 873–877 (1976). NB: The pK_a value was quoted from Eaton Labs product information literature.
1761	Dapsone ($C_{12}H_{12}N_2O_2S$) 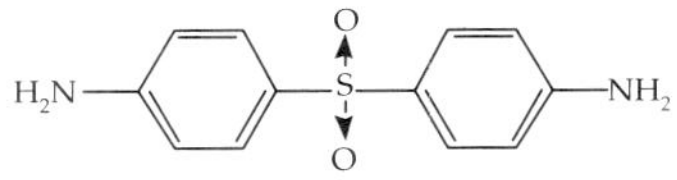	1.30 2.49	U U	+H +H			Craig; N&K; Chatten LG (ed.), *Pharmaceutical Chemistry*, vol. **1**, Dekker, New York, pp. 85–87 (1966); Bell, Roblin, 1942.
1762	Debrisoquin ($C_{10}H_{13}N_3$) 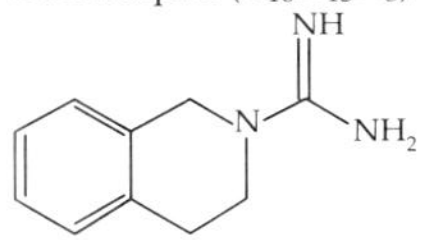	11.9	U	+H			Craig; N&K; W&G: Hengstmann JH, Falkner FC, Watson JT and Oates J, Quantitative determination of guanethidine and other guanido-containing drugs in biological fluids by gas chromatography with flame ionization detection and multiple ion detection, *Anal. Chem.*, **46**, 34–39 (1974).
1763	Debrisoquin	13.01±0.04	U	+H	potentio	H_2O $t = 25.0$ $I = 0.15$ (KCl)	Sirius Technical Application Notes, vol. **2**, p. 119 (1995). Sirius Analytical Instruments Ltd., Forest Row, East

(continued)

No.	Name	pK_a value(s)	Data quality	Ionization type	Method	Conditions t °C; I or c M	Comments and Reference(s)
							Sussex, RH18 5DW, UK. NB: Concentration of analyte, 13.3–14.0 mM. NB: Same value reported in ACD Labs database; Cited in Wan H, Holmen AG, Wang Y, Lindberg W, Englund M, Nagard MB and Thompson RA, High-throughput screening of pK_a values of pharmaceuticals by pressure-assisted capillary electrophoresis and mass spectrometry, *Rapid Commun. Mass Spectrom.*, **17**, 2639–2648 (2003). NB: Also gave a calculated value of 13.27.
1764	Dehydrocholic acid ($C_{24}H_{34}O_5$)	5.12	U	–H		H_2O $t = 20$ $c > CMC$	Craig; N&K; Johns WH and Bates TR, Quantification of the binding tendencies of cholestyramine II, *J. Pharm. Sci.*, **59**, 329–333 (1970).
1765	Dehydrocholic acid	4.91	U	–H		H_2O $t = 20$ $c < CMC$	Johns WH and Bates TR, Quantification of the binding tendencies of cholestyramine II, *J. Pharm. Sci.*, **59**, 329–333 (1970). NB: See Cholic acid.
1766	Demeclocycline	3.3 7.2 9.4	U U U	–H –H +H			Avery. NB: Assignment of pK_a values from Stephens CR, Murai K, Brunings KJ and Woodward RB, Acidity constants of the tetracycline antibiotics, *JACS*, **78**, 4155–4158 (1956), but see refs. 1 and 11 therein. See also Chlortetracycline.

1767	Demoxepam ($C_{15}H_{12}N_2O_2$) 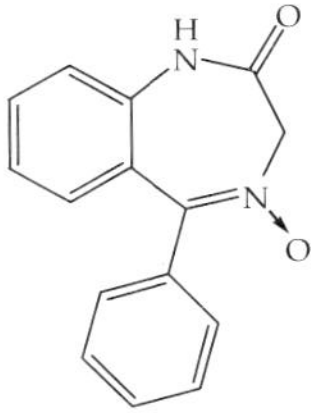	4.5 10.6	U U	+H +H			Craig
1768	Deoxycholic acid ($C_{24}H_{40}O_4$) 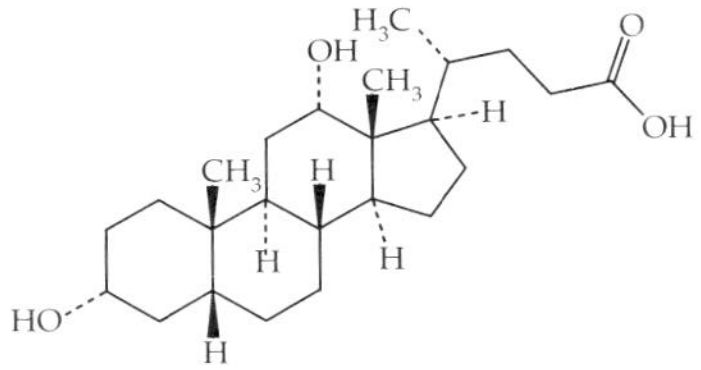	5.15±0.04	U	−H	potentio	H_2O $t = 20$ $c < 0.0045$ (CMC)	Johns WH and Bates TR., Quantification of the binding tendencies of cholestyramine I: effect of structure and added electrolytes on the binding of unconjugated and conjugated bile salt anions, *J. Pharm. Sci.*, **58**, 179–183 (1969). NB: These values were quoted from Ekwall P, Rosendahl T and Lofman N, Bile salt solutions. I. The dissociation constants of the cholic and deoxycholic acids, *Acta Chem. Scand.*, **11**, 590–598 (1957). The pK_a value becomes stable above 0.01 M.
1769	2′-Deoxyuridine ($C_9H_{12}N_2O_5$) 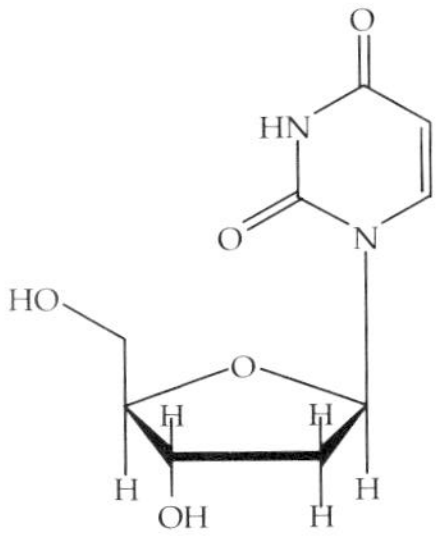	9.3	U	−H	spectro	H_2O $t = 25$	Nestler HJ and Garrett ER, Prediction of stability in pharmaceutical preparations XV, Kinetics of hydrolysis of 5-trifluoromethyl-2′-deoxyuridine, *J. Pharm. Sci.*, 1117–1125 (1968). Cited in Ritschel. NB: No further details reported. Also reported the following values (spectro unless noted): 5-trifluoromethyl-2′-deoxyuridine, 7.85; 5-carboxy-2′-deoxyuridine, 4.0 and 9.8; 5-trifluoromethyluracil, 7.4 and 12.6 (kinetic); 5-carboxyuracil, 4.20 and 9.10; 5-carboxyuracil, 4.25 (potentio) and 8.90 (potentio); uracil, 9.0 and >13.0.

(continued)

No.	Name	pK_a value(s)	Data quality	Ionization type	Method	Conditions t °C; I or c M	Comments and Reference(s)
1770	Deserpidine ($C_{32}H_{38}N_2O_8$) 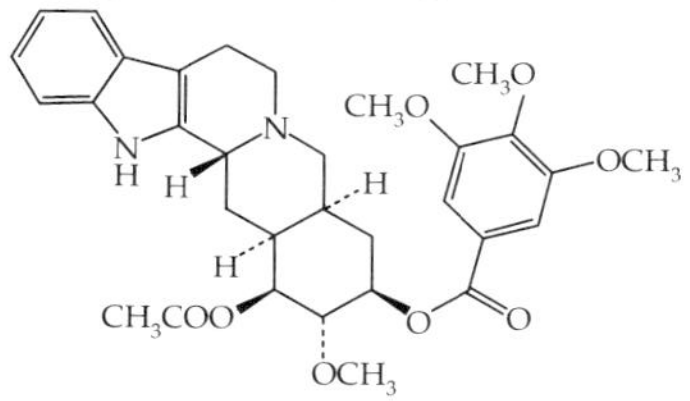	6.68	U	+H			Craig
1771	Deserpidine	6.68	U	+H		40% MeOH	N&K; Merck 9
1772	Desipramine	10.23	U	+H	potentio		Lombardo F, Obach RS, Shalaeva MY and Gao F, Prediction of human volume of distribution values for neutral and basic drugs. 2. Extended data set and leave-class-out statistics, *J. Med. Chem.*, **47**, 1242–1250 (2004); ref. not given: potentiometric titration.
1773	Desoxyephedrine (methamphetamine) ($C_{10}H_{15}N$) 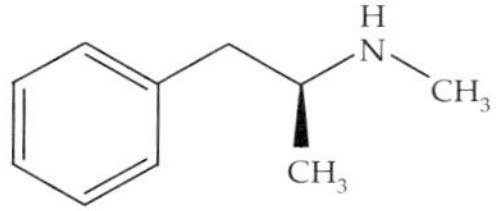	9.5	U	+H			Borodkin S and Yunker MH, Interaction of amine drugs with a polycarboxylic acid ion-exchange resin, *J. Pharm. Sci.*, **59**, 481–486 (1970). NB: This value was cited from unidentified literature.
1774	Dexamphetamine ($C_9H_{13}N$) 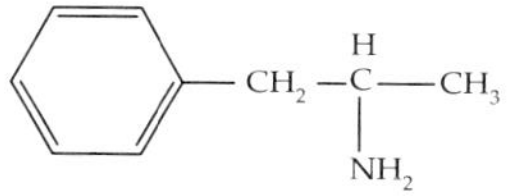	9.90	U	+H			Vree TB, Muskens ATJM and van Rossum JM, Some physicochemical properties of amphetamine and related drugs, *J. Pharm. Pharmacol.*, **21**, 774–775 (1969).

					Other sources: Kisbye J, *Pharm. Weekblad*, **93**, 206–215 (1958). Lewis GP, The importance of ionization in the activity of sympathomimetic amines, *Br. J. Pharmacol.*, **9**, 488–493 (1954).
1775	Dextromethorphan ($C_{18}H_{25}NO$) 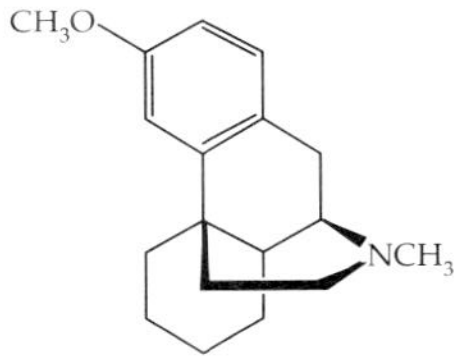	8.3	U	+H	Craig N&K; Borodkin S and Yunker MH, Interaction of amine drugs with a polycarboxylic acid ion-exchange resin, *J. Pharm. Sci.*, **59**, 481–486 (1970). Ritschel gave 8.25: ref. Garrett ER and Chemburkar PB. Evaluation, control and prediction of drug diffusion through polymeric membranes III, *J. Pharm. Sci.*, **57**, 1401–1409 (1968).
1776	Dextromoramide ($C_{25}H_{32}N_2O_2$) 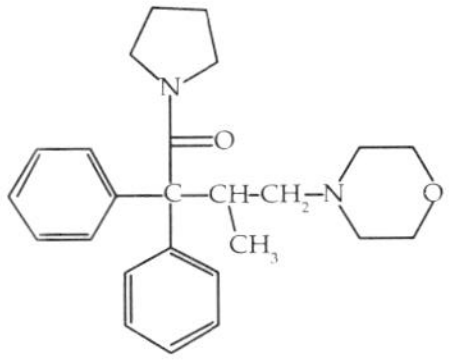	7.0	U	+H	Craig N&K; S&R
1777	Dextrose	12.1	U	−H	Sinko P (ed.), *Martin's Physical Pharmacy and Pharmaceutical Sciences*, 5th Edn., Lippincott, Williams and Wilkins, Baltimore MD (2006) Table 7–2.
1778	Diamorphine	7.83	U	+H	Beckett AH, Analgesics and their antagonists: Some steric and chemical considerations, *J. Pharm.*

(continued)

No.	Name	pK_a value(s)	Data quality	Ionization type	Method	Conditions t °C; I or c M	Comments and Reference(s)
							Pharmacol., **8**, 848–859 (1956); also cited in Clouet DH (ed.), *Narcotic Drugs Biochemical Pharmacology*, Plenum Press, New York, 52–53 (1971). NB: Quoted from Farmilo CG, Oestreicher PM and Levi L, Physical methods for the identification of narcotics. IB Common physical constants for identification of ninety-five narcotics and related compounds, *Bull. Narcotics*, UN Dept. Social Affairs vol. **6**, pp. 7–19 (1954).
1779	Diamorphine	7.83	U	+H			Barrett DA, Rutter N, Davis SS, An *in vitro* study of diamorphine permeation through premature human neonatal skin, *Pharm. Res.*, **10**(4), 583–587 (1993). NB: Value quoted without reference – probably the Farmilo value.
1780	Diatrizoic acid ($C_{11}H_9I_3N_2O_4$) 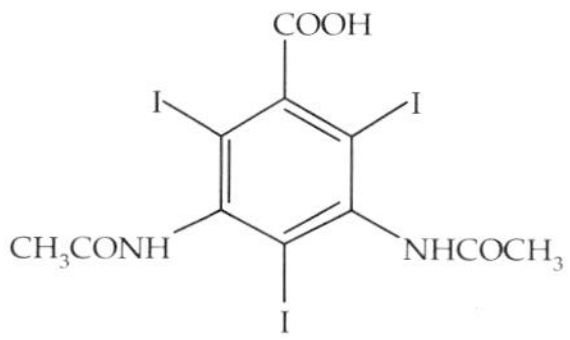	3.4	U	–H			Lerner HH, Diatrizoic acid, *APDS*, **4**, 137–167 (1975). NB: Cited Langecker H, Harwart A and Junkmann K, 3,5-Diacetamido-2,4,6-triiodobenzoic acid as an x-ray contrast medium, *Arch. Exp. Pathol. Pharmakol.*, **222**, 584–590 (1954).

COOH, I, I, CH_3CONH, $NHCOCH_3$, I

1781	Diazepam	3.4	U	+H	spectro		MacDonald A, Michaelis AF and Senkowski BZ, Diazepam, *APDS*, **1**, 90 (1972). NB: Cited Toome V, Hoffman-La Roche, personal communication.
1782	Diazepam	3.3	U	+H			Van der Kleijn E, Protein binding and lipophilic nature of ataractics of the meprobamate and diazepine group, *Arch. Int. Pharmacodyn.*, **179**, 225–250 (1969). NB: See Chlordiazepoxide for details.
1783	Diazoxide	8.5	U	−H			Craig; N&K; Anon.
1784	Dibucaine	−5.3	U	+H	spectro	aq. H_2SO_4	Padmanabhan GR, Dibucaine and Dibucaine Hydrochloride, *APDS*, **12**, 105–133 (1983).
		1.6	U	+H	spectro	1% EtOH	"The following pK_a values have been reported:
		8.31	U	+H	potentio		pK_{BH+} = 8.31 (potentiometric titration) pK_{BH2+} = 1.6 (absorption spectrophotometry) pK_{BH3+} = −5.3 (absorption spectrophotometry) BH3+, BH2+, and BH+ are respectively triply protonated, doubly protonated and mono protonated species. See Martucci JD, Schulman SG, *Anal. Chem. Acta*, **77**, 317–319 (1975)." NB: Martucci and Schulman also reported a value for pK_{a2} = 2.1 from fluorometric data. This value represents ionization in the first excited state.
1785	Dibucaine	8.5	U	+H			W&G: Truant AP and Takman B, Differential physical-chemical and neuropharmacologic properties of local anesthetic agents, *Anesth. Analg.*, **38**, 478–484 (1959). NB: Craig has 8.9 (+H).

(continued)

No.	Name	pK_a value(s)	Data quality	Ionization type	Method	Conditions t °C; I or c M	Comments and Reference(s)
1786	Dichlorphenamide ($C_6H_6Cl_2N_2O_4S_2$) SO$_2$NH$_2$ Cl SO$_2$NH$_2$ Cl	7.4 8.6	U U	–H –H			Craig N&K; Anon
1787	Diclofenac	4	U	–H			Adeyeye CM and Li P-K, Diclofenac sodium, *APDS*, **19**, 123–144 (1990). NB: See also: Moser P, Jakel K, Krupp P, Menasse R and Sallman A, Structure-activity relations of analogs of phenylbutazone, *Eur. J. Med. Chem.*, **10**, 613–617 (1975).
1788	Dicloxacillin	2.76	U	–H	potentio	H_2O $t = 37$ $I = 0.15$ (KCl)	Tsuji A, Kubo O, Miyamoto E and Yamana T, Physicochemical properties of β-lactam antibiotics: oil-water distribution, *J. Pharm. Sci.*, **66**, 1675–1679 (1977). NB: From extrapolation of pK_a' values (3.55–3.98) in EtOH-water mixtures (32.8–51.4%). Also reported in Craig; N&K.
1789	Dicumarol (bishydroxycoumarin) ($C_{19}H_{12}O_6$) O O O O OH OH	4.4 8.0	U U	–H –H	spectro	H_2O (extrap)	Cho MJ, Mitchell AG and Pernarowski M, Interaction of bishydroxycoumarin with human serum albumin, *J. Pharm. Sci.*, **60**, 196–200 (1971). NB: These apparent pK_a values were extrapolated from values in DMF-water mixtures. Cited MJ Cho, MSc Thesis, Univ. Br. Columbia, Vancouver, Canada (1970).

No.	Compound	pKa			Method	Solvent	Reference
1790	Diethazine ($C_{18}H_{22}N_2S$) 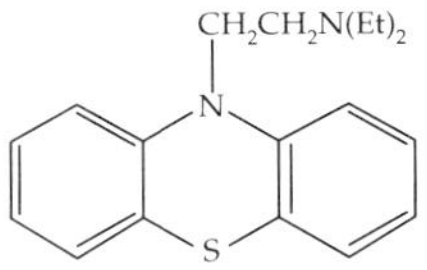 $CH_2CH_2N(Et)_2$, N, S	9.1	U	+H			Craig
1791	Diethylcarbamazine ($C_{10}H_{21}N_3O$) 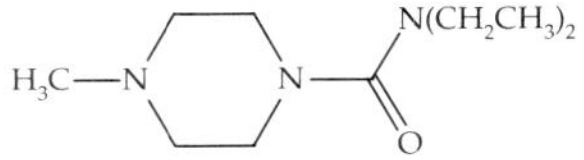 H_3C—N, N, $N(CH_2CH_3)_2$, O	7.7	U	+H			Craig
1792	Dihydrocodeinone 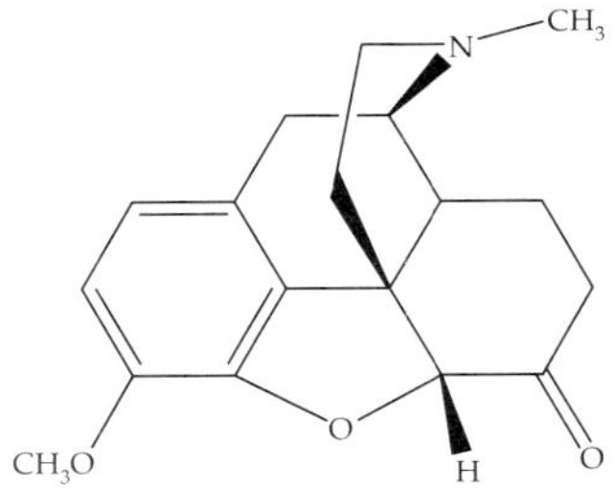 N, CH_3, CH_3O, O, H, O	6.61	U	+H	potentio	50% aq EtOH	Farmilo CG, Oestreicher PM and Levi L, Physical methods for the identification of narcotics. IB Common physical constants for identification of ninety-five narcotics and related compounds, *Bull. Narcotics*, UN Dept. Social Affairs vol. **6**, pp. 7–19 (1954). Cited in Clouet DH (ed.), *Narcotic Drugs Biochemical Pharmacology*, Plenum Press, New York, 52–53 (1971).
1793	Dihydroergotamine	8.0	U	+H			Ritschel gave 6.75 ± 0.03; this is the apparent value in 10% 7HPT. Ref. Maulding HV and Zoglio MA, Physical chemistry of ergot alkaloids and derivatives. I. Ionization constants of several medicinally active bases, *J. Pharm. Sci.*,

(continued)

No.	Name	pK_a value(s)	Data quality	Ionization type	Method	Conditions t °C; I or c M	Comments and Reference(s)
							59, 700–701 (1970). Notes the 8.0 value with a secondary reference to Robson JM and Stacey RS, *Recent advances in pharmacology*, 4th Edn., Little Brown and Co., Boston, p. 108 (1968).
1794	Dihydromorphine	8.55	U	+H	potentio	50% aq EtOH	Farmilo CG, Oestreicher PM and Levi L, Physical methods for the identification of narcotics. IB Common physical constants for identification of ninety-five narcotics and related compounds, *Bull. Narcotics*, UN Dept. Social Affairs vol. **6**, pp. 7–19 (1954); cited in Clouet DH (ed.), *Narcotic Drugs Biochemical Pharmacology*, Plenum Press, New York, 52–53 (1971).
1795	Dihydromorphinone	8.15	U	+H			Farmilo CG, Oestreicher PM and Levi L, Physical methods for the identification of narcotics. IB Common physical constants for identification of ninety-five narcotics and related compounds, *Bull. Narcotics*, UN Dept. Social Affairs, vol. **6**, pp. 7–19 (1954). Cited in Beckett AH, Analgesics and their antagonists. I., *J. Pharm. Pharmacol.*, **8**, 848–859 (1956); also cited in Clouet DH (ed.), *Narcotic Drugs Biochemical Pharmacology*, Plenum Press, New York, 52–53, (1971). NB: See Alphaprodine for details.

No.	Compound	pKa	Type	Ionization	Method		Reference / Comments
1796	Dihydrostreptomycin ($C_{21}H_{41}N_7O_{12}$)	7.8	U	+H			Craig NB: This value is most likely to be for the 2° amino group. There should be higher values for the guanidine residues.
1797	Dihydrostreptomycin	7.75	U	+H			N&K; Hoover
1798	Dilevolol ($C_{19}H_{24}N_2O_3$)	9.45	U	+H			Craig
1799	Diltiazem	7.70	U	+H			Craig
1800	Diltiazem	8.06	U	+H	potentio		Lombardo F, Obach RS, Shalaeva MY and Gao F, Prediction of human volume of distribution values for neutral and basic drugs. 2. Extended data set and leave-class-out statistics, *J. Med. Chem.*, **47**, 1242–1250 (2004); ref. not given: potentiometric titration.

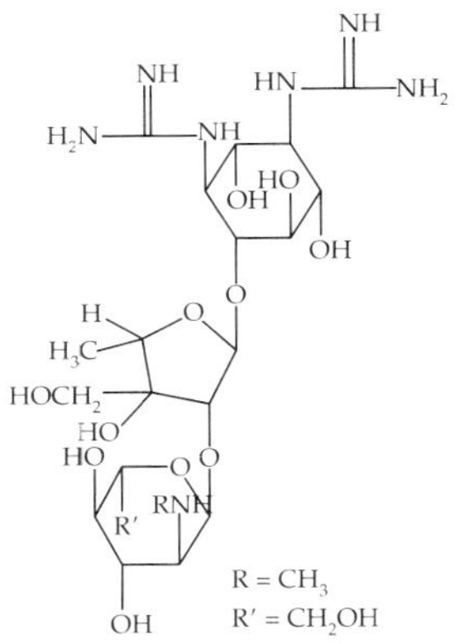

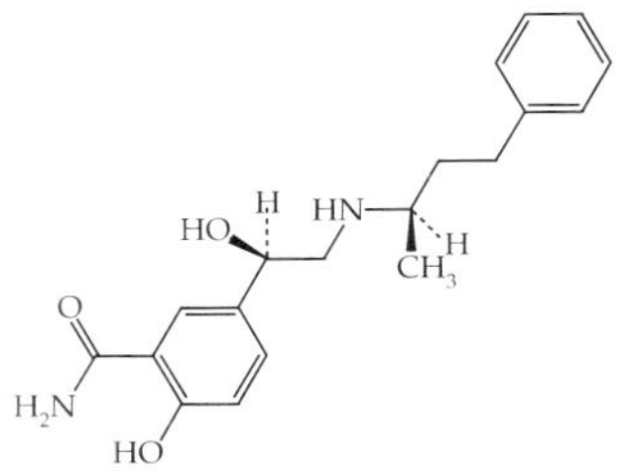

(continued)

No.	Name	pK_a value(s)	Data quality	Ionization type	Method	Conditions t °C; I or c M	Comments and Reference(s)
1801	Dimethadione ($C_5H_7NO_3$) 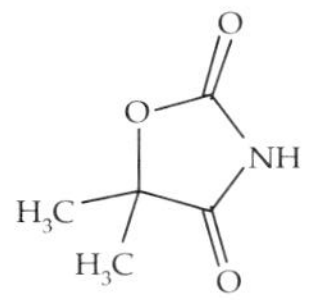	6.1	U	−H			Butler TC, The effects of N-methylation in 5,5-disubstituted derivatives of barbituric acid, hydantoin and 2,4-oxazolidinedione, *J. Am. Pharm. Assoc.*, **44**, 367–370 (1955). See also Craig.
1802	Dimethisoquin ($C_{17}H_{24}N_2O$) 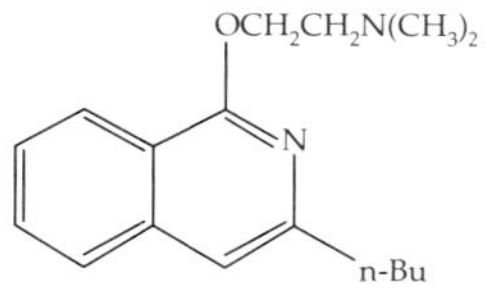	6.30	U	+H			Craig
1803	Dimethylhydantoin ($C_5H_8N_2O_2$) 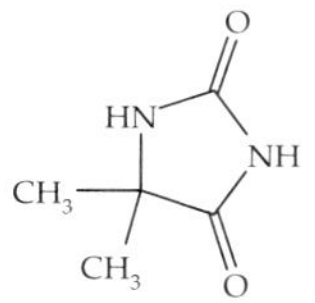	8.1	U	−H			Butler TC, The effects of N-methylation in 5,5-disubstituted derivatives of barbituric acid, hydantoin and 2,4-oxazolidinedione, *J. Am. Pharm. Assoc.*, **44**, 367–370 (1955). See also W&G.
1804	Dinoprostone (prostaglandin E_2) ($C_{20}H_{32}O_4$) 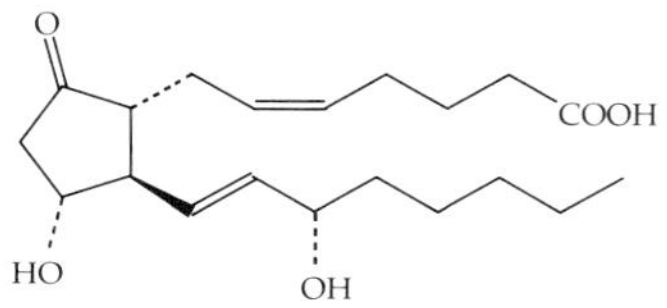	4.6	U	−H			Foye 3rd; see Azatadine; from McEvoy.

No.	Compound	pKa	Rating	Type	Method	Solvent	Reference / Notes
1805	Diperodon ($C_{22}H_{27}N_3O_4$) (piperidino)N–CH_2–CH($OCONHC_6H_5$)($CH_2OCONHC_6H_5$)	8.44	VU	+H		H_2O	Cohen JL, Diperodon, *APDS*, **6**, 99–112 (1977). NB: See also US Dispensatory, 27th Edn., (1973) p. 439. "Diperodon is a tertiary amine and is weakly basic. Aqueous solution of 1% diperodon hydrochloride have a pH of 5.1. Although the dissociation constant is not specifically reported in the literature a pK_a of 8.44 can be estimated from this information."
1806	Diphenhydramine	9.1	U	+H	potentio		Lombardo F, Obach RS, Shalaeva MY and Gao F, Prediction of human volume of distribution values for neutral and basic drugs. 2. Extended data set and leave-class-out statistics, *J. Med. Chem.*, **47**, 1242–1250 (2004); ref. not given: potentiometric titration.
1807	Diphenoxylate	7.1	U	+H	potentio		Hong DD, Diphenoxylate hydrochloride, *APDS*, **7**, 149–169 (1978). Cited Mead I, GD Searle & Co. (High Wycombe), personal communication. "The pK_a of diphenoxylate by the titrimetric method was found to be 7.1."
1808	Diphenoxylate	4.4	U	+H			N&K; Charles RL, Searle Labs, Chicago, personal communication. ACD gave an estimated value = 7.63±0.40; also mentions the N&K value of 4.4, but queries that value. Possibly this value was obtained in a mixed aqueous-organic solvent.
1809	Diphenylpyraline	8.9	U	+H			Craig. NB: This appears to be the Testa value (see no. 459).

(continued)

No.	Name	pK_a value(s)	Data quality	Ionization type	Method	Conditions t °C; I or c M	Comments and Reference(s)
1810	DL-Dipipanone ($C_{24}H_{31}NO$) 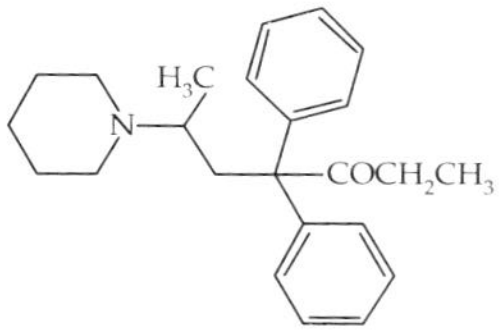	8.70	U	+H	potentio	50% aq EtOH	Farmilo CG, Oestreicher PM and Levi L, Physical methods for the identification of narcotics. IB Common physical constants for identification of ninety-five narcotics and related compounds, *Bull. Narcotics*, UN Dept. Social Affairs (1954) vol. **6**, pp. 7–19; cited in Clouet DH (ed.), *Narcotic Drugs Biochemical Pharmacology*, Plenum Press, New York, 52–53, (1971).
1811	DL-Dipipanone	9.08	U	+H			Taylor JF, Ph.D. Thesis, Univ. London (1968).
1812	Dipipanone	8.5	U	+H			Craig
1813	Dipivefrine ($C_{19}H_{29}NO_5$) 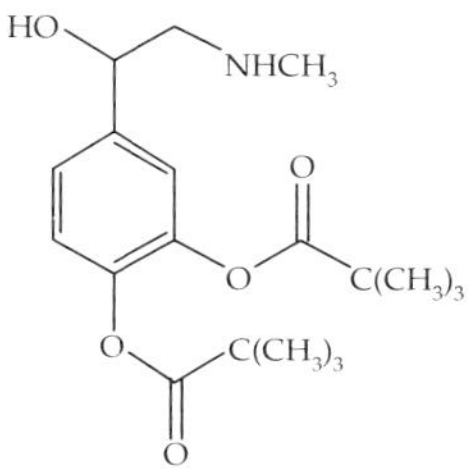	8.40	U	+H		RT	Wall GM and Fan TY, Dipivefrine hydrochloride, *APDS*, **22**, 229–262 (1993). NB: Alcon Laboratories, Inc., unpublished data on file.
1814	Dipyridamole ($C_{24}H_{40}N_8O_4$) 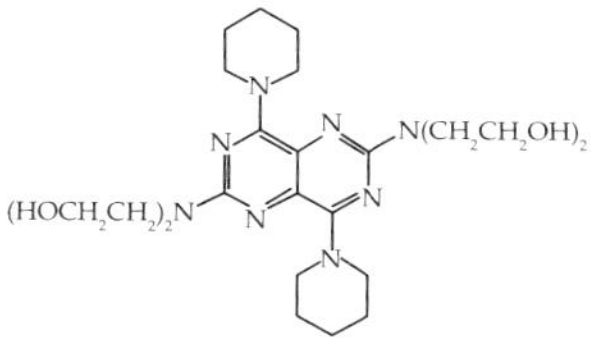	6.4	U	+H			Craig N&K; Ritschel gave 6.4; ref. Geigy Pharmaceuticals, Ardsley, New York 10502.

1815	Dobutamine ($C_{18}H_{23}NO_3$) 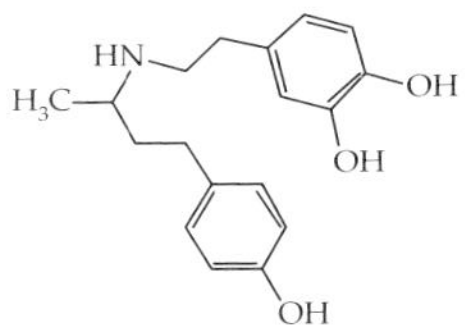	9.45	U	+H	DMF	Bishara RH and Long HB, Dobutamine hydrochloride, *APDS*, **8**, 139–158 (1979). NB: No reference, but probably by the potentiometric method.
1816	Domperidone ($C_{22}H_{24}ClN_5O_2$) 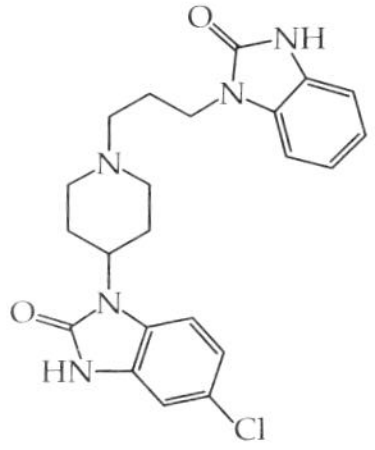	7.90	U	+H		Lombardo F, Obach RS, Shalaeva MY and Gao F, Prediction of human volume of distribution values for neutral and basic drugs. 2. Extended data set and leave-class-out statistics, *J. Med. Chem.*, **47**, 1242–1250 (2004). Ref. 286: Oliviera CR, Lima MCP, Carvalho CAM, Leysen JE and Carvalho AP, Partition coefficients of dopamine antagonists in brain membranes and liposomes, *Biochem. Pharmacol.*, **38**, 2113–2120 (1989). This reference states that the pK_a value was provided by Dr. JE Leysen, with no further details given.
1817	DOPA, L- (Levodopa) ($C_9H_{11}NO_4$)	2.3 9.0 10.2	U U U	−H −H, +H +H, −H		Lippold BC and Jaeger I, Stability and dissociation constants of L-dopa and alpha-L-methyldopa, *Arch. Pharm. Weinheim.*, **306**, 106–117 (1973). "A discussion of the oxidation of levodopa and methyldopa in alkaline solution is presented. The reaction rate increases with increasing supply of oxygen, increasing pH, but decreasing concentration of starting material. The pK_a values of levodopa are: 2.3, 9.0, 10.2, and 12.5, and of methyldopa: methyldopa 2.25, 9.0, 10.35, and 12.6."

(*continued*)

Appendix B *(continued)*

No.	Name	pK_a value(s)	Data quality	Ionization type	Method	Conditions t °C; I or c M	Comments and Reference(s)
1818	DOPA, L- (Levodopa)	2.31	U	−H	potentio		Gomez R, Hagel RB and MacMullan EA, Levodopa, *APDS*, **5**, 189–223 (1976).
		8.71	U	−H, +H			NB: Cited Gorton JE and Jameson RF, Complexes of doubly chelating ligands. I. Proton and copper(II) complexes of L-β-(3,4-dihydroxyphenyl)alanine (DOPA), *J. Chem. Soc* (A), 2615–2618 (1968).
		9.74	U	+H, −H			
		13.40	U	−H			
1819	Dorzolamide ($C_{10}H_{16}N_2O_4S_3$)	6.35	U	+H	potentio		Quint MP, Grove J and Thomas SM, Dorzolamide Hydrochloride, *APDS* **26**, 294–299 (1999). NB: No references given; data presumably obtained in house (Merck Sharp and Dohme, Riom, France; and Merck Research Labs, Rahway, NJ, USA), but see Brinzolamide. The assignments can be confirmed in part from the temperature dependence of the amine pK_a value = 0.019/K, which is higher than for typical anionic functional groups. See also Kamel K, Ogashiwa M, Ohsima A, Ohki Y, Takahashi M and Ishimaru S, *Iyakhun Kenkyu*, **25**, 6, 438 (1994) (ref. 13 in APDS 26 article on Brinzolamide. Assignments as claimed for Brinzolamide)
	O O CH$_3$ S S H SO$_2$NH$_2$ H NHCH$_2$CH$_3$	8.50	U	−H			

TABLE pH dependence of the equilibrium solubility of Form I dorzolamide

pH	Solubility (mg/mL)	pH	Solubility (mg/mL)	pH	Solubility (mg/mL)
4.97	29.79	5.74	30.06	8.00	3.59
5.18	33.35	6.09	13.21	8.50	5.33
5.35	35.92	7.00	3.74	9.00	11.71
5.56	40.00	7.50	3.19	9.50	40.10

1820	Doxepin	8.0	U	+H			N&K; Anon
1821	Doxorubicin	8.2	U	−H			Skovsgaard T and Nissen NI, Adriamycin, an antitumor antibiotic, *Dan. Med. Bull.*, **22**, 62–73 (1975). NB: See Foye, gave 8.2 and 10.2 (+H). See also Doxorubicin (below and nos. 469–470).
1822	Doxorubicin (Adriamycin)	8.25±0.60	U	−H	comp		ACD/pK_a estimate
		8.43±0.70	U	+H			
		11.9±0.4	U	−H			
		12.95±0.1	U	−H			
		13.8±0.70	U	−H			
1823	Doxycycline	3.5	U	−H			Craig; N&K; Jaffe JM, Colaizzi JL and Poust RI, Effect of altered urinary pH on tetracycline and doxycycline excretion in humans, *J. Pharmacokinet. Biopharm.*, **1**, 267–282 (1973).
		7.7	U	−H			
		9.5	U	+H			
1824	Doxylamine ($C_{17}H_{22}N_2O$) 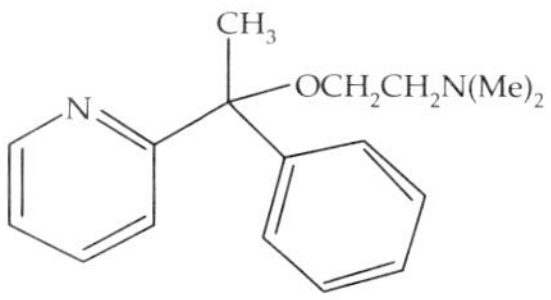	9.20	U	+H	potentio	H_2O $t = 25$ $c = 0.002–0.01$	Lordi NG and Christian JE, Physical properties and pharmacological activity: antihistaminics, *J. Am. Pharm. Ass., Sci. Edn.*, **45**, 300–305 (1956). See Chlorpheniramine (no. 1704) for details.
		4.4	U	+H			Craig; N&K
		9.2	U	+H			

(continued)

No.	Name	pK_a value(s)	Data quality	Ionization type	Method	Conditions t °C; I or c M	Comments and Reference(s)
1825	Droperidol ($C_{22}H_{22}FN_3O_2$)	7.64	U	+H	potentio		Janicki CA and Gilpin RK, Droperidol, *APDS*, **7**, 171–192 (1978). NB: Ref. Demoen P, Janssen Pharmaceutica, Beerse, Belgium, unpublished data. N&K; S&R
1826	Enalaprilat ($C_{18}H_{24}N_2O_5$)	2.30 3.39	U U	−H −H			Craig
		8.02	U	+H			
1827	Ephedrine	9.5	U	+H			Ritschel; Borodkin S and Yunker MH, Interaction of amine drugs with a polycarboxylic acid ion-exchange resin, *J. Pharm. Sci.*, **59**, 481–486 (1970).
1828	Ephedrine	9.6	U	+H			W&G: Schanker LS, Shore PA, Brodie BB and Hogben CAM, Absorption of drugs from the stomach. I. The rat, *JPET*, **120**, 528–539 (1957).
1829	Ephedrine	9.60	U	+H			Vree TB, Muskens ATJM and van Rossum JM, Some physicochemical properties of amphetamine and

							related drugs, *J. Pharm. Pharmacol.*, **21**, 774–775 (1969). Other sources: Kisbye J, *Pharm. Weekblad.*, **93**, 206–215 (1958); Brodie BB and Hogben CAM, Some physicochemical factors in drug action, *J. Pharm. Pharmacol.*, **9**, 345–380 (1957); Lewis GP, The importance of ionization in the activity of sympathomimetic amines *Br. J. Pharmacol.*, **9**, 488–493 (1954).
1830	Ergometrine ($C_{19}H_{23}N_3O_2$) HOCH$_2$ CH$_3$ O N H NH N CH$_3$	7.3	U	+H			Foye 3rd; see Azatadine; from N&K. N&K; S&R
1831	Ergonovine	6.8	U	+H			Craig; N&K; According to Merck 12, ergometrine = ergonovine.
1832	Ergotamine	6.40±0.09	U	+H	potentio	H_2O $t = 24.0$	Maulding HV and Zoglio MA, *J. Pharm. Sci.*, **59**, 700–701 (1970). Cited in Kreilgård B, Ergotamine Tartrate, *APDS*, **6**, 113–159 (1977). NB: The pK_a value was measured as a function of 7-hydroxypropyltheophylline concentration and extrapolated to zero [7HPT] (6.25 ± 0.04). The value was further corrected by addition of 0.15 ± 0.05 log unit, from comparison of the extrapolated pK_a value for methysergide and the corresponding value in the absence of 7HPT. It is not the apparent pK_a′ = 6.4 at 24° in 2% aqueous caffeine, as

(continued)

No.	Name	pK_a value(s)	Data quality	Ionization type	Method	Conditions t °C; I or c M	Comments and Reference(s)
							suggested by Kreilgård. NB: pK_a′ = 5.6 in 80% aqueous methylcellosolve has also been reported. (Hofmann A, Die Mutterkornalkaloide, F. Enke Verlag, Stuttgart, 1964)
1833	Ergotamine	6.3	U	+H	soly	H_2O	Lombardo F, Obach RS, Shalaeva MY and Gao F, Prediction of human volume of distribution values for neutral and basic drugs. 2. Extended data set and leave-class-out statistics, *J. Med. Chem.*, **47**, 1242–1250 (2004). Ref. 287: Kreilgård B, Ergotamine Tartrate, *APDS*, **6**, 113–159 (1977).
1834	Erythromycin	8.8	U	+H			Craig; N&K; Merck 9; Garrett ER, Heman-Ackah SM and Perry GL, Kinetics and mechanisms of action of drugs on microorganisms XI, *J. Pharm. Sci.*, **59**, 1449–1456 (1970) NB: No references or experimental details were cited.
1835	Erythromycin estolate ($C_{52}H_{97}NO_{18}S$) 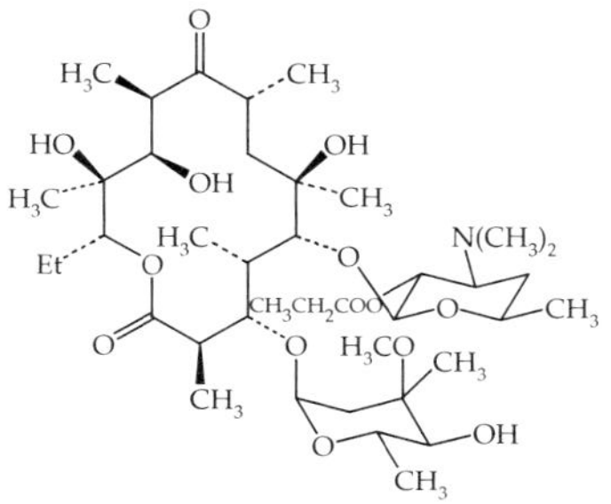	6.9	U	+H		66% aqueous DMF	Mann JM, Erythromycin estolate, *APDS*, **1**, 101–117 (1972). NB: Estolate = dodecyl sulfate salt.
1836	Erythromycin lactobionate (salt with lactobiono-δ-lactone) ($C_{37}H_{67}NO_{13}$)	8.8	U	+H	potentio	H_2O	McGuire JM, Bunch RL, Anderson RC, Boaz HE, Flynn EH, Powell HM and Smith JW, *Antibiot. Chemother.*, **2**, 281–283 (1952).

1837	Ethacrynic acid ($C_{13}H_{12}Cl_2O_4$) 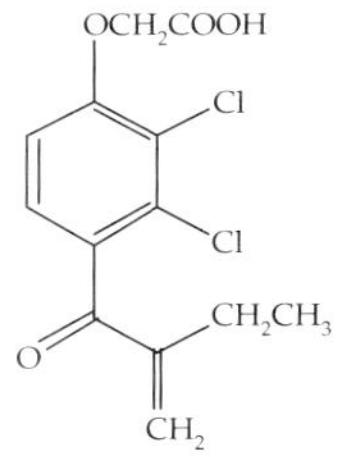	3.5		–H		Craig N&K; Merck 9 W&G: Merck 9
1838	Ethacrynic acid	2.80		–H		ACD Labs database
1839	Ethinylestradiol 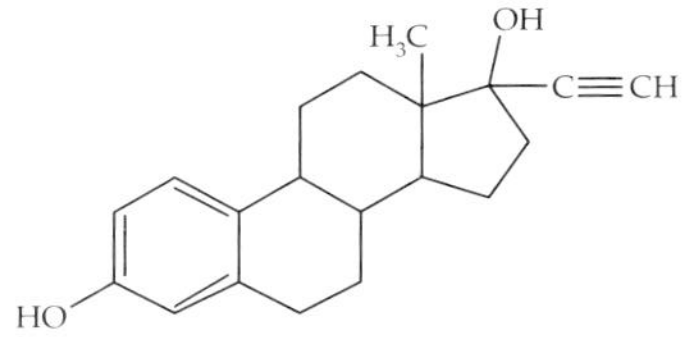	10.32	U	–H	soly/pH	Morvary J, Gyongysi J and Kiss L, Determination of the pK_a value of ethinyl estradiol on the basis of its solubility, *Acta Pharm. Hung.*, **50**, 243–247 (1980). NB: A gas-liquid chromatographic method was used to measure the very low solubilities of ethinyl estradiol in buffers of varying pH, to determine its pK_a value of 10.32. This was similar to the value obtained using a spectrophotometric method.
1840	Ethoheptazine ($C_{16}H_{23}NO_2$) 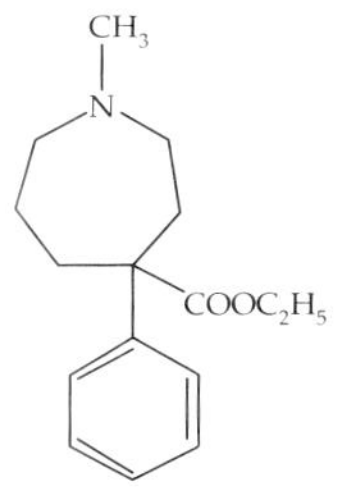	8.45	U	+H		Clouet DH (ed.), *Narcotic Drugs Biochemical Pharmacology*, Plenum Press, New York, 52–53 (1971); All data was cited from Farmilo CG, Oestreicher PM, Levi L, Physical methods for the identification of narcotics. IB Common physical constants for identification of ninety-five narcotics and related compounds, *Bull. Narcotics*, UN Dept. Social Affairs (1954) vol. **6**, pp. 7–19 (1954). Craig; N&K

(continued)

No.	Name	pK_a value(s)	Data quality	Ionization type	Method	Conditions t °C; I or c M	Comments and Reference(s)
1841	Ethopropazine	9.6	U	+H			Craig; N&K; Marshall PB, Some chemical and physical properties associated with histamine antagonism, *Br. J. Pharmacol.*, **10**, 270–278 (1955).
1842	Ethosuximide ($C_7H_{11}NO_2$) 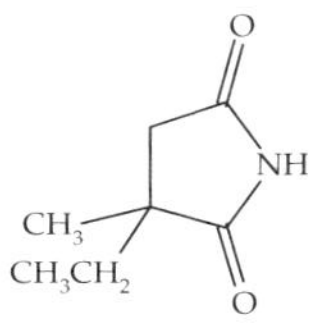	9.5	U	−H			Craig
1843	Ethosuximide	9.3	U	−H			N&K; Avery
1844	Ethylamphetamine ($C_{11}H_{17}N$)	10.23	U	+H			Vree TB, Muskens ATJM and van Rossum JM, Some physicochemical properties of amphetamine and related drugs, *J. Pharm. Pharmacol.*, **21**, 774–775 (1969).
1845	Ethyl biscoumacetate ($C_{10}H_{24}O_8$) 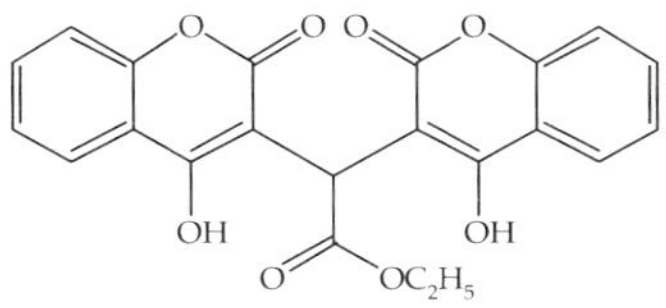	7.5	U	−H			Craig
1846	Ethyl biscoumacetate	3.1	U	−H			N&K; Avery
1847	Ethyl biscoumacetate	3.1	U	−H			W&G: Anton AH, A drug-induced change in the distribution and renal excretion of sulfonamides, *JPET*, **134**, 291–303 (1961). NB: The value was obtained from Dayton PG, personal communication.

1848	Ethylenediamine ($C_2H_8N_2$) 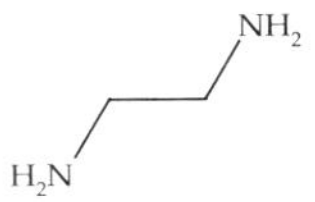	6.985 10.075	R R	+H +H	potentio	H_2O $t = 20.0$ $I = 0.07$–0.30 (pK_{a2}); 0.10–0.25 (pK_{a1})	Everett DH and Pinsent BRW, The dissociation constants of ethylenediammonium and hexamethylenediammonium ions from 0° to 60°, *Proc. Roy. Soc.*, Lond., **215A**, 416–429 (1952). Cited in Perrin Bases no. 104, ref. E36. NB: Used symmetrical hydrogen half cells with junction potentials. Raw data was extrapolated to zero ionic strength to give the thermodynamic values reported here. Numerous other values were reported. See also Bates RG, Amine buffers for pH control, *Ann. NY Acad. Sci.*, **92**, 341–356 (1961).
1849	2-Ethylmercapto-4-hydroxypyrimidine	7.01	U	−H?			Garrett ER and Weber DJ, Metal complexes of thiouracils I. Stability constants by potentiometric titration studies and structures of complexes, *J. Pharm. Sci.*, **59**, 1383–1398 (1970).
1850	Ethylmorphine	8.20	U	+H	potentio	50% aq EtOH	Farmilo CG, Oestreicher PM and Levi L, Physical methods for the identification of narcotics. IB Common physical constants for identification of ninety-five narcotics and related compounds, *Bull. Narcotics*, UN Dept. Social Affairs (1954) vol. **6**, pp. 7–19 cited in Clouet DH (ed.), *Narcotic Drugs Biochemical Pharmacology*, Plenum Press, New York, 52–53 (1971).
1851	Ethylmorphine	7.88	U	+H			N&K; Ritschel gave 7.88; ref, Parrott EL, Pharmaceutical Technology, Fundamental Pharmaceutics, Burgess Pub. Co, Minneaoplis, MN, p. 219 (1970).

(continued)

No.	Name	pK_a value(s)	Data quality	Ionization type	Method	Conditions t °C; I or c M	Comments and Reference(s)
1852	Ethylnorepinephrine ($C_{10}H_{15}NO_3$) 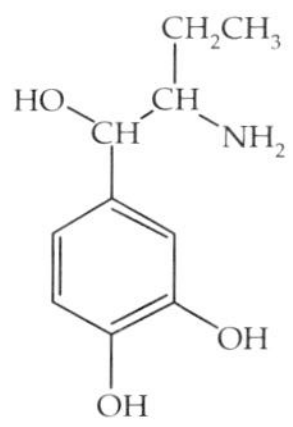	8.4	U	+H			Craig
1853	α-Ethylphenethylamine ($C_{10}H_{15}N$) 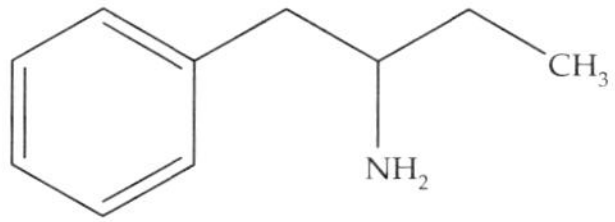	9.30	U	+H			Garrett ER and Chemburkar PB. Evaluation, control and prediction of drug diffusion through polymeric membranes III, *J. Pharm. Sci.*, **57**, 1401–1409 (1968).
1854	Ethylphenylhydantoin ($C_{11}H_{12}N_2O_2$) 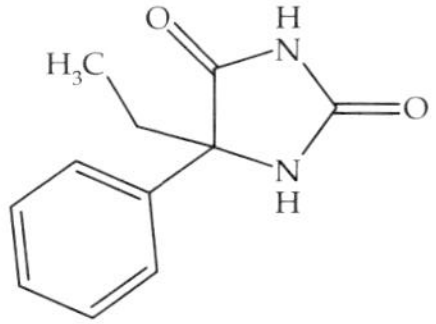	8.5	U	−H	spectro	H_2O RT I undefined	W&G: Butler TC, The effects of N-methylation in 5,5-disubstituted derivatives of barbituric acid, hydantoin, and 2,4-oxazolidinedione, *J. Am. Pharm. Assoc., Sci. Edn.*, **44**, 367–370 (1955). See Norparamethadione for additional details.
1855	Etidocaine	7.9	U	+H			Craig
1856	Etidocaine	7.7	U	+H			de Jong RH, Neural blockade by local anesthetics, *JAMA*, **238**, 1383–1385 (1977); cited in W&G.

1857	Etodolac ($C_{17}H_{21}NO_3$) 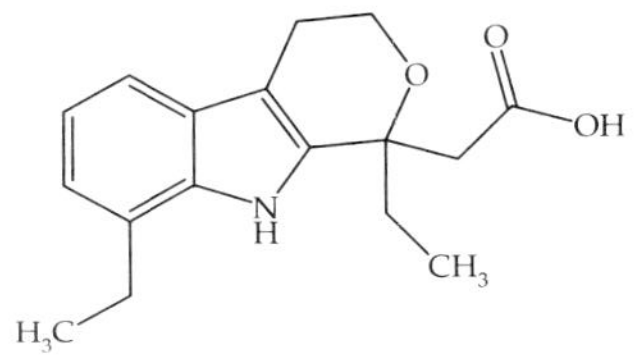	4.65	U	+H		Shah KP, Gumbhir-Shah K and Brittain HG, Etodolac, *APDS* **29**, 111 (2002). NB: No reference cited.
1858	Eugenol ($C_{10}H_{12}O_2$) 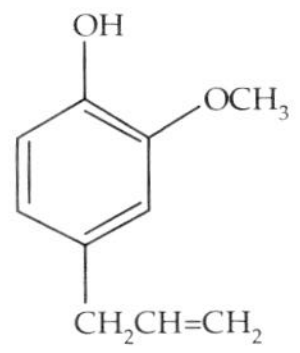	9.8	U	−H		Craig; Clarke
1859	Fencamfamine ($C_{15}H_{21}N$) 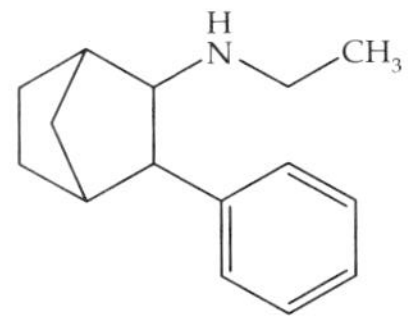	8.70	U	+H		Vree TB, Muskens ATJM and van Rossum JM, Some physicochemical properties of amphetamine and related drugs, *J. Pharm. Pharmacol.*, **21**, 774–775 (1969).
1860	Fenclofenac ($C_{14}H_{10}Cl_2O_3$) 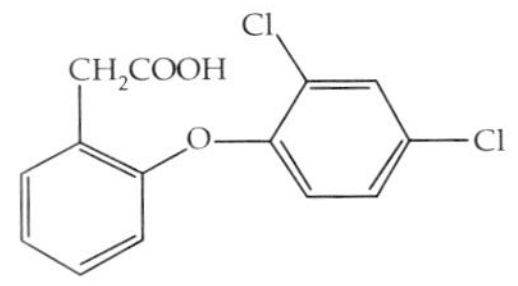	4.53	U	−H		Craig

(*continued*)

No.	Name	pK_a value(s)	Data quality	Ionization type	Method	Conditions t °C; I or c M	Comments and Reference(s)
1861	Fenfluramine ($C_{12}H_{16}F_3N$) 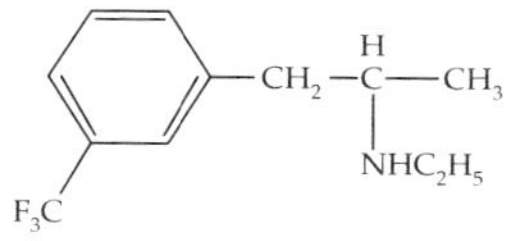	9.10	U	+H			Vree TB, Muskens ATJM and van Rossum JM, Some physicochemical properties of amphetamine and related drugs, *J. Pharm. Pharmacol.*, **21**, 774–775 (1969).
1862	Fenoprofen	4.5	U	−H		H_2O	Ward CK and Schirmer RE, Fenoprofen Calcium, *APDS*, **6**, 161–182 (1977). NB: No supporting reference. See also N&K; Anon.
		7.6	U	−H		66% aq. DMF	
1863	Fenoterol ($C_{17}H_{21}NO_4$) 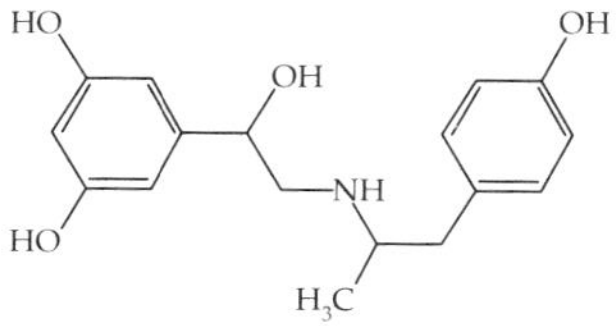	8.5 10.0	U U	+H −H			Craig
1864	Fenoterol	8.5 10.0	U U	+H −H			Al-Majed AA, Fenoterol Hydrobromide, *APDS* **27**, 43 (2000). Cited Clarke as source.
1865	Fentanyl	8.43	U	+H			Lombardo F, Obach RS, Shalaeva MY and Gao F, Prediction of human volume of distribution values for neutral and basic drugs. 2. Extended data set and leave-class-out

					statistics, *J. Med. Chem.*, **47**, 1242–1250 (2004); ref. 276: Meuldermans WEG, Hurkmans RMA, Heykants JJP, Plasma protein binding and distribution of fentanyl, sulfentanil, alfentanil and lofentanil in blood, *Arch. Int. Pharmacodyn.*, **257**, 4–19 (1982).
1866	Flecainide ($C_{17}H_{20}F_6N_2O_3$) 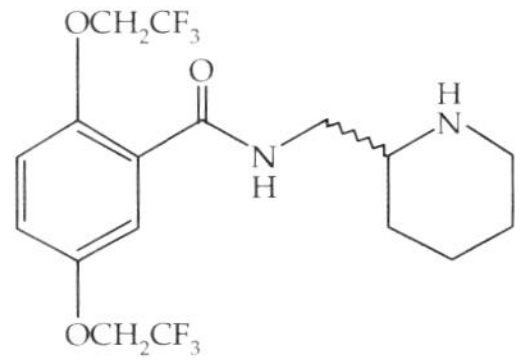	9.3	U	+H	Alessi-Severini S, Coutts RT, Jamali F and Pasutto FM, Flecainide, *APDS*, **21**, 169–195 (1992). Cited AFS Drug Information '88, ASHP, Bethesda, MD, 832–840 Lombardo F, Obach RS, Shalaeva MY and Gao F, Prediction of human volume of distribution values for neutral and basic drugs. 2. Extended data set and leave-class-out statistics, *J. Med. Chem.*, **47**, 1242–1250 (2004); ref. 288: Alessi-Severini S, Coutts RT, Jamali F and Pasutto FM, Flecainide, *APDS*, Brittain HG (ed.), AP Inc., NY, **21**, 169–195 (1992).
1867	Floxuridine ($C_9H_{11}FN_2O_5$) 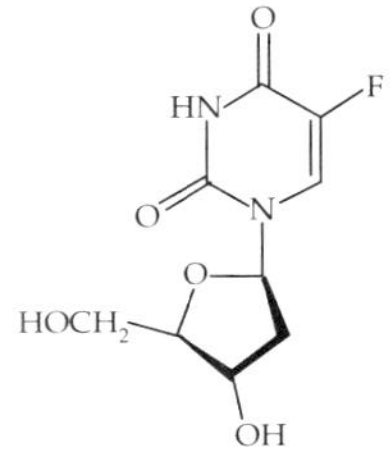	7.44	U	−H	Craig

(continued)

No.	Name	pK_a value(s)	Data quality	Ionization type	Method	Conditions t °C; I or c M	Comments and Reference(s)
1868	Flucloxacillin ($C_{19}H_{17}ClFN_2O_5S$) 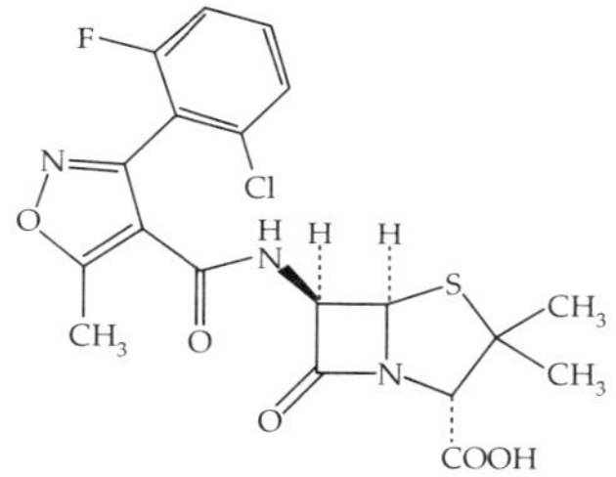	2.7	U	–H			Craig N&K; Avery NB: Also called Floxacillin.
1869	Flucytosine ($C_4H_4FN_3O$) 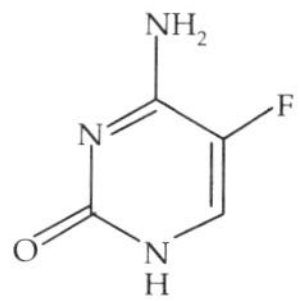	2.90 ± 0.05 10.71 ± 0.05	U U	+H –H	spectro	H_2O	Waysek EH and Johnson JH, Flucytosine, *APDS*, **5**, 115–138 (1976). Toome V, Hoffmann-La Roche Inc., unpublished data. N&K
1870	Flufenamic acid	3.9	U	–H	potentio		Abignente E and de Caprariis P, Flufenamic acid, *APDS*, **11**, 313–343 (1982).
		3.85	U	–H	soly/pH		"The pK_a of flufenamic acid was reported to be 3.9 by Aguiar and Fifelski (20) and 4.5 by Frey and El-Sayed (30). Terada *et al.* (24) have found a value of 3.85 using the pH-dependent solubility method . . .; this value is considerably different from the corresponding value obtained by potentiometric titration in 5–10% aqueous acetone (32)."

pK$_a$ values of flufenamic acid obtained by potentiometric titration

Solvent	pK$_a$	References
water	7.5	Jahn U, Wagner-Jauregg T, Wirkungsvergleich saurer Antiphlogistika im Bradykinin-, UV-Erythem- und Rattenpfotenödem-Test,. *Arzneim.-Forsch.*, **24**, 494–499 (1974). NB: See separate entry, no. 554.
50% aqueous ethanol	5.94	
80% aqueous 2-methoxyethanol	6.0	
75% aqueous methanol	5.75	Unterhalt B, *Arch. Pharm.*, **303**, 445–456 (1970).
Dioxane:water (2:1)	6.8	Lombardino JG, Otterness IG and Wiseman EH, *Arzneim.-Forsch.*, **25**, 1629–1635 (1975).

20. Aguiar AJ and Fifelski RJ, Effect of pH on the *in vitro* absorption of flufenamic acid, *J. Pharm. Sci.*, **55**, 1387–1391 (1966).
24. Terada H, Muraoka S and Fujita T, Structure-activity relationships of fenamic acids, *J. Med. Chem.*, **17**, 330–334 (1974). NB: See separate entry, no. 553.

(continued)

No.	Name	pK_a value(s)	Data quality	Ionization type	Method	Conditions t °C; I or c M	Comments and Reference(s)
							30. Frey HH and El-Sayed MA, Concentrations of acidic antiinflammatory drugs in gastric mucosa, *Arch. Int. Pharmacodyn. Ther.*, **230**, 300–308 (1977). 32. Terada H and Muraoka S, Physicochemical properties and uncoupling activity of 3′-substituted analogues of N-phenylanthranilic acid, *Mol. Pharmacol.*, **8**, 95–103 (1972). NB: See separate entry no. 552.
1871	Flumizole ($C_{18}H_{15}F_3N_2O_2$)	10.7	U	−H			Wiseman EH, McIlkenny HM and Bettis JW, Flumizole, a new non-steroidal antiinflammatory agent, *J. Pharm. Sci.*, **64**, 1469–1475 (1975). NB: Value quoted without references or experimental details. Craig
1872	Flunitrazepam	1.8	U	+H			Kaplan SA, Alexander K, Jack ML, Puglisi CV, DeSilva JAF, Lee TL and Wenfeld RE, Pharmacokinetic profiles of clonazepam in dog and

							human and of flunitrazepam in dog, *J. Pharm. Sci.*, **63**, 527–532 (1974). NB: Values given without reference or experimental support. See Clonazepam.
1873	2-(4-(p-Fluorobenzoyl)-piperidin-1-yl)-2′-acetonaphthone ($C_{24}H_{22}FNO_2$) 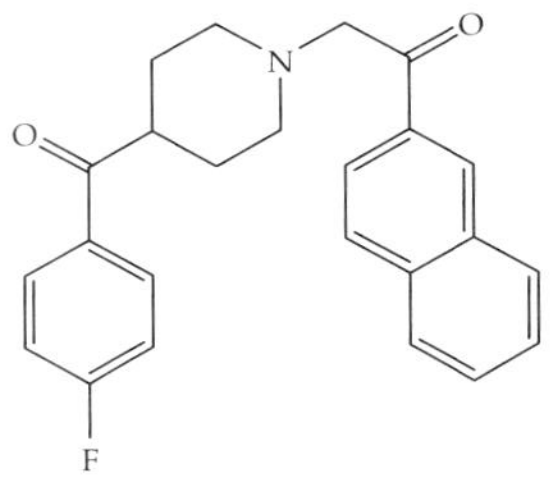	5.9	U	+H			Tokumura T, Tanaka T and Karibe N, Physicochemical properties of 2-(4-(p-fluoro-benzoyl)piperidin-1-yl)-2′-acetonaphthone hydrochloride and its gastrointestinal absorption in beagle dogs, *J. Pharm. Sci. Tech. Yakuzaigaku*, **54**, 95–102 (1994). "The physicochemical properties . . . of E-2001 (2-(4-(p-fluorobenzoyl)-piperidin-1-yl)-2′-acetonaphthone hydrochloride) were studied . . . . The pK_a and log P′ at pH 8 were 5.9 and 4.1, respectively. The solubility of the drug (0.1mcg/ml at pH 5) increased gradually with a decrease in pH and reached a value of 5 mg/ml at pH 3. . . ."
1874	Flupent(h)ixol	7.80	U	+H			Craig
1875	Fluphenazine enanthate ($C_{29}H_{38}F_3N_3O_2S$) 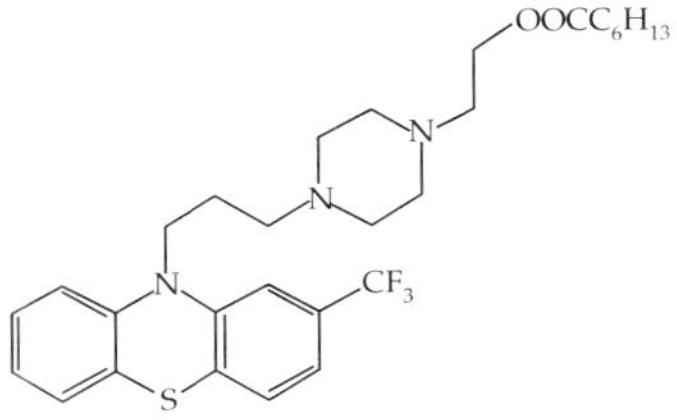	3.50 or 3.29 8.2 or 7.7	U VU	+H +H	potentio		Florey K, Fluphenazine Enanthate, *APDS*, **2**, 245–262 (1973). Cited Jacobson H, private communication. "The pK_{a1} value was determined by titration; the pK_{a2} value was estimated from the pH of the first equivalent point because fluphenazine enanthate precipitated at pH values slightly above this point." NB: The pK_{a2} value must be quite uncertain.

(*continued*)

Appendix B *(continued)*

No.	Name	pK_a value(s)	Data quality	Ionization type	Method	Conditions t °C; I or c M	Comments and Reference(s)
1876	Fluphenazine decanoate ($C_{32}H_{44}F_3N_3O_2S$) 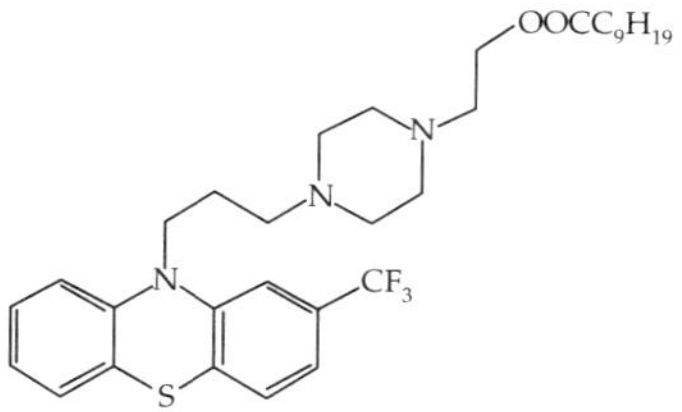						Clarke G, Fluphenazine Decanoate, *APDS*, **9**, 275–294 (1980). NB: The pK_a values were not actually reported in this source. They would be expected to be very similar to those reported for fluphenazine enanthate (3.4; 8.0).
1877	Flurazepam ($C_{21}H_{23}ClFN_3O$) 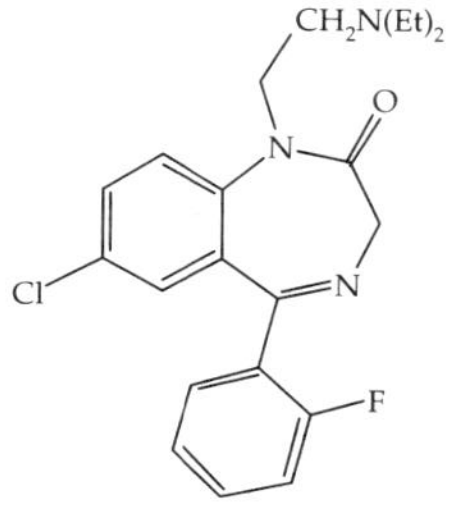	1.90 ± 0.05 8.16 ± 0.05	U U	+H −H	spectro	H_2O	Rudy BC and Senkowski BZ, Flurazepam hydrochloride, *APDS*, **3**, 307–331 (1974). Cited Toome V, Raymond G, Hoffman-La Roche Inc., unpublished data. NB: Apparent pK_{a2} = 7.0 ± 0.1 in 2-propanol:water (1:1) "In water, the trialkylamino type compounds are stronger bases, on the average, by 0.9 pK units. Therefore, the estimated pK_{a2} in water would be 7.9 which are in good agreement with that found spectrophotometrically."
1878	Flurazepam			+H			NB: See Flunitrazepam
1879	Flurbiprofen	3.80	U	−H		H_2O	ACDLabs database

No.	Compound	pKa	Data quality	Ionization type	Method	Conditions	Reference
1880	Fluvoxamine ($C_{15}H_{21}F_3N_2O_2$) 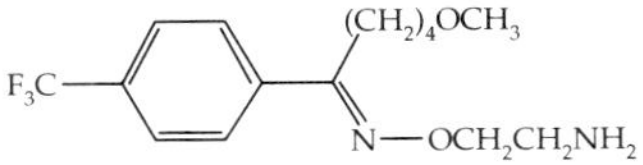	8.7	U	+H			Foda NH, Radwan MA and Al Deeb OA, Fluvoxamine Maleate, *APDS*, **24**, 165–208 (1996). NB: No pK_a reference was given.
1881	Folic acid	2.5	U	−H			Baker BR and Jordaan JH, Analogues of tetrahydrofolic acid XXVIII, *J. Pharm. Sci.*, **54**, 1740–1745 (1965). NB: Refers to Zakrzewski, Sigmund F, Relation between basicity of certain folate analogs and their affinities for folate reductase, *J. Biol. Chem.*, **238**, 4002–4004 (1963). This paper only gave data for the pterin nucleus.
1882	Fumaric acid ($C_4H_4O_4$) 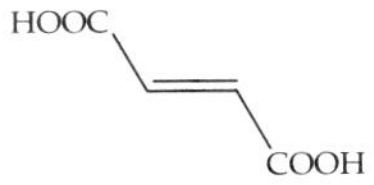	3.019 4.384	A A	−H −H	potentio	H_2O $t = 25.0 \pm 0.01$ N_2 atmos	German WL, Vogel AI, Jeffery GH, Thermodynamic primary and secondary dissociation constants of fumaric and maleic acids, *Phil. Mag.*, **22**, 790–800 (1936). NB: Used the quinhydrone electrode in cells with liquid junction potential. This electrode was necessary to avoid the reduction of the double bond that would otherwise have occurred if a hydrogen electrochemical cell had been used. See also W&G: Martin.
1883	Furaltadone ($C_{13}H_{16}F_3N_4O_6$) 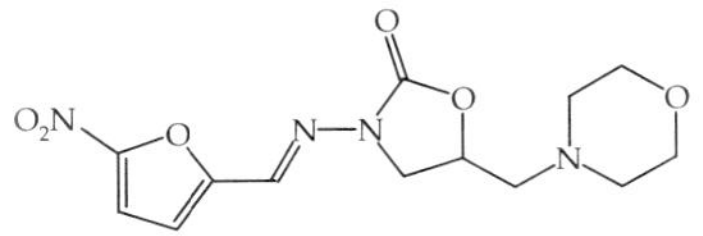	5.0	U	+H			W&G: Buzard JA, Conklin, JD and Buller RH, Lymphatic transport of selected nitrofuran derivatives in the dog, *Am. J. Physiol.*, **201**, 492–494 (1961).

(continued)

No.	Name	pK_a value(s)	Data quality	Ionization type	Method	Conditions t °C; I or c M	Comments and Reference(s)
1884	Furosemide (frusemide)	3.8	U	–H			Cruz JE, Maness DD and Yakatan GJ, Kinetics and mechanism of hydrolysis of furosemide, *Int. J. Pharm.*, **2**, 275–281 (1979). NB: Value given without references or experimental support. "An *in vitro* . . . study was performed to elucidate the . . . hydrolysis . . . of furosemide (I) as a function of pH and temperature. . . . below the reported pK_a of 3.8, the log K-pH profile indicated specific hydrogen ion catalysis on the undissociated species. . . ."
1885	Furosemide (frusemide)	4.7	U	–H			McCallister JB, Chin T-F and Lach JL, Diffuse reflectance studies of solid-solid interactions IV, *J. Pharm. Sci.*, **59**, 1286–1289 (1970). NB: The value was given without citation or experimental support.
1886	Fusidic acid ($C_{31}H_{48}O_6$)	5.35	U	–H			Craig N&K; Merck 9

1887	Galanthamine ($C_{17}H_{21}NO_3$)	8.32	U	+H			Lombardo F, Obach RS, Shalaeva MY and Gao F, Prediction of human volume of distribution values for neutral and basic drugs. 2. Extended data set and leave-class-out statistics, *J. Med. Chem.*, **47**, 1242–1250 (2004); ref. 289: Dictionary of Organic Compounds, CD-ROM, Version 10.1. Chapman and Hall/CRC, 2002.
1888	Gentamicin (C1: $C_{21}H_{43}N_5O_7$) (C2: $C_{20}H_{41}N_5O_7$) (C1a: $C_{19}H_{39}N_5O_7$) C_1 $R_1 = R_2 = CH_3$ C_2 $R_1 = CH_3$; $R_2 = H$ C_{1a} $R_1 = R_2 = H$	8.2	U	+H	potentio	66% DMF	Rosenkrantz B, Greco JR, Hoogerheide JG and Oden EM, Gentamicin Sulfate, *APDS*, **9**, 295–340. Cited Greco J, Rosenkrantz B, Schering-Plough Corp., unpublished results. N&K; Baldini JT, Schering Corp. Kenilworth NJ, Personal Communication, 1975.
		7.9	U	+H		H_2O	Craig; Done AK, *Emergency Med.*, **8**, 197–202 (1976). NB: Average values for 5 amino groups.
1889	Glibenclamide	6.5 ± 0.03	U	−H	soly/pH	H_2O	Crooks MJ and Brown KF, The binding of sulphonylureas to serum albumin, *J. Pharm. Pharmacol.*, **26**, 305–311 (1974). NB: Result stated to be identical to that of Hadju *et al.* (see no. 591).

(continued)

No.	Name	pK_a value(s)	Data quality	Ionization type	Method	Conditions t °C; I or c M	Comments and Reference(s)
1890	Glipizide ($C_{21}H_{27}N_5O_4S$) 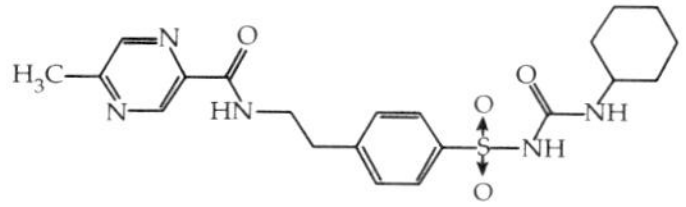	5.9	U	–H			Foye 3rd; see Azatadine; from McEvoy.
1891	D-Glucuronic acid ($C_6H_{10}O_7$) 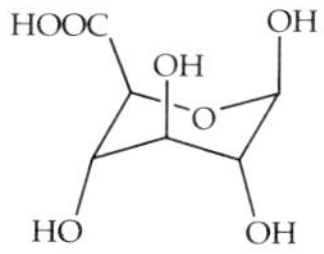	3.18	U	–H	potentio	H_2O $t = 20.0$	Hirsch P, The acid strength of glucuronic acid in comparison with that of oxycelluloses, *Rec. Trav. Chim. Belge.*, **71**, 999–1006 (1952). Cited in Kortum, ref. H36. Cited also in W&G, but with an incorrect reference.
1892	Glutarimide ($C_5H_7NO_2$) 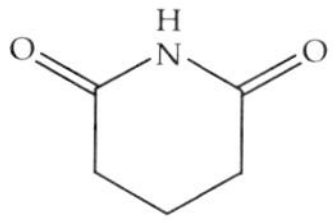	11.4	U	–H			W&G: Albert, Selective Toxicity, p. 281, (1968) reference is not correct.
1893	Glutethimide	9.2	U	–H			Agarwal SP and Blake MI, Determination of glutethimide, aminoglutethimide and bemegride by non-aqueous titration, *J. Pharm. Sci.*, **54**, 1668–1670 (1965). NB: The value was given without reference or supporting experimental data. See also Foye.
		11.2	U	–H	soly/pH	H_2O	Peinhardt G, Acidity and acidimetric titration of medically used glutaric acid imide, *Pharmazie*, **32**, 726–727 (1977) cited in CA 88:141587.

No.	Name	pKa	Method	Type	Conditions	Reference / Comments
1894	Glutethimide	4.518	U	–H	H_2O $t = 20.0$ $I = 0.1$	Aboul-Enein HY, Glutethimide, *APDS*, **5**, 139–187 (1976). ''Doornbos DA and deZeeuw RA, *Pharm. Weekblad*, 104, 233–251 (1969). Doornbos and DeZeeuw measured the ''proton lost'' dissociation constant K_2H and the ''proton gained'' K_3H dissociation constant for glutethimide by a previously described potentiometric titration method at 20°. K_2H and K_3H for glutethimide at ionic strength of $\mu = 0.10$ was found to be 4.518.'' NB: This value is inconsistent with other reported values for this compound (see no. 597) and is more likely to be the corresponding pK_b.
1895	Glyburide ($C_{23}H_{28}ClN_3O_5S$)	5.3	U	–H		N&K; Avery NB: See Glibenclamide
1896	Glycocholic acid	2.78	U	–H	H_2O $c < CMC$	Johns WH and Bates TR, Quantification of the binding tendencies of cholestyramine I: Effect of structure and added electrolytes on the binding of unconjugated and conjugated bile salt anions, *J. Pharm. Sci.*, **58**, 179–183 (1969). NB: These values were quoted from Ekwall P, Rosendahl T and Lofman N, Bile salt solutions. I. The dissociation constants of the cholic and deoxycholic acids, *Acta Chem. Scand.*, **11**, 590–598 (1957). They were measured at concentrations both above and below the critical micellar concentration range.
		4.35	U	–H	H_2O $c > CMC$	

(continued)

No.	Name	pK_a value(s)	Data quality	Ionization type	Method	Conditions t °C; I or c M	Comments and Reference(s)
1897	Glycodeoxycholic acid	2.46	U	–H		H_2O $c <$ CMC	Johns WH and Bates TR, Quantification of the binding tendencies of cholestyramine I: Effect of structure and added electrolytes on the binding of unconjugated and conjugated bile salt anions, *J. Pharm. Sci.*, **58**, 179–183 (1969). NB: These values were quoted from Ekwall P, Rosendahl T and Lofman N, Bile salt solutions. I. The dissociation constants of the cholic and deoxycholic acids, *Acta Chem. Scand.*, **11**, 590–598 (1957). They were measured at concentrations both above and below the critical micellar concentration range.
		3.98	U	–H		H_2O $c >$ CMC	
1898	Glycyclamide ($C_{14}H_{20}N_2O_3S$) 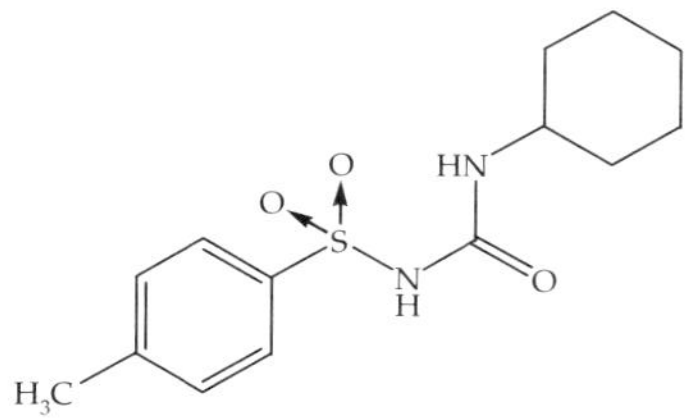	5.50	U	+H			Craig
1899	Guanabenz ($C_8H_8Cl_2N_4$) 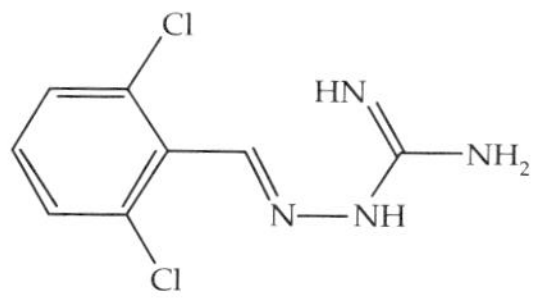	8.1	U	+H	spectro	40% EtOH	Shearer CM, Guanabenz Acetate, *APDS*, **15**, 319–336 (1986). Shearer CM, Wyeth Laboratories Inc., unpublished work. "By potentiometric titration [method of Parke TV, Davis WW, Use of apparent dissociation constants in qualitative organic analysis, *Anal. Chem.*, **26**, 642–645 (1954)] in 40% ethanol/water, pK_a = 8.1"

No.	Compound	pKa	Type	Mode	Method	Conditions	Reference / Comments
1900	Guanethidine ($C_{10}H_{22}N_4$) 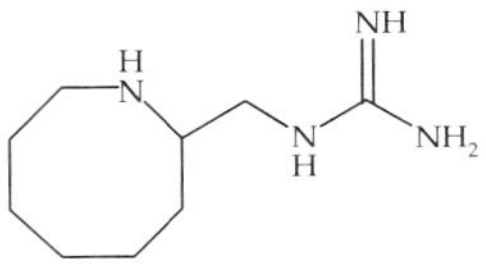	11.4	U	+H			Craig; N&K; Hengstmann JH, Falkner FC, Watson JT and Oates J, Quantitative determination of guanethidine and other guanido-containing drugs in biological fluids by gas chromatography with flame ionization detection and multiple ion detection, *Anal. Chem.*, **46**, 34–39 (1974). NB: See Bethanidine for further details.
1901	Guanethidine	8.3 11.4	U U	+H +H			W&G: Hengstmann JH, Falkner FC, Watson JT and Oates J, Quantitative determination of guanethidine and other guanido-containing drugs in biological fluids by gas chromatography with flame ionization detection and multiple ion detection, *Anal. Chem.*, **46**, 34–39 (1974). NB: See Bethanidine for further details.
1902	Guanidine (CH_5N_3) 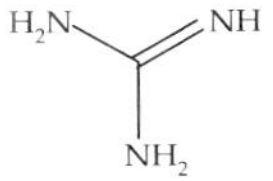	13.59 ± 0.05	U	+H	potentio	H_2O $t = 24.2$ $I = 2$	Hall NF, Sprinkle MR, *JACS*, **54**, 3469 (1932). Cited in Perrin Bases no. 3720, ref. H16. Used hydrogen/calomel electrodes in an unsymmetrical cell to measure buffer solutions of fixed concentration ratios. Ref. A69 (Angyal SJ and Warburton WK, Basic strengths of methylated guanidines, *JCS*, 2492–2494 (1951)). Used a glass electrode in an unsymmetrical cell, and gave $pK_a = 13.4$.

(continued)

Appendix B *(continued)*

No.	Name	pK_a value(s)	Data quality	Ionization type	Method	Conditions t °C; I or c M	Comments and Reference(s)
1903	Guanoxan ($C_{10}H_{13}N_3O_2$)	12.3	U	+H			Craig; N&K; Hengstmann JH, Falkner FC, Watson JT and Oates J, Quantitative determination of guanethidine and other guanido-containing drugs in biological fluids by gas chromatography with flame ionization detection and multiple ion detection, *Anal. Chem.*, **46**, 34–39 (1974). NB: See Bethanidine for further details.
1904	Haloperidol	8.3	U	+H	potentio	H_2O	Janicki CA and Ko CY, Haloperidol, *APDS*, **9**, 341–369 (1980). "The pK_a of haloperidol is 8.3 calculated by linear extrapolation using potentiometric titration in 15%, 25%, 35%, 45% methanol-water (v/v) with 0.005-N NaOH as titrant." N&K; Janssen PAJ, Westeringhe C van de, Jageneau AHM, *et al.*, Chemistry and pharmacology of CNS depressants related to 4-(4-hydroxy-4-phenylpiperidino)-butyrophenone. Part I, *J. Med. Pharm. Chem.*, **1**, 281–297 (1959).
1905	Haloperidol	8.3	U	+H			W&G: Janssen PAJ *et al.*, *J. Med. Pharm. Chem.*, **1**, 282–297 (1959).
1906	Haloperidol	8.65	U	+H	potentio		Lombardo F, Obach RS, Shalaeva MY and Gao F, Prediction of human volume of distribution values for neutral and basic drugs. 2. Extended data set and leave-class-out

							statistics, *J. Med. Chem.*, **47**, 1242–1250 (2004); ref. not given: potentiometric titration. NB: Oliviera CR, Lima MCP, Carvalho CAM, Leysen JE and Carvalho AP, Partition coefficients of dopamine antagonists in brain membranes and liposomes, *Biochem. Pharmacol.*, **38**, 2113–2120 (1989); gave a value of 8.66; refered to Leysen JE, Gommeren W, *J. Neurochem.*, **36**, 201–219 (1981); also Pauwels PJ, Leysen JE and Laduron PM, *Eur. J. Pharmacol.*, **124**, 291–298 (1986). Also referred to spiperone, pimozide, and domperidone.
1907	Harmol ($C_{12}H_{10}N_2O$) 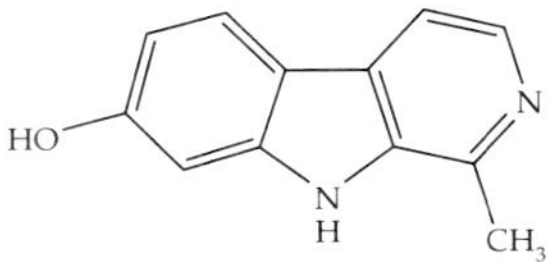	7.90	U	+H	spectro	H_2O	Balon-Almeida M, Munoz-Perez MA, Carmona-Guzman MC and Hidalgo-Toledo J, Ionization equilibria of harmol and harmalol in concentrated hydroxide solutions, *JCS Perk. II*, 1165–1167 (1988); Methodology is described in: Munoz-Perez MA, Carmona-Guzman MC, Hidalgo-Toledo J, Balon-Almeida M, Ionization equilibria of β-carbolines in concentrated hydroxide solutions, *JCS Perk. II*, 1573–1575 (1986).
		9.47	U	−H			
		15.75	U	−H			

Related cpds:

harmalol	8.62; 11.30; 16.01	harmaline	9.55; 15.39
harmine	7.45; 14.35	2,3-dimethylindole	15.57

(continued)

Appendix B *(continued)*

No.	Name	pK_a value(s)	Data quality	Ionization type	Method	Conditions t °C; I or c M	Comments and Reference(s)
							NB: The pK_{a1} values (phenolic) were reported by Tomas F, Zabala I and Olba A, Acid-base and tautomeric equilibria of harmol in the ground and first excited singlet states, *J. Photochem.*, **31**, 253–263 (1985); *ib.* **26**, 285–294 (1984); Douglas KT, Sharma RK, Walmsley JF and Hider RC, Ionization processes of some harmala alkaloids, *Mol. Pharmacol.*, **23**, 614–618 (1983).
1908	Hexylcaine ($C_{16}H_{23}NO_2$) 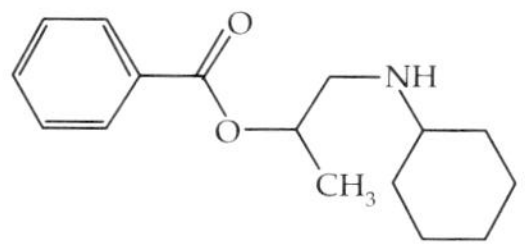	9.1	U	+H			W&G: Truant AP and Takman B, Differential physical-chemical and neuropharmacologic properties of local anesthetic agents, *Anesth. Analg.*, **38**, 478–484 (1959).
1909	Hexylcaine	9.1	U	+H			Craig
1910	Hexylresorcinol ($C_{12}H_{18}O_2$) 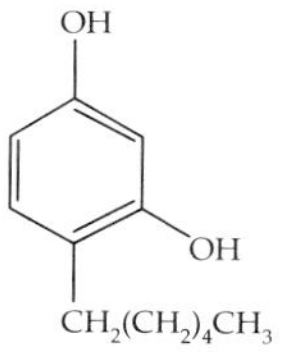	9.54	U	−H			Craig

1911	Hippuric acid ($C_9H_9NO_3$)	3.64	U	–H			N&K; Ritschel gave 3.64; ref. Parrott EL, Pharmaceutical Technology. W&G: Parrott EL and Saski W, *Exp. Pharm. Technol.*, Burgess, Minneapolis, MN, p. 255 (1965).
1912	Homatropine ($C_{16}H_{21}NO_3$)	9.9	U	+H		$t = 20.0$	Muhtadi FJ, Hassan MMA and Afify AFA, Homatropine hydrobromide, *APDS*, **16**, 245–290 (1987); cited Clarke, p. 660—This reference is incorrect—the entry for homatropine has no pK_a data. The reported value is given by: BPC, p. 414, although there is no primary reference.
1913	Homatropine	9.7	U	+H	spectro	$t = 23$	Schoorl N, Dissociation constants and titration exponents of several less common alkaloids, *Pharm. Weekblad*, **76**, 1497–1501 (1939); CA 34:1900. The study used the indicator spectrophotometric (colorimetric) method of Kolthoff. Also see N&K.

(continued)

Appendix B *(continued)*

No.	Name	pK_a value(s)	Data quality	Ionization type	Method	Conditions t °C; I or c M	Comments and Reference(s)
1914	Hycanthone ($C_{20}H_{24}N_2O_2S$) 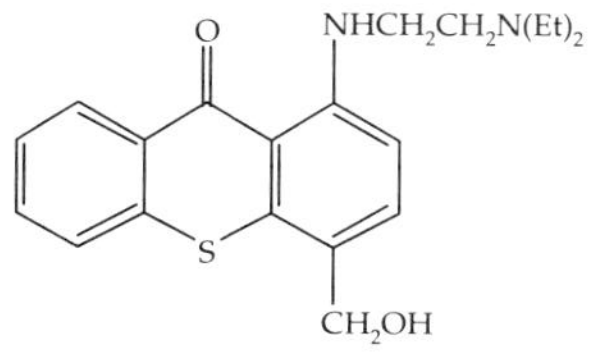	3.40	U	+H			Craig
1915	Hydantoin ($C_3H_4N_2O_2$) 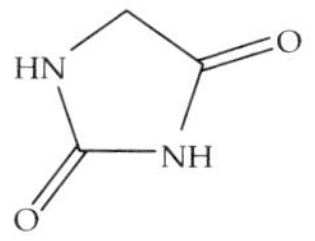	9.1	U	+H			W&G: Kortum *et al.*, 1961.
1916	Hydrochlorothiazide	7.0 9.2	U U	−H −H			N&K; Merck 9
1917	Hydrocodone ($C_{18}H_{21}NO_3$) 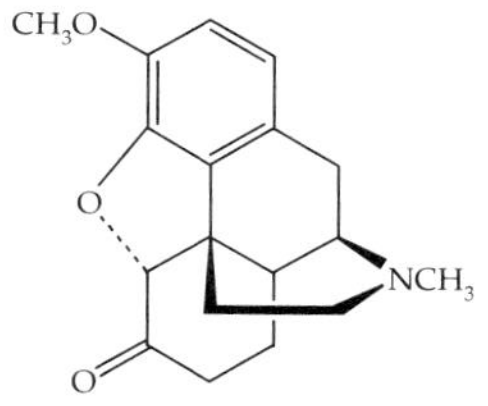	8.3	U	+H			Craig N&K; Ritschel gave 8.3; ref. Hager's Handbuch der Pharmazeutischen Praxis, vol. **I**, Allgemeiner Teil, Wirkstoffgruppen I, Springer-Verlag, p. 815 (1967).

1918	Hydromorphone ($C_{17}H_{19}NO_3$) 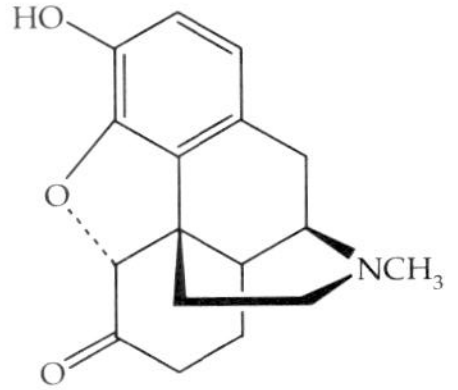	8.15	U	+H	Craig; N&K; Clouet DH (ed.), *Narcot. Drugs Biochem. Pharmacol.*, Plenum Press, New York, 52–53, (1971).
1919	Hydromorphone	7.8	U	+H	W&G cited Perrin Bases; this compound is not listed in either the 1965 initial publication or the 1972 supplement. Source not known.
1920	Hydroquinone	9.91	U	−H	Craig
		12.04	U	−H	
1921	4-Hydroxyamphetamine	9.6	U	+H	W&G; quoted Leffler EB, Spencer HM and Burger A, Dissociation constants of adrenergic amines, *JACS*, **73**, 2611–2613 (1951). NB: The real source of this value is not clear.
1922	4-Hydroxyamphetamine	8.25	U	+H	Beckett AH and Al-Sarraj S, The identification, properties and analysis of N-hydroxyamphetamine, a metabolite of amphetamine, *J. Pharm. Pharmacol.*, **25**, 328–334 (1973). NB: Value determined according to (method of) Leffler EB, Spencer HM and Burger A, Dissociation constants of adrenergic amines, *JACS*, **73**, 2611–2613 (1951).

(continued)

Appendix B *(continued)*

No.	Name	pK_a value(s)	Data quality	Ionization type	Method	Conditions t °C; I or c M	Comments and Reference(s)
1923	Hydroxyzine	2.13 7.13	U U	+H +H			N&K; Pardo A, Vivas S, Espana F and Fernandez-Alonso JI, Studies on the structure-activity relationship of tricyclic psychodrugs. III. Dissociation constants, *Afinidad*, **29**, 640–642 (1972); Persson B-A, Extraction of amines as complexes with inorganic anions, *Acta Pharm. Suecica*, **5**, 335–342 (1968).
1924	Hydroxyzine	1.8	U	+H			W&G: Tsau J and DeAngelis N, Hydroxyzine dihydrochloride, *APDS*, **7**, 319–341 (1978).
1925	Hyoscyamine	9.7	U	+H			Craig; Clarke, p. 676
1926	Hyoscyamine	9.3	U	+H			N&L; Ritschel gave 9.3 (apparent); ref. Robson JM and Stacey RS, *Recent advances in pharmacology*, 4th Edn., Little Brown and Co., Boston, p. 108 (1968).

No.	Name	pK_a value(s)	Data quality	Ionization type	Method	Conditions t°C; I or c M	Comments and Reference(s)
1927	Idoxuridine ($C_9H_{11}IN_2O_5$) 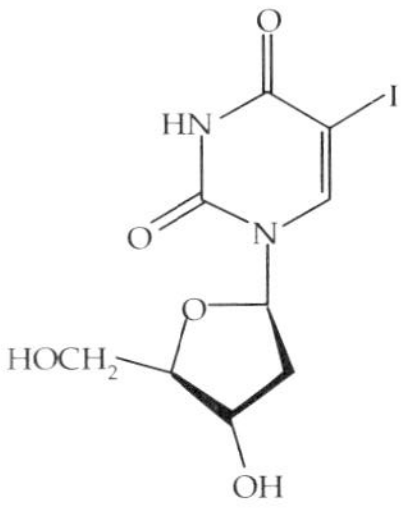	8.3	U	–H			Craig
1928	Idoxuridine	8.25	U	–H			N&K; Merck 9
1929	D-Indoprofen ($C_{17}H_{15}NO_3$) 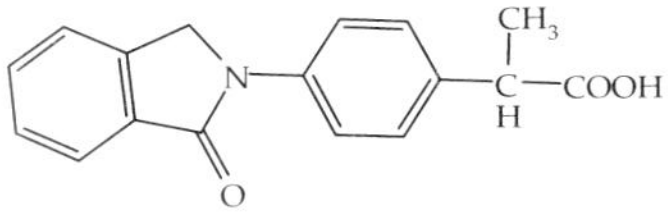	4.39	U	–H	Comp		ACD/pK_a Labs estimate
		5.8	U	–H			Craig
1930	Indoprofen	5.8	U	–H			Craig
1931	Indoprofen	5.8	U	–H			Fuccella LM, Conti F, Corvi G, Mandelli V, Randelli M and Stefanelli G, Double-blind study of the analgesic effect of indoprofen (K 4277), *Clin. Pharmacol. Ther.*, **17**, 277–283 (1975); cited in W&G
1932	Indoramin ($C_{22}H_{25}N_3O$) 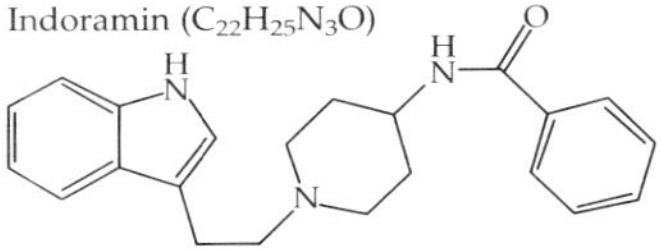	7.7	U	+H			Craig N&K; S&R

(continued)

No.	Name	pK_a value(s)	Data quality	Ionization type	Method	Conditions t°C; I or c M	Comments and Reference(s)
1933	Iocetamic acid ($C_{12}H_{13}I_3N_2O_3$) 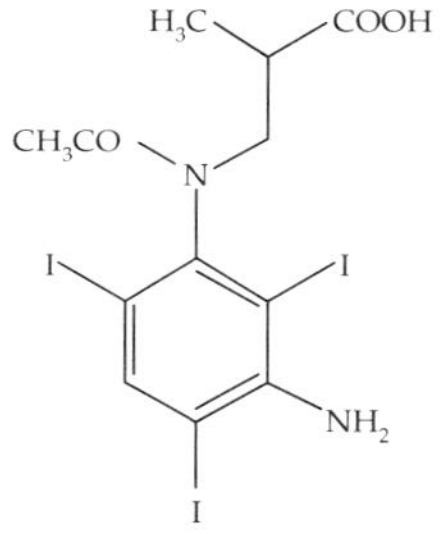	4.1 or 4.3	U	–H			McEvoy NB: Foye states that the value depends on optical isomer.
1934	Iodipamide	3.5	U	–H			Neudert W and Röpke H, The physico-chemical properties of the disodium salt of adipic acid bis(2,4,6-triiodo-3-carboxyanilide) and other triiodobenzene derivatives, *Chem. Ber.*, **87**, 659–667 (1954); cited in W&G.
1935	Iodoquinol ($C_9H_5I_2NO$) 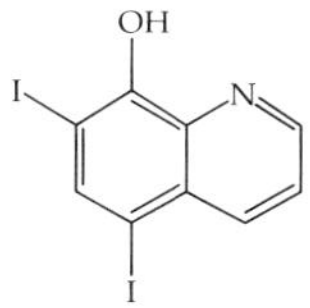	8.0	U	–H			Craig
1936	Iprindole ($C_{19}H_{28}N_2$) 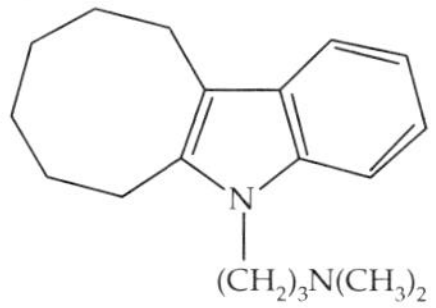	8.2	U	+H			Craig; N&K; S&R

1937	Ipronidazole ($C_7H_{11}N_3O_2$) 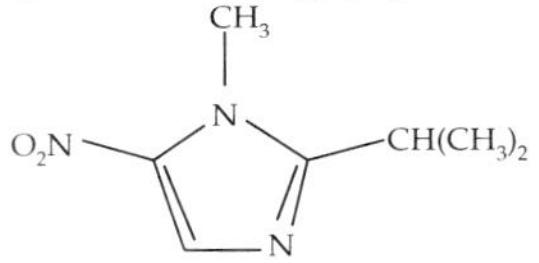	2.73	U	+H		Craig
1938	Isopropylamphetamine ($C_{12}H_{19}N$) 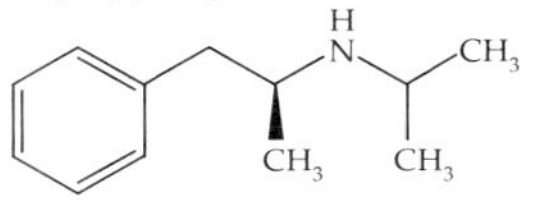	10.14	U	+H		Vree TB, Muskens ATJM and van Rossum JM, Some physicochemical properties of amphetamine and related drugs, *J. Pharm. Pharmacol.*, **21**, 774–775 (1969).
1939	Isoxsuprine ($C_{18}H_{23}NO_3$) 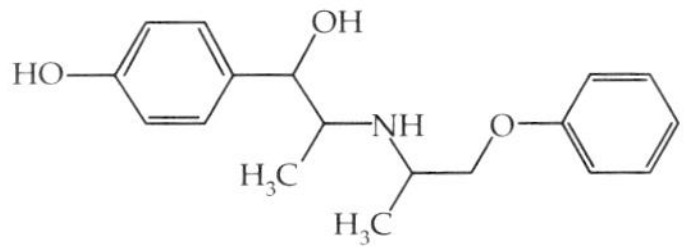	8.0 9.8	U U	+H −H		Belal F, Al-Badr A, Al-Majed AA and El-Subbagh HI, Isoxsuprine Hydrochloride, *APDS*, **26**, 370 (1999). NB: Cited Clarke as source.
1940	Isoxsuprine	8.0 9.8	U U	+H −H		Foye; N&K; Avery
1941	Lansoprazole ($C_{16}H_{14}F_3N_3O_2S$) 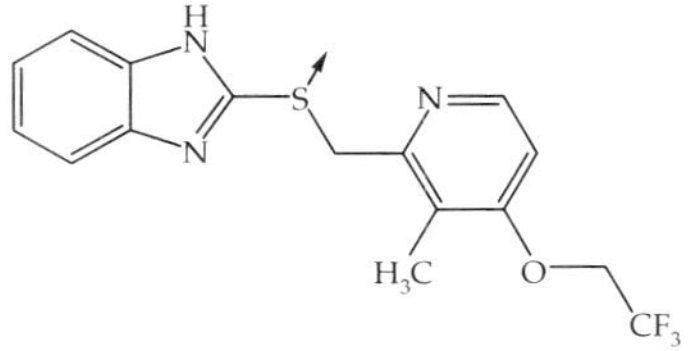	2.34 ± 0.37 3.53 ± 0.37 8.48 ± 0.30	VU VU VU	+H +H +H	comp	Al-Zehouri J, El-Subbagh HI and Al-Badr AA, Lansoprazole, *APDS*, **28**, 123 (2001). NB: Estimated using ACD PhysChem (Advanced Chemistry Development, Toronto, Canada).
1942	Levomethorphan ($C_{18}H_{25}NO$) 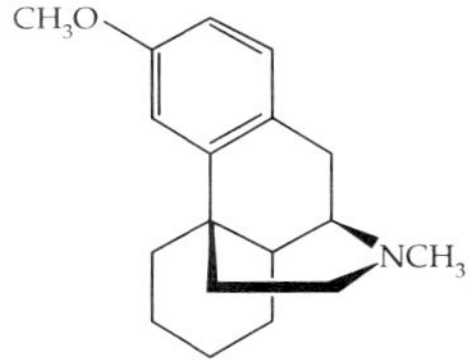	8.3	U	+H		Craig

(continued)

No.	Name	pK_a value(s)	Data quality	Ionization type	Method	Conditions t°C; I or c M	Comments and Reference(s)
1943	Levomoramide ($C_{25}H_{32}N_2O_2$) 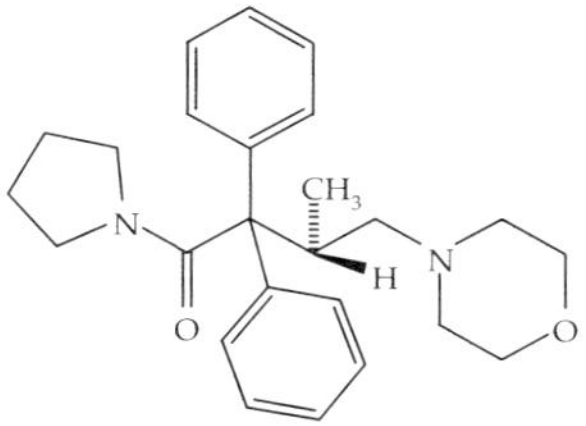	7.0	U	+H			Craig
1944	Levomoramide	6.60	U	+H	potentio	50% aq EtOH	Farmilo CG, Oestreicher PM and Levi L, Physical methods for the identification of narcotics. IB Common physical constants for identification of ninety-five narcotics and related compounds, *Bull. Narcotics, U.N. Dept. Social Affairs*, vol. 6, pp. 7–19 (1954); cited in Clouet DH (ed.), *Narcotic Drugs Biochemical Pharmacology*, Plenum Press, New York, 52–53 (1971).
1945	Liothyronine ($C_{15}H_{12}I_3NO_4$)	8.5	U	–H			Craig
		8.4	U	–H			W&G: Cited Smith RL, *Med. Chem.*, **2**, 477 (1964). This reference could not be identified.
		8.45	U	–H			N&K; Ritschel gave 8.45; ref, Smith Kline and French, Philadelphia, PA, 19101.
1946	Lithium Carbonate Li_2CO_3	6.38		–H			Stober HC, Lithium Carbonate, *APDS*, **15**, 367–391 (1986). NB: Cited Handbook of Chemistry and Physics, 66th Edn., CRC, Cleveland OH (1985) D163.
		10.25		–H			

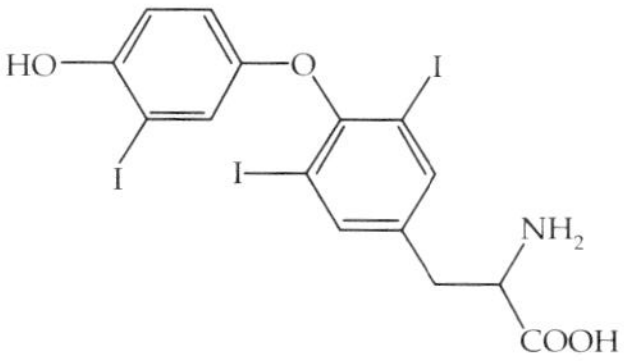

1947	Loxapine ($C_{18}H_{18}ClN_3O$) 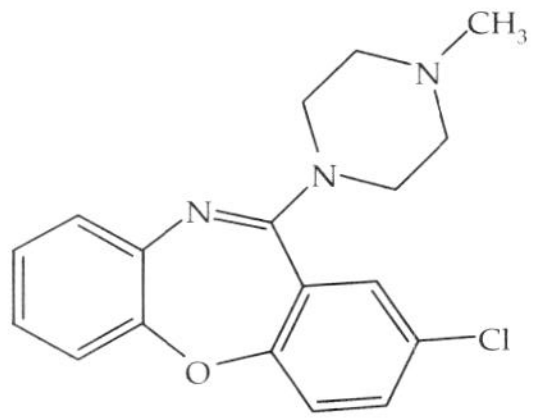	6.6	U	+H			Craig N&K; Anon
1948	Lysergide ($C_{20}H_{25}N_3O$) 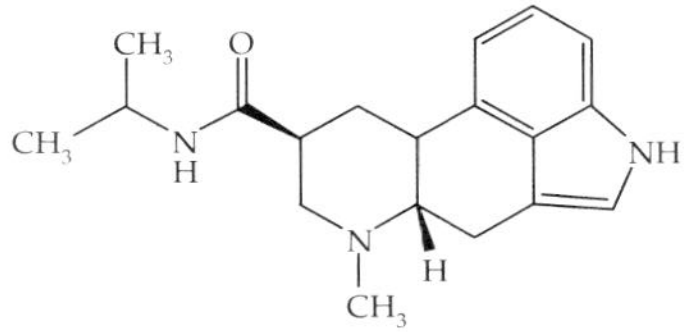	7.5	U	+H			Craig
1949	Mandelhydroxamic acid ($C_8H_9NO_3$) 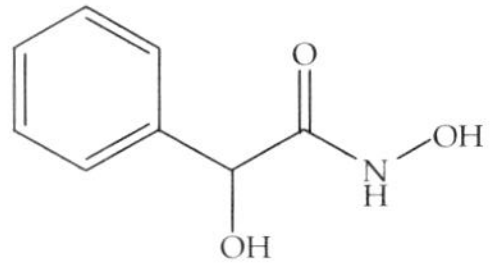	9.02	U	–H	potentio	$t = 25$	Jelikic M, Stankovic B and Dugandzic M, Spectrophotometric and potentiometric determination of the ionization constant of mandelhydroxamic acid, *Arch. Farm.*, **30,** 59–63 (1980).
		9.04	U	–H	spectro		
1950	Mandelic acid	3.37	U	–H			N&K; Merck 9
1951	Mandelic acid	3.8	U	–H			W&G: Parrott EL and Saski W, *Exp. Pharm. Technol.*, Burgess, Minneapolis, MN, p. 255 (1965).
1952	Maprotiline ($C_{20}H_{23}N$) 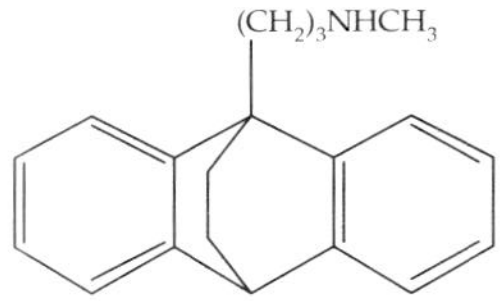	10.5 ± 0.2	U	+H		H_2O $t = 25.0$	Suh SK and Smith JB, Maprotiline hydrochloride, *APDS*, **15,** 393–426 (1986). Cited Jäkel K, Moser P, Stahl P, Ciba-Geigy Ltd., personal communication. Apparent pK_a = 9.0 ± 0.1 and 9.4 ± 0.1 (80% ethylene glycol monomethyl ether in water) (1, 6). 1. Stahl P, Ciba-Geigy Ltd., personal communication.

(continued)

No.	Name	pK_a value(s)	Data quality	Ionization type	Method	Conditions t°C; I or c M	Comments and Reference(s)
							6. Smith J, Hagman H, Mollica J, Ciba-Geigy Ltd., personal communication. NB: See also: Lombardo F, Obach RS, Shalaeva MY, Gao F, Prediction of human volume of distribution values for neutral and basic drugs. 2. Extended data set and leave-class-out statistics, *J. Med. Chem.*, **47,** 1242–1250 (2004); ref. 290: Suh SK, Smith JB, *APDS*, **15,** 393–426 (1986).
1953	Mazindol ($C_{16}H_{13}ClN_2O$) 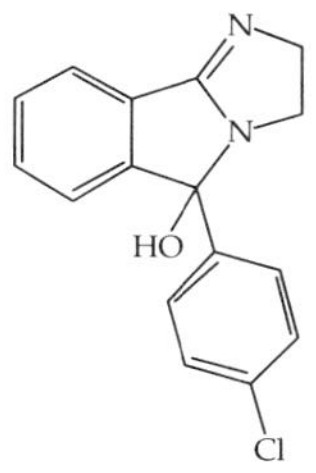	8.6	U	+H			Craig
1954	Mechlorethamine ($C_5H_{11}Cl_2N$) $CH_3N(CH_2CH_2Cl)_2$	6.43	U	+H			Craig; McEvoy
1955	Meclofenamic acid ($C_{14}H_{11}Cl_2NO_2$) 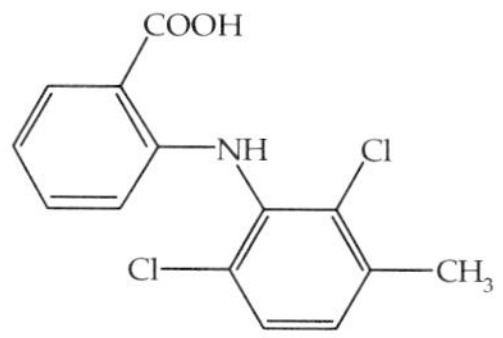	4.0	U	–H			McEvoy

1956	Mefloquine ($C_{17}H_{16}F_6N_2O$)	8.6	U	+H		H_2O	Mu JY, Israili ZH and Dayton PG, Disposition and metabolism of mefloquine hydrochloride (WR 142,490), a quinolinemethanol antimalarial, in the rat, *Drug Metab. Disp.*, **3,** 198–210 (1975). Cited in Lim P, Mefloquine hydrochloride, *APDS*, **14,** 157–179 (1985). NB: Value extrapolated from apparent pK_a values in aqueous EtOH: 8.54 (30% EtOH) 8.50 (50% EtOH) 8.42 (70% EtOH)
1957	Melphalan	2.5	U	−H		H_2O	Feyns LV, Melphalan, *APDS*, **13,** 265–297 (1984). NB: Evstratova KI, Goncharova NA and Solomko VY, *Farmatsiya* (Moscow), **17(4),** 33–36 (1968); CA 69:99338a also gave an apparent pK_a = 5.9 in 90% v/v acetone.
1958	Meperidine	8.72	U	+H	potentio	50% EtOH	Farmilo CG, Oestreicher PM and Levi L, Physical methods for the identification of narcotics. IB Common physical constants for identification of ninety-five narcotics and related compounds, *Bull. Narcotics*, U.N. Dept. Social Affairs, vol. 6, pp. 7–19 (1954). CA 48:69490. NB: Cited in Beckett AH, Analgesics and their antagonists. I., *J. Pharm. Pharmacol.*, **8,** 848–859 (1956); also cited in Clouet DH (ed.), *Narcotic Drugs Biochemical Pharmacology*, Plenum Press, New York, 52–53 (1971).
1959	Mephentermine	10.25	U	+H			Vree TB, Muskens ATJM and van Rossum JM, Some physicochemical properties of amphetamine and related drugs, *J. Pharm. Pharmacol.*, **21,** 774–775 (1969). NB: See Dexamphetamine for details.

(*continued*)

No.	Name	pK_a value(s)	Data quality	Ionization type	Method	Conditions t°C; I or c M	Comments and Reference(s)
1960	Mephenytoin ($C_{12}H_{14}N_2O_2$) 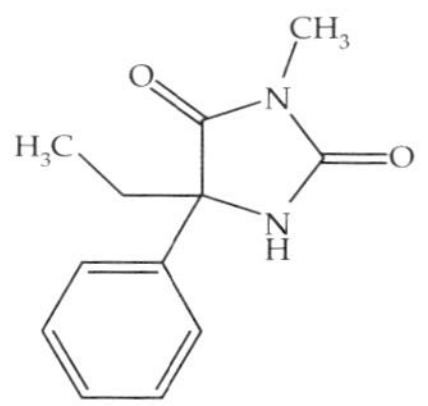	8.1	U	–H			W&G: Sandoz Pharmaceuticals product literature.
1961	Mepindolol ($C_{15}H_{22}N_2O_2$) 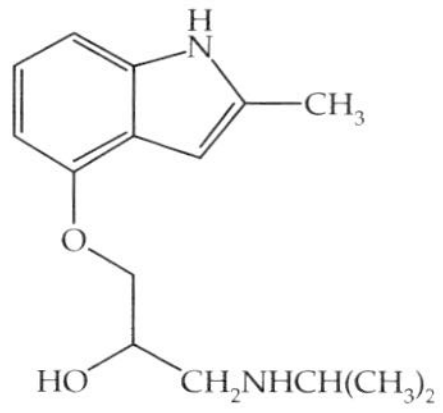	8.90	U	+H			Craig; Gugler R, Kreis L and Dengler HJ, Pharmacokinetics of a new β-adrenoceptor blocking agent, LF 17–895, in man, *Arzneim.-Forsch.*, **25,** 1067–1072 (1975).
1962	Mepivacaine	7.6	U	+H			de Jong RH, Neural blockade by local anesthetics, *JAMA*, **238,** 1383–1385 (1977); cited in W&G.
1963	Mercaptomerin ($C_{16}H_{27}HgNO_6S$) 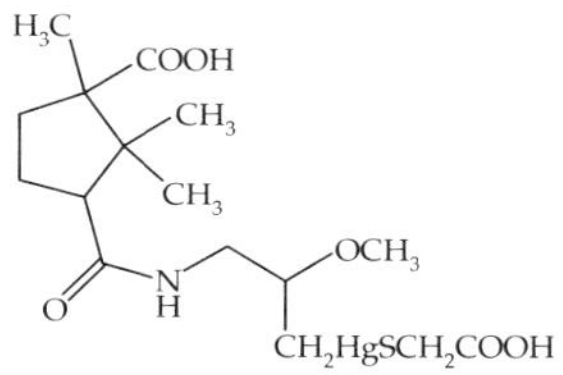	3.7 5.1	U U	–H –H			Craig; N&K; Anon

No.	Compound	pKa	Class	Type	Method	Conditions	Reference / Comments
1964	Mesalamine ($C_7H_7NO_3$) 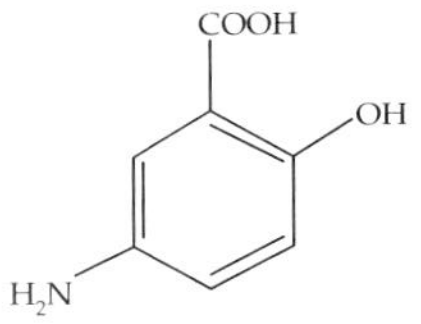	2.30 ± 0.09 5.69 ± 0.04	U U	−H +H	soly/pH spectro (λ = 298 nm)	H_2O $t = 25$	Dash AK and Brittain HG, Mesalamine, *APDS*, **25**, 209–242 (1998). "Following trends in the pH dependence of its ultraviolet spectrum, the dissociation constant of the amino group (pK_{a2}) in 5-aminosalicylic acid was determined to be 5.69. The dissociation constant of the carboxyl group pK_{a1} was obtained through studies of the compound solubility at pH values between 1.0 and 2.5, and was reported to be 2.30 [15] . . . The three dissociation constants . . . have been reported to be $pK_{a1} = 3.0$ (carboxylate group), $pK_{a2} = 6.0$ (amino group), and $pK_{a3} = 13.9$ (hydroxyl group) [13]."
		3.0 6.0 13.9	U U U	−H +H −H			13. Lund W (ed.), *The Pharmaceutical Codex*, 25th Edn., Pharmaceutical Press, London, 946–947 (1994). 15. French DL and Mauger JW, Evaluation of the physicochemical properties and dissolution characteristics of mesalamine, *Pharm. Res.*, **10**, 1285–1290 (1993).
1965	Mesna (2-mercapto-ethanesulfonic acid) ($C_2H_6O_3S_2$) $HSCH_2CH_2SO_3H$	9.1	U	−H		H_2O $t = 20$	Craig
1966	Metaproterenolol (orciprenaline)	8.8 11.8	U U	+H −H			N&K; Avery
1967	Metaraminol ($C_9H_{13}NO_2$) 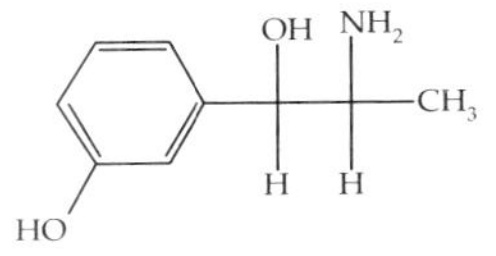	8.6	U	+H			Craig N&K; S&R

(continued)

No.	Name	pK_a value(s)	Data quality	Ionization type	Method	Conditions t°C; I or c M	Comments and Reference(s)
1968	Methacycline ($C_{22}H_{22}N_2O_8$) 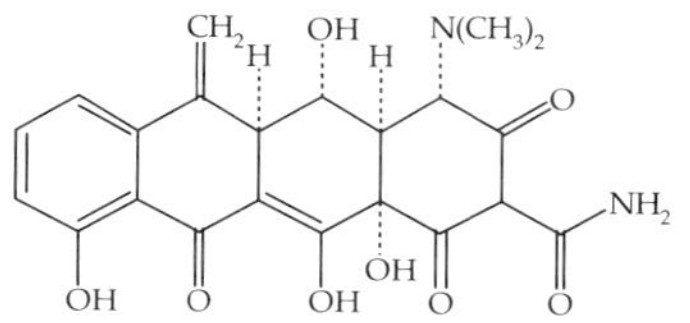	3.5 7.6 9.5	U U U	−H −H +H			Craig N&K; Avery
1969	Methadone ($C_{21}H_{27}NO$) 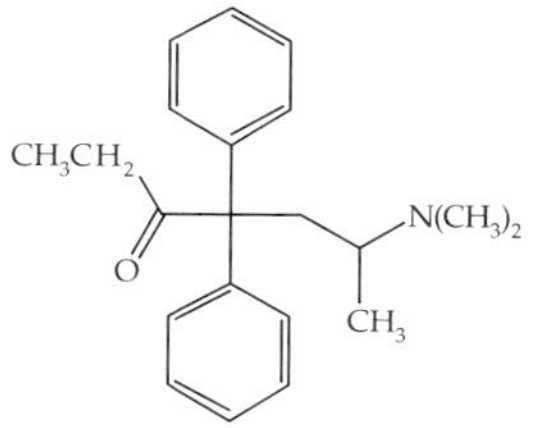	8.25	U	+H	potentio	EtOH/H_2O $t = 20.0$	Bishara, RH, Methadone hydrochloride, *APDS*, **3**, 365–439 (1974); see also *APDS*, **4**, 520; *APDS*, **9**, 601. "Levi *et al.* reported the pK_a of methadone, in water at 20 ° C, to be 8.25 (Levi L, Oestreicher PM, Farmilo CG, Nonaqueous titration of narcotics and alkaloids, *Bull. Narcotics*, **5**, 15–25 (1953))."
		8.3	U	+H			Persson BA and Schill G, Extraction of amines as complexes with inorganic anions, *Acta Pharm. Suecica*, **3**, 291–302 (1966). NB: No details are given for the measurement of what are described as "approximate acid dissociation constants."
1970	Methadone	10.12	U	+H	potentio	H_2O $t = 20.0$ I undefined	Marshall PB, Some chemical and physical properties associated with histamine antagonism, *Br. J. Pharmacol.*, **10**, 270–278 (1955). NB: Care taken to exclude CO_2. Corrections made for $[OH^-]$ concentrations in the Henderson-Hasselbalch equation. Values reported for numerous antihistamines.

1971	DL-Methadone	8.25	U	+H	potentio	50% aq EtOH	Farmilo CG, Oestreicher PM and Levi L, Physical methods for the identification of narcotics. IB Common physical constants for identification of ninety-five narcotics and related compounds, *Bull. Narcotics*, UN Dept. Social Affairs, vol. 6, pp. 7–19 (1954). CA 48:69490; NB: Cited in Beckett AH, Analgesics and their antagonists. I., *J. Pharm. Pharmacol.*, **8,** 848–859 (1956); also cited in Clouet DH (ed.), *Narcotic Drugs Biochemical Pharmacology*, Plenum Press, New York, 52–53 (1971). See Alphaprodine (no. 1549) for details.
1972	DL-Methadone	8.99	U	+H			Taylor JF, Ph.D. thesis, University of London (1968).
1973	Methamphetamine	10.1	U	+H			Craig
1974	Methamphetamine	9.5	U	+H			Borodkin S and Yunker MH, Interaction of amine drugs with a polycarboxylic acid ion-exchange resin, *J. Pharm. Sci.*, **59,** 481–486 (1970); cited in W&G.
1975	Methamphetamine	10.20	U	+H			Chatten LG and Harris LE, Relationship between $pK_b(H_2O)$ of organic compounds and $E_{1/2}$ values in several nonaqueous solvents, *Anal. Chem.*, **34,** 1495–1501 (1962). NB: The value was quoted from Kisbye J, *Pharm. Weekblad.*, **93,** 206–215 (1958).
1976	Methapyrilene ($C_{14}H_{19}N_3S$)	3.66 8.91	A U	+H +H	potentio	H_2O $t = 25$	Lordi NG and Christian JE, Physical properties and pharmacological activity: antihistaminics, *J. Am. Pharm. Assn., Sci. Edn.*, **45,** 300–305 (1956). NB: See Chlorpheniramine (no. 1704) for details.

$CH_2CH_2N(CH_3)_2$

(*continued*)

No.	Name	pK_a value(s)	Data quality	Ionization type	Method	Conditions t°C; I or c M	Comments and Reference(s)
1977	Methapyrilene	8.8	U	+H			Ritschel; Borodkin S and Yunker MH, Interaction of amine drugs with a polycarboxylic acid ion-exchange resin, *J. Pharm. Sci.*, **59**, 481–486 (1970).
1978	Methazolamide ($C_5H_8N_4O_3S_2$) 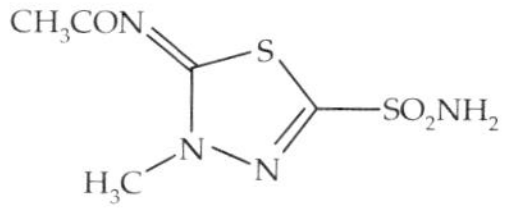	7.30	U	−H			Craig N&K; Merck 9 W&G; Merck 9
1979	Methdilazine ($C_{18}H_{20}N_2S$) 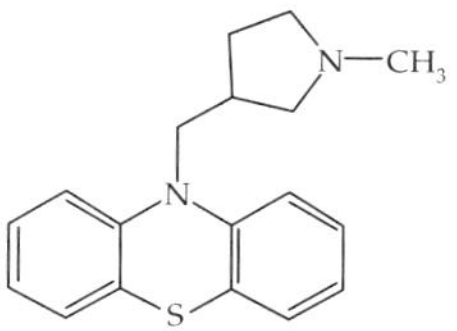	7.45	U	+H			Foye 3rd; see Azatadine. From N&K; also Hoover NB: See Pyrathiazine.
1980	Methenamine ($C_6H_{12}N_4$) 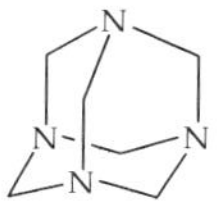	4.8	U	+H			McEvoy
1981	Methenamine	4.9	U	+H			W&G: Evstratova KI and Ivanova AI, Dissociation constants and methods for analysis of some organic bases, *Farmatsiya* (Moscow), **17(2)**, 41–45 (1968).
1982	Methenamine	6.2	U	+H			Perrin DD, Dempsey B and Serjeant E, pK_a *prediction for organic acids and bases*, Chapman and Hall, London, p. 34 (1981). NB: Also reported a predicted value for Methenamine of 6.0.

No.	Compound	pKa	Data quality	Type	Method	Conditions	Reference / Comments
1983	Methicillin	2.8	U	−H			Craig
1984	L-Methionine ($C_5H_{11}NO_2S$)	2.28 9.21	U U	−H +H			N&K; Merck 9
1985	Methopromazine (methoxypromazine) ($C_{18}H_{22}N_2OS$)	9.4	U	+H			W&G: Hulshoff and Perrin, 1976 NB: See Cyanopromazine for details.
1986	Methotrexate	3.76 4.83 5.60 ± 0.03	U U U	−H +H +H	spectro		Chamberlin AR, Cheung APK and Lim P, Methotrexate, *APDS*, **5**, 283–306, 1976. NB: Assigned to the diaminopteridinyl moiety by comparison with *p*-aminobenzoylglutamic acid (4.83 and 3.76) and 2,4-diaminopteridine (<0.5 and 5.32).
1987	Methotrexate	4.8 5.5	U U	+H +H			Liegler DG, Henderson ES, Hahn MA and Oliverio VT, Effect of organic acids on renal clearance of methotrexate in man, *Clin. Pharmacol. Ther.*, **10,** 849–857 (1969). Cited in W&G.
1988	*N*-Methyl-1-benzoyl ecgonine (NB: see comment).	8.65	U	+H	potentio	H_2O $t = 25$ $I < 0.002$	Krahl ME, Keltch AK and Clowes GHA, The role of changes in extracellular and intracellular hydrogen ion concentration in the action of local anesthetic bases, *JPET*, **68,** 330–350 (1940). Cited in Perrin Bases no. 2927 Ref. K58. This value is based on an apparent value for $pK_b' = 5.33$ at 25 °C with $pK_w = 14.00$ at $I = 0.00$ and $pK_w = 13.90$ at $I = 0.04$. NB: The name of the compound given here is the same as in the original reference and repeated by Perrin.

(continued)

No.	Name	pK_a value(s)	Data quality	Ionization type	Method	Conditions t°C; I or c M	Comments and Reference(s)
							However, this name is incorrect. Reference to the structural diagram in the original paper (Fig. 1, where there is also an error) shows that the structure intended is that of Cocaine, benzoyl methylecgonine ($C_{17}H_{21}NO_4$).
1989	Methylcaffeine ($C_9H_{12}N_4O_2$) 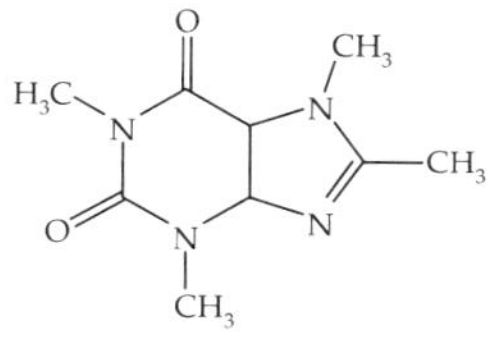	11.4 (pK_b) ($pK_a \sim 2.6$)	U	+H	potentio		Evstratova KI and Ivanova AI, Dissociation constants and methods for analysis of some organic bases, *Farmatsiya* (Moscow), **17(2)**, 41–45 (1968). NB: Cited in Perrin Bases Suppl no. 7481 Ref. E25.
1990	Methyldihydromorphinone (metopon)	8.08	U	+H	potentio	50% aq EtOH	Farmilo CG, Oestreicher PM and Levi L, Physical methods for the identification of narcotics. IB Common physical constants for identification of ninety-five narcotics and related compounds, *Bull. Narcotics*, UN Dept. Social Affairs, vol. 6, pp. 7–19 (1954); cited in Clouet DH (ed.), *Narcotic Drugs Biochemical Pharmacology*, Plenum Press, New York, 52–53 (1971).
1991	Methylephedrine ($C_{11}H_{17}NO$) 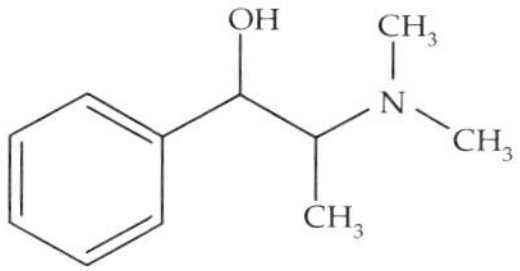	9.30	U	+H			Vree TB, Muskens ATJM and van Rossum JM, Some physicochemical properties of amphetamine and related drugs, *J. Pharm. Pharmacol.*, **21**, 774–775 (1969).

						Other sources: Kisbye J, *Pharm. Weekblad.*, **93,** 206–215 (1958). Leffler EB, Spencer HM, Burger A, Dissociation constants of adrenergic amines, *JACS*, **73,** 2611–2613 (1951). NB: There does not appear to be a structure corresponding to this name in the Leffler paper, nor is there a pK_b value.
1992	Methylethylamphetamine ($C_{12}H_{19}N$) 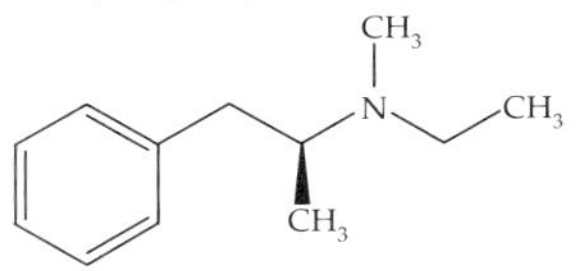	9.80	U	+H		Vree TB, Muskens ATJM and van Rossum JM, Some physicochemical properties of amphetamine and related drugs, *J. Pharm. Pharmacol.*, **21,** 774–775 (1969). NB: See Dexamphetamine for details.
1993	*N*-Methylglucamine	9.2	U	+H		W&G: Balasz L and Pungor E, Standard methylglucamine for acid titrations with high frequency end-point indication, *Mikrochim. Acta*, 309–313 (1962); CA 56:13524g.
1994	Methylisopropylamphetamine ($C_{12}H_{21}N$) 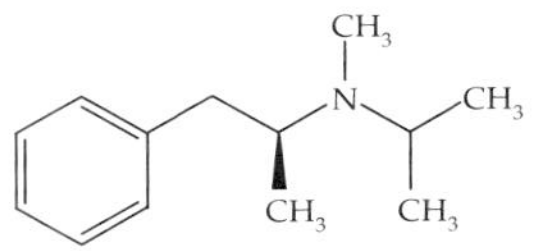	9.45	U	+H		Vree TB, Muskens ATJM and van Rossum JM, Some physicochemical properties of amphetamine and related drugs, *J. Pharm. Pharmacol.*, **21,** 774–775 (1969). NB: See Dexamphetamine for details.
1995	Methyl nicotinate ($C_7H_7NO_2$) 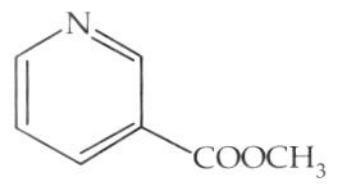	3.1	U	+H		Craig
1996	Methylparaben	8.4	U	–H	H_2O *t* undefined	W&G; Tammilehto S and Buchi J, *p*-Hydroxybenzoic acid esters (Nipagins). I. Physicochemical properties, *Pharm. Acta Helv.*, **43,** 726–738 (1968).

(*continued*)

No.	Name	pK_a value(s)	Data quality	Ionization type	Method	Conditions t°C; I or c M	Comments and Reference(s)
1997	α-Methylphenethylamine ($C_9H_{13}N$) 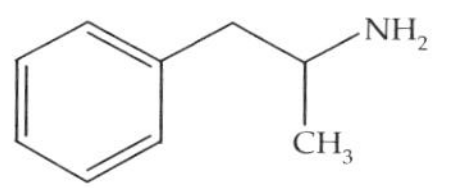	9.07	U	+H			Ritschel; Garrett ER and Chemburkar PB, Evaluation, control and prediction of drug diffusion through polymeric membranes III, *J. Pharm. Sci.*, **57,** 1401–1409 (1968).
1998	Methyl salicylate ($C_8H_8O_3$) 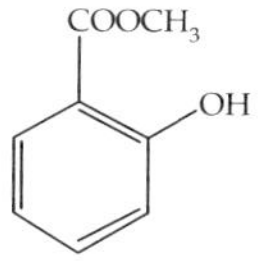	9.90	U	–H			Craig
1999	5-Methyl-2-thiouracil ($C_5H_6N_2OS$) 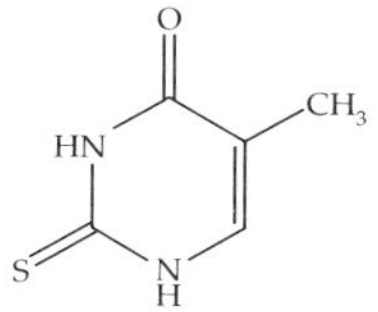	7.71	U	–H			Ritschel; Garrett ER and Weber DJ, Metal complexes of thiouracils I. Stability constants by potentiometric titration studies and structures of complexes, *J. Pharm. Sci.*, **59,** 1383–1398 (1970).
2000	6-Methylthiouracil ($C_5H_6N_2OS$) 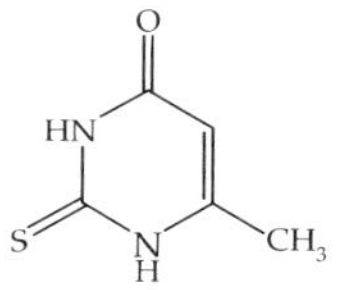	8.2	U	–H			Craig
2001	6-Methyl-2-thiouracil	7.73	U	–H			Ritschel; Garrett ER and Weber DJ, Metal complexes of thiouracils I. Stability constants by potentiometric titration studies and structures of complexes, *J. Pharm. Sci.*, **59,** 1383–1398 (1970).

2002	Methyprylon ($C_{10}H_{17}NO_2$) 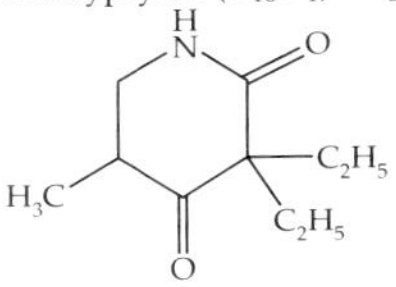	12.0	U	–H	spectro		Rudy BC and Senkowski BZ, Methyprylon, *APDS*, **2**, 363–382 (1973). Cited Lau E, Hoffmann-La Roche, Inc., unpublished data.
2003	Metolazone ($C_{16}H_{16}ClN_3O_3S$) 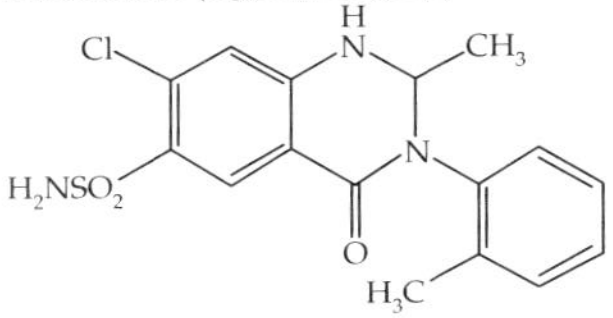	9.7	U	–H			Craig N&K; Avery
2004	Metopon ($C_{18}H_{21}NO_3$) 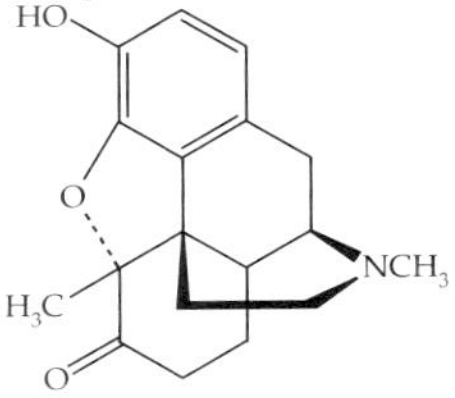	8.08	U	+H	potentio	50% aq EtOH	Farmilo CG, Oestreicher PM and Levi L, Physical methods for the identification of narcotics. IB Common physical constants for identification of ninety-five narcotics and related compounds, *Bull. Narcotics*, UN Dept. Social Affairs, vol. 6, pp. 7–19 (1954). CA 48:69490; NB: Cited in Beckett AH, Analgesics and their antagonists. I., *J. Pharm. Pharmacol.*, **8**, 848–859 (1956); See Alphaprodine for details.
2005	Metyrosine ($C_{10}H_{13}NO_3$) 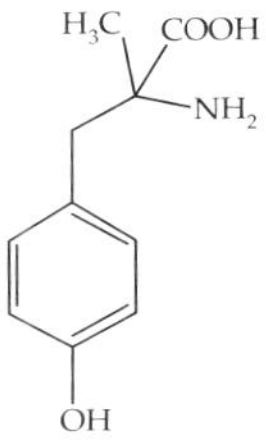	2.7 10.1	U U	–H –H			Foye

(continued)

No.	Name	pK_a value(s)	Data quality	Ionization type	Method	Conditions t°C; I or c M	Comments and Reference(s)
2006	Miconazole	6.5	U	+H			Beggs WH, Fungistatic activity of miconazole against Candida albicans in relation to pH and concentration of nonprotonated drug, *Mycoses*, **32(5)**, 239–244 (1989).
2007	Midazolam	6.2	U	+H			Craig
2008	Minocycline ($C_{23}H_{27}N_3O_7$)	2.8	U	−H			Craig
		5.0	U	−H			N&K; Avery
		7.8	U	−H			
		9.5	U	+H			
2009	Mitoxantrone ($C_{22}H_{28}N_4O_6$)	5.99	U	−H			Beijnen JH, Bult A and Underberg WJM, Mitoxantrone hydrochloride, *APDS*, **17**, 221–258, 1988. Cited Duchateau AMJA, *Pharm. Weekblad*, **122**, 286 (1987), however, this citation is unclear. "The mitoxantrone molecule contains several prototropic functions. The 1,4-hydroquinone moiety possesses acidic properties. The two nitrogen atoms attached to the tricyclic aromatic skeleton and the secondary nitrogens in the side chains can accept protons. No full characterization of the prototropic functions in mitoxantrone has been reported, so far. Two pK_a values, of 5.99 and 8.13, have been mentioned in the literature but they were not assigned to specific functions in the mitoxantrone molecule."
		8.13	U	+H			

2010	Morphine	8.02–8.05	U	+H	potentio	50% aq EtOH	Farmilo CG, Oestreicher PM and Levi L, Physical methods for the identification of narcotics. IB Common physical constants for identification of ninety-five narcotics and related compounds, *Bull. Narcotics*, UN Dept. Social Affairs, vol. 6, pp. 7–19 (1954); cited in Clouet DH (ed.), *Narcotic Drugs Biochemical Pharmacology*, Plenum Press, New York, 52–53 (1971).
2011	Morphine-N-oxide ($C_{17}H_{19}NO_4$) 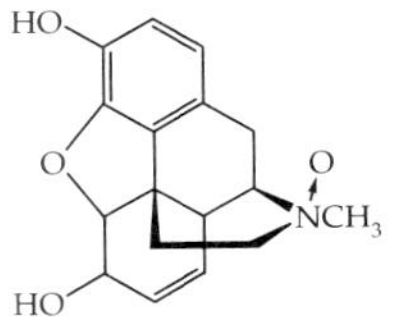	4.82	U	+H	potentio	50% aq EtOH	Farmilo CG, Oestreicher PM and Levi L, Physical methods for the identification of narcotics. IB Common physical constants for identification of ninety-five narcotics and related compounds, *Bull. Narcotics*, UN Dept. Social Affairs, vol. 6, pp. 7–19 (1954); cited in Clouet DH (ed.), *Narcotic Drugs Biochemical Pharmacology*, Plenum Press, New York, 52–53 (1971).
2012	Moxalactam ($C_{20}H_{20}N_6O_9S$) 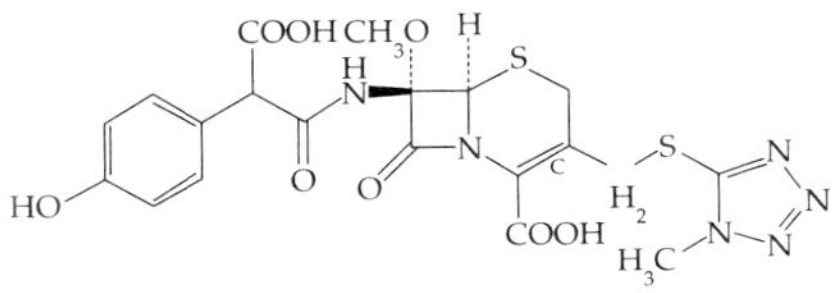	2.4 3.5 9.95	U U U	–H –H –H		H_2O	Lorenz LJ and Thomas PN, Moxalactam disodium, *APDS*, **13,** 305–329 (1984); no reference cited. NB: Gave other pK_a values in 66% DMF: COOH 1 4.9; COOH 2 6.1; Phenol 12.9
2013	Nabilone ($C_{24}H_{36}O_3$) 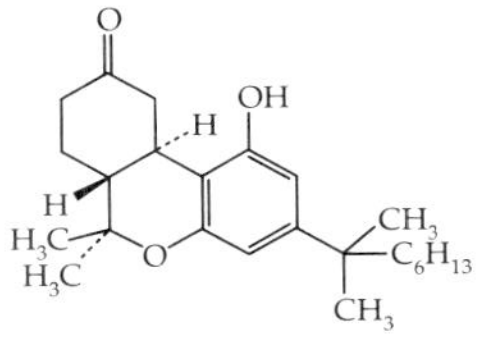	13.5	U	–H		66% DMF	Souter RW, Nabilone, *APDS*, **10,** 499–512 (1981); no ref. NB: This compound is a partly hindered phenol, however, semi-empirical MO models (Prankerd R, unpublished, 2005) do not indicate a major steric effect on the phenolic OH. The models indicate restriction in rotation around the phenolic C–O bond,

(*continued*)

No.	Name	pK_a value(s)	Data quality	Ionization type	Method	Conditions t°C; I or c M	Comments and Reference(s)
							but this should only increase the pK_a by about 0.3, due to the statistical effect. The very high observed value is likely to be a simple consequence of the partially aqueous medium. A spectrophotometric value in aqueous buffer is expected to be significantly closer to values typical for phenols, i.e., ~10.
2014	Nadolol ($C_{17}H_{27}O_4$) HO CH$_2$NHC(CH$_3$)$_3$ OH OH O	9.67	U	+H			Lombardo F, Obach RS, Shalaeva MY, Gao F, Prediction of human volume of distribution values for neutral and basic drugs. 2. Extended data set and leave-class-out statistics, *J. Med. Chem.*, **47,** 1242–1250 (2004). Ref. 275: Barbato F, Caliendo G, LaRotonda MI, Morrica P, Silipo C, Vittoria A, Relationships between octanol-water partition data, chromatographic indices and their dependence on pH in a set of beta-adrenoceptor blocking agents, *Farmaco*, **45,** 647–663 (1990). The same value was reported in Lien E, *US New Drug Digest: 1980–86*, Aurora, Nashville, 1987.
2015	Nadolol	9.4	U	+H			Hinderling PH, Schmidlin O and Seydel JK, Quantitative relationships between structure and pharmacokinetics of beta-adrenoceptor blocking agents in man, *J. Pharmacokin. Biopharm.*, **12,** 263–287 (1984).

2016	Nalbuphine ($C_{21}H_{27}NO_4$) 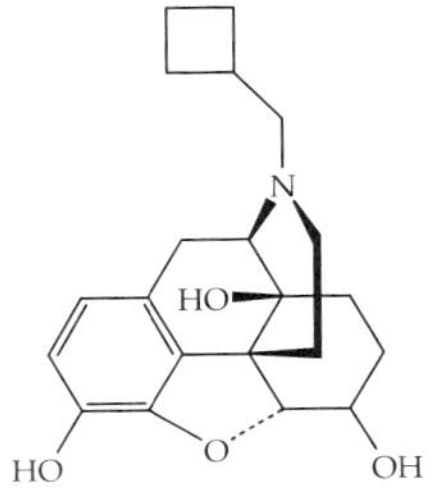	8.7 10.0	U U	+H –H			McEvoy
2017	Nalmefene ($C_{21}H_{25}NO_3$) 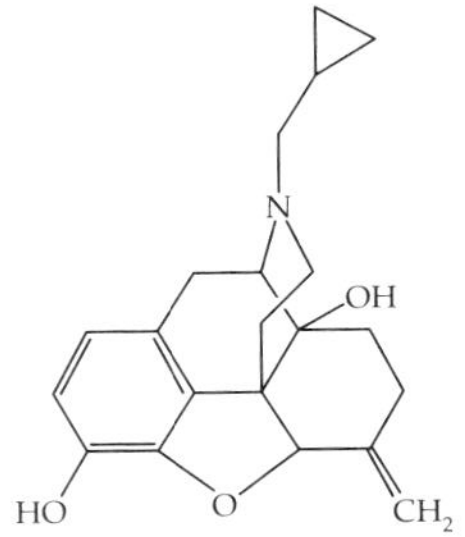	7.63	U	+H	potentio	H_2O	Brittain HG, Nalmefene Hydrochloride, *APDS*, **24**, 351–395 (1996). "Nalmefene hydrochloride contains a single ionizable group. Using an aqueous titration method, the pK_a of the compound was determined to be 7.63." NB: No reference was cited.
2018	Nalorphine	7.83	U	+H	potentio	50% aq EtOH	Farmilo CG, Oestreicher PM and Levi L, Physical methods for the identification of narcotics. IB Common physical constants for identification of ninety-five narcotics and related compounds, *Bull. Narcotics*, UN Dept. Social Affairs, vol. 6, pp. 7–19 (1954). CA 48:69490; Beckett AH, Analgesics and their antagonists. I., *J. Pharm. Pharmacol.*, **8,** 848–859 (1956); also cited in Clouet DH (ed.), *Narcotic Drugs Biochemical Pharmacology*, Plenum Press, New York, 52–53 (1971). NB: See Alphaprodine for details.

(continued)

No.	Name	pK_a value(s)	Data quality	Ionization type	Method	Conditions t°C; I or c M	Comments and Reference(s)
2019	Natamycin ($C_{33}H_{47}NO_{13}$) 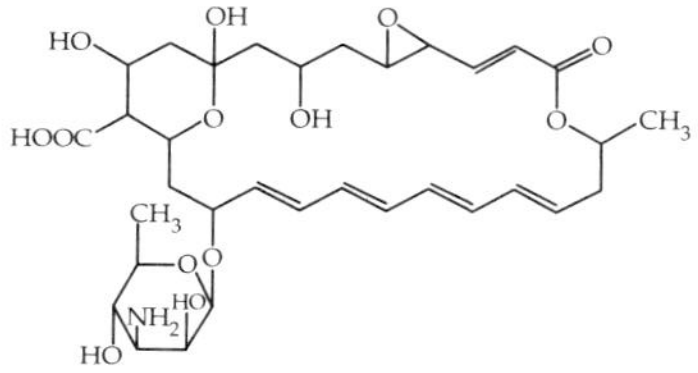	4.6 8.35	U U	−H +H		50% aq. MeOEtOH	Brik H, Natamycin, *APDS*, **10**, 513–557 (1981).
2020	Nefazodone ($C_{25}H_{32}ClN_5O_2$) 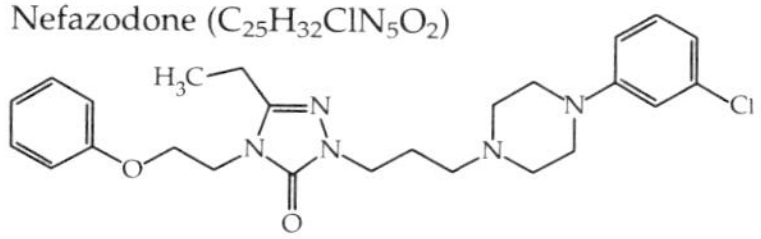	6.5	U	+H	potentio		Lombardo F, Obach RS, Shalaeva MY and Gao F, Prediction of human volume of distribution values for neutral and basic drugs. 2. Extended data set and leave-class-out statistics, *J. Med. Chem.*, **47,** 1242–1250 (2004). Ref. not given: potentiometric titration.
2021	Neostigmine ($C_{12}H_{19}N_2O_2$) 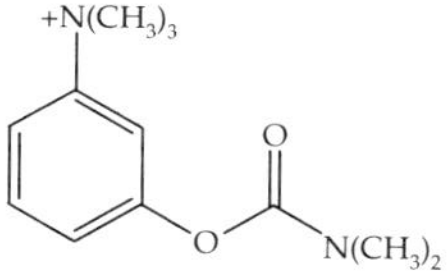	12.0	U	+H			Craig; N&K; Scott WE, Hoffman-LaRoche, personal communication, 1975. NB: This value appears to be either a false (mirage) constant, or just possibly a pK_b value for the carbamate.
2022	Nicorandil ($C_8H_9N_3O_4$) 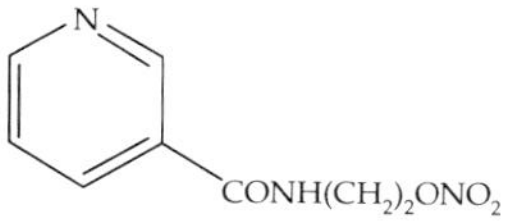		U	+H			Yanagisawa T and Hashimoto H and Taira N, Interaction of potassium channel openers and blockers in canine atrial muscle, *Br. J. Pharmacol.*, **97(3)**, 753–762 (1989). NB: This paper reported 'pK_a values', but the context showed that the equilibrium data referred to dissociation of the compound from pharmacological receptors. It should have a pK_a value similar to Nicotinamide (no. 887).

2023	Nicotine	8.5	U	+H			Ivey K and Triggs EJ, Absorption of nicotine by the human stomach and its effects on gastric ion fluxes and potential difference, *Am. J. Dig. Dis.*, **23,** 809–814 (1978). "Gastric absorption of nicotine (I), the effects of oral I intravenous I and cigarette smoking on ion fluxes and potential difference in the stomach of 13 healthy volunteers were studied. . . . It was concluded that I, a moderately strong [*sic*] base of pK_a 8.5, is best absorbed at alkaline pH in the undissociated unionized state. . . ."
2024	Nicotine	3.04 7.85	U U	+H +H			Ritschel: Parrott EL, Pharm. Technol., *Fund. Pharmaceut.* p. 220 (1970).
2025	Nicotine	3.10 8.01	U U	+H +H	potentio	H_2O $t = 25.0 \pm 0.1$	Barlow RB and Hamilton JT, Effects of pH on the activity of nicotine and nicotine mon-methiodide on the rat diaphragm preparation, *Br. J. Pharmacol.*, **18,** 543–549 (1962). The method was essentially measurement of the half-neutralization pH; it was described in *Br. J. Pharmacol.*, **18,** 543–549 (1962). The pK_a for the pyrrolidine nitrogen at 37 °C was estimated using a correction described by Albert (*Pharmacol. Rev.*, **4,** 136–167 (1952)): 8.01–12*0.022. NB: See nos. 878–879. Also reported pK_a values for numerous analogues.
		7.75	U	+H	comp	H_2O $t = 37$	
2026	Nicotine	8.07	U	+H			Ritschel; cited Beckett AH and Taylor JF, Blood concentration of pethidine and pentazocine in mother and infant at time of birth, *J. Pharm. Pharmacol.*, **19** (Suppl.), 50–52 (1967). NB: Gave the pK_a value without reference.

(*continued*)

No.	Name	pK_a value(s)	Data quality	Ionization type	Method	Conditions t°C; I or c M	Comments and Reference(s)
2027	Nicotine methiodide ($C_{11}H_{17}IN_2$) 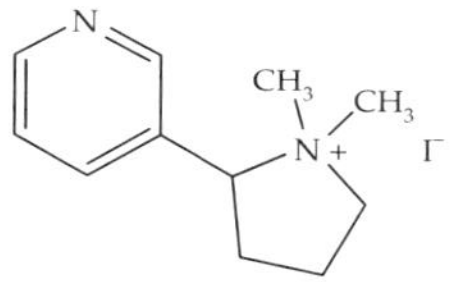	3.15	U	+H	potentio	H_2O $t = 25.0 \pm 0.1$	Barlow RB and Hamilton JT, Effects of pH on the activity of nicotine and nicotine mon-methiodide on the rat diaphragm preparation, *Br. J. Pharmacol.*, **18,** 543–549 (1962). See no. 2025 for details.
2028	Nitrofurantoin ($C_8H_6N_4O_5$) 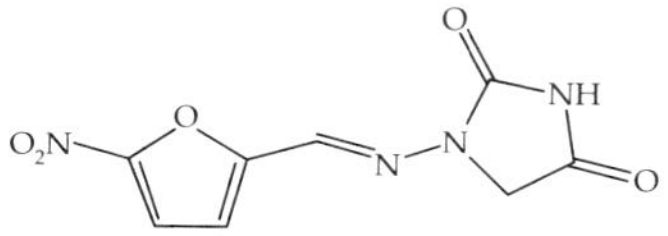	7.0	U	–H	potentio	H_2O $t = 25$ I undefined	Cadwallader DE and Jun HW, Nitrofurantoin, *APDS*, **5,** 345–373 (1976). "Michels (Norwich Pharmacal Co., personal communication) determined the pK_a of the free acid to be 7.0 using the method described by Stockton and Johnson (Stockton JR, Johnson CR, Dissociation constants of slightly soluble pharmaceuticals, *J. Am. Pharm. Assn.*, Sci. Edn., **33,** 383–384 (1944)). A pK_a of 7.2 is also reported for nitrofurantoin (*Merck Index*, 8th ed., Merck and Co., Rahway, NJ, p. 738 (1968))." NB: The Stockton and Johnson method is similar to the measurement of pH at half-neutralization, including extrapolation from mixed aqueous organic cosolvents, giving results that are quite uncertain. N&K quotes pK_a = 7.1, citing *APDS*, **5**. There should also be a pK_a value (+H) for the azomethine group.

No.	Compound	pKa	Data quality	Type	Method	Conditions	Reference
2029	Nitrofurantoin	7.2	U	−H			W&G: Buzard JA, Conklin, JD and Buller RH, Lymphatic transport of selected nitrofuran derivatives in the dog, *Am. J. Physiol.*, **201**, 492–494 (1961).
2030	Nizatidine ($C_{12}H_{21}N_5O_2S_2$) 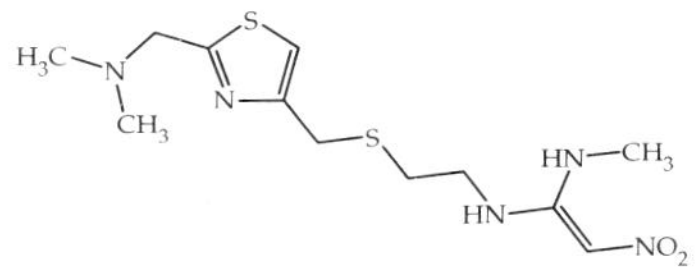	2.1 6.8	U U	+H +H		H_2O	Wozniak TJ and Nizatidine, *APDS*, **19**, 397–427 (1990). NB: No references were given
		6.3 8.4	U U	+H +H		DMF	
2031	Nizatidine	6.59	U	+H	potentio		Lombardo F, Obach RS, Shalaeva MY and Gao F, Prediction of human volume of distribution values for neutral and basic drugs. 2. Extended data set and leave-class-out statistics, *J. Med. Chem.*, **47**, 1242–1250 (2004). Ref. not given: potentiometric titration
		2.44 6.75	U U	+H +H	potentio	H_2O $t = 37$ $I = 0.15$ (KCl)	See Aspirin no. 96.
2032	Nordefrin (isoadrenaline) 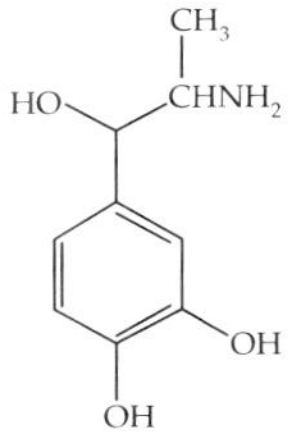	8.75 9.75	U U	+H −H			Craig
2033	L-Nordefrin (nordefrin) ($C_9H_{13}NO_3$)	8.55 9.75	U U	+H −H			Craig
2034	Norephedrine	9.55	U	+H			Vree TB, Muskens ATJM and van Rossum JM, Some physicochemical properties of amphetamine and related drugs, *J. Pharm. Pharmacol.*, **21**, 774–775 (1969). NB: See Dexamphetamine for details.

(continued)

No.	Name	pK_a value(s)	Data quality	Ionization type	Method	Conditions t°C; I or c M	Comments and Reference(s)
2035	Norepinephrine bitartrate (Levarterenol Bitartrate) ($C_8H_{11}NO_3.C_4H_6O_6$)	? 9.72 12 ? 10.3 13 ? 9.73 11.13	U U U U U U U U U	−H,+H +H,−H −H −H,+H +H,−H −H −H,+H +H,−H −H			Wilson TD, Levarterenol Bitartrate, *APDS*, **11,** 554–581 (1982). "No general agreement has been reached on the values of dissociation constants for levarterenol. ... While pK_{a3} results from the ionization of the second phenolic group, pK_{a1} and pK_{a2} as determined by titration procedures are assigned to the first phenolic and the ammonium ion or vice versa. It has been pointed out that the ionization of these two groups does not occur independently .... I is the cationic form, II is the neutral, III the zwitterionic, IV the monoanionic and V the dianionic form. The correct statements for the relation between the macro- and micro-ionization constants are: $K_{a1} = K_1 + K_2$ $1/K_{a2} = 1/K_3 + 1/K_4$
2036	Norketamine ($C_{12}H_{14}ClNO$) 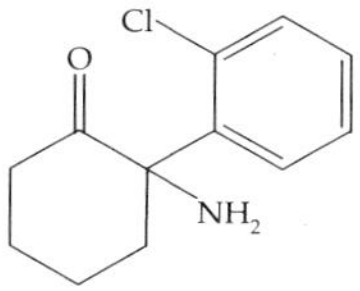	6.7	U	+H			W&G: Cohen ML and Trevor AJ, Cerebral accumulation of ketamine and the relation between metabolism of the drug and its pharmacological effects, *JPET*, **189,** 351–358 (1974).

2037	Normethadone ($C_{20}H_{25}NO$) 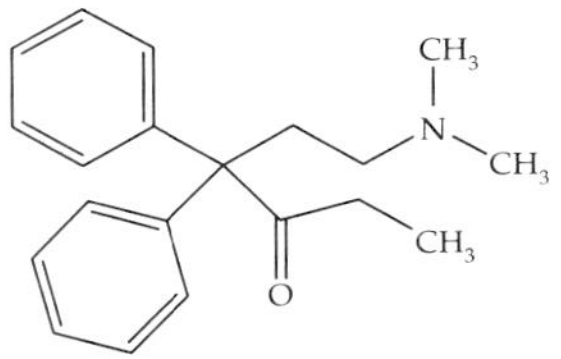	6.00	U	+H	potentio	50% aq EtOH	Farmilo CG, Oestreicher PM and Levi L, Physical methods for the identification of narcotics. IB Common physical constants for identification of ninety-five narcotics and related compounds, *Bull. Narcotics*, UN Dept. Social Affairs, vol. 6, pp. 7–19 (1954); cited in Clouet DH (ed.), *Narcotic Drugs Biochemical Pharmacology*, Plenum Press, New York, 52–53 (1971).
2038	Normethadone	8.14	U	+H			Taylor JF, Ph.D. Thesis, Univ London (1968).
2039	Norpseudoephedrine	9.40	U	+H			Vree TB, Muskens ATJM and van Rossum JM, Some physicochemical properties of amphetamine and related drugs, *J. Pharm. Pharmacol.*, **21,** 774–775 (1969). NB: See Dexamphetamine for details.
2040	Nortriptyline	10.1	U	+H	potentio		Lombardo F, Obach RS, Shalaeva MY and Gao F, Prediction of human volume of distribution values for neutral and basic drugs. 2. Extended data set and leave-class-out statistics, *J. Med. Chem.*, **47,** 1242–1250 (2004). Ref. not given: potentiometric titration.
2041	Noscapine (narcotine) ($C_{22}H_{23}NO_7$) 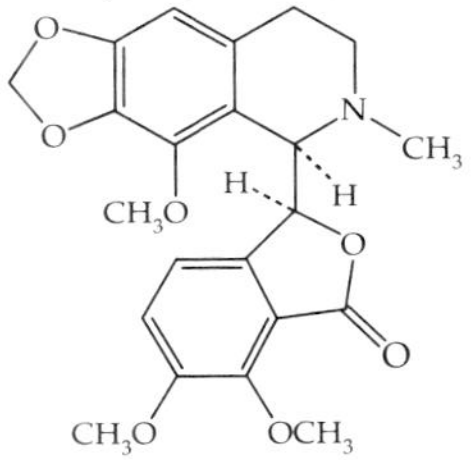	7.8	U	+H		H_2O	Al-Yahya MA and Hassan MMA, Noscapine, *APDS*, **11,** 407–456 (1982). Cited Merck 9, p. 872. This value is actually the pK_b, and probably came from Kolthoff. NB: The APDS monograph also quotes pK_a' = 4.85 (80% methylcellosolve) (Manske RHF, *The Alkaloids,* vol XII, Academic Press, NY, London, p. 396. 1970).

(*continued*)

No.	Name	pK_a value(s)	Data quality	Ionization type	Method	Conditions t°C; I or c M	Comments and Reference(s)
2042	Noscapine (narcotine)	6.2	U	+H		H_2O	W&G: Merck 9, p. 872. NB: This value is the pK_a, and has assumed measurement of the pK_b at 25 °C.
2043	Novobiocin ($C_{31}H_{36}NO_{11}$)	4.3	U	−H			Craig
		9.1	U	−H			N&K; Merck 9
							W&G; Merck 9
2044	Nystatin ($C_{47}H_{75}NO_{17}$)	5.12	U	−H	potentio	50% DMF	Michel GW, Nystatin, *APDS*, **6,** 341–421 (1977).
		8.89	U	+H			Ray-Johnson ML, Squibb International Dev. Lab., personal communication. Valenti V, The Squibb Institute for Medical Research, personal communication.

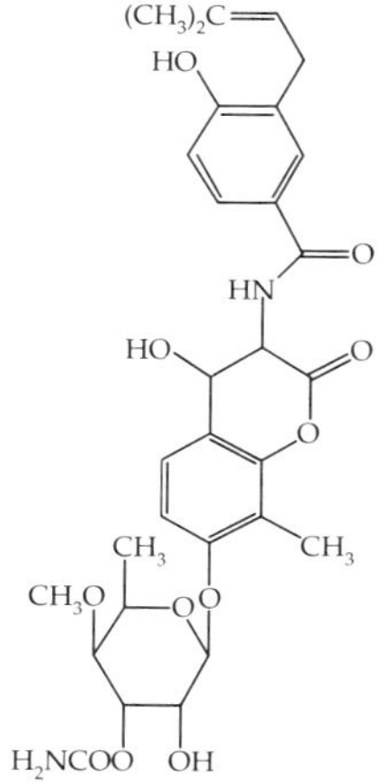

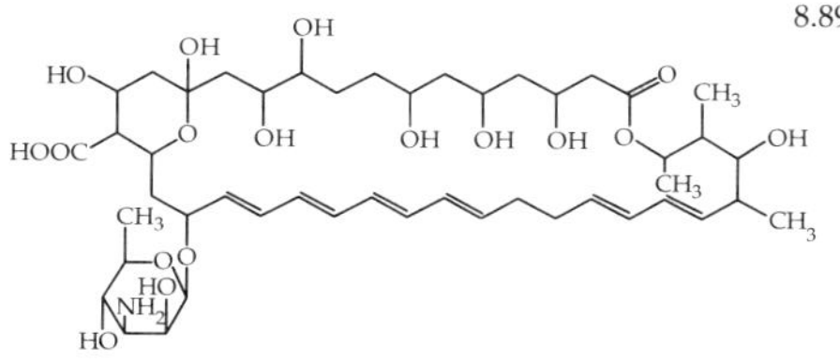

No.	Compound	pKa			Method	Conditions	Comments and Reference
		5.72 8.64	U U	−H +H	potentio	MeOH/2-methoxy-ethanol/H_2O	"Nystatin is an amphtoteric compound with a carboxyl and an amino function. Ray-Johnson determined the ionization constants of nystatin in a mixture of *N,N*-dimethylformamide/water (50:50) by direct titration and—following the general procedure of Albert and Serjeant—calculated the following pK_a values from the respective equilibrium constants: pK_1 (proton gained) = 5.12; pK_2 (proton lost) = 8.89 Valenti determined the ionization constants of nystatin in a solvent system composed of methanol, 2-methoxyethanol and water by potentiometric titration and established the following apparent pK_a values : pK_1 = 5.72; pK_2 = 8.64. The isoelectric point for nystatin in this system, calculated from the average of pK_1 and pK_2, was found to be at pH 7.18. There is, as yet, no experimental evidence to establish whether nystatin exists at the isoelectric point as a zwitterion or as an unionized molecule. Resolution of this question requires the examination of singly charged derivation of the antibiotic, such as an ester and/or suitable salt.
2045	Octopamine ($C_8H_{11}NO_2$) HO, CH_2NH_2, OH (structure)	8.88 9.53	U U	+H −H			Craig

(*continued*)

No.	Name	pK_a value(s)	Data quality	Ionization type	Method	Conditions t°C; I or c M	Comments and Reference(s)
2046	Ondansetron ($C_{18}H_9N_3O$) 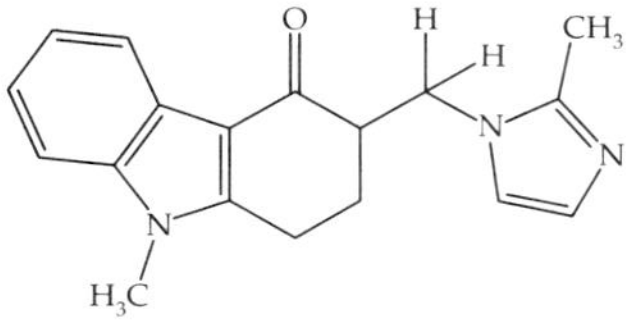	7.4	U	+H			Salem II, Lopez KJMR and Galan AC, Ondansetron Hydrochloride, *APDS*, **27**, 315 (2000). NB: No references cited.
2047	Ornidazole ($C_7H_{10}ClN_3O_3$) 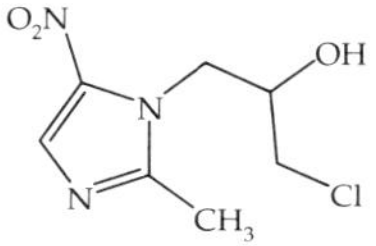	2.3	U	+H			Senkowsky BZ, *et al.*, undated personal communication cited by Schwartz DE and Jeunet F, Comparative pharmacokinetic studies of ornidazole and metronidazole in man, *Chemotherapy*, **22**, 19–29 (1976). NB: See Metronidazole.
2048	Orphenadrine ($C_{18}H_{23}NO$) 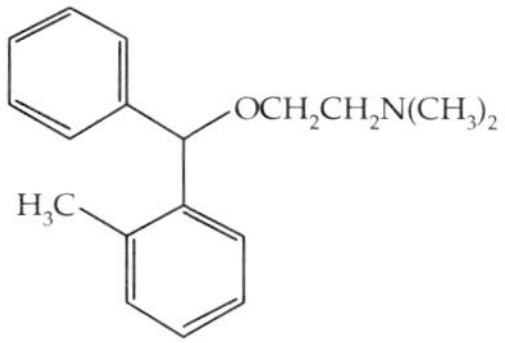	8.4	U	+H			Craig N&K; Avery
2049	Oxacillin ($C_{19}H_{19}N_3O_5S$) 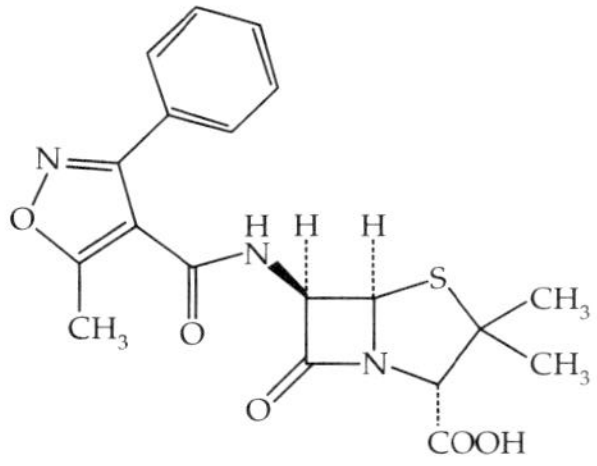	2.72	U	−H			Craig

2050	Oxacillin	2.84	U	−H	N&K; the reference cited therein is incorrect.
2051	Oxprenolol ($C_{15}H_{23}NO_3$) 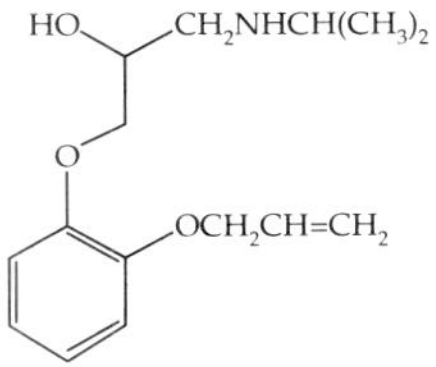 HO, $CH_2NHCH(CH_3)_2$, O, $OCH_2CH{=}CH_2$	9.5	U	+H	Hinderling PH, Schmidlin O and Seydel JK, Quantitative relationships between structure and pharmacokinetics of beta-adrenoceptor blocking agents in man, *J. Pharmacokin. Biopharm.*, **12,** 263–287 (1984).
2052	Oxybutynin ($C_{22}H_{31}NO_3$) 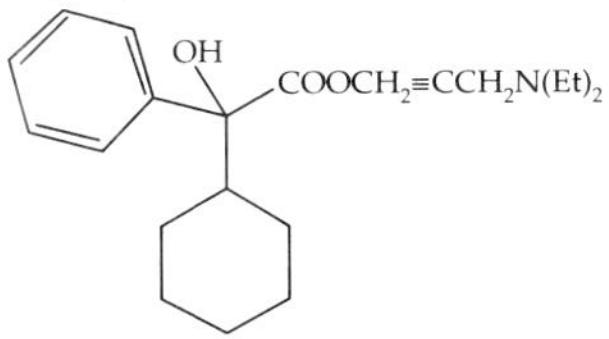 OH, $COOCH_2{\equiv}CCH_2N(Et)_2$	7.0	U	+H	McEvoy
2053	Oxycodone ($C_{18}H_{21}NO_4$) 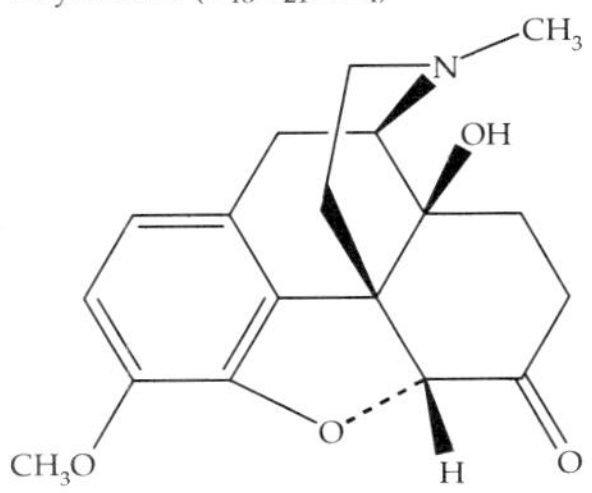 N, CH_3, OH, CH_3O, O, H, O	8.9	U	+H	Craig Bentley KW, *Chemistry of the Morphine Alkaloids*, Oxford (1954). N&K; Ritschel gave $pK_1 = 8.8–9.0$; ref. Endo Laboratories.

(continued)

No.	Name	pK_a value(s)	Data quality	Ionization type	Method	Conditions t°C; I or c M	Comments and Reference(s)
2054	Oxymorphone ($C_{17}H_{19}NO_4$) 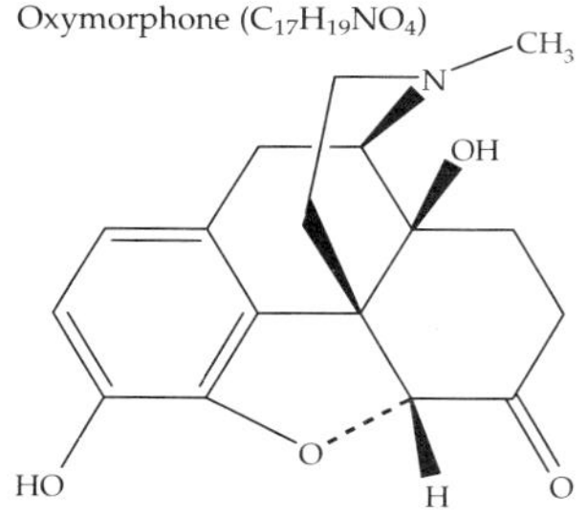	8.50 9.33	U U	+H −H			Craig N&K; Ritschel gave $pK_1 = 8.50$ and $pK_2 = 9.33$; ref. Endo Laboratories.
2055	Oxypurinol ($C_5H_4N_4O_2$) 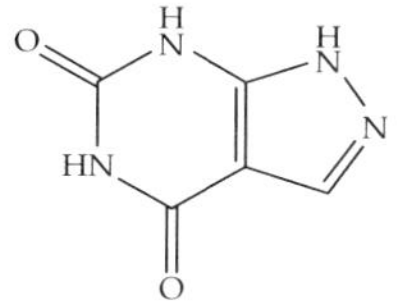	7.7	U	−H			Craig N&K; Avery
2056	Oxytocin ($C_{43}H_{66}N_{12}O_{12}S_2$) 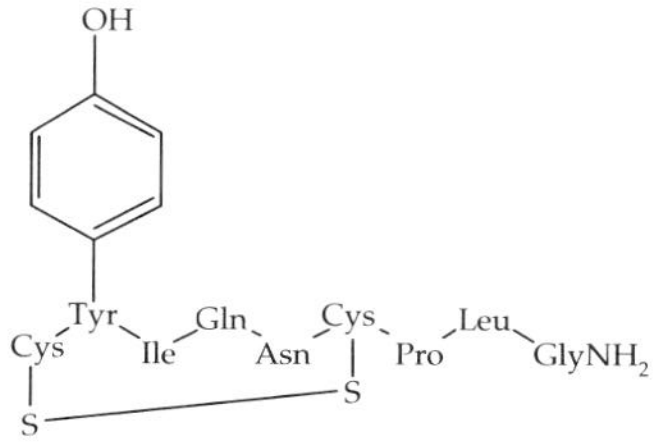	~ 6.1 ~ 10	VU VU	+H ($-NH_2$) −H (tyr)	comp		Nachtmann F, Krummen K, Maxl F and Riemer E, Oxytocin, *APDS*, **10,** 563–596 (1981). "Oxytocin is an amphoteric compound. Accordingly, the isoelectric point reported in the literature is at pH 7.7, consistent with the presence in the molecule of a free amino group and a free phenol group." NB: free amino group (pK_a ~6.1) on Cys, free phenol (pK_a ~10) on Tyr, according to ACD/pK_a Labs. Approximately consistent with the suggested pK_I.

No.	Compound	pKa	Data quality	Ionization	Method	Conditions	Reference / Notes
2057	Pamaquine (plasmoquin) ($C_{19}H_{29}N_3O$)	−1.33 3.48 10.2	U U U	+H +H +H	spectro	H_2O $t = 30$ $I = 0.1$	Craig; NB: This data appears to be that of Irvin and Irvin, 1948. See Appendix A, Quinine (no. 1219) for details.
2058	Pamaquine (plasmoquin)	8.7	U	+H			N&K; appears to be from Perrin Bases no. 1981. See also Perrin Bases nos. 1980 & 1982.
2059	Pamaquine (plasmoquin)	3.5 10.1	U U	+H +H			W&G: Christophers SR, Dissociation constants and solubilities of bases of anti-malarial compounds. I. Quinine. II. Atebrin, *Ann. Trop. Med. Parasitol.*, **31,** 43–69 (1937). CA 31:58602. NB: Results were reported as pK_b values.
2060	Pantoprazole ($C_{16}H_{15}F_2N_3O_4S$)	3.92 8.19	U U	+H −H			Badean AA, Nabulsi LN, Al-Omari MM, Daraghmeh NH, Ashour MK, Abdoh AM and Jaber AMY, Pantoprazole, *APDS*, **29,** 224 (2002). NB: Cited *Merck Index* 12.
2061	Paraxanthine ($C_7H_8N_4O_2$)	8.5	U	+H			N&K; Walther B, Carrupt P-A, El Tayar N and Testa B, 8-Substituted xanthines as phosphodiesterase inhibitors: Conformation-dependent lipophilicity and structure-activity relationships, *Helv. Chim. Acta*, **72,** 507–517 (1989). NB: See Xanthine Derivatives (no. 1513) for details.

(continued)

No.	Name	pK_a value(s)	Data quality	Ionization type	Method	Conditions t°C; I or c M	Comments and Reference(s)
2062	Pargyline ($C_{11}H_{13}N$) 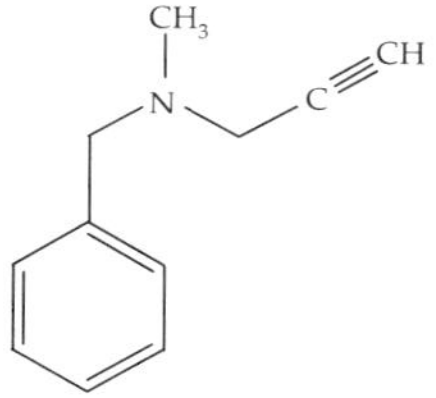	6.9	U	+H			Craig; N&K; Hoover
2063	Paroxetine ($C_{19}H_{20}FNO_3$) 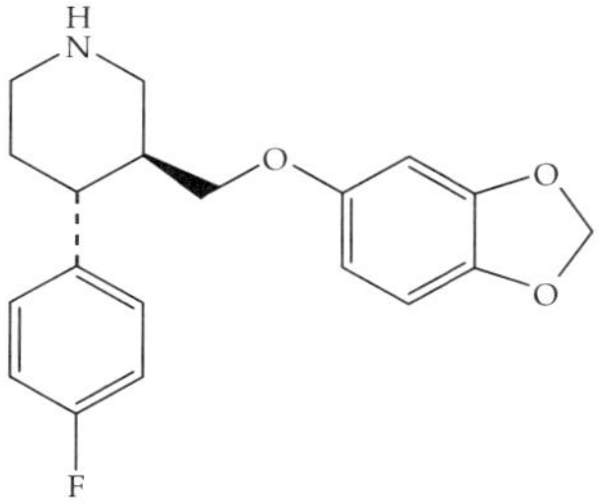	9.51	U	+H	CZE/pH		Lombardo F, Obach RS, Shalaeva MY and Gao F, Prediction of human volume of distribution values for neutral and basic drugs. 2. Extended data set and leave-class-out statistics, *J. Med. Chem.*, **47**, 1242–1250 (2004). Ref. not given: Capillary electrophoresis.
2064	Pemoline ($C_9H_8N_2O_2$) 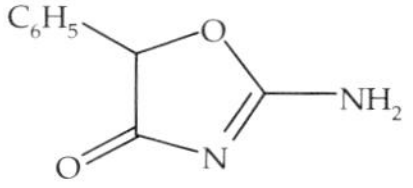	10.5	U	+H			Craig
2065	Pempidine ($C_{10}H_{21}N$)	11.25	U	+H		$t = 30$	Merck 13, no. 7207

2066	Penbutolol ($C_{18}H_{29}NO_2$) 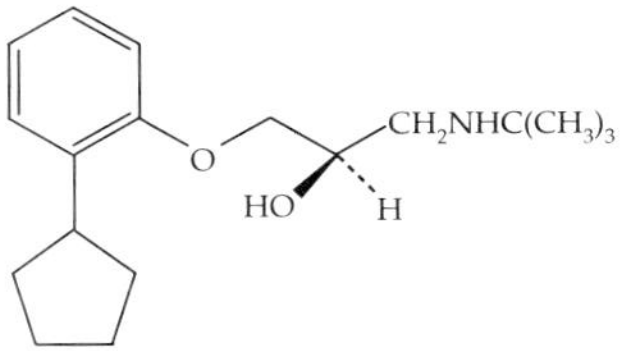	9.92 ± 0.06	U	+H	potentio	H_2O t = 25.0 I = 0.15 (KCl)	Sirius Technical Application Notes, vol. 2, p. 151 (1995). Sirius Analytical Instruments Ltd., Forest Row, East Sussex, RH18 5DW, UK. NB: From extrapolation to 0% MeOH of apparent pK_a (p$_s$$K_a$) data in 0–44 wt% MeOH by the Yasuda-Shedlovsky procedure. Concentration of analyte, 0.26–0.51 mM.
		9.26	U	+H		25–30% EtOH	Craig; also Merck 10; Hinderling PH, Schmidlin O and Seydel JK, Quantitative relationships between structure and pharmacokinetics of beta-adrenoceptor blocking agents in man, *J. Pharmacokin. Biopharm.*, **12,** 263–287 (1984); Hadju and Damm
2067	Pentamidine ($C_{19}H_{24}N_4O_2$) 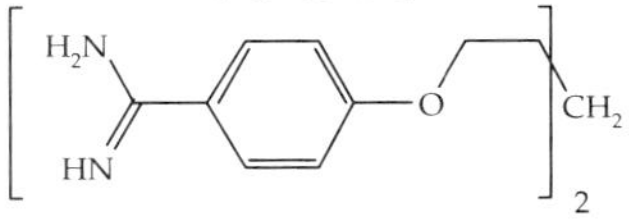	11.4	U	+H			Foye 3rd; see Azatadine from McEvoy.
2068	Pentazocine	8.76	U	+H			Clouet DH (ed.), *Narcotic drugs Biochemical Pharmacology*, Plenum, NY, 52–53 (1971); cited from Taylor JF, Ph.D. Thesis, University of London, (1968).
2069	Pentoxiphylline ($C_{13}H_{18}N_4O_3$) 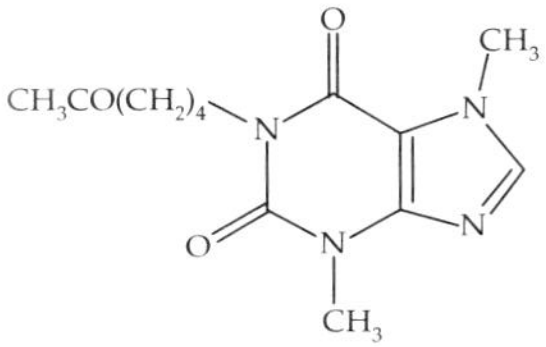	0.28	U	+H			Indrayanto G, Syahrani A, Moegihardjo, *et al.*, Pentoxifylline, *APDS*, **25,** 295–339 (1997). Cited Dollery, C, *Therapeutic Drugs*, Churchill Livingstone, Edinburgh, 50–52 (1991).
2070	Pentoxiphylline	0.3	U	+H			Craig

(continued)

No.	Name	pK_a value(s)	Data quality	Ionization type	Method	Conditions t°C; I or c M	Comments and Reference(s)
2071	Pergolide mesylate ($C_{19}H_{26}N_2S.CH_3SO_3H$) 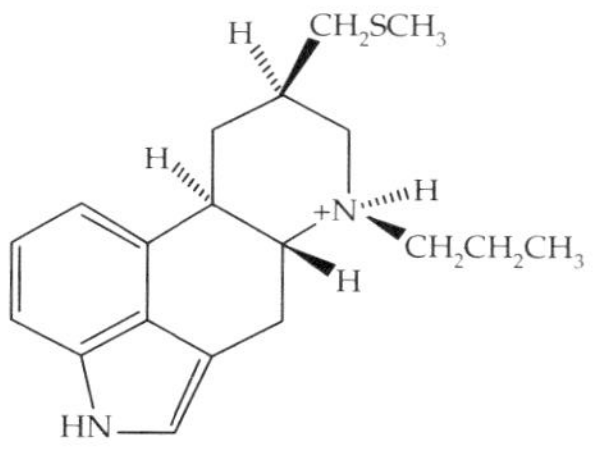	7.8	U	+H	potentio	DMF 66%	Sprankle DJ and Jensen EC, Pergolide Mesylate, *APDS*, **21**, 375–413 (1992). NB: No reference given. "The pK_a of pergolide mesylate in a 66% dimethylformamide solution was measured by potentiometric titration with 2 N potassium hydroxide. The pK_a of the secondary amine is 7.8."
2072	Perphenazine	3.70	U	+H			Craig; N&K
		7.8	U	+H			
2073	Phenacetin ($C_{10}H_{13}NO_2$) 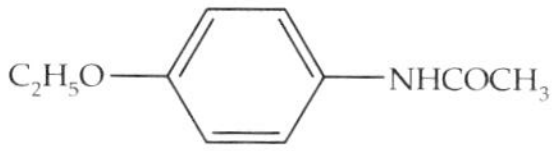	2.2	U	+H		H_2O	Craig
		3.5	U	+H		aqueous acetone	W&G: Evstratova KI, Goncharova NA and Solomko Vya, Dissociation constants of weak organic bases in acetone, *Farmatsiya* (Moscow), **17(4)**, 33–36 (1968).
2074	Phenadoxone ($C_{23}H_{29}NO_2$) 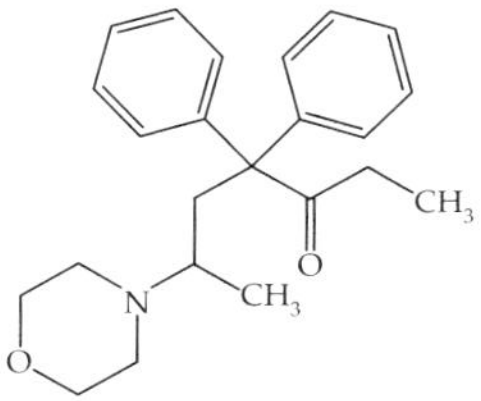	6.89	U	+H	potentio		Farmilo CG, Oestreicher PM and Levi L, Physical methods for the identification of narcotics. IB Common physical constants for identification of ninety-five narcotics and related compounds, *Bull. Narcotics*, UN Dept. Social Affairs, vol. 6, pp. 7–19 (1954). CA 48:69490; cited in Beckett AH, Analgesics and their antagonists. I., *J. Pharm. Pharmacol.*, **8**, 848–859 (1956). NB: See Alphaprodine for details.
2075	DL-phenadoxone	6.89	U	+H	potentio	50% aq EtOH	Farmilo CG, Oestreicher PM and Levi L, Physical methods for the identification of narcotics. IB

							Common physical constants for identification of ninety-five narcotics and related compounds, *Bull. Narcotics*, U.N. Dept. Social Affairs, vol. 6, pp. 7–19 (1954). CA 48:69490; cited in Clouet DH (ed.), *Narcotic Drugs Biochemical Pharmacology*, Plenum Press, New York, 52–53 (1971).
2076	DL-phenadoxone	6.75	U	+H			Taylor JF, Ph.D. Thesis, University of London (1968).
2077	Phencyclidine ($C_{17}H_{25}N$) 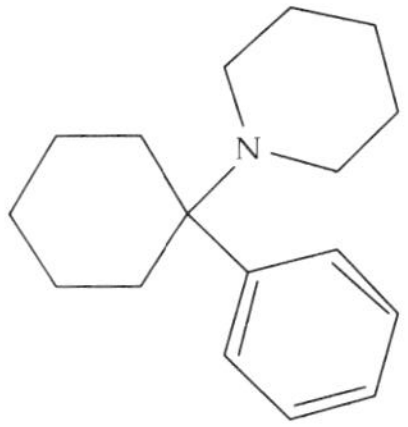	8.5	U	+H			Craig
2078	Phendimetrazine ($C_{12}H_{17}NO$) 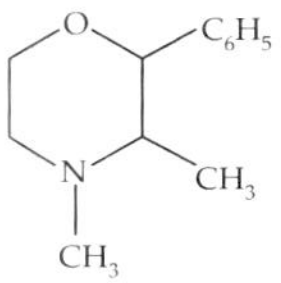	7.55	U	+H			Vree TB, Muskens ATJM and van Rossum JM, Some physicochemical properties of amphetamine and related drugs, *J. Pharm. Pharmacol.*, **21,** 774–775 (1969). NB: see Dexamphetamine for details.
2079	Phenethicillin	2.7	U	−H			Craig; N&K
2080	Phenindamine	8.29	U	+H	potentio	H_2O $t = 25$ $c = 0.002$ to 0.01	Craig; N&K; Lordi NG and Christian JE, Physical properties and pharmacological activity: Antihistaminics, *J. Am. Pharm. Ass., Sci. Edn.*, **45,** 300–305 (1956). See Chlorpheniramine (no. 1704) for details.
2081	Pheniramine	4.2	U	+H			Craig
		9.3	U	+H			Tolstoouhov AV, *Ionic interpretation of drug action in chemotherapeutic research,*
		9.27	U	+H			

(continued)

No.	Name	pK_a value(s)	Data quality	Ionization type	Method	Conditions t°C; I or c M	Comments and Reference(s)
							Chemical Rubber Publishing Co., NY, pp. 82–83 (1955); cited in Chatten LG (ed.), *Pharmaceutical Chemistry*, vol. 1, Dekker, New York, pp. 85–87 (1966). NB: See Chlorpheniramine (no. 1704) for comments noted previously about accuracy.
2082	Pheniramine	9.3	U	+H	potentio	H_2O $t = 25$ $c = 0.002$ to 0.01	W&G: Lordi NG and Christian JE, Physical properties and pharmacological activity: Antihistaminics, *J. Am. Pharm. Ass.*, Sci. Edn., **45,** 300–305 (1956). See Chlorpheniramine (no. 1704) for details.
2083	Pheniramine	9.27	U	+H			Foye 3rd; see Azatadine from N&K.
2084	Phenmetrazine ($C_{11}H_{15}NO$) 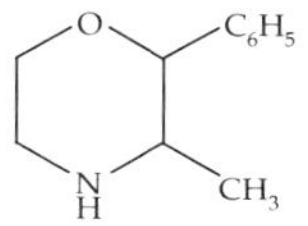	8.45	U	+H			Vree TB, Muskens ATJM and van Rossum JM, Some physicochemical properties of amphetamine and related drugs, *J. Pharm. Pharmacol.*, **21,** 774–775 (1969). NB: See Dexamphetamine for details.
2085	Phenmetrazine	8.4	U	+H			Craig
		8.5	U	+H			N&K; Vree TB, Muskens ATJM and van Rossum JM, Some physicochemical properties of amphetamine and related drugs, *J. Pharm. Pharmacol.*, **21,** 774–775 (1969); see Benzphetamine.
2086	Phenolsulphonphthalein (phenol red)	8.08	U	−H			Craig
		7.9	U	−H			N&K; Merck 9

2087	Phenomorphan ($C_{24}H_{29}NO$) 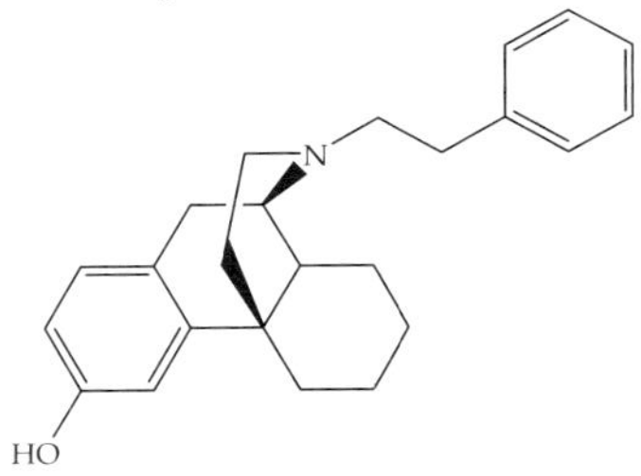 HO	7.30	U	+H	potentio	50% aq EtOH	Farmilo CG, Oestreicher PM and Levi L, Physical methods for the identification of narcotics. IB Common physical constants for identification of ninety-five narcotics and related compounds, *Bull. Narcotics*, UN Dept. Social Affairs, vol. 6, pp. 7–19 (1954); cited in Clouet DH (ed.), *Narcotic Drugs Biochemical Pharmacology*, Plenum Press, New York, 52–53 (1971).
2088	Phenoxybenzamine ($C_{18}H_{22}ClNO$) 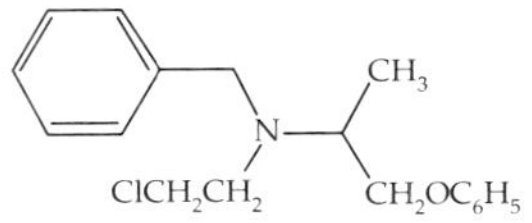 CH_3 N $ClCH_2CH_2$ $CH_2OC_6H_5$	4.4	U	+H			McEvoy
2089	Phenoxypropazine ($C_9H_{14}N_2O$) 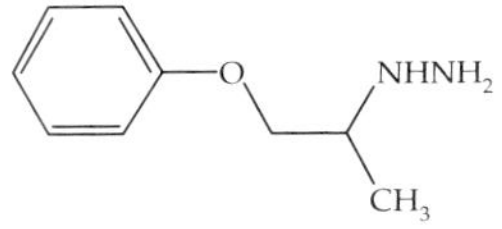 O $NHNH_2$ CH_3	6.9	U	+H			Craig N&K; Ritschel gave 6.9; ref. *Hager's Handbuch*, vol. II, *Wirkstoffgruppen* II, p. 407.
2090	Phentermine ($C_{10}H_{15}N$) 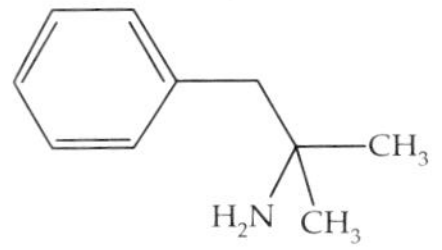 CH_3 H_2N CH_3	10.11	U	+H			Vree TB, Muskens ATJM and van Rossum JM, Some physicochemical properties of amphetamine and related drugs, *J. Pharm. Pharmacol.*, **21**, 774–775 (1969). NB: See Dexamphetamine for details.
2091	Phentermine	10.1	U	+H			Craig; N&K; Vree TB, Muskens ATJM and van Rossum JM, Some physicochemical properties of amphetamine and related drugs, *J. Pharm. Pharmacol.*, **21**, 774–775 (1969); see Benzphetamine.

(continued)

No.	Name	pK_a value(s)	Data quality	Ionization type	Method	Conditions t°C; I or c M	Comments and Reference(s)
2092	Phentolamine ($C_{17}H_{19}N_3O$) 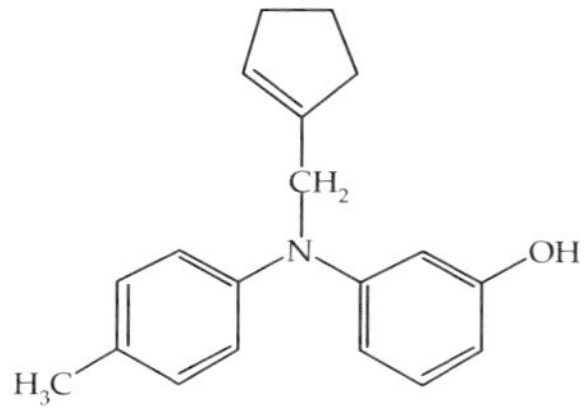	7.7	U	+H			Craig N&K; Ritschel gave 7.7 (apparent); ref., Robson JM and Stacey RS, *Recent advances in pharmacology*, 4th Edn., Little Brown and Co., Boston, p. 108 (1968).
2093	Phenylbutazone	4.89	U	–H		50% aqueous EtOH	Ali SL, Phenylbutazone, *APDS*, **11,** 483–518 (1982). This value is from Wallenfels K and Sund H, The mechanism of action of 3,5-dioxo-1,2-diphenyl-4-butylpyrazolidine, *Arzneim.-Forsch.*, **9,** 81–89 (1959). "Phenylbutazone is considered a carbon acid . and the pK_a value between 4.5–4.7 has also been given (7, 11). pK_a values in pure methanol, ethanol and water are given as 5.42, 5.76, and 5.07 respectively (8)." 7. Maulding HV and Zoglio MA, *J. Pharm. Sci.*, **60,** 309–311 (1971). 8. El-Fatatry HM, Sharafel-Deen MMK and Amer MM, *Pharmazie*, **34,** 155–157 (1979). 11. Girod E, Delley R and Häfliger F, *Helv. Chim. Acta*, **40,** 408–428 (1957) (see no. 1009).
2094	Phenylbutazone, isopropyl analogue	5.5	U	–H			Dayton PG, Berger L, Yu TF, Sicam LE, Landrau MA and Gutman AG, Relationship between pK_a and renal excretion of various phenylbutazone analogues, *Fed. Proc.*, **18,** 382 (1959).

No.	Compound	pKa	Class	Type	Method	Conditions	Reference / Comments
							NB: Conference abstract with no experimental details. Appears to largely repeat the data given under Phenylbutazone, no. 1018. Cited in W&G.
2095	Phenylephrine	8.9	U	−H, +H +H, −H			Wagner J, Grill H and Henschler D, Prodrugs of etilefrine: Synthesis and evaluation of 3′-(O-acyl) derivatives, *J. Pharm. Sci.*, **69**(12), 1423–1427 (1980). "Phenylethanolamines in solution represent a mixture of the uncharged form and ionic species (cation, anion, and zwitterion). At half-neutralization, these compounds are in apparent average pK_a value. The pK_a value is dependent on the substituent at the amino nitrogen and is increased from 8.67 (norfenefrine) to 8.9 (phenylephrine) or 9.0 (etilefrine) for the 3′-hydroxyphenylethanolamines when introducing a methyl or ethyl radical into the amino group."
2096	Phenylephrine	8.77 9.84	U U	−H, +H +H, −H			Gaglia CA Jr, Phenylephrine hydrochloride, *APDS*, **3,** 481–512, 1974. Cited Riegelman S, Strait LA and Fischer EZ, Acid dissociation constants of phenylalkanolamines, *J. Pharm. Sci.*, **51,** 129–133 (1962).
2097	Phenylethylamine	9.86	U	+H	potentio	H_2O $I = 0.1$	Lewis GP, The importance of ionization in the activity of sympathomimetic amines, *Br. J. Pharmacol.*, **9,** 488–493 (1954).
2098	Phenylethylamine	9.88	U	+H			Vree TB, Muskens ATJM and van Rossum JM, Some physicochemical properties of amphetamine and related drugs, *J. Pharm. Pharmacol.*, **21,** 774–775 (1969). NB: Cited

(continued)

No.	Name	pK_a value(s)	Data quality	Ionization type	Method	Conditions t°C; I or c M	Comments and Reference(s)
							Leffler EB, Spencer HM and Burger A, Dissociation constants of adrenergic amines, *JACS*, **73,** 2611–2613 (1951); this value does not correspond.
2099	DL-Phenyllactic acid ($C_9H_{10}O_3$) 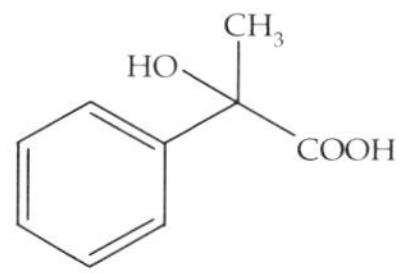	3.80	U	−H	potentio	H_2O $t = 25$	Randinitis EJ, Barr M, Wormser HC and Nagwekar JB, Kinetics of urinary excretion of D-(-)-mandelic acid and its homologs. I. Mutual inhibitory effect of D-(-)-mandelic acid and its certain homologs on their renal tubular secretion in rats, *J. Pharm. Sci.*, **59,** 806–812 (1970). NB: See Mandelic acid (no. 762) for details.
2100	Phenylpropanolamine	9.4	U	+H			Ritschel; Borodkin S and Yunker MH, Interaction of amine drugs with a polycarboxylic acid ion-exchange resin, *J. Pharm. Sci.*, **59,** 481–486 (1970).
2101	Phenylpropanolamine	9.4	U	+H			W&G: Lewis GP, The importance of ionization in the activity of sympathomimetic amines, *Br. J. Pharmacol.*, **9,** 488–493 (1954).
2102	Phenyltoloxamine ($C_{17}H_{21}NO$) 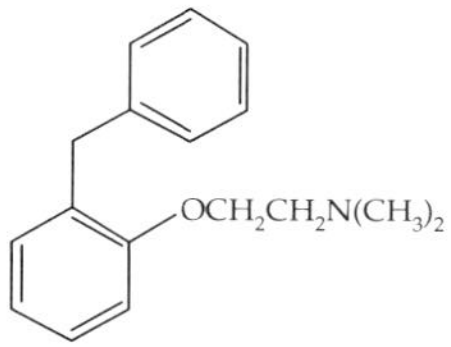	9.1	U	+H			Craig N&K; Ritschel gave 9.1; ref. Endo Laboratories, Garden City, NY, 11530.

2103	Pholcodine ($C_{23}H_{30}N_2O_4$) 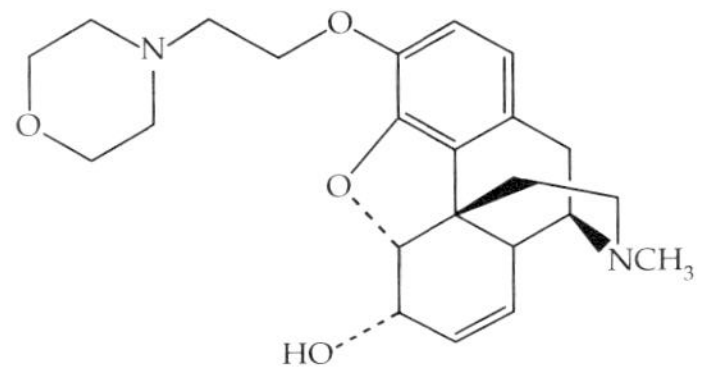	5.30	U	+H	potentio	50% aq EtOH	Farmilo CG, Oestreicher PM and Levi L, Physical methods for the identification of narcotics. IB Common physical constants for identification of ninety-five narcotics and related compounds, *Bull. Narcotics*, UN Dept. Social Affairs, vol. 6, pp. 7–19 (1954); cited in Clouet DH (ed.), *Narcotic Drugs Biochemical Pharmacology*, Plenum Press, New York, 52–53 (1971).
2104	Phthalimide ($C_8H_5NO_2$) 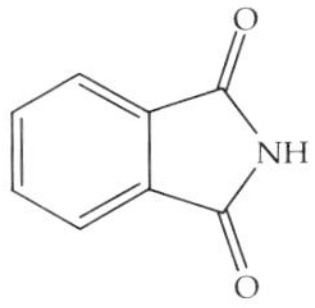	7.4	VU	−H	cond	H_2O $t = 25$ $c = 0.0022–0.0018$	Kendall J, Electrical conductivity and ionization constants of weak electrolytes in aqueous solution, *in* Washburn EW, Editor-in-Chief, *International Critical Tables*, vol. 6, McGraw-Hill, NY, 259–304 (1929). NB: Other value: $t = 18$, 1.86.
2105	Physostigmine salicylate	6.12 12.24	U U	+H +H			Muhtadi FJ and El-Hawary SS, Physostigmine Salicylate, *APDS*, **18,** 289–350 (1989). NB: These are clearly pK_b values.
2106	Piminodine ($C_{23}H_{30}N_2O_2$) 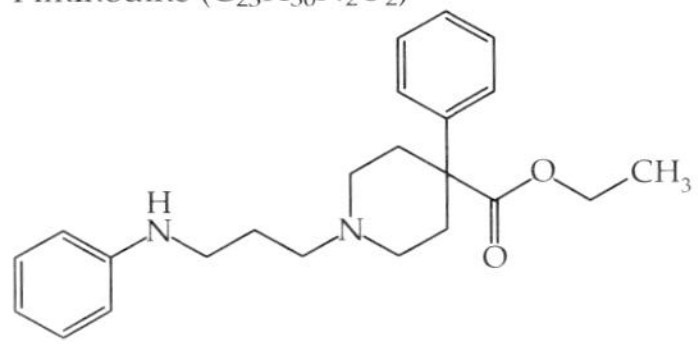	6.90	U	+H	potentio	50% aq EtOH	Clouet DH (ed.), *Narcotic Drugs Biochemical Pharmacology*, Plenum Press, New York, 52–53 (1971) cited from Farmilo CG, Oestreicher PM and Levi L, Physical methods for the identification of narcotics. IB Common physical constants for identification of ninety-five narcotics and related compounds, *Bull. Narcotics*, UN Dept. Social Affairs, vol. 6, pp. 7–19 (1954).

(continued)

No.	Name	pK_a value(s)	Data quality	Ionization type	Method	Conditions t°C; I or c M	Comments and Reference(s)
2107	Pimozide ($C_{28}H_{29}F_2N_3O$) 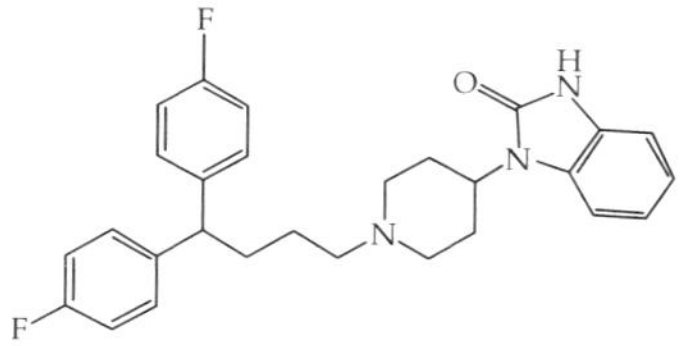	7.3 8.6	U U	+H +H			Craig
2108	Pimozide	7.32	U	+H			N&K; Merck 9
2109	Pinacidil ($C_{13}H_{19}N_5$) 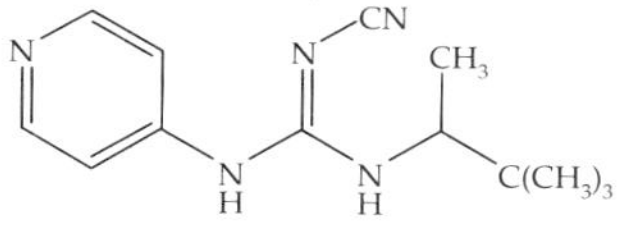		U	+H			Yanagisawa T, Hashimoto H and Taira N, Interaction of potassium channel openers and blockers in canine atrial muscle, *Br. J. Pharmacol.*, **97(3)**, 753–762 (1989). NB: This paper reported 'pK_a values', but the context showed that the data referred to dissociation of the compound from receptors. It should have pK_a values similar to 4-aminopyridine (pK_{a2} = 9.11; Perrin Bases no. 1027).
2110	Pinazepam ($C_{18}H_{13}ClN_2O$) 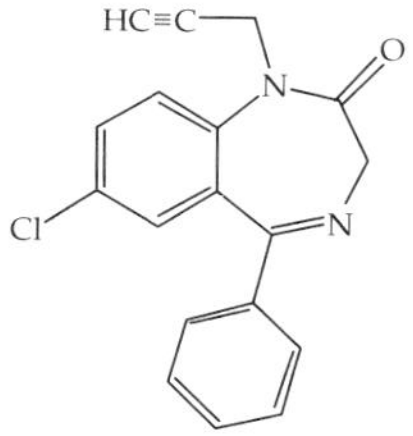	2.34	U	+H	spectro (λ = 282 nm)	H_2O *t* undefined *I* undefined	Filippi G and Trebbi A, Physicochemical profile of a new tranquilizer: Pinazepam, *Boll. Chim. Farm.*, **118**, 105–114 (1979). "The . . . physicochemical properties of pinazepam (Domar; *I*), are given. The compound has a pK_a of 2.34 and exhibits a UV absorption band at 238 and 288 nm. . . . ." NB: The final measurements were made in buffer solutions that were separated by steps of 0.3 pH unit. No activity corrections were made.

2111	Pindolol ($C_{14}H_{20}N_2O_2$) HO CH$_2$NHCH(CH$_3$)$_2$ 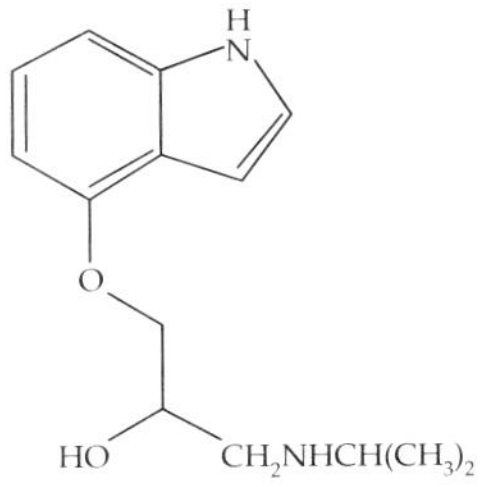	9.26	U	+H			Taylor EA, Jefferson D and Carroll JD, Turner P, Cerebrospinal fluid concentrations of propranolol, pindolol, and atenolol in man: evidence for central actions of beta-adrenoceptor antagonists, *Br. J. Clin. Pharmacol.*, **12,** 549–559 (1981). "The cerebrospinal fluid concentration of propranolol (lipid soluble) and pindolol (moderately lipid soluble) was proportional to the free plasma concentration .... It was confirmed that lipid solubility, pK_a, and the extent of plasma protein binding govern the extent and rate of penetration of a drug into cerebrospinal fluid." NB: No physicochemical data were measured in this study and the source of the pK_a values was not stated explicitly. However, log P values were reported from "ICI internal data", so presumably the pK_a values were the same.
2112	Pindolol	8.8	U	+H			Craig; N&K; Avery; Hinderling PH, Schmidlin O and Seydel JK, Quantitative relationships between structure and pharmacokinetics of beta-adrenoceptor blocking agents in man, *J. Pharmacokin. Biopharm.*, **12,** 263–287 (1984).
		9.54 ± 0.01	U	+H	potentio	H_2O $t = 25.0$ $I = 0.15$ (KCl)	Sirius Technical Application Notes, vol. 2, pp. 79–80 (1995). Sirius Analytical Instruments Ltd., Forest Row, East Sussex, RH18 5DW, UK. NB: Analyte concentration, 0.54–0.56 mM.

(continued)

Appendix B *(continued)*

No.	Name	pK_a value(s)	Data quality	Ionization type	Method	Conditions t°C; I or c M	Comments and Reference(s)
2113	Pindolol	9.54	U	+H			Lombardo F, Obach RS, Shalaeva MY and Gao F, Prediction of human volume of distribution values for neutral and basic drugs. 2. Extended data set and leave-class-out statistics, *J. Med. Chem.*, **47**, 1242–1250 (2004). Ref. 275: Barbato F, Caliendo G, LaRotonda MI, Morrica P, Silipo C and Vittoria A, Relationships between octanol-water partition data, chromatographic indices and their dependence on pH in a set of beta-adrenoceptor blocking agents, *Farmaco*, **45**, 647–663 (1990).
2114	Pipecuronium bromide (pipecurium bromide) ($C_{35}H_{62}Br_2N_4O_4$) 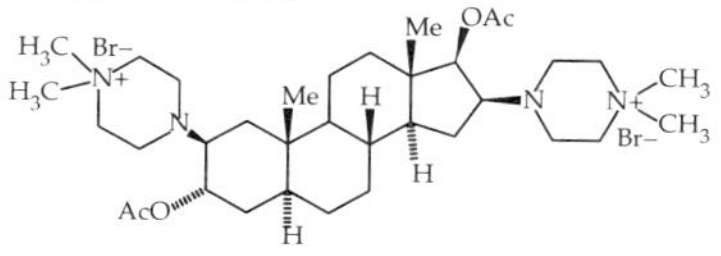	2.86 3.65	U U	+H +H	polarimetry		Bertha-Somodi Z and Pap Sziklary Z, Determination of the protonation constants of pipecuronium bromide on the basis of optical rotation as function of pH values, *Acta Pharm. Hung.*, **62**(3), 111–114 (1992). NB: Values assigned based on comparisons with the monopiperazino model compounds.
2115	Piperazine	5.7 10.0	U U	+H +H			W&G: Kolthoff IM, The dissociation constants, solubility product and titration of alkaloids, *Biochem. Z.*, **162**, 289–353 (1925).

No.	Compound	pKa			Method	Conditions	Reference / Comments
2116	6-Piperidino-4,4-diphenylheptan-3-one (dipipanone) ($C_{24}H_{31}NO$) 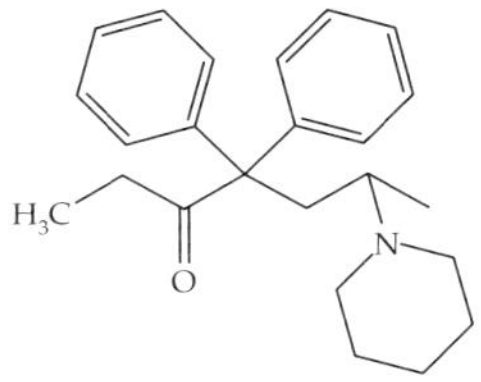	6.8	U	+H	potentio		Farmilo CG, Oestreicher PM and Levi L, Physical methods for the identification of narcotics. IB Common physical constants for identification of ninety-five narcotics and related compounds, *Bull. Narcotics*, UN Dept. Social Affairs, vol. 6, pp. 7–19 (1954). CA 48:69490; Beckett AH, Analgesics and their antagonists. I., *J. Pharm. Pharmacol.*, **8,** 848–859 (1956); See Alphaprodine for details. NB: See separate Dipipanone entry. This value is probably the pK_b value.
2117	Pipradol ($C_{18}H_{21}NO$) 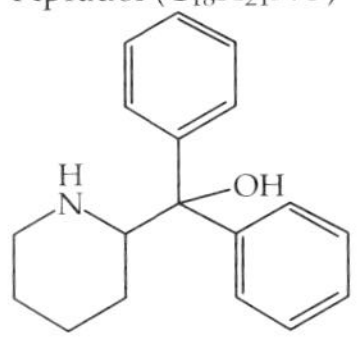	9.71	U	+H			Craig
2118	Pirbuterol	3.0 7.0 10.3	U U U	+H –H +H	potentio	H_2O $t = 21$ $c = 0.00025$	Bansal PC and Monkhouse DC, Stability of aqueous solutions of pirbuterol, *J. Pharm. Sci.*, **66,** 819–823 (1977). NB: Low concentrations indicate limited ability of the pH meter to discriminate changes in pH, however, careful exclusion of CO_2 should minimize this problem. The pK_a values were assigned by comparison with model compounds. Also reported values at 90 °C: 2.8, 7.1, 9.2. See also N&K; Craig.

(continued)

No.	Name	pK_a value(s)	Data quality	Ionization type	Method	Conditions t°C; I or c M	Comments and Reference(s)
2119	Pivampicillin ($C_{22}H_{29}N_3O_6S$) 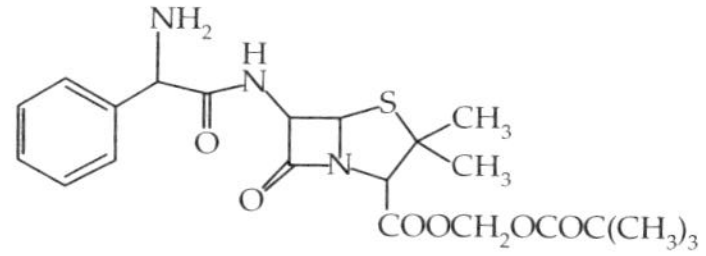	7.0	U	−H			Craig
2120	Pizotyline ($C_{19}H_{21}NS$) 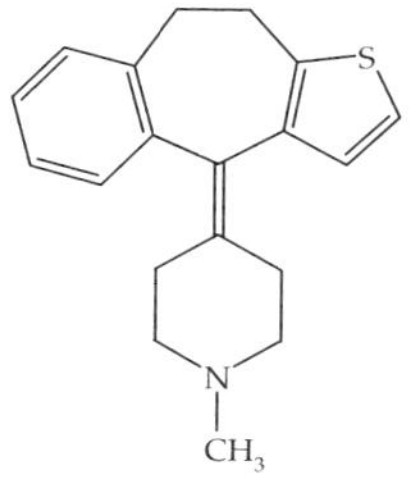	6.95	U	+H			N&K; Avery
2121	Polymyxin B–a mixture of polymyxin B_1 ($C_{56}H_{98}N_{16}O_{13}$) and polymyxin B_2 ($C_{55}H_{96}N_{16}O_{13}$)	8.9	U	+H			N&K; Avery NB: This pK_a value is for an average of five n-propylamine groups.
2122	Practolol ($C_{14}H_{22}N_2O_3$) 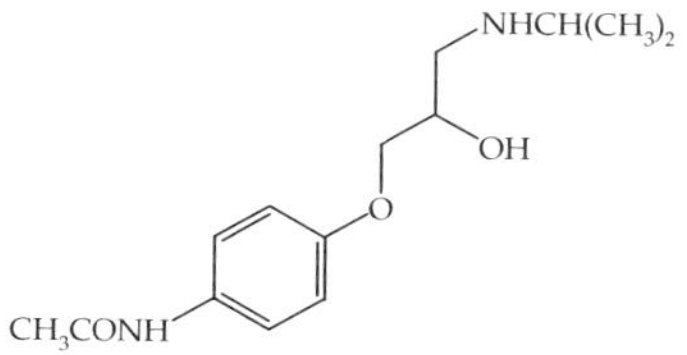	9.5	U	+H			Craig; N&K; Avery Hinderling PH, Schmidlin O and Seydel JK, Quantitative relationships between structure and pharmacokinetics of beta-adrenoceptor blocking agents in man, *J. Pharmacokin. Biopharm.*, **12**, 263–287 (1984).

2123	Pramoxine ($C_{17}H_{27}NO_3$) 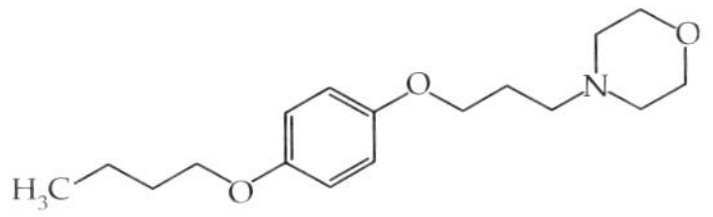	6.24	U	+H			Craig
2124	Prazepam ($C_{19}H_{17}ClN_2O$) 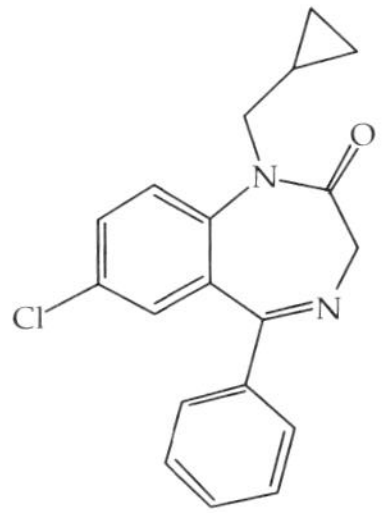	2.94	U	+H			Foye 3rd; see Azatadine from McEvoy.
2125	Prazepam	2.99	U	+H	spectro		Doi T, Okajima A, Ohkawa Y, Yoneda M and Nagai H, Physico-chemical properties and stabilities of prazepam, *Iyakuhin Kenkyu*, **9,** 205–215 (1978); cited in CA 88:141601a.
2126	Prazosin ($C_{19}H_{21}N_5O_4$) 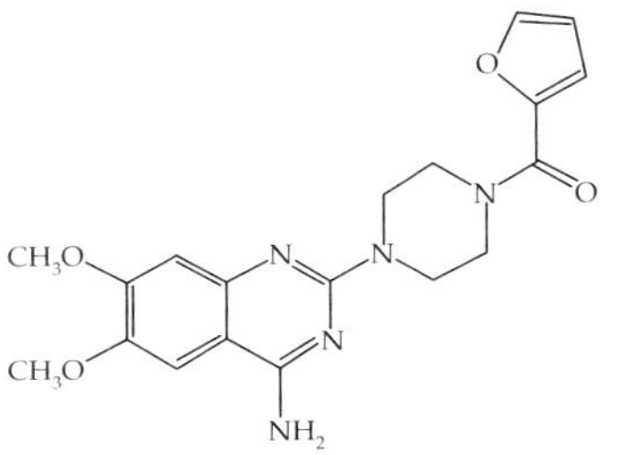	6.54	U	+H	potentio	50% EtOH	Kostek LJ, Prazosin, *APDS*, **18,** 351–378 (1989). NB: No reference given.
2127	Prazosin	6.5	U	+H			N&K; Anon

(continued)

No.	Name	pK_a value(s)	Data quality	Ionization type	Method	Conditions t°C; I or c M	Comments and Reference(s)
2128	Prenalterol ($C_{12}H_{19}NO_3$) 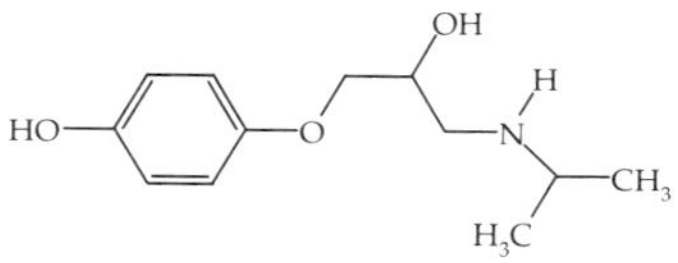	9.5 10.0	U U	−H, +H +H, −H			Craig
2129	Prilocaine ($C_{13}H_{20}N_2O$) 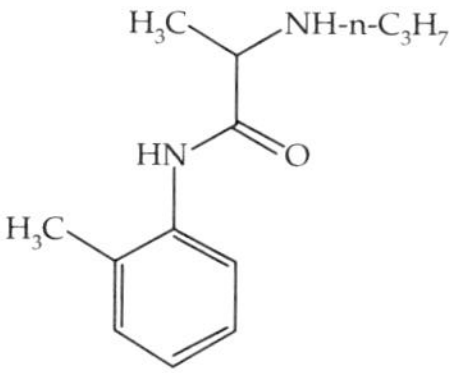	7.32 or 7.89	U	+H			Craig: Cited Lofgren N and Tegner C, Local anesthetics. XX. Synthesis of some α-monoalkylamino-2-methylpropionanilides. A new useful local anesthetic, *Acta Chem. Scand.*, **14,** 486–490 (1960). N&K; Ritschel gave 7.89; ref. Astra.
2130	Prilocaine	7.9	U	+H			de Jong RH, Neural blockade by local anesthetics, *JAMA*, **238,** 1383–1385 (1977); cited in W&G.
2131	Procainamide ($C_{13}H_{21}N_3O$) 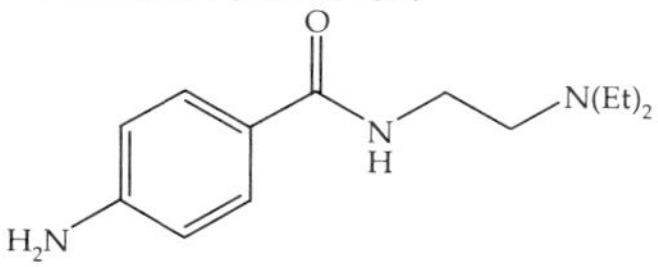	9.24 ± 0.10	U	+H	potentio		Poet RB and Kadin H, Procainamide hydrochloride, *APDS*, **4,** 333–383 (1975). Cited Jacobson H and Schaefer C, Squibb Institute, personal communication.
2132	Procainamide	9.2	U	+H		$t = 20$	Mian MS, El-Obeid HA and Al-Badr AA, Procainamide hydrochloride, *APDS*, **28,** 258 (2001). NB: Cited Clarke as source.

2133	Procaine	2.45 8.05	U U	+H +H	spectro	H_2O $t = 15$ $c \sim 0.005$ to 0.01	Kolthoff IM, The dissociation constants, solubility product and titration of alkaloids, *Biochem. Z.*, **162,** 289–353 (1925). Cited in Perrin Bases, no. 484 Ref. K47. See Aconitine for details.
		8.98	U	+H	potentio	H_2O $t = 25$ I undefined	Lordi NG and Christian JE, Physical properties and pharmacological activity: Antihistaminics, *J. Am. Pharm. Associ.*, Sci. Edn., **45,** 300–305 (1956). Cited in Perrin Bases, no. 484 Ref. L54. See Chlorpheniramine (no. 1704) for details.
		8.91	U	+H	potentio	H_2O $t = 25$ $c = 0.005$	Krahl ME, Keltch AK and Clowes GHA, The role of changes in extracellular and intracellular hydrogen ion concentration in the action of local anesthetic bases, *JPET*, **68,** 330–350 (1940). Cited in Perrin Bases, no. 484 Ref K58.
		8.11; 8.80	U	+H			Craig; Foye 3rd; see Azatadine gave N&K. N&K cited Chatten LG (ed.), *Pharmaceutical Chemistry*, vol. 1, Dekker, New York, pp. 85–87 (1966), which cited Martin, *Physical Pharmacy*, 1st Edn.
2134	Prochlorperazine	3.73 8.1	U U	+H +H			N&K; Green AL, Ionization constants and water solubilities of some aminoalkylphenothiazine tranquillizers and related compounds, *J. Pharm. Pharmacol.*, **19,** 10–16 (1967). NB: See Amitriptylline for details.
2135	Promazine	9.28	U	+H			Lombardo F, Obach RS, Shalaeva MY and Gao F, Prediction of human volume of distribution values for neutral and basic drugs. 2. Extended data set and leave-class-out statistics, *J. Med.*

(continued)

No.	Name	pK_a value(s)	Data quality	Ionization type	Method	Conditions t°C; I or c M	Comments and Reference(s)
							Chem., **47,** 1242–1250 (2004). Ref. 277: Mannhold R, Dross KP and Reffer RF, Drug lipophilicity in QSAR practice: I. A comparison of experimental with calculative approaches, *Quant. Struct.-Act. Relat.*, **9,** 21–28 (1990).
2136	Propafenone ($C_{21}H_{27}NO_3$) 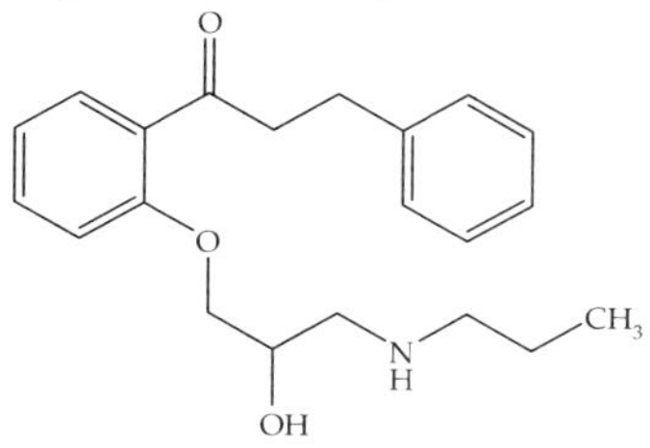	9.27	U	+H	potentio		Lombardo F, Obach RS, Shalaeva MY and Gao F, Prediction of human volume of distribution values for neutral and basic drugs. 2. Extended data set and leave-class-out statistics, *J. Med. Chem.*, **47,** 1242–1250 (2004). Ref. not given: potentiometric titration.
2137	Propicillin ($C_{18}H_{22}N_2O_5S$) 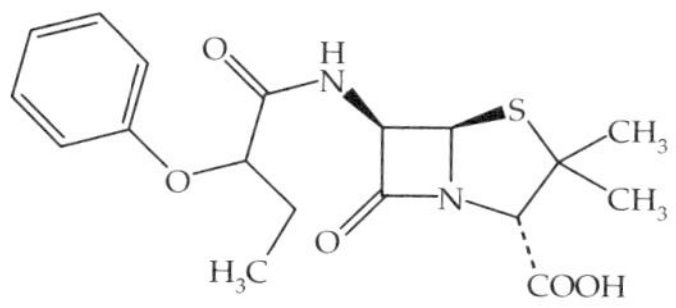	2.72 ± 0.04	A	−H	potentio	H_2O $t = 25$ $c = 0.01$	Rapson HDC and Bird AE, Ionization constants of some penicillins and of their alkaline and penicillinase hydrolysis products, *J. Pharm. Pharmacol.*, Suppl. 15, 222–231T (1963). NB: Potentiometric titrations used a glass electrode with an unsymmetrical cell and liquid junction potentials.
2138	Propiomazine ($C_{20}H_{24}N_2OS$) 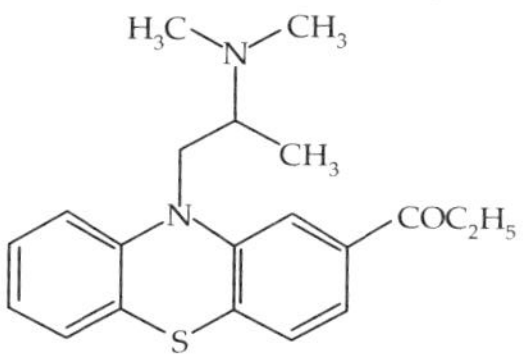	6.6	U	+H	potentio	H_2O	Crombie KB and Cullen LF, Propiomazine hydrochloride, *APDS*, **2,** 439–466 (1973). Crombie KB and Cullen LF, Wyeth Labs. Inc., unpublished results. "The pK_a for propiomazine has been determined potentiometrically to be 6.6 by aqueous titration with 0.1N sodium hydroxide."

2139	Propoxycaine ($C_{16}H_{26}N_2O_3$) 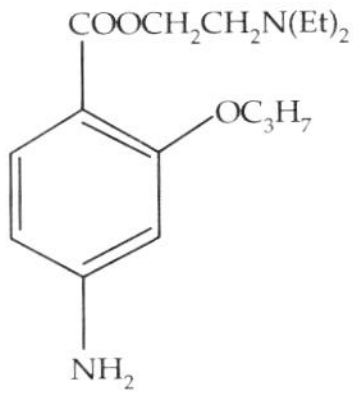	8.6	U	+H			Craig N&K; Hoover
2140	Propoxyphene ($C_{22}H_{29}NO_2$) 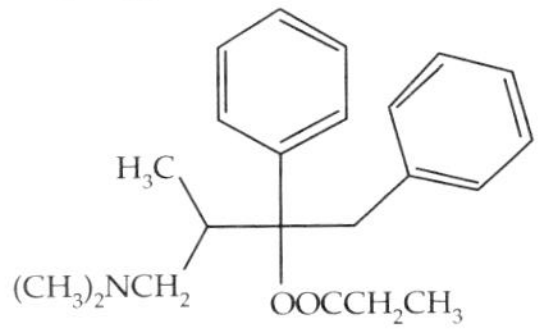	6.3	U	+H	potentio	50% aq EtOH	Craig N&K; Clouet, *Narcotic Drugs Biochemical Pharmacology* cited from Farmilo CG, Oestreicher PM and Levi L, Physical methods for the identification of narcotics. IB Common physical constants for identification of ninety-five narcotics and related compounds, *Bull. Narcotics*, UN Dept. Social Affairs, vol. 6, pp. 7–19 (1954).
2141	L-Propoxyphene (levopropoxyphene) ($C_{22}H_{29}NO_2$) 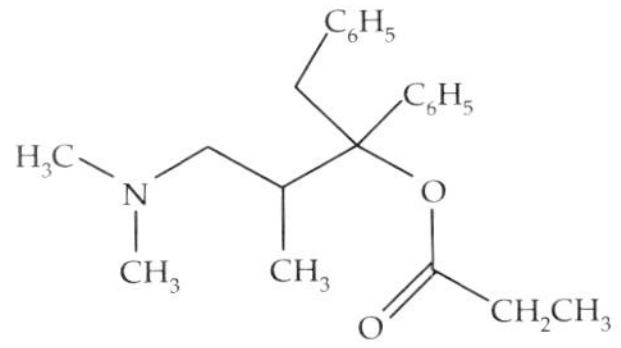	6.3	U	+H			Craig
2142	Propranolol	9.6	U	+H			Suzuki Y, Sugiyama Y, Saurada Y, Iga T and Hanano M, Assessment of the contribution of alpha-1-acid glycoprotein to the serum binding of basic drugs, *J. Pharm. Pharmacol.*, **37**, 712–717 (1985).

(continued)

Appendix B *(continued)*

No.	Name	pK_a value(s)	Data quality	Ionization type	Method	Conditions t°C; I or c M	Comments and Reference(s)
2143	Propranolol	9.45	U	+H			Lombardo F, Obach RS, Shalaeva MY and Gao F, Prediction of human volume of distribution values for neutral and basic drugs. 2. Extended data set and leave-class-out statistics, *J. Med. Chem.*, **47,** 1242–1250 (2004). Ref. 275: Barbato F, Caliendo G, LaRotonda MI, Morrica P, Silipo C and Vittoria A, Relationships between octanol-water partition data, chromatographic indices and their dependence on pH in a set of beta-adrenoceptor blocking agents, *Farmaco*, **45,** 647–663 (1990). See also: Hinderling PH, Schmidlin O and Seydel JK, Quantitative relationships between structure and pharmacokinetics of beta-adrenoceptor blocking agents in man, *J. Pharmacokin. Biopharm.*, **12,** 263–287 (1984). Taylor EA, Jefferson D, Carroll JD and Turner P, Cerebrospinal fluid concentrations of propranolol, pindolol, and atenolol in man: Evidence for central actions of beta-adrenoceptor antagonists, *Br. J. Clin. Pharmacol.*, **12,** 549–559 (1981). NB: See Pindolol for further details.

2144	Propylamphetamine ($C_{12}H_{19}N$)	9.98	U	+H		Vree TB, Muskens ATJM and van Rossum JM, Some physicochemical properties of amphetamine and related drugs, *J. Pharm. Pharmacol.*, **21,** 774–775 (1969). NB: see Dexamphetamine for details.
2145	Propylhexedrine	10.42	U	+H		Kisbye J, *Pharm. Weekblad.*, **93,** 206–215 (1958). Cited in: Craig; N&K; Chatten LG (ed.), *Pharmaceutical Chemistry* vol. 1, Dekker, New York, pp. 85–87 (1966), which gave $pK_b = 3.48 = pK_a = 10.52$, cited Chatten and Harris, *Anal. Chem.*, **34,** 1495–1501 (1962), but the value there is attributed to Kisbye J, *Pharm. Weekblad.*, **93,** 206–215 (1958) (also Kisbye J, *Dansk. Tiddshrift Farm.*, **32,** 174–186, 189–201 (1958). This value is definitely from Leffler EB, Spencer HM and Burger A, Dissociation constants of adrenergic amines, *JACS*, **73,** 2611–2613 (1951).
2146	Propylhexedrine	10.74	U	+H		Vree TB, Muskens ATJM and van Rossum JM, Some physicochemical properties of amphetamine and related drugs, *J. Pharm. Pharmacol.*, **21,** 774–775 (1969). NB: See Dexamphetamine for details.
2147	Propylparaben ($C_{10}H_{12}O_3$)	8.4	U	−H		W&G; Tammilehto S and Buchi J, p-Hydroxybenzoic acid esters (Nipagins). I. Physicochemical properties, *Pharm. Acta Helv.*, **43,** 726–738 (1968).

(continued)

No.	Name	pK_a value(s)	Data quality	Ionization type	Method	Conditions t°C; I or c M	Comments and Reference(s)
2148	6-n-Propyl-2-thiouracil ($C_7H_{10}N_2OS$) 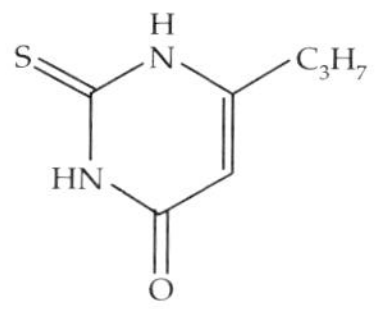	7.8; 8.3	U	−H			Craig; N&K; Garrett ER and Weber DJ, Metal complexes of thiouracils I. Stability constants by potentiometric titration studies and structures of complexes, *J. Pharm. Sci.*, **59,** 1383–1398 (1970). Ritschel also cited Garrett and Weber, giving 7.76.
2149	Prostaglandin E_1 ($C_{20}H_{34}O_5$) 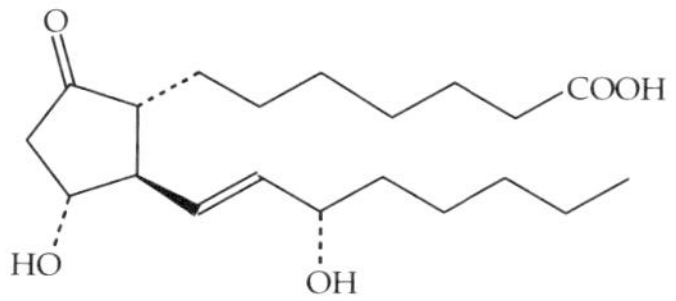	5.02	U	−H	??		Uekama K, Hirayama F, Tanaka H and Takematsu K, Partition behaviour and ion-pair formation of some prostaglandins, *Chem. Pharm. Bull.*, **26,** 3779–3784 (1978). NB: See separate entries for Dinoprost and Dinoprostone.
2150	Pseudoephedrine	9.7	U	+H			Ritschel; Borodkin S and Yunker MH, Interaction of amine drugs with a polycarboxylic acid ion-exchange resin, *J. Pharm. Sci.*, **59,** 481–486 (1970).
2151	Pseudoephedrine	9.86	U	+H			Vree TB, Muskens ATJM and van Rossum JM, Some physicochemical properties of amphetamine and related drugs, *J. Pharm. Pharmacol.*, **21,** 774–775 (1969). NB: See Dexamphetamine for details.
2152	Pyrazinamide ($C_5H_5N_3O$) 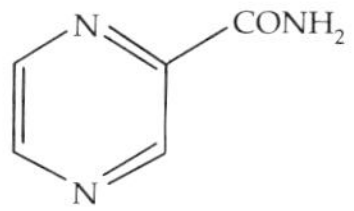	0.5	U	+H			Craig; N&K; 'Merck 9.
		−0.5	U	+H	spectro		Rogers EF, Leanza WJ, Becker HJ, Matzuk AR, O'Neill RC, Basso AJ, Stein GA, Solotorovsky M, Gregory FJ and Pfister K, Antitubercular diazine carboxamides, *Science*, **116,** 253–254 (1952). Cited in Perrin Bases no. 1494 Ref. R29. Used an undefined optical method.

No.	Compound	pKa	Class	Type	Method	Conditions	Comments/Reference
2153	Pyridine (C_5H_5N) 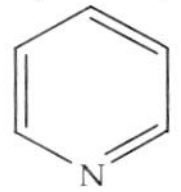	5.21 ± 0.05	A	+H	potentio	H_2O $t = 25$ $I \sim 0$	Perrin Bases no. 1018. NB: Numerous values reported over a temperature range 18 to 35 °C.
2154	Pyrilamine ($C_{17}H_{23}N_3O$) 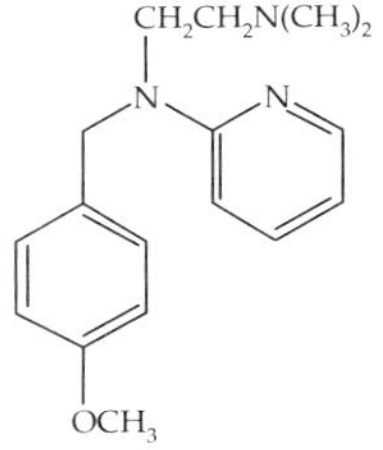	4.02 8.92	U U	+H +H	potentio	H_2O $t = 25$ $c = 0.002$ to 0.01	Craig; W&G: Cited Perrin Bases no. 1122.
2155	Pyrilamine	4.0 8.9	U U	+H +H			Persson BA and Schill G, Extraction of amines as complexes with inorganic anions, *Acta Pharm. Suecica*, **3,** 291–302 (1966). NB: No details are given for the measurement of what are described as "approximate acid dissociation constants."
2156	Pyrimethamine ($C_{12}H_{13}ClN_4$) 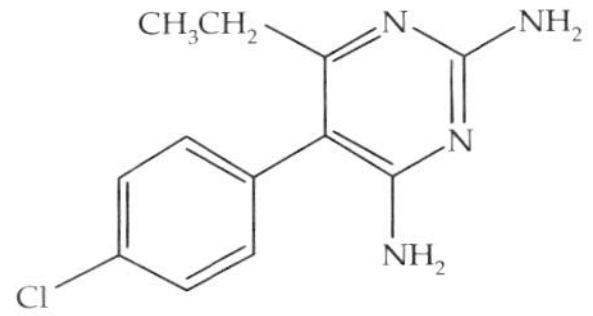	7	U	+H		$t = 20.0$	Loutfy MA and Aboul-Enein HY, Pyrimethamine, *APDS*, **12,** 463–479 (1983). BPC, p. 767
2157	Pyrimethamine	7.2	U	+H			N&K; Ritschel gave 7.2; ref. Burroughs-Wellcome, Tuckahoe, NY 10707.
2158	Pyrimethamine	7.2	U	+H			

(continued)

No.	Name	pK_a value(s)	Data quality	Ionization type	Method	Conditions t°C; I or c M	Comments and Reference(s)
							Anton AH, A drug-induced change in the distribution and renal excretion of sulfonamides, *JPET*, **134,** 291–303 (1961) . . . The value was obtained from Bases and Hitchings, unpublished observations.
2159	Pyrimethazine	9.4	U	+H			W&G: Sorby DL, Plein EM and Benmaman D, Adsorption of phenothiazine derivatives by solid adsorbents, *J. Pharm. Sci.*, **55,** 785–794 (1966). NB: See Chlorpromazine for details. This appears to be the same compound as pyrimethamine, but the discrepancy in the pK_a values is unexplained—possibly a difference in solvent.
2160	Pyrrobutamine ($C_{20}H_{22}ClN$) 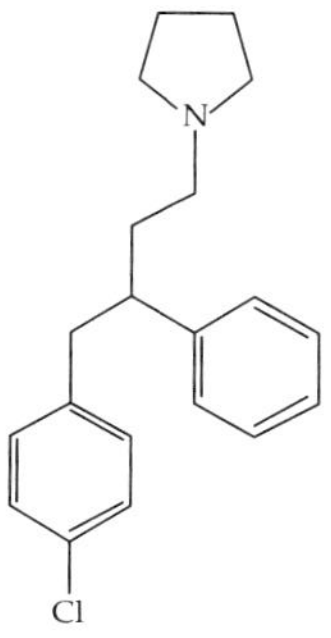	8.77	U	+H	potentio	H_2O $t = 25$ $c = 0.002$ to 0.01	Lordi NG and Christian JE, Physical properties and pharmacological activity: Antihistaminics, *J. Am. Pharm. Ass.*, Sci. Edn., **45,** 300–305 (1956). See Chlorpheniramine (no. 1704) for details. N&K; Craig

No.	Compound	pKa	Data quality	Ionization type	Method	Conditions	Comments / Reference
2161	Quinacrine	7.73	U	+H	spectro	H_2O	Irvin JL and Irvin EM, Apparent ionization exponents of homologs of quinacrine; electrostatic effects, *JACS*, **72,** 2743–2749 (1950). NB: Dilute (0.001 M) solutions were titrated, with resulting relatively low accuracy. Measurement of pK_{a2} involved extrapolation of glass electrode pH measurements from EtOH-water solutions. No treatment of activity effects. Cited in Craig; Perrin Bases no. 2679, ref. I7.
		7.69	U	+H	potentio	$t = 30$	
		10.18	U	+H	potentio	$I = 0.1$ (NaCl)	
2162	Quinacrine	8.0	U	+H			W&G: Perrin Bases
		10.2	U	+H			
2163	Quinacrine	7.73	U	+H			Lombardo F, Obach RS, Shalaeva MY and Gao F, Prediction of human volume of distribution values for neutral and basic drugs. 2. Extended data set and leave-class-out statistics, *J. Med. Chem.*, **47,** 1242–1250 (2004). Ref. 297: Irvin JL and Irvin EM, Apparent ionization exponents of homologs of quinacrine: Electrostatic effects, *JACS*, **72,** 2743–2749 (1950).
		10.2	U	+H			
2164	Quinethazone ($C_{10}H_{12}ClN_3O_3S$)	9.3	U	−H			Craig
		10.7	U	−H			N&K; Am Hosp Form Service.
2165	Quinidine	4.21	U	+H			N&K; cited Perrin Bases. NB: These are the values for Quinine, see Perrin Bases no. 2958.
		8.34	U	+H			

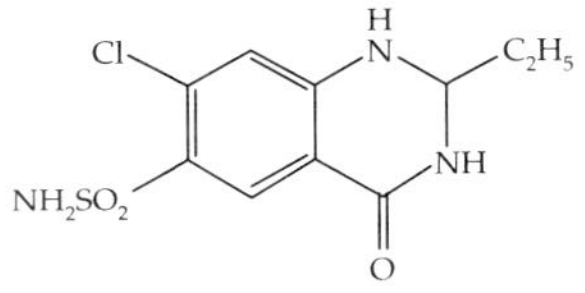

(continued)

No.	Name	pK_a value(s)	Data quality	Ionization type	Method	Conditions t°C; I or c M	Comments and Reference(s)
2166	Quinidine	8.05	U	+H			Lombardo F, Obach RS, Shalaeva MY and Gao F, Prediction of human volume of distribution values for neutral and basic drugs. 2. Extended data set and leave-class-out statistics, *J. Med. Chem.*, **47,** 1242–1250 (2004). Ref. 298: Tsai R-S, Carrupt PA, Testa B, Tayar, NE, Grunewald GL and Casy AF, Influence of stereochemical factors on the partition coefficient of diastereomers in a biphasic octan-1-ol/water system, *J. Chem. Res. (M)*, 1901–1920 (1993).
2167	quinolone–ciprofloxacin	6.00 8.80	U U	–H +H			Kitzes-Cohen R, Quinolones in CNS infections, *Quinolone Bull.*, **3,** 7–14 (1987).
2168	quinolone–enoxacin	6.00 8.50	U U	–H +H			Dalhoff A, A review of quinolone tissue pharmacokinetics, *in* Fernandes PB (ed.), *International Telesymposium on Quinolones*, JR Prous, Barcelona, pp. 277–312 (1989).
2169	quinolone–fleroxacin	5.70 8.00	U U	–H +H			Dalhoff A, A review of quinolone tissue pharmacokinetics, *in* Fernandes PB (ed.), *International Telesymposium on Quinolones*, JR Prous, Barcelona, pp. 277–312 (1989).
2170	quinolone–nalidixic acid	6.02	U	–H	spectro		Staroscik R and Sulkowska J, Acid-base equilibria of nalidixic acid, *Acta Pol. Pharm.*, **28,** 601–606 (1971).
2171	quinolone–nalidixic acid	6.12	U	–H	soly/pH		Staroscik R and Sulkowska J, Acid-base equilibria of nalidixic acid, *Acta Pol. Pharm.*, **28,** 601–606 (1971).

2172	quinolone–norfloxacin	6.40 8.70	U U	−H +H			Kitzes-Cohen R, Quinolones in CNS infections, *Quinolone Bull.*, **3**, 7–14 (1987).
2173	quinolone–norfloxacin	6.20 8.70	U U	−H +H			Stein G, Review of the bioavailability and pharmacokinetics of oral norfloxacin, *Am. J. Med.*, **82** (Suppl. 6B), 18–21 (1987).
2174	quinolone–ofloxacin	5.70 7.90	U U	−H +H			Kitzes-Cohen R, Quinolones in CNS infections, *Quinolone Bull.*, **3**, 7–14 (1987).
2175	Racemethorphan ($C_{18}H_{25}NO$) 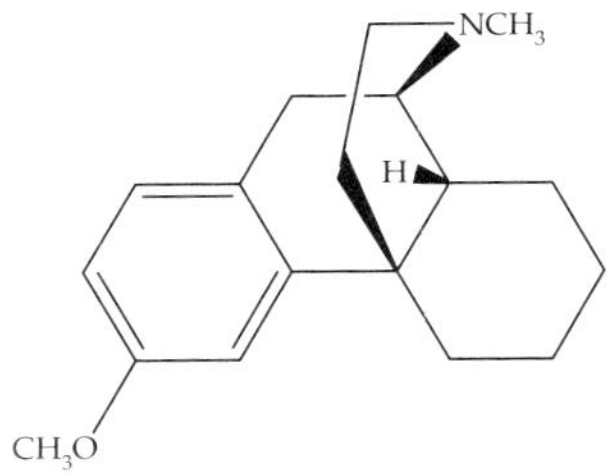	8.83	U	+H	potentio	50% aq EtOH	Farmilo CG, Oestreicher PM and Levi L, Physical methods for the identification of narcotics. IB Common physical constants for identification of ninety-five narcotics and related compounds, *Bull. Narcotics*, UN Dept. Social Affairs, vol. 6, pp. 7–19 (1954); cited in Clouet DH (ed.), *Narcotic Drugs Biochemical Pharmacology*, Plenum Press, New York, 52–53 (1971).
2176	Racemorphan (methorphinan) ($C_{17}H_{23}NO$) 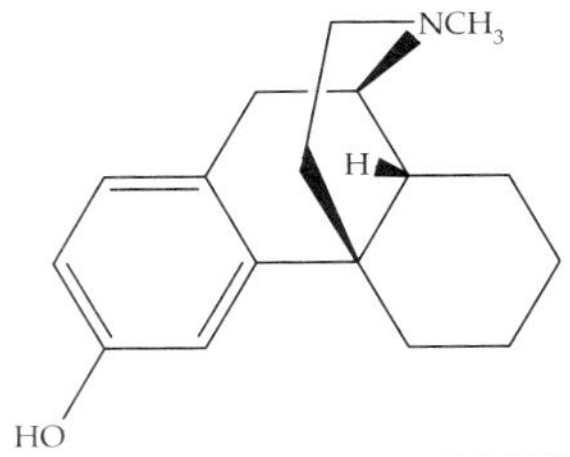	8.97	U	+H	potentio	50% aq EtOH	Farmilo CG, Oestreicher PM and Levi L, Physical methods for the identification of narcotics. IB Common physical constants for identification of ninety-five narcotics and related compounds, *Bull. Narcotics*, UN Dept. Social Affairs, vol. 6, pp. 7–19 (1954); cited in Clouet DH (ed.), *Narcotic Drugs Biochemical Pharmacology*, Plenum Press, New York, 52–53 (1971).

(continued)

No.	Name	pK_a value(s)	Data quality	Ionization type	Method	Conditions t°C; I or c M	Comments and Reference(s)
2177	Ranitidine ($C_{13}H_{22}N_4O_3S$) 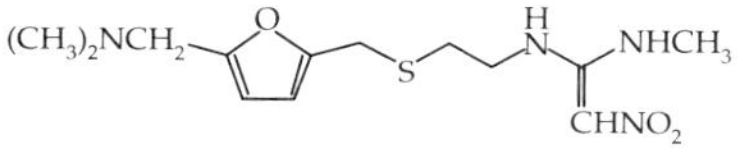	2.19 ± 0.04	U	+H	spectro		Hohnjec M, Kuftinec J, Malnar M, Skreblin M, Kajfez F, Nagl A and Blazevic N, Ranitidine, *APDS*, **15,** 533–561 (1986). Hohnjec M, Rendic S, Alebic-Kolbah T, Kajfez F, Blazevic N and Kuftinec J, *Acta Pharm. Jugosl.*, **31,** 131 (1981). NB: Albert and Serjeant (1984) says pK_a 2.3 (+H) and 8.2 (+H).
2178	Ranitidine	8.18	U	+H	potentio	H_2O $t = 25$ $I = 0.1$ (KCl)	Shankley NP, Black JW, Ganellin CR and Mitchell RC, *Br. J. Pharmacol.*, **94,** 264–274 (1988). "pK_a values were measured by MJ Graham (SK&F Ltd.) potentiometrically and corrected for 37° unless otherwise stated."
2179	Ranitidine	8.47	U	+H	potentio		Lombardo F, Obach RS, Shalaeva MY and Gao F, Prediction of human volume of distribution values for neutral and basic drugs. 2. Extended data set and leave-class-out statistics, *J. Med. Chem.*, **47,** 1242–1250 (2004). Ref. not given: potentiometric titration.
2180	Rescinnamine ($C_{35}H_{42}N_2O_9$) 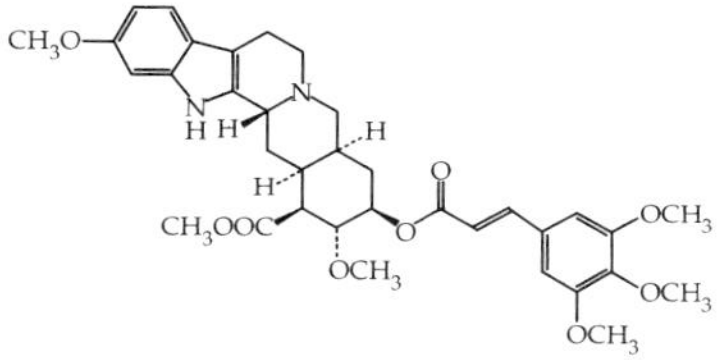	6.4	U	+H			McEvoy

No.	Compound	pKa	Method	Assignment	Reference
2181	Reserpine ($C_{33}H_{40}N_2O_9$) 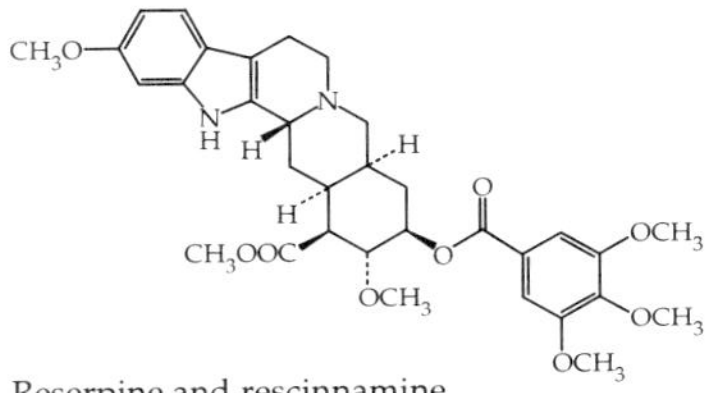	6.6	U	+H	Craig N&K; Chatten LG (ed.), *Pharmaceutical Chemistry*, vol. 1, Dekker, New York, pp. 85–87 (1966); Martin. Schirmer RE, Reserpine, *APDS*, **4,** 384–430 (1975) (no pK_a data).
2182	Reserpine and rescinnamine				Rojas JH, Extraction of ion pairs and distribution rates to the application of extraction in partition chromatography. Part 3. Reserpine and rescinnamine, *Rev. Colomb. Cienc. Quim. Farm.*, **4,** 35–42 (1985) CA 104:174462. "The determination of the pK_a values of reserpine and rescinnamine by spectrophotometric methods is described. The partition studies of the alkaloids at different concentrations of chloroform and either 0.1M sulfuric acid or 0.1M phosphoric acid are given."
2183	Rimoterol (rimiterol) ($C_{12}H_{17}NO_3$) 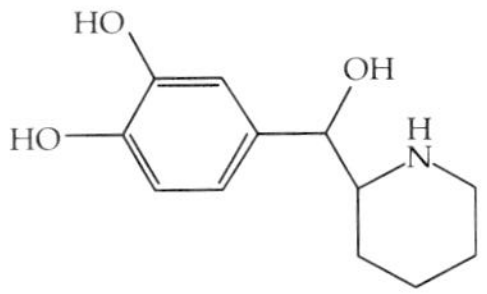	8.7 10.3	U U	+H, −H −H, +H	Craig

(continued)

No.	Name	pK_a value(s)	Data quality	Ionization type	Method	Conditions t°C; I or c M	Comments and Reference(s)
2184	Risperidone ($C_{23}H_{27}FN_4O_2$) 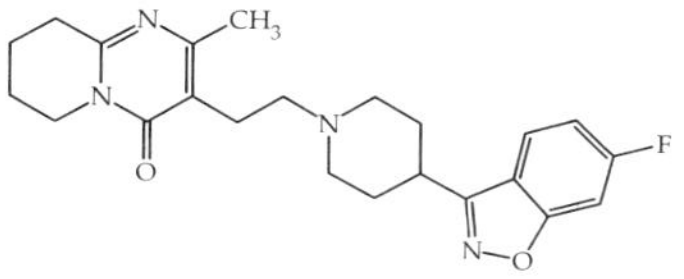	8.3	U	+H	CZE/pH		Lombardo F, Obach RS, Shalaeva MY and Gao F, Prediction of human volume of distribution values for neutral and basic drugs. 2. Extended data set and leave-class-out statistics, *J. Med. Chem.*, **47**, 1242–1250 (2004). Ref. not given: Capillary electrophoresis.
2185	Ritodrine ($C_{17}H_{21}NO_3$) 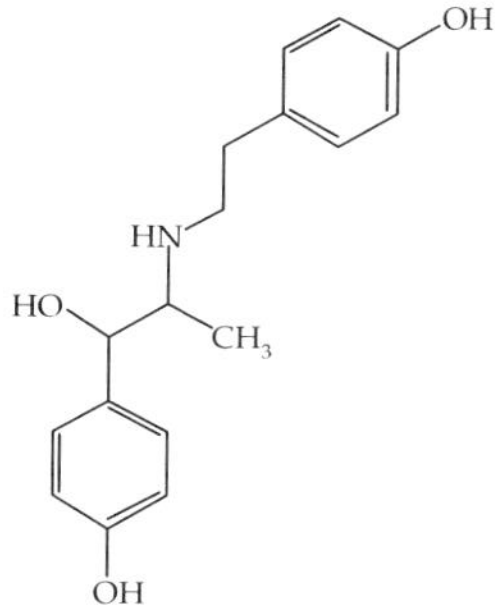	9.0	U	+H, −H			Foye 3rd; see Azatadine from McEvoy.
2186	Rivastigmine ($C_{14}H_{22}N_2O_2$) 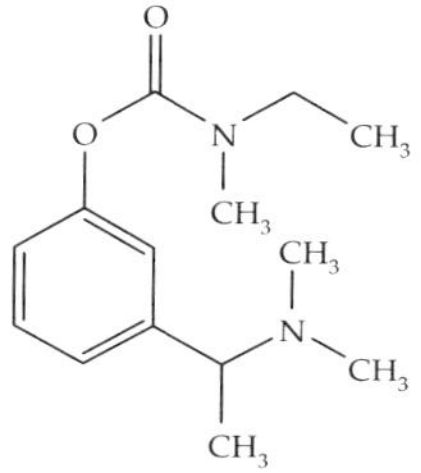	8.99	U	+H	CZE/pH		Lombardo F, Obach RS, Shalaeva MY and Gao F, Prediction of human volume of distribution values for neutral and basic drugs. 2. Extended data set and leave-class-out statistics, *J. Med. Chem.*, **47**, 1242–1250 (2004). Ref. not given: Capillary electrophoresis.

2187	Rolitetracycline ($C_{27}H_{33}N_3O_8$) 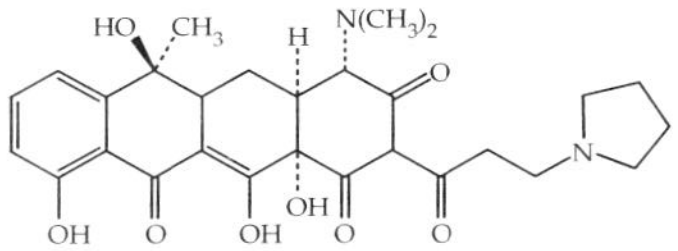	7.4	U	–H			Craig N&K; Avery NB: There should be three other pK_a values for Rolitetracycline.
2188	Rotoxamine (l-carbinoxamine) ($C_{16}H_{19}ClN_2O$) 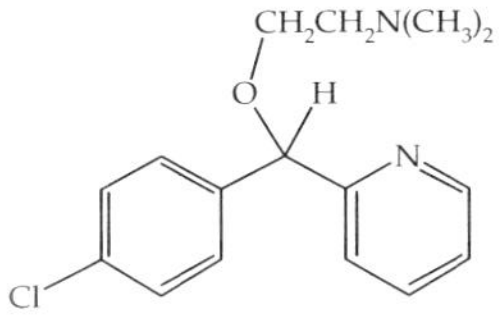	8.1	U	+H			Craig NB: Also known as Levocarbinoxamine.
2189	Saccharin ($C_7H_5NO_3S$) 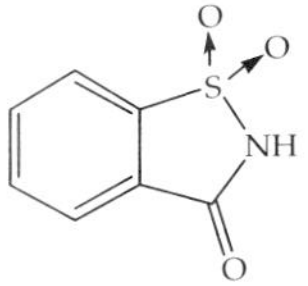	1.31	A	–H	spectro	H_2O $t = 25$ $I = 0.2$ (HCl-KCl)	Dawn H, Pitman IH, Higuchi T and Young S, N-chlorosaccharin as a possible chlorinating reagent: structure, chlorine potential, and stability in water and organic solvents, *J. Pharm. Sci.*, **59,** 955–959 (1970). Cited in Kortum, no. 3858, ref. D21.
		1.6	U	–H		H_2O $t = 18$	According to the Dawn *et al.* paper (1970), this value was obtained by Kolthoff (1925). Also cited in Zubair MU and Hassan MMA, Saccharin, *APDS*, **13,** 487–519 (1984). No reference cited. See also N&K; Martin; Ritschel.
2190	Salsalate ($C_{14}H_{10}O_5$) 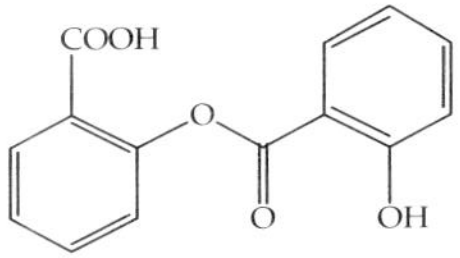	3.5 9.8	U U	–H –H			Craig

(continued)

No.	Name	pK_a value(s)	Data quality	Ionization type	Method	Conditions t°C; I or c M	Comments and Reference(s)
2191	Scopolamine	7.6	U	+H		$t = 23$	Muhtadi FJ and Hassan MMA, Scopolamine hydrobromide, *APDS*, **19**, 477–551 (1990). NB: Clarke, p. 674; also Craig.
2192	Serotonin ($C_{10}H_{12}N_2O$)	4.9	U	+H			Craig
		9.1	U	+H			N&K; Merck 9.
		9.8	U	–H			Perrin Bases ref R4: Rapaport MM, Green AA, Page IH, Serum vasoconstrictor (serotonin) IV. Isolation and characterization, *J. Biol. Chem.*, **176**, 1243–1251 (1948).
2193	Sildenafil ($C_{22}H_{30}N_6O_4S$) 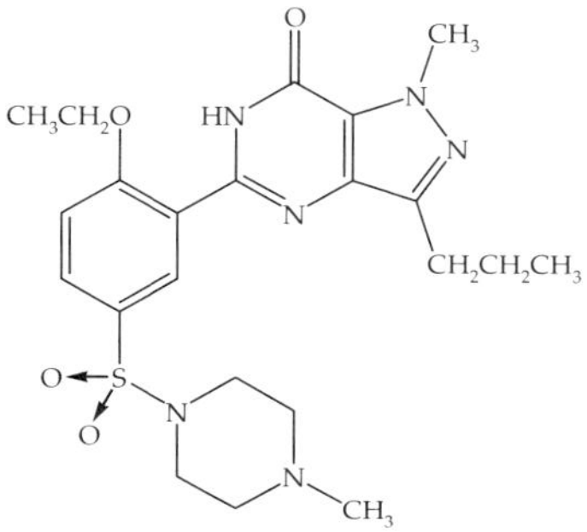	8.7	U	+H			Badwan AA, Nabusli L, Al-Omri MM, Daraghmeh N and Ashour M, Sildenafil, *APDS*, **27**, 352 (2000). "Sildenafil has a basic functional group, characterized by a pK_a value of 8.7 (NH-piperazine). In addition it has a weak acidic moiety (HN-amide) [4]." 4. Cooper JD, Muirhead DC, Taylor JE and Baker RP, *J. Chrom. Biomed. Sci. Appl.*, **701**, 87 (1997). NB: The very weakly acidic amide is confirmed by an increase in aqueous solubility at pH values between 9 and 12 (data in APDS 27).

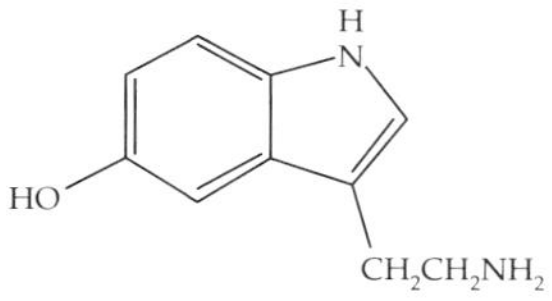

2194	Solasodine ($C_{27}H_{43}NO_2$)	6.31 (pK_b)	U	+H		EtOH (60%)	Indrayant G, Syahrani A, Sondakh R and Santosa MH, Solasodine, *APDS*, **24**, 487–522 (1996). NB: pK_b in 60% ethanol was 6.31. Conversion of this value to the corresponding pK_a would require pK_w for the 60% aqueous ethanol solvent system. Ref: Bentley KW and Kirby GW, *Elucidation of organic structure by physical and chemical methods*, 2nd Edn., *Interscience*, 665 (1972).
2195	Sorbitol	13.60	U	−H			Merck 10, p. 1248
2196	Sotalol	8.3	U	−H			Foster RT and Carr RA, Sotalol, *APDS*, **21,** 501–533 (1992). Cited W&G, which in turn cited Garrett and Schnelle, 1971 (see no. 1279). NB: Hinderling PH, Schmidlin O, Seydel JK, Quantitative relationships between structure and pharmacokinetics of beta-adrenoceptor blocking agents in man, *J. Pharmacokin. Biopharm.*, **12,** 263–287 (1984). See Sotalol, no. 1279.
		9.8	U	+H			
2197	Sotalol	9.76	U	+H			Lombardo F, Obach RS, Shalaeva MY, Gao F, Prediction of human volume of distribution values for neutral and basic drugs. 2. Extended data set and leave-class-out statistics, *J. Med. Chem.*, **47,** 1242–1250 (2004). Ref. 275 = Barbato F, Caliendo G, LaRotonda MI, Morrica P, Silipo C, Vittoria A, Relationships between octanol-water partition data, chromatographic indices and their dependence on pH in a set of beta-adrenoceptor blocking agents, *Farmaco*, **45,** 647–663 (1990).

(continued)

No.	Name	pK_a value(s)	Data quality	Ionization type	Method	Conditions t°C; I or c M	Comments and Reference(s)
2198	Sparteine ($C_{15}H_{26}N_2$)	4.80	U	+H			Craig
		11.96	U	+H			
2199	Spectinomycin ($C_{14}H_{24}N_2O_7$) 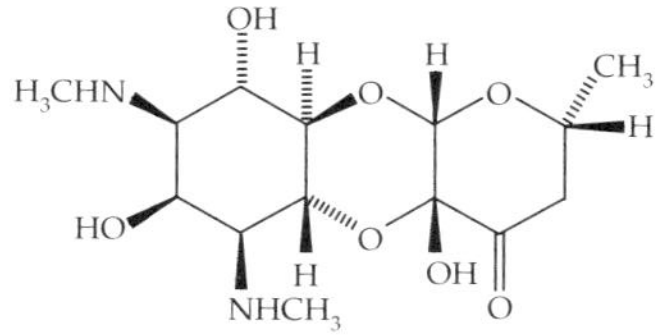	6.95	U	+H			Craig; N&K; Merck 9
		8.70	U	+H			
2200	Spiperone ($C_{23}H_{26}FN_3O_2$)	8.31	U	+H			Craig
		9.09	U	+H			

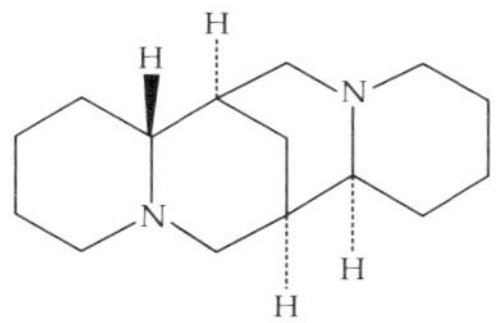

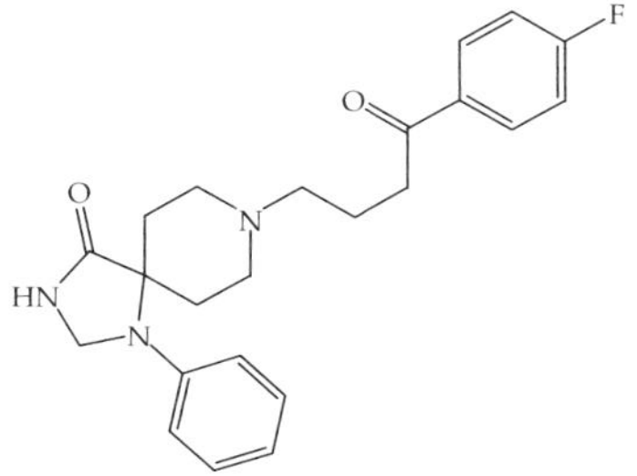

No.	Compound	pKa	Class	Type	Method	Conditions	Reference
2201	Streptozocin ($C_8H_{15}N_3O_7$) 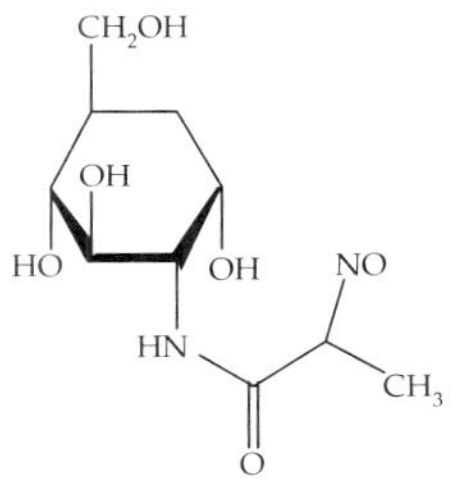	1.3	U	+H			McEvoy
2202	Strychnine	2.50	U	+H			Craig; N&K; Merck 9
		8.27	U	+H			Manske Alkaloids, vol. VI, pp. 179–195.
2203	Succinylsulfathiazole ($C_{13}H_{13}N_3O_5S_2$) 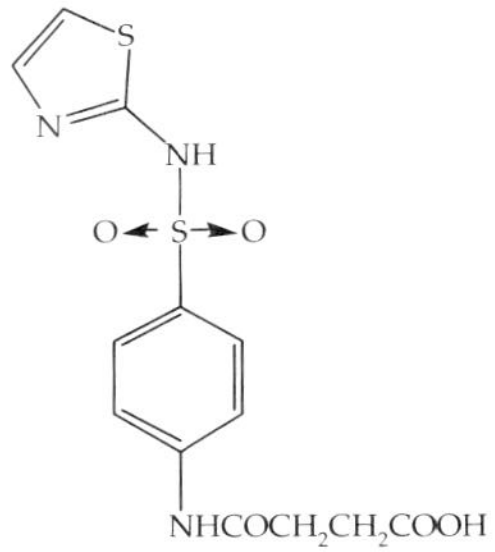	4.5	U	–H			Craig N&K; Hoover
2204	Sufentanil ($C_{22}H_{30}N_2O_2S$) 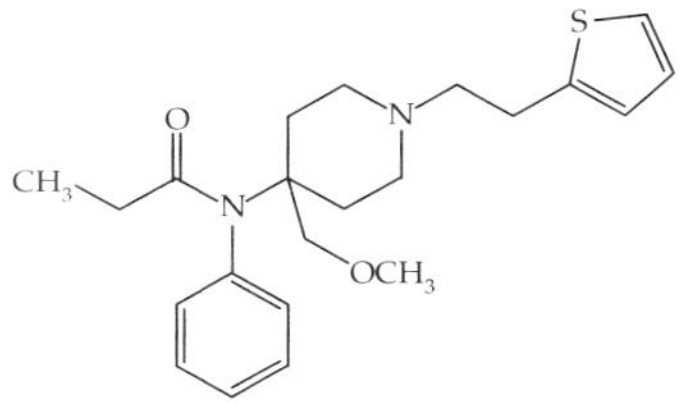	8.51	U	+H	soly/pH	$t = 35.0$	Roy SD and Flynn GL, Solubility behavior of narcotic analgesics in aqueous media: solubilities and dissociation constants of morphine, fentanyl, and sufentanil, *Pharm. Res.*, 6(2), 147–151 (1989). NB: See Fentanyl for details. NB: McEvoy gave 8.0.

(continued)

No.	Name	pK_a value(s)	Data quality	Ionization type	Method	Conditions t°C; I or c M	Comments and Reference(s)
2205	Sufentanil	7.85	U	+H	potentio		Lombardo F, Obach RS, Shalaeva MY and Gao F, Prediction of human volume of distribution values for neutral and basic drugs. 2. Extended data set and leave-class-out statistics, *J. Med. Chem.*, **47**, 1242–1250 (2004). Ref. 276: Meuldermans WEG, Hurkmans RMA, Heykants JJP, Plasma protein binding and distribution of fentanyl, sulfentanil, alfentanil and lofentanil in blood, *Arch. Int. Pharmacodyn.*, **257**, 4–19 (1982); Potentiometric titration.
2206	Sulfacarbamide	1.8 5.5	U U	+H −H			Ritschel: Struller T, Progress in sulphonamide research, *Progr. Drug Res.*, **12**, 389–457 (1968).
2207	Sulfacetamide	1.78 5.38	U U	+H −H		$t = 25$	N&K; Suzuki A, Higuchi WI Ho NFH, Theoretical model studies of drug absorption and transport in the gastrointestinal tract, *J. Pharm. Sci.*, **59**, 651–659 (1970). Craig; Clarke, p. 981; BPC, p. 862.
2208	Sulfacetamide	5.78	U	−H	potentio		Rieder J, Physicalisch-chemische und biologische untersuchungen an sulfonamiden, *Arzneim.-Forsch.*, **13**, 81–88 (1963). NB: See sulphanilamide for details.
2209	Sulfacetamide	5.38	U	−H		$t = 25$	Chatten LG (ed.), *Pharmaceutical Chemistry*, vol. 1, Dekker, New York, pp. 85–87 (1966); Bell, Roblin, *JACS*, 1942.

2210	Sulfachloropyridazine ($C_{10}H_9ClN_4O_2S$) 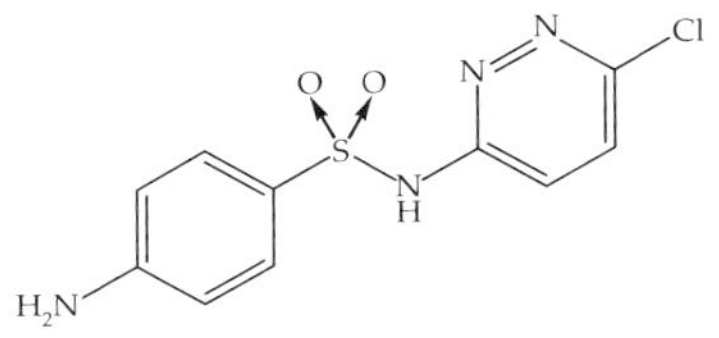	6.10	U	–H	potentio		Rieder J, Physicalisch-chemische und biologische untersuchungen an sulfonamiden, *Arzneim.-Forsch.*, **13,** 81–88 (1963). NB: See Sulphanilamide for details.
2211	Sulfachlorpyridazine	5.9	U	–H			Ritschel: Struller T, Progress in sulphonamide research, *Progr. Drug Res.*, **12,** 389–457 (1968).
2212	Sulfadiazine	6.52	U	–H	potentio		Rieder J, Physicalisch-chemische und biologische untersuchungen an sulfonamiden, *Arzneim.-Forsch.*, **13,** 81–88 (1963). NB: See Sulphanilamide for details.
2213	Sulfadiazine	6.37	U	–H	spectro	H_2O $t = 27 \pm 1$ $I = 0.2$	Ritschel: Yoshioka M, Hamamoto K, and Kubota T, Acid dissociation constants of sulfanilamides and substituent effects on the constants, *Yakugaku Zasshi*, **84,** 90–93 (1964).
2214	Sulfadiazine, N4-acetyl	6.34	U	–H	potentio		Rieder J, Physicalisch-chemische und biologische untersuchungen an sulfonamiden, *Arzneim.-Forsch.*, **13,** 81–88 (1963). NB: See Sulphanilamide for details.
2215	Sulfadimethoxine	2.02 6.70	U U	+H –H			Craig; N&K; Suzuki A, Higuchi WI, and Ho NFH, Theoretical model studies of drug absorption and transport in the gastrointestinal tract, *J. Pharm. Sci.*, **59,** 651–659. (1970).
2216	Sulfadimethoxine	6.32	U	–H	potentio		Rieder J, Physicalisch-chemische und biologische untersuchungen an sulfonamiden, *Arzneim.-Forsch.*, **13,** 81–88 (1963). NB: See Sulphanilamide for details.

(continued)

No.	Name	pK_a value(s)	Data quality	Ionization type	Method	Conditions t°C; I or c M	Comments and Reference(s)
2217	Sulfadimethoxine	5.94	U	–H	potentio	$t = 28$	Nakagaki M, Koga N and Terada H, Physicochemical studies on the binding of chemicals with proteins. I. The binding of several sulphonamides with serum albumin, *J. Pharm. Soc. Jpn.*, **83**, 586–590 (1963).
2218	Sulfadimethoxine	5.98	U	–H	spectro	H_2O $t = 27 \pm 1$ $I = 0.2$	Ritschel: Yoshioka M, Hamamoto K, and Kubota T, Acid dissociation constants of sulfanilamides and substituent effects on the constants, *Yakugaku Zasshi*, **84**, 90–93 (1964).
2219	Sulfadimethoxine, N4-acetyl ($C_{14}H_{16}N_4O_5S$) 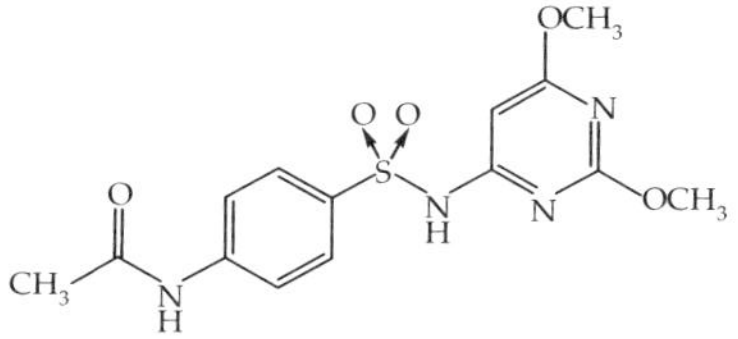	6.01	U	–H	potentio		Rieder J, Physicalisch-chemische und biologische untersuchungen an sulfonamiden, *Arzneim.-Forsch.*, **13**, 81–88 (1963). NB: See Sulphanilamide for details.
2220	Sulfadimethyloxazole ($C_{11}H_{13}N_3O_3S$) 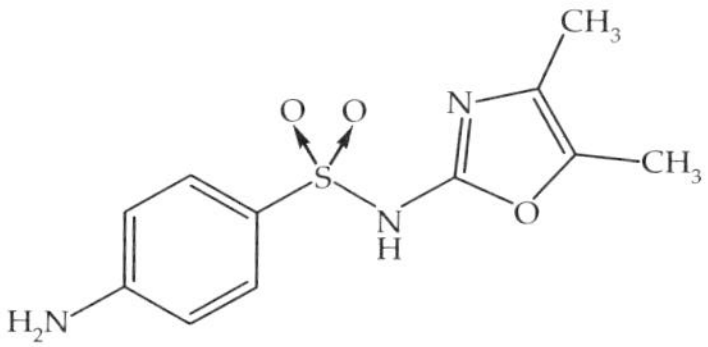	7.40	U	–H	potentio		Rieder J, Physicalisch-chemische und biologische untersuchungen an sulfonamiden, *Arzneim.-Forsch.*, **13**, 81–88 (1963). NB: See Sulphanilamide for details.
2221	Sulfadimidine	7.70	U	–H	potentio		Rieder J, Physicalisch-chemische und biologische untersuchungen an sulfonamiden, *Arzneim.-Forsch.*, **13**, 81–88 (1963). NB: See Sulphanilamide for details.

2222	Sulfadimidine	2.36	U	+H		H_2O (extrap) $t = 24 \pm 1$ $I \sim 0.002$	Chatten LG and Harris LE, Relationship between $pK_b(H_2O)$ of organic compounds and $E_{1/2}$ values in several nonaqueous solvents, *Anal. Chem.*, **34,** 1495–1501 (1962).
2223	Sulfadimidine	7.37	U	−H		$t = 25$	Chatten LG (ed.), *Pharmaceutical Chemistry*, vol. 1, Dekker, New York, pp. 85–87 (1966); Bell, Roblin, *JACS*, 1942.
2224	Sulfadimidine (sulfamethazine)	2.36 7.38	U U	+H −H			N&K; Suzuki A, Higuchi WI and Ho NFH, Theoretical model studies of drug absorption and transport in the gastrointestinal tract, *J. Pharm. Sci.*, **59,** 651–659 (1970).
2225	Sulfaethidole	1.93 5.60	U U	+H −H			N&K; Suzuki A, Higuchi WI and Ho NFH, Theoretical model studies of drug absorption and transport in the gastrointestinal tract, *J. Pharm. Sci.*, **59,** 651–659 (1970).
2226	Sulfaethidole	5.65	U	−H	potentio		Rieder J, Physicalisch-chemische und biologische untersuchungen an sulfonamiden, *Arzneim.-Forsch.*, **13,** 81–88 (1963). NB: See Sulphanilamide for details.
2227	Sulfafurazole	5.00	U	−H	potentio	$t = 28$	Nakagaki M, Koga N and Terada H, Physicochemical studies on the binding of chemicals with proteins. I. The binding of several sulphonamides with serum albumin, *J. Pharm. Soc. Jpn.*, **83,** 586–590 (1963).
2228	Sulfafurazole	4.79	U	−H	spectro	H_2O $t = 27 \pm 1$ $I = 0.2$	Ritschel: Yoshioka M, Hamamoto K, Kubota T, Acid dissociation constants of sulfanilamides and substituent effects on the constants, *Yakugaku Zasshi*, **84,** 90–93 (1964).

(continued)

No.	Name	pK_a value(s)	Data quality	Ionization type	Method	Conditions t°C; I or c M	Comments and Reference(s)
2229	Sulfaguanidine	2.75 12.05	U U	+H –H			Craig; N&K; Suzuki A, Higuchi WI and Ho NFH, Theoretical model studies of drug absorption and transport in the gastrointestinal tract, *J. Pharm. Sci.*, **59,** 651–659 (1970).
2230	Sulfamerazine	6.98	U	–H	potentio		Rieder J, Physicalisch-chemische und biologische untersuchungen an sulfonamiden, *Arzneim.-Forsch.*, **13,** 81–88 (1963). NB: See Sulphanilamide for details.
2231	Sulfamerazine	2.26 7.06	U U	+H –H			N&K; Suzuki A, Higuchi WI and Ho NFH, Theoretical model studies of drug absorption and transport in the gastrointestinal tract, *J. Pharm. Sci.*, **59,** 651–659 (1970).
2232	Sulfamerazine	7.06	U	–H			Chatten LG (ed.), *Pharmaceutical Chemistry*, vol. 1, Dekker, New York, pp. 85–87 (1966); Bell, Roblin, *JACS*, 1942.
2233	Sulfamerazine	2.08	U	+H		H_2O (extrap) $t = 24 \pm 1$ $I \sim 0.002$	Chatten LG and Harris LE, Relationship between pK_b (H_2O) of organic compounds and $E_{1/2}$ values in several nonaqueous solvents, *Anal. Chem.*, **34,** 1495–1501 (1962).
2234	Sulfamerazine	6.85	U	–H	spectro	H_2O $t = 27 \pm 1$ $I = 0.2$	Ritschel: Yoshioka M, Hamamoto K and Kubota T, Acid dissociation constants of sulfanilamides and substituent effects on the constants, *Yakugaku Zasshi*, **84,** 90–93 (1964).

No.	Compound	pKa			Method	Conditions	Reference
2235	Sulfamethizole ($C_9H_{10}N_4O_2S_2$)	2.20 5.45	U U	+H −H			Craig
2236	Sulfamethizole	2.00 5.45	U U	+H −H			N&K; Suzuki A, Higuchi WI and Ho NFH, Theoretical model studies of drug absorption and transport in the gastrointestinal tract, *J. Pharm. Sci.*, **59,** 651–659 (1970).
2237	Sulfamethizole	5.45	U	−H			Chatten LG (ed.), *Pharmaceutical Chemistry*, vol. 1, Dekker, New York, pp. 85–87 (1966); Merck 7.
2238	Sulfamethizole	5.22	U	−H	spectro	H_2O $t = 27 \pm 1$ $I = 0.2$	Ritschel: Yoshioka M, Hamamoto K and Kubota T, Acid dissociation constants of sulfanilamides and substituent effects on the constants, *Yakugaku Zasshi*, **84,** 90–93 (1964).
2239	Sulfamethoxazole ($C_{10}H_{11}N_3O_3S$) 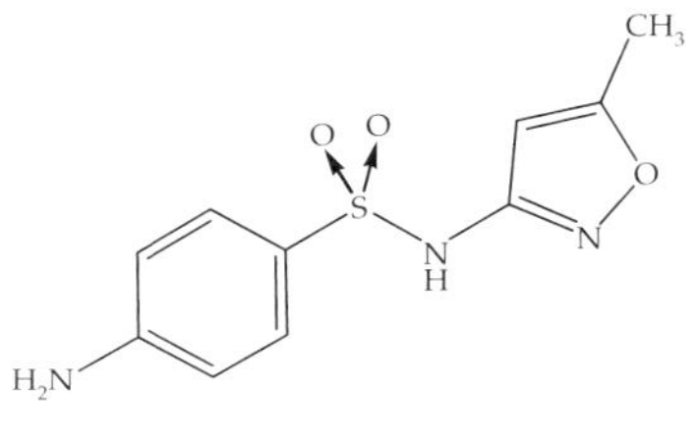	5.60 ± 0.05	A	−H	spectro, potentio		Rudy BC and Senkowski BZ, Sulfamethoxazole, *APDS*, **2,** 467–486 (1973). NB: Schmidli, Hoffmann-La Roche, Inc., unpublished data; Nakagaki M, Koga N and Terada H, Physicochemical studies on the binding of chemicals with proteins. I. The binding of several sulphonamides with serum albumin, *J. Pharm. Soc. Jpn.*, **83,** 586–590 (1963). "The pK_a of sulfamethoxazole has been determined spectrophotometrically to be 5.55 ± 0.05 and by titration of sulfamethoxazole in an excess of 0.1N HCl to be 5.63 ± 0.03. These values agree very well with the pK_a value of 5.60 at 25° reported by Nakagaki, Toga, and Terada."

(continued)

No.	Name	pK_a value(s)	Data quality	Ionization type	Method	Conditions t°C; I or c M	Comments and Reference(s)
2240	Sulfamethoxazole	5.72	U	–H	spectro	H_2O $t = 27 \pm 1$ $I = 0.2$	Ritschel: Yoshioka M, Hamamoto K and Kubota T, Acid dissociation constants of sulfanilamides and substituent effects on the constants, *Yakugaku Zasshi*, **84,** 90–93 (1964).
2241	Sulfamethoxazole	6.03	U	–H	potentio		Rieder J, Physicalisch-chemische und biologische untersuchungen an sulfonamiden, *Arzneim.-Forsch.*, **13,** 81–88 (1963). NB: See Sulphanilamide for details.
2242	Sulfamethoxazole, N4-acetyl ($C_{12}H_{13}N_3O_4S$) 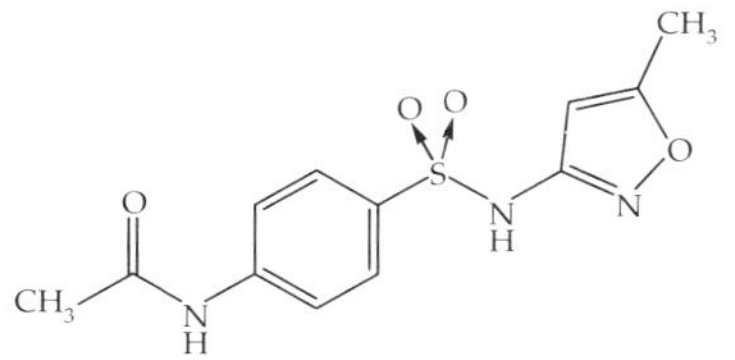	5.54	U	–H	potentio		Rieder J, Physicalisch-chemische und biologische untersuchungen an sulfonamiden, *Arzneim.-Forsch.*, **13,** 81–88 (1963). NB: See Sulphanilamide for details.
2243	Sulfamethoxydiazine ($C_{11}H_{12}N_4O_3S$) 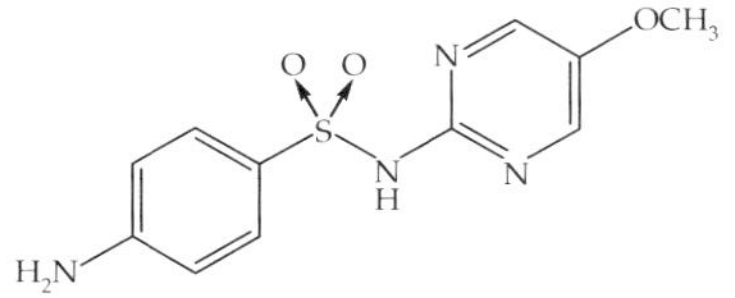	7.02	U	–H	potentio		Rieder J, Physicalisch-chemische und biologische untersuchungen an sulfonamiden, *Arzneim.-Forsch.*, **13,** 81–88 (1963). NB: See Sulphanilamide for details.
2244	Sulfamethoxypyridazine	7.20	U	–H	potentio		Rieder J, Physicalisch-chemische und biologische untersuchungen an sulfonamiden, *Arzneim.-Forsch.*, **13,** 81–88 (1963). NB: See Sulphanilamide for details.

2245	Sulfamethoxypyridazine	2.06 7.00	U U	+H –H			N&K; Suzuki A, Higuchi WI and Ho NFH, Theoretical model studies of drug absorption and transport in the gastrointestinal tract, *J. Pharm. Sci.*, **59,** 651–659 (1970).
2246	Sulfamethoxypyridazine	6.7	U	–H			Chatten LG (ed.), *Pharmaceutical Chemistry*, vol. 1, Dekker, New York, pp. 85–87 (1966); Merck 7.
2247	Sulfamethoxypyridazine	7.17	U	–H	spectro	H_2O $t = 27 \pm 1$ $I = 0.2$	Ritschel: Yoshioka M, Hamamoto K and Kubota T, Acid dissociation constants of sulfanilamides and substituent effects on the constants, *Yakugaku Zasshi*, **84,** 90–93 (1964).
2248	Sulfamethoxypyridazine, N4-acetyl	6.78	U	–H	potentio		Rieder J, Physicalisch-chemische und biologische untersuchungen an sulfonamiden, *Arzneim.-Forsch.*, **13,** 81–88 (1963). NB: See Sulphanilamide for details.
2249	Sulfamonomethoxine ($C_{11}H_{12}N_4O_2S$) 	5.94	U	–H	potentio		Rieder J, Physicalisch-chemische und biologische untersuchungen an sulfonamiden, *Arzneim.-Forsch.*, **13,** 81–88 (1963). NB: See Sulphanilamide for details.
2250	Sulfanilamide	10.08	U	–H	potentio		Rieder J, Physicalisch-chemische und biologische untersuchungen an sulfonamiden, *Arzneim.-Forsch.*, **13,** 81–88 (1963). NB: The pK_a potentiometric titrations were contracted out to Dr. B. Schmidli. No other experimental details were given.

(continued)

No.	Name	pK_a value(s)	Data quality	Ionization type	Method	Conditions t°C; I or c M	Comments and Reference(s)
2251	Sulfanilamide	2.08	U	+H		H_2O (extrap) $t = 24 \pm 1$ $I \sim 0.002$	Chatten LG and Harris LE, Relationship between $pK_b(H_2O)$ of organic compounds and $E_{1/2}$ values in several nonaqueous solvents, *Anal. Chem.*, **34,** 1495–1501 (1962).
2252	Sulfanilamide	10.4	U	–H			W&G: Brueckner AH, *Yale J. Biol. Med.*, **15,** 813 (1943).
2253	3-Sulphanilamido-4,5-dimethylpyrazole ($C_{11}H_{14}N_4O_2S$) 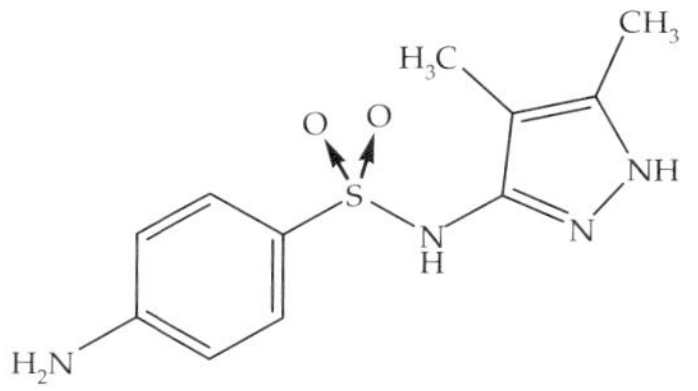	8.32	U	–H	potentio		Rieder J, Physicalisch-chemische und biologische untersuchungen an sulfonamiden, *Arzneim.-Forsch.*, **13,** 81–88 (1963). NB: See Sulphanilamide for details.
2254	2-Sulfanilamido-4,5,6-trimethoxypyrimidine ($C_{13}H_{16}N_4O_5S$) 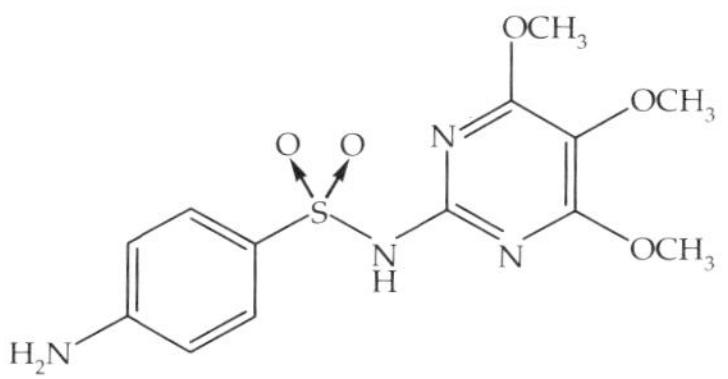	6.54	U	–H	potentio		Rieder J, Physicalisch-chemische und biologische untersuchungen an sulfonamiden, *Arzneim.-Forsch.*, **13,** 81–88 (1963). NB: See Sulphanilamide for details.
2255	Sulfanilic acid ($C_6H_7NO_3S$) 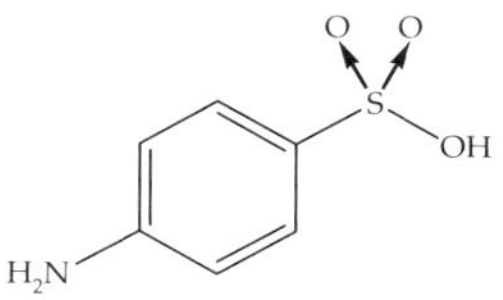	3.2	U	+H			W&G: Kolthoff IM and Stenger VA, *Volumetric Analysis*, vol. 1, 2nd Edn., Interscience, NY (1942).

2256	Sulfaperine (also called sulfamethyldiazine; 4-methylsulfadiazine)	6.77	U	−H	potentio		Rieder J, Physicalisch-chemische und biologische untersuchungen an sulfonamiden, *Arzneim.-Forsch.*, **13**, 81–88 (1963). NB: See Sulphanilamide for details.
2257	Sulfaphenazole	1.9 6.50	U U	+H −H			Craig; N&K; Suzuki A, Higuchi WI and Ho NFH, Theoretical model studies of drug absorption and transport in the gastrointestinal tract, *J. Pharm. Sci.*, **59**, 651–659 (1970).
2258	Sulfaphenazole	6.09	U	−H	potentio		Rieder J, Physicalisch-chemische und biologische untersuchungen an sulfonamiden, *Arzneim.-Forsch.*, **13**, 81–88 (1963). NB: See Sulphanilamide for details.
2259	Sulfaphenazole	5.89	U	−H	spectro	H_2O $t = 27 \pm 1$ $I = 0.2$	Ritschel: Yoshioka M, Hamamoto K and Kubota T, Acid dissociation constants of sulfanilamides and substituent effects on the constants, *Yakugaku Zasshi*, **84,** 90–93 (1964).
2260	Sulfaphenazole	5.80–6.13	U	−H			Elofsson R, Nilsson SO and Agren A, Complex formation between macromolecules and drugs IV, *Acta Pharm. Suec.*, **7,** 473–482 (1970). NB: This compound was not measured in the study but sourced from literature (ref. 8, Rieder J, Physicalisch-chemische und biologische untersuchungen an sulfonamiden, *Arzneim.-Forsch.*, **13,** 81–88 (1963); 9, Kruger-Thiemer, E and Bunger P, *Proc. Eur. Soc. Drug Toxicity.* vol. XI (1965) (Excerpta Medica Intl. Congress Series no. 97). See Sulphanilamide for details.

(*continued*)

No.	Name	pK_a value(s)	Data quality	Ionization type	Method	Conditions t°C; I or c M	Comments and Reference(s)
2261	Sulfapyridine	2.58 8.43	U U	+H −H			Craig; NB: This appears to be the Bell and Roblin data. See no. 1377. N&K; Suzuki A, Higuchi WI and Ho NFH, Theoretical model studies of drug absorption and transport in the gastrointestinal tract, *J. Pharm. Sci.*, **59**, 651–659 (1970).
2262	Sulfapyridine	8.56	U	−H	spectro	H_2O $t = 27 \pm 1$ $I = 0.2$	Ritschel: Yoshioka M, Hamamoto K and Kubota T, Acid dissociation constants of sulfanilamides and substituent effects on the constants, *Yakugaku Zasshi*, **84**, 90–93 (1964).
2263	Sulfathiazole	2.36 7.12	U U	+H −H			Craig; N&K; Suzuki A, Higuchi WI and Ho NFH, Theoretical model studies of drug absorption and transport in the gastrointestinal tract, *J. Pharm. Sci.*, **59**, 651–659 (1970). NB: Kapoor VK, Sulfathiazole, *APDS*, **22**, 389–430 (1993), citing Clarke, gave 7.1.
2264	Sulfathiazole	7.25	U	−H	potentio		Rieder J, Physicalisch-chemische und biologische untersuchungen an sulfonamiden, *Arzneim.-Forsch.*, **13**, 81–88 (1963). NB: See Sulphanilamide for details.
2265	Sulfathiazole	7.59	U	−H		$t = 25$	Chatten LG (ed.), *Pharmaceutical Chemistry*, vol. 1, Dekker, New York, pp. 85–87 (1966); Bell, Roblin, *JACS* 1942
2266	Sulfathiazole	7.00	U	−H	potentio	$t = 28$	Nakagaki M, Koga N and Terada H, Physicochemical studies on the binding of chemicals with proteins. I. The binding of several sulphonamides with serum albumin, *J. Pharm. Soc. Jpn.*, **83**, 586–590 (1963).

2267	Sulfathiazole	7.23	U	–H	spectro	H_2O $t = 27 \pm 1$ $I = 0.2$	Ritschel: Yoshioka M, Hamamoto K and Kubota T, Acid dissociation constants of sulfanilamides and substituent effects on the constants, *Yakugaku Zasshi*, **84,** 90–93 (1964).
2268	Sulfinpyrazone	2.8 3.25	U U	–H –H			Craig; N&K; Ritschel gave 2.8 only; ref Geigy Pharmaceuticals. NB: See Oxyphenbutazone.
2269	Sulfisomidine (2-Sulfanilamido-2, 4-dimethylpyrimidine)	7.25	U	–H	potentio		Rieder J, Physicalisch-chemische und biologische untersuchungen an sulfonamiden, *Arzneim.-Forsch.*, **13,** 81–88 (1963). NB: See Sulphanilamide for details.
2270	Sulfisomidine	2.36 7.5	U U	+H –H			N&K; Suzuki A, Higuchi WI and Ho NFH, Theoretical model studies of drug absorption and transport in the gastrointestinal tract, *J. Pharm. Sci.*, **59,** 651–659 (1970).
2271	Sulfisomidine	7.49	U	–H	spectro	H_2O $t = 27 \pm 1$ $I = 0.2$	Ritschel: Yoshioka M, Hamamoto K and Kubota T, Acid dissociation constants of sulfanilamides and substituent effects on the constants, *Yakugaku Zasshi*, **84,** 90–93 (1964).
2272	Sulfisomidine	7.57	U	–H	potentio		Rieder J, Physicalisch-chemische und biologische untersuchungen an sulfonamiden, *Arzneim.-Forsch.*, **13,** 81–88 (1963). NB: See Sulphanilamide for details.
2273	Sulfisomidine	7.17	U	–H	potentio	$t = 28$	Nakagaki M, Koga N and Terada H, Physicochemical studies on the binding of chemicals with proteins. I. The binding of several sulphonamides with serum albumin, *J. Pharm. Soc. Jpn.*, **83,** 586–590 (1963).

(continued)

No.	Name	pK_a value(s)	Data quality	Ionization type	Method	Conditions t°C; I or c M	Comments and Reference(s)
2274	Sulfisoxazole ($C_{11}H_{13}N_3O_3S$) 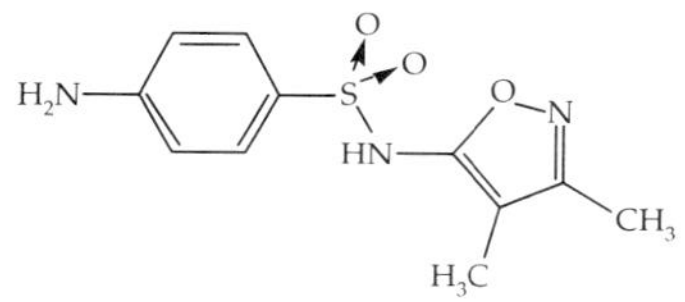	5.0	U	–H	spectro, potentio		Rudy BC and Senkowski BZ, Sulfisoxazole, *APDS*, **2,** 487–506 (1973). spectro: Motchane A, Hoffmann-La Roche, unpublished data. potentio: Rieder J, *Arzneim.-Forsch.*, **13,** 81–88 (1963); Nakagaki M, Koga N and Terada H, *J. Pharm. Soc. Jpn.*, **83,** 586 (1963).
2275	Sulfisoxazole	5.00	U	–H	potentio		Rieder J, Physicalisch-chemische und biologische untersuchungen an sulfonamiden, *Arzneim.-Forsch.*, **13,** 81–88 (1963). NB: See Sulphanilamide for details.
2276	Sulfisoxazole, N4-acetyl	4.72	U	–H	potentio		Rieder J, Physicalisch-chemische und biologische untersuchungen an sulfonamiden, *Arzneim.-Forsch.*, **13,** 81–88 (1963). NB: See Sulphanilamide for details.
2277	Sulthiame ($C_{10}H_{14}N_2O_4S_2$) 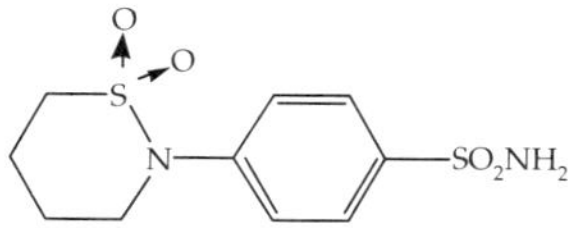	10.0	U	–H			Craig
2278	Sumatriptan ($C_{14}H_{21}N_3O_2S$) 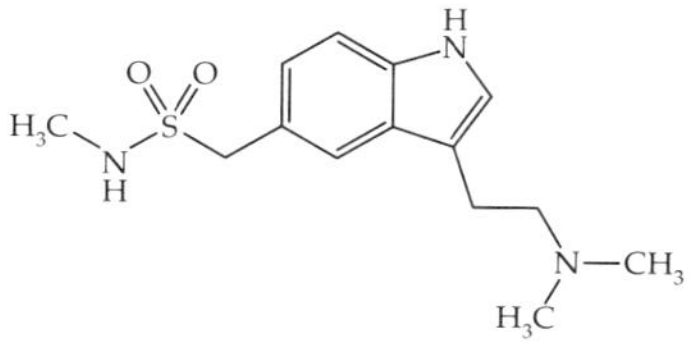	9.5	U	+H			Lombardo F, Obach RS, Shalaeva MY and Gao F, Prediction of human volume of distribution values for neutral and basic drugs. 2. Extended data set and leave-class-out statistics, *J. Med. Chem.*, **47,** 1242–1250 (2004). Ref. 300: O'Connor DO, Capel C, Rycroft W, Tattersall FD, Locker K, Sohal B, Graham MI and Evans DC. Influence of the physicochemistry on the brain penetration of the Triptans in rat. Poster presented at the XIV Course in Drug Research, June 5–6, 1997, Helsinki, Finland.

2279	p-Synephrine	9.3	U	−H			Craig
		10.2	U	+H			
2280	Tacrine ($C_{13}H_{14}N_2$)	9.8	U	+H			Lombardo F, Obach RS, Shalaeva MY and Gao F, Prediction of human volume of distribution values for neutral and basic drugs. 2. Extended data set and leave-class-out statistics, *J. Med. Chem.*, **47,** 1242–1250 (2004). Ref. 301: Desai MC, Thadeio PF, Lipinski CA, Liston DR, Spencer RW and Williams IH, Physical parameters for brain uptake: Optimizing logP, logD and pK_a of THA, *Bioorg. Med. Chem. Lett.*, **1,** 411–414 (1991).
2281	Tamoxifen	8.9	U	+H			McEvoy
2282	Tamoxifen	8.45	U	−H	potentio	H_2O $t = 25$	Bergström CAS, Strafford M, Lazorova L, Avdeef A, Luthman K and Artursson P, Absorption classification of oral drugs based on molecular surface properties, *J. Med. Chem.*, **46**(4), 558–570 (2003). NB: From extrapolation of aqueous-methanol mixtures to 0% methanol.
2283	Temazepam ($C_{16}H_{13}ClN_2O_2$)	1.6	U	+H			Craig
2284	Tenoxicam ($C_{13}H_{11}N_3O_4S_2$)	1.1	U	+H			Al-Obaid AM, Mian MS, Tenoxicam, *APDS*, **22,** 431–459 (1993). NB: The structure is probably better shown in the keto (carbon acid form). NB: No reference was given, although there was acknowledgement of assistance from Hoffman La-Roche with "product information and relevant literature."
		5.3	U	−H			

(continued)

No.	Name	pK_a value(s)	Data quality	Ionization type	Method	Conditions t°C; I or c M	Comments and Reference(s)
2285	Terbutaline	8.8 10.1 11.2	U U U	+H −H −H			Ahuja S and Ashman J, Terbutaline sulfate, *APDS*, **19,** 601–625 (1990). Ahuja S, personal communication (1976). NB: Same values in Craig and N&K.
2286	Terbutaline	8.67 ± 0.01 9.97 ± 0.01 11.02 ± 0.01	A U U	+H −H −H	potentio	H_2O $t = 25.0$ $I = 0.15$ (KCl)	Sirius Technical Application Notes, vol. 2, pp. 36–37 (1995). Sirius Analytical Instruments, Ltd., Forest Row, East Sussex, RH18 5DW, UK. NB: Concentration of analyte, 1.81.9 mM.
2287	Terfenadine	6.56	U	+H	potentio	H_2O (extrap) $t = 22$	Badwan AA, Al Kaysi HN, Owais LB, Salem MS and Arafat TA, Terfenadine, *APDS*, **19,** 627–662 (1990). NB: Titrations in methanol-water mixtures extrapolated to zero percent methanol by the Yasuda-Shedlovsky procedure, according to Newton DW, Murray WJ and Lovell MW, *J. Pharm. Sci.*, **71**(12), 1363–1366 (1982). Unpublished data obtained by M. Omari, Jordanian Pharmaceutical Manufacturing Co., Amman, Jordan. See also Australian National Drug Information Service, *Aust. J. Pharmacy*, **67,** 1077 (1986).
2288	Tetracaine	8.39	U	+H			Riaz M, Tetracaine, *APDS*, **18,** 379–411 (1989). NB: See Dittgen M and Jensch HP, Influence of the physicochemical properties of the drug on its release from acrylic films, *Acta Pharm. Jugo.*, **38**(4), 315–320 (1988). Also see Doyle TD and Proctor JB, Dermination of procaine and related local anaesthetics, *J. Assoc. Off. Anal. Chem.*, **58,** 88–92, 93–94 (1975); N&K; Anon.

No.	Compound	pK value	Assessment	Type	Method	Conditions	Reference
2289	Tetracaine	8.5	U	+H			W&G: Truant AP and Takman B, Differential physical-chemical and neuropharmacologic properties of local anesthetic agents, *Anesth. Analg.*, **38,** 478–484 (1959).
2290	Tetracycline	3.30 7.68 9.69 (10.7)	U U U U	−H −H +H −H		H_2O $t = 25.0$	Ali SL, Tetracycline hydrochloride, *APDS*, **13,** 597–645 (1984). pK_{a4} assigned to ionisation of the second proton of the extended phenolic-enolic system (see: Rigler NE, *et al.*, *Anal. Chem.*, **37,** 872–875 (1965)).
2291	Tetrahydrocannabinol ($C_{21}H_{30}O_2$)	10.6	U	−H			Albert A and Serjeant EP, *The determination of ionization constants*, 3rd Edn., Chapman and Hall, London (1984).
2292	Thebaine	8.15	U	+H	potentio	50% aq EtOH	Farmilo CG, Oestreicher PM and Levi L, Physical methods for the identification of narcotics. IB Common physical constants for identification of ninety-five narcotics and related compounds, *Bull. Narcotics*, UN Dept. Social Affairs, vol. 6, pp. 7–19 (1954); cited in Clouet DH (ed.), *Narcotic Drugs Biochemical Pharmacology*, Plenum Press, New York, 52–53 (1971).
2293	Thenyldiamine ($C_{14}H_{19}N_3S$)	3.94 8.93	U U	+H +H	potentio	H_2O $t = 25$ $c = 0.002$ to 0.01	Lordi NG and Christian JE, Physical properties and pharmacological activity: Antihistaminics, *J. Am. Pharm. Ass. Sci. Edn.*, **45,** 300–305 (1956). Cited in Perrin Bases no. 1124 ref. L54. NB: Used glass electrode in cell with liquid junctions. See Chlorpheniramine (no. 1704) for details.

(continued)

No.	Name	pK_a value(s)	Data quality	Ionization type	Method	Conditions t°C; I or c M	Comments and Reference(s)
2294	Thiabendazole ($C_{10}H_7N_3S$) 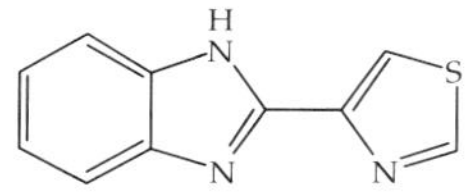	4.7	U	+H			McEvoy
2295	Thiamine ($C_{12}H_{17}N_4OS$) 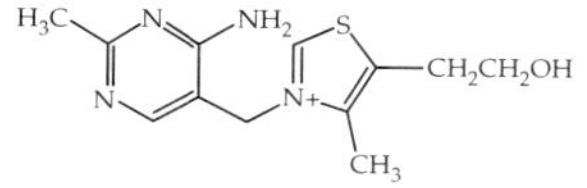	4.8 9.0	U U	+H +H			Carlin HS and Perkins AJ, Compatibilities of parenteral medications, *Am. J. Hosp. Pharm.*, **25,** 270–279 (1968).
2296	Thiamine	4.8	U	+H			Craig
2297	Thiamine	4.8 9.0	U U	+H +H			N&K; W&G: Carlin and Perkins data (see no. 2295).
2298	Thiamine	4.8	U	+H			Ritschel: Gupta VD, Cadwallader DE, Herman HB and Honigberg IL, Effect of pH and dye concentration on the extraction of a thiamine dye salt by an organic solvent, *J. Pharm. Sci.*, **57,** 1199–1202 (1968); cited $K_a = 1.58 \times 10^{-5}$; from Merck Index.
2299	Thiamine-O-monophosphate ($C_{12}H_{18}N_4O_4PS$) 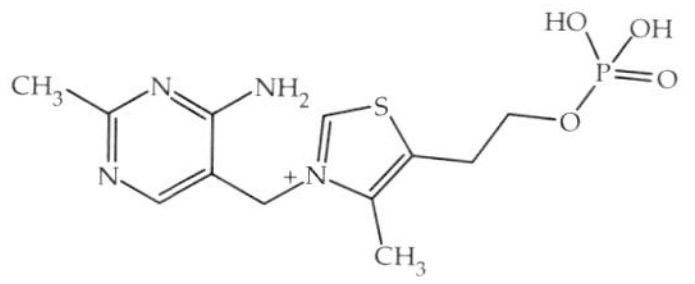	2.40 4.80 6.27 9.65 10.20	U U U U U	−H +H −H +H −H	potentio		Okamoto K, The alternating current polarography of thiamine O-monophosphate, *Bull. Chem. Soc. Jpn.*, **36,** 366–371 (1963). Cited in Perrin Bases, Suppl. No. 7798, ref. O3. NB: Polarographic behaviour was related to acid-base equilibria. No further details.
2300	Thioguanine ($C_5H_5N_5S$) 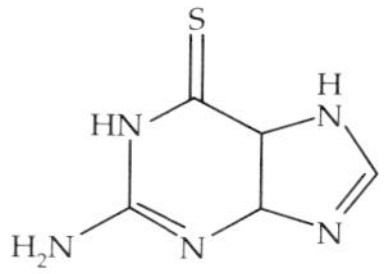	8.22	U	−H			Craig

2301	Thioridazine	9.5	U	+H		Abdel-Moety EM and Al-Rashood KA, Thioridazine and Thioridazine hydrochloride, *APDS*, **18,** 459–525 (1989). NB: See Remington, Foye and Green AL, Ionization constants and water solublities of some aminoalkylphenothiazine tranquilizers and related compounds, *J. Pharm. Pharmacol.*, **19,** 10–16 (1967). See Amitriptylline for details.
2302	Thiothixene ($C_{23}H_{29}N_3O_2S$)	7.67	U	+H		Craig
		7.97	U	+H		
2303	2-Thiouracil ($C_4H_4N_2OS$)	7.46	U	−H		N&K; Garrett ER and Weber DJ, Metal complexes of thiouracils I. Stability constants by potentiometric titration studies and structures of complexes, *J. Pharm. Sci.*, **59,** 1383–1398 (1970).
2304	Thonzylamine ($C_{16}H_{22}N_4O$)	2.17	U	+H		Craig
		8.96	U	+H		N&K; Chatten LG (ed.), *Pharmaceutical Chemistry*, vol. 1, Dekker, New York, pp. 85–87 (1966); Tolstoouhov (see Pheniramine, no. 2081).

(continued)

No.	Name	pK_a value(s)	Data quality	Ionization type	Method	Conditions t°C; I or c M	Comments and Reference(s)
2305	Thonzylamine	2.08 8.84	A U	+H +H	potentio	H_2O $t = 25$ $c = 0.002$ to 0.01	W&G: Lordi NG and Christian JE, Physical properties and pharmacological activity: Antihistaminics, *J. Am. Pharm. Ass. Sci. Edn.*, **45,** 300–305 (1956). See Chlorpheniramine (no. 1704) for details.
2306	Thyronine, L- ($C_{15}H_{12}I_3NO_4$) 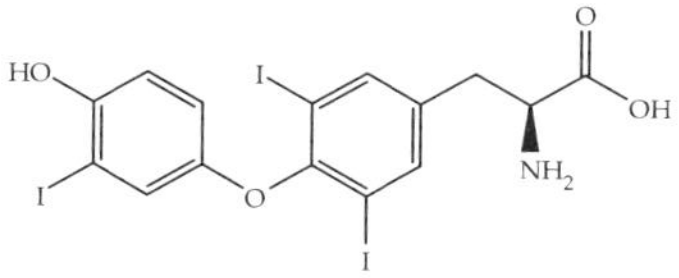	9.6	U	–H		H_2O	W&G: Smith RL, *Med. Chem.*, **2,** 477 (1964). NB: This reference could not be identified; see Liothyronine.
2307	Tiaprofenic acid (tiprofenic acid) ($C_{14}H_{12}O_3S$) 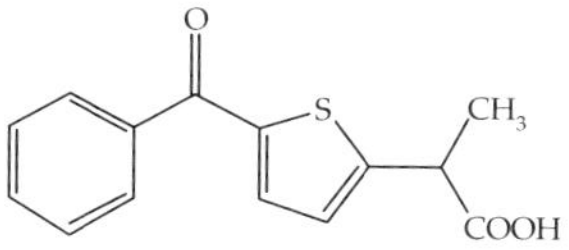	3.0	U	–H			Craig
2308	Ticarcillin ($C_{15}H_{16}N_2O_6S_2$)	2.89 ± 0.05 3.28 ± 0.04	U U	–H –H	potentio	H_2O $t = 25.0$ $I = 0.15$ (KCl)	Sirius Technical Application Notes, vol. 2, p. 109 (1995). Sirius Analytical Instruments, Ltd., Forest Row, East Sussex, RH18 5DW, UK.
		2.55 3.42	U U	–H –H			Craig; N&K; Anon.

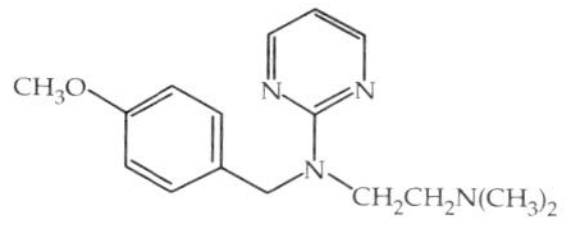

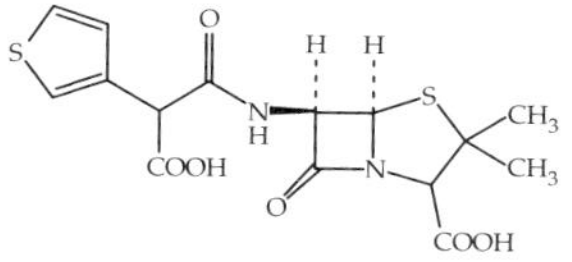

2309	Ticrynafen ($C_{13}H_8Cl_2O_4S$) 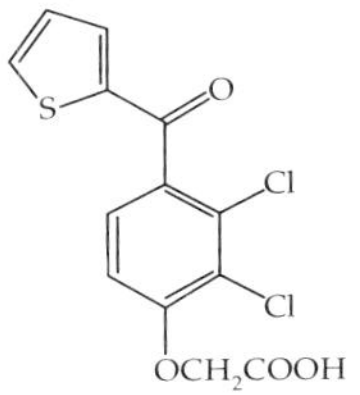	2.7	U	−H		Craig
2310	Timolol	8.8	U	+H		Lombardo F, Obach RS, Shalaeva MY and Gao F, Prediction of human volume of distribution values for neutral and basic drugs. 2. Extended data set and leave-class-out statistics, *J. Med. Chem.*, **47,** 1242–1250 (2004). Ref. 275: Barbato F, Caliendo G, LaRotonda MI, Morrica P, Silipo C and Vittoria A, Relationships between octanol-water partition data, chromatographic indices and their dependence on pH in a set of beta-adrenoceptor blocking agents, *Farmaco*, **45,** 647–663 (1990). See also Hinderling PH, Schmidlin O and Seydel JK, Quantitative relationships between structure and pharmacokinetics of beta-adrenoceptor blocking agents in man, *J. Pharmacokin. Biopharm.*, **12,** 263–287 (1984).
2311	Timoprazole ($C_{13}H_{11}N_3OS$) 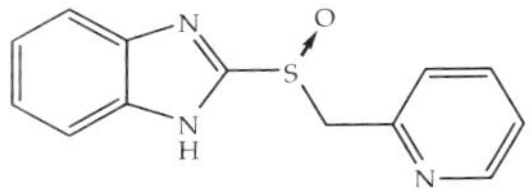	3.12 8.82	U U	+H +H		Craig

(continued)

No.	Name	pK_a value(s)	Data quality	Ionization type	Method	Conditions t°C; I or c M	Comments and Reference(s)
2312	Tiotidine ($C_{12}H_{18}N_6S_2$) 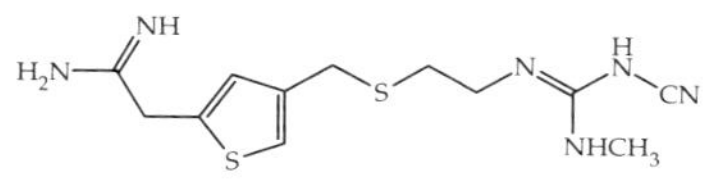	6.80	U	+H	potentio		Shankley NP, Black JW, Ganellin CR and Mitchell RC, *Br. J. Pharmacol.*, **94**, 264–274 (1988). NB: See Ranitidine for details. Also cited in Craig.
2313	Tobramycin ($C_{18}H_{37}N_5O_9$)	6.7	U	+H			Dash AK, Tobramycin, *APDS*, **24**, 579–613 (1996).
		8.3	U	+H			Albert A and Serjeant EP, *The determination of ionization constants*, 3rd Edn., Chapman and Hall, NY 174 (1984).
		9.9	U	+H			NB: Raymond GG and Born JL, An updated pK_a listing of medicinal compounds, *Drug Intel. Clin. Pharm.*, **20**, 683–686 (1986) reported four pK_a values (6.2, 7.4, 7.6, and 8.6).
2314	Tocainide	7.54	U	+H			Craig
2315	Tolamolol ($C_{20}H_{26}N_2O_4$) 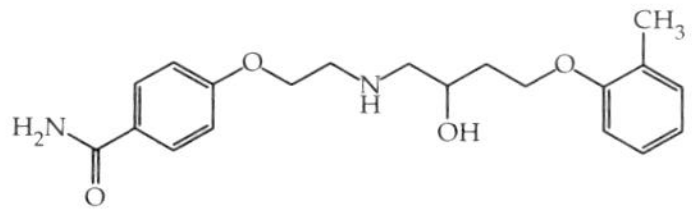	7.90	U	+H			Lombardo F, Obach RS, Shalaeva MY and Gao F, Prediction of human volume of distribution values for neutral and basic drugs. 2. Extended data set and leave-class-out statistics, *J. Med. Chem.*, **47**, 1242–1250 (2004). Ref. 275: Barbato F, Caliendo G, LaRotonda MI, Morrica P, Silipo C and Vittoria A, Relationships between octanol-water partition data, chromatographic indices and their dependence on pH in a set of beta-adrenoceptor blocking agents, *Farmaco*, **45**, 647–663 (1990).

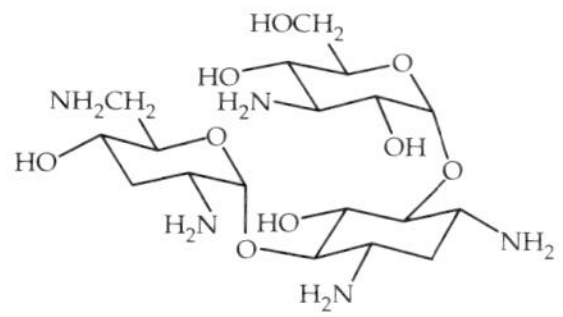

No.	Name	pKa			Method	Conditions	Reference
2316	Tolamolol	7.94	U	+H			Hinderling PH, Schmidlin O and Seydel JK, Quantitative relationships between structure and pharmacokinetics of beta-adrenoceptor blocking agents in man, *J. Pharmacokin. Biopharm.*, **12,** 263–287 (1984).
2317	Tolazamide	3.1	U	−H			Craig; N&K; Avery
		5.7	U	+H			
2318	Tolbutamide	5.43	U	−H	spectro	$t = 25$	Forist AA and Chulski T, pH-solubility relations for 1-butyl-3-(p-tolylsulfonyl)urea (Orinase) and its metabolite, 1-butyl-3-(p-carboxyphenylsulfonyl)urea, Metabolism, **5,** 807–812 (1956). NB: The metabolite was found to have $pK_a = 3.54$ at 37.5 °C by the solubility method. See also Glibenclamide for further discussion. See also Beyer WF, Jensen EH, Tolbutamide, *APDS*, **3,** 513–543 (1974).
		5.32	U	−H	soly/pH	$t = 37.5$	
2319	Tolterodine ($C_{22}H_{31}NO$) 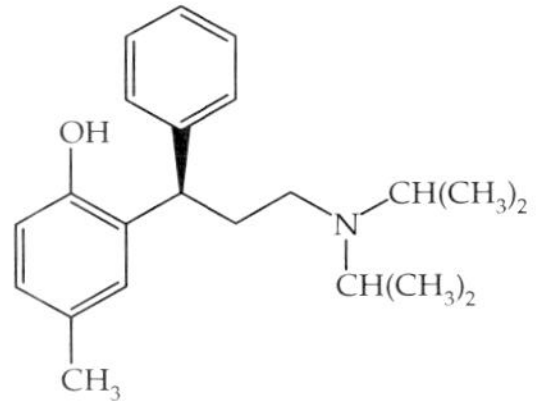	9.8	U	+H	potentio		Lombardo F, Obach RS, Shalaeva MY and Gao F, Prediction of human volume of distribution values for neutral and basic drugs. 2. Extended data set and leave-class-out statistics, *J. Med. Chem.*, **47,** 1242–1250 (2004). Ref. 302: Detrol® LA Capsules Monograph. Physician's Desk reference, 2001. On-line Version. Medical Economics Company, Montvale NJ, 2001; potentiometric titration.
2320	Tramazoline ($C_{13}H_{17}N_3$) 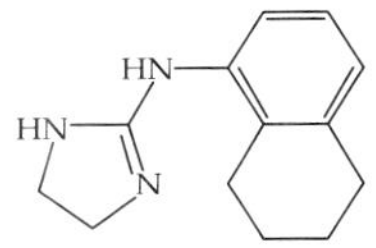	10.66	U	+H			Craig

(continued)

No.	Name	pK_a value(s)	Data quality	Ionization type	Method	Conditions t°C; I or c M	Comments and Reference(s)
2321	Tranexamic acid ($C_8H_{15}NO_2$)	4.3 10.6	U U	–H +H			N&K; S&R
2322	Tranylcypromine ($C_9H_{11}N$)	8.2	U	+H			Craig; N&K; Ritschel gave 8.2; ref. SKF NB: See also Abdel-Aleem H, El-Ashmawy MB, Belai F, El-Amam AA and Brittain HG, Tranylcypromine Sulfate, *APDS*, **25**, 501–533 (1997) (no specific reference).
2323	Trazodone	6.79	U	+H	potentio		Lombardo F, Obach RS, Shalaeva MY and Gao F, Prediction of human volume of distribution values for neutral and basic drugs. 2. Extended data set and leave-class-out statistics, *J. Med. Chem.*, **47**, 1242–1250 (2004). Ref. not given: Potentiometric titration.
2324	Triamterene ($C_{12}H_{11}N_7$)	6.2	U	+H			Craig; Clarke, p. 1037 N&K; Ritschel gave 6.2; ref. Smith, Kline & French.
2325	Trichloromethiazide	8.6	U	–H			Craig; Foye 3rd; see Azatadine from N&K; Anon.

2326	Trifluoperazine	4.10 8.36	U U	+H +H	potentio	Post A, Warren, RJ, Zarembo JE, Trifluoperazine hydrochloride, *APDS*, **9**, 544–579 (1980). "The apparent pK_{a1} and pK_{a2} have been determined using titrimetric and solubility measurements. As reported by many of these investigators, the determination of the pK_a of phenothiazines, in general, are difficult to obtain because of their poor water solubility. Thus the use of the term apparent pK_a. However, Green, using solubility measurements, did indeed confirm that apparent pK_{a2} for trifluoperazine is approximately 8.1–confirming the results obtained by titrimetric measurements."

pK_{a1}	pK_{a2}	Procedure (reference)
3.9	8.4	Titrimetric; Chatten LG and Harris LE, *Anal. Chem.*, **34**, 1495–1501 (1962).
3.9	8.1	Titrimetric; Murthy KS and Zografi G, *J. Pharm. Sci.*, **59**, 1281–1285 (1970).
4.10	8.36	Titrimetric; Sorby DL, Plein EM and Benmaman D, *J. Pharm. Sci.*, **55**, 785–794 (1966).
–	8.1	Solubility; Green AL, *J. Pharm. Pharmacol.*, **19**, 10–16 (1967). NB: See Amitriptylline for details.
–	8.3	TLC; Kraus L and Dumont E, *J. Chromatog.*, **56**, 159–162 (1971).

(continued)

No.	Name	pK_a value(s)	Data quality	Ionization type	Method	Conditions t°C; I or c M	Comments and Reference(s)
2327	Trifluoperazine	8.1	U	+H			Zografi G and Munshi MV, Effect of chemical modification on the surface activity of some phenothiazine derivatives, *J. Pharm. Sci.*, **59**, 819–822 (1970); cited in Ritschel.
2328	5-Trifluoromethyl-2′-deoxyuridine ($C_{10}H_{11}F_3N_2O_5$)	7.85	U	−H			Nestler HJ and Garrett ER, Prediction of stability in pharmaceutical preparations XV, Kinetics of hydrolysis of 5-trifluoromethyl-2′-deoxyuridine, *J. Pharm. Sci.*, 1117–1125 (1968). Cited in Ritschel.
2329	Triflupromazine	9.45	U	+H			Florey K, Triflupromazine hydrochloride, *APDS*, **2**, 523–550 (1973). NB: *APDS*, **5**, 557: Erratum. "The pK_a of 6.5 as listed in volume 2, p. 533 is in error. Reexamination by Dr. H. Jacobson gave a pK_a value of 9.45, in reasonable agreement with values determined by references 8 and 9.
2330	Triflupromazine	9.2	U	+H			Zografi G, Munshi MV, Effect of chemical modification on the surface activity of some phenothiazine derivatives, *J. Pharm. Sci.*, **59**, 819–822 (1970); cited in Ritschel.

2331	Trimeprazine (methylpromazine) ($C_{18}H_{22}N_2S$) 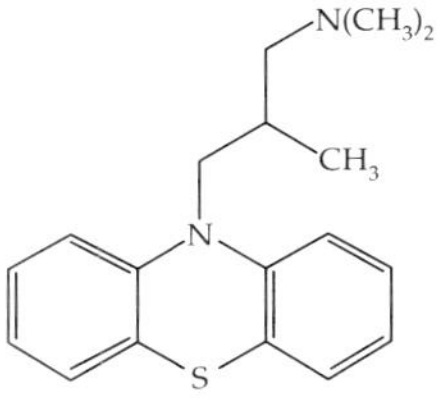	9.4	U	+H			W&G: Hulshoff and Perrin, 1976. NB: See Cyanopromazine for details.
2332	Trimeprazine	9.00	U	+H			Craig
2333	Trimethobenzamide ($C_{21}H_{28}N_2O_5$) 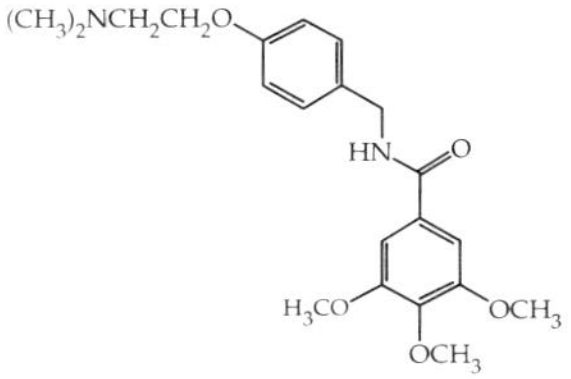	8.27 ± 0.03	U	+H	potentio	H_2O	Blessel KW, Rudy BC and Senkowski BZ, Trimethobenzamide hydrochloride, *APDS*, **2,** 551–570 (1973). NB: Cited Motchane A, Hoffmann-La Roche Inc., unpublished data.
2334	Trimethoprim	7.6	U	+H			Hansen I, Nielsen ML, Heerfordt L, Henriksen B and Bertelsen S, Trimethoprim in normal and pathological human lung tissue, *Chemotherapy*, **19**(4), 221–234 (1973). "The concentration of trimethoprim in normal and in pathological human lung tissue . . . was measured in 31 patients. . . .. The pK_a of 7.6 for trimethoprim implies that even small changes in the pH of the tissue in the acid direction will have a strong permeability-increasing effect on trimethoprim."
2335	Trimethoprim	7.2	U	+H			Kaplan SA, Weinfeld RE, Cotler S, Abruzzo CW and Alexander K, Pharmacokinetic profile of trimethoprim in dog and man, *J. Pharm. Sci.*, **59,** 358–363 (1970). NB: See also N&K.

(continued)

No.	Name	pK_a value(s)	Data quality	Ionization type	Method	Conditions t°C; I or c M	Comments and Reference(s)
2336	Trimethoprim	7.26	U	+H	potentio, CZE/pH		Lombardo F, Obach RS, Shalaeva MY and Gao F, Prediction of human volume of distribution values for neutral and basic drugs. 2. Extended data set and leave-class-out statistics, *J. Med. Chem.*, **47**, 1242–1250 (2004). Ref. not given: potentiometric titration; Capillary electrophoresis.
2337	Trimethoxy-3,4,5-phenylsulfonyl derivatives	7.1 to 7.6	U	+H			Foussard-Blanpin O, Uchida-Ernouf G, Moreau R and Adam Y, Central depressant activity of some derivatives with trimethoxy-3,4,5-phenylsulfonyl groups, *Ann. Pharm. Fr.*, **36,** 581–586 (1978). "The central depressant activity of 11 synthetic derivatives, differing from one another only by the nature of the chain linked to trimethoxy-3,4, 5-phenylsulfonyl was studied in mice. It was shown that the most active molecules are a hydroxysulfone with a primary alcohol grouping and a low partition coefficient in an octanol/water system; and 3 aminosulfones with a nitrogen atom, a part of a piperidine cycle or substituted by alkyls and with log P close to 2 and pK_a ranging from 7.1 to 7.6."

2338	Trimipramine ($C_{20}H_{26}N_2$) 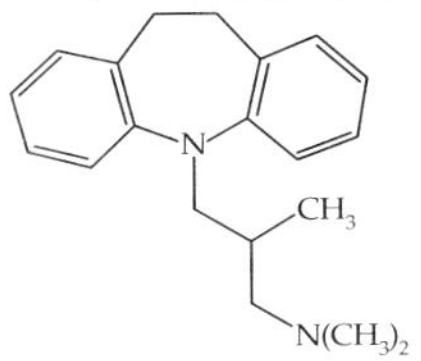	8.0	U	+H			McEvoy
2339	Trimipramine	9.24	U	+H	potentio		Lombardo F, Obach RS, Shalaeva MY and Gao F, Prediction of human volume of distribution values for neutral and basic drugs. 2. Extended data set and leave-class-out statistics, *J. Med. Chem.*, **47,** 1242–1250 (2004). Ref. not given: potentiometric titration.
2340	Tripelennamine	4.20 8.71	U U	+H +H			Craig; N&K; Chatten LG (ed.), *Pharmaceutical Chemistry*, vol. 1, Dekker, New York, pp. 85–87 (1966); Tolstoouhov.
2341	Tripelennamine	8.3	U	+H			Ritschel: Robson JM and Stacey RS, Recent advances in pharmacology, 4th Edn., Little Brown and Co., Boston, p. 108 (1968).
2342	Tripelennamine	3.92 8.96	A U	+H +H	potentio	H_2O $t = 25$ $c = 0.002$ to 0.01	W&G: Lordi NG and Christian JE, Physical properties and pharmacological activity: Antihistaminics, *J. Am. Pharm. Ass. Sci. Edn.*, **45,** 300–305 (1956). See Chlorpheniramine (no. 1704) for details.
2343	Triprolidine	3.6 9.3	U U	+H +H			Benezra SA and Yang C, Triprolidine hydrochloride, *APDS*, **8,** 509–528 (1979). Cited Morgan TA, Burroughs Wellcome Co., personal communication . NB: Craig gave only one value (6.4, +H).

(continued)

No.	Name	pK_a value(s)	Data quality	Ionization type	Method	Conditions t°C; I or c M	Comments and Reference(s)
2344	Triprolidine	~6.5	U	+H			DeAngelis RL, Kearney MF and Welch RM, Determination of triprolidine in human plasma by quantitative TLC, *J. Pharm. Sci.*, **66,** 841–843 (1977).
2345	Troleandomycin ($C_{41}H_{67}NO_{15}$) 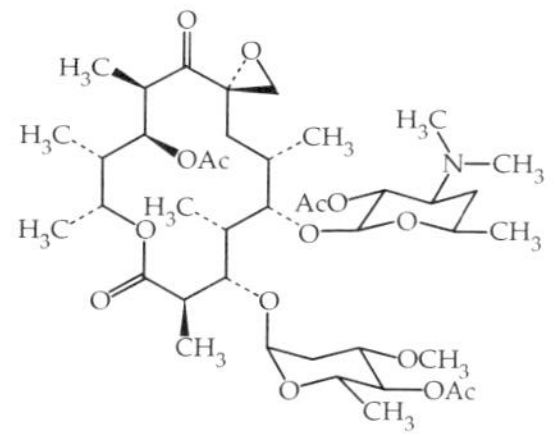	6.6	U	+H			Craig; N&K; Merck 9.
2346	Tromethamine	8.10	U	+H			Ritschel: Bruice TC and York JL, *JACS*, **83,** 1382–1387 (1961); cited in Brooke D, Guttman DE, Complex formation influence on reaction rate. IV. Studies on the kinetic behaviour of 3-carbethoxy-1-pyridinium cation, *J. Pharm. Sci.*, **57,** 1677–1684 (1968).
2347	Tropacocaine	4.32	U	+H			Ritschel: NB: This is a pK_b value. The pK_a value (assuming it is measured at 25 °C) = 9.69. Cited from *Merck* 8th Edn., p. 1083.
2348	Tropacocaine	9.68	U	+H			W&G: Kolthoff IM, The dissociation constants, solubility product and titration of alkaloids, *Biochem. Z.*, **162,** 289–353 (1925); Christophers SR, *Trans. Farad. Soc.*, **39,** 333–338 (1943).

2349	DL-Tropic acid	4.20	U	−H			Ritschel: Randinitis EJ, Barr M, Wormser HC and Nagwekar JB, Kinetics of urinary excretion of D-(-)-mandelic acid and its homologs. I. Mutual inhibitory effect of D-(-)-mandelic acid and its certain homologs on their renal tubular secretion in rats, *J. Pharm. Sci.*, **59**, 806–812 (1970). NB: See Mandelic acid (no. 762) for details.
2350	Tropicamide ($C_{17}H_{20}N_2O_2$) 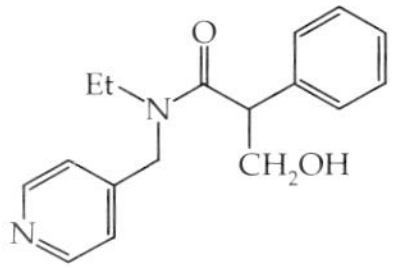	5.3	U	+H	potentio		Blessel KW, Rudy BC and Senkowski BZ, Tropicamide, *APDS*, **3**, 565–580 (1974). NB: No references given.
		5.2	U	+H	spectro		
2351	Tropicamide	5.25	U	+H	potentio		N&K; *APDS*, **3** (see no. 2350)
2352	Tyramine ($C_8H_{11}NO$) 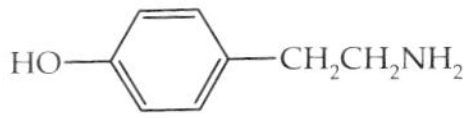	9.30	U	+H, −H			Craig
		10.9	U	−H, +H			
2353	Tyramine	9.5	U	+H, −H			Lewis GP, The importance of ionization in the activity of sympathomimetic amines, *Br. J. Pharmacol.*, **9**, 488–493 (1954); cited in W&G.
		10.8	U	−H, +H			
2354	Uracil ($C_4H_4N_2O_2$) 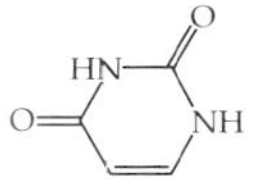	9.21 ± 0.01	U	−H	potentio	H_2O	Sirius Technical Application Notes, vol. 2, p. 6 (1995). Sirius Analytical Instruments Ltd., Forest Row, East Sussex, RH18 5DW, UK.
		13.28 ± 0.01	U	−H		$t = 25.0$	
						$I = 0.16$ (KCl)	
		9.0	U	−H			Ritschel: Nestler HJ and Garrett ER, Prediction of stability in pharmaceutical preparations XV, Kinetics of hydrolysis of 5-trifluoromethyl-2′-deoxyuridine, *J. Pharm. Sci.*, 1117–1125 (1968).
		13.0	U	−H			

(continued)

No.	Name	pK_a value(s)	Data quality	Ionization type	Method	Conditions t°C; I or c M	Comments and Reference(s)
2355	Urea (CH_4N_2O)	0.18	U	+H			McLean WM, Poland DM, Cohon MS, Penzotti SC and Mattocks AM, Effect of tris (hydroxymethyl) aminomethane on removal of urea by peritoneal dialysis, *J. Pharm. Sci.*, **56**, 1614–1621 (1967). Cited in W&G.
2356	Uric acid ($C_5H_4N_4O_3$)	5.47 ± 0.07	U	−H	potentio	H_2O $t = 25$	Bernoulli AL and Loebenstein A, Dissociation constants of uric acid, *Helv. Chim. Acta*, **23**, 245–247 (1940). Used a glass electrode with a salt bridge. Cited in Kortum, ref. B25.
		10.3	U	−H			See also W&G: White A, *et al.*, *Principles of Biochemistry*, McGraw-Hill, NY, p. 184 (1968).
2357	Venlafaxine ($C_{17}H_{27}NO_2$)	9.5	U	+H	potentio		Lombardo F, Obach RS, Shalaeva MY and Gao F, Prediction of human volume of distribution values for neutral and basic drugs. 2. Extended data set and leave-class-out statistics, *J. Med. Chem.*, **47**, 1242–1250 (2004). Ref. 292: Dallet P, Labat L, Richard M, Langlois MH and Dubost JP, A reversed phase HPLC method development for the separation of new antidepressants, *J. Liq. Chrom. Rel. Technol.*, **25**, 101–111 (2002): potentiometric titration.
2358	Verapamil	8.6	U	+H	potentio	H_2O	Chang ZL, Verapamil, *APDS*, **17**, 643–674 (1988). NB: No reference given. Extrapolated from MeOH-water mixtures. Solubility-pH data (Yunker M and Woodward S, Abbott Labs, personal communication) with the standard solubility-pH dependence equation gave a pK_a value of 9.06 ± 0.27.

2359	Verapamil	8.92	U	+H	CZE/pH	H_2O $t = 25$ $I = 0.1$	Lombardo F, Obach RS, Shalaeva MY and Gao F, Prediction of human volume of distribution values for neutral and basic drugs. 2. Extended data set and leave-class-out statistics, *J. Med. Chem.*, **47,** 1242–1250 (2004). Ref. 304: Hasegawa J, Fujita T, Hayashi Y, Iwamoto K and Watanabe J, pK_a determination of verapamil by liquid-liquid partition, *J. Pharm. Sci.*, **73,** 442–445 (1984). At 31 °C, the value was 8.79, and at 37 °C, the value was 8.68; 305: Ishihama Y, Oda Y and Asakawa N, Microscale determination of dissociation constants of multivalent pharmaceuticals by capillary electrophoresis, *J. Pharm. Sci.*, **83,** 1500–1507 (1994).
		8.92 ± 0.03	U	+H	partition	H_2O $t = 25.0 \pm 0.1$ $I = 0.0315$	
2360	Vidarabine ($C_{10}H_{13}N_5O_4$) NH$_2$ N N N N HOCH$_2$ O HO OH	3.55 11.4	A U	+H (N1) −H (OH)	potentio	H_2O $t = 20$ $I = 0.1$ (KCl)	Hong W, Chang T and Daly RE, Vidarabine, *APDS*, **15,** 647–672 (1986). NB: Craig gave 3.5, 12.5. 8. Sober HA (ed.), *Handbook of Biochemistry*, 2nd Edn., CRC, Cleveland OH (1970) J5. 9. Serjeant EP, Dempsey B, *Ionisation Constants of Organic Acids in Aqueous Solution*, IUPAC, Pergamon Press, Oxford, 561 (1979). 10. Martin RB and Mariam YH, Interactions between metal ions and nucleic bases, nucleosides, and nucleotides in solution, *Met. Ions Biol. Syst.*, **8,** 57–124 (1979). 11. Martell AE and Schwarzenbach J, *Helv. Chim. Acta*, **39,** 653–661 (1956). NB: Gave p$K_{a1} = 3.55 \pm 0.02$ for adenosine by potentiometry at $I = 0.1$ M (KCl) and $t = 20$ °C.

(continued)

No.	Name	pK_a value(s)	Data quality	Ionization type	Method	Conditions t°C; I or c M	Comments and Reference(s)
							12. Wallenfels K and Sund H, Mechanism of hydrogen transfer with pyridine nucleotides. III. Binary and ternary zinc complexes as models for enzyme-coenzyme bonds, *Biochem. Z.*, **329,** 41–47 (1957). 13. Gonnella NC, Nakanishi H, Holtwick JB, *et al.*, Studies of tautomers and protonation of adenine and its derivatives by nitrogen-15 nuclear magnetic resonance spectroscopy, *JACS*, **105,** 2050–2055 (1983). NB: Gave pK_{a1} = 4.1 ± 0.1 and pK_{a2} = 10.1 ± 0.2. Supports the current assignments.
2361	Viloxazine ($C_{13}H_{19}NO_3$) 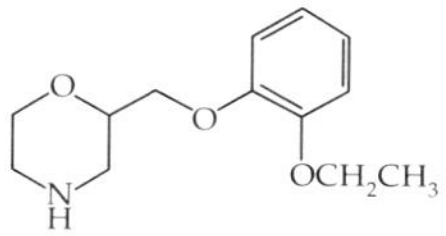	8.1	U	+H			Craig
2362	Vinblastine	5.0 7.0	U U	+H +H	potentio	66% DMF/ H_2O	Burns JH, Vinblastine sulfate, *APDS*, **1,** 450 (1972). NB: Beer CT, Cutts JH and Noble RL, US Pat. 3,097,137 July 9, 1963. The latest version of ACD/pK_a (ACD Labs, Toronto, Canada) gave the following values for vinblastine (applicable to water at 25 °C): 7.64 ± 0.60 (catharanthine 2° alicyclic nitrogen); 5.62 ± 0.70 (vindoline 3° alicyclic nitrogen); 1.23 ± 0.70 (vindoline 2,3-dihydroindole). The corresponding values for vincristine (applicable to water

						at 25 °C) are: 7.64 ± 0.60 (catharanthine 2° alicyclic nitrogen); 5.54 ± 0.70 (vindoline 3° alicyclic nitrogen); −2.95 ± 0.60 (catharanthine indole).
						N&K; Merck 9.
2363	Viomycin ($C_{25}H_{43}N_{13}O_{10}$) 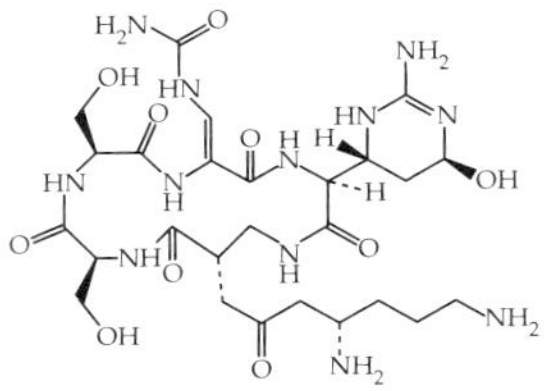	8.2 10.3	U U	+H +H, −H		ACD/pK_a suggests >14 for the guanidine, 10.3 for the terminal amino, 10.3 for the –OH adjacent to guanidine, 9.9 for the amide nitrogen included in the conjugated system, 8.8 for the 2° –NH_2 group. All other functionalities are predicted to be outside the range 0 to 14.
2364	Viomycin	2.8 5.87 13.4	U U U	? ? ?		Ritschel cited Dyer JR, Hayes HB, Miller EG, Jr, Chemistry of Viomycin, presented at the Third Interscience Conference on Antimicrobial Agents and Chemotherapy, Washington (Oct 28–30), 1963. NB: The 2.8 and 5.87 values do not correspond to any of the pK_a values estimated by ACD/pK_a, but may be pK_b values corresponding to the two amino groups. The value of 13.4 may be the pK_a for the guanidine function.
2365	Warfarin	5.0	U	−H	$t = 20.0$	Babhair SA, Tariq M and Al-Badr AA, Warfarin, *APDS*, **14,** 423–450 (1985). NB: BPC, p. 990. May be from O'Reilly RA, Nelson E and Levy G. Physicochemical and physiologic factors affecting the absorption of warfarin in man. *J. Pharm. Sci*, **55**, 435–437 (1966).

(continued)

Appendix B *(continued)*

No.	Name	pK_a value(s)	Data quality	Ionization type	Method	Conditions t°C; I or c M	Comments and Reference(s)
2366	Warfarin	5.5	U	−H			Robinson DS, Benjamin DM and McCormack JJ, Interaction of warfarin and nonsystemic gastrointestinal drugs, *Clin. Pharmacol. Ther.*, **12,** 491–495 (1971). "Interactions between warfarin and cholestyramine . . . were studied in 6 normal subjects. . . .. *In vitro* experiments demonstrated significant binding of warfarin to cholestyramine at a pH above its pK_a of 5.5."
2367	Xanthine ($C_5H_4N_4O_2$) 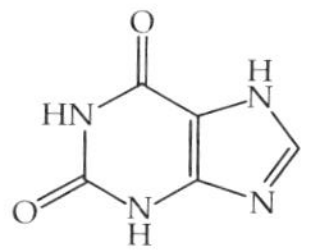	9.95	U	−H			Cohen JL and Connors KA, Stability and structure of some organic molecular complexes in aqueous solution, *J. Pharm. Sci.*, **59,** 1271–1276 (1970); cited in Ritschel.
2368	Zimeldine (zimelidine) ($C_{16}H_{17}BrN_2$) 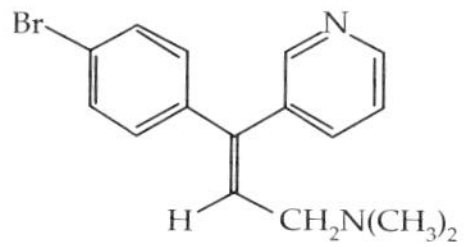	3.8 8.74	U U	+H +H			Craig

INDEX

ALPHABETICAL INDEX OF DRUGS OR DRUG RELATED COMPOUNDS FOUND IN PK_A DATABASE FILES

Appendix A – pK_a values found with significant data quality information
Appendix B – pK_a values found with little or no data quality information

Compound Name	*Appendix A*	*Appendix B*
A		
Acebutolol ($C_{18}H_{28}N_2O_4$)	1167	1523–1526
Acenocoumarin ($C_{19}H_{15}NO_6$)	637	
Acenocoumarol ($C_{19}H_{15}NO_6$)		1527
Acepromazine ($C_{19}H_{22}N_2OS$)	16	
7-Acetamidonitrazepam ($C_{17}H_{13}N_3O_4$)	205	
Acetaminophen ($C_8H_9NO_2$)	1–7	1528–1530
Acetanilide (C_8H_9NO)	8–9	1531
Acetarsone ($C_8H_{10}AsNO_5$)	10	
Acetazolamide ($C_4H_6N_4O_3S_2$)	11	1532–1535
Acetic acid ($C_2H_4O_2$)	12,252	
Acetohydroxamic acid ($C_2H_5NO_2$)		1536
N^2-Acetylacyclovir ($C_{10}H_{13}N_5O_4$)	20	
N-Acetylaminosalicylic acid ($C_9H_9NO_4$)	13	
3-Acetylazuloic acid ($C_{13}H_{10}O_3$)	115	
α-Acetylmethadol (Levomethadyl acetate) ($C_{23}H_{31}NO_2$)	14	1537
6-Acetylmorphine ($C_{19}H_{21}NO_4$)	15	
N-Acetylnorfloxacin ($C_{18}H_{20}FN_3O_4$)	1224	
N-(3′-Acetylphenyl)anthranilic acid ($C_{15}H_{13}NO_3$)	552–553	
Acetylpromazine (Acepromazine) ($C_{19}H_{22}N_2OS$)	16	
Aconitine ($C_{34}H_{47}NO_{11}$)	17	
Acridine derivatives	18–19	
Acridine orange ($C_{17}H_{19}N_3$)	18	
Acridine yellow ($C_{15}H_{15}N_3$)	18	
Acriquine	9	
Acyclovir ($C_8H_{11}N_5O_3$)	20–21	1538
Adenine ($C_5H_5N_5$)	22, 241	
Adenosine ($C_{10}H_{13}N_5O_4$)	23	1539
Adinazolam ($C_{19}H_{18}ClN_5$)	24	
Adrenaline ($C_9H_{13}NO_3$)	25, 497–502	

Adriamycin ($C_{27}H_{29}NO_{11}$)	469–470	1821–1822
Ajmaline ($C_{20}H_{26}N_2O_2$)		1540
Ala-Ile ($C_9H_{18}N_2O_3$)	453	
Ala-Leu ($C_9H_{18}N_2O_3$)	453	
Ala-Phe ($C_{12}H_{16}N_2O_3$)	453	
Albendazole sulphoxide ($C_{12}H_{15}N_3O_3S$)	26	
Albuterol (Salbutamol) ($C_{13}H_{21}NO_3$)	27, 1260	1541
Alclofenac ($C_{11}H_{11}ClO_3$)		1542–1543
Alfentanil ($C_{21}H_{32}N_6O_3$)		1544–1545
Alginic acids		1546
Allobarbital ($C_{10}H_{12}N_2O_3$)	158–160	
Allopurinol ($C_5H_4N_4O$)	96	1547–1548
Alphaprodine ($C_{16}H_{23}NO_2$)		1549
Alprenolol ($C_{15}H_{23}NO_2$)	28	1550
Altretamine (hexamethylmelamine) ($C_9H_{18}N_6$)		1551
Alypine (Amydricaine) ($C_{16}H_{26}N_2O_2$)	398	
Amantadine ($C_{10}H_{17}N$)	29–30	1552–1555
Amdinocillin (Mecillinam) ($C_{15}H_{23}N_3O_3S$)	31	1556
Amifloxacin ($C_{16}H_{19}FN_4O_3$)	1225–1226	
Amikacin ($C_{22}H_{43}N_5O_{13}$)		1557
Amiloride ($C_6H_8ClN_7O$)	32–33	1558
Aminacrine ($C_{13}H_{10}N_2$)		1559
Amino acid esters	34	
Amino acids		1560
2-Aminoacridine ($C_{13}H_{10}N_2$)	19	
3-Aminoacridine ($C_{13}H_{10}N_2$)	18	
9-Aminoacridine ($C_{13}H_{10}N_2$)	19	1559
4-Aminobenzoic acid ($C_7H_7NO_2$)	35–39	1561–1562
γ-Aminobutyric acid (GABA) ($C_4H_9NO_2$)	378	
ε-Aminocaproic acid ($C_6H_{13}NO_2$)		1563
1′-Aminocyclopentanecarboxamidopenicillin ($C_{14}H_{21}N_3O_4S$)	1511	
3-Amino-5-heptafluorobutyramido-1,2,4-triazole (Guanazole prodrug) ($C_6H_4F_7N_5O$)	602	
4-Aminohippuric acid ($C_9H_{10}N_2O_3$)		1564–1565
2-Amino-4-[4′-hydroxyphenyl]butane ($C_{10}H_{15}NO$)	41	
2-Amino-5-[4′-hydroxyphenyl]pentane ($C_{11}H_{17}NO$)	42	
1′-Amino-3′-methylcyclopentanecarboxamidopenicillin ($C_{15}H_{23}N_3O_4S$)	1512	
2-Amino-2-methyl-3-hydroxyoctane ($C_8H_{21}NO$)	43	
7-Aminonitrazepam ($C_{15}H_{12}N_4O_3$)	205	
6-Aminopenicillanic acid ($C_8H_{12}N_2O_3S$)	40	1566
Aminophenazone ($C_{13}H_{17}N_3O$)	46–47	1569
N-(3′-Aminophenyl)anthranilic acid ($C_{13}H_{12}N_2O_2$)	552–553	

2-Amino-4-phenylbutane ($C_{10}H_{15}N$)	44	
1-Amino-1-phenylethane ($C_8H_{11}N$)	973	
2-Amino-5-phenylpentane ($C_{11}H_{17}N$)	45	
Aminophylline ($C_{16}H_{24}N_{10}O_4$)		1567
Aminopterin ($C_{19}H_{20}N_8O_5$)		1568
Aminopyrine (aminophenazone) ($C_{13}H_{17}N_3O$)	46–47	1569
4-Aminoquinaldine ($C_9H_8N_2$)	48	
2-Aminoquinoline ($C_9H_8N_2$)	48	
4-Aminosalicyclic acid ($C_7H_7NO_3$)	49	1570
5-Aminosalicyclic acid ($C_7H_7NO_3$)	50	
Aminothiadiazole ($C_2H_3N_3S$)		1571
Amiodarone ($C_{25}H_{29}I_2NO_3$)	51–55	
Amitriptyline ($C_{20}H_{23}N$)	56–59	
Amizil	9	
Amlodipine ($C_{20}H_{25}ClN_2O_5$)		1572
Ammonia (H_3N)	60	
Amobarbital ($C_{11}H_{18}N_2O_3$)	145–148	
Amoxapine ($C_{17}H_{16}ClN_3O$)		1573
Amoxicillin ($C_{16}H_{19}N_3O_5S$)	61–66	1574
Amphetamine ($C_9H_{13}N$)	67–68	1575
Amphetamine, 4-hydroxy ($C_9H_{13}NO$)	69	
Amphetamine, 3-methoxy ($C_{10}H_{15}NO$)	70	
Amphetamine, 4-methoxy ($C_{10}H_{15}NO$)	71	
Amphotericin B ($C_{47}H_{73}NO_{17}$)		1576–1577
Ampicillin ($C_{16}H_{19}N_3O_4S$)	72–76	
Amydricaine (Alypine) ($C_{16}H_{26}N_2O_2$)	398	
Amylobarbitone ($C_{11}H_{18}N_2O_3$)	145–148	
Amylocaine ($C_{14}H_{21}NO_2$)	398	
n-Amylpenilloic acid ($C_{13}H_{24}N_2O_3S$)	77–78	
Anagrelide ($C_{10}H_7$ (Cl_2N_3O)	79	
Anhydro-4-epitetracycline ($C_{22}H_{22}N_2O_7$)	491–492	
Anhydrochlortetracycline ($C_{22}H_{21}ClN_2O_7$)	80	
Anileridine ($C_{22}H_{28}N_2O_2$)		1578
Anisindione ($C_{16}H_{12}O_3$)	81	
Antazoline ($C_{17}H_{19}N_3$)		1579–1581
Antifebrin (C_8H_9NO)	8–9	1531
Antipyrine ($C_{11}H_{12}N_2O$)	9	1582–1584
Apazone ($C_{16}H_{20}N_4O_2$)	108–109	
Apoatropine	652	
Apomorphine ($C_{17}H_{17}NO_2$)	82–84	
Apressine	9	
Aprindine ($C_{22}H_{30}N_2$)	85	
Aprobarbital ($C_{10}H_{14}N_2O_3$)	152–154	
Arecaidine ($C_7H_{11}NO_2$)	86	
Arecaidine methyl ester ($C_8H_{13}NO_2$)	87	

Arecoline ($C_8H_{13}NO_2$)	88–89	
Arsthinol ($C_{11}H_{14}AsNO_3S_2$)	10	
1-Arylpiperazine derivatives	90	
Ascorbic acid ($C_6H_8O_6$)	91–93	1585
Aspartame ($C_{14}H_{18}N_2O_5$)	94	
Aspirin ($C_9H_8O_4$)	95–97	1586–1587
Astemizole ($C_{28}H_{31}FN_4O$)	98	
Atenolol ($C_{14}H_{22}N_2O_3$)	99–102, 1167	1588–1590
Atomoxetine ($C_{17}H_{21}NO{\cdot}HCl$)		1591
Atomoxetine ($C_{17}H_{22}ClNO$)		1591
Atorvastatin ($C_{33}H_{35}FN_2O_5$)	103	
Atrazine ($C_8H_{14}ClN_5$)	104	1592
Atropine ($C_{17}H_{23}NO_3$)	9, 105–107	1593–1596
Atropine (non-aqueous titration) ($C_{17}H_{23}NO_3$)	194	
Azapropazone (apazone) ($C_{16}H_{20}N_4O_2$)	108–109	
Azatadine ($C_{20}H_{22}N_2$)		1597
Azathioprine ($C_9H_7N_7O_2S$)	110	1598
Azelastine ($C_{22}H_{24}ClN_3O$)	111–112	
Azithromycin ($C_{38}H_{72}N_2O_{12}$)	113	
Azlocillin ($C_{20}H_{23}N_5O_6S$)		1599
Aztreonam ($C_{13}H_{17}N_5O_8S_2$)	114	
Azuloic acid ($C_{11}H_8O_2$)	115	
B		
Bacampicillin ($C_{21}H_{27}N_3O_7S$)		1600
Baclofen ($C_{10}H_{12}ClNO_2$)		1601
Bamipine ($C_{19}H_{24}N_2$)	116	
Barbital ($C_8H_{12}N_2O_3$)	126–131	1602
Barbituric acid ($C_4H_4N_2O_3$)	117	
Barbituric acid, 5-allyl-5-ethyl ($C_9H_{12}N_2O_3$)	151	
Barbituric acid, 5-allyl-5-isopropyl (aprobarbital) ($C_{10}H_{14}N_2O_3$)	152–155	
Barbituric acid, 5-allyl-5-isobutyl ($C_{11}H_{16}N_2O_3$)	155–156	
Barbituric acid, 5-allyl-5-(1-methylbutyl) (secobarbital) ($C_{12}H_{18}N_2O_3$)	157	1608
Barbituric acid, 5-allyl-5-(1-methylbutyl)-2-thio (Thiamylal) ($C_{12}H_{18}N_2O_2S$)		1610
Barbituric acid, 5-allyl-5-(1-methylpent-2-ynyl)-*N*-methyl (methohexital) ($C_{14}H_{18}N_2O_3$)		1609
Barbituric acid, 5-allyl-5-(1-methylpropyl) (Talbutal) ($C_{11}H_{16}N_2O_3$)		1611
Barbituric acid, 5-t-butyl-5-(3-methylbut-2-enyl) ($C_{13}H_{20}N_2O_3$)	164	
Barbituric acid, 5-cyclohex-1′-enyl-1,5-dimethyl (hexobarbital) ($C_{12}H_{16}N_2O_3$)	125	

Barbituric acid, -1′,5-spiro(cyclobutane) ($C_7H_8N_2O_3$) 191
Barbituric acid, -1′,5-spiro(cyclohexane) ($C_9H_{12}N_2O_3$) 193
Barbituric acid, -1′,5-spiro(cyclopentane) ($C_8H_{10}N_2O_3$) 192
Barbituric acid, -1′,5-spiro(cyclopropane) ($C_6H_6N_2O_3$) 190
Barbituric acid, 5,5-diallyl (allobarbital) ($C_{10}H_{12}N_2O_3$) 158–160
Barbituric acid, 5,5-dibromo ($C_4H_2Br_2N_2O_3$) 118
Barbituric acid, 5,5-dichloro ($C_4H_2Cl_2N_2O_3$) 119
Barbituric acid, 5,5-diethyl (barbital) ($C_8H_{12}N_2O_3$) 126–131 1602
Barbituric acid, 5,5-diethyl-1-benzoyl ($C_{15}H_{16}N_2O_4$) 132
Barbituric acid, 5,5-diethyl-1-benzyl ($C_{15}H_{18}N_2O_3$) 132
Barbituric acid, 5,5-diethyl-1-(2-bromobenzoyl) ($C_{15}H_{15}BrN_2O_4$) 132
Barbituric acid, 5,5-diethyl-1-(3-bromobenzoyl) ($C_{15}H_{15}BrN_2O_4$) 132
Barbituric acid, 5,5-diethyl-1-(4-bromobenzoyl) ($C_{15}H_{15}BrN_2O_4$) 132
Barbituric acid, 5,5-diethyl-1-(4-chlorobenzyl) ($C_{15}H_{17}ClN_2O_3$) 132
Barbituric acid, 5,5-diethyl-1-(2-methoxybenzoyl) ($C_{16}H_{18}N_2O_5$) 132
Barbituric acid, 5,5-diethyl-1-(3-methoxybenzoyl) ($C_{16}H_{18}N_2O_5$) 132
Barbituric acid, 5,5-diethyl-1-(4-methoxybenzoyl) ($C_{16}H_{18}N_2O_5$) 132
Barbituric acid, 5,5-diethyl-1-methyl (metharbital) ($C_9H_{14}N_2O_3$) 132 1603–1604
Barbituric acid, 5,5-diethyl-1-(2-methylbenzoyl) ($C_{16}H_{18}N_2O_4$) 132
Barbituric acid, 5,5-diethyl-1-(3-methylbenzoyl) ($C_{16}H_{18}N_2O_4$) 132
Barbituric acid, 5,5-diethyl-1-(4-methylbenzoyl) ($C_{16}H_{18}N_2O_4$) 132
Barbituric acid, 5,5-diethyl-1-(4-nitrobenzyl) ($C_{15}H_{17}N_3O_5$) 132
Barbituric acid, 5,5-diethyl-1-(4-nitrophenyl) ($C_{14}H_{15}N_3O_5$) 132
Barbituric acid, 5,5-diethyl-1-phenyl ($C_{14}H_{16}N_2O_3$) 132
Barbituric acid, 5,5-dimethyl ($C_6H_8N_2O_3$) 120–122
Barbituric acid, 5,5-di-(3-methylbut-2-enyl) ($C_{14}H_{20}N_2O_3$) 165
Barbituric acid, 1,5-dimethyl-5-ethyl ($C_8H_{12}N_2O_3$) 133

Barbituric acid, 1,5-dimethyl-5-iso-propyl ($C_9H_{14}N_2O_3$)	133	
Barbituric acid, 5,5-diphenyl ($C_{16}H_{12}N_2O_3$)	189	
Barbituric acid, 5-ethyl-5-n-butyl ($C_{10}H_{16}N_2O_3$)	137, 155	
Barbituric acid, 5-ethyl-5-*iso*-butyl (butabarbital) ($C_{10}H_{16}N_2O_3$)	139	1605
Barbituric acid, 5-ethyl-5-*sec*-butyl ($C_{10}H_{16}N_2O_3$)	138	
Barbituric acid, 5-ethyl-5-(1,3-dimethylbutyl) ($C_{12}H_{20}N_2O_3$)	149	
Barbituric acid, 5-ethyl-5-methyl ($C_7H_{10}N_2O_3$)	123	
Barbituric acid, 5-ethyl-5-(1-methylbut-1-enyl) ($C_{11}H_{16}N_2O_3$)	166–167	
Barbituric acid, 5-ethyl-5-(1-methylbutyl) (pentobarbitone) ($C_{11}H_{18}N_2O_3$)	140–143, 155	
Barbituric acid, 5-ethyl-5-(1-methylbutyl)-2-thio (thiopentone) ($C_{11}H_{18}N_2O_2S$)	144, 1436	1606–1607
Barbituric acid, 5-ethyl-5-(3-methylbutyl) (amobarbital; amylobarbitone) ($C_{11}H_{18}N_2O_3$)	145–148	
Barbituric acid, 5-ethyl-5-(3-methylbut-2-enyl) ($C_{11}H_{16}N_2O_3$)	162	
Barbituric acid, 5-ethyl-5-(4-methylpent-1-en-2-yl) ($C_{12}H_{18}N_2O_3$)	168	
Barbituric acid, 5-ethyl-5-(3-nitrophenyl) ($C_{12}H_{11}N_3O_5$)	186–187	
Barbituric acid, 5-ethyl-5-(4-nitrophenyl) ($C_{12}H_{11}N_3O_5$)	188	
Barbituric acid, 5-ethyl-5-phenyl (Phenobarbital) ($C_{12}H_{12}N_2O_3$)	173–184	1612–1613
Barbituric acid, 5-ethyl-5-phenyl-1-benzoyl ($C_{19}H_{16}N_2O_4$)	185	
Barbituric acid, 5-ethyl-5-*iso*-propyl ($C_9H_{14}N_2O_3$)	134–136	
Barbituric acid, 1-methyl-5,5-diallyl ($C_{11}H_{14}N_2O_3$)	133	
Barbituric acid, 1-methyl-5,5-di-n-propyl ($C_{11}H_{18}N_2O_3$)	133	
Barbituric acid, 1-methyl-5-ethyl-5-butyl ($C_{11}H_{18}N_2O_3$)	133	
Barbituric acid, 1-methyl-5-ethyl-5-phenyl ($C_{13}H_{14}N_2O_3$)		1614
Barbituric acid, 5-methyl-5-(3-methyl-but-2-enyl) ($C_{10}H_{14}N_2O_3$)	161	
Barbituric acid, 5-methyl-5-(4-methylpent-1-en-2-yl) ($C_{11}H_{16}N_2O_3$)	169	
Barbituric acid, 5-methyl-5-phenyl ($C_{11}H_{10}N_2O_3$)	170–171	
Barbituric acid, 5-methyl-5-phenyl-1-benzoyl ($C_{18}H_{14}N_2O_4$)	172	

Barbituric acid, 1-methyl-5-propyl-5-iso-propyl ($C_{11}H_{18}N_2O_3$)	133	
Barbituric acid, 5-methyl-5-iso-propyl ($C_8H_{12}N_2O_3$)	124	
Barbituric acid, 5-*iso*-propyl-5-(3-methylbut-2-enyl) ($C_{12}H_{18}N_2O_3$)	163	
Barbituric acids, *N*-methylated	133	
Bases (non-aqueous titrations)	194	
Bemegride ($C_8H_{13}NO_2$)		1615
Bencyclane ($C_{19}H_{31}NO$)	195	
Bendazol	9	
Bendroflumethiazide ($C_{15}H_{14}F_3N_3O_4S_2$)	196–197	
Beneperidol ($C_{22}H_{24}FN_3O_2$)		1616
Benidipine ($C_{28}H_{31}N_3O_6$)	198	
Benperidol ($C_{22}H_{24}FN_3O_2$)		1616
Benzacine	9	
2-Benzenesulfanilamidopyrimidine ($C_{10}H_9N_3O_2S$)	199	
2-Benzenesulfonamidopyridine ($C_{11}H_9NO_2S$)	200	
Benzenesulfonamide ($C_6H_7NO_2S$)	201	
Benzenesulfonamide analogues	201	
1-(1,2-Benzisothiazol-3-yl)piperazine ($C_{11}H_{13}N_3S$)	90	
Benzocaine ($C_9H_{11}NO_2$)	202–203	1617–1619
Benzocaine (non-aqueous titration)	194	
1,4-Benzodiazepines	204	
1,4-Benzodiazepine metabolites	205	
Benzoic acid ($C_7H_6O_2$)	206–212, 252, 649, 874	
Benzoic acid, 4-amino ($C_7H_7NO_2$)	212	
Benzoic acid, 4-amino, diethylaminoethyl ester ($C_{13}H_{20}N_2O_2$)	212	
Benzoic acid, 4-bromo ($C_7H_5BrO_2$)	212	
Benzoic acid, 4-bromo, diethylaminoethyl ester ($C_{13}H_{18}BrNO_2$)	212	
Benzoic acid, 4-chloro ($C_7H_5ClO_2$)	212	
Benzoic acid, diethylaminoethyl ester ($C_{13}H_{19}NO_2$)	212	
Benzoic acid, 4-chloro, diethylaminoethyl ester ($C_{13}H_{18}ClNO_2$)	212	
Benzoic acid, 4-ethoxy ($C_9H_{10}O_2$)	212	
Benzoic acid, 4-ethoxy, diethylaminoethyl ester ($C_{15}H_{23}NO_2$)	212	
Benzoic acid, 4-*N*-ethylamino ($C_9H_{11}NO_2$)	212	
Benzoic acid, 4-(*N*-ethylamino), diethylaminoethyl ester ($C_{19}H_{24}N_2O_2$)	212	
Benzoic acid, 4-fluoro ($C_7H_5FO_2$)	212	

Benzoic acid, 4-fluoro, diethylaminoethyl ester ($C_{13}H_{18}FNO_2$)	212	
Benzoic acid, 4-hydroxy ($C_7H_6O_3$)	212	
Benzoic acid, 4-hydroxy, diethylaminoethyl ester ($C_{13}H_{19}NO_3$)	212	
Benzoic acid, 4-methyl ($C_8H_8O_2$)	212	
Benzoic acid, 4-methyl, diethylaminoethyl ester ($C_{14}H_{21}NO_2$)	212	
Benzoic acid, 4-nitro ($C_7H_5NO_4$)	212	
Benzoic acid, 4-nitro, diethylaminoethyl ester ($C_{13}H_{18}N_2O_4$)	212	
Benzoic acid, 4-substituted, diethylaminoethyl esters	212	
Benzoylecgonine ($C_{16}H_{19}NO_4$)	213	
Benzoylecgonine methyl ester (Cocaine) ($C_{17}H_{21}NO_4$)	214, 346–347	1739–1744, 1988
Benzoylhydrazine ($C_7H_8N_2O$)	699	
Benzphetamine ($C_{17}H_{21}N$)		1620
Benzquinamide ($C_{22}H_{32}N_2O_5$)	215	1621
Benztropine ($C_{21}H_{25}NO$)		1622
Benzylamphetamine ($C_{16}H_{19}N$)		1623
Benzyldesthiopenicillin	959	
D-(-)-Benzyllactic acid	762	
Benzylpenicillin (Penicillin G) ($C_{16}H_{18}N_2O_4S$)	217, 958–959	
Benzylpenicilloic acid ($C_{16}H_{20}N_2O_5S$)	218–219	
Benzylpenicilloic acid α-benzylamide ($C_{23}H_{25}N_3O_4S$)	220–221	
Benzylpenilloic acid ($C_{15}H_{20}N_2O_3S$)	222	
Betahistine ($C_8H_{12}N_2$)		1624
Betaprodine ($C_{16}H_{23}NO_2$)		1625
Betaxolol ($C_{18}H_{29}NO_3$)		1626
Bethanidine ($C_{10}H_{15}N_3$)		1627
Bevantolol	1167	
Biscoumacetic acid ($C_{20}H_{12}O_8$)		1628
Bishydroxycoumarin ($C_{19}H_{12}O_6$)		1789
Bisoprolol ($C_{18}H_{31}NO_4$)	223	
Boric acid (H_3BO_3)	224	
Brequinar ($C_{23}H_{15}F_2NO_2$)	225	
Brinzolamide ($C_{12}H_{21}N_3O_5S_3$)	226	
Bromazepam ($C_{14}H_{10}BrN_3O$)		1629–1630
3-Bromoazuloic acid ($C_{11}H_7BrO_2$)	115	
Bromocresol green ($C_{21}H_{14}Br_4O_5S$)	227	
Bromocriptine ($C_{32}H_{40}BrN_5O_5$)		1631–1632
Bromodiphenhydramine ($C_{17}H_{20}BrNO$)		1633

Bromophenol blue ($C_{19}H_{10}Br_4O_5S$)	228	
Bromothen ($C_{14}H_{18}BrN_3S$)		1634
8-Bromotheophylline ($C_7H_7BrN_4O_2$)	229	
Brompheniramine ($C_{16}H_{19}BrN_2$)	230	1635–1636
Brucine ($C_{23}H_{26}N_2O_4$)	231	1637
Bufuralol ($C_{16}H_{23}NO_2$)	1167	1638–1639
Bumetanide ($C_{17}H_{20}N_2O_5S$)	232–233	1640
Bunolol (levobunolol) ($C_{17}H_{25}NO_3$)	1167	1641–1642
Bupivacaine ($C_{18}H_{28}N_2O$)	234–236	
Buprenorphine ($C_{29}H_{41}NO_4$)	237–238	
Bupropion ($C_{13}H_{18}ClNO$)		1643
Burimamide ($C_9H_{16}N_4S$)		1644
Buspirone ($C_{21}H_{31}N_5O_2$)	239	
Butabarbital ($C_{10}H_{16}N_2O_3$)	139	1605
Butacaine ($C_{18}H_{30}N_2O_2$)		1645
Butaclamol ($C_{25}H_{31}NO$)		1646
Butamben ($C_{11}H_{15}NO_2$)		1647
Butanephrine ($C_{10}H_{15}NO_3$)	240	
Butorphanol ($C_{21}H_{29}NO_2$)		1648–1649
Butyl 4-aminobenzoate ($C_{11}H_{15}NO_2$)	38	
Butylated hydroxytoluene ($C_{15}H_{24}O$)		1650
Butylparaben ($C_{11}H_{14}O_3$)		1651–1652
Butyric acid ($C_4H_8O_2$)	378	
γ-Butyrobetaine (GBB) ($C_7H_{15}NO_2$)	378	
1-Butyryloxymethyl-5-fluorouracil	569	
C		
Caffeine ($C_8H_{10}N_4O_2$)	9, 241–243	1653
Camptothecin ($C_{20}H_{16}N_2O_4$)	244	1654
Candesartan cilexetil ($C_{33}H_{34}N_6O_6$)	245	
Cannabidiol ($C_{21}H_{30}O_2$)		1655
Capreomycin IA ($C_{25}H_{44}N_{14}O_8$)		1656
Capreomycin IB ($C_{25}H_{44}N_{14}O_7$)		1656
Captopril ($C_9H_{15}NO_3S$)	246	
Carbachol ($C_6H_{15}ClN_2O_2$)		1657
Carbenicillin ($C_{17}H_{18}N_2O_6S$)		1658
Carbenoxolone ($C_{34}H_{50}O_7$)	247–248	1659
Carbinoxamine ($C_{16}H_{19}ClN_2O$)	249	1660, 2188
β-Carboline alkaloids		1907
Carbomycin A (Magnamycin A) ($C_{42}H_{67}NO_{16}$)	250	
Carbomycin B ($C_{42}H_{67}NO_{15}$)	250	
Carbonic acid (CH_2O_3)	251	1661
5-Carboxy-2′-deoxyuridine ($C_{10}H_{12}N_2O_7$)		1769
Carboxylic acids	252	
5-Carboxyuracil ($C_5H_4N_2O_4$)		1769
Carbutamide ($C_{11}H_{17}N_3O_3S$)	253	

Carisoprodol ($C_{12}H_{24}N_2O_4$)		1662
Carnitine ($C_7H_{15}NO_3$)	378	
Carpindolol ($C_{19}H_{28}N_2O_4$)		1663
Cefaclor ($C_{15}H_{14}ClN_3O_4S$)		1664
Cefadroxil ($C_{16}H_{17}N_3O_5S$)	254	
Cefamandole ($C_{18}H_{18}N_6O_5S_2$)		1665
Cefazaflur ($C_{13}H_{13}F_3N_6O_4S_3$)	255	
Cefazolin ($C_{14}H_{14}N_8O_4S_3$)	256	1666–1669
Cefixime ($C_{16}H_{15}N_5O_7S_2$)		1670
Cefoperazone ($C_{25}H_{27}N_9O_8S_2$)		1671
Cefotaxime ($C_{16}H_{17}N_5O_7S_2$)	257	
Cefoxitin ($C_{16}H_{17}N_3O_7S_2$)		1672
Cefroxadine ($C_{16}H_{19}N_3O_5S$)	258	
Ceftazidime ($C_{22}H_{22}N_6O_7S_2$)		1673
Ceftizoxime ($C_{13}H_{13}N_5O_5S_2$)		1674
Ceftriaxone ($C_{18}H_{18}N_8O_7S_3$)		1675–1676
Cefuroxime ($C_{16}H_{16}N_4O_8S$)		1677
Celiprolol ($C_{20}H_{33}N_3O_4$)		1678
Cellulose acetate phthalate		1679
Cephacetrile ($C_{13}H_{13}N_3O_6S$)		1680
Cephalexin ($C_{16}H_{17}N_3O_4S$)	259–262	1681–1682
Cephaloglycin ($C_{18}H_{19}N_3O_6S$)	263–265	
Cephaloridine ($C_{19}H_{17}N_3O_4S_2$)	266–267	
Cephalosporin derivative ($C_{16}H_{17}N_3O_7S$)	268	
Cephalosporin, 5-amino-5-carboxyvaleramido, O-acetyl ($C_{17}H_{23}N_3O_8S$)	269	
Cephalosporin, 5-amino-5-carboxyvaleramido, O-carbamoyl ($C_{16}H_{22}N_4O_8S$)	269	
Cephalosporin, 5-amino-5-carboxyvaleramido, 7-methoxy, O-acetyl ($C_{18}H_{25}N_3O_9S$)	269	
Cephalosporin, 5-amino-5-carboxyvaleramido, 7-methoxy, O-carbamoyl ($C_{17}H_{24}N_4O_9S$)	269	
Cephalosporin C ($C_{16}H_{21}N_3O_8S$)	270	
Cephalosporin C, N-chloroacetyl ($C_{18}H_{22}ClN_3O_9S$)	270	
Cephalosporin C, deacetoxy ($C_{14}H_{19}N_3O_7S$)	270	
Cephalothin ($C_{16}H_{16}N_2O_6S_2$)	271–272	1683
Cephapirin ($C_{17}H_{17}N_3O_6S_2$)	273	1684
Cephradine ($C_{16}H_{15}N_5O_7S_2$)		1685
Cerivastatin ($C_{26}H_{34}FNO_5$)	274	
Chenodeoxycholic acid ($C_{24}H_{40}O_4$)	275	
Chenodeoxycholic acid derivatives	275	
Chenodiol ($C_{24}H_{40}O_4$)		1686
Chloral hydrate ($C_2H_3Cl_3O_2$)	276–277	
Chlorambucil ($C_{14}H_{19}Cl_2NO_2$)		1687–1688
Chlorcyclizine ($C_{18}H_{21}ClN_2$)	278	1689

Chlordiazepoxide ($C_{16}H_{14}ClN_3O$)	204, 279	1690–1691
Chlordiazepoxide lactam ($C_{15}H_{11}ClN_2O_2$)	205	
Chlorhexidine ($C_{22}H_{30}Cl_2N_{10}$)	282	1692
Chloridine	9	
Chlorimipramine ($C_{19}H_{23}ClN_2$)	280	
Chlormethiazole (C_6H_8ClNS)		1726–1728
Chloroacetazolamide ($C_4H_5ClN_4O_3S_2$)	11	
Chloroacetic acid ($C_2H_3ClO_2$)	252	
N-Chloroacetylcephalosporin C ($C_{18}H_{22}ClN_3O_9S$)	270	
N-Chloroacetyldeacetoxycephalosporin C ($C_{16}H_{20}ClN_3O_8S$)	270	
4-Chloroamphetamine ($C_9H_{12}ClN$)		1693
3-Chloroazuloic acid ($C_{11}H_7ClO_2$)	115	
Chlorocresol (C_7H_7ClO)		1694
(S)-6-Chloro-4-(cyclo-propylethynyl)-1,4-dihydro-4-(trifluoromethyl)-2H-3,1-benzoxazin-2-one (Efavirenz; DMP-266) ($C_{14}H_9ClF_3NO_2$)		1695
2-Chloro-2′,3′-dideoxy-adenosine (2-ClDDA) ($C_{10}H_{12}ClN_5O_2$)	281	
7-Chloro-4-epitetracycline ($C_{22}H_{23}ClN_2O_8$)	493–494	
Chlorophenols	see 231	
N^1-4-Chlorophenyl-N^5-alkylbiguanides	282	
N^1-4-Chlorophenyl-N^5-methylbiguanide ($C_9H_{12}ClN_5$)	282	
N^1-4-Chlorophenyl-N^5-ethylbiguanide ($C_{10}H_{14}ClN_5$)	282	
N^1-4-Chlorophenyl-N^5-propylbiguanide ($C_{11}H_{16}ClN_5$)	282	
N^1-4-Chlorophenyl-N^5-n-butylbiguanide ($C_{12}H_{18}ClN_5$)	282	
3-(4-Chlorophenyl)-5,6-dihydro-2-ethylimidazo [2,1-b]thiazole ($C_{13}H_{13}ClN_2S$)	283	
3-(4-Chlorophenyl)-2-ethyl-2,3,5,6-tetrahydroimidazo[2,1-b]thiazol-3-ol ($C_{13}H_{15}ClN_2OS$)	284	
1-(2-Chlorophenyl)piperazine ($C_{10}H_{13}ClN_2$)	90	
1-(3-Chlorophenyl)piperazine ($C_{10}H_{13}ClN_2$)	90	
1-(4-Chlorophenyl)piperazine ($C_{10}H_{13}ClN_2$)	90	
Chloroprocaine ($C_{13}H_{19}ClN_2O_2$)	741	
Chloroquine ($C_{18}H_{26}ClN_3$)	285–286	1696
Chloroquine phosphate ($C_{18}H_{26}ClN_3.H_3PO_4$)		1697
Chloroquine phosphate ($C_{18}H_{29}ClN_3O_4P$)		1697
Chlorothen ($C_{14}H_{18}ClN_3S$)		1698
8-Chlorotheophylline ($C_7H_7ClN_4O_2$)	287	1699–1701
Chlorothiazide ($C_7H_6ClN_3O_4S_2$)	288–289	1702–1703

Chlorpheniramine ($C_{16}H_{19}ClN_2$)	290	1704–1708
Chlorphentermine ($C_{10}H_{14}ClN$)		1709
Chlorpromazine ($C_{17}H_{19}ClN_2S$)	291–300, 984–985	1710
Chlorpromazine sulfoxide ($C_{17}H_{19}ClN_2OS$)		1711
Chlorpropamide ($C_{10}H_{13}ClN_2O_3S$)		1712–1713
Chlorprothixene ($C_{18}H_{18}ClNS$)		1714
Chlortetracycline ($C_{22}H_{23}ClN_2O_8$)	301–306	1715
Chlorthalidone ($C_{14}H_{11}ClN_2O_4S$)	307	1716
Chlorzoxazone ($C_7H_4ClNO_2$)		1717
Cholic acid ($C_{24}H_{40}O_5$)		1718
Chromonar ($C_{20}H_{27}NO_5$)		1719
Cimetidine ($C_{10}H_{16}N_6S$)	308–309	1720–1721
Cinchocaine ($C_{20}H_{29}N_3O_2$)	406–407	
Cinchonidine ($C_{19}H_{22}N_2O$)	310	
Cinchonine ($C_{19}H_{22}N_2O$)	311	1722–1723
Cinnamic acid ($C_9H_8O_2$)	252, 312	
Cinnarizine ($C_{26}H_{28}N_2$)	313	
Cinnopentazone ($C_{22}H_{22}N_2O_2$)	314	
Cinoxacin ($C_{12}H_{10}N_2O_5$)	315	
Ciprofloxacin ($C_{17}H_{18}FN_3O_3$)	316, 1227–1228	1724, 2167
Cisplatin degradation product (H_7ClN_2OPt)	317	
Citalopram ($C_{20}H_{21}FN_2O$)		1725
Citric acid ($C_6H_8O_7$)	318–319	
Clarithromycin ($C_{38}H_{69}NO_{13}$)	320–321	
Clindamycin ($C_{18}H_{33}ClN_2O_5S$)	322–325	
Clindamycin, 1′-demethyl-4′-pentyl ($C_{19}H_{35}ClN_2O_5S$)	325	
Clindamycin-2-palmitate ($C_{34}H_{63}ClN_2O_6S$)	326	
Clindamycin-2-phosphate ($C_{18}H_{35}ClN_2O_8SP$)	327	
Clioquinol (C_9H_5ClINO)	328–332	
Clofazimine ($C_{27}H_{22}Cl_2N_4$)	333–335	
Clofazimine analogue, des-iso-propyl ($C_{24}H_{16}Cl_2N_4$)	334	
Clofazimine analogue, N-diethylamino-2-ethyl ($C_{30}H_{29}Cl_2N_5$)	334	
Clofazimine analogue, N-diethylamino-4-(1-methylbutyl) ($C_{35}H_{41}Cl_2N_5$)	334	
Clofazimine analogue, N-diethylamino-3-propyl ($C_{31}H_{31}Cl_2N_5$)	334	
Clofazimine analogue, 4-piperidinylmethyl ($C_{33}H_{33}Cl_2N_5$)	334	
Clofazimine analogue, N-piperidinyl-3-propyl ($C_{35}H_{37}Cl_2N_5$)	334	

Clofazimine analogue, N-pyrrolidinyl-3-propyl ($C_{34}H_{35}Cl_2N_5$)	334	
Clofezone	336	
Clofluperol ($C_{22}H_{22}ClF_4NO_2$)	1273	
Clomethiazole (Chlormethiazole) (C_6H_8ClNS)		1726–1728
Clomipramine ($C_{19}H_{23}ClN_2$)		1729
Clonazepam ($C_{15}H_{10}ClN_3O_3$)		1730
Clonidine ($C_9H_9Cl_2N_3$)	337–338	1731–1734
Clopenthixol ($C_{22}H_{25}ClN_2OS$)	339	1735
Clorazepate ($C_{16}H_{11}ClN_2O_3$)		1736
Clorindione ($C_{15}H_9ClO_2$)	340	
Clotrimazole ($C_{22}H_{17}ClN_2$)	341	
Cloxacillin ($C_{19}H_{18}ClN_3O_5S$)	342, 962	
Clozapine ($C_{18}H_{19}ClN_4$)	343–344	1737–1738
Cobefrin (Nordefrin) ($C_9H_{13}NO_3$)	345, 906	2032–2033
Cocaine ($C_{17}H_{21}NO_4$)	214, 346–347	1739–1744, 1988
Codeine ($C_{18}H_{21}NO_3$)	9, 348–353	
Colchicine ($C_{22}H_{25}NO_6$)	354	1745–1746
Conhydrine ($C_8H_{17}NO$)	479	
Coniine ($C_8H_{17}N$)	479	
Coumermycin A_1 ($C_{55}H_{59}N_5O_{20}$)	355	
Coumetarol ($C_{21}H_{16}O_7$)	637	
Creatine ($C_4H_9N_3O_2$)	356	
Creatinine ($C_4H_7N_3O$)	357	
Cresol red ($C_{21}H_{18}O_5S$)	358	
Cromolyn ($C_{23}H_{16}O_{11}$)		1747
Cyanocobalamin ($C_{63}H_{88}CoN_{14}O_{14}P$)	359	
2-Cyanoguanidinophenytoin ($C_{16}H_{12}N_4O$)	360	
Cyanopromazine ($C_{18}H_{19}N_3S$)		1748
Cyclacillin ($C_{15}H_{23}N_3O_4S$)	361–362	1749
Cyclamic acid ($C_6H_{13}NO_3S$)	363	
Cyclazocine ($C_{18}H_{25}NO$)		1750
Cycliramine ($C_{18}H_{15}ClN_2$)	364	
Cyclizine ($C_{18}H_{22}N_2$)	365–366	1751
Cyclobarbital ($C_{12}H_{16}N_2O_3$)	150	
Cyclobenzaprine ($C_{20}H_{21}N$)		1752
α-Cyclodextrin ($C_{36}H_{60}O_{30}$)	367	
Cyclohexaamylose (α-cyclodextrin) ($C_{36}H_{60}O_{30}$)	367	
1-Cyclohexyl-2-aminopropane ($C_9H_{19}N$)	368	
Cyclopentamine ($C_9H_{19}N$)	369	1753
Cyclopentolate ($C_{17}H_{25}NO_3$)	370	
D-Cycloserine ($C_3H_6N_2O_2$)	371	1754
iso-Cyclosporin A ($C_{62}H_{111}N_{11}O_{12}$)	372	

Cyclothiazide ($C_{14}H_{16}ClN_3O_4S_2$)	197, 373–374	1755
Cyproheptadine ($C_{21}H_{21}N$)		1756
Cysteine ($C_3H_7NO_2$)	375, 955	
Cytarabine ($C_9H_{13}ClN_3O_5$)		1757
D		
Dacarbazine ($C_6H_{10}N_6O$)		1758
Dantrolene ($C_{14}H_{10}N_4O_5$)		1759–1760
Dapsone ($C_{12}H_{12}N_2O_2S$)	1395	1761
Daunorubicin ($C_{27}H_{29}NO_{10}$)	376–377	
Deacetoxycephalosporin C ($C_{14}H_{19}N_3O_7S$)	270	
Deacetoxycephalosporin C, N-chloroacetyl ($C_{16}H_{20}ClN_3O_8S$)	270	
Debrisoquin ($C_{10}H_{13}N_3$)		1762–1763
Decyl carnitine ($C_{17}H_{33}NO_4$)	378	
Deferoxamine ($C_{25}H_{48}N_6O_8$)	384	
Dehydrocholic acid ($C_{24}H_{34}O_5$)		1764–1765
6-Dehydroestrone ($C_{18}H_{20}O_2$)	379	
Demeclocycline (6-demethyl-7-chlorotetracycline) ($C_{21}H_{21}ClN_2O_8$)	380–381	1766
6-Demethyl-7-chlorotetracycline ($C_{21}H_{21}ClN_2O_8$)	380–381	1766
Demoxepam ($C_{15}H_{12}N_2O_2$)		1767
4-Deoxyacyclovir ($C_8H_{13}N_5O_2$)	20	
Deoxycholic acid ($C_{24}H_{40}O_4$)		1768
Deoxyepinephrine (Epinine) ($C_9H_{13}NO_2$)	382	
2′-Deoxyuridine ($C_9H_{12}N_2O_5$)		1769
Deprenyl ($C_{13}H_{17}N$)	1272	
Deramciclane ($C_{19}H_{31}NO$)	383	
Deserpidine ($C_{32}H_{38}N_2O_8$)		1770–1771
Desethylamiodarone ($C_{23}H_{25}I_2NO_3$)	53	
Desferrioxamine ($C_{25}H_{48}N_6O_8$)	384	
8-Desfluorolomefloxacin ($C_{17}H_{20}FN_3O_3$)	1229	
Desipramine ($C_{18}H_{22}N_2$)	385–387	1772
Desmethyldiazepam ($C_{15}H_{11}ClN_2O$)	205	
Desmethyldoxepin ($C_{18}H_{19}NO$)	388	
Desmethylpheniramine ($C_{15}H_{18}N_2$)	389	
Desmycosin ($C_{39}H_{65}NO_{14}$)	390	
Desmycarosylcarbomycin A ($C_{38}H_{72}N_2O_{12}$)	391	
Desoxyephedrine (methamphetamine) ($C_{10}H_{15}N$)		1773
Dexamethasone-21-phosphate ($C_{22}H_{30}FO_8P$)	392	
Dexamphetamine ($C_9H_{13}N$)	393	1774
Dextran	394	
Dextromethorphan ($C_{18}H_{25}NO$)		1775
Dextromoramide ($C_{25}H_{32}N_2O_2$)		1776

Dextrose (glucose) ($C_6H_{12}O_6$)	395–396, 596	1777
Dexverapamil ($C_{27}H_{38}N_2O_4$)	1495–1498	
3,5-Diacetamido-1,2,4-triazole (Guanazole prodrug) ($C_4H_9N_5O_2$)	602	
Diamorphine (heroin) ($C_{21}H_{23}NO_5$)	397–398	1778–1779
Dial ($C_{10}H_{12}N_2O_3$)	158–160	
4,4-Diaminodiphenylsulfone ($C_{12}H_{12}N_2O_2S$)	1395	
2,4-Diaminopyridine, 5-(2-bromo-4,5-methylenedioxybenzyl)-($C_{12}H_{11}BrN_4O_2$)	216	
2,4-Diaminopyridine, 5-(3,5-bis(dimethylamino)-4-methylbenzyl)- ($C_{16}H_{24}N_6$)	216	
2,4-Diaminopyridine, 5-(3,5-bis(methylamino)-4-methoxy-6-acetylbenzyl)- ($C_{18}H_{26}N_6O_2$)	216	
2,4-Diaminopyridine, 5-(2,6-dichloro-3,5-dimethoxybenzyl)- ($C_{13}H_{14}Cl_2N_4O_2$)	216	
2,4-Diaminopyridine, 5-(3,5-diethoxy-4-(2′-hydroxy-2′-propyl)benzyl)- ($C_{18}H_{26}N_4O_3$)	216	
2,4-Diaminopyridine, 5-(3,5-diethoxy-4-pyrrolidinylbenzyl)- ($C_{19}H_{23}N_5O_2$)	216	
2,4-Diaminopyridine, 5-(3,5-dihydroxy-4-methoxybenzyl)- ($C_{14}H_{18}N_4O_3$)	216	
2,4-Diaminopyridine, 5-(3,5-dimethoxy-4-aminobenzyl)- ($C_{13}H_{17}N_5O_2$)	216	
2,4-Diaminopyridine, 5-(3,4-dimethoxy-5-benzoylbenzyl)- ($C_{20}H_{20}N_4O_3$)	216	
2,4-Diaminopyridine, 5-(3,5-dimethoxy-4-(2-methoxyethyl)benzyl)- ($C_{16}H_{22}N_4O_3$)	216	
2,4-Diaminopyridine, 5-(2,4-dimethoxy-3-pyridylbenzyl)- ($C_{18}H_{19}N_5O_2$)	216	
2,4-Diaminopyridine, 5-(3,5-dimethoxy-4-pyridylbenzyl)- ($C_{18}H_{19}N_5O_2$)	216	
2,4-Diaminopyridine, 5-(3,5-dimethoxy-4-pyrrolidinylbenzyl)- ($C_{17}H_{19}N_5O_2$)	216	
2,4-Diaminopyridine, 5-(3,5-dimethoxy-4-thiomethylbenzyl)- ($C_{14}H_{18}N_4O_2S$)	216	
2,4-Diaminopyridine, 5-(3-methoxy-4,5-dihydroxybenzyl)- ($C_{12}H_{14}N_4O_3$)	216	
2,4-Diaminopyridine, 5-(3-methoxy-4,5-methylenedioxybenzyl)- ($C_{13}H_{14}N_4O_3$)	216	
2,4-Diaminopyridine, 5-(4,5-methylenedioxybenzyl)- ($C_{12}H_{12}N_4O_2$)	216	
2,4-Diaminopyridine, 5-substituted-benzyl derivatives ($C_{11}H_{12}N_4O_4$)	216	

2,4-Diaminopyridine, 5-(2,4,5-trimethoxybenzyl)- ($C_{14}H_{18}N_4O_3$)	216	
2,4-Diaminopyridine, 5-(3,4,5-trimethoxybenzyl)- ($C_{14}H_{18}N_4O_3$)	216	
Diatrizoic acid ($C_{11}H_9I_3N_2O_4$)		1780
Diazepam ($C_{16}H_{13}ClN_2O$)	204, 399–402	1781–1782
Diazoxide ($C_8H_7ClN_2O_2S$)	403	1783
3,5-Dibenzamido-1,2,4-triazole (Guanazole prodrug) ($C_{16}H_{13}N_5O_2$)	602	
Dibenzepine ($C_{18}H_{21}N_3O$)	404–405	
Dibucaine (Cinchocaine) ($C_{20}H_{29}N_3O_2$)	406–407, 741	1784–1785
Dibucaine O-methyl homologue ($C_{17}H_{23}N_3O_2$)	407	
Dibucaine O-ethyl homologue ($C_{18}H_{25}N_3O_2$)	407	
Dibucaine O-propyl homologue ($C_{19}H_{27}N_3O_2$)	407	
Dibucaine O-butyl homologue ($C_{20}H_{29}N_3O_2$)	407	
Dibucaine O-pentyl homologue ($C_{21}H_{31}N_3O_2$)	407	
Dibucaine O-hexyl homologue ($C_{22}H_{33}N_3O_2$)	407	
3,5-Dichlorophenol ($C_6H_4Cl_2O$)	408	
Dichlorphenamide ($C_6H_6Cl_2N_2O_4S_2$)		1786
Diclofenac ($C_{14}H_{11}Cl_2NO_2$)	409–413	1787
Dicloxacillin ($C_{19}H_{17}Cl_2N_3O_5S$)	414, 962	1788
Dicumarol (bishydroxycoumarin) ($C_{19}H_{12}O_6$)		1789
Dicyclomine ($C_{19}H_{35}NO_2$)	415	
Didanosine ($C_{10}H_{12}N_4O_3$)	416	
2′,3′-Dideoxyadenosine ($C_{10}H_{13}N_5O_2$)	417	
Didesmethylpheniramine ($C_{14}H_{16}N_2$)	418	
Diethazine ($C_{18}H_{22}N_2S$)	9	1790
Diethylamine ($C_4H_{11}N$)	9	
4-Diethylaminobenzoic acid ($C_{11}H_{15}NO_2$)	649	
4-Diethylaminosalicylic acid ($C_{11}H_{15}NO_3$)	649	
Diethylcarbamazine ($C_{10}H_{21}N_3O$)		1791
Diethylstilbestrol ($C_{18}H_{20}O_2$)	419	
Diethylsulfanilamide ($C_{10}H_{16}N_2O_2S$)	38	
Difloxacin ($C_{21}H_{19}F_2N_3O_3$)	1230	
Diflunisal ($C_{13}H_8F_2O_3$)	420–421	
α-(2,4-Difluorophenyl)-α-[1-(2-(2-pyridyl)-phenylethenyl)]-1H-1,2,4-triazole-1-ethanol (XD405, bis-mesylate salt) ($C_{13}H_8F_2O_3$)	422	
Dihydroarecaidine ($C_7H_{13}NO_2$)	423	
Dihydroarecaidine methyl ester ($C_8H_{15}NO_2$)	424	
Dihydrocodeine ($C_{18}H_{23}NO_3$)	425	
Dihydrocodeinone ($C_{18}H_{21}NO_4$)		1792
Dihydrodesoxycodeine ($C_{18}H_{23}NO_2$)	426	

Dihydrodesoxynorcodeine ($C_{17}H_{21}NO_2$)	427	
Dihydroequilin, 17α-($C_{18}H_{22}O_2$)	428	
Dihydroequilinen, 17β-($C_{18}H_{22}O_2$)	429	
Dihydroergocornine ($C_{31}H_{41}N_5O_5$)	430	
Dihydroergocriptine ($C_{32}H_{43}N_5O_5$)	431	
Dihydroergocristine ($C_{35}H_{41}N_5O_5$)	432	
Dihydroergonovine ($C_{19}H_{23}N_3O_2$)	433	
Dihydroergotamine ($C_{33}H_{37}N_5O_5$)	434	1793
Dihydroergotoxine	435	
Dihydrofolic acid ($C_{19}H_{21}N_7O_6$)	436	
Dihydromorphine ($C_{17}H_{21}NO_3$)	437	1794
Dihydromorphinone ($C_{17}H_{19}NO_4$)		1795
Dihydrostreptomycin ($C_{21}H_{41}N_7O_{12}$)		1796–1797
2,8-Dihydroxyadenine ($C_5H_5N_5O_2$)	438	
m,p-Dihydroxyphenylethylamine ($C_8H_{11}NO_2$)	1085	
m,p-Dihydroxyphenylethylamine, N-methyl ($C_9H_{13}NO_2$)	1086	
3,5-Di-iodo-L-tyrosine ($C_9H_9I_2NO_3$)	439	
Dilazep ($C_{31}H_{44}N_2O_{10}$)	440–441	
Dilevolol ($C_{19}H_{24}N_2O_3$)		1798
Diltiazem ($C_{22}H_{26}N_2O_4S$)	442	1799–1800
Dimedrol	9	
Dimethadione ($C_5H_7NO_3$)		1801
Dimethisoquin ($C_{17}H_{24}N_2O$)		1802
Dimethoxyamphetamine ($C_{11}H_{17}NO_2$)	443	
1-(3,4-Dimethoxyphenyl)-2-N-methylaminopropane ($C_{12}H_{19}NO_2$)	444	
4-Dimethylaminobenzoic acid ($C_9H_{11}NO_2$)	649	
4-Dimethylaminosalicylic acid ($C_9H_{11}NO_3$)	649	
Dimethylamphetamine ($C_{11}H_{17}N$)	447–449	
Dimethylhydantoin ($C_5H_8N_2O_2$)		1803
(3,4-Dimethyl-5-isoxazolyl)-4-amino-1,2-naphthoquinone ($C_{15}H_{12}N_2O_3$)	445	
Dimethyloxytetracycline ($C_{24}H_{28}N_2O_9$)	450	
Dimethylsulfanilamide ($C_8H_{12}N_2O_2S$)	38	
4′,6′-Dimethylsulfapyridine-1-oxide ($C_{13}H_{14}N_3O_3S$)	1381	
5,5-Dimethyl-3-(α,α,α, 4-tetrafluoro-*m*-tolyl) hydantoin ($C_{12}H_{10}F_4N_2O_2$)	446	
3,5-Di(p-nitrobenzamido)-1,2,4-triazole (Guanazole prodrug) ($C_{16}H_{11}N_7O_6$)	602	
Dinoprost (prostaglandin $F_{2\alpha}$) ($C_{20}H_{34}O_5$)	451	
Dinoprostone (prostaglandin E2) ($C_{20}H_{32}O_4$)		1804
1,2-Dioleoylphosphatidylethanolamine (DOPE) ($C_{41}H_{77}NO_8P$)	452	
Dionine	9	

Dipeptides	453, 483	
Tripeptides	483	
Diperodon ($C_{22}H_{27}N_3O_4$)		1805
Diphenhydramine ($C_{17}H_{21}NO$)	454–456	1806
Diphenhydramine analogues	455	
Diphenic acid ($C_{14}H_{10}O_4$)	457	
Diphenoxylate ($C_{30}H_{32}N_2O_2$)	458	1807–1808
Diphenylpyraline ($C_{19}H_{23}NO$)	459	1809
Dipipanone ($C_{24}H_{31}NO$)		1810–1812, 2116
Dipivefrine ($C_{19}H_{29}NO_5$)		1813
3,5-Dipropionamido-1,2,4-triazole (Guanazole prodrug) ($C_6H_{13}N_5O_2$)	602	
Dipyridamole ($C_{24}H_{40}N_8O_4$)		1814
Disopyramide ($C_{21}H_{29}N_3O$)	460–462	
3,5-Di(trifluoroacetamido)-1,2,4-triazole (Guanazole prodrug) ($C_4H_3F_6N_5O_2$)	602	
DMP-266 ($C_{14}H_9ClF_3NO_2$)		1695
DMP-777 (elastase inhibitor) ($C_{31}H_{40}N_4O_6$)	463	
Dobutamine ($C_{18}H_{23}NO_3$)		1815
Domperidone ($C_{22}H_{24}$ (ClN_5O_2)		1816
DOPA, L-(Levodopa) ($C_9H_{11}NO_4$)	464, 497	1817–1818
Dopamine ($C_8H_{11}NO_2$)	465, 497	
DOPE ($C_{41}H_{78}NO_8P$)	452	
Dorzolamide ($C_{10}H_{16}N_2O_4S_3$)		1819
Dosulepine ($C_{19}H_{21}NS$)	466	
Dothiepin (dosulepine) ($C_{19}H_{21}NS$)	466	
Doxepin ($C_{19}H_{21}NO$)	467–468	1820
Doxorubicin (adriamycin) ($C_{27}H_{29}NO_{11}$)	469–470	1821–1822
Doxycycline ($C_{22}H_{24}N_2O_8$)	471	1823
Doxylamine ($C_{17}H_{22}N_2O$)		1824
Droperidol ($C_{22}H_{22}FN_3O_2$)		1825

E

E2001 ($C_{24}H_{22}FNO_2$)		1873
Ebifuramin ($C_{19}H_{22}N_4O_7$)	472	
Ecgonine ($C_9H_{15}NO_3$)	473–474	
Ecgonine methyl ester ($C_{10}H_{17}NO_3$)	475	
Edetic acid (EDTA) ($C_{10}H_{16}N_2O_8$)	476	
Efavirenz ($C_{14}H_9ClF_3NO_2$)		1695
Elastase inhibitor (DMP-477) ($C_{31}H_{40}N_4O_6$)	477	
Emetine ($C_{29}H_{40}N_2O_4$)	478–480	
Enalapril ($C_{20}H_{28}N_2O_5$)	481–482	
Enalaprilat ($C_{18}H_{24}N_2O_5$)		1826
Enkephalin (met-enkephalin) ($C_{28}H_{37}N_5O_7$)	483	
Enoxacin ($C_{15}H_{17}FN_4O_3$)	1231	

Ephedrine ($C_{10}H_{15}NO$)	9, 484–490, 1178	1827–1829
Epianhydrotetracycline (anhydro-4-epitetracycline) ($C_{22}H_{22}N_2O_7$)	491–492	
Epichlortetracycline (7-chloro-4-epitetracycline) ($C_{22}H_{23}ClN_2O_8$)	493–494	
Epicillin ($C_{16}H_{21}N_3O_4S$)	495	
Epinastine ($C_{16}H_{15}N_3$)	496	
Epinephrine (adrenaline) ($C_9H_{13}NO_3$)	497–502	
Epinine ($C_9H_{13}NO_2$)	382	
Epiquinidine ($C_{20}H_{24}N_2O_2$)	1221	
Epiquinine ($C_{20}H_{24}N_2O_2$)	1221	
4-Epitetracycline ($C_{22}H_{24}N_2O_8$)	503	
Equilenin ($C_{18}H_{18}O_2$)	504	
Equilin ($C_{18}H_{20}O_2$)	505	
Ergometrine ($C_{19}H_{23}N_3O_2$)		1830
Ergonovine ($C_{19}H_{23}N_3O_2$)	506–507	1831
Ergostine ($C_{35}H_{39}N_5O_5$)	508–509	
Ergot alkaloids	see 506	
Ergotamine ($C_{33}H_{35}N_5O_5$)	510–511	1832–1833
Ergotaminine ($C_{33}H_{35}N_5O_5$)	512–513	
Erythromycin ($C_{37}H_{67}NO_{13}$)	514–519	1834
Erythromycin estolate ($C_{52}H_{97}NO_{18}S$)		1835
Erythromycin lactobionate ($C_{37}H_{67}NO_{13}$) (salt with lactobiono-δ-lactone)		1836
Erythromycyclamine ($C_{37}H_{70}N_2O_{12}$)	520	
Erythromycyclamine-11,12-carbonate ($C_{38}H_{68}N_2O_{13}$)	521	
Estazolam ($C_{16}H_{11}ClN_4$)	522	
Estradiol, 17α-($C_{18}H_{24}O_2$)	523	
Estradiol, 17β-($C_{18}H_{24}O_2$)	523	
Estriol ($C_{18}H_{24}O_3$)	524	
Estrone ($C_{18}H_{22}O_2$)	525	
Ethacrynic acid ($C_{13}H_{12}Cl_2O_4$)		1837–1838
Ethambutol ($C_{10}H_{24}N_2O_2$)	526	
Ethinylestradiol ($C_{20}H_{24}O_2$)	527–528	1839
Ethionamide ($C_8H_{10}N_2S$)	529	
Ethionine ($C_6H_{13}NO_2S$)	955	
Ethoheptazine ($C_{16}H_{23}NO_2$)		1840
Ethopropazine ($C_{19}H_{24}N_2S$)	530	1841
Ethosuximide ($C_7H_{11}NO_2$)		1842–1843
Ethoxazolamide ($C_9H_{10}N_2O_3S_2$)	531	
4-Ethoxybenzoic acid ($C_9H_{10}O_3$)	649	
4-Ethoxysalicylic acid ($C_9H_{10}O_4$)	649	
Ethoxzolamide ($C_9H_{10}N_2O_3S_2$)	531	

Ethyl 4-aminobenzoate ($C_9H_{11}NO_2$)	38	
Ethylamphetamine ($C_{11}H_{17}N$)		1844
α-Ethyl n-amylpenilloate	532–533	
Ethyl biscoumacetate ($C_{10}H_{24}O_8$)	637	1845–1847
Ethylenediamine ($C_2H_8N_2$)		1848
Ethylhydrocupreine ($C_{21}H_{28}N_2O_2$)	398	
2-Ethylmercapto-4-hydroxypyrimidine ($C_6H_8N_2OS$)		1849
Ethylmorphine ($C_{19}H_{23}NO_3$)	536	1850–1851
Ethylnorepinephrine ($C_{10}H_{15}NO_3$)		1852
α-Ethylphenethylamine ($C_{10}H_{15}N$)		1853
Ethylphenylephrine ($C_{10}H_{15}NO_2$)	538	
Ethylphenylhydantoin ($C_{11}H_{12}N_2O_2$)		1854
α-Ethylphenylpenilloate	534–535	
6′-Ethylsulfapyridine-1-oxide ($C_{13}H_{14}N_3O_3S$)	1381	
Etidocaine ($C_{17}H_{28}N_2O$)	537	1855–1856
Etilefrine (ethylphenylephrine) ($C_{10}H_{15}NO_2$)	538	
Etodolac ($C_{17}H_{21}NO_3$)		1857
Etomidate ($C_{14}H_{16}N_2O_2$)	539	
Etoposide ($C_{29}H_{32}O_{13}$)	540	
β-Eucaine ($C_{15}H_{21}NO_2$)	398, 541	
Eugenol ($C_{10}H_{12}O_2$)		1858
F		
Famotidine ($C_8H_{15}N_7O_2S_3$)	542–543	
Fenbufen ($C_{16}H_{14}O_3$)	544	
Fencamfamine ($C_{15}H_{21}N$)		1859
Fenclofenac ($C_{14}H_{10}Cl_2O_3$)		1860
Fendiline ($C_{23}H_{25}N$)	545	
Fenfluramine ($C_{12}H_{16}F_3N$)		1861
Fenoprofen ($C_{15}H_{14}O_3$)	546	1862
Fenoterol ($C_{17}H_{21}NO_4$)		1863–1864
Fentanyl ($C_{22}H_{28}N_2O$)	547	1865
Flavoxate ($C_{24}H_{25}NO_4$)	548	
Flecainide ($C_{17}H_{20}F_6N_2O_3$)		1866
Fleroxacin ($C_{17}H_{18}F_3N_3O_3$)	1232	
Floxuridine ($C_9H_{11}FN_2O_5$)		1867
Flucloxacillin ($C_{19}H_{17}ClFN_2O_5S$)	962	1868
Fluconazole ($C_{13}H_{12}F_2N_6O$)	549	
Flucytosine ($C_4H_4FN_3O$)		1869
Fludiazepam ($C_{16}H_{12}ClFN_2O$)	550	
Flufenamic acid ($C_{14}H_{10}F_3NO_2$)	551–554	1870
Flumequine ($C_{14}H_{12}FNO_3$)	555–557	
Flumethiazide ($C_8H_6F_3N_3O_4S_2$)	197, 558	
Flumizole ($C_{18}H_{15}F_3N_2O_2$)		1871
Flunitrazepam ($C_{16}H_{12}FN_3O_3$)	559–560	1872
Fluopromazine ($C_{18}H_{19}F_3N_2S$)	574	

Fluorescein ($C_{20}H_{12}O_5$)	561	
2-(4-(*p*-Fluorobenzoyl)piperidin-1-yl)-2′-acetonaphthone hydrochloride (E2001) ($C_{24}H_{22}FNO_2$)	1873	
8-Fluoro-norfloxacin ($C_{16}H_{17}F_2N_3O_3$)	1233	
8-Fluoro-pefloxacin ($C_{17}H_{19}F_2N_3O_3$)	1234	
1-(4-Fluorophenyl)piperazine ($C_{10}FH_{13}N_2$)	90	
1-(5-Fluoro-2-pyrimidinyl)piperazine ($C_{10}FH_{11}N_4$)	90	
5-Fluorouracil ($C_4H_3FN_2O_2$)	562	
5-Fluorouracil, 1-acetyloxymethyl ($C_7H_7FN_2O_4$)	569	
5-Fluorouracil, 3-acetyloxymethyl ($C_7H_7FN_2O_4$)	570	
5-Fluorouracil, 3-ethyloxycarbonyl ($C_7H_7FN_2O_4$	564	
5-Fluorouracil, 1-ethyloxycarbonyloxymethyl ($C_8H_9FN_2O_5$)	565	
5-Fluorouracil, 3-ethyloxycarbonyloxymethyl ($C_8H_9FN_2O_5$)	566	
5-Fluorouracil, 1-methyloxycarbonyl ($C_6H_5FN_2O_4$)	563	
5-Fluorouracil, 1-phenyloxycarbonyloxymethyl ($C_{12}H_9FN_2O_5$)	567	
5-Fluorouracil, 3-phenyloxycarbonyloxymethyl ($C_{12}H_9FN_2O_5$)	568	
Fluoxetine	96	
Flupenthixol ($C_{23}H_{25}F_3N_2OS$)	571	1874
Fluphenazine ($C_{22}H_{26}F_3N_3OS$)	572–573	
Fluphenazine enanthate ($C_{29}H_{38}F_3N_3O_2S$)		1875
Fluphenazine decanoate ($C_{32}H_{44}F_3N_3O_2S$)		1876
Flupromazine ($C_{18}H_{19}F_3N_2S$)	574	
Flurazepam ($C_{21}H_{23}ClFN_3O$)		1877–1878
Flurbiprofen ($C_{15}H_{13}FO_2$)	575	1879
Fluvoxamine ($C_{15}H_{21}F_3N_2O_2$)		1880
Folic acid ($C_{19}H_{19}N_7O_6$)	576	1881
Folic acid analogues	577	
Folinic acid (leucovorin) ($C_{20}H_{23}N_7O_7$)	578–579	
Formic acid (CH_2O_2)	580	
3-Formylazuloic acid ($C_{12}H_8O_3$)	115	
Frusemide ($C_{12}H_{11}ClN_2O_5S$)	581–586	1884–1885
Fumaric acid ($C_4H_4O_4$)		1882
Furaltadone ($C_{13}H_{16}F_3N_4O_6$)		1883
Furosemide (frusemide) ($C_{12}H_{11}ClN_2O_5S$)	581–586	1884–1885
Fusidic acid ($C_{31}H_{48}O_6$)		1886

G

GABA ($C_4H_9NO_2$)	378	
Galanthamine ($C_{17}H_{21}NO_3$)		1887
Gallic acid (3,4,5-trihydroxybenzoic acid) ($C_7H_6O_5$)	587	
GBB ($C_7H_{15}NO_2$)	378	

Gelatin	588	
Gentamicin C1 ($C_{21}H_{43}N_5O_7$)		1888
Gentamicin C1a ($C_{19}H_{39}N_5O_7$)		1888
Gentamicin C2 ($C_{20}H_{41}N_5O_7$)		1888
Glafenine ($C_{19}H_{17}ClN_2O_4$)	589	
Glibenclamide (glyburide) ($C_{23}H_{28}ClN_3O_5S$)	590–592	1889, 1895
Glipizide ($C_{21}H_{27}N_5O_4S$)		1890
Gluconic acid ($C_6H_{12}O_7$)	593	
D-Glucosamine ($C_6H_{13}NO_5$)	594	
D-Glucosamine-6-phosphate ($C_6H_{14}NO_8P$)	595	
Glucose ($C_6H_{12}O_6$)	395–396, 596	1777
D-Glucuronic acid ($C_6H_{10}O_7$)		1891
Glutarimide ($C_5H_7NO_2$)		1892
Glutethimide ($C_{13}H_{15}NO_2$)	597	1893–1894
Glyburide ($C_{23}H_{28}ClN_3O_5S$)	590–592	1889, 1895
Glycerol ($C_3H_8O_3$)	598	
Glycine xylidide ($C_{10}H_{14}N_2O$)	599	
Glycocholic acid ($C_{26}H_{43}NO_6$)	600	1896
Glycodeoxycholic acid ($C_{26}H_{43}NO_5$)	601	1897
Glycyclamide ($C_{14}H_{20}N_2O_3S$)		1898
Gly-Gly ($C_4H_8N_2O_3$)	453	
Gly-Phe ($C_{11}H_{14}N_2O_3$)	453	
Gly-Trp ($C_{13}H_{15}N_3O_3$)	453	
Guanabenz ($C_8H_8Cl_2N_4$)		1899
Guanazole prodrug (3-amino-5-heptafluorobutyramido-1,2,4-triazole) ($C_6H_4F_7N_5O$)	602	
Guanazole prodrug (3,5-diacetamido-1,2,4-triazole) ($C_4H_9N_5O_2$)	602	
Guanazole prodrug (3,5-dibenzamido-1,2,4-triazole) ($C_{16}H_{13}N_5O_2$)	602	
Guanazole prodrug (3,5-dipropionamido-1,2,4-triazole) ($C_6H_{13}N_5O_2$)	602	
Guanazole prodrug (3,5-di(*p*-nitrobenzamido)-1,2,4-triazole) ($C_{16}H_{11}N_7O_6$)	602	
Guanazole prodrug (3,5-di(trifluoroacetamido)-1,2,4-triazole) ($C_4H_3F_6N_5O_2$)	602	
Guanethidine ($C_{10}H_{22}N_4$)		1900–1901
Guanidine (CH_5N_3)		1902
Guanine ($C_5H_5N_5O$)	241	
Guanoxan ($C_{10}H_{13}N_3O_2$)		1903
Guluronic acid ($C_6H_{10}O_7$)		1546

H

Haloperidol ($C_{21}H_{23}ClFNO_2$)	603–604	1904–1906
Harmol ($C_{12}H_{10}N_2O$)		1907
Heliotridane ($C_8H_{15}N$)	605	
Heliotridene ($C_8H_{13}N$)	606	
n-Heptylpenicillin	959	
Heroin ($C_{21}H_{23}NO_5$)	397–398	1778–1779
Hexachlorophene ($C_{13}H_6Cl_6O_2$)	607–609	
Hexamethylmelamine ($C_9H_{18}N_6$)		1551
Hexetidine ($C_{21}H_{45}N_3$)	610–611	
Hexobarbital ($C_{12}H_{16}N_2O_3$)	125	
Hexobendine ($C_{30}H_{44}N_2O_{10}$)	612–613	
Hexylcaine ($C_{16}H_{23}NO_2$)		1908–1909
Hexylresorcinol ($C_{12}H_{18}O_2$)		1910
Hippuric acid ($C_9H_9NO_3$)		1911
Histamine ($C_5H_9N_3$)	614–615	
Histamine, diiodo ($C_5H_7I_2N_3$)	617	
Histamine, monoiodo ($C_5H_8IN_3$)	616	
Histapyrrodine ($C_{19}H_{24}N_2$)	618	
Histidine ($C_6H_9N_3O_2$)	1481	
HNB-1 ($C_{27}H_{26}NO_4$)	619	
HNB-5 ($C_{26}H_{25}N_2O_4$)	620	
Homatropine ($C_{16}H_{21}NO_3$)	398	1912–1913
Hycanthone ($C_{20}H_{24}N_2O_2S$)		1914
Hydantoin ($C_3H_4N_2O_2$)		1915
Hydralazine ($C_8H_8N_4$)	621–622	
Hydrochlorothiazide ($C_7H_8ClN_3O_4S_2$)	197, 623–626	1916
Hydrocodone ($C_{18}H_{21}NO_3$)	398	1917
Hydrocortisone hydrogen succinate ($C_{25}H_{34}O_8$)	627	
Hydrocortisone, imidazole-1-carboxylic acid prodrug ($C_{25}H_{32}N_2O_6$)	628	
Hydroflumethiazide ($C_8H_8F_3N_3O_4S_2$)	197, 629–631	
Hydrogen peroxide (H_2O_2)	632	
Hydromorphone ($C_{17}H_{19}NO_3$)	398	1918–1919
Hydroquinone ($C_6H_6O_2$)	633	1920
β-Hydroxy-γ-aminobutyric acid ($C_4H_9NO_3$)	378	
N-hydroxyamphetamine ($C_9H_{13}NO$)	634	
4-Hydroxyamphetamine (Paredrine) ($C_9H_{13}NO$)	635	1921–1922
2-Hydroxybenzoic acid ($C_7H_6O_3$)	252, 649, 1264–1269	
3-Hydroxybenzoic acid ($C_7H_6O_3$)	252	
4-Hydroxybenzoic acid ($C_7H_6O_3$)	252, 649	

p-Hydroxybenzylpenicillin ($C_{16}H_{18}N_2O_5S$)	959	
2-Hydroxycinnamic acid ($C_9H_8O_3$)	252	
10-Hydroxycodeine ($C_{18}H_{21}NO_4$)	636	
4-Hydroxycoumarin ($C_9H_6O_3$)	637	
4-Hydroxycoumarin derivatives	637	
14-cis-Hydroxydihydrocodeinone ($C_{18}H_{21}NO_4$)	638	
10-cis-Hydroxydihydrodesoxycodeine ($C_{18}H_{23}NO_3$)	639	
10-trans-Hydroxydihydrodesoxycodeine ($C_{18}H_{23}NO_3$)	640	
10-cis-Hydroxydihydronorcodeine ($C_{17}H_{21}NO_4$)	641	
10-trans-Hydroxydihydronorcodeine ($C_{17}H_{21}NO_4$)	642	
4-(2-Hydroxy-3-iso-propylaminopropoxy)indole ($C_{14}H_{20}N_2O_2$)	643	
3-Hydroxy-α-(methylamino) methylbenzenemethanol ($C_9H_{13}NO_2$)	644	
5-[1-Hydroxy-2-[(1-methylethyl)amino]-ethyl]-1,3-benzenediol (metaproterenol) ($C_{11}H_{17}NO_3$)	645	
4-Hydroxynorephedrine ($C_9H_{13}NO_2$)	647	
N-Hydroxyphentermine ($C_{10}H_{15}NO$)	646	
D-(-)-4-Hydroxy-4-phenylbutanoic acid	762	
m-Hydroxyphenylethylamine, 2-hydroxy ($C_8H_{11}NO_2$)	1081	
m-Hydroxyphenylethylamine, 2-hydroxy, *N*-methyl ($C_9H_{13}NO_2$)	1082	
p-Hydroxyphenylethylamine ($C_8H_{11}NO$)	1083	
p-Hydroxyphenylethylamine, *N*-methyl ($C_9H_{13}NO$)	1084	
p-Hydroxyphenylethylamine, 2-hydroxy ($C_8H_{11}NO_2$)	1087	
p-Hydroxyphenylethylamine, 2-hydroxy, *N*-methyl ($C_9H_{13}NO_2$)	1088	
7-(2-Hydroxypropyl)theophylline ($C_{10}H_{14}N_4O_3$)	648	
4-Hydroxysalicylic acid ($C_7H_6O_4$)	649	
Hydroxyzine ($C_{21}H_{27}ClN_2O_2$)	650–651	1923–1924
Hyoscine	652	
Hyoscyamine ($C_{17}H_{23}NO_3$)	652	1925–1926
Hypoxanthine ($C_5H_4N_4O$)	241, 653	

I

Ibuprofen ($C_{13}H_{18}O_2$)	654–664	
Idarubicin ($C_{26}H_{27}NO_9$)	376, 665	
Idoxuridine ($C_9H_{11}IN_2O_5$)	666	1927–1928
Imidazole ($C_3H_4N_2$)	667–668	
Imidazole, *N*-acetyl ($C_5H_6N_2O$)	669	
Imipenem ($C_{12}H_{17}N_3O_4S$)	670	
Imipramine ($C_{19}H_{24}N_2$)	671–674	
Indapamide ($C_{16}H_{16}$ (ClN_3O_3S)	675	

Indicators	676	
Indinavir ($C_{36}H_{47}N_5O_4$)	677	
Indomethacin ($C_{19}H_{16}ClNO_4$)	678–683	
D-Indoprofen ($C_{17}H_{15}NO_3$)		1929–1931
Indoramin ($C_{22}H_{25}N_3O$)		1932
Insulin, A21-aspartyl	684–685	
Iocetamic acid ($C_{12}H_{13}I_3N_2O_3$)		1933
Iodamide ($C_{12}H_{11}I_3N_2O_4$)	686–687	
Iodipamide ($C_{20}H_{14}I_6N_2O_6$)	688	1934
Iodoquinol ($C_9H_5I_2NO$)		1935
Iodoxamic acid ($C_{26}H_{26}I_6N_2O_{10}$)	689	
Iopamidol ($C_{17}H_{22}I_3N_3O_8$)	690	
Iopanoic acid ($C_{11}H_{12}I_3NO_2$)	691	
Iophenoxic acid ($C_{11}H_{11}I_3O_3$)	692	
Iprindole ($C_{19}H_{28}N_2$)		1936
Ipronidazole ($C_7H_{11}N_3O_2$)		1937
Irgafen ($C_{15}H_{16}N_2O_3S$)	1327	
Isocarboxazid ($C_{12}H_{13}N_3O_2$)	693	
Isochlorotetracycline ($C_{22}H_{23}ClN_2O_8$)	694–695	
Isolysergic acid ($C_{16}H_{16}N_2O_2$)	696	
DL-Isomethadone ($C_{21}H_{27}NO$)	697–698	
Isoniazid ($C_6H_7N_3O$)	699	
Isonicotinamide ($C_6H_6N_2O$)	699	
Isonicotinic acid ($C_6H_5NO_2$)	700, 881, 884	
Isonicotinic acid (non-aqueous titration) ($C_6H_5NO_2$)	194	
1-Isonicotinoyl-1-methylhydrazine ($C_7H_9N_3O$)	699	
1-Isonicotinoyl-2-methylhydrazine ($C_7H_9N_3O$)	699	
Isopilocarpine ($C_{11}H_{16}N_2O_2$)	701	
Isopropylamphetamine ($C_{12}H_{19}N$)		1938
Isopropylnorepinephrine ($C_{11}H_{17}NO_3$)	497	
Isoproterenol (DL-isoprenaline) ($C_{11}H_{17}NO_3$)	702–706	
Isopyridoxal ($C_8H_9NO_3$)	707	
Isopyridoxal-4-phosphate ($C_8H_{10}NO_6P$)	707	
Isoxsuprine ($C_{18}H_{23}NO_3$)		1939–1940
Itanoxone ($C_{17}H_{13}$ (ClO_3)	708	
J		
Juvenimicin ($C_{31}H_{51}NO_9$)	1258	
K		
Kanamycin A ($C_{18}H_{36}N_4O_{11}$)	709	
Ketamine ($C_{13}H_{16}ClNO$)	710–711	
Ketobemidone ($C_{15}H_{21}NO_2$)	712–713	
Ketoconazole ($C_{26}H_{28}Cl_2N_4O_4$)	714	
Ketoprofen ($C_{16}H_{14}O_3$)	715–716	
KHL 8430 ($C_{25}H_{29}NO_2$)	717	

L

Labetalol ($C_{19}H_{24}N_2O_3$)	718–720	
Lactic acid, D(−) ($C_3H_6O_3$)	721	
Lactic acid, L(+) ($C_3H_6O_3$)	722	
Lactic acid, (±) ($C_3H_6O_3$)	723–724	
Lansoprazole ($C_{16}H_{14}F_3N_3O_2S$)		1941
Lauric acid ($C_{12}H_{24}O_2$)	725	
Leu-His ($C_{12}H_{20}N_4O_3$)	453	
Leu-Phe ($C_{15}H_{22}N_2O_3$)	453	
Leu-Tyr ($C_{15}H_{22}N_2O_4$)	453	
Levallorphan tartrate ($C_{19}H_{25}NO.C_4H_6O_6$)	726	
Levallorphan tartrate ($C_{23}H_{31}NO_7$)	726	
Levallorphan ($C_{19}H_{25}NO$)	727–728	
Levarterenol (l-noradrenaline; l-norepinephrine) ($C_8H_{11}NO_3$)	729–731	
Levarterenol bitartrate ($C_8H_{11}NO_3.C_4H_6O_6$)		2035
Levarterenol bitartrate ($C_{12}H_{17}NO_9$)		2035
Levobunolol ($C_{17}H_{25}NO_3$)	1167	1641–1642
Levodopa ($C_9H_{11}NO_4$)	464, 732–733	1817–1818
Levomepromazine ($C_{19}H_{24}N_2OS$)	799–800, 992	
Levomethadyl acetate ($C_{23}H_{31}NO_2$)	14	
Levomethorphan ($C_{18}H_{25}NO$)		1942
Levomoramide ($C_{25}H_{32}N_2O_2$)		1943–1944
Levorphanol ($C_{17}H_{23}NO$)	734–736	
Levothyroxine ($C_{15}H_{11}I_4NO_4$)	1443	
Levulinic acid ($C_5H_8O_3$)	737	
Lidocaine (lignocaine; xylocaine) ($C_{14}H_{22}N_2O$)	738–746	
Lidocaine homologues	746	
Lignocaine ($C_{14}H_{22}N_2O$)	738–746	
Lincomycin ($C_{18}H_{34}N_2O_6S$)	747–748	
Linoleic acid ($C_{18}H_{32}O_2$)	749	
Liothyronine ($C_{15}H_{12}I_3NO_4$)		1945
Lisinopril ($C_{21}H_{31}N_3O_5$)	750–751	
Lisuride ($C_{20}H_{26}N_4O$)	752	
Lithium carbonate (CO_3Li_2)		1946
Lomefloxacin ($C_{17}H_{19}F_2N_3O_3$)	1235–1236	
Lonazolac ($C_{17}H_{13}ClN_2O_2$)	753	
Loperamide ($C_{29}H_{33}ClN_2O_2$)	754	
Lorazepam ($C_{15}H_{10}Cl_2N_2O_2$)	204, 755–756	
Loxapine ($C_{18}H_{18}ClN_3O$)		1947
Lysergic acid ($C_{16}H_{16}N_2O_2$)	757	

Lysergic acid derivatives	433	
Lysergide ($C_{20}H_{25}N_3O$)		1948
M		
Magnamycin A ($C_{42}H_{67}NO_{16}$)	250	
Maleic acid ($C_4H_4O_4$)	252, 758	
Malic acid ($C_4H_6O_5$)	759	
Malonic acid ($C_3H_4O_4$)	252, 760	
Mandelhydroxamic acid ($C_8H_9NO_3$)		1949
Mandelic acid ($C_8H_8O_3$)	761	1950–1951
D-(-)-Mandelic acid and analogues	762	
Mannitol ($C_6H_{14}O_6$)	763	
Mannuronic acid ($C_6H_{10}O_7$)		1546
Maprotiline ($C_{20}H_{23}N$)		1952
Mazindol ($C_{16}H_{13}ClN_2O$)		1953
Mebendazole ($C_{16}H_{13}ClN_3O_3$)	764	
Mebeverine ($C_{25}H_{35}NO_5$)	765	
Mebhydroline ($C_{19}H_{20}N_2$)	766	
Mecamylamine ($C_{11}H_{21}N$)	767	
Mechlorethamine ($C_5H_{11}Cl_2N$)		1954
Mecillinam ($C_{15}H_{23}N_3O_3S$)	31	1556
Meclizine ($C_{25}H_{27}ClN_2$)	768–769	
Meclofenamic acid ($C_{14}H_{11}Cl_2NO_2$)		1955
Medazepam ($C_{16}H_{15}ClN_2$)	204, 770–771	
Mefenamic acid ($C_{15}H_{15}NO_2$)	772–774	
Mefloquine ($C_{17}H_{16}F_6N_2O$)		1956
Melphalan ($C_{13}H_{18}Cl_2N_2O_2$)	775	1957
Mepazine (pecazine) ($C_{19}H_{22}N_2S$)	776, 952–953, 1002	
Meperidine (pethidine) ($C_{15}H_{21}NO_2$)	777, 968	1958
Mephentermine ($C_{11}H_{17}N$)	778–779	1959
Mephenytoin ($C_{12}H_{14}N_2O_2$)		1960
Mepindolol ($C_{15}H_{22}N_2O_2$)		1961
Mepivacaine ($C_{15}H_{22}N_2O$)	741, 780–781	1962
Mercaptomerin ($C_{16}H_{27}HgNO_6S$)		1963
6-Mercaptopurine ($C_5H_4N_4S$)	782–783	
Mesalamine ($C_7H_7NO_3$)		1964
Mesna (2-mercaptoethanesulfonic acid) ($C_2H_6O_3S_2$)		1965
Metaclazepam ($C_{18}H_{18}BrClN_2O$)	784	
Metaproterenol (orciprenaline) ($C_{11}H_{17}NO_3$)	645, 785–786	1966
Metaraminol ($C_9H_{13}NO_2$)		1967
Met-enkephalin ($C_{28}H_{37}N_5O_7$)	483	
Metformin ($C_4H_{11}N_5$)	787	

Methacycline ($C_{22}H_{22}N_2O_8$)		1968
Methadone ($C_{21}H_{27}NO$)		1969–1972
Methadone analogues	788	
Methamphetamine (methylamphetamine) ($C_{10}H_{15}N$)	789–790	1773, 1973–1975
Methamphetamine, 4-hydroxy ($C_{10}H_{15}NO$)	791	
Methamphetamine, *N*-(2-cyano)ethyl ($C_{13}H_{25}N_2$)	see 905	
Methaphenilene ($C_{15}H_{20}N_2S$)	792	
Methapyrilene ($C_{14}H_{19}N_3S$)		1976–1977
Methaqualone ($C_{16}H_{14}N_2O$)	793	
Metharbital ($C_9H_{14}N_2O_3$)	132	1603–1604
Methazolamide ($C_5H_8N_4O_3S_2$)		1978
Methdilazine ($C_{18}H_{20}N_2S$)		1979
Methenamine ($C_6H_{12}N_4$)		1980–1982
Methicillin ($C_{17}H_{20}N_2O_6S$)	794, 961	1983
Methionine ($C_5H_{11}NO_2S$)	955	1984
Methohexital ($C_{14}H_{18}N_2O_3$)		1609
Methopromazine (methoxypromazine) ($C_{18}H_{22}N_2OS$)		1985
D-Methorphan ($C_{18}H_{25}NO$)		1775
Methorphinan ($C_{17}H_{23}NO$)		2176
Methotrexate ($C_{20}H_{22}N_8O_5$)	795–798	1986–1987
Methotrexate esters	798	
Methotrimeprazine (levomepromazine) ($C_{19}H_{24}N_2OS$)	799–800, 992	
Methoxamine ($C_{11}H_{17}NO_3$)	801–802	
4-Methoxyamphetamine ($C_{10}H_{15}NO$)	803	
2-Methoxybenzoic acid ($C_8H_8O_3$)	252	
3-Methoxybenzoic acid ($C_8H_8O_3$)	252	
2-Methoxycinnamic acid ($C_{10}H_{10}O_3$)	252	
3-Methoxycinnamic acid ($C_{10}H_{10}O_3$)	252	
4-Methoxycinnamic acid ($C_{10}H_{10}O_3$)	252	
Methoxyphenamine ($C_{11}H_{17}NO$)	804	
1-(2-Methoxyphenyl)piperazine ($C_{11}H_{16}N_2O$)	90	
N-(3′-Methoxyphenyl)anthranilic acid ($C_{14}H_{13}NO_3$)	552–553	
Methoxypromazine ($C_{18}H_{22}N_2OS$)		1985
Methyclomethiazide (C_9H_{11} ($Cl_2N_3O_4S_2$)	805–806	
Methyclothiazide (C_9H_{11} ($Cl_2N_3O_4S_2$)	197	
DL-*N*-Methyladrenaline ($C_{10}H_{15}NO_3$)	807	
Methylamine (CH_5N)	9	
Methyl 4-aminobenzoate ($C_8H_9NO_2$)	38	
2-Methylamino-5-chlorobenzophenone ($C_{14}H_{12}ClNO$)	808	
Methylamphetamine ($C_{10}H_{15}N$)	789–790	1773, 1973–1975

N-Methyl-1-benzoyl ecgonine ($C_{17}H_{21}NO_4$)		1988
α-Methyl benzylpenilloate ($C_{16}H_{22}N_2O_3S$)	809	
Methylcaffeine ($C_9H_{12}N_4O_2$)	9	1989
Methyldihydromorphinone (metopon) ($C_{18}H_{21}NO_3$)		1990
5-Methyl-4,4-diphenyl-6-piperidino-3-hexanone ($C_{24}H_{31}NO$)	1126	
6-Methyl-4,4-diphenyl-6-piperidino-3-hexanone ($C_{24}H_{31}NO$)		1810–1812, 2116
L-α-Methyldopa ($C_{10}H_{13}NO_4$)	810–812	
D-α-Methyldopamine ($C_9H_{13}NO_2$)	813	
Methylenedioxyamphetamine (MDA) ($C_{12}H_{19}NO_2$)	814	
1-(3,4-Methylenedioxybenzyl)-4-(2-pyrimidinyl) piperazine ($C_{16}H_{18}N_4O_2$)	1128	
N-Methylephedrine ($C_{11}H_{17}NO$)	815, 1178	1991
Methylergonovine ($C_{20}H_{25}N_3O_2$)	816	
Methylethylamphetamine ($C_{12}H_{19}N$)		1992
N-Methylglucamine ($C_7H_{17}NO_5$)	817	1993
Methylhexaneamine ($C_7H_{17}N$)	818	
1-Methyl-1H-imidazole ($C_4H_6N_2$)	819	
Methylisopropylamphetamine ($C_{12}H_{21}N$)		1994
Methyl nicotinate ($C_7H_7NO_2$)		1995
Methylparaben ($C_8H_8O_3$)	820	1996
α-Methylphenethylamine ($C_9H_{13}N$)		1997
N-Methylphenethylamine ($C_9H_{13}N$)	821	
Methylphenidate ($C_{14}H_{19}NO_2$)	822–823	
N-(3′-Methylphenyl)anthranilic acid ($C_{14}H_{13}NO_2$)	552–553	
1-(2-Methylphenyl)piperazine ($C_{11}H_{16}N_2$)	90	
N-Methylpseudoephedrine ($C_{11}H_{17}NO$)	1178	
Methylprednisolone-21-phosphate ($C_{22}H_{31}O_8P$)	824	
Methylpromazine ($C_{18}H_{22}N_2S$)		2331–2332
Methyl salicylate ($C_8H_8O_3$)		1998
4-Methylsulfadiazine ($C_{11}H_{12}N_4O_2S$)	1373–1374	2256
3′-Methylsulfapyridine-1-oxide ($C_{12}H_{12}N_3O_3S$)	1381	
4′-Methylsulfapyridine-1-oxide ($C_{12}H_{12}N_3O_3S$)	1381	
5′-Methylsulfapyridine-1-oxide ($C_{12}H_{12}N_3O_3S$)	1381	
6′-Methylsulfapyridine-1-oxide ($C_{12}H_{12}N_3O_3S$)	1381	
5-Methyl-2-thiouracil ($C_5H_6N_2OS$)		1999
6-Methylthiouracil ($C_5H_6N_2OS$)		2000–2001
Methyprylon ($C_{10}H_{17}NO_2$)		2002
Methysergide ($C_{21}H_{27}N_3O_2$)	825–826	
Met-Leu ($C_{11}H_{22}N_2O_3S$)	453	
Metoclopramide ($C_{14}H_{22}ClN_3O_2$)	827–828	
Metolazone ($C_{16}H_{16}ClN_3O_3S$)		2003
Metopon (methyldihydromorphinone) ($C_{18}H_{21}NO_3$)		1990, 2004

Metoprolol ($C_{15}H_{25}NO_3$)	829–830, 1167	
Metronidazole ($C_6H_9N_3O_3$)	831–832	
Metyrosine ($C_{10}H_{13}NO_3$)		2005
Mexiletine ($C_{11}H_{17}NO$)	833–834	
Mezlocillin ($C_{21}H_{25}N_5O_8S_2$)	835	
Mianserin ($C_{18}H_{20}N_2$)	836	
Miconazole ($C_{18}H_{14}Cl_4N_2O$)	837	2006
Midazolam ($C_{18}H_{13}ClFN_3$)	838	2007
Minocycline ($C_{23}H_{27}N_3O_7$)		2008
Minoxidil ($C_9H_{15}N_5O$)	839	
Mirtazepine ($C_{17}H_{19}N_3$)	840	
Mitomycin C ($C_{15}H_{18}N_4O_5$)	841–844	
Mitoxantrone ($C_{22}H_{28}N_4O_6$)		2009
Molindone ($C_{16}H_{24}N_2O_2$)	845	
Monodesmethylpheniramine ($C_{15}H_{18}N_2$)	418	
Monoethylglycine xylidide ($C_{12}H_{18}N_2O$)	846	
Morizicine ($C_{22}H_{24}N_3O_4S$)	847	
Morphine ($C_{17}H_{19}NO_3$)	9, 848–857	2010
Morphine-3-glucuronide ($C_{23}H_{27}NO_9$)	858–859	
Morphine-6-glucuronide ($C_{23}H_{27}NO_9$)	860–861	
Morphine-*N*-oxide ($C_{17}H_{19}NO_4$)		2011
Moxalactam ($C_{20}H_{20}N_6O_9S$)		2012
Moxonidine	96	
Muroctasin (Romurtide) ($C_{43}H_{78}N_6O_{13}$)	862	
5-O-Mycaminosyltylonolide (OMT) ($C_{31}H_{51}NO_{10}$)	863	

N

Nabilone ($C_{24}H_{36}O_3$)		2013
Nadolol ($C_{17}H_{27}O_4$)	1167	2014–2015
Nafcillin ($C_{21}H_{22}N_2O_5S$)	864	
Nalbuphine ($C_{21}H_{27}NO_4$)		2016
Nalidixic acid ($C_{12}H_{12}N_2O_3$)	865–866, 1237–1238	
Nalmefene ($C_{21}H_{25}NO_3$)		2017
Nalorphine ($C_{19}H_{21}NO_3$)	867	2018
Naloxone ($C_{19}H_{21}NO_4$)	868	
Naltrexone ($C_{20}H_{23}NO_4$)	869	
Naphazoline ($C_{14}H_{14}N_2$)	870–871	
2-Naphthol ($C_{10}H_8O$)	252	
Naproxen ($C_{14}H_{14}O_3$)	872–874	
Narcotine ($C_{22}H_{23}NO_7$)	920–922	2041–2042
Natamycin ($C_{33}H_{47}NO_{13}$)		2019
Nebivolol ($C_{22}H_{25}F_2NO_4$)	875	
Nefazodone ($C_{25}H_{32}ClN_5O_2$)		2020

Neostigmine ($C_{12}H_{19}N_2O_2$)		2021
Neo-synephrine ($C_9H_{13}NO_2$)	1077	
Neurophysin I	876	
Niacinamide ($C_6H_6N_2O$)	877	
Nicorandil ($C_8H_9N_3O_4$)		2022
Nicotinamide (niacinamide) ($C_6H_6N_2O$)	877	
Nicotine ($C_{10}H_{14}N_2$)	878–880	2023–2026
Nicotine methiodide ($C_{11}H_{17}IN_2$)		2027
Nicotine analogues	880	
Nicotinic acid ($C_6H_5NO_2$)	881–883	
Nicotinic acid (non-aqueous titration) ($C_6H_5NO_2$)	194	
iso-Nicotinic acid ($C_6H_5NO_2$)	884	
Nicotinoylhydrazine ($C_6H_7N_3O$)	699	
Nifedipine ($C_{17}H_{18}N_2O_6$)	885–886	
Niflumic acid ($C_{13}H_9F_3N_2O_2$)	887–889	
Nikethamide ($C_{10}H_{14}N_2O$)	890	
Nimesulide ($C_{13}H_{12}N_2O_5S$)	891–893	
Nimetazepam ($C_{16}H_{13}N_3O_3$)	894	
Nitrazepam ($C_{15}H_{11}N_3O_3$)	204, 895–899	
Nitrazepam, 7-acetamido ($C_{17}H_{13}N_3O_4$)	205	
Nitrazepam, 7-amino ($C_{15}H_{12}N_4O_3$)	205	
3-Nitroazuloic acid ($C_{11}H_7NO_4$)	115	
2-Nitrobenzoic acid ($C_7H_5NO_4$)	252	
3-Nitrobenzoic acid ($C_7H_5NO_4$)	252	
4-Nitrobenzoic acid ($C_7H_5NO_4$)	252	
p-Nitrobenzoylhydrazine ($C_7H_7N_3O_3$)	699	
Nitrofurantoin ($C_8H_6N_4O_5$)		2028–2029
Nitrofurazone ($C_6H_6N_4O_4$)	900	
4-Nitrophenol ($C_6H_5NO_3$)	252	
8-Nitrotheophylline ($C_7H_7N_5O_4$)	901–902	1699
Nizatidine ($C_{12}H_{21}N_5O_2S_2$)	96	2030–2031
l-Noradrenaline ($C_8H_{11}NO_3$)	729–731, 908–910	
Norcarnitine ($C_6H_{13}NO_3$)	378	
Norcodeine ($C_{17}H_{19}NO_3$)	903–904	
Norcodeine, *N*-(2-cyano)ethyl ($C_{20}H_{22}N_2O_3$)	905	
Nordefrin (Cobefrin) ($C_9H_{13}NO_3$)	345, 906	2032–2033
Norephedrine ($C_9H_{13}NO$)	907	2034
l-Norepinephrine ($C_8H_{11}NO_3$)	497, 729–731, 908–910	
Norepinephrine bitartrate (levarterenol bitartrate) ($C_8H_{11}NO_3.C_4H_6O_6$)		2035

Norepinephrine bitartrate (levarterenol bitartrate) ($C_{12}H_{17}NO_9$)		2035
Norfenefrine ($C_8H_{11}NO_2$)	911	
Norfloxacin ($C_{16}H_{18}FN_3O_3$)	1239–1242	
Norfloxacin ethyl ester ($C_{18}H_{22}FN_3O_3$)	1243	
Norhyoscyamine ($C_{16}H_{21}NO_3$)	652, 912	
Norketamine ($C_{12}H_{14}ClNO$)		2036
Norketobemidone ($C_{14}H_{19}NO_2$)	913	
Normethadone ($C_{20}H_{25}NO$)		2037–2038
Normorphine ($C_{16}H_{17}NO_3$)	914–915	
Norparamethadione ($C_6H_9NO_3$)	916	
Norpseudoephedrine ($C_9H_{13}NO$)	917	2039
Nortrimethadione ($C_5H_7NO_3$)	918	
Nortriptyline ($C_{19}H_{21}N$)	919	2040
Noscapine (narcotine) ($C_{22}H_{23}NO_7$)	920–922	2041–2042
Novobiocin ($C_{31}H_{36}NO_{11}$)		2043
Novocaine	9	
Nystatin ($C_{47}H_{75}NO_{17}$)		2044
O		
Octodrine ($C_8H_{19}N$)	923	
Octopamine ($C_8H_{11}NO_2$)		2045
Ofloxacin ($C_{18}H_{20}FN_3O_4$)	1244–1245	
Olanzapine	96	
Oleandomycin ($C_{35}H_{61}NO_{12}$)	924–925	
Oleic acid ($C_{18}H_{34}O_2$)	926	
Omeprazole ($C_{17}H_{19}N_3O_3S$)	927	
OMT ($C_{31}H_{51}NO_{10}$)	863	
Ondansetron ($C_{18}H_9N_3O$)		2046
Opipramol ($C_{23}H_{29}N_3O$)	928	
Orbifloxacin ($C_{19}H_{20}F_3N_3O_3$)	1246	
Orciprenaline ($C_{11}H_{17}NO_3$)	785–786, 929	1966
Ornidazole ($C_7H_{10}ClN_3O_3$)		2047
Orphenadrine ($C_{18}H_{23}NO$)		2048
Oxacillin ($C_{19}H_{19}N_3O_5S$)	962	2049–2050
Oxamniquine ($C_{14}H_{21}N_3O_3$)	930	
Oxazepam ($C_{15}H_{11}ClN_2O_2$)	204, 931–933	
Oxprenolol ($C_{15}H_{23}NO_3$)	1167	2051
Oxybutynin ($C_{22}H_{31}NO_3$)		2052
Oxycodone ($C_{18}H_{21}NO_4$)	398, 934	2053
Oxymorphone ($C_{17}H_{19}NO_4$)		2054
Oxyphenbutazone ($C_{19}H_{20}N_2O_3$)	935–938	

Oxypurinol ($C_5H_4N_4O_2$)		2055
Oxytetracycline ($C_{22}H_{24}N_2O_9$)	939–945	
Oxytocin ($C_{43}H_{66}N_{12}O_{12}S_2$)		2056
P		
Pachycarpine	9	
Pamaquine (plasmoquin) ($C_{19}H_{29}N_3O$)		2057–2059
Pantoprazole ($C_{16}H_{15}F_2N_3O_4S$)		2060
Papaverine ($C_{20}H_{21}NO_4$)	9, 946–951	
Paracetamol ($C_8H_9NO_2$)	1–7	1528–1530
Paraxanthine ($C_7H_8N_4O_2$)		2061
Paredrine ($C_9H_{13}NO$)	635	
Pargyline ($C_{11}H_{13}N$)		2062
Paromomycin	96	
Paroxetine ($C_{19}H_{20}FNO_3$)		2063
Pecazine ($C_{19}H_{22}N_2S$)	776, 952–953, 1002	
Pefloxacin ($C_{17}H_{20}FN_3O_3$)	1247	
Pelargonic acid ($C_9H_{18}O_2$)	954	
Pemoline ($C_9H_8N_2O_2$)		2064
Pempidine ($C_{10}H_{21}N$)		2065
Penbutolol ($C_{18}H_{29}NO_2$)	1167	2066
Penicillamine ($C_5H_{11}NO_2S$)	375, 955–956	
Penicilloic acid ($C_8H_{14}N_2O_4S$)	957	
Penicillin G (benzylpenicillin) ($C_{16}H_{18}N_2O_4S$)	217, 958–959, 961	
Penicillin V (phenoxymethylpenicillin) ($C_{16}H_{18}N_2O_5S$)	960–962	
Pentamidine ($C_{19}H_{24}N_4O_2$)		2067
Pentazocine ($C_{19}H_{27}NO$)	963	2068
Pentobarbitone ($C_{11}H_{18}N_2O_3$)	140–143	
Pentostatin ($C_{11}H_{16}N_4O_4$)	964	
Pentoxiphylline ($C_{13}H_{18}N_4O_3$)		2069–2070
Pergolide mesylate ($C_{19}H_{26}N_2S.CH_3SO_3H$)		2071
Perhexilene ($C_{19}H_{35}N$)	965	
Perphenazine ($C_{21}H_{26}ClN_3OS$)	966–967	2072
Pethidine ($C_{15}H_{21}NO_2$)	777, 968	1958
Pharmorubicin ($C_{27}H_{29}NO_{11}$)	376, 969	
Phenacetin ($C_{10}H_{13}NO_2$)	9	2073
Phenadoxone ($C_{23}H_{29}NO_2$)		2074–2076
Phe-Gly ($C_{11}H_{14}N_2O_3$)	453	
Phe-Leu ($C_{15}H_{22}N_2O_3$)	453	
Phenazocine ($C_{22}H_{27}NO$)	970	
Phenazopyridine ($C_{11}H_{11}N_5$)	971	
Phencyclidine ($C_{17}H_{25}N$)		2077

Phendimetrazine ($C_{12}H_{17}NO$)		2078
Phenethicillin ($C_{17}H_{20}N_2O_5S$)	961, 972	2079
Phenoxypropylpenicillin ($C_{18}H_{22}N_2O_5S$)	1007	
α-Phenethylamine (1-amino-1-phenylethane) ($C_8H_{11}N$)	973	
Phenformin ($C_{10}H_{15}N_5$)	974	
Phenindamine ($C_{19}H_{19}N$)	975	2080
Phenindione ($C_{10}H_{15}O_2$)	976	
Pheniramine ($C_{16}H_{20}N_2$)	977	2081–2083
Phenmetrazine ($C_{11}H_{15}NO$)		2084–2085
Phenobarbital ($C_{12}H_{12}N_2O_3$)	173–184	1612–1613
Phenolphthalein ($C_{20}H_{14}O_4$)	978–979	
Phenol red ($C_{19}H_{14}O_5S$)	980	2086
Phenols	252	
Phenolsulphonphthalein (phenol red) ($C_{19}H_{14}O_5S$)	980	2086
Phenomorphan ($C_{24}H_{29}NO$)		2087
Phenothiazine ($C_{12}H_9NS$)	981	
Phenothiazine, 10-[*N*-(4-carbamoyl)piperidinyl] propyl-2-chloro- ($C_{21}H_{24}ClN_2OS$)	983	
Phenothiazine, 2-chloro-10-(3-dimethylaminopropyl)- ($C_{17}H_{19}ClN_2S$)	984–985	
Phenothiazine, 2-chloro-10-[*N*-(2-hydroxy)ethyl] piperazinylpropyl- ($C_{21}H_{22}ClN_3OS$)	986	
Phenothiazine, 2-chloro-10-[*N*-methyl] piperazinylpropyl- ($C_{20}H_{24}ClN_3S$)	987–988	
Phenothiazine, 2-chloro-10-[*N*-(2-propionyloxy) ethyl]piperazinylpropyl- ($C_{24}H_{26}ClN_3O_2S$)	989–990	
Phenothiazine derivatives	982	
Phenothiazine, 10-(2-diethylaminoethyl)- ($C_{18}H_{22}N_2S$)	991	
Phenothiazine, 10-(2-dimethylaminomethyl)propyl-2-methoxy- ($C_{19}H_{24}N_2OS$)	992	
Phenothiazine, 10-(2-dimethylaminopropyl)- ($C_{17}H_{20}N_2S$)	993–994	
Phenothiazine, 10-(3-dimethylaminopropyl)- ($C_{17}H_{20}N_2S$)	995	
Phenothiazine, 10-(3-dimethylaminopropyl)-2-trifluoromethyl- ($C_{18}H_{19}F_3N_2S$)	996–997	
Phenothiazine, 10-(*N*-methyl)piperazinylpropyl-2-trifluoromethyl- ($C_{21}H_{24}F_3N_3S$)	998–999	
Phenothiazine, 10-[2-(*N*-methyl)piperidinyl]ethyl-2-methylthio- ($C_{21}H_{22}N_2S_2$)	1000–1001	
Phenothiazine, 10-(*N*-methyl)piperidinyl]methyl-(mepazine) ($C_{19}H_{22}N_2S$)	1002	

Phenothiazine, 10-(*N*-pyrrolidinyl)ethyl- (pyrathiazine) ($C_{18}H_{20}N_2S$)	1003	
Phenoxyacetic acid ($C_8H_8O_3$)	1004	
Phenoxybenzamine ($C_{18}H_{22}ClNO$)		2088
Phenoxyethylpenicilloic acid ($C_{17}H_{22}N_2O_6S$)	1005	
Phenoxymethylpenicillin ($C_{16}H_{18}N_2O_5S$)	960–962	
Phenoxymethylpenicilloic acid ($C_{16}H_{20}N_2O_6S$)	1006	
Phenoxypropazine ($C_9H_{14}N_2O$)		2089
Phenoxypropylpenicilloic acid ($C_{18}H_{24}N_2O_6S$)	1007	
Phenprocoumon ($C_{18}H_{16}O_3$)	637, 1506	
Phentermine ($C_{10}H_{15}N$)		2090–2091
Phentolamine ($C_{17}H_{19}N_3O$)		2092
N-Phenylanthranillic acid ($C_{13}H_{11}NO_2$)	552, 1008	
Phenylbutazone ($C_{19}H_{20}N_2O_2$)	1009–1018	2093
Phenylbutazone analogs	1018	2094
5-Phenyl-1,3-dihydro-1,4-benzodiazepin-2-one ($C_{15}H_{12}N_2O$)	1019	
5-Phenyl-1,3-dihydro-1,4-benzodiazepin-2-one, 7-amino ($C_{15}H_{13}N_3O$)	1075	
5-Phenyl-1,3-dihydro-1,4-benzodiazepin-2-one, 2′-bromo-7-chloro ($C_{15}H_{10}BrClN_2O$)	1027	
5-Phenyl-1,3-dihydro-1,4-benzodiazepin-2-one, 1-[butane-2,4-diol]-2′-fluoro-7-chloro ($C_{19}H_{18}ClFN_2O_3$)	1070	
5-Phenyl-1,3-dihydro-1,4-benzodiazepin-2-one, 1-[butane-2,4-diol]-2′-fluoro-7-iodo ($C_{19}H_{18}FIN_2O_3$)	1066	
5-Phenyl-1,3-dihydro-1,4-benzodiazepin-2-one, 1-[butane-2,4-diol]-7-nitro ($C_{19}H_{19}N_3O_5$)	1076	
5-Phenyl-1,3-dihydro-1,4-benzodiazepin-2-one, 7-chloro ($C_{15}H_{11}ClN_2O$)	1040	
5-Phenyl-1,3-dihydro-1,4-benzodiazepin-2-one, 2′-chloro-7-nitro ($C_{15}H_{10}ClN_3O_3$)	1055	
5-Phenyl-1,3-dihydro-1,4-benzodiazepin-2-one, 7-cyano ($C_{16}H_{11}N_3O$)	1065	
5-Phenyl-1,3-dihydro-1,4-benzodiazepin-2-one, 2′,7-dichloro ($C_{15}H_{10}Cl_2N_2O$)	1028	
5-Phenyl-1,3-dihydro-1,4-benzodiazepin-2-one, 2′,6′-difluoro-7-chloro ($C_{15}H_9ClF_2N_2O$)	1050	
5-Phenyl-1,3-dihydro-1,4-benzodiazepin-2-one, 2′,6′-difluoro-8-chloro ($C_{15}H_{10}ClF_2N_2O$)	1044	
5-Phenyl-1,3-dihydro-1,4-benzodiazepin-2-one, 7-dimethylamino ($C_{17}H_{17}N_3O$)	1054	
5-Phenyl-1,3-dihydro-1,4-benzodiazepin-2-one, 1,2′-dimethyl-7-chloro ($C_{17}H_{15}ClN_2O$)	1037	

5-Phenyl-1,3-dihydro-1,4-benzodiazepin-2-one, 1,3-dimethyl-2′,7-dichloro ($C_{17}H_{14}Cl_2N_2O$) 1021
5-Phenyl-1,3-dihydro-1,4-benzodiazepin-2-one, 2′,7-dinitro ($C_{15}H_{10}N_4O_5$) 1069
5-Phenyl-1,3-dihydro-1,4-benzodiazepin-2-one, 1-ethyl-7-amino ($C_{17}H_{17}N_3O$) 1067
5-Phenyl-1,3-dihydro-1,4-benzodiazepin-2-one, 1-ethyl-7-chloro ($C_{17}H_{15}ClN_2O$) 1024
5-Phenyl-1,3-dihydro-1,4-benzodiazepin-2-one, 4′-fluoro ($C_{15}H_{11}FN_2O$) 1057
5-Phenyl-1,3-dihydro-1,4-benzodiazepin-2-one, 7-fluoro ($C_{15}H_{11}FN_2O$) 1058
5-Phenyl-1,3-dihydro-1,4-benzodiazepin-2-one, 2′-fluoro-7-acetyl ($C_{17}H_{13}FN_2O_2$) 1068
5-Phenyl-1,3-dihydro-1,4-benzodiazepin-2-one, 2′-fluoro-7-chloro ($C_{15}H_{10}ClFN_2O$) 1049
5-Phenyl-1,3-dihydro-1,4-benzodiazepin-2-one, 4′-fluoro-7-chloro ($C_{15}H_{10}ClF_2O$) 1043
5-Phenyl-1,3-dihydro-1,4-benzodiazepin-2-one, 2′-fluoro-7,8-dichloro ($C_{15}H_9Cl_2FN_2O$) 1020
5-Phenyl-1,3-dihydro-1,4-benzodiazepin-2-one, 2′-fluoro-7-ethyl ($C_{17}H_{15}FN_2O$) 1042
5-Phenyl-1,3-dihydro-1,4-benzodiazepin-2-one, 3-hydroxy-7-chloro ($C_{15}H_{11}ClN_2O_2$) 1060
5-Phenyl-1,3-dihydro-1,4-benzodiazepin-2-one, 3-hydroxy-2′,7-dichloro ($C_{15}H_{10}Cl_2N_2O_2$) 1056
5-Phenyl-1,3-dihydro-1,4-benzodiazepin-2-one, 7-methoxy ($C_{16}H_{14}N_2O_2$) 1059
5-Phenyl-1,3-dihydro-1,4-benzodiazepin-2-one, 2′-methoxy-7-chloro ($C_{16}H_{13}ClN_2O_2$) 1051
5-Phenyl-1,3-dihydro-1,4-benzodiazepin-2-one, 3′-methoxy-7-chloro ($C_{16}H_{13}ClN_2O_2$) 1030
5-Phenyl-1,3-dihydro-1,4-benzodiazepin-2-one, 1-methoxymethyl-7-amino ($C_{17}H_{17}N_3O_2$) 1072
5-Phenyl-1,3-dihydro-1,4-benzodiazepin-2-one, 1-methoxymethyl-2′-fluoro-7-amino ($C_{12}H_{16}FN_3O_2$) 1074
5-Phenyl-1,3-dihydro-1,4-benzodiazepin-2-one, 1-methoxymethyl-7-nitro ($C_{17}H_{16}N_3O_4$) 1064
5-Phenyl-1,3-dihydro-1,4-benzodiazepin-2-one, 1-methoxymethyl-7-chloro ($C_{17}H_{15}ClN_2O_2$) 1045
5-Phenyl-1,3-dihydro-1,4-benzodiazepin-2-one, 7-methyl ($C_{16}H_{14}N_2O$) 1052
5-Phenyl-1,3-dihydro-1,4-benzodiazepin-2-one, 1-methyl-7-amino ($C_{16}H_{15}N_3O$) 1071

5-Phenyl-1,3-dihydro-1,4-benzodiazepin-2-one, 1-methyl-7-chloro ($C_{16}H_{15}ClN_2O$) 1046
5-Phenyl-1,3-dihydro-1,4-benzodiazepin-2-one, 2′-methyl-7-chloro ($C_{16}H_{13}ClN_2O$) 1033
5-Phenyl-1,3-dihydro-1,4-benzodiazepin-2-one, 3-methyl-7-chloro ($C_{16}H_{13}ClN_2O$) 1023
5-Phenyl-1,3-dihydro-1,4-benzodiazepin-2-one, 1-methyl-4′-chloro-7-fluoro ($C_{16}H_{12}ClF_2O$) 1035
5-Phenyl-1,3-dihydro-1,4-benzodiazepin-2-one, 3-methyl-2′-chloro-7-nitro ($C_{16}H_{13}ClN_3O_3$) 1048
5-Phenyl-1,3-dihydro-1,4-benzodiazepin-2-one, 1-methyl-2′,7-dichloro ($C_{16}H_{12}Cl_2N_2O$) 1032
5-Phenyl-1,3-dihydro-1,4-benzodiazepin-2-one, 1-methyl-2′-fluoro-7-amino ($C_{16}H_{14}FN_3O$) 1073
5-Phenyl-1,3-dihydro-1,4-benzodiazepin-2-one, 1-methyl-2′-fluoro-7-chloro ($C_{16}H_{12}ClFN_2O$) 1047
5-Phenyl-1,3-dihydro-1,4-benzodiazepin-2-one, 1-methyl-2′-fluoro-7-iodo ($C_{16}H_{12}FIN_2O$) 1031
5-Phenyl-1,3-dihydro-1,4-benzodiazepin-2-one, 1-methyl-2′-fluoro-7-nitro ($C_{16}H_{13}FN_3O_3$) 1063
5-Phenyl-1,3-dihydro-1,4-benzodiazepin-2-one, 1-methyl-4′-methoxy-7-chloro ($C_{17}H_{15}ClN_2O_2$) 1029
5-Phenyl-1,3-dihydro-1,4-benzodiazepin-2-one, 1-methyl-7-nitro ($C_{16}H_{13}N_3O_3$) 1061
5-Phenyl-1,3-dihydro-1,4-benzodiazepin-2-one, 1-methyl-7-thiomethyl ($C_{17}H_{16}N_2OS$) 1039
5-Phenyl-1,3-dihydro-1,4-benzodiazepin-2-one, 1-methyl-2′,6′,7-trichloro ($C_{16}H_{11}Cl_3N_2O$) 1036
5-Phenyl-1,3-dihydro-1,4-benzodiazepin-2-one, 7-nitro ($C_{15}H_{12}N_3O_3$) 1062
5-Phenyl-1,3-dihydro-1,4-benzodiazepin-2-one, 7-thiomethyl ($C_{16}H_{14}N_2OS$) 1041
5-Phenyl-1,3-dihydro-1,4-benzodiazepin-2-one, 2′-thiomethyl-7-chloro ($C_{16}H_{13}ClN_2OS$) 1038
5-Phenyl-1,3-dihydro-1,4-benzodiazepin-2-one, 2′-trifluoromethyl ($C_{16}H_{11}F_3N_2O$) 1053
5-Phenyl-1,3-dihydro-1,4-benzodiazepin-2-one, 3′-trifluoromethyl ($C_{16}H_{11}F_3N_2O$) 1025
5-Phenyl-1,3-dihydro-1,4-benzodiazepin-2-one, 4′-trifluoromethyl ($C_{16}H_{11}F_3N_2O$) 1022
5-Phenyl-1,3-dihydro-1,4-benzodiazepin-2-one, 7-trifluoromethyl ($C_{16}H_{11}F_3N_2O$) 1034
5-Phenyl-1,3-dihydro-1,4-benzodiazepin-2-one, 2′-trifluoromethyl-7-chloro ($C_{16}H_{10}ClF_3N_2O$) 1026
Phenylephrine (Neo-synephrine) ($C_9H_{13}NO_2$) 1077 2095–2096

Phenylethylamine ($C_8H_{11}N$)	1078–1079	2097–2098
Phenylethylamine, *m*,p-dihydroxy ($C_8H_{11}NO_2$)	1085	
Phenylethylamine, *m*,*p*-dihydroxy, N-methyl ($C_9H_{13}NO_2$)	1086	
Phenylethylamine, 2-hydroxy ($C_8H_{11}NO$)	1080	
Phenylethylamine, 2-hydroxy, *m*-hydroxy ($C_8H_{11}NO_2$)	1081	
Phenylethylamine, 2-hydroxy, *m*-hydroxy, N-methyl ($C_9H_{13}NO_2$)	1082	
Phenylethylamine, 2-hydroxy, *p*-hydroxy ($C_8H_{11}NO_2$)	1087	
Phenylethylamine, 2-hydroxy, *p*-hydroxy, N-methyl ($C_9H_{13}NO_2$)	1088	
Phenylethylamine, 2-hydroxy, *N*-methyl ($C_9H_{13}NO$)	1089	
Phenylethylamine, *p*-hydroxy ($C_8H_{11}NO$)	1083	
Phenylethylamine, *p*-hydroxy, *N*-methyl ($C_9H_{13}NO$)	1084	
DL-Phenyllactic acid ($C_9H_{10}O_3$)		2099
3-Phenyloxycarbonyl-5-fluorouracil	564	
Phenylpenilloic acid ($C_{14}H_{17}N_2O_3S$)	1090–1091	
1-Phenylpiperazine ($C_{10}H_{14}N_2$)	90	
Phenylpropanolamine ($C_9H_{13}NO$)	1092–1094	2100–2101
Phenylpropylmethylamine (Vonedrine) ($C_{10}H_{15}N$)	1095–1096	
Phenyltoloxamine ($C_{17}H_{21}NO$)		2102
5-Phenylvaleric acid ($C_{10}H_{14}O_2$)	1097	
Phenyramidol ($C_{13}H_{14}N_2O$)	1098–1099	
Phenyramidol analogues	1099	
Phenytoin ($C_{15}H_{12}N_2O_2$)	1100–1104	
Phe-Phe ($C_{18}H_{20}N_2O_3$)	453	
Phe-Ser ($C_{12}H_{16}N_2O_4$)	453	
Phe-Tyr ($C_{18}H_{20}N_2O_4$)	453	
Pholcodine ($C_{23}H_{30}N_2O_4$)		2103
Phosphocreatine ($C_4H_{10}N_3O_5P$)	1105–1106	
2-Phthalic acid ($C_8H_6O_4$)	252, 1107	
Phthalimide ($C_8H_5NO_2$)		2104
Physostigmine ($C_{15}H_{21}N_3O_2$)	1109	
Physostigmine salicylate ($C_{22}H_{27}N_3O_5$)	1108	2105
Picolinic acid ($C_6H_5NO_2$)	881, 1110	
Picolinoylhydrazine ($C_6H_7N_3O$)	699	
Pilocarpine ($C_{11}H_{16}N_2O_2$)	9, 1111–1118	
Piminodine ($C_{23}H_{30}N_2O_2$)		2106
Pimozide ($C_{28}H_{29}F_2N_3O$)		2107–2108
Pinacidil ($C_{13}H_{19}N_5$)		2109
Pinazepam ($C_{18}H_{13}ClN_2O$)		2110
Pindolol ($C_{14}H_{20}N_2O_2$)		2111–2113

Pipamazine ($C_{21}H_{24}ClN_3OS$)	983, 1119	
Pipecuronium bromide (pipecurium bromide) ($C_{35}H_{62}Br_2N_4O_4$)		2114
Piperazine ($C_4H_{10}N_2$)	1120–1121	2115
Piperazine estrone sulfate ($C_{22}H_{32}N_2O_5S$)	1122	
Piperidine ($C_5H_{11}N$)	1123–1124	
6-Piperidino-4,4-diphenylheptan-3-one (dipipanone) ($C_{24}H_{31}NO$)		1810–1812, 2116
Piperine ($C_{17}H_{19}NO_3$)	1125	
DL-Pipidone (5-methyl-4,4-diphenyl-6-piperidino-3-hexanone)	1126	
Pipradol ($C_{18}H_{21}NO$)		2117
Pirbuterol ($C_{12}H_{20}N_2O_3$)	1127	2118
Piribedil (1-(3,4-methylenedioxybenzyl)-4-(2-pyrimidinyl)piperazine) ($C_{16}H_{18}N_4O_2$)	1128	
Piroxicam ($C_{15}H_{13}N_3O_4S$)	1129–1131	
(O-Pivaloyl)etilefrine ($C_{15}H_{23}NO_3$)	1132	
1-Pivaloyloxymethyl-5-fluorouracil	569	
Pivampicillin ($C_{22}H_{29}N_3O_6S$)		2119
Pizotyline ($C_{19}H_{21}NS$)		2120
Plasmoquin ($C_{19}H_{29}N_3O$)		2057–2059
Platyphylline	9	
Polymyxin B – a mixture of polymyxin B_1 ($C_{56}H_{98}N_{16}O_{13}$) and polymyxin B_2 ($C_{55}H_{96}N_{16}O_{13}$)	2121	
Polymyxin B_1 ($C_{56}H_{98}N_{16}O_{13}$)		2121
Polymyxin B_2 ($C_{55}H_{96}N_{16}O_{13}$)		2121
Polythiazide ($C_{11}H_{13}ClF_3N_3O_4S_3$)	197, 1133	
Porfiromycin ($C_{16}H_{20}N_4O_5$)	844, 1134	
Practolol ($C_{14}H_{22}N_2O_3$)		2122
Pralidoxime chloride ($C_7H_9ClN_2O$)	1135	
Pramoxine ($C_{17}H_{27}NO_3$)		2123
Prazepam ($C_{19}H_{17}ClN_2O$)		2124–2125
Prazosin ($C_{19}H_{21}N_5O_4$)		2126–2127
Prenalterol ($C_{12}H_{19}NO_3$)		2128
Prenylamine ($C_{24}H_{27}N$)	1136	
Prilocaine ($C_{13}H_{20}N_2O$)		2129–2130
Primaquine ($C_{15}H_{21}N_3O$)	1137	
Pristinamycin I_A (vernamycin Bα) ($C_{45}H_{54}N_8O_{10}$)	1138	
Probenecid ($C_{13}H_{19}NO_4S$)	1139	
Procainamide ($C_{13}H_{21}N_3O$)		2131–2132
Procaine ($C_{13}H_{20}N_2O_2$)	741, 1140–1142	2133
Procarbazine ($C_{12}H_{19}N_3O$)	1143	

Prochlorperazine ($C_{20}H_{24}ClN_3S$)	987–988, 1144–1147	2134
Proflavine ($C_{13}H_{11}N_3$)	18	
Progabide ($C_{17}H_{16}ClFN_2O_2$)	1148	
Promazine ($C_{17}H_{20}N_2S$)	995, 1149–1152	2135
Promedol	9	
Promethazine ($C_{17}H_{20}N_2S$)	1153–1156	
Propafenone ($C_{21}H_{27}NO_3$)		2136
Proparacaine ($C_{16}H_{26}N_2O_3$)	1157	
Propicillin ($C_{18}H_{22}N_2O_5S$)	961	2137
Propiomazine ($C_{20}H_{24}N_2OS$)		2138
Propionic acid ($C_3H_6O_2$)	252, 1158	
1-Propionyloxymethyl-5-fluorouracil	569	
Propofol ($C_{12}H_{18}O$)	1159	
Propoxycaine ($C_{16}H_{26}N_2O_3$)		2139
Propoxyphene ($C_{22}H_{29}NO_2$)		2140–2141
Propranolol ($C_{16}H_{21}NO_2$)	1160–1170	2142–2143
Propyl 4-aminobenzoate ($C_{10}H_{13}NO_2$)	38	
Propylamphetamine ($C_{12}H_{19}N$)		2144
Propylhexedrine ($C_{10}H_{21}N$)	1171	2145–2146
Propylparaben ($C_{10}H_{12}O_3$)		2147
6-n-Propyl-2-thiouracil ($C_7H_{10}N_2OS$)		2148
Prostaglandin E_1 ($C_{20}H_{34}O_5$)	1172	2149
Prostaglandin E_2 ($C_{20}H_{32}O_5$)	1173	1804
Prostaglandin $F_{2\alpha}$ ($C_{20}H_{34}O_5$)	451, 1174	
Pseudoecgonine ($C_9H_{15}NO_3$)	1175	
Pseudoecgonine, methyl ester ($C_{10}H_{17}NO_3$)	1176	
Pseudoephedrine ($C_{10}H_{15}NO$)	1177–1178	2150–2151
Pseudotropine ($C_8H_{15}NO$)	1179	
Pteridine, 2,4-diamino-6-formyl ($C_7H_6N_6O$)	1181	
Pteridine, 2,4-diamino-6-hydroxy ($C_6H_6N_6O$)	1182	
Pteridine, 2,4-diamino-6-methyl ($C_7H_8N_6$)	1180	
Pteridine, 2-amino-4-hydroxy-6-formyl ($C_7H_5N_5O_2$)	1183	
Pteridine, 2-amino-4-hydroxy-6-methyl ($C_7H_7N_5O$)	1184	
Purine ($C_5H_4N_4$)	241	2424
Purine derivatives	783	
Pyramidone	9	
Pyrathiazine ($C_{18}H_{20}N_2S$)	1003, 1185–1186	
Pyrazinamide ($C_5H_5N_3O$)		2152
Pyrazolic acid ($C_{17}H_{13}ClN_2O_2$)	1187	
Pyridine (C_5H_5N)	9	2153
Pyridoxal ($C_8H_9NO_3$)	1188–1190	

Pyridoxal, 3-methoxy ($C_9H_{11}NO_3$)	1191	
Pyridoxal, *N*-methyl ($C_9H_{12}NO_3$)	1192	
Pyridoxal-5-phosphate ($C_8H_{10}NO_6P$)	1193	
Pyridoxamine ($C_8H_{12}N_2O_2$)	1194–1197	
Pyridoxamine, 3-methoxy ($C_9H_{14}N_2O_2$)	1198	
Pyridoxamine, *N*-methyl ($C_9H_{15}N_2O_2$)	1199	
Pyridoxamine-5-phosphate ($C_8H_{13}N_2O_6P$)	1200–1201	
Pyridoxine ($C_8H_{11}NO_3$)	1202–1206	
1,3-bis[(2-Pyridyl)methyleneamino]guanidine ($C_{13}H_{13}N_7$)	1207	
Pyridylmethyl phosphate esters ($C_6H_8NO_4P$)	1208	
1-(2-Pyridyl)piperazine ($C_9H_{13}N_3$)	90	
Pyrilamine ($C_{17}H_{23}N_3O$)		2154–2155
Pyrimethamine ($C_{12}H_{13}ClN_4$)		2156–2158
Pyrimethazine ($C_{12}H_{13}ClN_4$)		2159
1-(2-Pyrimidinyl)piperazine ($C_8H_{12}N_4$)	90	
Pyrrobutamine ($C_{20}H_{22}ClN$)		2160
Pyrrolo[2,3-d]pyrimidine derivatives	1209	
Q		
Quinacrine ($C_{23}H_{30}ClN_3O$)	1210	2161–2163
Quinethazone ($C_{10}H_{12}ClN_3O_3S$)		2164
Quinidine ($C_{20}H_{24}N_2O_2$)	1211–1213, 1221	2165–2166
Quinine ($C_{20}H_{24}N_2O_2$)	9, 1214–1223	
1-(2-Quinolinyl)piperazine ($C_{13}H_{15}N_3$)	90	
Quinolone – *N*-acetylnorfloxacin ($C_{18}H_{20}FN_3O_4$)	1224	
Quinolone – amifloxacin ($C_{16}H_{19}FN_4O_3$)	1225–1226	
Quinolone – ciprofloxacin ($C_{17}H_{18}FN_3O_3$)	316, 1227–1228	1723, 2167
Quinolone – 8-desfluorolomefloxacin ($C_{17}H_{20}FN_3O_3$)	1229	
Quinolone – difloxacin ($C_{21}H_{19}F_2N_3O_3$)	1230	
Quinolone – enoxacin ($C_{15}H_{17}FN_4O_3$)	1231	2168
Quinolone – fleroxacin ($C_{17}H_{18}F_3N_3O_3$)	1232	2169
Quinolone – 8-fluoro-norfloxacin ($C_{16}H_{17}F_2N_3O_3$)	1233	
Quinolone – 8-fluoro-pefloxacin ($C_{17}H_{19}F_2N_3O_3$)	1234	
Quinolone – lomefloxacin ($C_{17}H_{19}F_2N_3O_3$)	1235–1236	
Quinolone – nalidixic acid ($C_{12}H_{12}N_2O_3$)	1237–1238	2170–2171
Quinolone – norfloxacin ($C_{16}H_{18}FN_3O_3$)	1239–1242	2172–2173
Quinolone – norfloxacin ethyl ester ($C_{18}H_{22}FN_3O_3$)	1243	
Quinolone – ofloxacin ($C_{18}H_{20}FN_3O_4$)	1244–1245	2174
Quinolone – orbifloxacin ($C_{19}H_{20}F_3N_3O_3$)	1246	
Quinolone – pefloxacin ($C_{17}H_{20}FN_3O_3$)	1247	
Quinolone – quinolone carboxylic acid ($C_{12}H_{11}NO_3$)	1248	
Quinolone – temafloxacin ($C_{21}H_{18}F_3N_3O_3$)	1249	

R

Racemethorphan ($C_{18}H_{25}NO$)		2175
Racemorphan (methorphinan) ($C_{17}H_{23}NO$)		2176
Ramipril ($C_{23}H_{32}N_2O_5$)	1250	
Ranitidine ($C_{13}H_{22}N_4O_3S$)		2177–2179
Remoxipride ($C_{16}H_{23}BrN_2O_3$)	1251	
Repromicin ($C_{38}H_{72}N_2O_{12}$)	1252	
Rescinnamine ($C_{35}H_{42}N_2O_9$)		2180, 2182
Reserpine ($C_{33}H_{40}N_2O_9$)		2181–2182
Riboflavine ($C_{17}H_{20}N_4O_6$)	1253–1255	
Rifampin ($C_{43}H_{58}N_4O_{12}$)	1256–1257	
Rifabutine	96	
Rimoterol (rimiterol) ($C_{12}H_{17}NO_3$)		2183
Risperidone ($C_{23}H_{27}FN_4O_2$)		2184
Ritodrine ($C_{17}H_{21}NO_3$)		2185
Rivastigmine ($C_{14}H_{22}N_2O_2$)		2186
Rolitetracycline ($C_{27}H_{33}N_3O_8$)		2187
Romurtide ($C_{43}H_{78}N_6O_{13}$)	862	
Rosaramicin (juvenimicin) ($C_{31}H_{51}NO_9$)	1258	
Rotoxamine (l-carbinoxamine) ($C_{16}H_{19}ClN_2O$)		2188
Roxithromycin ($C_{41}H_{76}N_2O_{15}$)	1259	

S

Saccharin ($C_7H_5NO_3S$)		2189
Salbutamol ($C_{13}H_{21}NO_3$)	27, 1260	1541
Salicylamide ($C_7H_7NO_2$)	1261–1263	
Salicylic acid (2-hydroxybenzoic acid) ($C_7H_6O_3$)	252, 649, 1264–1269	
Salicylic acid derivatives	1269	
Salsalate ($C_{14}H_{10}O_5$)		2190
Salsolidine	9	
Salsoline	9	
Sarcolysine II	9	
Scopolamine ($C_{17}H_{21}NO_4$)	398, 1270	2191
Secobarbital ($C_{12}H_{18}N_2O_3$)	157	1608
Seglitide ($C_{44}H_{56}N_8O_7$)	1271	
Selegiline (deprenyl) ($C_{13}H_{17}N$)	1272	
Seperidol (Clofluperol) ($C_{22}H_{22}ClF_4NO_2$)	1273	
Serotonin ($C_{10}H_{12}N_2O$)		2192
Sertraline ($C_{17}H_{17}Cl_2N$)	1274	
Ser-Leu ($C_9H_{18}N_2O_4$)	453	
Ser-Phe ($C_{12}H_{16}N_2O_4$)	453	
Sildenafil ($C_{22}H_{30}N_6O_4S$)		2193

Silybin ($C_{25}H_{22}O_{10}$); also Silycristin; Silydianin	1275	
Solasodine ($C_{27}H_{43}NO_2$)		2194
Soluflazine ($C_{28}H_{30}Cl_2FN_5O_2$)	1276	
Sorbic acid ($C_6H_8O_2$)	1277	
Sorbitol ($C_6H_{14}O_6$)	1278	2195
Sotalol ($C_{12}H_{20}N_2O_3S$)	1167, 1279	2196–2197
Sparteine ($C_{15}H_{26}N_2$)		2198
Spasmolytine	9	
Spectinomycin ($C_{14}H_{24}N_2O_7$)		2199
Spiperone ($C_{23}H_{26}FN_3O_2$)		2200
Stearic acid ($C_{18}H_{36}O_2$)	1280	
Streptomycin A ($C_{21}H_{39}N_7O_{12}$)	1281	
Streptomycin derivatives	1282	
Streptovitacin A degradation products	1283	
Streptozocin ($C_8H_{15}N_3O_7$)		2201
Strychnine ($C_{21}H_{22}N_2O_2$)	1284	2202
Succinic acid ($C_4H_6O_4$)	1285–1287	
Succinimide ($C_4H_5NO_2$)	1288	
Succinylsulfathiazole ($C_{13}H_{13}N_3O_5S_2$)		2203
Sucrose ($C_{12}H_{22}O_{11}$)	1289	
Sufentanil ($C_{22}H_{30}N_2O_2S$)		2204–2205
Sulfabenz ($C_{12}H_{12}N_2O_2S$)	1290	2207–2209
Sulfacarbamide ($C_7H_9N_3O_3S$)		2206
Sulfacetamide ($C_8H_{10}N_2O_3S$)	1291–1292	2207–2209
Sulfachloropyridazine ($C_{10}H_9ClN_4O_2S$)		2210–2211
Sulfadiazine ($C_{10}H_{10}N_4O_2S$)	1293–1300	2212–2213
Sulfadiazine, *N*[4]-acetyl ($C_{12}H_{12}N_4O_3S$)	1301	2214
Sulfadimethoxine ($C_{12}H_{14}N_4O_4S$)	1302	2215–2218
Sulfadimethoxine, *N*[4]-acetyl ($C_{14}H_{16}N_4O_5S$)		2219
Sulfadimethoxytriazine ($C_{11}H_{13}N_5O_4S$)	1303	
Sulfadimethyloxazole ($C_{11}H_{13}N_3O_3S$)		2220
Sulfadimidine ($C_{12}H_{14}N_4O_2S$)	1304–1308	2221–2224
Sulfaethidole ($C_{10}H_{12}N_4O_2S$)	1309	2225–2226
Sulfafurazole ($C_{11}H_{13}N_3O_3S$)	1310	2227–2228
Sulfaguanidine ($C_7H_{10}N_4O_2S$)	1311	2229
Sulfalene ($C_{11}H_{12}N_4O_3S$)	1312	
Sulfamerazine ($C_{11}H_{12}N_4O_2S$)	1313–1316	2230–2234
Sulfameter ($C_{11}H_{12}N_4O_3S$)	1323	2243
Sulfamethazine ($C_{12}H_{14}N_4O_2S$)	1304–1308	2221–2224
Sulfamethine ($C_{11}H_{12}N_4O_2S$)	1317	
Sulfamethizole (sulfamethylthiadiazole) ($C_9H_{10}N_4O_2S_2$)	1318–1320	2235–2238
Sulfamethomidine ($C_{12}H_{14}N_4O_3S$)	1322	
Sulfamethoxazole ($C_{10}H_{11}N_3O_3S$)		2239–2241
Sulfamethoxazole, *N*[4]-acetyl ($C_{12}H_{13}N_3O_4S$)		2242

Sulfamethoxydiazine (sulfameter) ($C_{11}H_{12}N_4O_3S$)	1323	2243
Sulfamethoxypyridazine ($C_{11}H_{12}N_4O_3S$)	1324–1325	2244–2247
Sulfamethoxypyridazine, N^4-acetyl ($C_{13}H_{14}N_4O_3S$)	1326	2248
Sulfamethyldiazine ($C_{11}H_{12}N_4O_2S$)	1373–1374	2256
Sulfamethylthiadiazole ($C_9H_{10}N_4O_2S_2$)	1318–1320	
Sulfamethythiazole, N^4-acetyl ($C_{11}H_{12}N_4O_3S_2$)	1321	
Sulfamilyl-3,4-xylamide (Irgafen) ($C_{15}H_{16}N_2O_3S$)	1327	
Sulfamonomethoxine ($C_{11}H_{12}N_4O_2S$)		2249
Sulfanilamide ($C_6H_8N_2O_2S$)	1328–1330	2250–2252
Sulfanilamide, N^1-acetyl ($C_8H_{10}N_2O_3S$)	1331	
Sulfanilamide, N^1-p-aminobenzoyl ($C_{13}H_{13}N_3O_3S$)	1332	
Sulfanilamide, N^1-benzoyl ($C_{13}H_{12}N_2O_3S$)	1333	
Sulfanilamide, N^1-ethylsulphonyl ($C_8H_{12}N_2O_4S_2$)	1334	
Sulfanilamide, N^1-p-aminophenyl ($C_{13}H_{15}N_3O_2S$)	1336	
Sulfanilamide, N^1-chloroacetyl ($C_8H_9ClN_2O_3S$)	1337	
Sulfanilamide, N^1,N^1-dimethyl ($C_8H_{12}N_2O_2S$)	1344	
Sulfanilamide, N^1-furfuryl ($C_{11}H_{12}N_2O_3S$)	1338	
Sulfanilamide, N^1-hydroxyethyl ($C_8H_{12}N_2O_3S$)	1345	
Sulfanilamide, N^1-methyl ($C_7H_{10}N_2O_2S$)	1339	
Sulfanilamide, N^1-phenyl ($C_{12}H_{12}N_2O_2S$)	1340	
Sulfanilamide, N^1-sulfanilyl ($C_{12}H_{13}N_3O_4S_2$)	1335	
Sulfanilamide, N^1-*o*-tolyl ($C_{13}H_{14}N_2O_2S$)	1341	
Sulfanilamide, N^1-*m*-tolyl ($C_{13}H_{14}N_2O_2S$)	1342	
Sulfanilamide, N^1-*p*-tolyl ($C_{13}H_{14}N_2O_2S$)	1343	
2-Sulfanilamido-5-aminopyridine ($C_{11}H_{12}N_4O_2S$)	1346	
5-Sulfanilamido-2-aminopyridine ($C_{11}H_{12}N_4O_2S$)	1347	
2-Sulfanilamido-4-aminopyrimidine ($C_{10}H_{11}N_5O_2S$)	1348	
2-Sulfanilamido-5-bromopyridine ($C_{11}H_{10}BrN_3O_2S$)	1349	
5-Sulfanilamido-2-bromopyridine ($C_{11}H_{10}BrN_3O_2S$)	1350	
5-Sulfanilamido-2-chloropyrimidine ($C_{10}H_9ClN_4O_2S$)	1351	
2-Sulfanilamido-2,4-dimethylpyrimidine ($C_{12}H_{14}N_4O_2S$)	1389–1390	
4-Sulfanilamido-2,6-dimethylpyrimidine ($C_{12}H_{14}N_4O_2S$)	1389–1390	
2-Sulfanilamido-4,6-dimethylpyrimidine ($C_{12}H_{14}N_4O_2S$)	1304–1308	
3-Sulfanilamido-4,5-dimethylpyrazole ($C_{11}H_{14}N_4O_2S$)		2253
2-Sulfanilamidoimidazole ($C_9H_{10}N_4O_2S$)	1352	
3-Sulfanilamido-4-methylfurazan ($C_9H_{10}N_4O_3S$)	1353	
5-Sulfanilamido-3-methylisoxazole ($C_{10}H_{11}N_3O_3S$)	1354	
2-Sulfanilamido-5-methyloxadiazole ($C_9H_{10}N_4O_3S$)	1355	

2-Sulfanilamido-4-methylpyrimidine ($C_{11}H_{12}N_4O_2S$)	1313–1316, 1373–1374	
2-Sulfanilamido-5-methylthiadiazole (sulfamethizole) ($C_9H_{10}N_4O_2S_2$)	1356	
2-Sulfanilamido-4-methylthiazole ($C_{10}H_{11}N_3O_2S_2$)	1357	
2-Sulfanilamido-oxazole ($C_9H_9N_3O_3S$)	1358	
2-Sulfanilamidopyrazine ($C_{10}H_{10}N_4O_2S$)	1359	
3-Sulfanilamidopyridazine ($C_{10}H_{10}N_4O_2S$)	1360	
3-Sulfanilamidopyridine ($C_{11}H_{11}N_3O_2S$)	1361	
4-Sulfanilamidopyrimidine ($C_{10}H_{10}N_4O_2S$)	1362	
5-Sulfanilamidopyrimidine ($C_{10}H_{10}N_4O_2S$)	1363	
2-Sulfanilamido-1,3,4-thiadiazole ($C_8H_8N_4O_2S_2$)	1364	
2-Sulfanilamidothiazole ($C_9H_9N_3O_2S_2$)	1385–1387	2263–2267
4-Sulfanilamido-1,2,4-triazole ($C_8H_9N_5O_2S$)	1365	
2-Sulfanilamido-4,5,6-trimethoxypyrimidine ($C_{13}H_{16}N_4O_5S$)		2254
Sulfanilic acid ($C_6H_7NO_3S$)		2255
Sulfanilylaminoguanidine ($C_7H_{11}N_5O_2S$)	1366	
Sulfanilylcyanamide ($C_7H_7N_3O_2S$)	1367	
Sulfanilylglycine ($C_7H_{10}N_2O_2S$)	1368	
Sulfanilylguanidine ($C_7H_{10}N_4O_2S$)	1369	
N^3-Sulfanilylmetanilamide ($C_{12}H_{13}N_3O_4S_2$)	1370	
N^4-Sulfanilylsulfanilamide ($C_{12}H_{13}N_3O_4S_2$)	1371	
Sulfanilylurea ($C_7H_9N_3O_3S$)	1372	
Sulfaperine (sulfamethyldiazine; 4-methylsulfadiazine) ($C_{11}H_{12}N_4O_2S$)	1373–1374	2256
Sulfaphenazole ($C_{15}H_{14}N_4O_2S$)	1375	2257–2260
Sulfapyridine ($C_{11}H_{11}N_3O_2S$)	1376–1380	2261–2262
Sulfapyridine, N^4-acetyl ($C_{13}H_{13}N_3O_3S$)	1383	
Sulfapyridine-1-oxide ($C_{11}H_{10}N_3O_3S$)	1381	
Sulfapyridine-1-oxide, N^4-acetyl, 6′-methyl ($C_{14}H_{14}N_3O_4S$)	1382	
Sulfapyridine-1-oxide, 4′,6′-dimethyl ($C_{13}H_{14}N_3O_3S$)	1381	
Sulfapyridine-1-oxide, 6′-ethyl ($C_{13}H_{14}N_3O_3S$)	1381	
Sulfapyridine-1-oxide, 3′-methyl ($C_{12}H_{12}N_3O_3S$)	1381	
Sulfapyridine-1-oxide, 4′-methyl ($C_{12}H_{12}N_3O_3S$)	1381	
Sulfapyridine-1-oxide, 5′-methyl ($C_{12}H_{12}N_3O_3S$)	1381	
Sulfapyridine-1-oxide, 6′-methyl ($C_{12}H_{12}N_3O_3S$)	1381	
Sulfasalazine ($C_{18}H_{14}N_4O_5S$)	1384	
Sulfathiazole (2-sulfanilamidothiazole) ($C_9H_9N_3O_2S_2$)	1385–1387	2263–2267
Sulfinpyrazone ($C_{23}H_{20}N_2O_3S$)	1388	2268
Sulfisomidine ($C_{12}H_{14}N_4O_2S$)	1389–1390	2269–2273
Sulfisoxazole ($C_{11}H_{13}N_3O_3S$)	1391–1392	2274–2275

Sulfisoxazole, N^4-acetyl ($C_{13}H_{15}N_3O_4S$)	1393	2276
Sulfonazole ($C_9H_9N_3O_2S_2$)	1394	
Sulfone, bis(4-aminophenyl) (Dapsone; 4,4-diaminodiphenylsulfone) ($C_{12}H_{12}N_2O_2S$)	1395	
Sulindac ($C_{20}H_{17}FO_3S$)	1396	
Sulpiride ($C_{15}H_{23}N_3O_4S$)	1397	
Sulthiame ($C_{10}H_{14}N_2O_4S_2$)		2277
Sumatriptan ($C_{14}H_{21}N_3O_2S$)		2278
Sympatol ($C_9H_{13}NO_2$)	1398–1399	
Synephrine (Sympatol) ($C_9H_{13}NO_2$)	1398–1399	2279

T

Tacrine ($C_{13}H_{14}N_2$)		2280
Talbutal ($C_{11}H_{16}N_2O_3$)		1611
Tamoxifen ($C_{26}H_{29}NO$)	1400	2281–2282
Tartaric acid ($C_4H_6O_6$)	1401	
Taurocholic acid ($C_{26}H_{45}NO_7S$)	1402	
Tecomine (tecomanine) ($C_{11}H_{17}NO$)	1403	
Temafloxacin ($C_{21}H_{18}F_3N_3O_3$)	1249	
Temazepam ($C_{16}H_{13}ClN_2O_2$)		2283
Teniposide ($C_{32}H_{32}O_{13}S$)	1404	
Tenoxicam ($C_{13}H_{11}N_3O_4S_2$)		2284
Terazosin ($C_{19}H_{25}N_5O_4$)	1405	
Terbinafine	96	
Terbutaline ($C_{12}H_{19}NO_3$)	1406–1408	2285–2286
Terfenadine ($C_{32}H_{41}NO_2$)	1409	2287
Testosterone, imidazole-1-carboxylic acid prodrug ($C_{23}H_{30}N_2O_3$)	1410	
Tetracaine ($C_{15}H_{24}N_2O_2$)	741, 1411–1412	2288–2289
Tetracycline ($C_{22}H_{24}N_2O_8$)	1413–1418	2290
Tetracycline methiodide ($C_{23}H_{27}IN_2O_8$)	1419	
Tetrahydrocannabinol ($C_{21}H_{30}O_2$)		2291
Tetrahydro-α-morphimethine ($C_{18}H_{25}NO_3$)	1420	
Tetrahydrozoline ($C_{13}H_{16}N_2$)	1421	
Th 1206 ($C_{12}H_{19}NO_3$)	1422	
THAM ($C_4H_{11}NO_3$)	194	2345–2346
Thebaine ($C_{19}H_{21}NO_3$)	1423	2292
Thenalidine ($C_{17}H_{22}N_2S$)	1424	
Thenyldiamine ($C_{14}H_{19}N_3S$)		2293
Theobromine ($C_7H_8N_4O_2$)	9, 479, 1425–1428	
Theophylline ($C_7H_8N_4O_2$)	9, 479, 1429–1434	
Thiabendazole ($C_{10}H_7N_3S$)		2294

Thiamine ($C_{12}H_{17}N_4OS$)		2295–2298
Thiamine-O-monophosphate ($C_{12}H_{18}N_4O_4PS$)		2299
Thiamylal ($C_{12}H_{18}N_2O_2S$)		1610
1-(2-Thiazolyl)piperazine ($C_7H_{11}N_3S$)	90	
Thioglucose ($C_6H_{12}O_5S$)	375	
Thioguanine ($C_5H_5N_5S$)		2300
Thiomalic acid ($C_4H_6O_4S$)	375, 1435	
Thiopentone ($C_{11}H_{18}N_2O_2S$)	144, 1436	1606–1607
Thiopropazate ($C_{23}H_{28}ClN_3O_2S$)	1437–1438	
Thioridazine ($C_{21}H_{26}N_2S_2$)	1439–1442	2301
Thiothixene ($C_{23}H_{29}N_3O_2S$)		2302
2-Thiouracil ($C_4H_4N_2OS$)		2303
Thonzylamine ($C_{16}H_{22}N_4O$)		2304–2305
L-Thyronine, ($C_{15}H_{12}I_3NO_4$)		2306
Thyroxine, L- (levothyroxine) ($C_{15}H_{11}I_4NO_4$)	1443	
Tiapamil ($C_{26}H_{37}NO_8S_2$)	1444	
Tiaprofenic acid (tiprofenic acid) ($C_{14}H_{12}O_3S$)		2307
Ticarcillin ($C_{15}H_{16}N_2O_6S_2$)		2308
Ticrynafen ($C_{13}H_8Cl_2O_4S$)		2309
Tienoxolol ($C_{21}H_{28}N_2O_5S$)	1445	
Tigloidine	652	
Tilmicosin ($C_{46}H_{80}N_2O_{13}$)	1446	
Timolol ($C_{13}H_{24}N_4O_3S$)	1167, 1447	2310
Timoprazole ($C_{13}H_{11}N_3OS$)		2311
Tiotidine ($C_{12}H_{18}N_6S_2$)		2312
Tixanox ($C_{15}H_{10}O_5S$)	1448	
Tobramycin ($C_{18}H_{37}N_5O_9$)		2313
Tocainide ($C_{11}H_{16}N_2O$)	1449	2314
Tolamolol ($C_{20}H_{26}N_2O_4$)		2315–2316
Tolazamide ($C_{14}H_{21}N_3O_3S$)	1450	2317
Tolazoline ($C_{10}H_{12}N_2$)	1451	
Tolbutamide ($C_{12}H_{18}N_2O_3S$)	1452–1453	2318
Tolmetin ($C_{15}H_{15}NO_3$)	1454	
Tolpropamide ($C_{18}H_{23}N$)	1455	
Tolterodine ($C_{22}H_{31}NO$)		2319
p-Toluic acid ($C_8H_8O_2$)	1456	
Tramazoline ($C_{13}H_{17}N_3$)		2320
Tranexamic acid ($C_8H_{15}NO_2$)		2321
Tranylcypromine ($C_9H_{11}N$)		2322
Trazodone ($C_{19}H_{22}ClN_5O$)	1457–1459	2323
Triamcinolone-16,17-acetonide-21-phosphate ($C_{24}H_{32}FO_9P$)	392	
Triamterene ($C_{12}H_{11}N_7$)		2324
Triazolam ($C_{17}H_{12}Cl_2N_4$)	1460	
Trichloromethiazide ($C_8H_8Cl_3N_3O_4S_2$)	197, 1461	2325
Triethylamine ($C_6H_{15}N$)	9	

Trifluoperazine ($C_{21}H_{24}F_3N_3S$)	998–999, 1462–1467	2326–2327
Trifluopromazine ($C_{18}H_{19}F_3N_2S$)	996–997, 1468–1469	2329–2330
5-Trifluoromethyl-2′-deoxyuridine ($C_{10}H_{11}F_3N_2O_5$)		1769, 2328
5-Trifluoromethyluracil ($C_5H_3F_3N_2O_2$)		1769
1-(3-Trifluoromethylphenyl)piperazine ($C_{11}H_{13}F_3N_2$)	90	
Trimeprazine (methylpromazine) ($C_{18}H_{22}N_2S$)		2331–2332
Trimethobenzamide ($C_{21}H_{28}N_2O_5$)		2333
Trimethoprim ($C_{14}H_{18}N_4O_{32}$)	1470–1471	2334–2336
Trimethoprim analogues	1471	
Trimethoxy-3,4,5-phenylsulfonyl derivatives		2337
Trimipramine ($C_{20}H_{26}N_2$)		2338–2339
Tripelennamine ($C_{16}H_{21}N_3$)	1472	2340–2342
Tripeptides	483	
Triprolidine ($C_{19}H_{22}N_2$)	1473	2343–2344
TRIS ($C_4H_{11}NO_3$)	1474	2346
TRIS (non-aqueous titration) ($C_4H_{11}NO_3$)	194	
Troleandomycin ($C_{41}H_{67}NO_{15}$)		2345
Tromethamine ($C_4H_{11}NO_3$)	194	2346
Tropacocaine ($C_{15}H_{19}NO_2$)	1475–1476	2347–2348
DL-Tropic acid ($C_9H_{10}O_3$)	1477	2349
Tropicamide ($C_{17}H_{20}N_2O_2$)		2350–2351
Tropine ($C_8H_{15}NO$)	1478–1479	
Trp-Gly ($C_{13}H_{15}N_3O_3$)	453	
Trp-Phe ($C_{20}H_{21}N_3O_3$)	453	
Tryptophan ($C_{11}H_{12}N_2O_2$)	1480–1481	
Tuaminoheptane ($C_7H_{17}N$)	1482	
Tylosin ($C_{46}H_{77}NO_{17}$)	1483	
Tyramine ($C_8H_{11}NO$)		2352–2353
Tyrosine ($C_9H_{11}NO_3$)	1484–1485	
U		
Uracil ($C_4H_4N_2O_2$)		1769, 2354
Urea (CH_4N_2O)		2355
Uric acid ($C_5H_4N_4O_3$)		2356
Urotropine	9	
Ursodeoxycholic acid (ursodiol) ($C_{24}H_{40}O_4$)	1486	
V		
Valeroidine	652	
Val-Gly ($C_7H_{14}N_2O_3$)	453	

Val-Tyr ($C_{14}H_{20}N_2O_4$)	453	
Valproic acid ($C_8H_{16}O_2$)	1487	
Vancomycin ($C_{66}H_{75}Cl_2N_9O_{24}$)	1488–1490	
Vanillic acid ($C_8H_8O_4$)	1491	
Vanillin ($C_8H_8O_3$)	1492	
Vanillin, *iso* ($C_8H_8O_3$)	1493	
Vanillin, *ortho* ($C_8H_8O_3$)	1494	
Venlafaxine ($C_{17}H_{27}NO_2$)		2357
Verapamil (dexverapamil) ($C_{27}H_{38}N_2O_4$)	1495–1498	2358–2359
Vernamycin Bα ($C_{45}H_{54}N_8O_{10}$)	1138	
Vidarabine ($C_{10}H_{13}N_5O_4$)		2360
Vinblastine (VLB; vincaleukoblastine) ($C_{46}H_{58}N_4O_9$)	1499	2362
Vincristine ($C_{46}H_{56}N_4O_{10}$)	1500–1501	2362
Vindesine ($C_{43}H_{55}N_5O_7$)	1502–1503	
Viomycin ($C_{25}H_{43}N_{13}O_{10}$)		2363–2364
Vonedrine ($C_{10}H_{15}N$)	1095–1096	
W		
Warfarin ($C_{19}H_{16}O_4$)	637, 1504–1508	2365–2366
Water (H_2O)	1509–1510	
WY- 4508 (1′-amino-cyclohexanecarboxamidopenicillin) ($C_{15}H_{23}N_3O_4S$)	1511	
WY- 7953 (1′-amino-cyclopentanecarboxamidopenicillin) ($C_{14}H_{21}N_3O_4S$)	1511	
WY- 8542 (1′-amino-3′-methylcyclopentanecarboxamidopenicillin) ($C_{15}H_{23}N_3O_4S$)	1512	
X		
Xanthine ($C_5H_4N_4O_2$)	241	2367
Xanthine derivatives	1513–1515	
Xipamide ($C_{15}H_{15}ClN_2O_4S$)	1516	
Xylocaine ($C_{14}H_{22}N_2O$)	738–746	
Xylometazoline ($C_{16}H_{24}N_2$)	1517	
Y		
Yohimbine ($C_{21}H_{26}N_2O_3$)	398, 479	
Z		
Zidovudine ($C_{10}H_{13}N_5O_4$)	1518	
Zileuton ($C_{11}H_{12}N_2O_2S$)	1519–1520	
Zimeldine (zimelidine) ($C_{16}H_{17}BrN_2$)		2368
Zomepirac ($C_{15}H_{14}ClNO_3$)	1521	
Zopiclone ($C_{17}H_{17}ClN_6O_3$)	1522	

CHEMICAL FORMULA INDEX OF DRUGS OR DRUG RELATED COMPOUNDS FOUND IN PK_a DATABASE FILES

Appendix A – pK_a values found with significant data quality information
Appendix B – pK_a values found with little or no data quality information

Chemical formula		*Appendix A*	*Appendix B*
C_1			
3-(4-Chlorophenyl)-2-ethyl-2,3,5,6-tetrahydroimidazo[2,1-b]thiazol-3-ol CHECK		284	
3-(4-Chlorophenyl)-5,6-dihydro-2-ethylimidazo[2,1-b]thiazole CHECK PAPER		283	
CH_2O_2	Formic acid	580	
CH_2O_3	Carbonic acid	251	1661
CH_4N_2O	Urea		2355
CH_5N	Methylamine	9	
CH_5N_3	Guanidine		1902
CO_3Li_2	Lithium carbonate		1946
C_2			
$C_2H_3ClO_2$	Chloroacetic acid	252	
$C_2H_3Cl_3O_2$	Chloral hydrate	276–277	
$C_2H_3N_3S$	Aminothiadiazole		1571
$C_2H_4O_2$	Acetic acid	12, 252	
$C_2H_5NO_2$	Acetohydroxamic acid		1536
$C_2H_6O_3S_2$	Mesna (2-mercaptoethanesulfonic acid)		1965
$C_2H_8N_2$	Ethylenediamine		1848
C_3			
$C_3H_4N_2$	Imidazole	667–668	
$C_3H_4N_2O_2$	Hydantoin		1915
$C_3H_4O_4$	Malonic acid	252, 760	
$C_3H_6N_2O_2$	D-Cycloserine	371	1754
$C_3H_6O_2$	Propionic acid	252, 1158	
$C_3H_6O_3$	Lactic acid, (±)	723–724	
$C_3H_6O_3$	Lactic acid, D(−)	721	
$C_3H_6O_3$	Lactic acid, L(+)	722	
$C_3H_7NO_2$	Cysteine	375, 955	
$C_3H_8O_3$	Glycerol	598	
C_4			
$C_4H_2Br_2N_2O_3$	Barbituric acid, 5,5-dibromo	118	
$C_4H_2Cl_2N_2O_3$	Barbituric acid, 5,5-dichloro	119	

$C_4H_3F_6N_5O_2$	3,5-Di(trifluoroacetamido)-1,2,4-triazole (Guanazole prodrug)	602	
$C_4H_3FN_2O_2$	5-Fluorouracil	562	
$C_4H_4FN_3O$	Flucytosine		1869
$C_4H_4N_2OS$	2-Thiouracil		2303
$C_4H_4N_2O_2$	Uracil		2354
$C_4H_4N_2O_3$	Barbituric acid	117	
$C_4H_4O_4$	Fumaric acid		1882
$C_4H_4O_4$	Maleic acid	252, 758	
$C_4H_5ClN_4O_3S_2$	Chloroacetazolamide	11	
$C_4H_5NO_2$	Succinimide	1288	
$C_4H_6N_2$	1-Methyl-1H-imidazole	819	
$C_4H_6N_4O_3S_2$	Acetazolamide	11	1532–1535
$C_4H_6O_4$	Succinic acid	1285–1287	
$C_4H_6O_4S$	Thiomalic acid	375, 1435	
$C_4H_6O_5$	Malic acid	759	
$C_4H_6O_6$	Tartaric acid	1401	
$C_4H_7N_3O$	Creatinine	357	
$C_4H_8N_2O_3$	Gly-Gly	453	
$C_4H_8O_2$	Butyric acid	378	
$C_4H_9N_3O_2$	Creatine	356	
$C_4H_9N_5O_2$	3,5-Diacetamido-1,2,4-triazole (Guanazole prodrug)	602	
$C_4H_9NO_2$	γ-Aminobutyric acid (GABA)	378	
$C_4H_9NO_3$	β-Hydroxy-γ-aminobutyric acid	378	
$C_4H_{10}N_2$	Piperazine	1120–1121	2115
$C_4H_{10}N_3O_5P$	Phosphocreatine	1105–1106	
$C_4H_{11}N$	Diethylamine	9	
$C_4H_{11}NO_3$	TRIS (non-aqueous titration)	194	
$C_4H_{11}NO_3$	Tromethamine (THAM, TRIS)	194	2345–2346
$C_4H_{11}N_5$	Metformin	787	
C_5			
$C_5H_3F_3N_2O_2$	5-Trifluoromethyluracil		1769
$C_5H_4N_2O_4$	5-Carboxyuracil		1769
$C_5H_4N_4$	Purine	241	2424
$C_5H_4N_4O$	Allopurinol		1547–1548
$C_5H_4N_4O$	Hypoxanthine	241, 653	
$C_5H_4N_4O_2$	Oxypurinol		2055
$C_5H_4N_4O_2$	Xanthine	241	2367
$C_5H_4N_4O_3$	Uric acid		2356
$C_5H_4N_4S$	6-Mercaptopurine	782–783	
C_5H_5N	Pyridine	9	2153
$C_5H_5N_3O$	Pyrazinamide		2152
$C_5H_5N_5$	Adenine	22, 241	

Formula	Compound		
$C_5H_5N_5O$	Guanine	241	
$C_5H_5N_5O_2$	2,8-Dihydroxyadenine	438	
$C_5H_5N_5S$	Thioguanine		2300
$C_5H_6N_2O$	Imidazole, N-acetyl	669	
$C_5H_6N_2OS$	5-Methyl-2-thiouracil		1999
$C_5H_6N_2OS$	6-Methylthiouracil		2000–2001
$C_5H_7I_2N_3$	Histamine, diiodo	617	
$C_5H_7NO_2$	Glutarimide		1892
$C_5H_7NO_3$	Dimethadione		1801
$C_5H_7NO_3$	Nortrimethadione	918	
$C_5H_8IN_3$	Histamine, monoiodo	616	
$C_5H_8N_2O_2$	Dimethylhydantoin		1803
$C_5H_8N_4O_3S_2$	Methazolamide		1978
$C_5H_8O_3$	Levulinic acid	737	
$C_5H_9N_3$	Histamine	614–615	
$C_5H_{11}Cl_2N$	Mechlorethamine		1954
$C_5H_{11}N$	Piperidine	1123–1124	
$C_5H_{11}NO_2S$	Methionine	955	1984
$C_5H_{11}NO_2S$	Penicillamine	375, 955–956	
C_6			
$C_6H_4Cl_2O$	3,5-Dichlorophenol	408	
$C_6H_4F_7N_5O$	3-Amino-5-heptafluorobutyramido-1,2,4-triazole (Guanazole prodrug)	602	
$C_6H_5FN_2O_4$	5-Fluorouracil, 1-methyloxycarbonyl	563	
$C_6H_5NO_2$	Isonicotinic acid (non-aqueous titration)	194	
$C_6H_5NO_2$	Isonicotinic acid	700, 881, 884	
$C_6H_5NO_2$	Nicotinic acid (non-aqueous titration)	194	
$C_6H_5NO_2$	Nicotinic acid	881	883
$C_6H_5NO_2$	Picolinic acid	881, 1110	
$C_6H_5NO_3$	4-Nitrophenol	252	
$C_6H_6Cl_2N_2O_4S_2$	Dichlorphenamide		1786
$C_6H_6N_2O$	Isonicotinamide	699	
$C_6H_6N_2O$	Nicotinamide (niacinamide)	877	
$C_6H_6N_2O_3$	Barbituric acid, -1′,5-spiroCyclopropane	190	
$C_6H_6N_4O_4$	Nitrofurazone	900	
$C_6H_6N_6O$	Pteridine, 2,4-diamino-6-hydroxy	1182	

$C_6H_6O_2$	Hydroquinone	633	1920
$C_6H_7NO_2S$	Benzenesulfonamide	201	
$C_6H_7NO_3S$	Sulfanilic acid		2255
$C_6H_7N_3O$	Isoniazid	699	
$C_6H_7N_3O$	Nicotinoylhydrazine	699	
$C_6H_7N_3O$	Picolinoylhydrazine	699	
C_6H_8ClNS	Clomethiazole (Chlormethiazole)		1726–1728
$C_6H_8ClN_7O$	Amiloride	32–33	1558
$C_6H_8NO_4P$	Pyridylmethyl phosphate esters	1208	
$C_6H_8N_2OS$	2-Ethylmercapto-4-hydroxypyrimidine		1849
$C_6H_8N_2O_2S$	Sulfanilamide	1328–1330	2250–2252
$C_6H_8N_2O_3$	Barbituric acid, 5,5-dimethyl	120–122	
$C_6H_8O_2$	Sorbic acid	1277	
$C_6H_8O_6$	Ascorbic acid	91–93	1585
$C_6H_8O_7$	Citric acid	318–319	
$C_6H_9N_3O_2$	Histidine	1481	
$C_6H_9N_3O_3$	Metronidazole	831–832	
$C_6H_9NO_3$	Norparamethadione	916	
$C_6H_{10}N_6O$	Dacarbazine		1758
$C_6H_{10}O_7$	D-Glucuronic acid		1891
$C_6H_{10}O_7$	Guluronic acid		1546
$C_6H_{10}O_7$	Mannuronic acid		1546
$C_6H_{12}N_4$	Methenamine		1980–1982
$C_6H_{12}O_5S$	Thioglucose	375	
$C_6H_{12}O_6$	Dextrose (glucose)	395–396, 596	1777
$C_6H_{12}O_7$	Gluconic acid	593	
$C_6H_{13}N_5O_2$	3,5-Dipropionamido-1,2,4-triazole (Guanazole prodrug)	602	
$C_6H_{13}NO_2$	ε-Aminocaproic acid		1563
$C_6H_{13}NO_2S$	Ethionine	955	
$C_6H_{13}NO_3$	Norcarnitine	378	
$C_6H_{13}NO_3S$	Cyclamic acid	363	
$C_6H_{13}NO_5$	D-Glucosamine	594	
$C_6H_{14}NO_8P$	D-Glucosamine-6-phosphate	595	
$C_6H_{14}O_6$	Mannitol	763	
$C_6H_{14}O_6$	Sorbitol	1278	2195
$C_6H_{15}ClN_2O_2$	Carbachol		1657
$C_6H_{15}N$	Triethylamine	9	
C_7			
$C_7H_4ClNO_2$	Chlorzoxazone		1717
$C_7H_5BrO_2$	Benzoic acid, 4-bromo	212	
$C_7H_5ClO_2$	Benzoic acid, 4-chloro	212	

$C_7H_5FO_2$	Benzoic acid, 4-fluoro	212	
$C_7H_5NO_3S$	Saccharin		2189
$C_7H_5NO_4$	2-Nitrobenzoic acid	252	
$C_7H_5NO_4$	3-Nitrobenzoic acid	252	
$C_7H_5NO_4$	4-Nitrobenzoic acid	252	
$C_7H_5NO_4$	Benzoic acid, 4-nitro	212	
$C_7H_5N_5O_2$	Pteridine, 2-amino-4-hydroxy-6-formyl	1183	
$C_7H_6ClN_3O_4S_2$	Chlorothiazide	288–289	1702–1703
$C_7H_6N_6O$	Pteridine, 2,4-diamino-6-formyl	1181	
$C_7H_6O_2$	Benzoic acid	206–212, 252, 649, 874	
$C_7H_6O_3$	2-Hydroxybenzoic acid	252, 649, 1264–1269	
$C_7H_6O_3$	3-Hydroxybenzoic acid	252	
$C_7H_6O_3$	4-Hydroxybenzoic acid	212, 252, 649	
$C_7H_6O_4$	4-Hydroxysalicylic acid	649	
$C_7H_6O_5$	Gallic acid (3,4,5-trihydroxybenzoic acid)	587	
$C_7H_7BrN_4O_2$	8-Bromotheophylline	229	
$C_7H_7ClN_4O_2$	8-Chlorotheophylline	287	1699–1701
C_7H_7ClO	Chlorocresol		1694
$C_7H_7FN_2O_4$	5-Fluorouracil, 1-acetyloxymethyl	569	
$C_7H_7FN_2O_4$	5-Fluorouracil, 3-acetyloxymethyl	570	
$C_7H_7FN_2O_4$	5-Fluorouracil, 3-ethyloxycarbonyl	564	
$C_7H_7N_3O_2S$	Sulfanilylcyanamide	1367	
$C_7H_7N_3O_3$	*p*-Nitrobenzoylhydrazine	699	
$C_7H_7N_5O$	Pteridine, 2-amino-4-hydroxy-6-methyl	1184	
$C_7H_7N_5O_4$	8-Nitrotheophylline	901–902	1699
$C_7H_7NO_2$	4-Aminobenzoic acid	35–39, 212	1561–1562
$C_7H_7NO_2$	Methyl nicotinate		1995
$C_7H_7NO_2$	Salicylamide	1261–1263	
$C_7H_7NO_3$	4-Aminosalicyclic acid	49	1570
$C_7H_7NO_3$	5-Aminosalicyclic acid	50	
$C_7H_7NO_3$	Mesalamine		1964
$C_7H_8ClN_3O_4S_2$	Hydrochlorothiazide	197, 623–626	1916
$C_7H_8N_2O$	Benzoylhydrazine	699	

$C_7H_8N_2O_3$	Barbituric acid, -1′,5-spiroCyclobutane	191	
$C_7H_8N_4O_2$	Paraxanthine		2061
$C_7H_8N_4O_2$	Theobromine	9, 479, 1425–1428	
$C_7H_8N_4O_2$	Theophylline	9, 479, 1429–1434	
$C_7H_8N_6$	Pteridine, 2,4-diamino-6-methyl	1180	
$C_7H_9ClN_2O$	Pralidoxime chloride	1135	
$C_7H_9N_3O$	1-Isonicotinoyl-1-methylhydrazine	699	
$C_7H_9N_3O$	1-Isonicotinoyl-2-methylhydrazine	699	
$C_7H_9N_3O_3S$	Sulfacarbamide, Sulfanilylurea	1372	2206
$C_7H_{10}ClN_3O_3$	Ornidazole		2047
$C_7H_{10}N_2OS$	6-n-Propyl-2-thiouracil		2148
$C_7H_{10}N_2O_2S$	Sulfanilamide, N^1-methyl	1339	
$C_7H_{10}N_2O_2S$	Sulfanilylglycine	1368	
$C_7H_{10}N_2O_3$	Barbituric acid, 5-ethyl-5-methyl	123	
$C_7H_{10}N_4O_2S$	Sulfaguanidine (Sulfanilylguanidine)	1311, 1369	2229
$C_7H_{11}N_3O_2$	Ipronidazole		1937
$C_7H_{11}N_3S$	1-(2-Thiazolyl)piperazine	90	
$C_7H_{11}N_5O_2S$	Sulfanilylaminoguanidine	1366	
$C_7H_{11}NO_2$	Arecaidine	86	
$C_7H_{11}NO_2$	Ethosuximide		1842–1843
$C_7H_{13}NO_2$	Dihydroarecaidine	423	
$C_7H_{14}N_2O_3$	Val-Gly	453	
$C_7H_{15}NO_2$	γ-Butyrobetaine (GBB)	378	
$C_7H_{15}NO_3$	Carnitine	378	
$C_7H_{17}N$	Methylhexaneamine	818	
$C_7H_{17}N$	Tuaminoheptane	1482	
$C_7H_{17}NO_5$	N-Methylglucamine	817	1993
C_8			
$C_8H_5NO_2$	Phthalimide		2104
$C_8H_6F_3N_3O_4S_2$	Flumethiazide	197, 558	
$C_8H_6N_4O_5$	Nitrofurantoin		2028–2029
$C_8H_6O_4$	2-Phthalic acid	252, 1107	
$C_8H_7ClN_2O_2S$	Diazoxide	403	1783
$C_8H_8Cl_2N_4$	Guanabenz		1899
$C_8H_8Cl_3N_3O_4S_2$	Trichloromethiazide	197, 1461	2325
$C_8H_8F_3N_3O_4S_2$	Hydroflumethiazide	197, 629–631	

$C_8H_8N_4$	Hydralazine	621–622	
$C_8H_8N_4O_2S_2$	2-Sulfanilamido-1,3,4-thiadiazole	1364	
$C_8H_8O_2$	Benzoic acid, 4-methyl	212, 1456	
$C_8H_8O_3$	2-Methoxybenzoic acid	252	
$C_8H_8O_3$	3-Methoxybenzoic acid	252	
$C_8H_8O_3$	Mandelic acid	761	1950–1951
$C_8H_8O_3$	Methyl salicylate		1998
$C_8H_8O_3$	Methylparaben	820	1996
$C_8H_8O_3$	Phenoxyacetic acid	1004	
$C_8H_8O_3$	Vanillin	1492	
$C_8H_8O_3$	Vanillin, *iso*	1493	
$C_8H_8O_3$	Vanillin, *ortho*	1494	
$C_8H_8O_4$	Vanillic acid	1491	
$C_8H_9ClN_2O_3S$	Sulfanilamide, N^1-chloroacetyl	1337	
$C_8H_9FN_2O_5$	5-Fluorouracil, 1-ethyloxycarbonyloxymethyl	565	
$C_8H_9FN_2O_5$	5-Fluorouracil, 3-ethyloxycarbonyloxymethyl	566	
C_8H_9NO	Acetanilide (Antifebrin)	8–9	1531
$C_8H_9NO_2$	Acetaminophen (Paracetamol)	1–7	1528–1530
$C_8H_9NO_2$	Methyl 4-aminobenzoate	38	
$C_8H_9NO_3$	Isopyridoxal	707	
$C_8H_9NO_3$	Mandelhydroxamic acid		1949
$C_8H_9NO_3$	Pyridoxal	1188–1190	
$C_8H_9N_3O_4$	Nicorandil		2022
$C_8H_9N_5O_2S$	4-Sulfanilamido-1,2,4-triazole	1365	
$C_8H_{10}AsNO_5$	Acetarsone	10	
$C_8H_{10}NO_6P$	Isopyridoxal-4-phosphate	707	
$C_8H_{10}NO_6P$	Pyridoxal-5-phosphate	1193	
$C_8H_{10}N_2O_3$	Barbituric acid, -1′,5-spiroCyclopentane	192	
$C_8H_{10}N_2O_3S$	Sulfacetamide	1291–1292	2207–2209
$C_8H_{10}N_2O_3S$	Sulfanilamide, N^1-acetyl	1331	
$C_8H_{10}N_2S$	Ethionamide	529	
$C_8H_{10}N_4O_2$	Caffeine	9, 241–243	1653
$C_8H_{11}N$	1-Amino-1-phenylethane	973	
$C_8H_{11}N$	Phenylethylamine	1078–1079	2097–2098
$C_8H_{11}N$	α-Phenethylamine (1-Amino-1-phenylethane)	973	
$C_8H_{11}NO$	Phenylethylamine, 2-hydroxy	1080	
$C_8H_{11}NO$	Phenylethylamine, p-hydroxy	1083	
$C_8H_{11}NO$	Tyramine	497	2352–2353
$C_8H_{11}NO_2$	Dopamine (*m,p*-Dihydroxyphenylethylamine)	465, 497, 1085	

$C_8H_{11}NO_2$	*m*-Hydroxyphenylethylamine, 2-hydroxy	1081	
$C_8H_{11}NO_2$	Norfenefrine	911	
$C_8H_{11}NO_2$	Octopamine		2045
$C_8H_{11}NO_2$	Phenylethylamine, 2-hydroxy, m-hydroxy	1081	
$C_8H_{11}NO_2$	Phenylethylamine, 2-hydroxy, p-hydroxy	1087	
$C_8H_{11}NO_2$	*p*-Hydroxyphenylethylamine, 2-hydroxy	1087	
$C_8H_{11}NO_3$	Levarterenol (l-noradrenaline; l-norepinephrine)	729–731	
$C_8H_{11}NO_3$	Pyridoxine	1202–1206	
$C_8H_{11}NO_3.C_4H_6O_6$	Norepinephrine bitartrate (Levarterenol bitartrate)		2035
$C_8H_{11}N_5O_3$	Acyclovir	20–21	1538
$C_8H_{12}N_2$	Betahistine		1624
$C_8H_{12}N_2O_2$	Pyridoxamine	1194–1197	
$C_8H_{12}N_2O_2S$	Dimethylsulfanilamide	38	
$C_8H_{12}N_2O_2S$	Sulfanilamide, N^1,N^1-dimethyl	1344	
$C_8H_{12}N_2O_3$	Barbital	126–131	1602
$C_8H_{12}N_2O_3$	Barbituric acid, 1,5-dimethyl-5-ethyl	133	
$C_8H_{12}N_2O_3$	Barbituric acid, 5,5-diethyl (barbital)	126–131	1602
$C_8H_{12}N_2O_3$	Barbituric acid, 5-methyl-5-isopropyl	124	
$C_8H_{12}N_2O_3S$	6-Aminopenicillanic acid	40	1566
$C_8H_{12}N_2O_3S$	Sulfanilamide, N^1-hydroxyethyl	1345	
$C_8H_{12}N_2O_4S_2$	Sulfanilamide, N^1-ethylsulphonyl	1334	
$C_8H_{12}N_4$	1-(2-Pyrimidinyl)piperazine	90	
$C_8H_{13}N$	Heliotridene	606	
$C_8H_{13}NO_2$	Arecaidine methyl ester	87	
$C_8H_{13}NO_2$	Arecoline	88–89	
$C_8H_{13}NO_2$	Bemegride		1615
$C_8H_{13}N_2O_6P$	Pyridoxamine-5-phosphate	1200–1201	
$C_8H_{13}N_5O_2$	4-Deoxyacyclovir	20	
$C_8H_{14}ClN_5$	Atrazine	104	1592
$C_8H_{14}N_2O_4S$	Penicilloic acid	957	
$C_8H_{15}N$	Heliotridane	605	
$C_8H_{15}NO$	Pseudotropine	1179	
$C_8H_{15}NO$	Tropine	1478–1479	
$C_8H_{15}NO_2$	Dihydroarecaidine methyl ester	424	
$C_8H_{15}NO_2$	Tranexamic acid		2321

$C_8H_{15}N_3O_7$	Streptozocin		2201
$C_8H_{15}N_7O_2S_3$	Famotidine	542–543	
$C_8H_{16}O_2$	Valproic acid	1487	
$C_8H_{19}N$	Octodrine	923	
$C_8H_{17}N$	Coniine	479	
$C_8H_{17}NO$	Conhydrine	479	
$C_8H_{21}NO$	2-Amino-2-methyl-3-hydroxyoctane	43	
C_9			
C_9H_5ClINO	Clioquinol	328–332	
$C_9H_5I_2NO$	Iodoquinol		1935
$C_9H_6O_3$	4-Hydroxycoumarin	637	
$C_9H_7N_7O_2S$	Azathioprine	110	1598
$C_9H_8N_2$	2-Aminoquinoline	48	
$C_9H_8N_2$	4-Aminoquinaldine	48	
$C_9H_8N_2O_2$	Pemoline		2064
$C_9H_8O_2$	Cinnamic acid	252, 312	
$C_9H_8O_3$	2-Hydroxycinnamic acid	252	
$C_9H_8O_4$	Aspirin	95–97	1586–1587
$C_9H_9Cl_2N_3$	Clonidine	337–338	1731–1734
$C_9H_9I_2NO_3$	3,5-Di-iodo-L-tyrosine	439	
$C_9H_9NO_3$	Hippuric acid		1911
$C_9H_9NO_4$	N-Acetylaminosalicylic acid	13	
$C_9H_9N_3O_2S_2$	2-Sulfanilamidothiazole	1385, 1387	2263–2267
$C_9H_9N_3O_2S_2$	Sulfathiazole (2-sulfanilamidothiazole)	1385–1387	2263–2267
$C_9H_9N_3O_2S_2$	Sulfonazole	1394	
$C_9H_9N_3O_3S$	2-Sulfanilamido-oxazole	1358	
$C_9H_{10}N_2O_3$	4-Aminohippuric acid		1564–1565
$C_9H_{10}N_2O_3S_2$	Ethoxazolamide (Ethoxzolamide)	531	
$C_9H_{10}N_4O_2S$	2-Sulfanilamidoimidazole	1352	
$C_9H_{10}N_4O_2S_2$	2-Sulfanilamido-5-methylthiadiazole (sulfamethizole; sulfamethylthiadiazole)	1318–1320, 1356	2235–2238
$C_9H_{10}N_4O_3S$	2-Sulfanilamido-5-methyloxadiazole	1355	
$C_9H_{10}N_4O_3S$	3-Sulfanilamido-4-methylfurazan	1353	
$C_9H_{10}O_2$	Benzoic acid, 4-ethoxy (4-Ethoxybenzoic acid)	212, 649	
$C_9H_{10}O_3$	DL-Phenyllactic acid		2099
$C_9H_{10}O_3$	DL-Tropic acid	1477	2349
$C_9H_{10}O_4$	4-Ethoxysalicylic acid	649	

$C_9H_{11}Cl_2N_3O_4S_2$	Methyclomethiazide (Methyclothiazide)	197, 805–806	
$C_9H_{11}FN_2O_5$	Floxuridine		1867
$C_9H_{11}IN_2O_5$	Idoxuridine	666	1927–1928
$C_9H_{11}N$	Tranylcypromine		2322
$C_9H_{11}NO_2$	Benzocaine	202–203	1617–1619
$C_9H_{11}NO_2$	Benzoic acid, 4-*N*-ethylamino	212	
$C_9H_{11}NO_2$	4-Dimethylaminobenzoic acid	37	
$C_9H_{11}NO_2$	Ethyl 4-aminobenzoate	38	
$C_9H_{11}NO_3$	4-Dimethylaminosalicylic acid	649	
$C_9H_{11}NO_3$	Pyridoxal, 3-methoxy	1191	
$C_9H_{11}NO_3$	Tyrosine	497, 1484–1485	
$C_9H_{11}NO_4$	DOPA, L- (Levodopa)	464, 497, 732–733	1817–1818
$C_9H_{12}ClN$	4-Chloroamphetamine		1693
$C_9H_{12}ClN_5$	N^1-4-Chlorophenyl-N^5-methylbiguanide	282	
$C_9H_{12}NO_3$	Pyridoxal, N-methyl	1192	
$C_9H_{12}N_2O_3$	Barbituric acid, -1′,5-spiro (cyclohexane)	193	
$C_9H_{12}N_2O_3$	Barbituric acid, 5-allyl-5-ethyl	151	
$C_9H_{12}N_2O_5$	2′-Deoxyuridine		1769
$C_9H_{12}N_4O_2$	Methylcaffeine	9	1989
$C_9H_{13}ClN_3O_5$	Cytarabine		1757
$C_9H_{13}N$	Amphetamine (Dexamphetamine)	67–68, 393	1575, 1774
$C_9H_{13}N$	N-Methylphenethylamine	821	
$C_9H_{13}N$	α-Methylphenethylamine		1997
$C_9H_{13}NO$	4-Hydroxyamphetamine (Paredrine)	635	1921–1922
$C_9H_{13}NO$	Amphetamine, 4-hydroxy	69	
$C_9H_{13}NO$	N-hydroxyamphetamine	634	
$C_9H_{13}NO$	Norephedrine	907	2034
$C_9H_{13}NO$	Norpseudoephedrine	917	2039
$C_9H_{13}NO$	Paredrine	635	
$C_9H_{13}NO$	Phenylethylamine, 2-hydroxy, N-methyl	1089	
$C_9H_{13}NO$	Phenylethylamine, p-hydroxy, N-methyl	1084	
$C_9H_{13}NO$	Phenylpropanolamine	1092–1094	2100–2101
$C_9H_{13}NO$	*p*-Hydroxyphenylethylamine, N-methyl	1084	
$C_9H_{13}NO_2$	3-Hydroxy-α-(methylamino) methylbenzenemethanol	644	

$C_9H_{13}NO_2$	4-Hydroxynorephedrine	647	
$C_9H_{13}NO_2$	Deoxyepinephrine (Epinine)	382	
$C_9H_{13}NO_2$	D-α-Methyldopamine	813	
$C_9H_{13}NO_2$	*m,p*-Dihydroxyphenylethylamine, *N*-methyl	1086	
$C_9H_{13}NO_2$	Metaraminol		1967
$C_9H_{13}NO_2$	*m*-Hydroxyphenylethylamine, 2-hydroxy, *N*-methyl	1082	
$C_9H_{13}NO_2$	Phenylephrine (Neo-synephrine)	1077	2095–2096
$C_9H_{13}NO_2$	Phenylethylamine, 2-hydroxy, *m*-hydroxy, *N*-methyl	1082	
$C_9H_{13}NO_2$	Phenylethylamine, 2-hydroxy, *p*-hydroxy, *N*-methyl	1088	
$C_9H_{13}NO_2$	Phenylethylamine, *m,p*-dihydroxy, *N*-methyl	1086	
$C_9H_{13}NO_2$	*p*-Hydroxyphenylethylamine, 2-hydroxy, *N*-methyl	1088	
$C_9H_{13}NO_2$	Synephrine (Sympatol)	1398–1399	2279
$C_9H_{13}NO_3$	Cobefrin (Nordefrin)	345, 906	2032–2033
$C_9H_{13}NO_3$	Epinephrine (Adrenaline)	25, 497–502	
$C_9H_{13}N_3$	1-(2-Pyridyl)piperazine	90	
$C_9H_{14}N_2O$	Phenoxypropazine		2089
$C_9H_{14}N_2O_2$	Pyridoxamine, 3-methoxy	1198	
$C_9H_{14}N_2O_3$	Barbituric acid, 1,5-dimethyl-5-iso-propyl	133	
$C_9H_{14}N_2O_3$	Barbituric acid, 5,5-diethyl-1-methyl (Metharbital)	132	1603–1604
$C_9H_{14}N_2O_3$	Barbituric acid, 5-ethyl-5-iso-propyl	134–136	
$C_9H_{15}NO_3$	Ecgonine	473–474	
$C_9H_{15}NO_3$	Pseudoecgonine	1175	
$C_9H_{15}NO_3S$	Captopril	246	
$C_9H_{15}N_2O_2$	Pyridoxamine, N-methyl	1199	
$C_9H_{15}N_5O$	Minoxidil	839	
$C_9H_{16}N_4S$	Burimamide		1644
$C_9H_{18}N_2O_3$	Ala-Ile	453	
$C_9H_{18}N_2O_3$	Ala-Leu	453	
$C_9H_{18}N_2O_4$	Ser-Leu	453	
$C_9H_{18}N_6$	Altretamine (Hexamethylmelamine)		1551
$C_9H_{18}O_2$	Pelargonic acid	954	
$C_9H_{19}N$	1-Cyclohexyl-2-aminopropane	368	
$C_9H_{19}N$	Cyclopentamine	369	1753

C_{10}

$C_{10}H_7Cl_2N_3O$	Anagrelide	79	
$C_{10}H_7N_3S$	Thiabendazole		2294
$C_{10}H_8O$	2-Naphthol	252	
$C_{10}H_9ClN_4O_2S$	5-Sulfanilamido-2-chloropyrimidine	1351	
$C_{10}H_9ClN_4O_2S$	Sulfachloropyridazine		2210–2211
$C_{10}H_9N_3O_2S$	2-Benzenesulfanil-amidopyrimidine	199	
$C_{10}H_{10}N_4O_2S$	2-Sulfanilamidopyrazine	1359	
$C_{10}H_{10}N_4O_2S$	3-Sulfanilamidopyridazine	1360	
$C_{10}H_{10}N_4O_2S$	4-Sulfanilamidopyrimidine	1362	
$C_{10}H_{10}N_4O_2S$	5-Sulfanilamidopyrimidine	1363	
$C_{10}H_{10}N_4O_2S$	Sulfadiazine	1293–1300	2212–2213
$C_{10}H_{10}O_3$	2-Methoxycinnamic acid	252	
$C_{10}H_{10}O_3$	3-Methoxycinnamic acid	252	
$C_{10}H_{10}O_3$	4-Methoxycinnamic acid	252	
$C_{10}H_{11}FN_4$	1-(5-Fluoro-2-pyrimidinyl) piperazine	90	
$C_{10}H_{11}F_3N_2O_5$	5-Trifluoromethyl-2′-deoxyuridine		1769, 2328
$C_{10}H_{11}N_3O_2S_2$	2-Sulfanilamido-4-methylthiazole	1357	
$C_{10}H_{11}N_3O_3S$	5-Sulfanilamido-3-methylisoxazole	1354	
$C_{10}H_{11}N_3O_3S$	Sulfamethoxazole		2239–2241
$C_{10}H_{11}N_5O_2S$	2-Sulfanilamido-4-aminopyrimidine	1348	
$C_{10}H_{12}ClNO_2$	Baclofen		1601
$C_{10}H_{12}ClN_3O_3S$	Quinethazone		2164
$C_{10}H_{12}ClN_5O_2$	2-Chloro-2′,3′-dideoxy-adenosine (2-ClDDA)	281	
$C_{10}H_{12}N_2$	Tolazoline	1451	
$C_{10}H_{12}N_2O$	Serotonin		2192
$C_{10}H_{12}N_2O_3$	Barbituric acid, 5,5-diallyl (Dial; Allobarbital)	158–160	
$C_{10}H_{12}N_2O_7$	5-Carboxy-2′-deoxyuridine		1769
$C_{10}H_{12}N_4O_2S$	Sulfaethidole	1309	2225–2226
$C_{10}H_{12}N_4O_3$	Didanosine	416	
$C_{10}H_{12}O_2$	Eugenol		1858
$C_{10}H_{12}O_3$	Propylparaben		2147
$C_{10}H_{13}ClN_2$	1-(2-Chlorophenyl)piperazine	90	
$C_{10}H_{13}ClN_2$	1-(3-Chlorophenyl)piperazine	90	
$C_{10}H_{13}ClN_2$	1-(4-Chlorophenyl)piperazine	90	
$C_{10}H_{13}ClN_2O_3S$	Chlorpropamide		1712–1713

$C_{10}H_{13}FN_2$	1-(4-Fluorophenyl)piperazine	90	
$C_{10}H_{13}NO_2$	Propyl 4-aminobenzoate	38	
$C_{10}H_{13}NO_2$	Phenacetin	9	2073
$C_{10}H_{13}NO_3$	Metyrosine		2005
$C_{10}H_{13}NO_4$	L-α-Methyldopa	810–812	
$C_{10}H_{13}N_3$	Debrisoquin		1762–1763
$C_{10}H_{13}N_3O_2$	Guanoxan		1903
$C_{10}H_{13}N_5O_2$	2′,3′-Dideoxyadenosine	417	
$C_{10}H_{13}N_5O_4$	Adenosine	23	1539
$C_{10}H_{13}N_5O_4$	N2-Acetylacyclovir	20	
$C_{10}H_{13}N_5O_4$	Vidarabine		2360
$C_{10}H_{13}N_5O_4$	Zidovudine	1518	
$C_{10}H_{14}ClN$	Chlorphentermine		1709
$C_{10}H_{14}ClN_5$	N^1-4-Chlorophenyl-N^5-ethylbiguanide	282	
$C_{10}H_{14}N_2$	1-Phenylpiperazine	90	
$C_{10}H_{14}N_2$	Nicotine	878–880	2023–2026
$C_{10}H_{14}N_2O$	Glycine xylidide	599	
$C_{10}H_{14}N_2O$	Nikethamide	890	
$C_{10}H_{14}N_2O_3$	Aprobarbital	152–154	
$C_{10}H_{14}N_2O_3$	Barbituric acid, 5-allyl-5-isopropyl (aprobarbital)	152–155	
$C_{10}H_{14}N_2O_3$	Barbituric acid, 5-methyl-5-(3-methyl-but-2-enyl)	161	
$C_{10}H_{14}N_2O_4S_2$	Sulthiame		2277
$C_{10}H_{14}N_4O_3$	7-(2-Hydroxypropyl) theophylline	648	
$C_{10}H_{14}O_2$	5-Phenylvaleric acid	1097	
$C_{10}H_{15}N$	2-Amino-4-phenylbutane	44	
$C_{10}H_{15}N$	Methamphetamine (Desoxyephedrine; methylamphetamine)	789–790	1773, 1973–1975
$C_{10}H_{15}N$	Phentermine		2090–2091
$C_{10}H_{15}N$	Phenylpropylmethylamine (Vonedrine)	1095–1096	
$C_{10}H_{15}N$	α-Ethylphenethylamine		1853
$C_{10}H_{15}NO$	2-Amino-4-[4′-hydroxyphenyl] butane	41	
$C_{10}H_{15}NO$	4-Methoxyamphetamine	803	
$C_{10}H_{15}NO$	Amphetamine, 3-methoxy	70	
$C_{10}H_{15}NO$	Amphetamine, 4-methoxy	71	
$C_{10}H_{15}NO$	Ephedrine	9, 484–490, 1178	1827–1829
$C_{10}H_{15}NO$	Methamphetamine, 4-hydroxy	791	
$C_{10}H_{15}NO$	N-Hydroxyphentermine	646	

$C_{10}H_{15}NO$	Pseudoephedrine	1177–1178	2150–2151
$C_{10}H_{15}NO_2$	Ethylphenylephrine	538	
$C_{10}H_{15}NO_2$	Etilefrine (ethylphenylephrine)	538	
$C_{10}H_{15}NO_3$	Butanephrine	240	
$C_{10}H_{15}NO_3$	DL-N-Methyladrenaline	807	
$C_{10}H_{15}NO_3$	Ethylnorepinephrine		1852
$C_{10}H_{15}N_3$	Bethanidine		1627
$C_{10}H_{15}N_5$	Phenformin	974	
$C_{10}H_{15}O_2$	Phenindione	976	
$C_{10}H_{16}N_2O_2S$	Diethylsulfanilamide	38	
$C_{10}H_{16}N_2O_3$	Barbituric acid, 5-ethyl-5-iso-butyl (butabarbital)	139	1605
$C_{10}H_{16}N_2O_3$	Barbituric acid, 5-ethyl-5-n-butyl	137, 155	
$C_{10}H_{16}N_2O_3$	Barbituric acid, 5-ethyl-5-sec-butyl	138	
$C_{10}H_{16}N_2O_3$	Butabarbital	139	1605
$C_{10}H_{16}N_2O_4S_3$	Dorzolamide		1819
$C_{10}H_{16}N_2O_8$	Edetic acid (EDTA)	476	
$C_{10}H_{16}N_6S$	Cimetidine	308–309	1720–1721
$C_{10}H_{17}N$	Amantadine	29–30	1552–1555
$C_{10}H_{17}NO_2$	Methyprylon		2002
$C_{10}H_{17}NO_3$	Ecgonine methyl ester	475	
$C_{10}H_{17}NO_3$	Pseudoecgonine, methyl ester	1176	
$C_{10}H_{21}N$	Pempidine		2065
$C_{10}H_{21}N$	Propylhexedrine	1171	2145–2146
$C_{10}H_{21}N_3O$	Diethylcarbamazine		1791
$C_{10}H_{22}N_4$	Guanethidine		1900–1901
$C_{10}H_{24}N_2O_2$	Ethambutol	526	
$C_{10}H_{24}O_8$	Ethyl biscoumacetate	637	1845–1847

C_{11}

$C_{11}H_7BrO_2$	3-Bromoazuloic acid	115	
$C_{11}H_7ClO_2$	3-Chloroazuloic acid	115	
$C_{11}H_7NO_4$	3-Nitroazuloic acid	115	
$C_{11}H_8O_2$	Azuloic acid	115	
$C_{11}H_9I_3N_2O_4$	Diatrizoic acid		1780
$C_{11}H_9NO_2S$	2-Benzenesulfonamidopyridine	200	
$C_{11}H_{10}BrN_3O_2S$	2-Sulfanilamido-5-bromopyridine	1349	
$C_{11}H_{10}BrN_3O_2S$	5-Sulfanilamido-2-bromopyridine	1350	
$C_{11}H_{10}N_2O_3$	Barbituric acid, 5-methyl-5-phenyl	170–171	
$C_{11}H_{10}N_3O_3S$	Sulfapyridine-1-oxide	1381	
$C_{11}H_{11}ClO_3$	Alclofenac		1542–1543
$C_{11}H_{11}I_3O_3$	Iophenoxic acid	692	

$C_{11}H_{11}N_3O_2S$	3-Sulfanilamidopyridine	1361	
$C_{11}H_{11}N_3O_2S$	Sulfapyridine	1376–1380	2261–2262
$C_{11}H_{11}N_5$	Phenazopyridine	971	
$C_{11}H_{12}I_3NO_2$	Iopanoic acid	691	
$C_{11}H_{12}N_2O$	Antipyrine	9	1582–1584
$C_{11}H_{12}N_2O_2$	Ethylphenylhydantoin		1854
$C_{11}H_{12}N_2O_2$	Tryptophan	1480–1481	
$C_{11}H_{12}N_2O_2S$	Zileuton	1519–1520	
$C_{11}H_{12}N_2O_3S$	Sulfanilamide, N[1]-furfuryl	1338	
$C_{11}H_{12}N_4O_2S$	2-Sulfanilamido-4-methylpyrimidine	1313–1316, 1373–1374	
$C_{11}H_{12}N_4O_2S$	2-Sulfanilamido-5-aminopyridine	1346	
$C_{11}H_{12}N_4O_2S$	4-Methylsulfadiazine	1373–1374	2256
$C_{11}H_{12}N_4O_2S$	5-Sulfanilamido-2-aminopyridine	1347	
$C_{11}H_{12}N_4O_2S$	Sulfamerazine	1313–1316	2230–2234
$C_{11}H_{12}N_4O_2S$	Sulfamethine	1317	
$C_{11}H_{12}N_4O_2S$	Sulfamethyldiazine	1373–1374	2256
$C_{11}H_{12}N_4O_2S$	Sulfamonomethoxine		2249
$C_{11}H_{12}N_4O_2S$	Sulfaperine (sulfamethyldiazine; 4-methylsulfadiazine)	1373–1374	2256
$C_{11}H_{12}N_4O_3S$	Sulfalene	1312	
$C_{11}H_{12}N_4O_3S$	Sulfamethoxydiazine (Sulfameter)	1323	2243
$C_{11}H_{12}N_4O_3S$	Sulfamethoxypyridazine	1324–1325	2244–2247
$C_{11}H_{12}N_4O_3S_2$	Sulfamethythiazole, N[4]-acetyl	1321	
$C_{11}H_{12}N_4O_4$	2,4-Diaminopyridine, 5-substituted-benzyl derivatives $C_{11}H_{13}ClF_3N_3O_4S_3$	216	
Polythiazide	197, 1133		
$C_{11}H_{13}F_3N_2$	1-(3-Trifluoromethylphenyl) piperazine	90	
$C_{11}H_{13}N$	Pargyline		2062
$C_{11}H_{13}N_3O_3S$	Sulfadimethyloxazole		2220
$C_{11}H_{13}N_3O_3S$	Sulfafurazole	1310	2227–2228
$C_{11}H_{13}N_3O_3S$	Sulfisoxazole	1391–1392	2274–2275
$C_{11}H_{13}N_3S$	1-(1,2-Benzisothiazol-3-yl) piperazine	90	
$C_{11}H_{13}N_5O_4S$	Sulfadimethoxytriazine	1303	
$C_{11}H_{14}AsNO_3S_2$	Arsthinol	10	
$C_{11}H_{14}N_2O_3$	Barbituric acid, 1-methyl-5, 5-diallyl	133	

$C_{11}H_{14}N_2O_3$	Gly-Phe	453	
$C_{11}H_{14}N_2O_3$	Phe-Gly	453	
$C_{11}H_{14}N_4O_2S$	3-Sulfanilamido-4,5-dimethylpyrazole		2253
$C_{11}H_{14}O_3$	Butylparaben		1651–1652
$C_{11}H_{15}NO$	Phenmetrazine		2084–2085
$C_{11}H_{15}NO_2$	Butamben		1647
$C_{11}H_{15}NO_2$	Butyl 4-aminobenzoate	38	
$C_{11}H_{15}NO_2$	4-Diethylaminobenzoic acid	649	
$C_{11}H_{15}NO_3$	4-Diethylaminosalicylic acid	649	
$C_{11}H_{15}NO_3$	Tyrosine ethyl ester	497, 1484–1485	
$C_{11}H_{16}ClN_5$	N^1-4-Chlorophenyl-N^5-propylbiguanide	282	
$C_{11}H_{16}N_2$	1-(2-Methylphenyl)piperazine	90	
$C_{11}H_{16}N_2O$	1-(2-Methoxyphenyl)piperazine	90	
$C_{11}H_{16}N_2O$	Tocainide	1449	2314
$C_{11}H_{16}N_2O_2$	Isopilocarpine	701	
$C_{11}H_{16}N_2O_2$	Pilocarpine	9, 1111–1118	
$C_{11}H_{16}N_2O_3$	Barbituric acid, 5-allyl-5-(1-methylpropyl) (Talbutal)		1610
$C_{11}H_{16}N_2O_3$	Barbituric acid, 5-allyl-5-isobutyl	155–156	
$C_{11}H_{16}N_2O_3$	Barbituric acid, 5-ethyl-5-(1-methylbut-1-enyl)	166–167	
$C_{11}H_{16}N_2O_3$	Barbituric acid, 5-ethyl-5-(3-methylbut-2-enyl)	162	
$C_{11}H_{16}N_2O_3$	Barbituric acid, 5-methyl-5-(4-methylpent-1-en-2-yl)	169	
$C_{11}H_{16}N_2O_3$	Talbutal		1610
$C_{11}H_{16}N_4O_4$	Pentostatin	964	
$C_{11}H_{17}IN_2$	Nicotine methiodide		2027
$C_{11}H_{17}N$	2-Amino-5-phenylpentane	45	
$C_{11}H_{17}N$	Dimethylamphetamine	447–449	
$C_{11}H_{17}N$	Ethylamphetamine		1844
$C_{11}H_{17}N$	Mephentermine	778–779	1959
$C_{11}H_{17}NO$	2-Amino-5-[4′-hydroxyphenyl] pentane	42	
$C_{11}H_{17}NO$	Methoxyphenamine	804	
$C_{11}H_{17}NO$	Mexiletine	833–834	
$C_{11}H_{17}NO$	N-Methylephedrine	815, 1178	1991
$C_{11}H_{17}NO$	N-Methylpseudoephedrine	1178	
$C_{11}H_{17}NO$	Tecomine (tecomanine)	1403	
$C_{11}H_{17}NO_2$	Dimethoxyamphetamine	443	

$C_{11}H_{17}NO_3$	5-[1-Hydroxy-2-[(1-methylethyl) amino]-ethyl]-1,3-benzenediol	645	
$C_{11}H_{17}NO_3$	Isopropylnorepinephrine	497	
$C_{11}H_{17}NO_3$	Isoproterenol (DL-isoprenaline)	702–706	
$C_{11}H_{17}NO_3$	Metaproterenol (orciprenaline)	645, 785–786	1966
$C_{11}H_{17}NO_3$	Methoxamine	801–802	
$C_{11}H_{17}NO_3$	Orciprenaline	785–786, 929	1966
$C_{11}H_{17}N_3O_3S$	Carbutamide	253	
$C_{11}H_{18}N_2O_2S$	Barbituric acid, 5-ethyl-5-(1-methylbutyl)-2-thio (Thiopental; Thiopentone)	144, 1436	1606–1607
$C_{11}H_{18}N_2O_3$	Barbituric acid, 1-methyl-5,5-di-n-propyl	133	
$C_{11}H_{18}N_2O_3$	Barbituric acid, 1-methyl-5-ethyl-5-butyl	133	
$C_{11}H_{18}N_2O_3$	Barbituric acid, 1-methyl-5-propyl-5-iso-propyl	133	
$C_{11}H_{18}N_2O_3$	Barbituric acid, 5-ethyl-5-(1-methylbutyl) (Pentobarbital; Pentobarbitone)	140–143, 155	
$C_{11}H_{18}N_2O_3$	Barbituric acid, 5-ethyl-5-(3-methylbutyl) (Amobarbital; Amylobarbitone)	145–148	
$C_{11}H_{21}N$	Mecamylamine	767	
$C_{11}H_{22}N_2O_3S$	Met-Leu	453	
C_{12}			
$C_{12}H_8O_3$	3-Formylazuloic acid	115	
$C_{12}H_9FN_2O_5$	5-Fluorouracil, 1-phenyloxy-carbonyloxymethyl	567	
$C_{12}H_9FN_2O_5$	5-Fluorouracil, 3-phenyloxy-carbonyloxymethyl	568	
$C_{12}H_9NS$	Phenothiazine	981	
$C_{12}H_{10}F_4N_2O_2$	5,5-Dimethyl-3-(α,α,α,4-tetrafluoro-*m*-tolyl) hydantoin	446	
$C_{12}H_{10}N_2O$	Harmol		1907
$C_{12}H_{10}N_2O_5$	Cinoxacin	315	
$C_{12}H_{11}BrN_4O_2$	2,4-Diaminopyridine, 5-(2-bromo-4,5-methylenedioxybenzyl)-	216	
$C_{12}H_{11}ClN_2O_5S$	Furosemide (Frusemide)	581–586	1884–1885
$C_{12}H_{11}I_3N_2O_4$	Iodamide	686–687	
$C_{12}H_{11}NO_3$	Quinolone carboxylic acid	1248	

$C_{12}H_{11}N_3O_5$	Barbituric acid, 5-ethyl-5-(3-nitrophenyl)	186–187	
$C_{12}H_{11}N_3O_5$	Barbituric acid, 5-ethyl-5-(4-nitrophenyl)	188	
$C_{12}H_{11}N_7$	Triamterene		2324
$C_{12}H_{12}N_2O_2S$	4,4-Diaminodiphenylsulfone (Dapsone)	1395	1761
$C_{12}H_{12}N_2O_2S$	Sulfabenz	1290	2207–2209
$C_{12}H_{12}N_2O_2S$	Sulfanilamide, N^1-phenyl	1340	
$C_{12}H_{12}N_2O_2S$	Sulfone, bis(4-aminophenyl) (Dapsone; 4,4-diaminodiphenylsulfone)	1395	
$C_{12}H_{12}N_2O_3$	Barbituric acid, 5-ethyl-5-phenyl (Phenobarbital; Phenobarbitone)	173–184	1612–1613
$C_{12}H_{12}N_2O_3$	Nalidixic acid	865–866, 1237–1238	2170–2171
$C_{12}H_{12}N_3O_3S$	3′-Methylsulfapyridine-1-oxide	1381	
$C_{12}H_{12}N_3O_3S$	4′-Methylsulfapyridine-1-oxide	1381	
$C_{12}H_{12}N_3O_3S$	5′-Methylsulfapyridine-1-oxide	1381	
$C_{12}H_{12}N_3O_3S$	6′-Methylsulfapyridine-1-oxide	1381	
$C_{12}H_{12}N_3O_3S$	Sulfapyridine-1-oxide, 3′-methyl	1381	
$C_{12}H_{12}N_3O_3S$	Sulfapyridine-1-oxide, 4′-methyl	1381	
$C_{12}H_{12}N_3O_3S$	Sulfapyridine-1-oxide, 5′-methyl	1381	
$C_{12}H_{12}N_3O_3S$	Sulfapyridine-1-oxide, 6′-methyl	1381	
$C_{12}H_{12}N_4O_2$	2,4-Diaminopyridine, 5-(4, 5-methylenedioxybenzyl)-	216	
$C_{12}H_{12}N_4O_3S$	Sulfadiazine, N^4-acetyl	1301	2214
$C_{12}H_{13}ClN_4$	Pyrimethamine		2156–2158
$C_{12}H_{13}ClN_4$	Pyrimethazine		2159
$C_{12}H_{13}I_3N_2O_3$	Iocetamic acid		1933
$C_{12}H_{13}N_3O_2$	Isocarboxazid	693	
$C_{12}H_{13}N_3O_4S$	Sulfamethoxazole, N^4-acetyl		2242
$C_{12}H_{13}N_3O_4S_2$	N^3-Sulfanilylmetanilamide	1370	
$C_{12}H_{13}N_3O_4S_2$	N^4-Sulfanilylsulfanilamide	1371	
$C_{12}H_{13}N_3O_4S_2$	Sulfanilamide, N^1-sulfanilyl	1335	
$C_{12}H_{14}ClNO$	Norketamine		2036
$C_{12}H_{14}N_2O_2$	Mephenytoin		1960
$C_{12}H_{14}N_4O_2S$	2-Sulfanilamido-2,4-dimethylpyrimidine	1389–1390	
$C_{12}H_{14}N_4O_2S$	2-Sulfanilamido-4,6-dimethylpyrimidine	1304–1308	
$C_{12}H_{14}N_4O_2S$	4-Sulfanilamido-2,6-dimethylpyrimidine	1389–1390	

$C_{12}H_{14}N_4O_2S$	Sulfadimidine (Sulfamethazine)	1304–1308	2221, 2224
$C_{12}H_{14}N_4O_2S$	Sulfisomidine	1389–1390	2269–2273
$C_{12}H_{14}N_4O_3$	2,4-Diaminopyridine, 5-(3-methoxy-4,5-dihydroxybenzyl)-	216	
$C_{12}H_{14}N_4O_3S$	Sulfamethomidine	1322	
$C_{12}H_{14}N_4O_4S$	Sulfadimethoxine	1302	2215–2218
$C_{12}H_{15}N_3O_3S$	Albendazole sulphoxide	26	
$C_{12}H_{16}FN_3O_2$	5-Phenyl-1,3-dihydro-1,4-benzodiazepin-2-one, 1-methoxymethyl-2′-fluoro-7-amino	1074	
$C_{12}H_{16}F_3N$	Fenfluramine		1861
$C_{12}H_{16}N_2O_3$	Ala-Phe	453	
$C_{12}H_{16}N_2O_3$	Barbituric acid, 5-cyclohex-1′-enyl-1,5-dimethyl (hexobarbital)	125	
$C_{12}H_{16}N_2O_3$	Cyclobarbital	150	
$C_{12}H_{16}N_2O_3$	Hexobarbital	125	
$C_{12}H_{16}N_2O_4$	Phe-Ser	453	
$C_{12}H_{16}N_2O_4$	Ser-Phe	453	
$C_{12}H_{17}NO$	Phendimetrazine		2078
$C_{12}H_{17}NO_3$	Rimoterol (rimiterol)		2183
$C_{12}H_{17}NO_9$	Levarterenol bitartrate		2035
$C_{12}H_{17}NO_9$	Norepinephrine bitartrate (levarterenol bitartrate)		2035
$C_{12}H_{17}N_3O_4S$	Imipenem	670	
$C_{12}H_{17}N_4OS$	Thiamine		2295–2298
$C_{12}H_{18}ClN_5$	N^1-4-Chlorophenyl-N^5-n-butylbiguanide	282	
$C_{12}H_{18}N_2O$	Monoethylglycine xylidide	846	
$C_{12}H_{18}N_2O_2S$	Barbituric acid, 5-allyl-5-(1-methylbutyl)-2-thio (Thiamylal)		1610
$C_{12}H_{18}N_2O_2S$	Thiamylal		1610
$C_{12}H_{18}N_2O_3$	Barbituric acid, 5-allyl-5-(1-methylbutyl) (secobarbital)	157	1608
$C_{12}H_{18}N_2O_3$	Barbituric acid, 5-ethyl-5-(4-methylpent-1-en-2-yl)	168	
$C_{12}H_{18}N_2O_3$	Barbituric acid, 5-iso-propyl-5-(3-methylbut-2-enyl)	163	
$C_{12}H_{18}N_2O_3$	Secobarbital	157	1608
$C_{12}H_{18}N_2O_3S$	Tolbutamide	1452–1453	2318
$C_{12}H_{18}N_4O_4PS$	Thiamine-O-monophosphate		2299
$C_{12}H_{18}N_6S_2$	Tiotidine		2312

$C_{12}H_{18}O$	Propofol	1159	
$C_{12}H_{18}O_2$	Hexylresorcinol		1910
$C_{12}H_{19}N$	Isopropylamphetamine		1938
$C_{12}H_{19}N$	Methylethylamphetamine		1992
$C_{12}H_{19}N$	Propylamphetamine		2144
$C_{12}H_{19}NO_2$	1-(3,4-Dimethoxyphenyl)-2-N-methylaminopropane	444	
$C_{12}H_{19}NO_2$	Methylenedioxyamphetamine (MDA)	814	
$C_{12}H_{19}NO_3$	Prenalterol		2128
$C_{12}H_{19}NO_3$	Terbutaline	1406–1408	2285–2286
$C_{12}H_{19}NO_3$	Th 1206	1422	
$C_{12}H_{19}N_2O_2$	Neostigmine		2021
$C_{12}H_{19}N_3O$	Procarbazine	1143	
$C_{12}H_{20}N_2O_3$	Barbituric acid, 5-ethyl-5-(1,3-dimethylbutyl)	149	
$C_{12}H_{20}N_2O_3$	Pirbuterol	1127	2118
$C_{12}H_{20}N_2O_3S$	Sotalol	1167, 1279	2196–2197
$C_{12}H_{20}N_4O_3$	Leu-His	453	
$C_{12}H_{21}N$	Methylisopropylamphetamine		1994
$C_{12}H_{21}N_3O_5S_3$	Brinzolamide	226	
$C_{12}H_{21}N_5O_2S_2$	Nizatidine		2030–2031
$C_{12}H_{22}O_{11}$	Sucrose	1289	
$C_{12}H_{24}N_2O_4$	Carisoprodol		1662
$C_{12}H_{24}O_2$	Lauric acid	725	
C_{13}			
$C_{13}H_6Cl_6O_2$	Hexachlorophene	607–609	
$C_{13}H_8Cl_2O_4S$	Ticrynafen		2309
$C_{13}H_8F_2O_3$	Diflunisal	420–421	
$C_{13}H_8F_2O_3$	α-(2,4-Difluorophenyl)-α-[1-(2-(2-pyridyl)-phenylethenyl)]-1H-1,2,4-triazole-1-ethanol (XD405, bis-mesylate salt)	422	
$C_{13}H_9F_3N_2O_2$	Niflumic acid	887–889	
$C_{13}H_{10}N_2$	2-Aminoacridine	19	
$C_{13}H_{10}N_2$	3-Aminoacridine	18	
$C_{13}H_{10}N_2$	9-Aminoacridine	19	1559
$C_{13}H_{10}N_2$	Aminacrine		1559
$C_{13}H_{10}O_3$	3-Acetylazuloic acid	115	
$C_{13}H_{11}NO_2$	N-Phenylanthranillic acid	552, 1008	
$C_{13}H_{11}N_3$	Proflavine	18	
$C_{13}H_{11}N_3O_4S_2$	Tenoxicam		2284
$C_{13}H_{11}N_3OS$	Timoprazole		2311
$C_{13}H_{12}Cl_2O_4$	Ethacrynic acid		1837–1838
$C_{13}H_{12}F_2N_6O$	Fluconazole	549	

$C_{13}H_{12}N_2O_2$	N-(3′-Aminophenyl)anthranilic acid	552–553	
$C_{13}H_{12}N_2O_3S$	Sulfanilamide, N^1-benzoyl	1333	
$C_{13}H_{12}N_2O_5S$	Nimesulide	891–893	
$C_{13}H_{13}ClN_2S$	3-(4-Chlorophenyl)-5,6-dihydro-2-ethylimidazo[2,1-b]thiazole	283	
$C_{13}H_{13}F_3N_6O_4S_3$	Cefazaflur	255	
$C_{13}H_{13}N_3O_3S$	Sulfanilamide, N^1-p-aminobenzoyl	1332	
$C_{13}H_{13}N_3O_3S$	Sulfapyridine, N^4-acetyl	1383	
$C_{13}H_{13}N_3O_5S_2$	Succinylsulfathiazole		2203
$C_{13}H_{13}N_3O_6S$	Cephacetrile		1680
$C_{13}H_{13}N_5O_5S_2$	Ceftizoxime		1674
$C_{13}H_{13}N_7$	1,3-bis[(2-Pyridyl) methyleneamino]guanidine	1207	
$C_{13}H_{14}Cl_2N_4O_2$	2,4-Diaminopyridine, 5-(2,6-dichloro-3,5-dimethoxybenzyl)-	216	
$C_{13}H_{14}N_2$	Tacrine		2280
$C_{13}H_{14}N_2O$	Phenyramidol	1098–1099	
$C_{13}H_{14}N_2O_2S$	Sulfanilamide, N^1-m-tolyl	1342	
$C_{13}H_{14}N_2O_2S$	Sulfanilamide, N^1-o-tolyl	1341	
$C_{13}H_{14}N_2O_2S$	Sulfanilamide, N^1-p-tolyl	1343	
$C_{13}H_{14}N_2O_3$	Barbituric acid, 1-methyl-5-ethyl-5-phenyl		1614
$C_{13}H_{14}N_3O_3S$	4′,6′-Dimethylsulfapyridine-1-oxide	1381	
$C_{13}H_{14}N_3O_3S$	6′-Ethylsulfapyridine-1-oxide	1381	
$C_{13}H_{14}N_3O_3S$	Sulfapyridine-1-oxide, 4′,6′-dimethyl	1381	
$C_{13}H_{14}N_3O_3S$	Sulfapyridine-1-oxide, 6′-ethyl	1381	
$C_{13}H_{14}N_4O_3$	2,4-Diaminopyridine, 5-(3-methoxy-4,5-methylenedioxybenzyl)-	216	
$C_{13}H_{14}N_4O_3S$	Sulfamethoxypyridazine, N^4-acetyl	1326	2248
$C_{13}H_{15}ClN_2OS$	3-(4-Chlorophenyl)-2-ethyl-2,3,5,6-tetrahydroimidazo[2,1-b]thiazol-3-ol	284	
$C_{13}H_{15}NO_2$	Glutethimide	597	1893–1894
$C_{13}H_{15}N_3$	1-(2-Quinolinyl)piperazine	90	
$C_{13}H_{15}N_3O_2S$	Sulfanilamide, N^1-p-aminophenyl	1336	
$C_{13}H_{15}N_3O_3$	Gly-Trp	453	
$C_{13}H_{15}N_3O_3$	Trp-Gly	453	

$C_{13}H_{15}N_3O_4S$	Sulfisoxazole, N^4-acetyl	1393	2276
$C_{13}H_{16}ClNO$	Ketamine	710–711	
$C_{13}H_{16}F_3N_4O_6$	Furaltadone		1883
$C_{13}H_{16}N_2$	Tetrahydrozoline	1421	
$C_{13}H_{16}N_4O_5S$	2-Sulfanilamido-4,5,6-trimethoxypyrimidine		2254
$C_{13}H_{17}N$	Selegiline (Deprenyl)	1272	
$C_{13}H_{17}N_3$	Tramazoline		2320
$C_{13}H_{17}N_3O$	Aminophenazone	46–47	1569
$C_{13}H_{17}N_3O$	Aminopyrine (aminophenazone)	46–47	1569
$C_{13}H_{17}N_5O_2$	2,4-Diaminopyridine, 5-(3,5-dimethoxy-4-aminobenzyl)-	216	
$C_{13}H_{17}N_5O_8S_2$	Aztreonam	114	
$C_{13}H_{18}BrNO_2$	Benzoic acid, 4-bromo, diethylaminoethyl ester	212	
$C_{13}H_{18}ClNO$	Bupropion		1643
$C_{13}H_{18}ClNO_2$	Benzoic acid, 4-chloro, diethylaminoethyl ester	212	
$C_{13}H_{18}Cl_2N_2O_2$	Melphalan	775	1957
$C_{13}H_{18}FNO_2$	Benzoic acid, 4-fluoro, diethylaminoethyl ester	212	
$C_{13}H_{18}N_2O_4$	Benzoic acid, 4-nitro, diethylaminoethyl ester	212	
$C_{13}H_{18}N_4O_3$	Pentoxiphylline		2069–2070
$C_{13}H_{18}O_2$	Ibuprofen	654–664	
$C_{13}H_{19}ClN_2O_2$	Chloroprocaine	741	
$C_{13}H_{19}NO_2$	Benzoic acid, diethylaminoethyl ester	212	
$C_{13}H_{19}NO_3$	Benzoic acid, 4-hydroxy, diethylaminoethyl ester	212	
$C_{13}H_{19}NO_4S$	Probenecid	1139	
$C_{13}H_{19}N_5$	Pinacidil		2109
$C_{13}H_{20}N_2O$	Prilocaine		2129–2130
$C_{13}H_{20}N_2O_2$	Benzoic acid, 4-amino, diethylaminoethyl ester (Procaine)	212, 741, 1140–1142	2133
$C_{13}H_{20}N_2O_3$	Barbituric acid, 5-t-butyl-5-(3-methylbut-2-enyl)	164	
$C_{13}H_{21}NO_3$	Albuterol (Salbutamol)	27, 1260	1541
$C_{13}H_{21}NO_3$	Salbutamol	27, 1260	1541
$C_{13}H_{21}N_3O$	Procainamide		2131–2132
$C_{13}H_{22}N_4O_3S$	Ranitidine		2177–2179
$C_{13}H_{24}N_2O_3S$	n-Amylpenilloic acid	77–78	
$C_{13}H_{24}N_4O_3S$	Timolol	1167, 1447	2310

$C_{13}H_{25}N_2$	Methamphetamine, *N*-(2-cyano) ethyl	905	
C_{14}			
$C_{14}H_9ClF_3NO_2$	(S)-6-Chloro-4-(cyclo-propylethynyl)-1,4-dihydro-4-(trifluoromethyl)-2H-3, 1-benzoxazin-2-one (Efavirenz; DMP-266)		1695
$C_{14}H_{10}BrN_3O$	Bromazepam		1629–1630
$C_{14}H_{10}Cl_2O_3$	Fenclofenac		1860
$C_{14}H_{10}F_3NO_2$	Flufenamic acid	551–554	1870
$C_{14}H_{10}N_4O_5$	Dantrolene		1759–1760
$C_{14}H_{10}O_4$	Diphenic acid	457	
$C_{14}H_{10}O_5$	Salsalate		2190
$C_{14}H_{11}ClN_2O_4S$	Chlorthalidone	307	1716
$C_{14}H_{11}Cl_2NO_2$	Diclofenac	409–413	1787
$C_{14}H_{11}Cl_2NO_2$	Meclofenamic acid		1955
$C_{14}H_{12}ClNO$	2-Methylamino-5-chlorobenzophenone	808	
$C_{14}H_{12}FNO_3$	Flumequine	555–557	
$C_{14}H_{12}O_3S$	Tiaprofenic acid (tiprofenic acid)		2307
$C_{14}H_{13}NO_2$	N-(3′-Methylphenyl)anthranilic acid	552–553	
$C_{14}H_{13}NO_3$	N-(3′-Methoxyphenyl) anthranilic acid	552–553	
$C_{14}H_{14}N_2$	Naphazoline	870–871	
$C_{14}H_{14}N_3O_4S$	Sulfapyridine-1-oxide, N^4-acetyl, 6′-methyl	1382	
$C_{14}H_{14}N_8O_4S_3$	Cefazolin	256	1666–1669
$C_{14}H_{14}O_3$	Naproxen	872–874	
$C_{14}H_{15}N_3O_5$	Barbituric acid, 5,5-diethyl-1-(4-nitrophenyl)	132	
$C_{14}H_{16}ClN_3O_4S_2$	Cyclothiazide	197, 373--374	1755
$C_{14}H_{16}N_2$	Didesmethylpheniramine	418	
$C_{14}H_{16}N_2O_2$	Etomidate	539	
$C_{14}H_{16}N_2O_3$	Barbituric acid, 5,5-diethyl-1-phenyl	132	
$C_{14}H_{16}N_4O_5S$	Sulfadimethoxine, N4-acetyl		2219
$C_{14}H_{17}N_2O_3S$	Phenylpenilloic acid	1090–1091	
$C_{14}H_{18}BrN_3S$	Bromothen		1634
$C_{14}H_{18}ClN_3S$	Chlorothen		1698
$C_{14}H_{18}N_2O_3$	Barbituric acid, 5-allyl-5-(1-methylpent-2-ynyl)-N-methyl (methohexital)		1609

$C_{14}H_{18}N_2O_5$	Aspartame	94	
$C_{14}H_{18}N_4O_2S$	2,4-Diaminopyridine, 5-(3, 5-dimethoxy-4-thiomethylbenzyl)-	216	
$C_{14}H_{18}N_4O_3$	2,4-Diaminopyridine, 5-(2,4, 5-trimethoxybenzyl)-	216	
$C_{14}H_{18}N_4O_3$	2,4-Diaminopyridine, 5-(3,4, 5-trimethoxybenzyl)-	216	
$C_{14}H_{18}N_4O_3$	2,4-Diaminopyridine, 5-(3,5-dihydroxy-4-methoxybenzyl)-	216	
$C_{14}H_{18}N_4O_3$	Trimethoprim	1470–1471	2334–2336
$C_{14}H_{19}Cl_2NO_2$	Chlorambucil		1687–1688
$C_{14}H_{19}NO_2$	Methylphenidate	822–823	
$C_{14}H_{19}NO_2$	Norketobemidone	913	
$C_{14}H_{19}N_3O_7S$	Cephalosporin C, deacetoxy	270	
*$C_{14}H_{19}N_3O_7S$	Deacetoxycephalosporin C	270	
$C_{14}H_{19}N_3S$	Methapyrilene		1976–1977
$C_{14}H_{19}N_3S$	Thenyldiamine		2293
$C_{14}H_{20}N_2O_2$	4-[2-Hydroxy-3-(iso-propylamino)propoxy]indole (Pindolol)	643	2111–2113
$C_{14}H_{20}N_2O_3$	Barbituric acid, 5,5-di-(3-methylbut-2-enyl)	165	
$C_{14}H_{20}N_2O_3S$	Glycyclamide		1898
$C_{14}H_{20}N_2O_4$	Val-Tyr	453	
$C_{14}H_{21}NO_2$	Amylocaine	398	
$C_{14}H_{21}NO_2$	Benzoic acid, 4-methyl, diethylaminoethyl ester	212	
$C_{14}H_{21}N_3O_2S$	Sumatriptan		2278
$C_{14}H_{21}N_3O_3$	Oxamniquine	930	
$C_{14}H_{21}N_3O_3S$	Tolazamide	1450	2317
$C_{14}H_{21}N_3O_4S$	1′-Aminocyclopentane-carboxamidopenicillin	1511	
$C_{14}H_{21}N_3O_4S$	WY- 7953 (1′-amino-cyclopentanecar-boxamidopenicillin)	1511	
$C_{14}H_{22}ClN_3O_2$	Metoclopramide	827–828	
$C_{14}H_{22}N_2O$	Lidocaine (lignocaine; xylocaine)	738–746	
$C_{14}H_{22}N_2O$	Lignocaine	738–746	
$C_{14}H_{22}N_2O$	Xylocaine	738–746	
$C_{14}H_{22}N_2O_2$	Rivastigmine		2186
$C_{14}H_{22}N_2O_3$	Atenolol	99–102, 1167	1588–1590
$C_{14}H_{22}N_2O_3$	Practolol		2122
$C_{14}H_{24}N_2O_7$	Spectinomycin		2199

C_{15}

$C_{15}H_9Cl_2FN_2O$	5-Phenyl-1,3-dihydro-1,4-benzodiazepin-2-one, 2′-fluoro-7,8-dichloro	1020
$C_{15}H_9ClF_2N_2O$	5-Phenyl-1,3-dihydro-1,4-benzodiazepin-2-one, 2′,6′-difluoro-7-chloro	1050
$C_{15}H_9ClO_2$	Clorindione	340
$C_{15}H_{10}BrClN_2O$	5-Phenyl-1,3-dihydro-1,4-benzodiazepin-2-one, 2′-bromo-7-chloro	1027
$C_{15}H_{10}ClFN_2O$	5-Phenyl-1,3-dihydro-1,4-benzodiazepin-2-one, 2′-fluoro-7-chloro	1049
$C_{15}H_{10}ClF_2N_2O$	5-Phenyl-1,3-dihydro-1,4-benzodiazepin-2-one, 2′,6′-difluoro-8-chloro	1044
$C_{15}H_{10}ClF_2O$	5-Phenyl-1,3-dihydro-1,4-benzodiazepin-2-one, 4′-fluoro-7-chloro	1043
$C_{15}H_{10}ClN_3O_3$	5-Phenyl-1,3-dihydro-1,4-benzodiazepin-2-one, 2′-chloro-7-nitro	1055
$C_{15}H_{10}ClN_3O_3$	Clonazepam	1730
$C_{15}H_{10}Cl_2N_2O$	5-Phenyl-1,3-dihydro-1,4-benzodiazepin-2-one, 2′,7-dichloro	1028
$C_{15}H_{10}Cl_2N_2O_2$	5-Phenyl-1,3-dihydro-1,4-benzodiazepin-2-one, 3-hydroxy-2′,7-dichloro	1056
$C_{15}H_{10}Cl_2N_2O_2$	Lorazepam	204, 755–756
$C_{15}H_{10}N_4O_5$	5-Phenyl-1,3-dihydro-1,4-benzodiazepin-2-one, 2′,7-dinitro	1069
$C_{15}H_{10}O_5S$	Tixanox	1448
$C_{15}H_{11}ClN_2O$	5-Phenyl-1,3-dihydro-1,4-benzodiazepin-2-one, 7-chloro	1040
$C_{15}H_{11}ClN_2O$	Desmethyldiazepam	205
$C_{15}H_{11}ClN_2O_2$	5-Phenyl-1,3-dihydro-1,4-benzodiazepin-2-one, 3-hydroxy-7-chloro	1060
$C_{15}H_{11}ClN_2O_2$	Chlordiazepoxide lactam	205
$C_{15}H_{11}ClN_2O_2$	Oxazepam	204, 931–933

$C_{15}H_{11}FN_2O$	5-Phenyl-1,3-dihydro-1,4-benzodiazepin-2-one, 4′-fluoro	1057	
$C_{15}H_{11}FN_2O$	5-Phenyl-1,3-dihydro-1,4-benzodiazepin-2-one, 7-fluoro	1058	
$C_{15}H_{11}I_4NO_4$	Thyroxine, L- (Levothyroxine)	1443	
$C_{15}H_{11}N_3O_3$	Nitrazepam	204, 895–899	
$C_{15}H_{12}I_3NO_4$	L-Thyronine (Liothyronine)		1945, 2306
$C_{15}H_{12}N_2O$	5-Phenyl-1,3-dihydro-1,4-benzodiazepin-2-one	1019	
$C_{15}H_{12}N_2O_2$	Demoxepam		1767
$C_{15}H_{12}N_2O_2$	Phenytoin	1100–1104	
$C_{15}H_{12}N_2O_3$	(3,4-Dimethyl-5-isoxazolyl)-4-amino-1,2-naphthoquinone	445	
$C_{15}H_{12}N_3O_3$	5-Phenyl-1,3-dihydro-1,4-benzodiazepin-2-one, 7-nitro	1062	
$C_{15}H_{12}N_4O_3$	7-Aminonitrazepam	205	
$C_{15}H_{12}N_4O_3$	Nitrazepam, 7-amino	205	
$C_{15}H_{13}FO_2$	Flurbiprofen	575	1879
$C_{15}H_{13}NO_3$	N-(3′-Acetylphenyl)anthranilic acid	552–553	
$C_{15}H_{13}N_3O$	5-Phenyl-1,3-dihydro-1,4-benzodiazepin-2-one, 7-amino	1075	
$C_{15}H_{13}N_3O_4S$	Piroxicam	1129–1131	
$C_{15}H_{14}ClNO_3$	Zomepirac	1521	
$C_{15}H_{14}ClN_3O_4S$	Cefaclor		1664
$C_{15}H_{14}F_3N_3O_4S_2$	Bendroflumethiazide	196–197	
$C_{15}H_{14}N_4O_2S$	Sulfaphenazole	1375	2257–2260
$C_{15}H_{14}O_3$	Fenoprofen	546	1862
$C_{15}H_{15}BrN_2O_4$	Barbituric acid, 5,5-diethyl-1-(2-bromobenzoyl)	132	
$C_{15}H_{15}BrN_2O_4$	Barbituric acid, 5,5-diethyl-1-(3-bromobenzoyl)	132	
$C_{15}H_{15}BrN_2O_4$	Barbituric acid, 5,5-diethyl-1-(4-bromobenzoyl)	132	
$C_{15}H_{15}ClN_2O_4S$	Xipamide	1516	
$C_{15}H_{15}NO_2$	Mefenamic acid	772–774	
$C_{15}H_{15}NO_3$	Tolmetin	1454	
$C_{15}H_{15}N_3$	Acridine yellow	18	
$C_{15}H_{16}N_2O_3S$	Sulfamilyl-3,4-xylamide (Irgafen)	1327	
$C_{15}H_{16}N_2O_4$	Barbituric acid, 5,5-diethyl-1-benzoyl	132	
$C_{15}H_{16}N_2O_6S_2$	Ticarcillin		2308

$C_{15}H_{17}ClN_2O_3$	Barbituric acid, 5,5-diethyl-1-(4-chlorobenzyl)	132	
$C_{15}H_{17}FN_4O_3$	Enoxacin	1231	2168
$C_{15}H_{17}N_3O_5$	Barbituric acid, 5,5-diethyl-1-(4-nitrobenzyl)	132	
$C_{15}H_{18}N_2$	Desmethylpheniramine (Monodesmethyl-pheniramine)	389, 418	
$C_{15}H_{18}N_2O_3$	Barbituric acid, 5,5-diethyl-1-benzyl	132	
$C_{15}H_{18}N_4O_5$	Mitomycin C	841–844	
$C_{15}H_{19}NO_2$	Tropacocaine	1475–1476	2347–2348
$C_{15}H_{20}N_2O_3S$	Benzylpenilloic acid	222	
$C_{15}H_{20}N_2S$	Methaphenilene	792	
$C_{15}H_{21}F_3N_2O_2$	Fluvoxamine		1880
$C_{15}H_{21}N$	Fencamfamine		1859
$C_{15}H_{21}NO_2$	Ketobemidone	712–713	
$C_{15}H_{21}NO_2$	Meperidine (Pethidine)	777, 968	1958
$C_{15}H_{21}NO_2$	β-Eucaine	398, 541	
$C_{15}H_{21}N_3O$	Primaquine	1137	
$C_{15}H_{21}N_3O_2$	Physostigmine	1109	
$C_{15}H_{22}N_2O$	Mepivacaine	741, 780–781	1962
$C_{15}H_{22}N_2O_2$	Mepindolol		1961
$C_{15}H_{22}N_2O_3$	Leu-Phe	453	
$C_{15}H_{22}N_2O_3$	Phe-Leu	453	
$C_{15}H_{22}N_2O_4$	Leu-Tyr	453	
$C_{15}H_{23}NO_2$	Alprenolol	28	1550
$C_{15}H_{23}NO_2$	Benzoic acid, 4-ethoxy, diethylaminoethyl ester	212	
$C_{15}H_{23}NO_3$	(O-Pivaloyl)etilefrine	1132	
$C_{15}H_{23}NO_3$	Oxprenolol	1167	2051
$C_{15}H_{23}N_3O_3S$	Amdinocillin (Mecillinam)	31	1556
$C_{15}H_{23}N_3O_4S$	1′-Amino-3′-methylcyclopentane-carboxamidopenicillin	1512	
$C_{15}H_{23}N_3O_4S$	Cyclacillin	361–362	1749
$C_{15}H_{23}N_3O_4S$	Sulpiride	1397	
$C_{15}H_{23}N_3O_4S$	WY- 4508 (1′-amino-cyclohexanecarbox-amidopenicillin)	1511	
$C_{15}H_{23}N_3O_4S$	WY- 8542 (1′-amino-3′-methylcyclopentanecarbox-amidopenicillin)	1512	

$C_{15}H_{24}N_2O_2$	Tetracaine	741, 1411–1412	2288–2289
$C_{15}H_{24}O$	Butylated hydroxytoluene		1650
$C_{15}H_{25}NO_3$	Metoprolol	829–830, 1167	
$C_{15}H_{26}N_2$	Sparteine		2198
C_{16}			
$C_{16}H_{10}ClF_3N_2O$	5-Phenyl-1,3-dihydro-1,4-benzodiazepin-2-one, 2′-trifluoromethyl-7-chloro	1026	
$C_{16}H_{11}ClN_2O_3$	Clorazepate		1736
$C_{16}H_{11}ClN_4$	Estazolam	522	
$C_{16}H_{11}Cl_3N_2O$	5-Phenyl-1,3-dihydro-1,4-benzodiazepin-2-one, 1-methyl-2′,6′,7-trichloro	1036	
$C_{16}H_{11}F_3N_2O$	5-Phenyl-1,3-dihydro-1,4-benzodiazepin-2-one, 2′-trifluoromethyl	1053	
$C_{16}H_{11}F_3N_2O$	5-Phenyl-1,3-dihydro-1,4-benzodiazepin-2-one, 3′-trifluoromethyl	1025	
$C_{16}H_{11}F_3N_2O$	5-Phenyl-1,3-dihydro-1,4-benzodiazepin-2-one, 4′-trifluoromethyl	1022	
$C_{16}H_{11}F_3N_2O$	5-Phenyl-1,3-dihydro-1,4-benzodiazepin-2-one, 7-trifluoromethyl	1034	
$C_{16}H_{11}N_3O$	5-Phenyl-1,3-dihydro-1,4-benzodiazepin-2-one, 7-cyano	1065	
$C_{16}H_{11}N_7O_6$	3,5-Di(p-nitrobenzamido)-1,2,4-triazole (Guanazole prodrug)	602	
$C_{16}H_{12}ClF_2O$	5-Phenyl-1,3-dihydro-1,4-benzodiazepin-2-one, 1-methyl-4′-chloro-7-fluoro	1035	
$C_{16}H_{12}ClFN_2O$	5-Phenyl-1,3-dihydro-1,4-benzodiazepin-2-one, 1-methyl-2′-fluoro-7-chloro	1047	
$C_{16}H_{12}ClFN_2O$	Fludiazepam	550	
$C_{16}H_{12}Cl_2N_2O$	5-Phenyl-1,3-dihydro-1,4-benzodiazepin-2-one, 1-methyl-2′,7-dichloro	1032	
$C_{16}H_{12}FIN_2O$	5-Phenyl-1,3-dihydro-1,4-benzodiazepin-2-one, 1-methyl-2′-fluoro-7-iodo	1031	
$C_{16}H_{12}FN_3O_3$	Flunitrazepam	559–560	1872

$C_{16}H_{12}N_2O_3$	Barbituric acid, 5,5-diphenyl	189	
$C_{16}H_{12}N_4O$	2-Cyanoguanidinophenytoin	360	
$C_{16}H_{12}O_3$	Anisindione	81	
$C_{16}H_{13}ClN_2O$	5-Phenyl-1,3-dihydro-1,4-benzodiazepin-2-one, 2′-methyl-7-chloro	1033	
$C_{16}H_{13}ClN_2O$	5-Phenyl-1,3-dihydro-1,4-benzodiazepin-2-one, 3-methyl-7-chloro	1023	
$C_{16}H_{13}ClN_2O$	Diazepam	204, 399–402	1781–1782
$C_{16}H_{13}ClN_2O$	Mazindol		1953
$C_{16}H_{13}ClN_2OS$	5-Phenyl-1,3-dihydro-1,4-benzodiazepin-2-one, 2′-thiomethyl-7-chloro	1038	
$C_{16}H_{13}ClN_2O_2$	Temazepam		2283
$C_{16}H_{13}ClN_2O_2$	5-Phenyl-1,3-dihydro-1,4-benzodiazepin-2-one, 2′-methoxy-7-chloro	1051	
$C_{16}H_{13}ClN_2O_2$	5-Phenyl-1,3-dihydro-1,4-benzodiazepin-2-one, 3′-methoxy-7-chloro	1030	
$C_{16}H_{13}ClN_3O_3$	Mebendazole	764	
$C_{16}H_{13}ClN_3O_3$	5-Phenyl-1,3-dihydro-1,4-benzodiazepin-2-one, 3-methyl-2′-chloro-7-nitro	1048	
$C_{16}H_{13}FN_3O_3$	5-Phenyl-1,3-dihydro-1,4-benzodiazepin-2-one, 1-methyl-2′-fluoro-7-nitro	1063	
$C_{16}H_{13}N_3O_3$	Nimetazepam	894	
$C_{16}H_{13}N_3O_3$	5-Phenyl-1,3-dihydro-1,4-benzodiazepin-2-one, 1-methyl-7-nitro	1061	
$C_{16}H_{13}N_5O_2$	3,5-Dibenzamido-1,2,4-triazole (Guanazole prodrug)	602	
$C_{16}H_{14}ClN_3O$	Chlordiazepoxide	204, 279	1690–1691
$C_{16}H_{14}FN_3O$	5-Phenyl-1,3-dihydro-1,4-benzodiazepin-2-one, 1-methyl-2′-fluoro-7-amino	1073	
$C_{16}H_{14}F_3N_3O_2S$	Lansoprazole		1941
$C_{16}H_{14}N_2O$	5-Phenyl-1,3-dihydro-1,4-benzodiazepin-2-one, 7-methyl	1052	
$C_{16}H_{14}N_2O$	Methaqualone	793	

$C_{16}H_{14}N_2OS$	5-Phenyl-1,3-dihydro-1,4-benzodiazepin-2-one, 7-thiomethyl	1041	
$C_{16}H_{14}N_2O_2$	5-Phenyl-1,3-dihydro-1,4-benzodiazepin-2-one, 7-methoxy	1059	
$C_{16}H_{14}O_3$	Fenbufen	544	
$C_{16}H_{14}O_3$	Ketoprofen	715–716	
$C_{16}H_{15}ClN_2$	Medazepam	204, 770–771	
$C_{16}H_{15}ClN_2O$	5-Phenyl-1,3-dihydro-1,4-benzodiazepin-2-one, 1-methyl-7-chloro	1046	
$C_{16}H_{15}F_2N_3O_4S$	Pantoprazole		2060
$C_{16}H_{15}N_3$	Epinastine	496	
$C_{16}H_{15}N_3O$	5-Phenyl-1,3-dihydro-1,4-benzodiazepin-2-one, 1-methyl-7-amino	1071	
$C_{16}H_{15}N_5O_7S_2$	Cefixime		1670
$C_{16}H_{15}N_5O_7S_2$	Cephradine		1685
$C_{16}H_{16}ClN_3O_3S$	Indapamide	675	
$C_{16}H_{16}ClN_3O_3S$	Metolazone		2003
$C_{16}H_{16}N_2O_2$	Isolysergic acid	696	
$C_{16}H_{16}N_2O_2$	Lysergic acid	757	
$C_{16}H_{16}N_2O_6S_2$	Cephalothin	271–272	1683
$C_{16}H_{16}N_4O_8S$	Cefuroxime		1677
$C_{16}H_{17}BrN_2$	Zimeldine (zimelidine)		2368
$C_{16}H_{17}F_2N_3O_3$	8-Fluoro-norfloxacin	1233	
$C_{16}H_{17}N_3O_4S$	Cephalexin	259–262	1681–1682
$C_{16}H_{17}N_3O_5S$	Cefadroxil	254	
$C_{16}H_{17}N_3O_7S$	Cephalosporin derivative	268	
$C_{16}H_{17}N_3O_7S_2$	Cefoxitin		1672
$C_{16}H_{17}N_5O_7S_2$	Cefotaxime	257	
$C_{16}H_{17}NO_3$	Normorphine	914–915	
$C_{16}H_{18}FN_3O_3$	Norfloxacin	1239–1242	2172–2173
$C_{16}H_{18}N_2O_4$	Barbituric acid, 5,5-diethyl-1-(2-methylbenzoyl)	132	
$C_{16}H_{18}N_2O_4$	Barbituric acid, 5,5-diethyl-1-(3-methylbenzoyl)	132	
$C_{16}H_{18}N_2O_4$	Barbituric acid, 5,5-diethyl-1-(4-methylbenzoyl)	132	
$C_{16}H_{18}N_2O_4S$	Benzylpenicillin (Penicillin G)	217, 958–959, 961	
$C_{16}H_{18}N_2O_5$	Barbituric acid, 5,5-diethyl-1-(2-methoxybenzoyl)	132	

$C_{16}H_{18}N_2O_5$	Barbituric acid, 5,5-diethyl-1-(3-methoxybenzoyl)	132	
$C_{16}H_{18}N_2O_5$	Barbituric acid, 5,5-diethyl-1-(4-methoxybenzoyl)	132	
$C_{16}H_{18}N_2O_5S$	*p*-Hydroxybenzylpenicillin	959	
$C_{16}H_{18}N_2O_5S$	Phenoxymethylpenicillin (Penicillin V)	960–962	
$C_{16}H_{18}N_4O_2$	1-(3,4-Methylenedioxybenzyl)-4-(2-pyrimidinyl)piperazine	1128	
$C_{16}H_{18}N_4O_2$	Piribedil (1-(3,4-methylenedioxybenzyl)-4-(2-pyrimidinyl)piperazine)	1128	
$C_{16}H_{19}BrN_2$	Brompheniramine	230	1635–1636
$C_{16}H_{19}ClN_2$	Chlorpheniramine	290	1704–1708
$C_{16}H_{19}ClN_2O$	Rotoxamine (l-Carbinoxamine)		2188
$C_{16}H_{19}FN_4O_3$	Amifloxacin	1225–1226	
$C_{16}H_{19}N$	Benzylamphetamine		1623
$C_{16}H_{19}NO_4$	Benzoylecgonine	213	
$C_{16}H_{19}N_3O_4S$	Ampicillin	72–76	
$C_{16}H_{19}N_3O_5S$	Amoxicillin	61–66	1574
$C_{16}H_{19}N_3O_5S$	Cefroxadine	258	
*$C_{16}H_{20}ClN_3O_8S$	Deacetoxycephalosporin C, *N*-chloroacetyl	270	
$C_{16}H_{20}ClN_3O_8S$	N-Chloroacetyldeacetoxycephalosporin C	270	
$C_{16}H_{20}N_2$	Pheniramine	977	2081–2083
$C_{16}H_{20}N_2O_5S$	Benzylpenicilloic acid	218–219	
$C_{16}H_{20}N_2O_6S$	Phenoxymethylpenicilloic acid	1006	
$C_{16}H_{20}N_4O_2$	Azapropazone (Apazone)	108–109	
$C_{16}H_{20}N_4O_5$	Porfiromycin	844, 1134	
$C_{16}H_{21}NO_2$	Propranolol	1160–1170	2142–2143
$C_{16}H_{21}NO_3$	Homatropine	398	1912–1913
$C_{16}H_{21}NO_3$	Norhyoscyamine	652, 912	
$C_{16}H_{21}N_3$	Tripelennamine	1472	2340–2342
$C_{16}H_{21}N_3O_4S$	Epicillin	495	
$C_{16}H_{21}N_3O_8S$	Cephalosporin C	270	
$C_{16}H_{22}N_2O_3S$	α-Methyl benzylpenilloate	809	
$C_{16}H_{22}N_4O$	Thonzylamine		2304–2305
$C_{16}H_{22}N_4O_3$	2,4-Diaminopyridine, 5-(3,5-dimethoxy-4-(2-methoxyethyl)benzyl)-	216	
$C_{16}H_{22}N_4O_8S$	Cephalosporin, 5-amino-5-carboxyvaleramido, O-carbamoyl	269	
$C_{16}H_{23}BrN_2O_3$	Remoxipride	1251	

$C_{16}H_{23}NO_2$	Alphaprodine		1549
$C_{16}H_{23}NO_2$	Betaprodine		1625
$C_{16}H_{23}NO_2$	Bufuralol	1167	1638–1639
$C_{16}H_{23}NO_2$	Ethoheptazine		1840
$C_{16}H_{23}NO_2$	Hexylcaine		1908–1909
$C_{16}H_{24}N_2$	Xylometazoline	1517	
$C_{16}H_{24}N_2O_2$	Molindone	845	
$C_{16}H_{24}N_6$	2,4-Diaminopyridine, 5-(3,5-bis (dimethylamino)-4-methylbenzyl)-	216	
$C_{16}H_{24}N_{10}O_4$	Aminophylline		1567
$C_{16}H_{26}N_2O_2$	Amydricaine (Alypine)	398	
$C_{16}H_{26}N_2O_3$	Proparacaine	1157	
$C_{16}H_{26}N_2O_3$	Propoxycaine		2139
$C_{16}H_{27}HgNO_6S$	Mercaptomerin		1963
C_{17}			
$C_{17}H_{12}Cl_2N_4$	Triazolam	1460	
$C_{17}H_{13}ClN_2O_2$	Lonazolac	753	
$C_{17}H_{13}ClN_2O_2$	Pyrazolic acid	1187	
$C_{17}H_{13}ClO_3$	Itanoxone	708	
$C_{17}H_{13}FN_2O_2$	5-Phenyl-1,3-dihydro-1,4-benzodiazepin-2-one, 2′-fluoro-7-acetyl	1068	
$C_{17}H_{13}N_3O_4$	7-Acetamidonitrazepam	205	
$C_{17}H_{13}N_3O_4$	Nitrazepam, 7-acetamido	205	
$C_{17}H_{14}Cl_2N_2O$	5-Phenyl-1,3-dihydro-1,4-benzodiazepin-2-one, 1,3-dimethyl-2′,7-dichloro	1021	
$C_{17}H_{15}ClN_2O$	5-Phenyl-1,3-dihydro-1,4-benzodiazepin-2-one, 1,2′-dimethyl-7-chloro	1037	
$C_{17}H_{15}ClN_2O$	5-Phenyl-1,3-dihydro-1,4-benzodiazepin-2-one, 1-ethyl-7-chloro	1024	
$C_{17}H_{15}ClN_2O_2$	5-Phenyl-1,3-dihydro-1,4-benzodiazepin-2-one, 1-methoxymethyl-7-chloro	1045	
$C_{17}H_{15}ClN_2O_2$	5-Phenyl-1,3-dihydro-1,4-benzodiazepin-2-one, 1-methyl-4′-methoxy-7-chloro	1029	
$C_{17}H_{15}FN_2O$	5-Phenyl-1,3-dihydro-1,4-benzodiazepin-2-one, 2′-fluoro-7-ethyl	1042	
$C_{17}H_{15}NO_3$	D-Indoprofen		1929–1931
$C_{17}H_{16}ClFN_2O_2$	Progabide	1148	

$C_{17}H_{16}ClN_3O$	Amoxapine		1573
$C_{17}H_{16}F_6N_2O$	Mefloquine		1956
$C_{17}H_{16}N_2OS$	5-Phenyl-1,3-dihydro-1,4-benzodiazepin-2-one, 1-methyl-7-thiomethyl	1039	
$C_{17}H_{16}N_3O_4$	5-Phenyl-1,3-dihydro-1,4-benzodiazepin-2-one, 1-methoxymethyl-7-nitro	1064	
$C_{17}H_{17}ClN_6O_3$	Zopiclone	1522	
$C_{17}H_{17}Cl_2N$	Sertraline	1274	
$C_{17}H_{17}NO_2$	Apomorphine	82–84	
$C_{17}H_{17}N_3O$	5-Phenyl-1,3-dihydro-1,4-benzodiazepin-2-one, 1-ethyl-7-amino	1067	
$C_{17}H_{17}N_3O$	5-Phenyl-1,3-dihydro-1,4-benzodiazepin-2-one, 7-dimethylamino	1054	
$C_{17}H_{17}N_3O_2$	5-Phenyl-1,3-dihydro-1,4-benzodiazepin-2-one, 1-methoxymethyl-7-amino	1072	
$C_{17}H_{17}N_3O_6S_2$	Cephapirin	273	1684
$C_{17}H_{18}FN_3O_3$	Ciprofloxacin	316, 1227–1228	1724, 2167
$C_{17}H_{18}F_3N_3O_3$	Fleroxacin	1232	2169
$C_{17}H_{18}N_2O_6$	Nifedipine	885–886	
$C_{17}H_{18}N_2O_6S$	Carbenicillin	962	1658
$C_{17}H_{19}ClN_2OS$	Chlorpromazine sulfoxide		1711
$C_{17}H_{19}ClN_2S$	Chlorpromazine	291–300, 984–985	1710
$C_{17}H_{19}ClN_2S$	Phenothiazine, 2-chloro-10-(3-dimethylaminopropyl)-	984–985	
$C_{17}H_{19}F_2N_3O_3$	8-Fluoro-pefloxacin	1234	
$C_{17}H_{19}F_2N_3O_3$	Lomefloxacin	1235–1236	
$C_{17}H_{19}NO_3$	Hydromorphone	398	1918–1919
$C_{17}H_{19}NO_3$	Morphine	9, 848–857	2010
$C_{17}H_{19}NO_3$	Norcodeine	903–904	
$C_{17}H_{19}NO_3$	Piperine	1125	
$C_{17}H_{19}NO_4$	Dihydromorphinone		1795
$C_{17}H_{19}NO_4$	Morphine-N-oxide		2011
$C_{17}H_{19}NO_4$	Oxymorphone		2054
$C_{17}H_{19}N_3$	Acridine orange	18	
$C_{17}H_{19}N_3$	Antazoline		1579–1581
$C_{17}H_{19}N_3$	Mirtazepine	840	
$C_{17}H_{19}N_3O$	Phentolamine		2092
$C_{17}H_{19}N_3O_3S$	Omeprazole	927	

$C_{17}H_{19}N_5O_2$	2,4-Diaminopyridine, 5-(3,5-dimethoxy-4-pyrrolidinylbenzyl)-	216	
$C_{17}H_{20}BrNO$	Bromodiphenhydramine		1633
$C_{17}H_{20}FN_3O_3$	8-Desfluorolomefloxacin	1229	
$C_{17}H_{20}FN_3O_3$	Pefloxacin	1247	
$C_{17}H_{20}F_6N_2O_3$	Flecainide		1866
$C_{17}H_{20}N_2O_2$	Tropicamide		2350–2351
$C_{17}H_{20}N_2O_5S$	Bumetanide	232–233	1640
$C_{17}H_{20}N_2O_5S$	Phenethicillin	961, 972	2079
$C_{17}H_{20}N_2O_6S$	Methicillin	794, 961	1983
$C_{17}H_{20}N_2S$	Phenothiazine, 10-(2-dimethylaminopropyl)-	993–994	
$C_{17}H_{20}N_2S$	Phenothiazine, 10-(3-dimethylaminopropyl)-	995	
$C_{17}H_{20}N_2S$	Promazine	995, 1149–1152	2135
$C_{17}H_{20}N_2S$	Promethazine	1153–1156	
$C_{17}H_{20}N_4O_6$	Riboflavine	1253–1255	
$C_{17}H_{21}N$	Benzphetamine		1620
$C_{17}H_{21}NO$	Diphenhydramine	454–456	1806
$C_{17}H_{21}NO$	Phenyltoloxamine		2102
$C_{17}H_{21}NO\cdot HCl$	Atomoxetine		1591
$C_{17}H_{21}NO_2$	Dihydrodesoxynorcodeine	427	
$C_{17}H_{21}NO_3$	Dihydromorphine	437	1794
$C_{17}H_{21}NO_3$	Etodolac		1857
$C_{17}H_{21}NO_3$	Galanthamine		1887
$C_{17}H_{21}NO_3$	Ritodrine		2185
$C_{17}H_{21}NO_4$	10-cis-Hydroxydihydronorcodeine	641	
$C_{17}H_{21}NO_4$	10-trans-Hydroxydihydronorcodeine	642	
$C_{17}H_{21}NO_4$	Benzoylecgonine methyl ester (Cocaine)	214, 346–347	1739–1744, 1988
$C_{17}H_{21}NO_4$	Fenoterol		1863–1864
$C_{17}H_{21}NO_4$	N-Methyl-1-benzoyl ecgonine		1988
$C_{17}H_{21}NO_4$	Scopolamine	398, 1270	2191
$C_{17}H_{22}ClNO$	Atomoxetine		1591
$C_{17}H_{22}I_3N_3O_8$	Iopamidol	690	
$C_{17}H_{22}N_2O$	Doxylamine		1824
$C_{17}H_{22}N_2O_6S$	Phenoxyethylpenicilloic acid	1005	
$C_{17}H_{22}N_2S$	Thenalidine	1424	
$C_{17}H_{23}NO$	Levorphanol	734–736	
$C_{17}H_{23}NO$	Racemorphan (Methorphinan)		2176
$C_{17}H_{23}NO_3$	Atropine (non-aqueous titration)	194	

$C_{17}H_{23}NO_3$	Atropine	9, 105–107	1593–1596
$C_{17}H_{23}NO_3$	Hyoscyamine	652	1925–1926
$C_{17}H_{23}N_3O$	Pyrilamine		2154–2155
$C_{17}H_{23}N_3O_2$	Dibucaine O-methyl homologue	407	
$C_{17}H_{23}N_3O_8S$	Cephalosporin, 5-amino-5-carboxyvaleramido, O-acetyl	269	
$C_{17}H_{24}N_2O$	Dimethisoquin		1802
$C_{17}H_{24}N_4O_9S$	Cephalosporin, 5-amino-5-carboxyvaleramido, 7-methoxy, O-carbamoyl	269	
$C_{17}H_{25}N$	Phencyclidine		2077
$C_{17}H_{25}NO_3$	Bunolol (levobunolol)	1167	1641–1642
$C_{17}H_{25}NO_3$	Cyclopentolate	370	
$C_{17}H_{25}NO_3$	Levobunolol	1167	1641–1642
$C_{17}H_{27}NO_2$	Venlafaxine		2357
$C_{17}H_{27}NO_3$	Pramoxine		2123
$C_{17}H_{27}O_4$	Nadolol	1167	2014–2015
$C_{17}H_{28}N_2O$	Etidocaine	537	1855–1856
$C_{17}H_{33}NO_4$	Decyl carnitine	378	
C_{18}			
$C_{18}H_9N_3O$	Ondansetron		2046
$C_{18}H_{13}ClFN_3$	Midazolam	838	2007
$C_{18}H_{13}ClN_2O$	Pinazepam		2110
$C_{18}H_{14}Cl_4N_2O$	Miconazole	837	2006
$C_{18}H_{14}N_2O_4$	Barbituric acid, 5-methyl-5-phenyl-1-benzoyl	172	
$C_{18}H_{14}N_4O_5S$	Sulfasalazine	1384	
$C_{18}H_{15}ClN_2$	Cycliramine	364	
$C_{18}H_{15}F_3N_2O_2$	Flumizole		1871
$C_{18}H_{16}O_3$	Phenprocoumon	637	
$C_{18}H_{18}BrClN_2O$	Metaclazepam	784	
$C_{18}H_{18}ClNS$	Chlorprothixene		1714
$C_{18}H_{18}ClN_3O$	Loxapine		1947
$C_{18}H_{18}N_6O_5S_2$	Cefamandole		1665
$C_{18}H_{18}N_8O_7S_3$	Ceftriaxone		1675–1676
$C_{18}H_{18}O_2$	Equilenin	504	
$C_{18}H_{19}ClN_4$	Clozapine	343–344	1737–1738
$C_{18}H_{19}F_3N_2S$	Fluopromazine	574	
$C_{18}H_{19}F_3N_2S$	Flupromazine	574	
$C_{18}H_{19}F_3N_2S$	Phenothiazine, 10-(3-dimethylaminopropyl)-2-trifluoromethyl-	996–997	
$C_{18}H_{19}F_3N_2S$	Trifluopromazine	996–997, 1468–1469	2329–2330

$C_{18}H_{19}NO$	Desmethyldoxepin	388	
$C_{18}H_{19}N_3O_6S$	Cephaloglycin	263–265	
$C_{18}H_{19}N_3S$	Cyanopromazine		1748
$C_{18}H_{19}N_5O_2$	2,4-Diaminopyridine, 5-(2,4-dimethoxy-3-pyridylbenzyl)-	216	
$C_{18}H_{19}N_5O_2$	2,4-Diaminopyridine, 5-(3,5-dimethoxy-4-pyridylbenzyl)-	216	
$C_{18}H_{20}FN_3O_4$	N-Acetylnorfloxacin	1224	
$C_{18}H_{20}FN_3O_4$	Ofloxacin	1244–1245	2174
$C_{18}H_{20}N_2$	Mianserin	836	
$C_{18}H_{20}N_2O_3$	Phe-Phe	453	
$C_{18}H_{20}N_2O_4$	Phe-Tyr	453	
$C_{18}H_{20}N_2S$	Methdilazine		1979
$C_{18}H_{20}N_2S$	Phenothiazine, 10-(N-pyrrolidinyl)ethyl- (Pyrathiazine)	1003, 1185–1186	
$C_{18}H_{20}O_2$	6-Dehydroestrone	379	
$C_{18}H_{20}O_2$	Diethylstilbestrol	419	
$C_{18}H_{20}O_2$	Equilin	505	
$C_{18}H_{21}ClN_2$	Chlorcyclizine	278	1689
$C_{18}H_{21}NO$	Pipradol		2117
$C_{18}H_{21}NO_3$	Codeine	9, 348–353	
$C_{18}H_{21}NO_3$	Hydrocodone	398	1917
$C_{18}H_{21}NO_3$	Methyldihydromorphinone (Metopon)		1990, 2004
$C_{18}H_{21}NO_4$	10-Hydroxycodeine	636	
$C_{18}H_{21}NO_4$	14-cis-Hydroxydihydrocodeinone	638	
$C_{18}H_{21}NO_4$	Dihydrocodeinone		1792
$C_{18}H_{21}NO_4$	Oxycodone	398, 934	2053
$C_{18}H_{21}N_3O$	Dibenzepine	404–405	
$C_{18}H_{22}ClN_3O_9S$	Cephalosporin C, N-chloroacetyl	270	
*$C_{18}H_{22}ClN_3O_9S$	N-Chloroacetylcephalosporin C	270	
$C_{18}H_{22}ClNO$	Phenoxybenzamine		2088
$C_{18}H_{22}FN_3O_3$	Norfloxacin ethyl ester	1243	
$C_{18}H_{22}N_2$	Cyclizine	365–366	1751
$C_{18}H_{22}N_2$	Desipramine	385–387	1772
$C_{18}H_{22}N_2OS$	Methopromazine (methoxypromazine)		1985
$C_{18}H_{22}N_2OS$	Methoxypromazine		1985
$C_{18}H_{22}N_2O_5S$	Propicillin	961	2137
$C_{18}H_{22}N_2S$	Diethazine	9	1790
$C_{18}H_{22}N_2S$	Phenothiazine, 10-(2-diethylaminoethyl)-	991	

$C_{18}H_{22}N_2S$	Trimeprazine (Methylpromazine)		2331–2332
$C_{18}H_{22}O_2$	Dihydroequilin, 17α-	428	
$C_{18}H_{22}O_2$	Dihydroequilinen, 17β-	429	
$C_{18}H_{22}O_2$	Estrone	525	
$C_{18}H_{23}N$	Tolpropamide	1455	
$C_{18}H_{23}NO$	Orphenadrine		2048
$C_{18}H_{23}NO_2$	Dihydrodesoxycodeine	426	
$C_{18}H_{23}NO_3$	10-cis-Hydroxydihydro-desoxycodeine	639	
$C_{18}H_{23}NO_3$	10-trans-Hydroxydihydro-desoxycodeine	640	
$C_{18}H_{23}NO_3$	Dihydrocodeine	425	
$C_{18}H_{23}NO_3$	Dobutamine		1815
$C_{18}H_{23}NO_3$	Isoxsuprine		1939–1940
$C_{18}H_{24}N_2O_5$	Enalaprilat		1826
$C_{18}H_{24}N_2O_6S$	Phenoxypropylpenicilloic acid	1007	
$C_{18}H_{24}O_2$	Estradiol, 17α-	523	
$C_{18}H_{24}O_2$	Estradiol, 17β-	523	
$C_{18}H_{24}O_3$	Estriol	524	
$C_{18}H_{25}NO$	Cyclazocine		1750
$C_{18}H_{25}NO$	Dextromethorphan		1775
$C_{18}H_{25}NO$	D-Methorphan		1775
$C_{18}H_{25}NO$	Levomethorphan		1942
$C_{18}H_{25}NO$	Racemethorphan		2175
$C_{18}H_{25}NO_3$	Tetrahydro-α-morphimethine	1420	
$C_{18}H_{25}N_3O_2$	Dibucaine O-ethyl homologue	407	
$C_{18}H_{25}N_3O_9S$	Cephalosporin, 5-amino-5-carboxyvaleramido, 7-methoxy, O-acetyl	269	
$C_{18}H_{26}ClN_3$	Chloroquine	285–286	1696
$C_{18}H_{26}ClN_3.H_3PO_4$	Chloroquine phosphate		1697
$C_{18}H_{26}N_4O_3$	2,4-Diaminopyridine, 5-(3,5-diethoxy-4-(2′-hydroxy-2′-propyl)benzyl)-	216	
$C_{18}H_{26}N_6O_2$	2,4-Diaminopyridine, 5-(3,5-bis (methylamino)-4-methoxy-6-acetylbenzyl)-	216	
$C_{18}H_{28}N_2O$	Bupivacaine	234–236	
$C_{18}H_{28}N_2O_4$	Acebutolol	1167	1523–1526
$C_{18}H_{29}ClN_3O_4P$	Chloroquine phosphate		1697
$C_{18}H_{29}NO_2$	Penbutolol	1167	2066
$C_{18}H_{29}NO_3$	Betaxolol		1626
$C_{18}H_{30}N_2O_2$	Butacaine		1645

$C_{18}H_{31}NO_4$	Bisoprolol	223	
$C_{18}H_{32}O_2$	Linoleic acid	749	
$C_{18}H_{33}ClN_2O_5S$	Clindamycin	322–325	
$C_{18}H_{34}N_2O_6S$	Lincomycin	747–748	
$C_{18}H_{34}O_2$	Oleic acid	926	
$C_{18}H_{35}ClN_2O_8SP$	Clindamycin-2-phosphate	327	
$C_{18}H_{36}N_4O_{11}$	Kanamycin A	709	
$C_{18}H_{36}O_2$	Stearic acid	1280	
$C_{18}H_{37}N_5O_9$	Tobramycin		2313
C_{19}			
$C_{19}H_{10}Br_4O_5S$	Bromophenol blue	228	
$C_{19}H_{12}O_6$	Dicumarol (Bishydroxycoumarin)		1789
$C_{19}H_{14}O_5S$	Phenolsulphonphthalein (Phenol red)	980	2086
$C_{19}H_{15}NO_6$	Acenocoumarin; Acenocoumarol	637	1527
$C_{19}H_{16}ClNO_4$	Indomethacin	678–683	
$C_{19}H_{16}N_2O_4$	Barbituric acid, 5-ethyl-5-phenyl-1-benzoyl	185	
$C_{19}H_{16}O_4$	Warfarin	637, 1504–1508	2365–2366
$C_{19}H_{17}ClFN_2O_5S$	Flucloxacillin	962	1868
$C_{19}H_{17}ClN_2O$	Prazepam		2124–2125
$C_{19}H_{17}ClN_2O_4$	Glafenine	589	
$C_{19}H_{17}Cl_2N_3O_5S$	Dicloxacillin	414, 962	1788
$C_{19}H_{17}N_3O_4S_2$	Cephaloridine	266–267	
$C_{19}H_{18}ClFN_2O_3$	5-Phenyl-1,3-dihydro-1,4-benzodiazepin-2-one, 1-[butane-2,4-diol]-2′-fluoro-7-chloro	1070	
$C_{19}H_{18}ClN_3O_5S$	Cloxacillin	342, 962	
$C_{19}H_{18}ClN_5$	Adinazolam	24	
$C_{19}H_{18}FIN_2O_3$	5-Phenyl-1,3-dihydro-1,4-benzodiazepin-2-one, 1-[butane-2,4-diol]-2′-fluoro-7-iodo	1066	
$C_{19}H_{19}N$	Phenindamine	975	2080
$C_{19}H_{19}N_3O_5$	5-Phenyl-1,3-dihydro-1,4-benzodiazepin-2-one, 1-[butane-2,4-diol]-7-nitro	1076	
$C_{19}H_{19}N_3O_5S$	Oxacillin	962	2049–2050
$C_{19}H_{19}N_7O_6$	Folic acid	576	1881

$C_{19}H_{20}F_3N_3O_3$	Orbifloxacin	1246	
$C_{19}H_{20}FNO_3$	Paroxetine		2063
$C_{19}H_{20}N_2$	Mebhydroline	766	
$C_{19}H_{20}N_2O_2$	Phenylbutazone	1009–1018	2093
$C_{19}H_{20}N_2O_3$	Oxyphenbutazone	935–938	
$C_{19}H_{20}N_8O_5$	Aminopterin		1568
$C_{19}H_{21}N$	Nortriptyline	919	2040
$C_{19}H_{21}NO$	Doxepin	467–468	1820
$C_{19}H_{21}NO_3$	Nalorphine	867	2018
$C_{19}H_{21}NO_3$	Thebaine	1423	2292
$C_{19}H_{21}NO_4$	6-Acetylmorphine	15	
$C_{19}H_{21}NO_4$	Naloxone	868	
$C_{19}H_{21}NS$	Dosulepine	466	
$C_{19}H_{21}NS$	Dothiepin (dosulepine)	466	
$C_{19}H_{21}NS$	Pizotyline		2120
$C_{19}H_{21}N_5O_4$	Prazosin		2126–2127
$C_{19}H_{21}N_7O_6$	Dihydrofolic acid	436	
$C_{19}H_{22}ClN_5O$	Trazodone	1457–1459	2323
$C_{19}H_{22}N_2$	Triprolidine	1473	2343–2344
$C_{19}H_{22}N_2O$	Cinchonidine	310	
$C_{19}H_{22}N_2O$	Cinchonine	311	1722–1723
$C_{19}H_{22}N_2OS$	Acepromazine	16	
$C_{19}H_{22}N_2OS$	Acetylpromazine (Acepromazine)	16	
$C_{19}H_{22}N_2S$	Phenothiazine, 10-(N-methyl) piperidinyl]methyl- (Mepazine; Pecazine)	776, 952–953, 1002	
$C_{19}H_{22}N_4O_7$	Ebifuramin	472	
$C_{19}H_{23}ClN_2$	Clomipramine (Chlorimipramine)	280	1729
$C_{19}H_{23}NO$	Diphenylpyraline	459	1809
$C_{19}H_{23}NO_3$	Ethylmorphine	536	1850–1851
$C_{19}H_{23}N_3O_2$	Ergometrine		1830
$C_{19}H_{23}N_3O_2$	Ergonovine	506–507	1831
$C_{19}H_{23}N_5O_2$	2,4-Diaminopyridine, 5-(3,5-diethoxy-4-pyrrolidinylbenzyl)-	216	
$C_{19}H_{24}N_2$	Bamipine	116	
$C_{19}H_{24}N_2$	Histapyrrodine	618	
$C_{19}H_{24}N_2$	Imipramine	671–674	
$C_{19}H_{24}N_2O_2$	Benzoic acid, 4-(N-ethylamino), diethylaminoethyl ester	212	
$C_{19}H_{24}N_2O_3$	Dilevolol		1798
$C_{19}H_{24}N_2O_3$	Labetalol	718–720	

$C_{19}H_{24}N_2OS$	Methotrimeprazine (Levomepromazine)	799–800, 992	
$C_{19}H_{24}N_2OS$	Phenothiazine, 10-(2-dimethylaminomethyl) propyl-2-methoxy-	992	
$C_{19}H_{24}N_2S$	Ethopropazine	530	1841
$C_{19}H_{24}N_4O_2$	Pentamidine		2067
$C_{19}H_{25}NO$	Levallorphan	727–728	
$C_{19}H_{25}NO.C_4H_6O_6$	Levallorphan tartrate	726	
$C_{19}H_{25}N_3O_2$	Dihydroergonovine	433	
$C_{19}H_{25}N_5O_4$	Terazosin	1405	
$C_{19}H_{26}N_2S.CH_3SO_3H$	Pergolide mesylate		2071
$C_{19}H_{27}NO$	Pentazocine	963	2068
$C_{19}H_{27}N_3O_2$	Dibucaine *O*-propyl homologue	407	
$C_{19}H_{28}N_2$	Iprindole		1936
$C_{19}H_{28}N_2O_4$	Carpindolol		1663
$C_{19}H_{29}NO_5$	Dipivefrine		1813
$C_{19}H_{29}N_3O$	Pamaquine (plasmoquin)		2057–2059
$C_{19}H_{31}NO$	Bencyclane	195	
$C_{19}H_{31}NO$	Deramciclane	383	
$C_{19}H_{35}ClN_2O_5S$	Clindamycin, 1′-demethyl-4′-pentyl	325	
$C_{19}H_{35}N$	Perhexilene	965	
$C_{19}H_{35}NO_2$	Dicyclomine	415	
$C_{19}H_{39}N_5O_7$	Gentamicin C1a		1888
C_{20}			
$C_{20}H_{12}O_5$	Fluorescein	561	
$C_{20}H_{12}O_8$	Biscoumacetic acid		1628
$C_{20}H_{14}I_6N_2O_6$	Iodipamide	688	1934
$C_{20}H_{14}O_4$	Phenolphthalein	978–979	
$C_{20}H_{16}N_2O_4$	Camptothecin	244	1654
$C_{20}H_{17}FO_3S$	Sulindac	1396	
$C_{20}H_{20}N_4O_3$	2,4-Diaminopyridine, 5-(3,4-dimethoxy-5-benzoylbenzyl)-	216	
$C_{20}H_{20}N_6O_9S$	Moxalactam		2012
$C_{20}H_{21}FN_2O$	Citalopram		1725
$C_{20}H_{21}N$	Cyclobenzaprine		1752
$C_{20}H_{21}NO_4$	Papaverine	9, 946–951	
$C_{20}H_{21}N_3O_3$	Trp-Phe	453	
$C_{20}H_{22}ClN$	Pyrrobutamine		2160
$C_{20}H_{22}N_2$	Azatadine		1597
$C_{20}H_{22}N_2O_3$	Norcodeine, *N*-(2-cyano)ethyl	905	
$C_{20}H_{22}N_8O_5$	Methotrexate	795–798	1986–1987

$C_{20}H_{23}N$	Amitriptyline	56–59	
$C_{20}H_{23}N$	Maprotiline		1952
$C_{20}H_{23}NO_4$	Naltrexone	869	
$C_{20}H_{23}N_5O_6S$	Azlocillin		1599
$C_{20}H_{23}N_7O_7$	Folinic acid (Leucovorin)	578–579	
$C_{20}H_{24}ClN_3S$	Phenothiazine, 2-chloro-10-[*N*-methyl]piperazinylpropyl- (Prochlorperazine)	987–988, 1144–1147	2134
$C_{20}H_{24}N_2OS$	Propiomazine		2138
$C_{20}H_{24}N_2O_2$	Epiquinidine	1221	
$C_{20}H_{24}N_2O_2$	Epiquinine	1221	
$C_{20}H_{24}N_2O_2$	Quinidine	1211–1213, 1221	2165–2166
$C_{20}H_{24}N_2O_2$	Quinine	9, 1214–1223	
$C_{20}H_{24}N_2O_2S$	Hycanthone		1914
$C_{20}H_{24}O_2$	Ethinylestradiol	527–528	1839
$C_{20}H_{25}ClN_2O_5$	Amlodipine		1572
$C_{20}H_{25}NO$	Normethadone		2037–2038
$C_{20}H_{25}N_3O$	Lysergide		1948
$C_{20}H_{25}N_3O_2$	Methylergonovine	816	
$C_{20}H_{26}N_2$	Trimipramine		2338–2339
$C_{20}H_{26}N_2O_2$	Ajmaline		1540
$C_{20}H_{26}N_2O_4$	Tolamolol		2315–2316
$C_{20}H_{26}N_4O$	Lisuride	752	
$C_{20}H_{27}NO_5$	Chromonar		1719
$C_{20}H_{28}N_2O_5$	Enalapril	481–482	
$C_{20}H_{29}N_3O_2$	Dibucaine (Cinchocaine)	406–407, 741	1784–1785
$C_{20}H_{29}N_3O_2$	Dibucaine O-butyl homologue	407	
$C_{20}H_{32}O_5$	Dinoprostone (Prostaglandin E_2)	1173	1804
$C_{20}H_{33}N_3O_4$	Celiprolol		1678
$C_{20}H_{34}O_5$	Dinoprost (Prostaglandin $F_{2\alpha}$)	451, 1174	
$C_{20}H_{34}O_5$	Prostaglandin E_1	1172	2149
$C_{20}H_{41}N_5O_7$	Gentamicin C2	1888	
C_{21}			
$C_{21}H_{14}Br_4O_5S$	Bromocresol green	227	
$C_{21}H_{16}O_7$	Coumetarol	637	
$C_{21}H_{18}F_3N_3O_3$	Temafloxacin	1249	
$C_{21}H_{18}O_5S$	Cresol red	358	
$C_{21}H_{19}F_2N_3O_3$	Difloxacin	1230	
$C_{21}H_{21}ClN_2O_8$	Demeclocycline (6-demethyl-7-chlorotetracycline)	380–381	1766
$C_{21}H_{21}N$	Cyproheptadine		1756

$C_{21}H_{22}ClN_3OS$	Phenothiazine, 2-chloro-10-[N-(2-hydroxy)ethyl]piperazinyl-propyl-	986	
$C_{21}H_{22}N_2O_2$	Strychnine	1284	2202
$C_{21}H_{22}N_2O_5S$	Nafcillin	864	
$C_{21}H_{22}N_2S_2$	Phenothiazine, 10-[2-(N-methyl) piperidinyl]ethyl-2-methylthio-	1000–1001	
$C_{21}H_{23}ClFNO_2$	Haloperidol	603–604	1904–1906
$C_{21}H_{23}ClFN_3O$	Flurazepam		1877–1878
$C_{21}H_{23}NO_5$	Diamorphine (heroin)	397–398	1778–1779
$C_{20}H_{32}ClN_3OS$	Phenothiazine, 10-[N-(4-carbamoyl) piperidinyl]propyl-2-chloro-	983	
$C_{20}H_{32}ClN_3OS$	Pipamazine	983, 1119	
$C_{21}H_{24}F_3N_3S)$	Phenothiazine, 10-(N-methyl) piperazinylpropyl-2-trifluoromethyl-(Trifluoperazine)	998–999, 1462–1467	2326–2327
$C_{21}H_{25}NO$	Benztropine		1622
$C_{21}H_{25}NO_3$	Nalmefene		2017
$C_{21}H_{25}N_5O_8S_2$	Mezlocillin	835	
$C_{21}H_{26}ClN_3OS$	Perphenazine	966–967	2072
$C_{21}H_{26}N_2O_3$	Yohimbine	398, 479	
$C_{21}H_{26}N_2S_2$	Thioridazine	1439–1442	2301
$C_{21}H_{27}ClN_2O_2$	Hydroxyzine	650–651	1923–1924
$C_{21}H_{27}NO$	DL-Isomethadone	697–698	
$C_{21}H_{27}NO$	Methadone		1969–1972
$C_{21}H_{27}NO_3$	Propafenone		2136
$C_{21}H_{27}NO_4$	Nalbuphine		2016
$C_{21}H_{27}N_3O_2$	Methysergide	825–826	
$C_{21}H_{27}N_3O_7S$	Bacampicillin		1600
$C_{21}H_{27}N_5O_4S$	Glipizide		1890
$C_{21}H_{28}N_2O_2$	Ethylhydrocupreine	398	
$C_{21}H_{28}N_2O_5$	Trimethobenzamide		2333
$C_{21}H_{28}N_2O_5S$	Tienoxolol	1445	
$C_{21}H_{29}NO_2$	Butorphanol		1648–1649
$C_{21}H_{29}N_3O$	Disopyramide	460–462	
$C_{21}H_{30}O_2$	Cannabidiol		1655
$C_{21}H_{30}O_2$	Tetrahydrocannabinol		2291
$C_{21}H_{31}N_3O_2$	Dibucaine *O*-pentyl homologue	407	
$C_{21}H_{31}N_3O_5$	Lisinopril	750–751	
$C_{21}H_{31}N_5O_2$	Buspirone	239	
$C_{21}H_{32}N_6O_3$	Alfentanil		1544–1545
$C_{21}H_{39}N_7O_{12}$	Streptomycin A	1281	

$C_{21}H_{41}N_7O_{12}$	Dihydrostreptomycin		1796–1797
$C_{21}H_{43}N_5O_7$	Gentamicin C1		1888
$C_{21}H_{45}N_3$	Hexetidine	610–611	
C_{22}			
$C_{22}H_{17}ClN_2$	Clotrimazole	341	
$C_{22}H_{21}ClN_2O_7$	Anhydrochlortetracycline	80	
$C_{22}H_{22}ClF_4NO_2$	Clofluperol	1273	
$C_{22}H_{22}ClF_4NO_2$	Seperidol (Clofluperol)	1273	
$C_{22}H_{22}FN_3O_2$	Droperidol		1825
$C_{22}H_{22}N_2O_2$	Cinnopentazone	314	
$C_{22}H_{22}N_2O_7$	Anhydro-4-epitetracycline	491–492	
$C_{22}H_{22}N_2O_7$	Epianhydrotetracycline (Anhydro-4-epitetracycline)	491–492	
$C_{22}H_{22}N_2O_8$	Methacycline		1968
$C_{22}H_{22}N_6O_7S_2$	Ceftazidime		1673
$C_{22}H_{23}ClN_2O_8$	Chlortetracycline	301–306	1715
$C_{22}H_{23}ClN_2O_8$	Epichlortetracycline (7-Chloro-4-epitetracycline)	493–494	
$C_{22}H_{23}ClN_2O_8$	Isochlorotetracycline	694–695	
$C_{22}H_{23}NO_7$	Noscapine (Narcotine)	920–922	2041–2042
$C_{22}H_{24}ClN_3O$	Azelastine	111–112	
$C_{22}H_{24}ClN_5O_2$	Domperidone		1816
$C_{22}H_{24}FN_3O_2$	Beneperidol		1616
$C_{22}H_{24}FN_3O_2$	Benperidol		1616
$C_{22}H_{24}N_2O_8$	4-Epitetracycline	503	
$C_{22}H_{24}N_2O_8$	Doxycycline	471	1823
$C_{22}H_{24}N_2O_8$	Tetracycline	1413–1418	2290
$C_{22}H_{24}N_2O_9$	Oxytetracycline	939–945	
$C_{22}H_{24}N_3O_4S$	Morizicine	847	
$C_{22}H_{25}ClN_2OS$	Clopenthixol	339	1735
$C_{22}H_{25}F_2NO_4$	Nebivolol	875	
$C_{22}H_{25}NO_6$	Colchicine	354	1745–1746
$C_{22}H_{25}N_3O$	Indoramin		1932
$C_{22}H_{26}F_3N_3OS$	Fluphenazine	572–573	
$C_{22}H_{26}N_2O_4S$	Diltiazem	442	1799–1800
$C_{22}H_{27}NO$	Phenazocine	970	
$C_{22}H_{27}N_3O_4$	Diperodon		1805
$C_{22}H_{27}N_3O_5$	Physostigmine salicylate	1108	2105
$C_{22}H_{28}N_2O$	Fentanyl	547	1865
$C_{22}H_{28}N_2O_2$	Anileridine		1578
$C_{22}H_{28}N_4O_6$	Mitoxantrone		2009
$C_{22}H_{29}NO_2$	Propoxyphene		2140–2141
$C_{22}H_{29}N_3O_6S$	Pivampicillin		2119
$C_{22}H_{30}Cl_2N_{10}$	Chlorhexidine	282	1692
$C_{22}H_{30}FO_8P$	Dexamethasone-21-phosphate	392	

$C_{22}H_{30}N_2$	Aprindine	85	
$C_{22}H_{30}N_2O_2S$	Sufentanil		2204–2205
$C_{22}H_{30}N_6O_4S$	Sildenafil		2193
$C_{22}H_{31}NO$	Tolterodine		2319
$C_{22}H_{31}NO_3$	Oxybutynin		2052
$C_{22}H_{31}O_8P$	Methylprednisolone-21-phosphate	824	
$C_{22}H_{32}N_2O_5$	Benzquinamide	215	1621
$C_{22}H_{32}N_2O_5S$	Piperazine estrone sulfate	1122	
$C_{22}H_{33}N_3O_2$	Dibucaine *O*-hexyl homologue	407	
$C_{22}H_{43}N_5O_{13}$	Amikacin		1557
C_{23}			
$C_{23}H_{15}F_2NO_2$	Brequinar	225	
$C_{23}H_{16}O_{11}$	Cromolyn		1747
$C_{23}H_{20}N_2O_3S$	Sulfinpyrazone	1388	2268
$C_{23}H_{25}F_3N_2OS$	Flupenthixol	571	1874
$C_{23}H_{25}I_2NO_3$	Desethylamiodarone	53	
$C_{23}H_{25}N$	Fendiline	545	
$C_{23}H_{25}N_3O_4S$	Benzylpenicilloic acid α-benzylamide	220–221	
$C_{23}H_{26}FN_3O_2$	Spiperone		2200
$C_{23}H_{26}N_2O_4$	Brucine	231	1637
$C_{23}H_{27}FN_4O_2$	Risperidone		2184
$C_{23}H_{27}IN_2O_8$	Tetracycline methiodide	1419	
$C_{23}H_{27}NO_9$	Morphine-3-glucuronide	858–859	
$C_{23}H_{27}NO_9$	Morphine-6-glucuronide	860–861	
$C_{23}H_{27}N_3O_7$	Minocycline		2008
$C_{23}H_{28}ClN_3O_2S$	Thiopropazate	1437–1438	
$C_{23}H_{28}ClN_3O_5S$	Glibenclamide (glyburide)	590–592	1889, 1895
$C_{23}H_{28}ClN_3O_5S$	Glyburide	590–592	1889, 1895
$C_{23}H_{29}NO_2$	Phenadoxone		2074–2076
$C_{23}H_{29}N_3O$	Opipramol	928	
$C_{23}H_{29}N_3O_2S$	Thiothixene		2302
$C_{23}H_{30}ClN_3O$	Quinacrine	1210	2161–2163
$C_{23}H_{30}N_2O_2$	Piminodine		2106
$C_{23}H_{30}N_2O_3$	Testosterone, imidazole-1-carboxylic acid prodrug	1410	
$C_{23}H_{30}N_2O_4$	Pholcodine		2103
$C_{23}H_{31}NO_2$	α-Acetylmethadol (Levomethadyl acetate)	14	1537
$C_{23}H_{31}NO_7$	Levallorphan tartrate	726	
$C_{23}H_{32}N_2O_5$	Ramipril	1250	

Formula	Name		
C_{24}			
$C_{24}H_{16}Cl_2N_4$	Clofazimine analogue, des-iso-propyl	334	
$C_{24}H_{22}FNO_2$	2-(4-(*p*-Fluorobenzoyl)piperidin-1-yl)-2′-acetonaphthone hydrochloride (E2001)	1873	
$C_{24}H_{22}FNO_2$	E2001		1873
$C_{24}H_{25}NO_4$	Flavoxate	548	
$C_{24}H_{26}ClN_3O_2S$	Phenothiazine, 2-chloro-10-[N-(2-propionyloxy)ethyl]-piperazinylpropyl-	989–990	
$C_{24}H_{27}N$	Prenylamine	1136	
$C_{24}H_{28}N_2O_9$	Dimethyloxytetracycline	450	
$C_{24}H_{29}NO$	Phenomorphan		2087
$C_{24}H_{31}NO$	6-Piperidino-4,4-diphenylheptan-3-one (Dipipanone)		1810–1812, 2116
$C_{24}H_{31}NO$	5-Methyl-4,4-diphenyl-6-piperidino-3-hexanone (DL-Pipidone)	1126	
$C_{24}H_{32}FO_9P$	Triamcinolone-16,17-acetonide-21-phosphate	392	
$C_{24}H_{34}O_5$	Dehydrocholic acid		1764–1765
$C_{24}H_{36}O_3$	Nabilone		2013
$C_{24}H_{40}N_8O_4$	Dipyridamole		1814
$C_{24}H_{40}O_4$	Chenodeoxycholic acid	275	
$C_{24}H_{40}O_4$	Chenodiol		1686
$C_{24}H_{40}O_4$	Deoxycholic acid		1768
$C_{24}H_{40}O_4$	Ursodeoxycholic acid (ursodiol)	1486	
$C_{24}H_{40}O_5$	Cholic acid		1718
C_{25}			
$C_{25}H_{22}O_{10}$	Silybin	1275	
$C_{25}H_{27}ClN_2$	Meclizine	768–769	
$C_{25}H_{27}N_9O_8S_2$	Cefoperazone		1671
$C_{25}H_{29}I_2NO_3$	Amiodarone	51–55	
$C_{25}H_{29}NO_2$	KHL 8430	717	
$C_{25}H_{31}NO$	Butaclamol		1646
$C_{25}H_{32}ClN_5O_2$	Nefazodone		2020
$C_{25}H_{32}N_2O_2$	Dextromoramide		1776
$C_{25}H_{32}N_2O_2$	Levomoramide		1943, 1944
$C_{25}H_{32}N_2O_6$	Hydrocortisone, imidazole-1-carboxylic acid prodrug	628	
$C_{25}H_{34}O_8$	Hydrocortisone hydrogen succinate	627	
$C_{25}H_{35}NO_5$	Mebeverine	765	

$C_{25}H_{43}N_{13}O_{10}$	Viomycin		2363–2364
$C_{25}H_{44}N_{14}O_7$	Capreomycin IB		1656
$C_{25}H_{44}N_{14}O_8$	Capreomycin IA		1656
$C_{25}H_{48}N_6O_8$	Desferrioxamine (deferoxamine)	384	
C_{26}			
$C_{26}H_{25}N_2O_4$	HNB-5	620	
$C_{26}H_{26}I_6N_2O_{10}$	Iodoxamic acid	689	
$C_{26}H_{27}NO_9$	Idarubicin	376, 665	
$C_{26}H_{28}Cl_2N_4O_4$	Ketoconazole	714	
$C_{26}H_{28}N_2$	Cinnarizine	313	
$C_{26}H_{29}NO$	Tamoxifen	1400	2281–2282
$C_{26}H_{34}FNO_5$	Cerivastatin	274	
$C_{26}H_{37}NO_8S_2$	Tiapamil	1444	
$C_{26}H_{43}NO_5$	Glycodeoxycholic acid	601	1897
$C_{26}H_{43}NO_6$	Glycocholic acid	600	1896
$C_{26}H_{45}NO_7S$	Taurocholic acid	1402	
C_{27}			
$C_{27}H_{22}Cl_2N_4$	Clofazimine	333–335	
$C_{27}H_{26}NO_4$	HNB-1	619	
$C_{27}H_{29}NO_{10}$	Daunorubicin	376–377	
$C_{27}H_{29}NO_{11}$	Doxorubicin (adriamycin)	469–470	1821–1822
$C_{27}H_{29}NO_{11}$	Pharmorubicin	376, 969	
$C_{27}H_{33}N_3O_8$	Rolitetracycline		2187
$C_{27}H_{38}N_2O_4$	Verapamil (dexverapamil)	1495–1498	2358–2359
$C_{27}H_{43}NO_2$	Solasodine		2194
C_{28}			
$C_{28}H_{29}F_2N_3O$	Pimozide		2107–2108
$C_{28}H_{30}Cl_2FN_5O_2$	Soluflazine	1276	
$C_{28}H_{31}FN_4O$	Astemizole	98	
$C_{28}H_{31}N_3O_6$	Benidipine	198	
$C_{28}H_{37}N_5O_7$	Enkephalin (met-enkephalin)	483	
$C_{27}H_{35}N_5O_7S$	Met-enkephalin	483	
C_{29}			
$C_{29}H_{32}O_{13}$	Etoposide	540	
$C_{29}H_{33}ClN_2O_2$	Loperamide	754	
$C_{29}H_{38}F_3N_3O_2S$	Fluphenazine enanthate		1875
$C_{29}H_{40}N_2O_4$	Emetine	478–480	
$C_{29}H_{41}NO_4$	Buprenorphine	237–238	
C_{30}			
$C_{30}H_{29}Cl_2N_5$	Clofazimine analogue, N-diethylamino-2-ethyl	334	
$C_{30}H_{32}N_2O_2$	Diphenoxylate	458	1807–1808
$C_{30}H_{44}N_2O_{10}$	Hexobendine	612–613	

Formula	Compound	Pages	Pages
C_{31}			
$C_{31}H_{31}Cl_2N_5$	Clofazimine analogue, *N*-diethylamino-3-propyl	334	
$C_{31}H_{36}NO_{11}$	Novobiocin		2043
$C_{31}H_{40}N_4O_6$	DMP-777 (elastase inhibitor)	463, 477	
$C_{31}H_{41}N_5O_5$	Dihydroergocornine	430	
$C_{31}H_{44}N_2O_{10}$	Dilazep	440–441	
$C_{31}H_{48}O_6$	Fusidic acid		1886
$C_{31}H_{51}NO_9$	Rosaramicin (juvenimicin)	1258	
$C_{31}H_{51}NO_{10}$	5-O-Mycaminosyltylonolide (OMT)	863	
C_{32}			
$C_{32}H_{32}O_{13}S$	Teniposide	1404	
$C_{32}H_{38}N_2O_8$	Deserpidine		1770–1771
$C_{32}H_{40}BrN_5O_5$	Bromocriptine		1631–1632
$C_{32}H_{41}NO_2$	Terfenadine	1409	2287
$C_{32}H_{43}N_5O_5$	Dihydroergocriptine	431	
$C_{32}H_{44}F_3N_3O_2S$	Fluphenazine decanoate		1876
C_{33}			
$C_{33}H_{33}Cl_2N_5$	Clofazimine analogue, 4-piperidinylmethyl	334	
$C_{33}H_{34}N_6O_6$	Candesartan cilexetil	245	
$C_{33}H_{35}FN_2O_5$	Atorvastatin	103	
$C_{33}H_{35}N_5O_5$	Ergotamine	510–511	1832–1833
$C_{33}H_{35}N_5O_5$	Ergotaminine	512–513	
$C_{33}H_{37}N_5O_5$	Dihydroergotamine	434	1793
$C_{33}H_{40}N_2O_9$	Reserpine		2181–2182
$C_{33}H_{47}NO_{13}$	Natamycin		2019
C_{34}			
$C_{34}H_{35}Cl_2N_5$	Clofazimine analogue, N-pyrrolidinyl-3-propyl	334	
$C_{34}H_{47}NO_{11}$	Aconitine	17	
$C_{34}H_{50}O_7$	Carbenoxolone	247–248	1659
$C_{34}H_{63}ClN_2O_6S$	Clindamycin-2-palmitate	326	
C_{35}			
$C_{35}H_{37}Cl_2N_5$	Clofazimine analogue, N-piperidinyl-3-propyl	334	
$C_{35}H_{39}N_5O_5$	Ergostine	508–509	
$C_{35}H_{41}Cl_2N_5$	Clofazimine analogue, N-diethylamino-4-(1-methylbutyl)	334	
$C_{35}H_{41}N_5O_5$	Dihydroergocristine	432	

$C_{35}H_{42}N_2O_9$	Rescinnamine		2180, 2182
$C_{35}H_{61}NO_{12}$	Oleandomycin	924–925	
$C_{35}H_{62}Br_2N_4O_4$	Pipecuronium bromide (Pipecurium bromide)		2114
C_{36}			
$C_{36}H_{47}N_5O_4$	Indinavir	677	
$C_{36}H_{60}O_{30}$	Cyclohexaamylose (α-Cyclodextrin)	367	
C_{37}			
$C_{37}H_{67}NO_{13}$	Erythromycin	514–519	1834
$C_{37}H_{67}NO_{13}\cdot C_{12}H_{22}O_{12}$	Erythromycin lactobionate (salt with lactobiono-δ-lactone)		1836
$C_{37}H_{70}N_2O_{12}$	Erythromycyclamine	520	
C_{38}			
$C_{38}H_{68}N_2O_{13}$	Erythromycyclamine-11,12-carbonate	521	
$C_{38}H_{69}NO_{13}$	Clarithromycin	320–321	
$C_{38}H_{72}N_2O_{12}$	Azithromycin	113	
$C_{38}H_{72}N_2O_{12}$	Desmycarosylcarbomycin A	391	
$C_{38}H_{72}N_2O_{12}$	Repromicin	1252	
C_{39}			
$C_{39}H_{65}NO_{14}$	Desmycosin	390	
C_{40-49}			
$C_{41}H_{67}NO_{15}$	Troleandomycin		2345
$C_{41}H_{76}N_2O_{15}$	Roxithromycin	1259	
$C_{41}H_{78}NO_8P$	1,2-Dioleoylphosphatidylethanolamine (DOPE)	452	
$C_{42}H_{67}NO_{15}$	Carbomycin B	250	
$C_{42}H_{67}NO_{16}$	Carbomycin A (Magnamycin A)	250	
$C_{43}H_{55}N_5O_7$	Vindesine	1502–1503	
$C_{43}H_{58}N_4O_{12}$	Rifampin	1256–1257	
$C_{43}H_{66}N_{12}O_{12}S_2$	Oxytocin		2056
$C_{43}H_{78}N_6O_{13}$	Muroctasin (Romurtide)	862	
$C_{44}H_{56}N_8O_7$	Seglitide	1271	
$C_{45}H_{54}N_8O_{10}$	Pristinamycin I_A (vernamycin Bα)	1138	
$C_{46}H_{56}N_4O_{10}$	Vincristine	1500–1501	2362
$C_{46}H_{58}N_4O_9$	Vinblastine (VLB; vincaleukoblastine)	1499	2362
$C_{46}H_{77}NO_{17}$	Tylosin	1483	
$C_{46}H_{80}N_2O_{13}$	Tilmicosin	1446	
$C_{47}H_{73}NO_{17}$	Amphotericin B		1576–1577

Formula	Name		
$C_{47}H_{75}NO_{17}$	Nystatin		2044
$C_{49}H_{89}NO_{25}$	Erythromycin lactobionate (salt with lactobiono-δ-lactone)		1836
C_{50-59}			
$C_{52}H_{97}NO_{18}S$	Erythromycin estolate		1835
$C_{55}H_{59}N_5O_{20}$	Coumermycin A_1	355	
$C_{55}H_{96}N_{16}O_{13}$	Polymyxin B_2		2121
$C_{56}H_{98}N_{16}O_{13}$	Polymyxin B_1		2121
C_{60-69}			
$C_{62}H_{111}N_{11}O_{12}$	*iso*-Cyclosporin A	372	
$C_{63}H_{88}CoN_{14}O_{14}P$	Cyanocobalamin	359	
$C_{66}H_{75}Cl_2N_9O_{24}$	Vancomycin	1488–1490	
H_{1-7}			
H_2O	Water	1509–1510	
H_2O_2	Hydrogen peroxide	632	
H_3BO_3	Boric acid	224	
H_3N	Ammonia	60	
H_7ClN_2OPt	Cisplatin degradation product	317	

SUBJECT INDEX

A

Accuracy and precision of pK_a values, physicochemical factors and, 8
  atmospheric CO_2 exclusion, 22–23
  chemical stability, 22
  experimental method, 9–15
  ionic strength, 21–22
  pH meters calibration with pH standards, 15–16
  solvent composition and polarity, 18–21
  temperature effects, 16–18
ACD/pK_a package versus measured pK_a values, 25
Acetic acid, pK_a values, 19
Acetonitrile, pK_a values, 19
4-aminobenzoic acid, 24
Artificial neural network (ANN) estimation methods, pK_a values, 14, 24, 26
Atmospheric CO_2, exclusion of, 22–23

B

Barbital. *See* 5,5-diethylbarbituric acid
Beckman Research models, 11
Benzoic acid, 5, 23, 28
Brønsted catalytic law, 24

C

Capillary electrophoresis (CE) studies, 5
Carbenoxolone, pK_a values, 28
Carboxylic acid group, pK_a values and, 19
CE methodology, pK_a screening and, 14–15
Chlorthalidone, 6, 28
Clofazimine, 6
Computer programs, pK_a values from, 24–26
Cyclopentane-1′,5-spiro derivative, 24

D

Data compilation, pK_a values and, 26–28
Davies modification, 21–22
Debye-Hückel equation, 8, 11
  Davies modification of, 21–22
  Guggenheim modification of, 10, 21
Dielectric constant (ε), 20
5,5-diethylbarbituric acid, 3, 5, 24, 26
  pK_a value, 23, 28
1,4-dioxane, pK_a values and, 19
DMSO solutions, 15

E

Electrochemical HPLC detectors, 15
Electromotive force (EMF) method, pK_a values and, 9, 12
Ephedrine, 5
Ethanol, pK_a values and, 19
5-ethyl-5-phenylbarbituric acid, 5

F

Famotidine, 6
Free energy change (ΔG^o), 26

G

Glibenclamide, 4, 6, 28
Guggenheim modification, 10, 21

H

Hammett and Taft relationships, 24
Henderson-Hasselbalch equation, 14, 22
HPLC assays, 27
HPLC-MS system, 16
Hydrolysis, of susceptible molecules, 22
Hydronium ion activity, 8, 12

I

Ibuprofen, 6, 28
International Union of Pure and Applied Chemistry (IUPAC), 2, 7, 15, 27
Ionic activity coefficient, 8
Ionic strength, 10–11, 14, 21–22
  pK_a values and, 6
Isonicotinic acid, 5, 12
5,5-di-isopropylbarbituric acid, 17
IUPAC pK_a compilations, 2–4

K

Kinetic methods, for pK_a determination, 13

L

Lidocaine, 6
Lignocaine, 28
Linear-free energy relationships (LFERs). *See* Hammett and Taft relationships

M

Mass spectrometry (MS), 14
Methanol, pK_a values and, 19
Methylcellosolve, pK_a values and, 19
Milli-Q water, 23
Molecular absorbance with ionization (dA/dpH method), 12
Morphine, 3

N

National Institute of Science and Technology (NIST), 7, 28
Nicotinic acid, 5
Nimesulide, 6, 28
N,N-dimethylacetamide, 19
N,N-dimethylformamide, 19
Noyes-Whitney equation, 19

P

pH
  definitions of, 8
  measurement methods, 9, 11–12, 15, 27
  meters calibration, pH standards and, 15–16
  and pK_a values, relationship between, 2, 7
Phenobarbital. *See* 5-ethyl-5-phenylbarbituric acid
Phenylbutazone, 6, 28
pK_a determination, evaluation methods
  accuracy and precision, 8
    atmospheric CO_2, exclusion of, 22–23
    chemical stability, 22
    experimental method, 9–15
    ionic strength, 21–22
    pH meters calibration with pH standards, 15–16
    solvent composition and polarity, 18–21
    temperature effects, 16–18
  data quality validation, 23
  inherent precision, 7–8
pK_a values, for pharmaceutical substances, 1
  from computer programs, 24–26
  data compilation, 26–28
  determination, evaluation methods for
    accuracy and precision, physicochemical factors and, 8–23
    data quality validation, 23
    inherent precision, 7–8
  IUPAC pK_a compilations, 2–4
  limitations of, 4–7
pK_b values, 5
Polarity and solvent composition, pK_a values and, 18
Potentiometric titration (weak acids and weak bases), 9–10
Propranolol, 6, 28
1,3-bis[(2-pyridyl)methyleneamino] guanidine, 12

R

Reversed phase-high performance liquid chromatography (RP-HPLC) column, 14

S

Sirius®, autotitration equipment, 11, 18
Sodium tetraborate (borax) solutions, 16
Solubility–pH dependence method, 10–15, 18–19, 22
Spectrophotometric method, 11, 18, 22–23
  pH measurements and, 12
Sulfamic acid group, pK_a values and, 19

T

Temperature, pK_a and pK_w values and, 16–18

U

US National Bureau of Standards (NBS), 7–9, 28

V

Valentiner-Lannung equation, 16–17
van't Hoff equation, 17

W

Water autoprotolysis constant (pK_w), 9

Y

Yasuda-Shedlovsky procedure, 3, 14, 20